Springer Proceedings in Mathematics & Statistics

Volume 50

For further volumes:
http://www.springer.com/series/10533

Springer Proceedings in Mathematics & Statistics

This book series features volumes composed of select contributions from workshops and conferences in all areas of current research in mathematics and statistics, including OR and optimization. In addition to an overall evaluation of the interest, scientific quality, and timeliness of each proposal at the hands of the publisher, individual contributions are all refereed to the high quality standards of leading journals in the field. Thus, this series provides the research community with well-edited, authoritative reports on developments in the most exciting areas of mathematical and statistical research today.

David H. Bailey • Heinz H. Bauschke
Peter Borwein • Frank Garvan • Michel Théra
Jon D. Vanderwerff • Henry Wolkowicz
Editors

Computational and Analytical Mathematics

In Honor of Jonathan Borwein's 60th Birthday

Springer

Editors
David H.Bailey
Lawrence Berkeley National Laboratory
Berkeley, CA, USA

Heinz H. Bauschke
Department of Mathematics
University of British Columbia
Kelowna, BC, Canada

Peter Borwein
Department of Mathematics
Simon Fraser University
Burnaby, BC, Canada

Frank Garvan
Department of Mathematics
University of Florida
Gainesville, FL, USA

Michel Théra
Department of Mathematics
Université de Limoges
Limoges Cedex, France

Jon D. Vanderwerff
Department of Mathematics
and Computer Science
La Sierra University
Riverside, CA, USA

Henry Wolkowicz
Faculty of Mathematics
University of Waterloo
Waterloo, ON, Canada

ISSN 2194-1009 ISSN 2194-1017 (electronic)
ISBN 978-1-4614-7620-7 ISBN 978-1-4614-7621-4 (eBook)
DOI 10.1007/978-1-4614-7621-4
Springer New York Heidelberg Dordrecht London

Library of Congress Control Number: 2013945306

Mathematics Subject Classifications (2010): 05A15, 05E05, 11A63, 11K16, 11K45, 11M41, 11R06, 26B25, 28Axx, 28A80, 28A25, 33C20, 37F10, 41A65, 46B10, 46B26, 46B50, 46B20, 46E30, 46N10, 46T99, 47H04, 47A06, 47B65, 47H05, 47H09, 47H10, 47J05, 47J25, 47J20, 47N10, 49J52, 49J53, 49K27, 49K40, 49M27, 49N15, 52A27, 52A41, 52B55, 54C15, 58C06, 58E17, 58E30, 65K05, 90C22, 90C25, 90C34, 90C46, 90C56, 91A05, 91B30, 91G10, 91G80

Printed on acid-free paper

Springer is part of Springer Science+Business Media (www.springer.com)

Preface

In the week of May 16–20, 2011, the workshop *Computational and Analytical Mathematics* (also known as *JonFest*) was held at the IRMACS Centre at Simon Fraser University in honour of Jonathan Michael Borwein's 60th birthday. It brought together nearly 100 experts from 14 countries.

Jon Borwein is one of the most productive Canadian researchers ever. His research interests are broad, ranging from analysis, computational mathematics, and optimization to experimental mathematics and number theory. He has authored or co-authored more than a dozen books and more than 300 papers.

Those who have had the fortune of collaborating with him as students or colleagues will testify to his immense knowledge, technical mastery, and deep intuition. He has been altering the life trajectories of many of his collaborators significantly and sometimes dramatically. His passion and relentless pursuit for useful and beautiful mathematics are extraordinary; the way he inspires and brings out the best in his students and collaborators is Steve Jobs-like!

This book brings together 31 carefully refereed research and review papers in the broad areas of Jon Borwein's interests. Most papers in this volume grew out of talks delivered at JonFest; however, some contributions are from experts who were unable to attend. Very sadly, one of the contributors, Richard Crandall, passed away in December 2012, before this book went into production.

We believe that the reader will find this book to be a delightful and valuable state-of-the-art account on some fascinating areas of Computational and Analytical Mathematics, ranging from Cantor fractals and strongly normal numbers to various algorithms in optimization and fixed point theory.

The editors thank the sponsors of JonFest—Interdisciplinary Research in the Mathematical and Computational Sciences (IRMACS) Centre at Simon Fraser University (SFU), Australian Mathematical Sciences Institute (AMSI), Mathematics of Information Technology and Complex Systems (MITACS), Pacific Institute for the Mathematical Sciences (PIMS), Fields Institute, and the Priority Research Centre for Computer-Assisted Research Mathematics and its Applications (CARMA)—for their financial and logistical support in hosting the workshop, and Pam Borghard and Veselin Jungic for their "on-site" help in the preparation and realization of the workshop at the IRMACS Centre.

We are very grateful to Dr. Hung Phan for his hard work and great help in the preparation of this volume which as a result not only is beautifully typeset but also exhibits a consistent structure. We also thank Ms. Elizabeth Loew from Springer for her help guiding this volume through production.

Finally, we thank the hardworking and dedicated referees who contributed crucially to the quality of this volume through their constructive and insightful reviews.

Berkeley, (USA)	David H. Bailey
Kelowna, (Canada)	Heinz H. Bauschke
Burnaby, (Canada)	Peter Borwein
Gainesville, (USA)	Frank Garvan
Limoges, (France)	Michel Théra
Riverside, (USA)	Jon D. Vanderwerff
Waterloo, (Canada)	Henry Wolkowicz

Contents

Contributors

David H. Bailey
Lawrence Berkeley National Laboratory, Berkeley, CA, USA,

Heinz H. Bauschke
Department of Mathematics, The University of British Columbia, Kelowna, BC, Canada

Adrian Belshaw
Department of Mathematics and Statistics, Capilano University, Sechelt, BC, Canada

Henri Bonnel
Université de la Nouvelle-Calédonie, ERIM, Nouméa Cédex, New Caledonia, France

Jonathan M. Borwein
CARMA, School of Mathematical and Physical Sciences, University of Newcastle, Newcastle, NSW, Australia
King Abdulaziz University, Jeddah 80200, Saudi Arabia

Peter Borwein
Department of Mathematics, Simon Fraser University, Burnaby, BC, Canada

Radu Ioan Boţ
Department of Mathematics, Chemnitz University of Technology, Chemnitz, Germany

David M. Bradley
Department of Mathematics and Statistics, University of Maine, Orono, ME, USA

Richard P. Brent
Mathematical Sciences Institute, Australian National University, Canberra, ACT, Australia

Luis M. Briceño-Arias
Department of Mathematics, Universidad de Chile, Santiago, Chile
Universidad Técnica Federico Santa María, Santiago, Chile

Regina S. Burachik
School of Mathematics and Statistics, University of South Australia, Mawson Lakes, SA, Australia

B. Cascales
Department of Mathematics, University of Murcia, Murcia, Spain

Tom Chappell
School of Mathematics and Physics, The University of Queensland, Brisbane, QLD, Australia

Yuen-Lam Cheung
Department of Combinatorics and Optimization, University of Waterloo, Waterloo, ON, Canada

Patrick L. Combettes
Laboratoire Jacques-Louis Lions – UMR CNRS 7598, UPMC Université Paris 06, Paris, France

Robert M. Corless
Department of Applied Mathematics, The University of Western Ontario, London, ON, Canada

Richard Crandall (1947–2012)

Ernö Robert Csetnek
Department of Mathematics, Chemnitz University of Technology, Chemnitz, Germany

Frank Deutsch
Department of Mathematics, The Pennsylvania State University, University Park, PA, USA

Asen L. Dontchev
University of Michigan and Mathematical Reviews, Ann Arbor, MI, USA

Mclean R. Edwards
Department of Mathematics, University of British Columbia, Vancouver, BC, Canada

Hélène Frankowska
CNRS, Institut de Mathématiques de Jussieu, Université Pierre et Marie Curie, Paris, France

J.R. Giles
School of Mathematical and Physical Sciences, The University of Newcastle, Callaghan, NSW, Australia

David Harvey
School of Mathematics and Statistics, University of New South Wales, Sydney, NSW, Australia

Hein Hundal
The Pennsylvania State University, Port Matilda, PA, USA

A.D. Ioffe
Department of Mathematics, Technion - Israel Institute of Technology, Haifa, Israel

P.S. Kenderov
Institute of Mathematics and Informatics, Bulgarian Academy of Sciences, Sofia, Bulgaria

Alexander Knecht
Department of Mathematics, La Sierra University, Riverside, CA, USA

W.M. Kozlowski
School of Mathematics and Statistics, University of New South Wales, Sydney, NSW, Australia

Alain Lascoux
CNRS, Institut Gaspard Monge, Université Paris-Est, Marne-la-Vallée, France

Piers W. Lawrence
Department of Applied Mathematics, The University of Western Ontario, London, ON, Canada

Yves Lucet
Department of Computer Science, University of British Columbia, Kelowna, BC, Canada

Victoria Martín-Márquez
Department of Mathematical Analysis, University of Seville, Seville, Spain

Marc Mazade
Université de Montpellier II, Montpellier, France

Jacqueline Morgan
Dipartimento di Matematica e Statistica and CSEF, Università di Napoli, Napoli, Italy

Dominikus Noll
Université Paul Sabatier, Institut de Mathématiques, Toulouse, France

J. Orihuela
Department of Mathematics, University of Murcia, Murcia, Spain

Gábor Pataki
Department of Statistics and Operations Research, University of North Carolina at Chapel Hill, Chapel Hill, NC, USA

Jean-Paul Penot
Laboratoire Jacques-Louis Lions, Université Pierre et Marie Curie, Paris Cedex 05, France

Simeon Reich
Department of Mathematics, The Technion - Israel Institute of Technology, Haifa, Israel

J.P. Revalski
Institute of Mathematics and Informatics, Bulgarian Academy of Sciences, Sofia, Bulgaria

Mathew Rogers
Department of Mathematics and Statistics, Université de Montréal, Montréal, QC, Canada

Aude Rondepierre
Institut de Mathématiques de Toulouse, INSA de Toulouse, Toulouse, France

M. Ruiz Galán
Department of Applied Mathematics, E. T. S. Ingeniería. Edificación., Granada, Spain

Shoham Sabach
Department of Mathematics, The Technion - Israel Institute of Technology, Haifa, Israel

Simon Schurr
spschurr@rogers.com

Brailey Sims
CARMA, School of Mathematical and Physical Sciences, The University of Newcastle, Callaghan, NSW, Australia

Lionel Thibault
Université de Montpellier II, Montpellier, France

Jon Vanderwerff
Department of Mathematics, La Sierra University, Riverside, CA, USA

Douglas E. Ward
Department of Mathematics, Miami University, Oxford, OH, USA

S. Ole Warnaar
School of Mathematics and Physics, The University of Queensland, Brisbane, QLD, Australia

Henry Wolkowicz
Department of Combinatorics and Optimization, University of Waterloo, Waterloo, ON, Canada

Stephen E. Wright
Department of Statistics, Miami University, Oxford, OH, USA

Liangjin Yao
CARMA, School of Mathematical and Physical Sciences, University of Newcastle, Newcastle, NSW, Australia

Boonrod Yuttanan
Department of Mathematics and Statistics, Prince of Songkla University, Songkhla, Thailand

Xia Zhou
Department of Mathematics, Zhejiang University, Hangzhou, Republic of China

Ludmil Zikatanov
Department of Mathematics, The Pennsylvania State University, University Park, PA, USA

Wadim Zudilin
School of Mathematical and Physical Sciences, The University of Newcastle, Callaghan, NSW, Australia

Chapter 1
Normal Numbers and Pseudorandom Generators

David H. Bailey and Jonathan M. Borwein

Abstract For an integer $b \geq 2$ a real number α is *b-normal* if, for all $m > 0$, every m-long string of digits in the base-b expansion of α appears, in the limit, with frequency b^{-m}. Although almost all reals in $[0, 1]$ are b-normal for every b, it has been rather difficult to exhibit explicit examples. No results whatsoever are known, one way or the other, for the class of "natural" mathematical constants, such as π, e, $\sqrt{2}$ and $\log 2$. In this paper, we summarize some previous normality results for a certain class of explicit reals and then show that a specific member of this class, while provably 2-normal, is provably *not* 6-normal. We then show that a practical and reasonably effective pseudorandom number generator can be defined based on the binary digits of this constant and conclude by sketching out some directions for further research.

Key words: Normal numbers • Stoneham numbers • Pseudorandom number generators

Mathematics Subject Classifications (2010): 11A63, 11K16, 11K45

COMMUNICATED BY HEINZ H. BAUSCHKE.

D.H. Bailey (✉)
Lawrence Berkeley National Laboratory, Berkeley, CA 94720, USA
e-mail: DHBailey@lbl.gov

J.M. Borwein
Centre for Computer Assisted Research Mathematics and its Applications (CARMA), University of Newcastle, Callaghan, NSW 2308, Australia

King Abdulaziz University, Jeddah 80200, Saudi Arabia
e-mail: jonathan.borwein@newcastle.edu.au

D.H. Bailey et al. (eds.), *Computational and Analytical Mathematics*, Springer Proceedings in Mathematics & Statistics 50, DOI 10.1007/978-1-4614-7621-4_1,

1.1 Introduction

For an integer $b \geq 2$ we say that a real number α is *b-normal* (or *normal base b*) if, for all $m > 0$, every m-long string of digits in the base-b expansion of α appears, in the limit, with frequency b^{-m}, or, in other words, with exactly the frequency one would expect if the digits appeared completely "at random." It follows from basic probability theory that, for any integer $b \geq 2$, almost all reals in the interval $(0,1)$ are b-normal. What's more, almost all reals in the unit interval are simultaneously b-normal for all integers $b \geq 2$.

Yet identifying even a single explicitly given real number that is b-normal for some b has proven frustratingly difficult. The first constant proven 10-normal was the Champernowne constant [7], namely $0.12345678910111213\ldots$, produced by concatenating the natural numbers in decimal format. This was extended to base-b normality (for base-b versions of the Champernowne constant). In 1946, Copeland and Erdös established that the concatenation of primes $0.23571113171923\ldots$ and also the concatenation of composites $0.46891012141516\ldots$, among others, are also 10-normal [8]. In general they proved:

Theorem 1.1 ([8]). *If $a_1, a_2, \cdots$ is an increasing sequence of integers such that for every $\theta < 1$ the number of a's up to N exceeds N^θ provided N is sufficiently large, then the infinite decimal*

$$0.a_1a_2a_3\cdots$$

is normal with respect to the base β in which these integers are expressed.

This clearly applies to the primes of the form $ak+c$ with a and c relatively prime in any given base and to the integers which are the sum of two squares (since every prime of the form $4k+1$ is included).

Some related results were established by Schmidt, including the following [15]. Write $p \sim q$ if there are positive integers r and s such that $p^r = q^s$. Then

Theorem 1.2. *If $p \sim q$, then any real number that is p-normal is also q-normal. However, if $p \not\sim q$, then there are uncountably many p-normal reals that are not q-normal.*

In a recent survey, Queffelec [14] described the above result and also presented the following, which he ascribed to Korobov:

Theorem 1.3. *Numbers of the form $\sum_k p^{-2^k} q^{-p^{2^k}}$, where p and q are relatively prime, are q-normal.*

Nonetheless, we are still completely in the dark as to the b-normality of "natural" constants of mathematics. Borel was the first to conjecture that *all* irrational algebraic numbers are b-normal for *every* integer $b \geq 2$. Yet not a single instance of this conjecture has ever been proven. We do not even know for certain whether or not the limiting frequency of zeroes in the binary expansion of $\sqrt{2}$ is one-half, although numerous large statistical analyses have failed to show any significant deviation from statistical normals. The same can be said for π and other basic constants, such as $e, \log 2$, and $\zeta(3)$. Clearly any result (one way or the other) for one of these constants would be a mathematical development of the first magnitude.

In the case of an algebraic number of degree d, it is now known that the number of ones in the binary expansion through bit position n must exceed $Cn^{1/d}$ for a positive number C (depending on the constant) and all sufficiently large n [4]. In particular, there must be at least $\sqrt{n}$ ones in the first n bits of $\sqrt{2}$. But this is clearly a relatively weak result, because, barring an enormous mathematical surprise, the correct limiting frequency of ones in the binary expansion of $\sqrt{2}$ is one-half.

In this paper, we briefly summarize some previously published normality results for a certain class of real constants, prove an interesting *non-normality* result, and then demonstrate how these normality results can be parlayed into producing a practical pseudorandom number generator. This generator can be implemented quite easily, is reasonably fast-running, and, in initial tests, seems to produce results of satisfactory "randomness." In addition, we show how all of this suggests a future direction to the long sought proof of normality for "natural" mathematical constants.

1.2 Normality of a Class of Generalized BBP-Type Constants

In [1], Richard Crandall and one of the present authors (Bailey) analyzed the class of constants

$$\alpha_{b,c}(r) = \sum_{k=1}^{\infty} \frac{1}{c^k b^{c^k + r_k}}, \tag{1.1}$$

where the integers $b > 1$ and $c > 1$ are co-prime, where r is any real in $[0,1]$, and where r_k is the kth binary digit of r. These constants qualify as "generalized BBP-type constants," because the nth base-b digit can be calculated directly, without needing to compute any of the first $n-1$ digits, by a simple and efficient algorithm similar to that first applied to π and $\log 2$ in the paper by Bailey et al. [3].

Bailey and Crandall were able to establish:

Theorem 1.4. *Every real constant of the class* (1.1) *is b-normal.*

Subsequently, Bailey and Misieurwicz were able to establish this same result (at least in a simple demonstrative case) via a much simpler argument, utilizing a "hot spot" lemma proven by ergodic theory techniques [2] (see also [5, p. 155]).

Fix integers b and c satisfying the above criteria, and let r and s be any reals in $[0,1]$. If $r \neq s$, then $\alpha_{b,c}(r) \neq \alpha_{b,c}(s)$, so that the class $A_{b,c} = \{\alpha_{b,c}(r),\ 0 \leq r \leq 1\}$ has uncountably many distinct elements (this was shown by Bailey and Crandall). However, it is not known whether the class $A_{b,c}$ contains any constants of mathematical significance, such as π or e.

In this paper we will focus on the constant $\alpha_{2,3}(0)$, which we will denote as α for short:

$$\begin{aligned}\alpha = \alpha_{2,3}(0) &= \sum_{k=1}^{\infty} \frac{1}{3^k 2^{3^k}} \\ &= 0.0418836808315029850712528986245716824260967584654857\ldots_{10} \\ &= 0.\mathtt{0AB8E38F684BDA12F684BF35BA781948B0FCD6E9E06522C3F35B}\ldots_{16}.\end{aligned} \tag{1.2}$$

Although its 2-normality follows from the results in either of the two papers mentioned above [1, 2], this particular constant was first proved 2-normal by Stoneham back in 1973 [16].

1.3 A Non-normality Result

It should be emphasized that just because a real constant is b-normal for some integer $b > 1$, it does not follow that it is c-normal for any other integer c, except in the case where $b^r = c^s$ for positive integers r and s (see Theorem 1.2). In other words, if a constant is 8-normal, it is clearly 16-normal (since base-16 digits can be written as four binary digits and base-8 digits can be written as three binary digits), but nothing can be said a priori about that constant's normality in any base that is not a power of two.

As mentioned above, there are very few normality results, and none is known for well-known constants of mathematics. But the same can be said about specific non-normality results, provided we exclude rationals (which repeat and thus are not normal) and examples, such as 1.0101000100000001... (i.e., ones appear in position 2^m), that are constructed specifically not to be normal but otherwise have relatively little mathematical interest (although Liouville's class of transcendental numbers is an exception). In particular, none of the well-known "natural" constants of mathematics have ever been proven *not* to be b-normal for some b. Indeed, such a result, say for $\pi, \log 2$, or $\sqrt{2}$, would be even more interesting than a proof of normality for that constant.

In that vein, here is an intriguing result regarding the α constant mentioned above:

Theorem 1.5. *α is* not *6-normal.*

1.3.1 Discussion

Let the notation $\{\cdot\}$ denote fractional part. Note that the base-6 digits immediately following position n in the base-6 expansion of α can be obtained by computing

Table 1.1 Base-6 expansion of α

0.
0130140430003334251130502130000001243555045432233011500243525320551352
3435410104300000000000000005141130054040555455303144250433435101241345
2351125142125134505503545015053522052044340452151505102411552500425130
0511244540010441311500324203032130000000000000000000000000000000000000
0000014212034311121452013525445342113412240220525301054204423552411055
4150155204350414555400310145303033532002534340401301240104453254343502
1420204324150255551010040433000455441145010313314511510144514123443342
3412400551313335045423530553151153501533452435450250055521453054234342
1530350125024205404135451231323245353031534552304115020154242121145201
5422225343403404505301233255344404431033324453321414150142334545424124
3203125340050134150245514404300000000000000000000000000000000000000000
0000000000000000000000000000000000000000000000000000000000000000000000
0000000000313350542444431111055534141052014540213412313001424333133115
...

$\{6^n\alpha\}$, which can be written as follows:

$$\{6^n\alpha\} = \left\{ \sum_{m=1}^{\lfloor \log_3 n \rfloor} 3^{n-m}2^{n-3^m} \right\} + \left\{ \sum_{m=\lfloor \log_3 n \rfloor+1}^{\infty} 3^{n-m}2^{n-3^m} \right\}. \tag{1.3}$$

Now note that the first portion of this expression is *zero*, since all terms of the summation are integers. That leaves the second expression.

Consider the case when $n = 3^m$, where $m \geq 1$ is an integer, and examine just the first term of the second summation. We see that this expression is

$$3^{3^m-(m+1)}2^{3^m-3^{m+1}} = 3^{3^m-m-1}2^{-2\cdot 3^m} = (3/4)^{3^m}/3^{m+1}. \tag{1.4}$$

We can generously bound the sum of all terms of the second summation by 1.00001 times this amount, for all $m \geq 1$, and by many times closer to unity for all $m \geq 2$. Thus we have

$$\{6^{3^m}\alpha\} \approx \frac{\left(\frac{3}{4}\right)^{3^m}}{3^{m+1}}, \tag{1.5}$$

and this approximation is as accurate as one wishes (in ratio) for all sufficiently large m.

Given the very small size of the expression $(3/4)^{3^m}/3^{m+1}$ for even moderate-sized m, it is clear the base-6 expansion will have very long stretches of zeroes beginning at positions $3^m + 1$. For example, by explicitly computing α to high precision, one can produce the counts of consecutive zeroes Z_m that immediately follow position 3^m in the base-6 expansion of α—see Tables 1.1 and 1.2.

Table 1.2 Counts Z_m of consecutive zeroes immediately following position 3^m in base-6 expansion of α

m	3^m	Z_m
1	3	1
2	9	3
3	27	6
4	81	16
5	243	42
6	729	121
7	2187	356
8	6561	1058
9	19683	3166
10	59049	9487

In total, there are 14,256 zeroes in these ten segments, which, including the last segment, span the first $59,049 + 9,487 = 68,536$ base-6 digits of α. In this tabulation we have of course ignored the many zeroes in the large "random" segments of the expansion. Thus the fraction of the first 68,536 digits that are zero is at least $14,256/68,536 = 0.20800747\ldots$, which is significantly more than the expected value $1/6 = 0.166666\ldots$.

A more careful analysis shows that this limiting ratio

$$\lim_{m\to\infty} \frac{\sum_{m\geq 1} Z_m}{3^m + Z_m} = \frac{3}{2} \cdot \frac{\log_6(4/3)}{1+\log_6(4/3)} \tag{1.6}$$

$$= \frac{1}{2}\log_2(4/3) = 0.2075187496\ldots \tag{1.7}$$

Complete details are given in the appendix. Also included in the appendix is a proof of this generalization of Theorem 1.5:

Theorem 1.6. *Given co-prime integers $b \geq 2$ and $c \geq 2$, the constant*

$$\alpha_{b,c} = \sum_{k\geq 1} 1/(c^k b^{c^k})$$

is not *bc-normal.*

These results thus constitute simple and concrete counter-examples to the question of whether normality in one base b implies normality in another base c (except in simple cases covered by the first part of Theorem 1.2). In particular, these results are explicit examples of part two of Theorem 1.2.

It is worth pointing out that Cassels proved that for almost all real x in the unit interval, x is 2-normal but not 3-normal, although he did not present any explicit example of such x [6]. Above we have presented an explicit real that is 2-normal but not 6-normal, which is almost but not quite such an example. Some related discussion is given in [13, 15, 17].

1.4 Alpha as a Pseudorandom Generator

The normality result for α (Theorem 1.4) suggests that the binary digits of α (certainly not its base-6 digits) could be used to fashion a practical pseudorandom number generator. Indeed, this was suggested in [1] and [5, p. 169–170]. We will show here how this can be done. The result is a generator that is both efficient on single-processor systems and also well suited for parallel processing: each processor can quickly and independently calculate the starting seed for its section of the resulting global sequence, which global sequence is the same as the sequence produced on a single-processor system (subject to some reasonable conditions). However, it is acknowledged that before such a generator is used in a "practical" application, it must be subjected to significant checking and testing. It should also be noted that just because a number is normal does not guarantee its suitability for pseudorandom generation (e.g., the convergence of the limiting frequencies might be very slow), although this particular scheme does appear to be reasonably well behaved.

1.4.1 Background

Define x_n to be the binary expansion of α starting with position $n+1$. Note that $x_n = \{2^n\alpha\}$, where $\{\cdot\}$ means the fractional part of the argument. First consider the case $n = 3^m$ for some integer m. In this case one can write

$$x_{3^m} = \{2^{3^m}\alpha\} = \left\{\sum_{k=1}^{m} \frac{2^{3^m-3^k}}{3^k}\right\} + \sum_{k=m+1}^{\infty} \frac{2^{3^m-3^k}}{3^k}. \tag{1.8}$$

Observe that the "tail" term (i.e., the second term) in this expression is exceedingly small once m is even moderately large—for example, when $m = 10$, this term is only about 10^{-35551}. This term will hereafter be abbreviated as ε_m. By expanding the first term, one obtains

$$x_{3^m} = \frac{(3^{m-1}2^{3^m-3} + 3^{m-2}2^{3^m-3^2} + \cdots + 3\cdot 2^{3^m-3^{m-1}} + 1) \bmod 3^m}{3^m} + \varepsilon_m. \tag{1.9}$$

The numerator is taken modulo 3^m, since only the remainder when divided by 3^m is of interest when finding the fractional part. By Euler's totient theorem, the next-to-last term in the numerator, when reduced modulo 3^m, is three. Similarly, it can be seen that every other term in the numerator, when reduced modulo 3^m, is equivalent to itself without the power-of-two part. In other words, the expression above reduces to

$$x_{3^m} = \frac{(3^{m-1} + 3^{m-2} + \cdots + 3 + 1) \bmod 3^m}{3^m} + \varepsilon_m \tag{1.10}$$

$$= \frac{3^m - 1}{2 \cdot 3^m} + \varepsilon_m = \frac{\lfloor 3^m/2 \rfloor}{3^m} + \varepsilon_m. \tag{1.11}$$

(The authors are indebted to Helaman Ferguson for a key idea in this proof.) More generally, for n that is not a power of three, one can write

$$x_n = \frac{(2^{n-3^m} \lfloor 3^m/2 \rfloor) \bmod 3^m}{3^m} + \varepsilon, \tag{1.12}$$

where m is chosen so that 3^m is the largest power of three less than or equal to n. In this case, one can be assured that $\varepsilon < 10^{-30}$ provided n is not within 100 of any power of three.

1.4.2 Algorithm

With this explicit expression in mind, an algorithm can be given for generating pseudorandom deviates, in the form of a sequence of IEEE 64-bit floating-point numbers in $(0,1)$. These deviates contain, in their mantissas, successive 53-bit segments of the binary expansion of α, beginning at some given starting position.

1.4.2.1 Initialization

First select a starting index a in the range

$$3^{33} + 100 = 5559060566555623 \leq a \leq 2^{53} = 9007199254740992. \tag{1.13}$$

The value of a can be thought of as the "seed" of the generator. Then calculate

$$z_0 = 2^{a-3^{33}} \cdot \lfloor 3^{33}/2 \rfloor \bmod 3^{33}. \tag{1.14}$$

1.4.2.2 Generate Iterates

Successive iterates of the generator can then be recursively computed by iterating

$$z_k = 2^{53} \cdot z_{k-1} \bmod 3^{33} \tag{1.15}$$

and then returning the values $z_k 3^{-33}$, which are 64-bit IEEE floating-point results in the unit interval.

1.4.2.3 Arithmetic

Several of the operations used in this scheme must be done with an accuracy of at least 106 mantissa bits. This can be done using "double-double" arithmetic. A double-double datum is represented by a pair of IEEE double-precision floating-point numbers: the first word is the closest 64-bit IEEE value to the double-double value, and the second word is the difference. Algorithms for performing basic double-double arithmetic algorithms, using only rounded 64-bit IEEE floating-point operations, are given in [9] or [5, p. 218–220]. These have been implemented in C++ and Fortran-90 double-double computation software packages, which include both basic-level arithmetic functions as well as common algebraic and transcendental functions, available from the first author's web site: http://crd.lbl.gov/~dhbailey/mpdist.

On the other hand, one could also use 128-bit integer or 128-bit IEEE floating-point arithmetic to do these operations, if these operations are available in hardware (software implementations tend to be relatively slow).

1.4.2.4 Implementation Details

The operation $2^{53} \cdot z_{k-1} \bmod 3^{33}$ can be performed efficiently as follows: (1) multiply 2^{53} by z_{k-1} (double times double yielding a double-double or 128-bit result); (2) multiply the result of step 1 (just the high-order portion will do) by 3^{-33} and take the greatest integer; (3) multiply the result of step 2 by 3^{33} (double times double yielding a double-double or 128-bit result); and (4) subtract the result of step 3 from the result of step 1 (using double-double or 128-bit arithmetic). It is possible that the result of step 2 might be one unit too high, or one too low, so that the result of step 4 may need to be adjusted accordingly: if it is negative, add 3^{33}; if it exceeds 3^{33}, subtract 3^{33}.

1.4.2.5 Exponentiation

The exponentiation required in the initialization may be done efficiently using the binary algorithm for exponentiation. This is merely the formal name for the observation that exponentiation can be economically performed by means of a factorization based on the binary expansion of the exponent. For example, one can write $3^{17} = ((((3^2)^2)^2)^2) \cdot 3$, thus producing the result in only five multiplications, instead of the usual 16. According to Knuth, this technique dates back at least to 200 BCE [10, p. 461]. In this application, the exponentiation result is required modulo a positive integer k. This can be done very efficiently by reducing modulo k the intermediate multiplication result at each step of the exponentiation algorithm. A formal statement of this scheme is as follows:

To compute $r = b^n \bmod k$, where r, b, n, and k are positive integers, first set t to be the largest power of two such that $t \leq n$, and set $r = 1$. Then

A: if $n \geq t$ then $r \leftarrow br \bmod k$; $n \leftarrow n - t$; endif
$t \leftarrow t/2$
if $t \geq 1$ then $r \leftarrow r^2 \bmod k$; go to A; endif

Note that the above algorithm is performed entirely with positive integers that do not exceed k^2 in size.

A full implementation of the entire pseudorandom scheme, which runs on any computer system with IEEE 64-bit arithmetic and a Fortran-90 compiler, can be obtained from the first author's web site: http://crd.lbl.gov/~dhbailey/mpdist. The code is straightforward and can easily be converted to other languages, such as C or Java.

1.4.3 Analysis

It can be seen from the above that the recursive sequence generating iterates, which contain successive 53-long segments of binary digits from the expansion of α, is nothing more than a special type of linear congruential pseudorandom number generator, a class that has been studied extensively by computer scientists and others [10, p. 10–26]. In other words, the binary digits of α are "locally" (within a range of indices spanned by successive powers of three) given by a linear congruential generator, with a modulus that is a large power of three.

This observation makes it an easy matter to determine the period P of the resulting generator [10, p. 17]: as specified above, $P = 2 \cdot 3^{32} \approx 3.706 \cdot 10^{15}$. Note, however, that the binary digits of the resulting sequence will match that of α only if $[a, a + 53n]$, where a is the starting index and n is the number of floating-point results generated, does not include a power of three or come within 100 of a power of three. If one can utilize 128-bit integer arithmetic, one could use a larger modulus, say 3^{40}, which would yield a period that is 2,187 times larger.

This scheme has one significant advantage over conventional linear congruential generators that use a power-of-two modulus: it cleanly avoids anomalies that sometimes arise in large scientific codes, when arrays with dimensions that are large powers of two are filled with pseudorandom data and then accessed both by row and by column (or plane), or which otherwise are accessed by large power-of-two data strides (as in a power-of-two FFT). This is because the pseudorandom data sequence accessed in this manner has a reduced period and thus may be not as "random" as desired. The usage of a modulus that is a large power of three is immune to these problems. The authors are not aware of any major scientific calculation that involves data access strides that are large powers of three.

1.4.4 Performance

As mentioned above, a Fortran-90 implementation of the scheme described above is available on the first author's web site. For comparison purposes, the conventional linear congruential generator

$$z_n = 5^{21} \cdot z_{n-1} \bmod 2^{53} \tag{1.16}$$

was implemented using the same software and programming style. These two codes were then tested on a 2.8 GHz Apple MacPro workstation, using the gfortran compiler (and running only on one of the eight cores). The program implementing the normal-number-based scheme required 3.553 s to generate an array of 100 million double-precision deviates. The conventional linear congruential system required essentially the same time.

By the way, the above program also is self-checking, in that it computes 100 million iterates using (1.15), then checks that the same value is produced by jumping ahead 100 million steps, by using formula (1.14). The present authors have used this program to check computational and data integrity on various computer systems. In at least one instance, the program disclosed intermittent memory errors.

1.4.5 Parallel Implementation

The scheme described above is very well suited for parallel processing, a trait not shared by a number of other commonly used pseudorandom schemes. Consider, for example, an implementation of the above pseudorandom scheme on a distributed memory system. Suppose that k is the processor number and p is the total number of processors used. Assume that a total of n pseudorandom deviates are to be generated, and assume that n is evenly divisible by p. Then each processor generates n/p results, with processor p using as a starting value $a + nk/p$. Note that each processor can quickly and independently generate its own value of z_0 by using formula (1.14).

In this way, the collective sequence generated by all processors coincides precisely with the sequence that is generated on a single-processor system. This feature is crucially important in parallel processing; permitting one can verify that a parallel program produces the same answers (to within reasonable numerical round-off error) as the single-processor version. It is also important, for the same reason, to permit one to compare results, say, between a run on 64 CPUs of a given system with one on 128 CPUs.

This scheme has been used to generate data for the fast Fourier transform (FFT) benchmark that is part of the benchmark suite for the high productivity computing systems (HPCS) program, funded by the US Defense Advanced Research Projects Agency (DARPA) and the US Department of Energy.

1.4.6 Variations

Some initial tests, conducted by Nelson Beebe of the University of Utah, found that if by chance one iterate is rather small, it will include as its trailing bits a few of the leading bits of the next result (this is a natural consequence of the construction). While the authors are not aware of any application for which this feature would have significant impact, it can be virtually eliminated by advancing the sequence by more than 53 bits—say by 64 bits—from iterate to iterate.

This can be done by simply altering formula (1.15) above to read

$$z_k = 2^{64} \cdot z_{k-1} \bmod 3^{33}. \tag{1.17}$$

This can be implemented as is, if one is using 128-bit integer or 128-bit IEEE floating-point arithmetic, but does not work correctly if one is using double-double arithmetic, because the product $2^{64} \cdot z_{k-1}$ could exceed 2^{106}, which is the maximum size of an integer that can be represented exactly as a double-double operand. When using double-double arithmetic, one can compute each iterate using the following:

$$z_k = 2^{11} \cdot (2^{53} \cdot z_{k-1} \bmod 3^{33}) \bmod 3^{33}. \tag{1.18}$$

Tests by the present authors, advancing 64 bits per result, showed no significant correlation to the leading bits of the next iterate. And, of course, the additional "skip" here could be more than 11; it could be any value up to 53.

Finally, there is no reason that other constants from this class could not also be used in a similar way. For example, a very similar generator could be constructed based on $\alpha_{2,5}$. One could also construct pseudorandom generators based on constants that are 3-normal or 5-normal, although one would lose the property that successive digits are precisely retained in consecutive computer words (which are based on binary arithmetic). The specific choice of multiplier and modulus can be made based on application requirements and the type of high-precision arithmetic that is available (e.g., double-double or 128-bit integer).

However, as we noted above, it is important to recognize that any proposed pseudorandom number generator, including this one, must be subjected to lengthy and rigorous testing [10–12]. Along this line, as noted above, generators of the general linear congruential family have problems, and it is not yet certain whether some variation or combination of generators in this class can be fashioned into a robust, reliable scheme that is both efficient and practical. But we do believe that these schemes are worthy of further study.

1.5 Conclusion and Directions for Further Work

In this paper, we have shown how the constant $\alpha = \sum_{n \geq 1} 1/(3^n 2^{3^n})$, which is provably 2-normal, is *not* 6-normal, as well as some generalizations. These results thus constitute simple and concrete counter-examples to the question of whether normality in one base b implies normality in another base c (except in simple cases covered by the first part of Theorem 1.2). In particular, these results are explicit examples of the second part of Theorem 1.2. We have also shown how a practical pseudorandom number generator can be constructed based on the binary digits of α, where each generated word consists of successive sections of its binary expansion.

Perhaps the most significant implication of the algorithm we have presented is not for its practical utility but instead for the insight it provides to the fundamental question of normality. In particular, the pseudorandom number construction implies that the digit expansions of one particular class of provably normal numbers consist of successive segments of exponentially growing length, and within each segment the digits are given by a specific type of linear congruential generator, with a period that also grows exponentially. From this perspective, the 2-normality of α is entirely plausible.

Now consider what this implies, say, for the normality of a constant such as $\log 2$. First recall the classical formula

$$\log 2 = \sum_{n=1}^{\infty} \frac{1}{n2^n}. \tag{1.19}$$

Thus, following the well-known BBP approach (see [3] or [5, Chap. 4]), we can write

$$\{2^d \log 2\} = \left\{ \sum_{n=1}^{d} \frac{2^{d-n} \bmod n}{n} \right\} + \left\{ \sum_{n=d+1}^{\infty} \frac{2^{d-n}}{n} \right\}. \tag{1.20}$$

This leads immediately to the BBP algorithm for computing the binary digits of $\log 2$ beginning after position d, since each term of the first summation can be computed very rapidly by means of the binary algorithm for exponentiation, and the second summation quickly converges.

But we can also view (1.20) for its insight on normality. Note that the binary expansion of $\log 2$ following position d can be seen as a sum of normalized linear congruential pseudorandom number generators, with periods (at least in some terms) that grow steadily with n (since the period of a linear congruential generator depends on the factorization of the modulus). But with increasing n, at least some terms will have prime moduli, resulting in relatively long periods. In fact, some will be primitive primes modulo two, which give the maximal period $(n-1)/2$. Note that the sum of normalized linear congruential generators can be rewritten as a single linear congruential generator. Thus it is plausible that the period of the sum of generators in the first portion of (1.20) increases without bound, resulting in a highly "random" expansion (although all of this needs to be worked out in detail).

We have attempted to develop these notions further, but so far we have not made a great deal of progress. But, at the least, this approach may be effective for constants such as

$$\beta = \sum_{n \in W}^{\infty} \frac{1}{n2^n}, \tag{1.21}$$

where W is the set of primitive primes modulo two, which as mentioned above give rise to maximal periods when used as a linear congruential modulus. Only time will tell.

Acknowledgements The first author is supported in part by the Director, Office of Computational and Technology Research, Division of Mathematical, Information, and Computational Sciences of the US Department of Energy, under contract number DE-AC02-05CH11231.

Appendix

Proof. $\alpha_{2,3}$ is not 6-normal.

Let Q_m be the base-6 expansion of $\alpha_{2,3}$ immediately following position 3^m (i.e., after the "decimal" point has been shifted to the right 3^m digits). We can write

$$\begin{aligned} Q_m &= 6^{3^m} \alpha_{2,3} \bmod 1 \\ &= \left(\sum_{k=1}^{m} 3^{3^m - k} 2^{3^m - 3^k} \right) \bmod 1 + \sum_{k=m+1}^{\infty} 3^{3^m - k} 2^{3^m - 3^k}. \end{aligned} \tag{1.22}$$

The first portion of this expression is zero, since all terms in the summation are integers. The small second portion is very accurately approximated by the first term of the series, namely $(3/4)^{3^m}/3^{m+1}$. In fact, for all $m \geq 1$,

$$\frac{(3/4)^{3^m}}{3^{m+1}} < Q_m < \frac{(3/4)^{3^m}}{3^{m+1}}(1 + 2 \cdot 10^{-6}). \tag{1.23}$$

Let $Z_m = \lfloor \log_6 1/Q_m \rfloor$ be the number of zeroes in the base-6 expansion of α that immediately follow position 3^m. Then for all $m \geq 1$, (1.23) can be rewritten

$$\begin{aligned} &3^m \log_6 \left(\frac{4}{3} \right) + (m+1) \log_6 3 - 2 \\ &< Z_m < 3^m \log_6 \left(\frac{4}{3} \right) + (m+1) \log_6 3. \end{aligned} \tag{1.24}$$

Now let F_m be the fraction of zeroes in the base-6 expansion of α up to position $3^m + Z_m$ (i.e., up to the end of the block of zeroes that immediately follows position 3^m). Clearly

$$F_m > \frac{\sum_{k=1}^{m} Z_k}{3^m + Z_m}, \tag{1.25}$$

since the numerator only counts zeroes in the long stretches. The summation in the numerator satisfies, for all sufficiently large m,

$$\sum_{k=1}^{m} Z_k > \frac{3}{2}\left(3^m - \frac{1}{3}\right)\log_6\left(\frac{4}{3}\right) + \frac{m(m+3)}{2}\log_6 3 - 2m$$
$$> \frac{3}{2}\cdot 3^m \log_6\left(\frac{4}{3}\right) - \frac{1}{2}\log_6\left(\frac{4}{3}\right) - 2m. \quad (1.26)$$

Now given any $\varepsilon > 0$, we can write, for all sufficiently large m,

$$F_m > \frac{\frac{3}{2}\cdot 3^m \log_6\left(\frac{4}{3}\right) - \frac{1}{2}\log_6\left(\frac{4}{3}\right) - 2m}{3^m + 3^m\log_6\left(\frac{4}{3}\right) + (m+1)\log_6 3}$$
$$= \frac{\frac{3}{2}\log_6\left(\frac{4}{3}\right) - \frac{1}{3^m}\left(\frac{1}{2}\log_6\left(\frac{4}{3}\right) + 2m\right)}{1 + \log_6\left(\frac{4}{3}\right) + \frac{(m+1)\log_6 3}{3^m}}$$
$$\geq \frac{\frac{3}{2}\log_6\left(\frac{4}{3}\right) - \varepsilon}{1 + \log_6\left(\frac{4}{3}\right) + \varepsilon} \geq \frac{1}{2}\log_2\left(\frac{4}{3}\right) - 2\varepsilon. \quad (1.27)$$

But $\beta = \frac{1}{2}\log_2(4/3)$ (which has numerical value $0.2075187496\ldots$) is clearly greater than $1/6$, since $(4/3)^3 = 64/27 > 2$. This means that infinitely often (namely, whenever $n = 3^m + Z_m$) the fraction of zeroes in the base-6 expansion of α up to position n exceeds $\frac{1}{2}(1/6 + \beta) > 1/6$. Thus α is not 6-normal. ■

Proof. Given co-prime integers $b \geq 2$ and $c \geq 2$, the constant $\alpha_{b,c} = \sum_{k\geq 1} 1/(c^k b^{c^k})$ is not bc-normal.

Let $Q_m(b,c)$ be the base-bc expansion of $\alpha_{b,c}$ immediately following position c^m. Then

$$Q_m(b,c) = (bc)^{c^m}\alpha_{b,c} \bmod 1$$
$$= \left(\sum_{k=1}^{m} c^{c^m - k} b^{c^m - c^k}\right) \bmod 1 + \sum_{k=m+1}^{\infty} c^{c^m - k} b^{c^m - c^k}. \quad (1.28)$$

As above, the first portion of this expression is zero, since all terms in the summation are integers, and the second portion is very accurately approximated by the first term of the series, namely $[\frac{c}{b(c-1)}]^{c^m}/c^{m+1}$. In fact, for any choice of b and c as above, and for all $m \geq 1$,

$$\frac{1}{c^{m+1}}\left[\frac{c}{b(c-1)}\right]^{c^m} < Q_m(b,c) < \frac{1}{c^{m+1}}\left[\frac{c}{b(c-1)}\right]^{c^m}\cdot(1 + 1/10). \quad (1.29)$$

Let $Z_m(b,c) = \lfloor \log_{bc} 1/Q_m(b,c) \rfloor$ be the number of zeroes that immediately follow position c^m. Then for all $m \geq 1$, (1.29) can be rewritten as

$$c^m \log_{bc}\left[\frac{b(c-1)}{c}\right] + (m+1)\log_{bc} c - 2$$

$$< Z_m(b,c) \; < \; c^m \log_{bc}\left[\frac{b(c-1)}{c}\right] + (m+1)\log_{bc} c. \tag{1.30}$$

Now let $F_m(b,c)$ be the fraction of zeroes up to position $c^m + Z_m(b,c)$. Clearly

$$F_m(b,c) > \frac{\sum_{k=1}^m Z_k(b,c)}{c^m + Z_m(b,c)}, \tag{1.31}$$

since the numerator only counts zeroes in the long stretches. The summation in the numerator of $F_m(b,c)$ satisfies

$$\sum_{k=1}^m Z_k(b,c) > \frac{c}{c-1}\left(c^m - \frac{1}{c}\right)\log_{bc}\left[\frac{b(c-1)}{c}\right] + \frac{m(m+3)}{2}\log_{bc} c - 2m$$

$$> \frac{c^{m+1}}{c-1}\log_{bc}\left[\frac{b(c-1)}{c}\right] - \frac{1}{c-1}\log_{bc}\left[\frac{b(c-1)}{c}\right] - 2m. \tag{1.32}$$

Thus given any $\varepsilon > 0$, we can write, for all sufficiently large m,

$$F_m(b,c) > \frac{\frac{c^{m+1}}{c-1}\log_{bc}\left[\frac{b(c-1)}{c}\right] - \frac{1}{c-1}\log_{bc}\left[\frac{b(c-1)}{c}\right] - 2m}{c^m + c^m\log_{bc}\left(\frac{b(c-1)}{c}\right) + (m+1)\log_{bc} c} \tag{1.33}$$

$$= \frac{\frac{c}{c-1}\log_{bc}\left[\frac{b(c-1)}{c}\right] - \frac{1}{c^m}\left(\frac{1}{c-1}\log_{bc}\left[\frac{b(c-1)}{c}\right] + 2m\right)}{1 + \log_{bc}\left[\frac{b(c-1)}{c}\right] + \frac{(m+1)\log_{bc} c}{c^m}}$$

$$\geq \frac{\frac{c}{c-1}\log_{bc}\left[\frac{b(c-1)}{c}\right] - \varepsilon}{1 + \log_{bc}\left[\frac{b(c-1)}{c}\right] + \varepsilon}$$

$$\geq \frac{c}{c-1} \cdot \frac{\log_{bc}\left[\frac{b(c-1)}{c}\right]}{1 + \log_{bc}\left[\frac{b(c-1)}{c}\right]} - 2\varepsilon$$

$$= T(b,c) - 2\varepsilon, \tag{1.34}$$

where

$$T(b,c) = \frac{c}{c-1} \cdot \frac{\log_{bc}\left[\frac{b(c-1)}{c}\right]}{1 + \log_{bc}\left[\frac{b(c-1)}{c}\right]}. \tag{1.35}$$

To establish the desired result that $T(b,c) > 1/(bc)$, first note that

$$T(b,c) > \frac{1}{2}\log_{bc}\left[\frac{b(c-1)}{c}\right] \geq \frac{1}{2}\log_{bc}\left(\frac{b}{2}\right). \tag{1.36}$$

Raise bc to the power of the right-hand side and also to the power $1/(bc)$. Then it suffices to demonstrate that

$$\frac{b}{2} > \left[(bc)^{1/(bc)}\right]^2. \tag{1.37}$$

The right-hand side is bounded above by $(e^{1/e})^2 = 2.0870652286\ldots$. Thus this inequality is clearly satisfied whenever $b \geq 5$.

If we also presume that $c \geq 5$, then by examining the middle of (1.36), it suffices to demonstrate that

$$\frac{1}{2}\log_{bc}\frac{4b}{5} > \frac{1}{bc} \tag{1.38}$$

or

$$\frac{4b}{5} > \left(e^{1/e}\right)^2. \tag{1.39}$$

But this is clearly satisfied whenever $b \geq 3$. For the case $b = 2$ and $c \geq 5$, we can write

$$T(b,c) = \frac{c}{c-1}\cdot\frac{\log_{2c}\left[\frac{2(c-1)}{c}\right]}{1+\log_{2c}\left[\frac{2(c-1)}{c}\right]} \geq \frac{\log_{2c}\left[\frac{2(c-1)}{c}\right]}{1+\log_{10}2}, \tag{1.40}$$

so by similar reasoning it suffices to demonstrate that

$$\frac{2(c-1)}{c} > \left(e^{1/e}\right)^{1+\log_{10}2} = 1.61384928833\ldots. \tag{1.41}$$

But this is clearly satisfied whenever $c \geq 6$.

The five remaining cases, namely $(2,3),(2,5),(3,2),(3,4),(4,3)$, are easily verified by explicitly computing numerical values of $T(b,c)$ using (1.35). As it turns out, the simple case that we worked out in detail above, namely $b = 2$ and $c = 3$, is the worst case, in the sense that for all other (b,c), the fraction $T(b,c)$ exceeds the natural frequency $1/(bc)$ by greater margins. ∎

References

1. Bailey, D.H., Crandall, R.E.: Random generators and normal numbers. Exp. Math. **11**(4), 527–546 (2002)
2. Bailey, D.H., Misiurewicz, M.: A strong hot spot theorem. Proc. Am. Math. Soc. **134**(9), 2495–2501 (2006)
3. Bailey, D.H., Borwein, P.B., Plouffe, S.: On the rapid computation of various polylogarithmic constants. Math. Comput. **66**(218), 903–913 (1997)
4. Bailey, D.H., Borwein, J.M., Crandall, R.E., Pomerance, C.: On the binary expansions of algebraic numbers. J. Number Theor. Bordeaux **16**, 487–518 (2004)
5. Borwein, J., Bailey, D.H.: Mathematics by Experiment: Plausible Reasoning in the 21st Century. AK Peters, Natick (2008)
6. Cassels, J.W.S.: On a problem of Steinhaus about normal numbers. Colloq. Math. **7**, 95–101 (1959)
7. Champernowne, D.G.: The construction of decimals normal in the scale of ten. J. London Math. Soc. **8**, 254–260 (1933)
8. Copeland, A.H., Erdös, P.: Note on normal numbers. Bull. Am. Math. Soc. **52**, 857–860 (1946)
9. Hida, Y., Li, X.S., Bailey, D.H.: Algorithms for quad-double precision floating point arithmetic. 15th IEEE Symposium on Computer Arithmetic. IEEE Computer Society, pp. 155–162. California University, Berkeley (2001)
10. Knuth, D.E.: The Art of Computer Programming, vol. 2. Addison-Wesley, Reading (1998)
11. L'Ecuyer, P.: Random number generation. In: Gentle, J.E., Haerdle, W., Mori, Y. (eds.) Handbook of Computational Statistics, Chap. II.2. Springer, Berlin (2004)
12. L'Ecuyer, P., Simard, R.: TestU01: A C Library for empirical testing of random number generators. ACM Trans. Math. Software **33**, 15 (2007)
13. Pollington, A.D.: The Hausdorff dimension of a set of normal numbers. Pacific J. Math. **95**, 193–204 (1981)
14. Queffelec, M.: Old and new results on normality. Lecture Notes – Monograph Series, Dynamics and Stochastics, vol. 48, pp. 225–236. Institute of Mathematical Statistics, Beachwood (2006)
15. Schmidt, W.: On normal numbers. Pacific J. Math. **10**, 661–672 (1960)
16. Stoneham, R.: On absolute (j, ε)-normality in the rational fractions with applications to normal numbers. Acta Arithmetica **22**, 277–286 (1973)
17. Weyl, H.: Über die Gleichverteilung von Zahlen mod. Eins. Math. Ann. **77**, 313–352 (1916)

Chapter 2
New Demiclosedness Principles for (Firmly) Nonexpansive Operators

Heinz H. Bauschke

Dedicated to Jonathan Borwein on the occasion of his 60th birthday

Abstract The demiclosedness principle is one of the key tools in nonlinear analysis and fixed point theory. In this note, this principle is extended and made more flexible by two mutually orthogonal affine subspaces. Versions for finitely many (firmly) nonexpansive operators are presented. As an application, a simple proof of the weak convergence of the Douglas-Rachford splitting algorithm is provided.

Key words: Demiclosedness principle • Douglas-Rachford algorithm • Firmly nonexpansive mapping • Maximal monotone operator • Nonexpansive mapping • Proximal algorithm • Resolvent • Splitting algorithm

Mathematics Subject Classifications (2010): Primary 47H05, 47H09; Secondary 47J25, 49M27, 65J15, 65K05, 65K15, 90C25

2.1 Introduction

Throughout this paper, we assume that

$$X \text{ is a real Hilbert space with inner product } \langle \cdot,\cdot \rangle \text{ and induced norm } \|\cdot\|. \tag{2.1}$$

COMMUNICATED BY MICHEL THÉRA.

H.H. Bauschke (✉)
Department of Mathematics, University of British Columbia, Kelowna, BC V1V 1V7, Canada
e-mail: heinz.bauschke@ubc.ca

D.H. Bailey et al. (eds.), *Computational and Analytical Mathematics*, Springer Proceedings in Mathematics & Statistics 50, DOI 10.1007/978-1-4614-7621-4_2,

We shall assume basic notation and results from fixed point theory and from monotone operator theory; see, e.g., [2, 4, 8, 15, 16, 20–22, 24]. The *graph* of a maximally monotone operator $A\colon X \rightrightarrows X$ is denoted by $\operatorname{gra} A$, its *resolvent* $(A+\mathrm{Id})^{-1}$ by J_A, its set of zeros by $\operatorname{zer} A = A^{-1}(0)$, and we set $R_A = 2J_A - \mathrm{Id}$, where Id is the identity operator. Weak convergence is indicated by $\rightharpoonup$.

Let $T: X \to X$. Recall that T is *firmly nonexpansive* if

$$(\forall x \in X)(\forall y \in X) \quad \|Tx - Ty\|^2 + \|(\mathrm{Id}-T)x - (\mathrm{Id}-T)y\|^2 \leq \|x-y\|^2. \tag{2.2}$$

It is well know that T is firmly nonexpansive if and only if $R = 2T - \mathrm{Id}$ is *nonexpansive*, i.e.,

$$(\forall x \in X)(\forall y \in X) \quad \|Rx - Ry\| \leq \|x-y\|. \tag{2.3}$$

Clearly, every firmly nonexpansive operator is nonexpansive. Building on work by Minty [19], Eckstein and Bertsekas [13] clearly linked firmly nonexpansive mappings to maximally monotone operators—the key result is the following: T is firmly nonexpansive if and only if $T = J_A$ for some maximally monotone operator A (namely, $T^{-1} - \mathrm{Id}$). This implies also a correspondence between maximally monotone operators and nonexpansive mappings (see [14, 17]). Thus, finding a zero of A is equivalent to finding a fixed point of J_A. Furthermore, the graph of any maximally monotone operator is beautifully described by the associated *Minty parametrization*:

$$\operatorname{gra} A = \big\{(J_A x, x - J_A x) \mid x \in X\big\}. \tag{2.4}$$

The most prominent example of firmly nonexpansive mappings are projectors, i.e., resolvents of normal cone operators associated with nonempty closed convex subsets of X. Despite being (firmly) nonexpansive and hence Lipschitz continuous, even projectors do not interact well with the weak topology as was first observed by Zarantonello [25]:

Example 2.1. Suppose that $X = \ell_2(\mathbb{N})$, set $C = \big\{x \in X \mid \|x\| \leq 1\big\}$, and denote the sequence of standard unit vectors in X by $(e_n)_{n\in\mathbb{N}}$. Set $(\forall n \in \mathbb{N})\ z_n = e_0 + e_n$. Then

$$z_n \rightharpoonup e_0 \quad \text{yet} \quad P_C z_n \rightharpoonup \tfrac{1}{\sqrt{2}} e_0 \neq e_0 = P_C e_0. \tag{2.5}$$

The following classical demiclosedness principle dates back to the 1960s and work by Browder [6]. It comes somewhat as a surprise in view of the previous example.

Fact 2.2 (Demiclosedness principle). Let S be a nonempty closed convex subset of X, let $T\colon S \to X$ be nonexpansive, let $(z_n)_{n\in\mathbb{N}}$ be a sequence in S converging weakly to z, and suppose that $z_n - Tz_n \to x$. Then $z - Tz = x$.

Remark 2.3. One might inquire whether or not the following even less restrictive demiclosedness principle holds:

$$\left.\begin{array}{r} z_n \rightharpoonup z \\ z_n - Tz_n \rightharpoonup x \end{array}\right\} \overset{?}{\Rightarrow} z - Tz = x. \tag{2.6}$$

However, this generalization is false: indeed, suppose that X, C, and $(z_n)_{n\in\mathbb{N}}$ are as in Example 2.1, and set $T = \mathrm{Id} - P_C$, which is (even firmly) nonexpansive. Then $z_n \rightharpoonup e_0$ and $z_n - Tz_n = P_C z_n \rightharpoonup \frac{1}{\sqrt{2}} e_0$ yet $e_0 - Te_0 = P_C e_0 = e_0 \neq \frac{1}{\sqrt{2}} e_0$.

The aim of this note is to provide new versions of the demiclosedness principle and illustrate their usefulness. The remainder of this paper is organized as follows. Section 2.2 presents new demiclosedness principles for one (firmly) nonexpansive operator. Multi-operator versions are provided in Sect. 2.3. The weak convergence of the Douglas-Rachford algorithm is rederived with a very transparent proof in Sect. 2.4.

2.2 Demiclosedness Principles

Fact 2.4 (Brezis). (See [5, Proposition 2.5 on P. 27], [23, Lemma 4], or [2, Corollary 20.49].) Let $A\colon X \rightrightarrows X$ be maximally monotone, let $(x,u) \in \operatorname{gra} A$, and let $(x_n,u_n)_{n\in\mathbb{N}}$ be a sequence in $X \times X$ such that $(x_n,u_n) \rightharpoonup (x,u)$ and $\overline{\lim}\langle x_n,u_n\rangle \leq \langle x,u\rangle$. Then $\langle x_n,u_n\rangle \to \langle x,u\rangle$ and $(x,u) \in \operatorname{gra} A$.

Theorem 2.5 (See also [2, Proposition 20.50]). *Let $A\colon X \rightrightarrows X$ be maximally monotone, let $(x,u) \in X \times X$, and let C and D be closed affine subspaces of X such that $D - D = (C - C)^\perp$. Furthermore, let $(x_n,u_n)_{n\in\mathbb{N}}$ be a sequence in* $\operatorname{gra} A$ *such that*

$$(x_n,u_n) \rightharpoonup (x,u) \quad \text{and} \quad (x_n,u_n) - P_{C\times D}(x_n,u_n) \to (0,0). \tag{2.7}$$

Then $(x,u) \in (C \times D) \cap \operatorname{gra} A$ and $\langle x_n,u_n\rangle \to \langle x,u\rangle$.

Proof. Set $V = C - C$, which is a closed linear subspace. Since $x_n - P_C x_n \to 0$, we have $P_C x_n \rightharpoonup x$ and thus $x \in C$. Likewise, $u \in D$ and hence

$$C = x + V \quad \text{and} \quad D = u + V^\perp. \tag{2.8}$$

It follows that

$$P_C\colon z \mapsto P_V z + P_{V^\perp} x \quad \text{and} \quad P_D\colon z \mapsto P_{V^\perp} z + P_V u. \tag{2.9}$$

Therefore, since P_V and $P_{V^\perp}$ are weakly continuous,

$$\langle x_n, u_n\rangle = \langle P_V x_n + P_{V^\perp} x_n, P_V u_n + P_{V^\perp} u_n\rangle \tag{2.10a}$$
$$= \langle P_V x_n, P_V u_n\rangle + \langle P_{V^\perp} x_n, P_{V^\perp} u_n\rangle \tag{2.10b}$$
$$= \langle P_V x_n, u_n - P_{V^\perp} u_n\rangle + \langle x_n - P_V x_n, P_{V^\perp} u_n\rangle \tag{2.10c}$$
$$= \langle P_V x_n, u_n - (P_D u_n - P_V u)\rangle \tag{2.10d}$$
$$+ \langle x_n - (P_C x_n - P_{V^\perp} x), P_{V^\perp} u_n\rangle \tag{2.10e}$$
$$= \langle P_V x_n, u_n - P_D u_n\rangle + \langle P_V x_n, P_V u\rangle \tag{2.10f}$$
$$+ \langle x_n - P_C x_n, P_{V^\perp} u_n\rangle + \langle P_{V^\perp} x, P_{V^\perp} u_n\rangle \tag{2.10g}$$
$$\to \langle P_V x, P_V u\rangle + \langle P_{V^\perp} x, P_{V^\perp} u\rangle \tag{2.10h}$$
$$= \langle x, u\rangle. \tag{2.10i}$$

The result now follows from Fact 2.4. ■

Remark 2.6. Theorem 2.5 generalizes [1, Theorem 2], which corresponds to the case when C is a closed linear subspace and $D = C^\perp$. A referee pointed out that Theorem 2.5 may be obtained from [1, Theorem 2] by a translation argument. However, the above proof of Theorem 2.5 is different and *much simpler* than the proof of [1, Theorem 2].

Corollary 2.7 (Firm Nonexpansiveness Principle). *Let $F: X \to X$ be firmly nonexpansive, let $(z_n)_{n\in\mathbb{N}}$ be a sequence in X such that $(z_n)_{n\in\mathbb{N}}$ converges weakly to $z \in X$, and suppose that $Fz_n \rightharpoonup x \in X$ and that C and D are closed affine subspaces of X such that $D - D = (C - C)^\perp$, $Fz_n - P_C F z_n \to 0$, and $(z_n - Fz_n) - P_D(z_n - Fz_n) \to 0$. Then $x \in C$, $z \in x + D$, and $x = Fz$.*

Proof. Set $A = F^{-1} - \mathrm{Id}$ so that $J_A = F$. By (2.4), A is maximally monotone and

$$(x_n, u_n)_{n\in\mathbb{N}} := (Fz_n, z_n - Fz_n)_{n\in\mathbb{N}} \tag{2.11}$$

is a sequence in gra$A$ that converges weakly to $(x, z - x)$. Thus, by Theorem 2.5, $x \in C$, $z - x \in D$, and $z - x \in Ax$. Therefore, $z \in x + Ax$, i.e., $x = J_A z = Fz$. ■

Corollary 2.8 (Nonexpansiveness Principle). *Let $T: X \to X$ be nonexpansive, let $(z_n)_{n\in\mathbb{N}}$ be a sequence in X such that $z_n \rightharpoonup z$, and suppose that $Tz_n \rightharpoonup y$ and that C and D are closed affine subspaces of X such that $D - D = (C - C)^\perp$, $z_n + Tz_n - P_C z_n - P_C T z_n \to 0$, and $z_n - Tz_n - P_D z_n - P_D(-Tz_n) \to 0$. Then $\frac{1}{2}z + \frac{1}{2}y \in C$, $\frac{1}{2}z - \frac{1}{2}y \in D$, and $y = Tz$.*

Proof. Set $F = \frac{1}{2}\mathrm{Id} + \frac{1}{2}T$, which is firmly nonexpansive. Then $Fz_n \rightharpoonup \frac{1}{2}z + \frac{1}{2}y =: x$. Since P_C is affine, we get

$$z_n + Tz_n - P_C z_n - P_C T z_n \to 0 \tag{2.12a}$$
$$\Leftrightarrow z_n + Tz_n - 2\big(\tfrac{1}{2}P_C z_n + \tfrac{1}{2}P_C T z_n\big) \to 0 \tag{2.12b}$$

$$\Leftrightarrow z_n + Tz_n - 2P_C\left(\tfrac{1}{2}z_n + \tfrac{1}{2}Tz_n\right) \to 0 \tag{2.12c}$$

$$\Leftrightarrow 2Fz_n - 2P_C Fz_n \to 0 \tag{2.12d}$$

$$\Leftrightarrow Fz_n - P_C Fz_n \to 0. \tag{2.12e}$$

Likewise, since $z_n - Fz_n = z_n - \frac{1}{2}z_n - \frac{1}{2}Tz_n = \frac{1}{2}z_n - \frac{1}{2}Tz_n$, we have

$$z_n - Tz_n - P_D z_n - P_D(-Tz_n) \to 0 \tag{2.13a}$$

$$\Leftrightarrow z_n - Tz_n - 2\left(\tfrac{1}{2}P_D z_n + \tfrac{1}{2}P_D(-Tz_n)\right) \to 0 \tag{2.13b}$$

$$\Leftrightarrow 2(z_n - Fz_n) - 2P_D\left(\tfrac{1}{2}z_n + \tfrac{1}{2}(-Tz_n)\right) \to 0 \tag{2.13c}$$

$$\Leftrightarrow z_n - Fz_n - P_D(z_n - Fz_n) \to 0. \tag{2.13d}$$

Thus, by Corollary 2.7, $x \in C$, $z \in x + D$, and $x = Fz$, i.e., $\frac{1}{2}z + \frac{1}{2}y \in C$, $z \in \frac{1}{2}z + \frac{1}{2}y + D$, and $\frac{1}{2}z + \frac{1}{2}y = Fz = \frac{1}{2}z + \frac{1}{2}Tz$, i.e., $\frac{1}{2}z + \frac{1}{2}y \in C$, $\frac{1}{2}z - \frac{1}{2}y \in D$, and $y = Tz$. ■

Corollary 2.9 (Classical Demiclosedness Principle). *Let S be a nonempty closed convex subset of X, let $T\colon S \to X$ be nonexpansive, let $(z_n)_{n\in\mathbb{N}}$ be a sequence in S converging weakly to z, and suppose that $z_n - Tz_n \to x$. Then $z - Tz = x$.*

Proof. We may and do assume that $S = X$ (otherwise, consider $T \circ P_S$ instead of T). Set $y = z - x$ and note that $Tz_n \rightharpoonup y$. Now set $C = X$ and $D = \{x/2\}$. Then $D - D = \{0\} = X^\perp = (X - X)^\perp = (D - D)^\perp$, $z_n + Tz_n - P_C z_n - P_C Tz_n \equiv 0$, and $z_n - Tz_n - P_D z_n - P_D(-Tz_n) = z_n - Tz_n - x/2 - x/2 \to 0$. Corollary 2.8 implies $y = Tz$, i.e., $z - x = Tz$. ■

2.3 Multi-operator Demiclosedness Principles

Set

$$I = \{1, 2, \ldots, m\}, \quad \text{where } m \text{ is an integer greater than or equal to 2.} \tag{2.14}$$

We shall work in the product Hilbert space

$$\mathbf{X} = X^I \tag{2.15}$$

with induced inner product $\langle \mathbf{x}, \mathbf{y}\rangle = \sum_{i\in I}\langle x_i, y_i\rangle$ and $\|\mathbf{x}\| = \sqrt{\sum_{i\in I}\|x_i\|^2}$, where $\mathbf{x} = (x_i)_{i\in I}$ and $\mathbf{y} = (y_i)_{i\in I}$ denote generic elements in $\mathbf{X}$.

We start with a multi-operator demiclosedness principle for firmly nonexpansive mappings, which we derive from the corresponding two-operator version

(Corollary 2.7). A referee pointed out that Theorem 2.10 is also equivalent to [1, Corollary 3] (see also [23, Lemma 5] for a Banach space extension of [1, Corollary 3]).

Theorem 2.10 (Multi-operator Demiclosedness Principle for Firmly Nonexpansive Operators). *Let $(F_i)_{i\in I}$ be a family of firmly nonexpansive operators on X, and let, for each $i \in I$, $(z_{i,n})_{n\in\mathbb{N}}$ be a sequence in X such that for all i and j in I,*

$$z_{i,n} \rightharpoonup z_i \text{ and } F_i z_{i,n} \rightharpoonup x, \tag{2.16a}$$

$$\sum_{i\in I}(z_{i,n} - F_i z_{i,n}) \to -mx + \sum_{i\in I} z_i, \tag{2.16b}$$

$$F_i z_{i,n} - F_j z_{j,n} \to 0. \tag{2.16c}$$

Then $F_i z_i = x$, for every $i \in I$.

Proof. Set $\mathbf{x} = (x)_{i\in I}$, $\mathbf{z} = (z_i)_{i\in I}$, $(\mathbf{z}_n) = (z_{i,n})_{n\in\mathbb{N}}$, and $\mathbf{C} = \{(y)_{i\in I} \mid y \in X\}$. Then $\mathbf{z}_n \rightharpoonup \mathbf{z}$ and $\mathbf{C}$ is a closed subspace of $\mathbf{X}$ with $\mathbf{C}^\perp = \{(y_i)_{i\in I} \mid \sum_{i\in I} y_i = 0\}$. Furthermore, we set $\mathbf{D} = \mathbf{z} - \mathbf{x} + \mathbf{C}^\perp$ so that $(\mathbf{C}-\mathbf{C})^\perp = \mathbf{C}^\perp = \mathbf{D} - \mathbf{D}$ and also $\mathbf{F}\colon (y_i)_{i\in I} \mapsto (Fy_i)_{i\in I}$. Then $\mathbf{F}$ is firmly nonexpansive on $\mathbf{X}$, and $\mathbf{F}\mathbf{z}_n \rightharpoonup \mathbf{x}$. Now (2.16c) implies

$$(\forall i \in I) \quad F_i z_{i,n} - \frac{1}{m}\sum_{j\in I} F_j z_{j,n} \to 0, \tag{2.17}$$

which—when viewed in $\mathbf{X}$—means that $\mathbf{F}\mathbf{z}_n - P_{\mathbf{C}}\mathbf{F}\mathbf{z}_n \to 0$. Similarly, using (2.16b),

$$\begin{aligned}
\mathbf{z}_n - \mathbf{F}\mathbf{z}_n - P_{\mathbf{D}}(\mathbf{z}_n - \mathbf{F}\mathbf{z}_n) &= \mathbf{z}_n - \mathbf{F}\mathbf{z}_n - P_{\mathbf{z}-\mathbf{x}+\mathbf{C}^\perp}(\mathbf{z}_n - \mathbf{F}\mathbf{z}_n) && (2.18a)\\
&= \mathbf{z}_n - \mathbf{F}\mathbf{z}_n - \big(\mathbf{z} - \mathbf{x} + P_{\mathbf{C}^\perp}\big(\mathbf{z}_n - \mathbf{F}\mathbf{z}_n - (\mathbf{z}-\mathbf{x})\big)\big) && (2.18b)\\
&= \big(\mathrm{Id} - P_{\mathbf{C}^\perp}\big)(\mathbf{z}_n - \mathbf{F}\mathbf{z}_n) - \big(\mathrm{Id} - P_{\mathbf{C}^\perp}\big)(\mathbf{z}-\mathbf{x}) && (2.18c)\\
&= P_{\mathbf{C}}(\mathbf{z}_n - \mathbf{F}\mathbf{z}_n) - P_{\mathbf{C}}(\mathbf{z}-\mathbf{x}) && (2.18d)\\
&= \Big(\tfrac{1}{m}\sum_{i\in I}\big(z_{i,n} - F_i z_{i,n} - z_i + x\big)\Big)_{j\in I} && (2.18e)\\
&\to 0. && (2.18f)
\end{aligned}$$

Therefore, by Corollary 2.7, $\mathbf{x} = \mathbf{F}\mathbf{z}$. ■

Theorem 2.11 (Multi-operator Demiclosedness Principle for Nonexpansive Operators). *Let $(T_i)_{i\in I}$ be a family of nonexpansive operators on X, and let, for each $i \in I$, $(x_{i,n})_{n\in\mathbb{N}}$ be a sequence in X such that for all i and j in I,*

$$z_{i,n} \rightharpoonup z_i \text{ and } T_i z_{i,n} \rightharpoonup y_i, \tag{2.19a}$$

$$\sum_{i\in I}\big(z_{i,n}-T_iz_{i,n}\big)\to\sum_{i\in I}\big(z_i-y_i\big) \tag{2.19b}$$

$$z_{i,n}-z_{j,n}+T_iz_{i,n}-T_jz_{j,n}\to 0. \tag{2.19c}$$

Then $T_iz_i=y_i$, *for each* $i\in I$.

Proof. Set $(\forall i\in I)$ $F_i=\frac{1}{2}\operatorname{Id}+\frac{1}{2}T_i$. Then F_i is firmly nonexpansive and $F_iz_{i,n}\rightharpoonup\frac{1}{2}z_i+\frac{1}{2}y_i$, for every $i\in I$. By (2.19c), $0\leftarrow 2F_iz_{i,n}-2F_jz_{j,n}=(z_{i,n}+T_iz_{i,n})-(z_{j,n}+T_jz_{j,n})\rightharpoonup(z_i+y_i)-(z_j+y_j)$, for all i and j in I. It follows that $x=\frac{1}{2}z_i+\frac{1}{2}y_i$ is *independent* of $i\in I$. Furthermore,

$$\sum_{i\in I}\big(z_{i,n}-F_iz_{i,n}\big)=\sum_{i\in I}\tfrac{1}{2}\big(z_{i,n}-T_iz_{i,n}\big) \tag{2.20a}$$

$$\to\sum_{i\in I}\tfrac{1}{2}\big(z_i-y_i\big) \tag{2.20b}$$

$$=\sum_{i\in I}\big(\tfrac{1}{2}z_i-\big(x-\tfrac{1}{2}z_i\big)\big) \tag{2.20c}$$

$$=-mx+\sum_{i\in I}z_i. \tag{2.20d}$$

Therefore, the conclusion follows from Theorem 2.10. ■

2.4 Application to Douglas-Rachford Splitting

In this section, we assume that A and B are maximally monotone operators on X such that

$$\operatorname{zer}(A+B)=(A+B)^{-1}(0)\neq\varnothing. \tag{2.21}$$

We set

$$T=\tfrac{1}{2}\operatorname{Id}+\tfrac{1}{2}R_BR_A=J_B(2J_A-\operatorname{Id})+(\operatorname{Id}-J_A), \tag{2.22}$$

which is the Douglas-Rachford splitting operator and where $R_A=2J_A-\operatorname{Id}$ and $R_B=2J_B-\operatorname{Id}$ are the "reflected resolvents" already considered in Sect. 2.1. (The term "reflected resolvent" is motivated by the fact that when J_A is a projection operator, then R_A is the corresponding reflection.) See [2, 10, 11] for further information on this algorithm and also [3] for some results for operators that are not maximally monotone. One has (see [10, Lemma 2.6(iii)] or [2, Proposition 25.1(ii)])

$$J_A\big(\operatorname{Fix}T\big)=\operatorname{zer}(A+B). \tag{2.23}$$

$$= z_{k_n} - J_A z_{k_n} + R_A z_{k_n} - J_B y_{k_n} \tag{2.30b}$$

$$= z_{k_n} - J_A z_{k_n} + y_{k_n} - J_B y_{k_n} \tag{2.30c}$$

$$\rightharpoonup z + y - 2x. \tag{2.30d}$$

To summarize,

$$(z_{k_n}, y_{k_n}) \rightharpoonup (z, y) \quad \text{and} \quad (J_A z_{k_n}, J_B y_{k_n}) \rightharpoonup (x, x), \tag{2.31a}$$

$$(z_{k_n} - J_A z_{k_n}) + (y_{k_n} - J_B y_{k_n}) \to -2x + z + y = 0, \tag{2.31b}$$

$$J_A z_{k_n} - J_B y_{k_n} \to 0. \tag{2.31c}$$

By Theorem 2.10, $J_A z = J_B y = x$. Hence $J_A z_{k_n} \rightharpoonup J_A z$. Since x was an arbitrary weak cluster point of the bounded sequence $(J_A z_n)_{n \in \mathbb{N}}$, we conclude that $J_A z_n \rightharpoonup J_A z$. ■

Motivated by a referee's comment, let us turn towards inexact iterations of T. The following result underlines the usefulness of the multi-operator demiclosedness principle.

Theorem 2.14. *Suppose that $(z_n)_{n \in \mathbb{N}}$ is a sequence in X such that $z_n - Tz_n \to 0$ and $z_n \rightharpoonup z$, where $z \in \operatorname{Fix} T$. Then $J_A z_n \rightharpoonup J_A z$.*

Proof. Argue exactly as in the proof of Fact 2.13. ■

We now present a prototypical result on inexact iterations; see [9–11, 13, 23] for many more results in this direction as well as [2] and also [12].

Corollary 2.15. *Suppose that $(z_n)_{n \in \mathbb{N}}$ and $(e_n)_{n \in \mathbb{N}}$ are sequences in X such that*

$$\sum_{n \in \mathbb{N}} \|e_n\| < +\infty \quad \text{and} \quad (\forall n \in \mathbb{N}) \quad z_{n+1} = e_n + Tz_n. \tag{2.32}$$

Then there exists $z \in \operatorname{Fix} T$ such that $z_n \rightharpoonup z$ and $J_A z_n \rightharpoonup J_A z$.

Proof. Combettes' [9, Proposition 4.2(ii)] yields $z_n - Tz_n \to 0$ while the existence of $z \in \operatorname{Fix} T$ such that $z_n \rightharpoonup z$ is guaranteed by his [9, Theorem 5.2(i)]. Now apply Theorem 2.14. ■

Unfortunately, the author is unaware of any existing actual numerical implementation guaranteeing summable errors; however, these theoretical results certainly increase confidence in the numerical stability of the Douglas-Rachford algorithm.

Acknowledgements HHB thanks Patrick Combettes and Jonathan Eckstein for their pertinent comments. HHB was partially supported by the Natural Sciences and Engineering Research Council of Canada and by the Canada Research Chair Program.

References

1. Bauschke, H.H.: A note on the paper by Eckstein and Svaiter on general projective splitting methods for sums of maximal monotone operators. SIAM J. Control Optim. **48**, 2513–2515 (2009)
2. Bauschke, H.H., Combettes, P.L.: Convex Analysis and Monotone Operator Theory in Hilbert Spaces. Springer, New York (2011)
3. Borwein, J.M., Sims, B.: The Douglas-Rachford algorithm in the absence of convexity. In: Bauschke, H.H., Burachik, R.S., Combettes, P.L., Elser, V., Luke, D.R., Wolkowicz, H. (eds.) Fixed-Point Algorithms for Inverse Problems in Science and Engineering, Springer Optimization and Its Applications, vol. 49, pp. 93–109. Springer, New York (2011)
4. Borwein, J.M., Vanderwerff, J.D.: Convex Functions. Cambridge University Press, Cambridge (2010)
5. Brézis, H.: Opérateurs Maximaux Monotones et Semi-Groupes de Contractions dans les Espaces de Hilbert. Elsevier, New York (1973)
6. Browder, F.E.: Semicontractive and semiaccretive nonlinear mappings in Banach spaces. Bull. Am. Math. Soc. **74**, 660–665 (1968)
7. Bruck, R.E., Reich, S.: Nonexpansive projections and resolvents of accretive operators in Banach spaces. Houston J. Math. **3**, 459–470 (1977)
8. Burachik, R.S., Iusem, A.N.: Set-Valued Mappings and Enlargements of Monotone Operators. Springer, New York (2008)
9. Combettes, P.L.: Quasi-Fejérian analysis of some optimization algorithms. In: Butnariu, D., Censor, Y., Reich, S. (eds.) Inherently Parallel Algorithms in Feasibility and Optimization and Their Applications, pp. 115–152. Elsevier, New York (2001)
10. Combettes, P.L.: Solving monotone inclusions via compositions of nonexpansive averaged operators. Optimization **53**, 475–504 (2004)
11. Combettes, P.L.: Iterative construction of the resolvent of a sum of maximal monotone operators. J. Convex Anal. **16**, 727–748 (2009)
12. Combettes, P.L., Svaiter, B.F.: Asymptotic behavior of alternating-direction method of multipliers (2011, preprint)
13. Eckstein, J., Bertsekas, D.P.: On the Douglas-Rachford splitting method and the proximal point algorithm for maximal monotone operators. Math. Program. **55**, 293–318 (1992)
14. Eckstein, J., Ferris, M.C.: Operator-splitting methods for monotone affine variational inequalities, with a parallel application to optimal control. Inform. J. Comput. **10**, 218–235 (1998)
15. Goebel, K., Kirk, W.A.: Topics in Metric Fixed Point Theory. Cambridge University Press, Cambridge (1990)
16. Goebel, K., Reich, S.: Uniform Convexity, Hyperbolic Geometry, and Nonexpansive Mappings. Marcel Dekker, New York (1984)
17. Lawrence, J., Spingarn, J.E.: On fixed points of nonexpansive piecewise isometric mappings. Proc. London Math. Soc. **55**, 605–624 (1987)
18. Lions, P.-L., Mercier, B.: Splitting algorithms for the sum of two nonlinear operators. SIAM J. Numer. Anal. **16**, 964–979 (1979)
19. Minty, G.J.: Monotone (nonlinear) operators in Hilbert spaces. Duke Math. J.**29**, 341–346 (1962)
20. Rockafellar, R.T., Wets, R.J-B.: Variational Analysis. Springer, New York (1998)
21. Simons, S.: Minimax and Monotonicity. Springer, Berlin (1998)
22. Simons, S.: From Hahn-Banach to Monotonicity. Springer, New York (2008)
23. Svaiter, B.F.: On weak convergence of the Douglas-Rachford method. SIAM J. Control Optim. **49**, 280–287 (2011)
24. Zălinescu, C.: Convex Analysis in General Vector Spaces. World Scientific Publishing, River Edge (2002)
25. Zarantonello, E.H.: Projections on convex sets in Hilbert space and spectral theory I. Projections on convex sets. In: Zarantonello, E.H. (ed.) Contributions to Nonlinear Functional Analysis, pp. 237–341. Academic Press, New York (1971)

Chapter 3
Champernowne's Number, Strong Normality, and the X Chromosome

Adrian Belshaw and Peter Borwein

This paper is dedicated to Jon Borwein in celebration of his 60th birthday

Abstract Champernowne's number is the best-known example of a normal number, but its digits are far from random. The sequence of nucleotides in the human X chromosome appears nonrandom in a similar way. We give a new asymptotic test of pseudorandomness, based on the law of the iterated logarithm; we call this new criterion "strong normality." We show that almost all numbers are strongly normal and that strong normality implies normality. However, Champernowne's number is not strongly normal. We adapt a method of Sierpiński to construct an example of a strongly normal number.

Key words: Champernowne's number • Law of the iterated logarithm • Normality of numbers • Random walks on digits of numbers • Random walks on nucleotide sequences

Mathematics Subject Classifications (2010): 11K16

COMMUNICATED BY FRANK GARVAN.

A. Belshaw (✉)
Department of Mathematics and Statistics, Capilano University, PO Box 1609
Sunshine Coast Campus 118C, Sechelt, BC V0N 3A0, Canada
e-mail: abelshaw@capilanou.ca

P. Borwein
Department of Mathematics, Simon Fraser University, 8888 University Drive,
Burnaby, BC V5A 1S6, Canada
e-mail: pborwein@sfu.ca

D.H. Bailey et al. (eds.), *Computational and Analytical Mathematics*, Springer Proceedings in Mathematics & Statistics 50, DOI 10.1007/978-1-4614-7621-4_3,

3.1 Normality

We can write a real number α in any integer base $r \geq 2$ as a sum of powers of the base:

$$\alpha = \sum_{j=-d}^{\infty} a_j r^{-j}.$$

The standard "decimal" notation is

$$\alpha = a_{-d}\, a_{-(d-1)} \cdots a_0 \,.\, a_1\, a_2 \,\cdots\,.$$

The sequence of digits $\{a_j\}$ gives the representation of α in the base r, and this representation is unique unless α is rational, in which case α may have two representations. (For example, in the base 10, $0.1 = 0.0999\cdots$.)

We call a subsequence of consecutive digits a *string*. The string may be finite or infinite; we call a finite string of t digits a *t-string*. An infinite string beginning in a specified position we call a *tail*, and we call a finite string beginning in a specified position a *block*.

A number α is *simply normal* in the base r if every 1-string in its base-r expansion occurs with an asymptotic frequency approaching $1/r$. That is, given the expansion $\{a_j\}$ of α in the base r, and letting $m_k(n)$ be the number of times that $a_j = k$ for $j \leq n$, we have

$$\lim_{n\to\infty} \frac{m_k(n)}{n} = \frac{1}{r}$$

for each $k \in \{0, 1, \ldots, r-1\}$. This is Borel's original definition [6].

A number is *normal* in the base r if every t-string in its base-r expansion occurs with a frequency approaching r^{-t}. Equivalently, a number is normal in the base r if it is simply normal in the base r^t for every positive integer t (see [6, 14, 17]).

A number is *absolutely normal* if it is normal in every base. Borel [6] showed that almost every real number is absolutely normal.

In 1933, Champernowne [8] produced the first concrete construction of a normal number. Champernowne's number is

$$\gamma_{10} = .1\ 2\ 3\ 4\ 5\ 6\ 7\ 8\ 9\ 10\ 11\ 12\ 13\ 14\ 15\ \cdots.$$

The number is written in the base 10, and its digits are obtained by concatenating the natural numbers written in the base 10. This number is likely the best-known example of a normal number.

Generally, the base-r Champernowne number is formed by concatenating the integers $1, 2, 3, \ldots$ in the base r. For example, the base-2 Champernowne number is written in the base 2 as

$$\gamma_2 = .1\ 10\ 11\ 100\ 101\ \cdots.$$

For any r, the base-r Champernowne number is normal in the base r. However, the question of its normality in any other base (not a power of r) is open. For example, it is not known whether the base-10 Champernowne number is normal in the base 2.

In 1917, Sierpiński [15] gave a construction of an absolutely normal number (in fact, one such number for each ε with $0 < \varepsilon \leq 1$). A computable version of this construction was given by Becher and Figueira [2].

Most fundamental irrational constants, such as $\sqrt{2}$, $\log 2$, π, and e, appear to be normal, and statistical tests done to date are consistent with the hypothesis that they are normal. (See, for example, Kanada on π [10] and Beyer, Metropolis and Neergard on irrational square roots [5].) However, there is no proof of the normality of any of these constants.

There is an extensive literature on normality in the sense of Borel. Introductions to the literature may be found in [4, 7].

3.2 Walks on the Digits of Numbers and on Chromosomes

In this section we graphically compare two walks on the digits of numbers with a walk on the values of the Liouville λ function and a walk on the nucleotides of the human X chromosome.

The walks are generated on a binary sequence of digits (Figs. 3.1 and 3.2) by converting each 0 in the sequence to -1 and then using digit pairs $(\pm 1, \pm 1)$ to walk $(\pm 1, \pm 1)$ in the plane. The colour or shading in the figures gives a rough indication of the number of steps taken in the walk. The values of the Liouville λ function (Fig. 3.3) are already ± 1.

There are four nucleotides in the X chromosome sequence, and each of the four is assigned one of the values $(\pm 1, \pm 1)$ to create a walk on the nucleotide sequence (Fig. 3.4). The nucleotide sequence is available on the UCSC Genome Browser [16].

A random walk on a million digits is expected to stay within roughly a thousand units of the origin, and this will be seen to hold for the walks on the digits of π and on the Liouville λ function values. On the other hand, the walks on the digits of Champernowne's number and on the X chromosome travel much farther than would be expected of a random walk.

The walk on the Liouville λ function moves away from the origin like $\sqrt{n}$, but it does not seem to move randomly near the origin. In fact, the positive values of λ first outweigh the negative values when $n = 906\,180\,359$ [12], which is not at all typical of a random walk.

3.3 Strong Normality

Mauduit and Sárközy [13] have shown that the digits of the base-2 Champernowne number γ_2 fail two tests of randomness. Dodge and Melfi [9] compared values of an

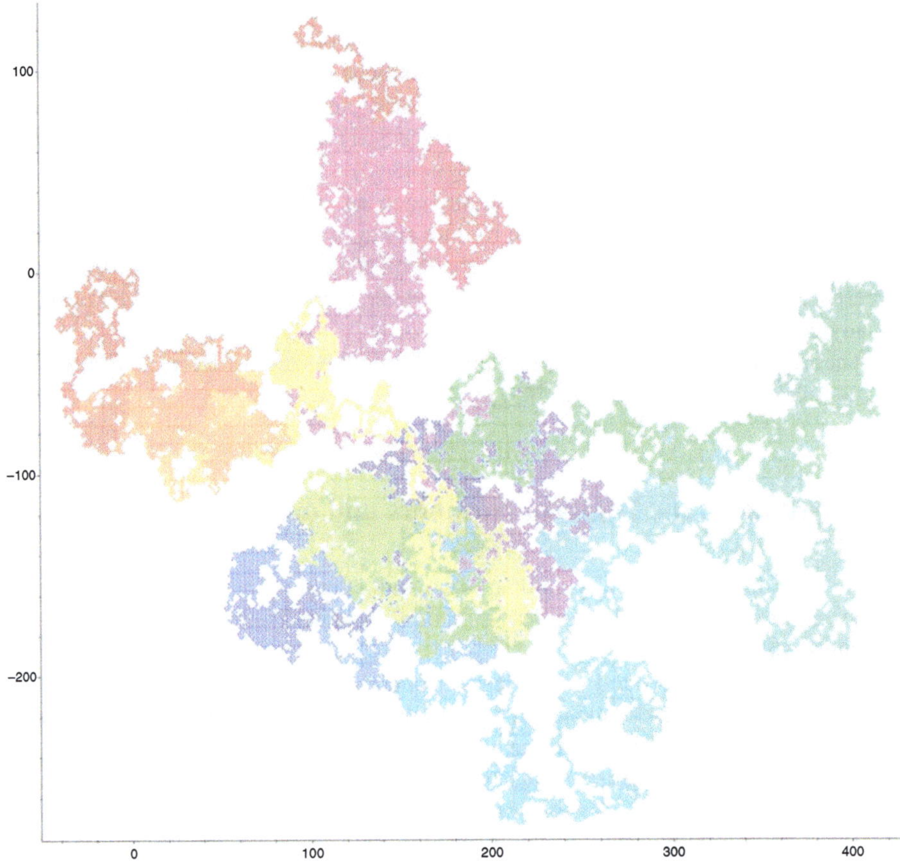

Fig. 3.1 A walk on 10^6 binary digits of π

autocorrelation function for Champernowne's number and π and found that π had the expected pseudorandom properties but that Champernowne's number did not.

Here we provide another test of pseudorandomness and show that it must be passed by almost all numbers. Our test is a simple one, in the spirit of Borel's test of normality, and Champernowne's number will be seen to fail the test.

If the digits of a real number α are chosen at random in the base r, the asymptotic frequency $m_k(n)/n$ of each 1-string approaches $1/r$ with probability 1. However, the *discrepancy* $m_k(n) - n/r$ does not approach any limit, but fluctuates with an expected value equal to the standard deviation $\sqrt{(r-1)n}/r$.

Kolmogorov's law of the iterated logarithm allows us to make a precise statement about the discrepancy of a random number. We use this to define our criterion.

Definition 3.1. For real α, and $m_k(n)$ as above, α is *simply strongly normal* in the base r if for each $k \in \{0,\ldots,r-1\}$

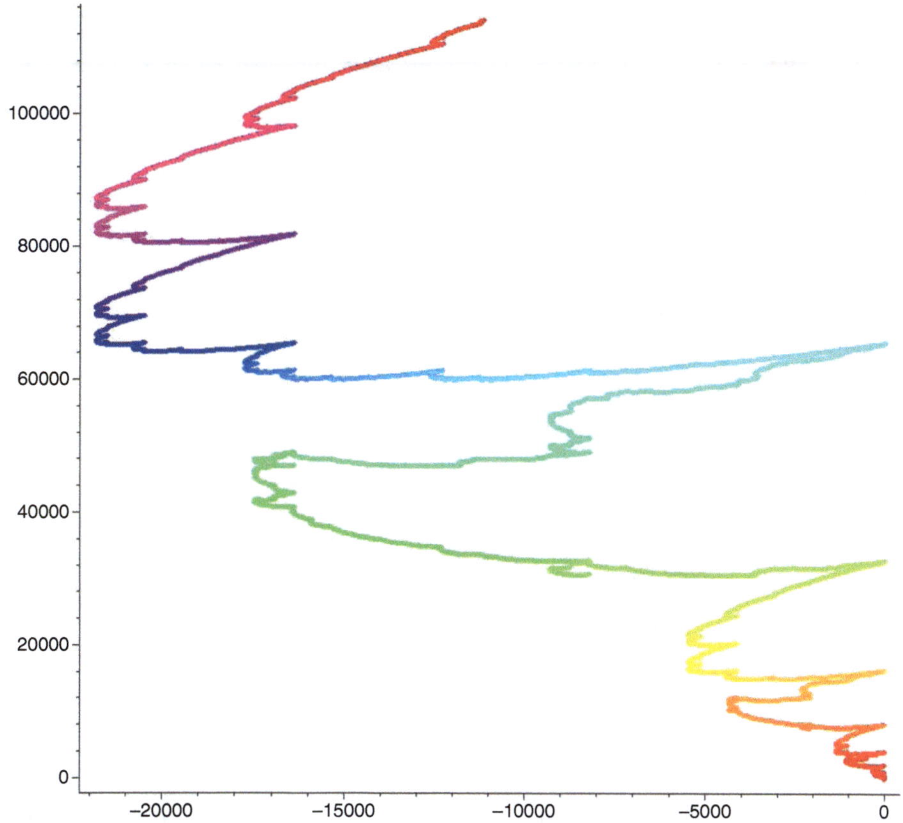

Fig. 3.2 A walk on 10^6 binary digits of the base-2 Champernowne number

$$\limsup_{n\to\infty} \frac{m_k(n) - \frac{n}{r}}{\frac{\sqrt{r-1}}{r}\sqrt{2n\log\log n}} = 1$$

and

$$\liminf_{n\to\infty} \frac{m_k(n) - \frac{n}{r}}{\frac{\sqrt{r-1}}{r}\sqrt{2n\log\log n}} = -1\,.$$

We make two further definitions analogous to the definitions of normality and absolute normality.

Definition 3.2. A number is *strongly normal* in the base r if it is simply strongly normal in each of the bases r^j, $j = 1, 2, 3, \ldots$.

Now let $z_0 \in X$ and define the sequence $(z_n)_{n\in\mathbb{N}}$ by

$$(\forall n \in \mathbb{N}) \quad z_{n+1} = Tz_n. \tag{2.24}$$

This sequence is very useful in determining a zero of $A+B$ as the next result illustrates.

Fact 2.12 (Lions–Mercier [18]). The sequence $(z_n)_{n\in\mathbb{N}}$ converges weakly to some point $z \in X$ such that $z \in \operatorname{Fix} T$ and $J_A z \in \operatorname{zer}(A+B)$. Moreover, the sequence $(J_A z_n)_{n\in\mathbb{N}}$ is bounded, and every weak cluster point of this sequence belongs to $\operatorname{zer}(A+B)$.

Since J_A is in general *not* sequentially weakly continuous (see Example 2.1), it is not obvious whether or not $J_A z_n \rightharpoonup J_A z$. However, recently Svaiter provided a relatively complicated proof that in fact weak convergence does hold. As an application, we rederive the most fundamental instance of his result with a considerably simpler and more conceptual proof.

Fact 2.13 (Svaiter [23]). The sequence $(J_A z_n)_{n\in\mathbb{N}}$ converges weakly to $J_A z$.

Proof. By Fact 2.12,

$$z_n \rightharpoonup z \in \operatorname{Fix} T. \tag{2.25}$$

Since J_A is (firmly) nonexpansive and $(z_n)_{n\in\mathbb{N}}$ is bounded, the sequence $(J_A z_n)_{n\in\mathbb{N}}$ is bounded as well. Let x be an arbitrary weak cluster point of $(J_A z_n)_{n\in\mathbb{N}}$, say

$$J_A z_{k_n} \rightharpoonup x \in \operatorname{zer}(A+B) \tag{2.26}$$

by Fact 2.12. Set $(\forall n \in \mathbb{N})\ y_n = R_A z_n$. Then

$$y_{k_n} \rightharpoonup y = 2x - z \in X. \tag{2.27}$$

Since the operator T is firmly nonexpansive and $\operatorname{Fix} T \neq \varnothing$, it follows from [7] that $z_n - Tz_n \to 0$ (i.e., T is "asymptotically regular"); thus,

$$J_A z_n - J_B y_n = z_n - Tz_n \to 0 \tag{2.28}$$

and hence

$$J_B y_{k_n} \rightharpoonup x. \tag{2.29}$$

Next,

$$0 \leftarrow J_A z_{k_n} - J_B y_{k_n} \tag{2.30a}$$

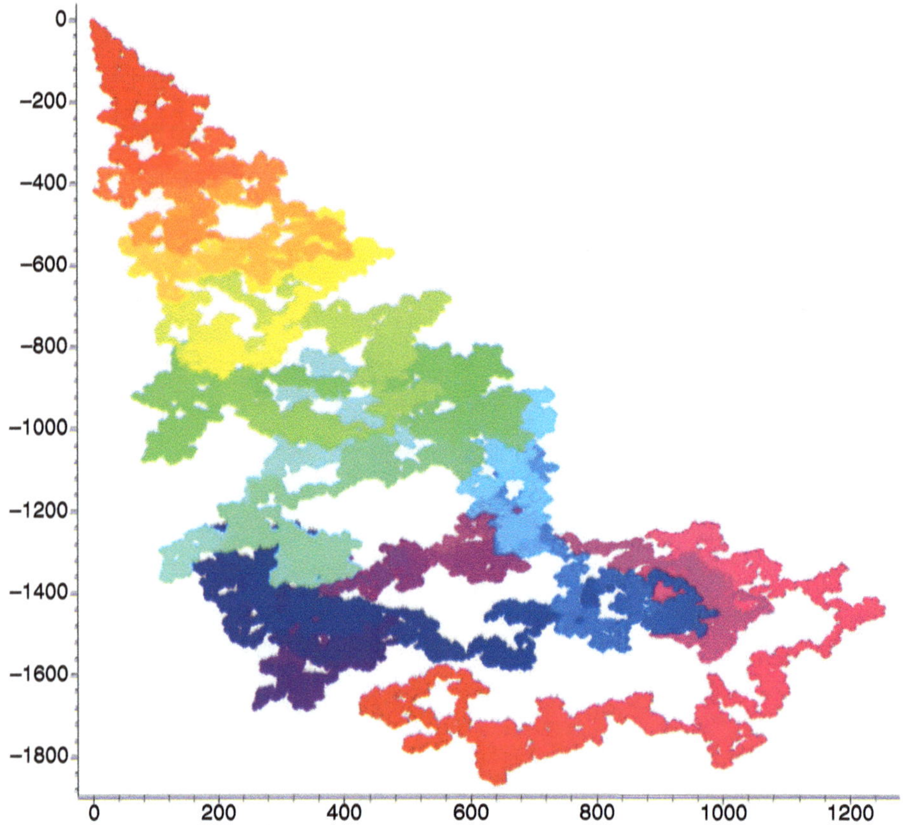

Fig. 3.3 A walk on 10^6 values of the Liouville λ function

Definition 3.3. A number is *absolutely strongly normal* if it is strongly normal in every base.

These definitions of strong normality are sharper than those given by one of the authors in [3].

3.4 Almost All Numbers Are Strongly Normal

Theorem 3.4. *Almost all numbers are simply strongly normal in any base r.*

Proof. Without loss of generality, we consider numbers in the interval $[0,1]$ and fix the integer base $r \geq 2$. We take Lebesgue measure to be our probability measure. For any k, $0 \leq k \leq r-1$, the ith digit of a randomly chosen number is k with probability r^{-1}. For $i \neq j$, the ith and jth digits are both k with probability r^{-2}, so the digits are pairwise independent.

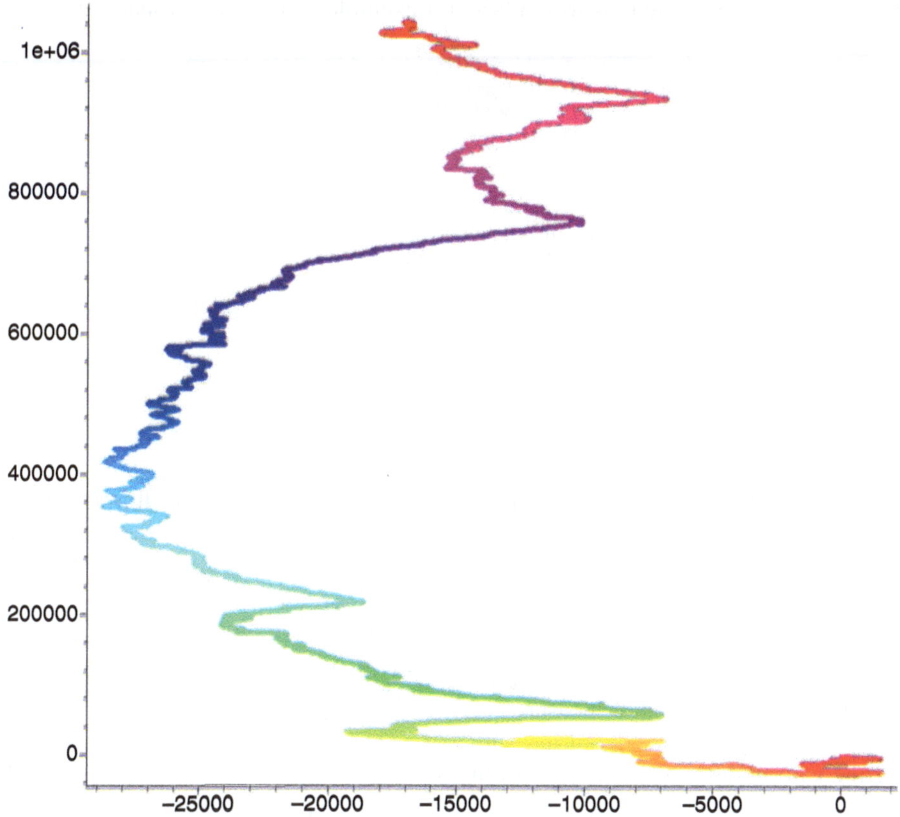

Fig. 3.4 A walk on the nucleotides of the human X chromosome

We define the sequence of random variables X_j by

$$X_j = \sqrt{r-1}$$

if the jth digit is k, with probability $\frac{1}{r}$, and

$$X_j = -\frac{1}{\sqrt{r-1}}$$

otherwise, with probability $\frac{r-1}{r}$.

Then the X_j form a sequence of independent identically distributed random variables with mean 0 and variance 1. Put

$$S_n = \sum_{j=1}^{n} X_j .$$

By the law of the iterated logarithm (see, for example, [11]), with probability 1,

$$\limsup_{n\to\infty} \frac{S_n}{\sqrt{2n\log\log n}} = 1 ,$$

and

$$\liminf_{n\to\infty} \frac{S_n}{\sqrt{2n\log\log n}} = -1 .$$

Now we note that if $m_k(n)$ is the number of occurrences of the digit k in the first n digits of our random number, then

$$S_n = m_k(n)\sqrt{r-1} - \frac{n - m_k(n)}{\sqrt{r-1}} .$$

Substituting this expression for S_n in the limits immediately above shows that the random number satisfies Definition 3.1 with probability 1. ■

This is easily extended.

Corollary 3.5. *Almost all numbers are strongly normal in any base r.*

Proof. By the theorem, the set of numbers in $[0,1]$ which fails to be simply strongly normal in the base r^j is of measure zero, for each j. The countable union of these sets of measure zero is also of measure zero. Therefore the set of numbers simply strongly normal in every base r^j is of measure 1. ■

The following corollary is proved in the same way as the last.

Corollary 3.6. *Almost all numbers are absolutely strongly normal.*

The results for $[0,1]$ are extended to $\mathbb{R}$ in the same way.

3.5 Champernowne's Number Is Not Strongly Normal

We begin by examining the digits of Champernowne's number in the base 2,

$$\gamma_2 = 0.1\ 10\ 11\ 100\ 101\ \cdots .$$

Each integer q, $2^{n-1} \leq q \leq 2^n - 1$, has an n-digit base-2 representation and so contributes an n-block to the expansion of γ_2. In each of these n-blocks, the first digit is 1. If we consider the remaining $n-1$ digits in each of these n-blocks, we see that every possible $(n-1)$-string occurs exactly once. The n-digit integers, concatenated, together contribute a block of length $n2^{n-1}$, and in this block, if we set aside the ones corresponding to the initial digit of each integer, the zeros and ones are equal in number. In the whole block there are $(n-1)2^{n-2}$ zeros and $(n-1)2^{n-2}+2^{n-1}$ ones. The excess of ones over zeros in the entire $(n2^{n-1})$-block is just equal to the number of integers, 2^{n-1}, contributing to the block.

As we concatenate the integers from 1 to 2^k-1, we write the first

$$N-1=\sum_{n=1}^{k} n2^{n-1}=(k-1)2^k+1$$

digits of γ_2. The excess of ones in the digits is

$$2^k-1.$$

The locally greatest excess of ones occurs at the first digit contributed by the integer 2^k, since each power of 2 is written as 1 followed by zeros. At this point the number of digits is $N=(k-1)2^k+2$ and the excess of ones is 2^k. That is, the actual number of ones in the first N digits is

$$m_1(N)=(k-2)2^{k-1}+1+2^k.$$

This gives

$$m_1(N)-\frac{N}{2}=2^{k-1}\,.$$

Thus, we have

$$\frac{m_1(N)-\frac{N}{2}}{N^{1/2+\varepsilon}}\geq\frac{2^{k-1}}{((k-1)2^k)^{1/2+\varepsilon}}.$$

For any sufficiently small positive ε, the right-hand expression is unbounded as $k\to\infty$. We have

$$\limsup_{N\to\infty}\frac{m_1(N)-\frac{N}{2}}{\frac{1}{2}\sqrt{2N\log\log N}}\geq\limsup_{N\to\infty}\frac{m_1(N)-\frac{N}{2}}{N^{1/2+\varepsilon}}=\infty\,.$$

We thus have:

Theorem 3.7. *The base-2 Champernowne number is not strongly normal in the base 2.*

One can show that Champernowne's number also fails the lower limit criterion. In fact, $m_1(N)-\dfrac{N}{2}>0$ for every N.

The theorem can be generalized to every Champernowne number, since there is a shortage of zeros in the base-r representation of the base-r Champernowne number. Each base-r Champernowne number fails to be strongly normal in the base r.

3.6 Strongly Normal Numbers Are Normal

Our definition of strong normality is strictly more stringent than Borel's definition of normality:

Theorem 3.8. *If a number α is simply strongly normal in the base r, then α is simply normal in the base r.*

Proof. It will suffice to show that if a number is not simply normal, then it cannot be simply strongly normal.

Let $m_k(n)$ be the number of occurrences of the 1-string k in the first n digits of the expansion of α in the base r, and suppose that α is not simply normal in the base r. This implies that for some k

$$\lim_{n\to\infty} \frac{rm_k(n)}{n} \neq 1.$$

Then there is some $Q > 1$ and infinitely many n_i such that either

$$rm_k(n_i) > Qn_i$$

or

$$rm_k(n_i) < \frac{n_i}{Q}.$$

If infinitely many n_i satisfy the former condition, then for these n_i,

$$m_k(n_i) - \frac{n_i}{r} > Q\frac{n_i}{r} - \frac{n_i}{r} = n_i P$$

where P is a positive constant.

Then for any $R > 0$,

$$\limsup_{n\to\infty} R\frac{m_k(n) - \frac{n}{r}}{\sqrt{2n\log\log n}} \geq \limsup_{n\to\infty} R\frac{nP}{\sqrt{2n\log\log n}} = \infty,$$

so α is not simply strongly normal.

On the other hand, if infinitely many n_i satisfy the latter condition, then for these n_i,

$$\frac{n_i}{r} - m_k(n_i) > \frac{n_i}{r} - \frac{n_i}{Qr} = n_i P,$$

and once again the constant P is positive. Now

$$\liminf_{n\to\infty} \frac{m_k(n) - \frac{n}{r}}{\sqrt{2n\log\log n}} = -\limsup_{n\to\infty} \frac{\frac{n}{r} - m_k(n)}{\sqrt{2n\log\log n}}$$

and so, in this case also, α fails to be simply strongly normal. ■

The general result is an immediate corollary.

Corollary 3.9. *If α is strongly normal in the base r, then α is normal in the base r.*

3.7 No Rational Number Is Simply Strongly Normal

In light of Theorem 3.8, it will suffice to show that no simply normal rational number can be simply strongly normal.

If α is rational and simply normal in the base r, then if we restrict ourselves to the first n digits in the repeating tail of the expansion, the frequency of any 1-string k is exactly n/r whenever n is a multiple of the length of the repeating string. The excess of occurrences of k can never exceed the constant number of times k occurs in the repeating string. Therefore, with $m_k(n)$ defined as in Sect. 3.3,

$$\limsup_{n\to\infty} \left(m_k(n) - \frac{n}{r}\right) = Q,$$

with Q a constant due in part to the initial non-repeating block and in part to the maximum excess in the tail.

But

$$\limsup_{n\to\infty} \frac{Q}{\sqrt{2n\log\log n}} = 0\,,$$

so α does not satisfy Definition 3.1.

3.8 Construction of an Absolutely Strongly Normal Number

To determine an absolutely strongly normal number, we modify Sierpiński's method of constructing an absolutely normal number [15]. We begin with an easy lemma.

Lemma 3.10. *Let $f(n)$ be a real-valued function of the first n base r digits of a number $\alpha \in [0,1]$, and suppose*

$$\mathbf{P}\left[\limsup_{n\to\infty} f(n) = 1\right] = 1$$

and

$$\mathbf{P}\left[\liminf_{n\to\infty} f(n) = -1\right] = 1 .$$

Given positive $\delta_1 > \delta_2 > \delta_3 > \cdots$, *and* $\varepsilon_1 > \varepsilon_2 > \varepsilon_3 > \cdots$, *we can find* $M_1 < M_2 < M_3 < \cdots$ *so that*

$$\mathbf{P}\left[\left|\sup_{M_i \le n < M_{i+1}} f(n) - 1\right| > \delta_i \quad \text{or} \quad \left|\inf_{M_i \le n < M_{i+1}} f(n) + 1\right| > \delta_i\right] < \varepsilon_i .$$

Notes. The function $f(n)$ depends on both n and α. The probability is the Lebesgue measure of the set of $\alpha \in [0,1]$ for which f satisfies the condition(s).

The lemma can easily be proved under more general assumptions.

Proof. For sufficiently large M,

$$\mathbf{P}\left[\sup_{n\ge M} f(n) > 1 + \delta_1\right] < \frac{\varepsilon_1}{4} \quad \text{and}$$

$$\mathbf{P}\left[\inf_{n\ge M} f(n) < -1 - \delta_1\right] < \frac{\varepsilon_1}{4} .$$

Set M_1 to be the least such M.
Now, as $M \to \infty$,

$$\mathbf{P}\left[\sup_{M_1 \le n < M} f(n) < 1 - \delta_1\right] \to 0 ,$$

and also

$$\mathbf{P}\left[\inf_{M_1 \le n < M} f(n) > -1 + \delta_1\right] \to 0 .$$

Thus, for sufficiently large M, these four conditions are satisfied:

$$\mathbf{P}\left[\sup_{M_1 \le n < M} f(n) < 1 - \delta_1\right] < \frac{\varepsilon_1}{4} ,$$

$$\mathbf{P}\left[\inf_{M_1 \le n < M} f(n) > -1 + \delta_1\right] < \frac{\varepsilon_1}{4} ,$$

$$\mathbf{P}\left[\sup_{n\ge M} f(n) > 1 + \delta_2\right] < \frac{\varepsilon_2}{4} ,$$

and

$$\mathbf{P}\left[\inf_{n\geq M} f(n) < -1-\delta_2\right] < \frac{\varepsilon_2}{4}\,.$$

We set M_2 to be the least $M > M_1$ satisfying all four conditions. Since

$$\mathbf{P}\left[\sup_{M_1\leq n<M_2} f(n) > 1+\delta_1\right] \leq \mathbf{P}\left[\sup_{n\geq M_1} f(n) > 1+\delta_1\right]$$

and

$$\mathbf{P}\left[\inf_{M_1\leq n<M_2} f(n) < -1-\delta_1\right] \leq \mathbf{P}\left[\inf_{n\geq M_1} f(n) < -1-\delta_1\right]\,,$$

we have

$$\mathbf{P}\left[\left|\sup_{M_1\leq n<M_2} f(n)-1\right| > \delta_1 \quad \text{or} \quad \left|\inf_{M_1\leq n<M_2} f(n)+1\right| > \delta_1\right] < \varepsilon_1\,.$$

We can continue in this way, recursively choosing $M_3, M_4, M_5, \ldots$ so that each M_i is the least satisfying the required conditions. ■

Now we fix an integer base $r \geq 2$ and a 1-string $k \in \{0,1,\ldots,r-1\}$. For each $\alpha \in [0,1]$, put

$$f(n) = f(\alpha,k,n) = \frac{m_k(n)-\dfrac{n}{r}}{\dfrac{\sqrt{r-1}}{r}\sqrt{2n\log\log n}}\,.$$

Here, as in Definition 3.1 of Sect. 3.3, $m_k(n)$ is the number of occurrences of k in the first n base r digits of α, and α is simply strongly normal in the base r if

$$\limsup_{n\to\infty} f(n) = 1$$

and

$$\liminf_{n\to\infty} f(n) = -1\,.$$

By Theorem 3.4, Sect. 3.4, these conditions hold with probability 1, so f satisfies the conditions of Lemma 3.10.

Now fix $0 < \varepsilon \leq 1$; set $\delta_i = \dfrac{1}{i}$ and $\varepsilon_i = \varepsilon_{r,i} = \dfrac{\varepsilon}{3\cdot 2^i r^3}$. These δ_i and ε_i also satisfy the conditions of Lemma 3.10.

We will construct a set $A_\varepsilon \subset [0,1]$, of measure less than 1, in such a way that every element of A_ε^C is absolutely strongly normal.

Let $M_1 < M_2 < M_3 < \cdots$ be determined as in the proof of Lemma 3.10, so that the conclusion of the lemma holds. We build a set $A_{r,i}$ containing those α for which the first M_{i+1} digits are, in a loose sense, far from simply strongly normal in the base r.

Around each $\alpha = .a_1a_2\cdots a_{M_{i+1}}\cdots$ such that

$$\left| \sup_{M_i \le n < M_{i+1}} f(n) - 1 \right| > \delta_i \tag{3.1}$$

or

$$\left| \inf_{M_i \le n < M_{i+1}} f(n) + 1 \right| > \delta_i \tag{3.2}$$

we construct an open interval containing α:

$$\left(\frac{a_1}{r} + \frac{a_2}{r^2} + \cdots + \frac{a_{M_{i+1}}}{r^{M_{i+1}}} - \frac{1}{r^{M_{i+1}}}, \frac{a_1}{r} + \frac{a_2}{r^2} + \cdots + \frac{a_{M_{i+1}}}{r^{M_{i+1}}} + \frac{2}{r^{M_{i+1}}} \right).$$

Let $A_{r,k,i}$ be the union of all the intervals constructed in this way. By our construction, the union of the closed intervals consisting of the numbers with initial digits $.a_1a_2\ldots a_{M_{i+1}}$ satisfying one of our two conditions (3.1) or (3.2) has measure less than ε_i, so, denoting Lebesgue measure by μ,

$$\mu\left(A_{r,k,i}\right) < 3\varepsilon_i = \frac{\varepsilon}{2^i r^3}.$$

In this way we construct $A_{r,k,i}$ for every base r and 1-string $k \in \{0,1,\ldots,r-1\}$. We let

$$A_\varepsilon = \bigcup_{r=2}^{\infty} \bigcup_{k=0}^{r-1} \bigcup_{i=1}^{\infty} A_{r,k,i},$$

so

$$\begin{aligned} \mu(A_\varepsilon) &\le \sum_{r=2}^{\infty} \sum_{k=0}^{r-1} \sum_{i=1}^{\infty} \mu\left(A_{r,k,i}\right) \\ &< \sum_{r=2}^{\infty} \sum_{k=0}^{r-1} \sum_{i=1}^{\infty} \frac{\varepsilon}{2^i r^3} \\ &= \left(\frac{\pi^2}{6} - 1 \right) \varepsilon. \end{aligned}$$

Let E_ε be the complement of A_ε in $[0,1]$. Since $\mu(A_\varepsilon) < 1$, E_ε is of positive measure. We claim that every element of E_ε is absolutely strongly normal.

For each base r and 1-string $k \in \{0,1,\ldots,r-1\}$, we have specified a set of integers $M_1 < M_2 < M_3 < \cdots$, depending on r and k. By our construction, if $\alpha \in E_\varepsilon$, then, recalling that f depends on α, we have

$$\left| \sup_{M_i \le n < M_{i+1}} f(n) - 1 \right| < \delta_i$$

and

$$\left| \inf_{M_i \le n < M_{i+1}} f(n) + 1 \right| < \delta_i$$

for every i. Clearly for this α, since $\delta_i \to 0$,

$$\limsup_{n \to \infty} f(n) = 1$$

and

$$\liminf_{n \to \infty} f(n) = -1 .$$

This is true for every k, so α is simply strongly normal to the base r, by Definition 3.1 (Sect. 3.3). Thus α is simply strongly normal to every base, and is therefore absolutely strongly normal by Definitions 3.2 and 3.3.

To specify an absolutely strongly normal number, we note that E_ε contains no interval, since, by Sect. 3.7, no rational number is simply strongly normal in any base. Since E_ε is bounded, $\inf E_\varepsilon$ is well defined; and $\inf E_\varepsilon \in E_\varepsilon$ since otherwise $\inf E_\varepsilon$ would be interior to some open interval of A_ε.

For example, $\inf E_1$ is a well-defined absolutely strongly normal number.

3.9 Further Questions

It should be possible to construct a computable absolutely strongly normal number by the method of Becher and Figueira [2].

We conjecture that such naturally occurring constants as the irrational numbers π, e, $\sqrt{2}$, and $\log 2$ are absolutely strongly normal.

On the other hand, we speculate that the binary Liouville λ number, created in the obvious way from the λ function values, may be normal but not strongly normal.

Bailey and Crandall [1] proved normality base 2 for an uncountable class of "generalized Stoneham constants," namely constants of the form

$$\alpha_{2,3}(r) = \sum_{k=0}^{\infty} \frac{1}{3^k 2^{3^k + r_k}},$$

where r_k is the kth binary digit of a real number r in the unit interval. This class of numbers may be a good place to look for examples of strong normality. However, new techniques may be required for this.

Acknowledgements Many thanks are due to Stephen Choi for his comments on the earlier ideas in [3] . We also give many thanks to Richard Lockhart for his help with ideas in probability. We are indebted to an anonymous referee for some extremely useful comments and criticisms.

References

1. Bailey, D.H., Crandall, R.E.: Random generators and normal numbers. Exp. Math. **11**(4), 527–546 (2003)
2. Becher, V., Figueira, S.: An example of a computable absolutely normal number. Theor. Comput. Sci. **270**, 947–958 (2002)
3. Belshaw, A.: On the normality of numbers. M.Sc. thesis, Simon Fraser University, Burnaby, BC (2005)
4. Berggren, L., Borwein, J., Borwein, P.: Pi: a Source Book, 3rd edn. Springer, New York (2004)
5. Beyer, W.A., Metropolis, N., Neergaard, J.R.: Statistical study of digits of some square roots of integers in various bases. Math. Comput. **24**, 455–473 (1970)
6. Borel, E.: Les probabilités dénombrables et leurs applications arithmétiques. Supplemento ai Rend. Circ. Mat. di Palermo **27**, 247–271 (1909)
7. Borwein, J., Bailey, D.: Mathematics by Experiment. A K Peters Ltd., Natick (2004)
8. Champernowne, D.G.: The construction of decimals normal in the scale of ten. J. London Math. Soc. **3**, 254–260 (1933)
9. Dodge, Y., Melfi, G.: On the reliability of random number generators. http://pictor.math.uqam.ca/~plouffe/articles/reliability.pdf
10. Kanada, Y.: Vectorization of multiple-precision arithmetic program and 201,326,395 decimal digits of π calculation. Supercomputing: Sci. Appl. **88**(II), pp. 117–128 (1988)
11. Laha, R.G., Rohatgi, V.K.: Probability Theory. Wiley, New York (1979)
12. Lehman, R.S.: On Liouville's function. Math. Comput. **14**, 311–320 (1960)
13. Mauduit, C., Sárközy, A.: On finite pseudorandom binary sequences II. The Champernowne, Rudin-Shapiro, and Thue-Morse sequences, a further construction. J. Number Theor. **73**, 256–276 (1998)
14. Niven, I., Zuckerman, H.S.: On the definition of normal numbers. Pacific J. Math. **1**, 103–109 (1951)
15. Sierpiński, W.: Démonstration élémentaire du théorème de M. Borel sur les nombres absolument normaux et détermination effective d'un tel nombre. Bull. Soc. Math. France **45**, 125–132 (1917)
16. UCSC Genome Browser. http://hgdownload.cse.ucsc.edu/goldenPath/hg19/chromosomes/
17. Wall, D.D.: Normal numbers. Ph.D. thesis, University of California, Berkeley, CA (1949)

Chapter 4
Optimality Conditions for Semivectorial Bilevel Convex Optimal Control Problems

Henri Bonnel and Jacqueline Morgan

Abstract We present optimality conditions for bilevel optimal control problems where the upper level is a scalar optimal control problem to be solved by a leader and the lower level is a multiobjective convex optimal control problem to be solved by several followers acting in a cooperative way inside the greatest coalition and choosing amongst efficient optimal controls. We deal with the so-called optimistic case, when the followers are assumed to choose the best choice for the leader amongst their best responses, as well with the so-called pessimistic case, when the best response chosen by the followers can be the worst choice for the leader. This paper continues the research initiated in Bonnel (SIAM J. Control Optim. **50**(6), 3224–3241, 2012) where existence results for these problems have been obtained.

Key words: Differential games • Multiobjective optimal control • Multiobjective optimization • Optimality conditions • Optimistic bilevel problem • Pareto (efficient) control set • Pessimistic bilevel problem • Semivectorial bilevel optimization • Stackelberg problem

---

COMMUNICATED BY MICHEL THÉRA.

H. Bonnel (✉)
Université de la Nouvelle-Calédonie, ERIM, BP. R4, F98851 Nouméa Cédex, New Caledonia, France
e-mail: bonnel@univ-nc.nc

J. Morgan
Department of Economy and Statistics & CSEF, University of Naples Federico II, Complesso Universitario di Monte S. Angelo, Via Cintia, 80126 Napoli, Italy
e-mail: morgan@unina.it

D.H. Bailey et al. (eds.), *Computational and Analytical Mathematics*, Springer Proceedings in Mathematics & Statistics 50, DOI 10.1007/978-1-4614-7621-4_4,

Mathematics Subject Classifications (2010): Primary 58E17; Secondary 49N99, 49N70, 49N10, 91A23, 90-08, 90C25, 90C29, 90C46, 90C48

4.1 Introduction

The aim of this paper is to obtain optimality conditions for the semivectorial bilevel optimal control problems introduced in [17] where existence results have been established.

Semivectorial bilevel optimal control problems are bilevel problems where the upper level corresponds to a scalar optimization problem and the lower level to a multiobjective optimal control problem. Multiobjective optimal control problems arise in many application areas where several conflicting objectives need to be considered. Minimizing several objective functionals leads to solutions such that none of the objective functional values can be improved further without deteriorating another. The set of all such solutions is referred to as efficient (also called Pareto optimal, noninferior, or nondominated) set of solutions (see, e.g. [38]). The lower level of the semivectorial bilevel optimal control problems can be associated to one player with p objective or to a "grand coalition" of a p-player "cooperative differential game", every player having its own objective and control function. We consider situations in which these p players react as "followers" to every decision imposed by a "leader" (who acts at the so-called upper level). The best reply correspondence of the followers being in general non-uniquely determined, the leader cannot predict the followers choice simply on the basis of his rational behaviour. So, the choice of the best strategy from the leader point of view depends of how the followers choose a strategy amongst his best responses. In this paper, we will consider two (extreme) possibilities:

1. The optimistic situation, when for every decision of the leader, the followers will choose a strategy amongst the efficient controls which minimizes the (scalar) objective of the leader; in this case the leader will choose a strategy which minimizes the best he can obtain amongst all the best responses of the followers:
2. The pessimistic situation, when the followers can choose amongst the efficient controls one which maximizes the (scalar) objective of the leader; in this case the leader will choose a strategy which minimizes the worst he could obtain amongst all the best responses of the followers.

The semivectorial bilevel control problems which model these two situations, and which will be described in the next section, include the following problems which have been intensively studied in the last decades, so we will give essentially a few earlier references:

- Optimizing a scalar-valued function over the efficient set associated to a multi-objective optimization (mathematical programming) problem (introduced in [47] and investigated in [8–13, 25–27, 33, 36, 37, 50] for a survey).

- Optimizing a scalar-valued function over an efficient control set associated to a multiobjective optimal control problem (introduced and investigated in [15], followed by [18])
- Semivectorial bilevel static problems (introduced and investigated in [16], followed by [3, 14, 22, 30, 31, 51], for the optimistic case)
- Stackelberg problems (introduced in [49] and investigated, e.g. in [6, 40, 43])
- Bilevel optimization problems (e.g. [24, 28, 29, 41, 44, 45] for an extensive bibliography)
- Stackelberg dynamic problems (introduced in [23, 48] and investigated, e.g. in [5, 6, 42, 45, 46], a book with an extensive bibliography)

In this paper, we rewrite the optimistic and pessimistic semivectorial bilevel control problems as bilevel problems where the lower level is a scalar optimization problem which admits a unique solution, using scalarization techniques as in [17]. So we are able to give optimality conditions for the lower level problem in the general case (supposing that the leader's controls are bounded) using Pontryagin maximum principle. This theoretically allows to obtain under suitable conditions the dependence of the optimal control on the leader's variables. However, this approach is very difficult to apply because one needs to solve a bilocal problem. That is why we consider the particular but important case when the followers' problem is linear-quadratic. In this case we show that using a resolvent matrix obtained from data, we can explicitly solve the bilocal problem and express the optimal control and the state as functions of leader's variables, and we show that these dependencies are continuously differentiable. Finally we present optimality conditions for the upper levels of the optimistic and pessimistic problems.

4.2 Preliminaries and Problem Statement

All the assumptions and notations considered in this section and introduced in [17] will be kept throughout this paper.

For the *leader* we denote by J_l the scalar objective, by u_l the control function and by $\mathscr{U}_l$ the set of admissible controls. For the followers we denote by $\mathbf{J_f} = (J_1, \ldots, J_p)$ the vector objective (p-scalar objectives) and by $\mathbf{u_f} = (u_1, \ldots, u_p)$ the control function whose values belong to the set $\mathbf{U_f} = U_1 \times \cdots \times U_p \subseteq \mathbb{R}^{m_f} = \mathbb{R}^{m_1} \times \cdots \times \mathbb{R}^{m_p}$. $\mathbf{U_f}$ is assumed to be nonempty, closed and convex, and $0 \in \mathbf{U_f}$. Real numbers t_0, T are fixed ($t_0 < T$) and represent respectively the initial time and an upper bound of the final time. The set of final time values $\mathscr{T} = [\underline{t}, \bar{t}] \subset]t_0, T[$, where $\underline{t} \leq \bar{t}$. The final time, denoted by $t_1 \in \mathscr{T}$, may be variable and it is decided by the leader; hence t_1 is fixed in the followers' problem. We assume that

$$\mathscr{U}_l \subset L_2^{m_l}([t_0, T]) \quad \text{is closed, nonempty and convex.} \tag{4.1}$$

For each fixed $(t_1,u_l) \in \mathscr{T} \times \mathscr{U}_l$, the followers have to solve the following parametric multiobjective control problem, called lower level problem:

$$(\mathbf{LL})_{(t_1,u_l)} \qquad \begin{cases} \mathbf{MIN}_{(\mathbf{u_f},x)} \ \mathbf{J_f}(t_1,u_l,\mathbf{u_f},x) \\ \text{subject to } (\mathbf{u_f},x) \text{ verifies (4.2)–(4.5)} \end{cases}$$

$$\mathbf{u_f}(t) \in \mathbf{U_f} \text{ a.e. on } [t_0,T], \ \ \mathbf{u_f}(t) = 0 \text{ a.e. on } [t_1,T], \tag{4.2}$$

$$\dot{x}(t) = A(t)x(t) + B_l(t)u_l(t) + \mathbf{B_f}(t)\mathbf{u_f}(t) \ \text{ a.e. on } [t_0,t_1], \tag{4.3}$$

$$x(t_0) = x_0, \tag{4.4}$$

$$x(t_1) \in \mathscr{F}, \tag{4.5}$$

where $A : [t_0,T] \to \mathbb{R}^{n\times n}$, $B_l : [t_0,T] \to \mathbb{R}^{n\times m_l}$ and $\mathbf{B_f} : [t_0,T] \to \mathbb{R}^{n\times m_f}$ are continuous matrix-valued functions and the control function $\mathbf{u_f} = (u_1,\dots,u_p) \in L_2^{m_f}([t_0,T]) = L_2^{m_1}([t_0,T]) \times \cdots \times L_2^{m_p}([t_0,T])$.

$L_2^m([t_0,T])$ stands for the usual Hilbert space of equivalence classes (two functions are equivalent iff they coincide a.e.) of (Lebesgue) measurable functions u from $[t_0,T]$ to $\mathbb{R}^m$, such that the function $t \mapsto u^T(t)u(t)$ is (Lebesgue) integrable over $[t_0,T]$ endowed with the norm $\|u\|_2 := \left(\int_{t_0}^T u^T(t)u(t)\mathrm{d}t\right)^{1/2}$. The target set $\mathscr{F} \subset \mathbb{R}^n$ is assumed to be closed, convex and nonempty.

The initial state $x_0 \in \mathbb{R}^n$ is specified.

For each $u = (t_1,u_l,\mathbf{u_f}) \in \mathscr{T} \times L_2^{m_l}([t_0,T]) \times L_2^{m_f}([t_0,T])$, under the above assumptions, there exists a unique solution (in the sense of Carathéodory) x_u of the Cauchy problem (4.3) and (4.4), and $x_u \in H_1^n([t_0,t_1])$. $H_1^n([t_0,t_1])$ stands for the Hilbert space of absolutely continuous functions from $[t_0,t_1]$ to $\mathbb{R}^n$ with derivative in $L_2^n([t_0,t_1])$ endowed with the norm $x \mapsto \|x\| := (\|\dot{x}\|_2^2 + \|x\|_2^2)^{1/2}$.

The *feasible set* $\mathscr{S}(t_1,u_l)$ for the problem $(\mathbf{LL})_{(t_1,u_l)}$ is defined in the following way:

$$\mathscr{S}(t_1,u_l) = \{(\mathbf{u_f},x) \in L_2^{m_f}([t_0,T]) \times H_1^n([t_0,t_1]) |\ (\mathbf{u_f},x) \ \text{ verifies relations (4.2)–(4.5)}\}. \tag{4.6}$$

Thus, problem $(\mathbf{LL})_{(t_1,u_l)}$ can be written as

$$(\mathbf{LL})_{(t_1,u_l)} \qquad \mathbf{MIN}_{(\mathbf{u_f},x)\in\mathscr{S}(t_1,u_l)} \ \mathbf{J_f}(t_1,u_l,\mathbf{u_f},x).$$

Next we give the following standard definitions.

Definition 4.1. For problem $(\mathbf{LL})_{(t_1,u_l)}$ the element $(\bar{\mathbf{u}}_\mathbf{f},\bar{x}) \in \mathscr{S}(t_1,u_l)$ is said to be

- *An efficient (or Pareto) control process* if there is no element $(\mathbf{u_f},x) \in \mathscr{S}(t_1,u_l)$ satisfying

$$\forall i \in \{1,\ldots,p\} \qquad J_i(t_1,u_l,\mathbf{u_f},x) \leq J_i(t_1,u_l,\bar{\mathbf{u}}_\mathbf{f},\bar{x})$$

and

$$\exists i_0 \in \{1,\ldots,p\} \qquad J_{i_0}(t_1,u_l,\mathbf{u_f},x) < J_{i_0}(t_1,u_l,\bar{\mathbf{u}}_\mathbf{f},\bar{x}).$$

- *A weakly efficient (or weakly Pareto) control process* if there is no element $(\mathbf{u_f},x) \in \mathscr{S}(t_1,u_l)$ satisfying

$$\forall i \in \{1,\ldots,p\} \qquad J_i(t_1,u_l,\mathbf{u_f},x) < J_i(t_1,u_l,\bar{\mathbf{u}}_\mathbf{f},\bar{x}).$$

- *A properly efficient (or properly Pareto) control process* (see [34] or [19, 38] for generalizations) if it is an efficient control process and there exists a real number $M > 0$ so that for every $i \in \{1,\ldots,p\}$ and every $(\mathbf{u_f},x) \in \mathscr{S}(t_1,u_l)$ with $J_i(t_1,u_l,\mathbf{u_f},x) < J_i(t_1,u_l,\bar{\mathbf{u}}_\mathbf{f},\bar{x})$ at least one $k \in \{1,\ldots,p\}$ exists with $J_k(t_1,u_l,\mathbf{u_f},x) > J_k(t_1,u_l,\bar{\mathbf{u}}_\mathbf{f},\bar{x})$ and

$$\frac{J_i(t_1,u_l,\bar{\mathbf{u}}_\mathbf{f},\bar{x}) - J_i(t_1,u_l,\mathbf{u_f},x)}{J_k(t_1,u_l,\mathbf{u_f},x) - J_k(t_1,u_l,\bar{\mathbf{u}}_\mathbf{f},\bar{x})} \leq M.$$

In the sequel the symbol $\sigma \in \{e, we, pe\}$ stands for "efficient" when $\sigma = e$, "weakly efficient" when $\sigma = we$ and "properly efficient" when $\sigma = pe$.

The set of all σ-control processes associated to problem $(\mathbf{LL})_{(t_1,u_l)}$ will be denoted by $\mathscr{P}_\sigma(t_1,u_l)$.

Finally we consider the following *semivectorial bilevel optimal control* problems:

$$(\text{OSVBC})_\sigma \qquad \min_{(t_1,u_l)\in\mathscr{T}\times\mathscr{U}_l} \; \min_{(\mathbf{u_f},x)\in\mathscr{P}_\sigma(t_1,u_l)} \; J_l(t_1,u_l,\mathbf{u_f},x)$$

called *optimistic semivectorial bilevel control problem* and

$$(\text{PSVBC})_\sigma \qquad \min_{(t_1,u_l)\in\mathscr{T}\times\mathscr{U}_l} \; \sup_{(\mathbf{u_f},x)\in\mathscr{P}_\sigma(t_1,u_l)} \; J_l(t_1,u_l,\mathbf{u_f},x)$$

called *pessimistic semivectorial bilevel control problem.*

Remark 4.2. Note that the terminal time t_1 is fixed for the lower level problem, but it is a decision variable for the leader. Of course, a particular case can be obtained when the terminal time t_1 is fixed for the leader too, i.e. when $\mathscr{T} = \{t_1\}$.

Remark 4.3. $(\mathbf{LL})_{(t_1,u_l)}$ may be also considered as the problem to be solved by the *grand coalition of a p-player cooperative differential game* (see [35] and its extensive references list) where the functional J_i and the control u_i represent the payoff and the control of the player number i, $i \in \{1,\ldots,p\}$. Then, our optimistic semivectorial bilevel problem corresponds to a strong Stackelberg problem in which, for any choice of (t_1,u_l), the leader can force the followers to choose amongst the σ-control processes one which minimizes the leader payoff. On the other hand, the pessimistic semivectorial bilevel problem corresponds to a weak Stackelberg problem in which, for any choice of the leader variables (t_1, u_l), the followers could choose amongst the σ-control processes one which is the worst for the leader.

We assume that for all $t_1 \in [t_0, T]$ and all $(u_l, \mathbf{u_f}, x) \in L_2^{m_l}([t_0,T]) \times L_2^{m_f}([t_0,T]) \times H_1^n([t_0,t_1])$, we have

$$J_l(t_1, u_l, \mathbf{u_f}, x) = \int_{t_0}^{t_1} f_l(t, u_l(t), \mathbf{u_f}(t), x(t))\mathrm{d}t,$$

and also, for all $i \in \{1, \dots, p\}$,

$$J_i(t_1, u_l, \mathbf{u_f}, x) = \psi_i(x(t_1)) + \int_{t_0}^{t_1} f_i(t, u_l(t), \mathbf{u_f}(t), x(t))\mathrm{d}t,$$

where, for all $i \in \{1, \dots, p\}$, the functions $\psi_i, \psi_l : \mathbb{R}^n \to \mathbb{R}$, $f_i, f_l : [t_0, T] \times \mathbb{R}^{m_l} \times \mathbb{R}^{m_f} \times \mathbb{R}^n \to \mathbb{R}$ verify the following *preliminary assumptions* :

$(\mathscr{PA})$

- ψ_i, f_i, f_l are continuously differentiable;
- there exist integrable functions $a_i, a_l : [t_0, T] \to \mathbb{R}$ and real numbers $b_i, b_l, c_i, c_l, d_i, d_l$, such that, for all $(t, u_l, \mathbf{u_f}, x) \in [t_0, T] \times \mathbb{R}^{m_l} \times \mathbb{R}^{m_f} \times \mathbb{R}^n$,
  $f_i(t, u_l, \mathbf{u_f}, x) \geqslant a_i(t) + b_i x^T x + c_i u_l^T u_l + d_i \mathbf{u_f}^T \mathbf{u_f}$,
  $f_l(t, u_l, \mathbf{u_f}, x) \geqslant a_l(t) + b_l x^T x + c_l u_l^T u_l + d_l \mathbf{u_f}^T \mathbf{u_f}$;
- ψ_i is a convex function;
- for each fixed $t \in [t_0, T]$, the function $f_i(t, \cdot, \cdot, \cdot)$ is convex on $\mathbb{R}^{m_l} \times \mathbb{R}^{m_f} \times \mathbb{R}^n$.

4.3 The Lower Level Problem

Let $t_1 \in \mathscr{T}$ be fixed, and let $\Phi : [t_0, t_1] \times [t_0, t_1] \to \mathbb{R}^{n \times n}$ be the matrix-valued function satisfying for each $s \in [t_0, t_1]$

$$\forall t \in [t_0, t_1] \qquad \frac{\partial \Phi}{\partial t}(t, s) = A(t)\Phi(t, s) \tag{4.7}$$

$$\Phi(s, s) = I_n \tag{4.8}$$

where I_n is the identity matrix.

Since, for each $(u_l, \mathbf{u_f}) \in L_2^{m_l}([t_0,T]) \times L_2^{m_f}([t_0,T])$, the unique solution $x_{(t_1, u_l, \mathbf{u_f})} \in H_1^n([t_0,t_1])$ of the Cauchy problem (4.3) and (4.4) is given by

$$\forall t \in [t_0, t_1] \quad x_{(t_1,u_l,\mathbf{u_f})}(t) = \Phi(t,t_0)x_0 + \int_{t_0}^{t} \Phi(t,s)(B_l(s)u_l(s) + \mathbf{B_f}(s)\mathbf{u_f}(s))\mathrm{d}s,$$

it is clear that the map $(u_l, \mathbf{u_f}) \mapsto x_{(t_1,u_l,\mathbf{u_f})}$ is affine from $L_2^{m_l}([t_0,T]) \times L_2^{m_f}([t_0,T])$ to $H_1^n([t_0,t_1])$. Moreover, using Cauchy–Schwartz inequality, we obtain easily that the map $(u_l, \mathbf{u_f}) \mapsto x_{(t_1,u_l,\mathbf{u_f})}$ is also continuous from $L_2^{m_l}([t_0,T]) \times L_2^{m_f}([t_0,T])$ to $H_1^n([t_0,t_1])$.

For each $i = 1, \ldots, p$, consider the functional

$$(u_l, \mathbf{u_f}) \mapsto \tilde{J}_i(t_1, u_l, \mathbf{u_f}) := J_i(t_1, u_l, \mathbf{u_f}, x_{(t_1,u_l,\mathbf{u_f})}). \tag{4.9}$$

Define also

$$(u_l, \mathbf{u_f}) \mapsto \tilde{J}_l(t_1, u_l, \mathbf{u_f}) := J_l(t_1, u_l, \mathbf{u_f}, x_{(t_1,u_l,\mathbf{u_f})}). \tag{4.10}$$

From [17, Lemmas 1 and 2] and the fact that $x_{(t_1,\cdot,\cdot)}$ is continuous and affine from $L_2^{m_l}([t_0,T]) \times L_2^{m_f}([t_0,T])$ to $H_1^n([t_0,t_1])$, we obtain the following.

Lemma 4.4. *For each $i = 1, \ldots, p$, the functional $\tilde{J}_i(t_1,\cdot,\cdot) : L_2^{m_l}([t_0,T]) \times L_2^{m_f}([t_0,T]) \to \mathbb{R} \cup \{+\infty\}$ is well defined, lower semicontinuous and convex.*

Also $\tilde{J}_l(t_1,\cdot,\cdot) : L_2^{m_l}([t_0,T]) \times L_2^{m_f}([t_0,T]) \to \mathbb{R} \cup \{+\infty\}$ is well defined and lower semicontinuous.

For each $(t_1, u_l) \in \mathscr{T} \times \mathscr{U}_l$ [see (4.1)], denote

$$\begin{aligned} \mathscr{U}_f(t_1,u_l) = \{&\mathbf{u_f} \in L_2^{m_f}([t_0,T]) |\ \mathbf{u_f}(t) \in \mathbf{U_f} \text{ a.e. on } [t_0,T], \\ &\mathbf{u_f}(t) = 0 \text{ a.e. on } [t_1,T],\ x_{(t_1,u_l,\mathbf{u_f})}(t_1) \in \mathscr{F}\}. \end{aligned} \tag{4.11}$$

For each $(t_1,u_l) \in \mathbb{R} \times L_2^{m_l}([t_0,T]) \setminus \mathscr{T} \times \mathscr{U}_l$ we put $\mathscr{U}_f(t_1,u_l) = \emptyset$. Thus $\mathscr{U}_f$ is a set-valued function $\mathscr{U}_f : \mathbb{R} \times L_2^{m_l}([t_0,T]) \rightrightarrows L_2^{m_f}([t_0,T])$.

Recall that

$$\text{dom}\,(\mathscr{U}_f) := \{(t_1,u_l) \in \mathbb{R} \times L_2^{m_l}([t_0,T]) |\ \mathscr{U}_f(t_1,u_l) \neq \emptyset\}$$

and

$$\text{Gr}\,(\mathscr{U}_f) = \{(t_1,u_l,\mathbf{u_f}) \in \mathbb{R} \times L_2^{m_l}([t_0,T]) \times L_2^{m_f}([t_0,T]) |\ \mathbf{u_f} \in \mathscr{U}_f(t_1,u_l)\}.$$

We will assume in the sequel that

$(\mathscr{H})$ $\quad\quad$ $\text{dom}\,(\mathscr{U}_f) = \mathscr{T} \times \mathscr{U}_l.$

Proposition 4.5. *Each of the following is a sufficient condition for $(\mathscr{H})$:*

(a) $\mathscr{F} = \mathbb{R}^n$.
(b) For each $t_1 \in \mathscr{T}$, the linear system

$$\dot{x}(t) = A(t)x(t) + \mathbf{B_f}(t)\mathbf{u_f}(t), x(t_0) = 0, \ \mathbf{u_f}(t) \in \mathbf{U_f} \text{ a.e. on } [t_0, t_1]$$

is controllable, i.e. for any $x_1 \in \mathbb{R}^n$*, there exists* $\mathbf{u_f} \in L_2^{m_f}([t_0,t_1])$ *such that* $\mathbf{u_f}(t) \in \mathbf{U_f}$ *a.e. on* $[t_0,t_1]$*, and the corresponding solution verifies* $x(t_1) = x_1$*.*

Proof. It is easy to adapt the proof given in [17, Proposition 1], where the initial condition is $x(t_0) = x_0$ (instead of $x(t_0) = 0$ as above). ■

It can be easily proved that $\mathscr{U}_f(t_1,u_l)$ is a convex subset of $L_2^{m_f}([t_0,T])$. Thus the problem $(\mathbf{LL})_{(t_1,u_l)}$ can be rewritten as a p-objective convex optimization problem:

$$(\mathbf{M})_{(t_1,u_l)} \qquad \begin{cases} \mathbf{MIN}_{\mathbf{u_f}} \ (\tilde{J}_1(t_1,u_l,\mathbf{u_f}),\ldots,\tilde{J}_p(t_1,u_l,\mathbf{u_f})) \\ \text{subject to } \ \mathbf{u_f} \in \mathscr{U}_f(t_1,u_l). \end{cases}$$

Definition 4.6. Let $\sigma \in \{e, we, pe\}$. An element $\mathbf{u_f} \in L_2^{m_f}([t_0,T])$ will be called *σ-control* of problem $(\mathbf{M})_{(t_1,u_l)}$ iff $(\mathbf{u_f}, x_{(t_1,u_l,\mathbf{u_f})})$ is a σ-control process of problem $(\mathbf{LL})_{(t_1,u_l)}$. We will denote $\mathscr{E}_\sigma(t_1,u_l)$ the set of all σ-controls of the p-objective optimization problem $(\mathbf{M})_{(t_1,u_l)}$.

Thus, using Lemma 4.4 and the well-known scalarization results from vector optimization [38, p. 302] we obtain the following.

Theorem 4.7 (see [17]). *Let* $(t_1,u_l) \in \mathscr{T} \times \mathscr{U}_l$ *and* $\hat{\mathbf{u}}_\mathbf{f} \in \mathscr{U}_f(t_1,u_l)$*, where* $\mathscr{U}_l$ *and* $\mathscr{U}_f$ *are given in* (4.1) *and* (4.11)*, respectively. The control process* $(\hat{\mathbf{u}}_\mathbf{f}, x_{(t_1,u_l,\hat{\mathbf{u}}_\mathbf{f})})$ *is weakly (resp. properly) efficient for problem* $(\mathbf{LL})_{(t_1,u_l)}$ *if and only if there exist nonnegative real numbers (resp. positive real numbers)* $\theta_1,\ldots,\theta_p$ *with* $\sum_{i=1}^p \theta_i = 1$ *such that* $\hat{\mathbf{u}}_\mathbf{f}$ *is an optimal control for the classical scalar optimal control problem:*

$$(\mathrm{S})_{(\theta_1,\ldots,\theta_p,t_1,u_l)} \qquad \begin{cases} \min\limits_{\mathbf{u_f}} \sum\limits_{i=1}^{p} \theta_i \tilde{J}_i(t_1,u_l,\mathbf{u_f}) \\ \text{subject to } \mathbf{u_f} \in \mathscr{U}_f(t_1,u_l). \end{cases}$$

In the sequel we need the following sets:

$$\Theta_\sigma = \begin{cases} \{(\theta_1,\ldots,\theta_p) \in]0,1[^p \mid \sum_{i=1}^p \theta_i = 1\} & \text{if } \sigma = pe \\ \\ \{(\theta_1,\ldots,\theta_p) \in [0,1]^p \mid \sum_{i=1}^p \theta_i = 1\} & \text{if } \sigma = we \end{cases} \tag{4.12}$$

and the following hypotheses:

$$H_\sigma(t_1): \begin{cases} (\exists i \in \{1,\ldots,p\}) \ (\forall (t,v,x) \in [t_0,t_1] \times \mathbb{R}^{m_l} \times \mathbb{R}^n) \\ \mathbf{u_f} \mapsto f_i(t,v,\mathbf{u_f},x) \ \text{ is strictly convex on } \mathbb{R}^m & \text{if } \sigma = pe \\ \\ (\forall i \in \{1,\ldots,p\}) \ (\forall (t,v,x) \in [t_0,t_1] \times \mathbb{R}^{m_l} \times \mathbb{R}^n) \\ \mathbf{u_f} \mapsto f_i(t,v,\mathbf{u_f},x) \ \text{ is strictly convex on } \mathbb{R}^m & \text{if } \sigma = we \end{cases}$$

and

$$(Hc)_\sigma : \begin{cases} \forall i \in \{1,\dots,p\} : \ \psi_i \geqslant 0, b_i = c_i = 0, \ d_i \geqslant 0, \ \sum_{j=1}^p d_j > 0 & \text{if } \sigma = pe \\ \forall i \in \{1,\dots,p\} : \ \psi_i \geqslant 0, b_i = c_i = 0, \ d_i > 0 & \text{if } \sigma = we, \end{cases}$$

where b_i, c_i, d_i have been introduced in the preliminary assumptions $(\mathscr{PA})$.

Theorem 4.8 (see [17]). *Let $\sigma \in \{we, pe\}$ and $(t_1, u_l) \in \mathscr{T} \times \mathscr{U}_l$. Assume that $H_\sigma(t_1)$ holds. Moreover, suppose that at least one of the following hypotheses holds:*

(i) $\mathbf{U_f}$ *is bounded.*
(ii) $(Hc)_\sigma$.

Then, for each $\theta = (\theta_1, \dots, \theta_p) \in \Theta_\sigma$, there exists a unique optimal control $\mathbf{u_f}(\theta, t_1, u_l, \cdot) \in \mathscr{U}_f(t_1, u_l)$ of the scalar problem $(S)_{(\theta, t_1, u_l)}$.

It is obvious that according to Theorem 4.7, $\mathbf{u_f}(\theta, t_1, u_l, \cdot)$ is a σ-control for multiobjective problem $(\mathbf{M})_{(t_1, u_l)}$. Moreover, Theorem 4.7 implies also that for each σ-control $\mathbf{u_f} \in \mathscr{U}_f(t_1, u_l)$ of the multiobjective problem $(\mathbf{M})_{(t_1, u_l)}$, there exists $\theta \in \Theta_\sigma$ such that $\mathbf{u_f}$ is the unique optimal control of the scalar problem $(S)_{(\theta, t_1, u_l)}$.

Thus we can state the following.

Corollary 4.9. *Let $(t_1, u_l) \in \mathscr{T} \times \mathscr{U}_l$. Under the hypotheses of Theorem 4.8 we have that the correspondence $\theta \mapsto \mathbf{u_f}(\theta, t_1, u_l, \cdot)$ is a surjection from Θ_σ to the set $\mathscr{E}_\sigma(t_1, u_l)$.*

In the sequel we will keep all the hypotheses of Theorem 4.8 in addition to the preliminary assumptions $(\mathscr{PA})$.

4.4 Equivalent Formulations of Problems $(\mathrm{OSVBC})_\sigma$ and $(\mathrm{PSVBC})_\sigma$

Consider, for each $(\theta, t_1, u_l) \in \Theta_\sigma \times \mathscr{T} \times \mathscr{U}_l \subset \mathbb{R}^p \times \mathbb{R} \times L_2^{m_l}([t_0, T])$, the function $F(\theta, t_1, u_l, \cdot) : \mathscr{U}_f(t_1, u_l) \to \mathbb{R}$ defined by

$$\forall \mathbf{u_f} \in \mathscr{U}_f(t_1, u_l) \qquad F(\theta, t_1, u_l, \mathbf{u_f}) := \sum_{i=1}^p \theta_i \tilde{J}_i(t_1, u_l, \mathbf{u_f}),$$

where $\mathscr{U}_f(t_1, u_l)$ and $\tilde{J}_i$ are given respectively in (4.11) and (4.9).

Note that problem $(\mathrm{OSVBC})_\sigma$ can be written equivalently as an optimistic semivectorial bilevel optimization problem:

$$(\mathrm{OSVB})_\sigma \qquad \min_{(t_1, u_l) \in \mathscr{T} \times \mathscr{U}_l} \ \min_{\mathbf{u_f} \in \mathscr{E}_\sigma(t_1, u_l)} \tilde{J}_l(t_1, u_l, \mathbf{u_f}).$$

According to Theorem 4.8, for each $(\theta, t_1, u_l) \in \Theta_\sigma \times \mathscr{T} \times \mathscr{U}_l$, there exists a unique minimizer $\mathbf{u_f}(\theta, t_1, u_l, \cdot) \in \mathscr{U}_f(t_1, u_l)$ of $F(\theta, t_1, u_l, \cdot)$ over $\mathscr{U}_f(t_1, u_l)$. According to Corollary 4.9, for each $(t_1, u_l) \in \mathscr{T} \times \mathscr{U}_l$, we have

$$\mathscr{E}_\sigma(t_1, u_l) = \bigcup_{\theta \in \Theta_\sigma} \{\mathbf{u_f}(\theta, t_1, u_l, \cdot)\}. \tag{4.13}$$

Then we obviously have the following.

Proposition 4.10 (see [17]). *Problem* $(\mathrm{OSVB})_\sigma$ *is equivalent to the problem*

$$\min_{(t_1,u_l) \in \mathscr{T} \times \mathscr{U}_l} \min_{\theta \in \Theta_\sigma} \tilde{J}_l(t_1, u_l, \mathbf{u_f}(\theta, t_1, u_l, \cdot)).$$

Thus, the optimistic semivectorial problem $(\mathrm{OSVB})_\sigma$ can be rewritten as an optimistic bilevel optimization problem (also called strong Stackelberg problem):

$$(\mathrm{OB})_\sigma \quad \begin{cases} \min\limits_{(t_1,u_l) \in \mathscr{T} \times \mathscr{U}_l} \min\limits_{\theta \in \Theta_\sigma} \tilde{J}_l(t_1, u_l, \mathbf{u_f}(\theta, t_1, u_l, \cdot)) \\ \text{where } \mathbf{u_f}(\theta, t_1, u_l, \cdot) \text{ is the unique minimizer to the problem} \\ (\mathrm{S})_{(\theta, t_1, u_l)}: \quad \min\limits_{\mathbf{u_f} \in \mathscr{U}_f(t_1, u_l)} F(\theta, t_1, u_l, \mathbf{u_f}). \end{cases}$$

Here the *upper and lower levels are given by scalar optimization problems and the lower level admits a unique solution.*

In the same way the pessimistic semivectorial problem can be rewritten as a pessimistic bilevel optimization problem (leading to a so-called weak Stackelberg problem; see [20] where this terminology was introduced).

Proposition 4.11 (see [17]). *Problem* $(\mathrm{PSVBC})_\sigma$ *is equivalent to the problem*

$$\min_{(t_1,u_l) \in \mathscr{T} \times \mathscr{U}_l} \sup_{\theta \in \Theta_\sigma} \tilde{J}_l(t_1, u_l, \mathbf{u_f}(\theta, t_1, u_l, \cdot)).$$

Finally, we can rewrite that problem as

$$(\mathrm{PB})_\sigma \quad \begin{cases} \min\limits_{(t_1,u_l) \in \mathscr{T} \times \mathscr{U}_l} \sup\limits_{\theta \in \Theta_\sigma} \tilde{J}_l(t_1, u_l, \mathbf{u_f}(\theta, t_1, u_l, \cdot)) \\ \text{where } \mathbf{u_f}(\theta, t_1, u_l, \cdot) \text{ is the unique minimizer of the problem} \\ (S)_{(\theta, t_1, u_l)}: \quad \min\limits_{\mathbf{u_f} \in \mathscr{U}_f(t_1, u_l)} F(\theta, t_1, u_l, \mathbf{u_f}). \end{cases}$$

4.5 Necessary and Sufficient Conditions for the Scalarized Lower Level Problem

Let $(t_1, u_l) \in \mathscr{T} \times \mathscr{U}_l$ and $\theta = (\theta_1, \ldots, \theta_p) \in \Theta_\sigma$ be given. The scalarized problem $(\mathrm{S})_{(\theta,t_1,u_l)}$ can be written as

$$\begin{aligned}
\min_{(\mathbf{u_f},x) \in L_2^{m_f}([t_0,T]) \times H_1^n([t_0,t_1])} \quad & \left[\sum_{i=1}^{p} \theta_i \psi_i(x(t_1)) + \int_{t_0}^{t_1} \left(\sum_{i=1}^{p} \theta_i f_i(t, u_l(t), \mathbf{u_f}(t), x(t)) \right) dt \right] \\
\text{s.t.} \quad & \mathbf{u_f}(t) \in \mathbf{U_f} \text{ a.e. on } [t_0,T], \quad \mathbf{u_f}(t) = 0 \text{ a.e. on } [t_1,T], \\
& \dot{x}(t) = A(t)x(t) + B_l(t)u_l(t) + \mathbf{B_f}(t)\mathbf{u_f}(t) \ \text{ a.e. on } [t_0,t_1] \\
& x(t_0) = x_0 \\
& x(t_1) \in \mathscr{F}.
\end{aligned}$$

Let $H : [t_0,t_1] \times \mathbb{R}^{m_l} \times \mathbb{R}^{m_f} \times \mathbb{R}^n \times \mathbb{R} \times \mathbb{R}^n \to \mathbb{R}$ be the Hamilton-Pontryagin function associated to this control problem (see, e.g. [2] or [39]) defined by

$$H(t, u_l, \mathbf{u_f}, x, \lambda_0, \lambda) = \lambda^T \left(A(t)x + B_l(t)u_l + \mathbf{B_f}(t)\mathbf{u_f} \right) - \lambda_0 \sum_{i=1}^{p} \theta_i f_i(t, u_l, \mathbf{u_f}, x).$$

Let $\lambda(\cdot) = (\lambda_1(\cdot), \ldots, \lambda_n(\cdot)) \in W_{1,\infty}^n([t_0,t_1])$ be the adjoint function, where $W_{1,\infty}^n([t_0, t_1])$ is the Banach space of absolutely continuous functions from $[t_0,t_1]$ to $\mathbb{R}^n$ having derivative in the Banach space $L_\infty^n([t_0,t_1])$ of essentially bounded measurable functions (see, e.g. [21] for details).

Since we use L_2 controls, and the Pontryagin maximum principle usually uses controls in L_∞, we will consider two particular situations in order to be able to get necessary and sufficient conditions for problem $(\mathrm{S})_{(\theta,t_1,u_l)}$, as stated below.

4.5.1 *The Case When $\mathbf{U_f}$ Is Bounded and $\mathscr{U}_l \subset L_\infty^{m_l}([t_0,T]) \cap L_2^{m_l}([t_0,T])$*

In this subsection we assume the set $\mathbf{U_f}$ is bounded (and closed, convex with nonempty interior) and the leader's controls are essentially bounded, i.e. $\mathscr{U}_l \subset L_\infty^{m_l}([t_0,T]) \cap L_2^{m_l}([t_0,T])$. Also, suppose the target set $\mathscr{F} = \{x \in \mathbb{R}^n | \, Gx = a\}$, where the matrix $G \in \mathbb{R}^{k \times n}$, and $a \in \mathbb{R}^k$ are given. Moreover we assume that $\operatorname{rank}(G) = k > 0$. However the results presented in this subsection are also valid when $\mathscr{F} = \mathbb{R}^n$ by taking $G = 0$, $a = 0$.

We obtain the following.

Theorem 4.12 (Necessary conditions). *Let* $(\mathbf{u_f}_*, x_*) \in L_2^{m_f}([t_0,T]) \times H_1^n([t_0,t_1])$ *be an optimal control process for problem* $(S)_{(\theta,t_1,u_l)}$. *Then there exist* $\lambda(\cdot) \in W_{1,\infty}^n([t_0,t_1])$, *a nonnegative real number* λ_0 *and a vector* $v \in \mathbb{R}^k$ *with* $(\lambda(\cdot), \lambda_0, v) \neq 0$ *such that*

$$\dot{\lambda}^T(t) = -\lambda^T(t)A(t) + \lambda_0 \sum_{i=1}^{p} \theta_i \frac{\partial f_i}{\partial x}(t, u_l(t), \mathbf{u_f}_*(t), x_*(t)), \text{ a.e. on } [t_0,t_1] \quad (4.14)$$

$$\lambda^T(t_1) = -\lambda_0 \sum_{i=1}^{p} \theta_i \frac{\partial \psi_i}{\partial x}(x_*(t_1)) + v^T G, \quad (4.15)$$

and, for almost all $t \in [t_0,t_1]$,

$$H(t, u_l(t), \mathbf{u_f}_*(t), x_*(t), \lambda_0, \lambda(t)) = \max_{\mathbf{v_f} \in \mathbf{U_f}} H(t, u_l(t), \mathbf{v_f}, x_*(t), \lambda_0, \lambda(t)). \quad (4.16)$$

Moreover, if the linearized system

$$\dot{x}(t) = A(t)x(t) + \mathbf{B_f}(t)\mathbf{u_f}(t) \text{ a.e. on } [t_0,t_1] \quad (4.17)$$

$$x(t_0) = 0 \quad (4.18)$$

is controllable,[1] *then we can take above* $\lambda_0 = 1$.

Sufficient conditions. *Let* $(x_*, \mathbf{u_f}_*) \in H_1^n([t_0,t_1]) \times L_2^{m_f}([t_0,T])$ *verifying* (4.2)–(4.5). *If there exist* $\lambda(\cdot) \in W_{1,\infty}^n([t_0,t_1])$ *and* $v \in \mathbb{R}^k$ *such that* (4.14)–(4.16) *are verified with* $\lambda_0 = 1$, *then* $(x_*, \mathbf{u_f}_*)$ *is an optimal control process for problem* $(S)_{(\theta,t_1,u_l)}$.

Proof. Since $\mathbf{U_f}$ is bounded, $\{\mathbf{u_f}(\cdot) \in L_2^{m_f}([t_0,T]) | \mathbf{u_f}(t) \in \mathbf{U_f}\} \subset L_\infty^{m_f}([t_0,T])$. For the same reason $u_l(\cdot) \in L_\infty^{m_l}([t_0,t_1])$. Thus we have $\mathbf{u_f}_* \in L_\infty^{m_f}([t_0,T])$; hence $x_* \in W_{1,\infty}^n([t_0,t_1])$ and $\lambda(\cdot) \in W_{1,\infty}^n([t_0,t_1])$. Therefore we can apply [39, Theorem 5.19] to obtain the first part (necessary conditions). Note that [39, Theorem 5.19] is stated for autonomous systems, but the same proof apply for non-autonomous systems.

For the second part (sufficiency conditions) we can use [39, Theorem 5.22] which also holds for non-autonomous systems with the same proof. ■

Remark 4.13. Since $\mathbf{U_f}$ is convex and closed and H is concave w.r.t. $\mathbf{u_f}$, relation (4.16) can equivalently be written as a variational inequality:

$$\forall \mathbf{v_f} \in \mathbf{U_f} \quad \Big(\lambda^T(t)\mathbf{B_f}(t) - \lambda_0 \sum_{i=1}^{p} \theta_i \frac{\partial f_i}{\partial \mathbf{u_f}}(t, u_l(t), \mathbf{u_f}_*(t), x_*(t))\Big)(\mathbf{v_f} - \mathbf{u_f}_*(t)) \leq 0$$

$$\text{a.e. on } [t_0,t_1].$$

[1]If A and $\mathbf{B_f}$ do not depend on t, it is well known that this system is controllable if, and only if, $\operatorname{rank}(\mathbf{B_f}, A\mathbf{B_f}, A^2\mathbf{B_f}, \ldots, A^{n-1}\mathbf{B_f}) = n$.

Finally, we can conclude the following.

Corollary 4.14. *Let $(t_1, u_l) \in \mathscr{U}_l$, and let $\theta \in \Theta_\sigma$. Assume that the linearized system* (4.17) *and* (4.18) *is controllable. Let $\mathbf{u_f} \in L_2^{m_f}([t_0, T])$. Then $\mathbf{u_f}(\cdot) = \mathbf{u_f}(\theta, t_1, u_l, \cdot)$ (i.e. $\mathbf{u_f}$ is the unique optimal control for problem $S_{(\theta,t_1,u_l)}$ presented in Theorem 4.8) if, and only if, there exists $\big(x(\cdot), \lambda(\cdot), v\big) \in H_1^n([t_0,t_1]) \times W_{1,\infty}^n([t_0,t_1]) \times \mathbb{R}^k$ such that*

$$\mathbf{u_f}(t) \in \mathbf{U_f} \text{ a.e. on } [t_0, T], \quad \mathbf{u_f}(t) = 0 \text{ a.e. on } [t_1, T], \tag{4.19}$$

$$\dot{x}(t) = A(t)x(t) + B_l(t)\,u_l(t) + \mathbf{B_f}(t)\mathbf{u_f}(t) \text{ a.e. on } [t_0, t_1], \tag{4.20}$$

$$x(t_0) = x_0, \tag{4.21}$$

$$Gx(t_1) = a, \tag{4.22}$$

$$\dot{\lambda}^T(t) = -\lambda^T(t)A(t) + \sum_{i=1}^{p} \theta_i \frac{\partial f_i}{\partial x}(t, u_l(t), \mathbf{u_f}(t), x(t)) \text{ a.e. on } [t_0, t_1], \tag{4.23}$$

$$\lambda^T(t_1) = -\sum_{i=1}^{p} \theta_i \frac{\partial \psi_i}{\partial x}(x(t_1)) + v^T G, \tag{4.24}$$

and, for almost all $t \in [t_0, t_1]$,

$$\forall \mathbf{v_f} \in \mathbf{U_f} \qquad \Big(\lambda^T(t)\mathbf{B_f}(t) - \sum_{i=1}^{p} \theta_i \frac{\partial f_i}{\partial \mathbf{u_f}}(t, u_l(t), \mathbf{u_f}(t), x_*(t))\Big)(\mathbf{v_f} - \mathbf{u_f}(t)) \leq 0. \tag{4.25}$$

4.5.2 The Case $\mathbf{U_f} = \mathbb{R}^{m_f}$: The Followers Problem Is Linear-Quadratic; Explicit Expressions of $\mathbf{u_f}(\theta, t_1, u_l, \cdot)$ and $x_{(t_1, u_l, \mathbf{u_f}(\theta, t_1, u_l, \cdot))}$

In this subsection we consider the case when $\mathbf{U_f} = \mathbb{R}^{m_f}$, $\mathscr{U}_l$ is an arbitrary closed, convex set with nonempty interior in $L_2^{m_l}([t_0, T])$ and the endpoint is free, i.e. the target set $\mathscr{F} = \mathbb{R}^n$. The objectives of the followers are quadratic, i.e. for $i = 1, \ldots, p$, and $(t, u_l, \mathbf{u_f}, x) \in [t_0, T] \times \mathbb{R}^{m_l} \times \mathbb{R}^{m_f} \times \mathbb{R}^n$

$$f_i(t, u_l, \mathbf{u_f}, x) = x^T Q_i(t) x + \mathbf{u_f}^T R_i(t) \mathbf{u_f},$$

where $Q_i(\cdot) : [t_0, T] \to \mathbb{R}^{n \times n}$ and $R_i(\cdot) : [t_0, T] \to \mathbb{R}^{m_f \times m_f}$ are continuous positive semidefinite matrix-valued functions.

Also

$$\psi_i(x) = x^T Q_i^f x,$$

where Q_i^f is a symmetric positive semidefinite matrix.

Moreover we make the following assumption:

$$(\mathrm{HLQP})_\sigma: \quad \begin{cases} \forall (i,t) \in \{1,\dots,p\} \times [t_0,T] \quad R_i(t) > 0 & \text{if } \sigma = we, \\ (\exists i \in \{1,\dots,p\})\,(\forall t \in [t_0,T]) \quad R_i(t) > 0 & \text{if } \sigma = pe. \end{cases}$$

Note that this particular choice of f_i and ψ_i agrees with all the assumptions $(\mathscr{P}\mathscr{A})$.
Let us denote

$$Q(\theta,\cdot) = \sum_{i=1}^{p} \theta_i Q_i(\cdot); \quad R(\theta,\cdot) = \sum_{i=1}^{p} \theta_i R_i(\cdot); \quad Q^f(\theta) = \sum_{i=1}^{p} \theta_i Q_i^f.$$

Thus, the scalarized problem $(\mathrm{S})_{(\theta,t_1,u_l)}$ becomes the linear-quadratic problem

$$(\mathrm{LQP}) \quad \begin{cases} \min \left(x(t_1)^T Q^f(\theta) x(t_1) + \int_{t_0}^{t_1} (x(t)^T Q(\theta,t) x(t) + \mathbf{u_f}(t)^T R(\theta,t) \mathbf{u_f}(t)) \mathrm{d}t \right) \\ \text{s.t.} \quad \dot{x}(t) = A(t)x(t) + \mathbf{B_f}(t)\mathbf{u_f}(t) + B_l(t)u_l(t) \quad \text{a.e. on } [t_0,t_1], \\ \qquad x(t_0) = x_0. \end{cases}$$

We have the following result which is probably known also for L_2 controls, but we will present a proof for the sake of completeness.

Theorem 4.15. *Let $(x_*(\cdot),\mathbf{u_{f*}}(\cdot)) \in H_1^n([t_0,t_1]) \times L_2^{m_f}([t_0,t_1])$ verify the differential system and the initial condition for problem (LQP). Then the control process $(x_*(\cdot),\mathbf{u_{f*}}(\cdot))$ is optimal for problem (LQP) if, and only if, there exists a function $\lambda(\cdot) \in H_1^n([t_0,t_1])$ such that*

$$\dot{\lambda}^T(t) = -\lambda^T(t)A(t) - x_*^T(t)Q(\theta,t) \quad \text{a.e. on } [t_0,t_1], \tag{4.26}$$

$$\lambda^T(t_1) = x_*^T(t_1)Q^f(\theta), \tag{4.27}$$

$$\mathbf{u_{f*}}(t) = -R^{-1}(\theta,t)\mathbf{B_f}^T(t)\lambda(t) \quad \text{a.e. on } [t_0,t_1]. \tag{4.28}$$

Proof. Assume that $\lambda(\cdot) \in H_1^n([t_0,t_1])$ verifies (4.26)–(4.28). Let $(x,\mathbf{u_f}) \in H_1^n([t_0,t_1]) \times L_2^{m_f}([t_0,t_1])$ verify the differential system and the initial condition for problem (LQP). We have for almost all $t \in [t_0,t_1]$

$$\begin{aligned} \frac{\mathrm{d}}{\mathrm{d}t}\left(\lambda^T(t)(x(t)-x_*(t))\right) &= \dot{\lambda}^T(t)(x(t)-x_*(t)) + \lambda^T(t)(\dot{x}(t)-\dot{x}_*(t)) \\ &= -(\lambda^T(t)A(t) + x_*^T(t)Q(\theta,t))(x(t)-x_*(t)) \\ &\quad + \lambda^T(t)\left(A(t)(x(t)-x_*(t)) + \mathbf{B_f}(t)(\mathbf{u_f}(t)-\mathbf{u_{f*}}(t))\right) \\ &= -x_*^T(t)Q(\theta,t)(x(t)-x_*(t)) - \mathbf{u_{f*}}^T(t)R(\theta,t)(\mathbf{u_f}(t)-\mathbf{u_{f*}}(t)). \end{aligned}$$

With the initial condition for $x(\cdot), x_*(\cdot)$ and final condition for $\lambda(\cdot)$ we get by integration

$$x_*^T(t_1)Q^f(\theta)(x(t_1)-x_*(t_1)) = -\int_{t_0}^{t_1}\Big(x_*^T(t)Q(\theta,t)(x(t)-x_*(t)) + \mathbf{u_{f*}}^T(t)R(\theta,t)(\mathbf{u_f}(t)-\mathbf{u_{f*}}(t))\Big)\mathrm{d}t. \tag{4.29}$$

Denote

$$J(x(\cdot),\mathbf{u_f}(\cdot)) = \Big(x(t_1)^T Q^f(\theta)x(t_1) + \int_{t_0}^{t_1}(x(t)^T Q(\theta,t)x(t) + \mathbf{u_f}(t)^T R(\theta,t)\mathbf{u_f}(t))\mathrm{d}t\Big).$$

For any symmetric positive semidefinite matrix P and for all vectors v, v_*, we obviously have

$$v^T P v - v_*^T P v_* \geq 2v_*^T P(v - v_*).$$

Therefore

$$\begin{aligned} J(x(\cdot),\mathbf{u_f}(\cdot)) - J(x_*(\cdot),\mathbf{u_{f*}}(\cdot)) \geq 2\Big[&x_*^T(t_1)Q^f(\theta)(x(t_1)-x_*(t_1)) \\ &+ \int_{t_0}^{t_1}\Big(x_*^T(t)Q(\theta,t)(x(t)-x_*(t)) \\ &+ \mathbf{u_{f*}}^T(t)R(\theta,t)(\mathbf{u_f}(t)-\mathbf{u_{f*}}(t))\Big)\mathrm{d}t\Big]. \end{aligned}$$

From (4.29) the last expression is zero; hence $J(x(\cdot),u(\cdot)) - J(x_*(\cdot),\mathbf{u_{f*}}(\cdot)) \geq 0$. Thus $(x_*(\cdot),\mathbf{u_{f*}}(\cdot))$ is an optimal control process for problem (SQP).

Conversely, let $(x_*(\cdot),\mathbf{u_{f*}}(\cdot)) \in H_1^n([t_0,t_1]) \times L_2^{m_f}([t_0,t_1])$ be a solution of (LQP) (which exists and is unique according to Theorem 4.8). Let $\lambda(\cdot) \in H_1^n([t_0,t_1])$ be the solution of the linear system (4.26) verifying the final condition (4.27). For any $\mathbf{u_f}(\cdot) \in L_2^{m_f}([t_0,t_1])$, denoting by $x(\cdot)$ the corresponding solution of the differential system and the initial condition for problem (LQP), we have (using a similar calculus as before)

$$\lambda^T(t_1)(x(t_1)-x_*(t_1)) = -\int_{t_0}^{t_1}\Big(x_*^T(t)Q(\theta,t)(x(t)-x_*(t)) + \lambda^T(t)\mathbf{B_f}(t)(\mathbf{u_f}(t)-\mathbf{u_{f*}}(t))\Big)\mathrm{d}t.$$

On the other, using the fact that the directional derivative of J at the optimal point $(x_*(\cdot),\mathbf{u_{f*}}(\cdot))$ in the direction $(x(\cdot),\mathbf{u_f}(\cdot)) - (x_*(\cdot),\mathbf{u_{f*}}(\cdot))$ is positive we have

$$x_*^T(t_1)Q^f(\theta)(x(t_1)-x_*(t_1))+\int_{t_0}^{t_1}(x_*^T(t)Q(\theta,t)(x(t)-x_*(t))$$
$$+\mathbf{u_{f*}}^T(t)R(\theta,t)(\mathbf{u_f}(t)-\mathbf{u_{f*}}(t)))dt\geq 0.$$

Finally we obtain

$$\int_{t_0}^{t_1}(\lambda^T(t)\mathbf{B_f}(t)-\mathbf{u_{f*}}^T(t)R(\theta,t))(\mathbf{u_f}(t)-\mathbf{u_{f*}}(t)))dt\leq 0.$$

Since $\mathbf{u_f}(\cdot)$ can be arbitrarily chosen in $L_2^{m_f}([t_0,t_1])$, we obtain that (4.28) is satisfied. ■

Next we will show that, in the linear-quadratic case, it is possible to compute explicitly the optimal control and state as a function of the parameters θ, t_1, u_l by means of a $2n\times 2n$ resolvent matrix of a linear differential system based on data. This fact will allow us to find explicit optimality conditions for our bilevel problems.

Recall that $\mathbf{u_f}(\theta,t_1,u_l,\cdot)$ denotes the unique optimal control of the scalarized problem $(S)_{(\theta,t_1,u_l)}$. The corresponding unique state and adjoint state (verifying Theorem 4.15) will be denoted by $x(\theta,t_1,u_l,\cdot)$ and $\lambda(\theta,t_1,u_l,\cdot)$.

To be more precise, the functions $x(\theta,t_1,u_l,\cdot)$ and $\lambda(\theta,t_1,u_l,\cdot)$ verify the following boundary linear problem:

$$\frac{\partial x}{\partial t}(\theta,t_1,u_l,t)=A(t)x(\theta,t_1,u_l,t)-\mathbf{B_f}(t)R^{-1}(\theta,t)\mathbf{B_f}(t)^T\lambda(\theta,t_1,u_l,t)$$
$$+B_l(t)u_l(t)\quad \text{a.e. on } [t_0,t_1], \tag{4.30}$$

$$\frac{\partial \lambda}{\partial t}(\theta,t_1,u_l,t)=-A(t)^T\lambda(\theta,t_1,u_l,t)-Q(\theta,t)x(\theta,t_1,u_l,t)\quad \text{a.e. on } [t_0,t_1], \tag{4.31}$$

$$x(\theta,t_1,u_l,t_0)=x_0, \tag{4.32}$$

$$\lambda(\theta,t_1,u_l,t_1)=Q^f(\theta)x(\theta,t_1,u_l,t_1) \tag{4.33}$$

and

$$\mathbf{u_f}(\theta,t_1,u_l,t)=-R^{-1}(\theta,t)\mathbf{B_f}^T(t)\lambda(\theta,t_1,u_l,t)\quad \text{a.e. on } [t_0,t_1]. \tag{4.34}$$

Given $t_1\in\mathscr{T}$ and $\theta\in\Theta_\sigma$, consider the matrix-valued function $P(\theta,t_1,\cdot):[t_0,t_1]\to\mathbb{R}^{n\times n}$ which, under our hypotheses about matrices $Q^f(\theta)$, $Q(\theta,t)$, $R(\theta,t)$, is the unique continuously differentiable solution (see, e.g. [1]) of the Riccati matrix differential equation (RMDE) on $[t_0,t_1]$:

$$\frac{\partial P}{\partial t}(\theta,t_1,t)=-A(t)^TP(\theta,t_1,t)-P(\theta,t_1,t)A(t)-Q(\theta,t)$$
$$+P(\theta,t_1,t)\mathbf{B_f}(t)R(\theta,t)^{-1}\mathbf{B_f}(t)^TP(\theta,t_1,t)$$

satisfying the final time condition

$$P(\theta,t_1,t_1) = Q^f(\theta). \tag{4.35}$$

Moreover, $P(\theta,t_1,t)$ is a symmetric positive definite matrix for each t.

Following [18] we can express P in terms of a resolvent matrix depending directly on data. Thus consider for all $(\theta,t) \in \Theta_\sigma \times [t_0,t_1]$ the $2n \times 2n$ matrix which defines the linear system (4.30) and (4.31)

$$L(\theta,t) = \begin{pmatrix} A(t) & -\mathbf{B_f}(t)R^{-1}(\theta,t)\mathbf{B_f}^T(t) \\ -Q(\theta,t) & -A^T(t) \end{pmatrix}.$$

The proof of the following result can be found in [18].

Proposition 4.16. *Let $\Psi(\theta,\cdot,\cdot)$ be the resolvent (or state transition) matrix associated to the linear differential system defined by $L(\theta,t)$, i.e. for each $s \in [t_0,T]$, $\Psi(\theta,\cdot,s)$ satisfies the Cauchy problem:*

$$\frac{\partial \Psi}{\partial t}(\theta,t,s) = L(\theta,t)\Psi(\theta,t,s), \;\; t \in [t_0,T], \quad \Psi(\theta,s,s) = I_{2n}.$$

Let us divide the matrix $\Psi(\theta,t,s)$ into four $n \times n$ blocks:

$$\Psi(\theta,t,s) = \begin{pmatrix} \Psi_{11}(\theta,t,s) & \Psi_{12}(\theta,t,s) \\ \Psi_{21}(\theta,t,s) & \Psi_{22}(\theta,t,s) \end{pmatrix}.$$

Then, for all $t \in [t_0,t_1]$, the matrix $[\Psi_{11}(\theta,t,t_1) + \Psi_{12}(\theta,t,t_1)Q^f(\theta)]$ is invertible and

$$P(\theta,t_1,t) = \left[\Psi_{21}(\theta,t,t_1) + \Psi_{22}(\theta,t,t_1)Q^f(\theta)\right]\left[\Psi_{11}(\theta,t,t_1) + \Psi_{12}(\theta,t,t_1)Q^f(\theta)\right]^{-1}. \tag{4.36}$$

Next, let us denote by $\xi(\theta,t_1,u_l,\cdot) \in H_1^n([t_0,t_1])$ the unique solution of the following linear Cauchy problem:

$$\frac{\partial \xi}{\partial t}(\theta,t_1,u_l,t) = \left(-A(t)^T + P(\theta,t_1,t)\mathbf{B_f}(t)R^{-1}(\theta,t)\mathbf{B_f}(t)\right)\xi(\theta,t_1,u_l,t)$$
$$- P(\theta,t_1,t)B_l(t)u_l(t) \quad \text{a.e. on } [t_0,t_1], \tag{4.37}$$
$$\xi(\theta,t_1,u_l,t_1) = 0. \tag{4.38}$$

Lemma 4.17. *For all $t \in [t_0,t_1]$ we have*

$$\lambda(\theta,t_1,u_l,t) = P(\theta,t_1,t)x(\theta,t_1,u_l,t) + \xi(\theta,t_1,u_l,t). \tag{4.39}$$

Proof. Computing the derivative $\frac{\partial}{\partial t}\Big(\lambda(\theta,t_1,u_l,t) - P(\theta,t_1,t)x(\theta,t_1,u_l,t) - \xi(\theta,t_1,u_l,t)\Big)$ and then, using (4.30)–(4.33), (RMDE), (4.35), (4.37), and (4.38), the result follows easily. ■

Denote by $\Xi(\theta,t_1,\cdot,\cdot)$ the resolvent matrix associated to (4.37), i.e. for all $(\theta,t_1,s) \in \Theta_\sigma \times \mathscr{T} \times [t_0,T]$

$$\frac{\partial \Xi}{\partial t}(\theta,t_1,t,s) = \big(-A(t)^T + P(\theta,t_1,t)\mathbf{B}_\mathbf{f}(t)R^{-1}(\theta,t)\mathbf{B}_\mathbf{f}(t)\big)\Xi(\theta,t_1,t,s),\ t \in [t_0,T] \tag{4.40}$$

$$\Xi(\theta,t_1,s,s) = I_n. \tag{4.41}$$

Based on this we are able to solve the boundary problem (4.30)–(4.33) in terms of data.

Corollary 4.18. *For all* $(\theta,t_1,u_l) \in \Theta_\sigma \times \mathscr{T} \times L_2^{m_l}([t_0,T])$ *and for all* $t \in [t_0,t_1]$ *we have*

$$\begin{pmatrix} x(\theta,t_1,u_l,t) \\ \lambda(\theta,t_1,u_l,t) \end{pmatrix} = \Psi(\theta,t,t_0)\begin{pmatrix} x_0 \\ P(\theta,t_1,t_0)x_0 + \xi(\theta,t_1,u_l,t_0) \end{pmatrix} + \int_{t_0}^{t} \Psi(\theta,t,s)\begin{pmatrix} B_l(s)u_l(s) \\ 0 \end{pmatrix} \mathrm{d}s,$$

where

$$\xi(\theta,t_1,u_l,t_0) = \int_{t_0}^{t_1} \Xi(\theta,t_1,t_0,s)P(\theta,t_1,s)B_l(s)u_l(s)\mathrm{d}s.$$

Remark 4.19. The right-hand side member in the formulas giving $x(\theta,t_1,u_l,t)$ and $\lambda(\theta,t_1,u_l,t)$ in Corollary 4.18 is defined for all $(t_1,t) \in]t_0,T[\times[t_0,T]$ (and not only for $(t_1,t) \in \mathscr{T} \times [t_0,t_1]$) and for all θ belonging to an open convex set Ω with $\Theta_\sigma \subseteq \Omega$. Indeed, the formulas in Corollary 4.18 have a meaning as long as $R(\theta,t) > 0$.

When $\sigma = pe$, by $(\text{HLQP})_{pe}$ it is obvious that we can take $\Omega = \mathbb{R}^p_{++}$.

When $\sigma = we$, the continuous function $[t_0,T] \times \mathbb{R}^{m_f} \ni (t,\mathbf{u}_f) \mapsto \mathbf{u}_\mathbf{f}{}^T R_i(t)\mathbf{u}_\mathbf{f}$ attains its minimum value, say α_i, on the compact set $[t_0,T] \times \mathbb{S}$, where $\mathbb{S}$ is the unit sphere in $\mathbb{R}^{m_f}$, $i = 1,\ldots,p$. According to $(\text{HLQP})_{we}$ we have $\alpha_i > 0$ for all i. Then, it is easy to see that we can take

$$\Omega = \{\theta \in \mathbb{R}^p | \sum_{i=1}^{p} \theta_i\alpha_i > 0\}.$$

We will extend the functions $x(\cdot,\cdot,\cdot,\cdot)$ and $\lambda(\cdot,\cdot,\cdot,\cdot)$ based on these formulas as continuous functions from $\Omega\times]t_0,T[\times L_2^{m_l}([t_0,T])\times[t_0,T]$ to $\mathbb{R}^n$. Moreover, based on (4.34), we will extend also the function $\mathbf{u_f}(\cdot,\cdot,\cdot,\cdot)$ as a continuous function from $\Omega\times]t_0,T[\times L_2^{m_l}([t_0,T])\times[t_0,T]$ to $\mathbb{R}^{m_f}$. These extensions are necessary further in order to obtain optimality conditions for the upper level.

Using the differentiability with respect to parameters of a differential equation and some straightforward computation we have the following.

Proposition 4.20. *The resolvent $\Psi(\cdot,\cdot,\cdot)$ is continuously differentiable on $\Omega\times[t_0,T]\times[t_0,T]$. We have the following formulas for all $(\theta,t,s)\in\Omega\times[t_0,T]\times[t_0,T]$ and $i=1,\ldots,p$:*

$$\frac{\partial\Psi}{\partial\theta_i}(\theta,t,s)=\int_s^t\Psi(\theta,t,\tau)\frac{\partial L}{\partial\theta_i}(\theta,\tau)\Psi(\theta,\tau,s)\mathrm{d}\tau,\quad\text{where}\tag{4.42}$$

$$\frac{\partial L}{\partial\theta_i}(\theta,t)=\begin{pmatrix}0 & \mathbf{B_f}(t)R^{-1}(\theta,t)R_i(t)R^{-1}(\theta,t)\mathbf{B_f}(t)^T\\ -Q_i(t) & 0\end{pmatrix},\tag{4.43}$$

$$\frac{\partial\Psi}{\partial s}(\theta,t,s)=-\Psi(\theta,t,s)L(\theta,s).\tag{4.44}$$

By (4.36) and the previous proposition we obtain immediately the following.

Proposition 4.21. *The matrix-valued function $P(\cdot,\cdot,\cdot)$ is continuously differentiable on $\Omega\times[t_0,T]\times[t_0,T]$ and verifies the following formulas:*

$$\begin{aligned}\frac{\partial P}{\partial\theta_i}(\theta,t_1,t)=&\Big[\frac{\partial\Psi_{21}}{\partial\theta_i}(\theta,t,t_1)+\frac{\partial\Psi_{22}}{\partial\theta_i}(\theta,t,t_1)Q^f(\theta)+\Psi_{22}(\theta,t,t_1)Q_i^f\Big]\\&\times\Big[\Psi_{11}(\theta,t,t_1)+\Psi_{12}(\theta,t,t_1)Q^f(\theta)\Big]^{-1}\\&-\Big[\Psi_{21}(\theta,t,t_1)+\Psi_{22}(\theta,t,t_1)Q^f(\theta)\Big]\Big[\Psi_{11}(\theta,t,t_1)+\Psi_{12}(\theta,t,t_1)Q^f(\theta)\Big]^{-1}\\&\times\Big[\frac{\partial\Psi_{11}}{\partial\theta_i}(\theta,t,t_1)+\frac{\partial\Psi_{12}}{\partial\theta_i}(\theta,t,t_1)Q^f(\theta)+\Psi_{12}(\theta,t,t_1)Q_i^f\Big]\\&\times\Big[\Psi_{11}(\theta,t,t_1)+\Psi_{12}(\theta,t,t_1)Q^f(\theta)\Big]^{-1}\end{aligned}\tag{4.45}$$

and

$$\frac{\partial\Psi}{\partial\theta_i}(\theta,t,s)=\begin{pmatrix}\frac{\partial\Psi_{11}}{\partial\theta_i}(\theta,t,s) & \frac{\partial\Psi_{12}}{\partial\theta_i}(\theta,t,s)\\ \frac{\partial\Psi_{21}}{\partial\theta_i}(\theta,t,s) & \frac{\partial\Psi_{22}}{\partial\theta_i}(\theta,t,s)\end{pmatrix}.$$

Using an analogue calculus we obtain

$$\begin{aligned}
&\frac{\partial P}{\partial t_1}(\theta,t_1,t)\\
&=\Big[\frac{\partial \Psi_{21}}{\partial t_1}(\theta,t,t_1)+\frac{\partial \Psi_{22}}{\partial t_1}(\theta,t,t_1)Q^f(\theta)\Big]\Big[\Psi_{11}(\theta,t,t_1)+\Psi_{12}(\theta,t,t_1)Q^f(\theta)\Big]^{-1}\\
&\quad-\Big[\Psi_{21}(\theta,t,t_1)+\Psi_{22}(\theta,t,t_1)Q^f(\theta)\Big]\Big[\Psi_{11}(\theta,t,t_1)+\Psi_{12}(\theta,t,t_1)Q^f(\theta)\Big]^{-1}\\
&\quad\times\Big[\frac{\partial \Psi_{11}}{\partial t_1}(\theta,t,t_1)+\frac{\partial \Psi_{12}}{\partial t_1}(\theta,t,t_1)Q^f(\theta)\Big]\Big[\Psi_{11}(\theta,t,t_1)+\Psi_{12}(\theta,t,t_1)Q^f(\theta)\Big]^{-1}.
\end{aligned} \tag{4.46}$$

The computation of $\frac{\partial \Psi_{ij}}{\partial t_1}(\theta,t,t_1)$ *can be obtained using* (4.44)*:*

$$\begin{pmatrix} \frac{\partial \Psi_{11}}{\partial t_1}(\theta,t,t_1) & \frac{\partial \Psi_{12}}{\partial t_1}(\theta,t,t_1) \\ \\ \frac{\partial \Psi_{21}}{\partial t_1}(\theta,t,t_1) & \frac{\partial \Psi_{22}}{\partial t_1}(\theta,t,t_1) \end{pmatrix} = -\begin{pmatrix} \Psi_{11}(\theta,t,t_1) & \Psi_{12}(\theta,t,t_1) \\ \\ \Psi_{21}(\theta,t,t_1) & \Psi_{22}(\theta,t,t_1) \end{pmatrix} L(\theta,t_1). \tag{4.47}$$

Proposition 4.22. *The resolvent* $\Xi(\cdot,\cdot,\cdot,\cdot)$ *is continuously differentiable on* $\Omega\times[t_0,T]\times[t_0,T]$*, and denoting*

$$\mathscr{A}(\theta,t_1,t):=-A(t)^T+P(\theta,t_1,t)\mathbf{B_f}(t)R^{-1}(\theta,t)\mathbf{B_f}(t), \tag{4.48}$$

we have

$$\frac{\partial \Xi}{\partial \theta_i}(\theta,t_1,t,s)=\int_s^t \Xi(\theta,t_1,t,\tau)\frac{\partial \mathscr{A}}{\partial \theta_i}(\theta,t_1,\tau)\Xi(\theta,t_1,\tau,s)\mathrm{d}\tau, \tag{4.49}$$

$$\frac{\partial \Xi}{\partial t_1}(\theta,t_1,t,s)=\int_s^t \Xi(\theta,t_1,t,\tau)\frac{\partial \mathscr{A}}{\partial t_1}(\theta,t_1,\tau)\Xi(\theta,t_1,\tau,s)\mathrm{d}\tau, \tag{4.50}$$

$$\frac{\partial \Xi}{\partial s}(\theta,t_1,t,s)=-\Xi(\theta,t_1,t,s)\mathscr{A}(\theta,t_1,s). \tag{4.51}$$

The computation of the partial derivatives of $\mathscr{A}(\theta,t_1,t)$ *can be obtained using* (4.36)*, Proposition 4.21 and the obvious formulas:*

$$\frac{\partial}{\partial \theta_i}R^{-1}(\theta,t)=-R^{-1}(\theta,t)R_i(t)R^{-1}(\theta,t).$$

Proposition 4.23. *For all* $(\theta,t_1)\in\Omega\times]t_0,T[$*, the maps* $u_l\mapsto x(\theta,t_1,u_l,\cdot)$*,* $u_l\mapsto\lambda(\theta,t_1,u_l,\cdot)$*, respectively,* $u_l\mapsto\mathbf{u_f}(\theta,t_1,u_l,\cdot)$ *are affine and continuous from* $L_2^{m_l}([t_0,T])$ *to* $H_1^n([t_0,t_1])$*, respectively, from* $L_2^{m_l}([t_0,T])$ *to* $L_2^{m_f}([t_0,T])$*. Therefore they are* *continuously Fréchet differentiable on* $L_2^{m_l}([t_0,T])$ *and, for any* $u_l \in L_2^{m_l}([t_0,t_1])$, *their Fréchet differentials (which are linear continuous maps from* $L_2^{m_l}([t_0,T])$ *to* $H_1^n([t_0,t_1])$ *and, respectively, from* $L_2^{m_l}([t_0,T])$ *to* $L_2^{m_f}([t_0,T])$*) verify for all* $h \in L_2^{m_l}([t_0,T])$ *and for all* $t \in [t_0,t_1]$*:*

$$\frac{\partial}{\partial u_l} x(\theta,t_1,u_l,t)\cdot h = \Psi_{12}(\theta,t,t_0)\int_{t_0}^{t_1} \Xi(\theta,t_1,t_0,s)P(\theta,t_1,s)B_l(s)h(s)\mathrm{d}s + \int_{t_0}^{t} \Psi_{11}(\theta,t,s)B_l(s)h(s)\mathrm{d}s \tag{4.52}$$

$$\frac{\partial}{\partial u_l} \lambda(\theta,t_1,u_l,t)\cdot h = \Psi_{22}(\theta,t,t_0)\int_{t_0}^{t_1} \Xi(\theta,t_1,t_0,s)P(\theta,t_1,s)B_l(s)h(s)\mathrm{d}s + \int_{t_0}^{t} \Psi_{21}(\theta,t,s)B_l(s)h(s)\mathrm{d}s \tag{4.53}$$

$$\frac{\partial}{\partial u_l} \mathbf{u_f}(\theta,t_1,u_l,t)\cdot h = -R^{-1}(\theta,t)\mathbf{B_f}(t)^T \frac{\partial}{\partial u_l}\lambda(\theta,t_1,u_l,t)\cdot h,. \tag{4.54}$$

Proof. It is easy to see from Corollary 4.18 and (4.30) and (4.31) that the maps $u_l \mapsto x(\theta,t_1,u_l,\cdot)$ and $u_l \mapsto \lambda(\theta,t_1,u_l,\cdot)$ are affine and continuous from $L_2^{m_l}([t_0,T])$ to $H_1^n([t_0,t_1])$; hence (4.52) and (4.53) hold. Then, by (4.34), we obtain that the map $u_l \mapsto \mathbf{u_f}(\theta,t_1,u_l,\cdot)$ from $L_2^{m_l}([t_0,T])$ to $L_2^{m_f}([t_0,T])$ is affine and continuous and we get (4.54). ∎

Theorem 4.24 (Regularity of $\mathbf{u_f}(\cdot,\cdot,\cdot,\cdot)$ and $x(\cdot,\cdot,\cdot,\cdot)$).

1. *The functions* $\mathbf{u_f}(\cdot,\cdot,\cdot,\cdot) : \Omega\times]t_0,T[\times L_2^{m_l}([t_0,T])\times[t_0,T]\to\mathbb{R}^{m_f}$ *and* $x(\cdot,\cdot,\cdot,\cdot) : \Omega\times]t_0,T[\times L_2^{m_l}([t_0,T])\times[t_0,T]\to\mathbb{R}^n$ *are continuous.*
2. *The function* $(\theta,t_1,u_l)\mapsto \mathbf{u_f}(\theta,t_1,u_l,\cdot)$ *from* $\Omega\times]t_0,T[\times L_2^{m_l}([t_0,T])$ *to* $L_2^{m_f}([t_0,T])$ *is continuous as well as the function* $(\theta,t_1,u_l)\mapsto x(\theta,t_1,u_l,\cdot)$ *from* $\Omega\times]t_0,T[\times L_2^{m_l}([t_0,T])$ *to* $L_2^n([t_0,T])$.
3. *For each fixed* $(\bar\theta,\bar t_1,\bar u_l)\in\Omega\times]t_0,T[\times L_2^{m_l}([t_0,T])$*:*
   - *The function* $\theta\mapsto \mathbf{u_f}(\theta,\bar t_1,\bar u_l,\cdot)$ *from* Ω *to* $L_2^{m_f}([t_0,T])$ *and the function*[2] $\theta\mapsto x(\theta,\bar t_1,\bar u_l,\cdot)$ *from* Ω *to* $L_2^n([t_0,T])$ *are continuously Fréchet differentiable on* Ω.
   - *The function* $u_l\mapsto \mathbf{u_f}(\bar\theta,\bar t_1,u_l,\cdot)$ *from* $L_2^{m_l}([t_0,T])$ *to* $L_2^{m_f}([t_0,T])$ *and the function* $u_l\mapsto x(\bar\theta,\bar t_1,u_l,\cdot)$ *from* $L_2^{m_l}([t_0,T])$ *to* $H_1^n([t_0,T])$ *are continuously Fréchet differentiable.*
   - *The functions* $t_1\mapsto \mathbf{u_f}(\bar\theta,t_1,\bar u_l,\cdot)$ *from* $]t_0,T[$ *to* $L_2^{m_f}([t_0,T])$ *and* $t_1\mapsto x(\bar\theta,t_1,\bar u_l,\cdot)$ *from* $]t_0,T[$ *to* $L_2^n([t_0,T])$ *are a.e. differentiable on* $]t_0,T[$*, and for almost all* $t_1\in]t_0,T[$, $\frac{\partial \mathbf{u_f}}{\partial t_1}(\bar\theta,\bar t_1,\bar u_l,\cdot)\in L_2^{m_f}([t_0,T])$ *and* $\frac{\partial x}{\partial t_1}(\bar\theta,\bar t_1,\bar u_l,\cdot)\in L_2^n([t_0,T])$.

[2]Note that the embedding $H_1^n([t_0,T])\subset L_2^n([t_0,T])$ is continuous.

Moreover, for each $t_1 \in]t_0,T[$ such that $\bar{u}_l$ is continuous[3] *at t_1, these functions are differentiable in t_1.*

4. *The functions $\mathbf{u_f}(\cdot,\cdot,\cdot,\cdot)$, $x(\cdot,\cdot,\cdot,\cdot)$ and their partial derivatives can be explicitly represented as functions of data (supposing we are able to compute the resolvent matrices Ψ and Ξ).*

Proof. By Corollary 4.18, Remark 4.19 and Propositions 4.20–4.23, we obtain points 1 and 4.

To prove point 2 we will use the fact that, by Corollary 4.18, we can write

$$x(\theta,t_1,u_l,t) = \alpha(\theta,t_1,t) + \int_{t_0}^{T} X(\theta,t_1,t,s)u_l(s)\mathrm{d}s,$$

where

$$\alpha(\theta,t_1,t) = \big(\Psi_{11}(\theta,t,t_0) + \Psi_{12}P(\theta,t_1,t_0)\big)x_0$$

and $X(\theta,t_1,t,s)$ is described later in relations (4.61) and (4.63). Obviously α : $\Omega\times]t_0,T[\times[t_0,T] \to \mathbb{R}^n$ is a continuous function, and for each $s \in [t_0,T]$, $X(\cdot,\cdot,\cdot,s)$ is continuous on $\Omega\times]t_0,T[\times[t_0,T] \to \mathbb{R}^{n\times m_l}$, and, for each $(\theta,t_1,t) \in \Omega\times]t_0,T[\times[t_0,T]$, $X(\theta,t_1,t,\cdot) \in L_2^{n\times m_l}([t_0,T])$.

We obtain easily that the function $(\theta,t_1) \mapsto \alpha(\theta,t_1,\cdot)$ is continuous from $\Omega\times]t_0,T[$ to $\mathscr{C}([t_0,T];\mathbb{R}^n)$, where $\mathscr{C}([t_0,T];\mathbb{R}^n)$ is the Banach space of continuous functions on $[t_0,T]$ with values in $\mathbb{R}^n$ endowed with the uniform convergence norm.

Since the embedding $\mathscr{C}([t_0,T];\mathbb{R}^n) \subset L_2^n([t_0,T])$ is continuous, we obtain that the function $(\theta,t_1) \mapsto \alpha(\theta,t_1,\cdot)$ is continuous from $\Omega\times]t_0,T[$ to $L_2^n([t_0,T])$.

Also, using Lebesgue's dominated convergence theorem, we obtain easily that the function $(\theta,t_1,t) \mapsto X(\theta,t_1,t,\cdot)$ is continuous from $\Omega\times]t_0,T[\times[t_0,T]$ to $L_2^{n\times m_l}([t_0,T])$. Denoting $y(\theta,t_1,u_l,t) = \int_{t_0}^{T} X(\theta,t_1,t,s)u_l(s)\mathrm{d}s$, and writing

$$\begin{aligned} y(\theta',t_1',u_l',t) - y(\theta,t_1,u_l,t) &= \big(y(\theta',t_1',u_l',t) - y(\theta',t_1',u_l,t)\big) \\ &\quad + \big(y(\theta',t_1',u_l,t) - y(\theta,t_1,u_l,t)\big), \end{aligned}$$

we obtain that

$$\begin{aligned} |y(\theta',t_1',u_l',t) - y(\theta,t_1,u_l,t)| &\le \|X(\theta',t_1',t,\cdot)\|_2 \cdot \|u_l' - u_l\|_2 \\ &\quad + \|X(\theta',t_1',t,\cdot) - X(\theta,t_1,t,\cdot)\|_2 \cdot \|u_l\|_2 \end{aligned}$$

which finally prove the continuity of the function $(\theta,t_1,u_l) \mapsto x(\theta,t_1,u_l,\cdot)$ from $\Omega\times]t_0,T[\times L_2^{m_l}([t_0,T])$ to $L_2^n([t_0,T])$.

---

[3]In the sense that there exists a function $\tilde{u}_l$ continuous at t_1 and $\bar{u}_l(t) = \tilde{u}_l(t)$ a.e. on $[t_0,T]$. Note that by Lusin's theorem, we can find measurable sets of arbitrarily small positive measure and such functions $\tilde{u}_l$ which are continuous on the complement of those sets.

With similar arguments we can prove the continuity of the function $(\theta, t_1, u_l) \mapsto \mathbf{u_f}(\theta, t_1, u_l, \cdot)$ from $\Omega \times]t_0, T[\times L_2^{m_l}([t_0, T])$ to $L_2^{m_f}([t_0, T])$ and point 3. ■

4.6 Optimality Conditions for the Upper Level, i.e. for Problems $(\mathbf{OB})_\sigma$ and $(\mathbf{PB})_\sigma$

In this section we will restrain to the case considered in Sect. 4.5.2. Moreover we will suppose that $\mathscr{U}_l$ is the closed ball

$$\mathscr{U}_l = \left\{ u_l \in L_2^{m_l}([t_0, T]) \mid \|u_l\|_2 \leq R \right\}, \tag{4.55}$$

where R is a strictly positive real.

4.6.1 *The Optimistic Bilevel Problem*

We begin with some preliminary results in order to obtain an existence result when $\mathbf{U_f}$ is not assumed to be bounded, so we cannot apply the results obtained in [17]. We could adapt the proof given in [17], but we will give direct proofs for the sake of completeness.

Lemma 4.25. *Let X and Y be arbitrary sets and let $J : X \times Y \to \mathbb{R} \cup \{+\infty\}$ such that, for each $x \in X$, the set* argmin $J(x, \cdot)$ *is nonempty. Then the problems*

$$\min_{(x,y) \in X \times Y} J(x,y) \tag{4.56}$$

and

$$\min_{x \in X} \min_{y \in Y} J(x,y) \tag{4.57}$$

are equivalent, i.e. problem (4.56) *is solvable if and only if problem* (4.57) *is solvable. In this case the solution sets coincide as well as the minimal values.*

Proof. Let $(\hat{x}, \hat{y}) \in X \times Y$ be a solution for problem (4.56), i.e. $(\hat{x}, \hat{y}) \in \operatorname{argmin} J(\cdot, \cdot)$. Then, for each $x \in X$, we have obviously $J(\hat{x}, \hat{y}) = \min_{y \in Y} J(\hat{x}, y) \leq \min_{y \in Y} J(x, y)$; hence $J(\hat{x}, \hat{y}) = \min_{x \in X} \min_{y \in Y} J(x,y)$, and $(\hat{x}, \hat{y})$ is a solution for problem (4.57).

Conversely, let $(\bar{x}, \bar{y})$ be a solution for problem (4.57). This means that, for all $x \in X$ and $y' \in \operatorname{argmin} J(x, \cdot)$, we have we have $J(\bar{x}, \bar{y}) \leq J(x, y') = \min_{y \in Y} J(x, y)$; hence for all $(x,y) \in X \times Y$, we have $J(\bar{x}, \bar{y}) \leq J(x,y)$. Therefore $(\bar{x}, \bar{y})$ is a solution for problem (4.56). ■

Lemma 4.26. *Let $X = X' \times X''$ where X' is a compact metric space, X'' is a closed bounded convex set in a reflexive Banach space $\mathscr{X}''$ and let Y be a compact metric space. Let $J : X \times Y \to \mathbb{R} \cup \{+\infty\}$ be a lower semicontinuous function on the topological product space $X' \times (X'', s) \times Y$, where s denotes the topology on X'' induced by the strong topology of $\mathscr{X}''$. Suppose that $J(x', \cdot, y)$ is convex for each fixed $(x', y) \in X' \times Y$.*

Then the hypotheses of Lemma 4.25 are fulfilled, and $\operatorname{argmin} J(\cdot,\cdot,\cdot) \neq \emptyset$.

Proof. 1. From Banach–Alaoglu–Kakutani theorem, X'' is compact for the weak topology of $\mathscr{X}''$ denoted w. Thus $X \times Y = (X' \times X'') \times Y$ is compact in the topological product space $[X' \times (\mathscr{X}'', w)] \times Y$. Let us show that J is sequentially lower semicontinuous on $[X' \times (X'', w_{X''})] \times Y$, where $w_{X''}$ stands for the topology on X'' induced by the weak topology of $\mathscr{X}''$. Indeed, for any real α, let us denote

$$SL_\alpha = \{(x', x'', y) \in X' \times X'' \times Y | J(x', x'', y) \leq \alpha\}.$$

Since J is lower semicontinuous on $X' \times (X'', s) \times Y$ we have that SL_α is closed in $X' \times (X'', s) \times Y$. Consider now a sequence $((x'_k, x''_k, y_k))_k$ in SL_α convergent to some (x', x'', y) in $X' \times (\mathscr{X}'', w) \times Y$. Since (x''_k) converges weakly to x'', by Mazur's lemma [32, p. 6], there is a sequence $(\bar{x}''_k)$ converging to x'' in (X'', s) such that, for any k, $\bar{x}''_k$ is a convex combination of x''_k's. Then, by the convexity of X'' and of $J(x'_k, \cdot, y_k)$, we have $\bar{x}''_k \in X''$ and

$$J(x'_k, \bar{x}''_k, y_k) \leq J(x'_k, x''_k, y_k) \leq \alpha.$$

Thus $(x'_k, \bar{x}''_k, y_k) \in SL_\alpha$ and $(x'_k, \bar{x}''_k, y_k)$ converges to (x', x'', y) in $X' \times (X'', s) \times Y$; hence $(x', x'', y) \in SL_\alpha$. Therefore SL_α is sequentially closed in $X' \times (\mathscr{X}'', w) \times Y$; hence J is sequentially lower semicontinuous on $X' \times (\mathscr{X}'', w) \times Y$. Finally, by Weierstrass' theorem, we obtain that $\operatorname{argmin} J(\cdot,\cdot,\cdot) \neq \emptyset$.

Let now $x = (x', x'') \in X = X' \times X''$ be fixed. Since Y is compact and $J(x, \cdot)$ is lower semicontinuous on Y, we obtain from Weierstrass' theorem that argmin $J(x, \cdot) \neq \emptyset$. ■

Let $\hat{J}_l : \Omega \times]t_0, T[\times \mathscr{U}_l \to \mathbb{R} \cup \{+\infty\}$ be defined by

$$\hat{J}_l(\theta, t_1, u_l) := \tilde{J}_l(t_1, u_l, \mathbf{u_f}(\theta, t_1, u_l, \cdot)) = J_l(t_1, u_l, \mathbf{u_f}(\theta, t_1, u_l, \cdot), x(\theta, t_1, u_l, \cdot)). \tag{4.58}$$

Theorem 4.27. *In addition to hypotheses $(\mathscr{PA})$ we suppose that, for each $t \in [t_0, T]$, $f_l(t, \cdot, \cdot, \cdot)$ is a convex function.*

Moreover we suppose the following hypothesis:

$$\text{(Hf)} \quad \begin{cases} \text{there is some } \alpha \in \mathrm{L}_\infty([\mathrm{t}_0, \mathrm{T}]) \text{ and some real constant } \beta \text{ such that,} \\ \text{for almost all } \mathrm{t} \in [\mathrm{t}_0, \mathrm{T}], \text{ and for all } (\mathrm{u}_l, \mathbf{u_f}, \mathrm{x}) \in \mathbb{R}^{\mathrm{m}_l} \times \mathbb{R}^{\mathrm{m_f}} \times \mathbb{R}^{\mathrm{n}}, \\ \quad |\nabla_{(u_l, \mathbf{u_f}, x)} f_l(t, u_l, \mathbf{u_f}, x)| \leq \alpha(t) + \beta |(u_l, \mathbf{u_f}, x)|. \end{cases} \tag{4.59}$$

Then problem $(OB)_{we}$ has at least one solution and it is equivalent to the problem

$$(P_l) \qquad \min_{(\theta,t_1,u_l)\in\Theta_{we}\times\mathscr{T}\times\mathscr{U}_l} \hat{J}_l(\theta,t_1,u_l).$$

Proof. We will show that all the hypotheses of Lemma 4.26 are fulfilled (denoting $X'=\mathscr{T}$, $X''=\mathscr{U}_l, Y=\Theta_{we}$, $\mathscr{X}''=L_2^{m_l}([t_0,T])$, $x'=t_1$, $x''=u_l, y=\theta$, $J(x',x'',y)=\hat{J}_l(\theta,t_1,u_l)$), and then the conclusion follows from Lemma 4.25.

$\mathscr{U}_l$ is (strongly) closed, bounded and convex in $L_2^{m_l}([t_0,T])$; $\mathscr{T}$ and Θ_{we} are compact. For fixed $(t_1,\theta)\in\mathscr{T}\times\Theta_{we}$, the function $\hat{J}_l(\theta,\cdot,t_1)$ is convex since, for any $t\in[t_0,T]$, the function $f_l(t,\cdot,\cdot,\cdot)$ is convex, and $u_l\mapsto\mathbf{u_f}(\theta,t_1,u_l,\cdot)$, $u_l\mapsto x(\theta,t_1,u_l,\cdot)$ are affine functions by Proposition 4.23.

To finish the proof it is sufficient to show that $\hat{J}_l$ is lower semicontinuous on $\Theta_{we}\times\mathscr{T}\times\mathscr{U}_l$, where $\mathscr{U}_l$ is endowed with the topology induced by the strong topology of $L_2^{m_l}([t_0,T])$. Let $(\theta^k,t_1^k,u_l^k)_k$ be a sequence in $\Theta_{we}\times\mathscr{T}\times\mathscr{U}_l$ which converges (strongly) to an element $(\bar{\theta},\bar{t}_1,\bar{u}_l)$. Since $\Theta_{we}\times\mathscr{T}\times\mathscr{U}_l$ is closed we have $(\bar{\theta},\bar{t}_1,\bar{u}_l)\in\Theta_{we}\times\mathscr{T}\times\mathscr{U}_l$.

We obtain from Lemma 4.4, Theorem 4.24 and (4.58) that, for each fixed $t_1\in\mathscr{T}$, the function $\hat{J}_l(\cdot,t_1,\cdot)$ is lower semicontinuous. On the other hand we have

$$\hat{J}_l(\theta^k,t_1^k,u_l^k)=\hat{J}_l(\theta^k,\bar{t}_1,u_l^k)+(\hat{J}_l(\theta^k,t_1^k,u_l^k)-\hat{J}_l(\theta^k,\bar{t}_1,u_l^k)),$$

and the term $(\hat{J}_l(\theta^k,t_1^k,u_l^k)-\hat{J}_l(\theta^k,\bar{t}_1,u_l^k))$ tends to 0 as $k\to+\infty$. Indeed,

$$\begin{aligned}\hat{J}_l(\theta^k,t_1^k,u_l^k)-\hat{J}_l(\theta^k,\bar{t}_1,u_l^k)=&\int_{t_0}^{t_1^k} f_l(t,u_l^k(t),\mathbf{u_f}(\theta^k,t_1^k,u_l^k,t),x(\theta^k,t_1^k,u_l^k,t))\mathrm{d}t\\&-\int_{t_0}^{\bar{t}_1} f_l(t,u_l^k(t),\mathbf{u_f}(\theta^k,\bar{t}_1,u_l^k,t),x(\theta^k,\bar{t}_1,u_l^k,t))\mathrm{d}t.\end{aligned} \tag{4.60}$$

Since the sequence (u_l^k) is bounded in $L_2^{m_l}([t_0,T])$, by (Hf) and Theorem 4.24 there is a constant $M>0$, such that, for all $k\in\mathbb{N}$ and almost all $t\in[t_0,T]$,

$$|f_l(t,u_l^k(t),\mathbf{u_f}(\theta^k,t_1^k,u_l^k,t),x(\theta^k,t_1^k,u_l^k,t))|\le M$$

and

$$|f_l(t,u_l^k(t),\mathbf{u_f}(\theta^k,\bar{t}_1,u_l^k,t),x(\theta^k,\bar{t}_1,u_l^k,t))|\le M.$$

Finally, let us show that both integrals in (4.60) have the same limit as $k\to+\infty$, which is $\int_{t_0}^{\bar{t}_1} f_l(t,\bar{u}_l(t),\mathbf{u_f}(\bar{\theta},\bar{t}_1,\bar{u}_l,t),x(\bar{\theta},\bar{t}_1,\bar{u}_l,t))\mathrm{d}t$. To do this it is sufficient to prove that these convergences hold for a subsequence. Since (u_l^k) converges in

$L_2^{m_l}([t_0,T])$, there exists a subsequence $(u_l^{k'})_{k'}$, such that $(u_l^{k'}(t))_{k'}$ converges to $\bar{u}_l(t)$ a.e. on $[t_0,T]$. Then, we can apply Lebesgue's dominated convergence theorem to obtain the last claim.

Therefore, using the fact that for each $t_1 \in \mathscr{T}$ the function $\hat{J}_l(\cdot,t_1,\cdot)$ is lower semicontinuous, we obtain

$$\lim_{k\to+\infty} \hat{J}_l(\theta^k,t_1^k,u_l^k) = \lim_{k\to+\infty} \hat{J}_l(\theta^k,\bar{t}_1,u_l^k) \geq \hat{J}_l(\bar{\theta},\bar{t}_1,\bar{u}_l). \qquad \blacksquare$$

We denote $(f_l)'_{u_l}(\cdot,\cdot,\cdot,\cdot) : [t_0,T] \times \mathbb{R}^{m_l} \times \mathbb{R}^{m_f} \times \mathbb{R}^n \to \mathbb{R}^{m_l}$, $(f_l)'_{\mathbf{u_f}}(\cdot,\cdot,\cdot,\cdot) : [t_0,T] \times \mathbb{R}^{m_l} \times \mathbb{R}^{m_f} \times \mathbb{R}^n \to \mathbb{R}^{m_f}$, $(f_l)'_x(\cdot,\cdot,\cdot,\cdot) : [t_0,T] \times \mathbb{R}^{m_l} \times \mathbb{R}^{m_f} \times \mathbb{R}^n \to \mathbb{R}^n$ the partial derivatives of f_l with respect to the variables located on the second, third and fourth position, respectively.

Also, let us denote for all $(\theta,t_1,t,s) \in \Omega\times]t_0,T[\times[t_0,T]\times[t_0,T]$,

$$X(\theta,t_1,t,s) = \Big[\chi_{[t_0,t_1]}(s)\Psi_{12}(\theta,t,t_0)\Xi(\theta,t_1,t_0,s)P(\theta,t_1,s) + \chi_{[t_0,t]}(s)\Psi_{11}(\theta,t,s)\Big]B_l(s) \tag{4.61}$$

$$Y(\theta,t_1,t,s) = -R^{-1}(\theta,t)\mathbf{B_f}(t)^T\Big[\chi_{[t_0,t_1]}(s)\Psi_{22}(\theta,t,t_0)\Xi(\theta,t_1,t_0,s)P(\theta,t_1,s) + \chi_{[t_0,t]}(s)\Psi_{21}(\theta,t,s)\Big]B_l(s), \tag{4.62}$$

where $\chi_{[t_0,t]} : [t_0,T] \to \mathbb{R}$ is the characteristic function

$$\chi_{[t_0,t]}(s) = \begin{cases} 1 & \text{if } s \in [t_0,t], \\ 0 & \text{otherwise.} \end{cases} \tag{4.63}$$

Thus, formulas (4.52), (4.54) become

$$\frac{\partial}{\partial u_l}x(\theta,t_1,u_l,\cdot)\cdot h = \int_{t_0}^{T} X(\theta,t_1,\cdot,s)h(s)\mathrm{d}s, \tag{4.64}$$

$$\frac{\partial}{\partial u_l}\mathbf{u_f}(\theta,t_1,u_l,\cdot)\cdot h = \int_{t_0}^{T} Y(\theta,t_1,\cdot,s)h(s)\mathrm{d}s. \tag{4.65}$$

Next result is necessary to ensure the differentiability of $\hat{J}_l$.

Lemma 4.28. *Suppose that f_l satisfies the hypothesis (Hf) given in Theorem 4.27, in addition to the hypothesis $(\mathscr{PA})$. Then, for each fixed $t_1 \in]t_0,T[$, the functional $\hat{J}_l(\cdot,t_1,\cdot) : \Omega \times L_2^{m_l}([t_0,T]) \to \mathbb{R}$ is well defined and continuously Fréchet differentiable. Its partial derivatives with respect to θ_i, $i=1,\dots,p$ are given by*

$$\frac{\partial \hat{f}_l}{\partial \theta_i}(\theta,t_1,u_l) = \int_{t_0}^{t_1} (f_l)'_{\mathbf{u_f}}(t,u_l(t),\mathbf{u_f}(\theta,t_1,u_l,t),x(\theta,t_1,u_l,t))^T \frac{\partial \mathbf{u_f}}{\partial \theta_i}(\theta,t_1,u_l,t)\mathrm{d}t$$
$$+ \int_{t_0}^{t_1} (f_l)'_x(t,u_l(t),\mathbf{u_f}(\theta,t_1,u_l,t),x(\theta,t_1,u_l,t))^T \frac{\partial x}{\partial \theta_i}(\theta,t_1,u_l,t)\mathrm{d}t. \quad (4.66)$$

Its partial Fréchet gradient with respect to u_l at (θ,t_1,u_l) is given, for almost all $s \in [t_0,t_1]$, by[4]

$$\nabla_{u_l}\hat{J}_l(\theta,t_1,u_l)(s) = (f_l)'_{u_l}(s,u_l(s),\mathbf{u_f}(\theta,t_1,u_l,s),x(\theta,t_1,u_l,s))$$
$$+ \int_{t_0}^{T} L^T(\theta,t_1,t,s)(f_l)'_{\mathbf{u_f}}(t,u_l(t),\mathbf{u_f}(\theta,t_1,u_l,t),x(\theta,t_1,u_l,t))\mathrm{d}t$$
$$+ \int_{t_0}^{T} X^T(\theta,t_1,t,s)(f_l)'_x(t,u_l(t),\mathbf{u_f}(\theta,t_1,u_l,t),x(\theta,t_1,u_l,t))\mathrm{d}t. \quad (4.67)$$

Moreover, for each fixed $(\theta,u_l) \in \Omega \times L_2^{m_l}([t_0,T])$, the function $\hat{J}_l(\theta,\cdot,u_l) \in H_1([t_0,T])$, and for almost all $t_1 \in]t_0,T[$, its derivative is given by

$$\frac{\partial \hat{J}_l}{\partial t_1}(\theta,t_1,u_l) = f_l(t_1,u_l(t_1),\mathbf{u_f}(\theta,t_1,u_l,t_1),x(\theta,t_1,u_l,t_1))$$
$$+ \int_{t_0}^{t_1} (f_l)'_{\mathbf{u_f}}(t,u_l(t),\mathbf{u_f}(\theta,t_1,u_l,t),x(\theta,t_1,u_l,t))^T \frac{\partial \mathbf{u_f}}{\partial t_1}(\theta,t_1,u_l,t)\mathrm{d}t$$
$$+ \int_{t_0}^{t_1} (f_l)'_x(t,u_l(t),\mathbf{u_f}(\theta,t_1,u_l,t),x(\theta,t_1,u_l,t))^T \frac{\partial x}{\partial t_1}(\theta,t_1,u_l,t)\mathrm{d}t. \quad (4.68)$$

In particular, at each point t_1 such that u_l is continuous at t_1 (see footnote 3), the real-valued function $t \mapsto \hat{J}_l(\theta,t,u_l)$ is differentiable.

Proof. By [4, Example 2, p. 20] we have that the functional $J_l(t_1,\cdot,\cdot,\cdot) : L_2^{m_l}([t_0,T]) \times L_2^{m_f}([t_0,T]) \times H_1^n([t_0,T]) \to \mathbb{R}$ is well defined and is continuously Fréchet differentiable for each fixed $t_1 \in]t_0,T[$. Moreover, its partial derivatives satisfy, for all $(t_1,u_l,\mathbf{u_f},x) \in]t_0,T[\times L_2^{m_l}([t_0,T]) \times L_2^{m_f}([t_0,T]) \times H_1^n([t_0,T])$, the following equations:

$$\frac{\partial J_l}{\partial u_l}(t_1,u_l,\mathbf{u_f},x)\cdot v = \int_{t_0}^{t_1} (f_l)'_{u_l}(t,u_l(t),\mathbf{u_f}(t),x(t))^T v(t)\mathrm{d}t \quad \forall v \in L_2^{m_l}([t_0,T]),$$

---

[4] We identify the Hilbert space $L_2^{m_l}([t_0,T])$ with its dual according to Riesz-Fréchet theorem; hence $\nabla_{u_l}\hat{J}_l(\theta,t_1,u_l) \in L_2^{m_l}([t_0,T])$ (see, e.g. [7, p. 38]).

$$\frac{\partial J_l}{\partial \mathbf{u_f}}(t_1,u_l,\mathbf{u_f},x)\cdot w = \int_{t_0}^{t_1} (f_l)'_{\mathbf{u_f}}(t,u_l(t),\mathbf{u_f}(t),x(t))^T w(t)\mathrm{d}t \quad \forall w \in L_2^{m_f}([t_0,T]),$$

$$\frac{\partial J_l}{\partial x}(t_1,u_l,\mathbf{u_f},x)\cdot z = \int_{t_0}^{t_1} (f_l)'_{x}(t,u_l(t),\mathbf{u_f}(t),x(t))^T z(t)\mathrm{d}t \quad \forall z \in H_1^{n}([t_0,T]).$$

Also, for each fixed $(u_l,\mathbf{u_f},x) \in L_2^{m_l}([t_0,T]) \times L_2^{m_f}([t_0,T]) \times H_1^n([t_0,T])$ and for almost all $t_1 \in]t_0,T]$,

$$\frac{\partial J_l}{\partial t_1}(t_1,u_l,\mathbf{u_f},x) = f_l(t_1,u_l(t_1),\mathbf{u_f}(t_1),x(t_1)).$$

Let us identify, using Riesz-Fréchet theorem, the Hilbert spaces $L_2^{m_l}([t_0,T])$, $L_2^{m_f}([t_0,T])$ and $L_2^n([t_0,T])$ with their duals, and do not identify $H_1^n([t_0,T])$ with its dual $H_1^n([t_0,T])^*$. Based on the fact that (see [21, pp. 81–82] for details)

$$H_1^n([t_0,T]) \subset L_2^n([t_0,T]) \equiv L_2^n([t_0,T])^* \subset H_1^n([t_0,T])^*$$

and both embeddings are continuous and dense, and the duality product between $H_1^n([t_0,T])$ and $H_1^n([t_0,T])^*$ coincide with the inner product in $L_2^n([t_0,T])$ on $H_1^n([t_0,T]) \times L_2^n([t_0,T])$, we have that the Fréchet gradients $\nabla_{u_l} J_l(t_1,u_l,\mathbf{u_f},x) \in L_2^{m_l}([t_0,T])$, $\nabla_{\mathbf{u_f}} J_l(t_1,u_l,\mathbf{u_f},x) \in L_2^{m_f}([t_0,T])$ and $\nabla_x J_l(t_1,u_l,\mathbf{u_f},x) \in L_2^n([t_0,T])$ are given for almost all $t \in [t_0,T]$ by

$$\nabla_{u_l} J_l(t_1,u_l,\mathbf{u_f},x)(t) = \begin{cases} (f_l)'_{u_l}(t,u_l(t),\mathbf{u_f}(t),x(t)), \text{ if } t \in [t_0,t_1], \\ 0, \text{ if } t \in]t_1,T], \end{cases}$$

$$\nabla_{\mathbf{u_f}} J_l(t_1,u_l,\mathbf{u_f},x)(t) = \begin{cases} (f_l)'_{\mathbf{u_f}}(t,u_l(t),\mathbf{u_f}(t),x(t)), \text{ if } t \in [t_0,t_1], \\ 0, \text{ if } t \in]t_1,T], \end{cases}$$

$$\nabla_{x} J_l(t_1,u_l,\mathbf{u_f},x)(t) = \begin{cases} (f_l)'_{x}(t,u_l(t),\mathbf{u_f}(t),x(t)), \text{ if } t \in [t_0,t_1], \\ 0, \text{ if } t \in]t_1,T], \end{cases}$$

Now, using the chain rule in (4.58), we obtain immediately (4.66) and (4.68) and also

$$\begin{aligned} \nabla_{u_l} \hat{J}_l(\theta,t_1,u_l)(t) =& (f_l)'_{u_l}(t,u_l(t),\mathbf{u_f}(\theta,t_1,u_l,t),x(\theta,t_1,u_l,t)) \\ &+ \left(\frac{\partial}{\partial u_l}\mathbf{u_f}(\theta,t_1,u_l,\cdot)\right)^* (f_l)'_{\mathbf{u_f}}(t,u_l(t),\mathbf{u_f}(\theta,t_1,u_l,t),x(\theta,t_1,u_l,t)) \\ &+ \left(\frac{\partial}{\partial u_l}x(\theta,t_1,u_l,\cdot)\right)^* (f_l)'_{x}(t,u_l(t),\mathbf{u_f}(\theta,t_1,u_l,t),x(\theta,t_1,u_l,t)), \end{aligned} \tag{4.69}$$

and, for almost all $t \in]t_1,T]$, $\nabla_{u_l}\hat{J}_l(\theta,t_1,u_l)(t) = 0$, where M^* stands for the adjoint operator of a linear continuous operator M between two Hilbert spaces.

Fix $(\theta,t_1,u_l) \in \Omega\times]t_0,T[\times L_2^{m_l}([t_0,T])$. Since the embedding $H_1^n([t_0,T]) \subset L_2^n([t_0,T])$ is continuous, we can consider the partial Fréchet derivative $\frac{\partial}{\partial u_l}x(\theta,t_t,u_l,\cdot)$ as a linear continuous operator from $L_2^{m_l}([t_0,T])$ to $L_2^n([t_0,T])$. Denote $\langle\cdot,\cdot\rangle_n$ the inner product in $L_2^n([t_0,T])$. For all $h \in L_2^{m_l}([t_0,T])$, $k \in L_2^n([t_0,T])$ we have

$$\begin{aligned}\langle\frac{\partial}{\partial u_l}x(\theta,t_t,u_l,\cdot)h,k\rangle_n &= \int_{t_0}^{T} k^T(t)\left(\int_{t_0}^{T} X(\theta,t_1,t,s)h(s)\mathrm{d}s\right)\mathrm{d}t\\ &= \int_{t_0}^{T} h^T(s)\left(\int_{t_0}^{T} X^T(\theta,t_1,t,s)k(t)\mathrm{d}t\right)\mathrm{d}s\\ &= \langle h,\left(\frac{\partial}{\partial u_l}x(\theta,t_t,u_l,\cdot)\right)^* k\rangle_{m_l};\end{aligned}$$

hence

$$\left(\frac{\partial}{\partial u_l}x(\theta,t_t,u_l,\cdot)\right)^* \cdot k = \int_{t_0}^{T} X^T(\theta,t_1,t,\cdot)k(t)\mathrm{d}t. \tag{4.70}$$

In the same way we get for all $k \in L_2^{m_f}([t_0,T])$

$$\left(\frac{\partial}{\partial u_l}\mathbf{u_f}(\theta,t_t,u_l,\cdot)\right)^* \cdot k = \int_{t_0}^{T} Y^T(\theta,t_1,t,\cdot)k(t)\mathrm{d}t. \tag{4.71}$$

Finally (4.67) follows from (4.69). ■

Theorem 4.29 (First-order necessary conditions when the final time is fixed, i.e. $\mathscr{T} = \{t_1\}$**).** *Suppose that* $\mathscr{T} = \{t_1\}$, *and* f_l *satisfies hypotheses* $(\mathscr{PA})$, *(Hf), and* $f_l(t,\cdot,\cdot,\cdot)$ *is convex for all* $t \in [t_0,T]$.

Let $(\bar{\theta},\bar{u}_l) \in \Theta_{we} \times \mathscr{U}_l$ *solve* $(\mathrm{OB})_{we}$. *Then there are nonnegative real numbers* $\mu, l_1,\ldots,l_p$ *and a real number* ν *such that*

$$\nabla_{u_l}\hat{J}_l(\bar{\theta},t_1,\bar{u}_l)(t) + \mu\bar{u}_l(t) = 0 \qquad \text{a.e. on } [t_0,T], \tag{4.72}$$

$$\frac{\partial \hat{J}_l}{\partial \theta_i}(\bar{\theta},t_1,\bar{u}_l) - l_i + \nu = 0, \qquad i = 1,\ldots,p, \tag{4.73}$$

$$\mu(\|\bar{u}_l\|_2 - R) = 0, \tag{4.74}$$

$$l_i\bar{\theta}_i = 0, \qquad i = 1,\ldots,p, \tag{4.75}$$

and of course

$$\sum_{i=1}^{p} \bar{\theta}_i = 1, \tag{4.76}$$

$$\|\bar{u}_l\|_2 \le R, \quad \bar{\theta}_i \ge 0, \qquad i = 1, \dots, p. \tag{4.77}$$

Remark 4.30. According to (4.67), equation (4.72) is a Fredholm integral equation in the unknown $\bar{u}_l$ (linear if $f_l(t,\cdot,\cdot,\cdot)$ is quadratic, case which satisfies hypothesis (Hf)), depending on $2p+1$ parameters (μ and $\bar{\theta}_i$). Assuming that we are able to solve this integral equation, (4.73)–(4.76) represent a nonlinear system with $2p+2$ equations and $2p+2$ unknowns μ, ν, θ_i, l_i. A similar remark applies to the next theorem.

Theorem 4.31 (First-order necessary conditions when the final time $t_1 \in \mathscr{T} = [\underline{t}, \bar{t}] \subset]t_0, T[$). *Suppose that f_l satisfies hypotheses $(\mathscr{PA})$, (Hf) and $f_l(t,\cdot,\cdot,\cdot)$ is convex for all $t \in [t_0, T]$.*

Let $(\bar{t}_1, \bar{\theta}, \bar{u}_l) \in \mathscr{T} \times \Theta_{we} \times \mathscr{U}_l$ solve $(OB)_{we}$. Suppose that $\bar{u}_l$ is continuous at $\bar{t}_1$ (see footnote 3). Then there are nonnegative real numbers $\mu, l_1, \dots, l_p, l_{p+1}, l_{p+2}$ and a real number ν such that

$$\nabla_{u_l} \hat{f}_l(\bar{\theta}, t_1, \bar{u}_l)(t) + \mu \bar{u}_l(t) = 0 \qquad \text{a.e. on } [t_0, T], \tag{4.78}$$

$$\frac{\partial \hat{f}_l}{\partial \theta_i}(\bar{\theta}, t_1, \bar{u}_l) - l_i + \nu = 0, \qquad i = 1, \dots, p, \tag{4.79}$$

$$\frac{\partial \hat{f}_l}{\partial t_1}(\bar{\theta}, t_1, \bar{u}_l) - l_{p+1} + l_{p+2} = 0, \tag{4.80}$$

$$\mu(\|\bar{u}_l\|_2 - R) = 0, \tag{4.81}$$

$$l_i \bar{\theta}_i = 0, \qquad i = 1, \dots, p, \tag{4.82}$$

$$l_{p+1}(\bar{t}_1 - \underline{t}) = 0, \tag{4.83}$$

$$l_{p+2}(\bar{t} - \bar{t}_1) = 0, \tag{4.84}$$

and of course

$$\sum_{i=1}^{p} \bar{\theta}_i = 1, \tag{4.85}$$

$$\|\bar{u}_l\|_2 \le R, \quad \bar{\theta}_i \ge 0, \qquad i = 1, \dots, p. \tag{4.86}$$

The proof of Theorems 4.29 and 4.31 is a direct application of the generalized Lagrange multiplier rule under Kurcyusz–Robinson–Zowe regularity condition (see [39, Theorem 5.3]) and is based on Theorem 4.27 and on Lemma 4.28.

4.6.2 The Pessimistic Bilevel Problem

In this section we assume that $f_l(t,\cdot,\cdot,\cdot)$ is quadratic, i.e. for all $(t,u_l,\mathbf{u_f},x) \in [t_0,T]\times\mathbb{R}^{m_l}\times\mathbb{R}^{m_f}\times\mathbb{R}^n$,

$$f_l(t,u_l,\mathbf{u_f},x) = u_l^T S_l(t)u_l + \mathbf{u_f}^T R_l(t)\mathbf{u_f} + x^T Q_l(t)x, \tag{4.87}$$

where $S_l(\cdot), R_l(\cdot), Q_l(\cdot)$ are continuous symmetric matrix-valued functions. Note that this function satisfies hypotheses ($\mathscr{P}\mathscr{A}$) and (Hf).

According to [4, Example 3, p. 14] the functional $J_l(t_1,\cdot,\cdot,\cdot) : L_2^{m_l}([t_0,T]) \times L_2^{m_f}([t_0,T]) \times H_1^n([t_0,T])\times$ is well defined and continuous. Therefore, by Theorem 4.24, the functional $\hat{J}_l(\cdot,\cdot,\cdot)$ has finite values and is continuous on $\Theta_{we}\times\mathscr{T}\times\mathscr{U}_l$.

Moreover, since Θ_{we} is compact, the pessimistic problem $(\mathrm{PB})_{we}$ can be written as

$$\min_{(t_1,u_l)\in\mathscr{T}\times\mathscr{U}_l}\ \max_{\theta\in\Theta_{we}} \hat{J}_l(\theta,t_1,u_l).$$

Theorem 4.32 (First-order necessary conditions when the final time is fixed, i.e. $\mathscr{T}=\{t_1\}$). *Suppose that $\mathscr{T}=\{t_1\}$.*

Let $(\bar\theta,\bar u_l)\in\Theta_{we}\times\mathscr{U}_l$ solve $(\mathrm{PB})_{we}$. Then there are nonnegative real numbers $\mu, l_1,\dots,l_p$ and a real number ν such that

$$\nabla_{u_l}\hat{J}_l(\bar\theta,t_1,\bar u_l)(t) + \mu\bar u_l(t) = 0 \qquad \text{a.e. on } [t_0,T], \tag{4.88}$$

$$\frac{\partial\hat{J}_l}{\partial\theta_i}(\bar\theta,t_1,\bar u_l) + l_i + \nu = 0, \qquad i=1,\dots,p, \tag{4.89}$$

$$\mu(\|\bar u_l\|_2 - R) = 0, \tag{4.90}$$

$$l_i\bar\theta_i = 0, \qquad i=1,\dots,p, \tag{4.91}$$

and of course

$$\sum_{i=1}^{p}\bar\theta_i = 1, \tag{4.92}$$

$$\|\bar u_l\|_2 \le R, \quad \bar\theta_i \ge 0, \qquad i=1,\dots,p. \tag{4.93}$$

Proof. We have that $\bar\theta$ is a maximizer of $\hat{J}_l(\cdot,t_1,\bar u_l)$ over Θ_{we}. By Karush–Kuhn–Tucker theorem, since on Θ_{we} the linear independence of gradients of active constraints holds (hence Mangasarian–Fromowitz regularity condition holds), and based on Lemma 4.28, we obtain that there are nonnegative reals $l_1,\dots,l_p$ and a real ν such that (4.89) and (4.91) hold and of course (4.92) and (4.93).

Moreover, $\bar{u}_l$ is a minimizer of $\hat{J}_l(\bar{\theta},t_1,\cdot)$ over the ball $\mathscr{U}_l$. By the generalized Lagrange multiplier rule under Kurcyusz-Robinson-Zowe regularity condition (see [39, Theorem 5.3]), and based on Lemma 4.28, we obtain (4.88) and (4.90). ■

Theorem 4.33 (First-order necessary conditions when the final time $t_1 \in \mathscr{T} = [\underline{t},\bar{t}] \subset]t_0,T[$**).** *Let* $(\bar{t}_1,\bar{\theta},\bar{u}_l) \in \mathscr{T} \times \Theta_{we} \times \mathscr{U}_l$ *solve (PB)*$_{we}$*. Suppose that* $\bar{u}_l$ *is continuous at* $\bar{t}_1$ *(see footnote 3). Then there are nonnegative real numbers* $\mu, l_1,\dots,l_p,l_{p+1},l_{p+2}$ *and a real number* ν *such that*

$$\nabla_{u_l}\hat{J}_l(\bar{\theta},t_1,\bar{u}_l)(t) + \mu\bar{u}_l(t) = 0 \qquad \text{a.e. on } [t_0,T], \tag{4.94}$$

$$\frac{\partial \hat{J}_l}{\partial \theta_i}(\bar{\theta},t_1,\bar{u}_l) + l_i + \nu = 0, \qquad i = 1,\dots,p, \tag{4.95}$$

$$\frac{\partial \hat{J}_l}{\partial t_1}(\bar{\theta},t_1,\bar{u}_l) - l_{p+1} + l_{p+2} = 0, \tag{4.96}$$

$$\mu(\|\bar{u}_l\|_2 - R) = 0, \tag{4.97}$$

$$l_i\bar{\theta}_i = 0, \qquad i = 1,\dots,p, \tag{4.98}$$

$$l_{p+1}(\bar{t}_1 - \underline{t}) = 0, \tag{4.99}$$

$$l_{p+2}(\bar{t} - \bar{t}_1) = 0, \tag{4.100}$$

and of course

$$\sum_{i=1}^{p} \bar{\theta}_i = 1, \tag{4.101}$$

$$\|\bar{u}_l\|_2 \le R, \quad \bar{\theta}_i \ge 0, \qquad i = 1,\dots,p. \tag{4.102}$$

The proof is identical to the proof of Theorem 4.32.

Remark 4.34. A similar comment as in Remark 4.30 can be done for the last two theorems. Moreover, in this case the computation of the partial derivatives and gradients in Lemma 4.28 is simplified since, by (4.87), we have

$$(f_l)'_{u_l}(t,u_l,\mathbf{u_f},x) = 2u_l^T S_l(t),$$

$$(f_l)'_{\mathbf{u_f}}(t,u_l,\mathbf{u_f},x) = 2\mathbf{u_f}^T R_l(t),$$

$$(f_l)'_x(t,u_l,\mathbf{u_f},x) = x^T Q_l(t).$$

Acknowledgements Henri Bonnel thanks the University of Naples Federico II and CSEF for its support and the Department of Mathematics and Statistics for its hospitality.

References

1. Abou-Kandil, H., Freiling, G., Ionescu, V., Jank, G.: Matrix Riccati Equations in Control and Systems Theory. Birkhauser, Basel (2003)
2. Alexeev, V.M., Tikhomirov, V.M., Fomin, S.V.: Optimal Control. Plenum, New York (1987)
3. Ankhili, Z., Mansouri, A.: An exact penalty on bilevel programs with linear vector optimization lower level. European J. Oper. Res. **197**, 36–41 (2009)
4. Aubin, J.-P., Ekeland, I.: Applied Nonlinear Analysis. Wiley, New York (1984)
5. Bagchi, A.: Stackelberg Differential Games in Economic Models. Lecture Notes in Control and Information Sciences, vol. 64. Springer, Berlin (1984)
6. Basar, T., Olsder, G.J.: Dynamic Noncooperative Game Theory, 2nd edn. Academic, London/New York (1995)
7. Bauschke, H.H., Combettes, P.L.: Convex Analysis and Monotone Operator Theory in Hilbert Spaces. CMS Books in Mathematics. Springer, New York (2010)
8. Benson, H.P.: Optimization over the efficient set. J. Math. Anal. Appl. **98**, 562–580 (1984)
9. Benson, H.P.: A finite, non-adjacent extreme point search algorithm for optimization over the efficient set. J. Optim. Theory Appl. **73**, 47–64 (1992)
10. Bolintinéanu, S.: Minimization of a quasi-concave function over an efficient set. Math. Program. **61**, 89–110 (1993)
11. Bolintinéanu, S.: Necessary conditions for nonlinear suboptimization over the weakly-efficient set. J. Optim. Theory Appl. **78**, 579–598 (1993)
12. Bolintinéanu, S.: Optimality conditions for minimization over the (weakly or properly) efficient set. J. Math. Anal. Appl. **173**(2), 523–541 (1993)
13. Bolintinéanu, S., El Maghri, M.: Pénalisation dans l'optimisation sur l'ensemble faiblement efficient. RAIRO Oper. Res. **31**(3), 295–310 (1997)
14. Bonnel, H.: Optimality conditions for the semivectorial bilevel optimization problem. Pacific J. Optim. **2**(3), 447–468 (2006)
15. Bonnel, H., Kaya, C.Y.: Optimization over the efficient set in multiobjective convex optimal control problems. J. Optim. Theory Appl. **147**(1), 93–112 (2010)
16. Bonnel, H., Morgan, J.: Semivectorial bilevel optimization problem: Penalty Approach. J. Optim. Theory Appl. **131**(3), 365–382 (2006)
17. Bonnel, H., Morgan, J.: Semivectorial bilevel convex optimal control problems: existence results. SIAM J. Control Optim. **50**(6), 3224–3241 (2012)
18. Bonnel, H., Pham, N.S.: Nonsmooth optimization over the (weakly or properly) pareto set of a linear-quadratic multi-objective control problem: Explicit Optimality Conditions. J. Ind. Manag. Optim. **7**(4), 789–809 (2011)
19. Borwein, J.: Proper efficient points for maximizations with respect to cones. SIAM J. Control Optim. **15**(1), 57–63 (1977)
20. Breton, M., Alj, A., Haurie, A.: Sequential Stackelberg Equilibrium in Two-person Games. J. Optim. Theory Appl. **59**, 71–97 (1988)
21. Brezis, H.: Analyse fonctionnelle : théorie et applications. Dunod, Paris (1999)
22. Calvete, H., Galé, C.: On linear bilevel problems with multiple objectives at the lower level. Omega **39**, 33–40 (2011)(Elsevier)
23. Chen, T., Cruz Jr., J.B.: Stackelberg solution for two person games with biased information patterns. IEEE Trans. Automatic Control **17**, 791–798 (1972)
24. Colson, B., Marcotte, P., Savard, G.: An overview of bilevel optimization. Ann. Oper. Res. **153**, 235–256 (2007)
25. Craven, B.D.: Aspects of multicriteria optimization. Recent Prospects in Mathematical Programming. Gordon and Breach, Philadelphia (1991)
26. Dauer, J.P.: Optimization over the efficient set using an active constraint approach. Z. Oper. Res. **35**, 185–195 (1991)
27. Dauer, J.P., Fosnaugh, T.A.: Optimization over the efficient set. J. Global Optim. **7**, 261–277 (1995)

28. Dempe, S.: Foundations of Bilevel Programming. Kluwer Academic Publishers, Dordrecht (2002)
29. Dempe, S.: Annotated bibliography on bilevel programming and mathematical programs with equilibrium constraints. Optimization **52**, 333–359 (2003)
30. Dempe, S., Gadhi, N., Zemkoho, A.B.: New optimality conditions for the semivectorial bilevel optimization problem. J. Optim. Theory Appl. **157**, 54–74 (2013)
31. Eichfelder, G.: Multiobjective bilevel optimization. Math. Program. Ser. A **123**, 419–449 (2010)
32. Ekeland, I., Témam, R.: Convex Analysis and Variational Problems. Classics in Applied Mathematics, vol. 28. SIAM, Philadelphia (1999)
33. Fülöp, J.: A cutting plane algorithm for linear optimization over the efficient set. Generalized Convexity, Lecture notes in Economics and Mathematical System, vol. 405, pp. 374–385. Springer, Berlin (1994)
34. Geoffrion, A.M.: Proper efficiency and the theory of vector maximization. J. Math. Anal. Appl. **22**, 618–630 (1968)
35. Haurie, A.: A Historical Perspective on Cooperative Differential Games. Advances in dynamic games and applications (Maastricht, 1998). Ann. Internat. Soc. Dynam. Games, Part I **6**, 19–29 (2001)
36. Horst, R., Thoai, N.V.: Maximizing a concave function over the efficient or weakly-efficient set. European J. Oper. Res. **117**, 239–252 (1999)
37. Horst, R., Thoai, N.V., Yamamoto, Y., Zenke, D.: On Optimization over the Efficient Set in Linear Multicriteria Programming. J. Optim. Theory Appl. **134**, 433–443 (2007)
38. Jahn, J.: Vector Optimization. Springer, Berlin (2004)
39. Jahn, J.: Introduction to the Theory of Nonlinear Optimization. Springer, Berlin (2007)
40. Lignola, M.B., Morgan, J.: Topological Existence and Stability for Stackelberg Problems. J. Optim. Theory Appl. **84**, 575–596 (1995)
41. Lignola, M.B., Morgan, J.: Stability of regularized bilevel programming problems. J. Optim. Theory Appl. **93**, 575–596 (1997)
42. Loridan, P., Morgan, J.: Approximation of the Stackelberg problem and applications in control theory. Proceedings of the Vth IFAC Workshop on Control Applications of Non Linear Programming and Optimization, Capri, 1985. Pergamon Press, Oxford (1986)
43. Loridan, P., Morgan, J.: A theoretical approximation scheme for Stackelberg problems. J. Optim. Theory Appl. **61**, 95–110 (1989)
44. Loridan, P., Morgan, J.: New results on approximate solutions in two-level optimization. Optimization **20**, 819–836 (1989)
45. Morgan, J.: Constrained well-posed two-level optimization problems. In: Clarke, F., Dem'yanov, V.F., Giannessi, F. (eds.) Nonsmooth Optimization and Related Topics. Ettore Majorana International Sciences Series, pp. 307–326. Plenum Press, New York (1989)
46. Morgan, J.: Existence for Hierarchical differential non-zero sum games with coupled constraints. Workshop of the International Society of Game Theory and Applications. Sils-Maria, Switzerland (1997)
47. Philip, J.: Algorithms for the vector maximization problem. Math. Program. **2**, 207–229 (1972)
48. Simaan, M., Cruz, J.B., Jr.: On the Stackelberg strategy in nonzero-sum games. J. Optim. Theory Appl. **11**(5), 533–555 (1973)
49. Von Stackelberg, H.: The Theory of the Market Economy. Oxford University Press, Oxford (1952)
50. Yamamoto, Y.: Optimization over the efficient set: overview. J. Global Optim. **22**, 285–317 (2002)
51. Zheng, Y., Wan, Z.: A solution method for semivectorial bilevel programming problem via penalty method. J. Appl. Math. Comput. **37**, 207–219 (2011)

Chapter 5
Monotone Operators Without Enlargements

Jonathan M. Borwein, Regina S. Burachik, and Liangjin Yao

Abstract Enlargements have proven to be useful tools for studying maximally monotone mappings. It is therefore natural to ask in which cases the enlargement does not change the original mapping. Svaiter has recently characterized non-enlargeable operators in reflexive Banach spaces and has also given some partial results in the nonreflexive case. In the present paper, we provide another characterization of non-enlargeable operators in nonreflexive Banach spaces under a closedness assumption on the graph. Furthermore, and still for general Banach spaces, we present a new proof of the maximality of the sum of two maximally monotone linear relations. We also present a new proof of the maximality of the sum of a maximally monotone linear relation and a normal cone operator when the domain of the linear relation intersects the interior of the domain of the normal cone.

Key words: Adjoint • Fenchel conjugate • Fitzpatrick function • Linear relation • Maximally monotone operator • Monotone operator • Multifunction • Normal cone operator • Non-enlargeable operator • Operator of type (FPV) • Partial inf-convolution • Set-valued operator

---

COMMUNICATED BY HEINZ H. BAUSCHKE.

J.M. Borwein
CARMA, University of Newcastle, Newcastle, NSW 2308, Australia

King Abdulaziz University, Jeddah 80200, Saudi Arabia
e-mail: jonathan.borwein@newcastle.edu.au

R.S. Burachik
School of Mathematics and Statistics, University of South Australia,
Mawson Lakes, SA 5095, Australia
e-mail: regina.burachik@unisa.edu.au

L. Yao (✉)
CARMA, University of Newcastle, Newcastle, NSW 2308, Australia
e-mail: liangjin.yao@gmail.com

D.H. Bailey et al. (eds.), *Computational and Analytical Mathematics*, Springer Proceedings in Mathematics & Statistics 50, DOI 10.1007/978-1-4614-7621-4_5,

Mathematics Subject Classifications (2010): 47A06, 47H05, 47B65, 47N10, 90C25

5.1 Introduction

Maximally monotone operators have proven to be a significant class of objects in both modern optimization and functional analysis. They extend both the concept of subdifferentials of convex functions, as well as that of a positive semi-definite function. Their study in the context of Banach spaces, and in particular nonreflexive ones, arises naturally in the theory of partial differential equations, equilibrium problems, and variational inequalities. For a detailed study of these operators, see, e.g., [12–14], or the books [3, 15, 20, 27, 32–34, 46, 47].

A useful tool for studying or proving properties of a maximally monotone operator A is the concept of the "enlargement of A". A main example of this usefulness is Rockafellar's proof of maximality of the subdifferential of a convex function (Fact 5.3 below), which uses the concept of ε-subdifferential. The latter is an enlargement of the subdifferential introduced in [18].

Broadly speaking, an enlargement is a multifunction which approximates the original maximally monotone operator in a convenient way. Another useful way to study a maximally monotone operator is by associating to it a convex function called the Fitzpatrick function. The latter was introduced by Fitzpatrick in [22] and its connection with enlargements, as shown in [21], is contained in (5.4) below. Enlargements of positive sets in SSDB spaces (see [34, Sect. 21]) have recently been studied in [16].

Our first aim in the present paper is to provide further characterizations of maximally monotone operators which are not enlargeable, in the setting of possibly nonreflexive Banach spaces (see Sect. 5.4). In other words, in which cases the enlargement does not change the graph of a maximally monotone mapping defined in a Banach space. We address this issue Corollary 5.28, under a closedness assumption on the graph of the operator.

Our other aim is to use the Fitzpatrick function to derive new results which establish the maximality of the sum of two maximally monotone operators in nonreflexive spaces (see Sect. 5.5). First, we provide a different proof of the maximality of the sum of two maximally monotone linear relations. Second, we provide a proof of the maximality of the sum of a maximally monotone linear relation and a normal cone operator when the domain of the operator intersects the interior of the domain of the normal cone.

5.2 Technical Preliminaries

Throughout this paper, X is a real Banach space with norm $\|\cdot\|$, and X^* is the continuous dual of X. The spaces X and X^* are paired by the duality pairing, denoted as $\langle\cdot,\cdot\rangle$. The space X is identified with its canonical image in the bidual space X^{**}. Furthermore, $X\times X^*$ and $(X\times X^*)^* := X^*\times X^{**}$ are paired via $\langle(x,x^*),(y^*,y^{**})\rangle := \langle x,y^*\rangle + \langle x^*,y^{**}\rangle$, where $(x,x^*)\in X\times X^*$ and $(y^*,y^{**})\in X^*\times X^{**}$.

Let $A\colon X\rightrightarrows X^*$ be a *set-valued operator* (also known as a multifunction) from X to X^*, i.e., for every $x\in X$, $Ax\subseteq X^*$, and let $\operatorname{gra}A := \{(x,x^*)\in X\times X^* \mid x^*\in Ax\}$ be the *graph* of A. The *domain* of A is $\operatorname{dom}A := \{x\in X \mid Ax\neq\varnothing\}$, and $\operatorname{ran}A := A(X)$ for the *range* of A. Recall that A is *monotone* if

$$\langle x-y,x^*-y^*\rangle \geq 0,\quad \forall(x,x^*)\in\operatorname{gra}A\ \forall(y,y^*)\in\operatorname{gra}A, \tag{5.1}$$

and *maximally monotone* if A is monotone and A has no proper monotone extension (in the sense of graph inclusion). Let $A: X\rightrightarrows X^*$ be monotone and $(x,x^*)\in X\times X^*$. We say (x,x^*) is *monotonically related to* $\operatorname{gra}A$ if

$$\langle x-y,x^*-y^*\rangle \geq 0,\quad \forall(y,y^*)\in\operatorname{gra}A.$$

Let $A: X\rightrightarrows X^*$ be maximally monotone. We say A is *of type (FPV)* if for every open convex set $U\subseteq X$ such that $U\cap\operatorname{dom}A\neq\varnothing$, the implication

$$x\in U \text{and}\,(x,x^*)\,\text{is monotonically related to}\ \operatorname{gra}A\cap U\times X^* \Rightarrow (x,x^*)\in\operatorname{gra}A$$

holds. Maximally monotone operators of type (FPV) are relevant primarily in the context of nonreflexive Banach spaces. Indeed, it follows from [34, Theorem 44.1] and a well-known result from [30] (or from [34, Theorems 38.4 and 39.1]) that every maximally monotone operator defined in a reflexive Banach space is of type (FPV). As mentioned in [34, Sect. 44], an example of a maximally monotone operator which is not of type (FPV) has not been found yet.

Let $A: X\rightrightarrows X^*$ be monotone such that $\operatorname{gra}A\neq\varnothing$. The *Fitzpatrick function* associated with A is defined by

$$F_A\colon X\times X^*\to\,]-\infty,+\infty]: (x,x^*)\mapsto \sup_{(a,a^*)\in\operatorname{gra}A}\big(\langle x,a^*\rangle+\langle a,x^*\rangle-\langle a,a^*\rangle\big).$$

When A is maximally monotone, a fundamental property of the Fitzpatrick function F_A (see Fact 5.5) is that

$$F_A(x,x^*)\geq\langle x,x^*\rangle \text{ for all } (x,x^*)\in X\times X^*, \tag{5.2}$$

$$F_A(x,x^*)=\langle x,x^*\rangle \text{ for all } (x,x^*)\in\operatorname{gra}A. \tag{5.3}$$

Hence, for a fixed $\varepsilon \geq 0$, the set of pairs (x,x^*) for which $F_A(x,x^*) \leq \langle x,x^*\rangle + \varepsilon$ contains the graph of A. This motivates the definition of enlargement of A for a general monotone mapping A, which is as follows.

Let $\varepsilon \geq 0$. We define $A_\varepsilon : X \rightrightarrows X^*$ by

$$\begin{aligned}\operatorname{gra} A_\varepsilon &:= \left\{(x,x^*) \in X \times X^* \mid \langle x^* - y^*, x - y\rangle \geq -\varepsilon,\ \forall (y,y^*) \in \operatorname{gra} A\right\} \\ &= \left\{(x,x^*) \in X \times X^* \mid F_A(x,x^*) \leq \langle x,x^*\rangle + \varepsilon\right\}. \end{aligned} \tag{5.4}$$

Let $A : X \rightrightarrows X^*$ be monotone. We say A is *enlargeable* if $\operatorname{gra} A \subsetneqq \operatorname{gra} A_\varepsilon$ for some $\varepsilon \geq 0$, and A is *non-enlargeable* if $\operatorname{gra} A = \operatorname{gra} A_\varepsilon$ for every $\varepsilon \geq 0$. Lemma 23.1 in [34, 36] proves that if a proper and convex function verifies (5.2), then the set of all pairs (x,x^*) at which (5.3) holds is a monotone set. Therefore, if A is non-enlargeable then it must be maximally monotone. As the referee has pointed out another proof is as follows: if A is non-enlargeable then $A = A_0$ and hence A is maximally monotone.

We adopt the notation used in the books [15, Chap. 2] and [12, 33, 34]. Given a subset C of X, $\operatorname{int} C$ is the *interior* of C, $\overline{C}$ is the *norm closure* of C. The *support function* of C, written as σ_C, is defined by $\sigma_C(x^*) := \sup_{c\in C}\langle c,x^*\rangle$. The *indicator function* of C, written as ι_C, is defined at $x \in X$ by

$$\iota_C(x) := \begin{cases} 0, & \text{if } x \in C; \\ +\infty, & \text{otherwise.} \end{cases} \tag{5.5}$$

For every $x \in X$, the *normal cone operator* of C at x is defined by $N_C(x) := \{x^* \in X^* \mid \sup_{c\in C}\langle c - x, x^*\rangle \leq 0\}$, if $x \in C$; and $N_C(x) := \varnothing$, if $x \notin C$. The *closed unit ball* is $B_X := \{x \in X \mid \|x\| \leq 1\}$, and $\mathbb{N} := \{1,2,3,\ldots\}$.

If Z is a real Banach space with dual Z^* and a set $S \subseteq Z$, we denote $S^\perp$ by $S^\perp := \{z^* \in Z^* \mid \langle z^*, s\rangle = 0, \quad \forall s \in S\}$. The *adjoint* of an operator A, written A^*, is defined by

$$\operatorname{gra} A^* := \left\{(x^{**},x^*) \in X^{**} \times X^* \mid (x^*, -x^{**}) \in (\operatorname{gra} A)^\perp\right\}.$$

We will be interested in monotone operators which are *linear relations*, i.e., such that $\operatorname{gra} A$ is a linear subspace. Note that in this situation, A^* is also a linear relation. Moreover, A is *symmetric* if $\operatorname{gra} A \subseteq \operatorname{gra} A^*$. Equivalently, for all $(x,x^*), (y,y^*) \in \operatorname{gra} A$ it holds that

$$\langle x,y^*\rangle = \langle y,x^*\rangle. \tag{5.6}$$

We say that a linear relation A is *skew* if $\operatorname{gra} A \subseteq \operatorname{gra}(-A^*)$. Equivalently, for all $(x,x^*) \in \operatorname{gra} A$ we have

$$\langle x,x^*\rangle = 0. \tag{5.7}$$

We define the *symmetric part* a of A via

$$A_+ := \tfrac{1}{2}A + \tfrac{1}{2}A^*. \tag{5.8}$$

It is easy to check that A_+ is symmetric.

Let $f\colon X \to\,]-\infty,+\infty]$. Then $\operatorname{dom} f := f^{-1}(\mathbb{R})$ is the *domain* of f, and $f^*\colon X^* \to [-\infty,+\infty] : x^* \mapsto \sup_{x\in X}(\langle x,x^*\rangle - f(x))$ is the *Fenchel conjugate* of f. We denote by $\overline{f}$ the lower semicontinuous hull of f. We say that f is proper if $\operatorname{dom} f \neq \varnothing$. Let f be proper. The *subdifferential* of f is defined by

$$\partial f\colon X \rightrightarrows X^*\colon x \mapsto \{x^* \in X^* \mid (\forall y \in X)\ \langle y-x,x^*\rangle + f(x) \leq f(y)\}.$$

For $\varepsilon \geq 0$, the *ε-subdifferential* of f is defined by

$$\partial_\varepsilon f\colon X \rightrightarrows X^*\colon x \mapsto \{x^* \in X^* \mid (\forall y \in X)\ \langle y-x,x^*\rangle + f(x) \leq f(y) + \varepsilon\}.$$

Note that $\partial f = \partial_0 f$. Given $x \in X$, we say $\partial_\varepsilon \iota_C(x)$ is the *ε-normal set* of C at x (see [24]).

Relatedly, we say A is of Brønsted–Rockafellar (BR) type [15, 34] if whenever $(x,x^*) \in X \times X^*$, $\alpha,\beta > 0$ while

$$\inf_{(a,a^*)\in \operatorname{gra} A} \langle x-a,x^*-a^*\rangle > -\alpha\beta$$

then there exists $(b,b^*) \in \operatorname{gra} A$ such that $\|x-b\| < \alpha$, $\|x^*-b^*\| < \beta$. The name is motivated by the celebrated theorem of Brønsted and Rockafellar [15,34] which can be stated now as saying that all closed convex subgradients are of type (BR).

Let $g\colon X \to\,]-\infty,+\infty]$. The *inf-convolution* of f and g, $f\Box g$, is defined by

$$f\Box g : x \to \inf_{y\in X}\left[f(y) + g(x-y)\right].$$

Let Y be another real Banach space. We set $P_X : X \times Y \to X\colon (x,y) \mapsto x$. We denote $\operatorname{Id} : X \to X$ by the *identity mapping*.

Let $F_1, F_2\colon X \times Y \to\,]-\infty,+\infty]$. Then the *partial inf-convolution* $F_1\Box_2 F_2$ is the function defined on $X \times Y$ by

$$F_1\Box_2 F_2\colon (x,y) \mapsto \inf_{v\in Y}\left[F_1(x,y-v) + F_2(x,v)\right]. \tag{5.9}$$

5.3 Auxiliary Results

We collect in this section some facts we will use later on. These facts involve convex functions, maximally monotone operators, and Fitzpatrick functions.

Fact 5.1 (See [27, Propositions 3.3 and 1.11]). Let $f : X \to \left]-\infty,+\infty\right]$ be a lower semicontinuous convex and $\operatorname{int}\operatorname{dom} f \neq \varnothing$. Then f is continuous on $\operatorname{int}\operatorname{dom} f$ and $\partial f(x) \neq \varnothing$ for every $x \in \operatorname{int}\operatorname{dom} f$.

Fact 5.2 (Rockafellar). (See [29, Theorem 3(a)], [34, Corollary 10.3], or [46, Theorem 2.8.7(iii)].) Let $f, g : X \to \left]-\infty,+\infty\right]$ be proper convex functions. Assume that there exists a point $x_0 \in \operatorname{dom} f \cap \operatorname{dom} g$ such that g is continuous at x_0. Then for every $z^* \in X^*$, there exists $y^* \in X^*$ such that

$$(f+g)^*(z^*) = f^*(y^*) + g^*(z^* - y^*). \tag{5.10}$$

Fact 5.3 (Rockafellar). (See [31, Theorem A], [46, Theorem 3.2.8], [34, Theorem 18.7] or [25, Theorem 2.1].) Let $f : X \to \left]-\infty,+\infty\right]$ be a proper lower semicontinuous convex function. Then ∂f is maximally monotone.

Fact 5.4 (Attouch-Brezis). (See [1, Theorem 1.1] or [34, Remark 15.2].) Let $f, g : X \to \left]-\infty,+\infty\right]$ be proper lower semicontinuous and convex. Assume that

$$\bigcup_{\lambda>0} \lambda \left[\operatorname{dom} f - \operatorname{dom} g\right] \text{ is a closed subspace of } X.$$

Then

$$(f+g)^*(z^*) = \min_{y^* \in X^*} \left[f^*(y^*) + g^*(z^* - y^*)\right], \quad \forall z^* \in X^*.$$

Fact 5.3 above relates a convex function with maximal monotonicity. Fitzpatrick functions go in the opposite way: from maximally monotone operators to convex functions.

Fact 5.5 (Fitzpatrick). (See [22, Corollary 3.9] and [12, 15].) Let $A \colon X \rightrightarrows X^*$ be maximally monotone. Then for every $(x,x^*) \in X \times X^*$, the inequality $\langle x, x^* \rangle \leq F_A(x,x^*)$ is true, and the equality holds if and only if $(x,x^*) \in \operatorname{gra} A$.

It was pointed out in [34, Problem 31.3] that it is unknown whether $\overline{\operatorname{dom} A}$ is necessarily convex when A is maximally monotone and X is not reflexive. When A is of type (FPV), the question was answered positively by using F_A.

Fact 5.6 (Simons). (See [34, Theorem 44.2].) Let $A : X \rightrightarrows X^*$ be maximally monotone of type (FPV). Then $\overline{\operatorname{dom} A} = \overline{P_X\left[\operatorname{dom} F_A\right]}$ and $\overline{\operatorname{dom} A}$ is convex.

We observe that when A is of type (FPV) then also $\operatorname{dom} A_\varepsilon$ has convex closure.

Remark 5.7. Let A be of type (FPV) and fix $\varepsilon \geq 0$. Then by (5.4), Facts 5.5 and 5.6, we have $\operatorname{dom} A \subseteq \operatorname{dom} A_\varepsilon \subseteq P_X\left[\operatorname{dom} F_A\right] \subseteq \overline{\operatorname{dom} A}$. Thus we obtain

$$\overline{\operatorname{dom} A} = \overline{\left[\operatorname{dom} A_\varepsilon\right]} = \overline{P_X\left[\operatorname{dom} F_A\right]},$$

and this set is convex because $\overline{\operatorname{dom} F_A}$ is convex. As a result, for every A of type (FPV) it holds that $\overline{\operatorname{dom} A} = \overline{[\operatorname{dom} A_\varepsilon]}$ and this set is convex.

We recall below some necessary conditions for a maximally monotone operator to be of type (FPV).

Fact 5.8 (Simons). (See [34, Theorem 46.1].) Let $A : X \rightrightarrows X^*$ be a maximally monotone linear relation. Then A is of type (FPV).

Fact 5.9 (Fitzpatrick-Phelps and Verona–Verona). (See [23, Corollary 3.4], [38, Corollary 4] or [34, Theorem 48.4(d)].) Let $f : X \to]-\infty, +\infty]$ be proper, lower semicontinuous, and convex. Then ∂f is of type (FPV).

Fact 5.10 (See [45, Corollary 3.3]). Let $A : X \rightrightarrows X^*$ be a maximally monotone linear relation, and $f : X \to]-\infty, +\infty]$ be a proper lower semicontinuous convex function with $\operatorname{dom} A \cap \operatorname{int} \operatorname{dom} \partial f \neq \varnothing$. Then $A + \partial f$ is of type (FPV).

Fact 5.11 (Phelps-Simons). (See [28, Corollary 2.6 and Proposition 3.2(h)].) Let $A: X \to X^*$ be monotone and linear. Then A is maximally monotone and continuous.

Fact 5.12 (See [7, Theorem 4.2] or [26, Lemma 1.5]). Let $A : X \rightrightarrows X^*$ be maximally monotone such that $\operatorname{gra} A$ is convex. Then $\operatorname{gra} A$ is affine.

Remark 5.13. In [42, Proposition 5(ii)], it was shown that Fact 5.12 can be extended to a locally convex space.

Fact 5.14 (Simons). (See [34, Lemma 19.7 and Sect. 22].) Let $A : X \rightrightarrows X^*$ be a monotone operator such that $\operatorname{gra} A$ is convex with $\operatorname{gra} A \neq \varnothing$. Then the function

$$g: X \times X^* \to]-\infty, +\infty] : (x, x^*) \mapsto \langle x, x^* \rangle + \iota_{\operatorname{gra} A}(x, x^*) \tag{5.11}$$

is proper and convex.

Fact 5.15 (See [40, Theorem 3.4 and Corollary 5.6], or [34, Theorem 24.1(b)]). Let $A, B : X \rightrightarrows X^*$ be maximally monotone operators. Assume that

$$\bigcup_{\lambda>0} \lambda \left[P_X(\operatorname{dom} F_A) - P_X(\operatorname{dom} F_B)\right] \text{ is a closed subspace.}$$

If

$$F_{A+B} \geq \langle \cdot, \cdot \rangle \text{ on } \quad X \times X^*, \tag{5.12}$$

then $A + B$ is maximally monotone.

Definition 5.16 (Fitzpatrick family). Let $A: X \rightrightarrows X^*$ be maximally monotone. The associated *Fitzpatrick family* $\mathscr{F}_A$ consists of all functions $F: X \times X^* \to]-\infty, +\infty]$ that are lower semicontinuous and convex, and that satisfy $F \geq \langle \cdot, \cdot \rangle$, and $F = \langle \cdot, \cdot \rangle$ on $\operatorname{gra} A$.

Fact 5.17 (Fitzpatrick). (See [22, Theorem 3.10] or [21].) Let $A\colon X \rightrightarrows X^*$ be maximally monotone. Then for every $(x,x^*) \in X \times X^*$,

$$F_A(x,x^*) = \min\{F(x,x^*) \mid F \in \mathscr{F}_A\}.$$

Corollary 5.18. *Let $A\colon X \rightrightarrows X^*$ be a maximally monotone operator such that* gra$A$ *is convex. Then for every $(x,x^*) \in X \times X^*$,*

$$F_A(x,x^*) = \min\{F(x,x^*) \mid F \in \mathscr{F}_A\} \quad \textit{and} \quad g(x,x^*) = \max\{F(x,x^*) \mid F \in \mathscr{F}_A\},$$

where $g := \langle\cdot,\cdot\rangle + \iota_{\operatorname{gra}A}$.

Proof. Apply Facts 5.14 and 5.17. ■

Fact 5.19 (See [34, Lemma 23.9], or [4, Proposition 4.2]). Let $A, B\colon X \rightrightarrows X^*$ be monotone operators and $\operatorname{dom}A \cap \operatorname{dom}B \neq \varnothing$. Then $F_{A+B} \leq F_A \Box_2 F_B$.

Let X,Y be two real Banach spaces and let $h : X \times Y \to \left]-\infty,+\infty\right]$ be a convex function. We say that h is *separable* if there exist convex functions $h_1 : X \to \left]-\infty,+\infty\right]$ and $h_2 : Y \to \left]-\infty,+\infty\right]$ such that $h(x,y) = h_1(x) + h_2(y)$. This situation is denoted as $h = h_1 \oplus h_2$. We recall below some cases in which the Fitzpatrick function is separable.

Fact 5.20 (See [2, Corollary 5.9] or [11, Fact 4.1]). Let C be a nonempty closed convex subset of X. Then $F_{N_C} = \iota_C \oplus \iota_C^*$.

Fact 5.21 (See [2, Theorem 5.3]). Let $f : X \to \left]-\infty,+\infty\right]$ be a proper lower semicontinuous sublinear function. Then $F_{\partial f} = f \oplus f^*$ and $\mathscr{F}_{\partial f} = \{f \oplus f^*\}$.

Remark 5.22. Let f be as in Fact 5.21, then

$$\begin{aligned} \operatorname{gra}(\partial f)_\varepsilon &= \{(x,x^*) \in X \times X^* \mid f(x) + f^*(x^*) \leq \langle x,x^*\rangle + \varepsilon\} \\ &= \operatorname{gra}\partial_\varepsilon f, \quad \forall \varepsilon \geq 0. \end{aligned} \tag{5.13}$$

Fact 5.23 (Svaiter). (See [37, p. 312].) Let $A\colon X \rightrightarrows X^*$ be maximally monotone. Then A is non-enlargeable if and only if $\operatorname{gra}A = \operatorname{dom}F_A$ and then $\operatorname{gra}A$ is convex.

It is immediate from the definitions that:

Fact 5.24. Every non-enlargeable maximally monotone operator is of type (BR).

Fact 5.21 and the subsequent remark refer to a case in which all enlargements of A coincide, or, equivalently, the Fitzpatrick family is a singleton. It is natural to deduce that a non-enlargeable operator will also have a single element in its Fitzpatrick family.

Corollary 5.25. *Let $A\colon X \rightrightarrows X^*$ be maximally monotone. Then A is non-enlargeable if and only if $F_A = \iota_{\operatorname{gra}A} + \langle\cdot,\cdot\rangle$ and hence $\mathscr{F}_A = \{\iota_{\operatorname{gra}A} + \langle\cdot,\cdot\rangle\}$.*

Proof. "$\Rightarrow$": By Fact 5.23, we have gra$A$ is convex. By Facts 5.5 and 5.23, we have $F_A = \iota_{\text{gra}A} + \langle\cdot,\cdot\rangle$. Then by Corollary 5.18, $\mathscr{F}_A = \{\iota_{\text{gra}A} + \langle\cdot,\cdot\rangle\}$. "$\Leftarrow$": Apply directly Fact 5.23. ■

Remark 5.26. The condition that $\mathscr{F}_A$ is singleton does not guarantee that gra$A$ is convex. For example, let $f : X \to\ ]-\infty,+\infty]$ be a proper lower semicontinuous sublinear function. Then by Fact 5.21, $\mathscr{F}_A$ is singleton but gra∂f is not necessarily convex.

5.4 Non-enlargeable Monotone Linear Relations

We begin with a basic characterization.

Theorem 5.27. *Let $A\colon X \rightrightarrows X^*$ be a maximally monotone linear relation such that* gra$A$ *is weak×weak* closed. Then A is non-enlargeable if and only if* $\text{gra}(-A^*) \cap X \times X^* \subseteq \text{gra}A$. *In this situation, we have that* $\langle x,x^*\rangle = 0, \forall (x,x^*) \in \text{gra}(-A^*) \cap X \times X^*$.

Proof. "$\Rightarrow$": By Corollary 5.25,

$$F_A = \iota_{\text{gra}A} + \langle\cdot,\cdot\rangle. \tag{5.14}$$

Let $(x,x^*) \in \text{gra}(-A^*) \cap X \times X^*$. Then we have

$$\begin{aligned} F_A(x,x^*) &= \sup_{(a,a^*)\in\text{gra}A} \{\langle a^*,x\rangle + \langle a,x^*\rangle - \langle a,a^*\rangle\} \\ &= \sup_{(a,a^*)\in\text{gra}A} \{-\langle a,a^*\rangle\} \\ &= 0. \end{aligned} \tag{5.15}$$

Then by (5.15), $(x,x^*) \in \text{gra}A$ and $\langle x,x^*\rangle = 0$. Hence $\text{gra}(-A^*) \cap X \times X^* \subseteq \text{gra}A$.

"$\Leftarrow$": By the assumption that gra$A$ is weak×weak* closed, we have

$$[\text{gra}(-A^*) \cap X \times X^*]^\perp \cap X^* \times X = \left[(\text{gra}A^{-1})^\perp \cap X \times X^*\right]^\perp \cap X^* \times X = \text{gra}A^{-1}. \tag{5.16}$$

By [37, Lemma 2.1(2)], we have

$$\langle z,z^*\rangle = 0, \quad \forall (z,z^*) \in \text{gra}(-A^*) \cap X \times X^*. \tag{5.17}$$

Hence $A^*|_X$ is skew. Let $(x,x^*) \in X \times X^*$. Then by (5.17), we have

$$\begin{aligned}
F_A(x,x^*) &= \sup_{(a,a^*)\in \operatorname{gra} A} \left\{\langle x,a^*\rangle + \langle x^*,a\rangle - \langle a,a^*\rangle\right\} \\
&\geq \sup_{(a,a^*)\in \operatorname{gra}(-A^*)\cap X\times X^*} \left\{\langle x,a^*\rangle + \langle x^*,a\rangle - \langle a,a^*\rangle\right\} \\
&= \sup_{(a,a^*)\in \operatorname{gra}(-A^*)\cap X\times X^*} \left\{\langle x,a^*\rangle + \langle x^*,a\rangle\right\} \\
&= \iota_{\left(\operatorname{gra}(-A^*)\cap X\times X^*\right)^{\perp}\cap X^*\times X}(x^*,x) \\
&= \iota_{\operatorname{gra} A}(x,x^*) \quad \text{[by (5.16)]}.
\end{aligned} \tag{5.18}$$

Hence by Fact 5.5

$$F_A(x,x^*) = \langle x,x^*\rangle + \iota_{\operatorname{gra} A}(x,x^*). \tag{5.19}$$

Hence by Corollary 5.25, A is non-enlargeable. ■

The following corollary, which holds in a general Banach space, provides a characterization of non-enlargeable operators under a closedness assumption on the graph. A characterization of non-enlargeable linear operators for reflexive spaces (in which the closure assumption is hidden) was established by Svaiter in [37, Theorem 2.5].

Corollary 5.28. *Let* $A\colon X \rightrightarrows X^*$ *be maximally monotone and suppose that* $\operatorname{gra} A$ *is weak×weak* closed. Select* $(a,a^*) \in \operatorname{gra} A$ *and set* $\operatorname{gra}\tilde{A} := \operatorname{gra} A - \{(a,a^*)\}$. *Then* A *is non-enlargeable if and only if* $\operatorname{gra} A$ *is convex and* $\operatorname{gra}(-\tilde{A}^*)\cap X\times X^* \subseteq \operatorname{gra}\tilde{A}$. *In particular,* $\langle x,x^*\rangle = 0, \forall (x,x^*) \in \operatorname{gra}\tilde{A}^*\cap X\times X^*$.

Proof. "⇒": By the assumption that A is non-enlargeable, so is $\tilde{A}$. By Fact 5.23, $\operatorname{gra} A$ is convex and then $\operatorname{gra} A$ is affine by Fact 5.12. Thus $\tilde{A}$ is a linear relation. Now we can apply Theorem 5.27 to $\tilde{A}$. "⇐": Apply Fact 5.12 and Theorem 5.27 directly. ■

Remark 5.29. We cannot remove the condition that "$\operatorname{gra} A$ is convex" in Corollary 5.28. For example, let $X=\mathbb{R}^n$ with the Euclidean norm. Suppose that $f := \|\cdot\|$. Then ∂f is maximally monotone by Fact 5.3, and hence $\operatorname{gra}\partial f$ is weak×weak* closed. Now we show that

$$\operatorname{gra}(\partial f)^* = \{(0,0)\}. \tag{5.20}$$

Note that

$$\partial f(x) = \begin{cases} B_X, & \text{if } x=0; \\ \{\frac{x}{\|x\|}\}, & \text{otherwise.} \end{cases} \tag{5.21}$$

Let $(z,z^*) \in \operatorname{gra}(\partial f)^*$. By (5.21), we have $(0,B_X) \subseteq \operatorname{gra}\partial f$ and thus

$$\langle -z, B_X \rangle = 0. \tag{5.22}$$

Thus $z = 0$. Hence

$$\langle z^*, a \rangle = 0, \quad \forall a \in \operatorname{dom} \partial f. \tag{5.23}$$

Since $\operatorname{dom} \partial f = X$, $z^* = 0$ by (5.23). Hence $(z, z^*) = (0,0)$ and thus (5.20) holds. By (5.20), $\operatorname{gra} -(\partial f)^* \subseteq \operatorname{gra} \partial f$. However, $\operatorname{gra} \partial f$ is not convex. Indeed, let $e_k = (0, \ldots, 0, 1, 0, \cdots, 0)$: the kth entry is 1 and the others are 0. Take

$$a = \frac{e_1 - e_2}{\sqrt{2}} \quad \text{and} \quad b = \frac{e_2 - e_3}{\sqrt{2}}.$$

Then $(a,a) \in \operatorname{gra} \partial f$ and $(b,b) \in \operatorname{gra} \partial f$ by (5.21), but

$$\frac{1}{2}(a,a) + \frac{1}{2}(b,b) \notin \operatorname{gra} \partial f.$$

Hence ∂f is enlargeable by Fact 5.23.

In the case of a skew operator we can be more exacting.

Corollary 5.30. *Let $A\colon X \rightrightarrows X^*$ be a maximally monotone and skew operator and $\varepsilon \geq 0$. Then*

(i) $\operatorname{gra} A_\varepsilon = \{(x, x^*) \in \operatorname{gra}(-A^*) \cap X \times X^* \mid \langle x, x^* \rangle \geq -\varepsilon\}$.
(ii) A is non-enlargeable if and only if $\operatorname{gra} A = \operatorname{gra}(-A^*) \cap X \times X^*$.
(iii) A is non-enlargeable if and only if $\operatorname{dom} A = \operatorname{dom} A^* \cap X$.
(iv) Assume that X is reflexive. Then $F_{A^*} = \iota_{\operatorname{gra} A^*} + \langle \cdot, \cdot \rangle$ *and hence A^* is non-enlargeable.*

Proof.

(i) By [10, Lemma 3.1], we have

$$F_A = \iota_{\operatorname{gra}(-A^*) \cap X \times X^*}. \tag{5.24}$$

Hence $(x, x^*) \in \operatorname{gra} A_\varepsilon$ if and only if $F_A(x, x^*) \leq \langle x, x^* \rangle + \varepsilon$. This yields $(x, x^*) \in \operatorname{gra}(-A^*) \cap X \times X^*$ and $0 \leq \langle x, x^* \rangle + \varepsilon$.
(ii) From Fact 5.23 we have that $\operatorname{dom} F_A = \operatorname{gra} A$. The claim now follows by combining the latter with (5.24).
(iii) For "$\Rightarrow$": use (ii). "$\Leftarrow$": Since A is skew, we have $\operatorname{gra}(-A^*) \cap X \times X^* \supseteq \operatorname{gra} A$. Using this and (ii), it suffices to show that $\operatorname{gra}(-A^*) \cap X \times X^* \subseteq \operatorname{gra} A$. Let $(x, x^*) \in \operatorname{gra}(-A^*) \cap X \times X^*$. By the assumption, $x \in \operatorname{dom} A$. Let $y^* \in Ax$. Note that $\langle x, -x^* \rangle = \langle x, y^* \rangle = 0$, where the first equality follows from the definition of A^* and the second one from the fact that A is skew. In this case we claim that

(x,x^*) is monotonically related to gra$A$. Indeed, let $(a,a^*) \in \operatorname{gra}A$. Since A is skew we have $\langle a,a^*\rangle = 0$. Thus

$$\langle x-a,x^*-a^*\rangle = \langle x,x^*\rangle - \langle (x^*,x),(a,a^*)\rangle + \langle a,a^*\rangle = 0$$

since $(x^*,x) \in (\operatorname{gra}A)^{\perp}$ and $\langle x,x^*\rangle = \langle a,a^*\rangle = 0$. Hence (x,x^*) is monotonically related to gra$A$. By maximality we conclude $(x,x^*) \in \operatorname{gra}A$. Hence $\operatorname{gra}(-A^*) \cap X \times X^* \subseteq \operatorname{gra}A$.

(iv) Now assume that X is reflexive. By [17, Theorem 2] (or see [35, 43]), A^* is maximally monotone. Since $\operatorname{gra}A \subseteq \operatorname{gra}(-A^*)$ we deduce that $\operatorname{gra}(-A^{**}) = \operatorname{gra}(-A) \subseteq \operatorname{gra}A^*$. The latter inclusion and Theorem 5.27 applied to the operator A^* yield A^* non-enlargeable. The conclusion now follows by applying Corollary 5.25 to A^*.

■

5.4.1 Limiting Examples and Remarks

It is possible for a non-enlargeable maximally monotone operator to be non-skew. This is the case for the operator A^* in Example 5.33.

Example 5.31. Let $A\colon X \rightrightarrows X^*$ be a non-enlargeable maximally monotone operator. By Fact 5.23 and Fact 5.12, gra$A$ is affine. Let $f : X \to \left]-\infty,+\infty\right]$ be a proper lower semicontinuous convex function with $\operatorname{dom}A \cap \operatorname{int}\operatorname{dom}\partial f \neq \varnothing$ such that $\operatorname{dom}A \cap \operatorname{dom}\partial f$ is not an affine set. By Fact 5.10, $A+\partial f$ is maximally monotone. Since $\operatorname{gra}(A+\partial f)$ is not affine, $A+\partial f$ is enlargeable.

The operator in the following example was studied in detail in [9].

Fact 5.32. Suppose that $X = \ell^2$, and that $A : \ell^2 \rightrightarrows \ell^2$ is given by

$$Ax := \frac{\left(\sum_{i<n} x_i - \sum_{i>n} x_i\right)_{n\in\mathbb{N}}}{2} = \left(\sum_{i<n} x_i + \tfrac{1}{2}x_n\right)_{n\in\mathbb{N}}, \quad \forall x = (x_n)_{n\in\mathbb{N}} \in \operatorname{dom}A, \tag{5.25}$$

where $\operatorname{dom}A := \left\{x := (x_n)_{n\in\mathbb{N}} \in \ell^2 \mid \sum_{i\geq 1} x_i = 0, \left(\sum_{i\leq n} x_i\right)_{n\in\mathbb{N}} \in \ell^2\right\}$ and $\sum_{i<1} x_i := 0$. Now [9, Propositions 3.6] states that

$$A^*x = \left(\tfrac{1}{2}x_n + \sum_{i>n} x_i\right)_{n\in\mathbb{N}}, \tag{5.26}$$

where

$$x = (x_n)_{n\in\mathbb{N}} \in \operatorname{dom}A^* = \left\{x = (x_n)_{n\in\mathbb{N}} \in \ell^2 \;\middle|\; \left(\sum_{i>n} x_i\right)_{n\in\mathbb{N}} \in \ell^2\right\}.$$

Then A is an at most single-valued linear relation such that the following hold (proofs of all claims are in brackets):

(i) A is maximally monotone and skew ([9, Propositions 3.5 and 3.2]).
(ii) A^* is maximally monotone but not skew ([9, Theorem 3.9 and Proposition 3.6]).
(iii) $\operatorname{dom} A$ is dense in ℓ^2 ([28, Theorem 2.5]), and $\operatorname{dom} A \subsetneqq \operatorname{dom} A^*$ ([9, Proposition 3.6]).
(iv) $\langle A^* x, x\rangle = \frac{1}{2}s^2, \quad \forall x = (x_n)_{n\in\mathbb{N}} \in \operatorname{dom} A^*$ with $s := \sum_{i\geq 1} x_i$ ([9, Proposition 3.7]).

Example 5.33. Suppose that X and A are as in Fact 5.32. Then A is enlargeable but A^* is non-enlargeable and is not skew. Moreover,

$$\operatorname{gra} A_\varepsilon = \Big\{(x,x^*) \in \operatorname{gra}(-A^*) \mid \Big|\sum_{i\geq 1} x_i\Big| \leq \sqrt{2\varepsilon},\ x = (x_n)_{n\in\mathbb{N}}\Big\},$$

where $\varepsilon \geq 0$.

Proof. By Corollary 5.30(iii) and Fact 5.32(iii), A must be enlargeable. For the second claim, note that $X = \ell^2$ is reflexive, and hence by Fact 5.32(i) and Corollary 5.30(iv), for every skew operator we must have A^* non-enlargeable. For the last statement, apply Corollary 5.30(i) and Fact 5.32(iv) directly to obtain $\operatorname{gra} A_\varepsilon$. ■

Example 5.34. Let C be a nonempty closed convex subset of X and $\varepsilon \geq 0$. Then

$$\operatorname{gra}(N_C)_\varepsilon = \big\{(x,x^*) \in C \times X^* \mid \sigma_C(x^*) \leq \langle x, x^*\rangle + \varepsilon\big\}.$$

Moreover, $(N_C)_\varepsilon = \partial_\varepsilon \iota_C$. Therefore, (for every $x \in X$) $(N_C)_\varepsilon(x)$ is the ε−normal set of C at x .

Proof. By Fact 5.20, we have

$$\begin{aligned}(x,x^*) \in \operatorname{gra}\,(N_C)_\varepsilon &\Leftrightarrow F_{N_C}(x,x^*) = \iota_C(x) + \sigma_C(x^*) \leq \langle x,x^*\rangle + \varepsilon \qquad (5.27)\\ &\Leftrightarrow x \in C,\ \sigma_C(x^*) \leq \langle x,x^*\rangle + \varepsilon.\end{aligned}$$

By (5.27) and [46, Theorem 2.4.2(ii)], $(N_C)_\varepsilon = \partial_\varepsilon \iota_C$. Hence (for every $x \in X$) $(N_C)_\varepsilon(x)$ is the ε−normal set of C at x. ■

Example 5.35. Let $f(x) := \|x\|$, $\forall x \in X$ and $\varepsilon \geq 0$. Then

$$\operatorname{gra}(\partial f)_\varepsilon = \big\{(x,x^*) \in X \times B_{X^*} \mid \|x\| \leq \langle x,x^*\rangle + \varepsilon\big\}.$$

In particular, $(\partial f)_\varepsilon(0) = B_{X^*}$.

Proof. Note that f is sublinear, and hence by Fact 5.21 and Remark 5.22 we can write

$$\begin{aligned}(x,x^*) \in \operatorname{gra}(\partial f)_\varepsilon &\Leftrightarrow F_{\partial f}(x,x^*) = f(x) + f^*(x^*) \leq \langle x,x^*\rangle + \varepsilon \quad \text{[by (5.13)]}\\ &\Leftrightarrow \|x\| + \iota_{B_{X^*}}(x^*) \leq \langle x,x^*\rangle + \varepsilon \quad \text{(by [46, Corollary 2.4.16])}\\ &\Leftrightarrow x^* \in B_{X^*},\ \|x\| \leq \langle x,x^*\rangle + \varepsilon.\end{aligned}$$

Hence $(\partial f)_\varepsilon(0) = B_{X^*}$. ■

Example 5.36. Let $p > 1$ and $f(x) := \frac{1}{p}\|x\|^p$, $\forall x \in X$. Then

$$(\partial f)_\varepsilon(0) = p^{\frac{1}{p}}(q\varepsilon)^{\frac{1}{q}} B_{X^*},$$

where $\frac{1}{p} + \frac{1}{q} = 1$ and $\varepsilon \geq 0$.

Proof. We have

$$\begin{aligned}x^* \in (\partial f)_\varepsilon(0) &\Leftrightarrow \langle x^* - y^*, -y\rangle \geq -\varepsilon, \quad \forall y^* \in \partial f(y)\\ &\Leftrightarrow \langle x^*, -y\rangle + \|y\|^p \geq -\varepsilon, \quad \forall y \in X\\ &\Leftrightarrow \langle x^*, y\rangle - \|y\|^p \leq \varepsilon, \quad \forall y \in X\\ &\Leftrightarrow p \sup_{y\in X}\left[\langle \tfrac{1}{p}x^*, y\rangle - \tfrac{1}{p}\|y\|^p\right] \leq \varepsilon\\ &\Leftrightarrow p \cdot \tfrac{1}{q}\|\tfrac{1}{p}x^*\|^q \leq \varepsilon\\ &\Leftrightarrow \|x^*\|^q \leq q\varepsilon p^{q-1} = q\varepsilon p^{\frac{q}{p}}\\ &\Leftrightarrow x^* \in p^{\frac{1}{p}}(q\varepsilon)^{\frac{1}{q}} B_{X^*}.\end{aligned}$$

■

5.4.2 Applications of Fitzpatrick's Last Function

For a monotone linear operator $A\colon X \to X^*$ it will be very useful to define the following quadratic function (which is actually a special case of *Fitzpatrick's last function* [15] for the linear relation A):

$$q_A\colon x \mapsto \tfrac{1}{2}\langle x, Ax\rangle.$$

Then $q_A = q_{A_+}$. We shall use the well-known fact (see, e.g., [28]) that

$$\nabla q_A = A_+, \tag{5.28}$$

where the gradient operator ∇ is understood in the Gâteaux sense.

The next result was first given in [8, Proposition 2.2] for a reflexive space. The proof is easily adapted to a general Banach space.

Fact 5.37. Let $A\colon X \to X^*$ be linear continuous, symmetric, and monotone. Then

$$\big(\forall (x,x^*) \in X \times X^*\big) \quad q_A^*(x^* + Ax) = q_A(x) + \langle x, x^*\rangle + q_A^*(x^*) \tag{5.29}$$

and $q_A^* \circ A = q_A$.

The next result was first proven in [4, Proposition 3.7(iv)] and [5, Theorem 2.3(i)] in Hilbert space. We now extend it to a general Banach space.

Proposition 5.38. *Let $A\colon X \to X^*$ be linear and monotone. Then*

$$F_A(x,x^*) = 2q_{A_+}^*(\tfrac{1}{2}x^* + \tfrac{1}{2}A^*x) = \tfrac{1}{2}q_{A_+}^*(x^* + A^*x), \quad \forall (x,x^*) \in X \times X, \tag{5.30}$$

and $\operatorname{ran} A_+ \subseteq \operatorname{dom} \partial q_{A_+}^* \subseteq \operatorname{dom} q_{A_+}^* \subseteq \overline{\operatorname{ran} A_+}$. *If* $\operatorname{ran} A_+$ *is closed, then* $\operatorname{dom} q_{A_+}^* = \operatorname{dom} \partial q_{A_+}^* = \operatorname{ran} A_+$.

Proof. By Fact 5.11, $\operatorname{dom} A^* \cap X = X$, so for every $x, y \in X$ we have $x, y \in \operatorname{dom} A^* \cap \operatorname{dom} A$. The latter fact and the definition of A^* yield $\langle y, A^*x\rangle = \langle x, Ay\rangle$. Hence for every $(x,x^*) \in X \times X^*$,

$$\begin{aligned} F_A(x,x^*) &= \sup_{y\in X} \langle x, Ay\rangle + \langle y, x^*\rangle - \langle y, Ay\rangle \\ &= 2\sup_{y\in X} \langle y, \tfrac{1}{2}x^* + \tfrac{1}{2}A^*x\rangle - q_{A_+}(y) \\ &= 2q_{A_+}^*(\tfrac{1}{2}x^* + \tfrac{1}{2}A^*x) \\ &= \tfrac{1}{2}q_{A_+}^*(x^* + A^*x), \end{aligned} \tag{5.31}$$

where we also used the fact that $q_A = q_{A_+}$ in the second equality. The third equality follows from the definition of Fenchel conjugate. By [46, Proposition 2.4.4(iv)],

$$\operatorname{ran} \partial q_{A_+} \subseteq \operatorname{dom} \partial q_{A_+}^* \tag{5.32}$$

By (5.28), $\operatorname{ran} \partial q_{A_+} = \operatorname{ran} A_+$. Then by (5.32),

$$\operatorname{ran} A_+ \subseteq \operatorname{dom} \partial q_{A_+}^* \subseteq \operatorname{dom} q_{A_+}^* \tag{5.33}$$

Then by the Brøndsted–Rockafellar Theorem (see [46, Theorem 3.1.2]),

$$\operatorname{ran} A_+ \subseteq \operatorname{dom} \partial q_{A_+}^* \subseteq \operatorname{dom} q_{A_+}^* \subseteq \overline{\operatorname{ran} A_+}.$$

Hence, under the assumption that $\operatorname{ran} A_+$ is closed, we have $\operatorname{ran} A_+ = \operatorname{dom} \partial q_{A_+}^* = \operatorname{dom} q_{A_+}^*$. ■

We can now apply the last proposition to obtain a formula for the enlargement of a single-valued operator.

Proposition 5.39 (Enlargement of a monotone linear operator). *Let $A : X \to X^*$ be a linear and monotone operator, and $\varepsilon \geq 0$. Then*

$$A_\varepsilon(x) = \left\{Ax + z^* \mid q_A^*(z^*) \leq 2\varepsilon\right\}, \quad \forall x \in X. \tag{5.34}$$

Moreover, A is non-enlargeable if and only if A is skew.

Proof. Fix $x \in X$, $z^* \in X^*$ and $x^* = Ax + z^*$. Then by Proposition 5.38 and Fact 5.37,

$$\begin{aligned}
x^* \in A_\varepsilon(x) &\Leftrightarrow F_A(x, Ax + z^*) \leq \langle x, Ax + z^* \rangle + \varepsilon \\
&\Leftrightarrow \tfrac{1}{2} q_{A_+}^*(Ax + z^* + A^* x) \leq \langle x, Ax + z^* \rangle + \varepsilon \\
&\Leftrightarrow \tfrac{1}{2} q_{A_+}^*(A_+(2x) + z^*) \leq \langle x, Ax + z^* \rangle + \varepsilon \\
&\Leftrightarrow \tfrac{1}{2}\left[q_{A_+}^*(z^*) + 2\langle x, z^* \rangle + 2\langle x, Ax \rangle\right] \leq \langle x, Ax + z^* \rangle + \varepsilon \\
&\Leftrightarrow q_A^*(z^*) \leq 2\varepsilon,
\end{aligned}$$

where we also used in the last equivalence the fact that $q_A = q_{A_+}$. Now we show the second statement. By Fact 5.11, $\operatorname{dom} A^* \cap X = X$. Then by Theorem 5.27 and Corollary 5.30(iii), we have A is non-enlargeable if and only if A is skew. ■

A result similar to Corollary 5.40 below was proved in [19, Proposition 2.2] in reflexive space. Their proof still requires the constraint that $\operatorname{ran}(A + A^*)$ is closed.

Corollary 5.40. *Let $A : X \to X^*$ be a linear and monotone operator such that $\operatorname{ran}(A + A^*)$ is closed. Then*

$$A_\varepsilon(x) = \left\{Ax + (A + A^*)z \mid q_A(z) \leq \tfrac{1}{2}\varepsilon\right\}, \quad \forall x \in X.$$

Proof. By Fact 5.11, A is continuous and $\operatorname{dom} A^* \cap X = X$. Proposition 5.39 yields

$$x^* \in A_\varepsilon(x) \Leftrightarrow x^* = Ax + z^*, \; q_A^*(z^*) \leq 2\varepsilon. \tag{5.35}$$

In particular, $z^* \in \operatorname{dom} q_A^*$. Since $\operatorname{ran}(A_+)$ is closed, Proposition 5.38 yields

$$\operatorname{ran}(A_+) = \operatorname{ran}(A + A^*) = \operatorname{dom} q_{A_+}^* = \operatorname{dom} q_A^*.$$

The above expression and the fact that $z^* \in \operatorname{dom} q_A^*$ implies that there exists $z \in X$ such that $z^* = (A + A^*)z$. Note also that (by Fact 5.37)

$$q_A^*(z^*) = q_{A_+}^*(z^*) = q_{A_+}^*(A_+(2z)) = q_{A_+}(2z) = 4q_A(z),$$

where we used Fact 5.37 in the last equality. Using this in (5.35) gives

$$x^* \in A_\varepsilon(x) \Leftrightarrow x^* = Ax + (A + A^*)z, \; 4q_A(z) \leq 2\varepsilon$$

$$\Leftrightarrow x^* = Ax + (A + A^*)z,\ q_A(z) \le \tfrac{1}{2}\varepsilon,$$

establishing the claim. ■

We conclude the section with two examples.

Example 5.41 (Rotation). Assume that X is the Euclidean plane $\mathbb{R}^2$, let $\theta \in \left[0, \frac{\pi}{2}\right]$, and set

$$A := \begin{pmatrix} \cos\theta & -\sin\theta \\ \sin\theta & \cos\theta \end{pmatrix}. \tag{5.36}$$

Then for every $(\varepsilon, x) \in \mathbb{R}_+ \times \mathbb{R}^2$,

$$A_\varepsilon(x) = \left\{Ax + v \mid v \in 2\sqrt{(\cos\theta)\varepsilon}\, B_X\right\}. \tag{5.37}$$

Proof. We consider two cases.

Case 1: $\theta = \frac{\pi}{2}$.
Then A is skew operator. By Corollary 5.30, $A_\varepsilon = A$ and hence (5.37) holds.

Case 2: $\theta \in \left[0, \frac{\pi}{2}\right[$.
Let $x \in \mathbb{R}^2$. Note that $\frac{A+A^*}{2} = (\cos\theta)\,\mathrm{Id}$, $q_A = \frac{\cos\theta}{2}\|\cdot\|^2$. Then by Corollary 5.40,

$$A_\varepsilon(x) = \left\{Ax + 2(\cos\theta)z \mid q_A(z) = \tfrac{\cos\theta}{2}\|z\|^2 \le \tfrac{1}{2}\varepsilon\right\}.$$

Thus,

$$A_\varepsilon(x) = \left\{Ax + v \mid \|v\| \le 2\sqrt{(\cos\theta)\varepsilon}\right\} = \left\{Ax + v \mid v \in 2\sqrt{(\cos\theta)\varepsilon}\, B_X\right\}.$$

■

Example 5.42 (Identity). Assume that X is a Hilbert space, and $A := \mathrm{Id}$. Let $\varepsilon \ge 0$. Then

$$\operatorname{gra} A_\varepsilon = \left\{(x, x^*) \in X \times X \mid x^* \in x + 2\sqrt{\varepsilon} B_X\right\}.$$

Proof. By [4, Example 3.10], we have

$$\begin{aligned}(x, x^*) \in \operatorname{gra} A_\varepsilon &\Leftrightarrow \tfrac{1}{4}\|x + x^*\|^2 \le \langle x, x^*\rangle + \varepsilon \\ &\Leftrightarrow \tfrac{1}{4}\|x - x^*\|^2 \le \varepsilon \\ &\Leftrightarrow \|x - x^*\| \le 2\sqrt{\varepsilon} \\ &\Leftrightarrow x^* \in x + 2\sqrt{\varepsilon} B_X.\end{aligned}$$

■

5.5 Sums of Operators

The conclusion of the lemma below has been established for reflexive Banach spaces in [7, Lemma 5.8]. Our proof for a general Banach space assumes the operators to be of type (FPV) and follows closely that of [7, Lemma 5.8].

Lemma 5.43. *Let $A,B\colon X \rightrightarrows X^*$ be maximally monotone of type (FPV), and suppose that $\bigcup_{\lambda>0}\lambda\,[\operatorname{dom}A-\operatorname{dom}B]$ is a closed subspace of X. Then we have*

$$\bigcup_{\lambda>0}\lambda\,[\operatorname{dom}A-\operatorname{dom}B]=\bigcup_{\lambda>0}\lambda\,[P_X\operatorname{dom}F_A-P_X\operatorname{dom}F_B].$$

Proof. By Facts 5.5 and 5.6, we have

$$\begin{aligned}\bigcup_{\lambda>0}\lambda\,[\operatorname{dom}A-\operatorname{dom}B]&\subseteq\bigcup_{\lambda>0}\lambda\,[P_X\operatorname{dom}F_A-P_X\operatorname{dom}F_B]\subseteq\bigcup_{\lambda>0}\lambda\,\left[\overline{\operatorname{dom}A}-\overline{\operatorname{dom}B}\right]\\&\subseteq\bigcup_{\lambda>0}\lambda\,\overline{[\operatorname{dom}A-\operatorname{dom}B]}\subseteq\overline{\bigcup_{\lambda>0}\lambda\,[\operatorname{dom}A-\operatorname{dom}B]}\\&=\bigcup_{\lambda>0}\lambda\,[\operatorname{dom}A-\operatorname{dom}B]\quad\text{(by the assumption).}\end{aligned}$$

■

Corollary 5.44. *Let $A,B\colon X \rightrightarrows X^*$ be maximally monotone linear relations, and suppose that* $\operatorname{dom}A-\operatorname{dom}B$ *is a closed subspace. Then*

$$[\operatorname{dom}A-\operatorname{dom}B]=\bigcup_{\lambda>0}\lambda\,[P_X\operatorname{dom}F_A-P_X\operatorname{dom}F_B].$$

Proof. Directly apply Fact 5.8 and Lemma 5.43. ■

Corollary 5.45. *Let $A\colon X \rightrightarrows X^*$ be a maximally monotone linear relation and let $C\subseteq X$ be a nonempty and closed convex set. Assume that $\bigcup_{\lambda>0}\lambda\,[\operatorname{dom}A-C]$ is a closed subspace. Then*

$$\bigcup_{\lambda>0}\lambda\,\left[P_X\operatorname{dom}F_A-P_X\operatorname{dom}F_{N_C}\right]=\bigcup_{\lambda>0}\lambda\,[\operatorname{dom}A-C].$$

Proof. Let $B=N_C$. Then apply directly Facts 5.8, 5.9 and Lemma 5.43. ■

Theorem 5.46 below was proved in [7, Theorem 5.10] for a reflexive space. We extend it to a general Banach space.

Theorem 5.46 (Fitzpatrick function of the sum). *Let $A,B\colon X \rightrightarrows X^*$ be maximally monotone linear relations, and suppose that* $\operatorname{dom}A-\operatorname{dom}B$ *is closed. Then*

$$F_{A+B} = F_A \Box_2 F_B,$$

and the partial infimal convolution is exact everywhere.

Proof. Let $(z,z^*) \in X \times X^*$. By Fact 5.19, it suffices to show that there exists $v^* \in X^*$ such that

$$F_{A+B}(z,z^*) \geq F_A(z,z^*-v^*) + F_B(z,v^*). \tag{5.38}$$

If $(z,z^*) \notin \operatorname{dom} F_{A+B}$, clearly, (5.38) holds.

Now assume that $(z,z^*) \in \operatorname{dom} F_{A+B}$. Then

$$\begin{aligned} &F_{A+B}(z,z^*) \\ &= \sup_{\{x,x^*,y^*\}} \left[\langle x,z^*\rangle + \langle z,x^*\rangle - \langle x,x^*\rangle + \langle z-x,y^*\rangle - \iota_{\operatorname{gra} A}(x,x^*) - \iota_{\operatorname{gra} B}(x,y^*)\right]. \end{aligned} \tag{5.39}$$

Let $Y = X^*$ and define $F, K : X \times X^* \times Y \to \left]-\infty,+\infty\right]$ respectively by

$$\begin{aligned} F &: (x,x^*,y^*) \in X \times X^* \times Y \to \langle x,x^*\rangle + \iota_{\operatorname{gra} A}(x,x^*) \\ K &: (x,x^*,y^*) \in X \times X^* \times Y \to \langle x,y^*\rangle + \iota_{\operatorname{gra} B}(x,y^*) \end{aligned}$$

Then by (5.39),

$$F_{A+B}(z,z^*) = (F+K)^*(z^*,z,z) \tag{5.40}$$

By Fact 5.14 and the assumptions, F and K are proper lower semicontinuous and convex. The definitions of F and K yield

$$\operatorname{dom} F - \operatorname{dom} K = [\operatorname{dom} A - \operatorname{dom} B] \times X^* \times Y, \quad \text{which is a closed subspace.}$$

Thus by Fact 5.4 and (5.40), there exists $(z_0^*, z_0^{**}, z_1^{**}) \in X^* \times X^{**} \times Y^*$ such that

$$\begin{aligned} F_{A+B}(z,z^*) &= F^*(z^*-z_0^*, z-z_0^{**}, z-z_1^{**}) + K^*(z_0^*, z_0^{**}, z_1^{**}) \\ &= F^*(z^*-z_0^*, z, 0) + K^*(z_0^*, 0, z) \quad (\text{by } (z,z^*) \in \operatorname{dom} F_{A+B}) \\ &= F_A(z,z^*-z_0^*) + F_B(z,z_0^*). \end{aligned}$$

Thus (5.38) holds by taking $v^* = z_0^*$ and hence $F_{A+B} = F_A \Box_2 F_B$. ∎

The next result was first obtained by Voisei in [39] while Simons gave a different proof in [34, Theorem 46.3]. We are now in position to provide a third approach.

Theorem 5.47. *Let $A, B \colon X \rightrightarrows X^*$ be maximally monotone linear relations, and suppose that* $\operatorname{dom} A - \operatorname{dom} B$ *is closed. Then $A+B$ is maximally monotone.*

Proof. By Fact 5.5, we have that $F_A \geq \langle\cdot,\cdot\rangle$ and $F_B \geq \langle\cdot,\cdot\rangle$. Using now Theorem 5.46 and (5.9) implies that $F_{A+B} \geq \langle\cdot,\cdot\rangle$. Combining the last inequality with Corollary 5.44 and Fact 5.15, we conclude that $A+B$ is maximally monotone. ■

Theorem 5.48. *Let $A,B\colon X \rightrightarrows X^*$ be maximally monotone linear relations, and suppose that* $\operatorname{dom} A - \operatorname{dom} B$ *is closed. Assume that A and B are non-enlargeable. Then*

$$F_{A+B} = \iota_{\operatorname{gra}(A+B)} + \langle\cdot,\cdot\rangle$$

and hence $A+B$ is non-enlargeable.

Proof. By Corollary 5.25, we have

$$F_A = \iota_{\operatorname{gra} A} + \langle\cdot,\cdot\rangle \quad \text{and} \quad F_B = \iota_{\operatorname{gra} B} + \langle\cdot,\cdot\rangle. \tag{5.41}$$

Let $(x,x^*) \in X \times X^*$. Then by (5.41) and Theorem 5.46, we have

$$\begin{aligned} F_{A+B}(x,x^*) &= \min_{y^* \in X^*} \left\{ \iota_{\operatorname{gra} A}(x,x^*-y^*) + \langle x^*-y^*,x\rangle + \iota_{\operatorname{gra} B}(x,y^*) + \langle y^*,x\rangle \right\} \\ &= \iota_{\operatorname{gra}(A+B)}(x,x^*) + \langle x^*,x\rangle. \end{aligned}$$

By Theorem 5.47 we have that $A+B$ is maximally monotone. Now we can apply Corollary 5.25 to $A+B$ to conclude that $A+B$ is non-enlargeable. ■

The proof of Theorem 5.49 in part follows that of [6, Theorem 3.1].

Theorem 5.49. *Let $A : X \rightrightarrows X^*$ be a maximally monotone linear relation. Suppose C is a nonempty closed convex subset of X, and that* $\operatorname{dom} A \cap \operatorname{int} C \neq \varnothing$. *Then $F_{A+N_C} = F_A \Box_2 F_{N_C}$, and the partial infimal convolution is exact everywhere.*

Proof. Let $(z,z^*) \in X \times X^*$. By Fact 5.19, it suffices to show that there exists $v^* \in X^*$ such that

$$F_{A+N_C}(z,z^*) \geq F_A(z,v^*) + F_{N_C}(z,z^*-v^*). \tag{5.42}$$

If $(z,z^*) \notin \operatorname{dom} F_{A+N_C}$, clearly, (5.42) holds.

Now assume that

$$(z,z^*) \in \operatorname{dom} F_{A+N_C}. \tag{5.43}$$

By Facts 5.10 and 5.6,

$$P_X\left[\operatorname{dom} F_{A+N_C}\right] \subseteq \overline{\left[\operatorname{dom}(A+N_C)\right]} \subseteq C.$$

Thus, by (5.43), we have

$$z \in C. \tag{5.44}$$

Set

$$g\colon X\times X^*\to\,]-\infty,+\infty]: (x,x^*)\mapsto\langle x,x^*\rangle+\iota_{\operatorname{gra}A}(x,x^*). \tag{5.45}$$

By Fact 5.14, g is convex. Hence,

$$h=g+\iota_{C\times X^*} \tag{5.46}$$

is convex as well. Let

$$c_0\in\operatorname{dom}A\cap\operatorname{int}C, \tag{5.47}$$

and let $c_0^*\in Ac_0$. Then $(c_0,c_0^*)\in\operatorname{gra}A\cap(\operatorname{int}C\times X^*)=\operatorname{dom}g\cap\operatorname{int}\operatorname{dom}\iota_{C\times X^*}$. Let us compute $F_{A+N_C}(z,z^*)$. As in (5.39) we can write

$$\begin{aligned}
&F_{A+N_C}(z,z^*)\\
&=\sup_{(x,x^*,c^*)}\left[\langle x,z^*\rangle+\langle z,x^*\rangle-\langle x,x^*\rangle+\langle z-x,c^*\rangle-\iota_{\operatorname{gra}A}(x,x^*)-\iota_{\operatorname{gra}N_C}(x,c^*)\right]\\
&\geq\sup_{(x,x^*)}\left[\langle x,z^*\rangle+\langle z,x^*\rangle-\langle x,x^*\rangle-\iota_{\operatorname{gra}A}(x,x^*)-\iota_{C\times X^*}(x,x^*)\right]\\
&=\sup_{(x,x^*)}\left[\langle x,z^*\rangle+\langle z,x^*\rangle-h(x,x^*)\right]\\
&=h^*(z^*,z),
\end{aligned}$$

where we took $c^*=0$ in the inequality. By Fact 5.1, $\iota_{C\times X^*}$ is continuous at $(c_0,c_0^*)\in\operatorname{int}\operatorname{dom}\iota_{C\times X^*}$. Since $(c_0,c_0^*)\in\operatorname{dom}g\cap\operatorname{int}\operatorname{dom}\iota_{C\times X^*}$ we can use Fact 5.2 to conclude the existence of $(y^*,y^{**})\in X^*\times X^{**}$ such that

$$\begin{aligned}
h^*(z^*,z)&=g^*(y^*,y^{**})+\iota_{C\times X^*}^*(z^*-y^*,z-y^{**})\\
&=g^*(y^*,y^{**})+\iota_C^*(z^*-y^*)+\iota_{\{0\}}(z-y^{**}).
\end{aligned} \tag{5.48}$$

Then by (5.43) and (5.48) we must have $z=y^{**}$. Thus by (5.48) and the definition of g we have

$$\begin{aligned}
&F_{A+N_C}(z,z^*)\geq g^*(y^*,z)+\iota_C^*(z^*-y^*)=F_A(z,y^*)+\iota_C^*(z^*-y^*)\\
&=F_A(z,y^*)+\iota_C^*(z^*-y^*)+\iota_C(z)\quad\text{[by (5.44)]}\\
&=F_A(z,y^*)+F_{N_C}(z,z^*-y^*)\quad\text{[by Fact 5.20]}.
\end{aligned}$$

Hence (5.42) holds by taking $v^*=y^*$ and thus $F_{A+N_C}=F_A\Box_2F_{N_C}$. ■

We decode the prior result as follows:

Corollary 5.50 (Normal cone). *Let $A : X \rightrightarrows X^*$ be a maximally monotone linear relation. Suppose C is a nonempty closed convex subset of X, and that* $\operatorname{dom} A \cap \operatorname{int} C \neq \varnothing$. *Then $A + N_C$ is maximally monotone.*

Proof. By Fact 5.5, we have that $F_A \geq \langle\cdot,\cdot\rangle$ and $F_{N_C} \geq \langle\cdot,\cdot\rangle$. Using now Theorem 5.49 and (5.9) implies that $F_{A+N_C} \geq \langle\cdot,\cdot\rangle$. Combining the last inequality with Corollary 5.44 and Fact 5.15, we conclude that $A + N_C$ is maximally monotone. ■

Remark 5.51. Corollary 5.50 was first established in [6, Theorem 3.1]. See [41,44, 45] for generalizations.

To conclude we revisit a quite subtle example. All statements in the fact below have been proved in [10, Example 4.1 and Theorem 3.6(vii)].

Fact 5.52. Consider $X := c_0$, with norm $\|\cdot\|_\infty$ so that $X^* = \ell^1$ with norm $\|\cdot\|_1$, and $X^{**} = \ell^\infty$ with second dual norm $\|\cdot\|_*$. Fix $\alpha := (\alpha_n)_{n\in\mathbb{N}} \in \ell^\infty$ with $\limsup \alpha_n \neq 0$, and define $A_\alpha : \ell^1 \to \ell^\infty$ by

$$(A_\alpha x^*)_n := \alpha_n^2 x_n^* + 2\sum_{i>n} \alpha_n \alpha_i x_i^*, \quad \forall x^* = (x_n^*)_{n\in\mathbb{N}} \in \ell^1. \tag{5.49}$$

Finally, let $T_\alpha : c_0 \rightrightarrows X^*$ be defined by

$$\begin{aligned}\operatorname{gra} T_\alpha &:= \{(-A_\alpha x^*, x^*) \mid x^* \in X^*, \langle \alpha, x^*\rangle = 0\}\\ &= \Big\{\big((-\sum_{i>n} \alpha_n \alpha_i x_i^* + \sum_{i<n} \alpha_n \alpha_i x_i^*)_n, x^*\big) \mid x^* \in X^*, \langle \alpha, x^*\rangle = 0\Big\}.\end{aligned} \tag{5.50}$$

Then

(i) $\langle A_\alpha x^*, x^*\rangle = \langle \alpha, x^*\rangle^2, \quad \forall x^* = (x_n^*)_{n\in\mathbb{N}} \in \ell^1$ and so (5.50) is well defined.
(ii) A_α is a maximally monotone operator on ℓ^1.
(iii) T_α is a maximally monotone and skew operator on c_0.
(iv) $F_{T_\alpha} = \iota_C$, where $C := \{(-A_\alpha x^*, x^*) \mid x^* \in X^*\}$.

This set of affairs allows us to show the following:

Example 5.53. Let $X = c_0, A_\alpha, C$, and T_α be defined as in Fact 5.52. Then $T_\alpha : c_0 \rightrightarrows \ell^1$ is a maximally monotone enlargeable skew linear relation. Indeed

$$\operatorname{gra}(T_\alpha + N_{B_X})_\varepsilon = \Big\{(-A_\alpha x^*, z^*) \in B_X \times X^* \mid x^* \in X, \|z^* - x^*\|_1 \leq \langle -A_\alpha x^*, z^*\rangle + \varepsilon\Big\}.$$

Proof. From (5.50), we have that $\operatorname{gra} T_\alpha \subsetneqq C$ therefore Fact 5.52(iv) yields $F_{T_\alpha} \neq \iota_{\operatorname{gra} T_\alpha} + \langle\cdot,\cdot\rangle$. Using now Fact 5.52(iii) and Corollary 5.25, we conclude that T_α is enlargeable.

Now we determine $\operatorname{gra}(T_\alpha + N_{B_X})_\varepsilon$. By Fact 5.52(iii), Theorem 5.49, and (5.4), we have

$$\begin{aligned}
&(z,z^*) \in \text{gra}(T_\alpha + N_{B_X})_\varepsilon \\
&\Leftrightarrow F_{T_\alpha} \Box_2 F_{N_{B_X}}(z,z^*) \le \langle z,z^*\rangle + \varepsilon \\
&\Leftrightarrow F_{T_\alpha}(z,x^*) + \iota_{B_X}(z) + \iota^*_{B_X}(z^*-x^*) \le \langle z,z^*\rangle + \varepsilon,\ \exists x^* \in X^* \quad \text{(by Fact 5.20)} \\
&\Leftrightarrow z \in B_X,\ \iota_C(z,x^*) + \|z^*-x^*\|_1 \le \langle z,z^*\rangle + \varepsilon,\ \exists x^* \in X^* \text{ (by Fact 5.52(iv))} \\
&\Leftrightarrow z = -A_\alpha x^* \in B_X,\ \|z^*-x^*\|_1 \le \langle z,z^*\rangle + \varepsilon,\ \exists x^* \in X^* \\
&\Leftrightarrow z = -A_\alpha x^* \in B_X,\ \|z^*-x^*\|_1 \le \langle -A_\alpha x^*,z^*\rangle + \varepsilon,\ \exists x^* \in X^*.
\end{aligned}$$

This is the desired result. ■

Acknowledgements The authors thank Dr. Heinz Bauschke and Dr. Xianfu Wang for their valuable discussions and comments. The authors thank an anonymous referee for his/her careful reading and pertinent comments. Jonathan Borwein was partially supported by the Australian Research Council. The third author thanks CARMA at the University of Newcastle and the School of Mathematics and Statistics of University of South Australia for the support of his visit to Australia, which started this research.

References

1. Attouch, H., Brézis, H.: Duality for the sum of convex functions in general Banach spaces. In: Barroso, J.A. (ed.) Aspects of Mathematics and Its Applications, pp. 125–133. Elsevier, North-Holland (1986)
2. Bartz, S., Bauschke, H.H., Borwein, J.M., Reich, S., Wang, X.: Fitzpatrick functions, cyclic monotonicity and Rockafellar's antiderivative. Nonlinear Anal. **66**, 1198–1223 (2007)
3. Bauschke, H.H., Combettes, P.L.: Convex Analysis and Monotone Operator Theory in Hilbert Spaces. Springer, New York (2011)
4. Bauschke, H.H., McLaren, D.A., Sendov, H.S.: Fitzpatrick functions: inequalities, examples and remarks on a problem by S. Fitzpatrick. J. Convex Anal. **13**, 499–523 (2006)
5. Bauschke, H.H., Borwein, J.M., Wang, X.: Fitzpatrick functions and continuous linear monotone operators. SIAM J. Optim. **18**, 789–809 (2007)
6. Bauschke, H.H., Wang, X., Yao, L.: An answer to S. Simons' question on the maximal monotonicity of the sum of a maximal monotone linear operator and a normal cone operator. Set-Valued Var. Anal. **17**, 195–201 (2009)
7. Bauschke, H.H., Wang, X., Yao, L.: Monotone linear relations: maximality and Fitzpatrick functions. J. Convex Anal. **16**, 673–686 (2009)
8. Bauschke, H.H., Wang, X., Yao, L.: Autoconjugate representers for linear monotone operators. Math. Program. Ser. B **123**, 5–24 (2010)
9. Bauschke, H.H., Wang, X., Yao, L.: Examples of discontinuous maximal monotone linear operators and the solution to a recent problem posed by B.F. Svaiter. J. Math. Anal. Appl. **370**, 224–241 (2010)
10. Bauschke, H.H., Borwein, J.M., Wang, X., Yao, L.: Construction of pathological maximally monotone operators on non-reflexive Banach spaces. Set-Valued Var. Anal., **20**(3), 387–415 (2012). DOI: 10.1007/s11228-012-0209-0. Available at http://arxiv.org/abs/1108.1463
11. Bauschke, H.H., Borwein, J.M., Wang, X., Yao, L.: Monotone operators and "bigger conjugate" functions. J. Convex Anal., **20**, 143–155 (2013)

12. Borwein, J.M.: Maximal monotonicity via convex analysis. J. Convex Anal. **13**, 561–586 (2006)
13. Borwein, J.M.: Maximality of sums of two maximal monotone operators in general Banach space. Proc. Am. Math. Soc. **135**, 3917–3924 (2007)
14. Borwein, J.M.: Fifty years of maximal monotonicity. Optim. Lett. **4**, 473–490 (2010)
15. Borwein, J.M., Vanderwerff, J.D.: Convex Functions. Cambridge University Press, Cambridge (2010)
16. Boţ, R.I., Csetnek, E.R.: Enlargements of positive sets. J. Math. Anal. Appl. **356**, 328–337 (2009)
17. Brézis, H., Browder, F.E.: Linear maximal monotone operators and singular nonlinear integral equations of Hammerstein type. In: Nonlinear Analysis (collection of Papers in Honor of Erich H. Rothe), pp. 31–42. Academic, New York (1978)
18. Brøndsted, A., Rockafellar, R.T.: On the subdifferentiability of convex functions. Proc. Am. Math. Soc. **16**, 605–611 (1965)
19. Burachik, R.S., Iusem, A.N.: On non-enlargeable and fully enlargeable monotone operators. J. Convex Anal. **13**, 603–622 (2006)
20. Burachik, R.S., Iusem, A.N.: Set-Valued Mappings and Enlargements of Monotone Operators. Springer, New York (2008)
21. Burachik, R.S., Svaiter, B.F.: Maximal monotone operators, convex functions and a special family of enlargements. Set-Valued Anal. **10**, 297–316 (2002)
22. Fitzpatrick, S.: Representing monotone operators by convex functions. In: Workshop/Miniconference on Functional Analysis and Optimization (Canberra 1988). Proceedings of the Centre for Mathematical Analysis, Australian National University, vol. 20, pp. 59–65, Canberra, Australia, 1988
23. Fitzpatrick, S.P., Phelps, R.R.: Some properties of maximal monotone operators on nonreflexive Banach spaces. Set-Valued Anal. **3**, 51–69 (1995)
24. Hiriart-Urruty, J.-B., Moussaoui, M., Seeger, A., Volle, M.: Subdifferential calculus without qualification conditions, using approximate subdifferentials: a survey. Nonlinear Anal. **24**, 1727–1754 (1995)
25. Marques Alves, M., Svaiter, B.F.: A new proof for maximal monotonicity of subdifferential operators. J. Convex Anal. **15**, 345–348 (2008)
26. Marques Alves, M., Svaiter, B.F.: Maximal monotone operators with a unique extension to the bidual. J. Convex Anal. **16**, 409–421 (2009)
27. Phelps, R.R.:: Convex functions, Monotone Operators and Differentiability, 2nd edn. Springer, New york (1993)
28. Phelps, R.R., Simons, S.: Unbounded linear monotone operators on nonreflexive Banach spaces. J. Convex Anal. **5**, 303–328 (1998)
29. Rockafellar, R.T.: Extension of Fenchel's duality theorem for convex functions. Duke Math. J. **33**, 81–89 (1966)
30. Rockafellar, R.T.: On the maximal monotonicity of sums on nonlinear monotone operators. Trans. Am. Math. Soc. **149**, 75–88 (1970)
31. Rockafellar, R.T.: On the maximal monotonicity of subdifferential mappings. Pacific J. Math. **33**, 209–216 (1970)
32. Rockafellar, R.T., Wets, R.J-B.: Variational Analysis, 3rd edn. Springer, New York (2009)
33. Simons, S.: Minimax and Monotonicity. Springer, Berlin (1998)
34. Simons, S.: From Hahn-Banach to Monotonicity. Springer, New York (2008)
35. Simons, S.: A Brezis-Browder theorem for SSDB spaces. http://arxiv.org/abs/1004.4251v3 (2010)
36. Svaiter, B.F.: A family of enlargements of maximal monotone operators. Set-Valued Anal. **8**, 311–328 (2000)
37. Svaiter, B.F.: Non-enlargeable operators and self-cancelling operators. J. Convex Anal. **17**, 309–320 (2010)
38. Verona, A., Verona, M.E.: Regular maximal monotone operators. Set-Valued Anal. **6**, 303–312 (1998)

39. Voisei, M.D.: The sum theorem for linear maximal monotone operators. Math. Sci. Res. J. **10**, 83–85 (2006)
40. Voisei, M.D.: The sum and chain rules for maximal monotone operators. Set-Valued Anal. **16**, 461–476 (2008)
41. Voisei, M.D.: A Sum Theorem for (FPV) operators and normal cones. J. Math. Anal. Appl. **371**, 661–664 (2010)
42. Voisei, M.D., Zălinescu, C.: Linear monotone subspaces of locally convex spaces. Set-Valued Var. Anal. **18**, 29–55 (2010)
43. Yao, L.: The Brézis-Browder Theorem revisited and properties of Fitzpatrick functions of order *n*. Fixed-Point Algorithms for Inverse Problems in Science and Engineering (Banff 2009). Springer Optimization and Its Applications, vol. 49, pp. 391–402. Springer, New York (2011)
44. Yao, L.: The sum of a maximal monotone operator of type (FPV) and a maximal monotone operator with full domain is maximally monotone. Nonlinear Anal. **74**, 6144–6152 (2011)
45. Yao, L.: The sum of a maximally monotone linear relation and the subdifferential of a proper lower semicontinuous convex function is maximally monotone. Set-Valued Var. Anal., **20**, 155–167 (2012)
46. Zălinescu, C.: Convex Analysis in General Vector Spaces. World Scientific Publishing, River Edge (2002)
47. Zeidler, E.: Nonlinear Functional Analysis and Its Application, Part II/B. Nonlinear Monotone Operators. Springer, New York (1990)

Chapter 6
A Brøndsted–Rockafellar Theorem for Diagonal Subdifferential Operators

Radu Ioan Boţ and Ernö Robert Csetnek

Dedicated to Jonathan Borwein on the occasion of his 60th birthday

Abstract In this note we give a Brøndsted–Rockafellar Theorem for diagonal subdifferential operators in Banach spaces. To this end we apply an Ekeland-type variational principle for monotone bifunctions.

Key words: Brøndsted–Rockafellar Theorem • Ekeland variational principle • Diagonal subdifferential operator • Monotone bifunction • Subdifferential

Mathematics Subject Classifications (2010): Primary 58E30; Secondary 90C25

6.1 Introduction

Throughout this paper X denotes a real Banach space and X^* its topological dual space endowed with the dual norm. Since there is no danger of confusion, we use $\|\cdot\|$ as notation for the norms of both spaces X and X^*. We denote by $\langle x^*, x\rangle$ the value of the linear and continuous functional $x^* \in X^*$ at $x \in X$.

A function $f : X \to \overline{\mathbb{R}} := \mathbb{R} \cup \{\pm\infty\}$ is called proper if the set $\operatorname{dom} f := \{x \in X : f(x) < +\infty\}$, called *effective domain* of f, is nonempty and $f(x) > -\infty$ for all $x \in X$. We consider also the *epigraph* of f, which is the set $\operatorname{epi} f = \{(x,r) \in X \times \mathbb{R} :$

COMMUNICATED BY HEINZ H. BAUSCHKE.

R.I. Boţ (✉) • E.R. Csetnek
Department of Mathematics, Chemnitz University of Technology,
D-09107 Chemnitz, Germany
e-mail: radu.bot@mathematik.tu-chemnitz.de; robert.csetnek@mathematik.tu-chemnitz.de

D.H. Bailey et al. (eds.), *Computational and Analytical Mathematics*, Springer Proceedings in Mathematics & Statistics 50, DOI 10.1007/978-1-4614-7621-4_6,

$f(x) \leq r\}$. For a set $C \subseteq X$, let $\delta_C : X \to \overline{\mathbb{R}}$ be its *indicator function*, which is the function taking the values 0 on C and $+\infty$ otherwise.

The (convex) subdifferential of f at an element $x \in X$ such that $f(x) \in \mathbb{R}$ is defined as $\partial f(x) := \{x^* \in X^* : f(y) - f(x) \geq \langle x^*, y - x\rangle \ \forall y \in X\}$, while in case $f(x) \notin \mathbb{R}$ one takes by convention $\partial f(x) := \emptyset$. For every $\varepsilon \geq 0$, the *ε-subdifferential* of f, defined as $\partial_\varepsilon f(x) = \{x^* \in X^* : f(y) - f(x) \geq \langle x^*, y - x\rangle - \varepsilon \ \forall y \in X\}$ for $x \in X$ such that $f(x) \in \mathbb{R}$, and $\partial_\varepsilon f(x) := \emptyset$ otherwise, represents an enlargement of its subdifferential. Let us notice that in contrast to the classical subdifferential, the ε-subdifferential of a proper, convex, and lower semicontinuous function at each point of its effective domain is in general a nonempty set, provided that $\varepsilon > 0$ (cf. [12, Proposition 3.15]; see also [15, Theorem 2.4.4(iii)]).

For $\varepsilon \geq 0$, the *ε-normal set* of C at $x \in X$ is defined by $N_C^\varepsilon(x) := \partial_\varepsilon \delta_C(x)$, that is, $N_C^\varepsilon(x) = \{x^* \in X^* : \langle x^*, y - x\rangle \leq \varepsilon \ \forall y \in C\}$ if $x \in C$, and $N_C^\varepsilon(x) = \emptyset$ otherwise. The *normal cone* of the set C at $x \in X$ is $N_C(x) := N_C^0(x)$, that is, $N_C(x) = \{x^* \in X^* : \langle x^*, y - x\rangle \leq 0 \ \forall y \in C\}$ if $x \in C$, and $N_C(x) = \emptyset$ otherwise.

For the following characterizations of the ε-subdifferential via the ε-normal set, we refer, for instance, to [13] (the extension from finite to infinite dimensional spaces is straightforward). If $x \in X$ is such that $f(x) \in \mathbb{R}$, then for all $\varepsilon \geq 0$ it holds $x^* \in \partial_\varepsilon f(x)$ if and only if $(x^*, -1) \in N_{\operatorname{epi} f}^\varepsilon(x, f(x))$. Moreover, for $r \in \mathbb{R}$ with $f(x) \leq r$, the relation $(x^*, -1) \in N_{\operatorname{epi} f}(x, r)$ implies $r = f(x)$. Furthermore, if $(x^*, -s) \in N_{\operatorname{epi} f}(x, r)$, then $s \geq 0$ and, if, additionally, $s \neq 0$, then $r = f(x)$ and $(1/s)x^* \in \partial f(x)$.

The celebrated Brøndsted–Rockafellar Theorem [6], which we recall as follows, emphasizes the fact that the ε-subdifferential of a proper, convex, and lower semicontinuous function can be seen as an approximation of its subdifferential.

Theorem 6.1 (Brøndsted–Rockafellar Theorem [6]). *Let $f : X \to \overline{\mathbb{R}}$ be a proper, convex, and lower semicontinuous function and $x_0 \in \operatorname{dom} f$. Take $\varepsilon > 0$ and $x_0^* \in \partial_\varepsilon f(x_0)$. Then for all $\lambda > 0$, there exist $x \in X$ and $x^* \in X^*$ such that*

$$x^* \in \partial f(x), \ \|x - x_0\| \leq \frac{\varepsilon}{\lambda} \text{ and } \|x^* - x_0^*\| \leq \lambda.$$

Let us mention that a method for proving this result is by applying the Ekeland variational principle (see [12, Theorem 3.17]). For a more elaborated version of Theorem 6.1, we refer the interested reader to a result given by Borwein in [4] (see, also, [15, Theorem 3.1.1].

The aim of this note is to provide a Brøndsted–Rockafellar Theorem for so-called diagonal subdifferential operators. These are set-valued operators $A^F : X \rightrightarrows X^*$ defined by (see [1, 5, 8–10])

$$A^F(x) = \begin{cases} \{x^* \in X^* : F(x,y) - F(x,x) \geq \langle x^*, y - x\rangle \ \forall y \in C\}, & \text{if } x \in C, \\ \emptyset, & \text{otherwise,} \end{cases}$$

where C is a nonempty subset of X and $F : C \times C \to \mathbb{R}$ is a so-called bifunction. The term *diagonal subdifferential operator* is justified by the formula $A^F(x) = \partial(F(x,\cdot)+\delta_C)(x)$ for all $x \in X$.

Bifunctions have been intensively studied in connection with equilibrium problems since the publication of the seminal work of Blum and Oettli [3] and, recently, in the context of diagonal subdifferential operators, when characterizing properties like local boundedness [1], monotonicity, and maximal monotonicity in both reflexive [8, 9] and nonreflexive Banach spaces [5, 10].

A further operator of the same type, which has been considered in the literature, is ${}^F\!A : X \rightrightarrows X^*$, defined by

$$
{}^F\!A(x) = \begin{cases} \{x^* \in X^* : F(x,x) - F(y,x) \geq \langle x^*, y-x\rangle \ \forall y \in C\}, & \text{if } x \in C, \\ \emptyset, & \text{otherwise.} \end{cases}
$$

Notice that when F is *monotone*, namely, $F(x,y)+F(y,x) \leq 0$ for all $x,y \in C$ (see [3]) and $F(x,x) = 0$ for all $x \in C$, then $A^F(x) \subseteq {}^F\!A(x)$ for all $x \in C$. Furthermore, if C is convex and closed, $F(x,x) = 0$, $F(x,\cdot)$ is convex and $F(\cdot,y)$ is upper hemicontinuous, i.e., upper semicontinuous along segments, for all $x,y \in C$, then ${}^F\!A(x) \subseteq A^F(x)$ for all $x \in C$ (cf. [5, Lemma 5]). Under these hypotheses one can transfer properties from ${}^F\!A$ to A^F and vice versa.

In the following we will concentrate ourselves on A^F and consider, in analogy to the definition of the ε-subdifferential, what we call to be the *ε-diagonal subdifferential operator* of F, $A^F_\varepsilon : X \rightrightarrows X^*$, defined by

$$
A^F_\varepsilon(x) = \begin{cases} \{x^* \in X^* : F(x,y) - F(x,x) \geq \langle x^*, y-x\rangle - \varepsilon \ \forall y \in C\}, & \text{if } x \in C, \\ \emptyset, & \text{otherwise.} \end{cases}
$$

If C is a nonempty, convex, and closed set and $x \in C$ is such that $F(x,\cdot)$ is convex and lower semicontinuous, then $A^F_\varepsilon(x) \neq \emptyset$ for all $\varepsilon > 0$.

The main result of this paper is represented by a Brøndsted–Rockafellar Theorem for the diagonal subdifferential operator A^F, the proof of which relies on the Ekeland variational principle for bifunctions given in [2].

For a generalization of the Brøndsted–Rockafellar Theorem for maximal monotone operators, we refer to [14, Theorem 29.9], whereby, as pointed out in [14, pp. 152–153], this result holds only in reflexive Banach spaces. Later, a special formulation of this theorem in the nonreflexive case was given in [11].

In contrast to this, our approach does not rely on the maximal monotonicity of the diagonal subdifferential operator, while the result holds in general Banach spaces. We present also some consequences of the given Brøndsted–Rockafellar theorem concerning the density of the domain of diagonal subdifferential operators. We close the note by showing that a Brøndsted–Rockafellar-type theorem for subdifferential operators can be obtained as a particular case of our main result.

6.2 A Brøndsted–Rockafellar Theorem

The following Ekeland variational principle for bifunctions was given in [2]. Although this result was stated there in Euclidian spaces, it is valid in general Banach spaces, too.

Theorem 6.2. *Assume that C is nonempty, convex, and closed set and $f : C \times C \to \mathbb{R}$ satisfies:*

(i) $f(x,\cdot)$ *is lower bounded and lower semicontinuous for every* $x \in C$.
(ii) $f(x,x) = 0$ *for every* $x \in C$.
(iii) $f(x,y) + f(y,z) \geq f(x,z)$ *for every* $x,y,z \in C$.

Then, for every $\varepsilon > 0$ *and for every* $x_0 \in C$, *there exists* $\overline{x} \in C$ *such that*

$$f(x_0,\overline{x}) + \varepsilon\|x_0 - \overline{x}\| \leq 0$$

and

$$f(\overline{x},x) + \varepsilon\|\overline{x} - x\| > 0 \;\forall x \in C,\; x \neq \overline{x}.$$

Remark 6.3. By taking $z = x$, the assumptions (iii) and (ii) in the above theorem imply that $f(x,y) + f(y,x) \geq 0$ for all $x,y \in C$, which means that $-f$ is monotone.

Theorem 6.2 will be an essential ingredient in proving the following Brøndsted–Rockafellar Theorem for diagonal subdifferential operators.

Theorem 6.4. *Assume that C is a nonempty, convex, and closed set and $F : C \times C \to \mathbb{R}$ satisfies:*

(i) $F(x,\cdot)$ *is a convex and lower semicontinuous function for every* $x \in C$.
(ii) $F(x,x) = 0$ *for every* $x \in C$.
(iii) $F(x,y) + F(y,z) \geq F(x,z)$ *for every* $x,y,z \in C$.

Take $\varepsilon > 0$, $x_0 \in C$ *and* $x_0^* \in A_\varepsilon^F(x_0)$. *Then for all* $\lambda > 0$, *there exist* $x^* \in X^*$ *and* $x \in C$ *such that*

$$x^* \in A^F(x),\; \|x - x_0\| \leq \frac{\varepsilon}{\lambda} \text{ and } \|x^* - x_0^*\| \leq \lambda.$$

Proof. We fix $\varepsilon > 0$, $x_0 \in C$ and $x_0^* \in A_\varepsilon^F(x_0)$. According to the definition of the operator A_ε^F, we have

$$F(x_0,y) \geq \langle x_0^*, y - x_0\rangle - \varepsilon \;\forall y \in C. \tag{6.1}$$

Let us define the bifunction $f : C \times C \to \mathbb{R}$ by

$$f(x,y) = F(x,y) - \langle x_0^*, y - x\rangle \text{ for all } (x,y) \in C \times C.$$

We want to apply Theorem 6.2 to f and show to this aim that the assumptions (i)–(iii) in Theorem 6.2 are verified. Indeed, the lower semicontinuity of the function $f(x,\cdot)$ and the relation $f(x,x)=0$, for all $x \in C$, are inherited from the corresponding properties of F. One can easily see that (iii) is fulfilled, too: for $x,y,z \in C$ it holds

$$f(x,y)+f(y,z) = F(x,y)+F(y,z) - \langle x_0^*, z-x\rangle \geq F(x,z) - \langle x_0^*, z-x\rangle = f(x,z).$$

It remains to prove that $f(x,\cdot)$ is lower bounded for all $x \in C$. Take an arbitrary $x \in C$. By using (6.1) we get for all $y \in C$

$$f(x,y) \geq f(x_0,y) - f(x_0,x) = F(x_0,y) - \langle x_0^*, y-x_0\rangle - f(x_0,x) \geq -\varepsilon - f(x_0,x)$$

and the desired property follows.

Take now $\lambda > 0$. A direct application of Theorem 6.2 guarantees the existence of $\overline{x} \in C$ such that

$$f(x_0,\overline{x}) + \lambda\|x_0-\overline{x}\| \leq 0 \tag{6.2}$$

and

$$f(\overline{x},x) + \lambda\|\overline{x}-x\| > 0 \;\forall x \in C,\; x \neq \overline{x}. \tag{6.3}$$

From (6.2) we obtain

$$F(x_0,\overline{x}) - \langle x_0^*, \overline{x}-x_0\rangle + \lambda\|x_0-\overline{x}\| \leq 0,$$

which combined with (6.1) ensures

$$\lambda\|x_0-\overline{x}\| \leq \langle x_0^*, \overline{x}-x_0\rangle - F(x_0,\overline{x}) \leq \varepsilon,$$

hence $\|x_0-\overline{x}\| \leq \frac{\varepsilon}{\lambda}$.

Further, notice that (6.3) implies

$$0 \in \partial\big(f(\overline{x},\cdot) + \delta_C + \lambda\|\overline{x}-\cdot\|\big)(\overline{x}).$$

Since the functions in the above statement are convex and $\|\overline{x}-\cdot\|$ is continuous, we obtain via the subdifferential sum formula (cf. [15, Theorem 2.8.7])

$$0 \in \partial\big(f(\overline{x},\cdot)+\delta_C\big)(\overline{x}) + \partial\big(\lambda\|\overline{x}-\cdot\|\big)(\overline{x}). \tag{6.4}$$

Taking into account the definition of the bifunction f, we get (cf. [15, Theorem 2.4.2(vi)]) $\partial\big(f(\overline{x},\cdot)+\delta_C\big)(\overline{x}) = \partial\big(F(\overline{x},\cdot)+\delta_C\big)(\overline{x}) - x_0^* = A^F(\overline{x}) - x_0^*$. Moreover, $\partial\big(\lambda\|\overline{x}-\cdot\|\big)(\overline{x}) = \lambda B_{X^*}$, where B_{X^*} denotes the closed unit ball of the dual space X^* (see, for instance, [15, Corollary 2.4.16]). Hence, (6.4) is nothing else than

$$0 \in A^F(\overline{x}) - x_0^* + \lambda B_{X^*},$$

from which we conclude that there exists $x^* \in A^F(\overline{x})$ with $\|x^* - x_0^*\| \le \lambda$ and the proof is complete. ■

For a similar result like the one given in Theorem 6.4, but formulated in reflexive Banach spaces and by assuming (Blum–Oettli) maximal monotonicity for the bifunction F (see [3] for the definition of this notion), we refer the reader to [7, Theorem 1.1].

A direct consequence of the above Brøndsted–Rockafellar Theorem is the following result concerning the density of $D(A^F)$ in C, where $D(A^F) = \{x \in X : A^F(x) \neq \emptyset\}$ is the *domain* of the operator A^F.

Corollary 6.5. *Assume that the hypotheses of Theorem 6.4 are fulfilled. Then* $\overline{D(A^F)} = C$, *hence* $\overline{D(A^F)}$ *is a convex set.*

Proof. The implication $D(A^F) \subseteq C$ is obvious. Take now an arbitrary $x_0 \in C$. For all $n \in \mathbb{N}$, we have that $A^F_{1/n}(x_0) \neq \emptyset$, hence we can choose $x_n^* \in A^F_{1/n}(x_0)$. Theorem 6.4 guarantees the existence of $u_n^* \in X^*$ and $u_n \in C$ such that

$$u_n^* \in A^F(u_n),\ \|u_n - x_0\| \le \sqrt{1/n} \text{ and } \|u_n^* - x_n^*\| \le \sqrt{1/n} \text{ for all } n \in \mathbb{N}.$$

Since $u_n \in D(A^F)$ for all $n \in \mathbb{N}$, we get from above that $x_0 \in \overline{D(A^F)}$. ■

Remark 6.6. Similar statements to the one in Corollary 6.5 were furnished in [8, Sect. 4] in reflexive Banach spaces and by assuming maximal monotonicity for A^F.

Let us show how Theorem 6.4 can be used in order to derive the classical Brøndsted–Rockafellar theorem for the subdifferential operator in case the domain of the function is closed.

Corollary 6.7. *Let* $f : X \to \overline{\mathbb{R}}$ *be a proper, convex, and lower semicontinuous function such that* $\operatorname{dom} f$ *is closed. Take* $x_0 \in \operatorname{dom} f$, $\varepsilon > 0$ *and* $x_0^* \in \partial_\varepsilon f(x_0)$. *Then for all* $\lambda > 0$, *there exist* $x^* \in X^*$ *and* $x \in X$ *such that*

$$x^* \in \partial f(x),\ \|x - x_0\| \le \frac{\varepsilon}{\lambda} \text{ and } \|x^* - x_0^*\| \le \lambda.$$

Proof. The result follows by applying Theorem 6.4 for $C = \operatorname{dom} f$ and the bifunction $F : \operatorname{dom} f \times \operatorname{dom} f \to \mathbb{R}$ defined by $F(x,y) = f(y) - f(x)$. ■

The restriction "$\operatorname{dom} f$ closed" comes from the fact that in Theorems 6.2 and 6.4 the set C is assumed to be a closed set. In the following Brøndsted–Rockafellar-type Theorem for subdifferential operators, which we obtain as a consequence of Corollary 6.7, we abandon this assumption.

Corollary 6.8. *Let* $f : X \to \overline{\mathbb{R}}$ *be a proper, convex, and lower semicontinuous function. Take* $x_0 \in \operatorname{dom} f$, $\varepsilon > 0$ *and* $x_0^* \in \partial_\varepsilon f(x_0)$. *Then for all* $\lambda > 0$, *there exist* $x^* \in X^*$ *and* $x \in X$ *such that*

$$x^* \in \partial f(x),\ \|x - x_0\| \leq \varepsilon\left(\frac{1}{\lambda} + 1\right) \text{ and } \|x^* - x_0^*\| \leq \lambda.$$

Proof. Take $x_0 \in \operatorname{dom} f$, $\varepsilon > 0$, $x_0^* \in \partial_\varepsilon f(x_0)$ and $\lambda > 0$. We consider $X \times \mathbb{R}$ endowed with the norm defined for all $(x,r) \in X \times \mathbb{R}$ as being $\|(x,r)\| = (\|x\|^2 + r^2)^{1/2}$. We divide the proof in two steps.

(I) Consider the case $x_0^* = 0$. We have $0 \in \partial_\varepsilon f(x_0)$, hence $(0,-1) \in N^\varepsilon_{\operatorname{epi} f}(x_0, f(x_0)) = \partial_\varepsilon \delta_{\operatorname{epi} f}(x_0, f(x_0))$. By applying Corollary 6.7 for the function $\delta_{\operatorname{epi} f}$ and $\lambda := \lambda/(\lambda+1)$, we obtain the existence of $(x,r) \in \operatorname{epi} f$ and $(x^*,-s) \in \partial \delta_{\operatorname{epi} f}(x,r) = N_{\operatorname{epi} f}(x,r)$ such that

$$\|(x,r) - (x_0, f(x_0))\| \leq \varepsilon \frac{1+\lambda}{\lambda} \text{ and } \|(x^*,-s) - (0,-1)\| \leq \frac{\lambda}{1+\lambda}.$$

From here, it follows

$$\|x - x_0\| \leq \varepsilon/\lambda + \varepsilon, s \geq 0, \|x^*\| \leq \frac{\lambda}{1+\lambda} \text{ and } |s-1| \leq \frac{\lambda}{1+\lambda}.$$

The last inequality ensures $0 < \frac{1}{1+\lambda} \leq s$, hence $r = f(x)$ and $(1/s)x^* \in \partial f(x)$. Moreover, $\|(1/s)x^*\| \leq \frac{\lambda}{1+\lambda} \cdot (1+\lambda) = \lambda$.

(II) Let us consider now the general case, when $x_0^* \in \partial_\varepsilon f(x_0)$ is an arbitrary element. Define the function $g : X \to \overline{\mathbb{R}}$, $g(x) = f(x) - \langle x_0^*, x\rangle$, for all $x \in X$. Notice that $\partial_\alpha g(x) = \partial_\alpha f(x) - x_0^*$ for all $\alpha \geq 0$, hence the condition $x_0^* \in \partial_\varepsilon f(x_0)$ guarantees $0 \in \partial_\varepsilon g(x_0)$. Applying the statement obtained in the first part of the proof for g, we obtain that there exist $x^* \in X^*$ and $x \in X$ such that

$$x^* \in \partial g(x),\ \|x - x_0\| \leq \varepsilon\left(\frac{1}{\lambda} + 1\right) \text{ and } \|x^*\| \leq \lambda.$$

Thus, $x^* + x_0^* \in \partial f(x)$, $\|x - x_0\| \leq \varepsilon\left(\frac{1}{\lambda} + 1\right)$ and $\|(x^* + x_0^*) - x_0^*\| = \|x^*\| \leq \lambda$; hence, the proof is complete.

■

The bounds in Corollary 6.8 differ from the ones in Theorem 6.1, nevertheless, by taking $\lambda = \sqrt{\varepsilon}$, they become $\sqrt{\varepsilon} + \varepsilon$ and, respectively, $\sqrt{\varepsilon}$, and allow one to derive (by letting $\varepsilon \searrow 0$) the classical density result regarding the domain of the subdifferential.

However, it remains an open question if Theorem 6.1 can be deduced from Theorem 6.4.

Acknowledgements Research partially supported by DFG (German Research Foundation), project BO 2516/4-1.

References

1. Alizadeh, M.H., Hadjisavvas, N.: Local boundedness of monotone bifunctions. J. Global Optim. **53**(2), 231–241 (2012). doi:10.1007/s10898-011-9677-2
2. Bianchi, M., Kassay, G., Pini, R.: Existence of equilibria via Ekeland's principle. J. Math. Anal. Appl. **305**, 502–512 (2005)
3. Blum, E., Oettli, W.: From optimization and variational inequalities to equilibrium problems. Math. Student **63**, 123–145 (1994)
4. Borwein, J.M.: A note on ε-subgradients and maximal monotonicity. Pacific J. Math. **103**, 307–314 (1982)
5. Boţ, R.I., Grad, S.-M.: Approaching the maximal monotonicity of bifunctions via representative functions. J. Convex Anal. **19**, (2012)
6. Brøndsted, A., Rockafellar, R.T.: On the subdifferentiability of convex functions. Proc. Am. Math. Soc. **16**, 605–611 (1965)
7. Chbani, Z., Riahi, H.: Variational principles for monotone operators and maximal bifunctions. Serdica Math. J. **29**, 159–166, (2003)
8. Hadjisavvas, N., Khatibzadeh, H.: Maximal monotonicity of bifunctions. Optimization **59**, 147–160 (2010)
9. Iusem, A.N.: On the maximal monotonicity of diagonal subdifferential operators. J. Convex Anal. **18**, 489–503 (2011)
10. Iusem, A.N., Svaiter, B.F.: On diagonal subdifferential operators in nonreflexive Banach spaces. Set-Valued Var. Anal. **20**, 1–14 (2012)
11. Marques Alves, M., Svaiter, B.F.: Brønsted-Rockafellar property and maximality of monotone operators representable by convex functions in non-reflexive Banach spaces. J. Convex Anal. **15**, 693–706 (2008)
12. Phelps, R.R.: Convex Functions, Monotone Operators and Differentiability, 2nd edn. Lecture Notes in Mathematics, vol. 1364, Springer, Berlin (1993)
13. Rockafellar, R.T., Wets, R.J.-B.: Variational Analysis: Grundlehren der Mathematischen Wissenschaften (Fundamental Principles of Mathematical Sciences), vol. 317. Springer, Berlin (1998)
14. Simons, S.: From Hahn-Banach to Monotonicity. Springer, Berlin (2008)
15. Zălinescu, C.: Convex Analysis in General Vector Spaces. World Scientific, Singapore (2002)

Chapter 7
A q-Analog of Euler's Reduction Formula for the Double Zeta Function

David M. Bradley and Xia Zhou

Abstract The double zeta function is a function of two arguments defined by a double Dirichlet series and was first studied by Euler in response to a letter from Goldbach in 1742. By calculating many examples, Euler inferred a closed-form evaluation of the double zeta function in terms of values of the Riemann zeta function, in the case when the two arguments are positive integers with opposite parity. Here, we establish a q-analog of Euler's evaluation. That is, we state and prove a 1-parameter generalization that reduces to Euler's evaluation in the limit as the parameter q tends to 1.

Key words: Euler sums • Multiple harmonic series • Multiple zeta values • q-series, Lambert series • q-analog

Mathematics Subject Classifications (2010): Primary 11M41; Secondary 11M06, 05A30, 33E20, 30B50

COMMUNICATED BY HEINZ H. BAUSCHKE.

D.M. Bradley (✉)
Department of Mathematics and Statistics, University of Maine, 5752 Neville Hall, Orono, ME 04469-5752, USA
e-mail: dbradley@member.ams.org

X. Zhou
Department of Mathematics, Zhejiang University, Hangzhou 310027, Republic of China
e-mail: xiazhou0821@hotmail.com

D.H. Bailey et al. (eds.), *Computational and Analytical Mathematics*, Springer Proceedings in Mathematics & Statistics 50, DOI 10.1007/978-1-4614-7621-4_7,

7.1 Introduction

The double zeta function is defined by

$$\zeta(s,t) := \sum_{n=1}^{\infty} \frac{1}{n^s} \sum_{k=1}^{n-1} \frac{1}{k^t}, \qquad \Re(s) > 1, \qquad \Re(s+t) > 2. \tag{7.1}$$

The sums (7.1), and more generally those of the form

$$\zeta(s_1, s_2, \ldots, s_m) := \sum_{k_1 > k_2 > \cdots > k_m > 0} \prod_{j=1}^{m} \frac{1}{k_j^{s_j}}, \quad \sum_{j=1}^{n} \Re(s_j) > n, \quad n = 1, 2, \ldots, m, \tag{7.2}$$

have attracted increasing attention in recent years; see, e.g., [2–5, 7–10, 12, 15, 20]. The survey articles [6, 16, 22, 25] provide an extensive list of references. In (7.2) the sum is over all positive integers $k_1, \ldots, k_m$ satisfying the indicated inequalities. Note that with positive integer arguments, $s_1 > 1$ is necessary and sufficient for convergence. As is now customary, we refer to the parameter m in (7.2) as the depth. Of course (7.2) reduces to the familiar Riemann zeta function when the depth $m = 1$.

The problem of evaluating sums of the form (7.1) with integers $s > 1$ and $t > 0$ seems to have been first proposed in a letter from Goldbach to Euler [18] in 1742. (See also [17, 19] and [1, p. 253].) Calculating several examples led Euler to infer a closed-form evaluation of the double zeta function in terms of values of the Riemann zeta function, in the case when the two arguments have opposite parity. Euler's evaluation can be expressed as follows. Let $s-1$ and $t-1$ be positive integers with opposite parity (i.e., $s+t$ is odd) and let $2h = \max(s,t)$. Then

$$\begin{aligned} \zeta(s,t) &= (-1)^{s+1} \sum_{k=1}^{h} \left[\binom{s+t-2k-1}{t-1} + \binom{s+t-2k-1}{s-1} \right] \zeta(2k)\zeta(s+t-2k) \\ &\quad + \frac{1}{2}\big((1+(-1)^s\big)\zeta(s)\zeta(t) + \frac{1}{2}\left[(-1)^s \binom{s+t}{s} - 1\right] \zeta(s+t). \end{aligned} \tag{7.3}$$

If we interpret $\zeta(1) = 0$, then Euler's formula (7.3) gives true results also when $t = 1$ and s is even, but this case is subsumed by another formula of Euler, namely

$$2\zeta(s,1) = s\zeta(s+1) - \sum_{k=2}^{s-1} \zeta(k)\zeta(s+1-k), \tag{7.4}$$

which is valid for all integers $s > 1$.

The evaluations (7.3) and (7.4) are both examples of *reduction* formulas, since they both give a closed-form evaluation of a sum of depth 2 in terms of sums of depth 1. More generally (see, e.g., [7, 8]) a reduction formula expresses an instance of (7.2) in terms of lower depth sums.

With the general goal of gaining a more complete understanding of the myriad relations satisfied by the multiple zeta functions (7.2) in mind, a q-analog of (7.2) was introduced in [11] and independently in [21, 23] as

$$\zeta[s_1, s_2, \ldots, s_m] := \sum_{k_1 > k_2 > \cdots > k_m > 0} \prod_{j=1}^{m} \frac{q^{(s_j-1)k_j}}{[k_j]_q^{s_j}}, \tag{7.5}$$

where $0 < q < 1$ and for any integer k,

$$[k]_q := \frac{1-q^k}{1-q}.$$

Observe that we now have

$$\zeta(s_1, \ldots, s_m) = \lim_{q \to 1} \zeta[s_1, \ldots, s_m],$$

so that (7.5) represents a generalization of (7.2). The papers [11–14] consider values of the multiple q-zeta functions (7.5) and establish several infinite classes of relations satisfied by them. In particular, the following q-analog of (7.4) was established.

Theorem 7.1 (Corollary 8 of [11]). *Let $s-1$ be a positive integer. Then*

$$2\zeta[s,1] = s\zeta[s+1] + (1-q)(s-2)\zeta[s] - \sum_{k=2}^{s-1} \zeta[k]\zeta[s+1-k].$$

Here, we continue this general program of study by establishing a q-analog of Euler's reduction formula (7.3). Throughout the remainder of this paper, s and t denote positive integers with additional restrictions noted where needed, and q is real with $0 < q < 1$.

7.2 q-Analog of Euler's Reduction Formula

Throughout this section, we assume $s > 1$. We've seen that $\zeta[s,t]$ as given by (7.5) is a q-analog of $\zeta(s,t)$ in (7.1). Here, we introduce additional q-analogs of $\zeta(s,t)$ by defining

$$\zeta_1[s,t] = \zeta_1[s,t;q] := (-1)^t \sum_{u>v>0} \frac{q^{(s-1)u+(t-1)(-v)}}{[u]_q^s[-v]_q^t} = \sum_{u>v>0} \frac{q^{(s-1)u+v}}{[u]_q^s[v]_q^t}$$

and

$$\zeta_2[s,t] = \zeta_2[s,t;q] := (-1)^s \sum_{u>v>0} \frac{q^{(s-1)(-u)+(t-1)v}}{[-u]_q^s[v]_q^t} = \sum_{u>v>0} \frac{q^{u+(t-1)v}}{[u]_q^s[v]_q^t}$$

$$= q^{s+t}\zeta_1[s,t;1/q].$$

Let

$$\zeta_-[s] := \sum_{n=1}^{\infty} \frac{q^{(s-1)(-n)}}{[-n]_q^s} = (-1)^s \sum_{n=1}^{\infty} \frac{q^n}{[n]_q^s}$$

and for convenience, put

$$\zeta_\pm[s] := \zeta[s] + \zeta_-[s] = \sum_{0\neq n\in\mathbf{Z}} \frac{q^{(s-1)n}}{[n]_q^s} = (-1)^s \sum_{0\neq n\in\mathbf{Z}} \frac{q^n}{[n]_q^s}.$$

Note that if $s-1$ is a positive integer and $n \neq 0$, then

$$\frac{q^n}{[n]_q^s} = \frac{q^n}{[n]_q^2}\left(1 - q + \frac{q^n}{[n]_q}\right)^{s-2} = \sum_{k=0}^{s-2} \binom{s-2}{k} (1-q)^k \frac{q^{(s-1-k)n}}{[n]_q^{s-k}}$$

and so

$$\zeta_-[s] = (-1)^s \sum_{k=0}^{s-2} \binom{s-2}{k} (1-q)^k \zeta[s-k] \tag{7.6}$$

and

$$\zeta_\pm[s] = \left(1 + (-1)^s\right)\zeta[s] + (-1)^s \sum_{k=1}^{s-2} \binom{s-2}{k} (1-q)^k \zeta[s-k] \tag{7.7}$$

are expressible in terms of values of the q-Riemann zeta function, i.e., (7.2) with $m=1$. Finally, as in [13], let

$$\varphi[s] := \sum_{n=1}^{\infty} \frac{(n-1)q^{(s-1)n}}{[n]_q^s} = \sum_{n=1}^{\infty} \frac{nq^{(s-1)n}}{[n]_q^s} - \zeta[s].$$

We also employ the notation [24]

$$\binom{z}{a,\, b} := \binom{z}{a}\binom{z-a}{b} = \binom{z}{b}\binom{z-b}{a} = \binom{z}{a+b}\frac{(a+b)!}{a!b!}$$

for the trinomial coefficient, in which a, b are nonnegative integers and which reduces to $z!/a!b!(z-a-b)!$ if z is an integer not less than $a+b$. We can now state our main result.

Theorem 7.2 (q-Analog of Euler's double zeta reduction). *Let $s-1$ and $t-1$ be positive integers, and let $0<q<1$. Then*

$$
\begin{aligned}
&(-1)^t\zeta_1[s,t]-(-1)^s\zeta_2[s,t]\\
&=\sum_{a=0}^{s-2}\sum_{b=0}^{s-2-a}\binom{a+t-1}{a,b}(1-q)^b(\zeta_\pm[s-a-b]\zeta[a+t]\\
&-\zeta[s+t-b]-(1-q)\zeta[s+t-b-1])\\
&+\sum_{a=0}^{t-2}\sum_{b=0}^{t-2-a}\binom{a+s-1}{a,b}(1-q)^b(\zeta_\pm[t-a-b]\zeta[a+s]\\
&-\zeta[s+t-b]-(1-q)\zeta[s+t-b-1])\\
&-\sum_{j=1}^{\min(s,t)}\binom{s+t-j-1}{s-j,t-j}(1-q)^{j-1}(2\zeta[s+t-j+1]-(1-q)\varphi[s+t-j])\\
&-\zeta_\pm[s]\zeta[t]+(-1)^s\sum_{k=0}^{s-1}\binom{s-1}{k}(1-q)^k\zeta[s+t-k].
\end{aligned}
$$

Corollary 7.3 (Euler's double zeta reduction). *Let $s-1$ and $t-1$ be positive integers with opposite parity, and let $2h=\max(s,t)$. Then* (7.3) *holds.*

Corollary 7.4. *Let $s-1$ and $t-1$ be positive integers with like parity, and let $2h=\max(s,t)$. Then*

$$
\begin{aligned}
&2\sum_{k=1}^{h}\left[\binom{s+t-2k-1}{s-1}+\binom{s+t-2k-1}{t-1}\right]\zeta(2k)\zeta(s+t-2k)\\
&\qquad=(1+(-1)^s)\zeta(s)\zeta(t)+\left[\binom{s+t}{t}-(-1)^s\right]\zeta(s+t).
\end{aligned}
$$

Proof. Let $q\to 1$ in Theorem 7.2. With the obvious notation

$$
\zeta_\pm(s):=\lim_{q\to 1}\zeta_\pm[s]=\sum_{0\neq n\in\mathbf{Z}}\frac{1}{n^s}=(1+(-1)^s)\zeta(s),
$$

we find that

$$
\begin{aligned}
(-1)^t\zeta(s,t)-(-1)^s\zeta(s,t)&=\sum_{a=0}^{s-2}\binom{a+t-1}{a}(\zeta_\pm(s-a)\zeta(a+t)-\zeta(s+t))\\
&+\sum_{a=0}^{t-2}\binom{a+s-1}{a}(\zeta_\pm(t-a)\zeta(a+s)-\zeta(s+t))\\
&-2\binom{s+t-2}{s-1}\zeta(s+t)-\zeta_\pm(s)\zeta(t)+(-1)^s\zeta(s+t).
\end{aligned}
$$

Since

$$\sum_{a=0}^{s-2}\binom{a+t-1}{a}=\binom{s+t-2}{t} \quad\text{and}\quad \sum_{a=0}^{t-2}\binom{a+s-1}{a}=\binom{s+t-2}{s},$$

and

$$\binom{s+t-2}{t}+\binom{s+t-2}{s}+2\binom{s+t-2}{s-1}=\binom{s+t}{t},$$

it follows that

$$\begin{aligned}
&(-1)^t\zeta(s,t)-(-1)^s\zeta(s,t)\\
&=\sum_{a=0}^{s-2}\binom{a+t-1}{a}\zeta_\pm(s-a)\zeta(a+t)+\sum_{a=0}^{t-2}\binom{a+s-1}{a}\zeta_\pm(t-a)\zeta(a+s)\\
&\quad-\left[\binom{s+t-2}{t}+\binom{s+t-2}{s}+2\binom{s+t-2}{s-1}-(-1)^s\right]\zeta(s+t)-\zeta_\pm(s)\zeta(t)\\
&=\sum_{j=2}^{s}\binom{s+t-j-1}{t-1}\zeta_\pm(j)\zeta(s+t-j)+\sum_{j=2}^{t}\binom{s+t-j-1}{s-1}\zeta_\pm(j)\zeta(s+t-j)\\
&\quad-\left[\binom{s+t}{t}-(-1)^s\right]\zeta(s+t)-\zeta_\pm(s)\zeta(t)\\
&=2\sum_{k=1}^{s/2}\binom{s+t-2k-1}{t-1}\zeta(2k)\zeta(s+t-2k)\\
&\quad+2\sum_{k=1}^{t/2}\binom{s+t-2k-1}{s-1}\zeta(2k)\zeta(s+t-2k)\\
&\quad-(1+(-1)^s)\zeta(s)\zeta(t)-\left[\binom{s+t}{t}-(-1)^s\right]\zeta(s+t).
\end{aligned}\tag{7.8}$$

Since the binomial coefficients vanish if k exceeds the indicated range of summation above, we can replace the two sums by a single sum on k ranging from 1 up to h. If s and t have opposite parity, multiply both sides by $(-1)^t=(-1)^{s+1}$ and divide each term by 2 to complete the proof of Corollary 7.3. For Corollary 7.4, note that if s and t have like parity, then the left hand side of (7.8) vanishes. ■

7.3 Proof of Theorem 7.2

The key ingredient is the following partial fraction decomposition.

Lemma 7.5 (cf. Lemma 3.1 of [13] and Lemma 1 of [24]). *If s and t are positive integers, and u and v are non-zero real numbers such that $u+v\neq 0$, then*

$$\frac{1}{[u]_q^s[v]_q^t}=\sum_{a=0}^{s-1}\sum_{b=0}^{s-1-a}\binom{a+t-1}{a,b}\frac{(1-q)^b q^{(t-1-b)u+av}}{[u]_q^{s-a-b}[u+v]_q^{a+t}}$$
$$+\sum_{a=0}^{t-1}\sum_{b=0}^{t-1-a}\binom{a+s-1}{a,b}\frac{(1-q)^b q^{au+(s-1-b)v}}{[v]_q^{t-a-b}[u+v]_q^{a+s}}$$
$$-\sum_{j=1}^{\min(s,t)}\binom{s+t-j-1}{s-j,t-j}\frac{(1-q)^j q^{(t-j)u+(s-j)v}}{[u+v]_q^{s+t-j}}.$$

Proof. As in [13], let x and y be non-zero real numbers such that $x+y+(q-1)xy\neq 0$. Apply the partial differential operator

$$\frac{1}{(r-1)!}\left(-\frac{\partial}{\partial x}\right)^{r-1}\frac{1}{(s-1)!}\left(-\frac{\partial}{\partial y}\right)^{s-1}$$

to both sides of the identity

$$\frac{1}{xy}=\frac{1}{x+y+(q-1)xy}\left(\frac{1}{x}+\frac{1}{y}+q-1\right);$$

then let $x=[u]_q$, $y=[v]_q$ and observe that $x+y+(q-1)xy=[u+v]_q$. ■

We now proceed with the proof of Theorem 7.2. First, multiply both sides of Lemma 7.5 by $q^{(s-1)u+(t-1)v}$ to obtain

$$\frac{q^{(s-1)u}q^{(t-1)v}}{[u]_q^s[v]_q^t}=\sum_{a=0}^{s-1}\sum_{b=0}^{s-1-a}\binom{a+t-1}{a,b}\frac{(1-q)^b q^{(s-a-b-1)u}q^{(a+t-1)(u+v)}}{[u]_q^{s-a-b}[u+v]_q^{a+t}}$$
$$+\sum_{a=0}^{t-1}\sum_{b=0}^{t-1-a}\binom{a+s-1}{a,b}\frac{(1-q)^b q^{(t-a-b-1)v}q^{(a+s-1)(u+v)}}{[v]_q^{t-a-b}[u+v]_q^{a+s}}$$
$$-\sum_{j=1}^{\min(s,t)}\binom{s+t-j-1}{s-j,t-j}\frac{(1-q)^j q^{(s+t-j-1)(u+v)}}{[u+v]_q^{s+t-j}}.$$

After replacing u by $u-v$ and v by $-v$, we find that

$$\frac{q^{(s-1)(u+v)}q^{(t-1)(-v)}}{[u+v]_q^s[-v]_q^t}=\sum_{a=0}^{s-1}\sum_{b=0}^{s-1-a}\binom{a+t-1}{a,b}\frac{(1-q)^b q^{(s-a-b-1)(u+v)}q^{(a+t-1)u}}{[u+v]_q^{s-a-b}[u]_q^{a+t}}$$
$$+\sum_{a=0}^{t-1}\sum_{b=0}^{t-1-a}\binom{a+s-1}{a,b}\frac{(1-q)^b q^{(t-a-b-1)(-v)}q^{(a+s-1)u}}{[-v]_q^{t-a-b}[u]_q^{a+s}}$$

$$
- \sum_{j=1}^{\min(s,t)} \binom{s+t-j-1}{s-j,t-j} \frac{(1-q)^j q^{(s+t-j-1)u}}{[u]_q^{s+t-j}}. \tag{7.9}
$$

We'd like to sum (7.9) over all ordered pairs of positive integers (u,v), but we must exercise some care in doing so since some of the terms on the right hand side may diverge. The difficulty can be circumvented by judiciously combining the troublesome terms before summing. To this end, observe that

$$
\begin{aligned}
&\sum_{a=0}^{s-1} \binom{a+t-1}{a,s-1-a} \frac{(1-q)^{s-1-a} q^{(a+t-1)u}}{[u+v]_q [u]_q^{a+t}} \\
&+\sum_{a=0}^{t-1} \binom{a+s-1}{a,t-1-a} \frac{(1-q)^{t-1-a} q^{(a+s-1)u}}{[-v]_q [u]_q^{a+s}} \\
&-\sum_{j=1}^{\min(s,t)} \binom{s+t-j-1}{s-j,t-j} \frac{(1-q)^j q^{(s+t-j-1)u}}{[u]_q^{s+t-j}} \\
&= \sum_{j=1}^{s} \binom{s+t-j-1}{s-j,j-1} \frac{(1-q)^{j-1} q^{(s+t-j-1)u}}{[u+v]_q [u]_q^{s+t-j}} \\
&+\sum_{j=1}^{t} \binom{s+t-j-1}{t-j,j-1} \frac{(1-q)^{j-1} q^{(s+t-j-1)u}}{[-v]_q [u]_q^{s+t-j}} \\
&-\sum_{j=1}^{\min(s,t)} \binom{s+t-j-1}{s-j,t-j} \frac{(1-q)^j q^{(s+t-j-1)u}}{[u]_q^{s+t-j}} \\
&= \left(\frac{1}{[u+v]_q} + \frac{1}{[-v]_q} - (1-q) \right) \sum_{j=1}^{\min(s,t)} \binom{s+t-j-1}{s-j,t-j} \frac{(1-q)^{j-1} q^{(s+t-j-1)u}}{[u]_q^{s+t-j}} \\
&= \left(\frac{1}{[u+v]_q} - \frac{1}{[v]_q} \right) \sum_{j=1}^{\min(s,t)} \binom{s+t-j-1}{s-j,t-j} \frac{(1-q)^{j-1} q^{(s+t-j-1)u}}{[u]_q^{s+t-j}},
\end{aligned} \tag{7.10}
$$

where we have used the fact that

$$
\binom{s+t-j-1}{s-j,j-1} = \binom{s+t-j-1}{t-j,j-1} = \binom{s+t-j-1}{s-j,t-j}
$$

vanishes if $j > \min(s,t)$. Substituting (7.10) into (7.9) yields

$$
\frac{q^{(s-1)(u+v)} q^{(t-1)(-v)}}{[u+v]_q^s [-v]_q^t}
$$

$$= \sum_{a=0}^{s-2} \sum_{b=0}^{s-2-a} \binom{a+t-1}{a,b} \frac{(1-q)^b q^{(s-a-b-1)(u+v)} q^{(a+t-1)u}}{[u+v]_q^{s-a-b} [u]_q^{a+t}}$$

$$+ \sum_{a=0}^{t-2} \sum_{b=0}^{t-2-a} \binom{a+s-1}{a,b} \frac{(1-q)^b q^{(t-a-b-1)(-v)} q^{(a+s-1)u}}{[-v]_q^{t-a-b} [u]_q^{a+s}}$$

$$- \left(\frac{1}{[v]_q} - \frac{1}{[u+v]_q} \right) \sum_{j=1}^{\min(s,t)} \binom{s+t-j-1}{s-j,t-j} \frac{(1-q)^{j-1} q^{(s+t-j-1)u}}{[u]_q^{s+t-j}}. \quad (7.11)$$

Now assume that $s > 1$. Then

$$\sum_{u,v=1}^{\infty} \left(\frac{1}{[v]_q} - \frac{1}{[u+v]_q} \right) \frac{q^{(s+t-j-1)u}}{[u]_q^{s+t-j}} = \sum_{n=1}^{\infty} \frac{q^{(s+t-j-1)n}}{[n]_q^{s+t-j}} \sum_{k>n} \left(\frac{1}{[k-n]_q} - \frac{1}{[k]_q} \right).$$

Recalling that $0 < q < 1$, we evaluate the telescoping sum

$$\begin{aligned} \sum_{k>n} \left(\frac{1}{[k-n]_q} - \frac{1}{[k]_q} \right) &= \lim_{N \to \infty} \sum_{k=n+1}^{n+N} \left(\frac{1}{[k-n]_q} - \frac{1}{[k]_q} \right) \\ &= \lim_{N \to \infty} \sum_{k=1}^{n} \left(\frac{1}{[k]_q} - \frac{1}{[N+k]_q} \right) \\ &= (q-1)n + \sum_{k=1}^{n} \frac{1}{[k]_q}, \end{aligned}$$

so that

$$\sum_{u,v=1}^{\infty} \left(\frac{1}{[v]_q} - \frac{1}{[u+v]_q} \right) \frac{q^{(s+t-j-1)u}}{[u]_q^{s+t-j}}$$

$$= (q-1)\big(\varphi[s+t-j] + \zeta[s+t-j]\big) + \sum_{n=1}^{\infty} \frac{q^{(s+t-j-1)n}}{[n]_q^{s+t-j}} \sum_{k=1}^{n} \frac{1}{[k]_q}.$$

But this last double sum evaluates as

$$\begin{aligned} &\sum_{n=1}^{\infty} \frac{q^{(s+t-j-1)n}}{[n]_q^{s+t-j}} \sum_{k=1}^{n} \frac{1}{[k]_q} \\ &= \sum_{n>k>0} \frac{q^{(s+t-j-1)n}}{[n]_q^{s+t-j} [k]_q} + \sum_{n=1}^{\infty} \frac{q^{(s+t-j-1)n}}{[n]_q^{s+t-j+1}} \\ &= \zeta[s+t-j,1] + \sum_{n=1}^{\infty} \frac{q^{(s+t-j)n}}{[n]_q^{s+t-j+1}} + \sum_{n=1}^{\infty} \frac{q^{(s+t-j-1)n} - q^{(s+t-j)n}}{[n]_q^{s+t-j+1}} \end{aligned}$$

$$= \zeta[s+t-j,1] + \zeta[s+t-j+1] + (1-q)\sum_{n=1}^{\infty}\left(\frac{1-q^n}{1-q}\right)\frac{q^{(s+t-j-1)n}}{[n]_q^{s+t-j+1}}$$

$$= \zeta[s+t-j,1] + \zeta[s+t-j+1] + (1-q)\sum_{n=1}^{\infty}\frac{q^{(s+t-j-1)n}}{[n]_q^{s+t-j}}$$

$$= \zeta[s+t-j,1] + \zeta[s+t-j+1] + (1-q)\zeta[s+t-j].$$

It follows that

$$\sum_{u,v=1}^{\infty}\left(\frac{1}{[v]_q} - \frac{1}{[u+v]_q}\right)\frac{q^{(s+t-j-1)u}}{[u]_q^{s+t-j}}$$
$$= \zeta[s+t-j,1] + \zeta[s+t-j+1] + (q-1)\varphi[s+t-j].$$

Consequently, summing (7.11) over all ordered pairs of positive integers (u,v) yields

$$\begin{aligned}(-1)^t\zeta_1[s,t] &= \sum_{a=0}^{s-2}\sum_{b=0}^{s-2-a}\binom{a+t-1}{a,b}(1-q)^b\zeta[s-a-b,a+t]\\ &\quad+\sum_{a=0}^{t-2}\sum_{b=0}^{t-2-a}\binom{a+s-1}{a,b}(1-q)^b\zeta_-[t-a-b]\zeta[a+s]\\ &\quad-\sum_{j=1}^{\min(s,t)}\binom{s+t-j-1}{s-j,t-j}(1-q)^{j-1}\\ &\quad\times\big(\zeta[s+t-j,1]+\zeta[s+t-j+1]-(1-q)\varphi[s+t-j]\big).\end{aligned} \tag{7.12}$$

Now assume also that $t>1$. For each pair of integers (a,b) with $0\le a\le s-1$, $0\le b\le s-2-a$, we apply the q-stuffle multiplication rule [11, Equation (2.2)] in the form

$$\begin{aligned}\zeta[s-a-b]\zeta[a+t] &= \zeta[s-a-b,a+t]+\zeta[a+t,s-a-b]\\ &\quad+\zeta[s+t-b]+(1-q)\zeta[s+t-b-1],\end{aligned}$$

substituting for $\zeta[s-a-b,a+t]$ in (7.12). Thus, we find that

$$\begin{aligned}(-1)^t\zeta_1[s,t] = \sum_{a=0}^{s-2}\sum_{b=0}^{s-2-a}\binom{a+t-1}{a,b}(1-q)^b\big(&\zeta[s-a-b]\zeta[a+t]-\zeta[s+t-b]\\ &-(1-q)\zeta[s+t-b-1]-\zeta[a+t,s-a-b]\big)\end{aligned}$$

$$+\sum_{a=0}^{t-2}\sum_{b=0}^{t-2-a}\binom{a+s-1}{a,b}(1-q)^b\zeta_-[t-a-b]\zeta[a+s]$$
$$-\sum_{j=1}^{\min(s,t)}\binom{s+t-j-1}{s-j,t-j}(1-q)^{j-1}$$
$$\times\big(\zeta[s+t-j,1]+\zeta[s+t-j+1]-(1-q)\varphi[s+t-j]\big).$$

The sum of $\zeta[s+t-j,1]$ over j can be combined with the double sum of $\zeta[a+t,s-a-b]$ over a and b by extending the range of the latter to include the value $b=s-1-a$. Doing this yields

$$(-1)^t\zeta_1[s,t]=\sum_{a=0}^{s-2}\sum_{b=0}^{s-2-a}\binom{a+t-1}{a,b}(1-q)^b\big(\zeta[s-a-b]\zeta[a+t]$$
$$-\zeta[s+t-b]-(1-q)\zeta[s+t-b-1]\big)$$
$$+\sum_{a=0}^{t-2}\sum_{b=0}^{t-2-a}\binom{a+s-1}{a,b}(1-q)^b\zeta_-[t-a-b]\zeta[a+s]$$
$$-\sum_{j=1}^{\min(s,t)}\binom{s+t-j-1}{s-j,t-j}(1-q)^{j-1}$$
$$\times\big(\zeta[s+t-j+1]-(1-q)\varphi[s+t-j]\big)$$
$$-\sum_{a=0}^{s-1}\sum_{b=0}^{s-1-a}\binom{a+t-1}{a,b}(1-q)^b\zeta[t+a,s-a-b].$$

It follows that for integers $s>1$ and $t>1$,

$$(-1)^s\zeta_1[t,s]+(-1)^t\zeta_1[s,t]$$
$$=\sum_{a=0}^{s-2}\sum_{b=0}^{s-2-a}\binom{a+t-1}{a,b}(1-q)^b\big(\zeta_\pm[s-a-b]\zeta[a+t]$$
$$-\zeta[s+t-b]-(1-q)\zeta[s+t-b-1]\big)$$
$$+\sum_{a=0}^{t-2}\sum_{b=0}^{t-2-a}\binom{a+s-1}{a,b}(1-q)^b\big(\zeta_\pm[t-a-b]\zeta[a+s]$$
$$-\zeta[s+t-b]-(1-q)\zeta[s+t-b-1]\big)$$
$$-2\sum_{j=1}^{\min(s,t)}\binom{s+t-j-1}{s-j,t-j}(1-q)^{j-1}\big(\zeta[s+t-j+1]-(1-q)\varphi[s+t-j]\big)$$

$$-\sum_{a=0}^{s-1}\sum_{b=0}^{s-1-a}\binom{a+t-1}{a,b}(1-q)^b\zeta[t+a,s-a-b]$$
$$-\sum_{a=0}^{t-1}\sum_{b=0}^{t-1-a}\binom{a+s-1}{a,b}(1-q)^b\zeta[s+a,t-a-b]. \tag{7.13}$$

By Theorem 2.1 of [13],

$$\zeta[s]\zeta[t]=\sum_{a=0}^{s-1}\sum_{b=0}^{s-1-a}\binom{a+t-1}{a,b}(1-q)^b\zeta[t+a,s-a-b]$$
$$+\sum_{a=0}^{t-1}\sum_{b=0}^{t-1-a}\binom{a+s-1}{a,b}(1-q)^b\zeta[s+a,t-a-b]$$
$$-\sum_{j=1}^{\min(s,t)}\binom{s+t-j-1}{s-j,t-j}(1-q)^j\varphi[s+t-j].$$

We use this latter decomposition formula to eliminate the last two sums of double q-zeta values in (7.13), obtaining

$$(-1)^s\zeta_1[t,s]+(-1)^t\zeta_1[s,t]+\zeta[s]\zeta[t]$$
$$=\sum_{a=0}^{s-2}\sum_{b=0}^{s-2-a}\binom{a+t-1}{a,b}(1-q)^b(\zeta_\pm[s-a-b]\zeta[a+t]$$
$$-\zeta[s+t-b]-(1-q)\zeta[s+t-b-1])$$
$$+\sum_{a=0}^{t-2}\sum_{b=0}^{t-2-a}\binom{a+s-1}{a,b}(1-q)^b(\zeta_\pm[t-a-b]\zeta[a+s]$$
$$-\zeta[s+t-b]-(1-q)\zeta[s+t-b-1])$$
$$-\sum_{j=1}^{\min(s,t)}\binom{s+t-j-1}{s-j,t-j}(1-q)^{j-1}$$
$$\times\big(2\zeta[s+t-j+1]-(1-q)\varphi[s+t-j]\big). \tag{7.14}$$

But

$$\zeta_-[s]\zeta[t]=(-1)^s\sum_{u=1}^{\infty}\frac{q^u}{[u]_q^s}\sum_{v=1}^{\infty}\frac{q^{(t-1)v}}{[v]_q^t}$$
$$=(-1)^s\sum_{u>v>0}\frac{q^{u+(t-1)v}}{[u]_q^s[v]_q^t}+(-1)^s\sum_{v>u>0}\frac{q^{(t-1)v+u}}{[v]_q^t[u]_q^s}+(-1)^s\sum_{v=1}^{\infty}\frac{q^{tv}}{[v]_q^{s+t}}.$$

Since

$$\frac{q^{tv}}{[v]_q^{s+t}} = \frac{q^{tv}}{[v]_q^{t+1}}\left(1-q+\frac{q^v}{[v]_q}\right)^{s-1} = \sum_{k=0}^{s-1}\binom{s-1}{k}\frac{(1-q)^k q^{(s+t-k-1)v}}{[v]_q^{s+t-k}},$$

it follows that

$$\sum_{v=1}^{\infty}\frac{q^{tv}}{[v]_q^{s+t}} = \sum_{k=0}^{s-1}\binom{s-1}{k}(1-q)^k\zeta[s+t-k],$$

and therefore

$$\zeta_-[s]\zeta[t] = (-1)^s\zeta_2[s,t]+(-1)^s\zeta_1[t,s]+(-1)^s\sum_{k=0}^{s-1}\binom{s-1}{k}(1-q)^k\zeta[s+t-k].$$

We now use this formula to substitute the initial $(-1)^s\zeta_1[t,s]$ term in (7.14) to complete the proof.

References

1. Berndt, B.: Ramanujan's Notebooks, Part I. Springer, New York (1985)
2. Borwein, J.M., Bradley, D.M., Broadhurst, D.J.: Evaluations of k-fold Euler/Zagier sums: a compendium of results for arbitrary k. Electron. J. Combin. **4**(2), #R5 (1997) (Wilf Festschrift)
3. Borwein, J.M., Bradley, D.M., Broadhurst, D.J., Lisoněk, P.: Combinatorial aspects of multiple zeta values. Electron. J. Combin. **5**(1), #R38 (1998)
4. Borwein, J.M., Bradley, D.M., Broadhurst, D.J., Lisoněk, P.: Special values of multiple polylogarithms. Trans. Am. Math. Soc. **353**(3), 907–941 (2001)
5. Borwein, D., Borwein, J.M., Bradley, D.M.: Parametric Euler sum identities. J. Math. Anal. Appl. **316**(1), 328–338 (2006). doi:10.1016/j.jmaa.2005.04.040
6. Bowman, D., Bradley, D.M.: Multiple polylogarithms: a brief survey. In: Berndt, B.C., Ono, K. (eds.) Proceedings of a Conference on q-Series with Applications to Combinatorics, Number Theory and Physics. Contemporary Mathematics, vol. 291, pp. 71–92. American Mathematical Society, Providence (2001)
7. Bowman, D., Bradley, D.M.: The algebra and combinatorics of shuffles and multiple zeta values. J. Combin. Theory, Ser. A **97**(1), 43–61 (2002)
8. Bowman, D., Bradley, D.M.: Resolution of some open problems concerning multiple zeta evaluations of arbitrary depth. Compositio Math. **139**(1), 85–100 (2003). doi:10.1023/B:COMP:0000005036.52387.da
9. Bowman, D., Bradley, D.M., Ryoo, J.: Some multi-set inclusions associated with shuffle convolutions and multiple zeta values. European J. Combin. **24**(1), 121–127 (2003)
10. Bradley, D.M.: Partition identities for the multiple zeta function. In: Aoki, T., Kanemitsu, S., Nakahara, M., Ohno, Y. (eds.) Zeta Functions, Topology, and Quantum Physics, Developments in Mathematics, vol. 14, pp. 19–29. Springer, New York (2005)
11. Bradley, D.M.: Multiple q-zeta values. J. Algebra **283**(2), 752–798 (2005). doi:10.1016/j.jalgebra.2004.09.017
12. Bradley, D.M.: Duality for finite multiple harmonic q-series. Discrete Math. **300**(1–3), 44–56 (2005). doi:10.1016/j.disc.2005.06.008 [MR 2170113] (2006m:05019)

13. Bradley, D.M.: A q-analog of Euler's decomposition formula for the double zeta function. Int. J. Math. Math. Sci. **2005**(21), 3453–3458 (2005). doi:10.1155/IJMMS.2005.3453 [MR 2206867] (2006k:11174)
14. Bradley, D.M.: On the sum formula for multiple q-zeta values. Rocky Mountain J. Math. **37**(5), 1427–1434 (2007)
15. Broadhurst, D.J., Kreimer, D.: Association of multiple zeta values with positive knots via Feynman diagrams up to 9 loops. Phys. Lett. B **393**(3–4), 403–412 (1997)
16. Cartier, P.: Fonctions polylogarithmes, nombres polyzêtas et groupes pro-unipotents. Astérisque **282**(viii), 137–173 (2002)
17. Euler, L.: Meditationes circa singulare serierum genus. Novi Comm. Acad. Sci. Petropol. **20**, 140–186 (1775). Reprinted in Opera Omnia, ser. I, **15**, 217–267, B.G. Teubner, Berlin (1927)
18. Euler, L.: Briefwechsel, vol. 1. Birhäuser, Basel (1975)
19. Euler, L., Goldbach, C.: Briefwechsel, pp. 1729–1764. Akademie, Berlin (1965)
20. Le, T.Q.T., Murakami, J.: Kontsevich's integral for the Homfly polynomial and relations between values of multiple zeta functions. Topology Appl. **62**(2), 193–206 (1995)
21. Okuda, J., Takeyama, Y.: On relations for the multiple q-zeta values. Ramanujan J. **14**(3), 379–387 (2007)
22. Waldschmidt, M.: Valeurs zêta multiples: une introduction. J. Théor. Nombres Bordeaux **12**(2), 581–595 (2000)
23. Zhao, J.: Multiple q-zeta functions and multiple q-polylogarithms. Ramanujan J. **14**(2), 189–221 (2007)
24. Zhou, X., Cai, T., Bradley, D.M.: Signed q-analogs of Tornheim's double series. Proc. Am. Math. Soc. **136**(8), 2689–2698 (2008)
25. Zudilin, V.V.: Algebraic relations for multiple zeta values (Russian). Uspekhi Mat. Nauk **58**(1), 3–32 (2003). Translation in Russian Math. Surveys **58**(1), 1–29 (2003)

Chapter 8
Fast Computation of Bernoulli, Tangent and Secant Numbers

Richard P. Brent and David Harvey

Abstract We consider the computation of Bernoulli, Tangent (zag), and Secant (zig or Euler) numbers. In particular, we give asymptotically fast algorithms for computing the first n such numbers $O(n^2(\log n)^{2+o(1)})$. We also give very short in-place algorithms for computing the first n Tangent or Secant numbers in $O(n^2)$ integer operations. These algorithms are extremely simple and fast for moderate values of n. They are faster and use less space than the algorithms of Atkinson (for Tangent and Secant numbers) and Akiyama and Tanigawa (for Bernoulli numbers).

Key words: Bernoulli number • Tangent number • Secant number • Asymptotically fast algorithm • In-place algorithm

Mathematics Subject Classifications (2010): Primary 05A15; Secondary 11B68, 11B83, 11-04, 11Y55, 11Y60, 65-04, 68R05

COMMUNICATED BY DAVID H. BAILEY.

R.P. Brent (✉)
Mathematical Sciences Institute, Australian National University, Canberra, ACT 0200, Australia
e-mail: Tangent@rpbrent.com

D. Harvey
School of Mathematics and Statistics, University of New South Wales, Sydney, NSW 2052, Australia
e-mail: D.Harvey@unsw.edu.au

D.H. Bailey et al. (eds.), *Computational and Analytical Mathematics*, Springer Proceedings in Mathematics & Statistics 50, DOI 10.1007/978-1-4614-7621-4_8,

8.1 Introduction

The *Bernoulli numbers* are rational numbers B_n defined by the generating function

$$\sum_{n\geq 0} B_n \frac{z^n}{n!} = \frac{z}{\exp(z)-1}. \tag{8.1}$$

Bernoulli numbers are of interest in number theory and are related to special values of the Riemann zeta function (see Sect. 8.2). They also occur as coefficients in the Euler–Maclaurin formula, so are relevant to high-precision computation of special functions [7, Sect. 4.5].

It is sometimes convenient to consider *scaled* Bernoulli numbers

$$C_n = \frac{B_{2n}}{(2n)!}, \tag{8.2}$$

with generating function

$$\sum_{n\geq 0} C_n z^{2n} = \frac{z/2}{\tanh(z/2)}. \tag{8.3}$$

The generating functions (8.1) and (8.3) only differ by the single term $B_1 z$, since the other odd terms vanish.

The *Tangent numbers* T_n and *Secant numbers* S_n are defined by

$$\sum_{n>0} T_n \frac{z^{2n-1}}{(2n-1)!} = \tan z, \quad \sum_{n\geq 0} S_n \frac{z^{2n}}{(2n)!} = \sec z. \tag{8.4}$$

In this paper, which is based on an a talk given by the first author at a workshop held to mark Jonathan Borwein's sixtieth birthday, we consider some algorithms for computing Bernoulli, Tangent, and Secant numbers. For background, combinatorial interpretations, and references, see Abramowitz and Stegun [1, Chap. 23] (where the notation differs from ours, e.g. $(-1)^n E_{2n}$ is used for our S_n), and Sloane's [27] sequences A000367, A000182, and A000364.

Let $M(n)$ be the number of bit-operations required for n-bit integer multiplication. The Schönhage–Strassen algorithm [25] gives $M(n) = O(n \log n \log\log n)$, and Fürer [17] has recently given an improved bound $M(n) = O(n(\log n)2^{\log^* n})$. For simplicity we merely assume that $M(n) = O(n(\log n)^{1+o(1)})$, where the $o(1)$ term depends on the precise algorithm used for multiplication. For example, if the Schönhage–Strassen algorithm is used, then the $o(1)$ term can be replaced by $\log\log\log n / \log\log n$.

In Sects. 8.2 and 8.3 we mention some relevant and generally well-known facts concerning Bernoulli, Tangent, and Secant numbers.

Recently, Harvey [20] showed that the *single* number B_n can be computed in $O(n^2(\log n)^{2+o(1)})$ bit-operations using a modular algorithm. In this paper we show that *all* the Bernoulli numbers $B_0, \ldots, B_n$ can be computed with the same complexity bound (and similarly for Secant and Tangent numbers).

In Sect. 8.4 we give a relatively simple algorithm that achieves the slightly weaker bound $O(n^2(\log n)^{3+o(1)})$. In Sect. 8.5 we describe the improvement to $O(n^2(\log n)^{2+o(1)})$. The idea is similar to that espoused by Steel [29], although we reduce the problem to division rather than multiplication. It is an open question whether the *single* number B_{2n} can be computed in $o(n^2)$ bit-operations.

In Sect. 8.6 we give very short in-place algorithms for computing the first n Secant or Tangent numbers using $O(n^2)$ integer operations. These algorithms are extremely simple and fast for moderate values of n (say $n \le 1000$), although asymptotically not as fast as the algorithms given in Sects. 8.4 and 8.5. Bernoulli numbers can easily be deduced from the corresponding Tangent numbers using the relation (8.14) below.

8.2 Bernoulli Numbers

From the generating function (8.1) it is easy to see that the B_n are rational numbers, with $B_{2n+1} = 0$ if $n > 0$. The first few nonzero B_n are $B_0 = 1$, $B_1 = -1/2$, $B_2 = 1/6$, $B_4 = -1/30$, $B_6 = 1/42$, $B_8 = -1/30$, $B_{10} = 5/66$, $B_{12} = -691/2730$, and $B_{14} = 7/6$.

The denominators of the Bernoulli numbers are given by the *Von Staudt–Clausen Theorem* [12, 28], which states that

$$B'_{2n} := B_{2n} + \sum_{(p-1)|2n} \frac{1}{p} \in \mathbb{Z}.$$

Here the sum is over all primes p for which $p-1$ divides $2n$.

Since the "correction" $B'_{2n} - B_{2n}$ is easy to compute, it might be convenient in a program to store the integers B'_{2n} instead of the rational numbers B_{2n} or C_n.

Euler found that the Riemann zeta-function for even non-negative integer arguments can be expressed in terms of Bernoulli numbers—the relation is

$$(-1)^{n-1}\frac{B_{2n}}{(2n)!} = \frac{2\zeta(2n)}{(2\pi)^{2n}}. \tag{8.5}$$

Since $\zeta(2n) = 1 + O(4^{-n})$ as $n \to +\infty$, we see that

$$|B_{2n}| \sim \frac{2\,(2n)!}{(2\pi)^{2n}}.$$

From Stirling's approximation to $(2n)!$, the number of bits in the integer part of B_{2n} is $2n\lg n + O(n)$ (we write lg for $\log_2$). Thus, it takes $\Omega(n^2 \log n)$ space to store $B_1, \ldots, B_n$. We cannot expect any algorithm to compute $B_1, \ldots, B_n$ in fewer than $\Omega(n^2 \log n)$ bit-operations.

Another connection between the Bernoulli numbers and the Riemann zeta-function is the identity

$$\frac{B_{n+1}}{n+1} = -\zeta(-n) \tag{8.6}$$

for $n \in \mathbb{Z}$, $n \geq 1$. This follows from (8.5) and the functional equation for the zeta-function or directly from a contour integral representation of the zeta-function [31].

From the generating function (8.1), multiplying both sides by $\exp(z) - 1$ and equating coefficients of z, we obtain the recurrence

$$\sum_{j=0}^{k} \binom{k+1}{j} B_j = 0 \text{ for } k > 0. \tag{8.7}$$

This recurrence has traditionally been used to compute $B_0, \ldots, B_{2n}$ with $O(n^2)$ arithmetic operations, for example, in [22]. However, this is unsatisfactory if floating-point numbers are used, because the recurrence is *numerically unstable*: the relative error in the computed B_{2n} is of order $4^n \varepsilon$ if the floating-point arithmetic has precision ε, i.e., $\lg(1/\varepsilon)$ bits.

Let C_n be defined by (8.2). Then, multiplying each side of (8.3) by $\sinh(z/2)/(z/2)$ and equating coefficients gives the recurrence

$$\sum_{j=0}^{k} \frac{C_j}{(2k+1-2j)!\,4^{k-j}} = \frac{1}{(2k)!\,4^k}. \tag{8.8}$$

Using this recurrence to evaluate $C_0, C_1, \ldots, C_n$, the relative error in the computed C_n is only $O(n^2 \varepsilon)$, which is satisfactory from a numerical point of view.

Equation (8.5) can be used in several ways to compute Bernoulli numbers. If we want just one Bernoulli number B_{2n} then $\zeta(2n)$ on the right-hand side of (8.5) can be evaluated to sufficient accuracy using the Euler product: this is the "zeta-function" algorithm for computing Bernoulli numbers mentioned (with several references to earlier work) by Harvey [20]. On the other hand, if we want several Bernoulli numbers, then we can use the generating function

$$\frac{\pi z}{\tanh(\pi z)} = -2 \sum_{k=0}^{\infty} (-1)^k \zeta(2k) z^{2k}, \tag{8.9}$$

computing the coefficients of z^{2k}, $k \leq n$, to sufficient accuracy, as mentioned in [3, 8, 9]. This is similar to the fast algorithm that we describe in Sect. 8.4. The similarity can be seen more clearly if we replace πz by z in (8.9), giving

$$\frac{z}{\tanh(z)} = -2 \sum_{k=0}^{\infty} (-1)^k \frac{\zeta(2k)}{\pi^{2k}} z^{2k}, \tag{8.10}$$

since it is the rational number $\zeta(2n)/\pi^{2n}$ that we need in order to compute B_{2n} from (8.5). In fact, it is easy to see that (8.10) is equivalent to (8.3).

There is a vast literature on Bernoulli, Tangent, and Secant numbers. For example, the bibliography of Dilcher and Slavutskii [15] contains more than 2,000 items. Thus, we do not attempt to give a complete list of references to related work. However, we briefly mention the problem of computing *irregular primes* [10, 11], which are odd primes p such that p divides the class number of the pth cyclotomic field. The algorithms that we present in Sects. 8.4 and 8.5 below are not suitable for this task because they take too much memory. It is much more space-efficient to use a modular algorithm where the computations are performed modulo a single prime (or maybe the product of a small number of primes), as in [10, 11, 14, 20]. Space can also be saved by the technique of "multisectioning", which is described by Crandall [13, Sect. 3.2] and Hare [19].

8.3 Tangent and Secant Numbers

The *Tangent numbers* T_n $(n > 0)$ (also called *zag* numbers) are defined by

$$\sum_{n>0} T_n \frac{z^{2n-1}}{(2n-1)!} = \tan z = \frac{\sin z}{\cos z}.$$

Similarly, the *Secant numbers* S_n $(n \geq 0)$ (also called *Euler* or *zig* numbers) are defined by

$$\sum_{n\geq 0} S_n \frac{z^{2n}}{(2n)!} = \sec z = \frac{1}{\cos z}.$$

Unlike the Bernoulli numbers, the Tangent and Secant numbers are positive integers. Because $\tan z$ and $\sec z$ have poles at $z = \pi/2$, we expect T_n to grow roughly like $(2n-1)!(2/\pi)^n$ and S_n like $(2n)!(2/\pi)^n$. To obtain more precise estimates, let

$$\zeta_0(s) = (1-2^{-s})\zeta(s) = 1 + 3^{-s} + 5^{-s} + \cdots$$

be the *odd* zeta-function. Then

$$\frac{T_n}{(2n-1)!} = \frac{2^{2n+1}\zeta_0(2n)}{\pi^{2n}} \sim \frac{2^{2n+1}}{\pi^{2n}} \tag{8.11}$$

(this can be proved in the same way as Euler's relation (8.5) for the Bernoulli numbers). We also have [1, (23.2.22)]

$$\frac{S_n}{(2n)!} = \frac{2^{2n+2}\beta(2n+1)}{\pi^{2n+1}} \sim \frac{2^{2n+2}}{\pi^{2n+1}}, \tag{8.12}$$

where

$$\beta(s) = \sum_{j=0}^{\infty} (-1)^j (2j+1)^{-s}. \tag{8.13}$$

From (8.5) and (8.11), we see that

$$T_n = (-1)^{n-1} 2^{2n} (2^{2n} - 1) \frac{B_{2n}}{2n}. \tag{8.14}$$

This can also be proved directly, without involving the zeta-function, by using the identity

$$\tan z = \frac{1}{\tan z} - \frac{2}{\tan(2z)}.$$

Since $T_n \in \mathbb{Z}$, it follows from (8.14) that the odd primes in the denominator of B_{2n} must divide $2^{2n} - 1$. This is compatible with the Von Staudt–Clausen theorem, since $(p-1)|2n$ implies $p|(2^{2n}-1)$ by Fermat's little theorem.

T_n has about $4n$ more bits than $\lceil B_{2n} \rceil$, but both have $2n\lg n + O(n)$ bits, so asymptotically there is not much difference between the sizes of T_n and $\lceil B_{2n} \rceil$. Thus, if our aim is to compute B_{2n}, we do not lose much by first computing T_n, and this may be more convenient since $T_n \in \mathbb{Z}$, $B_{2n} \in \mathbb{Q}$.

8.4 A Fast Algorithm for Bernoulli Numbers

Harvey [20] showed how B_n could be computed exactly, using a modular algorithm and the Chinese remainder theorem, in $O(n^2(\log n)^{2+o(1)})$ bit-operations. The same complexity can be obtained using (8.5) and the Euler product for the zeta-function (see the discussion in Harvey [20, Sect. 1]).

In this section we show how to compute *all* of $B_0, \ldots, B_n$ with almost the same complexity bound (only larger by a factor $O(\log n)$). In Sect. 8.5 we give an even faster algorithm, which avoids the $O(\log n)$ factor.

Let $A(z) = a_0 + a_1 z + a_2 z^2 + \cdots$ be a power series with coefficients in $\mathbb{R}$, with $a_0 \neq 0$. Let $B(z) = b_0 + b_1 z + \cdots$ be the *reciprocal* power series, so $A(z)B(z) = 1$. Using the FFT, we can multiply polynomials of degree $n-1$ with $O(n \log n)$ real operations. Using Newton's method [24, 26], we can compute $b_0, \ldots, b_{n-1}$ with the *same* complexity $O(n \log n)$, up to a constant factor.

Taking $A(z) = (\exp(z) - 1)/z$ and working with N-bit floating-point numbers, where $N = n\lg(n) + O(n)$, we get $B_0, \ldots, B_n$ to sufficient accuracy to deduce the exact (rational) result. (Alternatively, use (8.3) to avoid computing the terms with odd subscripts, since these vanish except for B_1.) The work involved is $O(n \log n)$ floating-point operations, each of which can be done with N-bit accuracy in $O(n(\log n)^{2+o(1)})$ bit-operations. Thus, overall we get $B_0, \ldots, B_n$

with $O(n^2(\log n)^{3+o(1)})$ bit-operations. Similarly for Secant and Tangent numbers. We omit a precise specification of N and a detailed error analysis of the algorithm, since it is improved in the following section.

8.5 A Faster Algorithm for Tangent and Bernoulli Numbers

To improve the algorithm of Sect. 8.4 for Bernoulli numbers, we use the "Kronecker–Schönhage trick" [7, Sect. 1.9]. Instead of working with power series $A(z)$ (or polynomials, which can be regarded as truncated power series), we work with binary numbers $A(z)$ where z is a suitable (negative) power of 2.

The idea is to compute a single real number $\mathscr{A}$ which is defined in such a way that the numbers that we want to compute are encoded in the binary representation of $\mathscr{A}$. For example, consider the series

$$\sum_{k>0} k^2 z^k = \frac{z(1+z)}{(1-z)^3}, \quad |z| < 1.$$

The right-hand side is an easily computed rational function of z, say $A(z)$. We use decimal rather than binary for expository purposes. With $z = 10^{-3}$ we easily find

$$A(10^{-3}) = \frac{1001000}{997002999} = 0.00\underline{1}\,00\underline{4}\,00\underline{9}\,0\underline{16}\,0\underline{25}\,0\underline{36}\,0\underline{49}\,0\underline{64}\,0\underline{81}\,\underline{100}\cdots$$

Thus, if we are interested in the finite sequence of squares $(1^2, 2^2, 3^2, \ldots, 10^2)$, it is sufficient to compute $\mathscr{A} = A(10^{-3})$ correctly rounded to 30 decimal places, and we can then "read off" the squares from the decimal representation of $\mathscr{A}$.

Of course, this example is purely for illustrative purposes, because it is easy to compute the sequence of squares directly. However, we use the same idea to compute Tangent numbers. Suppose we want the first n Tangent numbers $(T_1, T_2, \ldots, T_n)$. The generating function

$$\tan z = \sum_{k\geq 1} T_k \frac{z^{2k-1}}{(2k-1)!}$$

gives us almost what we need, but not quite, because the coefficients are rationals, not integers. Instead, consider

$$(2n-1)!\tan z = \sum_{k=1}^{n} T'_{k,n} z^{2k-1} + R_n(z), \tag{8.15}$$

where

$$T'_{k,n} = \frac{(2n-1)!}{(2k-1)!} T_k \tag{8.16}$$

is an integer for $1 \le k \le n$, and

$$R_n(z) = \sum_{k=n+1}^{\infty} T'_{k,n} z^{2k-1} = (2n-1)! \sum_{k=n+1}^{\infty} T_k \frac{z^{2k-1}}{(2k-1)!} \tag{8.17}$$

is a remainder term which is small if z is sufficiently small. Thus, choosing $z = 2^{-p}$ with p sufficiently large, the first $2np$ binary places of $(2n-1)! \tan z$ define $T'_{1,n}, T'_{2,n}, \ldots, T'_{n,n}$. Once we have computed $T'_{1,n}, T'_{2,n}, \ldots, T'_{n,n}$ it is easy to deduce $T_1, T_2, \ldots, T_n$ from

$$T_k = \frac{T'_{k,n}}{(2n-1)!/(2k-1)!}.$$

For this idea to work, two conditions must be satisfied. First, we need

$$0 \le T'_{k,n} < 1/z^2 = 2^{2p}, \quad 1 \le k \le n, \tag{8.18}$$

so we can read off the $T'_{k,n}$ from the binary representation of $(2n-1)! \tan z$. Since we have a good asymptotic estimate for T_k, it is not hard to choose p sufficiently large for this condition to hold.

Second, we need the remainder term $R_n(z)$ to be sufficiently small that it does not influence the estimation of $T'_{n,n}$. A sufficient condition is

$$0 \le R_n(z) < z^{2n-1}. \tag{8.19}$$

Choosing z sufficiently small (i.e., p sufficiently large) guarantees that condition (8.19) holds, since $R_n(z)$ is $O(z^{2n+1})$ as $z \to 0$ with n fixed.

Lemmas 8.3 and 8.4 below give sufficient conditions for (8.18) and (8.19) to hold.

Lemma 8.1.

$$\frac{T_k}{(2k-1)!} \le \left(\frac{2}{\pi}\right)^{2(k-1)} \quad \text{for } k \ge 1.$$

Proof. From (8.11),

$$\frac{T_k}{(2k-1)!} = 2\left(\frac{2}{\pi}\right)^{2k} \zeta_0(2k) \le 2\left(\frac{2}{\pi}\right)^{2k} \zeta_0(2) \le \left(\frac{2}{\pi}\right)^{2k} \frac{\pi^2}{4} = \left(\frac{2}{\pi}\right)^{2(k-1)}.$$

■

Lemma 8.2. $(2n-1)! \le n^{2n-1}$ *for* $n \ge 1$.

Proof.

$$(2n-1)! = n \prod_{j=1}^{n-1}(n-j)(n+j) = n \prod_{j=1}^{n-1}(n^2-j^2) \le n^{2n-1}$$

with equality iff $n = 1$. ∎

Lemma 8.3. *If* $k \ge 1$, $n \ge 2$, $p = \lceil n\lg(n)\rceil$, $z = 2^{-p}$, *and* $T'_{k,n}$ *is as in* (8.16), *then* $z \le n^{-n}$ *and* $T'_{k,n} < 1/z^2$.

Proof. We have $z = 2^{-p} = 2^{-\lceil n\lg(n)\rceil} \le 2^{-n\lg(n)} = n^{-n}$, which proves the first part of the Lemma.

Assume $k \ge 1$ and $n \ge 2$. From Lemma 8.1, we have

$$T'_{k,n} \le (2n-1)!\left(\frac{2}{\pi}\right)^{2(k-1)} \le (2n-1)!,$$

and from Lemma 8.2 it follows that

$$T'_{k,n} \le n^{2n-1} < n^{2n}.$$

From the first part of the Lemma, $n^{2n} \le 1/z^2$, so the second part follows. ∎

Lemma 8.4. *If* $n \ge 2$, $p = \lceil n\lg(n)\rceil$, $z = 2^{-p}$, *and* $R_n(z)$ *is as defined in* (8.17), *then* $0 < R_n(z) < 0.1\,z^{2n-1}$.

Proof. Since all the terms in the sum defining $R_n(z)$ are positive, it is immediate that $R_n(z) > 0$. Since $n \ge 2$, we have $p \ge 2$ and $z \le 1/4$. Now, using Lemma 8.1,

$$\begin{aligned}
R_n(z) &= \sum_{k=n+1}^{\infty} T'_{k,n} z^{2k-1} \\
&\le (2n-1)! \sum_{k=n+1}^{\infty} \left(\frac{2}{\pi}\right)^{2(k-1)} z^{2k-1} \\
&\le (2n-1)! \left(\frac{2}{\pi}\right)^{2n} z^{2n+1} \left(1 + \left(\frac{2z}{\pi}\right)^2 + \left(\frac{2z}{\pi}\right)^4 + \cdots\right) \\
&\le (2n-1)! \left(\frac{2}{\pi}\right)^{2n} z^{2n+1} \Bigg/ \left(1 - \left(\frac{2z}{\pi}\right)^2\right).
\end{aligned}$$

Since $z \le 1/4$, we have $1/(1-(2z/\pi)^2) < 1.026$. Also, from Lemma 8.2, $(2n-1)! \le n^{2n-1}$. Thus, we have

Input: integer $n \geq 2$
Output: Tangent numbers $T_1, \ldots, T_n$ and (optional) Bernoulli numbers $B_2, B_4, \ldots, B_{2n}$
$p \leftarrow \lceil n \lg(n) \rceil$
$z \leftarrow 2^{-p}$
$S \leftarrow \sum_{0 \leq k < n} (-1)^k z^{2k+1} \times (2n)!/(2k+1)!$
$C \leftarrow \sum_{0 \leq k < n} (-1)^k z^{2k} \times (2n)!/(2k)!$
$V \leftarrow \lfloor z^{1-2n} \times (2n-1)! \times S/C \rceil$ (here $\lfloor x \rceil$ means round x to nearest integer)
Extract $T'_{k,n} = T_k (2n-1)!/(2k-1)!$, $1 \leq k \leq n$, from the binary representation of V
$T_k \leftarrow T'_{k,n} \times (2k-1)!/(2n-1)!$, $k = n, n-1, \ldots, 1$
$B_{2k} \leftarrow (-1)^{k-1} (k \times T_k / 2^{2k-1})/(2^{2k}-1)$, $k = 1, 2, \ldots, n$ (optional)
return $T_1, T_2, \ldots, T_n$ and (optional) $B_2, B_4, \ldots, B_{2n}$

Fig. 8.1 Algorithm FastTangentNumbers (also optionally computes Bernoulli numbers)

$$\frac{R_n(z)}{z^{2n-1}} < 1.026 n^{2n-1} \left(\frac{2}{\pi}\right)^{2n} z^2.$$

Now $z^2 \leq n^{-2n}$ from the first part of Lemma 8.3, so

$$\frac{R_n(z)}{z^{2n-1}} < \frac{1.026}{n} \left(\frac{2}{\pi}\right)^{2n}. \tag{8.20}$$

The right-hand side is a monotonic decreasing function of n, so is bounded above by its value when $n = 2$, giving $R_n(z)/z^{2n-1} < 0.1$. ■

A high-level description of the resulting Algorithm FastTangentNumbers is given in Fig. 8.1. The algorithm computes the Tangent numbers $T_1, T_2, \ldots, T_n$ using the Kronecker–Schönhage trick as described above, and deduces the Bernoulli numbers $B_2, B_4, \ldots, B_{2n}$ from the relation (8.14).

In order to achieve the best complexity, the algorithm must be implemented carefully using binary arithmetic. The computations of S (an approximation to $(2n)! \sin z$) and C (an approximation to $(2n)! \cos z$) involve computing ratios of factorials such as $(2n)!/(2k)!$, where $0 \leq k \leq n$. This can be done in time $O(n^2 (\log n)^2)$ by a straightforward algorithm. The N-bit division to compute S/C (an approximation to $\tan z$) can be done in time $O(N \log(N) \log\log(N))$ by the Schönhage–Strassen algorithm combined with Newton's method [7, Sect. 4.2.2]. Here it is sufficient to take $N = 2np + 2 = 2n^2 \lg(n) + O(n)$. Note that

$$V = \sum_{k=1}^{n} 2^{2(n-k)p} T'_{k,n} \tag{8.21}$$

is just the finite sum in (8.15) scaled by z^{1-2n} (a power of two), and the integers $T'_{k,n}$ can simply be "read off" from the binary representation of V in n blocks of $2p$ consecutive bits. The $T'_{k,n}$ can then be scaled by ratios of factorials in time $O(n^2 (\log n)^{2+o(1)})$ to give the Tangent numbers $T_1, T_2, \ldots, T_n$.

The correctness of the computed Tangent numbers follows from Lemmas 8.3 and 8.4, apart from possible errors introduced by S/C being only an approximation to $\tan(z)$. Lemma 8.5 shows that this error is sufficiently small.

Lemma 8.5. *Suppose that $n \geq 2$, z, S and C as in Algorithm FastTangentNumbers. Then*

$$z^{1-2n}(2n-1)!\left|\frac{S}{C} - \tan z\right| < 0.02. \tag{8.22}$$

Proof. We use the inequality

$$\left|\frac{A}{B} - \frac{A'}{B'}\right| \leq \frac{|A|\cdot|B-B'| + |B|\cdot|A-A'|}{|B|\cdot|B'|}. \tag{8.23}$$

Take $A = \sin z$, $B = \cos z$, $A' = S/(2n)!$, $B' = C/(2n)!$ in (8.23). Since $n \geq 2$ we have $0 < z \leq 1/4$. Then $|A| = |\sin z| < z$. Also, $|B| = |\cos z| > 31/32$ from the Taylor series $\cos z = 1 - z^2/2 + \cdots$, which has terms of alternating sign and decreasing magnitude. By similar arguments, $|B'| \geq 31/32$, $|B-B'| < z^{2n}/(2n)!$, and $|A-A'| < z^{2n+1}/(2n+1)!$. Combining these inequalities and using (8.23), we obtain

$$\left|\frac{S}{C} - \tan z\right| < \frac{6\cdot 32\cdot 32\, z^{2n+1}}{5\cdot 31\cdot 31\,(2n)!} < \frac{1.28\, z^{2n+1}}{(2n)!}.$$

Multiplying both sides by $z^{1-2n}(2n-1)!$ and using $1.28z^2/(2n) \leq 0.02$, we obtain the inequality (8.22). This completes the proof of Lemma 8.5. ∎

In view of the constant 0.02 in (8.22) and the constant 0.1 in Lemma 8.4, the effect of all sources of error in computing $z^{1-2n}(2n-1)!\tan z$ is at most $0.12 < 1/2$, which is too small to change the computed integer V, that is to say, the computed V is indeed given by (8.21).

The computation of the Bernoulli numbers $B_2, B_4, \ldots, B_{2n}$ from $T_1, \ldots, T_n$, is straightforward (details depending on exactly how rational numbers are to be represented). The entire computation takes time

$$O(N(\log N)^{1+o(1)}) = O(n^2(\log n)^{2+o(1)}).$$

Thus, we have proved:

Theorem 8.6. *The Tangent numbers $T_1, \ldots, T_n$ and Bernoulli numbers $B_2, B_4, \ldots, B_{2n}$ can be computed in $O(n^2(\log n)^{2+o(1)})$ bit-operations using $O(n^2 \log n)$ space.*

A small modification of the above can be used to compute the Secant numbers $S_0, S_1, \ldots, S_n$ in $O(n^2(\log n)^{2+o(1)})$ bit-operations and $O(n^2 \log n)$ space. The bound on Tangent numbers given by Lemma 8.1 can be replaced by the bound

$$\frac{S_n}{(2n)!} \le 2\left(\frac{2}{\pi}\right)^{2n+1}$$

which follows from (8.12) since $\beta(2n+1) < 1$.

We remark that an efficient implementation of Algorithm FastTangentNumbers in a high-level language such as Sage [30] or Magma [5] is nontrivial, because it requires access to the internal binary representation of high-precision integers. Everything can be done using (implicitly scaled) integer arithmetic—there is no need for floating-point—but for the sake of clarity we did not include the scaling in Fig. 8.1. If floating-point arithmetic is used, a precision of N bits is sufficient, where $N = 2np + 2$.

Comparing our Algorithm FastTangentNumbers with Harvey's modular algorithm [20], we see that there is a space-time trade-off: Harvey's algorithm uses less space (by a factor of order n) to compute a single B_n, but more time (again by a factor of order n) to compute all of $B_1, \ldots, B_n$. Harvey's algorithm has better locality and is readily parallelizable.

In the following section we give much simpler algorithms which are fast enough for most practical purposes and are based on three-term recurrence relations.

8.6 Algorithms Based on Three-Term Recurrences

Akiyama and Tanigawa [21] gave an algorithm for computing Bernoulli numbers based on a three-term recurrence. However, it is only useful for exact computations, since it is numerically unstable if applied using floating-point arithmetic. It is faster to use a stable recurrence for computing Tangent numbers and then deduce the Bernoulli numbers from (8.14).

8.6.1 Bernoulli and Tangent Numbers

We now give a stable three-term recurrence and corresponding in-place algorithm for computing Tangent numbers. The algorithm is perfectly stable since all operations are on positive integers and there is no cancellation. Also, it involves less arithmetic than the Akiyama–Tanigawa algorithm. This is partly because the operations are on integers rather than rationals and partly because there are fewer operations since we take advantage of zeros.

Bernoulli numbers can be computed using Algorithm TangentNumbers and the relation (8.14). The time required for the application of (8.14) is negligible.

The recurrence (8.24) that we use was given by Buckholtz and Knuth [23], but they did not give our in-place Algorithm TangentNumbers explicitly. Related recurrences with applications to parallel computation were considered by Hare [19].

Fig. 8.2 Algorithm TangentNumbers

Input: positive integer n
Output: Tangent numbers $T_1, \ldots, T_n$

$T_1 \leftarrow 1$
for k **from** 2 **to** n
  $T_k \leftarrow (k-1)T_{k-1}$
for k **from** 2 **to** n
  for j **from** k **to** n
    $T_j \leftarrow (j-k)T_{j-1} + (j-k+2)T_j$
return $T_1, T_2, \ldots, T_n$.

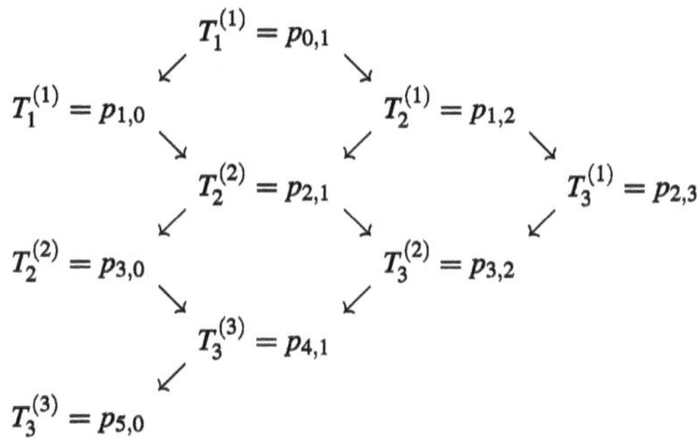

Fig. 8.3 Dataflow in algorithm TangentNumbers for $n = 3$

Write $t = \tan x$, $D = \mathrm{d}/\mathrm{d}x$, so $Dt = 1 + t^2$ and $D(t^n) = nt^{n-1}(1+t^2)$ for $n \geq 1$. It is clear that $D^n t$ is a polynomial in t, say $P_n(t)$. Write $P_n(t) = \sum_{j \geq 0} p_{n,j} t^j$. Then $\deg(P_n) = n + 1$ and, from the formula for $D(t^n)$,

$$p_{n,j} = (j-1)p_{n-1,j-1} + (j+1)p_{n-1,j+1}. \tag{8.24}$$

We are interested in $T_k = (\mathrm{d}/\mathrm{d}x)^{2k-1} \tan x|_{x=0} = P_{2k-1}(0) = p_{2k-1,0}$, which can be computed from the recurrence (8.24) in $O(k^2)$ operations using the obvious boundary conditions. We save work by noticing that $p_{n,j} = 0$ if $n + j$ is even. The resulting algorithm is given in Fig. 8.2.

The first **for** loop initializes $T_k = p_{k-1,k} = (k-1)!$. The variable T_k is then used to store $p_{k,k-1}, p_{k+1,k-2}, \ldots, p_{2k-2,1}, p_{2k-1,0}$ at successive iterations of the second **for** loop. Thus, when the algorithm terminates, $T_k = p_{2k-1,0}$, as expected.

The process in the case $n = 3$ is illustrated in Fig. 8.3, where $T_k^{(m)}$ denotes the value of the variable T_k at successive iterations $m = 1, 2, \ldots, n$. It is instructive to compare a similar figure for the Akiyama–Tanigawa algorithm in [21].

Algorithm TangentNumbers takes $\Theta(n^2)$ operations on positive integers. The integers T_n have $O(n \log n)$ bits, other integers have $O(\log n)$ bits. Thus, the overall complexity is $O(n^3(\log n)^{1+o(1)})$ bit-operations, or $O(n^3 \log n)$ word-operations if n fits in a single word.

The algorithm is not optimal, but it is good in practice for moderate values of n, and much simpler than asymptotically faster algorithms such as those described in Sects. 8.4 and 8.5. For example, using a straightforward Magma implementation of Algorithm TangentNumbers, we computed the first 1,000 Tangent numbers in 1.50 s

Fig. 8.4 Algorithm SecantNumbers

Input: positive integer n
Output: Secant numbers $S_0, S_1 \ldots, S_n$
  $S_0 \leftarrow 1$
  for k **from** 1 **to** n
    $S_k \leftarrow kS_{k-1}$
  for k **from** 1 **to** n
    for j **from** $k+1$ **to** n
      $S_j \leftarrow (j-k)S_{j-1} + (j-k+1)S_j$
  return $S_0, S_1, \ldots, S_n$.

on a 2.26 GHz Intel Core 2 Duo. For comparison, it takes 1.92 s for a single N-bit division computing T in Algorithm FastTangentNumbers (where $N = 19{,}931{,}568$ corresponds to $n = 1{,}000$). Thus, we expect the crossover point where Algorithm FastTangentNumbers actually becomes faster to be slightly larger than $n = 1{,}000$ (but dependent on implementation details).

8.6.2 Secant Numbers

A similar algorithm may be used to compute Secant numbers. Let $s = \sec x$, $t = \tan x$, and $D = \mathrm{d}/\mathrm{d}x$. Then $Ds = st$, $D^2 s = s(1+2t^2)$, and in general $D^n s = sQ_n(t)$, where $Q_n(t)$ is a polynomial of degree n in t. The Secant numbers are given by $S_k = Q_{2k}(0)$. Let $Q_n(t) = \sum_{k\geq 0} q_{n,k} t^k$. From

$$D(st^k) = st^{k+1} + kst^{k-1}(1+t^2)$$

we obtain the three-term recurrence

$$q_{n+1,k} = kq_{n,k-1} + (k+1)q_{n,k+1} \text{ for } 1 \leq k \leq n. \tag{8.25}$$

By avoiding the computation of terms $q_{n,k}$ that are known to be zero ($n+k$ odd), and ordering the computation in a manner analogous to that used for Algorithm TangentNumbers, we obtain Algorithm SecantNumbers (see Fig. 8.4), which computes the Secant numbers in place using non-negative integer arithmetic.

8.6.3 Comparison with Atkinson's Algorithm

Atkinson [2] gave an elegant algorithm for computing both the Tangent numbers $T_1, T_2, \ldots, T_n$ and the Secant numbers $S_0, S_1, \ldots, S_n$ using a "Pascal's triangle" style of algorithm that only involves additions of non-negative integers. Since a triangle with $2n+1$ rows in involved, Atkinson's algorithm requires $2n^2 + O(n)$

integer additions. This can be compared with $n^2/2+O(n)$ additions and $n^2+O(n)$ multiplications (by small integers) for our Algorithm TangentNumbers, and similarly for Algorithm SecantNumbers.

Thus, we might expect Atkinson's algorithm to be slower than Algorithm TangentNumbers. Computational experiments confirm this. With $n = 1,000$, Algorithm TangentNumbers programmed in Magma takes 1.50 s on a 2.26 GHz Intel Core 2 Duo, algorithm SecantNumbers also takes 1.50 s, and Atkinson's algorithm takes 4.51 s. Thus, even if both Tangent and Secant numbers are required, Atkinson's algorithm is slightly slower. It also requires about twice as much memory.

Acknowledgements We thank Jon Borwein for encouraging the belief that high-precision computations are useful in "experimental" mathematics [4], e.g. in the PSLQ algorithm [16]. Ben F. "Tex" Logan, Jr. (unpublished observation, mentioned in [18, Sect. 6.5]) suggested the use of Tangent numbers to compute Bernoulli numbers. Christian Reinsch (about 1979, acknowledged in [6], Personal communication to R.P. Brent) pointed out the numerical instability of the recurrence (8.7) and suggested the use of the numerically stable recurrence (8.8). Christopher Heckman kindly drew our attention to Atkinson's algorithm [2]. We thank Paul Zimmermann for his comments. Some of the material presented here is drawn from the recent book *Modern Computer Arithmetic* [7] (and as-yet-unpublished solutions to exercises in the book). In particular, see [7, Sect. 4.7.2 and Exercises 4.35–4.41]. Finally, we thank David Bailey, Richard Crandall, and two anonymous referees for suggestions and pointers to additional references.

Note added in proof Recently the second author [A subquadratic algorithm for computing the n-th Bernoulli number, arXiv:1209.0533, to appear in Mathematics of Computation] has given an improved algorithm for the computation of a single Bernoulli number. The new algorithm reduces the exponent from $2+o(1)$ to $4/3+o(1)$.

References

1. Abramowitz, M., Stegun, I.A.: Handbook of Mathematical Functions. Dover (1973)
2. Atkinson, M.D.: How to compute the series expansions of sec x and tan x. Am. Math. Monthly **93**, 387–389 (1986)
3. Bailey, D.H., Borwein, J.M., Crandall, R.E.: On the Khintchine constant. Math. Comput. **66**, 417–431 (1997)
4. Borwein, J.M., Corless, R.M.: Emerging tools for experimental mathematics. Am. Math. Mon. **106**, 899–909 (1999)
5. Bosma, W., Cannon, J., Playoust, C.: The Magma algebra system. I: the user language. J. Symbolic Comput. **24**, 235–265 (1997)
6. Brent, R.P.: Unrestricted algorithms for elementary and special functions. Information Processing, vol. 80, pp. 613–619. North-Holland, Amsterdam (1980). arXiv:1004.3621v1
7. Brent, R.P., Zimmermann, P.: Modern Computer Arithmetic. Cambridge University Press, Cambridge (2010). arXiv:1004.4710v1
8. Buhler, J., Crandall, R., Sompolski, R.: Irregular primes to one million. Math. Comput. **59**, 717–722 (1992)
9. Buhler, J., Crandall, R., Ernvall, R., Metsänkylä, T.: Irregular primes to four million. Math. Comput. **61**, 151–153 (1993)
10. Buhler, J., Crandall, R., Ernvall, R., Metsänkylä, T., Shokrollahi, M.A.: Irregular primes and cyclotomic invariants to twelve million. J. Symbolic Comput. **31**, 89–96 (2001)

11. Buhler, J., Harvey, D.: Irregular primes to 163 million. Math. Comput. **80**, 2435–2444 (2011)
12. Clausen, T.: Theorem. Astron. Nachr. **17**, 351–352 (1840)
13. Crandall, R.E.: Topics in Advanced Scientific Computation. Springer, New York (1996)
14. Crandall, R.E., Pomerance, C.: Prime Numbers: A Computational Perspective. Springer, New York (2001)
15. Dilcher, K., Slavutskii, I.Sh.: A Bibliography of Bernoulli Numbers (last updated March 3, 2007). http://www.mscs.dal.ca/%7Edilcher/bernoulli.html.
16. Ferguson, H.R.P., Bailey, D.H., Arno, S.: Analysis of PSLQ, an integer relation finding algorithm. Math. Comput. **68**, 351–369 (1999)
17. Fürer, M.: Faster integer multiplication. Proceedings of 39th Annual ACM Symposium on Theory of Computing (STOC), pp. 57–66. ACM, San Diego (2007)
18. Graham, R.L., Knuth, D.E., Patashnik, O.: Concrete Mathematics, 3rd edn. Addison-Wesley, Reading (1994)
19. Hare, K.: Multisectioning, rational poly-exponential functions and parallel computation. M.Sc. thesis, Department of Mathematics and Statistics, Simon Fraser University, Canada (2002)
20. Harvey, D.: A multimodular algorithm for computing Bernoulli numbers. Math. Comput. **79**, 2361–2370 (2010)
21. Kaneko, M.: The Akiyama–Tanigawa algorithm for Bernoulli numbers. J. Integer Seq. **3**, Article 00.2.9, 6 (2000). http://www.cs.uwaterloo.ca/journals/JIS/
22. Knuth, D.E.: Euler's constant to 1271 places. Math. Comput. **16**, 275–281 (1962)
23. Knuth, D.E., Buckholtz, T.J.: Computation of Tangent, Euler, and Bernoulli numbers. Math. Comput. **21**, 663–688 (1967)
24. Kung, H.T.: On computing reciprocals of power series. Numer. Math. **22**, 341–348 (1974)
25. Schönhage, A., Strassen, V.: Schnelle Multiplikation großer Zahlen. Computing **7**, 281–292 (1971)
26. Sieveking, M.: An algorithm for division of power series. Computing **10**, 153–156 (1972)
27. Sloane, N.J.A.: The on-line encyclopedia of integer sequences. http://oeis.org
28. Von Staudt, K.G.C.: Beweis eines Lehrsatzes, die Bernoullischen Zahlen betreffend. J. Reine Angew. Math. **21**, 372–374 (1840). http://gdz.sub.uni-goettingen.de
29. Steel, A.: Reduce everything to multiplication. Presented at Computing by the Numbers: Algorithms, Precision and Complexity. Workshop for Richard Brent's 60th Birthday, Berlin, 2006. http://www.mathematik.hu-berlin.de/%7Egaggle/EVENTS/2006/BRENT60/
30. Stein, W. et al.: Sage. http://www.sagemath.org/
31. Titchmarsh, E.C.: The Theory of the Riemann Zeta-Function, 2nd edn (revised by D. R. Heath-Brown). Clarendon Press, Oxford (1986)

Chapter 9
Monotone Operator Methods for Nash Equilibria in Non-potential Games

Luis M. Briceño-Arias and Patrick L. Combettes

Abstract We observe that a significant class of Nash equilibrium problems in non-potential games can be associated with monotone inclusion problems. We propose splitting techniques to solve such problems and establish their convergence. Applications to generalized Nash equilibria, zero-sum games, and cyclic proximation problems are demonstrated.

Key words: Monotone operator • Nash equilibrium • Potential game • Proximal algorithm • Splitting method • Zero-sum game

Mathematics Subject Classifications (2010): Primary 91A05, 49M27, 47H05; Secondary 90C25

9.1 Problem Statement

Consider a game with $m \geq 2$ players indexed by $i \in \{1,\dots,m\}$. The strategy x_i of the ith player lies in a real Hilbert space $\mathcal{H}_i$ and the problem is to find $x_1 \in \mathcal{H}_1,\dots,x_m \in \mathcal{H}_m$ such that

---

Communicated by Heinz H. Bauschke.

L.M. Briceño-Arias
Department of Mathematics, Universidad de Chile, Center for Mathematical Modeling, CNRS–UMI 2807, Santiago, Chile, and Universidad Técnica Federico Santa María, Santiago, Chile
e-mail: lbriceno@dim.uchile.cl

P.L. Combettes
Laboratoire Jacques-Louis Lions – UMR CNRS 7598, UPMC Université Paris 06, 4, Place Jussieu 75005 Paris, France
e-mail: plc@math.jussieu.fr

D.H. Bailey et al. (eds.), *Computational and Analytical Mathematics*, Springer Proceedings in Mathematics & Statistics 50, DOI 10.1007/978-1-4614-7621-4_9,

$$(\forall i \in \{1,\dots,m\}) \quad x_i \in \underset{x\in\mathscr{H}_i}{\operatorname{Argmin}} \boldsymbol{f}(x_1,\dots,x_{i-1},x,x_{i+1},\dots,x_m) + \boldsymbol{g}_i(x_1,\dots,x_{i-1},x,x_{i+1},\dots,x_m), \tag{9.1}$$

where $(\boldsymbol{g}_i)_{1\leq i\leq m}$ represents the individual penalty of player i depending on the strategies of all players and $\boldsymbol{f}$ is a convex penalty which is common to all players and models the collective discomfort of the group. At this level of generality, no reliable method exists for solving (9.1) and some hypotheses are required. In this paper we focus on the following setting.

Problem 9.1. Let $m \geq 2$ be an integer and let $\boldsymbol{f}\colon \mathscr{H}_1 \oplus \cdots \oplus \mathscr{H}_m \to \left]-\infty,+\infty\right]$ be a proper lower semicontinuous convex function. For every $i \in \{1,\dots,m\}$, let $\boldsymbol{g}_i\colon \mathscr{H}_1 \oplus \cdots \oplus \mathscr{H}_m \to \left]-\infty,+\infty\right]$ be such that, for every $x_1 \in \mathscr{H}_1,\dots,x_m \in \mathscr{H}_m$, the function $x \mapsto \boldsymbol{g}_i(x_1,\dots,x_{i-1},x,x_{i+1},\dots,x_m)$ is convex and differentiable on $\mathscr{H}_i$, and denote by $\nabla_i \boldsymbol{g}_i(x_1,\dots,x_m)$ its derivative at x_i. Moreover,

$$\big(\forall (x_1,\dots,x_m) \in \mathscr{H}_1 \oplus \cdots \oplus \mathscr{H}_m\big)\big(\forall (y_1,\dots,y_m) \in \mathscr{H}_1 \oplus \cdots \oplus \mathscr{H}_m\big) \quad \sum_{i=1}^{m} \langle \nabla_i \boldsymbol{g}_i(x_1,\dots,x_m) - \nabla_i \boldsymbol{g}_i(y_1,\dots,y_m) \mid x_i - y_i\rangle \geq 0. \tag{9.2}$$

The problem is to find $x_1 \in \mathscr{H}_1, \dots, x_m \in \mathscr{H}_m$ such that

$$\begin{cases} x_1 \in \underset{x\in\mathscr{H}_1}{\operatorname{Argmin}}\; \boldsymbol{f}(x,x_2,\dots,x_m) + \boldsymbol{g}_1(x,x_2,\dots,x_m) \\ \quad\vdots \\ x_m \in \underset{x\in\mathscr{H}_m}{\operatorname{Argmin}}\; \boldsymbol{f}(x_1,\dots,x_{m-1},x) + \boldsymbol{g}_m(x_1,\dots,x_{m-1},x). \end{cases} \tag{9.3}$$

In the special case when, for every $i \in \{1,\dots,m\}$, $\boldsymbol{g}_i = \boldsymbol{g}$ is convex, Problem 9.1 amounts to finding a Nash equilibrium of a potential game, i.e., a game in which the penalty of every player can be represented by a common potential $\boldsymbol{f}+\boldsymbol{g}$ [14]. Hence, Nash equilibria can be found by solving

$$\underset{x_1\in\mathscr{H}_1,\dots,x_m\in\mathscr{H}_m}{\operatorname{minimize}} \quad \boldsymbol{f}(x_1,\dots,x_m) + \boldsymbol{g}(x_1,\dots,x_m). \tag{9.4}$$

Thus, the problem reduces to the minimization of the sum of two convex functions on the Hilbert space $\mathscr{H}_1 \oplus \cdots \oplus \mathscr{H}_m$ and various methods are available to tackle it under suitable assumptions (see for instance [5, Chap. 27]). On the other hand, in the particular case when $\boldsymbol{f}$ is separable, a review of methods for solving (9.3) is provided in [8]. In this paper we address the more challenging non-potential setting, in which the functions $(\boldsymbol{g}_i)_{1\leq i\leq m}$ need not be identical nor convex, but they must satisfy (9.2), and $\boldsymbol{f}$ need not be separable. Let us note that (9.2) actually implies, for every $i \in \{1,\dots,m\}$, the convexity of $\boldsymbol{g}_i$ with respect to its ith variable.

Our methodology consists of using monotone operator splitting techniques for solving an auxiliary monotone inclusion, the solutions of which are Nash equilibria of Problem 9.1. In Sect. 9.2 we review the notation and background material needed subsequently. In Sect. 9.3 we introduce the auxiliary monotone inclusion problem and provide conditions ensuring the existence of solutions to the auxiliary problem. We also propose two methods for solving Problem 9.1 and establish their convergence. Finally, in Sect. 9.4, the proposed methods are applied to the construction of generalized Nash equilibria, to zero-sum games, and to cyclic proximation problems.

9.2 Notation and Background

Throughout this paper, $\mathcal{H}$, $\mathcal{G}$, and $(\mathcal{H}_i)_{1\leq i\leq m}$ are real Hilbert spaces. For convenience, their scalar products are all denoted by $\langle \cdot \mid \cdot\rangle$ and the associated norms by $\|\cdot\|$. Let $A\colon \mathcal{H}\to 2^{\mathcal{H}}$ be a set-valued operator. The domain of A is

$$\operatorname{dom} A = \big\{x\in\mathcal{H} \bigm| Ax\neq\varnothing\big\}, \tag{9.5}$$

the set of zeros of A is

$$\operatorname{zer} A = \big\{x\in\mathcal{H} \bigm| 0\in Ax\big\}, \tag{9.6}$$

the graph of A is

$$\operatorname{gra} A = \big\{(x,u)\in\mathcal{H}\times\mathcal{H} \bigm| u\in Ax\big\}, \tag{9.7}$$

the range of A is

$$\operatorname{ran} A = \big\{u\in\mathcal{H} \bigm| (\exists x\in\mathcal{H})\ u\in Ax\big\}, \tag{9.8}$$

the inverse of A is the set-valued operator

$$A^{-1}\colon \mathcal{H}\to 2^{\mathcal{H}}\colon u\mapsto \big\{x\in\mathcal{H} \bigm| u\in Ax\big\}, \tag{9.9}$$

and the resolvent of A is

$$J_A = (\operatorname{Id}+A)^{-1}. \tag{9.10}$$

In addition, A is monotone if

$$(\forall(x,y)\in\mathcal{H}\times\mathcal{H})(\forall(u,v)\in Ax\times Ay)\quad \langle x-y \mid u-v\rangle\geq 0 \tag{9.11}$$

and it is maximally monotone if, furthermore, every monotone operator $B\colon \mathcal{H}\to 2^{\mathcal{H}}$ such that $\operatorname{gra} A\subset \operatorname{gra} B$ coincides with A.

We denote by $\Gamma_0(\mathcal{H})$ the class of lower semicontinuous convex functions $\varphi\colon \mathcal{H} \to]-\infty,+\infty]$ which are proper in the sense that $\operatorname{dom}\varphi = \{x \in \mathcal{H} \mid \varphi(x) < +\infty\} \neq \varnothing$. Let $\varphi \in \Gamma_0(\mathcal{H})$. The proximity operator of φ is

$$\operatorname{prox}_\varphi\colon \mathcal{H} \to \mathcal{H}\colon x \mapsto \underset{y\in\mathcal{H}}{\operatorname{argmin}}\ \varphi(y) + \frac{1}{2}\|x-y\|^2, \tag{9.12}$$

and the subdifferential of φ is the maximally monotone operator

$$\partial\varphi\colon \mathcal{H} \to 2^{\mathcal{H}}\colon x \mapsto \big\{u \in \mathcal{H} \mid (\forall y \in \mathcal{H})\ \langle y-x \mid u\rangle + \varphi(x) \leq \varphi(y)\big\}. \tag{9.13}$$

We have

$$\underset{x\in\mathcal{H}}{\operatorname{Argmin}}\ \varphi(x) = \operatorname{zer}\partial\varphi \quad \text{and} \quad \operatorname{prox}_\varphi = J_{\partial\varphi}. \tag{9.14}$$

Let $\beta \in]0,+\infty[$. An operator $T\colon \mathcal{H} \to \mathcal{H}$ is β-cocoercive (or βT is firmly nonexpansive) if

$$(\forall x \in \mathcal{H})(\forall y \in \mathcal{H}) \quad \langle x-y \mid Tx-Ty\rangle \geq \beta\|Tx-Ty\|^2, \tag{9.15}$$

which implies that it is monotone and β^{-1}–Lipschitzian. Let C be a nonempty convex subset of $\mathcal{H}$. The indicator function of C is

$$\iota_C\colon \mathcal{H} \to]-\infty,+\infty]\colon x \mapsto \begin{cases} 0, & \text{if } x \in C;\\ +\infty, & \text{if } x \notin C \end{cases} \tag{9.16}$$

and $\partial\iota_C = N_C$ is the normal cone operator of C, i.e.,

$$N_C\colon \mathcal{H} \to 2^{\mathcal{H}}\colon x \mapsto \begin{cases} \big\{u \in \mathcal{H} \mid (\forall y \in C)\ \langle y-x \mid u\rangle \leq 0\big\}, & \text{if } x \in C;\\ \varnothing, & \text{otherwise.} \end{cases} \tag{9.17}$$

If C is closed, for every $x \in \mathcal{H}$, there exists a unique point $P_C x \in C$ such that $\|x - P_C x\| = \inf_{y\in C}\|x-y\|$; $P_C x$ is called the projection of x onto C and we have $P_C = \operatorname{prox}_{\iota_C}$. In addition, the symbols $\rightharpoonup$ and $\to$ denote respectively weak and strong convergence. For a detailed account of the tools described above, see [5].

9.3 Model, Algorithms, and Convergence

We investigate an auxiliary monotone inclusion problem, the solutions of which are Nash equilibria of Problem 9.1 and propose two splitting methods to solve it. Both involve the proximity operator prox_f, which can be computed explicitly in several instances [5, 7]. We henceforth denote by $\mathscr{H}$ the direct sum of the Hilbert spaces $(\mathscr{H}_i)_{1\leq i\leq m}$, i.e., the product space $\mathscr{H}_1 \times \cdots \times \mathscr{H}_m$ equipped with the scalar product

$$\langle\langle \cdot \mid \cdot \rangle\rangle \colon \big((x_i)_{1\leq i\leq m}, (y_i)_{1\leq i\leq m}\big) \mapsto \sum_{i=1}^m \langle x_i \mid y_i \rangle. \tag{9.18}$$

We denote the associated norm by $|||\cdot|||$, a generic element of $\boldsymbol{\mathscr{H}}$ by $\boldsymbol{x} = (x_i)_{1\leq i\leq m}$, and the identity operator on $\boldsymbol{\mathscr{H}}$ by $\mathbf{Id}$.

9.3.1 A Monotone Inclusion Model

With the notation and hypotheses of Problem 9.1, let us set

$$\boldsymbol{A} = \partial \boldsymbol{f} \quad \text{and} \quad \boldsymbol{B} \colon \boldsymbol{\mathscr{H}} \to \boldsymbol{\mathscr{H}} \colon \boldsymbol{x} \mapsto \big(\nabla_1 \boldsymbol{g}_1(\boldsymbol{x}), \ldots, \nabla_m \boldsymbol{g}_m(\boldsymbol{x})\big). \tag{9.19}$$

We consider the inclusion problem

$$\text{find} \quad \boldsymbol{x} \in \operatorname{zer}(\boldsymbol{A}+\boldsymbol{B}). \tag{9.20}$$

Since $\boldsymbol{f} \in \Gamma_0(\boldsymbol{\mathscr{H}})$, $\boldsymbol{A}$ is maximally monotone. On the other hand, it follows from (9.2) that $\boldsymbol{B}$ is monotone. The following result establishes a connection between the monotone inclusion problem (9.20) and Problem 9.1.

Proposition 9.2. *Using the notation and hypotheses of Problem 9.1, let* $\boldsymbol{A}$ *and* $\boldsymbol{B}$ *be as in* (9.19)*. Then every point in* $\operatorname{zer}(\boldsymbol{A}+\boldsymbol{B})$ *is a solution to Problem 9.1.*

Proof. Suppose that $\operatorname{zer}(\boldsymbol{A}+\boldsymbol{B}) \neq \varnothing$ and let $(x_1,\ldots,x_m) \in \boldsymbol{\mathscr{H}}$. Then [5, Proposition 16.6] asserts that

$$\boldsymbol{A}(x_1,\ldots,x_m) \subset \partial\big(\boldsymbol{f}(\cdot,x_2,\ldots,x_m)\big)(x_1) \times \cdots \times \partial\big(\boldsymbol{f}(x_1,\ldots,x_{m-1},\cdot)\big)(x_m). \tag{9.21}$$

Hence, since $\operatorname{dom}\boldsymbol{g}_1(\cdot,x_2,\ldots,x_m) = \mathscr{H}_1$, ..., $\operatorname{dom}\boldsymbol{g}_m(x_1,\ldots,x_{m-1},\cdot) = \mathscr{H}_m$, we derive from (9.19), (9.14), and [5, Corollary 16.38(iii)] that

$$\begin{aligned}
&(x_1,\ldots,x_m) \in \operatorname{zer}(\boldsymbol{A}+\boldsymbol{B}) \\
&\quad\Leftrightarrow\quad -\boldsymbol{B}(x_1,\ldots,x_m) \in \boldsymbol{A}(x_1,\ldots,x_m) \\
&\quad\Rightarrow\quad \begin{cases} -\nabla_1 \boldsymbol{g}_1(x_1,\ldots,x_m) \in \partial\big(\boldsymbol{f}(\cdot,x_2,\ldots,x_m)\big)(x_1) \\ \qquad\vdots \\ -\nabla_m \boldsymbol{g}_m(x_1,\ldots,x_m) \in \partial\big(\boldsymbol{f}(x_1,\ldots,x_{m-1},\cdot)\big)(x_m) \end{cases} \\
&\quad\Leftrightarrow\quad (x_1,\ldots,x_m) \text{ solves Problem 9.1,}
\end{aligned} \tag{9.22}$$

which yields the result. ∎

Proposition 9.2 asserts that we can solve Problem 9.1 by solving (9.20), provided that the latter has solutions. The following result provides instances in which this property is satisfied. First, we need the following definitions (see [5, Chaps. 21–24]):

Let $A\colon \mathcal{H} \to 2^{\mathcal{H}}$ be monotone. Then A is 3^* monotone if $\operatorname{dom} A \times \operatorname{ran} A \subset \operatorname{dom} F_A$, where

$$F_A\colon \mathcal{H} \times \mathcal{H} \to \left]-\infty,+\infty\right] : (x,u) \mapsto \langle x \mid u\rangle - \inf_{(y,v)\in \operatorname{gra} A} \langle x-y \mid u-v\rangle. \tag{9.23}$$

On the other hand, A is uniformly monotone if there exists an increasing function $\phi\colon \left[0,+\infty\right[\to [0,+\infty]$ vanishing only at 0 such that

$$\big(\forall (x,y) \in \mathcal{H} \times \mathcal{H}\big)\big(\forall (u,v) \in Ax \times Ay\big) \quad \langle x-y \mid u-v\rangle \geq \phi(\|x-y\|). \tag{9.24}$$

A function $\varphi \in \Gamma_0(\mathcal{H})$ is uniformly convex if there exists an increasing function $\phi\colon \left[0,+\infty\right[\to [0,+\infty]$ vanishing only at 0 such that

$$\begin{aligned}&(\forall (x,y) \in \operatorname{dom}\varphi \times \operatorname{dom}\varphi)(\forall \alpha \in \left]0,1\right[)\\ &\qquad \varphi(\alpha x + (1-\alpha)y) + \alpha(1-\alpha)\phi(\|x-y\|) \leq \alpha\varphi(x) + (1-\alpha)\varphi(y).\end{aligned} \tag{9.25}$$

The function ϕ in (9.24) and (9.25) is called the modulus of uniform monotonicity and of uniform convexity, respectively, and it is said to be supercoercive if $\lim_{t\to+\infty} \phi(t)/t = +\infty$.

Proposition 9.3. *With the notation and hypotheses of Problem 9.1, let* $\boldsymbol{B}$ *be as in* (9.19)*. Suppose that* $\boldsymbol{B}$ *is maximally monotone and that one of the following holds:*

(i) $\lim_{|||\boldsymbol{x}|||\to+\infty} \inf |||\partial\boldsymbol{f}(\boldsymbol{x}) + \boldsymbol{B}\boldsymbol{x}||| = +\infty$.
(ii) $\partial\boldsymbol{f} + \boldsymbol{B}$ *is uniformly monotone with a supercoercive modulus.*
(iii) $(\operatorname{dom}\partial\boldsymbol{f}) \cap \operatorname{dom}\boldsymbol{B}$ *is bounded.*
(iv) $\boldsymbol{f} = \iota_{\boldsymbol{C}}$*, where* $\boldsymbol{C}$ *is a nonempty closed convex bounded subset of* $\boldsymbol{\mathcal{H}}$.
(v) $\boldsymbol{f}$ *is uniformly convex with a supercoercive modulus.*
(vi) $\boldsymbol{B}$ *is* 3^* *monotone, and* $\partial\boldsymbol{f}$ *or* $\boldsymbol{B}$ *is surjective.*
(vii) $\boldsymbol{B}$ *is uniformly monotone with a supercoercive modulus.*
(viii) $\boldsymbol{B}$ *is linear and bounded, there exists* $\beta \in \left]0,+\infty\right[$ *such that* $\boldsymbol{B}$ *is* β*–cocoercive, and* $\partial\boldsymbol{f}$ *or* $\boldsymbol{B}$ *is surjective.*

Then $\operatorname{zer}(\partial\boldsymbol{f} + \boldsymbol{B}) \neq \varnothing$*. In addition, if (ii), (v), or (vii) holds,* $\operatorname{zer}(\partial\boldsymbol{f} + \boldsymbol{B})$ *is a singleton.*

Proof. First note that, for every $(x_i)_{1\leq i\leq m} \in \mathcal{H}$, $\operatorname{dom}\nabla_1 g_1(\cdot,x_2,\dots,x_m) = \mathcal{H}_1,\dots,$ $\operatorname{dom}\nabla_m g_m(x_1,\dots,x_{m-1},\cdot) = \mathcal{H}_m$. Hence, it follows from (9.19) that $\operatorname{dom}\boldsymbol{B} = \mathcal{H}$ and, therefore, from [5, Corollary 24.4(i)] that $\partial\boldsymbol{f}+\boldsymbol{B}$ is maximally monotone. In addition, it follows from [5, Example 24.9] that $\partial\boldsymbol{f}$ is 3* monotone.

(i) This follows from [5, Corollary 21.20].
(ii) This follows from [5, Corollary 23.37(i)].
(iii) Since $\operatorname{dom}(\partial\boldsymbol{f}+\boldsymbol{B}) = (\operatorname{dom}\partial\boldsymbol{f})\cap\operatorname{dom}\boldsymbol{B}$, the result follows from [5, Proposition 23.36(iii)].
(iv)⇒(iii) $\boldsymbol{f} = \iota_{\boldsymbol{C}} \in \Gamma_0(\mathcal{H})$ and $\operatorname{dom}\partial\boldsymbol{f} = \boldsymbol{C}$ is bounded.
(v)⇒(ii) It follows from (9.19) and [5, Example 22.3(iii)] that $\partial\boldsymbol{f}$ is uniformly monotone. Hence, $\partial\boldsymbol{f}+\boldsymbol{B}$ is uniformly monotone.
(vi) This follows from [5, Corollary 24.22(ii)].
(vii)⇒(ii) Clear.
(viii)⇒(vi) This follows from [5, Proposition 24.12].

Finally, the uniqueness of a zero of $\partial\boldsymbol{f}+\boldsymbol{B}$ in cases (ii), (v), and (vii) follows from the strict monotonicity of $\partial\boldsymbol{f}+\boldsymbol{B}$. ■

9.3.2 Forward–Backward–Forward Algorithm

Our first method for solving Problem 9.1 is derived from an algorithm proposed in [6], which itself is a variant of a method proposed in [16].

Theorem 9.4. *In Problem 9.1, suppose that there exist* $(z_1,\dots,z_m) \in \mathcal{H}$ *such that*

$$-\big(\nabla_1 g_1(z_1,\dots,z_m),\dots,\nabla_m g_m(z_1,\dots,z_m)\big) \in \partial\boldsymbol{f}(z_1,\dots,z_m) \tag{9.26}$$

and $\chi \in \left]0,+\infty\right[$ *such that*

$$(\forall(x_1,\dots,x_m)\in\mathcal{H})(\forall(y_1,\dots,y_m)\in\mathcal{H})$$
$$\sum_{i=1}^m \|\nabla_i g_i(x_1,\dots,x_m)-\nabla_i g_i(y_1,\dots,y_m)\|^2 \leq \chi^2\sum_{i=1}^m\|x_i-y_i\|^2. \tag{9.27}$$

Let $\varepsilon \in \left]0,1/(\chi+1)\right[$ *and let* $(\gamma_n)_{n\in\mathbb{N}}$ *be a sequence in* $[\varepsilon,(1-\varepsilon)/\chi]$. *Moreover, for every* $i\in\{1,\dots,m\}$, *let* $x_{i,0}\in\mathcal{H}_i$, *and let* $(a_{i,n})_{n\in\mathbb{N}}$, $(b_{i,n})_{n\in\mathbb{N}}$, *and* $(c_{i,n})_{n\in\mathbb{N}}$ *be absolutely summable sequences in* $\mathcal{H}_i$. *Now consider the following routine:*

$$(\forall n \in \mathbb{N}) \quad \left\lfloor \begin{array}{l} \textit{for } i = 1,\dots,m \\ \left\lfloor y_{i,n} = x_{i,n} - \gamma_n(\nabla_i \boldsymbol{g}_i(x_{1,n},\dots,x_{m,n}) + a_{i,n}) \right. \\ (p_{1,n},\dots,p_{m,n}) = \operatorname{prox}_{\gamma_n \boldsymbol{f}}(y_{1,n},\dots,y_{m,n}) + (b_{1,n},\dots,b_{m,n}) \\ \textit{for } i = 1,\dots,m \\ \left\lfloor \begin{array}{l} q_{i,n} = p_{i,n} - \gamma_n(\nabla_i \boldsymbol{g}_i(p_{1,n},\dots,p_{m,n}) + c_{i,n}) \\ x_{i,n+1} = x_{i,n} - y_{i,n} + q_{i,n}. \end{array} \right. \end{array} \right. \tag{9.28}$$

Then there exists a solution $(\overline{x}_1,\dots,\overline{x}_m)$ to Problem 9.1 such that, for every $i \in \{1,\dots,m\}$, $x_{i,n} \rightharpoonup \overline{x}_i$ and $p_{i,n} \rightharpoonup \overline{x}_i$.

Proof. Let $\boldsymbol{A}$ and $\boldsymbol{B}$ be defined as (9.19). Then (9.26) yields $\operatorname{zer}(\boldsymbol{A}+\boldsymbol{B}) \neq \varnothing$, and, for every $\gamma \in \,]0,+\infty[$, (9.14) yields $J_{\gamma \boldsymbol{A}} = \operatorname{prox}_{\gamma \boldsymbol{f}}$. In addition, we deduce from (9.2) and (9.27) that $\boldsymbol{B}$ is monotone and χ-Lipschitzian. Now set

$$(\forall n \in \mathbb{N}) \quad \begin{cases} \boldsymbol{x}_n = (x_{1,n},\dots,x_{m,n}) \\ \boldsymbol{y}_n = (y_{1,n},\dots,y_{m,n}) \\ \boldsymbol{p}_n = (p_{1,n},\dots,p_{m,n}) \\ \boldsymbol{q}_n = (q_{1,n},\dots,q_{m,n}) \end{cases} \tag{9.29}$$

and

$$(\forall n \in \mathbb{N}) \quad \begin{cases} \boldsymbol{a}_n = (a_{1,n},\dots,a_{m,n}) \\ \boldsymbol{b}_n = (b_{1,n},\dots,b_{m,n}) \\ \boldsymbol{c}_n = (c_{1,n},\dots,c_{m,n}). \end{cases} \tag{9.30}$$

Then (9.28) is equivalent to

$$(\forall n \in \mathbb{N}) \quad \left\lfloor \begin{array}{l} \boldsymbol{y}_n = \boldsymbol{x}_n - \gamma_n(\boldsymbol{B}\boldsymbol{x}_n + \boldsymbol{a}_n) \\ \boldsymbol{p}_n = J_{\gamma_n \boldsymbol{A}} \boldsymbol{y}_n + \boldsymbol{b}_n \\ \boldsymbol{q}_n = \boldsymbol{p}_n - \gamma_n(\boldsymbol{B}\boldsymbol{p}_n + \boldsymbol{c}_n) \\ \boldsymbol{x}_{n+1} = \boldsymbol{x}_n - \boldsymbol{y}_n + \boldsymbol{q}_n. \end{array} \right. \tag{9.31}$$

Thus, the result follows from [6, Theorem 2.5(ii)] and Proposition 9.2. ∎

Note that two (forward) gradient steps involving the individual penalties $(\boldsymbol{g}_i)_{1 \le i \le m}$ and one (backward) proximal step involving the common penalty $\boldsymbol{f}$ are required at each iteration of (9.28).

9.3.3 Forward–Backward Algorithm

Our second method for solving Problem 9.1 is somewhat simpler than (9.28) but requires stronger hypotheses on $(\boldsymbol{g}_i)_{1\leq i\leq m}$. This method is an application of the forward–backward splitting algorithm (see [3, 9] and the references therein for background).

Theorem 9.5. *In Problem 9.1, suppose that there exist* $(z_1,\dots,z_m)\in\boldsymbol{\mathcal{H}}$ *such that*

$$-\big(\nabla_1\boldsymbol{g}_1(z_1,\dots,z_m),\dots,\nabla_m\boldsymbol{g}_m(z_1,\dots,z_m)\big)\in\partial\boldsymbol{f}(z_1,\dots,z_m) \tag{9.32}$$

and $\chi\in\,]0,+\infty[$ *such that*

$$\begin{aligned}&(\forall(x_1,\dots,x_m)\in\boldsymbol{\mathcal{H}})(\forall(y_1,\dots,y_m)\in\boldsymbol{\mathcal{H}})\\ &\quad\sum_{i=1}^m\langle\nabla_i\boldsymbol{g}_i(x_1,\dots,x_m)-\nabla_i\boldsymbol{g}_i(y_1,\dots,y_m)\mid x_i-y_i\rangle\\ &\qquad\qquad\geq\frac{1}{\chi}\sum_{i=1}^m\|\nabla_i\boldsymbol{g}_i(x_1,\dots,x_m)-\nabla_i\boldsymbol{g}_i(y_1,\dots,y_m)\|^2.\end{aligned} \tag{9.33}$$

Let $\varepsilon\in\,]0,2/(\chi+1)[$ *and let* $(\gamma_n)_{n\in\mathbb{N}}$ *be a sequence in* $[\varepsilon,(2-\varepsilon)/\chi]$. *Moreover, for every* $i\in\{1,\dots,m\}$, *let* $x_{i,0}\in\mathcal{H}_i$, *and let* $(a_{i,n})_{n\in\mathbb{N}}$ *and* $(b_{i,n})_{n\in\mathbb{N}}$ *be absolutely summable sequences in* $\mathcal{H}_i$. *Now consider the following routine:*

$$(\forall n\in\mathbb{N})\quad\left\lfloor\begin{array}{l}\textit{for } i=1,\dots,m\\ \lfloor\, y_{i,n}=x_{i,n}-\gamma_n(\nabla_i\boldsymbol{g}_i(x_{1,n},\dots,x_{m,n})+a_{i,n})\\ (x_{1,n+1},\dots,x_{m,n+1})=\operatorname{prox}_{\gamma_n\boldsymbol{f}}(y_{1,n},\dots,y_{m,n})+(b_{1,n},\dots,b_{m,n}).\end{array}\right. \tag{9.34}$$

Then there exists a solution $(\overline{x}_1,\dots,\overline{x}_m)$ *to Problem 9.1 such that, for every* $i\in\{1,\dots,m\}$, $x_{i,n}\rightharpoonup\overline{x}_i$ *and* $\nabla_i\boldsymbol{g}_i(x_{1,n},\dots,x_{m,n})\to\nabla_i\boldsymbol{g}_i(\overline{x}_1,\dots,\overline{x}_m)$.

Proof. If we define $\boldsymbol{A}$ and $\boldsymbol{B}$ as in (9.19), (9.32) is equivalent to $\operatorname{zer}(\boldsymbol{A}+\boldsymbol{B})\neq\varnothing$, and it follows from (9.33) that $\boldsymbol{B}$ is χ^{-1}–cocoercive. Moreover, (9.34) can be recast as

$$(\forall n\in\mathbb{N})\quad\left\lfloor\begin{array}{l}\boldsymbol{y}_n=\boldsymbol{x}_n-\gamma_n(\boldsymbol{B}\boldsymbol{x}_n+\boldsymbol{a}_n)\\ \boldsymbol{x}_{n+1}=J_{\gamma_n\boldsymbol{A}}\boldsymbol{y}_n+\boldsymbol{b}_n.\end{array}\right. \tag{9.35}$$

The result hence follows from Proposition 9.2 and [3, Theorem 2.8(i) and (ii)]. ∎

As illustrated in the following example, Theorem 9.5 imposes more restrictions on $(\boldsymbol{g}_i)_{1\leq i\leq m}$. However, unlike the forward–backward–forward algorithm used in Sect. 9.3.2, it employs only one forward step at each iteration. In addition, this method allows for larger gradient steps since the sequence $(\gamma_n)_{n\in\mathbb{N}}$ lies in $]0,2/\chi[$, as opposed to $]0,1/\chi[$ in Theorem 9.4.

Example 9.6. In Problem 9.1, set $m = 2$, let $L\colon \mathcal{H}_1 \to \mathcal{H}_2$ be linear and bounded, and set

$$\begin{cases} \boldsymbol{g}_1\colon (x_1,x_2) \mapsto \langle Lx_1 \mid x_2\rangle \\ \boldsymbol{g}_2\colon (x_1,x_2) \mapsto -\langle Lx_1 \mid x_2\rangle. \end{cases} \tag{9.36}$$

It is readily checked that all the assumptions of Problem 9.1 are satisfied, as well as (9.27) with $\chi = \|L\|$. However, (9.33) does not hold since

$$\begin{aligned} &(\forall (x_1,x_2) \in \mathcal{H}_1 \oplus \mathcal{H}_2)(\forall (y_1,y_2) \in \mathcal{H}_1 \oplus \mathcal{H}_2) \\ &\qquad \langle \nabla_1 \boldsymbol{g}_1(x_1,x_2) - \nabla_1 \boldsymbol{g}_1(y_1,y_2) \mid x_1 - y_1\rangle \\ &\qquad +\langle \nabla_2 \boldsymbol{g}_2(x_1,x_2) - \nabla_2 \boldsymbol{g}_2(y_1,y_2) \mid x_2 - y_2\rangle = 0. \end{aligned} \tag{9.37}$$

9.4 Applications

The previous results can be used to solve a wide variety of instances of Problem 9.1. We discuss several examples.

9.4.1 Saddle Functions and Zero-Sum Games

We consider an instance of Problem 9.1 with $m = 2$ players whose individual penalties $\boldsymbol{g}_1$ and $\boldsymbol{g}_2$ are saddle functions.

Example 9.7. Let $\chi \in \left]0,+\infty\right[$, let $\boldsymbol{f} \in \Gamma_0(\mathcal{H}_1 \oplus \mathcal{H}_2)$, and let $\boldsymbol{\mathcal{L}}\colon \mathcal{H}_1 \oplus \mathcal{H}_2 \to \mathbb{R}$ be a differentiable function with a χ-Lipschitzian gradient such that, for every $x_1 \in \mathcal{H}_1$, $\boldsymbol{\mathcal{L}}(x_1,\cdot)$ is concave and, for every $x_2 \in \mathcal{H}_2$, $\boldsymbol{\mathcal{L}}(\cdot,x_2)$ is convex. The problem is to find $x_1 \in \mathcal{H}_1$ and $x_2 \in \mathcal{H}_2$ such that

$$\begin{cases} x_1 \in \underset{x \in \mathcal{H}_1}{\operatorname{Argmin}}\; \boldsymbol{f}(x,x_2) + \boldsymbol{\mathcal{L}}(x,x_2) \\ x_2 \in \underset{x \in \mathcal{H}_2}{\operatorname{Argmin}}\; \boldsymbol{f}(x_1,x) - \boldsymbol{\mathcal{L}}(x_1,x). \end{cases} \tag{9.38}$$

Proposition 9.8. *In Example 9.7, suppose that there exists $(z_1,z_2) \in \mathcal{H}_1 \oplus \mathcal{H}_2$ such that*

$$\big(-\nabla_1 \boldsymbol{\mathcal{L}}(z_1,z_2), \nabla_2 \boldsymbol{\mathcal{L}}(z_1,z_2)\big) \in \partial \boldsymbol{f}(z_1,z_2). \tag{9.39}$$

Let $\varepsilon \in \left]0,1/(\chi+1)\right[$ and let $(\gamma_n)_{n\in\mathbb{N}}$ be a sequence in $[\varepsilon,(1-\varepsilon)/\chi]$. Moreover, let $(x_{1,0},x_{2,0}) \in \mathcal{H}_1 \oplus \mathcal{H}_2$, let $(a_{1,n})_{n\in\mathbb{N}}$, $(b_{1,n})_{n\in\mathbb{N}}$, and $(c_{1,n})_{n\in\mathbb{N}}$ be absolutely *summable sequences in* $\mathcal{H}_1$, *and let* $(a_{2,n})_{n\in\mathbb{N}}$, $(b_{2,n})_{n\in\mathbb{N}}$, *and* $(c_{2,n})_{n\in\mathbb{N}}$ *be absolutely summable sequences in* $\mathcal{H}_2$. *Now consider the following routine:*

$$(\forall n \in \mathbb{N}) \quad \left\lfloor \begin{array}{l} y_{1,n} = x_{1,n} - \gamma_n(\nabla_1 \mathscr{L}(x_{1,n}, x_{2,n}) + a_{1,n}) \\ y_{2,n} = x_{2,n} + \gamma_n(\nabla_2 \mathscr{L}(x_{1,n}, x_{2,n}) + a_{2,n}) \\ (p_{1,n}, p_{2,n}) = \mathrm{prox}_{\gamma_n f}(y_{1,n}, y_{2,n}) + (b_{1,n}, b_{2,n}) \\ q_{1,n} = p_{1,n} - \gamma_n(\nabla_1 \mathscr{L}(p_{1,n}, p_{2,n}) + c_{1,n}) \\ q_{2,n} = p_{2,n} + \gamma_n(\nabla_2 \mathscr{L}(p_{1,n}, p_{2,n}) + c_{2,n}) \\ x_{1,n+1} = x_{1,n} - y_{1,n} + q_{1,n} \\ x_{2,n+1} = x_{2,n} - y_{2,n} + q_{2,n}. \end{array} \right. \tag{9.40}$$

Then there exists a solution $(\overline{x}_1, \overline{x}_2)$ *to Example 9.7 such that* $x_{1,n} \rightharpoonup \overline{x}_1$, $p_{1,n} \rightharpoonup \overline{x}_1$, $x_{2,n} \rightharpoonup \overline{x}_2$, *and* $p_{2,n} \rightharpoonup \overline{x}_2$.

Proof. Example 9.7 corresponds to the particular instance of Problem 9.1 in which $m = 2$, $\boldsymbol{g}_1 = \mathscr{L}$, and $\boldsymbol{g}_2 = -\mathscr{L}$. Indeed, it follows from [15, Theorem 1] that the operator

$$(x_1, x_2) \mapsto \big(\nabla_1 \mathscr{L}(x_1, x_2), -\nabla_2 \mathscr{L}(x_1, x_2)\big) \tag{9.41}$$

is monotone in $\mathcal{H}_1 \oplus \mathcal{H}_2$ and, hence, (9.2) holds. In addition, (9.39) implies (9.26) and, since $\nabla \mathscr{L}$ is χ-Lipschitzian, (9.27) holds. Altogether, since (9.28) reduces to (9.40), the result follows from Theorem 9.4. ∎

Next, we examine an application of Proposition 9.8 to 2-player finite zero-sum games.

Example 9.9. We consider a 2-player finite zero-sum game (for complements and background on finite games, see [17]). Let S_1 be the finite set of pure strategies of player 1, with cardinality N_1, and let

$$C_1 = \left\{ (\xi_j)_{1 \leq j \leq N_1} \in [0,1]^{N_1} \;\middle|\; \sum_{j=1}^{N_1} \xi_j = 1 \right\} \tag{9.42}$$

be his set of mixed strategies (S_2, N_2, and C_2 are defined likewise). Moreover, let L be an $N_1 \times N_2$ real cost matrix such that

$$(\exists z_1 \in C_1)(\exists z_2 \in C_2) \quad -Lz_2 \in N_{C_1} z_1 \quad \text{and} \quad L^\top z_1 \in N_{C_2} z_2. \tag{9.43}$$

The problem is to

$$\text{find} \quad x_1 \in \mathbb{R}^{N_1} \quad \text{and} \quad x_2 \in \mathbb{R}^{N_2} \quad \text{such that} \quad \begin{cases} x_1 \in \underset{x \in C_1}{\mathrm{Argmin}}\; x^\top L x_2 \\ x_2 \in \underset{x \in C_2}{\mathrm{Argmax}}\; x_1^\top L x. \end{cases} \tag{9.44}$$

Since the penalty function of player 1 is $(x_1,x_2) \mapsto x_1^\top L x_2$ and the penalty function of player 2 is $(x_1,x_2) \mapsto -x_1^\top L x_2$, (9.44) is a zero-sum game. It corresponds to the particular instance of Example 9.7 in which $\mathcal{H}_1 = \mathbb{R}^{N_1}$, $\mathcal{H}_2 = \mathbb{R}^{N_2}$, $\boldsymbol{f}\colon (x_1,x_2) \mapsto \iota_{C_1}(x_1) + \iota_{C_2}(x_2)$, and $\boldsymbol{\mathcal{L}}\colon (x_1,x_2) \mapsto x_1^\top L x_2$. Indeed, since C_1 and C_2 are nonempty closed convex sets, $\boldsymbol{f} \in \Gamma_0(\mathcal{H}_1 \oplus \mathcal{H}_2)$. Moreover, $x_1 \mapsto \boldsymbol{\mathcal{L}}(x_1,x_2)$ and $x_2 \mapsto -\boldsymbol{\mathcal{L}}(x_1,x_2)$ are convex, and $\nabla\boldsymbol{\mathcal{L}}\colon (x_1,x_2) \mapsto (Lx_2, L^\top x_1)$ is linear and bounded, with $\|\nabla\boldsymbol{\mathcal{L}}\| = \|L\|$. In addition, for every $\gamma \in \left]0,+\infty\right[$, $\operatorname{prox}_{\gamma\boldsymbol{f}} = (P_{C_1}, P_{C_2})$ [5, Proposition 23.30]. Hence, (9.40) reduces to (we set the error terms to zero for simplicity)

$$
(\forall n \in \mathbb{N}) \quad
\left\lfloor
\begin{array}{l}
y_{1,n} = x_{1,n} - \gamma_n L x_{2,n} \\
y_{2,n} = x_{2,n} + \gamma_n L^\top x_{1,n} \\
p_{1,n} = P_{C_1} y_{1,n} \\
p_{2,n} = P_{C_2} y_{2,n} \\
q_{1,n} = p_{1,n} - \gamma_n L p_{2,n} \\
q_{2,n} = p_{2,n} + \gamma_n L^\top p_{1,n} \\
x_{1,n+1} = x_{1,n} - y_{1,n} + q_{1,n} \\
x_{2,n+1} = x_{2,n} - y_{2,n} + q_{2,n},
\end{array}
\right.
\tag{9.45}
$$

where $(\gamma_n)_{n\in\mathbb{N}}$ is a sequence in $\left[\varepsilon, \frac{1-\varepsilon}{\|L\|}\right]$ for some arbitrary $\varepsilon \in \left]0, \frac{1}{\|L\|+1}\right[$. Since $\partial\boldsymbol{f}\colon (x_1,x_2) \mapsto N_{C_1}x_1 \times N_{C_2}x_2$, (9.43) yields (9.39). Altogether, Proposition 9.8 asserts that the sequence $(x_{1,n},x_{2,n})_{n\in\mathbb{N}}$ generated by (9.45) converges to $(\overline{x}_1,\overline{x}_2) \in \mathbb{R}^{N_1} \times \mathbb{R}^{N_2}$, such that $(\overline{x}_1,\overline{x}_2)$ is a solution to (9.44).

9.4.2 Generalized Nash Equilibria

We consider the particular case of Problem 9.1 in which $\boldsymbol{f}$ is the indicator function of a closed convex subset of $\boldsymbol{\mathcal{H}} = \mathcal{H}_1 \oplus \cdots \oplus \mathcal{H}_m$.

Example 9.10. Let $\boldsymbol{C} \subset \boldsymbol{\mathcal{H}}$ be a nonempty closed convex set and, for every $i \in \{1,\ldots,m\}$, let $\boldsymbol{g}_i\colon \boldsymbol{\mathcal{H}} \to \left]-\infty,+\infty\right]$ be a function which is differentiable with respect to its ith variable. Suppose that

$$
(\forall(x_1,\ldots,x_m) \in \boldsymbol{\mathcal{H}})\,(\forall(y_1,\ldots,y_m) \in \boldsymbol{\mathcal{H}})
\quad \sum_{i=1}^{m} \langle \nabla_i \boldsymbol{g}_i(x_1,\ldots,x_m) - \nabla_i \boldsymbol{g}_i(y_1,\ldots,y_m) \mid x_i - y_i \rangle \geq 0 \tag{9.46}
$$

and set

$$
(\forall(x_1,\ldots,x_m) \in \boldsymbol{\mathcal{H}})
$$

$$\begin{cases} Q_1(x_2,\ldots,x_m) = \{x \in \mathcal{H}_1 \mid (x,x_2,\ldots,x_m) \in C\} \\ \quad\vdots \\ Q_m(x_1,\ldots,x_{m-1}) = \{x \in \mathcal{H}_m \mid (x_1,\ldots,x_{m-1},x) \in C\}. \end{cases} \tag{9.47}$$

The problem is to find $x_1 \in \mathcal{H}_1, \ldots, x_m \in \mathcal{H}_m$ such that

$$\begin{cases} x_1 \in \operatorname*{Argmin}_{x \in Q_1(x_2,\ldots,x_m)} g_1(x,x_2,\ldots,x_m) \\ \quad\vdots \\ x_m \in \operatorname*{Argmin}_{x \in Q_m(x_1,\ldots,x_{m-1})} g_m(x_1,\ldots,x_{m-1},x). \end{cases} \tag{9.48}$$

The solutions to Example 9.10 are called generalized Nash equilibria [11], social equilibria [10], or equilibria of abstract economies [1], and their existence has been studied in [1, 10]. We deduce from Proposition 9.2 that we can find a solution to Example 9.10 by solving a variational inequality in $\mathcal{H}$, provided the latter has solutions. This observation is also made in [11], which investigates a Euclidean setting in which additional smoothness properties are imposed on $(g_i)_{1\leq i\leq m}$. An alternative approach for solving Example 9.10 in Euclidean spaces is also proposed in [13] with stronger differentiability properties on $(g_i)_{1\leq i\leq m}$ and a monotonicity assumption of the form (9.46). However, the convergence of the method is not guaranteed. Below we derive from Sect. 9.3.2 a weakly convergent method for solving Example 9.10.

Proposition 9.11. *In Example 9.10, suppose that there exist $(z_1,\ldots,z_m) \in \mathcal{H}$ such that*

$$-\big(\nabla_1 g_1(z_1,\ldots,z_m),\ldots,\nabla_m g_m(z_1,\ldots,z_m)\big) \in N_C(z_1,\ldots,z_m) \tag{9.49}$$

and $\chi \in \,]0,+\infty[$ such that

$$(\forall (x_1,\ldots,x_m) \in \mathcal{H})(\forall (y_1,\ldots,y_m) \in \mathcal{H})$$
$$\sum_{i=1}^{m} \|\nabla_i g_i(x_1,\ldots,x_m) - \nabla_i g_i(y_1,\ldots,y_m)\|^2 \leq \chi^2 \sum_{i=1}^{m} \|x_i - y_i\|^2. \tag{9.50}$$

Let $\varepsilon \in \,]0,1/(\chi+1)[$ and let $(\gamma_n)_{n\in\mathbb{N}}$ be a sequence in $[\varepsilon,(1-\varepsilon)/\chi]$. Moreover, for every $i \in \{1,\ldots,m\}$, let $x_{i,0} \in \mathcal{H}_i$, and let $(a_{i,n})_{n\in\mathbb{N}}$, $(b_{i,n})_{n\in\mathbb{N}}$, and $(c_{i,n})_{n\in\mathbb{N}}$ be absolutely summable sequences in $\mathcal{H}_i$. Now consider the following routine:

$$(\forall n \in \mathbb{N}) \quad \left\lfloor \begin{array}{l} \text{for } i = 1, \ldots, m \\ \left\lfloor y_{i,n} = x_{i,n} - \gamma_n(\nabla_i \boldsymbol{g}_i(x_{1,n}, \ldots, x_{m,n}) + a_{i,n}) \right. \\ (p_{1,n}, \ldots, p_{m,n}) = P_C(y_{1,n}, \ldots, y_{m,n}) + (b_{1,n}, \ldots, b_{m,n}) \\ \text{for } i = 1, \ldots, m \\ \left\lfloor \begin{array}{l} q_{i,n} = p_{i,n} - \gamma_n(\nabla_i \boldsymbol{g}_i(p_{1,n}, \ldots, p_{m,n}) + c_{i,n}) \\ x_{i,n+1} = x_{i,n} - y_{i,n} + q_{i,n}. \end{array} \right. \end{array} \right. \tag{9.51}$$

Then there exists a solution $(\overline{x}_1, \ldots, \overline{x}_m)$ *to Example 9.10 such that, for every* $i \in \{1, \ldots, m\}$, $x_{i,n} \rightharpoonup \overline{x}_i$ *and* $p_{i,n} \rightharpoonup \overline{x}_i$.

Proof. Example 9.10 corresponds to the particular instance of Problem 9.1 in which $\boldsymbol{f} = \iota_C$. Since $P_C = \operatorname{prox}_{\boldsymbol{f}}$, the result follows from Theorem 9.4. ∎

9.4.3 Cyclic Proximation Problem

We consider the following problem in $\boldsymbol{\mathcal{H}} = \mathcal{H}_1 \oplus \cdots \oplus \mathcal{H}_m$.

Example 9.12. Let $\mathcal{G}$ be a real Hilbert space, let $\boldsymbol{f} \in \Gamma_0(\boldsymbol{\mathcal{H}})$, and, for every $i \in \{1, \ldots, m\}$, let $L_i \colon \mathcal{H}_i \to \mathcal{G}$ be a bounded linear operator. The problem is to find $x_1 \in \mathcal{H}_1, \ldots, x_m \in \mathcal{H}_m$ such that

$$\begin{cases} x_1 & \in \underset{x \in \mathcal{H}_1}{\operatorname{Argmin}}\, \boldsymbol{f}(x, x_2, \ldots, x_m) + \dfrac{1}{2}\|L_1 x - L_2 x_2\|^2 \\ x_2 & \in \underset{x \in \mathcal{H}_2}{\operatorname{Argmin}}\, \boldsymbol{f}(x_1, x, \ldots, x_m) + \dfrac{1}{2}\|L_2 x - L_3 x_3\|^2 \\ & \vdots \\ x_m & \in \underset{x \in \mathcal{H}_m}{\operatorname{Argmin}}\, \boldsymbol{f}(x_1, \ldots, x_{m-1}, x) + \dfrac{1}{2}\|L_m x - L_1 x_1\|^2. \end{cases} \tag{9.52}$$

For every $i \in \{1, \ldots, m\}$, the individual penalty function of player i models his desire to keep some linear transformation L_i of his strategy close to some linear transformation of that of the next player $i+1$. In the particular case when $\boldsymbol{f} \colon (x_i)_{1 \le i \le m} \mapsto \sum_{i=1}^m f_i(x_i)$, a similar formulation is studied in [2, Sect. 3.1], where an algorithm is proposed for solving (9.52). However, each step of the algorithm involves the proximity operator of a sum of convex functions, which is extremely difficult to implement numerically. The method described below circumvents this difficulty.

Proposition 9.13. *In Example 9.12, suppose that there exists* $(z_1, \ldots, z_m) \in \boldsymbol{\mathcal{H}}$ *such that*

$$\big(L_1^*(L_2 z_2 - L_1 z_1), \ldots, L_m^*(L_1 z_1 - L_m z_m)\big) \in \partial \boldsymbol{f}(z_1, \ldots, z_m). \tag{9.53}$$

Set $\chi = 2\max_{1\leq i\leq m}\|L_i\|^2$, *let* $\varepsilon \in \,]0,2/(\chi+1)[$, *and let* $(\gamma_n)_{n\in\mathbb{N}}$ *be a sequence in* $[\varepsilon,(2-\varepsilon)/\chi]$. *For every* $i\in\{1,\dots,m\}$, *let* $x_{i,0}\in\mathcal{H}_i$, *and let* $(a_{i,n})_{n\in\mathbb{N}}$ *and* $(b_{i,n})_{n\in\mathbb{N}}$ *be absolutely summable sequences in* $\mathcal{H}_i$. *Now set* $L_{m+1}=L_1$, *for every* $n\in\mathbb{N}$, *set* $x_{m+1,n}=x_{1,n}$, *and consider the following routine:*

$$(\forall n\in\mathbb{N})\quad \left\lfloor \begin{array}{l} \text{for } i=1,\dots,m \\ \lfloor\, y_{i,n}=x_{i,n}-\gamma_n\big(L_i^*(L_ix_{i,n}-L_{i+1}x_{i+1,n})+a_{i,n}\big) \\ (x_{1,n+1},\dots,x_{m,n+1})=\operatorname{prox}_{\gamma_n f}(y_{1,n},\dots,y_{m,n})+(b_{1,n},\dots,b_{m,n}). \end{array}\right. \tag{9.54}$$

Then there exists a solution $(\overline{x}_1,\dots,\overline{x}_m)$ *to Example 9.12 such that, for every* $i\in\{1,\dots,m\}$, $x_{i,n}\rightharpoonup\overline{x}_i$ *and* $L_i^*\big(L_i(x_{i,n}-\overline{x}_i)-L_{i+1}(x_{i+1,n}-\overline{x}_{i+1})\big)\to 0$.

Proof. Note that Example 9.12 corresponds to the particular instance of Problem 9.1 in which, for every $i\in\{1,\dots,m\}$, $\boldsymbol{g}_i\colon (x_i)_{1\leq i\leq m}\mapsto \|L_ix_i-L_{i+1}x_{i+1}\|^2/2$, where we set $x_{m+1}=x_1$. Indeed, since

$$(\forall (x_1,\dots,x_m)\in\boldsymbol{\mathcal{H}})\quad \begin{cases} \nabla_1\boldsymbol{g}_1(x_1,\dots,x_m) &= L_1^*(L_1x_1-L_2x_2)\\ &\ \ \vdots\\ \nabla_m\boldsymbol{g}_m(x_1,\dots,x_m) &= L_m^*(L_mx_m-L_1x_1), \end{cases} \tag{9.55}$$

the operator $(x_i)_{1\leq i\leq m}\mapsto(\nabla_i\boldsymbol{g}_i(x_1,\dots,x_m))_{1\leq i\leq m}$ is linear and bounded. Thus, for every $(x_1,\dots,x_m)\in\boldsymbol{\mathcal{H}}$,

$$\begin{aligned} &\sum_{i=1}^m\langle\nabla_i\boldsymbol{g}_i(x_1,\dots,x_m)\mid x_i\rangle\\ &\qquad=\sum_{i=1}^m\langle L_i^*(L_ix_i-L_{i+1}x_{i+1})\mid x_i\rangle\\ &\qquad=\sum_{i=1}^m\langle L_ix_i-L_{i+1}x_{i+1}\mid L_ix_i\rangle\\ &\qquad=\sum_{i=1}^m\|L_ix_i\|^2-\sum_{i=1}^m\langle L_{i+1}x_{i+1}\mid L_ix_i\rangle\\ &\qquad=\frac12\sum_{i=1}^m\|L_ix_i\|^2+\frac12\sum_{i=1}^m\|L_{i+1}x_{i+1}\|^2-\sum_{i=1}^m\langle L_{i+1}x_{i+1}\mid L_ix_i\rangle\\ &\qquad=\sum_{i=1}^m\frac12\|L_ix_i-L_{i+1}x_{i+1}\|^2\\ &\qquad=\sum_{i=1}^m\frac{1}{2\|L_i\|^2}\|L_i\|^2\|L_ix_i-L_{i+1}x_{i+1}\|^2 \end{aligned}$$

$$\geq \chi^{-1} \sum_{i=1}^{m} \|L_i^*(L_i x_i - L_{i+1} x_{i+1})\|^2$$

$$= \chi^{-1} \sum_{i=1}^{m} \|\nabla_i \boldsymbol{g}_i(x_1, \ldots, x_m)\|^2, \tag{9.56}$$

and, hence, (9.33) and (9.2) hold. In addition, (9.53) yields (9.32). Altogether, since (9.34) reduces to (9.54), the result follows from Theorem 9.5. ∎

We present below an application of Proposition 9.13 to cyclic proximation problems and, in particular, to cyclic projection problems.

Example 9.14. We apply Example 9.12 to cyclic evaluations of proximity operators. For every $i \in \{1,\ldots,m\}$, let $\mathcal{H}_i = \mathcal{H}$, let $f_i \in \Gamma_0(\mathcal{H})$, let $L_i = \mathrm{Id}$, and set $\boldsymbol{f}\colon (x_i)_{1\leq i\leq m} \mapsto \sum_{i=1}^m f_i(x_i)$. In view of (9.12), Example 9.12 reduces to finding $x_1 \in \mathcal{H}, \ldots, x_m \in \mathcal{H}$ such that

$$\begin{cases} x_1 = \operatorname{prox}_{f_1} x_2 \\ x_2 = \operatorname{prox}_{f_2} x_3 \\ \quad\vdots \\ x_m = \operatorname{prox}_{f_m} x_1. \end{cases} \tag{9.57}$$

It is assumed that (9.57) has at least one solution. Since $\operatorname{prox}_{\boldsymbol{f}}\colon (x_i)_{1\leq i\leq m} \mapsto (\operatorname{prox}_{f_i} x_i)_{1\leq i\leq m}$ [5, Proposition 23.30], (9.54) becomes (we set errors to zero for simplicity)

$$(\forall n \in \mathbb{N}) \quad \left\lfloor \begin{array}{l} \text{for } i = 1,\ldots,m \\ \lfloor x_{i,n+1} = \operatorname{prox}_{\gamma_n f_i}\big((1-\gamma_n)x_{i,n} + \gamma_n x_{i+1,n}\big), \end{array} \right. \tag{9.58}$$

where $(x_{i,0})_{1\leq i\leq m} \in \mathcal{H}^m$ and $(\gamma_n)_{n\in\mathbb{N}}$ is a sequence in $[\varepsilon, 1-\varepsilon]$ for some arbitrary $\varepsilon \in \left]0, 1/2\right[$. Proposition 9.13 asserts that the sequences $(x_{1,n})_{n\in\mathbb{N}}, \ldots, (x_{m,n})_{n\in\mathbb{N}}$ generated by (9.58) converge weakly to points $\overline{x}_1 \in \mathcal{H}, \ldots, \overline{x}_m \in \mathcal{H}$, respectively, such that $(\overline{x}_1, \ldots, \overline{x}_m)$ is a solution to (9.57).

In the particular case when for every $i \in \{1,\ldots,m\}$, $f_i = \iota_{C_i}$, a solution of (9.57) represents a cycle of points in $C_1, \ldots, C_m$. It can be interpreted as a Nash equilibrium of the game in which, for every $i \in \{1,\ldots,m\}$, the strategies of player i belong to C_i and its penalty function is $(x_i)_{1\leq i\leq m} \mapsto \|x_i - x_{i+1}\|^2$, that is, player i wants to have strategies as close as possible to the strategies of player $i+1$. Such schemes go back at least to [12]. It has recently been proved [4] that, in this case, if $m > 2$, the cycles are not minimizers of any potential, from which we infer that this problem cannot be reduced to a potential game. Note that (9.58) becomes

$$(\forall n \in \mathbb{N}) \quad \left\lfloor \begin{array}{l} \text{for } i = 1,\ldots,m \\ \lfloor x_{i,n+1} = P_{C_i}\big((1-\gamma_n)x_{i,n} + \gamma_n x_{i+1,n}\big), \end{array} \right. \tag{9.59}$$

and the sequences $(x_{1,n})_{n\in\mathbb{N}},\ldots,(x_{m,n})_{n\in\mathbb{N}}$ thus generated converge weakly to points $\overline{x}_1 \in \mathscr{H},\ldots,\overline{x}_m \in \mathscr{H}$, respectively, such that $(\overline{x}_1,\ldots,\overline{x}_m)$ is a cycle. The existence of cycles has been proved in [12] when one of the sets $C_1,\ldots,C_m$ is bounded. Thus, (9.59) is an alternative parallel algorithm to the method of successive projections [12].

Acknowledgements This work was supported by the Agence Nationale de la Recherche under grant ANR-08-BLAN-0294-02.

References

1. Arrow, K.J., Debreu, G.: Existence of an equilibrium for a competitive economy. Econometrica **22**, 265–290 (1954)
2. Attouch, H., Bolte, J., Redont, P., Soubeyran, A.: Alternating proximal algorithms for weakly coupled convex minimization problems: Applications to dynamical games and PDE's. J. Convex Anal. **15**, 485–506 (2008)
3. Attouch, H., Briceño-Arias, L.M., Combettes, P.L.: A parallel splitting method for coupled monotone inclusions. SIAM J. Control Optim. **48**, 3246–3270 (2010)
4. Baillon, J.-B., Combettes, P.L., Cominetti, R.: There is no variational characterization of the cycles in the method of periodic projections. J. Func. Anal. **262**, 400–408 (2012)
5. Bauschke, H.H., Combettes, P.L.: Convex Analysis and Monotone Operator Theory in Hilbert Spaces. Springer, New York (2011)
6. Briceño-Arias, L.M., Combettes, P.L.: A monotone+skew splitting model for composite monotone inclusions in duality. SIAM J. Optim. **21**, 1230–1250 (2011)
7. Briceño-Arias, L.M., Combettes, P.L., Pesquet, J.-C., Pustelnik, N.: Proximal algorithms for multicomponent image processing. J. Math. Imaging Vision **41**, 3–22 (2011)
8. Cohen, G.: Nash equilibria: gradient and decomposition algorithms. Large Scale Syst. **12**, 173–184 (1987)
9. Combettes, P.L.: Solving monotone inclusions via compositions of nonexpansive averaged operators. Optimization **53**, 475–504 (2004)
10. Debreu, G.: A social equilibrium existence theorem. Proc. Natl. Acad. Sci. USA **38**, 886–893 (1952)
11. Facchinei, F., Kanzow, C.: Generalized Nash equilibrium problems. Ann. Oper. Res. **175**, 177–211 (2010)
12. Gubin, L.G., Polyak, B.T., Raik, E.V.: The method of projections for finding the common point of convex sets. Comput. Math. Math. Phys. **7**, 1–24 (1967)
13. Von Heusinger, A., Kanzow, C.: Relaxation methods for generalized Nash equilibrium problems with inexact line search. J. Optim. Theory Appl. **143**, 159–183 (2009)
14. Monderer, D., Shapley, L.S.: Potential games. Games Econom. Behav. **14**, 124–143 (1996)
15. Rockafellar, R.T.: Monotone operators associated with saddle-functions and minimax problems. In: Browder, F.E. (ed.) Nonlinear Functional Analysis, Part 1. Proceedings of Symposium on Pure Mathematics, vol. 18, pp. 241–250. American Mathematical Society, Providence (1970)
16. Tseng, P.: A modified forward-backward splitting method for maximal monotone mappings. SIAM J. Control Optim. **38**, 431–446 (2000)
17. Weibull, J.W.: Evolutionary Game Theory. MIT Press, Cambridge (1995)

Chapter 10
Compactness, Optimality, and Risk

B. Cascales, J. Orihuela, and M. Ruiz Galán

Dedicated to Jonathan Borwein on the occasion of his 60th birthday

Abstract This is a survey about one of the most important achievements in optimization in Banach space theory, namely, James' weak compactness theorem, its relatives, and its applications. We present here a good number of topics related to James' weak compactness theorem and try to keep the technicalities needed as simple as possible: Simons' inequality is our preferred tool. Besides the expected applications to measures of weak noncompactness, compactness with respect to boundaries, size of sets of norm-attaining functionals, etc., we also exhibit other very recent developments in the area. In particular we deal with functions and their level sets to study a new Simons' inequality on unbounded sets that appear as the epigraph of some fixed function f. Applications to variational problems for f and to risk measures associated with its Fenchel conjugate f^* are studied.

Key words: Compactness • I-generation • Measure of non-weak compactness • Nonattaining functionals • Optimization • Reflexivity • Risk measures • Simons' inequality • Variational problems

---

COMMUNICATED BY JON D. VANDERWERFF.

B. Cascales (✉) • J. Orihuela
Department of Mathematics,University of Murcia,Campus Espinardo 15, 30100 Murcia, Spain
e-mail: beca@um.es; joseori@um.es

M. Ruiz Galán
Department of Applied Mathematics, E. T. S. Ingeniería. Edificación., c/ Severo Ochoa s/n, 1871 Granada, Spain
e-mail: mruiz@ugr.es

D.H. Bailey et al. (eds.), *Computational and Analytical Mathematics*, Springer Proceedings in Mathematics & Statistics 50, DOI 10.1007/978-1-4614-7621-4_10,

Mathematics Subject Classifications (2010): 46B10, 46B26, 46B50, 46E30, 46N10, 47J20, 49J52, 91B30, 91G10, 91G80

10.1 Introduction

In 1957 James proved that a separable Banach space is reflexive whenever each continuous and linear functional on it attains its supremum on the unit ball; see [82, Theorem 3]. This result was generalized in 1964 to the nonseparable case in [83, Theorem 5]: in what follows we will refer to it as *James' reflexivity theorem*. More generally (and we shall refer to it as to *James' weak compactness theorem*), the following characterization of weak compactness was obtained in [84, Theorem 5]:

Theorem 10.1 (James). *A weakly closed and bounded subset A of a real Banach space is weakly compact if, and only if, every continuous and linear functional attains its supremum on A.*

This central result in Functional Analysis can be extended to complete locally convex spaces, as shown in [84, Theorem 6]. Note that it is not valid in the absence of completeness, as seen in [86]. Since a complex Banach space can be considered naturally as a real Banach space with the same weak topology, James' weak compactness theorem is easily transferred to the complex case. Nonetheless, and because of the strongly real nature of the optimization assumption, the setting for this survey will be that of real Banach spaces.

We refer to [53, 81, 85] for different characterizations of weak compactness.

James' weak compactness theorem has two important peculiarities. The first one is that it has plenty of direct applications as well as it implies a number of important theorems in the setting of Banach spaces. Regarding the latter, we can say that this result is a sort of metatheorem within Functional Analysis. Thus, for instance, the Krein–Šmulian theorem (i.e., the closed convex hull of a weakly compact subset of a Banach space is weakly compact) or the Milman–Pettis theorem (i.e., every uniformly convex Banach space is reflexive) straightforwardly follows from it. Also, the Eberlein–Šmulian theorem, that states that a nonempty subset A of a Banach space E is relatively weakly compact in E if, and only if, it is relatively weakly countably compact in E, can be easily derived from James' weak compactness theorem. Indeed, assume that A is relatively weakly countably compact in E and for a given continuous and linear functional x^* on E, let $\{x_n\}_{n\geq 1}$ be a sequence in A satisfying

$$\lim_n x^*(x_n) = \sup_A x^* \in (-\infty, \infty].$$

If $x_0 \in E$ is a w-cluster point of the sequence $\{x_n\}_{n\geq 1}$, then

$$\sup_A x^* = x^*(x_0) < \infty.$$

The boundedness of A follows from the Banach–Steinhaus theorem, and that A is relatively weakly compact is then a consequence of James' weak compactness theorem.

The second singularity regarding James' weak compactness theorem is that this result not only has attracted the attention of many researchers due to the huge number of its different applications, but also that several authors in the last decades tried to find a reasonable simple proof for it. This search has produced plenty of new important techniques in the area.

Pryce, in [125], simplified the proof of James' weak compactness theorem by using two basic ideas. The first one was to use the Eberlein–Grothendieck double-limit condition (see, for instance, [53, pp. 11–18] or [135, Theorem 28.36]) that states that a bounded subset A of a Banach space E is relatively weakly compact if, and only if,

$$\lim_m \lim_n x_m^*(x_n) = \lim_n \lim_m x_m^*(x_n) \tag{10.1}$$

for all sequences $\{x_n\}_{n\geq 1}$ in A and all bounded sequences $\{x_m^*\}_{m\geq 1}$ in E^* for which the above iterated limits do exist. Pryce's second idea was to use the following diagonal argument.

Lemma 10.2 (Pryce). *Let X be a nonempty set, $\{f_n\}_{n\geq 1}$ a uniformly bounded sequence in $\ell^\infty(X)$, and D a separable subset of $\ell^\infty(X)$. Then there exists a subsequence $\{f_{n_k}\}_{k\geq 1}$ of $\{f_n\}_{n\geq 1}$ such that*

$$\sup_X \left(f - \limsup_k f_{n_k} \right) = \sup_X \left(f - \liminf_k f_{n_k} \right),$$

for every $f \in D$.

We should stress here that from the lemma above it follows that for any further subsequence $\{f_{n_{k_j}}\}_{j\geq 1}$ of $\{f_{n_k}\}_{k\geq 1}$ we also have

$$\sup_X \left(f - \limsup_j f_{n_{k_j}} \right) = \sup_X \left(f - \liminf_j f_{n_{k_j}} \right),$$

for every $f \in D$. With the above tools, Pryce's proof of James' weak compactness theorem is done by contradiction: if a weakly closed and bounded subset A of a Banach space E is not weakly compact, then there exist sequences $\{x_n\}_{n\geq 1}$ and $\{x_m^*\}_{m\geq 1}$ for which (10.1) does not hold. Lemma 10.2 applied to $\{x_m^*\}_{m\geq 1}$ helped Pryce to derive the existence of a continuous linear functional that does not attain its supremum on A. In the text by Holmes [81, Theorem 19.A], one can find Pryce's proof for Banach spaces whose dual unit ball is w^*-sequentially compact: Pryce's original arguments are simplified in this case.

In 1972 Simons gave another simpler proof of James' weak compactness theorem in [137]. The proof by Simons uses an *ad hoc* minimax theorem (with optimization and convexity hypotheses) that follows from a diagonal argument different from that of Pryce above, together with a deep result known henceforth as *Simons' inequality* (see [136, Lemma 2]) that we recall immediately below.

Lemma 10.3 (Simons). *Let $\{f_n\}_{n\geq 1}$ be a uniformly bounded sequence in $\ell^\infty(X)$ and let W be its convex hull. If Y is a subset of X with the property that for every sequence of nonnegative numbers $\{\lambda_n\}_{n\geq 1}$ with $\sum_{n=1}^{\infty}\lambda_n = 1$ there exists $y \in Y$ such that*

$$\sum_{n=1}^{\infty}\lambda_n f_n(y) = \sup\left\{\sum_{n=1}^{\infty}\lambda_n f_n(x) : x \in X\right\},$$

then

$$\inf\left\{\sup_X g : g \in W\right\} \leq \sup_{y\in Y}\left\{\limsup_n f_n(y)\right\}.$$

A converse minimax theorem (see [137, Theorem 15]) (see also [139, Theorem 5.6] and [133, Lemma 18]) provides an easier proof of James' weak compactness theorem and a minimax characterization of weak compactness.

A different proof of James' weak compactness theorem, and even simpler than that in [84], was stated by James himself in [87]. He took into account ideas coming from Simons' inequality in his new proof. The result proved is: *A separable Banach space E is reflexive if, and only if, there exists $\theta \in (0,1)$ such that for every sequence $\{x_n^*\}_{n\geq 1}$ in the unit ball of its dual space, either $\{x_n^*\}_{n\geq 1}$ is not weak*-null or*

$$\inf_{x^*\in C}\|x^*\| < \theta,$$

where C is the convex hull of $\{x_n^*\colon n \geq 1\}$—the characterization of weak compact subsets of a separable Banach spaces is easily guessed by analogy. If the assumption of separability on E is dropped, a similar characterization is obtained, but perturbing the functionals in the convex hull of $\{x_n^*\colon\ n \geq 1\}$ by functionals in the annihilator of a nonreflexive separable subspace X of E: *E is reflexive if, and only if, there exists $\theta \in (0,1)$ such that for each subspace X of E and for every sequence $\{x_n^*\}_{n\geq 1}$ in the unit ball of the dual space of E, either $\{x_n^*\}_{n\geq 1}$ is not null for the topology in E^* of pointwise convergence on X or*

$$\inf_{x^*\in C,\ w\in X^\perp}\|x^* - w\| < \theta,$$

with C being the convex hull of $\{x_n^*\colon n \geq 1\}$.

It should be noted that the new conditions that characterize reflexivity above imply in fact that every continuous and linear functional attains the norm.

In 1974 De Wilde [152] stated yet another proof of James' weak compactness theorem, that basically uses as main tools the diagonal argument of Pryce and the ideas of Simons in [136] together with the Eberlein–Grothendieck double-limit condition.

More recently, Morillon [111] has given a different proof of James' reflexivity theorem, based on a previous result by her [112, Theorem 3.9] establishing, on the one hand, James' reflexivity theorem for spaces with a w^*-block compact dual unit ball by means of Simons' inequality and Rosenthal's ℓ^1-theorem, and extending, on the other hand, the proof to the general case with an adaptation of a result of Hagler and Jonhson [72]. Along with these ideas another proof of James' reflexivity theorem has been given by Kalenda in [92]. Very recently, Pfitzner has gone a step further using the ideas above to solve the so-called *boundary problem* of Godefroy, [59, Question 2]—see Sect. 10.4, giving yet another approach to James' weak compactness theorem [122].

Another approach to James' reflexivity theorem in the separable case is due to Rodé [129], by using his form of the minimax theorem in the setting of the so-called "superconvex analysis." Let us also point out that for separable Banach spaces, the proof in [45, Theorem I.3.2], directly deduced from the Simons inequality, can be considered an easy one. A completely different proof using Bishop–Phelps and Krein–Milman theorems is due to Fonf, Lindenstrauss, and Phelps [56, Theorem 5.9], and an alternative approach is due to Moors [108, Theorem 4]. Nevertheless, the combinatorial principles involved (known in the literature as the (I)-formula) are equivalent to Simons' inequality; see [93, Lemma 2.1 and Remark 2.2] and [35, Theorem 2.2]. We refer the interested reader to the papers by Kalenda [92, 93], where other proofs for James' reflexivity theorem using (I)-envelopes in some special cases can be found.

The leitmotif in this survey is Simons' inequality, which is used, to a large extent, as the main tool for proving the results, most of them self-contained and different from the original ones. Section 10.2 is devoted to the discussion of a generalization of the Simons inequality, where the uniform boundedness condition is relaxed, together with its natural consequences as unbounded sup-limsup's and Rainwater–Simons' theorems. The first part of Sect. 10.3 is devoted to providing a proof of James' weak compactness theorem that, going back to the work of James, explicitly supplies nonattaining functionals in the absence of weak compactness; in the second part of Sect. 10.3 we study several measures of weak noncompactness and we introduce a new one that is very close to Simons' inequality. Section 10.4 deals with the study of boundaries in Banach spaces and some deep related results, that can be viewed as extensions of James' weak compactness theorem. Other extensions of James' weak compactness theorem are presented in Sect. 10.5, where we mainly focus our attention on those of perturbed nature, which have found some applications in mathematical finance and variational analysis, as seen in Sect. 10.6.

Let us note that each section of this paper concludes with a selected open problem.

10.1.1 Notation and Terminology

Most of our notation and terminology are standard, otherwise it is either explained here or when needed: unexplained concepts and terminology can be found in our standard references for Banach spaces [45, 49, 90] and topology [48, 95]. By letters E, K, T, X, etc. we denote sets and sometimes topological spaces. Our topological spaces are assumed to be completely regular.

All vector spaces E that we consider in this paper are assumed to be real. Frequently, E denotes a normed space endowed with a norm $\|\cdot\|$, and E^* stands for its dual space. Given a subset S of a vector space, we write $\operatorname{conv}(S)$ and $\operatorname{span}(S)$ to denote, respectively, the convex and the linear hull of S. If S is a subset of E^*, then $\sigma(E,S)$ denotes the weakest topology for E that makes each member of S continuous, or equivalently, the topology of pointwise convergence on S. Dually, if S is a subset of E, then $\sigma(E^*,S)$ is the topology for E^* of pointwise convergence on S. In particular, $\sigma(E,E^*)$ and $\sigma(E^*,E)$ are the weak (denoted by w) and weak* (denoted by w^*) topologies, respectively. Of course, $\sigma(E,S)$ is always a locally convex topology, that is, Hausdorff if, and only if, $E^* = \overline{\operatorname{span} S}^{w^*}$ (and similarly for $\sigma(E^*,S)$). Given $x^* \in E^*$ and $x \in E$, we write $\langle x^*, x\rangle$ and $x^*(x)$ for the evaluation of x^* at x. If $x \in E$ and $\delta > 0$, we denote by $B(x,\delta)$ (resp. $B[x,\delta]$) the open (resp. closed) ball centered at x of radius δ: we will simplify our notation and just write $B_E := B[0,1]$; the unit sphere $\{x \in E : \|x\| = 1\}$ will be denoted by S_E. Given a nonempty set X and $f \in \mathbb{R}^X$, we write

$$S_X(f) := \sup_{x\in X} f(x) \in (-\infty, \infty].$$

$\ell^\infty(X)$ stands for the Banach space of real-valued bounded functions defined on X, endowed with the supremum norm $S_X(|\cdot|)$.

10.2 Simons' Inequality for Pointwise Bounded Subsets of $\mathbb{R}^X$

The main goal of this section is to derive a generalized version of Simons' inequality, Theorem 10.5, in a pointwise bounded setting, as opposed to the usual uniform bounded context. As a consequence, we derive an unbounded version of the so-called Rainwater–Simons theorem, Corollary 10.7, that will provide us with some generalizations of James' weak compactness theorem, as well as new developments and applications in Sects. 10.5 and 10.6. In addition, the aforementioned result will allow us to present the state of the art of a number of issues related to boundaries in Banach spaces in Sect. 10.4.

The inequality presented in Lemma 10.3, as Simons himself says in [136], is inspired by some of James' and Pryce's arguments in [84, 125] and contains the essence of the proof of James' weak compactness theorem in the separable case.

As mentioned in the Introduction, James included later the novel contribution of Simons in his proof in [87]. We refer to [45, 61] for some applications of Simons' inequality, to [35, 44, 99, 114] for proper extensions, and to [115] for a slightly different proof.

Given a pointwise bounded sequence $\{f_n\}_{n\geq 1}$ in $\mathbb{R}^X$, we define

$$\mathrm{co}_{\sigma_p}\{f_n\colon n\geq 1\} := \left\{\sum_{n=1}^{\infty}\lambda_n f_n\colon \lambda_n\geq 0 \text{ for every } n\geq 1 \text{ and } \sum_{n=1}^{\infty}\lambda_n = 1\right\},$$

where a function of the form $\sum_{n=1}^{\infty}\lambda_n f_n \in \mathbb{R}^X$ is obviously defined by

$$\left(\sum_{n=1}^{\infty}\lambda_n f_n\right)(x) := \sum_{n=1}^{\infty}\lambda_n f_n(x)$$

for every $x\in X$.

Instead of presenting the results of Simons in [136, 138], we adapt them to a pointwise but not necessarily uniformly bounded framework. This adaptation allows us to extend the original results of Simons and provides new applications, as we show below.

The next result follows by arguing as in the "Additive Diagonal Lemma" in [138]. Hereafter, any sum $\sum_{n=1}^{0}\cdots$ is understood to be 0.

Lemma 10.4. *If $\{f_n\}_{n\geq 1}$ is a pointwise bounded sequence in $\mathbb{R}^X$ and $\varepsilon>0$, then for every $m\geq 1$ there exists $g_m\in \mathrm{co}_{\sigma_p}\{f_n\colon n\geq m\}$ such that*

$$S_X\left(\sum_{n=1}^{m-1}\frac{g_n}{2^n}\right)\leq\left(1-\frac{1}{2^{m-1}}\right)S_X\left(\sum_{n=1}^{\infty}\frac{g_n}{2^n}\right)+\frac{\varepsilon}{2^{m-1}}.$$

Proof. It suffices to choose inductively, for each $m\geq 1$, $g_m\in \mathrm{co}_{\sigma_p}\{f_n\colon n\geq m\}$ satisfying

$$S_X\left(\sum_{n=1}^{m-1}\frac{g_n}{2^n}+\frac{g_m}{2^{m-1}}\right)\leq \inf_{g\in \mathrm{co}_{\sigma_p}\{f_n\colon n\geq m\}} S_X\left(\sum_{n=1}^{m-1}\frac{g_n}{2^n}+\frac{g}{2^{m-1}}\right)+\frac{2\varepsilon}{4^m}. \tag{10.2}$$

The existence of such g_m follows from the easy fact that

$$\inf_{g\in \mathrm{co}_{\sigma_p}\{f_n\colon n\geq m\}} S_X(g) > -\infty,$$

according with the pointwise boundedness of our sequence $\{f_n\}_{n\geq 1}$. Since

$$2^{m-1}\sum_{n=m}^{\infty}\frac{g_n}{2^n}\in \mathrm{co}_{\sigma_p}\{f_n\colon n\geq m\},$$

then inequality (10.2) implies

$$S_X\left(\left(\sum_{n=1}^{m-1}\frac{g_n}{2^n}\right)+\frac{g_m}{2^{m-1}}\right)\le S_X\left(\sum_{n=1}^{\infty}\frac{g_n}{2^n}\right)+\frac{2\varepsilon}{4^m}. \tag{10.3}$$

From the equality

$$\sum_{n=1}^{m-1}\frac{g_n}{2^n}=\sum_{k=1}^{m-1}\frac{1}{2^{m-k}}\left(\left(\sum_{n=1}^{k-1}\frac{g_n}{2^n}\right)+\frac{g_k}{2^{k-1}}\right),$$

and the help of (10.3) we finally derive that

$$\begin{aligned}S_X\left(\sum_{n=1}^{m-1}\frac{g_n}{2^n}\right)&\le\sum_{k=1}^{m-1}\frac{1}{2^{m-k}}S_X\left(\left(\sum_{n=1}^{k-1}\frac{g_n}{2^n}\right)+\frac{g_k}{2^{k-1}}\right)\\&\le\sum_{k=1}^{m-1}\frac{1}{2^{m-k}}\left(S_X\left(\sum_{n=1}^{\infty}\frac{g_n}{2^n}\right)+\frac{2\varepsilon}{4^k}\right)\\&=\left(1-\frac{1}{2^{m-1}}\right)S_X\left(\sum_{n=1}^{\infty}\frac{g_n}{2^n}\right)+\left(1-\frac{1}{2^{m-1}}\right)\frac{2\varepsilon}{2^m}\\&\le\left(1-\frac{1}{2^{m-1}}\right)S_X\left(\sum_{n=1}^{\infty}\frac{g_n}{2^n}\right)+\frac{\varepsilon}{2^{m-1}},\end{aligned}$$

and the proof is over. ■

We now arrive at the announced extension of Simons' inequality. Unlike the original work [136], we only assume pointwise boundedness of the sequence $\{f_n\}_{n\geq 1}$. Let us also emphasize that the extension of Simons' inequality stated in [114] is a particular case of the following non uniform version:

Theorem 10.5 (Simons' inequality in $\mathbb{R}^X$). *Let X be a nonempty set, let $\{f_n\}_{n\geq 1}$ be a pointwise bounded sequence in $\mathbb{R}^X$, and let Y be a subset of X such that*

$$\text{for every } g\in \mathrm{co}_{\sigma_p}\{f_n\colon n\geq 1\} \text{ there exists } y\in Y \text{ with } g(y)=S_X(g).$$

Then

$$\inf_{g\in \mathrm{co}_{\sigma_p}\{f_n\colon n\geq 1\}} S_X(g)\le S_Y\left(\limsup_n f_n\right).$$

Proof. It suffices to prove that for every $\varepsilon>0$ there exist $y\in Y$ and $g\in \mathrm{co}_{\sigma_p}\{f_n\colon n\geq 1\}$ such that

$$S_X(g)-\varepsilon\le\limsup_n f_n(y).$$

Fix $\varepsilon > 0$. Then Lemma 10.4 provides us with a sequence $\{g_m\}_{m\geq 1}$ in $\mathbb{R}^X$ such that for every $m \geq 1$, $g_m \in \mathrm{co}_{\sigma_p}\{f_n \colon n \geq m\}$ and

$$S_X\left(\sum_{n=1}^{m-1} \frac{g_n}{2^n}\right) \leq \left(1 - \frac{1}{2^{m-1}}\right) S_X\left(\sum_{n=1}^{\infty} \frac{g_n}{2^n}\right) + \frac{\varepsilon}{2^{m-1}}. \tag{10.4}$$

Let us write $g := \sum_{n=1}^{\infty} \frac{g_n}{2^n} \in \mathrm{co}_{\sigma_p}\{f_n \colon n \geq 1\}$. Then by hypothesis there exists $y \in Y$ with

$$g(y) = S_X(g), \tag{10.5}$$

and so it follows from (10.4) and (10.5) that given $m \geq 1$,

$$\begin{aligned}\left(1 - \frac{1}{2^{m-1}}\right) g(y) + \frac{\varepsilon}{2^{m-1}} &\geq S_X\left(\sum_{n=1}^{m-1} \frac{g_n}{2^n}\right)\\ &\geq \sum_{n=1}^{m-1} \frac{g_n(y)}{2^n}\\ &= g(y) - \sum_{n=m}^{\infty} \frac{g_n(y)}{2^n}.\end{aligned}$$

Therefore,

$$\inf_{m\geq 1} 2^{m-1} \sum_{n=m}^{\infty} \frac{g_n(y)}{2^n} \geq g(y) - \varepsilon. \tag{10.6}$$

Since for every $m \geq 1$ we have $2^{m-1}\sum_{n=m}^{\infty} 2^n = 1$, we conclude that

$$\sup_{n\geq m} f_n(y) \geq 2^{m-1} \sum_{n=m}^{\infty} \frac{g_n(y)}{2^n}.$$

Now, with this last inequality in mind together with (10.5) and (10.6), we arrive at

$$\begin{aligned}\limsup_n f_n(y) &= \inf_{m\geq 1} \sup_{n\geq m} f_n(y)\\ &\geq \inf_{m\geq 1} 2^{m-1} \sum_{n=m}^{\infty} \frac{g_n(y)}{2^n}\\ &\geq g(y) - \varepsilon\\ &= S_X(g) - \varepsilon,\end{aligned}$$

as was to be shown. ∎

Both in the original version of Simons' inequality and in the previous one, a uniform behavior follows from a pointwise one, resembling Mazur's theorem for continuous functions when X is a compact topological space; see [146, Sect. 3, p. 14]. Indeed, it turns out that Simons' inequality tell us that

$$\inf\{\|g\|_\infty : g \in \mathrm{co}\{f_n : n \geq 1\}\} = 0,$$

whenever a uniformly bounded sequence of continuous functions $\{f_n\}_{n\geq 1}$ pointwise converges to zero on a compact space X.

As a consequence of the above version of Simons' inequality we deduce the following generalization of the sup-limsup theorem of Simons [136, Theorem 3] (see also [133, Theorem 7]). This result has recently been stated in [119, Corollary 1], but using the tools in [133].

Corollary 10.6 (Simons' sup-limsup theorem in $\mathbb{R}^X$). *Let X be a nonempty set, let $\{f_n\}_{n\geq 1}$ be a pointwise bounded sequence in $\mathbb{R}^X$, and let Y be a subset of X such that*

$$\text{for every } g \in \mathrm{co}_{\sigma_p}\{f_n : n \geq 1\} \text{ there exists } y \in Y \text{ with } g(y) = S_X(g).$$

Then

$$S_X\left(\limsup_n f_n\right) = S_Y\left(\limsup_n f_n\right).$$

Proof. Let us assume, arguing by reductio ad absurdum, that there exists $x_0 \in X$ such that

$$\limsup_n f_n(x_0) > S_Y\left(\limsup_n f_n\right).$$

We assume then, passing to a subsequence if necessary, that

$$\inf_{n\geq 1} f_n(x_0) > S_Y\left(\limsup_n f_n\right).$$

In particular,

$$\inf_{g\in \mathrm{co}_{\sigma_p}\{f_n : n\geq 1\}} g(x_0) > S_Y\left(\limsup_n f_n\right),$$

and then, by applying Theorem 10.5, we arrive at

$$\begin{aligned} S_Y\left(\limsup_n f_n\right) &\geq \inf_{g\in \mathrm{co}_{\sigma_p}\{f_n : n\geq 1\}} S_X(g) \\ &\geq \inf_{g\in \mathrm{co}_{\sigma_p}\{f_n : n\geq 1\}} g(x_0) \\ &> S_Y\left(\limsup_n f_n\right), \end{aligned}$$

a contradiction. ∎

In the Banach space framework we obtain the sup-limsup's type result below, which also generalizes the so-called Rainwater–Simons theorem; see [136, Corollary 11] (see also [138, Sup-limsup Theorem], [101, Theorem 5.1] and [116, Theorem 2.2], the recent extension [108, Corollary 3], and for some related results [75]). It is a direct consequence of the Simons sup-limsup theorem in $\mathbb{R}^X$, Corollary 10.6, as in the uniform setting; see [51, Theorem 3.134]. In particular it generalizes the Rainwater theorem [127], which asserts that a sequence $\{x_n\}_{n\geq 1}$ in a Banach space E is weakly null if it is bounded and for each extreme point e^* of B_{E^*},

$$\lim_n e^*(x_n) = 0.$$

Given a bounded sequence $\{x_n\}_{n\geq 1}$ in a Banach space E, we define

$$\mathrm{co}_\sigma\{x_n\colon n\geq 1\} := \left\{ \sum_{n=1}^{\infty} \lambda_n x_n\colon \text{ for all } n\geq 1,\ \lambda_n \geq 0 \text{ and } \sum_{n=1}^{\infty}\lambda_n = 1\right\}$$

Note that the above series are clearly norm-convergent and that

$$\mathrm{co}_\sigma\{x_n\colon n\geq 1\} = \mathrm{co}_{\sigma_p}\{x_n\colon n\geq 1\}$$

when for the second set we look at the x_n's as functions defined on B_{E^*}.

Corollary 10.7 (Unbounded Rainwater–Simons' theorem). *If E is a Banach space, C is a subset of E^*, B is a nonempty subset of C, and $\{x_n\}_{n\geq 1}$ is a bounded sequence in E such that*

$$\textit{for every } x\in \mathrm{co}_\sigma\{x_n :\ n\geq 1\} \textit{ there exists } b^*\in B \textit{ with } b^*(x) = S_C(x),$$

then

$$S_B\left(\limsup_n x_n\right) = S_C\left(\limsup_n x_n\right).$$

As a consequence

$$\sigma(E,B)\text{-}\lim_n x_n = 0 \ \Rightarrow\ \sigma(E,C)\text{-}\lim_n x_n = 0.$$

The unbounded Rainwater–Simons theorem (or the Simons inequality in $\mathbb{R}^X$) not only gives as special cases those classical results that follow from Simons's inequality (some of them are discussed here, besides the already mentioned [45,61]), but it also provides new applications whose discussion we delay until the next sections. We only remark here that Moors has recently obtained a particular case of the unbounded Rainwater–Simons theorem (see [108, Corollary 1]), which leads him to a proof of James' weak compactness theorem for Banach spaces whose dual unit ball is w^*-sequentially compact.

A very interesting consequence of Simons' inequality in the bounded case is the (I)-formula (10.7) of Fonf and Lindenstrauss; see [35, 55]:

Corollary 10.8 (Fonf–Lindenstrauss' theorem). *Let E be a Banach space, B a bounded subset of E^* such that for every $x \in E$ there exists some $b_0^* \in B$ satisfying $b_0^*(x) = \sup_{b^* \in B} b^*(x)$. Then we have that, for every covering $B \subset \bigcup_{n=1}^\infty D_n$ by an increasing sequence of w^*-closed convex subsets $D_n \subset \overline{\mathrm{co}(B)}^{w^*}$, the following equality holds true:*

$$\overline{\bigcup_{n=1}^\infty D_n}^{\|\cdot\|} = \overline{\mathrm{co}(B)}^{w^*}. \tag{10.7}$$

Proof. Here is the proof given in [35, Theorem 2.2]. We proceed by contradiction assuming that there exists $z_0^* \in \overline{\mathrm{co}(B)}^{w^*}$ such that $z_0^* \notin \overline{\bigcup_{n=1}^\infty D_n}^{\|\cdot\|}$. Fix $\delta > 0$ such that

$$B[z_0^*, \delta] \cap D_n = \emptyset, \text{ for every } n \geq 1.$$

The separation theorem in (E^*, w^*), when applied to the w^*-compact set $B[0, \delta]$ and the w^*-closed set $D_n - z_0^*$, provides us with a norm-one $x_n \in E$ and $\alpha_n \in \mathbb{R}$ such that

$$\inf_{v^* \in B[0,\delta]} x_n(v^*) > \alpha_n > \sup_{y^* \in D_n} x_n(y^*) - x_n(z_0^*).$$

But

$$-\delta = \inf_{v^* \in B[0,\delta]} x_n(v^*),$$

and consequently the sequence $\{x_n\}_{n \geq 1}$ in B_E satisfies

$$x_n(z_0^*) - \delta > x_n(y^*) \tag{10.8}$$

for each $n \geq 1$ and $y^* \in D_n$. Fix a w^*-cluster point $x^{**} \in B_{E^{**}}$ of the sequence $\{x_n\}_{n \geq 1}$ and let $\{x_{n_k}\}_{k \geq 1}$ be a subsequence of $\{x_n\}_{n \geq 1}$ such that $x^{**}(z_0^*) = \lim_k x_{n_k}(z_0^*)$. We can and do assume that for every $k \geq 1$,

$$x_{n_k}(z_0^*) > x^{**}(z_0^*) - \frac{\delta}{2}. \tag{10.9}$$

Since $B \subset \bigcup_{n=1}^\infty D_n$ and $\{D_n\}_{n \geq 1}$ is an increasing sequence of sets, given $b^* \in B$ there exists $k_0 \geq 1$ such that $b^* \in D_{n_k}$ for each $k \geq k_0$. Now inequality (10.8) yields

$$x^{**}(z_0^*) - \delta \geq \limsup_k x_{n_k}(b^*), \quad \text{for every } b^* \in B, \tag{10.10}$$

and, on the other hand, inequality (10.9) implies that

$$w(z_0^*) \geq x^{**}(z_0^*) - \frac{\delta}{2}, \quad \text{for every } w \in \mathrm{co}_\sigma\{x_{n_k} : k \geq 1\}. \tag{10.11}$$

Now Theorem 10.5 can be applied to the sequence $\{x_{n_k}\}_{k\geq 1}$, to deduce

$$\begin{aligned} x^{**}(z_0^*) - \delta &\overset{(10.10)}{\geq} \sup_{b^* \in B} \limsup_k x_{n_k}(b^*) \geq \\ &\geq \inf\left\{ \sup\{w(z^*) : z^* \in \overline{\mathrm{co}(B)}^{w^*}, w \in \mathrm{co}_\sigma\{x_{n_k} : k \in \mathbb{N}\}\}\right\} \\ &\geq \inf\left\{w(z_0^*) : w \in \mathrm{co}_\sigma\{x_{n_k} : k \in \mathbb{N}\}\right\} \overset{(10.11)}{\geq} x^{**}(z_0^*) - \frac{\delta}{2}. \end{aligned}$$

From the inequalities above we obtain $0 \geq \delta$, which is a contradiction that completes the proof. ■

To conclude this section, let us emphasize that in [35, Theorem 2.2] the equivalence between Simons' inequality, the sup-limsup theorem of Simons, and the (I)-formula of Fonf and Lindenstrauss was established in the bounded case. However, in the unbounded case we propose the following question:

Question 10.9. Are the unbounded versions of Simons' inequality and sup-limsup theorem of Simons equivalent to some kind of I-formula for the unbounded case?

10.3 Nonattaining Functionals

This section is devoted to describe how to obtain nonattaining functionals in the absence of weak compactness. Simons' inequality provides us a first way of doing it in a wide class of Banach spaces, which includes those whose dual unit balls are w^*-sequentially compact. We introduce a new measure of weak noncompactness, tightly connected with Simons' inequality, and we relate it with recent quantification results of classical theorems about weakly compact sets.

When Simons' inequality in $l^\infty(\mathbb{N})$ holds for a w^*-null sequence $\{x_n^*\}_{n\geq 1}$ in a dual Banach space E^*, it follows that the origin belongs to the norm-closed convex hull of the sequence, $\overline{\mathrm{co}\{x_n^* : n \geq 1\}}^{\|\cdot\|}$. Therefore every time we have a w^*-null sequence $\{x_n^*\}_{n\geq 1}$ with $0 \notin \overline{\mathrm{co}\{x_n^* : n \geq 1\}}^{\|\cdot\|}$ we will have some $x_0^* \in \mathrm{co}_\sigma\{x_n^* : n \geq 1\}$ such that x_0^* does not attain its supremum on B_E.

We note that just Simons' inequality, or its equivalent sup-limsup theorem, provides us with the tools to give a simple proof of James' weak compactness theorem for a wide class of Banach spaces. We first recall the following concept:

Definition 10.10. Let E be a vector space, and let $\{x_n\}_{n\geq 1}$ and $\{y_n\}_{n\geq 1}$ be sequences in E. We say that $\{y_n\}_{n\geq 1}$ is a *convex block sequence* of $\{x_n\}_{n\geq 1}$ if for a certain sequence of nonempty finite subsets of integers $\{F_n\}_{n\geq 1}$ with

$$\max F_1 < \min F_2 \leq \max F_2 < \min F_3 \leq \cdots \leq \max F_n < \min F_{n+1} \leq \cdots$$

and adequate sets of positive numbers $\{\lambda_i^n : i \in F_n\} \subset (0,1]$ we have that

$$\sum_{i\in F_n} \lambda_i^n = 1 \quad \text{and} \quad y_n = \sum_{i\in F_n} \lambda_i^n x_i.$$

For a Banach space E, its dual unit ball B_{E^*} is said to be *w^*-convex block compact* provided that each sequence $\{x_n^*\}_{n\geq 1}$ in B_{E^*} has a convex block w^*-convergent sequence.

It is clear that if the dual unit ball B_{E^*} of a Banach space E is w^*-sequentially compact, then it is w^*-convex block compact. This happens, for example, when E is a weakly Lindelöf determined (in short, WLD) Banach space; see [74]. Let us emphasize that both kinds of compactness do not coincide. Indeed, on the one hand, an example of a Banach space with a non w^*-sequentially compact dual unit ball and not containing $\ell^1(\mathbb{N})$ is presented in [73]. On the other hand, it is proved in [24] that if a Banach space E does not contain an isomorphic copy of $\ell^1(\mathbb{N})$, then B_{E^*} is w^*-convex block compact. This last result was extended for spaces not containing an isomorphic copy of $\ell^1(\mathbb{R})$ under Martin Axiom and the negation of the Continuum hypothesis in [80].

For a bounded sequence $\{x_n^*\}_{n\geq 1}$ in a dual Banach space E^*, we denote by $L_{E^*}\{x_n^*\}$ the set of all cluster points of the given sequence in the w^*-topology, and when no confusion arises, we just write $L\{x_n^*\}$.

Lemma 10.11. *Suppose that E is a Banach space, $\{x_n\}_{n\geq 1}$ is a bounded sequence in E and x_0^{**} in E^{**} is a w^*-cluster point of $\{x_n\}_{n\geq 1}$ with $d(x_0^{**},E) > 0$. Then for every α with $d(x_0^{**},E) > \alpha > 0$ there exists a sequence $\{x_n^*\}_{n\geq 1}$ in B_{E^*} such that*

$$\langle x_n^*, x_0^{**}\rangle > \alpha \tag{10.12}$$

whenever $n \geq 1$, and

$$\langle x_0^*, x_0^{**}\rangle = 0 \tag{10.13}$$

for any $x_0^* \in L\{x_n^*\}$.

Proof. The Hahn–Banach theorem applies to provide us with $x^{***} \in B_{E^{***}}$ satisfying $x^{***}_{|E} = 0$ and $x^{***}(x_0^{**}) = d(x_0^{**},E)$. For every $n \geq 1$ the set

$$V_n := \{y^{***} \in E^{***} : y^{***}(x_0^{**}) > \alpha,\ |y^{***}(x_i)| \leq \frac{1}{n},\ i = 1,2,\ldots,n\}$$

is a w^*-open neighborhood of x^{***}, and therefore, by Goldstein's theorem, we can pick up $x_n^* \in B_{E^*} \cap V_n$. The sequence $\{x_n^*\}_{n\geq 1}$ clearly satisfies

$$\lim_n \langle x_n^*, x_p \rangle = 0, \quad \text{for all } p \in \mathbb{N},$$

and for each $n \geq 1$

$$\langle x_n^*, x_0^{**} \rangle > \alpha.$$

Fix an arbitrary $x_0^* \in L\{x_n^*\}$. For every $p \geq 1$ we have that

$$\langle x_0^*, x_p \rangle = 0,$$

and thus

$$\langle x_0^*, x_0^{**} \rangle = 0,$$

because $x_0^{**} \in \overline{\{x_p : p = 1, 2, \cdots\}}^{w^*}$. ■

Theorem 10.12. *Let E be a Banach space with a w^*-convex block compact dual unit ball. If a bounded subset A of E is not weakly relatively compact, then there exists a sequence of linear functionals $\{y_n^*\}_{n\geq 1} \subset B_{E^*}$ with a w^*-limit point y_0^*, and some $g^* \in \mathrm{co}_\sigma\{y_n^* : n \geq 1\}$, such that $g^* - y_0^*$ does not attain its supremum on A.*

Proof. Assume that A is not weakly relatively compact, which in view of the Eberlein–Šmulian theorem is equivalent to the existence of a sequence $\{x_n\}_{n\geq 1}$ in A and a w^*-cluster point $x_0^{**} \in E^{**} \setminus E$ of it. Then Lemma 10.11 applies to provide us with a sequence $\{x_n^*\}_{n\geq 1}$ in B_{E^*} and $\alpha > 0$ satisfying (10.12) and (10.13).

Let $\{y_n^*\}_{n\geq 1}$ be a convex block sequence of $\{x_n^*\}_{n\geq 1}$ and let $y_0^* \in B_{E^*}$ such that $w^*\text{-}\lim_n y_n^* = y_0^*$. It is clear that (10.12) and (10.13) are valid when replacing $\{x_n^*\}_{n\geq 1}$ and x_0^* with $\{y_n^*\}_{n\geq 1}$ and y_0^*, respectively. Then

$$\begin{aligned} S_{\overline{A}^{w^*}}\left(\limsup_n (y_n^* - y_0^*)\right) &\geq \limsup_n (y_n^* - y_0^*)(x_0^{**}) \\ &\geq \alpha \\ &> 0 \\ &= S_A\left(\limsup_n (y_n^* - y_0^*)\right), \end{aligned}$$

so in view of the Rainwater–Simons theorem, Corollary 10.7, there exists $g^* \in \mathrm{co}_\sigma\{y_n^* : n \geq 1\}$ such that $g^* - y_0^*$ does not attain its supremum on A, as announced. ■

In Sect. 10.5.2 we shall show a nonlinear extension of this result, with the use of the (necessarily unbounded) Rainwater–Simons theorem, Corollary 10.7. For the space $\ell^1(\mathbb{N})$, James constructed in [82] a continuous linear functional $g : \ell^1(\mathbb{N}) \to \mathbb{R}$

such that g can be extended to $\hat{g} \in E^*$ on any Banach space E containing $\ell^1(\mathbb{N})$, but $\hat{g}$ does not attain its supremum on B_E. Rosenthal's $\ell^1(\mathbb{N})$-theorem, together with Theorem 10.12, provides another approach for James' reflexivity theorem. These ideas, developed by Morillon in [111], are the basis for new approaches to the weak compactness theorem of James, as the very successful one due to Pfitzner in [122].

We now deal with the general version of Theorem 10.12, that is, James' weak compactness theorem with no additional assumptions on the Banach space. If E is a Banach space and A is a bounded subset of E, we denote by $\|\cdot\|_A$ the seminorm on the dual space E^* given by the Minkowski functional of its polar set, *i.e.*, the seminorm of uniform convergence on the set A. If $A = -A$, given a bounded sequence $\{x_n^*\}_{n\geq 1}$ in E^* and $h^* \in L\{x_n^*\}$, Simons' inequality for the sequence $\{x_n^* - h^*\}_{n\geq 1}$ in $\ell^\infty(A)$ reads as follows: Under the assumption that every element in $\mathrm{co}_{\sigma_p}\{x_n^* - h^* : n \geq 1\}$ attains its supremum on A,

$$\mathrm{dist}_{\|\cdot\|_A}(h^*, \mathrm{co}\{x_n^* : n \geq 1\}) \leq S_A\left(\limsup_n x_n^* - h^*\right).$$

Therefore,

$$\mathrm{dist}_{\|\cdot\|_A}(L\{x_n^*\}, \mathrm{co}\{x_n^* : n \geq 1\}) \leq \inf_{h^* \in L\{x_n^*\}} S_A\left(\limsup_n x_n^* - h^*\right).$$

We state the following characterization:

Proposition 10.13. *Let A be a bounded subset of a Banach space E. Then A is weakly relatively compact if, and only if, for every bounded sequence $\{x_n^*\}_{n\geq 1}$ in E^* we have*

$$\mathrm{dist}_{\|\cdot\|_A}(L\{x_n^*\}, \mathrm{co}\{x_n^* : n \geq 1\}) = 0. \tag{10.14}$$

Proof. We first prove that if A is weakly relatively compact then equality (10.14) holds for any bounded sequence $\{x_n^*\}_{n\geq 1}$ in E^*. To this end, we note that, since $\overline{\mathrm{co}(A)}^{\|\cdot\|}$ is weakly compact by the Krein–Šmulian theorem, the seminorm $\|\cdot\|_A = \|\cdot\|_{\overline{\mathrm{co}(A)}^{\|\cdot\|}}$ is continuous for the Mackey topology $\mu(E^*, E)$. Hence we have the inclusions

$$L\{x_n^*\} \subset \overline{\mathrm{co}\{x_n^* : n \geq 1\}}^{w^*} = \overline{\mathrm{co}\{x_n^* : n \geq 1\}}^{\mu(E^*,E)} \subset \overline{\mathrm{co}\{x_n^* : n \geq 1\}}^{\|\cdot\|_A}$$

that clearly explain the validity of (10.14).

To prove the converse we will show that if A is not weakly relatively compact in E, then there exists a sequence $\{x_n^*\}_{n\geq 1} \subset B_{E^*}$ such that

$$\mathrm{dist}_{\|\cdot\|_A}(L\{x_n^*\}, \mathrm{co}\{x_n^* : n \geq 1\}) > 0.$$

Let us assume that A is not relatively weakly compact in E. Then the Eberlein–Šmulian theorem guarantees the existence of a sequence $\{x_n\}_{n\geq 1}$ in A with a w^*-cluster point $x_0^{**} \in E^{**} \setminus E$. If $d(x_0^{**}, E) > \alpha > 0$, an appeal to Lemma 10.11 provides us with a sequence $\{x_n^*\}_{n\geq 1}$ in B_{E^*} satisfying

$$\langle x_n^*, x_0^{**} \rangle > \alpha$$

whenever $n \geq 1$ and

$$\langle x_0^*, x_0^{**} \rangle = 0$$

for any $x_0^* \in L\{x_n^*\}$. Therefore we have that

$$\| \sum_{i=1}^{n} \lambda_i x_{n_i}^* - x_0^* \|_A \geq \left\langle \sum_{i=1}^{n} \lambda_i x_{n_i}^* - x_0^*, x_0^{**} \right\rangle > \alpha$$

for any convex combination $\sum_{i=1}^{n} \lambda_i x_{n_i}^*$, and consequently

$$\text{dist}_{\|\cdot\|_A}(L\{x_n^*\}, \text{co}\{x_n^* : n \geq 1\}) \geq \alpha > 0, \tag{10.15}$$

and the proof is over. ■

Pryce's diagonal procedure is used in the proof of the following result:

Proposition 10.14. *Let E be a Banach space, A a bounded subset of E with $A = -A$, $\{x_n^*\}_{n\geq 1}$ a bounded sequence in the dual space E^* and D its norm-closed linear span in E^*. Then there exists a subsequence $\{x_{n_k}^*\}_{k\geq 1}$ of $\{x_n^*\}_{n\geq 1}$ such that*

$$S_A\left(x^* - \liminf_k x_{n_k}^*\right) = S_A\left(x^* - \limsup_k x_{n_k}^*\right) = \text{dist}_{\|\cdot\|_A}(x^*, L\{x_{n_k}^*\}) \tag{10.16}$$

for all $x^* \in D$.

Proof. Lemma 10.2 implies the existence of a subsequence $\{x_{n_k}^*\}_{k\geq 1}$ of $\{x_n^*\}_{n\geq 1}$ such that

$$S_A\left(x^* - \liminf_k x_{n_k}^*\right) = S_A\left(x^* - \limsup_k x_{n_k}^*\right)$$

for all $x^* \in D$. Since for any $h^* \in L\{x_{n_k}^*\}$ we have

$$\liminf_k x_{n_k}^*(a) \leq h^*(a) \leq \limsup_k x_{n_k}^*(a)$$

for all $a \in A$, it follows that

$$S_A\left(x^* - \liminf_k x_{n_k}^*\right) = \|x^* - h^*\|_A = S_A\left(x^* - \limsup_k x_{n_k}^*\right).$$

Therefore

$$S_A\left(x^* - \liminf_k x^*_{n_k}\right) = S_A\left(x^* - \limsup_k x^*_{n_k}\right) = \mathrm{dist}_{\|\cdot\|_A}(x^*, L\{x^*_{n_k}\})$$

for all $x^* \in D$, and the proof is finished. ■

Equality (10.16) will be in general the source to look for nonattaining linear functionals whenever we have

$$\mathrm{dist}_{\|\cdot\|_A}(L\{x^*_{n_k}\}, \mathrm{co}\{x^*_{n_k} : k \geq 1\}) > 0,$$

which means, in view of Proposition 10.13, whenever A is a nonrelatively weakly compact subset of E. Until now all such constructions depend on this fact, which is called the *technique of the undetermined function*. The next result is so far the most general perturbed version for the existence of nonattaining functionals; see [133, Corollary 8]:

Theorem 10.15. *Let X be a nonempty set, $\{h_j\}_{j\geq 1}$ a bounded sequence in $\ell^\infty(X)$, $\varphi \in \ell^\infty(X)$ with $\varphi \geq 0$ and $\delta > 0$ such that*

$$S_X\left(h - \limsup_j h_j - \varphi\right) = S_X\left(h - \liminf_j h_j - \varphi\right) \geq \delta,$$

whenever $h \in \mathrm{co}_\sigma\{h_j : j \geq 1\}$. Then there exists a sequence $\{g_i\}_{i\geq 1}$ in $\ell^\infty(X)$ with

$$g_i \in co_\sigma\{h_j : j \geq i\}, \quad \text{for all } i \geq 1,$$

and there exists $g_0 \in \mathrm{co}_\sigma\{g_i : i \geq 1\}$ such that for all $g \in \ell^\infty(X)$ with

$$\liminf_i g_i \leq g \leq \limsup_i g_i \quad \text{on } X,$$

the function $g_0 - g - \varphi$ does not attain its supremum on X.

The proof given in [133] for the above result involves an adaptation of the additive diagonal lemma we have used for Simons' inequality in $\mathbb{R}^X$, Theorem 10.5. Let us include here a proof for the following consequence, that was stated first in this way by James in [87, Theorems 2 and 4].

Theorem 10.16 (James). *Let A be a nonempty bounded subset of a Banach space E which is not weakly relatively compact. Then there exist a sequence $\{g^*_n\}_{n\geq 1}$ in B_{E^*} and some $g_0 \in \mathrm{co}_\sigma\{g^*_n : n \geq 1\}$ such that, for every $h \in \ell^\infty(A)$ with*

$$\liminf_n g^*_n \leq h \leq \limsup_n g^*_n \quad \text{on } A,$$

we have that $g_0 - h$ does not attain its supremum on A.

Proof. Without loss of generality we can assume that A is convex and that $A = -A$. Proposition 10.13 gives us a sequence $\{x_n^*\}_{n\geq 1}$ in B_{E^*} such that $\text{dist}_{\|\cdot\|_A}(L\{x_n^*\}, \text{co}\{x_n^* : n \geq 1\}) > 0$. By Proposition 10.14 there exists a subsequence $\{x_{n_k}^*\}_{k\geq 1}$ of $\{x_n^*\}_{n\geq 1}$ that verifies the hypothesis of Theorem 10.15 with $\varphi = 0$. So we find a sequence $\{g_n^*\}_{n\geq 1}$ with $g_n^* \in \text{co}_\sigma\{x_{n_k}^* : k \geq n\}$, for every $n \in \mathbb{N}$, and $g_0 \in \text{co}_\sigma\{g_n^* : n \geq 1\}$ such that $g_0 - h$ does not attain its supremum on A, where h is any function in $\ell^\infty(A)$ with $\liminf_n g_n^* \leq h \leq \limsup_n g_n^*$ on A. ■

In particular we have seen how to construct linear functionals $g_0 - g$ that do not attain their supremum on A, whenever g is a w^*-cluster point of the sequence $\{g_n^*\}_{n\geq 1}$ in B_{E^*}.

We finish this section with a short visit to the so-called *measures of weak noncompactness* in Banach spaces: the relationship of these measures with the techniques already presented in this survey will be plain clear when progressing in our discussion below.

We refer the interested reader to [14, 105], where measures of weak noncompactness are axiomatically defined. A measure of weak noncompactness is a nonnegative function μ defined on the family $\mathscr{M}_E$ of bounded subsets of a Banach space E, with the following properties:

(i) $\mu(A) = 0$ if, and only if, A is weakly relatively compact in E
(ii) If $A \subset B$ then $\mu(A) \leq \mu(B)$
(iii) $\mu(\text{conv}(A)) = \mu(A)$
(iv) $\mu(A \cup B) = \max\{\mu(A), \mu(B)\}$
(v) $\mu(A + B) \leq \mu(A) + \mu(B)$
(vi) $\mu(\lambda A) = |\lambda|\mu(A)$

Inspired by Proposition 10.13, we introduce the following:

Definition 10.17. For a bounded subset A of a Banach space E, $\sigma(A)$ stands for the quantity

$$\sup_{\{x_n^*\}_{n\geq 1}\subset B_{E^*}} \text{dist}_{\|\cdot\|_A}(L\{x_n^*\}, \text{co}\{x_n^* : n \geq 1\}).$$

Observe that σ satisfies properties (i), (ii), (iii), (iv), and (vi), and therefore σ can be considered as a measure of weak noncompactness. Beyond the formalities we will refer *in general* to measures of weak noncompactness to quantities as above fulfilling property (i) and sometimes a few of the others. These measures of noncompactness or weak noncompactness have been successfully applied to the study of compactness, operator theory, differential equations, and integral equations; see, for instance, [10–12, 20, 33, 36, 50, 64, 66, 68, 103–105].

The next definition collects several measures of weak noncompactness that appeared in the aforementioned literature. If A and B are nonempty subsets of E^{**}, then $d(A,B)$ denotes the *usual* inf *distance* (associated to the bidual norm) between A and B, and the *Hausdorff nonsymmetrized distance* from A to B is defined by

$$\hat{\mathrm{d}}(A,B) = \sup\{d(a,B) : a \in A\}.$$

Notice that $\hat{\mathrm{d}}(A,B)$ can be different from $\hat{\mathrm{d}}(B,A)$, and that $\max\{\hat{\mathrm{d}}(A,B), \hat{\mathrm{d}}(B,A)\}$ is the *Hausdorff distance* between A and B. Notice further that $\hat{\mathrm{d}}(A,B) = 0$ if, and only if, $A \subset \overline{B}$ (norm-closure) and that

$$\hat{\mathrm{d}}(A,B) = \inf\{\varepsilon > 0 : A \subset B + \varepsilon B_{E^{**}}\}.$$

Definition 10.18. Given a bounded subset A of a Banach space E we define

$$\omega(A) := \inf\{\varepsilon > 0 : A \subset K_\varepsilon + \varepsilon B_E \text{ and } K_\varepsilon \subset E \text{ is } w\text{-compact}\},$$

$$\gamma(A) := \sup\{|\lim_n \lim_m x_m^*(x_n) - \lim_m \lim_n x_m^*(x_n)| : \{x_m^*\}_{m\geq 1} \subset B_{E^*}, \{x_n\}_{n\geq 1} \subset A\},$$

assuming the involved limits exist,

$$\mathrm{ck_E}(A) := \sup_{\{x_n\}_{n\geq 1} \subset A} d(L_{E^{**}}\{x_n\}, E),$$

$$\mathrm{k}(A) := \hat{\mathrm{d}}(\overline{A}^{w^*}, E) = \sup_{x^{**} \in \overline{A}^{w^*}} d(x^{**}, E),$$

and

$$\mathrm{Ja_E}(A) := \inf\{\varepsilon > 0 : \text{ for every } x^* \in E^*, \text{ there exists } x^{**} \in \overline{A}^{w^*}$$
$$\text{such that } x^{**}(x^*) = S_A(x^*) \text{ and } d(x^{**}, E) \leq \varepsilon\}.$$

The function ω was introduced by de Blasi [20] as a measure of weak noncompactness that is somehow the counterpart for the weak topology of the classical Kuratowski measure of norm noncompactness. Properties for γ can be found in [11, 12, 33, 50, 105] and for $\mathrm{ck_E}$ in [11]—note that $\mathrm{ck_E}$ is denoted as ck in that paper. The quantity k has been used in [11, 33, 50, 64]. A thorough study for $\mathrm{Ja_E}$ has been done in [36] to prove, amongst other things, a quantitative version of James' weak compactness theorem, whose statement is presented as part of Theorem 10.19 bellow. This theorem tells us that *all classical approaches used to study weak compactness in Banach spaces (Tychonoff's theorem, Eberlein–Šmulian's theorem, Eberlein–Grothendieck double-limit criterion, and James' weak compactness theorem) are qualitatively and quantitatively equivalent.*

Theorem 10.19. *For any bounded subset A of a Banach space E the following inequalities hold true:*

$$\begin{array}{ccccccccc} & & & & \sigma(A) & \leq & 2\omega(A) & & \\ & & & & r & & \text{VI} & & \\ \frac{1}{2}\gamma(A) & \leq & \mathrm{Ja}_{\mathrm{E}}(A) & \leq & \mathrm{ck}_E(A) & \leq & k(A) & \leq & \gamma(A). \end{array} \tag{10.17}$$

Moreover, for any $x^{**} \in \overline{A}^{w^*}$ *there exists a sequence* $\{x_n\}_{n\geq 1}$ *in* A *such that*

$$\|x^{**} - y^{**}\| \leq \gamma(A) \tag{10.18}$$

for any w^**-cluster point* y^{**} *of* $\{x_n\}_{n\geq 1}$ *in* E^{**}.

Furthermore, A *is weakly relatively compact in* E *if, and only if, one (equivalently, all) of the numbers* $\gamma(A), \mathrm{Ja}_{\mathrm{E}}(A), \mathrm{ck}_E(A), \mathrm{k}(A), \sigma(A)$, *and* $\omega(A)$ *is zero.*

Proof. A full proof with references to prior work for the inequalities

$$\frac{1}{2}\gamma(A) \leq \mathrm{ck}_E(A) \leq k(A) \leq \gamma(A) \leq 2\omega(A)$$

and (10.18) is provided in [11, Theorem 2.3]. The inequalities

$$\frac{1}{2}\gamma(A) \leq \mathrm{Ja}_{\mathrm{E}}(A) \leq \mathrm{ck}_E(A)$$

are established in Theorem 3.1 and Proposition 2.2 of [36].

To prove $\mathrm{ck}_E(A) \leq \sigma(A)$ we proceed as follows. If $0 = ck_E(A)$, the inequality is clear. Assume that $0 < \mathrm{ck}_E(A)$ and take an arbitrary $0 < \alpha < \mathrm{ck}_E(A)$. By the very definition of $\mathrm{ck}_E(A)$ there exist a sequence $\{x_n\}_{n\geq 1}$ in A and a w^*-cluster point $x_0^{**} \in E^{**}$ with $d(x_0^{**}, E) > \alpha > 0$. If we read now the second part of the proof of Proposition 10.13, we end up producing a sequence $\{x_n^*\}_{n\geq 1}$ in B_{E^*} that according to inequality (10.15) satisfies

$$\mathrm{dist}_{\|\cdot\|_A}(L\{x_n^*\}, \mathrm{co}\{x_n^* : n \geq 1\}) \geq \alpha.$$

Since α with $0 < \alpha < \mathrm{ck}_E(A)$ is arbitrary, the above inequality yields $\mathrm{ck}_E(A) \leq \sigma(A)$.

To complete the chain of inequalities we establish $\sigma(A) \leq 2\omega(A)$. Let $\omega(A) < \varepsilon$ and take a weakly compact subset K_ε of E such that $A \subset K_\varepsilon + \varepsilon B_E$. This inclusion leads to the inequality

$$\|\cdot\|_A \leq \|\cdot\|_{K_\varepsilon} + \varepsilon\|\cdot\|. \tag{10.19}$$

Fix an arbitrary sequence $\{x_n^*\}_{n\geq 1}$ in B_{E^*} and now take a w^*-cluster point $x_0^* \in L\{x_n^*\}$. Since K_ε is weakly compact we know that $x_0^* \in \overline{\mathrm{co}\{x_n^* : n \geq 1\}}^{\|\cdot\|_{K_\varepsilon}}$. Hence, for an arbitrary $\eta > 0$, we can find a convex combination $\sum_{i=1}^n \lambda_i x_{n_i}^*$ with $\|x_0^* - \sum_{i=1}^n \lambda_i x_{n_i}^*\|_{K_\varepsilon} < \eta$. Thus, inequality (10.19) allows us to conclude that

$$\operatorname{dist}_{\|\cdot\|_A}(L\{x_n^*\}, \operatorname{co}\{x_n^* : n \geq 1\}) \leq \left\| x_0^* - \sum_{i=1}^{n} \lambda_i x_{n_i}^* \right\|_A$$

$$\leq \left\| x_0^* - \sum_{i=1}^{n} \lambda_i x_{n_i}^* \right\|_{K_\varepsilon} + \varepsilon \left\| x_0^* - \sum_{i=1}^{n} \lambda_i x_{n_i}^* \right\| \leq \eta + 2\varepsilon.$$

Since ε, η and $\{x_n^*\}_{n\geq 1}$ are arbitrary, we conclude $\sigma(A) \leq 2\omega(A)$.

Finally, recall a well-known result of Grothendieck [46, Lemma 2, p. 227] stating that $\omega(A) = 0$ if, and only if, A is weakly relatively compact in E. Observe that, as a consequence of (10.17), one of the numbers $\gamma(A), \mathrm{Ja}_E(A), \mathrm{ck}_E(A), \mathrm{k}(A)$ is zero if, and only if, all of them are zero. Clearly, $\mathrm{k}(A) = 0$ if, and only if, $\overline{A}^{w^*} \subset E$, that is equivalent to the fact that A is weakly relatively compact by Tychonoff's theorem. To establish $\sigma(A) = 0$ if, and only if, A is weakly relatively compact either use Proposition 10.13 or the comments above for ω and ck_E, together with the inequalities $\mathrm{ck}_E(A) \leq \sigma(A) \leq \omega(A)$. The proof is over. ∎

It is worth noticing that the inequalities

$$\mathrm{ck}_E(A) \leq k(A) \leq 2\,\mathrm{ck}_E(A),$$

that follow from (10.17), offer a quantitative version (and imply) of the Eberlein–Šmulian theorem saying that weakly relatively countably compact sets in Banach spaces are weakly relatively compact. Note also that (10.18) implies that points in the weak closure of a weakly relatively compact set of a Banach space are reachable by weakly convergent sequences from within the set (summing up, the inequalities are a *quantitative* version of the angelicity of weakly compact sets in Banach spaces; see Definition 10.20). In a different order of ideas the inequality

$$\frac{1}{2}\gamma(A) \leq \mathrm{Ja}_E(A) \tag{10.20}$$

implies James' weak compactness theorem, Theorem 10.1, and since $\mathrm{Ja}_E(A) \leq \mathrm{ck}_E(A)$ as well, we therefore know that James' weak compactness theorem can be derived and implies the other classical results about weak compactness in Banach spaces. We should mention that the proof of inequality (10.20) in [36, Theorem 3.1] follows the arguments by Pryce in [125] suitably adapted and strengthened for the occasion: assuming that $0 < r < \gamma(A)$, two sequences $\{x_n\}_{n\geq 1} \subset A$ and $\{x_m^*\}_{m\geq 1} \subset B_{E^*}$ are produced satisfying

$$\lim_m \lim_n x_m^*(x_n) - \lim_n \lim_m x_m^*(x_n) > r.$$

Then Lemma 10.2 is applied to the sequence $\{x_m^*\}_{m\geq 1}$, and after some twisting and fine adjustments in Pryce's original arguments, for arbitrary $0 < r' < r$ a sequence $\{g_n^*\}_{n\geq 1}$ in B_{E^*} and $g_0 \in \mathrm{co}_\sigma\{g_n^* :\ n \geq 1\}$ are produced with the property that for any w^*-cluster point $h \in B_{E^*}$ of $\{g_n^*\}_{n\geq 1}$, if $x^{**} \in \overline{A}^{w^*}$ is such that

$$x^{**}(g_0 - h) = S_A(g_0 - h)$$

then $d(x^{**},E) \geq \frac{1}{2}r'$. Since $0 < r < \gamma(A)$ and $r' \in (0,r)$ are arbitrary the inequality (10.20) follows. Of course, $g_0 - h \in E^*$ does not attain its supremum on A but we moreover know how far from E in $\overline{A}^{w^*}$ we need to go in order that $g_0 - h$ might attain it: compare with Theorem 10.16.

The aforementioned references contain examples showing when the inequalities in (10.17) are sharp, as well as sufficient conditions of when the inequalities become equalities. An example of the latter is given in the theorem below, where we use the notion of angelic space that follows.

Definition 10.20 (Fremlin). A regular topological space T is *angelic* if every relatively countably compact subset A of T is relatively compact and its closure $\overline{A}$ is made up of the limits of sequences from A.

In angelic spaces the different concepts of compactness and relative compactness coincide: the (relatively) countably compact, (relatively) compact, and (relatively) sequentially compact subsets are the same, as seen in [53]. Examples of angelic spaces include $C(K)$ endowed with the topology $t_p(K)$ of pointwise convergence on a countably compact space K ([71, 96]) and all Banach spaces in their weak topologies. Another class of angelic spaces are dual spaces of weakly countably K-determined Banach spaces, endowed with their w^*-topology [117].

Theorem 10.21 ([36, Theorem 6.1]). *Let E be a Banach space such that (B_{E^*}, w^*) is angelic. Then for any bounded subset A of E we have*

$$\frac{1}{2}\gamma(A) \leq \gamma_0(A) = \mathrm{Ja}_E(A) = \mathrm{ck}_E(A) = \mathrm{k}(A) \leq \gamma(A),$$

where

$$\gamma_0(A) := \sup\{|\lim_i \lim_j x_i^*(x_j)| : \{x_j\}_{j\geq 1} \subset A, \{x_i^*\}_{i\geq 1} \subset B_{E^*}, x_i^* \xrightarrow{w^*} 0\}.$$

A moment of thought and the help of Riesz's lemma suffice to conclude that for the unit ball B_E we have that

$$k(B_E) = \sup_{x^{**} \in B_{E^{**}}} d(x^{**},E) \in \{0,1\}.$$

Reflexivity of E is equivalent to $k(B_E) = 0$ and non reflexivity to $k(B_E) = 1$. Note then that, when (B_{E^*}, w^*) is angelic, reflexivity of E is equivalent to $\mathrm{Ja}_E(B_E) = 0$, and non reflexivity to $\mathrm{Ja}_E(B_E) = 1$. In other words, James' reflexivity theorem can be strengthened to: *If there exists $0 < \varepsilon < 1$ such that for every $x^* \in E^*$ there exists $x^{**} \in B_{E^{**}}$ with $d(x^{**},E) \leq \varepsilon$ and $S_{B_E}(x^*) = x^{**}(x^*)$, then E is reflexive.* Indeed, the above comments provide a proof of this result when (B_{E^*}, w^*) is angelic; for the general case we refer to [69].

With regard to convex hulls, the quantities in Theorem 10.19 behave quite differently. Indeed, if A is a bounded set of a Banach space E, then the following statements hold:

$$\gamma(\mathrm{co}(A)) = \gamma(A), \quad \mathrm{Ja}_{\mathrm{E}}(\mathrm{co}(A)) \leq \mathrm{Ja}_{\mathrm{E}}(A);$$
$$\mathrm{ck}_E(\mathrm{co}(A)) \leq 2\,\mathrm{ck}_E(A), \quad \mathrm{k}(\mathrm{co}(A)) \leq 2\mathrm{k}(A);$$
$$\sigma(\mathrm{co}(A)) = \sigma(A), \quad \omega(\mathrm{co}(A)) = \omega(A).$$

Constant 2 for ck_E and k is sharp, [36, 64, 68], and it is unknown if Ja_{E} might really decrease when passing to convex hulls. The equality $\gamma(A) = \gamma(\mathrm{co}(A))$ is a bit delicate and has been established in [33, 50].

Last, but not least, we present yet another measure of weak noncompactness inspired by James' ideas in [85]. Following [105], for a given bounded sequence $\{x_n\}_{n\geq 1}$ in a Banach space, we define

$$\mathrm{csep}(\{x_n\}_{n\geq 1}) := \inf\{\|u_1 - u_2\| : (u_1, u_2) \in \mathrm{scc}(\{x_n\}_{n\geq 1})\},$$

where

$$\mathrm{scc}(\{x_n\}_{n\geq 1}) := \{(u_1, u_2) : u_1 \in \mathrm{conv}\{x_i\}_{1\leq i\leq m}, u_2 \in \mathrm{conv}\{x_i\}_{i\geq m+1}, m \in \mathbb{N}\}.$$

Definition 10.22 ([105, Definition 2.2]). If A is a bounded subset of a Banach space, we define

$$\alpha(A) := \sup\{\mathrm{csep}(\{x_n\}_{n\geq 1}) : \{x_n\}_{n\geq 1} \subset A\}.$$

It is proved in [105] that the relationship of α with the measures of weak noncompactness already presented are given by the formulas:

$$\alpha(A) = \sup\left\{d(x^{**}, \mathrm{conv}\{x_n : n \geq 1\} : \{x_n\}_{n\geq 1} \subset A,\, x^{**} \in L_{E^{**}}\{x_n\}\right\}$$

and

$$\gamma(A) = \alpha(\mathrm{conv}(A)).$$

For the measure of weak noncompactness σ introduced in Definition 10.17, and in view of Theorem 10.19, the following question naturally arises:

Question 10.23. With regard to the measure of weak noncompactness σ, are the derived estimates sharp? Is it equivalent to the others (except ω)?

10.4 Boundaries

Given a w^*-compact subset C of E^*, a *boundary* for C is a subset B of C with the property that

for every $x \in E$ there exists some $b^* \in B$ such that $b^*(x) = \sup\{c^*(x) : c^* \in C\}$.

Note that if C is moreover convex, then the Hahn–Banach theorem shows that $\overline{\mathrm{co}(B)}^{w^*} = C$. In addition, the set $\mathrm{ext}(C)$ of the extreme points of C is a boundary for C, thanks to Bauer's maximum principle (see [53, p. 6]), and therefore also satisfies $C = \overline{\mathrm{co}(\mathrm{ext}(C))}^{w^*}$. Note that Milman's theorem [46, Corollary IX.4] tells us that $\mathrm{ext}(C) \subset \overline{B}^{w^*}$. Nonetheless, in general, boundaries can be disjoint of the set of extreme points as the following example shows: let Γ be a uncountable set and consider $\big(\ell^1(\Gamma), \|\cdot\|_1\big)$ and

$$B := \Big\{(x_\gamma)_{\gamma\in\Gamma} : x_\gamma \in \{-1,0,1\} \text{ and } \{\gamma \in \Gamma : x_\gamma \neq 0\} \text{ is countable}\Big\}.$$

A moment of thought suffices to conclude that B is a boundary for the dual unit ball $B_{\ell^\infty(\Gamma)}$ that is clearly disjoint from $\mathrm{ext}\big(B_{\ell^\infty(\Gamma)}\big)$; see [136, Example 7].

If B is a boundary for B_{E^*}, we will say that B is a *boundary for E*.

Two problems regarding boundaries in Banach spaces have attracted the attention of a good number of authors during the years, namely:

The study of *strong* boundaries. The goal here is to find conditions under which a boundary B for the w^*-compact convex C is *strong*, i.e., $\overline{\mathrm{co}(B)}^{\|\cdot\|} = C$.

The boundary problem. Let E be a Banach space, let B be a boundary for E, and let A be a bounded and $\sigma(E,B)$-compact subset of E. Is A weakly compact? (Godefroy, [59, Question V.2]).

At first glance, the two questions above may look unrelated. They are not. Indeed, on the one hand, the boundary problem has an easy and positive answer for all strong boundaries B in B_{E^*}. On the other hand, many studies about strong boundaries and several partial answers to the boundary problem use Simons' inequality as a tool. Regarding strong boundaries, the following references are a good source for information [34, 35, 39, 45, 51, 55, 56, 59, 61, 77, 78, 88, 123, 130, 148]. At the end of this section we will provide some recent results on strong boundaries.

Let us start by considering the boundary problem. It has been recently solved in full generality in the paper [122]. It is interesting to recall the old roots and the long history of the problem.

The first result that provided a partial positive result to the boundary problem (before its formulation as such a question) was the following characterization of weak compactness in continuous function spaces, due to Grothendieck; see [71, Théorème 5]:

Theorem 10.24. *If K is a Hausdorff and compact topological space and A is a subset of $C(K)$, then A is weakly compact if, and only if, it is bounded and compact for the topology of the pointwise convergence on K.*

More generally, Theorem 10.24 was generalized by Bourgain and Talagrand [25, Théorème 1] in the following terms:

Theorem 10.25. *Let E be a Banach space, $B = \mathrm{ext}(B_{E^*})$ and let A be a bounded and $\sigma(E,B)$-compact subset of E. Then A is weakly compact.*

Note that the result of Bourgain and Talagrand is far from being a full solution to the boundary problem, because as presented above there are examples of boundaries of Banach spaces that do not contain any extreme point.

Bearing in mind the Rainwater–Simons theorem, Corollary 10.7, it is easy to give another partial solution to the boundary problem.

Corollary 10.26. *For any separable Banach space E and any boundary for E, the boundary problem has positive answer.*

Proof. Let B be a boundary for E and let A be a bounded and $\sigma(E,B)$-compact subset of E. Since E is separable, the unit ball (B_{E^*},w^*) is metrizable and separable. It follows that B is w^*-separable. Take D a countable and w^*-dense subset of B. The topology $\sigma(E,D)$ is then Hausdorff, metrizable, and coarser than $\sigma(E,B)$. Consequently we obtain that $\sigma(E,D)$ and $\sigma(E,B)$ coincide when restricted to A and we conclude that $(A,\sigma(E,B))$ is sequentially compact. An application of Corollary 10.7 taking into account the Eberlein–Šmulian theorem gives us that A is weakly compact, which concludes the proof. ∎

A first approach to the next result appears implicitly in [136, Theorem 5]. Using the ideas of Pryce in [125] and those of Rodé on the so-called "superconvex analysis" in [129], Konig formulated it in [101, Theorem 5.2, p. 104]. We present here our approach based on the criteria given by Theorem 10.15.

Theorem 10.27. *Let E be a Banach space and $B(\subset B_{E^*})$ a boundary for E. If A is a bounded convex subset of E such that for every sequence $\{a_n\}_{n\geq 1}$ in A there exists $z\in E$ such that*

$$\liminf_n \langle a_n, b^*\rangle \leq \langle z, b^*\rangle \leq \limsup_n \langle a_n, b^*\rangle \tag{10.21}$$

for every $b^*\in B$, then A is weakly relatively compact.

Proof. Let us proceed by contradiction and assume that A is not weakly relatively compact in E. Then the Eberlein–Šmulian theorem says that there exists a sequence $\{a_n\}_{n\geq 1}\subset A$ without weak cluster points in E. According to Pryce's diagonal argument, Lemma 10.2, we can and do assume that

$$\begin{aligned} S_B\left(a - \liminf_n a_n\right) &= S_B\left(a - \liminf_k a_{n_k}\right) \\ &= S_B\left(a - \limsup_k a_{n_k}\right) \\ &= S_B\left(a - \limsup_n a_n\right) \end{aligned}$$

for every $a\in \mathrm{co}_\sigma\{a_n : n\geq 1\}$ and every subsequence of integers $n_1 < n_2 < \cdots$.

Let us fix $x_0\in E$ such that for every $b^*\in B$

$$\liminf\langle a_n, b^*\rangle \leq \langle x_0, b^*\rangle \leq \limsup\langle a_n, b^*\rangle.$$

Keeping in mind that A is w^*-relatively compact in E^{**}, we know that $\{a_n\}_{n\geq 1}$ has a w^*-cluster point $x_0^{**} \in E^{**} \setminus E$. Let us fix $h^* \in B_{E^*}$ and $\xi \in \mathbb{R}$ such that

$$h^*(x_0) < \xi < h^*(x_0^{**}).$$

Since $h^*(x_0^{**})$ is a cluster point of the sequence $\{h^*(a_n)\}_{n\geq 1}$, then there exists a subsequence $\{a_{n_k}\}_{k\geq 1}$ of $\{a_n\}_{n\geq 1}$ such that $h^*(a_{n_k}) > \xi$ for every $k \geq 1$. Thus we also have $h^*(a) \geq \xi$ for every $a \in \mathrm{co}_\sigma\{a_{n_k} : k \geq 1\}$. Consequently we have that

$$\begin{aligned} S_B\left(a - \liminf_n a_n\right) &= S_B\left(a - \liminf_k a_{n_k}\right) = S_B\left(a - \limsup_k a_{n_k}\right) \\ &= S_B\left(a - \limsup_n a_n\right) = S_B\left(a - x_0\right) = S_{B_{E^*}}\left(a - x_0\right) \\ &\geq h^*(a) - h^*(x_0) \geq \xi - h^*(x_0) > 0 \end{aligned}$$

for every $a \in \mathrm{co}_\sigma\{a_{n_k} : k \geq 1\}$. We can apply now Theorem 10.15 with $X := B$, $\varphi = 0$ and $\{h_j\}_{j\geq 1}$ being $\{a_{n_k}\}_{k\geq 1}$ to get a sequence $\{y_i\}_{i\geq 1}$ such that for all $i \geq 1$, $y_i \in \mathrm{co}_\sigma\{a_{n_j} : j \geq i\}$, together with some $y_0 \in \mathrm{co}_\sigma\{y_i : i \geq 1\}$, in such a way that $y_0 - y$ does not attain its supremum on B for any y with

$$\liminf_i y_i(b^*) \leq y(b^*) \leq \limsup_i y_i(b^*), \quad \text{for all } b^* \in B.$$

Given $i \geq 1$, since $y_i \in \overline{\mathrm{co}}^{\|\cdot\|}\{a_{n_j} : j \geq i\}$ we can pick up $z_i \in \mathrm{co}\{a_{n_j} : j \geq i\}$ with $\|y_i - z_i\|_\infty < 2^{-i}$. Note that the convexity of A implies $z_i \in A$ for every $i \geq 1$. But our hypothesis provide us with some $z \in E$ such that

$$\liminf_i y_i(b^*) = \liminf_i z_i(b^*) \leq z(b^*) \leq \limsup_i z_i(b^*) = \limsup_i y_i(b^*)$$

for every $b^* \in B$. Thus we have that $y_0 - z \in E$ does not attain its norm on B, which contradicts that B is a boundary for E and the proof is over. ■

The following result straightforwardly follows from Theorem 10.27.

Theorem 10.28. *Let E be a Banach space and $B(\subset B_{E^*})$ a boundary for E. If A is a convex bounded and $\sigma(E,B)$-relatively countably compact subset of E, then it is weakly relatively compact.*

Proof. It suffices to note that if A is $\sigma(E,B)$-relatively countably compact in E, then for any given sequence $\{a_n\}_{n\geq 1}$ in A and each $\sigma(E,B)$-cluster point $z \in E$ of it, z satisfies the inequalities in (10.21). Then Theorem 10.27 applies and the proof is over. ■

A different proof for Theorem 10.28, even in a more general setting, can be found in [53, Corollary 3, p. 78]: the arguments for this proof go back to the construction of norm-nonattaining functionals in Pryce's proof of James' weak compactness theorem. A different proof by Godefroy appeared in [60, Proposition II.21] (this proof has been rewritten in [51, Theorem 3.140]).

Theorem 10.28 opens another door for positive answers to the boundary problem as long as for the given boundary $B(\subset B_{E^*})$ for E and the norm-bounded $\sigma(E,B)$-compact set $A(\subset E)$ we have that $\overline{\mathrm{co}(A)}^{\sigma(E,B)} \subset E$ is $\sigma(E,B)$-compact. In other words, the boundary problem would have a positive answer subject to the locally convex space $(E,\sigma(E,B))$ satisfies Krein–Šmulian's property just mentioned. Note though, that the classical Krein–Šmulian theorem only works for locally convex topologies in between the weak and the norm-topology of E and that $\sigma(E,B)$ can be strictly coarser than the weak topology, [102, Sect. 24]. Positive results along this direction were established in [30–32].

Recall that a subset B of B_{E^*} is said to be *norming* (resp. 1-*norming*) if

$$|||x||| = \sup\{|b^*(x)| : b^* \in B\}$$

is a norm in E equivalent (resp. equal) to the original norm of E. Particularly, if $B(\subset B_{E^*})$ is a boundary for E then B is 1-norming.

The three results that follow are set up to address the boundary problem from the point of view of the existence of isomorphic copies of the basis of $\ell^1(\mathbb{R})$. A proof for these results can be found in [32] (see also [30]).

Theorem 10.29 (Krein–Šmulian type result). *Let E be a Banach space and let B be a norming subset of B_{E^*}. If E does not contain an isomorphic copy of $\ell^1(\mathbb{R})$, then the $\sigma(E,B)$-closed convex hull of every bounded $\sigma(E,B)$-relatively compact subset of X is $\sigma(E,B)$-compact.*

Corollary 10.30. *Let E be a Banach space which does not contain an isomorphic copy of $\ell^1(\mathbb{R})$ and let $B(\subset B_{E^*})$ be a boundary for E. Then, every bounded $\sigma(E,B)$-compact subset of E is weakly compact.*

Theorem 10.31. *Let E be a Banach, $B(\subset B_{E^*})$ a boundary for E and let A be a bounded subset of E. Then, the following statements are equivalent:*

(i) A is weakly compact.
(ii) A is $\sigma(E,B)$-compact and does not contain a family $(x_\alpha)_{\alpha\in\mathbb{R}}$ equivalent to the usual basis of $\ell^1(\mathbb{R})$.

Note that Theorems 10.29 and 10.28 straightforwardly imply Corollary 10.30. Theorem 10.29 is of interest by itself. The original proof for this result in [32] uses techniques of Pettis integration together with fine subtleties about independent families of sets in the sense of Rosenthal. Other proofs are available as for instance in [30, 67], where it is established that if for the Banach space E the Krein–Šmulian property in Theorem 10.29 holds true for any norming set $B(\subset B_{E^*})$ then E cannot contain isomorphically $\ell^1(\mathbb{R})$ (see also [21] for related results).

It is worth mentioning a few things about the class of Banach spaces not containing isomorphic copies of $\ell^1(\mathbb{R})$. Good references for this class of Banach spaces are [79, 106, 144]. On the one hand, a Banach space E does not contain isomorphically $\ell^1(\mathbb{R})$ if, and only if, $\ell^\infty(\mathbb{N})$ is not a quotient of E, [120, Lemma 4.2]. On the other hand, E does not admit $\ell^\infty(\mathbb{N})$ as a quotient if, and only if, the dual unit ball (B_{E^*}, w^*) does not contain a homeomorphic copy of the Stone-Čech compactification of the natural numbers, $\beta\mathbb{N}$, [144]. In particular each one of the following classes of Banach spaces are made up of spaces which do not contain isomorphically $\ell^1(\mathbb{R})$:

(a) Banach spaces with a weak*-sequentially compact dual unit ball
(b) Banach spaces which are Lindelöf for their weak topologies, or more in general, Banach spaces with the property $(\mathscr{C})$ of Corson

Recall that E has *property* $(\mathscr{C})$ (see [124]), if every family of convex closed subsets of it with empty intersection has a countable subfamily with empty intersection.

Finally, the positive answer to the boundary problem due to Pfitzner (see [122, Theorem 9]) is formulated as follows:

Theorem 10.32 (Pfitzner). *Let A be a bounded set in a Banach space E and let $B(\subset E^*)$ be a boundary of a w^*-compact subset C of E^*. If A is $\sigma(E,B)$-countably compact then A is $\sigma(E,C)$-sequentially compact. In particular, if B is a boundary for E, then a bounded subset of E is weakly compact if, and only if, it is $\sigma(E,B)$-compact.*

In the proof of this fine result, Pfitzner does a localized analysis on A that goes beyond Theorem 10.31 and involves the quantitative version of Rosenthal's ℓ^1-theorem in [17], Simons' inequality, and a modification of a result of Hagler and Johnson in [72].

Although Theorem 10.32 answers in full generality the boundary problem a few open problems still remain. For instance, it is unknown if given a boundary B $(\subset B_{E^*})$ for E, the topology $\sigma(E,B)$ is angelic on bounded subsets of E. A few comments are needed here. We first note that since in angelic spaces compact subsets are sequentially compact, [53], when $\sigma(E,B)$ is angelic on bounded subsets of E, a positive answer to the boundary problem is easily given as a consequence of Rainwater–Simons' theorem, Corollary 10.7—see Corollary 10.26 as illustration. In general it is not true that $(E,\sigma(E,B))$ is angelic; see [141, Theorem 1.1(b)]: an L^1-predual E is constructed together with a $\sigma(E,\text{ext}(B_{E^*})$-countably compact set $A \subset E$ for which not every point $x \in \overline{A}^{\sigma(E,\text{ext}(B_{E^*})}$ is the $\sigma(E,\text{ext}(B_{E^*})$-limit of a sequence in A (see also [110]). Nonetheless there are cases where angelicity of $\sigma(E,B)$ (or $\sigma(E,B)$ on bounded sets) is known, and therefore for these cases a stronger positive answer to the boundary problem is provided. One of this cases is presented in [25] where it is proved that for any Banach space E the topology $\sigma(E,\text{ext}(B_{E^*}))$ is angelic on bounded sets—compared with [141, Theorem 1.1(b)]. Two more of these positive cases are presented below in Theorems 10.35 and 10.36.

The proof of Theorem 10.35 needs the two lemmas that follow. The first one (see [30, Lemma 4.5]) that implicitly appears in a particular case in [29] can be considered as a kind of strong version of an "Angelic Lemma" in the spirit of [53, Lemma in p. 28].

Lemma 10.33. *Let X be a nonempty set and τ, $\mathfrak{T}$ two Hausdorff topologies on X such that (X,τ) is regular and $(X,\mathfrak{T})$ is angelic. Assume that for every sequence $\{x_n\}_{n\geq 1}$ in X with a τ-cluster point $x \in X$, x is $\mathfrak{T}$-cluster point of $\{x_n\}_{n\geq 1}$. The following assertions hold true:*

(i) If L is a τ-relatively countably compact subset of X, then L is $\mathfrak{T}$-relatively compact.
(ii) If L is a τ-compact subset of X, then L is $\mathfrak{T}$- compact.
(iii) (X,τ) is an angelic space.

The lemma below (see [29, Lemma 1] and [30, Lemma 4.7]) evokes properties of the real-compactification (also called the *repletion*) of a topological space, cf. [53, Sect. 4.6].

Lemma 10.34. *Let K be a compact space and $B(\subset B_{C(K)^*})$ a boundary for the Banach space $(C(K), \|\cdot\|_\infty)$. If $\{f_n\}_{n\geq 1}$ is an arbitrary sequence in $C(K)$ and $x \in K$, then there exists $\mu \in B$ such that*

$$f_n(x) = \int_K f_n \mathrm{d}\mu$$

for every $n \geq 1$.

Proof. If we define the continuous function $g : K \to [0,1]$ by the expression

$$g(t) := 1 - \sum_{n=1}^{\infty} \frac{1}{2^n} \frac{|f_n(t) - f_n(x)|}{1 + |f_n(t) - f_n(x)|} \qquad (t \in K),$$

then

$$F := \bigcap_{n=1}^{\infty} \{y \in K : f_n(y) = f_n(x)\} = \{y \in K : g(y) = 1 = \|g\|_\infty\}. \tag{10.22}$$

Since B is a boundary, there exists $\mu \in B$ such that $\int_K g \mathrm{d}\mu = 1$. So we arrive at

$$1 = \|\mu\| = |\mu|(K) \geq \int_K g \mathrm{d}|\mu| \geq \int_K g \mathrm{d}\mu = 1, \tag{10.23}$$

in other words,

$$0 = |\mu|(K) - \int_K g \mathrm{d}|\mu| = \int_K (1-g) \mathrm{d}|\mu|.$$

Since $1 - g \geq 0$ we obtain $|\mu|(\{y \in K : 1 - g(y) > 0\}) = 0$, that is $|\mu|(K \setminus F) = 0$. Therefore, for every $n \in \mathbb{N}$, we have

$$\int_K f_n \mathrm{d}\mu = \int_F f_n \mathrm{d}\mu = \int_F f_n(x) \mathrm{d}\mu = f_n(x)$$

because $\mu(F) = \int_F g\mathrm{d}\mu = \int_K g\mathrm{d}\mu = 1$ by the equalities (10.22) and (10.23) (note that μ is actually a probability!). ■

We are ready to proof the next result that appeared in [29, 30]:

Theorem 10.35. *Let K be a compact space and $B(\subset B_{C(K)^*})$ a boundary for the Banach space $(C(K), \|\cdot\|_\infty)$. Then the following statements hold true:*

(i) $(C(K), \sigma(C(K),B))$ is angelic.
(ii) If a subset A of $C(K)$ is $\sigma(C(K),B)$-relatively countably compact in $C(K)$, then A is $\sigma(C(K),B)$-relatively sequentially compact.
(iii) If A is a norm-bounded and $\sigma(C(K),B)$-compact subset of $C(K)$, then A is weakly compact.

Proof. Let us fix the notation $X := C(K)$, $\tau := \sigma(C(K),B)$ and $\mathfrak{T} := t_p(K)$ the topology of pointwise convergence on $C(K)$. Then Lemma 10.34 implies that the hypotheses in Lemma 10.33 are fulfilled. On the one hand, let $\{f_n\}_{n\geq 1}$ be a sequence in $C(K)$ that has τ-cluster point $f_0 \in C(K)$ and take an arbitrary $\mathfrak{T}$-open neighborhood of f_0

$$V(f_0, x_1, x_2, \ldots, x_m, \varepsilon) := \{g \in C(K) : \sup_{1\leq i\leq m} |g(x_i) - f_0(x_i)| < \varepsilon\},$$

with $\varepsilon > 0$, $x_1, x_2, \ldots, x_m \in K$. Use Lemma 10.34 to pick $\mu_i \in B$ associated to each x_i and the sequence $\{f_n\}_{n\geq 1} \cup \{f_0\}$, $1 \leq i \leq m$. Since $\{f_n\}_{n\geq 1}$ visits frequently the τ-open neighborhood of f_0

$$V(f_0, \mu_1, \mu_2, \ldots, \mu_m, \varepsilon) := \left\{g \in C(K) : \sup_{1\leq i\leq m} \left| \int_K g\mathrm{d}\mu_i - \int_K f_0 \mathrm{d}\mu_i \right| < \varepsilon \right\},$$

we conclude that $\{f_n\}_{n\geq 1}$ visits frequently $V(f_0, x_1, x_2, \ldots, x_m, \varepsilon)$, hence f_0 is also a $\mathfrak{T}$-cluster point of $\{f_n\}_{n\geq 1}$. On the other hand, the space $(C(K), t_p(K))$ is angelic, [71, 96] (see also [53]). Therefore $(C(K), \sigma(C(K),B))$ is angelic by Lemma 10.33 that explains (i). Since in angelic spaces relatively countably compactness implies relatively sequentially compactness, statement (ii) follows from (i). Finally (iii) follows from (ii) and the Rainwater–Simons theorem, Corollary 10.7—we have no need here for the general solution given in Theorem 10.32 for the boundary problem. ■

Given a topological space X we denote by $C_b(X)$ the Banach space of bounded continuous real-valued functions on X endowed with the supremum norm $\|\cdot\|_\infty$. $\mathscr{M}(X)$ stands for the dual space $(C_b(X), \|\cdot\|_\infty)^*$, for which we adopt the Alexandroff representation as the space of finite, finitely additive and zero-set regular Baire measures on X [150, Theorem 6].

The following result was published in [31]:

Theorem 10.36. *Let E be a Banach space whose dual unit ball B_{E^*} is w^*-angelic and let B be a subset of B_{E^*}:*

(i) If B is norming and A is a bounded and $\sigma(E,B)$-relatively countably compact subset of E, then $\overline{\mathrm{co}(A)}^{\sigma(E,B)}$ is $\sigma(E,B)$-compact.
(ii) If B if a boundary for E, then every bounded $\sigma(E,B)$-relatively countably compact subset of E is weakly relatively compact. Therefore the topology $\sigma(E,B)$ is angelic on bounded sets of E.

Proof. It is clear that (ii) follows from (i) when taking into account Theorem 10.28.

Here is a proof for (i). We note first that is not restrictive to assume that B is 1-norming and in this case $\overline{\mathrm{co}(B)}^{w^*} = B_{E^*}$. Consider $X := \overline{A}^{\sigma(E,B)}$ endowed with the topology induced by $\sigma(E,B)$. Now we will state that every Baire probability μ on X has a barycenter x_μ in X. Since A is $\sigma(E,B)$-relatively countably compact, every $\sigma(E,B)$-continuous real function on X is bounded, which means that X is a pseudocompact space. For pseudocompact spaces X, the space $\mathscr{M}(X)$ is made up of countably additive measures defined on the Baire σ-field $\mathscr{B}a$ of X, [58] and [150, Theorem 21]. Take a Baire probability μ on X and $x^* \in B_{E^*}$. On the one hand, since (B_{E^*}, w^*) is angelic, for every $x^* \in B_{E^*}$ there exists a sequence in $\mathrm{co}(B)$ that w^*-converges to x^*, and therefore $x^*|_X$ is $\mathscr{B}a$-measurable. On the other hand, X is norm-bounded and thus $x^*|_X$ is also bounded, hence μ-integrable. Since $x^* \in E^*$ is arbitrary, for the given μ we can consider the linear functional $T_\mu : E^* \to \mathbb{R}$ given for each $x^* \in E^*$ by the formula

$$T_\mu(x^*) := \int_X x^*|_X \mathrm{d}\mu.$$

We claim that $T_\mu|_{B_{E^*}}$ is w^*-continuous. To this end it is enough to prove that for any subset C of B_{E^*} we have that

$$T_\mu(\overline{C}^{w^*}) \subset \overline{T_\mu(C)}. \tag{10.24}$$

Take $y^* \in \overline{C}^{w^*}$ and use the angelicity of (B_{E^*}, w^*) to pick up a sequence $\{y_n^*\}_{n\geq 1}$ in C with $y^* = w^*\text{-}\lim_n y_n^*$; in particular we have that considered as functions, the sequence $\{y_n^*|_X\}_{n\geq 1}$ converges pointwise to $y^*|_X$ and it is uniformly bounded on X. The Lebesgue convergence theorem gives us that $T_\mu(y^*) = \lim_n T_\mu(y_n^*)$ and this proves (10.24). Now Grothendieck's completeness theorem, [102, Sect. 21.9.4], applies to conclude the existence of an element x_μ in E such that $T_\mu(x^*) = x^*(x_\mu)$ for every $x^* \in E^*$. x_μ is the barycenter of μ that we are looking for. Now we define the map $\phi : \mu \to x_\mu$ from the $\sigma(\mathscr{M}(X), C_b(X))$-compact convex set $\mathscr{P}(X)$ of all Baire probabilities on X into E. It is easy to prove that ϕ is $\sigma(\mathscr{M}(X), C_b(X))$-to-$\sigma(E,B)$ continuous and its range $\phi(\mathscr{P}(X))$ is a $\sigma(E,B)$-compact convex set that contains X. The proof is concluded. ∎

A particular class of angelic compact spaces is that of the Corson compact spaces: a compact space K is said to be *Corson compact* if for some set Γ it is (homeomorphic to) a compact subset of $[0,1]^{\Gamma}$ such that for every $x=(x(\gamma))$ in K the set $\{\gamma : x(\gamma) \neq 0\}$ is countable; see [40]. If we assume that (B_{E^*}, w^*) is Corson compact, techniques of Radon–Nikodým compact spaces introduced in [113] can be used to prove that (i) in Theorem 10.36 can be completed by proving that A is also $\sigma(E,B)$-relatively sequentially compact. Let us remark that many Banach spaces have w^*-angelic dual unit ball as for instance the weakly compactly generated or more general the weakly countably K-determined Banach spaces; see [117, 145].

We finish this section with a few brief comments regarding strong boundaries. If B is a norm-separable boundary for a w^*-compact subset C in E^*, then B is a *strong boundary* of C, in the sense that C is the norm-closed convex hull of B. This result was first stated in [130], and later, with techniques based on (I)-generation in [55,56]—note that it straightforwardly follows from Corollary 10.8. If the boundary B is weakly Lindelöf it is an open problem to know if it is strong. When B is weakly Lindelöf determined, the angelic character of $C_p((B,w))$ (see [117]) tells us that every $x^{**} \in B_{E^{**}}$ is the pointwise limit of a sequence of elements in B_E and Simons' inequality implies that B is a strong boundary (see [59, Theorem I.2]). If C is a w^*-compact and weakly Lindelöf subset of E^* we also have that every boundary of C is strong (see [34, Theorem 5.7]). For separable Banach spaces E without isomorphic copies of $\ell^1(\mathbb{N})$ we also have that every boundary of any w^*-compact set is a strong boundary [59]. In the nonseparable case the same is true if the boundary is assumed to be w^*-K-analytic as established in the result below that can be found in [35, Theorem 5.6]:

Theorem 10.37. *A Banach space E does not contain isomorphic copies of $\ell^1(\mathbb{N})$ if, and only if, each w^*-K-analytic boundary of any w^*-compact subset C of E^* is strong.*

In particular, w^*-analytic boundaries are always strong boundaries in the former situation. We note that recently Theorem 10.37 has been extended to w^*-K-countably determined boundaries in [65]. In a different order of ideas, let us remark here that the sup-limsup theorem can be extended to more general functions in this situation; see [35, Theorem 5.9]:

Theorem 10.38. *Let E be a Banach space without isomorphic copies of $\ell^1(\mathbb{N})$, C a w^*-compact subset in E^* and B a boundary of C. Let $\{z_n^{**}\}_{n\geq 1}$ be a sequence in E^{**} such that for all $n \geq 1$, $z_n^{**} = w^*\text{-}\lim_m z_m^n$, for some $\{z_m^n\}_{m\geq 1} \subset E$. Then we have*

$$\sup_{b^* \in B} \{\limsup_n z_n^{**}(b^*)\} = \sup_{x^* \in C} \{\limsup_n z_n^{**}(x^*)\}.$$

When the boundary is built up by using a measurable map, it is always strong.

Theorem 10.39. *Let E be a Banach space, and let C be a w^*-compact subset of E^*. Assume that $f : E \to C$ is a norm-to-norm Borel map such that $\langle x, f(x)\rangle = S_C(x)$ for every $x \in E$. Then*

$$\overline{\mathrm{co}(f(X))}^{\|\cdot\|} = C.$$

Proof. Cascales et al. [34, Corollary 2.7] says that we are in conditions to apply [35, Theorem 4.3] to get the conclusion. ■

Borel maps between complete metric spaces send separable sets to separable ones; see [142, Theorem 4.3.8]. This fact implies that a w^*-compact set C as in Theorem 10.39 is going to be fragmented by the norm of E^*. Indeed, for every separable subspace S of E, we have that $f(S)$ is a separable boundary of the w^*-compact set $C_{|S}(\subset S^*)$, thus $C_{|S} = \overline{\mathrm{co} f(S)_{|S}}^{\|\cdot\|_{S^*}}$ is a separable subset of S^*, and therefore C is fragmented by the norm of E^*; see [113]. If $C = B_{E^*}$ the space E must be an Asplund space. With these results in mind, strong boundaries of an Asplund space are characterized in terms of the following concept, introduced in [35]. A subset C of the dual of a Banach space E is said to be *finitely self-predictable* if there is a map $\xi : \mathscr{F}_E \longrightarrow \mathscr{F}_{\mathrm{co}(C)}$ from the family of all finite subsets of E into the family of all finite subsets of $\mathrm{co}(C)$ such that for each increasing sequence $\{\sigma_n\}_{n\geq 1}$ in $\mathscr{F}_E$ with

$$\Sigma = \bigcup_{n=1}^{\infty} \sigma_n, \qquad D = \bigcup_{n=1}^{\infty} \xi(\sigma_n),$$

we have that

$$C_{|\Sigma} \subset \overline{\mathrm{co}}^{\|\cdot\|}(D_{|\Sigma}).$$

The characterization of strong boundaries in Asplund spaces is stated in the following terms; see [35, Theorem 3.9]:

Theorem 10.40. *For a boundary B of an Asplund space, B is a strong boundary if, and only if, it is finitely self-predictable.*

In particular, Asplund spaces are those Banach spaces for which the above equivalence holds; see [35, Theorem 3.10]. A procedure for generating finitely self-predictable subsets is also provided in [35, Corollary 4.4], as the range of σ-fragmented selectors (see [88] for the definition) of the duality mapping, which leads to another characterization of Asplund spaces; see [35, Corollary 4.5].

In a different order of ideas, the paper [94] contains a good number of interesting results of how to transfer topological properties from a boundary B of C to the whole set C (in particular fragmentability) as well as how to embed a Haar system in an analytic boundary of a separable non-Asplund space. Other results about w^*-K-analytic boundaries not containing isomorphic copies of the basis of $\ell^1(\mathbb{R})$ can be found in [65]—see also Theorem 10.31.

We finish this section with the following open question:

Question 10.41. Let E be a Banach space and B a boundary of it. Is $\sigma(E,B)$ an angelic topology on bounded sets of E?

10.5 Extensions of James' Weak Compactness Theorem

Since its appearance, James' weak compactness theorem has become the subject of much interest for many researchers. As discussed in the Introduction, one of the concerns about it has been to obtain proofs which are simpler than the original one. Another, and we deal with it in this section, is to generalize it, which in particular has led to new applications that we will show in Sect. 10.6. Clearly the commented developments on boundaries represent a first group of results along these lines. The other extensions that we present fall into two kind of results. On the one hand, we can have those that for a Banach space E guarantee reflexivity, whenever the set $\mathrm{NA}(E)$ of the continuous and linear functionals that attain their norms,

$$\mathrm{NA}(E) := \{x^* \in E^* : \text{ there exists } x_0 \in B_E \text{ such that } x^*(x_0) = \|x^*\|\},$$

is large enough. On the other hand, we have James' type results but considering more general optimization problems.

10.5.1 Size of the Set of Norm Attaining Functionals

Roughly speaking, the basic question we are concerned with here is whether the reflexivity of a Banach space E follows from the fact that the set of norm-attaining functionals $\mathrm{NA}(E)$ is not small in some sense. Most of these results are based on a suitable meaning for being topologically big.

With regard to the norm-topology, the concrete question is to know whether a Banach space E is reflexive provided that the set $\mathrm{NA}(E)$ has nonempty norm-interior. The space $\ell^1(\mathbb{N})$ shows that the answer is negative, and in addition it is easily proven in [3, Corollary 2] that every Banach space admits an equivalent norm for which the set of norm-attaining functionals has nonempty norm-interior. For this very reason we cannot assume an isomorphic hypothesis on the space when studying the question above. Some geometric properties have been considered. Before collecting some results in this direction, let us say something more from the isomorphic point of view. In 1950 Klee proved that a Banach space E is reflexive provided that for every space isomorphic to E, each functional attains its norm [100]. Latter, in 1999 Namioka asked whether a Banach space E is reflexive whenever the set $\mathrm{NA}(X)$ has nonempty norm-interior for each Banach space X isomorphic with E. In [1, Theorem 1.3], Acosta and Kadets provided a positive answer (see also [2]).

In order to state the known results for the norm-topology, let us recall that a Banach space E has the *Mazur intersection property* when each bounded, closed, and convex subset of E is an intersection of closed balls ([107]). This is the case of a space with a Fréchet differentiable norm ([45, Proposition II.4.5]). Another different geometric condition is this one: a Banach space E is *weakly Hahn–Banach smooth* if each $x^* \in \mathrm{NA}(E)$ has a unique Hahn–Banach extension to E^{**}. It is clear that if E

is *very smooth* (its duality mapping is single valued and norm-to-weak continuous [140]), then it is weakly Hahn–Banach smooth. Examples of very smooth spaces are those with a Fréchet differentiable norm and those which are an M-ideal in its bidual [76, 151]—for instance c_0 or the space of compact operators on ℓ^2. The following statement, shown in [89, Proposition 3.3] and [4, Theorem 1], provides a first generalization of James' reflexivity theorem for the above classes of Banach spaces:

Theorem 10.42. *Suppose that E is a Banach space that has the Mazur intersection property or is weakly Hahn–Banach smooth. Then E is reflexive if, and only if,* $\mathrm{NA}(E)$ *has nonempty norm-interior.*

The above result is a consequence of James' reflexivity theorem applied to an adequate renorming, in the Mazur intersection property case, and of the Simons inequality after a sequential reduction, for weakly Hahn–Banach smooth spaces.

Note that Theorem 10.42 fails when the space is *smooth* (norm Gâteaux differentiable). Indeed, any separable Banach space is isomorphic to another smooth Banach space whose set of norm-attaining functional has nonempty norm-interior; see [3, Proposition 9].

For some concrete Banach spaces we can say something better. For instance, the sequence space c_0 satisfies that the set $\mathrm{NA}(c_0)$ is of the first Baire category, since it is nothing more than the subset of sequences in $\ell^1(\mathbb{N})$ with finite support. Bourgain and Stegall generalized it for any separable Banach space whose unit ball is not dentable. As a matter of fact, they established the following result in [26, Theorem 3.5.5]:

Theorem 10.43. *If E is a Banach space and C is a closed, bounded, and convex subset of E that is separable and nondentable, then the set of functionals in* E^* *that attain their supremum on C is of the first Baire category in* E^*.

When C is the unit ball of the continuous function space on a infinite Hausdorff and compact topological space K, Kenderov, Moors, and Sciffer proved in [97] that $\mathrm{NA}(C(K))$ is also of the first Baire category. However we do not know whether or not Theorem 10.43 is valid if C is nonseparable. However, Moors has provided us (private communication) with the proof of the following unpublished result which follows from Lemma 4.3 in [109]: Suppose that a Banach space E admits an equivalent weakly midpoint LUR norm and that E has the Namioka property, i.e., every weakly continuous mapping acting from a Baire space into E is densely norm continuous. Then every closed, bounded, and convex subset C of E for which the set of functionals in E^* attaining their supremum on C is of the second Baire category in E^* has at least one strongly exposed point. In particular, C is dentable.

Now we present a group of results whose hypotheses involve the weak topology of the dual space. Jiménez-Sevilla and Moreno showed a series of results, from which we emphasize the following consequence of Simons' inequality [89, Proposition 3.10]:

Theorem 10.44. *Let E be a separable Banach space such that the set* $\mathrm{NA}(E) \cap S_{E^*}$ *has nonempty relative weak interior in S_{E^*}. Then E is reflexive.*

Regarding the w^*-topology in the dual space, the first result was obtained, also applying Simons' inequality, by Deville, Godefroy, and Saint Raymond [41, Lemma 11] and is the version for the w^*-topology of the preceding theorem. Later, an adequate use of James' reflexivity theorem for a renorming of the original space implies the same assertion, but removing the separability assumption [89, Proposition 3.2]:

Theorem 10.45. *A Banach space is reflexive if, and only if, the set of norm-one norm-attaining functionals contains a nonempty relative w^*-open subset of its unit sphere.*

This result has been improved for a certain class of Banach spaces, for instance, for *Grothendieck spaces, i.e.,* those Banach spaces for which the sequential convergence in its dual space for the w-topology is equal to that of the w^*-topology. It is clear that any reflexive space is a Grothendieck space and the converse is true when the space does not contain $\ell^1(\mathbb{N})$; see [63, 149]. Moreover, the Eberlein–Šmulian theorem guarantees that a Banach space with a w^*-sequentially compact dual unit ball is reflexive whenever is a Grothendieck space.

Theorem 10.46. *If E is a Banach space E that is not Grothendieck, then* $\mathrm{NA}(E)$ *is not a w^*-G_δ subset of E^*.*

This result has been stated in [1, Theorem 2.5], although it previously appeared in [41, Theorem 3] for separable spaces. Finally, a characterization of the reflexivity in terms of the w^*-topology, and once again by means of the Simons inequality but with other kind of assumptions, was obtained in [6, Theorem 1]:

Theorem 10.47. *Assume that E is a Banach space that does not contain $\ell^1(\mathbb{N})$ and that for some $r > 0$*

$$B_{E^*} = \overline{\mathrm{co}}^{w^*}\{x^* \in S_{E^*} : x^* + rB_{E^*} \subset \mathrm{NA}(E)\}.$$

Then E is reflexive.

A similar result is stated in [6, Proposition 4], but replacing the assumption of non containing $\ell^1(\mathbb{N})$ with that of the norm of the space is not *rough, i.e.,* there exists $\varepsilon > 0$ such that for all $x \in E$

$$\limsup_{h \to 0} \frac{\|x+h\| + \|x-h\| - 2\|x\|}{\|h\|} \geq \varepsilon.$$

Here we have emphasized some extensions of James' reflexivity theorem in connection to the size of the set of norm-attaining functionals, but there are other ways of measuring such size. For example, one can look for linear subspaces into $\mathrm{NA}(E)$.

The first of these results was obtained by Petunin and Plichko in [121]. To motivate it, let us observe that for a dual space $E = F^*$ we have that F is a closed and w^*-dense subspace of E^* with $F \subset \mathrm{NA}(E)$. Their result deals with the converse:

Theorem 10.48. *A separable Banach space E is isometric to a dual space provided that there exists a Banach space F which is w^*-dense in E^* and satisfies $F \subset \mathrm{NA}(E)$.*

There are some recent results that provide conditions implying that the set of norm-attaining functionals contains an infinite-dimensional linear subspace. See [9,15,57] and the references therein. For instance, in [57] the following renorming result is stated:

Theorem 10.49. *Every Banach space that admits an infinite-dimensional separable quotient is isomorphic to another Banach space whose set of norm-attaining functionals contains an infinite-dimensional linear subspace.*

However, some questions still remain to be studied. For instance, whether for every infinite-dimensional Banach space E, the set $\mathrm{NA}(E)$ contains a linear subspace of dimension 2 is an irritating open problem, posed in [15, Question 2.24].

10.5.2 Optimizing Other Kind of Functions

In the past several years, some extensions of James' weak compactness theorem appeared. A common thing for these results is that the optimization condition—each continuous and linear functional attains its supremum on a weakly closed and bounded subset of the space—is replaced by another one: the objective function is more general. We present some of them here, when considering either polynomials or perturbed functionals.

For a Banach space E and $n \geq 1$, let us consider the space $\mathscr{P}(^nE)$ of all continuous n-homogeneous polynomials on E, endowed with its usual sup norm. Recall that a polynomial in $\mathscr{P}(^nE)$ *attains the norm* when the supremum defining its norm is a maximum. It is clear that if for some n each polynomial in $\mathscr{P}(^nE)$ attains its norm, then every functional attains the norm and thus James' reflexivity theorem implies the reflexivity of E. So the polynomial version of James' reflexivity theorem should be stated in terms of a subset of $\mathscr{P}(^nE)$. This is done in the following characterization (see [131, Theorem 2]), when dealing with weak compactness of a bounded, closed, and convex subset of E:

Theorem 10.50. *A bounded, closed, and convex subset A of a Banach space E is weakly compact if, and only if, there exist $n \geq 1$ and $x_1^*, \dots, x_n^* \in E^*$ such that for all $x^* \in E^*$, the absolute value of the continuous $(n+1)$-homogeneous polynomial*

$$x \mapsto x_1^*(x) \cdots x_n^*(x) x^*(x), \quad (x \in E),$$

when restricted to A, attains its supremum and

$$A \not\subset \cup_{j=1}^{n} \ker x_j^*.$$

Similar results for symmetric multilinear forms, including some improved versions for the case $A = B_E$, can be found in [8, 131].

A related question to that of "norm attaining" (or "sup attaining") is that of "numerical radius attaining." More specifically, the *numerical radius* of a continuous and linear operator $T : E \longrightarrow E$ is the real number $v(T)$ given by

$$v(T) := \sup\{|x^*Tx| : (x,x^*) \in \Pi(E)\},$$

where $\Pi(E) := \{(x,x^*) \in S_E \times S_{E^*} : x^*(x) = 1\}$ and such an operator T is said to *attain the numerical radius* when there exists $(x_0,x_0^*) \in \Pi(E)$ with $|x_0^*Tx_0| = v(T)$.

The following sufficient condition for reflexivity was stated in [5, Theorem 1] (see also [132, Corollary 3.5] for a more general statement about weak compactness), and was obtained by applying the minimax theorem [137, Theorem 5].

Theorem 10.51. *A Banach space such that every rank-one operator on it attains its numerical radius is reflexive.*

Surprisingly enough, the easy-to-prove part in the classical James' reflexivity theorem does not hold. Indeed, a Banach space is finite dimensional if, and only if, in any equivalent norm each rank-one operator attains its numerical radius, as seen in [5, Example] and [7, Theorem 7].

However, the James type result that seems to be more applied nowadays (see Sect. 10.6) is a perturbed version: there exists a fixed function $f : E \longrightarrow \mathbb{R} \cup \{\infty\}$ such that

$$\text{for every } x^* \in E^*, \quad x^* - f \text{ attains its supremum on } E.$$

Let us note that this optimization condition generalizes that in the classical James' weak compactness theorem. Indeed, $x^* \in E^*$ attains its supremum on the set $A (\subset E)$ if, and only if, $x^* - \delta_A$ attains its supremum on E, where δ_A denotes the *indicator function* of A defined as

$$\delta_A(x) := \begin{cases} 0, & \text{if } x \in A \\ \infty, & \text{otherwise} \end{cases}.$$

The first result along these lines was stated in [27, 52] by Calvert and Fitzpatrick.

Theorem 10.52. *A Banach space is reflexive whenever its dual space coincides with the range of the subdifferential of an extended real-valued coercive, convex, and lower semicontinuous function whose effective domain has nonempty norm-interior.*

The erratum [27] makes [52] more difficult to follow, since the main addendum requires to correct non-written proofs of some statements in [52], which are adapted from [84]. A complete and more general approach was presented in Theorems 2, 5 and 7 of [118].

Let us point out that, for a Banach space E and a proper function $f : E \longrightarrow \mathbb{R} \cup \{\infty\}$, *coercive* means

$$\lim_{\|x\| \to \infty} \frac{f(x)}{\|x\|} = \infty,$$

and that the *effective domain* of f, $\operatorname{dom}(f)$, is the set of those $x \in E$ with $f(x)$ finite.

Taking into account that for a function $f : E \longrightarrow \mathbb{R} \cup \{\infty\}$ which is *proper* ($\operatorname{dom}(f) \neq \emptyset$), and $x \in \operatorname{dom}(f)$, we have that the subdifferential of f at x is given by

$$\partial f(x) = \{x^* \in E^* : \ x^* - f \text{ attains its supremum on } E \text{ at } x\},$$

then the surjectivity assumption in Calvert and Fitzpatrick's theorem is once again a perturbed optimization result.

Another perturbed version of James' weak compactness theorem, different from the preceding one, was established in [133, Theorem 16] as a consequence of a minimax result [133, Theorem 14]. In order to state that minimax theorem, generalizing [137, Theorem 14], the authors used the ideas of Pryce in Lemma 10.2 and a refinement of the arguments in [138]. Such a perturbed theorem reads as follows in the Banach space framework:

Theorem 10.53. *Let A be a weakly closed subset of a Banach space E for which there exists $\psi \in \ell^\infty(A)$ such that*

$$\text{for each } x^* \in E^*, \ x^*|_A - \psi \text{ attains its supremum.}$$

Then A is weakly compact.

Here the perturbation f (defined on the whole E) is given by

$$f(x) := \begin{cases} \psi(x), & \text{if } x \in A \\ \infty, & \text{for } x \in E \setminus A \end{cases}.$$

The second named author in this survey obtained another perturbed James type result in the class of separable Banach spaces. This result was motivated by financial applications, and once again, it was proved by applying adequately Simons' inequality. Its proof was included in the Appendix of [91]:

Theorem 10.54. *Suppose that E is a separable Banach space and that $f : E \longrightarrow \mathbb{R} \cup \{\infty\}$ is a proper function whose effective domain is bounded and such that*

$$\text{for each } x^* \in E^*, \quad x^* - f \text{ attains its supremum on } E.$$

Then for every $c \in \mathbb{R}$ the sublevel set $f^{-1}((-\infty,c])$ is weakly compact.

In the preceding versions of the weak compactness theorem of James, the perturbation functions are coercive. Recently, the following characterization has been developed in [118, Theorem 5]:

Theorem 10.55. *Let E be a Banach space and suppose that $f : E \longrightarrow \mathbb{R} \cup \{+\infty\}$ is a proper, coercive, and weakly lower semicontinuous function. Then*

$$\text{for all } x^* \in E^*, \quad x^* - f \text{ attains its supremum on } E$$

if, and only if,

$$\text{for each } c \in \mathbb{R}, \text{ the sublevel set } f^{-1}((-\infty,c]) \text{ is weakly compact.}$$

The proof makes use of the perturbed technique of the undetermined function as explained in Theorem 10.15.

Let us also emphasize that there are previous topological results along the lines of Theorem 10.55; see [23, Theorems 2.1 and 2.4].

Since for any reflexive Banach space E the proper, noncoercive, and weakly lower semicontinuous function $f = \|\cdot\|$ satisfies that for every $c \in \mathbb{R}$ the sublevel set $f^{-1}((-\infty,c])$ is weakly compact, although $\partial f(E) = B_{E^*}$, then the coercivity cannot be dropped in one direction of the former theorem. Nevertheless, for the converse implication, Saint Raymond has just obtained the nice theorem that follows, [134, Theorem 11]:

Theorem 10.56 (Saint Raymond). *If E is a Banach space and $f : E \longrightarrow \mathbb{R} \cup \{\infty\}$ is a proper weakly lower semicontinuous function such that for every $x^* \in E^*$, $x^* - f$ attains its supremum, then for each $c \in \mathbb{R}$, the sublevel set $f^{-1}((-\infty,c])$ is weakly compact.*

Remark 10.57. The fact that for a proper function $f : E \longrightarrow \mathbb{R} \cup \{\infty\}$ with $\partial f(E) = E^*$ all its sublevel sets are relatively weakly compact can be straightforwardly derived from Theorem 10.56. To see it, replace f with the proper weakly lower semicontinuous function $\tilde{f} : E \longrightarrow \mathbb{R} \cup \{\infty\}$ defined for every $x \in E$ as

$$\tilde{f}(x) := \inf\{t \in \mathbb{R} \colon \; (x,t) \in \overline{\operatorname{epi}(f)}^{\sigma(E\times\mathbb{R}, E^*\times\mathbb{R})}\},$$

where $\operatorname{epi}(f)$ is the *epigraph* of f, that is,

$$\operatorname{epi}(f) := \{(x,t) \in E \times \mathbb{R} \colon \; f(x) \le t\}.$$

Furthermore, when $\operatorname{dom}(f)$ has nonempty norm-interior, we have that E is reflexive as a consequence of the Baire Category theorem.

Note that Theorem 10.56 provides an answer to the problem posed in [27]: given a Banach space E and a convex and lower semicontinuous function $f : E \longrightarrow \mathbb{R} \cup \{\infty\}$

whose effective domain has nonempty norm-interior, is it true that the surjectivity of its subdifferential is equivalent to the reflexivity of E and the fact that for all $x^* \in E^*$, the function $x^* - f$ is bounded above?

On the other hand, Bauschke proved that each real infinite-dimensional reflexive Banach space E has a proper, convex, and lower semicontinuous function $f : E \longrightarrow \mathbb{R} \cup \{+\infty\}$ such that

$$\text{for each } x^* \in E^*, \quad x^* - f \text{ is bounded above,}$$

but f is not coercive; see [16, Theorem 3.6]. From here it follows that $\partial f(E) = E^*$, as seen in [118, Theorem 3]. Thus Theorem 10.56 properly extends one direction of Theorem 10.55.

Now let us show how Saint Raymond's result, Theorem 10.56, following the ideas in [118, Corollary 5], has some consequences for multivalued mappings. Let us recall that given a Banach space E and a multivalued operator $\Phi : E \longrightarrow 2^{E^*}$, the *domain* of Φ is the subset of E

$$\mathrm{D}(\Phi) := \{x \in E : \Phi(x) \text{ is nonempty}\},$$

and its *range* is the subset of E^*

$$\Phi(E) := \{x^* \in E^* : \text{there exists } x \in E \text{ with } x^* \in \Phi(x)\}.$$

In addition, Φ is said to be *monotone* if

$$\inf_{\substack{x,y \in \mathrm{D}(\Phi) \\ x^* \in \Phi(x),\ y^* \in \Phi(y)}} \langle x^* - y^*, x - y \rangle \geq 0,$$

and *cyclically monotone* when the inequality

$$\sum_{j=1}^{n} \langle x_j^*, x_j - x_{j-1} \rangle \geq 0$$

holds, whenever $n \geq 2$, $x_0, x_1, \ldots, x_n \in \mathrm{D}(\Phi)$ with $x_0 = x_n$ and for $j = 1, \ldots, n$, $x_j^* \in \Phi(x_j)$.

If Φ is a cyclically monotone operator then there exists a proper and convex function $f : E \longrightarrow \mathbb{R} \cup \{+\infty\}$ such that for every $x \in E$,

$$\Phi(x) \subset \partial f(x),$$

see [128, Theorem 1], and so Theorem 10.56 leads to the following James' type result for cyclically monotone operators:

Corollary 10.58. *Let E be a Banach space and let $\Phi : E \longrightarrow 2^{E^*}$ be a cyclically monotone operator such that $\mathrm{D}(\Phi)$ has nonempty norm-interior and*

$$\Phi(E) = E^*.$$

Then E is reflexive.

Note that this result does not provide a satisfactory answer to the following open problem, posed in [52]: Assume that E is a Banach space and $\Phi : E \longrightarrow 2^{E^*}$ is a monotone operator such that $\mathrm{D}(\Phi)$ has nonempty interior and $\Phi(E) = E^*$. Is E reflexive?

To conclude this section we provide a proof of Theorem 10.56 for the wide class of Banach spaces with w^*-convex block compact dual unit balls, which easily follows from the unbounded Rainwater–Simons theorem, Corollary 10.7; see [119, Theorem 4]. The following lemma produces the sequence needed to apply it:

Lemma 10.59. *Suppose that the dual unit ball of E is w^*-convex block compact and that A is a nonempty, bounded subset of E. Then A is weakly relatively compact if, and only if, each w^*-null sequence in E^* is also $\sigma(E^*, \overline{A}^{w^*})$-null.*

Proof. If A is weakly relatively compact, then we have $A = \overline{A}^{w^*}$ and the conclusion follows. According to Proposition 10.13, to see the reverse implication we have to check the validity of the identity

$$\mathrm{dist}_{\|\cdot\|_A}(L\{x_n^*\}, \mathrm{co}\{x_n^* : n \geq 1\}) = 0 \tag{10.25}$$

for every bounded sequence $\{x_n^*\}_{n\geq 1}$ in E^*. Thus, let us fix $\{x_n^*\}_{n\geq 1}$ a bounded sequence in B_{E^*}. Since B_{E^*} is w^*-convex block compact, there exist a block sequence $\{y_n^*\}_{n\geq 1}$ of $\{x_n^*\}_{n\geq 1}$ and an $x_0^* \in B_{E^*}$ such that

$$w^*\text{-}\lim_n y_n^* = x_0^*.$$

Then, by assumption, $\{y_n^*\}_{n\geq 1}$ also converges to x_0^* pointwise on $\overline{A}^{w^*} \subset E^{**}$. Mazur's theorem applied to the sequence of continuous functions $\{y_n^*\}_{n\geq 1}$ restricted to the w^*-compact space $\overline{A}^{w^*}$ tells us that

$$0 = \mathrm{dist}_{\|\cdot\|_{\overline{A}^{w^*}}}(x_0^*, \mathrm{co}\{y_n^* : n \geq 1\}) = \mathrm{dist}_{\|\cdot\|_A}(x_0^*, \mathrm{co}\{x_n^* : n \geq 1\}) \geq 0,$$

It is not difficult to check that $x_0^* \in L\{x_n^*\}$ and (10.25) is proved, and we have concluded the proof. ∎

Following [119], we present the next proof of Theorem 10.56 for the class of Banach spaces with w^*-convex block compact dual unit balls:

Theorem 10.60. *Let E be a Banach space whose dual unit ball is w^*-convex block compact and let $f : E \longrightarrow \mathbb{R} \cup \{+\infty\}$ be a proper map such that*

$$\textit{for all } x^* \in E^*, \quad x^* - f \textit{ attains its supremum on } E.$$

Then

for every $c \in \mathbb{R}$, the sublevel set $f^{-1}((-\infty, c])$ is weakly relatively compact.

Proof. We first claim that for every $(x^*, \lambda) \in E^* \times \mathbb{R}$ with $\lambda < 0$, there exists $x_0 \in E$ with $f(x_0) < +\infty$ and such that

$$\sup\{(x^*, \lambda)(x,t) : (x,t) \in \operatorname{epi}(f)\} = x^*(x_0) - \lambda f(x_0). \tag{10.26}$$

In fact, the optimization problem

$$\sup_{x \in E}\{\langle x, x^*\rangle - f(x)\} \tag{10.27}$$

may be rewritten as

$$\sup_{(x,t)\in \operatorname{epi}(f)} \{(x^*, -1), (x,t)\} \tag{10.28}$$

and the supremum in (10.27) is attained if, and only if, the supremum in (10.28) is attained.

Let us fix $c \in \mathbb{R}$ and assume that $A := f^{-1}((-\infty, c])$ is nonempty. The uniform boundedness principle and the optimization assumption on f imply that A is bounded. In order to obtain the relative weak compactness of A we apply Lemma 10.59. Thus, let us consider a w^*-null sequence $\{x_n^*\}_{n\geq 1}$ in E^* and let us show that it is also $\sigma(E^*, \overline{A}^{w^*})$-null.

It follows from the unbounded Rainwater–Simons theorem, Corollary 10.7, taking the Banach space $E^* \times \mathbb{R}$,

$$B := \operatorname{epi}(f) \subset C := \overline{\operatorname{epi}(f)}^{\sigma(E^{**}\times\mathbb{R}, E^*\times\mathbb{R})}$$

and the bounded sequence

$$\left\{\left(x_n^*, -\frac{1}{n}\right)\right\}_{n\geq 1},$$

that

$$\sigma(E^* \times \mathbb{R}, B)\text{-}\lim_n \left(x_n^*, -\frac{1}{n}\right) = \sigma(E^* \times \mathbb{R}, C)\text{-}\lim_n \left(x_n^*, -\frac{1}{n}\right),$$

But $w^*\text{-}\lim_{n\geq 1} x_n^* = 0$, so we have that

$$\sigma(E^* \times \mathbb{R}, C)\text{-}\lim_n \left(x_n^*, -\frac{1}{n}\right) = 0.$$

As a consequence, since $A \times \{c\} \subset B$, then $\overline{A}^{w^*} \times \{c\} \subset C$, and so

$$\sigma(E^*, \overline{A}^{w^*})\text{-}\lim_n x_n^* = 0,$$

as announced. ■

Theorem 10.60 was first presented at the meeting Analysis, Stochastics, and Applications, held at Viena in July 2010, to celebrate Walter Schachermayer's 60th Birthday; see

http://www.mat.univie.ac.at/$\sim$anstap10/slides/Orihuela.pdf,

where the conjecture of its validity for any Banach space was considered. Later on, in the Workshop on Computational and Analytical Mathematics in honor of Jonathan Borwein's 60th Birthday, held at Vancouver in May 2011; see

http://conferences.irmacs.sfu.ca/jonfest2011/,

Theorem 10.60 and its application Theorem 10.65 were discussed too. Both results can be found published by the second and third named authors of this survey in the paper [119]. In September 2011 we were informed by J. Saint Raymond that he had independently obtained Theorem 10.60 without any restriction on the Banach space E in [134]: Saint Raymond's proof is based upon a clever and nontrivial reduction to the classical James' weak compactness theorem instead of dealing with unbounded sup-limsup results as presented here, as well as in [119]. Nevertheless, our approach contains classical James' result without using it inside the proof, together with the generalizations of Simons' inequalities for unbounded sets in Sect. 10.2.

The proof of Theorem 10.60 has been obtained by means of elementary techniques for Banach spaces with a w^*-convex block compact dual unit ball, in particular for the separable ones. For this very reason, an easy reduction to the separable case would provide us with a basic proof of the theorem. In that direction, we suggest the following question:

Question 10.61. Let E be a Banach space, $\rho : E^* \times E^* \longrightarrow [0,\infty)$ a pseudometric on E^* for pointwise convergence on a countable set $A(\subset B_{E^{**}})$, where

$$A = A_0 \cup \{x_0^{**}\}, A_0 \subset E, x_0^{**} \in \overline{A_0}^{w^*}.$$

Given $\{x_n^*\}_{n\geq 1}$ a sequence in B_{E^*} such that

$$\sigma(E^*, A_0)\text{-}\lim_n x_n^* = 0,$$

is it possible to find a sequence $\{y_n^*\}_{n\geq 1}$ in E^* with

$$w^*\text{-}\lim_n y_n^* = 0$$

and

$$\lim_n \rho(x_n^*, y_n^*) = 0?$$

10.6 Applications to Convex Analysis and Finance

Since its publication, the applicability of James' weak compactness theorem has been steady. As mentioned in the Introduction, James' weak compactness theorem implies almost straightforwardly a number of important results in Functional Analysis. In this section we focus on some consequences of Theorem 10.56, which have been recently obtained from Theorems 10.55 and 10.60 in the areas of finance and variational analysis. But before describing them, a bit of history on known applications of the theorem of weak compactness of James.

It is in 1968 when appeared the first work mentioning application: in [147] it was proved that a quasi-complete locally convex space-valued measure always has a relatively weakly compact range. On the other hand, Dieudonné [47] gave an example of a Banach space for which the Peano theorem about the existence of solutions to ordinary differential equations fails. Then Cellina [37] stated, with the aid of James' reflexivity theorem, that a Banach space is reflexive provided that the Peano theorem holds true for it. Later, Godunov [62] proved that indeed the space is finite dimensional. In [13] one can find some related results to the failure of Peano's theorem in an infinite dimensional Banach space, as a consequence of James' reflexivity theorem. Finally, let us emphasize the well-known fact (see, for instance, [22, Theorem 2.2.5]) that the completeness of a metric space is equivalent to the validity of the famous Ekeland variational principle. In [143] a characterization of the reflexivity of a normed space is established, also in terms of the Ekeland variational principle, and making use once again of James' reflexivity theorem.

10.6.1 Nonlinear Variational Problems

Our goal is to deal with some consequences of Theorem 10.56 for nonlinear variational problems, following the ideas in [118, Sect. 4]. For this very reason, let us first recall that variational equations are the standard setting to studying and obtaining weak solutions for large portion of differential problems. Such variational equations, in the presence of symmetry, turn into variational problems for which one has to deduce the existence of a minimum. We prove that this kind of result, always stated in the reflexive context, only make sense for this class of Banach spaces.

To be more precise, let us evoke the so-called *main theorem on convex minimum problems* (see, for instance, [153, Theorem 25E, p. 516]), which is a straightforward consequence of the classical theorem of Weierstrass (continuous functions defined on a compact space attain their minimum): in a reflexive Banach space E the sub-differential of every proper, coercive, convex, and lower semicontinuous function $f : E \longrightarrow \mathbb{R} \cup \{+\infty\}$ is onto, that is, for each $x^* \in E^*$, the optimization problem

$$\text{find } x_0 \in E \text{ such that } f(x_0) - x^*(x_0) = \inf_{x \in E} (f(x) - x^*(x)) \tag{10.29}$$

admits a solution. This result guarantees the solvability of nonlinear variational equations derived from the weak formulation of a wide range of boundary value problems. For instance, given $1 < p < \infty$, a positive integer N, and a bounded open subset Ω of $\mathbb{R}^N$, let E be the reflexive Sobolev space $W_0^{1,p}(\Omega)$ and consider the coercive, convex, and continuous function $f : E \longrightarrow \mathbb{R}$ defined by

$$f(x) := \frac{1}{p}\int_\Omega |\nabla x|^p d\lambda \qquad (x \in E),$$

where $|\cdot|$ is the Euclidean norm. By the main theorem on convex minimum problems we have $\partial f(E) = E^*$. But taking into account that the p-laplacian operator $\triangle_p$, defined for each $x \in E$ as

$$\triangle_p(x) := \operatorname{div}\left(|\nabla x|^{p-2}\nabla x\right),$$

satisfies that for all $x \in E$

$$\partial f(x) = \{-\triangle_p x\}$$

(see [98, Proposition 6.1]), then given any $h^* \in E^*$, the nonlinear boundary value problem

$$\begin{cases} -\triangle_p x &= h^* \quad \text{in } \Omega \\ \quad\;\; x &= 0 \quad \text{on } \partial\Omega \end{cases}$$

admits a weak solution $x \in E$.

We conclude this subsection by applying Theorem 10.56 (see also Remark 10.57) to show that the adequate setting for dealing with some common variational problems, as p-laplacian above, is that of the reflexive spaces. To properly frame the result it is convenient to recall some usual notions. For a Banach space E, an operator $\Phi : E \longrightarrow E^*$ is said to be *strongly monotone* if

$$\inf_{\substack{x,y\in E \\ x\neq y}} \frac{\langle \Phi(x) - \Phi(y), x-y\rangle}{\|x-y\|^2} > 0,$$

hemicontinuous if for all $x,y,z \in E$, the function

$$t \in [0,1] \mapsto (\Phi(x+ty))(z) \in \mathbb{R}$$

is continuous, *bounded* when the image under Φ of a bounded set is also bounded, and *coercive* whenever the function

$$x \in E \mapsto (\Phi(x))(x) \in \mathbb{R}$$

is coercive. The result below appears in [28, Corollary 2.101] and it includes as a special case the celebrated Lax–Milgram theorem:

Proposition 10.62. *If E is a reflexive Banach space and $\Phi : E \longrightarrow E^*$ is a monotone, hemicontinuous, bounded, and coercive operator, then Φ is surjective.*

This result applies to several problems in nonlinear variational analysis, including one of its most popular particular cases: in a real reflexive Banach space E, given $x_0^* \in E^*$, the equation

$$\text{find } x \in E \text{ such that } \Phi(x) = x_0^*$$

admits a unique solution, whenever $\Phi : E \longrightarrow E^*$ is a Lipschitz continuous and strongly monotone operator. We refer to [70, Example 3.51] for usual applications.

When Φ is symmetric, that is,

$$\text{for every } x, y \in E, \quad \langle \Phi(x), y\rangle = \langle \Phi(y)), x\rangle,$$

the equation $\Phi(x) = x_0^*$ leads to the nonlinear optimization problem involving the function

$$f(x) := \frac{1}{2}(\Phi(x))(x), \qquad x \in E.$$

As a consequence of Theorem 10.56, or more specifically of Remark 10.57, the natural context for Proposition 10.62, at least with symmetry, is the reflexive one, as shown in the next corollary whose proof is completely analogous to that of [118, Corollary 3]:

Corollary 10.63. *A Banach space E is reflexive, provided there exists a monotone, symmetric, and surjective operator $\Phi : E \longrightarrow E^*$.*

10.6.2 Mathematical Finance

We now turn our attention to some recent applications of James' weak compactness theorem in mathematical finance. Let us fix a probability space $(\Omega, \mathscr{F}, \mathbb{P})$ together with $\mathscr{X}$, a linear space of functions in $\mathbb{R}^\Omega$ that contains the constant functions. We assume here that $(\Omega, \mathscr{F}, \mathbb{P})$ is atomless, although in practice this is not a restriction, since the property of being atomless is equivalent to the fact that we can define a random variable on $(\Omega, \mathscr{F}, \mathbb{P})$ that has a continuous distribution function. The space $\mathscr{X}$ will describe all possible financial positions $X : \Omega \longrightarrow \mathbb{R}$, where $X(\omega)$ is the discounted net worth of the position at the end of the trading period if the scenario $\omega \in \Omega$ is realized. The problem of quantifying the risk of a financial position $X \in \mathscr{X}$ is modeled with functions $\rho : \mathscr{X} \longrightarrow \mathbb{R}$ that satisfy:

(i) *Monotonicity*: if $X \leq Y$, then $\rho(X) \geq \rho(Y)$.
(ii) *Cash invariance*: if $m \in \mathbb{R}$ then $\rho(X+m) = \rho(X) - m$.

Such a function ρ is called a *monetary measure of risk* (see Chapter 4 in [54]). When ρ is also a convex function, then it is called a *convex measure of risk*. In many occasions we have $\mathscr{X} = \mathbb{L}^\infty(\Omega, \mathscr{F}, \mathbb{P})$, and it is important to have results for representing the risk measure as

$$\rho(X) = \sup_{Y \in \mathbb{L}^1(\Omega, \mathscr{F}, \mathbb{P})} \{\mathbb{E}[Y \cdot X] - \rho^*(Y)\}. \tag{10.30}$$

Here ρ^* is the Fenchel–Legendre conjugate of ρ, that is, for every $Y \in (\mathbb{L}^\infty(\Omega, \mathscr{F}, \mathbb{P}))^*$,

$$\rho^*(Y) = \sup_{X \in \mathbb{L}^\infty(\Omega, \mathscr{F}, \mathbb{P})} \{\langle Y, X\rangle - \rho(X)\}.$$

To have this representation is equivalent to have the so-called *Fatou property*, *i.e.*, for any bounded sequence $\{X_n\}_{n\geq 1}$ that converges pointwise almost surely (shortly, a.s) to some X,

$$\rho(X) \leq \liminf_n \rho(X_n)$$

(see [54, Theorem 4.31]). A natural question is whether the supremum (10.30) is attained. In general the answer is no, as it is shown by the essential supremum map on $\mathbb{L}^\infty(\Omega, \mathscr{F}, \mathbb{P})$; see [54, Example 4.36]. The representation formula (10.30) with a maximum instead of a supremum has been studied by Delbaen (see [42, Theorems 8 and 9]) (see also [54, Corollary 4.35]) in the case of coherent risk measures, that is, the convex ones that also are positively homogeneous. The fact that the order continuity of ρ is equivalent to the supremum becoming a maximum, that is, for every $X \in \mathbb{L}^\infty(\Omega, \mathscr{F}, \mathbb{P})$:

$$\rho(X) = \max_{Y \in \mathbb{L}^1(\Omega, \mathscr{F}, \mathbb{P})} \{\mathbb{E}[Y \cdot X] - \rho^*(Y)\},$$

for an arbitrary convex risk measure ρ, is the statement of the so-called Jouini–Schachermayer–Touzi theorem in [42, Theorem 2] (see also [91, Theorem 5.2] for the original reference). Let us remark that order sequential continuity for a map ρ in $\mathbb{L}^\infty(\Omega, \mathscr{F}, \mathbb{P})$ is equivalent to have

$$\lim_n \rho(X_n) = \rho(X),$$

whenever $\{X_n\}_{n\geq 1}$ is a bounded sequence in $\mathbb{L}^\infty$ pointwise a.s. convergent to X. Indeed, it is said that a map $\rho : \mathbb{L}^\infty(\Omega, \mathscr{F}, \mathbb{P}) \longrightarrow \mathbb{R} \cup \{+\infty\}$ verifies the *Lebesgue property* provided that it is sequentially order continuous. The precise statement is the following one:

Theorem 10.64 (Jouini, Schachermayer, and Touzi). *Let $\rho : \mathbb{L}^\infty(\Omega,\mathscr{F},\mathbb{P}) \longrightarrow \mathbb{R}$ be a convex risk measure with the Fatou property, and let $\rho^* : (\mathbb{L}^\infty(\Omega,\mathscr{F},\mathbb{P}))^* \longrightarrow [0,+\infty]$ be its Fenchel–Legendre conjugate. The following are equivalent:*

(i) For every $c \in \mathbb{R}$, $\{Y \in \mathbb{L}^1(\Omega,\mathscr{F},\mathbb{P}) : \rho^*(Y) \le c\}$ is a weakly compact subset of $\mathbb{L}^1(\Omega,\mathscr{F},\mathbb{P})$.

(ii) For every $X \in \mathbb{L}^\infty(\Omega,\mathscr{F},\mathbb{P})$, the supremum in the equality

$$\rho(X) = \sup_{Y\in\mathbb{L}^1(\Omega,\mathscr{F},\mathbb{P})} \{\mathbb{E}[XY] - \rho^*(Y)\}$$

is attained.

(iii) For every bounded sequence $\{X_n\}_{n\ge1}$ in $\mathbb{L}^\infty(\Omega,\mathscr{F},\mathbb{P})$ tending a.s. to $X \in \mathbb{L}^\infty(\Omega,\mathscr{F},\mathbb{P})$, we have

$$\lim_n \rho(X_n) = \rho(X).$$

The proof of this result required compactness arguments of the perturbed James type and it was based on Theorem 10.54; see [91, Theorem A.1]. In [42] this result is already presented as a generalization of James' weak compactness theorem. Let us observe that we can apply Theorem 10.60 for $f = \rho^*$ to obtain the proof for the main implication (ii) $\Rightarrow$ (i) above. Indeed, $\mathbb{L}^1(\Omega,\mathscr{F},\mathbb{P})$ is weakly compactly generated and so its dual ball is w^*-sequentially compact.

Delbaen gave a different approach for Theorem 10.64. His proof is valid for nonseparable $\mathbb{L}^1(\Omega,\mathscr{F},\mathbb{P})$ spaces, and it is based in a homogenization trick to reduce the matter to a direct application of the classical James' weak compactness theorem, as well as the Dunford–Pettis theorem characterizing weakly compact sets in $\mathbb{L}^1(\Omega,\mathscr{F},\mathbb{P})$.

For our next application let us recall that a *Young function* Ψ is an even, convex function $\Psi : E \to [0,+\infty]$ with the properties:

1. $\Psi(0) = 0$
2. $\lim_{x\to\infty} \Psi(x) = +\infty$
3. $\Psi < +\infty$ in a neighborhood of 0

The Orlicz space L^Ψ is defined as

$$L^\Psi(\Omega,\mathscr{F},\mathbb{P}) := \{X \in L^0(\Omega,\mathscr{F},\mathbb{P}) : \text{there exists } \alpha > 0 \text{ with } \mathrm{e}_\mathbb{P}[\Psi(\alpha X)] < +\infty\},$$

and we consider the Luxembourg norm on it:

$$N_\Psi(X) := \inf\left\{c > 0 : \mathrm{e}_\mathbb{P}\left[\Psi\left(\frac{1}{c}X\right)\right] \le 1\right\}, \qquad (X \in L^\Psi(\Omega,\mathscr{F},\mathbb{P})).$$

With the usual pointwise lattice operations, $L^\Psi(\Omega,\mathscr{F},\mathbb{P})$ is a Banach lattice and we have the inclusions

$$L^\infty(\Omega,\mathscr{F},\mathbb{P}) \subset L^\Psi(\Omega,\mathscr{F},\mathbb{P}) \subset L^1(\Omega,\mathscr{F},\mathbb{P}).$$

Moreover, $(L^\Psi)^* = L^{\Psi^*} \oplus G$ where G is the singular band and L^{Ψ^*} is the order continuous band identified with the Orlicz space L^{Ψ^*}, where

$$\Psi^*(y) := \sup_{x\in\mathbb{R}}\{yx - \Psi(x)\}$$

is the Young function conjugate to Ψ, [126].

Risk measures defined on $L^\Psi(\Omega,\mathscr{F},\mathbb{P})$ and their robust representation are of interest in mathematical finance too. Delbaen has recently proved that a risk measure defined on $\mathbb{L}^\infty(\Omega,\mathscr{F},\mathbb{P})$ finitely extends to an Orlicz space if, and only if, it verifies the equivalent conditions of Theorem 10.64; see [43, Sect. 4.16]. Theorem 10.64 is extended to Orlicz spaces in [119, Theorem 1].

Theorem 10.65 (Lebesgue risk measures in Orlicz spaces). *Let Ψ be a Young function with finite conjugate Ψ^* and let*

$$\alpha : (\mathbb{L}^\Psi(\Omega,\mathscr{F},\mathbb{P}))^* \to \mathbb{R}\cup\{+\infty\}$$

be a $\sigma((\mathbb{L}^\Psi)^*,\mathbb{L}^\Psi)$-lower semicontinuous penalty function representing a finite monetary risk measure ρ as

$$\rho(X) = \sup_{Y\in\mathbb{M}^{\Psi^*}}\{-\mathbb{E}[X\cdot Y] - \alpha(Y)\}.$$

The following are equivalent:

(i) For each $c\in\mathbb{R}$, $\alpha^{-1}((-\infty,c])$ is a weakly compact subset of $\mathbb{M}^{\Psi^*}(\Omega,\mathscr{F},\mathbb{P})$.
(ii) For every $X\in\mathbb{L}^\Psi(\Omega,\mathscr{F},\mathbb{P})$, the supremum in the equality

$$\rho(X) = \sup_{Y\in\mathbb{M}^{\Psi^*}}\{-\mathbb{E}[X\cdot Y] - \alpha(Y)\}$$

is attained.
(iii) ρ is order sequentially continuous.

Let us notice that order sequential continuity for a map ρ in $\mathbb{L}^\Psi$ is equivalent to having

$$\lim_n \rho(X_n) = \rho(X)$$

whenever (X_n) is a sequence in L^Ψ a.s. convergent to X and bounded by some $Z\in L^\Psi$, i.e., $|X_n|\le Z$ for all $n\in\mathbb{N}$. For that reason it is also said that a map $\rho : L^\Psi \to (-\infty,+\infty]$ verifies the Lebesgue property whenever it is sequentially order continuous. Orlicz spaces provide a general framework of Banach lattices for applications in mathematical finance, for a general picture see [18, 19, 38].

Noncoercive growing conditions for penalty functions in the Orlicz case have been studied in [38]. More precisely, let us recall that a Young function Φ verifies the Δ_2 condition if there exist $t_0 > 0$ and $K > 0$ such that for every $t > t_0$

$$\Phi(2t) \leq K\Phi(t).$$

In addition, the Orlicz heart M^Ψ is the Morse subspace of all $X \in L^\Psi$ such that for every $\beta > 0$

$$e_{\mathbb{P}}[\Psi(\beta X)] < +\infty.$$

In [38, Theorem 4.5] it is proved that a risk measure ρ, defined by a penalty function α, is finite on the Morse subspace $\mathbb{M}^\Psi \subset L^\Psi$ if, and only if, α satisfies the growing condition

$$\alpha(Y) \geq a + b\|Y\|_{\Psi^*}$$

for all $Y \in \mathbb{L}^{\Psi^*}$, and fixed numbers a, b with $b > 0$. Theorem 10.60 can be applied for $f = \rho^*$ because the spaces involved in the representation formulas have w^*-sequentially compact dual balls.

When Ψ is a Young function such that either Ψ or its conjugate verify the Δ_2 condition we have the following result for the risk measures studied by Cheredito and Li in [38]:

Corollary 10.66 ([119], Corollaries 6 and 7). *Let Ψ be a Young with finite conjugate Ψ^* and such that either Ψ or Ψ^* verify the Δ_2 condition. Let $\rho : \mathbb{L}^\Psi(\Omega, \mathscr{F}, \mathbb{P}) \to \mathbb{R}$ be a finite convex risk measure with the Fatou property, and*

$$\rho^* : \mathbb{L}^{\Psi^*}(\Omega, \mathscr{F}, \mathbb{P}) \to \mathbb{R} \cup \{+\infty\}$$

its Fenchel–Legendre conjugate defined on the dual space. The following are equivalent:

(i) For every $c \in \mathbb{R}$, $(\rho^*)^{-1}((-\infty, c])$ is a weakly compact subset of $\mathbb{M}^{\Psi^*}(\Omega, \mathscr{F}, \mathbb{P})$.
(ii) For every $X \in \mathbb{L}^\Psi(\Omega, \mathscr{F}, \mathbb{P})$, the supremum in the equality

$$\rho(X) = \sup_{Y \in (\mathbb{M}^{\Psi^*})^+, e(Y)=1} \{-\mathbb{E}[X \cdot Y] - \rho^*(-Y)\}$$

is attained.
(iii) ρ is sequentially order continuous.
(iv) $\lim_n \rho(X_n) = \rho(X)$ whenever $X_n \nearrow X$ in $\mathbb{L}^\Psi$.
(v) $\operatorname{dom}(\rho^*) \subset \mathbb{M}^{\Psi^*}$.

We conclude this section with the following question:

Question 10.67. Does Corollary 10.63 remain valid in absence of symmetry?

Acknowledgements *To finish our contribution let us remark we are very grateful to the anonymous referee who highly improved the redaction of our paper.*
B. Cascales and J. Orihuela's research was partially supported by MTM2008-05396 and MTM2011-25377/MTM Fondos FEDER; Fundación Sénaca 08848/PI/08, CARM; and that of M. Ruiz Galán by Junta de Andalucía grant FQM359.

References

1. Acosta, M.D., Kadets, V.: A characterization of reflexivity. Math. Ann. **349**, 577–588 (2011)
2. Acosta, M.D., Montesinos, V.: On a problem of Namioka on norm-attaining functionals. Math. Z. **256**, 295–300 (2007)
3. Acosta, M.D., Ruiz Galán, M.: New characterizations of the reflexivity in terms of the set of norm attaining functionals. Canad. Math. Bull. **41**, 279–289 (1998)
4. Acosta, M.D., Ruiz Galán, M.: Norm attaining operators and reflexivity. Rend. Circ. Mat. Palermo **56**, 171–177 (1998)
5. Acosta, M.D., Ruiz Galán, M.: A version of James' theorem for numerical radius. Bull. London Math. Soc. **31**, 67–74 (1999)
6. Acosta, M.D., Becerra Guerrero, J., Ruiz Galán, M.: Dual spaces generated by the interior of the set of norm attaining functionals. Studia Math. **149**, 175–183 (2002)
7. Acosta, M.D., Becerra Guerrero, J., Ruiz Galán, M.: Numerical-radius-attaining polynomials. Quart. J. Math. **54**, 1–10 (2003)
8. Acosta, M.D., Becerra Guerrero, J., Ruiz Galán, M.: James type results for polynomials and symmetric multilinear forms. Ark. Mat. **42**, 1–11 (2004)
9. Acosta, M.D., Aizpuru, A., Aron, R., García-Pacheco, F.J.: Functionals that do not attain their norm. Bull. Belg. Math. Soc. Simon Stevin **14**, 407–418 (2007)
10. Angosto, C.: Distance to function spaces. Ph.D. thesis, Universidad de Murcia (2007)
11. Angosto, C., Cascales, B.: Measures of weak noncompactness in Banach spaces. Topology Appl. **156**, 1412–1421 (2009)
12. Astala, K., Tylli, H.O.: Seminorms related to weak compactness and to Tauberian operators. Math. Proc. Cambridge Philos. Soc. **107**, 367–375 (1990)
13. Azagra, D., Dobrowolski, T.: Smooth negligibility of compact sets in infinite-dimensional Banach spaces, with applications. Math. Ann. **312**, 445–463 (1998)
14. Banaś, J., Martinón, A.: Measures of weak noncompactness in Banach sequence spaces. Portugal. Math. **52**, 131–138 (1995)
15. Bandyopadhyay, P., Godefroy, G.: Linear structures in the set of norm-attaining functionals on a Banach space. J. Convex Anal. **13**, 489–497 (2006)
16. Bauschke, H.H., Borwein, J.M., Combettes, P.L.: Essential smoothness, essential strict convexity, and Legendre functions in Banach spaces. Commun. Contemp. Math. **3**, 615–647. (2001)
17. Behrends, E.: New proofs of Rosenthal's ℓ^1−theorem and the Josefson–Nissenzweig theorem. Bull. Pol. Acad. Sci. Math. **43**, 283–295 (1996)
18. Biagini, S., Fritelli, M.: A unified framework for utility maximization problems: an Orlicz space approach. Ann. Appl. Prob. **18**, 929–966 (2008)
19. Biagini, S., Fritelli, M.: On the extension of the Namioka–Klee theorem and on the fatou property for risk measures. Optimality and Risk–Modern Trends in Mathematical Finance: The Kabanov Festschrift, pp. 1–28. Springer, Berlin (2009)
20. De Blasi, F.S.: On a property of the unit sphere in a Banach space. Colloq. Math. **65**, 333–343 (1992)
21. Bonet, J., Cascales, B.: Noncomplete Mackey topologies on Banach spaces. Bull. Aust. Math. Soc. **81**, 409–413 (2010)

22. Borwein, J.M., Zhu, Q.J.: Techniques of Variational Analysis. CMS Books in Mathematics, vol. 20. Springer, New York (2005)
23. Borwein, J.M., Cheng, L., Fabián, M., Revalski, J.P.: A one perturbation variational principle and applications. Set-Valued Anal. **12**, 49–60 (2004)
24. Bourgain, J.: La propiété de Radon Nikodym. Publications Mathématiques de l'Université Pierre et Marie Curie, **36** (1979)
25. Bourgain, J., Talagrand, M.: Compacité extrêmale. Proc. Am. Math. Soc. **80**, 68–70 (1980)
26. Bourgin, R.D.: Geometric Aspects of Convex Sets with the Radon–Nikodym Property. Lecture Notes in Mathematics, vol. 993. Springer, Berlin (1983)
27. Calvert, B., Fitzpatrick, S.: Erratum: In a nonreflexive space the subdifferential is not onto. Math. Zeitschrift **235**, 627 (2000)
28. Carl, S., Le, V.K., Motreanu, D.: Nonsmooth Variational Problems and Their Inequalities: Comparison Principles and Applications. Springer Monographs in Mathematics. Springer, New York (2007)
29. Cascales, B., Godefroy, G.: Angelicity and the boundary problem. Mathematika **45**, 105–112 (1998)
30. Cascales, B., Shvydkoy, R.: On the Krein–Šmulian theorem for weaker topologies. Illinois J. Math. **47**, 957–976 (2003)
31. Cascales, B., Vera, G.: Topologies weaker than the weak topology of a Banach space. J. Math. Anal. Appl. **182**, 41–68 (1994)
32. Cascales, B., Manjabacas, G., Vera, G.: A Krein–Šmulian type result in Banach spaces. Quart. J. Math. Oxford **48**, 161–167 (1997)
33. Cascales, B., Marciszesky, W., Raja, M.: Distance to spaces of continuous functions. Topology Appl. **153**, 2303–2319 (2006)
34. Cascales, B., Muñoz, M., Orihuela, J.: James boundaries and σ-fragmented selectors. Studia Math. **188**, 97–122 (2008)
35. Cascales, B., Fonf, V.P., Orihuela, J., Troyanski, S.: Boundaries of Asplund spaces. J. Funct. Anal. **259**, 1346–1368 (2010)
36. Cascales, B., Kalenda, O.F.K., Spurný, J.: A quantitative version of James's compactness theorem. Proc. Roy. Edinburgh Soc. **35**(2), 369–386 (2012)
37. Cellina, A.: On the nonexistence of solutions of differential equations in nonreflexive spaces. Bull. Am. Math. Soc. **78**, 1069–1072 (1972)
38. Cheridito, P., Li, T.: Risks measures on Orlicz hearts. Math. Finance **19**, 189–214 (2009)
39. Contreras, M., Payá, R.: On upper semicontinuity of duality mappings. Proc. Am. Math. Soc. **121**, 451–459 (1994)
40. Corson, H.H.: Normality in subsets of product spaces. Am. J. Math. **81**, 785–796 (1959)
41. Debs, G., Godefroy, G., Saint Raymond, J.: Topological properties of the set of norm-attaining linear functionals. Canad. J. Math. **47**, 318–329 (1995)
42. Delbaen, F.: Differentiability properties of utility functions. In: Delbaen, F. et al. (eds.) Optimality and Risk–Modern Trends in Mathematical Finance, pp. 39–48. Springer, New York (2009)
43. Delbaen, F.: Draft: Monetary utility functions. Lectures Notes in preparation, CSFI 3, Osaka University Press (2012)
44. Deville, R., Finet, C.: An extension of Simons' inequality and applications. Revista Matemática Complutense **14**, 95–104 (2001)
45. Deville, R., Godefroy, G., Zizler, V.: Smoothness and renormings in Banach spaces. Pitman Monographs and Surveys in Pure and Applied Mathematics, vol. 64. Longman Scientific and Technical, New York (1993)
46. Diestel, J.: Sequences and series in Banach spaces. Graduate Texts in Mathematics, vol. 92. Springer, New York (1984)
47. Dieudonné, J.: Deux exemples singuliérs d'équations différentielles. Acta Sci. Math. Szeged **12**, Leopoldo Fejér et Frererico Riesz LXX annos natis dedicatus, pars B, pp. 38–40 (1950)

48. Engelking, R.: General Topology. Translated from the Polish by the author. Monografie Matematyczne, Tom 60 (Mathematical Monographs, vol. 60). PWN–Polish Scientific Publishers, Warsaw (1977)
49. Fabian, M., Habala, P., Hájek, P., Montesinos, V., Pelant, J., Zizler, V.: Functional Analysis and Infinite-Dimensional Geometry. CMS Books in Mathematics, vol. 8. Springer, New York (2001)
50. Fabian, M., Hájek, P., Montesinos, V., Zizler, V.: A quantitative version of Krein's theorem. Rev. Mat. Iberoamericana **21**, 237–248 (2005)
51. Fabian, M., Habala, P., Hájek, P., Montesinos, V., Zizler, V.: Banach Space Theory: The Basis for Linear and Nonlinear Analysis. CMS Books in Mathematics. Springer, New York (2011)
52. Fitzpatrick, S., Calvert, B.: In a nonreflexive space the subdifferential is not onto. Math. Zeitschhrift **189**, 555–560 (1985)
53. Floret, K.: Weakly Compact Sets. Lecture Notes in Mathematics, vol. 801. Springer, Berlin (1980)
54. Föllmer, H., Schied, A.: Stochastic Finance. An Introduction in Discrete Time, 2nd edn. de Gruyter Studies in Mathematics vol. 27. Walter de Gruyter, Berlin (2004)
55. Fonf, V.P., Lindenstrauss, J.: Boundaries and generation of convex sets. Israel J. Math. **136**, 157–172 (2003)
56. Fonf, V.P., Lindenstrauss, J., Phelps, R.R.: Infinite-dimensional convexity. In: Johnson,W.B., Lindenstrauss, J. (eds.) Handbook of Banach Spaces, vol. I, pp. 599–670. Elsevier, North-Holland (2001)
57. García-Pacheco, F.J., Puglisi, D.: Lineability of functionals and operators. Studia Math. **201**, 37–47 (2010)
58. Glicksberg, I.: The representation of functionals by integrals. Duke Math. J. **19**, 253–261 (1952)
59. Godefroy, G.: Boundaries of convex sets and interpolation sets. Math. Ann. **277**, 173–184 (1987)
60. Godefroy, G.: Five lectures in geometry of Banach spaces. In: Seminar on Functional Analysis, 1987 (Murcia, 1987). Notas de Matemtica, vol. 1, pp. 9–67. University of Murcia, Murcia (1988)
61. Godefroy, G.: Some applications of Simons' inequality. Serdica Math. J. **26**, 59–78 (2000)
62. Godunov, A.N.: The Peano theorem in Banach spaces. Funkcional. Anal. i Priložen **9**, 59–60 (1974)
63. González, M., Gutiérrez, J.M.: Polynomial Grothendieck properties. Glasgow Math. J. **37**, 211–219 (1995)
64. Granero, A.S.: An extension of the Krein–Šmulian theorem. Rev. Mat. Iberoam. **22**, 93–110 (2006)
65. Granero, A.S., Hernández, J.M: On James boundaries in dual Banach spaces. J. Funct. Anal. **263**, 429–447 (2012)
66. Granero, A.S., Sánchez, M.: Convexity, compactness and distances. In: Methods in Banach Space Theory. London Mathematical Society Lecture Note Series, vol. 337, pp. 215–237. Cambridge University Press, Cambridge (2006)
67. Granero, A.S., Sánchez, M.: The class of universally Krein–Šmulian Banach spaces. Bull. Lond. Math. Soc. **39**, 529–540 (2007)
68. Granero, A.S., Hájek, P., Santalucía, V.M.: Convexity and w^*-compactness in Banach spaces. Math. Ann. **328**, 625–631 (2004)
69. Granero, A.S., Hernández, J.M., Pfitzner, H.: The distance $\mathrm{dist}(\mathscr{B},X)$ when $\mathscr{B}$ is a boundary of $B(X^{**})$. Proc. Am. Math. Soc. **139**, 1095–1098 (2011)
70. Grossmann, C., Roos, H., Stynes, M.: Numerical Treatment of Partial Differential Equations (Universitex). Springer, Berlin (2007)
71. Grothendieck, A.: Critères de compacité dans les espaces fonctionnels generaux. Am. J. Math. **74**, 168–186 (1952)
72. Hagler, J., Johnson, W.B.: On Banach spaces whose dual balls are not weak* sequentially compact. Israel J. Math. **28**, 235–330 (1977)

73. Hagler, J., Odell, E.: A Banach space not containing ℓ_1 whose dual ball is not weak* sequentially compact. Illinois J. Math. **22**, 290–294 (1978)
74. Hájek, P., Montesinos, V., Vanderwerff, J., Zizler, V.: Biorthogonal Systems in Banach Spaces. CMS Books in Mathematics, vol. 26. Springer, New York (2008)
75. Hardtke, J.: Rainwater–Simons type convergence theorems for generalized convergence methods. Acta Comm. Univ. Tartu. Math. **14**, 65–74 (2010)
76. Harmand, P., Werner, D., Werner, W.: M-Ideals in Banach Spaces and Banach Algebras. Lecture Notes in Mathematics, vol. 1547. Springer, Berlin (1993)
77. Haydon, R.: An extreme point criterion for separability of a dual Banach space, and a new proof of a theorem of Corson. Quart. J. Math. Oxford **27**, 379–385 (1976)
78. Haydon, R.: Some more characterizations of Banach spaces containing l_1. Math. Proc. Cambridge Philos. Soc. **80**, 269–276 (1976)
79. Haydon, R.: On Banach spaces which contain $l^1(\tau)$ and types of measures on compact spaces. Israel J. Math. **28**, 313–324 (1977)
80. Haydon, R., Levy, M., Odell, E.: On sequences without weak* convergent convex block subsequences. Proc. Am. Mat. Soc. **100**, 94–98 (1987)
81. Holmes, R.: Geometric Functional Analysis and Its Applications. Graduate Texts in Mathematics, vol. 24. Springer, New York (1975)
82. James, R.C.: Reflexivity and the supremum of linear functionals. Ann. Math. **66**, 159–169 (1957)
83. James, R.C.: Characterizations of reflexivity. Studia Math. **23**, 205–216 (1964)
84. James, R.C.: Weakly compact sets. Trans. Am. Math. Soc. **113**, 129–140 (1964)
85. James, R.C.: Weak compactness and reflexivity. Israel J. Math. **2**, 101–119 (1964)
86. James, R.C.: A counterexample for a sup theorem in normed spaces. Israel J. Math. **9**, 511–512 (1971)
87. James, R.C.: Reflexivity and the Sup of Linear Functionals. Israel J. Math. **13**, 289–300 (1972)
88. Jayne, J.E., Orihuela, J., Pallarés, A.J., Vera, G.: σ−fragmentability of multivalued maps and selection theorems. J. Funct. Anal. **117**, 243–273 (1993)
89. Jiménez-Sevilla, M., Moreno, J.P.: A note on norm attaining functionals. Proc. Am. Math. Soc. **126**, 1989–1997 (1998)
90. Johnson, W.B., Lindenstrauss, J.: Basic concepts in the geometry of Banach spaces. In: Handbook of the Geometry of Banach Spaces, vol. I, pp. 1–84. Elsevier, North–Holland (2001)
91. Jouini, E., Schachermayer, W., Touzi, N.: Law invariant risk measures have the Fatou property. Adv. Math. Econ. **9**, 49–71 (2006)
92. Kalenda, O.: (I)-envelopes of unit balls and James' characterization of reflexivity. Studia Math. **182**, 29–40 (2007)
93. Kalenda, O.: (I)-envelopes of closed convex sets in Banach spaces. Israel J. Math. **162**, 157–181 (2007)
94. Kalenda, O., Spurny, J.: Boundaries of compact convex sets and fragmentability. J. Funct. Anal. **256**, 865–880 (2009)
95. Kelley, J.L.: General Topology. Reprint of the 1955 edition (Van Nostrand, Toronto). Graduate Texts in Mathematics, vol. 27. Springer, New York (1975)
96. Kelley, J.L., Namioka, I.: Linear topological spaces. With the collaboration of W. F. Donoghue, Jr., Kenneth R. Lucas, B. J. Pettis, Ebbe Thue Poulsen, G. Baley Price, Wendy Robertson, W. R. Scott, and Kennan T. Smith. Second corrected printing. Graduate Texts in Math. Springer-Verlag, New York (1976)
97. Kenderov, P.S., Moors, W.B., Sciffer, S.: Norm attaining functionals on $C(T)$. Proc. Am. Math. Soc. **126**, 153–157 (1998)
98. Kenmochi, N.: Monotonicity and compactness methods for nonlinear variational inequalities. In: Chipot, M. (ed.) Handbook of Differential Equations: Stationary Partial Differential Equations, vol. 4. Elsevier, Amsterdam (2007)

99. Kivisoo, K., Oja, E.: Extension of Simons' inequality. Proc. Am. Math. Soc. **133**, 3485–3496 (2005)
100. Klee, V.L.: Some characterizations of reflexivity. Revista de Ciencias **52**, 15–23 (1950)
101. König, H.: Theory and applications of superconvex spaces. In: Nagel, R., Schlotterbeck, U., Wolff, M.P.H. (eds.) Aspects of Positivity in Functional Analysis. North–Holland (1986)
102. Köthe, G.: Topological vector spaces. I. Translated from the German by D. J. H. Garling. Die Grundlehren der Mathematischen Wissenschaften, Band 159. Springer, New York (1969)
103. Kryczka, A.: Quantitative approach to weak noncompactness in the polygon interpolation method. Bull. Austral. Math. Soc. **69**, 49–62 (2004)
104. Kryczka, A., Prus, S.: Measure of weak noncompactness under complex interpolation. Studia Math. **147**, 89–102 (2001)
105. Kryczka, A., Prus, S., Szczepanik, M.: Measure of weak noncompactness and real interpolation of operators. Bull. Austral. Math. Soc. **62**, 389–401 (2000)
106. Lacey, H.E.: The isometric theory of classical Banach spaces. Die Grundlehren der Mathematischen Wissenschaften, Band 208. Springer, New York (1974)
107. Mazur, S.: Über schwache konvergenz in den Raumen ℓ_p. Studia Math. **4**, 128–133 (1933)
108. Moors, W.B.: An elementary proof of James' characterization of weak compactness. Bull. Aust. Math. Soc. **84**, 98–102 (2011)
109. Moors, W.B., Giles, J.R.: Generic continuity of minimal set-valued mappings. J. Austral. Math. Soc. Ser. A **63**, 238–262 (1997)
110. Moors, W.B., Reznichenko, E.A.: Separable subspaces of affine function spaces on convex compact sets. Topology Appl. **155**, 1306–1322 (2008)
111. Morillon, M.: A new proof of James' sup theorem. Extracta Math. **20**, 261–271 (2005)
112. Morillon, M.: James sequences and dependent choices. Math. Log. Quart. **51**, 171–186 (2005)
113. Namioka, I.: Radon–Nikodým compact spaces and fragmentability. Mathematika **34**, 258–281 (1987)
114. Neumann, M.: Varianten zum Konvergenzsatz von Simons und Anwendungen in der Choquet theorie. Arch. Math. **28**, 182–192 (1977)
115. Oja, E.: A proof of the Simons inequality. Acta Comment. Univ. Tartu. Math. **2**, 27–28 (1998)
116. Oja, E.: Geometry of Banach spaces having shrinking approximations of the identity. Trans. Am. Math. Soc. **352**, 2801–2823 (2000)
117. Orihuela, J.: Pointwise compactness in spaces of continuous functions. J. London Math. Soc. **36**, 143–152 (1987)
118. Orihuela, J., Ruiz Galán, M.: A coercive James's weak compactness theorem and nonlinear variational problems. Nonlinear Anal. **75**, 598–611 (2012)
119. Orihuela, J., Ruiz Galán, M.: Lebesgue property for convex risk measures on Orlicz spaces. Math. Finan. Econ. **6**, 15–35 (2012)
120. Pełczyński, A.: On Banach spaces containing $L_1(\mu)$. Studia Math. **30**, 231–246 (1968)
121. Petunin, J.I., Plichko, A.N.: Some properties of the set of functionals that attain a supremum on the unit sphere. Ukrain. Math. Zh. **26**, 102–106 (1974)
122. Pfitzner, H.: Boundaries for Banach spaces determine weak compactness. Invent. Math. **182**, 585–604 (2010)
123. Plichko, A.: On sequential property of Banach spaces and duality (2003, preprint)
124. Pol, R.: On a question of H.H. Corson and some related problems. Fund. Math. **109**, 143–154 (1980)
125. Pryce, J.D.: Weak compactness in locally convex spaces. Proc. Am. Math. Soc. **17**, 148–155 (1966)
126. Rao, M.M., Ren, Z.D.: Theory of Orlicz Spaces. Marcel Dekker, New York (1991)
127. Rainwater, J.: Weak convergence of bounded sequences. Proc. Am. Math. Soc. **14**, 999 (1963)
128. Rockafellar, R.T.: Characterization of the subdifferentials of convex functions. Pacific J. Math. **17**, 497–510 (1966)
129. Rodé, G.: Ein Grenzwertsatz für stützende Funktionen auf superkonvexen Räumen. Dissertation, Universität des Saarlandes, Saarbrücken (1977)
130. Rodé, G.: Superkonvexität und schwache Kompaktheit. Arch. Math. **36**, 62–72 (1981)

131. Ruiz Galán, M.: Polynomials, symmetric multilinear forms and weak compactness. Positivity **8**, 297–304 (2004)
132. Ruiz Galán, M.: Convex numerical radius. J. Math. Anal. Appl. **361**, 481–491 (2010)
133. Ruiz Galán, M., Simons, S.: A new minimax theorem and a perturbed James's theorem. Bull. Austral. Math. Soc. **66**, 43–56 (2002)
134. Saint Raymond, J.: Characterizing convex functions on a reflexive Banach space. Mediterranean J. Math. **10**(2), 927–940 (2013)
135. Schechter, E.: Handbook of Analysis and Its Foundations. Academic, San Diego (1997)
136. Simons, S.: A convergence theorem with boundary. Pacific J. Math. **40**, 703–708 (1972)
137. Simons, S.: Maximinimax, minimax, and antiminimax theorems and a result of R.C. James. Pacific J. Math. **40**, 709–718 (1972)
138. Simons, S.: An eigenvector proof of Fatou's lemma for continuous functions. Math. Intelligencer **17**, 67–70 (1995)
139. Simons, S.: From Hahn–Banach to Monotonicity. Lecture Notes in Mathematics, vol. 1693. Springer, New York (2008)
140. Smith, M.A., Sullivan, F.: Extremely smooth Banach spaces. In: Baker, J., Cleaver, C., Diestel, J. (eds.) Banach Spaces of Analytic Functions (Proc. Conf. Kent, Ohio, 1976). Lecture Notes in Mathematics, vol. 604, pp. 125–137. Springer, Berlin (1977)
141. Spurný, J.: The boundary problem for L^1-preduals. Illinois J. Math. **52**, 1183–1193 (2008)
142. Srivastava, S.M.: A Course on Borel Sets. Graduate Texts in Mathematics, vol. 180. Springer, New York (1998)
143. Suzuki, T.: Characterizations of reflexivity and compactness via the strong Ekeland variational principle. Nonlinear Anal. **72**, 2204–2209 (2010)
144. Talagrand, M.: Sur les espaces de Banach contenant $l^1(\tau)$. Israel J. Math. **40**, 324–330 (1982)
145. Talagrand, M.: Pettis Integral and Measure Theory. Memoirs of the American Mathematical Society, vol. 307. American Mathematical Society, Providence (1984)
146. Todorcevic, S.: Topics in Topology. Lecture Notes in Math. **1652**. Springer (1997)
147. Tweddle, I.: Weak compactness in locally convex spaces. Glasgow Math. J. **9**, 123–127 (1968)
148. Valdivia, M.: Boundaries of convex sets. Rev. Real Acad. Cienc. Exact. Fís. Natur. Madrid **87**, 177–183 (1993)
149. Valdivia, M.: Fréchet spaces with no subspaces isomorphic to ℓ_1. Math. Japon. **38**, 397–411 (1993)
150. Varadarajan, S.: Measures on topological spaces. Am. Math. Soc. Transl. **48**, 161–228 (1965)
151. Werner, D.: New classes of Banach spaces which are M-ideals in their biduals. Math. Proc. Cambridge Philos. Soc. **111**, 337–354 (1992)
152. De Wilde, M.: Pointwise compactness in spaces of functions and R. C. James theorem. Math. Ann. **208**, 33–47 (1974)
153. Zeidler, E.: Nonlinear Functional Analysis and Its Applications II/B Nonlinear Monotone Operators. Springer, New York (1990)

Chapter 11
Logarithmic and Complex Constant Term Identities

Tom Chappell, Alain Lascoux, S. Ole Warnaar, and Wadim Zudilin

To Jon

Abstract In recent work on the representation theory of vertex algebras related to the Virasoro minimal models $M(2,p)$, Adamović and Milas discovered logarithmic analogues of (special cases of) the famous Dyson and Morris constant term identities. In this paper we show how the identities of Adamović and Milas arise naturally by differentiating as-yet-conjectural complex analogues of the constant term identities of Dyson and Morris. We also discuss the existence of complex and logarithmic constant term identities for arbitrary root systems, and in particular prove such identities for the root system G_2.

Key words: Constant term identities • Jon's birthday • Perfect matchings • Pfaffians • Root systems

Mathematics Subject Classifications (2010): Primary 05E05; Secondary 05A05, 05A10, 05A19, 33C20.

COMMUNICATED BY DAVID H. BAILEY.

T. Chappell • S.O. Warnaar
School of Mathematics and Physics, The University of Queensland,
Brisbane, QLD 4072, Australia
e-mail: thomas.chappell@uqconnect.edu.au; o.warnaar@maths.uq.edu.au

A. Lascoux
CNRS, Institut Gaspard Monge, Université Paris-Est, Marne-la-Vallée, France
e-mail: al@univ-mlv.fr

W. Zudilin (✉)
School of Mathematical and Physical Sciences, The University of Newcastle,
Callaghan, NSW 2308, Australia
e-mail: wadim.zudilin@newcastle.edu.au

D.H. Bailey et al. (eds.), *Computational and Analytical Mathematics*, Springer Proceedings in Mathematics & Statistics 50, DOI 10.1007/978-1-4614-7621-4_11,

11.1 Jonathan Borwein

Jon Borwein is known for his love of mathematical *constants*. We hope this paper will spark his interest in *constant terms*.

11.2 Constant Term Identities

The study of constant term identities originated in Dyson's famous 1962 paper *Statistical theory of the energy levels of complex systems* [9]. In this paper Dyson conjectured that for $a_1,\dots,a_n$ nonnegative integers,

$$\mathrm{CT} \prod_{1\le i\ne j\le n} \Big(1-\frac{x_i}{x_j}\Big)^{a_i} = \frac{(a_1+a_2+\cdots+a_n)!}{a_1!a_2!\cdots a_n!}, \tag{11.1}$$

where $\mathrm{CT} f(X)$ stands for the constant term of the Laurent polynomial (or possibly Laurent series) $f(X) = f(x_1,\dots,x_n)$. Dyson's conjecture was almost instantly proved by Gunson and Wilson [14, 36]. In a very elegant proof, published several years later [13], Good showed that (11.1) is a direct consequence of Lagrange interpolation applied to $f(X) = 1$.

In 1982 Macdonald generalised the equal-parameter case of Dyson's ex-conjecture, i.e.,

$$\mathrm{CT} \prod_{1\le i\ne j\le n} \Big(1-\frac{x_i}{x_j}\Big)^{k} = \frac{(kn)!}{(k!)^n}, \tag{11.2}$$

to all irreducible, reduced root systems; here (11.2) corresponds to the root system A_{n-1}. Adopting standard notation and terminology—see [17] or the next section—Macdonald conjectured that [25]

$$\mathrm{CT} \prod_{\alpha\in\Phi} (1-\mathrm{e}^{\alpha})^k = \prod_{i=1}^{r} \binom{kd_i}{k}, \tag{11.3}$$

where Φ is one of the root systems $\mathrm{A}_{n-1},\mathrm{B}_n,\mathrm{C}_n,\mathrm{D}_n,\mathrm{E}_6,\mathrm{E}_7,\mathrm{E}_8,\mathrm{F}_4,\mathrm{G}_2$ of rank r and $d_1,\dots,d_r$ are the degrees of its fundamental invariants. For $k=1$ the Macdonald conjectures are an easy consequence of Weyl's denominator formula

$$\sum_{w\in W} \mathrm{sgn}(w)\,\mathrm{e}^{w(\rho)-\rho} = \prod_{\alpha>0} (1-\mathrm{e}^{-\alpha})$$

(where W is the Weyl group of Φ and ρ the Weyl vector), and for $\mathrm{B}_n,\mathrm{C}_n,\mathrm{D}_n$ but k general they follow from the Selberg integral. The first uniform proof of (11.3)—based on hypergeometric shift operators—was given by Opdam in 1989 [24].

In his Ph.D. thesis [27] Morris used the Selberg integral to prove a generalisation of (11.2), now commonly referred to as the Morris or Macdonald–Morris constant term identity:

$$\mathrm{CT}\left[\prod_{i=1}^{n}(1-x_i)^a\Big(1-\frac{1}{x_i}\Big)^b\prod_{1\le i\ne j\le n}\Big(1-\frac{x_i}{x_j}\Big)^k\right]=\prod_{i=0}^{n-1}\frac{(a+b+ik)!((i+1)k)!}{(a+ik)!(b+ik)!k!}, \tag{11.4}$$

where a and b are arbitrary nonnegative integers.

In their recent representation-theoretic work on W-algebra extensions of the $M(2,p)$ minimal models of conformal field theory [1, 2], Adamović and Milas discovered a novel type of constant term identities, which they termed *logarithmic constant term identities*. Before stating the results of Adamović and Milas, some more notation is needed.

Let $(a)_m=a(a+1)\cdots(a+m-1)$ denote the usual Pochhammer symbol or rising factorial, and let u be either a formal or complex variable. Then the (generalised) binomial coefficient $\binom{u}{m}$ is defined as

$$\binom{u}{m}=(-1)^m\frac{(-u)_m}{m!}.$$

We now interpret $(1-x)^u$ and $\log(1-x)$ as the (formal) power series

$$(1-x)^u=\sum_{m=0}^{\infty}(-x)^m\binom{u}{m} \tag{11.5}$$

and

$$\log(1-x)=-\sum_{m=1}^{\infty}\frac{x^m}{m}=\frac{\mathrm{d}}{\mathrm{d}u}(1-x)^u\Big|_{u=0}.$$

Finally, for $X=(x_1,\dots,x_n)$, we define the Vandermonde product

$$\Delta(X)=\prod_{1\le i<j\le n}(x_i-x_j).$$

One of the discoveries of Adamović and Milas is the following beautiful logarithmic analogue of the equal-parameter case (11.2) of Dyson's identity.

Conjecture 11.1 ([1, Conjecture A.12]). For n an odd positive integer and k a nonnegative integer define $m:=(n-1)/2$ and $K:=2k+1$. Then

$$\mathrm{CT}\left[\Delta(X)\prod_{i=1}^{n}x_i^{-m}\prod_{i=1}^{m}\log\Big(1-\frac{x_{2i}}{x_{2i-1}}\Big)\prod_{1\le i\ne j\le n}\Big(1-\frac{x_i}{x_j}\Big)^k\right]=\frac{(nK)!!}{n!!(K!!)^n}. \tag{11.6}$$

We remark that the kernel on the left is a Laurent series in X of (total) degree 0. Moreover, in the absence of the term $\prod_{i=1}^{m}\log(1-x_{2i}/x_{2i-1})$ the kernel is a skew-symmetric Laurent polynomial which therefore has a vanishing constant term. Using identities for harmonic numbers, Adamović and Milas proved (11.6) for $n=3$; see [1, Corollary 11.11].

Another result of Adamović and Milas, first conjectured in [1, Conjecture 10.3] (and proved for $n=3$ in (the second) Theorem 1.1 of that paper, see page 3925) and subsequently proved in [2, Theorem 1.4], is the following Morris-type logarithmic constant term identity.

Theorem 11.2. *With the same notation as above,*

$$\operatorname{CT}\left[\Delta(X)\prod_{i=1}^{n}x_i^{2-(k+1)(n+1)}(1-x_i)^a\prod_{i=1}^{m}\log\left(1-\frac{x_{2i}}{x_{2i-1}}\right)\prod_{1\le i\neq j\le n}\left(1-\frac{x_i}{x_j}\right)^k\right]$$
$$=c_{nk}\prod_{i=0}^{n-1}\binom{a+Ki/2}{(m+1)K-1},\qquad(11.7)$$

where a is an indeterminate, c_{nk} a nonzero constant, and

$$c_{3,k}=\frac{(3K)!(k!)^3}{6(3k+1)!(K!)^3}\binom{3K-1}{2K-1}^{-1}\binom{5K/2-1}{2K-1}^{-1}.\qquad(11.8)$$

As we shall see later, the above can be generalised to include an additional free parameter resulting in a logarithmic constant term identity more closely resembling Morris' identity; see (11.9) below.

The work of Adamović and Milas raises the following obvious questions:

1. Can any of the methods of proof of the classical constant term identities, see, e.g., [7, 8, 11–15, 19–21, 24, 30–32, 36–40], be utilised to prove the logarithmic counterparts?
2. Do more of Macdonald's identities (11.3) admit logarithmic analogues?
3. All of the classical constant term identities have q-analogues [16, 18, 25, 27]. Do such q-analogues also exist in the logarithmic setting?

As to the first and third questions, we can be disappointingly short; we have not been able to successfully apply any of the known methods of proof of constant term identities to also prove Conjecture 11.1, and attempts to find q-analogues have been equally unsuccessful. (In fact, we now believe q-analogues do not exist.)

As to the second question, we have found a very appealing explanation—itself based on further conjectures!—of the logarithmic constant term identities of Adamović and Milas. They arise by differentiating a complex version of Morris' constant term identity. Although such complex constant term identities are conjectured to exist for other root systems as well—this is actually proved in the case G_2—it seems that only for A_{2n} and G_2 these complex identities imply elegant logarithmic identities.

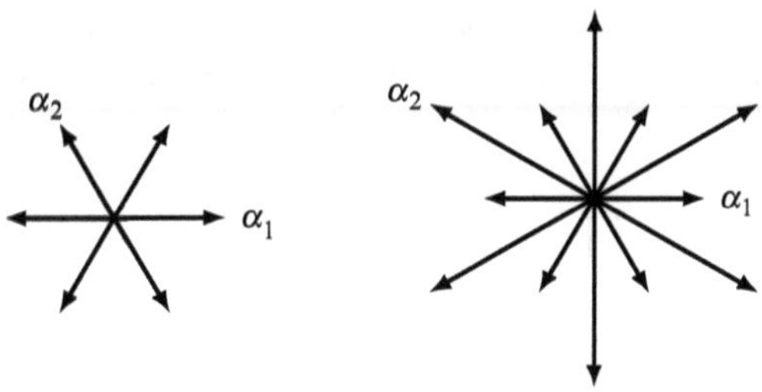

Fig. 11.1 The root systems A_2 (*left*) and G_2 (*right*) with $\Delta = \{\alpha_1, \alpha_2\}$

The remainder of this paper is organised as follows. In the next section we introduce some standard notation related to root systems. Then, in Sect. 11.4, we study certain sign functions and prove a related Pfaffian identity needed subsequently. In Sect. 11.5, we conjecture a complex analogue of the Morris constant term identity (11.4) for n odd and prove this for $n = 3$ using Zeilberger's method of creative telescoping [4, 28]. In Sect. 11.6 we show that the complex Morris identity implies the following logarithmic analogue of (11.4).

Theorem 11.3 (Logarithmic Morris constant term identity). *With the same notation as in Conjecture 11.1 and conditional on the complex Morris constant term identity* (11.24) *to hold, we have*

$$\operatorname{CT}\left[\Delta(X)\prod_{i=1}^{n} x_i^{-m}(1-x_i)^a\left(1-\frac{1}{x_i}\right)^b\prod_{i=1}^{m}\log\left(1-\frac{x_{2i}}{x_{2i-1}}\right)\prod_{1\le i\ne j\le n}\left(1-\frac{x_i}{x_j}\right)^k\right]$$
$$=\frac{1}{n!!}\prod_{i=0}^{n-1}\frac{(2a+2b+iK)!!((i+1)K)!!}{(2a+iK)!!(2b+iK)!!K!!}, \tag{11.9}$$

where a, b are nonnegative integers.

In Sect. 11.7 we prove complex as well as logarithmic analogues of (11.3) for the root system G_2, and finally, in Sect. 11.8, we briefly discuss the classical roots systems B_n, C_n and D_n.

11.3 Preliminaries on Root Systems and Constant Terms

In the final two sections of this paper we consider root systems of types other than A, and below we briefly recall some standard notation concerning root systems and constant term identities. For more details we refer the reader to [17, 25].

Let Φ be an irreducible, reduced root system in a real Euclidean space E with bilinear symmetric form $(\cdot,\cdot)$. Fix a base Δ of Φ and denote by Φ^+ the set of positive roots. Write $\alpha > 0$ if $\alpha \in \Phi^+$. The Weyl vector ρ is defined as half the sum of the positive roots: $\rho = \frac{1}{2}\sum_{\alpha>0}\alpha$. The height $\operatorname{ht}(\beta)$ of the root β is given by $\mathrm{ht}(\beta) = (\beta, \rho)$. Let r be the rank of Φ (that is, the dimension of E). Then the degrees $1 < d_1 \le d_2 \le \cdots \le d_r$ of the fundamental invariants of Φ are uniquely determined by

$$\prod_{i\ge 1} \frac{1-t^{d_i}}{1-t} = \prod_{\alpha>0} \frac{1-t^{\mathrm{ht}(\alpha)+1}}{1-t^{\mathrm{ht}(\alpha)}}.$$

For example, in the standard representation of the root system A_{n-1},

$$E = \{(x_1,\dots,x_n) \in \mathbb{R}^n : x_1 + \cdots + x_n = 0\}, \tag{11.10}$$

$$\Phi = \{\varepsilon_i - \varepsilon_j : 1 \le i \ne j \le n\}$$

and

$$\Delta = \{\alpha_1,\dots,\alpha_{n-1}\} = \{\varepsilon_i - \varepsilon_{i+1} : 1 \le i \le n-1\},$$

where ε_i denotes the ith standard unit vector in $\mathbb{R}^n$. Since $\mathrm{ht}(\varepsilon_i - \varepsilon_j) = j - i$,

$$\prod_{\alpha>0} \frac{1-t^{\mathrm{ht}(\alpha)+1}}{1-t^{\mathrm{ht}(\alpha)}} = \prod_{1\le i<j\le n} \frac{1-t^{j-i+1}}{1-t^{j-i}} = \prod_{i=1}^{n} \frac{1-t^i}{1-t}.$$

The degrees of A_{n-1} are thus $\{2,3,\dots,n\}$, and the A_{n-1} case of (11.3) is readily seen to be (11.2).

As a second example we consider the root system G_2 which is made up of two copies of A_2—one scaled. E is (11.10) for $n = 3$, and the canonical choice of simple roots is given by

$$\alpha_1 = \varepsilon_1 - \varepsilon_2 \quad \text{and} \quad \alpha_2 = 2\varepsilon_2 - \varepsilon_1 - \varepsilon_3.$$

The following additional four roots complete the set of positive root Φ^+:

$$\begin{aligned}
\alpha_1 + \alpha_2 &= \varepsilon_2 - \varepsilon_3, \\
2\alpha_1 + \alpha_2 &= \varepsilon_1 - \varepsilon_3, \\
3\alpha_1 + \alpha_2 &= 2\varepsilon_1 - \varepsilon_2 - \varepsilon_3, \\
3\alpha_1 + 2\alpha_2 &= \varepsilon_1 + \varepsilon_2 - 2\varepsilon_3.
\end{aligned}$$

The degrees of G_2 are now easily found to be $\{2,6\}$ and, after the identification $(\mathrm{e}^{\varepsilon_1}, \mathrm{e}^{\varepsilon_2}, \mathrm{e}^{\varepsilon_2}) = (x,y,z)$, the constant term identity (11.3) becomes

$$\begin{aligned}
&\mathrm{CT}\Big[\Big(1-\frac{x^2}{yz}\Big)^k\Big(1-\frac{y^2}{xz}\Big)^k\Big(1-\frac{z^2}{xy}\Big)^k\Big(1-\frac{yz}{x^2}\Big)^k\Big(1-\frac{xz}{y^2}\Big)^k\Big(1-\frac{xy}{z^2}\Big)^k \\
&\quad\times\Big(1-\frac{x}{y}\Big)^k\Big(1-\frac{x}{z}\Big)^k\Big(1-\frac{y}{x}\Big)^k\Big(1-\frac{y}{z}\Big)^k\Big(1-\frac{z}{x}\Big)^k\Big(1-\frac{z}{y}\Big)^k\Big] = \binom{2k}{k}\binom{6k}{k}.
\end{aligned} \tag{11.11}$$

This was first proved, in independent work, by Habsieger and Zeilberger [15, 38], who both utilised the A_2 case of Morris' constant term identity (11.4). They in fact proved a (q-analogue of a) slightly more general result related to another conjecture of Macdonald we discuss next.

Macdonald's (ex-)conjecture (11.3) may be generalised by replacing the exponent k on the left by k_α, where k_α depends only on the length of the root α, i.e., $k_\alpha = k_\beta$ if $\|\alpha\| = \|\beta\|$, where $\|\cdot\| := (\cdot,\cdot)^{1/2}$. If $\alpha^\vee = 2\alpha/\|\alpha\|^2$ is the coroot corresponding to α and $\rho_k = \frac{1}{2}\sum_{\alpha>0} k_\alpha \alpha$, then Macdonald's generalisation of (11.3) is

$$\mathrm{CT} \prod_{\alpha\in\Phi} (1-\mathrm{e}^\alpha)^{k_\alpha} = \prod_{\alpha>0} \frac{|(\rho_k,\alpha^\vee)+k_\alpha|!}{|(\rho_k,\alpha^\vee)|!}. \tag{11.12}$$

If k_α is independent of α, i.e., $k_\alpha = k$, then $\rho_k = k\rho$ and the above right-hand side may be simplified to that of (11.3).

As an example of (11.12) we consider the full Habsieger–Zeilberger theorem for G_2 [15, 38].

Theorem 11.4. *Let Φ_s and Φ_l denote the set of short and long roots of G_2, respectively. Then*

$$\mathrm{CT} \prod_{\alpha\in\Phi_l} (1-\mathrm{e}^\alpha)^k \prod_{\alpha\in\Phi_s} (1-\mathrm{e}^\alpha)^m = \frac{(3k+3m)!(3k)!(2k)!(2m)!}{(3k+2m)!(2k+m)!(k+m)!k!k!m!}. \tag{11.13}$$

Note that for $k=0$ or $m=0$ this yields (11.2) for $n=3$. As we shall see in Sect. 11.7, it is the above identity, not its equal-parameter case (11.11), that admits a logarithmic analogue.

11.4 The Signatures τ_{ij}

In our discussion of complex and logarithmic constant term identities in Sects. 11.5–11.8, an important role is played by certain signatures τ_{ij}. For the convenience of the reader, in this section we have collected all relevant facts about the τ_{ij}.

For a fixed odd positive integer n and $m := (n-1)/2$ define τ_{ij} for $1 \le i < j \le n$ by

$$\tau_{ij} = \begin{cases} 1 & \text{if } j \le m+i, \\ -1 & \text{if } j > m+i, \end{cases} \tag{11.14}$$

and extend this to all $1 \le i, j \le n$ by setting $\tau_{ij} = -\tau_{ji}$. Assuming that $1 \le i < n$ we have

$$\tau_{in} = \chi(n \le m+i) - \chi(n > m+i),$$

where $\chi(\text{true}) = 1$ and $\chi(\text{false}) = 0$. Since $n - m = m + 1$, this is the same as

$$\tau_{in} = \chi(i > m) - \chi(i \leq m) = -\tau_{1,i+1} = \tau_{i+1,1}.$$

For $1 \leq i, j < n$ we clearly also have $\tau_{ij} = \tau_{i+1,j+1}$. Hence the matrix

$$\mathrm{T} := (\tau_{ij})_{1 \leq i,j \leq n} \tag{11.15}$$

is a *skew-symmetric circulant matrix*. For example, for $n = 5$,

$$\mathrm{T} = \begin{pmatrix} 0 & 1 & 1 & -1 & -1 \\ -1 & 0 & 1 & 1 & -1 \\ -1 & -1 & 0 & 1 & 1 \\ 1 & -1 & -1 & 0 & 1 \\ 1 & 1 & -1 & -1 & 0 \end{pmatrix}.$$

We note that all of the row sums (and column sums) of the above matrix are zero. Because T is a circulant matrix, to verify this property holds for all (odd) n, we only need to verify this for the first row:

$$\sum_{j=1}^{n} \tau_{1j} = \sum_{j=2}^{m+1} 1 - \sum_{j=m+2}^{n} 1 = m - (n - m - 1) = m - m = 0.$$

By the skew symmetry, the vanishing of the row sums may also be stated as follows.

Lemma 11.5. *For* $1 \leq i \leq n$,

$$\sum_{j=1}^{i-1} \tau_{ji} = \sum_{j=i+1}^{n} \tau_{ij}.$$

A property of the signatures τ_{ij}, which will be important in our later discussions, can be neatly clarified by having recourse to Pfaffians.

By a *perfect matching* (or 1-factor) on $[n+1] := \{1, 2, \ldots, n+1\}$ we mean a graph on the vertex set $[n+1]$ such that each vertex has degree one; see, e.g., [6, 35]. If in a perfect matching π the vertices $i<j$ are connected by an edge we say that $(i,j)\in\pi$. Two edges (i,j) and (k,l) of π are said to be crossing if $i<k<j<l$ or $k<i<l<j$. The crossing number $c(i,j)$ of the edge $(i,j) \in \pi$ is the number of edges crossed by (i,j), and the crossing number $c(\pi)$ is the total number of pairs of crossing edges: $c(\pi)=\frac{1}{2}\sum_{(i,j)\in\pi} c(i,j)$. We can embed perfect matching in the xy-plane, such that (i) the vertex labelled i occurs at the point $(i,0)$, and (ii) the edges (i,j) and (k,l) intersect exactly once if they are crossing and do not intersect if they are non-crossing. For example, the perfect matching $\{(1,3),(2,7),(4,5),(6,8)\}$ corresponds to

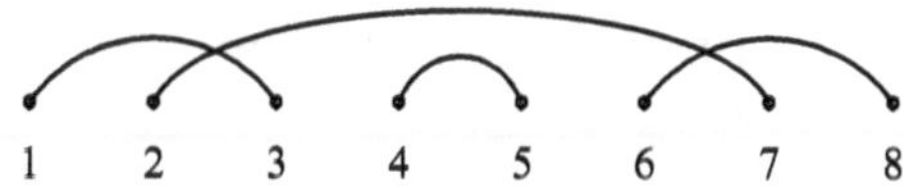

and has crossing number 2 ($c(4,5)=0$, $c(1,3)=c(6,8)=1$ and $c(2,7)=2$).

The *Pfaffian* of a $(2N)\times(2N)$ skew-symmetric matrix A is defined as [6, 22, 23, 35]:

$$\mathrm{Pf}(A) := \sum_{\pi}(-1)^{c(\pi)} \prod_{(i,j)\in\pi} A_{ij}. \tag{11.16}$$

After these preliminaries on perfect matching and Pfaffians we now form a second skew-symmetric matrix, closely related to T. First we extend the τ_{ij} to $1\le i,j\le n+1$ by setting $\tau_{i,n+1}=b_i$. We then define the $(n+1)\times(n+1)$ skew-symmetric matrix $Q(a,b)=(Q_{ij}(a,b))_{1\le i,j\le n+1}$, where $a=(a_1,\dots,a_{n+1})$ and $b=(b_1,\dots,b_n)$, as follows:

$$Q_{ij}(a,b)=\tau_{ij}a_ia_j \qquad \text{for } 1\le i<j\le n+1. \tag{11.17}$$

For example, for $n=5$,

$$Q(a,b)=\begin{pmatrix} 0 & a_1a_2 & a_1a_3 & -a_1a_4 & -a_1a_5 & a_1a_6b_1\\ -a_2a_1 & 0 & a_2a_3 & a_2a_4 & -a_2a_5 & a_2a_6b_2\\ -a_3a_1 & -a_3a_2 & 0 & a_3a_4 & a_3a_5 & a_3a_6b_3\\ a_4a_1 & -a_4a_2 & -a_4a_3 & 0 & a_4a_5 & a_4a_6b_4\\ a_5a_1 & a_5a_2 & -a_5a_3 & -a_5a_4 & 0 & a_5a_6b_5\\ -a_6a_1b_1 & -a_6a_2b_2 & -a_6a_3b_3 & -a_6a_4b_4 & -a_6a_5b_5 & 0 \end{pmatrix}.$$

Note that T is the submatrix of $Q\big((1^{n+1}),b\big)$ obtained by deleting the last row and column.

Proposition 11.6. *We have*

$$\mathrm{Pf}\big(Q(a,b)\big)=(-1)^{\binom{m}{2}}a_1a_2\cdots a_{n+1}(b_1+b_2+\cdots+b_n).$$

Proof. The main point of our proof below is to exploit a cyclic symmetry of the terms contributing to $\mathrm{Pf}\big(Q(a,b)\big)$. This reduces the computation of the Pfaffian to that of a sub-Pfaffian of lower order.

Let $S(\pi;a,b)$ denote the summand of $\mathrm{Pf}\big(Q(a,b)\big)$, that is,

$$\mathrm{Pf}\big(Q(a,b)\big)=\sum_{\pi}S(\pi;a,b) \qquad \text{with} \quad S(\pi;a,b)=(-1)^{c(\pi)}\prod_{(i,j)\in\pi}Q_{ij}(a,b).$$

From the definition (11.17) of $Q_{ij}(a,b)$ and the fact that π is a perfect matching on $[n+1]$,

$$S(\pi;a,b)=(-1)^{c(\pi)}\prod_{(i,j)\in\pi} a_i a_j \tau_{ij}=(-1)^{c(\pi)}a_1\cdots a_{n+1}\prod_{(i,j)\in\pi}\tau_{ij}. \tag{11.18}$$

We now observe that $S(\pi;a,b)$ is, up to a cyclic permutation of b, invariant under the permutation w given by $(1,2,3,\dots,n,n+1)\mapsto(n,1,2,\dots,n-1,n+1)$. To see this, denote by π' the image of π under w. For example, the image of the perfect matching given on the previous page is

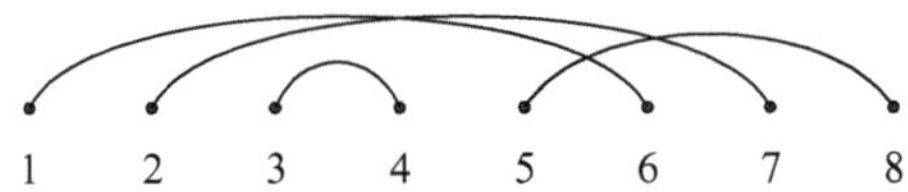

Under the permutation w, all edges not containing the vertices 1 or $n+1$ are shifted one unit to the left: $(i,j)\mapsto(i-1,j-1)$. For the edge $(1,j)$ containing vertex 1 we have:

(i) If $j\le n$ then $(1,j)\mapsto(j-1,n)$. This also implies that the edge $(j',n+1)$ ($j'\ge 2$) containing vertex $n+1$ maps to $(j'-1,n+1)$.
(ii) If $j=n+1$ then $(1,j)=(1,n+1)\mapsto(n,n+1)=(j-1,n+1)$.

First we consider (i). If we remove the edge $(1,j)$ from π and carry out w, then the number of crossings of its image is exactly that of π. Hence we only need to focus on the edge $(1,j)$ and its image under w. In π the edge $(1,j)$ has crossing number $c(1,j)\equiv j \pmod 2$, while the edge $(j-1,n)$ in π' has crossing number $c(j-1,n)\equiv n-j\equiv j+1 \pmod 2$. Hence $(-1)^{c(\pi)}=-(-1)^{c(\pi')}$. Since $\tau_{ij}=\tau_{i-1,j-1}$ (for $2\le i<j\le n$) and $\tau_{1,j}=-\tau_{j-1,n}$ it thus follows that π and π' have the same sign. Finally we note that under w, $b_i=\tau_{i,n+1}\mapsto\tau_{i-1,n+1}=b_{i-1}$ (since $i\ne 1$). We thus conclude that

$$S\big(\pi;a,(b_1,\dots,b_n)\big)\mapsto S\big(\pi';a,(b_2,\dots,b_n,b_1)\big), \tag{11.19}$$

where we note that both sides depend on a single $b_i(\ne b_1)$ only. For example, the perfect matching in the above two figures correspond to

$$\begin{aligned}&S\big((1,3),(2,7),(4,5),(6,8);a,(b_1,\dots,b_7)\big)\\&\quad=(-1)^2\cdot a_1a_3\cdot(-a_2a_7)\cdot a_4a_5\cdot a_6a_8b_6=-a_1\cdots a_8b_6\end{aligned}$$

and

$$\begin{aligned}&S\big((1,6),(2,7),(3,4),(5,8);a,(b_1,\dots,b_7)\big)\\&\quad=(-1)^3\cdot(-a_1a_6)\cdot(-a_2a_7)\cdot a_3a_4\times a_5a_8b_5=-a_1\cdots a_8b_5.\end{aligned}$$

The case (ii) is even simpler; the edge $(1,n+1)$ in π and its image $(n,n+1)$ in π' both have crossing number 0. The crossing numbers of all other edges do not change by a global shift of one unit to the right, so that $c(\pi) = c(\pi')$:

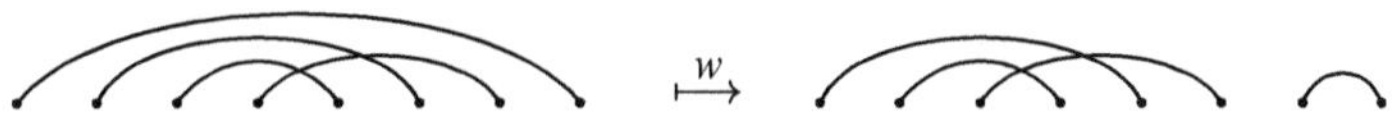

Moreover, $\tau_{ij} = \tau_{i-1,j-1}$ (for $2 \le i < j \le n$) so that π and π' again have the same sign. Finally, from $b_1 = \tau_{1,n+1} \mapsto \tau_{n,n+1} = b_n$, it follows that once again (11.19) holds, where this time both sides depend only on b_1.

From (11.19) it follows that the Pfaffian $\mathrm{Pf}\big(Q(a,b)\big)$ is symmetric under cyclic permutations of the b_i. But since the Pfaffian, viewed as a function of b, has degree 1 it thus follows [see also (11.18)] that

$$\mathrm{Pf}\big(Q(a,b)\big) = Ca_1 \cdots a_{n+1}(b_1 + \cdots + b_n)$$

for some yet-unknown constant C. We shall determine C by computing the coefficient of b_n of $\mathrm{Pf}\big(Q((1^{n+1}),b\big)$, which is equal to the Pfaffian of the $(2m) \times (2m)$ submatrix M of T obtained by deleting its last row and column.

We recall the property $\mathrm{Pf}(M) = \mathrm{Pf}(U^t M U)$ of Pfaffians, where U is a unipotent triangular matrix [35]. Choosing the nonzero entries of the $(2m) \times (2m)$ matrix U to be $U_{ii} = 1$ for $i = 1, \dots, 2m$, and $U_{i,i+m} = 1$ for $i = 1, \dots, m$, one transforms M into

$$\left(\begin{array}{c|c} M' & I \\ \hline -I & \varnothing \end{array}\right),$$

where M' is the upper-left $m \times m$ submatrix of M and I is the $m \times m$ identity matrix. The Pfaffian of the above matrix, and hence that of M, is exactly (cf. [35]) $(-1)^{\binom{m}{2}} \det(I) = (-1)^{\binom{m}{2}}$. This, in turn, implies that $C = (-1)^{\binom{m}{2}}$, and the required formula follows. ∎

Remark 11.7. By a slight modification of the above proof the following more general Pfaffian results. Let

$$Q_{ij}(X,a,b) := \tau_{ij} a_i a_j (x_i + x_j) \qquad \text{for } 1 \le i < j \le n$$

and

$$Q_{i,n+1}(X,a,b) := \tau_{i,n+1} a_i a_{n+1} = a_i a_{n+1} b_i \qquad \text{for } 1 \le i \le n,$$

and use this to form the $(n+1) \times (n+1)$ skew-symmetric matrix $Q(X,a,b)$.

Then

$$\mathrm{Pf}\big(Q(X,a,b)\big) = 2^{m-1}(-1)^{\binom{m}{2}} a_1 a_2 \cdots a_{n+1} \sum_{i=1}^{n} b_i (x_{i+1} \cdots x_{i+m} + x_{i+m+1} \cdots x_{i+n-1}),$$

where $x_{i+n} := x_i$ for $i \geq 1$. For $X = (1/2, \ldots, 1/2)$ this yields Proposition 11.6.

11.5 The Complex Morris Constant Term Identity

Thanks to Lemma 11.5,

$$\begin{aligned}
\prod_{1\leq i\neq j\leq n} \Big(1 - \frac{x_i}{x_j}\Big) &= \prod_{1\leq i<j\leq n} \Big(-\frac{x_j}{x_i}\Big)^{\tau_{ij}} \Big(1 - \Big(\frac{x_i}{x_j}\Big)^{\tau_{ij}}\Big)^2 \\
&= (-1)^{\binom{n}{2}} \prod_{i=1}^{n} x_i^{\sum_{j=1}^{i-1}\tau_{ji} - \sum_{j=i+1}^{n}\tau_{ij}} \prod_{1\leq i<j\leq n} \Big(1 - \Big(\frac{x_i}{x_j}\Big)^{\tau_{ij}}\Big)^2 \\
&= (-1)^m \prod_{1\leq i<j\leq n} \Big(1 - \Big(\frac{x_i}{x_j}\Big)^{\tau_{ij}}\Big)^2. \qquad (11.20)
\end{aligned}$$

For odd values of n Morris' constant term identity (11.4) can thus be rewritten in the equivalent form

$$\begin{aligned}
&\mathrm{CT}\Bigg[\prod_{i=1}^{n}(1-x_i)^a\Big(1-\frac{1}{x_i}\Big)^b \prod_{1\leq i<j\leq n}\Big(1-\Big(\frac{x_i}{x_j}\Big)^{\tau_{ij}}\Big)^{2k}\Bigg] \\
&\qquad = (-1)^{km}\prod_{i=0}^{n-1}\frac{(a+b+ik)!((i+1)k)!}{(a+ik)!(b+ik)!k!}. \qquad (11.21)
\end{aligned}$$

The crucial point about this rewriting is that in the product

$$\prod_{1\leq i<j\leq n}\Big(1-\Big(\frac{x_i}{x_j}\Big)^{\tau_{ij}}\Big)^{2k}$$

each of the variables $x_1, x_2, \ldots, x_n$ occurs exactly m times in one of the numerators and m times in one of the denominators. For example,

$$\prod_{1\leq i<j\leq 3}\Big(1-\Big(\frac{x_i}{x_j}\Big)^{\tau_{ij}}\Big)^{2k} = \Big(1-\frac{x_1}{x_2}\Big)^{2k}\Big(1-\frac{x_2}{x_3}\Big)^{2k}\Big(1-\frac{x_3}{x_1}\Big)^{2k}.$$

Obviously, for n even such a rewriting is not possible.

We are now interested in the question as to what happens when $2k$ is replaced by an arbitrary complex variable u. For $n=3$ we will later prove the following proposition.

Proposition 11.8. *For a,b nonnegative integers and* $\mathrm{Re}(1+\frac{3}{2}u)>0$,

$$\mathrm{CT}\Big[(1-x)^a(1-y)^a(1-z)^a\Big(1-\frac{1}{x}\Big)^b\Big(1-\frac{1}{y}\Big)^b\Big(1-\frac{1}{z}\Big)^b \times\Big(1-\frac{x}{y}\Big)^u\Big(1-\frac{y}{z}\Big)^u\Big(1-\frac{z}{x}\Big)^u\Big] = \cos\big(\tfrac{1}{2}\pi u\big)\frac{\Gamma(1+\frac{3}{2}u)}{\Gamma^3(1+\frac{1}{2}u)}\prod_{i=0}^{2}\frac{(1+\frac{1}{2}iu)_{a+b}}{(1+\frac{1}{2}iu)_a(1+\frac{1}{2}iu)_b}. \tag{11.22}$$

As follows from its proof, a slightly more general result in fact holds. Using $(z)_{n+m}=(z)_n(z+n)_m$ and $(1-x)^a(1-x^{-1})^b=(-x)^{-b}(1-x)^{a+b}$, then replacing $a\mapsto a-b$, and finally using $(z-b)_b=(-1)^b(1-z)_b$, the identity (11.22) can also be stated as

$$[x^by^bz^b]\Big[(1-x)^a(1-y)^a(1-z)^a\Big(1-\frac{x}{y}\Big)^u\Big(1-\frac{y}{z}\Big)^u\Big(1-\frac{z}{x}\Big)^u\Big] = \cos\big(\tfrac{1}{2}\pi u\big)\frac{\Gamma(1+\frac{3}{2}u)}{\Gamma^3(1+\frac{1}{2}u)}\prod_{i=0}^{2}\frac{(-a-\frac{1}{2}iu)_b}{(1+\frac{1}{2}iu)_b}, \tag{11.23}$$

where $[X^c]f(X)$ (with $X^c=x_1^{c_1}\cdots x_n^{c_n}$) denotes the coefficient of X^c in $f(X)$. This alternative form of (11.22) is true for all $a,u\in\mathbb{C}$ such that $\mathrm{Re}(1+\frac{3}{2}u)>0$.

In view of Proposition 11.8 it seems reasonable to make the following more general conjecture.

Conjecture 11.9 (Complex Morris constant term identity). Let n be an odd positive integer, a,b nonnegative integers and $u\in\mathbb{C}$ such that $\mathrm{Re}(1+\frac{1}{2}nu)>0$. Then there exists a polynomial $P_n(x)$, independent of a and b, such that $P_n(0)=1/(n-2)!!$, $P_n(1)=1$, and

$$\mathrm{CT}\Big[\prod_{i=1}^{n}(1-x_i)^a\Big(1-\frac{1}{x_i}\Big)^b\prod_{1\le i<j\le n}\Big(1-\Big(\frac{x_i}{x_j}\Big)^{\tau_{ij}}\Big)^u\Big] = x^mP_n(x^2)\frac{\Gamma(1+\frac{1}{2}nu)}{\Gamma^n(1+\frac{1}{2}u)}\prod_{i=0}^{n-1}\frac{(1+\frac{1}{2}iu)_{a+b}}{(1+\frac{1}{2}iu)_a(1+\frac{1}{2}iu)_b}, \tag{11.24}$$

where $x=x(u):=\cos\big(\frac{1}{2}\pi u\big)$ and $m:=(n-1)/2$.

Note that for u an odd positive integer the kernel on the left of (11.24) is a skew-symmetric function, so that its constant term trivially vanishes. When u is an even integer, say $2k$ then $x = \cos(\pi k) = (-1)^k$ so that $x^m P_n(x^2) = (-1)^{km} P_n(1) = (-1)^{km}$ in accordance with (11.21). Similar to the case $n = 3$, in the form

$$\big[(x_1 \cdots x_n)^b\big] \left[\mathrm{CT}\left[\prod_{i=1}^{n}(1-x_i)^a \prod_{1\le i<j\le n}\left(1-\left(\frac{x_i}{x_j}\right)^{\tau_{ij}}\right)^u\right]\right.$$

$$= x^m P_n(x^2) \frac{\Gamma(1+\frac{1}{2}nu)}{\Gamma^n(1+\frac{1}{2}u)} \prod_{i=0}^{n-1} \frac{(-a-\frac{1}{2}iu)_b}{(1+\frac{1}{2}iu)_b},$$

Conjecture 11.9 should hold for all $a \in \mathbb{C}$.

For $n = 1$ the left-side of (11.24) does not depend on u so that $P_1(x) = 1$. Moreover, from Proposition 11.8, it follows that also $P_3(x) = 1$. Extensive numerical computations leave little doubt that the next two instances of $P_n(x)$ are given by

$$P_5(x) = \frac{1}{3}(1+2x)$$

$$P_7(x) = \frac{1}{45}(3+26x-16x^2+32x^3).$$

Conjecturally, we also have $\deg(P_n(x)) = \binom{m}{2}$ and

$$P_n'(0) = 2\binom{m}{2}\frac{2n-1}{9(n-2)!!}$$

$$P_n'(1) = \frac{2}{3}\binom{m}{2}, \qquad P_n''(1) = \frac{2}{45}\binom{m}{3}(19m+23),$$

but beyond this we know very little about $P_n(x)$.

To conclude our discussion of the polynomials $P_n(x)$ we note that if $z_i = z_i(u) := \cos(i\pi u)$, then

$$P_5\big(x^2(u)\big) = \frac{1}{3}(2+z_1)$$

$$P_7\big(x^2(u)\big) = \frac{1}{45}(20+20z_1+4z_2+z_3),$$

suggesting that the coefficients of z_i admit a combinatorial interpretation.

As will be shown in the next section, the complex Morris constant term identity (11.24) implies the logarithmic Morris constant term identity (11.9), and the only properties of $P_n(x)$ that are essential in the proof are $P_n(0) = 1/(n-2)!!$ and $P_n(1) = 1$.

To conclude this section we give a proof of Proposition 11.8. The reader unfamiliar with the basic setup of the method of creative telescoping is advised to consult the text [28].

Proof of Proposition 11.8. Instead of proving (11.22) we establish the slightly more general (11.23).

By a sixfold use of the binomial expansion (11.5), the constant term identity (11.23) can be written as the following combinatorial sum:

$$\sum_{m_0,m_1,m_2=0}^{\infty} \prod_{i=0}^{2} (-1)^{m_i} \binom{u}{m_i} \binom{a}{b+m_i-m_{i+1}} = \cos\left(\tfrac{1}{2}\pi u\right) \frac{\Gamma(1+\frac{3}{2}u)}{\Gamma^3(1+\frac{1}{2}u)} \prod_{i=0}^{2} \frac{(-a-\frac{1}{2}iu)_b}{(1+\frac{1}{2}iu)_b},$$

where $m_3 := m_0$ and where $a,u \in \mathbb{C}$ such that $\mathrm{Re}(1+\frac{3}{2}u) > 0$ and b is a nonnegative integer. If we denote the summand of this identity by $f_b(\frac{1}{2}u,-1-a;m)$ where $m := (m_0,m_1,m_2)$, then we need to prove that

$$F_b(u,v) := \sum_{m\in\mathbb{Z}^3} f_b(u,v;m) = \cos\left(\pi u\right) \frac{\Gamma(1+3u)}{\Gamma^3(1+u)} \prod_{i=0}^{2} \frac{(1+v-iu)_b}{(1+iu)_b}, \tag{11.25}$$

for $\mathrm{Re}(1+3u) > 0$.

In our working below we suppress the dependence of the various functions on the variables u and v. In particular we write F_b and $f_b(m)$ for $F_b(u,v)$ and $f_b(u,v;m)$.

The function $f_0(m)$ vanishes unless $m_0 = m_1 = m_2$. Hence

$$F_0 = \sum_{m=0}^{\infty} (-1)^m \binom{2u}{m}^3 = {}_3F_2\left[\begin{matrix}-2u,-2u,-2u\\1,1\end{matrix};1\right],$$

where we adopt standard notation for (generalised) hypergeometric series; see, e.g., [3,5]. The ${}_3F_2$ series is summable by the $2a = b = c = -2u$ case of Dixon's sum [3, Eq. (2.2.11)]

$${}_3F_2\left[\begin{matrix}2a,b,c\\1+2a-b,1+2a-c\end{matrix};1\right] = \frac{\Gamma(1+a)\Gamma(1+2a-b)\Gamma(1+2a-c)\Gamma(1+a-b-c)}{\Gamma(1+2a)\Gamma(1+a-b)\Gamma(1+a-c)\Gamma(1+2a-b-c)} \tag{11.26}$$

for $\mathrm{Re}(1+a-b-c) > 0$. As a result,

$$F_0 = \frac{\Gamma(1-u)\Gamma(1+u)}{\Gamma(1-2u)\Gamma(1+2u)} \cdot \frac{\Gamma(1+3u)}{\Gamma^3(1+u)} = \cos(\pi u) \frac{\Gamma(1+3u)}{\Gamma^3(1+u)},$$

proving the $b = 0$ instance of (11.25).

In the remainder we assume that $b \geq 1$.

Let $\mathscr{C}$ be the generator of the cyclic group C_3 acting on m as $\mathscr{C}(m) = (m_2,m_0,m_1)$. With the help of the multivariable Zeilberger algorithm [4], one discovers the (humanly verifiable) rational function identity

$$t_b(m)\prod_{i=0}^{2}(b+iu)+\prod_{i=0}^{2}(b+v-iu)$$
$$=\sum_{i=0}^{2}\Big(r_b(\boldsymbol{e}_1+\mathscr{C}^i(m))\,s_b(\mathscr{C}^i(m))+r_b(\mathscr{C}^i(m))\Big), \qquad (11.27)$$

where

$$r_b(m)=-\frac{m_0(b+v+m_2-m_0)}{6(b+m_1-m_2)(b+m_2-m_0)}$$
$$\times\big((2b+v)(3b^2+3bv+2uv)+2(m_1-m_2)(3b^2+3bv+v^2-uv)\big),$$
$$s_b(m)=-\frac{f_{b-1}(\boldsymbol{e}_1+m)}{f_{b-1}(m)}=\frac{(2u-m_0)(b+v+m_0-m_1)(b+m_2-m_0-1)}{(1+m_0)(b+m_0-m_1)(b+v+m_2-m_0-1)},$$
$$t_b(m)=-\frac{f_b(m)}{f_{b-1}(m)}=\prod_{i=0}^{2}\frac{b+v+m_i-m_{i+1}}{b+m_i-m_{i+1}},$$

and $e_1+m:=(1+m_0,m_1,m_2)$. If we multiply (11.27) by $-f_{b-1}(m)$ and use that $f_b(m)=f_b(\mathscr{C}^i(m))$ we find that

$$f_b(m)\prod_{i=0}^{2}(b+iu)-f_{b-1}(m)\prod_{i=0}^{2}(b+v-iu)$$
$$=\sum_{i=0}^{2}\Big[r_b(e_1+\mathscr{C}^i(m))\,f_{b-1}(e_1+\mathscr{C}^i(m))-r_b(\mathscr{C}^i(m))\,f_{b-1}(\mathscr{C}^i(m))\Big].$$

Summing this over $m\in\mathbb{Z}^3$ the right-hand side telescopes to zero, resulting in

$$F_b=F_{b-1}\prod_{i=0}^{2}\frac{(b+v-iu)}{(b+iu)}.$$

By b-fold iteration this yields

$$F_b=F_0\prod_{i=0}^{2}\frac{(1+v-iu)_b}{(1+iu)_b}. \qquad \blacksquare$$

11.6 The Logarithmic Morris Constant Term Identity

This section contains three parts. In the first very short part, we present an integral analogue of the logarithmic Morris constant term identity. This integral may be viewed as a logarithmic version of the well-known Morris integral. The second and third, more substantial parts, contain, respectively, a proof of Theorem 11.3 and, exploiting some further results of Adamović and Milas, a strengthening of this theorem.

11.6.1 A Logarithmic Morris Integral

By a repeated use of Cauchy's integral formula, constant term identities such as (11.4) or (11.9) may be recast in the form of multiple integral evaluations. In the case of (11.4) this leads to the well-known Morris integral [10,27]

$$\begin{aligned}\int\limits_{[-\frac{1}{2}\pi,\frac{1}{2}\pi]^n} &\prod_{i=1}^{n} \mathrm{e}^{\mathrm{i}(a-b)\theta_i}\sin^{a+b}(\theta_i) \prod_{1\le i<j\le n} \sin^{2k}(\theta_i-\theta_j)\,\mathrm{d}\theta_1\cdots\mathrm{d}\theta_n \\ &= \big(B_{k,n}(a,b)\big)^n \prod_{i=0}^{n-1}\frac{(a+b+ik)!((i+1)k)!}{(a+ik)!(b+ik)!k!},\end{aligned}$$

where $B_{k,n}(a,b) = \pi\,\mathrm{i}^{a-b}2^{-k(n-1)-a-b}$. The Morris integral may be shown to be a simple consequence of the Selberg integral [10,29]. Thanks to (11.9) we now have a logarithmic analogue of the Morris integral as follows:

$$\begin{aligned}\int\limits_{[-\frac{1}{2}\pi,\frac{1}{2}\pi]^n} &\prod_{i=1}^{n} \mathrm{e}^{\mathrm{i}(a-b)\theta_i}\sin^{a+b}(\theta_i) \prod_{i=1}^{m} \log\big(1-\mathrm{e}^{2\mathrm{i}(\theta_{2i}-\theta_{2i-1})}\big) \\ &\times \prod_{1\le i<j\le n} \sin^{K}(\theta_i-\theta_j)\,\mathrm{d}\theta_1\cdots\mathrm{d}\theta_n \\ &= \big(C_{k,n}(a,b)\big)^n \frac{1}{n!!}\prod_{i=0}^{n-1}\frac{(2a+2b+iK)!!((i+1)K)!!}{(2a+iK)!!(2b+iK)!!K!!},\end{aligned}$$

where $C_{k,n}(a,b) = \pi\,\mathrm{i}^{a-b-m}2^{-Km-a-b}$. Unfortunately, this cannot be rewritten further in a form that one could truly call a logarithmic Selberg integral.

11.6.2 Proof of Theorem 11.3

In this subsection we prove that the logarithmic Morris constant term identity (11.9) is nothing but the mth derivative of the complex Morris constant term identity (11.24) evaluated at $u = K := 2k+1$.

To set things up we first prepare a technical lemma. For $\mathfrak{S}_n$ the symmetric group on n letters and $w \in \mathfrak{S}_n$, we denote by $\operatorname{sgn}(w)$ the signature of the permutation w; see, e.g., [26]. The identity permutation in $\mathfrak{S}_n$ will be written as $\mathbb{I}$.

Lemma 11.10. *For n an odd integer, set $m := (n-1)/2$. Let t_{ij} for $1 \le i < j \le n+1$ be a collection of signatures (i.e., each t_{ij} is either $+1$ or -1) such that $t_{i,n+1} = 1$ and $\tilde{Q}$ a skew-symmetric matrix with entries $\tilde{Q}_{ij} = t_{ij}$ for $1 \le i < j \le n+1$.*

If $f(X)$ is a skew-symmetric polynomial in $X = (x_1, \dots, x_n)$, $g(z)$ a Laurent polynomial or Laurent series in the scalar variable z, and $g_{ij}(X) := g((x_i/x_j)^{t_{ij}})$, then the following statements hold:

1. *For $w \in \mathfrak{S}_n$, denote $g(w;X) := \prod_{k=1}^{m} g(x_{w_{2k-1}}/x_{w_{2k}})$. Then*
$$\operatorname{CT}\left[f(X)g(w;X)\right] = \operatorname{sgn}(w)\operatorname{CT}\left[f(X)g(\mathbb{I};X)\right].$$
2. *For π a perfect matching on $[n+1]$,*
$$\sum_{\pi}\operatorname{CT}\Bigg[f(X)\prod_{\substack{(i,j)\in\pi\\ j\neq n+1}} g_{ij}(X)\Bigg] = \operatorname{Pf}(\tilde{Q})\operatorname{CT}\left[f(X)g(\mathbb{I};X)\right]. \tag{11.28}$$

We will be needing a special case of this lemma corresponding to $t_{ij} = \tau_{ij}$ for $1 \le i,j \le n$, with the τ_{ij} defined in (11.14). Then the matrix $\tilde{Q}$ coincides with $Q\big((1^{n+1}),(1^n)\big)$ of (11.17), so that by Lemma 11.6, $\operatorname{Pf}(\tilde{Q}) = (-1)^{\binom{m}{2}}n$. We summarise this in the following corollary.

Corollary 11.11. *If in Lemma 11.10 we specialise $t_{ij} = \tau_{ij}$ for $1 \le i < j \le n$, then*
$$\sum_{\pi}\operatorname{CT}\Bigg[f(X)\prod_{\substack{(i,j)\in\pi\\ j\neq n+1}} g_{ij}(X)\Bigg] = (-1)^{\binom{m}{2}}n\operatorname{CT}\left[f(X)g(\mathbb{I};X)\right]. \tag{11.29}$$

Proof of Lemma 11.10.

1. According to the "Stanton–Stembridge trick" [33, 34, 39],
$$\operatorname{CT}\left[h(X)\right] = \operatorname{CT}\left[w\big(h(X)\big)\right] \quad \text{for } w \in \mathfrak{S}_n,$$
where $w(h(X))$ is shorthand for $h(x_{w_1}, \dots, x_{w_n})$.
For our particular choice of h, the skew-symmetric factor $f(X)$ produces the claimed sign.
2. A permutation $w \in \mathfrak{S}_n$ may be interpreted as a signed perfect matching $(-1)^{d(w)}(w_1,w_2)\cdots(w_{n-2},w_{n-1})(w_n,w_{n+1})$, where $d(w)$ counts the number $|\{k \le m : w_{2k-1} > w_{2k}\}|$. By claim (11.10), the left-hand side of (11.28) is a multiple of $\operatorname{CT}\left[f(X)g(\mathbb{I};X)\right]$; the factor is exactly the sum $\sum_{\pi}(-1)^{c(\pi)}\prod t_{ij}$, in which one recognises the Pfaffian of $\tilde{Q}$. ■

Conditional proof of (11.9). Suppressing the a and b dependence, denote the left- and right-hand sides of (11.24) by $L_n(u)$ and $R_n(u)$, respectively. We then wish to show that (11.9) is identical to

$$L_n^{(m)}(K) = R_n^{(m)}(K).$$

Let us first consider the right-hand side, which we write as $R_n(u) = p_n(u)r_n(u)$, where

$$p_n(u) = x^m P_n(x^2), \qquad x = x(u) = \cos\left(\tfrac{1}{2}\pi u\right)$$

and

$$r_n(u) = \frac{\Gamma(1+\frac{1}{2}nu)}{\Gamma^n(1+\frac{1}{2}u)} \prod_{i=0}^{n-1} \frac{(1+\frac{1}{2}iu)_{a+b}}{(1+\frac{1}{2}iu)_a(1+\frac{1}{2}iu)_b}. \tag{11.30}$$

Since $x(K) = 0$, it follows that for $0 \le j \le m$,

$$p_n^{(j)}(K) = (-1)^{km+m} \left(\frac{\pi}{2}\right)^m \frac{m!}{(n-2)!!} \delta_{jm}. \tag{11.31}$$

Therefore, since $r_n(u)$ is m times differentiable at $u = K$,

$$R_n^{(m)}(K) = p_n^{(m)}(K) r_n(K). \tag{11.32}$$

By the functional equation for the gamma function

$$\Gamma(1+\tfrac{1}{2}N) = \begin{cases} N!!2^{-N/2}\sqrt{\pi/2} & \text{if } N > 0 \text{ is odd,} \\ N!!2^{-N/2} & \text{if } N \ge 0 \text{ is even,} \end{cases} \tag{11.33}$$

and, consequently,

$$(1+\tfrac{1}{2}N)_a = \frac{(N+2a)!!}{2^a N!!} \tag{11.34}$$

for any nonnegative integer N. Applying these formulae to (11.30) with $u = K$, we find that

$$r_n(K) = \left(\frac{2}{\pi}\right)^m \prod_{i=0}^{n-1} \frac{(2a+2b+iK)!!((i+1)K)!!}{(2a+iK)!!(2b+iK)!!K!!}.$$

Combined with (11.31) and (11.32) this implies

$$R_n^{(m)}(K) = (-1)^{(k+1)m} \frac{m!}{(n-2)!!} \prod_{i=0}^{n-1} \frac{(2a+2b+iK)!!((i+1)K)!!}{(2a+iK)!!(2b+iK)!!K!!}. \tag{11.35}$$

Next we focus on the calculation of $L_n^{(m)}(K)$. To keep all equations in check we define

$$f_{ab}(X) := \prod_{i=1}^{n}(1-x_i)^a\Big(1-\frac{1}{x_i}\Big)^b.$$

and

$$F_{ab}(X) := \Delta(X)\prod_{i=1}^{n} x_i^{-m}(1-x_i)^a\Big(1-\frac{1}{x_i}\Big)^b \prod_{1\le i\ne j\le n}\Big(1-\frac{x_i}{x_j}\Big)^k. \tag{11.36}$$

Let $i := (i_1,\dots,i_m)$ and $j := (j_1,\dots,j_m)$. Then, by a straightforward application of the product rule,

$$L_n^{(m)}(u) = \sum_{1\le i_1<j_1\le n}\cdots\sum_{1\le i_m<j_m\le n} L_{n;i,j}(u),$$

where

$$L_{n;i,j}(u) = \mathrm{CT}\bigg[f_{ab}(X)\prod_{1\le i<j\le n}\Big(1-\Big(\frac{x_i}{x_j}\Big)^{\tau_{ij}}\Big)^u\prod_{\ell=1}^{m}\log\Big(1-\Big(\frac{x_{i_\ell}}{x_{j_\ell}}\Big)^{\tau_{i_\ell j_\ell}}\Big)\bigg].$$

For $u=K$ the kernel without the product over logarithms is a skew-symmetric function in X, so that $L_{n;i,j}(K)=0$ if there exists a pair of variables, say x_r and x_s, that does not occur in the product of logarithms. In other words, $L_{n;i,j}(K)$ vanishes unless all of the $2m=n-1$ entries of i and j are distinct:

$$L_n^{(m)}(K) = \sum \mathrm{CT}\bigg[f_{ab}(X)\prod_{1\le i<j\le n}\Big(1-\Big(\frac{x_i}{x_j}\Big)^{\tau_{ij}}\Big)^K\prod_{\ell=1}^{m}\log\Big(1-\Big(\frac{x_{i_\ell}}{x_{j_\ell}}\Big)^{\tau_{i_\ell j_\ell}}\Big)\bigg],$$

where the sum is over $1\le i_\ell<j_\ell\le n$ for $1\le\ell\le m$ such that all of $i_1,\dots,i_m$, $j_1,\dots,j_m$ are distinct. By (11.20) and

$$\prod_{1\le i<j\le n}\Big(1-\Big(\frac{x_i}{x_j}\Big)^{\tau_{ij}}\Big) = (-1)^{\binom{m}{2}}\Delta(X)\prod_{i=1}^{n}x_i^{-m}$$

this can be simplified to

$$L_n^{(m)}(K) = (-1)^{km+\binom{m}{2}}\sum\mathrm{CT}\bigg[F_{ab}(X)\prod_{\ell=1}^{m}\log\Big(1-\Big(\frac{x_{i_\ell}}{x_{j_\ell}}\Big)^{\tau_{i_\ell j_\ell}}\Big)\bigg].$$

Using the $\mathfrak{S}_m$ symmetry of the product over the logarithmic terms, this reduces further to

$$L_n^{(m)}(K) = (-1)^{km+\binom{m}{2}} m! \sum \mathrm{CT}\left[F_{ab}(X) \prod_{\ell=1}^{m} \log\left(1 - \left(\frac{x_{i_\ell}}{x_{j_\ell}}\right)^{\tau_{i_\ell j_\ell}}\right)\right],$$

where $1 \le i_\ell < j_\ell \le n$ for $1 \le \ell \le m$ such that $i_1 < i_2 < \cdots < i_m$ and all of $i_1, \dots, i_m, j_1, \dots, j_m$ are pairwise distinct.

For the term in the summand corresponding to i, j there is exactly one integer ℓ in $[n]$ such that $\ell \notin i, j$. Pair this integer with $n+1$ to form the edge $(\ell, n+1)$ in a perfect matching on $[n+1]$. The other edges of this perfect matching are given by the m distinct pairs $(i_1, j_1), \dots, (i_m, j_m)$. Hence

$$L_n^{(m)}(K) = (-1)^{km+\binom{m}{2}} m! \sum_{\pi} \mathrm{CT}\left[F_{ab}(X) \prod_{\substack{(i,j)\in\pi \\ j\neq n+1}} \log\left(1 - \left(\frac{x_i}{x_j}\right)^{\tau_{ij}}\right)\right].$$

Since $F_{ab}(X)$ is a skew-symmetric function (it is the product of a symmetric function times the skew-symmetric Vandermonde product $\Delta(X)$) we are in a position to apply Corollary 11.11. Thus

$$L_n^{(m)}(K) = (-1)^{km} n m! \,\mathrm{CT}\left[F_{ab}(X) \prod_{i=1}^{m} \log\left(1 - \frac{x_{2i-1}}{x_{2i}}\right)\right].$$

Finally we replace $X \mapsto X^{-1}$ using $F_{ab}(X^{-1}) = (-1)^m F_{ba}(X)$ and use the symmetry in a and b to find

$$L_n^{(m)}(K) = (-1)^{(k+1)m} n m! \,\mathrm{CT}\left[F_{ab}(X) \prod_{i=1}^{m} \log\left(1 - \frac{x_{2i}}{x_{2i-1}}\right)\right].$$

Equating this with (11.35) completes the proof of (11.9). ■

11.6.3 A Strengthening of Theorem 11.3

As will be described in more detail below, using some further results of Adamović and Milas, it follows that the logarithmic Morris constant term identity (11.9) holds provided it holds for $a = b = 0$, i.e., provided the logarithmic analogue (11.2) of Dyson's identity holds. The proof of Theorem 11.3 given in the previous subsection implies that the latter follows from what could be termed the complex analogue of Dyson's identity, i.e., the $a = b = 0$ case of (11.24):

$$\mathrm{CT}\left[\prod_{i=1}^{n} \prod_{1\le l<j\le n} \left(1 - \left(\frac{x_i}{x_j}\right)^{\tau_{ij}}\right)^{u}\right] = x^m P_n(x^2) \frac{\Gamma(1+\frac{1}{2}nu)}{\Gamma^n(1+\frac{1}{2}u)}. \tag{11.37}$$

As a consequence of all this, Theorem 11.3 can be strengthened as follows.

Theorem 11.12 (Logarithmic Morris constant term identity, strong version). *The complex Dyson constant term identity* (11.37) *implies the logarithmic Morris constant term identity.*

To justify this claim, let $e_r(X)$ for $r = 0, 1, \ldots, n$ denote the rth elementary symmetric function. The $e_r(X)$ have generating function [26]

$$\sum_{r=0}^{n} z^r e_r(X) = \prod_{i=1}^{n} (1 + zx_i). \tag{11.38}$$

Recalling definition (11.36) of F_{ab}, we now define $f_r(a) = f_r(a, b, k, n)$ by

$$f_r(a) = \mathrm{CT}\left[(-1)^r e_r(X) G_{ab}(X)\right],$$

where

$$G_{ab}(X) = F_{ab}(X) \prod_{i=1}^{m} \log\left(1 - \frac{x_{2i}}{x_{2i-1}}\right).$$

In the following b may be viewed as a formal or complex variable, but a must be taken to be an integer.

From (11.38) with $z = -1$ it follows that

$$\sum_{r=0}^{n} f_r(a) = \mathrm{CT}\left[G_{a+1,b}(X)\right] = f_0(a+1). \tag{11.39}$$

According to [2, Theorem 7.1] (translated into the notation of this paper) we also have

$$(n-r)(2b+rK) f_r(a) = (r+1)(2a+2+(n-r-1)K) f_{r+1}(a), \tag{11.40}$$

where we recall that $K := 2k+1$. Iterating this recursion yields

$$f_r(a) = f_0(a) \binom{n}{r} \prod_{i=0}^{r-1} \frac{2b+iK}{2a+2+(n-i-1)K}.$$

Summing both sides over r and using (11.39) leads to

$$f_0(a+1) = f_0(a)\, {}_2F_1\left[\begin{matrix} -n, 2b/K \\ 1-n-(2a+2)/K \end{matrix}; 1\right].$$

The ${}_2F_1$ series sums to $((2a+2b+2)/K)_n / ((2a+2)K)_n$ by the Chu–Vandermonde sum [3, Corollary 2.2.3]. Therefore,

$$f_0(a+1) = f_0(a) \prod_{i=0}^{n-1} \frac{2a+2b+2+iK}{2a+2+iK}.$$

This functional equation can be solved to finally yield

$$f_0(a) = f_0(0) \prod_{i=0}^{n-1} \frac{(2a+2b+iK)!!(iK)!!}{(2b+iK)!!(2a+iK)!!}.$$

To summarise the above computations, we have established that

$$\mathrm{CT}\big[G_{ab}(X)\big] = \mathrm{CT}\big[G_{0,b}(X)\big] \prod_{i=0}^{n-1} \frac{(2a+2b+iK)!!(iK)!!}{(2b+iK)!!(2a+iK)!!}.$$

But since $G_{0,0}(X)$ is homogeneous of degree 0 it follows that

$$\mathrm{CT}\big[G_{0,b}(X)\big] = \mathrm{CT}\big[G_{0,0}(X)\big],$$

so that indeed the logarithmic Morris constant term identity is implied by its $a = b = 0$ case.

We finally remark that it seems highly plausible that the recurrence (11.40) has the following analogue for the complex Morris identity (enhanced by the term $(-1)^r e_r(X)$ in the kernel):

$$(n-r)(2b+ru)f_r(a) = (r+1)(2a+2+(n-r-1)u)f_{r+1}(a).$$

However, the fact that for general complex u the kernel is not a skew-symmetric function seems to prevent the proof of [2, Theorem 7.1] carrying over to the complex case in a straightforward manner.

11.7 The Root System G_2

In this section we prove complex and logarithmic analogues of the Habsieger–Zeilberger identity (11.13).

Theorem 11.13 (Complex G_2 constant term identity). *For $u, v \in \mathbb{C}$ such that $\mathrm{Re}(1+\frac{3}{2}u) > 0$ and $\mathrm{Re}(1+\frac{3}{2}(u+v)) > 0$,*

$$\begin{aligned}&\mathrm{CT}\left[\left(1-\frac{yz}{x^2}\right)^u\left(1-\frac{xz}{y^2}\right)^u\left(1-\frac{xy}{z^2}\right)^u\left(1-\frac{x}{y}\right)^v\left(1-\frac{y}{z}\right)^v\left(1-\frac{z}{x}\right)^v\right]\\ &= \frac{\cos\left(\frac{1}{2}\pi u\right)\cos\left(\frac{1}{2}\pi v\right)\Gamma(1+\frac{3}{2}(u+v))\Gamma(1+\frac{3}{2}u)\Gamma(1+u)\Gamma(1+v)}{\Gamma(1+\frac{3}{2}u+v)\Gamma(1+u+\frac{1}{2}v)\Gamma(1+\frac{1}{2}(u+v))\Gamma^2(1+\frac{1}{2}u)\Gamma(1+\frac{1}{2}v)}. \qquad (11.41)\end{aligned}$$

Proof. We adopt the method of proof employed by Habsieger and Zeilberger [15, 38] in their proof of Theorem 11.4.

If $A(x,y,z;a,u)$ denotes the kernel on the left of the complex Morris identity (11.23) for $n=3$, and if and $G(x,y,z;u,v)$ denotes the kernel on the left of (11.41), then

$$G(x,y,z;u,v) = A(x/y,y/z,z/x,v,u).$$

Therefore,

$$\begin{aligned}
\mathrm{CT}\, G(x,y,z;u,v) &= \mathrm{CT}\, A(x/y,y/z,z/x;v,u) \\
&= \mathrm{CT}\, A(x,y,z;v,u)\big|_{xyz=1} \\
&= \sum_{b=0}^{\infty} \big[x^b y^b z^b\big] A(x,y,z;v,u) \\
&= \cos\left(\tfrac{1}{2}\pi u\right) \frac{\Gamma(1+\frac{3}{2}u)}{\Gamma^3(1+\frac{1}{2}u)}\, {}_3F_2\left[\begin{matrix} -v,-\frac{1}{2}u-v,-u-v \\ 1+\frac{1}{2}u,1+u \end{matrix};1\right],
\end{aligned}$$

where the last equality follows from (11.23). Summing the ${}_3F_2$ series by Dixon's sum (11.26) with $(2a,b,c) \mapsto (-v,-\frac{1}{2}u-v,-u-v)$ completes the proof. ∎

Just as for the root system A_{n-1}, we can use the complex G_2 constant term identity to prove a logarithmic analogue of (11.13).

Theorem 11.14. *Assume the representation of the* G_2 *root system as given in Sect. 11.3, and let* Φ_s^+ *and* Φ_l^+ *denote the set of positive short and positive long roots, respectively. Define*

$$G(K,M) = \frac{1}{3} \frac{(3K+3M)!!(3K)!!(2K)!!(2M)!!}{(3K+2M)!!(2K+M)!!(K+M)!!K!!K!!M!!}.$$

Then for k,m *nonnegative integers,*

$$\mathrm{CT}\left[\mathrm{e}^{-3\alpha_1-2\alpha_2}\log(1-\mathrm{e}^{\alpha_2}) \prod_{\alpha\in\Phi_l^+}(1-\mathrm{e}^{\alpha}) \prod_{\alpha\in\Phi_l}(1-\mathrm{e}^{\alpha})^k \prod_{\alpha\in\Phi_s}(1-\mathrm{e}^{\alpha})^m\right] = G(K,M),$$

where $(K,M) := (2k+1,2m)$, *and*

$$\mathrm{CT}\left[\mathrm{e}^{-2\alpha_1-\alpha_2}\log(1-\mathrm{e}^{\alpha_1}) \prod_{\alpha\in\Phi_s^+}(1-\mathrm{e}^{\alpha}) \prod_{\alpha\in\Phi_l}(1-\mathrm{e}^{\alpha})^k \prod_{\alpha\in\Phi_s}(1-\mathrm{e}^{\alpha})^m\right] = G(K,M),$$

where $(K,M) := (2k,2m+1)$.

We can more explicitly write the kernels of the two logarithmic G_2 constant term identities as

$$\begin{aligned}&\frac{z^2}{xy}\Big(1-\frac{x^2}{yz}\Big)\Big(1-\frac{y^2}{xz}\Big)\Big(1-\frac{xy}{z^2}\Big)\log\Big(1-\frac{y^2}{xz}\Big)\\&\quad\times\bigg(\Big(1-\frac{x^2}{yz}\Big)\Big(1-\frac{y^2}{xz}\Big)\Big(1-\frac{z^2}{xy}\Big)\Big(1-\frac{yz}{x^2}\Big)\Big(1-\frac{xz}{y^2}\Big)\Big(1-\frac{xy}{z^2}\Big)\bigg)^k\\&\quad\times\bigg(\Big(1-\frac{x}{y}\Big)\Big(1-\frac{x}{z}\Big)\Big(1-\frac{y}{x}\Big)\Big(1-\frac{y}{z}\Big)\Big(1-\frac{z}{x}\Big)\Big(1-\frac{z}{y}\Big)\bigg)^m\end{aligned}$$

and

$$\begin{aligned}&\frac{z}{x}\Big(1-\frac{x}{y}\Big)\Big(1-\frac{y}{z}\Big)\Big(1-\frac{x}{z}\Big)\log\Big(1-\frac{x}{y}\Big)\\&\quad\times\bigg(\Big(1-\frac{x^2}{yz}\Big)\Big(1-\frac{y^2}{xz}\Big)\Big(1-\frac{z^2}{xy}\Big)\Big(1-\frac{yz}{x^2}\Big)\Big(1-\frac{xz}{y^2}\Big)\Big(1-\frac{xy}{z^2}\Big)\bigg)^k\\&\quad\times\bigg(\Big(1-\frac{x}{y}\Big)\Big(1-\frac{x}{z}\Big)\Big(1-\frac{y}{x}\Big)\Big(1-\frac{y}{z}\Big)\Big(1-\frac{z}{x}\Big)\Big(1-\frac{z}{y}\Big)\bigg)^m,\end{aligned}$$

respectively.

Proof of Theorem 11.14. If we differentiate (11.41) with respect to u, use the cyclic symmetry in (x,y,z) of the kernel on the left and finally set $u=2k+1=K$, we get

$$\begin{aligned}&3\,\mathrm{CT}\bigg[\log\Big(1-\frac{xz}{y^2}\Big)\Big(1-\frac{yz}{x^2}\Big)^K\Big(1-\frac{xz}{y^2}\Big)^K\Big(1-\frac{xy}{z^2}\Big)^K\Big(1-\frac{x}{y}\Big)^v\Big(1-\frac{y}{z}\Big)^v\Big(1-\frac{z}{x}\Big)^v\bigg]\\&\quad=\frac{\pi}{2}\,\frac{(-1)^{k+1}\cos\big(\frac{1}{2}\pi v\big)\Gamma(1+\frac{3}{2}(K+v))\Gamma(1+\frac{3}{2}K)\Gamma(1+K)\Gamma(1+v)}{\Gamma(1+\frac{3}{2}K+v)\Gamma(1+K+\frac{1}{2}v)\Gamma(1+\frac{1}{2}(K+v))\Gamma^2(1+\frac{1}{2}K)\Gamma(1+\frac{1}{2}v)}.\end{aligned}$$

Setting $v=2m=M$ and carrying out some simplifications using (11.33) and (11.34) completes the proof of the first claim.

In much the same way, if we differentiate (11.41) with respect to v, use the cyclic symmetry in (x,y,z) and set $v=2m+1=M$, we get

$$\begin{aligned}&3\,\mathrm{CT}\bigg[\log\Big(1-\frac{x}{y}\Big)\Big(1-\frac{yz}{x^2}\Big)^u\Big(1-\frac{xz}{y^2}\Big)^u\Big(1-\frac{xy}{z^2}\Big)^u\Big(1-\frac{x}{y}\Big)^M\Big(1-\frac{y}{z}\Big)^M\Big(1-\frac{z}{x}\Big)^M\bigg]\\&\quad=\frac{\pi}{2}\,\frac{(-1)^{m+1}\cos\big(\frac{1}{2}\pi u\big)\Gamma(1+\frac{3}{2}(u+M))\Gamma(1+\frac{3}{2}u)\Gamma(1+u)\Gamma(1+M)}{\Gamma(1+\frac{3}{2}u+M)\Gamma(1+u+\frac{1}{2}M)\Gamma(1+\frac{1}{2}(u+M))\Gamma^2(1+\frac{1}{2}u)\Gamma(1+\frac{1}{2}M)}.\end{aligned}$$

Setting $u=2k=K$ yields the second claim. ■

11.8 Other Root Systems

Although further root systems admit complex analogues of the Macdonald constant term identities (11.3) or (11.12), it seems the existence of elegant logarithmic identities is restricted to A_{2n} and G_2. To see why this is so, we will discuss the root systems B_n, C_n and D_n. In order to treat all three simultaneously, it will be convenient to consider the more general *non-reduced* root system BC_n. With ε_i again denoting the ith standard unit vector in $\mathbb{R}^n$, this root system is given by

$$\Phi = \{\pm\varepsilon_i:\ 1 \le i \le n\} \cup \{\pm 2\varepsilon_i:\ 1 \le i \le n\} \cup \{\pm\varepsilon_i \pm \varepsilon_j:\ 1 \le i < j \le n\}.$$

Using the Selberg integral, Macdonald proved that [25]

$$\begin{aligned} &\mathrm{CT}\Bigg[\prod_{i=1}^{n}(1-x_i^{\pm})^a(1-x_i^{\pm 2})^b \prod_{1\le i<j\le n}(1-x_i^{\pm}x_j^{\pm})^k\Bigg] \\ &\quad = \prod_{i=0}^{n-1} \frac{(k+ik)!(2a+2b+2ik)!(2b+2ik)!}{k!(a+b+ik)!(b+ik)!(a+2b+(n+i-1)k)!}, \end{aligned} \tag{11.42}$$

where a,b,k are nonnegative integers and where we have adopted the standard shorthand notation $(1-x^{\pm}) := (1-x)(1-1/x)$, $(1-x^{\pm 2}) := (1-x^2)(1-1/x^2)$, $(1-x^{\pm}y^{\pm}) := (1-xy)(1-x/y)(1-y/x)(1-1/xy)$. For $b=0$ the above identity is the B_n case of (11.12), for $a=0$ it is the C_n case of (11.12) and for $a=b=0$ it is the D_n case of (11.12) [and also (11.3)].

A first task in finding a complex analogue of (11.42) is to fix signatures τ_{ij} and σ_{ij} for $1 \le i,j \le n$ such that

$$\prod_{1\le i<j\le n}(1-x_i^{\pm}x_j^{\pm}) = \prod_{1\le i<j\le n}\Big(1-(x_ix_j)^{\sigma_{ij}}\Big)^2\Big(1-\Big(\frac{x_i}{x_j}\Big)^{\tau_{ij}}\Big)^2. \tag{11.43}$$

This would allow the rewriting of (11.42) as

$$\begin{aligned} &\mathrm{CT}\Bigg[\prod_{i=1}^{n}(1-x_i^{\pm})^a(1-x_i^{\pm 2})^b \prod_{1\le i<j\le n}\Big(1-(x_ix_j)^{\sigma_{ij}}\Big)^{2k}\Big(1-\Big(\frac{x_i}{x_j}\Big)^{\tau_{ij}}\Big)^{2k}\Bigg] \\ &\quad = \prod_{i=0}^{n-1} \frac{(k+ik)!(2a+2b+2ik)!(2b+2ik)!}{k!(a+b+ik)!(b+ik)!(a+2b+(n+i-1)k)!}, \end{aligned} \tag{11.44}$$

after which $2k$ can be replaced by the complex variable u.

In the following we abbreviate (11.43) as $L(X) = R_{\sigma\tau}(X)$. In order to satisfy this equation, we note that for an arbitrary choice of the signatures σ_{ij} and τ_{ij},

$$
\begin{aligned}
L(X) &= \prod_{1\le i<j\le n} \left(1-(x_ix_j)^{\pm\sigma_{ij}}\right)\left(1-\left(\frac{x_i}{x_j}\right)^{\pm\tau_{ij}}\right) \\
&= \prod_{1\le i<j\le n} (x_ix_j)^{-\sigma_{ij}}\left(\frac{x_i}{x_j}\right)^{-\tau_{ij}}\left(1-(x_ix_j)^{\sigma_{ij}}\right)^2\left(1-\left(\frac{x_i}{x_j}\right)^{\tau_{ij}}\right)^2 \\
&= R_{\sigma\tau}(X)\prod_{i=1}^{n} x_i^{-\sum_{j>i}(\sigma_{ij}+\tau_{ij})-\sum_{j<i}(\sigma_{ji}-\tau_{ji})}.
\end{aligned}
$$

We must therefore fix the σ_{ij} and τ_{ij} such that

$$\sum_{j=i+1}^{n}(\sigma_{ij}+\tau_{ij})+\sum_{j=1}^{i-1}(\sigma_{ji}-\tau_{ji})=0 \tag{11.45}$$

for all $1\le i\le n$. If we sum this over all i, this gives

$$0=\sum_{1\le i<j\le n}\sigma_{ij}\equiv\binom{n}{2}\pmod 2.$$

We thus conclude that a necessary condition for (11.45), and hence (11.43), to hold is that $n\equiv 0,1 \pmod 4$. As we shall show next it is also a sufficient condition, as there are many solutions to (11.45) for the above two congruence classes.

Lemma 11.15. *For $n\equiv 1 \pmod 4$ define $m:=(n-1)/2$ and $p:=m/2$. If we choose τ_{ij} as in* (11.14) *and σ_{ij}, $1\le i<j\le n$, as*

$$\sigma_{ij}=\begin{cases}-1 & \text{if } p<j-i\le 3p,\\ 1 & \text{otherwise,}\end{cases} \tag{11.46}$$

then (11.45)*, and thus* (11.43)*, is satisfied.*

We can extend the definition of σ_{ij} to all $1\le i,j\le n$ by setting $\sigma_{ij}=-\sigma_{ji}$. Then the matrix $\Sigma=(\sigma_{ij})_{1\le i,j\le n}$ is a skew-symmetric Toeplitz matrix. For example, for $n=5$, the above choice for the σ_{ij} generates

$$\Sigma=\begin{pmatrix}0&1&-1&-1&1\\-1&0&1&-1&-1\\1&-1&0&1&-1\\1&1&-1&0&1\\-1&1&1&-1&0\end{pmatrix}.$$

Proof of Lemma 11.15. Note that by Lemma 11.5 we only need to prove that

$$\sum_{j=i+1}^{n}\sigma_{ij}+\sum_{j=1}^{i-1}\sigma_{ji}=0.$$

If for $1 \le j \le i-1$ we define $\sigma_{i,n+j} := -\sigma_{ij} = \sigma_{ji}$ then this becomes

$$\sum_{j=i+1}^{n+i-1} \sigma_{ij} = 0. \tag{11.47}$$

We now observe that $\sigma_{i+1,j+1} = \sigma_{i,j}$. For $j < n$ or $j > n$ this follows immediately from (11.46). For $j = n$ it follows from $\sigma_{1,i+1} = \sigma_{i,n}$, which again follows from (11.46) since $p < n-i \le 3p$ is equivalent to $p < i \le 3p$. Thanks to $\sigma_{i+1,j+1} = \sigma_{i,j}$ we only need to check (11.47) for $i = 1$. Then

$$\sum_{j=2}^{n} \sigma_{ij} = \sum_{j=2}^{p+1} 1 - \sum_{j=p+2}^{3p+1} 1 + \sum_{j=3p+2}^{n} 1 = n - 4p - 1 = 0. \qquad \blacksquare$$

Lemma 11.16. *For $n \equiv 0 \pmod 4$ define $m := (n-2)/2$. If we choose τ_{ij} as in* (11.14) *and σ_{ij} as*

$$\sigma_{ij} = \begin{cases} 1 & \text{if } i+j \text{ is even or } i+j = m+2, \\ -1 & \text{if } i+j \text{ is odd and } i+j \neq m+2, \end{cases}$$

then (11.45)*, and thus* (11.43)*, is satisfied.*

Proof. By a simple modification of Lemma 11.5 it follows that for n even and $m = (n-2)/2$,

$$\sum_{j=i+1}^{n} \tau_{ij} - \sum_{j=1}^{i-1} \tau_{ji} = \begin{cases} -1 & \text{if } 1 \le i \le m+1, \\ 1 & \text{if } m+1 < i < n. \end{cases}$$

Hence we must show that

$$\sum_{j=i+1}^{n} \sigma_{ij} + \sum_{j=1}^{i-1} \sigma_{ji} = \begin{cases} 1 & \text{if } 1 \le i \le m+1, \\ -1 & \text{if } m+1 < i < n. \end{cases}$$

But this is obvious. The sum on the left is over $n-1$ terms, with one more odd $i+j$ then even $i+j$. Hence, without the exceptional condition on $i+j = m+2$, the sum would always be -1. To have $i+j = m+2$ as part of one of the two sums we must have $i \le m+1$, in which case one -1 is changed to a $+1$ leading to a sum of $+1$ instead of -1. $\blacksquare$

Lemmas 11.15 and 11.16 backed up with numerical data for $n = 4$ and $n = 5$ suggest the following generalisation of (11.44).

Conjecture 11.17 (Complex BC_n constant term identity). Let $n \equiv \zeta \pmod 4$ where $\zeta = 0, 1$, and let $u \in \mathbb{C}$ such that $\min\{\mathrm{Re}(1+2b+(n-1)u), \mathrm{Re}(1+\frac{1}{2}nu)\} > 0$. Assume that τ_{ij} and σ_{ij} for $1 \le i < j \le n$ are signatures satisfying (11.45). Then there exists a polynomial $P_n(x)$, independent of a and b, such that $P_n(1) = 1$ and

$$\begin{aligned} &\mathrm{CT}\Bigg[\prod_{i=1}^{n}(1-x_i^{\pm})^a(1-x_i^{\pm 2})^b \prod_{1\le i<j\le n}\Big(1-(x_ix_j)^{\sigma_{ij}}\Big)^u\Big(1-\Big(\frac{x_i}{x_j}\Big)^{\tau_{ij}}\Big)^u\Bigg] \\ &= x^{n-\zeta}P_n(x^2)\frac{\Gamma(1+\frac{1}{2}nu)}{\Gamma(1+\frac{1}{2}(n-1)u)\Gamma^n(1+\frac{1}{2}u)}\prod_{i=1}^{n-1}\frac{\Gamma(1+iu)}{\Gamma(1+(i-\frac{1}{2})u)} \\ &\quad\times\prod_{i=0}^{n-1}\frac{(\frac{1}{2}+\frac{1}{2}iu)_{a+b}(\frac{1}{2}+\frac{1}{2}iu)_b}{(1+\frac{1}{2}(n+i-1)u)_{a+2b}}, \end{aligned} \tag{11.48}$$

where $x = x(u) := \cos\left(\frac{1}{2}\pi u\right)$. Trivially, $P_1(x) = 1$. Conjecturally,

$$P_4(x) = 1 \quad \text{and} \quad P_5(x) = \frac{1}{15}(3+4x+8x^2).$$

We note that the D_4 case of the conjecture, i.e., $a = b = 0$ and $n = 4$, is equivalent to the following new hypergeometric multisum identity

$$\sum\prod_{1\le i<j\le 4}(-1)^{k_{ij}}\binom{u}{k_{ij}}\binom{u}{m_{ij}} = \cos^4(\tfrac{1}{2}\pi u)\frac{\Gamma(1+u)\Gamma^2(1+2u)\Gamma(1+3u)}{\Gamma^5(1+\frac{1}{2}u)\Gamma^2(1+\frac{3}{2}u)\Gamma(1+\frac{5}{2}u)},$$

where the sum is over $\{k_{ij}\}_{1\le i<j\le 4}$ and $\{m_{ij}\}_{1\le i<j\le 4}$ subject to the constraints

$$\begin{aligned} k_{12}-k_{13}-k_{14}+m_{12}+m_{13}-m_{14} &= 0, \\ k_{12}-k_{23}+k_{24}-m_{12}+m_{23}-m_{24} &= 0, \\ k_{13}-k_{23}+k_{34}+m_{13}-m_{23}-m_{34} &= 0, \\ k_{14}+k_{24}-k_{34}-m_{14}+m_{24}-m_{34} &= 0, \end{aligned}$$

or, equivalently,

$$\sum_{\substack{1\le i<j\le 4\\ i=p \text{ or } j=p}}(\tau_{ij}k_{ij}+\sigma_{ij}m_{ij}) = 0 \qquad \text{for } 1\le p\le 4.$$

Unfortunately, from the point of view of logarithmic constant term identities, (11.48) is not good news. On the right-hand side the exponent $n-\zeta$ of x is too high relative to the rank n of the root system. (Compare with $m = (n-1)/2$ versus $n-1$ for A_{n-1}.) If we write (11.48) as $L_n(u) = R_n(u)$ and define $K := 2k+1$, then due to the factor $x^{n-\zeta}$, $R_n^{(j)}(K) = 0$ for all $1 \le j < n-\zeta$. Much like the Morris case, $R_n^{(n-\zeta)}(K)$ yields a ratio of products of double factorials:

$$R_n^{(n-\zeta)}(K) = (n-\zeta)!P_n(0)\frac{(nK)!!}{((n-1)K)!!(K!!)^n}\prod_{i=1}^{n-1}\frac{(2iK)!!}{((2i-1)K)!!}$$
$$\times\prod_{i=0}^{n-1}\frac{(2b+iK-1)!!(2a+2b+iK-1)!!((n+i-1)K)!!}{(2a+4b+(n+i-1)K)!!(iK-1)!!(iK-1)!!}.$$

However, if we differentiate $L_n(u)$ as many as $n-\zeta$ times, a large number of different types of logarithmic terms give a nonvanishing contribution to $L_n^{(n-\zeta)}(K)$—unlike type A where only terms with the same functional form (corresponding to perfect matchings) survive the specialisation $u = K$. For example, for $n = 4$ terms such as

$$\log^3\Big(1-\frac{x_1}{x_2}\Big)\log\Big(1-\frac{1}{x_2x_3}\Big),$$
$$\log^2\Big(1-\frac{x_1}{x_2}\Big)\log(1-x_1x_2)\log\Big(1-\frac{1}{x_2x_3}\Big),$$
$$\log^2\Big(1-\frac{x_1}{x_2}\Big)\log^2(1-x_3x_4),$$

and many similar such terms, all give nonvanishing contributions.

Acknowledgements The authors have greatly benefited from Tony Guttmann's and Vivien Challis' expertise in numerical computations. The authors also thank Antun Milas and the anonymous referee for very helpful remarks, leading to the inclusion of Sect. 11.6.3. OW and WZ are supported by the Australian Research Council.

References

1. Adamović, D., Milas, A.: On W-algebras associated to $(2,p)$ minimal models and their representations. Int. Math. Res. Not. IMRN **20**, 3896–3934 (2010)
2. Adamović, D., Milas, A.: On W-algebra extensions of $(2,p)$ minimal models: $p > 3$. J. Algebra **344**, 313–332 (2011)
3. Andrews, G.E., Askey, R., Roy, R.: Special Functions, Encyclopedia of Mathematics and Its Applications, vol. 71. Cambridge University Press, Cambridge (1999)
4. Apagodu, M., Zeilberger, D.: Multi-variable Zeilberger and Almkvist–Zeilberger algorithms and the sharpening of Wilf–Zeilberger theory. Adv. Appl. Math. **37**, 139–152 (2006). http://www.math.rutgers.edu/~zeilberg/mamarim/mamarimhtml/multiZ.html
5. Bailey, W.N.: Generalized Hypergeometric Series, Cambridge Tracts in Mathematics and Mathematical Physics, vol. 32. Stechert-Hafner, Inc., New York (1964)
6. Bressoud, D.M.: Proofs and Confirmations—The Story of the Alternating Sign Matrix Conjecture. Cambridge University Press, Cambridge (1999)

7. Bressoud, D.M., Goulden, I.P.: Constant term identities extending the q-Dyson theorem. Trans. Amer. Math. Soc. **291**, 203–228 (1985)
8. Cherednik, I.: Double affine Hecke algebras and Macdonald's conjectures. Ann. Math. (2) **141**, 191–216 (1995)
9. Dyson, F.J.: Statistical theory of the energy levels of complex systems I. J. Math. Phys. **3**, 140–156 (1962)
10. Forrester, P.J., Warnaar, S.O.: The importance of the Selberg integral. Bull. Amer. Math. Soc. (N.S.) **45**, 489–534 (2008)
11. Gessel, I.M., Xin, G.: A short proof of the Zeilberger–Bressoud q-Dyson theorem. Proc. Amer. Math. Soc. **134**, 2179–2187 (2006)
12. Gessel, I.M., Lv, L., Xin, G., Zhou, Y.: A unified elementary approach to the Dyson, Morris, Aomoto, and Forrester constant term identities. J. Combin. Theory Ser. A **115**, 1417–1435 (2008)
13. Good, I.J.: Short proof of a conjecture by Dyson. J. Math. Phys. **11**, 1884 (1970)
14. Gunson, J.: unpublished
15. Habsieger, L.: La q-conjecture de Macdonald–Morris pour G_2. C. R. Acad. Sci. Paris Sér. I Math. **303**, 211–213 (1986)
16. Habsieger, L.: Une q-intégrale de Selberg et Askey. SIAM J. Math. Anal. **19**, 1475–1489 (1988)
17. Humphreys, J.E.: Introduction to Lie Algebras and Representation Theory. Graduate Texts in Mathematics, vol. 9. Springer, New York (1978)
18. Kadell, K.W.J.: A proof of Askey's conjectured q-analogue of Selberg's integral and a conjecture of Morris. SIAM J. Math. Anal. **19**, 969–986 (1988)
19. Kadell, K.W.J.: Aomoto's machine and the Dyson constant term identity. Methods Appl. Anal. **5**, 335–350 (1998)
20. Kaneko, J.: Forrester's conjectured constant term identity II. Ann. Comb. **6**, 383–397 (2002)
21. Károlyi, G., Nagy, Z.L.: A short proof of Andrews' q-Dyson conjecture. Proc. Amer. Math. Soc., so appear
22. Knuth, D.E.: Overlapping Pfaffians. Electron. J. Combin. **3**, 13 (1996) (Research Paper 5)
23. Krattenthaler, C.: Advanced determinant calculus. Sém. Lothar. Combin. **42**, 67 (1999) (Art. B42q)
24. Opdam, E.M.: Some applications of hypergeometric shift operators. Invent. Math. **98**, 1–18 (1989)
25. Macdonald, I.G.: Some conjectures for root systems. SIAM J. Math. Anal. **13**, 988–1007 (1982)
26. Macdonald, I.G.: Symmetric Functions and Hall Polynomials, 2nd edn. Oxford University Press, New York (1995)
27. Morris, W.G.: Constant term identities for finite and affine root systems: conjectures and theorems. Ph.D. Thesis, University of Wisconsin-Madison (1982)
28. Petkovšek, M., Wilf, H.S., Zeilberger, D.: $A = B$. A. K. Peters, Ltd., Wellesley (1996)
29. Selberg, A.: Bemerkninger om et multipelt integral. Norske Mat. Tidsskr. **26**, 71–78 (1944)
30. Sills, A.V.: Disturbing the Dyson conjecture, in a generally GOOD way. J. Combin. Theory Ser. A **113**, 1368–1380 (2006)
31. Sills, A.V.: Disturbing the q-Dyson conjecture. In: Tapas in Experimental Mathematics. Contemporary Mathematics, vol. 457, pp. 265–271. American Mathematical Society, Providence (2008)
32. Sills, A.V., Zeilberger, D.: Disturbing the Dyson conjecture (in a GOOD way). Experiment. Math. **15**, 187–191 (2006)
33. Stanton, D.: Sign variations of the Macdonald identities. SIAM J. Math. Anal. **17**, 1454–1460 (1986)
34. Stembridge, J.R.: A short proof of Macdonald's conjecture for the root systems of type A. Proc. Amer. Math. Soc. **102**, 777–786 (1988)
35. Stembridge, J.R.: Nonintersecting paths, Pfaffians, and plane partitions. Adv. Math. **83**, 96–131 (1990)

36. Wilson, K.: Proof of a conjecture of Dyson. J. Math. Phys. **3**, 1040–1043 (1962)
37. Zeilberger, D.: A combinatorial proof of Dyson's conjecture. Discrete Math. **41**, 317–321 (1982)
38. Zeilberger, D.: A proof of the G_2 case of Macdonald's root system–Dyson conjecture. SIAM J. Math. Anal. **18**, 880–883 (1987)
39. Zeilberger, D.: A Stembridge–Stanton style elementary proof of the Habsieger–Kadell q-Morris identity. Discrete Math. **79**, 313–322 (1990)
40. Zeilberger, D., Bressoud, D.M.: A proof of Andrews' q-Dyson conjecture. Discrete Math. **54**, 201–224 (1985)

Chapter 12
Preprocessing and Regularization for Degenerate Semidefinite Programs

Yuen-Lam Cheung, Simon Schurr, and Henry Wolkowicz

Abstract This paper presents a backward stable preprocessing technique for (nearly) ill-posed semidefinite programming, SDP, problems, i.e., programs for which the Slater constraint qualification (SCQ), the existence of strictly feasible points, (nearly) fails. Current popular algorithms for semidefinite programming rely on *primal-dual interior-point, p-d i-p,* methods. These algorithms require the SCQ for both the primal and dual problems. This assumption guarantees the existence of Lagrange multipliers, well-posedness of the problem, and stability of algorithms. However, there are many instances of SDPs where the SCQ fails or *nearly* fails. Our backward stable preprocessing technique is based on applying the Borwein–Wolkowicz facial reduction process to find a finite number, k, of *rank-revealing orthogonal rotations* of the problem. After an appropriate truncation, this results in a smaller, well-posed, *nearby* problem that satisfies the Robinson constraint qualification, and one that can be solved by standard SDP solvers. The case $k = 1$ is of particular interest and is characterized by strict complementarity of an auxiliary problem.

Key words: Backward stability • Degeneracy • Preprocessing • Semidefinite programming • Strong duality

Mathematics Subject Classifications (2010): Primary, 49K40, 90C22; Secondary, 65K10, 90C25, 90C46.

COMMUNICATED BY HEINZ H. BAUSCHKE.

Y.-L. Cheung • S. Schurr • H. Wolkowicz (✉)
Department of Combinatorics and Optimization, University of Waterloo, Waterloo, ON N2L 3G1, Canada
e-mail: yl2cheun@uwaterloo.ca; spschurr@rogers.com; hwolkowicz@uwaterloo.ca

D.H. Bailey et al. (eds.), *Computational and Analytical Mathematics*, Springer Proceedings in Mathematics & Statistics 50, DOI 10.1007/978-1-4614-7621-4_12,

12.1 Introduction

The aim of this paper is to develop a backward stable preprocessing technique to handle (nearly) ill-posed semidefinite programming, SDP, problems, i.e., programs for which the Slater constraint qualification (Slater CQ or SCQ), the existence of strictly feasible points, (nearly) fails. The technique is based on applying the Borwein–Wolkowicz *facial reduction* process [11, 12] to find a finite number k of *rank-revealing orthogonal rotation* steps. Each step is based on solving an auxiliary problem (AP) where it and its dual satisfy the Slater CQ. After an appropriate truncation, this results in a smaller, well-posed, *nearby* problem for which the Robinson constraint qualification (RCQ) [52] holds; and one that can be solved by standard SDP solvers. In addition, the case $k = 1$ is of particular interest and is characterized by strict complementarity of the (AP).

In particular, we study SDPs of the following form:

$$\text{(P)} \qquad v_P := \sup_y \{b^T y : \mathscr{A}^* y \preceq C\}, \tag{12.1}$$

where the optimal value v_P is finite, $b \in \mathbb{R}^m$, $C \in \mathbb{S}^n$, and $\mathscr{A} : \mathbb{S}^n \to \mathbb{R}^m$ is an onto linear transformation from the space $\mathbb{S}^n$ of $n \times n$ real symmetric matrices to $\mathbb{R}^m$. The adjoint of $\mathscr{A}$ is $\mathscr{A}^* y = \sum_{i=1}^m y_i A_i$, where $A_i \in \mathbb{S}^n, i = 1, \ldots, m$. The symbol $\preceq$ denotes the Löwner partial order induced by the cone $\mathbb{S}^n_+$ of positive semidefinite matrices, i.e., $\mathscr{A}^* y \preceq C$ if and only if $C - \mathscr{A}^* y \in \mathbb{S}^n_+$. (Note that the cone optimization problem (12.1) is commonly used as the dual problem in the SDP literature, though it is often the primal in the linear matrix inequality (LMI) literature, e.g., [13].) If (P) is *strictly feasible*, then one can use standard solution techniques; if (P) is *strongly infeasible*, then one can set $v_P = -\infty$, e.g., [38,43,47,62,65]. If neither of these two feasibility conditions can be verified, then we apply our preprocessing technique that finds a rotation of the problem that is akin to *rank-revealing* matrix rotations. (See e.g., [58,59] for equivalent matrix results.) This rotation finds an equivalent (nearly) block diagonal problem which allows for simple strong dualization by solving only the most significant block of (P) for which the Slater CQ holds. This is equivalent to restricting the original problem to a face of $\mathbb{S}^n_+$, i.e., the preprocessing can be considered as a *facial reduction* of (P). Moreover, it provides a *backward stable* approach for solving (P) when it is feasible and the SCQ fails; and it solves a nearby problem when (P) is *weakly infeasible*.

The Lagrangian dual to (12.1) is

$$\text{(D)} \qquad v_D := \inf_X \{\langle C, X\rangle : \mathscr{A}(X) = b, X \succeq 0\}, \tag{12.2}$$

where $\langle C, X\rangle := \operatorname{trace} CX = \sum_{ij} C_{ij} X_{ij}$ denotes the trace inner product of the symmetric matrices C and X and $\mathscr{A}(X) = (\langle A_i, X\rangle) \in \mathbb{R}^m$. Weak duality $v_D \geq v_P$ follows easily. The usual constraint qualification (CQ) used for (P) is SCQ, i.e., strict feasibility $\mathscr{A}^* y \prec C$ (or $C - \mathscr{A}^* y \in \mathbb{S}^n_{++}$, the cone of positive definite matrices). If we assume the Slater CQ holds and the primal optimal value is finite, then strong duality holds, i.e., we have a zero duality gap and attainment of the dual optimal value. Strong duality results for (12.1) without any constraint qualification are given in [10–12, 48, 49, 72], and more recently in [50, 66]. Related closure conditions appear in [44]; and, properties of problems where strong duality fails appear in [45].

General surveys on SDP are in, e.g., [4, 63, 68, 75]. Further general results on SDP appear in the recent survey [31].

Many popular algorithms for (P) are based on Newton's method and a *primal-dual interior-point, p-d i-p,* approach, e.g., the codes (latest at the URLs in the citations) CSDP, SeDuMi, SDPT3, SDPA [9, 60, 67, 76]; see also the

SDP URL: www-user.tu-chemnitz.de/~helmberg/sdp_software.html.

To find the search direction, these algorithms apply symmetrization in combination with block elimination to find the Newton search direction. The symmetrization and elimination steps both result in ill-conditioned linear systems, even for well conditioned SDP problems, e.g., [19, 73]. And, these methods are very susceptible to numerical difficulties and high iteration counts in the case when SCQ nearly fails; see, e.g., [21–24]. Our aim in this paper is to provide a stable regularization process based on orthogonal rotations for problems where strict feasibility (nearly) fails. Related papers on regularization are, e.g., [30, 39]; and papers on high accuracy solutions for algorithms SDPA-GMP,-QD,-DD are, e.g., [77]. In addition, a popular approach uses a self-dual embedding, e.g., [16, 17]. This approach results in SCQ holding by using homogenization and increasing the number of variables. In contrast, our approach reduces the size of the problem in a preprocessing step in order to guarantee SCQ.

12.1.1 Outline

We continue in Sect. 12.1.2 with preliminary notation and results for cone programming. In Sect. 12.2 we recall the history and outline the similarities and differences of what facial reduction means first for linear programming (LP), and then for ordinary convex programming (CP), and finally for SDP, which has elements from both LP and CP. Instances and applications where the SCQ fails are given in Sect. 12.2.3.1. Then, Sect. 12.3 presents the theoretical background and tools needed for the facial reduction algorithm for SDP. This includes results on strong duality in Sect. 12.3.1; and, various theorems of the alternative, with cones having both nonempty and empty interior, are given in Sect. 12.3.2. A stable auxiliary problem (12.18) for identifying the minimal face containing the feasible set is presented and studied in Sect. 12.3.3; see, e.g., Theorem 12.13. In particular, we relate the question of transforming the unstable problem of finding the minimal face to the existence of a primal-dual optimal pair satisfying strict complementarity and to the number of steps in the facial reduction. See Remark 12.12 and Sect. 12.3.5. The resulting information from the auxiliary problem for problems where SCQ (nearly) fails is given in Theorem 12.17 and Propositions 12.18, 12.19. This information can be used to construct equivalent problems. In particular, a rank-revealing rotation is used in Sect. 12.3.4 to yield two equivalent problems that are useful in sensitivity analysis, see Theorem 12.22. In particular, this shows the backward stability with respect to perturbations in the parameter β in the definition of the cone T_β for the problem. Truncating the (near) singular blocks to zero yields two smaller equivalent, regularized problems in Sect. 12.3.4.1.

The facial reduction is studied in Sect. 12.4. An outline of the facial reduction using a rank-revealing rotation process is given in Sect. 12.4.1. Backward stability results are presented in Sect. 12.4.2.

Preliminary numerical tests, as well as a technique for generating instances with a finite duality gap useful for numerical tests, are given in Sect. 12.5. Concluding remarks appear in Sect. 12.6.

12.1.2 Preliminary Definitions

Let $(\mathscr{V}, \langle\cdot,\cdot\rangle_{\mathscr{V}})$ be a finite-dimensional inner product space and K be a (closed) *convex cone* in $\mathscr{V}$, i.e., $\lambda K \subseteq K, \forall \lambda \geq 0$, and $K + K \subseteq K$. K is *pointed* if $K \cap (-K) = \{0\}$; K is *proper* if K is pointed and $\operatorname{int} K \neq \emptyset$; the *polar* or *dual cone* of K is $K^* := \{\phi : \langle \phi, k \rangle \geq 0, \forall k \in K\}$. We denote by $\preceq_K$ the partial order with respect to K. That is, $x_1 \preceq_K x_2$ means that $x_2 - x_1 \in K$. We also write $x_1 \prec_K x_2$ to mean that $x_2 - x_1 \in \operatorname{int} K$. In particular with $\mathscr{V} = \mathbb{S}^n$, $K = \mathbb{S}^n_+$ yields the partial order induced by the cone of positive semidefinite matrices in $\mathbb{S}^n$, i.e., the so-called Löwner partial order. We denote this simply with $X \preceq Y$ for $Y - X \in \mathbb{S}^n_+$. $\operatorname{cone}(S)$ denotes the convex cone generated by the set S. In particular, for any nonzero vector x, the *ray generated by* x is defined by $\operatorname{cone}(x)$. The ray generated by $s \in K$ is called an *extreme ray* if $0 \preceq_K u \preceq_K s$ implies that $u \in \operatorname{cone}(s)$. The subset $F \subseteq K$ is a *face of the cone* K, denoted $F \trianglelefteq K$, if

$$(s \in F, 0 \preceq_K u \preceq_K s) \implies (\operatorname{cone}(u) \subseteq F). \tag{12.3}$$

Equivalently, $F \trianglelefteq K$ if F is a cone and $\left(x, y \in K, \frac{1}{2}(x+y) \in F\right) \implies (\{x,y\} \subseteq F)$. If $F \trianglelefteq K$ but is not equal to K, we write $F \triangleleft K$. If $\{0\} \neq F \triangleleft K$, then F is a *proper face* of K. For $S \subseteq K$, we let face(S) denote the smallest face of K that contains S. A face $F \trianglelefteq K$ is an *exposed face* if it is the intersection of K with a hyperplane. The cone K is *facially exposed* if every face $F \trianglelefteq K$ is exposed. If $F \trianglelefteq K$, then the *conjugate face* is $F^c := K^* \cap \{F\}^\perp$. Note that the conjugate face F^c is *exposed* using any $s \in \operatorname{relint} F$ (where $\operatorname{relint} S$ denotes the *relative interior* of the set S), i.e., $F^c = K^* \cap \{s\}^\perp, \forall s \in \operatorname{relint} F$. In addition, note that $\mathbb{S}^n_+$ is self-dual (i.e., $(\mathbb{S}^n_+)^* = \mathbb{S}^n_+$) and is facially exposed.

For the general conic programming problem, the constraint linear transformation $\mathscr{A} : \mathscr{V} \to \mathscr{W}$ maps between two Euclidean spaces. The adjoint of $\mathscr{A}$ is denoted by $\mathscr{A}^* : \mathscr{W} \to \mathscr{V}$, and the Moore–Penrose generalized inverse of $\mathscr{A}$ is denoted by $\mathscr{A}^\dagger : \mathscr{W} \to \mathscr{V}$.

A linear conic program may take the form

$$(\mathrm{P_{conic}}) \qquad v_P^{\mathrm{conic}} = \sup_y \{\langle b, y\rangle \ : \ C - \mathscr{A}^* y \succeq_K 0\}, \tag{12.4}$$

with $b \in \mathscr{W}$ and $C \in \mathscr{V}$. Its dual is given by

$$(\mathrm{D_{conic}}) \qquad v_D^{\mathrm{conic}} = \inf_X \{\langle C, X\rangle \ : \mathscr{A}(X) = b, X \succeq_{K^*} 0\}. \tag{12.5}$$

Note that the RCQ is said to hold for the linear conic program $(\mathrm{P_{conic}})$ if $0 \in \mathrm{int}(C - \mathscr{A}^*(\mathbb{R}^m) - \mathbb{S}^n_+)$; see [53]. As pointed out in [61], the Robinson CQ is equivalent to the Mangasarian–Fromovitz constraint qualification in the case of conventional nonlinear programming. Also, it is easy to see that the Slater CQ, strict feasibility, implies RCQ.

Denote the feasible solution and slack sets of (12.4) and (12.5) by $\mathscr{F}_P = \mathscr{F}_P^y = \{y : \ \mathscr{A}^* y \preceq_K C\}$, $\mathscr{F}_P^Z = \{Z : \ Z = C - \mathscr{A}^* y \succeq_K 0\}$, and $\mathscr{F}_D = \{X : \ \mathscr{A}(X) = b, \ X \succeq_{K^*} 0\}$, respectively. The *minimal face* of (12.4) is the intersection of all faces of K containing the feasible slack vectors:

$$f_P = f_P^Z := \mathrm{face}(C - \mathscr{A}^*(\mathscr{F}_P)) = \cap\{H \trianglelefteq K \ : \ C - \mathscr{A}^*(\mathscr{F}_P) \subseteq H\}.$$

Here, $\mathscr{A}^*(\mathscr{F}_P)$ is the linear image of the set $\mathscr{F}_P$ under $\mathscr{A}^*$.

We continue with the notation specifically for $\mathscr{V} = \mathbb{S}^n$, $K = \mathbb{S}^n_+$, and $\mathscr{W} = \mathbb{R}^m$. Then (12.4) [respectively, (12.5)] is the same as (12.1) [respectively, (12.2)]. We let e_i denote the ith unit vector, and $E_{ij} := \frac{1}{\sqrt{2}}(e_i e_j^T + e_j e_i^T)$ are the unit matrices in $\mathbb{S}^n$. For specific $A_i \in \mathbb{S}^n, i = 1, \ldots, m$, we let $\|\mathscr{A}\|_2$ denote the spectral norm of $\mathscr{A}$ and define the Frobenius norm (Hilbert–Schmidt norm) of $\mathscr{A}$ as $\|\mathscr{A}\|_F := \sqrt{\sum_{i=1}^m \|A_i\|_F^2}$.

Unless stated otherwise, all vector norms are assumed to be 2-norm, and all matrix norms in this paper are Frobenius norms. Then, e.g., [32, Chap. 5], for any $X \in \mathbb{S}^n$,

$$\|\mathscr{A}(X)\|_2 \leq \|\mathscr{A}\|_2 \|X\|_F \leq \|\mathscr{A}\|_F \|X\|_F. \tag{12.6}$$

We summarize our assumptions in the following.

Assumption 12.1. $\mathscr{F}_P \neq \emptyset$; $\mathscr{A}$ is onto.

12.2 Framework for Regularization/Preprocessing

The case of preprocessing for linear programming is well known. The situation for general convex programming is not. We now outline the preprocessing and facial reduction for the cases of linear programming (LP); ordinary convex programming (CP); and SDP. We include details on motivation involving numerical stability and convergence for algorithms. In all three cases, the facial reduction can be regarded as a Robinson-type regularization procedure.

12.2.1 *The Case of Linear Programming, LP*

Preprocessing is essential for LP, in particular for the application of interior-point methods. Suppose that the constraint in (12.4) is $\mathscr{A}^*y \preceq_K c$ with $K = \mathbb{R}^n_+$, the nonnegative orthant, i.e., it is equivalent to the elementwise inequality $A^T y \le c, c \in \mathbb{R}^n$, with the (full row rank) matrix A being $m \times n$. Then ($\mathrm{P}_{\text{conic}}$) and ($\mathrm{D}_{\text{conic}}$) form the standard primal-dual LP pair. Preprocessing is an essential step in algorithms for solving LP, e.g., [20, 27, 35]. In particular, interior-point methods require strictly feasible points for both the primal and dual LPs. Under the assumption that $\mathscr{F}_P \neq \emptyset$, lack of strict feasibility for the primal is equivalent to the existence of an unbounded set of dual optimal solutions. This results in convergence problems, since current primal-dual interior-point methods follow the *central path* and converge to the analytic center of the optimal set. From a standard Farkas' lemma argument, we know that the Slater CQ, the existence of a strictly feasible point $A^T\hat{y} < c$, holds if and only if

$$\text{the system } \boxed{0 \neq d \ge 0, Ad = 0, c^T d = 0} \text{ is inconsistent.} \tag{12.7}$$

In fact, after a permutation of columns if needed, we can partition both A, c as

$$A = \begin{bmatrix} A^< & A^= \end{bmatrix}, \text{ with } A^= \text{ size } m \times t, \quad c = \begin{pmatrix} c^< \\ c^= \end{pmatrix},$$

so that we have

$$A^{<T}\hat{y} < c^<, \quad A^{=T}\hat{y} = c^=, \text{ for some } \hat{y} \in \mathbb{R}^m, \qquad \text{and } A^T y \le c \implies A^{=T} y = c^=,$$

i.e., the constraints $A^{=T} y \le c^=$ are the *implicit equality constraints*, with indices given in

$$\mathscr{P} := \{1, \dots, n\}, \quad \mathscr{P}^< := \{1, \dots, n-t\}, \quad \mathscr{P}^= := \{n-t+1, \dots, n\}.$$

Moreover, the indices for $c^=$ (and columns of $A^=$) correspond to the indices in a *maximal positive* solution d in (12.7); and, the nonnegative linear dependence in (12.7) implies that there are redundant implicit equality constraints that we can discard, yielding the smaller $(A^=_{\bar{R}})^T y = c^=_{\bar{R}}$ with $A^=_{\bar{R}}$ full column rank. Therefore, an equivalent problem to ($\mathrm{P_{conic}}$) is

$$(\mathrm{P_{reg}}) \qquad v_P := \max\{b^T y \,:\, A^{<T} y \le c^<,\ A^{=T}_{\bar{R}} y = c^=_{\bar{R}}\}. \qquad (12.8)$$

And this LP satisfies the RCQ; see Corollary 12.17, Item 2, below. In this case RCQ is equivalent to the Mangasarian–Fromovitz constraint qualification (MFCQ), i.e., there exists a feasible $\hat{y}$ which satisfies the inequality constraints strictly, $A^{<T}\hat{y} < c^<$, and the matrix $A^=$ for the equality constraints is full row rank; see, e.g., [8, 40]. The MFCQ characterizes stability with respect to right-hand side perturbations and is equivalent to having a compact set of dual optimal solutions. Thus, recognizing and changing the implicit equality constraints to equality constraints and removing redundant equality constraints provides a simple *regularization of LP*.

Let f_P denote the minimal face of the LP. Then note that we can rewrite the constraint as

$$A^T y \preceq_{f_P} c, \quad \text{with } f_P := \{z \in \mathbb{R}^n_+ : z_i = 0, i \in \mathscr{P}^=\}.$$

Therefore, rewriting the constraint using the minimal face provides a regularization for LP. This is followed by discarding redundant equality constraints to obtain the MFCQ. This reduces the number of constraints and thus the dimension of the dual variables. Finally, the dimension of the problem can be further reduced by eliminating the equality constraints completely using the nullspace representation. However, this last step can result in loss of sparsity and is usually not done.

We can similarly use a theorem of the alternative to recognize failure of strict feasibility in the dual, i.e., the (in)consistency of the system $0 \neq A^T v \geq 0, b^T v = 0$. This corresponds to identifying which variables x_i are identically zero on the feasible set. The regularization then simply discards these variables along with the corresponding columns of A, c.

12.2.2 The Case of Ordinary Convex Programming, CP

We now move from LP to nonlinear convex programming. We consider the *ordinary convex program (CP)*

$$(\mathrm{CP}) \qquad v_{\mathrm{CP}} := \sup\{b^T y \,:\, g(y) \le 0\}, \qquad (12.9)$$

where $g(y) = (g_i(y)) \in \mathbb{R}^n$ and $g_i : \mathbb{R}^m \to \mathbb{R}$ are convex functions, for all i. (Without loss of generality, we let the objective function $f(y) = b^T y$ be linear. This can always be achieved by replacing a concave objective function with a new variable $\sup t$, and adding a new constraint $-f(y) \le -t$.) The quadratic programming case has been well studied [28,41]. Some preprocessing results for the general CP case are known, e.g., [15]. However, preprocessing for general CP is not as well known as for LP. In fact (see [6]) as for LP there is a set of *implicit equality constraints for CP*, i.e., we can partition the constraint index set $\mathscr{P} = \{1,\dots,n\}$ into two sets:

$$\mathscr{P}^= = \{i \in \mathscr{P} : y \text{ feasible } \implies g_i(y) = 0\}, \quad \mathscr{P}^< = \mathscr{P} \setminus \mathscr{P}^=. \tag{12.10}$$

Therefore, as above for LP, we can rewrite the constraints in CP using the minimal face f_P to get $g(y) \preceq_{f_P} 0$. However, this is not a true convex program since the new equality constraints are not affine. However, surprisingly the corresponding feasible set for the implicit equality constraints is convex, e.g., [6]. We include the result and a proof for completeness.

Lemma 12.2. *Let the convex program (CP) be given, and let $\mathscr{P}^=$ be defined as in* (12.10)*. Then the set $\mathscr{F}^= := \{y : g_i(y) = 0, \forall i \in \mathscr{P}^=\}$ satisfies*

$$\mathscr{F}^= = \{y : g_i(y) \le 0, \forall i \in \mathscr{P}^=\},$$

and thus is a convex set.

Proof. Let $g^=(y) = (g_i(y))_{i \in \mathscr{P}^=}$ and $g^<(y) = (g_i(y))_{i \in \mathscr{P}^<}$. By definition of $\mathscr{P}^<$, there exists a feasible $\hat{y} \in \mathscr{F}$ with $g^<(\hat{y}) < 0$; and, suppose that there exists $\bar{y}$ with $g^=(\bar{y}) \le 0$, and $g_{i_0}(\bar{y}) < 0$, for some $i_0 \in \mathscr{P}^=$. Then for small $\alpha > 0$ the point $y_\alpha := \alpha\hat{y} + (1-\alpha)\bar{y} \in \mathscr{F}$ and $g_{i_0}(y_\alpha) < 0$. This contradicts the definition of $\mathscr{P}^=$. ■

This means that we can regularize CP by replacing the implicit equality constraints as follows:

$$(\text{CP}_{\text{reg}}) \qquad v_{\text{CP}} := \sup\{b^T y : g^<(y) \le 0, y \in \mathscr{F}^=\}. \tag{12.11}$$

The generalized Slater CQ holds for the *regularized convex program* (CP_{reg}). Let

$$\phi(\lambda) = \sup_{y \in \mathscr{F}^=} b^T y - \lambda^T g^<(y)$$

denote the *regularized dual functional for CP*. Then strong duality holds for CP with the *regularized dual program*, i.e.,

$$\begin{aligned} v_{\text{CP}} = v_{CPD} &:= \inf_{\lambda \ge 0} \phi(\lambda) \\ &= \phi(\lambda^*), \end{aligned}$$

for some (dual optimal) $\lambda^* \geq 0$. The Karush–Kuhn–Tucker (KKT) optimality conditions applied to (12.11) imply that

$$\begin{gathered} y^* \text{ is optimal for } \mathrm{CP_{reg}} \\ \text{if and only if} \end{gathered}$$

$$\begin{cases} y^* \in \mathscr{F} & \text{(primal feasibility)} \\ b - \nabla g^{<}(y^*)\lambda^* \in (\mathscr{F}^{=} - y^*)^*, \text{ for some } \lambda^* \geq 0 & \text{(dual feasibility)} \\ g^{<}(y^*)^T \lambda^* = 0 & \text{(complementary slackness)} \end{cases}$$

This differs from the standard KKT conditions in that we need the polar set

$$(\mathscr{F}^{=} - y^*)^* = \overline{\mathrm{cone}\,(\mathscr{F}^{=} - y^*)}^* = (D^{=}(y^*))^*, \tag{12.12}$$

where $D^{=}(y^*)$ denotes the *cone of directions of constancy* of the implicit equality constraints $\mathscr{P}^{=}$, e.g., [6]. Thus we need to be able to find this cone numerically; see [71]. A backward stable algorithm for the cone of directions of constancy is presented in [37].

Note that a convex function f is faithfully convex if f is affine on a line segment only if it is affine on the whole line containing that segment; see [54]. Analytic convex functions are faithfully convex, as are strictly convex functions . For faithfully convex functions, the set $\mathscr{F}^{=}$ is an affine manifold, $\mathscr{F}^{=} = \{y : Vy = V\hat{y}\}$, where $\hat{y} \in \mathscr{F}$ is feasible, and the nullspace of the matrix V gives the intersection of the cones of directions of constancy $D^{=}$. Without loss of generality, let V be chosen full row rank. Then in this case we can rewrite the regularized problem as

$$(\mathrm{CP_{reg}}) \qquad v_{\mathrm{CP}} := \sup\{b^T y \ : g^{<}(y) \leq 0, Vy = V\hat{y}\}, \tag{12.13}$$

which is a convex program for which the MFCQ holds. Thus by identifying the implicit equalities and replacing them with the linear equalities that represent the cone of directions of constancy, we obtain the regularized convex program. If we let $g^R(y) = \begin{pmatrix} g^{<}(y) \\ Vy - V\hat{y} \end{pmatrix}$, then writing the constraint $g(y) \leq 0$ using g^R and the minimal cone f_P as $g^R(y) \preceq_{f_P} 0$ results in the regularized CP for which MFCQ holds.

12.2.3 *The Case of Semidefinite Programming, SDP*

Finally, we consider our case of interest, the SDP given in (12.1). In this case, the cone for the constraint partial order is $\mathbb{S}^n_+$, a *nonpolyhedral* cone. Thus we have elements of both LP and CP. Significant preprocessing is not done in current public domain SDP codes. Theoretical results are known (see, e.g., [34]) for results on redundant constraints using a probabilistic approach. However [10], the notion of minimal face can be used to regularize SDP. Surprisingly, the above result for LP in (12.8) holds. A regularized problem for (P) for which strong duality holds has constraints of the form $\mathscr{A}^*y \preceq_{f_P} C$ without the need for an extra polar set as in (12.12) that is used in the CP case, i.e., changing the cone for the partial order regularizes the problem. However, as in the LP case where we had to discard redundant implicit equality constraints, extra work has to be done to ensure that the RCQ holds. The details for the facial reduction now follow in Sect. 12.3. An equivalent regularized problem is presented in Corollary 12.24, i.e., rather than a permutation of columns needed in the LP case, we perform a rotation of the problem constraint matrices, and then we get a similar division of the constraints as in (12.8); and, setting the implicit equality constraints to equality results in a regularized problem for which the RCQ holds.

12.2.3.1 Instances Where the Slater CQ Fails for SDP

Instances where SCQ fails for CP are given in [6]. It is known that the SCQ holds generically for SDP, e.g., [3]. However, there are surprisingly many SDPs that arise from relaxations of hard combinatorial problems where SCQ fails. In addition, there are many instances where the structure of the problems allows for exact facial reduction. This was shown for the quadratic assignment problem in [80] and for the graph partitioning problem in [74]. For these two instances, the barycenter of the feasible set is found explicitly and then used to project the problem onto the minimal face; thus we simultaneously regularize and simplify the problems. In general, the affine hull of the feasible solutions of the SDP are found and used to find Slater points. This is formalized and generalized in [64, 65]. In particular, SDP relaxations that arise from problems with matrix variables that have 0, 1 constraints along with row and column constraints result in SDP relaxations where the Slater CQ fails.

Important applications occur in the facial reduction algorithm for sensor network localization and molecular conformation problems given in [36]. Cliques in the graph result in corresponding dimension reduction of the minimal face of the problem resulting in efficient and accurate solution techniques. Another instance is the SDP relaxation of the side chain positioning problem studied in [14]. Further applications that exploit the failure of the Slater CQ for SDP relaxations appear in, e.g., [1, 2, 5, 69].

12.3 Theory

We now present the theoretical tools that are needed for the facial reduction algorithm for SDP. This includes the well-known results for strong duality, the theorems of the alternative to identify strict feasibility, and, in addition, a stable subproblem to apply the theorems of the alternative. Note that we use K to represent the cone $\mathbb{S}^n_+$ to emphasize that many of the results hold for more general closed convex cones.

12.3.1 Strong Duality for Cone Optimization

We first summarize some results on *strong duality* for the conic convex program in the form (12.4). Strong duality for (12.4) means that there is a *zero duality gap*, $v_P^{\text{conic}} = v_D^{\text{conic}}$, and the dual optimal value v_D (12.5) is attained. However, it is easy to construct examples where strong duality fails; see, e.g., [45,49,75] and Sect. 12.5 below.

It is well known that for a finite-dimensional LP, strong duality fails only if the primal problem and/or its dual is infeasible. In fact, in LP both problems are feasible and both of the optimal values are attained (and equal) if, and only if, the optimal value of one of the problems is finite. In general (conic) convex optimization, the situation is more complicated, since the underlying cones in the primal and dual optimization problems need not be polyhedral. Consequently, even if a primal problem and its dual are feasible, a nonzero duality gap and/or non-attainment of the optimal values may ensue unless some *constraint qualification* holds; see, e.g., [7, 55]. More specific examples for our cone situations appear in, e.g., [38], [51, Sect. 3.2], and [63, Sect. 4].

Failure of strong duality is problematic, since many classes of p-d i-p algorithms require not only that a primal-dual pair of problems possess a zero duality gap, but also that the (generalized) Slater CQ holds for both primal and dual, i.e., that strict feasibility holds for both problems. In [10–12], an equivalent *strongly dualized primal problem* corresponding to (12.4), given by

$$\text{(SP)} \qquad v_{SP}^{\text{conic}} := \sup\{\langle b,y\rangle \,:\, \mathscr{A}^* y \preceq_{f_P} C\}, \tag{12.14}$$

where $f_P \trianglelefteq K$ is the minimal face of K containing the feasible region of (12.4), is considered. The equivalence is in the sense that the feasible set is unchanged

$$\mathscr{A}^* y \preceq_K C \iff \mathscr{A}^* y \preceq_{f_P} C.$$

This means that for any face F we have

$$f_P \trianglelefteq F \trianglelefteq K \implies \{\mathscr{A}^* y \preceq_K C \iff \mathscr{A}^* y \preceq_F C\}.$$

The Lagrangian dual of (12.14) is given by

$$\text{(DSP)} \qquad v_{DSP}^{\text{conic}} := \inf\{\langle C,X\rangle \,:\, \mathscr{A}(X) = b,\, X \succeq_{f_P^*} 0\}. \tag{12.15}$$

We note that the linearity of the constraint means that an equality set of the type in (12.12) is not needed.

Theorem 12.3 ([10]). *Suppose that the optimal value v_P^{conic} in* (12.4) *is finite. Then strong duality holds for the pair* (12.14) *and* (12.15)*, or equivalently, for the pair* (12.4) *and* (12.15)*; i.e., $v_P^{\text{conic}} = v_{SP}^{\text{conic}} = v_{DSP}^{\text{conic}}$ and the dual optimal value v_{DSP}^{conic} is attained.*

12.3.2 Theorems of the Alternative

In this section, we state some theorems of the alternative for the Slater CQ of the conic convex program (12.4), which are essential to our reduction process. We first recall the notion of recession direction [for the dual (12.5)] and its relationship with the minimal face of the primal feasible region.

Definition 12.4. The convex cone of *recession directions* for (12.5) is

$$\mathscr{R}_{\mathrm{D}} := \{D \in \mathscr{V} : \mathscr{A}(D) = 0,\ \langle C, D\rangle = 0,\ D \succeq_{K^*} 0\}. \tag{12.16}$$

The cone $\mathscr{R}_{\mathrm{D}}$ consists of feasible directions for the homogeneous problem along which the dual objective function is constant.

Lemma 12.5. *Suppose that the feasible set $\mathscr{F}_P \neq \emptyset$ for* (12.4)*, and let $0 \neq D \in \mathscr{R}_{\mathrm{D}}$. Then the minimal face of* (12.4) *satisfies*

$$f_P \trianglelefteq K \cap \{D\}^{\perp} \triangleleft K.$$

Proof. We have

$$0 = \langle C, D\rangle - \langle \mathscr{F}_P, \mathscr{A}(D)\rangle = \langle C - \mathscr{A}^*(\mathscr{F}_P), D\rangle.$$

Hence $C - \mathscr{A}^*(\mathscr{F}_P) \subseteq \{D\}^{\perp} \cap K$, which is a face of K. It follows that $f_P \subseteq \{D\}^{\perp} \cap K$. The required result now follows from the fact that f_P is (by definition) a face of K, and D is nonzero. ■

Lemma 12.5 indicates that if we are able to find an element $D \in \mathscr{R}_{\mathrm{D}} \backslash \{0\}$, then D gives us a smaller face of K that contains $\mathscr{F}_P^Z$. The following lemma shows that the existence of such a direction D is *equivalent* to the failure of the Slater CQ for a feasible program (12.4). The lemma specializes [12, Theorem 7.1] and forms the basis of our reduction process.

Lemma 12.6 ([12]). *Suppose that* $\operatorname{int} K \neq \emptyset$ *and $\mathscr{F}_P \neq \emptyset$. Then exactly one of the following two systems is consistent:*

1. $\mathscr{A}(D) = 0$, $\langle C, D\rangle = 0$, *and* $0 \neq D \succeq_{K^*} 0$ ($\mathscr{R}_{\mathrm{D}} \backslash \{0\}$)
2. $\mathscr{A}^* y \prec_K C$ *(Slater CQ)*

Proof. Suppose that D satisfies the system in Item 1. Then for all $y \in \mathscr{F}_P$, we have $\langle C - \mathscr{A}^* y, D\rangle = \langle C, D\rangle - \langle y, (\mathscr{A}(D))\rangle = 0$. Hence $\mathscr{F}_P^Z \subseteq K \cap \{D\}^{\perp}$. But $\{D\}^{\perp} \cap \operatorname{int} K = \emptyset$ as $0 \neq D \succeq_{K^*} 0$. This implies that the Slater CQ (as in Item 2) fails.

Conversely, suppose that the Slater CQ in Item 2 fails. We have $\operatorname{int} K \neq \emptyset$ and

$$0 \notin (\mathscr{A}^*(\mathbb{R}^m) - C) + \operatorname{int} K.$$

Therefore, we can find $D \neq 0$ to separate the open set $(\mathscr{A}^*(\mathbb{R}^m) - C) + \operatorname{int} K$ from 0. Hence we have

$$\langle D, Z \rangle \geq \langle D, C - \mathscr{A}^* y \rangle,$$

for all $Z \in K$ and $y \in \mathscr{W}$. This implies that $D \in K^*$ and $\langle D, C \rangle \leq \langle D, \mathscr{A}^* y \rangle$, for all $y \in \mathscr{W}$. This implies that $\langle \mathscr{A}(D), y \rangle = 0$ for all $y \in \mathscr{W}$; hence $\mathscr{A}(D) = 0$. To see that $\langle C, D \rangle = 0$, fix any $\hat{y} \in \mathscr{F}_P$. Then $0 \geq \langle D, C \rangle = \langle D, C - \mathscr{A}^* \hat{y} \rangle \geq 0$, so $\langle D, C \rangle = 0$. ■

We have an equivalent characterization for the generalized Slater CQ for the dual problem. This can be used to extend our results to ($\mathrm{D_{conic}}$) .

Corollary 12.7. *Suppose that* $\operatorname{int} K^* \neq \emptyset$ *and* $\mathscr{F}_D \neq \emptyset$. *Then exactly one of the following two systems is consistent:*

1. $0 \neq \mathscr{A}^* v \succeq_K 0$, *and* $\langle b, v \rangle = 0$.
2. $\mathscr{A}(X) = b, X \succ_{K^*} 0$ *(generalized Slater CQ).*

Proof. Let $\mathscr{K}$ be a one-one linear transformation with range $\mathscr{R}(\mathscr{K}) = \mathscr{N}(\mathscr{A})$, and let $\hat{X}$ satisfy $\mathscr{A}(\hat{X}) = b$. Then, Item 2 is consistent if, and only if, there exists $\hat{u}$ such that $X = \hat{X} - \mathscr{K}\hat{u} \succ_{K^*} 0$. This is equivalent to $\mathscr{K}\hat{u} \prec_{K^*} \hat{X}$. Therefore, $\mathscr{K}, \hat{X}$ play the roles of $\mathscr{A}^*, C$, respectively, in Lemma 12.6. Therefore, an alternative system is $\mathscr{K}^*(Z) = 0, 0 \neq Z \succeq_K 0$, and $\langle \hat{X}, Z \rangle = 0$. Since $\mathscr{N}(\mathscr{K}^*) = \mathscr{R}(\mathscr{A}^*)$, this is equivalent to $0 \neq Z = \mathscr{A}^* v \succeq_K 0$, and $\langle \hat{X}, Z \rangle = 0$, or $0 \neq \mathscr{A}^* v \succeq_K 0$, and $\langle b, v \rangle = 0$. ■

We can extend Lemma 12.6 to problems with additional equality constraints.

Corollary 12.8. *Consider the modification of the primal* (12.4) *obtained by adding equality constraints:*

$$(P_B) \qquad v_{P_B} := \sup\{\langle b, y \rangle \ : \mathscr{A}^* y \preceq_K C, \mathscr{B} y = f\}, \tag{12.17}$$

where $\mathscr{B} : \mathscr{W} \to \mathscr{W}'$ *is an onto linear transformation. Assume that* $\operatorname{int} K \neq \emptyset$ *and* (P_B) *is feasible. Let* $\bar{C} = C - \mathscr{A}^* \mathscr{B}^\dagger f$. *Then exactly one of the following two systems is consistent:*

1. $\mathscr{A}(D) + \mathscr{B}^* v = 0$, $\langle \bar{C}, D \rangle = 0$, $0 \neq D \succeq_{K^*} 0$.
2. $\mathscr{A}^* y \prec_K C$, $\mathscr{B} y = f$.

Proof. Let $\bar{y} = \mathscr{B}^\dagger f$ be the particular solution (of minimum norm) of $\mathscr{B} y = f$. Since $\mathscr{B}$ is onto, we conclude that $\mathscr{B} y = f$ if, and only if, $y = \bar{y} + \mathscr{C}^* v$, for some v, where the range of the linear transformation $\mathscr{C}^*$ is equal to the nullspace of $\mathscr{B}$. We can now substitute for y and obtain the equivalent constraint $\mathscr{A}^*(\bar{y} + \mathscr{C}^* v) \preceq_K C$; equivalently we get $\mathscr{A}^* \mathscr{C}^* v \preceq_K C - \mathscr{A}^* \bar{y}$. Therefore, Item 2 holds at $y = \hat{y} = \bar{y} + \mathscr{C}^* \hat{v}$, for some $\hat{v}$, if, and only if, $\mathscr{A}^* \mathscr{C}^* \hat{v} \prec_K C - \mathscr{A}^* \bar{y}$. The result now follows immediately from Lemma 12.6 by equating the linear transformation $\mathscr{A}^* \mathscr{C}^*$ with $\mathscr{A}^*$ and the right-hand side $C - \mathscr{A}^* \bar{y}$ with C. Then the system in Item 1 in Lemma 12.6 becomes $\mathscr{C}(\mathscr{A}(D)) = 0, \langle (C - \mathscr{A}^* \bar{y}), D \rangle = 0$. The result follows since the nullspace of $\mathscr{C}$ is equal to the range of $\mathscr{B}^*$. ■

We can also extend Lemma 12.6 to the important case where $\operatorname{int} K = \emptyset$. This occurs at each iteration of the facial reduction.

Corollary 12.9. *Suppose that* $\operatorname{int} K = \emptyset$, $\mathscr{F}_P \neq \emptyset$, *and* $C \in \operatorname{span}(K)$. *Then the linear manifold*

$$\mathbb{S}_y := \{y \in \mathscr{W} : C - \mathscr{A}^* y \in \operatorname{span}(K)\}$$

is a subspace. Moreover, let $\mathscr{P}$ *be a one-one linear transformation with*

$$\mathscr{R}(\mathscr{P}) = (\mathscr{A}^*)^\dagger \operatorname{span}(K).$$

Then exactly one of the following two systems is consistent:

1. $\mathscr{P}^* \mathscr{A}(D) = 0$, $\langle C, D \rangle = 0$, $D \in \operatorname{span}(K)$, *and* $0 \neq D \succeq_{K^*} 0$.
2. $C - \mathscr{A}^* y \in \operatorname{relint} K$.

Proof. Since $C \in \operatorname{span}(K) = K - K$, we get that $0 \in \mathbb{S}_y$, i.e., $\mathbb{S}_y$ is a subspace.

Let $\mathscr{T}$ denote an onto linear transformation acting on $\mathscr{V}$ such that the nullspace $\mathscr{N}(\mathscr{T}) = \operatorname{span}(K)^\perp$, and $\mathscr{T}^*$ is a partial isometry, i.e., $\mathscr{T}^* = \mathscr{T}^\dagger$. Therefore, $\mathscr{T}$ is one-to-one and is onto $\operatorname{span}(K)$. Then

$$\begin{aligned}
\mathscr{A}^* y \preceq_K C &\iff \mathscr{A}^* y \preceq_K C \text{ and } \mathscr{A}^* y \in \operatorname{span}(K), && \text{since } C \in K - K\\
&\iff (\mathscr{A}^* \mathscr{P}) w \preceq_K C,\ y = \mathscr{P} w, \text{ for some } w, && \text{by definition of } \mathscr{P}\\
&\iff (\mathscr{T}\mathscr{A}^* \mathscr{P}) w \preceq_{\mathscr{T}(K)} \mathscr{T}(C),\ y = \mathscr{T} w, \text{ for some } w, && \text{by definition of } \mathscr{T},
\end{aligned}$$

i.e., (12.1) is equivalent to

$$v_P := \sup\{\langle \mathscr{P}^* b, w \rangle : (\mathscr{T}\mathscr{A}^* \mathscr{P}) w \preceq_{\mathscr{T}(K)} \mathscr{T}(C)\}.$$

The corresponding dual is

$$v_D := \inf\left\{\langle \mathscr{T}(C), D \rangle : \mathscr{P}^* \mathscr{A} \mathscr{T}^*(D) = \mathscr{P}^* b,\ D \succeq_{(\mathscr{T}(K))^*} 0\right\}.$$

By construction, $\operatorname{int} \mathscr{T}(K) \neq \emptyset$, so we may apply Lemma 12.6. We conclude that exactly one of the following two systems is consistent:

1. $\mathscr{P}^* \mathscr{A} \mathscr{T}^*(D) = 0$, $0 \neq D \succeq_{(\mathscr{T}(K))^*} 0$, and $\langle \mathscr{T}(C), D \rangle = 0$.
2. $(\mathscr{T}\mathscr{A}^* \mathscr{P}) w \prec_{\mathscr{T}(K)} \mathscr{T}(D)$ (Slater CQ).

The required result follows, since we can now identify $\mathscr{T}^*(D)$ with $D \in \operatorname{span}(K)$, and $\mathscr{T}(C)$ with C. ■

Remark 12.10. Ideally, we would like to find $\hat{D} \in \operatorname{relint}\left(\mathscr{F}_P^Z\right)^c = \operatorname{relint}((C + \mathscr{R}(\mathscr{A}^*)) \cap K)^c$, since then we have found the minimal face $f_P = \{\hat{D}\}^\perp \cap K$. This is difficult to do numerically. Instead, Lemma 12.6 compromises and finds a point in a larger set $D \in \left(\mathscr{N}(\mathscr{A}) \cap \{C\}^\perp \cap K^*\right) \setminus \{0\}$. This allows for the reduction of $K \leftarrow$

$K \cap \{D\}^\perp$. Repeating to find another D is difficult without the subspace reduction using $\mathscr{P}$ in Corollary 12.9. This emphasizes the importance of the minimal subspace form reduction as an aid to the minimal cone reduction, [66].

A similar argument applies to the regularization of the dual as given in Corollary 12.7. Let $\mathscr{F}_D = (\hat{X} + \mathscr{N}(\mathscr{A})) \cap K^*$, where $\mathscr{A}(\hat{X}) = b$. We note that a compromise to finding $\hat{Z} \in \operatorname{relint}(\mathscr{F}_P^z)^c = \operatorname{relint}((\hat{X} + \mathscr{N}(\mathscr{A})) \cap K^*)^c$, $f_D = \{\hat{Z}\}^\perp \cap K^*$ is finding $Z \in (\mathscr{R}(\mathscr{A}^*) \cap \{\hat{X}\}^\perp \cap K) \backslash \{0\}$, where $0 = \langle Z, \hat{X} \rangle = \langle \mathscr{A}^* v, \hat{X} \rangle = \langle v, b \rangle$.

12.3.3 Stable Auxiliary Subproblem

From this section on we restrict the application of facial reduction to the SDP in (12.1). (Note that the notion of auxiliary problem as well as Theorems 12.13 and 12.17, below, apply to the more general conic convex program (12.4).) Each iteration of the facial reduction algorithm involves two steps. First, we apply Lemma 12.6 and find a point D in the relative interior of the recession cone $\mathscr{R}_{\mathrm{D}}$. Then, we project onto the span of the conjugate face $\{D\}^\perp \cap \mathbb{S}^n_+ \supseteq f_P$. This yields a smaller dimensional equivalent problem. The first step to find D is well suited for interior-point algorithms if we can formulate a suitable conic optimization problem. We now formulate and present the properties of a stable auxiliary problem for finding D. The following is well known, e.g., [42, Theorems 10.4.1, 10.4.7].

Theorem 12.11. *If the (generalized) Slater CQ holds for both primal problem* (12.1) *and dual problem* (12.2)*, then as the barrier parameter $\mu \to 0^+$, the primal-dual central path converges to a point $(\hat{X}, \hat{y}, \hat{Z})$, where $\hat{Z} = C - \mathscr{A}^*\hat{y}$, such that $\hat{X}$ is in the relative interior of the set of optimal solutions of* (12.2) *and $(\hat{y}, \hat{Z})$ is in the relative interior of the set of optimal solutions of* (12.1).

Remark 12.12. Many polynomial time algorithms for SDP assume that the Newton search directions can be calculated accurately. However, difficulties can arise in calculating accurate search directions if the corresponding Jacobians become increasingly ill-conditioned. This is the case in most of the current implementations of interior-point methods due to symmetrization and block elimination steps; see, e.g., [19]. In addition, the ill-conditioning arises if the Jacobian of the optimality conditions is not full rank at the optimal solution, as is the case if strict complementarity fails for the SDP. This key question is discussed further in Sect. 12.3.5, below.

According to Theorem 12.11, if we can formulate a pair of auxiliary primal-dual cone optimization problems, each with generalized Slater points such that the relative interior of $\mathscr{R}_{\mathrm{D}}$ coincides with the relative interior of the optimal solution set of one of our auxiliary problems, then we can design an interior-point algorithm for the auxiliary primal-dual pair, making sure that the iterates of our algorithm stay close to the central path (as they approach the optimal solution set) and generate our desired $X \in \operatorname{relint}\mathscr{R}_{\mathrm{D}}$.

This is precisely what we accomplish next. In the special case of $K = \mathbb{S}^n_+$, this corresponds to finding maximum rank feasible solutions for the underlying auxiliary SDPs, since the relative interiors of the faces are characterized by their maximal rank elements.

Define the linear transformation $\mathscr{A}_C : \mathbb{S}^n \to \mathbb{R}^{m+1}$ by

$$\mathscr{A}_C(D) = \begin{pmatrix} \mathscr{A}(D) \\ \langle C, D\rangle \end{pmatrix},$$

This presents a homogenized form of the constraint of (12.1) and combines the two constraints in Lemma 12.6, Item 1. Now consider the following conic optimization problem, which we shall henceforth refer to as the *auxiliary problem*:

$$(AP) \qquad \begin{array}{rll} val_P^{aux} := & \min\limits_{\delta,D} & \delta \\ & \text{s.t.} & \|\mathscr{A}_C(D)\| \le \delta \\ & & \langle \frac{1}{\sqrt{n}} I, D\rangle = 1 \\ & & D \succeq 0. \end{array} \tag{12.18}$$

This auxiliary problem is related to the study of the distances to infeasibility in, e.g., [46]. The Lagrangian dual of (12.18) is

$$\begin{aligned} &\sup_{W\succeq 0, \begin{pmatrix}\beta\\ u\end{pmatrix}\succeq_{\mathscr{Q}} 0} \inf_{\delta,D} \delta + \gamma\left(1 - \left\langle D, \frac{1}{\sqrt{n}} I\right\rangle\right) - \langle W, D\rangle - \left\langle \begin{pmatrix}\beta\\ u\end{pmatrix}, \begin{pmatrix}\delta\\ \mathscr{A}_C(D)\end{pmatrix}\right\rangle \\ &= \sup_{W\succeq 0, \begin{pmatrix}\beta\\ u\end{pmatrix}\succeq_{\mathscr{Q}} 0} \inf_{\delta,D} \delta(1-\beta) - \left\langle D, \mathscr{A}_C^* u + \gamma \frac{1}{\sqrt{n}} I + W\right\rangle + \gamma, \end{aligned} \tag{12.19}$$

where $\mathscr{Q} := \left\{ \begin{pmatrix}\beta\\ u\end{pmatrix} \in \mathbb{R}^{m+2} : \|u\| \le \beta \right\}$ refers to the second-order cone. Since the inner infimum of (12.19) is unconstrained, we get the following equivalent dual:

$$(DAP) \qquad \begin{array}{rll} val_D^{aux} := & \sup\limits_{\gamma,u,W} & \gamma \\ & \text{s.t.} & \mathscr{A}_C^* u + \gamma \frac{1}{\sqrt{n}} I + W = 0 \\ & & \|u\| \le 1 \\ & & W \succeq 0. \end{array} \tag{12.20}$$

A strictly feasible primal-dual point for (12.18) and (12.20) is given by

$$D = \frac{1}{\sqrt{n}} I,\ \delta > \left\| \mathscr{A}_C \left(\frac{1}{\sqrt{n}} I \right) \right\|, \quad \text{and} \quad \gamma = -1,\ u = 0,\ W = \frac{1}{\sqrt{n}} I, \tag{12.21}$$

showing that the generalized Slater CQ holds for the pair (12.18)–(12.20).

Observe that the complexity of solving (12.18) is essentially that of solving the original dual (12.2). Recalling that if a path-following interior-point method is applied to solve (12.18), one arrives at a point in the relative interior of the set of optimal solutions, a primal optimal solution (δ^*, D^*) obtained is such that D^* is of maximum rank.

12.3.3.1 Auxiliary Problem Information for Minimal Face of $\mathscr{F}_P^Z$

This section outlines some useful information that the auxiliary problem provides. Theoretically, in the case when the Slater CQ (nearly) fails for (12.1), the auxiliary problem provides a more refined description of the feasible region, as Theorem 12.13 shows. Computationally, the auxiliary problem gives a measure of how close the feasible region of (12.1) is to being a subset of a face of the cone of positive semidefinite matrices, as shown by: (i) the cosine-angle upper bound (near orthogonality) of the feasible set with the conjugate face given in Theorem 12.17; (ii) the cosine-angle lower bound (closeness) of the feasible set with a proper face of $\mathbb{S}_+^n$ in Proposition 12.18; and (iii) the near common block singularity bound for all the feasible slacks obtained after an appropriate orthogonal rotation, in Corollary 12.19.

We first illustrate the stability of the auxiliary problem and show how a primal-dual solution can be used to obtain useful information about the original pair of conic problems.

Theorem 12.13. *The primal-dual pair of problems* (12.18) *and* (12.20) *satisfy the generalized Slater CQ, both have optimal solutions, and their (nonnegative) optimal values are equal. Moreover, letting* (δ^*, D^*) *be an optimal solution of* (12.18)*, the following holds under the assumption that* $\mathscr{F}_P \neq \emptyset$*:*

1. *If* $\delta^* = 0$ *and* $D^* \succ 0$*, then the Slater CQ fails for* (12.1) *but the generalized Slater CQ holds for* (12.2)*. In fact, the primal minimal face and the only primal feasible (hence optimal) solution are*

$$f_P = \{0\}, \quad y^* = (\mathscr{A}^*)^\dagger(C).$$

2. *If* $\delta^* = 0$ *and* $D^* \not\succ 0$*, then the Slater CQ fails for* (12.1) *and the minimal face satisfies*

$$f_P \trianglelefteq \mathbb{S}_+^n \cap \{D^*\}^\perp \triangleleft \mathbb{S}_+^n. \tag{12.22}$$

3. *If* $\delta^* > 0$*, then the Slater CQ holds for* (12.1)*.*

Proof. A strictly feasible pair for (12.18)–(12.20) is given in (12.21). Hence by strong duality both problems have equal optimal values and both values are attained.

1. Suppose that $\delta^* = 0$ and $D^* \succ 0$. It follows that $\mathscr{A}_C(D^*) = 0$ and $D^* \neq 0$. It follows from Lemma 12.5 that

$$f_P \trianglelefteq \mathbb{S}^n_+ \cap \{D^*\}^\perp = \{0\}.$$

   Hence all feasible points for (12.1) satisfy $C - \mathscr{A}^* y = 0$. Since $\mathscr{A}$ is onto, we conclude that the unique solution of this linear system is $y = (\mathscr{A}^*)^\dagger(C)$.

   Since $\mathscr{A}$ is onto, there exists $\bar{X}$ such that $\mathscr{A}(\bar{X}) = b$. Thus, for every $t \geq 0$, $\mathscr{A}(\bar{X} + tD^*) = b$, and for t large enough, $\bar{X} + tD^* \succ 0$. Therefore, the generalized Slater CQ holds for (12.2).
2. The result follows from Lemma 12.5.
3. If $\delta^* > 0$, then $\mathscr{R}_D = \{0\}$, where $\mathscr{R}_D$ was defined in (12.16). It follows from Lemma 12.6 that the Slater CQ holds for (12.1). ■

Remark 12.14. Theorem 12.13 shows that if the primal problem (12.1) is feasible, then by definition of (AP) as in (12.18), $\delta^* = 0$ if, and only if, $\mathscr{A}_C$ has a right singular vector D such that $D \succeq 0$ and the corresponding singular value is zero, i.e., we could replace (AP) with $\min\{\|\mathscr{A}_C(D)\| : \|D\| = 1, D \succeq 0\}$. Therefore, we could solve (AP) using a basis for the nullspace of $\mathscr{A}_C$, e.g., using an onto linear function $\mathscr{N}_{\mathscr{A}_C}$ on $\mathbb{S}^n$ that satisfies $\mathscr{R}(\mathscr{N}^*_{\mathscr{A}_C}) = \mathscr{N}(\mathscr{A}_C)$, and an approach based on maximizing the smallest eigenvalue:

$$\delta \approx \sup_y \left\{ \lambda_{\min}(\mathscr{N}^*_{\mathscr{A}_C} y) : \operatorname{trace}(\mathscr{N}^*_{\mathscr{A}_C} y) = 1, \|y\| \leq 1 \right\},$$

so, in the case when $\delta^* = 0$, both (AP) and (DAP) can be seen as a max-min eigenvalue problem (subject to a bound and a linear constraint).

Finding $0 \neq D \succeq 0$ that solves $\mathscr{A}_C(D) = 0$ is also equivalent to the SDP:

$$\begin{aligned} &\inf_D \ \|D\| \\ &\text{s.t. } \mathscr{A}_C(D) = 0, \ \langle I, D \rangle = \sqrt{n}, \ D \succeq 0, \end{aligned} \tag{12.23}$$

a program for which the Slater CQ generally fails. (See Item 2 of Theorem 12.13.) This suggests that the problem of finding the recession direction $0 \neq D \succeq 0$ that certifies a failure for (12.1) to satisfy the Slater CQ may be a difficult problem.

One may detect whether the Slater CQ fails for the dual (12.2) using the auxiliary problem (12.18) and its dual (12.20).

Proposition 12.15. *Assume that* (12.2) *is feasible, i.e., there exists* $\hat{X} \in \mathbb{S}^n_+$ *such that* $\mathscr{A}(\hat{X}) = b$. *Then we have that* X *is feasible for* (12.2) *if and only if*

$$X = \hat{X} + \mathscr{N}^*_{\mathscr{A}} y \succeq 0,$$

where $\mathcal{N}_{\mathscr{A}} : \mathbb{S}^n \to \mathbb{R}^{n(n+1)/2-m}$ *is an onto linear transformation such that* $\mathscr{R}(\mathcal{N}_{\mathscr{A}}^*) = \mathcal{N}(\mathscr{A})$*. Then the corresponding auxiliary problem*

$$\inf_{\delta,D} \delta \quad \text{s.t.} \quad \left\| \begin{pmatrix} \mathcal{N}_{\mathscr{A}}(D) \\ \langle \hat{X}, D \rangle \end{pmatrix} \right\| \leq \delta, \ \langle I, D \rangle = \sqrt{n}, \ D \succeq 0$$

certifies either that (12.2) *satisfies the Slater CQ or that* 0 *is the only feasible slack of* (12.2) *or detects a smaller face of* $\mathbb{S}^n_+$ *containing* $\mathscr{F}_D$.

The results in Proposition 12.15 follows directly from the corresponding results for the primal problem (12.1). An alternative form of the auxiliary problem for (12.2) can be defined using the theorem of the alternative in Corollary 12.7.

Proposition 12.16. *Assume that* (12.2) *is feasible. The dual auxiliary problem*

$$\sup_{v,\lambda} \lambda \quad \text{s.t.} \quad (\mathscr{A}(I))^T v = 1, \ b^T v = 0, \ \mathscr{A}^* v \succeq \lambda I \tag{12.24}$$

determines if (12.2) *satisfies the Slater CQ. The dual of* (12.24) *is given by*

$$\inf_{\mu,\Omega} \mu_2 \quad \text{s.t.} \quad \langle I, \Omega \rangle = 1, \ \mathscr{A}(\Omega) - \mu_1 \mathscr{A}(I) - \mu_2 b = 0, \ \Omega \succeq 0, \tag{12.25}$$

and the following hold under the assumption that (12.2) *is feasible:*

1. *If* (12.24) *is infeasible, then* (12.2) *must satisfy the Slater CQ.*
2. *If* (12.24) *is feasible, then both* (12.24) *and* (12.25) *satisfy the Slater CQ. Moreover, the Slater CQ holds for* (12.2) *if and only if the optimal value of* (12.24) *is negative.*
3. *If* (v^*, λ^*) *is an optimal solution of* (12.24) *with* $\lambda^* \geq 0$*, then* $\mathscr{F}_D \subseteq \mathbb{S}^n_+ \cap \{\mathscr{A}^* v^*\}^\perp \lhd \mathbb{S}^n_+$.

   Since X feasible for (12.2) *implies that*

$$\langle \mathscr{A}^* v^*, X \rangle = (v^*)^T (\mathscr{A}(X)) = (v^*)^T b = 0,$$

   we conclude that $\mathscr{F}_D \subseteq \mathbb{S}^n_+ \cap \{\mathscr{A}^* v^*\}^\perp \lhd \mathbb{S}^n_+$*. Therefore, if* (12.2) *fails the Slater CQ, then, by solving* (12.24)*, we can obtain a proper face of* $\mathbb{S}^n_+$ *that contains the feasible region* $\mathscr{F}_D$ *of* (12.2).

Proof. The Lagrangian of (12.24) is given by

$$\begin{aligned} L(v, \lambda, \mu, \Omega) &= \lambda + \mu_1(1 - (\mathscr{A}(I)^T v)) + \mu_2(-b^T v) + \langle \Omega, \mathscr{A}^* v - \lambda I \rangle \\ &= \lambda(1 - \langle I, \Omega \rangle) + v^T(\mathscr{A}(\Omega) - \mu_1 \mathscr{A}(I) - \mu_2 b) + \mu_2. \end{aligned}$$

This yields the dual program (12.25).

If (12.24) is infeasible, then we must have $b \neq 0$ and $\mathscr{A}(I) = kb$ for some $k \in \mathbb{R}$. If $k > 0$, then $k^{-1}I$ is a Slater point for (12.2). If $k = 0$, then $\mathscr{A}(\hat{X} + \lambda I) = b$ and $\hat{X} + \lambda I \succ 0$ for any $\hat{X}$ satisfying $\mathscr{A}(\hat{X}) = b$ and sufficiently large $\lambda > 0$. If $k < 0$, then $\mathscr{A}(2\hat{X} + k^{-1}I) = b$ for $\hat{X} \succeq 0$ satisfying $\mathscr{A}(\hat{X}) = b$; and we have $2\hat{X} + k^{-1}I \succ 0$.

If (12.24) is feasible, i.e., if there exists $\hat{v}$ such that $(\mathscr{A}(I))^T v = 1$ and $b^T \hat{v} = 0$, then

$$(\hat{v}, \hat{\lambda}) = \left(\hat{v}, \hat{\lambda} = \lambda_{\min}(\mathscr{A}^*\hat{v}) - 1\right), \quad (\hat{\mu}, \hat{\Omega}) = \left(\begin{pmatrix} 1/n \\ 0 \end{pmatrix}, \frac{1}{n}I\right)$$

is strictly feasible for (12.24) and (12.25), respectively.

Let (v^*, λ^*) be an optimal solution of (12.25). If $\lambda^* \leq 0$, then for any $v \in \mathbb{R}^m$ with $\mathscr{A}^*y \succeq 0$ and $b^T v = 0$, v cannot be feasible for (12.24) so $\langle I, \mathscr{A}^*v \rangle \leq 0$. This implies that $\mathscr{A}^*v = 0$. By Corollary 12.7, the Slater CQ holds for (12.2). If $\lambda^* > 0$, then v^* certifies that the Slater CQ fails for (12.2), again by Corollary 12.7. ∎

The next result shows that δ^* from (AP) is a measure of how close the Slater CQ is to failing.

Theorem 12.17. *Let (δ^*, D^*) denote an optimal solution of the auxiliary problem (12.18). Then δ^* bounds how far the feasible primal slacks $Z = C - \mathscr{A}^*y \succeq 0$ are from orthogonality to D^*:*

$$0 \leq \sup_{0 \preceq Z = C - \mathscr{A}^*y \neq 0} \frac{\langle D^*, Z \rangle}{\|D^*\| \|Z\|} \leq \alpha(\mathscr{A}, C) := \begin{cases} \dfrac{\delta^*}{\sigma_{\min}(\mathscr{A})} & \text{if } C \in \mathscr{R}(\mathscr{A}^*), \\ \dfrac{\delta^*}{\sigma_{\min}(\mathscr{A}_C)} & \text{if } C \notin \mathscr{R}(\mathscr{A}^*). \end{cases} \tag{12.26}$$

Proof. Since $\langle \frac{1}{\sqrt{n}} I, D^* \rangle = 1$, we get

$$\|D^*\| \geq \frac{\left\langle \frac{1}{\sqrt{n}} I, D^* \right\rangle}{\|\frac{1}{\sqrt{n}} I\|} = \frac{1}{\frac{1}{\sqrt{n}}\|I\|} = 1.$$

If $C = \mathscr{A}^* y_C$ for some $y_C \in \mathbb{R}^m$, then for any $Z = C - \mathscr{A}^*y \succeq 0$,

$$\begin{aligned} \cos\theta_{D^*,Z} := \frac{\langle D^*, C - \mathscr{A}^*y \rangle}{\|D^*\| \|C - \mathscr{A}^*y\|} &\leq \frac{\langle \mathscr{A}(D^*), y_C - y \rangle}{\|\mathscr{A}^*(y_C - y)\|} \\ &\leq \frac{\|\mathscr{A}(D^*)\| \; \|y_C - y\|}{\sigma_{\min}(\mathscr{A}^*) \, \|y_C - y\|} \\ &\leq \frac{\delta^*}{\sigma_{\min}(\mathscr{A})}. \end{aligned}$$

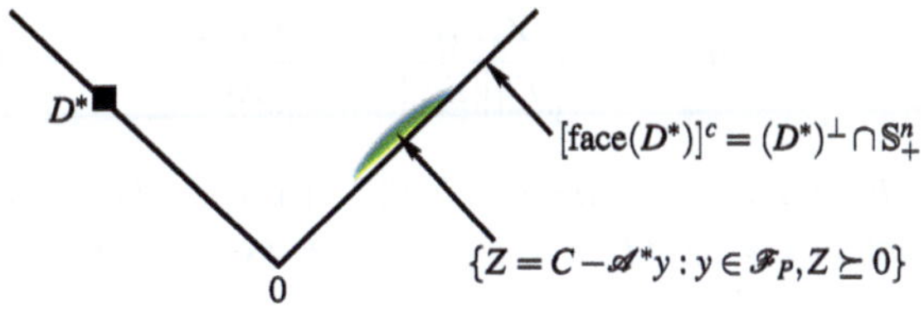

Fig. 12.1 Minimal face; $0 < \delta^* \ll 1$

If $C \notin \mathscr{R}(\mathscr{A}^*)$, then by Assumption 12.1, $\mathscr{A}_C$ is onto so $\langle D^*, C - \mathscr{A}^* y\rangle = \left\langle \mathscr{A}_C(D^*), \begin{pmatrix} -y \\ 1 \end{pmatrix} \right\rangle$ implies that $0 \preceq C - \mathscr{A}^* y \neq 0, \forall y \in \mathscr{F}_P$. Therefore the cosine of the angle $\theta_{D^*,Z}$ between D^* and $Z = C - \mathscr{A}^* y \succeq 0$ is bounded by

$$\cos \theta_{D^*,Z} = \frac{\langle D^*, C - \mathscr{A}^* y\rangle}{\|D^*\| \|C - \mathscr{A}^* y\|} \leq \frac{\left\langle \mathscr{A}_C(D^*), \begin{pmatrix} -y \\ 1 \end{pmatrix} \right\rangle}{\left\| \mathscr{A}_C^* \begin{pmatrix} -y \\ 1 \end{pmatrix} \right\|}$$
$$\leq \frac{\|\mathscr{A}_C(D^*)\| \left\| \begin{pmatrix} -y \\ 1 \end{pmatrix} \right\|}{\sigma_{\min}(\mathscr{A}_C) \left\| \begin{pmatrix} -y \\ 1 \end{pmatrix} \right\|}$$
$$= \frac{\delta^*}{\sigma_{\min}(\mathscr{A}_C)}.$$

■

Theorem 12.17 provides a lower bound for the angle and distance between feasible slack vectors and the vector D^* on the boundary of $\mathbb{S}^n_+$. For our purposes, the theorem is only useful when $\alpha(\mathscr{A}, C)$ is small. Given that $\delta^* = \|\mathscr{A}_C(D^*)\|$, we see that the lower bound is independent of simple scaling of $\mathscr{A}_C$, though not necessarily independent of the conditioning of $\mathscr{A}_C$. Thus, δ^* provides qualitative information about both the conditioning of $\mathscr{A}_C$ and the distance to infeasibility.

We now strengthen the result in Theorem 12.17 by using more information from D^*. In applications we expect to choose the partitions of U and D^* to satisfy $\lambda_{\min}(D_+) >> \lambda_{\max}(D_\varepsilon)$ (Fig. 12.1).

Proposition 12.18. *Let (δ^*, D^*) denote an optimal solution of the auxiliary problem (12.18), and let*

$$D^* = \begin{bmatrix} P & Q \end{bmatrix} \begin{bmatrix} D_+ & 0 \\ 0 & D_\varepsilon \end{bmatrix} \begin{bmatrix} P & Q \end{bmatrix}^T, \tag{12.27}$$

with $U = \begin{bmatrix} P & Q \end{bmatrix}$ orthogonal, and $D_+ \succ 0$.

Let $0 \neq Z := C - \mathscr{A}^* y \succeq 0$ and $Z_Q := QQ^T Z QQ^T$. Then Z_Q is the closest point in $\mathscr{R}(Q \cdot Q^T) \cap \mathbb{S}^n_+$ to Z; and, the cosine of the angle θ_{Z,Z_Q} between Z and the face $\mathscr{R}(Q \cdot Q^T) \cap \mathbb{S}^n_+$ satisfies

$$\cos\theta_{Z,Z_Q} := \frac{\langle Z,Z_Q\rangle}{\|Z\|\|Z_Q\|} = \frac{\|Q^TZQ\|}{\|Z\|} \geq 1-\alpha(\mathscr{A},C)\frac{\|D^*\|}{\lambda_{\min}(D_+)}, \tag{12.28}$$

where $\alpha(\mathscr{A},C)$ is defined in (12.26)*. Thus the angle between any feasible slack and the face $\mathscr{R}(Q\cdot Q^T)\cap\mathbb{S}^n_+$ cannot be too large in the sense that*

$$\inf_{0\neq Z=C-\mathscr{A}^*y\succeq 0} \cos\theta_{Z,Z_Q} \geq 1-\alpha(\mathscr{A},C)\frac{\|D^*\|}{\lambda_{\min}(D_+)}.$$

Moreover, the normalized distance to the face is bounded as in

$$\|Z-Z_Q\|^2 \leq 2\|Z\|^2\left[\alpha(\mathscr{A},C)\frac{\|D^*\|}{\lambda_{\min}(D_+)}\right]. \tag{12.29}$$

Proof. Since $Z\succeq 0$, we have $Q^TZQ\in \operatorname{argmin}_{W\succeq 0}\|Z-QWQ^T\|$. This shows that $Z_Q := QQ^TZQQ^T$ is the closest point in $\mathscr{R}(Q\cdot Q^T)\cap\mathbb{S}^n_+$ to Z. The expression for the angle in (12.28) follows using

$$\frac{\langle Z,Z_Q\rangle}{\|Z\|\|Z_Q\|} = \frac{\|Q^TZQ\|^2}{\|Z\|\|Q^TZQ\|} = \frac{\|Q^TZQ\|}{\|Z\|}. \tag{12.30}$$

From Theorem 12.17, we see that $0\neq Z=C-\mathscr{A}^*y\succeq 0$ implies that $\left\langle \frac{1}{\|Z\|}Z,D^*\right\rangle \leq \alpha(\mathscr{A},C)\|D^*\|$. Therefore, the optimal value of the following optimization problem provides a lower bound on the quantity in (12.30):

$$\begin{array}{rll} \gamma_0 := & \min\limits_Z & \|Q^TZQ\| \\ & \text{s.t.} & \langle Z,D^*\rangle \leq \alpha(\mathscr{A},C)\|D^*\| \\ & & \|Z\|^2=1, \quad Z\succeq 0. \end{array} \tag{12.31}$$

Since $\langle Z,D^*\rangle = \langle P^TZP,D_+\rangle + \langle Q^TZQ,D_\varepsilon\rangle \geq \langle P^TZP,D_+\rangle$ whenever $Z\succeq 0$, we have

$$\begin{array}{rll} \gamma_0 \geq \gamma := & \min\limits_Z & \|Q^TZQ\| \\ & \text{s.t.} & \langle P^TZP,D_+\rangle \leq \alpha(\mathscr{A},C)\|D^*\| \\ & & \|Z\|^2=1, \quad Z\succeq 0. \end{array} \tag{12.32}$$

It is possible to find the optimal value γ of (12.32). After the orthogonal rotation

$$Z = \begin{bmatrix} P & Q\end{bmatrix}\begin{bmatrix} S & V \\ V^T & W\end{bmatrix}\begin{bmatrix} P & Q\end{bmatrix}^T = PSP^T+PVQ^T+QV^TP^T+QWQ^T,$$

where $S\in\mathbb{S}^{n-\bar{n}}_+$, $W\in\mathbb{S}^{\bar{n}}_+$ and $V\in\mathbb{R}^{(n-\bar{n})\times\bar{n}}$, (12.32) can be rewritten as

$$\begin{array}{rll} \gamma = \min\limits_{S,V,W} & \|W\| & \\ \text{s.t.} & \langle S, D_+ \rangle \leq \alpha(\mathscr{A}, C)\|D^*\| & \\ & \|S\|^2 + 2\|V\|^2 + \|W\|^2 = 1 & \\ & \begin{bmatrix} S & V \\ V^T & W \end{bmatrix} \in \mathbb{S}^n_+. & \end{array} \tag{12.33}$$

Since

$$\|V\|^2 \leq \|S\|\|W\| \tag{12.34}$$

holds whenever $\begin{bmatrix} S & V \\ V^T & W \end{bmatrix} \succeq 0$, we have that $(\|S\| + \|W\|)^2 \geq \|S\|^2 + 2\|V\|^2 + \|W\|^2$. This yields

$$\begin{array}{rll} \gamma \geq \bar{\gamma} := \min_{S,V,W} & \|W\| \\ \text{s.t.} & \langle S, D_+ \rangle \leq \alpha(\mathscr{A}, C)\|D^*\| \\ & \|S\| + \|W\| \geq 1 \\ & S \succeq 0,\ W \succeq 0. \end{array} \qquad \begin{array}{rll} \bar{\gamma} \geq \min\limits_{S} & 1 - \|S\| \\ \text{s.t.} & \langle S, D_+ \rangle \leq \alpha(\mathscr{A}, C)\|D^*\| \\ & S \succeq 0 \end{array} \tag{12.35}$$

Since $\lambda_{\min}(D_+)\|S\| \leq \langle S, D_+ \rangle \leq \alpha(\mathscr{A}, C)\|D^*\|$, we see that the objective value of the last optimization problem in (12.35) is bounded below by $1 - \alpha(\mathscr{A}, C)\|D^*\|/\lambda_{\min}(D_+)$. Now let u be a normalized eigenvector of D_+ corresponding to its smallest eigenvalue $\lambda_{\min}(D_+)$. Then $S^* = \frac{\alpha(\mathscr{A}, C)\|D^*\|}{\lambda_{\min}(D_+)} uu^T$ solves the last optimization problem in (12.35), with corresponding optimal value $1 - \frac{\alpha(\mathscr{A}, C)\|D^*\|}{\lambda_{\min}(D_+)}$.

Let $\beta := \min\left\{ \frac{\alpha(\mathscr{A}, C)\|D^*\|}{\lambda_{\min}(D_+)}, 1 \right\}$. Then $\gamma \geq 1 - \beta$. Also,

$$\begin{bmatrix} S & V \\ V^T & W \end{bmatrix} := \begin{pmatrix} \sqrt{\beta} u \\ \sqrt{1-\beta} e_1 \end{pmatrix} \begin{pmatrix} \sqrt{\beta} u \\ \sqrt{1-\beta} e_1 \end{pmatrix}^T = \begin{bmatrix} \beta uu^T & \sqrt{\beta(1-\beta)} u e_1^T \\ \sqrt{\beta(1-\beta)} e_1 u^T & (1-\beta) e_1 e_1^T \end{bmatrix} \in \mathbb{S}^n_+.$$

Therefore (S, V, W) is feasible for (12.33) and attains an objective value $1 - \beta$. This shows that $\gamma = 1 - \beta$ and proves (12.28).

The last claim (12.29) follows immediately from

$$\begin{aligned} \|Z - Z_Q\|^2 &= \|Z\|^2 \left(1 - \frac{\|Q^T Z Q\|^2}{\|Z\|^2} \right) \\ &\leq \|Z\|^2 \left[1 - \left(1 - \alpha(\mathscr{A}, C) \frac{\|D^*\|}{\lambda_{\min}(D_+)} \right)^2 \right] \\ &\leq 2\|Z\|^2 \alpha(\mathscr{A}, C) \frac{\|D^*\|}{\lambda_{\min}(D_+)}. \end{aligned}$$

■

These results are related to the extreme angles between vectors in a cone studied in [29, 33]. Moreover, it is related to the distances to infeasibility in, e.g., [46], in which the distance to infeasibility is shown to provide backward and forward error bounds.

We now see that we can use the rotation $U = \begin{bmatrix} P \ Q \end{bmatrix}$ obtained from the diagonalization of the optimal D^* in the auxiliary problem (12.18) to reveal *nearness to infeasibility*, as discussed in, e.g., [46]. Or, in our approach, this reveals nearness to a facial decomposition. We use the following results to bound the size of certain blocks of a feasible slack Z.

Corollary 12.19. *Let (δ^*, D^*) denote an optimal solution of the auxiliary problem (12.18), as in Theorem 12.17, and let*

$$D^* = \begin{bmatrix} P \ Q \end{bmatrix} \begin{bmatrix} D_+ & 0 \\ 0 & D_\varepsilon \end{bmatrix} \begin{bmatrix} P \ Q \end{bmatrix}^T, \tag{12.36}$$

with $U = \begin{bmatrix} P \ Q \end{bmatrix}$ orthogonal, and $D_+ \succ 0$. Then for any feasible slack $0 \neq Z = C - \mathscr{A}^* y \succeq 0$, we have

$$\operatorname{trace} P^T Z P \leq \alpha(\mathscr{A}, C) \frac{\|D^*\|}{\lambda_{\min}(D_+)} \|Z\|, \tag{12.37}$$

where $\alpha(\mathscr{A}, C)$ is defined in (12.26).

Proof. Since

$$\begin{aligned} \langle D^*, Z \rangle &= \left\langle \begin{bmatrix} D_+ & 0 \\ 0 & D_\varepsilon \end{bmatrix}, \begin{bmatrix} P^T Z P & P^T Z Q \\ Q^T Z P & Q^T Z Q \end{bmatrix} \right\rangle \\ &= \langle D_+, P^T Z P \rangle + \langle D_\varepsilon, Q^T Z Q \rangle \\ &\geq \langle D_+, P^T Z P \rangle \\ &\geq \lambda_{\min}(D_+) \operatorname{trace} P^T Z P, \end{aligned} \tag{12.38}$$

the claim follows from Theorem 12.17. ■

Remark 12.20. We now summarize the information available from a solution of the auxiliary problem, with optima $\delta^* \geq 0, D^* \not\succ 0$. We let $0 \neq Z = C - \mathscr{A}^* y \succeq 0$ denote a feasible slack. In particular, we emphasize the information obtained from the rotation $U^T Z U$ using the orthogonal U that block diagonalizes D^* and from the *closest* point $Z_Q = QQ^T Z QQ^T$. We note that replacing all feasible Z with the *projected* Z_Q provides a nearby problem for the backward stability argument. Alternatively, we can view the nearby problem by projecting the data $A_i \leftarrow QQ^T A_i QQ^T, \forall i$, $C \leftarrow QQ^T C QQ^T$.

1. From (12.26) in Theorem 12.17, we get a lower bound on the angle (upper bound on the cosine of the angle):

$$\cos \theta_{D^*, Z} = \frac{\langle D^*, Z \rangle}{\|D^*\| \|Z\|} \leq \alpha(\mathscr{A}, C).$$

2. In Proposition 12.18 with orthogonal $U = \begin{bmatrix} P \ Q \end{bmatrix}$, we get upper bounds on the angle between a feasible slack and the face defined using $Q \cdot Q^T$ and on the normalized distance to the face:

$$\cos \theta_{Z,Z_Q} := \frac{\langle Z, Z_Q \rangle}{\|Z\| \|Z_Q\|} = \frac{\|Q^T Z Q\|}{\|Z\|} \geq 1 - \alpha(\mathscr{A}, C) \frac{\|D^*\|}{\lambda_{\min}(D_+)}.$$

$$\|Z - Z_Q\|^2 \leq 2\|Z\|^2 \left[\alpha(\mathscr{A}, C) \frac{\|D^*\|}{\lambda_{\min}(D_+)} \right].$$

3. After the rotation using the orthogonal U, the $(1,1)$ principal block is bounded as

$$\operatorname{trace} P^T Z P \leq \alpha(\mathscr{A}, C) \frac{\|D^*\|}{\lambda_{\min}(D_+)} \|Z\|.$$

12.3.4 Rank-Revealing Rotation and Equivalent Problems

We may use the results from Theorem 12.17 and Corollary 12.19 to get two *rotated* optimization problems equivalent to (12.1). The equivalent problems indicate that, in the case when δ^* is sufficiently small, it is possible to reduce the dimension of the problem and get a *nearby* problem that helps in the facial reduction. The two equivalent formulations can be used to illustrate backward stability with respect to a perturbation of the cone $\mathbb{S}^n_+$.

First we need to find a suitable shift of C to allow a proper facial projection. This is used in Theorem 12.22, below.

Lemma 12.21. *Let $\delta^*, D^*, U = \begin{bmatrix} P \ Q \end{bmatrix}, D_+, D_\varepsilon$ be defined as in the hypothesis of Corollary 12.19. Let $(y_Q, W_Q) \in \mathbb{R}^m \times \mathbb{S}^{\bar{n}}$ be the best least squares solution to the equation $QWQ^T + \mathscr{A}^* y = C$, that is, (y_Q, W_Q) is the optimal solution of minimum norm to the linear least squares problem*

$$\min_{y,W} \frac{1}{2} \|C - (QWQ^T + \mathscr{A}^* y)\|^2. \tag{12.39}$$

Let $C_Q := QW_Q Q^T$ and $C_{\mathrm{res}} := C - (C_Q + \mathscr{A}^* y_Q)$. Then

$$Q^T C_{\mathrm{res}} Q = 0, \quad \text{and} \quad \mathscr{A}(C_{\mathrm{res}}) = 0. \tag{12.40}$$

Moreover, if $\delta^* = 0$, then for any feasible solution y of (12.1), we get

$$C - \mathscr{A}^* y \in \mathscr{R}(Q \cdot Q^T), \tag{12.41}$$

and further $(y, Q^T(C - \mathscr{A}^* y)Q)$ is an optimal solution of (12.39), whose optimal value is zero.

Proof. Let $\Omega(y,W) := \frac{1}{2}\|C-(QWQ^T+\mathscr{A}^*y)\|^2$. Since

$$\begin{aligned}\Omega(y,W) =& \frac{1}{2}\|C\|^2+\frac{1}{2}\|\mathscr{A}^*y\|^2+\frac{1}{2}\|W\|^2+\left\langle QWQ^T,\mathscr{A}^*y\right\rangle \\ &-\left\langle Q^TCQ,W\right\rangle-\langle\mathscr{A}(C),y\rangle,\end{aligned}$$

we have (y_Q,W_Q) solves (12.39) if, and only if,

$$\nabla_y\Omega = \mathscr{A}\left(QWQ^T-(C-\mathscr{A}^*y)\right)=0, \tag{12.42}$$

$$\text{and}\quad \nabla_w\Omega = W-\left[Q^T(C-\mathscr{A}^*y)Q\right]=0. \tag{12.43}$$

Then (12.40) follows immediately by substitution.

If $\delta^*=0$, then $\langle D^*,A_i\rangle=0$ for $i=1,\dots,m$ and $\langle D^*,C\rangle=0$. Hence, for any $y\in\mathbb{R}^m$,

$$\langle D_+,P^T(C-\mathscr{A}^*y)P\rangle+\langle D_\varepsilon,Q^T(C-\mathscr{A}^*y)Q\rangle=\langle D^*,C-\mathscr{A}^*y\rangle=0.$$

If $C-\mathscr{A}^*y\succeq 0$, then we must have $P^T(C-\mathscr{A}^*y)P=0$ (as $D_+\succ 0$), and so $P^T(C-\mathscr{A}^*y)Q=0$. Hence

$$\begin{aligned}C-\mathscr{A}^*y &= UU^T(C-\mathscr{A}^*y)UU^T \\ &= U\left[P\ Q\right]^T(C-\mathscr{A}^*y)\left[P\ Q\right]U^T \\ &= QQ^T(C-\mathscr{A}^*y)QQ^T,\end{aligned}$$

i.e., we conclude (12.41) holds.

The last statement now follows from substituting $W=Q^T(C-\mathscr{A}^*y)Q$ in (12.39). ■

We can now use the rotation from Corollary 12.19 with a shift of C (to $C_{\text{res}}+C_Q=C-\mathscr{A}^*y_Q$) to get two equivalent problems to (P). This emphasizes that when δ^* is *small*, then the auxiliary problem reveals a block structure with one principal block and three *small/negligible* blocks. If δ is small, then β in the following Theorem 12.22 is *small*. Then fixing $\beta=0$ results in a nearby problem to (P) that illustrates backward stability of the facial reduction.

Theorem 12.22. *Let $\delta^*,D^*,U=\left[P\ Q\right],D_+,D_\varepsilon$ be defined as in the hypothesis of Corollary 12.19, and let $y_Q,W_Q,C_Q,C_{\text{res}}$ be defined as in Lemma 12.21. Define the scalar*

$$\beta := \alpha(\mathscr{A},C)\frac{\|D^*\|}{\lambda_{\min}(D_+)}, \tag{12.44}$$

and the convex cone $T_\beta \subseteq \mathbb{S}^n_+$ *partitioned appropriately as in* (12.36),

$$T_\beta := \left\{ Z = \begin{bmatrix} A & B \\ B^T & C \end{bmatrix} \in \mathbb{S}^n_+ : \operatorname{trace} A \le \beta \operatorname{trace} Z \right\}. \tag{12.45}$$

Then we get the following two equivalent programs to (P) in (12.1)*:*

1. *Using the rotation U and the cone* T_β*,*

$$v_P = \sup_y \left\{ b^T y : \begin{bmatrix} P^T Z P & P^T Z Q \\ Q^T Z P & Q^T Z Q \end{bmatrix} \succeq_{T_\beta} 0, Z = C - \mathscr{A}^* y \right\}; \tag{12.46}$$

2. *Using* (y_Q, W_Q)*,*

$$v_P = b^T y_Q + \sup_y \left\{ b^T y : \begin{bmatrix} P^T Z P & P^T Z Q \\ Q^T Z P & Q^T Z Q \end{bmatrix} \succeq_{T_\beta} 0, Z = C_{\text{res}} + C_Q - \mathscr{A}^* y \right\}. \tag{12.47}$$

Proof. From Corollary 12.19,

$$\mathscr{F}_P = \left\{ y : \begin{bmatrix} P^T Z P & P^T Z Q \\ Q^T Z P & Q^T Z Q \end{bmatrix} \succeq_{T_\beta} 0, Z = C - \mathscr{A}^* y \right\}. \tag{12.48}$$

Hence the equivalence of (12.1) with (12.46) follows.

For (12.47), first note that for any $y \in \mathbb{R}^m$,

$$Z := C_{\text{res}} + C_Q - \mathscr{A}^* y = C - \mathscr{A}^*(y + y_Q),$$

so $Z \succeq 0$ if and only if $y + y_Q \in \mathscr{F}_P$, if and only if $Z \in T_\beta$. Hence

$$\mathscr{F}_P = y_Q + \left\{ y : \begin{bmatrix} P^T Z P & P^T Z Q \\ Q^T Z P & Q^T Z Q \end{bmatrix} \succeq_{T_\beta} 0, Z = C_{\text{res}} + Q W_Q Q^T - \mathscr{A}^* y \right\}, \tag{12.49}$$

and (12.47) follows. ∎

Remark 12.23. As mentioned above, Theorem 12.22 illustrates the backward stability of the facial reduction. It is difficult to state this precisely due to the shifts done and the changes to the constraints in the algorithm. For simplicity, we just discuss one iteration. The original problem (P) is equivalent to the problem in (12.46). Therefore, a facial reduction step can be applied to the original problem or equivalently to (12.46). We then perturb this problem in (12.46) by setting $\beta = 0$. The algorithm applied to this nearby problem with exact arithmetic will result in the same step.

12.3.4.1 Reduction to Two Smaller Problems

Following the results from Theorems 12.13 and 12.22, we focus on the case where $\delta^* = 0$ and $\mathscr{R}_D \cap \mathbb{S}^n_{++} = \emptyset$. In this case we get a proper face $Q\mathbb{S}^{\bar{n}}_+ Q^T \lhd \mathbb{S}^n_+$. We obtain two different equivalent formulations of the problem by restricting to this smaller face. In the first case, we stay in the same dimension for the domain variable y but decrease the constraint space and include equality constraints. In the second case, we eliminate the equality constraints and move to a smaller dimensional space for y. We first see that when we have found the minimal face, then we obtain an equivalent regularized problem as was done for LP in Sect. 12.2.1.

Corollary 12.24. *Suppose that the minimal face f_P of (P) is found using the orthogonal $U = \begin{bmatrix} P_{\text{fin}} & Q_{\text{fin}} \end{bmatrix}$, so that $f_P = Q_{\text{fin}} \mathbb{S}^r_+ Q^T_{\text{fin}}$, $0 < r < n$. Then an equivalent problem to (P) is*

$$(P_{PQ,reg}) \qquad \begin{aligned} v_P = \sup\ & b^T y \\ \text{s.t. } & Q^T_{\text{fin}}(\mathscr{A}^* y) Q_{\text{fin}} \preceq Q^T_{\text{fin}} C Q_{\text{fin}} \\ & \mathscr{A}^*_{\text{fin}} y \qquad\quad = \mathscr{A}^*_{\text{fin}} y_{Q_{\text{fin}}}, \end{aligned} \tag{12.50}$$

where $(y_{Q_{\text{fin}}}, W_{Q_{\text{fin}}})$ solves the least squares problem $\min_{y,W} \|C - (\mathscr{A}^* y + Q_{\text{fin}} W Q^T_{\text{fin}})\|$, and $\mathscr{A}^*_{\text{fin}} : \mathbb{R}^m \to \mathbb{R}^t$ is a full rank (onto) representation of the linear transformation

$$y \mapsto \begin{bmatrix} P^T_{\text{fin}}(\mathscr{A}^* y) P_{\text{fin}} \\ Q^T_{\text{fin}}(\mathscr{A}^* y) P_{\text{fin}} \end{bmatrix}.$$

Moreover, $(P_{PQ,reg})$ is regularized, i.e., the RCQ holds.

Proof. The result follows immediately from Theorem 12.22, since the definition of the minimal face implies that there exists a feasible $\hat{y}$ which satisfies the constraints in (12.50). The new equality constraint is constructed to be full rank and not change the feasible set. ■

Alternatively, we now reduce (12.1) to an equivalent problem over a spectrahedron in a lower dimension using the spectral decomposition of D^*.

Proposition 12.25. *Let the notation and hypotheses in Theorem 12.22 hold with $\delta^* = 0$ and $D^* = \begin{bmatrix} P & Q \end{bmatrix} \begin{bmatrix} D_+ & 0 \\ 0 & 0 \end{bmatrix} \begin{bmatrix} P^T \\ Q^T \end{bmatrix}$, where $\begin{bmatrix} P & Q \end{bmatrix}$ is orthogonal, $Q \in \mathbb{R}^{n \times \bar{n}}$ and $D_+ \succ 0$. Then*

$$\begin{aligned} v_P = \sup \{ b^T y : \ & Q^T (C - \mathscr{A}^* y) Q \succeq 0, \\ & P^T (\mathscr{A}^* y) P = P^T (\mathscr{A}^* y_Q) P, \\ & Q^T (\mathscr{A}^* y) P = Q^T (\mathscr{A}^* y_Q) P \ \}. \end{aligned} \tag{12.51}$$

Moreover:

1. *If* $\mathscr{R}(Q \cdot Q^T) \cap \mathscr{R}(\mathscr{A}^*) = \{0\}$, *then for any* $y_1, y_2 \in \mathscr{F}_P$, $b^T y_1 = b^T y_2 = v_P$.
2. *If* $\mathscr{R}(Q \cdot Q^T) \cap \mathscr{R}(\mathscr{A}^*) \neq \{0\}$, *and if, for some* $\bar{m} > 0$, $\mathscr{P} : \mathbb{R}^{\bar{m}} \to \mathbb{R}^m$ *is an injective linear map such that* $\mathscr{R}(\mathscr{A}^* \mathscr{P}) = \mathscr{R}(\mathscr{A}^*) \cap \mathscr{R}(Q \cdot Q^T)$, *then we have*

$$v_P = b^T y_Q + \sup_v \left\{ (\mathscr{P}^* b)^T v : W_Q - Q^T (\mathscr{A}^* \mathscr{P} v) Q \succeq 0 \right\}. \tag{12.52}$$

And, if v^* *is an optimal solution of* (12.52), *then* $y^* = y_Q + \mathscr{P} v^*$ *is an optimal solution of* (12.1).

Proof. Since $\delta^* = 0$, from Lemma 12.21 we have that $C = C_Q + \mathscr{A}^* y_Q, C_Q = Q W_Q Q^T$, for some $y_Q \in \mathbb{R}^m$ and $W_Q \in \mathbb{S}^{\bar{n}}$. Hence by (12.48),

$$\begin{aligned} \mathscr{F}_P &= \{ y \in \mathbb{R}^m : Q^T (C - \mathscr{A}^* y) Q \succeq 0, P^T (C - \mathscr{A}^* y) P = 0, Q^T (C - \mathscr{A}^* y) P = 0 \} \\ &= \{ y \in \mathbb{R}^m : Q^T (C - \mathscr{A}^* y) Q \succeq 0, P^T (\mathscr{A}^* (y - y_Q)) P = 0, Q^T (\mathscr{A}^* (y - y_Q)) P = 0 \}, \end{aligned} \tag{12.53}$$

and (12.51) follows:

1. Since $C - \mathscr{A}^* y \in \mathscr{R}(Q \cdot Q^T), \forall y \in \mathscr{F}_P$, we get $\mathscr{A}^*(y_2 - y_1) = (C - \mathscr{A}^* y_1) - (C - \mathscr{A}^* y_2) \in \mathscr{R}(Q \cdot Q^T) \cap \mathscr{R}(\mathscr{A}^*) = \{0\}$. Given that $\mathscr{A}$ is onto, we get $b = \mathscr{A}(\hat{X})$, for some $\hat{X} \in \mathbb{S}^n$, and

$$b^T (y_2 - y_1) = \langle \hat{X}, \mathscr{A}^* (y_2 - y_1) \rangle = 0.$$

2. From (12.53),

$$\begin{aligned} \mathscr{F}_P &= y_Q + \{ y : W_Q - Q^T (\mathscr{A}^* y) Q \succeq 0, P^T (\mathscr{A}^* y) P = 0, Q^T (\mathscr{A}^* y) P = 0 \} \\ &= y_Q + \{ y : W_Q - Q^T (\mathscr{A}^* y) Q \succeq 0, \mathscr{A}^* y \in \mathscr{R}(Q \cdot Q^T) \} \\ &= y_Q + \{ \mathscr{P} v : W_Q - Q^T (\mathscr{A}^* \mathscr{P} v) Q \succeq 0 \}, \end{aligned}$$

   the last equality follows from the choice of $\mathscr{P}$. Therefore, (12.52) follows, and if v^* is an optimal solution of (12.52), then $y_Q + \mathscr{P} v^*$ is an optimal solution of (12.1).

■

Next we establish the existence of the operator $\mathscr{P}$ mentioned in Proposition 12.25.

Proposition 12.26. *For any* $n \times n$ *orthogonal matrix* $U = \begin{bmatrix} P & Q \end{bmatrix}$ *and any surjective linear operator* $\mathscr{A} : \mathbb{S}^n \to \mathbb{R}^m$ *with* $\bar{m} := \dim(\mathscr{R}(\mathscr{A}^*) \cap \mathscr{R}(Q \cdot Q^T)) > 0$, *there exists a one-one linear transformation* $\mathscr{P} : \mathbb{R}^{\bar{m}} \to \mathbb{R}^m$ *that satisfies*

$$\mathscr{R}(\mathscr{A}^*\mathscr{P}) = \mathscr{R}(Q\cdot Q^T)\cap\mathscr{R}(\mathscr{A}^*), \tag{12.54}$$

$$\mathscr{R}(\mathscr{P}) = \mathscr{N}\left(P^T(\mathscr{A}^*\cdot)P\right)\cap\mathscr{N}\left(P^T(\mathscr{A}^*\cdot)Q\right). \tag{12.55}$$

Moreover, $\bar{\mathscr{A}}:\mathbb{S}^{\bar{n}}\to\mathbb{R}^{\bar{m}}$ *is defined by*

$$\bar{\mathscr{A}}^*(\cdot) := Q^T\left(\mathscr{A}^*\mathscr{P}(\cdot)\right)Q$$

is onto.

Proof. Recall that for any matrix $X\in\mathbb{S}^n$

$$X = UU^TXUU^T = PP^TXPP^T + PP^TXQQ^T + QQ^TXPP^T + QQ^TXQQ^T.$$

Moreover, $P^TQ=0$. Therefore, $X\in\mathscr{R}(Q\cdot Q^T)$ implies $P^TXP=0$ and $P^TXQ=0$. Conversely, $P^TXP=0$ and $P^TXQ=0$ implies $X=QQ^TXQQ^T$. Therefore $X\in\mathscr{R}(Q\cdot Q^T)$ if, and only if, $P^TXP=0$ and $P^TXQ=0$.

For any $y\in\mathbb{R}^m$, $\mathscr{A}^*y\in\mathscr{R}(Q\cdot Q^T)$ if, and only if,

$$\sum_{i=1}^m(P^TA_iP)y_i = 0 \quad\text{and}\quad \sum_{i=1}^m(P^TA_iQ)y_i = 0,$$

which holds if, and only if, $y\in\operatorname{span}\{\beta\}$, where $\beta:=\{y_1,\dots,y_{\bar{m}}\}$ is a basis of the linear subspace

$$\left\{y:\sum_{i=1}^m(P^TA_iP)y_i=0\right\}\cap\left\{y:\sum_{i=1}^m(P^TA_iQ)y_i=0\right\}$$
$$= \mathscr{N}\left(P^T(\mathscr{A}^*\cdot)P\right)\cap\mathscr{N}\left(P^T(\mathscr{A}^*\cdot)Q\right).$$

Now define $\mathscr{P}:\mathbb{R}^{\bar{m}}\to\mathbb{R}^m$ by

$$\mathscr{P}v = \sum_{i=1}^{\bar{m}} v_iy_i \quad \text{for } \lambda\in\mathbb{R}^{\bar{m}}.$$

Then, by definition of $\mathscr{P}$, we have

$$\mathscr{R}(\mathscr{A}^*\mathscr{P}) = \mathscr{R}(Q\cdot Q^T)\cap\mathscr{R}(\mathscr{A}^*)$$
$$\text{and}\quad \mathscr{R}(\mathscr{P}) = \mathscr{N}\left(P^T(\mathscr{A}^*\cdot)P\right)\cap\mathscr{N}\left(P^T(\mathscr{A}^*\cdot)Q\right).$$

The onto property of $\bar{\mathscr{A}}$ follows from (12.54) and the fact that both $\mathscr{P},\mathscr{A}^*$ are one-one. Note that if $\bar{\mathscr{A}}^*v=0$, noting that $\mathscr{A}^*\mathscr{P}v=QWQ^T$ for some $W\in\mathbb{S}^{\bar{n}}$ by (12.54), we have that $w=0$ so $\mathscr{A}^*\mathscr{P}v=0$. Since both $\mathscr{A}^*$ and $\mathscr{P}$ injective, we have that $v=0$. ■

12.3.5 LP, SDP, and the Role of Strict Complementarity

The (near) loss of the Slater CQ results in both theoretical and numerical difficulties, e.g., [46]. In addition, both theoretical and numerical difficulties arise from the loss of strict complementarity, [70]. The connection between strong duality, the Slater CQ, and strict complementarity is seen through the notion of complementarity partitions [66]. We now see that this plays a key role in the stability and in determining the number of steps k for the facial reduction. In particular, we see that $k = 1$ is characterized by strict complementary slackness and therefore results in a stable formulation.

Definition 12.27. The pair of faces $F_1 \unlhd K, F_2 \unlhd K^*$ form a *complementarity partition of* K, K^* if $F_1 \subseteq (F_2)^c$. (Equivalently, $F_2 \subseteq (F_1)^c$.) The partition is *proper* if both F_1 and F_2 are proper faces. The partition is *strict* if $(F_1)^c = F_2$ or $(F_2)^c = F_1$.

We now see the importance of this notion for the facial reduction.

Theorem 12.28. *Let $\delta^* = 0, D^* \succeq 0$ be the optimum of (AP) with dual optimum (γ^*, u^*, W^*). Then the following are equivalent:*

1. *If $D^* = \begin{bmatrix} P & Q \end{bmatrix} \begin{bmatrix} D_+ & 0 \\ 0 & 0 \end{bmatrix} \begin{bmatrix} P^T \\ Q^T \end{bmatrix}$ is a maximal rank element of $\mathscr{R}_D$, where $\begin{bmatrix} P & Q \end{bmatrix}$ is orthogonal, $Q \in \mathbb{R}^{n \times \bar{n}}$ and $D_+ \succ 0$, then the reduced problem in* (12.52) *using D^* satisfies the Slater CQ; only one step of facial reduction is needed.*
2. *Strict complementarity holds for (AP); that is, the primal-dual optimal solution pair $(0, D^*), (0, u^*, W^*)$ for* (12.18) *and* (12.20) *satisfy* $\operatorname{rank}(D^*) + \operatorname{rank}(W^*) = n$.
3. *The faces of $\mathbb{S}^n_+$ defined by*

$$f^0_{aux,P} := \text{face}\left(\{D \in \mathbb{S}^n : \mathscr{A}(D) = 0,\ \langle C, D\rangle = 0,\ D \succeq 0\}\right)$$

$$f^0_{aux,D} := \text{face}\left(\left\{W \in \mathbb{S}^n : W = \mathscr{A}_C^* z \succeq 0,\ \textit{for some } z \in \mathbb{R}^{\bar{m}+1}\right\}\right)$$

form a strict complementarity partition of $\mathbb{S}^n_+$.

Proof. (1) $\iff$ (2): If (12.52) satisfies the Slater CQ, then there exists $\tilde{v} \in \mathbb{R}^{\bar{m}}$ such that $W_Q - \bar{\mathscr{A}}^* \tilde{v} \succ 0$. This implies that $\tilde{Z} := Q(W_Q - \bar{\mathscr{A}}^* \tilde{v})Q^T$ is of rank $\bar{n}$. Moreover,

$$0 \preceq \tilde{Z} = QW_Q Q - \mathscr{A}^* \mathscr{P} \tilde{v} = C - \mathscr{A}^*(y_Q + \mathscr{P}\tilde{v}) = \mathscr{A}_C^* \begin{pmatrix} -(y_Q + \mathscr{P}\tilde{v}) \\ 1 \end{pmatrix}.$$

Hence, letting

$$\tilde{u} = \frac{\begin{pmatrix} y_Q + \mathscr{P}\tilde{v} \\ -1 \end{pmatrix}}{\left\| \begin{pmatrix} y_Q + \mathscr{P}\tilde{v} \\ -1 \end{pmatrix} \right\|} \quad \text{and} \quad \tilde{W} = \frac{1}{\left\| \begin{pmatrix} y_Q + \mathscr{P}\tilde{v} \\ -1 \end{pmatrix} \right\|} \tilde{Z},$$

we have that $(0,\tilde{u},\tilde{W})$ is an optimal solution of (12.20). Since $\operatorname{rank}(D^*)+\operatorname{rank}(\tilde{W}) = (n-\bar{n})+\bar{n}=n$, we get that strict complementarity holds.

Conversely, suppose that strict complementarity holds for (AP), and let D^* be a maximum rank optimal solution as described in the hypothesis of Item 1. Then there exists an optimal solution $(0,u^*,W^*)$ for (12.20) such that $\operatorname{rank}(W^*)=\bar{n}$. By complementary slackness, $0=\langle D^*,W^*\rangle = \langle D_+,P^TW^*P\rangle$, so $W^*\in\mathscr{R}(Q\cdot Q^T)$ and $Q^TW^*Q\succ 0$. Let $u^*=\begin{pmatrix}\tilde{y}\\-\tilde{\alpha}\end{pmatrix}$, so

$$W^*=\tilde{\alpha}C-\mathscr{A}^*\tilde{y}=\tilde{\alpha}C_Q-\mathscr{A}^*(\tilde{y}-\tilde{\alpha}y_Q).$$

Since $W^*,C_Q\in\mathscr{R}(Q\cdot Q^T)$ implies that $\mathscr{A}^*(\tilde{y}-\tilde{\alpha}y_Q)=\mathscr{A}^*\mathscr{P}\tilde{v}$ for some $\tilde{v}\in\mathbb{R}^{\bar{m}}$, we get

$$0\prec Q^TW^*Q=\tilde{\alpha}\bar{C}-\bar{\mathscr{A}}^*\tilde{v}.$$

Without loss of generality, we may assume that $\tilde{\alpha}=\pm 1$ or 0. If $\tilde{\alpha}=1$, then $\bar{C}-\bar{\mathscr{A}}^*\tilde{v}\succ 0$ is a Slater point for (12.52). Consider the remaining two cases. Since (12.1) is assumed to be feasible, the equivalent program (12.52) is also feasible so there exists $\hat{v}$ such that $\bar{C}-\bar{\mathscr{A}}^*\hat{v}\succeq 0$. If $\tilde{\alpha}=0$, then $\bar{C}-\bar{\mathscr{A}}^*(\hat{v}+\tilde{v})\succ 0$. If $\tilde{\alpha}=-1$, then $\bar{C}-\bar{\mathscr{A}}^*(2\hat{v}+\tilde{v})\succ 0$. Hence (12.52) satisfies the Slater CQ.

(2) $\Longleftrightarrow$ (3): Notice that $f^0_{aux,P}$ and $f^0_{aux,D}$ are the minimal faces of $\mathbb{S}^n_+$ containing the optimal slacks of (12.18) and (12.20), respectively, and that $f^0_{aux,P}, f^0_{aux,D}$ form a complementarity partition of $\mathbb{S}^n_+=(\mathbb{S}^n_+)^*$. The complementarity partition is strict if and only if there exist primal-dual optimal slacks D^* and W^* such that $\operatorname{rank}(D^*)+\operatorname{rank}(W^*)=n$. Hence (2) and (3) are equivalent. ■

In the special case where the Slater CQ fails and (12.1) is a linear program (and, more generally, the special case of optimizing over an arbitrary polyhedral cone; see, e.g., [56, 57, 78, 79]), we see that one single iteration of facial reduction yields a reduced problem that satisfies the Slater CQ.

Corollary 12.29. *Assume that the optimal value of (AP) equals zero, with D^* being a maximum rank optimal solution of (AP). If $A_i=\operatorname{Diag}(a_i)$ for some $a_i\in\mathbb{R}^n$, for $i=1,\dots,m$, and $C=\operatorname{Diag}(c)$, for some $c\in\mathbb{R}^n$, then the reduced problem* (12.52) *satisfies the Slater CQ.*

Proof. In this diagonal case, the SDP is equivalent to an LP. The Goldman–Tucker theorem [25] implies that there exists a required optimal primal-dual pair for (12.18) and (12.20) that satisfies strict complementarity, so Item 2 in Theorem 12.28 holds. By Theorem 12.28, the reduced problem (12.52) satisfies the Slater CQ. ■

12.4 Facial Reduction

We now study facial reduction for (P) and its sensitivity analysis.

12.4.1 Two Types

We first outline two algorithms for facial reduction that find the minimal face f_P of (P). Both are based on solving the auxiliary problem and applying Lemma 12.6. The first algorithm repeatedly finds a face F containing the minimal face and then projects the problem into $F - F$, thus reducing both the size of the constraints and the dimension of the variables till finally obtaining the Slater CQ. The second algorithm also repeatedly finds F; but then it identifies the implicit equality constraints till eventually obtaining MFCQ.

12.4.1.1 Dimension Reduction and Regularization for the Slater CQ

Suppose that Slater's CQ fails for our given input $\mathscr{A} : \mathbb{S}^n \to \mathbb{R}^m$, $C \in \mathbb{S}^n$, i.e., the minimal face $f_P \lhd F := \mathbb{S}^n_+$. Our procedure consists of a finite number of repetitions of the following two steps that begin with $k = n$.

1. We first identify $0 \neq D \in (f_P)^c$ using the auxiliary problem (12.18). This means that $f_P \unlhd F \leftarrow \left(\mathbb{S}^k_+ \cap \{D\}^\perp\right)$ and the interior of this new face F is empty.
2. We then project the problem (P) into $\text{span}(F)$. Thus we reduce the dimension of the variables and size of the constraints of our problem; the new cone satisfies $\text{int}\, F \neq \emptyset$. We set $k \leftarrow \dim(F)$.[1]

Therefore, in the case that $\text{int}\, F = \emptyset$, we need to obtain an equivalent problem to (P) in the subspace $\text{span}(F) = F - F$. One essential step is finding a subspace intersection. We can apply the algorithm in, e.g., [26, Thm 12.4.2]. In particular, by abuse of notation, let H_1, H_2 be matrices with orthonormal columns representing the orthonormal bases of the subspaces $\mathscr{H}_1, \mathscr{H}_2$, respectively. Then we need only find a singular value decomposition $H_1^T H_2 = U \Sigma V^T$ and find which singular vectors correspond to singular values $\Sigma_{ii}, i = 1, \ldots, r$, (close to) 1. Then both $H_1 U(:, 1:r)$ and $H_2 V(:, 1:r)$ provide matrices whose ranges yield the intersection. The cone $\mathbb{S}^n_+$ possesses a "self-replicating" structure. Therefore we choose an isometry $\mathscr{I}$ so that $\mathscr{I}(\mathbb{S}^n_+ \cap (F - F))$ is a smaller dimensional PSD cone $\mathbb{S}^r_+$.

Algorithm 12.1 outlines one iteration of facial reduction. The output returns an equivalent problem $(\bar{\mathscr{A}}, \bar{b}, \bar{C})$ on a smaller face of $\mathbb{S}^n_+$ that contains the set of feasible

[1] Note that for numerical stability and well-posedness, it is essential that there exists Lagrange multipliers and that $\text{int}\, F \neq \emptyset$. Regularization involves finding both a minimal face and a minimal subspace; see [66].

Algorithm 12.1: One iteration of facial reduction

Input: $\mathscr{A} : \mathbb{S}^n \to \mathbb{R}^m, b \in \mathbb{R}^m, C \in \mathbb{S}^n$;
Obtain an optimal solution (δ^*, D^*) of (AP)
if $\delta^* > 0$, **then**
  STOP; Slater CQ holds for $(\mathscr{A}, b, C)$.
else
  if $D^* \succ 0$, **then**
    STOP; generalized Slater CQ holds for $(\mathscr{A}, b, C)$ (see Theorem 12.13);
  else
    Obtain eigenvalue decomposition $D^* = \begin{bmatrix} P & Q \end{bmatrix} \begin{bmatrix} D_+ & 0 \\ 0 & 0 \end{bmatrix} \begin{bmatrix} P^T \\ Q^T \end{bmatrix}$ as described in Proposition 12.25, with $Q \in \mathbb{R}^{n \times \bar{n}}$;
    if $\mathscr{R}(Q \cdot Q^T) \cap \mathscr{R}(\mathscr{A}^*) = \{0\}$, **then**
      STOP; all feasible solutions of $\sup_y \{b^T y : C - \mathscr{A}^* y \succeq 0\}$ are optimal.
    else
      find $\bar{m}$, $\mathscr{P} : \mathbb{R}^{\bar{m}} \to \mathbb{R}^m$ satisfying the conditions in Proposition 12.25;
      solve (12.39) for (y_Q, W_Q);
      $\bar{C} \leftarrow W_Q$;
      $\bar{b} \leftarrow \mathscr{P}^* b$;
      $\bar{\mathscr{A}}^* \leftarrow Q^T(\mathscr{A}^* \mathscr{P}(\cdot))Q$;
      Output: $\bar{\mathscr{A}} : \mathbb{S}^{\bar{n}} \to \mathbb{R}^{\bar{m}}, \bar{b} \in \mathbb{R}^{\bar{m}}, \bar{C} \in \mathbb{S}^{\bar{n}}; y_Q \in \mathbb{R}^m, \mathscr{P} : \mathbb{R}^{\bar{m}} \to \mathbb{R}^m$;
    end if
  end if
end if

slacks $\mathscr{F}_P^Z$; and, we also obtain the linear transformation $\mathscr{P}$ and point y_Q, which are needed for recovering an optimal solution of the original problem (P). (See Proposition 12.25.)

Two numerical aspects arising in Algorithm 12.1 need to be considered. The first issue concerns the determination of rank(D^*). In practice, the spectral decomposition of D^* would be of the form

$$D^* = \begin{bmatrix} P & Q \end{bmatrix} \begin{bmatrix} D_+ & 0 \\ 0 & D_\varepsilon \end{bmatrix} \begin{bmatrix} P^T \\ Q^T \end{bmatrix} \text{ with } D_\varepsilon \approx 0, \quad \text{instead of } \ D^* = \begin{bmatrix} P & Q \end{bmatrix} \begin{bmatrix} D_+ & 0 \\ 0 & 0 \end{bmatrix} \begin{bmatrix} P^T \\ Q^T \end{bmatrix}.$$

We need to decide which of the eigenvalues of D^* are small enough so that they can be safely rounded down to zero. This is important for the determination of Q, which gives the smaller face $\mathscr{R}(Q \cdot Q^T) \cap \mathbb{S}^n_+$ containing the feasible region $\mathscr{F}_P^Z$. The partitioning of D^* can be done by using similar techniques as in the determination of numerical rank. Assuming that $\lambda_1(D^*) \geq \lambda_2(D^*) \geq \cdots \geq \lambda_n(D^*) \geq 0$, the *numerical rank* rank(D^*, ε) of D^* with respect to a zero tolerance $\varepsilon > 0$ is defined via

$$\lambda_{\text{rank}(D^*,\varepsilon)}(D^*) > \varepsilon \geq \lambda_{\text{rank}(D^*,\varepsilon)+1}(D^*).$$

In implementing Algorithm 12.1, to determine the partitioning of D^*, we use the numerical rank with respect to $\frac{\varepsilon\|D^*\|}{\sqrt{n}}$ where $\varepsilon \in (0,1)$ is fixed: take $r = \operatorname{rank}\left(D^*, \frac{\varepsilon\|D^*\|}{\sqrt{n}}\right)$,

$$D_+ = \operatorname{Diag}(\lambda_1(D^*),\dots,\lambda_r(D^*)), \quad D_\varepsilon = \operatorname{Diag}(\lambda_{r+1}(D^*),\dots,\lambda_n(D^*)),$$

and partition $\begin{bmatrix} P & Q \end{bmatrix}$ accordingly. Then

$$\lambda_{\min}(D_+) > \frac{\varepsilon\|D^*\|}{\sqrt{n}} \geq \lambda_{\max}(D_\varepsilon) \implies \|D_\varepsilon\| \leq \varepsilon\|D^*\|.$$

Also,

$$\frac{\|D_\varepsilon\|^2}{\|D_+\|^2} = \frac{\|D_\varepsilon\|^2}{\|D^*\|^2 - \|D_\varepsilon\|^2} \leq \frac{\varepsilon^2\|D^*\|^2}{(1-\varepsilon^2)\|D^*\|^2} = \frac{1}{\varepsilon^{-2}-1} \tag{12.56}$$

that is, D_ε is negligible comparing with D_+.

The second issue is the computation of intersection of subspaces, $\mathscr{R}(Q \cdot Q^T) \cap \mathscr{R}(\mathscr{A}^*)$ (and in particular, finding one-one map $\mathscr{P}$ such that $\mathscr{R}(\mathscr{A}^*\mathscr{P}) = \mathscr{R}(Q \cdot Q^T) \cap \mathscr{R}(\mathscr{A}^*)$). This can be done using the following result on subspace intersection.

Theorem 12.30 ([26], Sect. 12.4.3). *Given $Q \in \mathbb{R}^{n\times\bar{n}}$ of full rank and onto linear map $\mathscr{A} : \mathbb{S}^n \to \mathbb{R}^m$, there exist $U_1^{\rm sp},\dots,U^{\rm sp}_{\min\{m,\bar{n}^2\}}, V_1^{\rm sp},\dots,V^{\rm sp}_{\min\{m,\bar{n}^2\}} \in \mathbb{S}^n$ such that*

$$\begin{aligned}
\sigma_1^{\rm sp} := \left\langle U_1^{\rm sp}, V_1^{\rm sp}\right\rangle &= \max\left\{\langle U,V\rangle : \|U\| = 1 = \|V\|,\ U \in \mathscr{R}(Q\cdot Q^T),\ V \in \mathscr{R}(\mathscr{A}^*)\right\},\\
\sigma_k^{\rm sp} := \left\langle U_k^{\rm sp}, V_k^{\rm sp}\right\rangle &= \max\left\{\langle U,V\rangle : \|U\| = 1 = \|V\|,\ U \in \mathscr{R}(Q\cdot Q^T),\ V \in \mathscr{R}(\mathscr{A}^*),\right.\\
&\qquad\left.\left\langle U, U_i^{\rm sp}\right\rangle = 0 = \left\langle V, V_i^{\rm sp}\right\rangle,\ \forall i = 1,\dots,k-1\right\},
\end{aligned} \tag{12.57}$$

for $k = 2,\dots,\min\{m,\bar{n}^2\}$, and $1 \geq \sigma_1^{\rm sp} \geq \sigma_2^{\rm sp} \geq \dots \geq \sigma^{\rm sp}_{\min\{m,\bar{n}^2\}} \geq 0$. Suppose that

$$\sigma_1^{\rm sp} = \dots = \sigma_{\bar{m}}^{\rm sp} = 1 > \sigma_{\bar{m}+1}^{\rm sp} \geq \dots \geq \sigma^{\rm sp}_{\min\{\bar{n},m\}}, \tag{12.58}$$

then

$$\mathscr{R}(Q\cdot Q^T) \cap \mathscr{R}(\mathscr{A}^*) = \operatorname{span}\left(U_1^{\rm sp},\dots,U_{\bar{m}}^{\rm sp}\right) = \operatorname{span}\left(V_1^{\rm sp},\dots,V_{\bar{m}}^{\rm sp}\right), \tag{12.59}$$

and $\mathscr{P} : \mathbb{R}^{\bar{m}} \to \mathbb{R}^m$ defined by $\mathscr{P}v = \sum_{i=1}^{\bar{m}} v_i y_i^{\rm sp}$ for $v \in \mathbb{R}^{\bar{m}}$, where $\mathscr{A}^* y_i^{\rm sp} = V_i^{\rm sp}$ for $i = 1,\dots,\bar{m}$, is one-one linear and satisfies $\mathscr{R}(\mathscr{A}^*\mathscr{P}) = \mathscr{R}(Q\cdot Q^T) \cap \mathscr{R}(\mathscr{A}^*)$.

In practice, we do not get $\sigma_i^{\rm sp} = 1$ (for $i = 1,\dots,\bar{m}$) exactly. For a fixed tolerance $\varepsilon^{\rm sp} \geq 0$, suppose that

$$1 \geq \sigma_1^{\rm sp} \geq \dots \geq \sigma_{\bar{m}}^{\rm sp} \geq 1 - \varepsilon^{\rm sp} > \sigma_{\bar{m}+1}^{\rm sp} \geq \dots \geq \sigma^{\rm sp}_{\min\{\bar{n},m\}} \geq 0. \tag{12.60}$$

Then we would take the approximation

$$\mathscr{R}(Q\cdot Q^T)\cap\mathscr{R}(\mathscr{A}^*)\approx\operatorname{span}\left(U_1^{\mathrm{sp}},\ldots,U_{\bar{m}}^{\mathrm{sp}}\right)\approx\operatorname{span}\left(V_1^{\mathrm{sp}},\ldots,V_{\bar{m}}^{\mathrm{sp}}\right). \tag{12.61}$$

Observe that with the chosen tolerance $\varepsilon^{\mathrm{sp}}$, we have that the cosines of the principal angles between $\mathscr{R}(Q\cdot Q^T)$ and $\operatorname{span}\left(V_1^{\mathrm{sp}},\ldots,V_{\bar{m}}^{\mathrm{sp}}\right)$ is no less than $1-\varepsilon^{\mathrm{sp}}$; in particular, $\|U_k^{\mathrm{sp}}-V_k^{\mathrm{sp}}\|^2\le 2\varepsilon^{\mathrm{sp}}$ and $\|Q^TV_k^{\mathrm{sp}}Q\|\ge\sigma_k^{\mathrm{sp}}\ge 1-\varepsilon^{\mathrm{sp}}$ for $k=1,\ldots,\bar{m}$.

Remark 12.31. Using $V_1^{\mathrm{sp}},\ldots,V_{\min\{m,\bar{n}^2\}}^{\mathrm{sp}}$ from Theorem 12.30, we may replace $A_1,\ldots,A_m$ by $V_1^{\mathrm{sp}},\ldots,V_m^{\mathrm{sp}}$ (which may require extending $V_1^{\mathrm{sp}},\ldots,V_{\min\{m,\bar{n}^2\}}^{\mathrm{sp}}$ to a basis of $\mathscr{R}(\mathscr{A}^*)$, if $m>\bar{n}^2$).

If the subspace intersection is exact (as in (12.58) and (12.59) in Theorem 12.30), then $\mathscr{R}(Q\cdot Q^T)\cap\mathscr{R}(\mathscr{A}^*)=\operatorname{span}(A_1,\ldots,A_{\bar{m}})$ would hold. If the intersection is inexact (as in (12.60) and (12.61)), then we may replace $\mathscr{A}$ by $\breve{\mathscr{A}}:\mathbb{S}^n\to\mathbb{R}^m$, defined by

$$\breve{A}_i=\begin{cases}U_i^{\mathrm{sp}} & \text{if } i=1,\ldots,\bar{m},\\ V_i^{\mathrm{sp}} & \text{if } i=\bar{m}+1,\ldots,m,\end{cases}$$

which is a perturbation of $\mathscr{A}$ with $\|\mathscr{A}^*-\breve{\mathscr{A}}^*\|_F=\sqrt{\sum_{i=1}^{\bar{m}}\|U_i^{\mathrm{sp}}-V_i^{\mathrm{sp}}\|^2}\le\sqrt{2\bar{m}\varepsilon^{\mathrm{sp}}}$. Then $\mathscr{R}(Q\cdot Q^T)\cap\mathscr{R}(\breve{\mathscr{A}}^*)=\operatorname{span}(\breve{A}_1,\ldots,\breve{A}_{\bar{m}})$ because $\breve{A}_i\in\mathscr{R}(Q\cdot Q^T)\cap\mathscr{R}(\breve{\mathscr{A}}^*)$ for $i=1,\ldots,\bar{m}$ and

$$\begin{aligned}
&\max_{U,V}\Big\{\langle U,V\rangle : U\in\mathscr{R}(Q\cdot Q^T),\|U\|=1,V\in\mathscr{R}(\breve{\mathscr{A}}^*),\|V\|=1,\\
&\qquad\qquad \left\langle U,U_j^{\mathrm{sp}}\right\rangle=0=\left\langle V,U_j^{\mathrm{sp}}\right\rangle\ \forall j=1,\ldots,\bar{m},\Big\}\\
&\le\max_{U,y}\left\{\left\langle U,\sum_{i=1}^{\bar{m}}y_jU_j^{\mathrm{sp}}+\sum_{i=\bar{m}+1}^{m}y_jV_j^{\mathrm{sp}}\right\rangle : U\in\mathscr{R}(Q\cdot Q^T),\|U\|=1,\|y\|=1,\right.\\
&\qquad\qquad \left.\left\langle U,U_j^{\mathrm{sp}}\right\rangle=0\ \forall j=1,\ldots,\bar{m},\right\}\\
&=\max_{U,y}\left\{\left\langle U,\sum_{i=\bar{m}+1}^{m}y_jV_j^{\mathrm{sp}}\right\rangle : U\in\mathscr{R}(Q\cdot Q^T),\|U\|=1,\|y\|=1,\right.\\
&\qquad\qquad\qquad\qquad \left.\left\langle U,U_j^{\mathrm{sp}}\right\rangle=0\ \forall j=1,\ldots,\bar{m},\right\}\\
&=\sigma_{\bar{m}+1}^{\mathrm{sp}}<1-\varepsilon^{\mathrm{sp}}<1.
\end{aligned}$$

To increase the robustness of the computation of $\mathscr{R}(Q\cdot Q^T)\cap\mathscr{R}(\mathscr{A}^*)$ in deciding whether σ_i^{sp} is 1 or not, we may follow similar treatment in [18] where one decides which singular values are zero by checking the ratios between successive small singular values.

Algorithm 12.2: Preprocessing for (AP)

Input: $A_1,\ldots,A_m,A_{m+1} := C \in \mathbb{S}^n$;
Output: δ^*, $P \in \mathbb{R}^{n\times(n-\bar{n})}$, $D_+ \in \mathbb{S}^{n-\bar{n}}$ satisfying $D_+ \succ 0$; (so $D^* = PD_+P^T$);
if *one of the* A_i *(*$i \in \{1,\ldots,m+1\}$*) is definite* **then**
  STOP; (12.62) does not have a solution.
else
  if *some of the* $A = \begin{bmatrix} U & \tilde{U} \end{bmatrix} \begin{bmatrix} \hat{D} & 0 \\ 0 & 0 \end{bmatrix} \begin{bmatrix} U^T \\ \tilde{U}^T \end{bmatrix} \in \{A_i : i = 1,\ldots,m+1\}$ *satisfies* $\hat{D} \succ 0$, **then**
    reduce the size using $A_i \leftarrow \tilde{U}^T A_i \tilde{U}, \forall i$;
  else
    if $\exists 0 \neq V \in \mathbb{R}^{n\times r}$ *such that* $A_iV = 0$ *for all* $i = 1,\ldots,m+1$, **then**
      We get $\langle A_i, VV^T\rangle = 0\ \forall i = 1,\ldots,m+1$;
      $\delta^* = 0, D^* = VV^T$ solves (AP); STOP;
    else
      Use an SDP solver to solve (AP).
    end if
  end if
end if

12.4.1.2 Implicit Equality Constraints and Regularization for MFCQ

The second algorithm for facial reduction involves repeated use of two steps again:

1. We repeat step 1 in Sect. 12.4.1.1 and use (AP) to find the face F.
2. We then find the implicit equality constraints and ensure that they are linearly independent, see Corollary 12.24 and Proposition 12.25.

12.4.1.3 Preprocessing for the Auxiliary Problem

We can take advantage of the fact that eigenvalue-eigenvector calculations are efficient and accurate to obtain a more accurate optimal solution (δ^*, D^*) of (AP), i.e., to decide whether the linear system

$$\langle A_i, D\rangle = 0 \ \forall i = 1,\ldots,m+1 \quad (\text{where } A_{m+1} := C), \quad 0 \neq D \succeq 0 \tag{12.62}$$

has a solution, we can use Algorithm 12.2 as a preprocessor for Algorithm 12.1.

More precisely, Algorithm 12.2 tries to find a solution D^* satisfying (12.62) without using an SDP solver. It attempts to find a vector v in the nullspace of all the A_i, and then sets $D^* = vv^T$. In addition, any semidefinite A_i allows a reduction to a smaller dimensional space.

12.4.2 Backward Stability of One Iteration of Facial Reduction

We now provide the details for one iteration of the main algorithm, see Theorem 12.38. Algorithm 12.1 involves many nontrivial subroutines, each of which would introduce some numerical errors. First we need to obtain an optimal solution (δ^*, D^*) of (AP); in practice we can only get an approximate optimal solution, as δ^* is never exactly zero, and we decide whether the true value of δ^* is zero when the computed value is only close to zero. Second we need to obtain the eigenvalue decomposition of D^*. There comes the issue of determining which of the nearly zero eigenvalues are indeed zero. (Since (AP) is not solved exactly, the approximate solution D^* would have eigenvalues that are positive but close to zero.) Finally, the subspace intersection $\mathscr{R}(Q\cdot Q^T)\cap\mathscr{R}(\mathscr{A}^*)$ (for finding $\bar{m}$ and $\mathscr{P}$) can only be computed approximately via a singular value decomposition, because in practice we would take singular vectors corresponding to singular values that are approximately (but not exactly) 1.

It is important that Algorithm 12.1 is robust against such numerical issues arising from the subroutines. We show that Algorithm 12.1 is backward stable (with respect to these three categories of numerical errors), i.e., for any given input $(\mathscr{A}, b, c)$, there exists $(\tilde{\mathscr{A}}, \tilde{b}, \tilde{C}) \approx (\mathscr{A}, b, C)$ such that the computed result of Algorithm 12.1 applied on $(\mathscr{A}, b, C)$ is equal to the exact result of the same algorithm applied on $(\tilde{\mathscr{A}}, \tilde{b}, \tilde{C})$ (when (AP) is solved exactly and the subspace intersection is determined exactly).

We first show that $\|C_{\text{res}}\|$ is relatively small, given a small $\alpha(\mathscr{A}, C)$.

Lemma 12.32. *Let y_Q, C_Q, C_{res} be defined as in Lemma 12.21. Then the norm of C_{res} is small in the sense that*

$$\|C_{\text{res}}\| \leq \sqrt{2}\left[\frac{\|D^*\|}{\lambda_{\min}(D_+)}\alpha(\mathscr{A}, C)\right]^{1/2}\left(\min_{Z=C-\mathscr{A}^*y\succeq 0}\|Z\|\cdot\right). \tag{12.63}$$

Proof. By optimality, for any $y \in \mathscr{F}_p$,

$$\|C_{\text{res}}\| \leq \min_W \|C - \mathscr{A}^*y - QWQ^T\| = \|Z - QQ^TZQQ^T\|,$$

where $Z := C - \mathscr{A}^*y$. Therefore (12.63) follows from Proposition 12.18. ■

The following technical results shows the relationship between the quantity $\min_{\|y\|=1}\|\mathscr{A}^*y\|^2 - \|Q^T(\mathscr{A}^*y)Q\|^2$ and the cosine of the smallest principal angle between $\mathscr{R}(\mathscr{A}^*)$ and $\mathscr{R}(Q\cdot Q^T)$, defined in (12.57).

Lemma 12.33. *Let $Q \in \mathbb{R}^{n\times\bar{n}}$ satisfy $Q^TQ = I_{\bar{n}}$. Then*

$$\tau := \min_{\|y\|=1}\left\{\|\mathscr{A}^*y\|^2 - \|Q^T(\mathscr{A}^*y)Q\|^2\right\} \geq \left(1 - (\sigma_1^{\text{sp}})^2\right)\sigma_{\min}(\mathscr{A}^*)^2 \geq 0, \tag{12.64}$$

where σ_1^{sp} is defined in (12.57). *Moreover,*

$$\tau = 0 \iff \sigma_1^{\text{sp}} = 1 \iff \mathscr{R}(Q\cdot Q^T)\cap\mathscr{R}(\mathscr{A}^*) \neq \{0\}. \tag{12.65}$$

Proof. By definition of σ_1^{sp},

$$\begin{aligned}
&\max_V \left\{ \max_{\|U\|=1, U\in\mathscr{R}(Q\cdot Q^T)} \langle U,V\rangle : \|V\|=1, V\in\mathscr{R}(\mathscr{A}^*) \right\} \\
\geq &\max_{\|U\|=1, U\in\mathscr{R}(Q\cdot Q^T)} \langle U,V_1^{\mathrm{sp}}\rangle \quad \geq \quad \langle U_1^{\mathrm{sp}},V_1^{\mathrm{sp}}\rangle \quad = \quad \sigma_1^{\mathrm{sp}} \\
\geq &\max_V \left\{ \max_{\|U\|=1, U\in\mathscr{R}(Q\cdot Q^T)} \langle U,V\rangle : \|V\|=1, V\in\mathscr{R}(\mathscr{A}^*) \right\},
\end{aligned}$$

so equality holds throughout, implying that

$$\begin{aligned}
\sigma_1^{\mathrm{sp}} &= \max_V \left\{ \max_{\|U\|=1, U\in\mathscr{R}(Q\cdot Q^T)} \langle U,V\rangle : \|V\|=1, V\in\mathscr{R}(\mathscr{A}^*) \right\} \\
&= \max_y \left\{ \max_{\|W\|=1} \langle QWQ^T, \mathscr{A}^*y\rangle : \|\mathscr{A}^*y\|=1 \right\} \\
&= \max_y \left\{ \|Q^T(\mathscr{A}^*y)Q\| : \|\mathscr{A}^*y\|=1 \right\}.
\end{aligned}$$

Obviously, $\|\mathscr{A}^*y\|=1$ implies that the orthogonal projection $QQ^T(\mathscr{A}^*y)QQ^T$ onto $\mathscr{R}(Q\cdot Q^T)$ is of norm no larger than one:

$$\|Q^T(\mathscr{A}^*y)Q\| = \|QQ^T(\mathscr{A}^*y)QQ^T\| \leq \|\mathscr{A}^*y\| = 1. \tag{12.66}$$

Hence $\sigma_1^{\mathrm{sp}} \in [0,1]$. In addition, equality holds in (12.66) if and only if $\mathscr{A}^*y \in \mathscr{R}(Q\cdot Q^T)$, hence

$$\sigma_1^{\mathrm{sp}} = 1 \iff \mathscr{R}(\mathscr{A}^*) \cap \mathscr{R}(Q\cdot Q^T) \neq \{0\}. \tag{12.67}$$

Whenever $\|y\|=1$, $\|\mathscr{A}^*y\| \geq \sigma_{\min}(\mathscr{A}^*)$. Hence

$$\begin{aligned}
\tau &= \min_y \left\{ \|\mathscr{A}^*y\|^2 - \|Q^T(\mathscr{A}^*y)Q\|^2 : \|y\|=1 \right\} \\
&= \sigma_{\min}(\mathscr{A}^*)^2 \min_y \left\{ \|\mathscr{A}^*y\|^2 - \|Q^T(\mathscr{A}^*y)Q\|^2 : \|y\| = \frac{1}{\sigma_{\min}(\mathscr{A}^*)} \right\} \\
&\geq \sigma_{\min}(\mathscr{A}^*)^2 \min_y \left\{ \|\mathscr{A}^*y\|^2 - \|Q^T(\mathscr{A}^*y)Q\|^2 : \|\mathscr{A}^*y\| \geq 1 \right\} \\
&= \sigma_{\min}(\mathscr{A}^*)^2 \min_y \left\{ \|\mathscr{A}^*y\|^2 - \|Q^T(\mathscr{A}^*y)Q\|^2 : \|\mathscr{A}^*y\| = 1 \right\} \\
&= \sigma_{\min}(\mathscr{A}^*)^2 \left(1 - \max_y \left\{ \|Q^T(\mathscr{A}^*y)Q\|^2 : \|\mathscr{A}^*y\| = 1 \right\} \right) \\
&= \sigma_{\min}(\mathscr{A}^*)^2 \left(1 - (\sigma_1^{\mathrm{sp}})^2 \right).
\end{aligned}$$

This together with $\sigma_1^{\mathrm{sp}} \in [0,1]$ proves (12.64). If $\tau = 0$, then $\sigma_1^{\mathrm{sp}} = 1$ since $\sigma_{\min}(\mathscr{A}^*) > 0$. Then (12.67) implies that $\mathscr{R}(\mathscr{A}^*) \cap \mathscr{R}(Q \cdot Q^T) \neq \{0\}$. Conversely, if $\mathscr{R}(\mathscr{A}^*) \cap \mathscr{R}(Q \cdot Q^T) \neq \{0\}$, then there exists $\hat{y}$ such that $\|\hat{y}\| = 1$ and $\mathscr{A}^*\hat{y} \in \mathscr{R}(Q \cdot Q^T)$. This implies that

$$0 \leq \tau \leq \|\mathscr{A}^*\hat{y}\|^2 - \|Q^T(\mathscr{A}^*\hat{y})Q\|^2 = 0,$$

so $\tau = 0$. This together with (12.67) proves the second claim (12.65). ∎

Next we prove that two classes of matrices are positive semidefinite and show their eigenvalue bounds, which will be useful in the backward stability result.

Lemma 12.34. *Suppose $A_1, \ldots, A_m, D^* \in \mathbb{S}^n$. Then the matrix $\hat{M} \in \mathbb{S}^m$ defined by*

$$\hat{M}_{ij} = \langle A_i, D^* \rangle \langle A_j, D^* \rangle \quad (i,j = 1, \ldots, m)$$

is positive semidefinite. Moreover, the largest eigenvalue $\lambda_{\max}(\hat{M}) \leq \sum_{i=1}^m \langle A_i, D^* \rangle^2$.

Proof. For any $y \in \mathbb{R}^m$,

$$y^T \hat{M} y = \sum_{i,j=1}^m \langle A_i, D^* \rangle \langle A_j, D^* \rangle y_i y_j = \left(\sum_{i=1}^m \langle A_i, D^* \rangle y_i \right)^2.$$

Hence $\hat{M}$ is positive semidefinite. Moreover, by the Cauchy Schwarz inequality we have

$$y^T \hat{M} y = \left(\sum_{i=1}^m \langle A_i, D^* \rangle y_i \right)^2 \leq \left(\sum_{i=1}^m \langle A_i, D^* \rangle^2 \right) \|y\|_2^2.$$

Hence $\lambda_{\max}(\hat{M}) \leq \sum_{i=1}^m \langle A_i, D^* \rangle^2$. ∎

Lemma 12.35. *Suppose $A_1, \ldots, A_m \in \mathbb{S}^n$ and $Q \in \mathbb{R}^{n \times \bar{n}}$ has orthonormal columns. Then the matrix $M \in \mathbb{S}^m$ defined by*

$$M_{ij} = \langle A_i, A_j \rangle - \langle Q^T A_i Q, Q^T A_j Q \rangle, \quad i,j = 1, \ldots, m,$$

is positive semidefinite, with the smallest eigenvalue $\lambda_{\min}(M) \geq \tau$, where τ is defined in (12.64).

Proof. For any $y \in \mathbb{R}^m$, we have

$$\begin{aligned} y^T M y &= \sum_{i,j=1}^m \langle y_i A_i, y_j A_j \rangle - \langle y_i Q^T A_i Q, y_j Q^T A_j Q \rangle \\ &= \|\mathscr{A}^* y\|^2 - \left\|Q^T(\mathscr{A}^* y)Q\right\|^2 \geq \tau \|y\|^2. \end{aligned}$$

Hence $M \in \mathbb{S}_+^m$ and $\lambda_{\min}(M) \geq \tau$. ∎

The following lemma shows that when nonnegative δ^* is approximately zero and $D^* = PD_+P^T + QD_\varepsilon Q^T \approx PD_+P^T$ with $D_+ \succ 0$, under a mild assumption (12.70) it is possible to find a linear operator $\hat{\mathscr{A}}$ "near" $\mathscr{A}$ such that we can take the following approximation:

$$\delta^* \leftarrow 0, \quad D^* \leftarrow PD_+P^T, \quad \mathscr{A}^* \leftarrow \hat{\mathscr{A}}^*,$$

and we maintain that $\hat{\mathscr{A}}(PD_+P^T) = 0$ and $\mathscr{R}(Q \cdot Q^T) \cap \mathscr{R}(\mathscr{A}^*) = \mathscr{R}(Q \cdot Q^T) \cap \mathscr{R}(\hat{\mathscr{A}}^*)$.

Lemma 12.36. *Let $\mathscr{A} : \mathbb{S}^n \to \mathbb{R}^m : X \mapsto (\langle A_i, X \rangle)$ be onto. Let $D^* = \begin{bmatrix} P & Q \end{bmatrix} \begin{bmatrix} D_+ & 0 \\ 0 & D_\varepsilon \end{bmatrix} \begin{bmatrix} P^T \\ Q^T \end{bmatrix} \in \mathbb{S}^n_+$, where $\begin{bmatrix} P & Q \end{bmatrix} \in \mathbb{R}^{n \times n}$ is an orthogonal matrix, $D_+ \succ 0$ and $D_\varepsilon \succeq 0$. Suppose that*

$$\mathscr{R}(Q \cdot Q^T) \cap \mathscr{R}(\mathscr{A}^*) = \operatorname{span}(A_1, \ldots, A_{\bar{m}}), \tag{12.68}$$

for some $\bar{m} \in \{1, \ldots, m\}$. Then

$$\min_{\|y\|=1, y \in \mathbb{R}^{m-\bar{m}}} \left\{ \left\| \sum_{i=1}^{m-\bar{m}} y_i A_{\bar{m}+i} \right\|^2 - \left\| \sum_{i=1}^{m-\bar{m}} y_i Q^T A_{\bar{m}+i} Q \right\|^2 \right\} > 0. \tag{12.69}$$

Assume that

$$\begin{aligned} &\min_{\|y\|=1, y \in \mathbb{R}^{m-\bar{m}}} \left\{ \left\| \sum_{i=1}^{m-\bar{m}} y_i A_{\bar{m}+i} \right\|^2 - \left\| \sum_{i=1}^{m-\bar{m}} y_i Q^T A_{\bar{m}+i} Q \right\|^2 \right\} \\ &\quad > \frac{2}{\|D_+\|^2} \left(\|\mathscr{A}(D^*)\|^2 + \|D_\varepsilon\|^2 \sum_{i=\bar{m}+1}^{m} \|A_i\|^2 \right). \end{aligned} \tag{12.70}$$

Define $\tilde{A}_i$ to be the projection of A_i on $\{PD_+P^T\}^\perp$:

$$\tilde{A}_i := A_i - \frac{\langle A_i, PD_+P^T \rangle}{\langle D_+, D_+ \rangle} PD_+P^T, \quad \forall i = 1, \ldots, m. \tag{12.71}$$

Then

$$\mathscr{R}(Q \cdot Q^T) \cap \mathscr{R}(\tilde{\mathscr{A}}^*) = \mathscr{R}(Q \cdot Q^T) \cap \mathscr{R}(\mathscr{A}^*). \tag{12.72}$$

Proof. We first prove the strict inequality (12.69). First observe that since

$$\left\| \sum_{i=1}^{m-\bar{m}} y_i A_{\bar{m}+i} \right\|^2 - \left\| \sum_{i=1}^{m-\bar{m}} y_i Q^T A_{\bar{m}+i} Q \right\|^2 = \left\| \sum_{i=1}^{m-\bar{m}} y_i (A_{\bar{m}+i} - QQ^T A_{\bar{m}+i} QQ^T) \right\|^2 \geq 0,$$

the optimal value is always nonnegative. Let $\bar{y}$ solve the minimization problem in (12.69). If $\left\|\sum_{i=1}^{m-\bar{m}} \bar{y}_i A_{\bar{m}+i}\right\|^2 - \left\|\sum_{i=1}^{m-\bar{m}} \bar{y}_i Q^T A_{\bar{m}+i} Q\right\|^2 = 0$, then

$$0 \neq \sum_{i=1}^{m-\bar{m}} \bar{y}_i A_{\bar{m}+i} \in \mathscr{R}(Q \cdot Q^T) \cap \mathscr{R}(\mathscr{A}^*) = \operatorname{span}(A_1, \ldots, A_{\bar{m}}),$$

which is absurd since $A_1, \ldots, A_m$ are linearly independent.

Now we prove (12.72). Observe that for $j = 1, \ldots, \bar{m}$, $A_j \in \mathscr{R}(Q \cdot Q^T)$ so $\langle A_j, PD_+P^T \rangle = 0$, which implies that $\tilde{A}_j = A_j$. Moreover,

$$\operatorname{span}(A_1, \ldots, A_{\bar{m}}) \subseteq \mathscr{R}(Q \cdot Q^T) \cap \mathscr{R}(\tilde{A}^*).$$

Conversely, suppose that $B := \tilde{\mathscr{A}}^* y \in \mathscr{R}(Q \cdot Q^T)$. Since $\tilde{A}_j = A_j \in \mathscr{R}(Q \cdot Q^T)$ for $j = 1, \ldots, \bar{m}$,

$$B = QQ^T BQQ^T \implies \sum_{j=\bar{m}+1}^{m} y_j(\tilde{A}_j - QQ^T \tilde{A}_j QQ^T) = 0$$

We show that $y_{\bar{m}+1} = \cdots = y_m = 0$. In fact, since $Q^T(PD_+P^T)Q = 0$, $\sum_{j=\bar{m}+1}^{m} y_j (\tilde{A}_j - QQ^T \tilde{A}_j QQ^T) = 0$ implies

$$\sum_{j=\bar{m}+1}^{m} y_j QQ^T A_j QQ^T = \sum_{j=\bar{m}+1}^{m} y_j A_j - \left(\sum_{j=\bar{m}+1}^{m} \frac{\langle A_j, PD_+P^T \rangle}{\langle D_+, D_+ \rangle} y_j \right) PD_+P^T.$$

For $i = \bar{m}+1, \ldots, m$, taking inner product on both sides with A_i,

$$\sum_{j=\bar{m}+1}^{m} \langle Q^T A_i Q, Q^T A_j Q \rangle y_j = \sum_{j=\bar{m}+1}^{m} \langle A_i, A_j \rangle y_j - \sum_{j=\bar{m}+1}^{m} \frac{\langle A_i, PD_+P^T \rangle \langle A_j, PD_+P^T \rangle}{\langle D_+, D_+ \rangle} y_j,$$

which holds if, and only if,

$$(M - \tilde{M}) \begin{pmatrix} y_{\bar{m}+1} \\ \vdots \\ y_m \end{pmatrix} = 0, \tag{12.73}$$

where $M, \tilde{M} \in \mathbb{S}^{m-\bar{m}}$ are defined by

$$M_{(i-\bar{m}),(j-\bar{m})} = \langle A_i, A_j \rangle - \langle Q^T A_i Q, Q^T A_j Q \rangle,$$
$$\tilde{M}_{(i-\bar{m}),(j-\bar{m})} = \frac{\langle A_i, PD_+P^T \rangle \langle A_j, PD_+P^T \rangle}{\langle D_+, D_+ \rangle}, \quad \forall i, j = \bar{m}+1, \ldots, m.$$

We show that (12.73) implies that $y_{\bar{m}+1} = \cdots = y_m = 0$ by proving that $M - \tilde{M}$ is indeed positive definite. By Lemmas 12.34 and 12.35,

$$\begin{aligned}
\lambda_{\min}(M-\tilde{M}) &\geq \lambda_{\min}(M) - \lambda_{\max}(\tilde{M}) \\
&\geq \min_{\|y\|=1} \left\{ \left\| \sum_{i=1}^{m-\bar{m}} y_i A_{\bar{m}+i} \right\|^2 - \left\| \sum_{i=1}^{m-\bar{m}} y_i Q^T A_{\bar{m}+i} Q \right\|^2 \right\} - \frac{\sum_{i=\bar{m}+1}^{m} \langle A_i, PD_+P^T \rangle^2}{\langle D_+, D_+ \rangle}.
\end{aligned}$$

To see that $\lambda_{\min}(M-\tilde{M}) > 0$, note that since $D^* = PD_+P^T + QD_\varepsilon Q^T$, for all i,

$$\begin{aligned}
\left|\langle A_i, PD_+P^T \rangle\right| &\leq \left|\langle A_i, D^* \rangle\right| + \left|\langle A_i, QD_\varepsilon Q^T \rangle\right| \\
&\leq |\langle A_i, D^* \rangle| + \|A_i\| \|QD_\varepsilon Q^T\| \\
&= |\langle A_i, D^* \rangle| + \|A_i\| \|D_\varepsilon\| \\
&\leq \sqrt{2} \left(|\langle A_i, D^* \rangle|^2 + \|A_i\|^2 \|D_\varepsilon\|^2 \right)^{1/2}.
\end{aligned}$$

Hence

$$\begin{aligned}
\sum_{i=\bar{m}+1}^{m} \left|\langle A_i, PD_+P^T \rangle\right|^2 &\leq 2 \sum_{i=\bar{m}+1}^{m} \left(|\langle A_i, D^* \rangle|^2 + \|A_i\|^2 \|D_\varepsilon\|^2 \right) \\
&\leq 2\|\mathscr{A}(D^*)\|^2 + 2\|D_\varepsilon\|^2 \sum_{i=\bar{m}+1}^{m} \|A_i\|^2,
\end{aligned}$$

and that $\lambda_{\min}(M-\tilde{M}) > 0$ follows from the assumption (12.70). This implies that $y_{\bar{m}+1} = \cdots = y_m = 0$. Therefore $B = \sum_{i=1}^{\bar{m}} y_i \tilde{A}_i$, and by (12.68)

$$\mathscr{R}(Q \cdot Q^T) \cap \mathscr{R}(\tilde{\mathscr{A}}^*) = \operatorname{span}(A_1, \ldots, A_{\bar{m}}) = \mathscr{R}(Q \cdot Q^T) \cap \mathscr{R}(\mathscr{A}^*).$$

■

Remark 12.37. We make a remark about the assumption (12.70) in Lemma 12.36. We argue that the right-hand side expression

$$\frac{2}{\|D_+\|^2} \left(\|\mathscr{A}(D^*)\|^2 + \|D_\varepsilon\|^2 \sum_{i=\bar{m}+1}^{m} \|A_i\|^2 \right)$$

is close to zero (when $\delta^* \approx 0$ and when D_ε is chosen appropriately). Assume that the spectral decomposition of D^* is partitioned as described in Sect. 12.4.1.1. Then (since $\|D_\varepsilon\| \leq \varepsilon \|D^*\|$)

$$\frac{2}{\|D_+\|^2} \|\mathscr{A}(D^*)\|^2 \leq \frac{2(\delta^*)^2}{\|D^*\|^2 - \|D_\varepsilon\|^2} \leq \frac{2(\delta^*)^2}{\|D^*\|^2 - \varepsilon^2 \|D^*\|^2} \leq \frac{2n(\delta^*)^2}{1-\varepsilon^2}$$

and

$$\frac{2\|D_\varepsilon\|^2}{\|D_+\|^2} \sum_{i=\bar{m}+1}^{m} \|A_i\|^2 \le \frac{2\varepsilon^2}{1-\varepsilon^2} \sum_{i=\bar{m}+1}^{m} \|A_i\|^2.$$

Therefore as long as ε and δ^* are small enough (taking into account n and $\sum_{i=\bar{m}+1}^{m} \|A_i\|^2$), then the right-hand side of (12.70) would be close to zero.

Here we provide the backward stability result for one step of the facial reduction algorithm. That is, we show that the smaller problem obtained from one step of facial reduction with $\delta^* \ge 0$ is equivalent to applying facial reduction exactly to an SDP instance "nearby" to the original SDP instance.

Theorem 12.38. *Suppose* $\mathscr{A} : \mathbb{S}^n \to \mathbb{R}^m$, $b \in \mathbb{R}^m$, *and* $C \in \mathbb{S}^n$ *are given so that* (12.1) *is feasible and Algorithm 12.1 returns* (δ^*, D^*), *with* $0 \le \delta^* \approx 0$ *and spectral decomposition* $D^* = \begin{bmatrix} P & Q \end{bmatrix} \begin{bmatrix} D_+ & 0 \\ 0 & D_\varepsilon \end{bmatrix} \begin{bmatrix} P^T \\ Q^T \end{bmatrix}$, *and* $(\bar{\mathscr{A}}, \bar{b}, \bar{C}, y_Q, \mathscr{P})$. *In addition, assume that*

$$\mathscr{P} : \mathbb{R}^{\bar{m}} \to \mathbb{R}^m : v \mapsto \begin{pmatrix} v \\ 0 \end{pmatrix}, \quad \text{so } \mathscr{R}(\mathscr{A}^* \mathscr{P}) = \operatorname{span}(A_1, \ldots, A_{\bar{m}}).$$

Assume also that (12.70) *holds. For* $i = 1, \ldots, m$, *define* $\tilde{A}_i \in \mathbb{S}^n$ *as in* (12.71), *and* $\tilde{\mathscr{A}}^* y := \sum_{i=1}^{m} y_i \tilde{A}_i$. *Let* $\tilde{C} = \tilde{\mathscr{A}}^* y_Q + Q\bar{C}Q^T$. *Then* $(\bar{\mathscr{A}}, \bar{b}, \bar{C})$ *is the exact output of Algorithm 12.1 applied on* $(\tilde{\mathscr{A}}, b, \tilde{C})$, *that is, the following hold:*

1. $\tilde{\mathscr{A}}_{\tilde{C}}(PD_+P^T) = \begin{pmatrix} \tilde{\mathscr{A}}(PD_+P^T) \\ \langle \tilde{C}, PD_+P^T \rangle \end{pmatrix} = 0$,
2. $(y_Q, \bar{C})$ *solves*

$$\min_{y,Q} \frac{1}{2} \left\| \tilde{\mathscr{A}}^* y + QWQ^T - \tilde{C} \right\|^2. \tag{12.74}$$

3. $\mathscr{R}(\tilde{\mathscr{A}}^* \mathscr{P}) = \mathscr{R}(Q \cdot Q^T) \cap \mathscr{R}(\tilde{\mathscr{A}}^*)$.

Moreover, $(\tilde{\mathscr{A}}, b, \tilde{C})$ *is close to* $(\mathscr{A}, b, C)$ *in the sense that*

$$\sum_{i=1}^{m} \|A_i - \tilde{A}_i\|^2 \le \frac{2}{\|D_+\|^2} \left((\delta^*)^2 + \|D_\varepsilon\|^2 \sum_{i=1}^{m} \|A_i\|^2 \right), \tag{12.75}$$

$$\begin{aligned} \|C - \tilde{C}\| \le{}& \frac{\sqrt{2}}{\|D_+\|} \left((\delta^*)^2 + \|D_\varepsilon\|^2 \sum_{i=1}^{m} \|A_i\|^2 \right)^{1/2} \|y_Q\| \\ &+ \sqrt{2} \left[\frac{\|D^*\|}{\lambda_{\min}(D_+)} \alpha(\mathscr{A}, C) \right]^{1/2} \left(\min_{Z = C - \mathscr{A}^* y \succeq 0} \|Z\| \right), \end{aligned} \tag{12.76}$$

where $\alpha(\mathscr{A}, c)$ *is defined in* (12.26).

Proof. First we show that $(\bar{\mathscr{A}}, \bar{b}, \bar{C})$ is the exact output of Algorithm 12.1 applied on $(\tilde{\mathscr{A}}, b, \tilde{C})$:

1. For $i = 1, \ldots, m$, by definition of $\tilde{A}_i$ in (12.71), we have $\langle \tilde{A}_i, PD_+P^T \rangle = 0$. Hence $\tilde{\mathscr{A}}(PD_+P^T) = 0$. Also, $\langle \tilde{C}, PD_+P^T \rangle = y_Q^T(\tilde{\mathscr{A}}(PD_+P^T)) + \langle \bar{C}, Q^T(PD_+P^T)Q \rangle = 0$.
2. By definition, $\tilde{C} - \tilde{\mathscr{A}}^* y_Q - Q\bar{C}Q^T = 0$, so $(y_Q, \bar{C})$ solves the least squares problem (12.74).
3. Given (12.70), we have that

$$\mathscr{R}(Q \cdot Q^T) \cap \mathscr{R}(\tilde{\mathscr{A}}^*) = \mathscr{R}(Q \cdot Q^T) \cap \mathscr{R}(\mathscr{A}^*) = \mathscr{R}(A_1, \ldots, A_{\bar{m}})$$
$$= \mathscr{R}(\tilde{A}_1, \ldots, \tilde{A}_{\bar{m}}) = \mathscr{R}(\tilde{\mathscr{A}}^* \mathscr{P}).$$

The results (12.75) and (12.76) follow easily:

$$\sum_{i=1}^{m} \|A_i - \tilde{A}_i\|^2 = \sum_{i=1}^{m} \frac{|\langle A_i, PD_+P^T \rangle|^2}{\|D_+\|^2} \le \sum_{i=1}^{m} \frac{2|\langle A_i, D^* \rangle|^2 + 2\|A_i\|^2 \|D_\varepsilon\|^2}{\|D_+\|^2}$$
$$\le \frac{2}{\|D_+\|^2} \left((\delta^*)^2 + \|D_\varepsilon\|^2 \sum_{i=1}^{m} \|A_i\|^2 \right),$$

and

$$\begin{aligned}
\|C - \tilde{C}\| &\le \|\mathscr{A}^* y_Q - \tilde{\mathscr{A}}^* y_Q\| + \|C_{\text{res}}\| \\
&\le \sum_{i=1}^{m} |(y_Q)_i| \|A_i - \tilde{A}_i\| + \|C_{\text{res}}\| \\
&\le \|y_Q\| \left(\sum_{i=1}^{m} \|A_i - \tilde{A}_i\|^2 \right)^{1/2} + \|C_{\text{res}}\| \\
&\le \frac{\sqrt{2}}{\|D_+\|} \left((\delta^*)^2 + \|D_\varepsilon\|^2 \sum_{i=1}^{m} \|A_i\|^2 \right)^{1/2} \|y_Q\| \\
&\quad + \sqrt{2} \left[\frac{\|D^*\|}{\lambda_{\min}(D_+)} \alpha(\mathscr{A}, C) \right]^{1/2} \left(\min_{Z = C - \mathscr{A}^* y \succeq 0} \|Z\| \right),
\end{aligned}$$

from (12.75) and (12.63). ■

12.5 Test Problem Descriptions

12.5.1 *Worst-Case Instance*

From Tunçel [65], we consider the following *worst-case* problem instance in the sense that for $n \geq 3$, the facial reduction process in Algorithm 12.1 requires $n-1$ steps to obtain the minimal face. Let $b = e_2 \in \mathbb{R}^n$, $C = 0$, and $\mathscr{A} : \mathbb{S}^n_+ \to \mathbb{R}^n$ be defined by

$$A_1 = e_1 e_1^T,\ A_2 = e_1 e_2^T + e_2 e_1^T,\ A_i = e_{i-1} e_{i-1}^T + e_1 e_i^T + e_i e_1^T \text{ for } i = 3, \ldots, n.$$

It is easy to see that

$$\mathscr{F}_P^Z = \left\{C - \mathscr{A}^* y \in \mathbb{S}^n_+ : y \in \mathbb{R}^n\right\} = \left\{\mu e_1 e_1^T : \mu \geq 0\right\},$$

(so $\mathscr{F}_P^Z$ has empty interior) and

$$\sup\{b^T y : C - \mathscr{A}^* y \succeq 0\} = \sup\{y_2 : -\mathscr{A}^* y = \mu e_1 e_1^T, \mu \geq 0\} = 0,$$

which is attained by any feasible solution.

Now consider the auxiliary problem

$$\min \|\mathscr{A}_C(D)\| = \left[D_{11}^2 + 4D_{12}^2 + \sum_{i=3}^{n} (D_{i-1,i-1} + 2D_{1i})\right]^{1/2} \quad \text{s.t. } \langle D, I \rangle = \sqrt{n},\ D \succeq 0.$$

An optimal solution is $D^* = \sqrt{n} e_n e_n^T$, which attains objective value zero. It is easy to see this is the only solution. More precisely, any solution D attaining objective value 0 must satisfy $D_{11} = 0$, and by the positive semidefiniteness constraint $D_{1,i} = 0$ for $i = 2, \ldots, n$ and so $D_{ii} = 0$ for $i = 2, \ldots, n-1$. So D_{nn} is the only nonzero entry and must equal $\sqrt{n}$ by the linear constraint $\langle D, I \rangle = \sqrt{n}$. Therefore, Q from Proposition 12.18 must have $n-1$ columns, implying that the reduced problem is in $\mathbb{S}^{n-1}$. Theoretically, each facial reduction step via the auxiliary problem can only reduce the dimension by one. Moreover, after each reduction step, we get the same SDP with n reduced by one. Hence it would take $n-1$ facial reduction steps before a reduced problem with strictly feasible solutions is found. This realizes the result in [12] on the upper bound of the number of facial reduction steps needed.

12.5.2 *Generating Instances with Finite Nonzero Duality Gaps*

In this section we give a procedure for generating SDP instances with finite nonzero duality gaps. The algorithm is due to the results in [66, 70].

Algorithm 12.3: Generating SDP instance that has a finite nonzero duality gap

1 Input: problem dimensions m, n; desired duality gap g;

2 Output: linear map $\mathscr{A}:\mathbb{S}^n\to\mathbb{R}^m$, $b\in\mathbb{R}^m$, $C\in\mathbb{S}^n$ such that the corresponding primal dual pair (12.1)–(12.2) has a finite nonzero duality gap;

1. Pick any positive integer r_1,r_3 that satisfy $r_1+r_3+1=n$, and any positive integer $p\le r_3$.
2. Choose $A_i\succeq 0$ for $i=1,\dots,p$ so that $\dim(\text{face}(\{A_i:i=1,\dots,p\}))=r_3$. Specifically, choose $A_1,\dots,A_p$ so that

$$\text{face}(\{A_i:1,\dots,p\})=\begin{bmatrix}0&0&0\\0&0&0\\0&0&\mathbb{S}^{r_3}_+\end{bmatrix}. \tag{12.77}$$

3. Choose $A_{p+1},\dots,A_m$ of the form

$$A_i=\begin{bmatrix}0&0&(A_i)_{13}\\0&(A_i)_{22}&*\\(A_i)_{13}^T&*&*\end{bmatrix},$$

where an asterisk denotes a block having arbitrary elements, such that $(A_{p+1})_{13},\dots,(A_m)_{13}$ are linearly independent, and $(A_i)_{22}\succ 0$ for some $i\in\{p+1,\dots,m\}$.

4. Pick

$$\bar{X}=\begin{bmatrix}0&0&0\\0&\sqrt{g}&0\\0&0&0\end{bmatrix}. \tag{12.78}$$

5. Take $b=\mathscr{A}(\bar{X})$, $C=\bar{X}$.

Finite nonzero duality gaps and strict complementarity are closely tied together for cone optimization problems; using the concept of a *complementarity partition*, we can generate instances that fail to have strict complementarity; these in turn can be used to generate instances with finite nonzero duality gaps. See [66, 70].

Theorem 12.39. *Given any positive integers n, $m\le n(n+1)/2$ and any $g>0$ as input for Algorithm 12.3, the following statements hold for the primal-dual pair* (12.1)–(12.2) *corresponding to the output data from Algorithm 12.3:*

1. *Both* (12.1) *and* (12.2) *are feasible.*
2. *All primal feasible points are optimal and $v_P=0$.*
3. *All dual feasible point are optimal and $v_D=g>0$.*

It follows that (12.1) *and* (12.2) *possess a finite positive duality gap.*

Proof. Consider the primal problem (12.1). Equation (12.1) is feasible because $C:=\bar{X}$ given in (12.78) is positive semidefinite. Note that by definition of $\mathscr{A}$ in Algorithm 12.3, for any $y\in\mathbb{R}^m$,

$$C-\sum_{i=1}^{p} y_i A_i = \begin{bmatrix} 0 & 0 & 0 \\ 0 & \sqrt{g} & 0 \\ 0 & 0 & * \end{bmatrix} \text{ and } -\sum_{i=p+1}^{m} y_i A_i = \begin{bmatrix} 0 & 0 & * \\ 0 & * & * \\ * & * & * \end{bmatrix},$$

so if $y \in \mathbb{R}^m$ satisfies $Z := C - \mathscr{A}^* y \succeq 0$, then $\sum_{i=p+1}^{m} y_i A_i = 0$ must hold. This implies $\sum_{i=p+1}^{m} y_i (A_i)_{13} = 0$. Since $(A_{p+1})_{13}, \dots, (A_m)_{13}$ are linearly independent, we must have $y_{p+1} = \cdots = y_m = 0$. Consequently, if y is feasible for (12.1), then

$$\mathscr{A}^* y = \begin{bmatrix} 0 & 0 & 0 \\ 0 & 0 & 0 \\ 0 & 0 & -Z_{33} \end{bmatrix}$$

for some $Z_{33} \succeq 0$. The corresponding objective value in (12.1) is given by

$$b^T y = \langle \bar{X}, \mathscr{A}^* y \rangle = 0.$$

This shows that the objective value of (12.1) is constant over the feasible region. Hence $v_P = 0$, and all primal feasible solutions are optimal.

Consider the dual problem (12.2). By the choice of b, $\bar{X} \succeq 0$ is a feasible solution, so (12.2) is feasible too. From (12.77), we have that $b_1 = \cdots = b_p = 0$. Let $X \succeq 0$ be feasible for (12.1). Then $\langle A_i, X \rangle = b_i = 0$ for $i = 1, \dots, p$, implying that the (3,3) block of X must be zero by (12.77), so

$$X = \begin{bmatrix} * & * & 0 \\ * & * & 0 \\ 0 & 0 & 0 \end{bmatrix}.$$

Since $\alpha = (A_j)_{22} > 0$ for some $j \in \{p+1, \dots, m\}$, we have that

$$\alpha X_{22} = \langle A_j, X \rangle = \langle A_j, \bar{X} \rangle = \alpha \sqrt{g},$$

so $X_{22} = \sqrt{g}$ and $\langle C, X \rangle = g$. Therefore the objective value of (12.2) is constant and equals $g > 0$ over the feasible region, and all feasible solutions are optimal. ∎

12.5.3 Numerical Results

Table 12.1 shows a comparison of solving SDP instances *with* versus *without* facial reduction. Examples 1 through 9 are specially generated problems available online at the URL for this paper.[2] In particular: Example 3 has a positive duality gap, $v_P = 0 < v_D = 1$; for Example 4, the dual is infeasible; in Example 5, the Slater CQ holds; Examples 9a, 9b are instances of the worst-case problems presented

[2]orion.math.uwaterloo.ca/~hwolkowi/henry/reports/ABSTRACTS.html.

Table 12.1 Comparisons with/without facial reduction

Name	n	m	True primal optimal value	True dual optimal value	Primal optimal value *with* facial reduction	Primal optimal value *without* facial reduction
Example 1	3	2	0	0	0	−6.30238e−016
Example 2	3	2	0	1	0	+0.570395
Example 3	3	4	0	0	0	+6.91452e−005
Example 4	3	3	0	Infeasible	0	+Inf
Example 5	10	5	*	*	+5.02950e+02	+5.02950e+02
Example 6	6	8	1	1	+1	+1
Example 7	5	3	0	0	0	−2.76307e−012
Example 9a	20	20	0	Infeasible	0	Inf
Example 9b	100	100	0	Infeasible	0	Inf
RandGen1	10	5	0	1.4509	+1.5914e−015	+1.16729e−012
RandGen2	100	67	0	5.5288e+003	+1.1056e−010	NaN
RandGen4	200	140	0	2.6168e+004	+1.02803e−009	NaN
RandGen5	120	45	0	0.0381	−5.47393e−015	−1.63758e−015
RandGen6	320	140	0	2.5869e+005	+5.9077e−025	NaN
RandGen7	40	27	0	168.5226	−5.2203e−029	+5.64118e−011
RandGen8	60	40	0	4.1908	−2.03227e−029	NaN
RandGen9	60	40	0	61.0780	+5.61602e−015	−3.52291e−012
RandGen10	180	100	0	5.1461e+004	+2.47204e−010	NaN
RandGen11	255	150	0	4.6639e+004	+7.71685e−010	NaN

in Sect. 12.5.1. The remaining instances RandGen1–RandGen11 are generated randomly with most of them having a finite positive duality gap, as described in Sect. 12.5.2. These instances generically require only one iteration of facial reduction. The software package SeDuMi is used to solve the SDPs that arise.

One general observation is that, if the instance has primal-dual optimal solutions and has zero duality gap, SeDuMi is able to find the optimal solutions. However, if the instance has finite nonzero duality gaps, and if the instance is not too small, SeDuMi is unable to compute any solution, and returns NaN.

SeDuMi, based on self-dual embedding, embeds the input primal-dual pair into a larger SDP that satisfies the Slater CQ [16]. Theoretically, the lack of the Slater CQ in a given primal-dual pair is not an issue for SeDuMi. It is not known what exactly causes problem on SeDuMi when handling instances where a nonzero duality gap is present.

12.6 Conclusions and Future Work

In this paper we have presented a preprocessing technique for SDP problems where the Slater CQ (nearly) fails. This is based on solving a stable auxiliary problem that approximately identifies the minimal face for (P). We have included a backward

11. Borwein, J.M., Wolkowicz, H.: Facial reduction for a cone-convex programming problem. J. Austral. Math. Soc. Ser. A **30**(3), 369–380 (1980/1981)
12. Borwein, J.M., Wolkowicz, H.: Regularizing the abstract convex program. J. Math. Anal. Appl. **83**(2), 495–530 (1981)
13. Boyd, S., Balakrishnan, V., Feron, E., El Ghaoui, L.: Control system analysis and synthesis via linear matrix inequalities. In: Proceedings of the American Control Conference, pp. 2147–2154 (1993)
14. Burkowski, F., Cheung, Y-L., Wolkowicz, H.: Efficient use of semidefinite programming for selection of rotamers in protein conformations. Technical Report, p. 30, University of Waterloo, Waterloo, Ontario (2012)
15. Caron, R.J., Boneh, A., Boneh, S.: Redundancy. In: Advances in Sensitivity Analysis and Parametric Programming. International Series in Operations Research & Management Science, vol. 6, pp. 13.1–13.41. Kluwer Academic Publishers, Boston (1997)
16. De Klerk, E.: Interior point methods for semidefinite programming. Ph.D. Thesis, Delft University (1997)
17. De Klerk, E.: Aspects of Semidefinite Programming: Interior Point Algorithms and Selected Applications. Applied Optimization Series. Kluwer Academic, Boston (2002)
18. Demmel, J., Kågström, B.: The generalized Schur decomposition of an arbitrary pencil $A - \lambda B$; robust software with error bounds and applications II: software and applications. ACM Trans. Math. Soft. **19**(2), 175–201 (1993)
19. Doan, X.V., Kruk, S., Wolkowicz, H.: A robust algorithm for semidefinite programming. Optim. Methods Soft. **27**(4–5), 667–693 (2012)
20. Fourer, R., Gay, D.M.: Experience with a primal presolve algorithm. In: Large Scale Optimization (Gainesville, FL, 1993), pp. 135–154. Kluwer Academic Publishers, Dordrecht (1994)
21. Freund, R.M.: Complexity of an algorithm for finding an approximate solution of a semi-definite program with no regularity assumption. Technical Report OR 302-94, MIT, Cambridge (1994)
22. Freund, R.M.: Complexity of convex optimization using geometry-based measures and a reference point. Math. Program. Ser. A **99**(2), 197–221 (2004)
23. Freund, R.M., Vera, J.R.: Some characterizations and properties of the "distance to ill-posedness" and the condition measure of a conic linear system. Technical report, MIT, Cambridge (1997)
24. Freund, R.M., Ordóñez, F., Toh, K.C.: Behavioral measures and their correlation with IPM iteration counts on semi-definite programming problems. USC-ISE Working Paper #2005-02, MIT (2005). www-rcf.usc.edu/~fordon/
25. Goldman, A.J., Tucker, A.W.: Theory of linear programming. In: Linear Inequalities and Related Systems. Annals of Mathematics Studies, vol. 38, pp. 53–97. Princeton University Press, Princeton (1956)
26. Golub, G.H., Van Loan, C.F.: Matrix Computations, 3rd edn. Johns Hopkins University Press, Baltimore (1996)
27. Gondzio, J.: Presolve analysis of linear programs prior to applying an interior point method. Informs J. Comput. **9**(1), 73–91 (1997)
28. Gould, N.I.M., Toint, Ph.L.: Preprocessing for quadratic programming. Math. Program. Ser. B **100**(1), 95–132 (2004)
29. Gourion, D., Seeger, A.: Critical angles in polyhedral convex cones: numerical and statistical considerations. Math. Program. **123**(1), 173–198 (2010)
30. Gruber, G., Rendl, F.: Computational experience with ill-posed problems in semidefinite programming. Comput. Optim. Appl. **21**(2), 201–212 (2002)
31. Hiriart-Urruty, J-B., Malick, J.: A fresh variational-analysis look at the positive semidefinite matrices world. Technical Report, University of Tolouse, Toulouse, France (2010)
32. Horn, R.A., Johnson, C.R.: Matrix Analysis (Corrected reprint of the 1985 original). Cambridge University Press, Cambridge (1990)
33. Iusem, A., Seeger, A.: Searching for critical angles in a convex cone. Math. Program. Ser. B **120**(1), 3–25 (2009)

34. Jibrin, S.: Redundancy in semidefinite programming. Ph.D. Thesis. Carleton University, Ottawa, Ontario, Canada (1997)
35. Karwan, M.H., Lotfi, V., Telgen, J., Zionts, S.: Redundancy in Mathematical Programming. Springer, New York (1983)
36. Krislock, N., Wolkowicz, H.: Explicit sensor network localization using semidefinite representations and facial reductions. SIAM J. Optim. **20**(5), 2679–2708 (2010)
37. Lamoureux, M., Wolkowicz, H.: Numerical decomposition of a convex function. J. Optim. Theory Appl. **47**(1), 51–64 (1985)
38. Luo, Z-Q., Sturm, J.F., Zhang, S.: Conic convex programming and self-dual embedding. Optim. Methods Soft. **14**(3), 169–218 (2000)
39. Malick, J., Povh, J., Rendl, F., Wiegele, A.: Regularization methods for semidefinite programming. SIAM J. Optim. **20**(1), 336–356 (2009)
40. Mangasarian, O.L., Fromovitz, S.: The Fritz John necessary optimality conditions in the presence of equality and inequality constraints. J. Math. Anal. Appl. **17**, 37–47 (1967)
41. Mészáros, C., Suhl, U.H.: Advanced preprocessing techniques for linear and quadratic programming. OR Spectrum **25**, 575–595 (2003). doi: 10.1007/s00291-003-0130-x
42. Monteiro, R.D.C., Todd, M.J.: Path-following methods. In: Handbook of Semidefinite Programming, pp. 267–306. Kluwer Academic Publishers, Boston (2000)
43. Nesterov, Y.E., Todd, M.J., Ye, Y.: Infeasible-start primal-dual methods and infeasibility detectors for nonlinear programming problems. Math. Program. Ser. A **84**(2), 227–267 (1999)
44. Pataki, G.: On the closedness of the linear image of a closed convex cone. Math. Oper. Res. **32**(2), 395–412 (2007)
45. Pataki, G.: Bad semidefinite programs: they all look the same. Technical Report, Department of Operations Research, University of North Carolina, Chapel Hill (2011)
46. Peña, J., Renegar, J.: Computing approximate solutions for convex conic systems of constraints. Math. Program. Ser. A **87**(3), 351–383 (2000)
47. Pólik, I., Terlaky, T.: New stopping criteria for detecting infeasibility in conic optimization. Optim. Lett. **3**(2), 187–198 (2009)
48. Ramana, M.V.: An algorithmic analysis of multiquadratic and semidefinite programming problems. Ph.D. Thesis, Johns Hopkins University, Baltimore (1993)
49. Ramana, M.V.: An exact duality theory for semidefinite programming and its complexity implications. Math. Program. **77**(2), 129–162 (1997)
50. Ramana, M.V., Tunçel, L., Wolkowicz, H.: Strong duality for semidefinite programming. SIAM J. Optim. **7**(3), 641–662 (1997)
51. Renegar, J.: A Mathematical View of Interior-Point Methods in Convex Optimization. MPS/SIAM Series on Optimization. SIAM, Philadelphia (2001)
52. Robinson, S.M.: Stability theory for systems of inequalities I: Linear systems. SIAM J. Numer. Anal. **12**, 754–769 (1975)
53. Robinson, S.M.: First order conditions for general nonlinear optimization. SIAM J. Appl. Math. **30**(4), 597–607 (1976)
54. Rockafellar, R.T: Some convex programs whose duals are linearly constrained. In: Nonlinear Programming (Proceedings of a Symposium, University of Wisconsin, Madison, Wisconsin, 1970), pp. 293–322. Academic, New York (1970)
55. Rockafellar, R.T.: Convex Analysis (Reprint of the 1970 original) Princeton Landmarks in Mathematics, Princeton Paperbacks. Princeton University Press, Princeton (1997)
56. Shapiro, A.: On duality theory of conic linear problems. In: Semi-Infinite Programming (Alicante, 1999). Nonconvex Optim. Appl. vol. 57, pp. 135–165. Kluwer Academic Publishers, Dordrecht (2001)
57. Shapiro, A., Nemirovskii, A.: Duality of linear conic problems. Technical Report, School of Industrial and Systems Engineering, Georgia Institute of Technology, Atlanta, Georgia (2003)
58. Stewart, G.W.: Rank degeneracy. SIAM J. Sci. Stat. Comput. **5**(2), 403–413 (1984)
59. Stewart, G.W.: Determining rank in the presence of error. In: Linear Algebra for Large Scale and Real-Time Applications (Leuven, 1992). NATO Advanced Science Institute Series E: Applied Sciences, vol. 232, pp. 275–291. Kluwer Academic Publishers, Dordrecht (1993)

60. Sturm, J.F.: Using SeDuMi 1.02, a MATLAB toolbox for optimization over symmetric cones. Optim. Methods Soft. **11/12**(1–4), 625–653 (1999). sedumi.ie.lehigh.edu.
61. Sun, D.: The strong second-order sufficient condition and constraint nondegeneracy in nonlinear semidefinite programming and their implications. Math. Oper. Res. **31**(4), 761–776 (2006)
62. Todd, M.J., Ye, Y.: Approximate Farkas lemmas and stopping rules for iterative infeasible-point algorithms for linear programming. Math. Program. Ser. A **81**(1), 1–21 (1998)
63. Todd, M.J.: Semidefinite programming. Acta Numerica **10**, 515–560 (2001)
64. Tunçel, L.: On the Slater condition for the SDP relaxations of nonconvex sets. Oper. Res. Lett. **29**(4), 181–186 (2001)
65. Tunçel, L.: Polyhedral and Semidefinite Programming Methods in Combinatorial Optimization. Fields Institute Monographs, vol. 27. American Mathematical Society, Providence (2010)
66. Tunçel, L., Wolkowicz, H.: Strong duality and minimal representations for cone optimization. Comput. Optim. Appl. **53**(2),619–648 (2012)
67. Tütüncü, R.H., Toh, K.C., Todd, M.J.: Solving semidefinite-quadratic-linear programs using SDPT3. Math. Program. Ser. B **95**(2), 189–217 (2003). www.math.nus.edu.sg/~mattohkc/sdpt3.html.
68. Vandenberghe, L., Boyd, S.: Semidefinite programming. SIAM Rev. **38**(1), 49–95 (1996)
69. Waki, H., Kim, S., Kojima, M., Muramatsu, M.: Sums of squares and semidefinite program relaxations for polynomial optimization problems with structured sparsity. SIAM J. Optim. **17**(1), 218–242 (2006)
70. Wei, H., Wolkowicz, H.: Generating and solving hard instances in semidefinite programming. Math. Program. **125**(1), 31–45 (2010)
71. Wolkowicz, H.: Calculating the cone of directions of constancy. J. Optim. Theory Appl. **25**(3), 451–457 (1978)
72. Wolkowicz, H.: Some applications of optimization in matrix theory. Linear Algebra Appl. **40**, 101–118 (1981)
73. Wolkowicz, H.: Solving semidefinite programs using preconditioned conjugate gradients. Optim. Methods Soft. **19**(6), 653–672 (2004)
74. Wolkowicz, H., Zhao, Q.: Semidefinite programming relaxations for the graph partitioning problem. Discrete Appl. Math. **96/97**, 461–479 (1999) (Selected for the special Editors' Choice, Edition 1999)
75. Wolkowicz, H., Saigal, R., Vandenberghe, L. (eds.): Handbook of Semidefinite Programming: Theory, Algorithms, and Applications. International Series in Operations Research & Management Science, vol. 27. Kluwer Academic Publishers, Boston (2000)
76. Yamashita, M., Fujisawa, K., Kojima, M.: Implementation and evaluation of SDPA 6.0 (semidefinite programming algorithm 6.0). Optim. Methods Soft. **18**(4), 491–505 (2003). sdpa.indsys.chuo-u.ac.jp/sdpa/
77. Yamashita, M., Fujisawa, K., Nakata, K., Nakata, M., Fukuda, M., Kobayashi, K., Goto, K.: A high-performance software package for semidefinite programs: Sdpa7. Technical Report, Department of Information Sciences, Tokyo Institute of Technology, Tokyo, Japan (2010)
78. Zălinescu, C: On zero duality gap and the Farkas lemma for conic programming. Math. Oper. Res. **33**(4), 991–1001 (2008)
79. Zălinescu, C.: On duality gap in linear conic problems. Technical Report, University of Alexandru Ioan Cusa, Iasi, Romania (2010)
80. Zhao, Q., Karisch, S.E., Rendl, F., Wolkowicz, H.: Semidefinite programming relaxations for the quadratic assignment problem. Semidefinite programming and interior-point approaches for combinatorial optimization problems (Fields Institute, Toronto, ON, 1996). J. Comb. Optim. **2**(1), 71–109 (1998)

Chapter 13
The Largest Roots of the Mandelbrot Polynomials

Robert M. Corless and Piers W. Lawrence

Abstract This paper gives some details of the experimental discovery and partial proof of a simple asymptotic development for the largest magnitude roots of the Mandelbrot polynomials defined by $p_0(z) = 1$ and $p_{n+1}(z) = zp_n^2(z) + 1$.

Key words: Asymptotic expansion • Eigenvalues • Mandelbrot polynomials • Periodic points • Polynomial zeros

Mathematics Subject Classifications (2010): Primary 37F10; Secondary 37F45, 41A60.

13.1 Background, Experiments, and Results

In the paper [5] we undertook to use the Mandelbrot polynomials, which satisfy $p_0(z) = 1$ and

$$p_{n+1}(z) = zp_n^2(z) + 1\,, \tag{13.1}$$

as a family of test examples for a general class of eigenvalue methods for finding roots of polynomials. More, one of us (PWL) invented an interesting recursively constructed zero-one matrix family whose eigenvalues were the roots of $p_n(z)$ and which allowed the computation of all $2^{20} - 1 = 1{,}048{,}575$ roots of $p_{20}(z)$.

COMMUNICATED BY HEINZ H. BAUSCHKE.

R.M. Corless (✉) • P.W. Lawrence
Department of Applied Mathematics, The University of Western Ontario,
College Room 272, London, ON N6A 5B7, Canada
e-mail: rcorless@uwo.ca; plawren@uwo.ca

D.H. Bailey et al. (eds.), *Computational and Analytical Mathematics*, Springer Proceedings in Mathematics & Statistics 50, DOI 10.1007/978-1-4614-7621-4_13,

When another of us (RMC) was presenting these results at JonFest DownUnder, in Newcastle, Neil Calkin asked about the root of largest magnitude. That question sparked this paper, and we thank Neil for the discussions that took place at the conference.

We begin with the well-known observation that the largest root is quite close to, but slightly closer to zero than, -2. The classical approach to find a root, given an initial guess, is Newton's method. For that, we need derivatives: obviously, $p_0'(z) = 0$, and

$$p_{n+1}'(z) = p_n^2(z) + 2zp_n(z)p_n'(z)\,. \tag{13.2}$$

Notice also that $p_0(-2) = 1$ but $p_1(-2) = -2 \cdot 1^2 + 1 = -1$ and thereafter $p_{n+1}(-2) = -2\cdot(-1)^2+1 = -1$. Thus, all first derivatives $p_k'(-2)$ are known from

$$\begin{aligned} p_{n+1}'(-2) &= (-1)^2 + 2\cdot(-2)(-1)p_n'(-2) \\ &= 4p_n'(-2) + 1\,, \end{aligned} \tag{13.3}$$

which is easily solved to give

$$p_n'(-2) = \frac{4^n - 1}{3}\,. \tag{13.4}$$

That the derivatives are all integers also follows from the definition, as it is easily seen that the coefficients of $p_k(z)$ in the monomial basis are positive integers.

The Newton estimate (which is not quite right, as we will see very soon) is thus

$$z_k \doteq -2 + \frac{3}{4^k - 1}\,. \tag{13.5}$$

In one sense this is quite successful. For large k this suggests that the largest magnitude zero is close to -2, which it is. But in another sense, it is not very successful; the error is in fact $O(4^{-k})$, not smaller (even though the initial guess is already $O(4^{-k})$ accurate, so this is not an improvement), and taking yet another Newton step hardly improves this estimate at all! Indeed Newton's method converges initially only very slowly from here. Of course the problem is growth in the higher derivatives. The Newton estimate is based on the expansion

$$p_k(-2+\varepsilon) = p_k(-2) + p_k'(-2)\varepsilon + \frac{1}{2}p_k''(-2)\varepsilon^2 + \cdots \tag{13.6}$$

neglecting the (usually benign) terms of $O(\varepsilon^2)$. However, here $p_k'(-2) = (4^k - 1)/3$ and $\varepsilon = 3/(4^k - 1)$, so while ε is very small, and ε^2 is smaller, we really should check $p_k''(-2)$.

Taking the second derivative is simple: With $p_0''(z) = 0$ and

$$p_{n+1}''(z) = 4p_n(z)p_n'(z) + 2z(p'(z))^2 + 2zp_n(z)p_n''(z)\,, \tag{13.7}$$

we can compute all values of $p_k''(-2)$. At $z=-2$, $p_k(-2)=-1$ and $p_k'(-2)=(4^k-1)/3$; therefore the recurrence for the second derivatives is

$$p_{n+1}''(-2) = -4\left(\frac{4^n-1}{3}\right) - 4\left(\frac{4^n-1}{3}\right)^2 + 4p_n''(-2), \qquad (13.8)$$

which is nearly as easy to solve as the first one. One can use Maple's `rsolve`, as we did, to find

$$p_k''(-2) = -\frac{1}{27}4^{2k} + \left(\frac{1}{3} - \frac{k}{9}\right)4^k - \frac{8}{27}. \qquad (13.9)$$

Now the problem with Newton's method becomes apparent: This is $O(\varepsilon^{-2})$, therefore we cannot neglect the $O(\varepsilon^2)$ term!

In a fit of enthusiasm we compute a few more derivatives:

$$p_k'''(-2) = \frac{1}{15}\varepsilon^{-3} + O(\varepsilon^{-2}), \qquad (13.10)$$

$$p_k^{(iv)}(-2) = -\frac{1}{105}\varepsilon^{-4} + O(\varepsilon^{-3}), \qquad (13.11)$$

and so on, giving (to $O(\varepsilon)$)

$$0 \underset{\text{wishful}}{=} p_n(-2+\varepsilon) = -1 + 1 - \frac{1}{3\cdot 2!} + \frac{1}{15\cdot 3!} - \frac{1}{105\cdot 4!} + \cdots \qquad (13.12)$$

which is tantalizing, but wrong.

But what if, instead of using Newton to move from -2 to $-2+\varepsilon$, we instead moved to $-2+\alpha\varepsilon$? Could we find a useful α from the series?

This would give

$$0 = p_n(-2+\alpha\varepsilon) = -1 + \alpha - \frac{\alpha^2}{3\cdot 2!} + \frac{\alpha^3}{15\cdot 3!} - \frac{\alpha^4}{105\cdot 4!} + \cdots. \qquad (13.13)$$

Now the natural thing to do is to ask the on-line encyclopedia of integer sequences [6] (OEIS) if it knows those integers. It does! (And in fact they are easy: they are the coefficients of $-\cos\sqrt{2\alpha} = -1 + \alpha - \frac{\alpha^2}{6} + \frac{\alpha^3}{90} - \cdots$.) This is the first really big break and in essence leads to everything that follows.

To make things explicit we present, in tidier form,

Fact 1:

$$p_k\left(-2 + \frac{3}{2}\cdot\theta^2\cdot 4^{-k}\right) = -\cos\theta + O(4^{-k}). \qquad (13.14)$$

Table 13.1 Numerical verification of Eq. (13.15): comparison of the prediction of the formula with the numerically computed largest magnitude root (computed by the method of [5])

k	$\log_4(z-z_{k,0})$
1	-1.87
2	-3.09
3	-4.76
4	-6.52
5	-8.34
6	-10.19
7	-12.07
8	-13.97
9	-15.87
10	-17.79
11	-19.72
12	-21.65
13	-23.59
14	-25.54
15	-27.48
16	-29.43
17	-31.39
18	-33.35
19	-35.31
20	-37.27

The error improves as k increases, being approximately $O(4^{-2k})$, as claimed

This suggests that the largest magnitude zero of $p_k(z)$ begins (with $\theta = \pi/2$) (Table 13.1):

Fact 2:

$$z = -2 + \frac{3}{8}\pi^2 \cdot 4^{-k} + O(4^{-2k})\,. \tag{13.15}$$

We numerically verified this to high precision, as the reader may do for themselves: choose your favourite multiple-precision arithmetic, write a recursive program to compute $p_k(z)$ given z and k (don't, of course, expand into the monomial basis—just use the recurrence relation itself), and choose a large value of k (say $k = 30$), evaluate the value of z given above to high precision, and then evaluate $p_{30}(z)$. Its value should be comparable to 4^{-30}, indeed $\approx -1.23 \cdot 10^{-17}$, while both $p_{29}(z)$ and $p_{31}(z)$ are order 1.

Greatly encouraged, we go back to the recurrence relations for $p_k^{(\ell)}(-2)$ to look at the higher-order terms. Indeed we can make progress there, too, which we do not describe in all its false starts and missteps here; but the $\left(\frac{1}{3} - \frac{k}{9}\right)4^k$ term in $p_k''(-2)$, which correctly leads us to the conjecture

$$p_k\left(-2+\frac{3}{2}\theta^2 4^{-k}\right) = -\cos\theta + (\tilde{a}(\theta)k+\tilde{b}(\theta))4^{-k} + O(4^{-2k}), \qquad (13.16)$$

allows similar use of the OEIS with computer-generated series terms to succeed. Instead of detailing that experimental approach, we start over, with more rigour.

13.2 What's Really True

We now proceed with a partial proof containing a kind of analytical discovery of what $\tilde{a}(\theta)$ and $\tilde{b}(\theta)$ are. Consider the basic iteration (13.1), and suppose, as a sort of inductive step, that $p_k\left(-2+\frac{3}{2}\theta^2 4^{-k}\right)$ is as given in Eq. (13.16). It is easy to see that $p_1(z) = z+1$ is quite close to $-\cos\theta$ when $z = -2+3\theta^2/2\cdot 4^{-1}$ on, say, $0 \le \theta \le \sqrt{30}/3 \doteq 1.8257$, for example, so that we may begin the approximations already when $k = 1$. Now consider the case with a fixed but unspecified k.

Then it must be true that

$$\begin{aligned} &p_{k+1}\left(-2+\frac{3}{2}\theta^2 4^{-k-1}\right) \\ &= \left(-2+\frac{3}{8}\theta^2 4^{-k}\right) p_k^2\left(-2+\frac{3}{2}\left(\frac{\theta}{2}\right)^2 4^{-k}\right)+1 \\ &= \left(-2+\frac{3}{8}\theta^2 4^{-k}\right)\left(-\cos\frac{\theta}{2}+\left(\tilde{a}\left(\frac{\theta}{2}\right)k+\tilde{b}\left(\frac{\theta}{2}\right)\right)4^{-k}\right)^2+1, \end{aligned} \qquad (13.17)$$

which is supposed to equal

$$-\cos\theta + \left(\tilde{a}(\theta)(k+1)+\tilde{b}(\theta)\right)4^{-k-1}+\cdots. \qquad (13.18)$$

When we expand the right-hand side out, we get

$$\begin{aligned} &\left(-2+\frac{3}{8}\theta^2 4^{-k}\right)\left(\cos^2\frac{\theta}{2}-2\cos\frac{\theta}{2}\left(\tilde{a}\left(\frac{\theta}{2}\right)k+\tilde{b}\left(\frac{\theta}{2}\right)\right)4^{-k}+\cdots\right)+1 \\ &= 1-2\cos^2\frac{\theta}{2}+\left(\frac{3}{8}\theta^2\cos^2\frac{\theta}{2}+4\cos\frac{\theta}{2}\left(\tilde{a}\left(\frac{\theta}{2}\right)k+\tilde{b}\left(\frac{\theta}{2}\right)\right)\right)4^{-k}+\cdots \end{aligned} \qquad (13.19)$$

and we are delighted to see that $1-2\cos^2\frac{\theta}{2}$ becomes $-\cos\theta$, as desired.

The other terms are a little more complicated: equating terms multiplied by $k4^{-k}$ we get

$$\frac{\tilde{a}(\theta)}{4} = 4\cos\frac{\theta}{2}\cdot\tilde{a}\left(\frac{\theta}{2}\right). \qquad (13.20)$$

Now, taking advantage of our earlier experimental work, we merely verify that

$$\tilde{a}(\theta) = K\theta^3 \sin\theta \tag{13.21}$$

solves this, for any K:

$$K\theta^3 \sin\theta = 16\cos\frac{\theta}{2}\cdot K\cdot\left(\frac{\theta}{2}\right)^3 \sin\frac{\theta}{2}. \tag{13.22}$$

This is a linear homogeneous functional equation and has a unique solution [1]. Therefore $\tilde{a}(\theta)$ has been identified.

Now consider $\tilde{b}(\theta)$. This must satisfy

$$\frac{1}{4}(\tilde{b}(\theta)+\tilde{a}(\theta)) = \frac{3}{8}\theta^2\cos^2\frac{\theta}{2} + 4\cos\frac{\theta}{2}\tilde{b}\left(\frac{\theta}{2}\right) \tag{13.23}$$

or

$$\tilde{b}(\theta) = -K\theta^3\sin\theta + \frac{3}{2}\theta^2\cos^2\frac{\theta}{2} + 16\cos\frac{\theta}{2}\cdot\tilde{b}\left(\frac{\theta}{2}\right). \tag{13.24}$$

This can be simplified as follows. Put $\theta = 0$: $\tilde{b}(0) = -0+0+16\cdot\tilde{b}(0)$ or $\tilde{b}(0) = 0$. Similarly $\lim_{\theta\to 0}\left(\frac{\tilde{b}(\theta)}{\theta}\right) = 0$. Thus put $\tilde{b}(\theta) = \theta^2 b(\theta)$. Then

$$\theta^2 b(\theta) = -K\theta^3\sin\theta + \frac{3}{2}\theta^2\cos^2\frac{\theta}{2} + 16\cos\frac{\theta}{2}\cdot\left(\frac{\theta}{2}\right)^2 b\left(\frac{\theta}{2}\right) \tag{13.25}$$

or

$$b(\theta) = -K\theta\sin\theta + \frac{3}{2}\cos^2\frac{\theta}{2} + 4\cos\frac{\theta}{2}b\left(\frac{\theta}{2}\right). \tag{13.26}$$

Now as $\theta \to 0$, if $b(\theta)$ is continuous, we must have

$$b(0) = \frac{3}{2} + 4b(0) \tag{13.27}$$

or

$$b(0) = -\frac{1}{2}. \tag{13.28}$$

Trying a power series solution

$$b(\theta) = -\frac{1}{2} + \sum_{\ell\geq 1} b_\ell\cdot\theta^{2\ell} \tag{13.29}$$

(the functional equation is even, so it makes sense to look for an even solution), we have

$$(-\frac{1}{2}+b_1\theta^2+\cdots) = -K\theta^2+\frac{3}{2}\left(1-\frac{(\theta/2)^2}{2}+\cdots\right)^2$$
$$+4\left(1-\frac{(\theta/2)^2}{2}+\cdots\right)\left(-\frac{1}{2}+b_1\frac{\theta^2}{4}+\cdots\right) \quad (13.30)$$

and the terms containing b_1 drop out, leaving

$$0 = \left(-K-\frac{3}{8}+4\cdot\frac{1}{16}\right)\theta^2+\cdots$$
$$= \left(-K-\frac{1}{8}\right)\theta^2+\cdots \quad (13.31)$$

and thus we can only solve for $b(\theta)$ if $K=-\frac{1}{8}$, that is, this requires

$$\tilde{a}(\theta) = \theta^3 a(\theta) = \frac{\theta^3\sin\theta}{8}. \quad (13.32)$$

Thereafter, equating coefficients places no restriction on b_1 but requires each coefficient to be constrained as follows:

$$b_2 = \frac{1}{144}-\frac{b_1}{3!}, \quad (13.33)$$

$$b_3 = -\frac{1}{5{,}400}+\frac{b_1}{5!}, \quad (13.34)$$

$$b_4 = -\frac{11}{25{,}401{,}600}-\frac{b_1}{7!}, \quad (13.35)$$

and so on. The OEIS does not recognize these numbers (neither did we, but a little more work later pays off, as we will see).

Experimentally, by computing $p_{80}(-2+\frac{3}{2}\theta^2 4^{-80})$ to a ridiculous number of places, at $\theta = 10^{-50}$, adding $\cos\theta + 4^{-80}(\frac{1}{8}\theta^3\sin\theta\cdot 80+\frac{\theta^2}{2})$, we find that $b_1=\frac{3}{8}$, to beyond reasonable doubt. This, then, requires the sequence of b_k to start

$$\left[-\frac{1}{2},\frac{3}{8},-\frac{1}{18},\frac{127}{43{,}200},-\frac{1{,}901}{25{,}401{,}600},\cdots\right]. \quad (13.36)$$

This essentially completes our construction (we return to $3/8$ in a moment).

What we have proved is, if for all θ we have

$$p_k(-2+\frac{3}{2}\theta^2 4^{-k}) = -\cos\theta+\left(-\frac{1}{8}\theta^3\sin\theta\cdot k+\theta^2 b(\theta)\right)4^{-k}+O(4^{-2k}), \quad (13.37)$$

where $b(\theta) = -\frac{1}{2} + \frac{3}{8}\theta^2 - \frac{1}{18}\theta^4 + \frac{127}{43{,}200}\theta^6 + \cdots$ satisfies

$$b(\theta) = 4\cos\frac{\theta}{2} b\left(\frac{\theta}{2}\right) + \frac{1}{8}\theta\sin\theta + \frac{3}{2}\cos^2\frac{\theta}{2}, \tag{13.38}$$

then it follows that

$$\begin{aligned} & p_{k+1}(-2+\frac{3}{2}\theta^2 4^{-k-1}) \\ &\quad = -\cos\theta + \left(-\frac{1}{8}\theta^3\sin\theta\cdot(k+1) + \theta^2 b(\theta)\right)4^{-k-1} + O(4^{-2k-2}). \end{aligned} \tag{13.39}$$

13.3 The Mysterious 3/8 or $\frac{1}{2!}b''(0)$

By numerical evidence (high-precision finite differences on computer values of $p_k(z)$ for large values of k and values of z very near -2) we identified $b''(0) = \frac{3}{4}$. Later, we found the functional equation (13.38) but noted that while this equation requires $b(0) = -1/2$ and $b'(0) = 0$ all by itself, it leaves free a multiple of the homogeneous solution, namely $K\theta\sin\theta = K(\theta^2 - \theta^4/3! + \theta^6/5! - \cdots)$ and so cannot by itself determine $b''(0)$. What helps is a boundary condition, applied at $\theta = 0$, for all k. Using Eq. (13.37) we find by differentiating *four* times with respect to θ that

$$\frac{d^4}{d\theta^4} p_k(-2+\frac{3}{2}\theta^2 4^{-k})\Big|_{\theta=0} = -1 + 4^{-k}\cdot(-3k + 12b''(0)) + O(4^{-2k}). \tag{13.40}$$

But we already know how to differentiate $p_k(z)$ and so, using the chain rule,

$$\frac{d}{d\theta} p_k(\xi) = p_k'(\xi)\cdot(3\cdot 4^{-k}), \theta \tag{13.41}$$

$$\frac{d^2}{d\theta^2} p_k(\xi) = p_k'(\xi)\cdot(3\cdot 4^{-k}) + p_k''(\xi)\cdot(3\cdot 4^{-k})^2\theta^2, \tag{13.42}$$

$$\frac{d^3}{d\theta^3} p_k(\xi) = 3\cdot p_k''(\xi)\cdot(3\cdot 4^{-k})^2\theta + p_k'''(\xi)\cdot(3\cdot 4^{-k})^3\theta^3, \tag{13.43}$$

$$\frac{d^4}{d\theta^4} p_k(\xi) = 3\cdot p_k''(\xi)\cdot(3\cdot 4^{-k})^2 + 6\cdot p_k'''(\xi)\cdot(3\cdot 4^{-k})^3\theta^2 + p_k^{(iv)}(\xi)(3\cdot 4^{-k})^4\theta^4, \tag{13.44}$$

where, of course, $\xi = -2 + \frac{3}{2}\theta^2 4^{-k}$.

Therefore, at $\theta = 0$,

$$\left.\frac{d^4}{d\theta^4} p_k(\xi)\right|_{\theta=0} = 27 \cdot 4^{-2k} p_k''(-2)\,. \tag{13.45}$$

We have already worked out $p_k''(-2)$ which is

$$p_k''(-2) = -\frac{1}{27} 4^{2k} + \left(\frac{1}{3} - \frac{k}{9}\right) 4^k - \frac{8}{27}\,, \tag{13.46}$$

so $27 \cdot 4^{-2k} p_k''(-2) = -1 + (9 - 3k)4^{-k} - 8 \cdot 4^{-2k}$ forcing, for compliance, the equality of $-1 + (12b''(0) - 3k)4^{-k} + O(4^{-2k})$.

This in turn requires

$$12b''(0) = 9 \tag{13.47}$$

or

$$b''(0) = \frac{3}{4}\,. \tag{13.48}$$

This retrospective identification of $b''(0)$ fixes the choice of $K \cdot \theta \sin \theta$ and thereby $b(\theta)$. It seems to us remarkable that this non-standard initial condition suffices.

13.4 The Taylor Series for $b(\theta)$

Define the numbers b_n by $b_0 = -1/2$, $b_1 = 3/8$, and for $n \geq 2$ by

$$b_n = \frac{1}{2^{2n} - 4} \left[4 \sum_{\ell=1}^{n} \frac{(-1)^\ell}{(2\ell)!} b_{n-\ell} + \frac{3}{4} \frac{(-1)^n 2^{2n}}{(2n)!} + \frac{(-1)^{n-1} 2^{2n-1}}{4(2n-1)!} \right]. \tag{13.49}$$

This recurrence relation gives a sequence starting

$$\left[-\frac{1}{2}, \frac{3}{8}, -\frac{1}{18}, \frac{127}{43{,}200}, -\frac{1{,}901}{25{,}401{,}600}, \cdots \right]. \tag{13.50}$$

Lemma 13.1. *If* $b_n = \frac{(-1)^{n-1} a_n}{(2n-2)!}$, *then* $|a_n| < 1$, *for* $n \geq 1$.

Proof. If $n = 1$, $b_1 = \frac{3}{8} = \frac{(-1)^{1-1} \cdot a_1}{0!}$ so $a_1 = \frac{3}{8} < 1$.

error analysis and some preliminary tests that successfully solve problems where the CQ fails and also problems that have a duality gap. The optimal value of our (AP) has significance as a measure of *nearness to infeasibility*.

Though our stable (AP) satisfied both the primal and dual generalized Slater CQ, high accuracy solutions were difficult to obtain for unstructured general problems. (AP) is equivalent to the underdetermined linear least squares problem

$$\min \|\mathscr{A}_C(D)\|_2^2 \quad \text{s.t.} \quad \langle I, D\rangle = \sqrt{n}, \quad D \succeq 0, \tag{12.79}$$

which is known to be difficult to solve. High accuracy solutions are essential in performing a proper facial reduction.

Extensions of some of our results can be made to general conic convex programming, in which case the partial orderings in (12.1) and (12.2) are induced by a proper closed convex cone K and the dual cone K^*, respectively.

Acknowledgements Research of the first and the third authors is supported by The Natural Sciences and Engineering Research Council of Canada and by TATA Consultancy Services. Research of the second author is supported by The Natural Sciences and Engineering Research Council of Canada. The authors thank the referee as well as Gábor Pataki for their helpful comments and suggestions.

References

1. Alfakih, A., Khandani, A., Wolkowicz, H.: Solving Euclidean distance matrix completion problems via semidefinite programming. Computational optimization–a tribute to Olvi Mangasarian, Part I. Comput. Optim. Appl. **12**(1–3), 13–30 (1999)
2. Alipanahi, B., Krislock, N., Ghodsi, A.: Manifold learning by semidefinite facial reduction. Technical Report Submitted to Machine Learning Journal, University of Waterloo, Waterloo, Ontario (2010)
3. Alizadeh, F., Haeberly, J.-P.A., Overton, M.L.: Complementarity and nondegeneracy in semidefinite programming. Math. Program. **77**, 111–128 (1997)
4. Anjos, A.F., Lasserre, J.B. (eds.): Handbook on Semidefinite, Conic and Polynomial Optimization. International Series in Operations Research & Management Science. Springer, New York (2011)
5. Anjos, M.F., Wolkowicz, H.: Strengthened semidefinite relaxations via a second lifting for the Max-Cut problem. Foundations of heuristics in combinatorial optimization. Discrete Appl. Math. **119**(1–2), 79–106 (2002)
6. Ben-Israel, A., Ben-Tal, A., Zlobec, S.: Optimality in Nonlinear Programming: A Feasible Directions Approach. A Wiley-Interscience Publication, New York (1981)
7. Ben-Israel, A., Charnes, A., Kortanek, K.: Duality and asymptotic solvability over cones. Bull. Amer. Math. Soc. **75**(2), 318–324 (1969)
8. Bonnans, J.F., Shapiro, A.: Perturbation Analysis of Optimization Problems. Springer Series in Operations Research. Springer, New York (2000)
9. Borchers, B.: CSDP, a C library for semidefinite programming. Optim. Methods Soft. **11/12**(1–4), 613–623 (1999). projects.coin-or.org/Csdp
10. Borwein, J.M., Wolkowicz, H.: Characterization of optimality for the abstract convex program with finite-dimensional range. J. Austral. Math. Soc. Ser. A **30**(4), 390–411 (1980/1981)

Assume $|a_j| < 1$ for $1 \le j \le k-1$. Then

$$b_k = \frac{(-1)^{k-1}a_k}{(2k-2)!} = \frac{1}{2^{k-4}-4}\left[4\sum_{j=1}^{k-1}\frac{(-1)^j(-1)^{k-j-1}a_{k-j}}{(2j)!(2(k-j)-2)!} \right.$$
$$\left. +4\frac{(-1)^k(-1/2)}{(2k)!} + \frac{3}{4}\frac{(-1)^k 2^{2k}}{(2k)!} + \frac{(-1)^{k-1}2^{2k-1}}{4(2k-1)!} \right] \tag{13.51}$$

multiplying through by $(-1)^{k-1}(2k-2)!$,

$$a_k = \frac{1}{2^{2k}-4}\left[4\sum_{j=1}^{k-1}\frac{(2k-2)!a_{k-j}}{(2j)!(2(k-j)-2)!} + \frac{2}{(2k)(2k-1)} \right.$$
$$\left. + \frac{2^{2k-3}}{(2k-1)} - \frac{2\cdot 2^{2k-2}}{(2k)(2k-1)} \right]. \tag{13.52}$$

The triangle inequality gives

$$|a_k| < \frac{4}{4^k-4}\sum_{j=1}^{k-1}\frac{(2k-2)!}{(2j)!(2(k-j)-2)!} + \left|\frac{2^{2k-3}}{2k-1} + \frac{2-3\cdot 2^{2k-2}}{(2k)(2k-1)}\right|\frac{1}{4^k-4}. \tag{13.53}$$

Using Maple, we find that

$$\sum_{j=1}^{k-1}\frac{(2k-2)!}{(2j)!(2(k-j)-2)!} = \frac{1}{2}4^{k-1} - 1, \tag{13.54}$$

so for $k \ge 2$

$$|a_k| < \frac{1}{4^{k-1}-1}\cdot\left(\frac{1}{2}4^{k-1}-1\right) + \frac{1}{4^k-4}\left|\frac{2-3\cdot 4^{k-1}}{(2k)(2k-1)}\right| + \frac{\frac{1}{2}4^{k-1}}{(2k-1)(4^k-4)}. \tag{13.55}$$

The last term is less than $\frac{1}{6}\frac{1}{(4-4^{-k})} < \frac{1}{12}$, and the second last term is less than $\frac{2}{4^k}\frac{3\cdot 4^{k-1}}{4\cdot 3} = \frac{1}{8}$, so

$$|a_k| < \frac{1}{2} + \frac{1}{8} + \frac{1}{12} < 1. \tag{13.56}$$

■

Corollary 13.2. *The series*

$$b(\theta) = -\frac{1}{2} + \frac{3}{8}\theta^2 + \sum_{k\geq 2} b_k \theta^{2k} \tag{13.57}$$

defines an entire function.

Remark 13.3. Solving the equation

$$b(\theta) = 4\cos\frac{\theta}{2}\cdot b\left(\frac{\theta}{2}\right) + \frac{1}{8}\theta\sin\theta + \frac{3}{2}\cos^2\frac{\theta}{2} \tag{13.58}$$

subject to $b''(0) = 3/4$, in series, gives the recurrence relation (13.49). The process is tedious but straightforward.

Theorem 13.4. *Suppose $n = 2^\ell q$ and q is odd. Then*

$$b(n\pi) = \frac{2^{2\ell}-1}{2}. \tag{13.59}$$

Proof.

$$b(q\pi) = 4\cos\left(\frac{q\pi}{2}\right)b\left(\frac{q\pi}{2}\right) + \frac{1}{8}(q\pi)\sin(q\pi) + \frac{3}{2}\cos^2\left(\frac{q\pi}{2}\right) = 0 \tag{13.60}$$

since $\cos\left(\frac{q\pi}{2}\right) = \sin(q\pi) = 0$.

Then this is $\ell = 0$, the base of the induction. Suppose the theorem is true for $\ell = n$. Consider

$$b(2^{n+1}q\pi) = 4\cos(2^n q\pi)b(2^n q\pi) + \frac{1}{8}(2^{n+1}q\pi)\sin(2^{n+1}q\pi) + \frac{3}{2}\cos^2(2^n q\pi) \tag{13.61}$$

$$= \begin{cases} -4\cdot 0 + \frac{3}{2} & \text{if } n = 0 \\ 4\cdot\left(\frac{2^{2n}-1}{2}\right) + \frac{3}{2} & \text{if } n > 0 \end{cases} \tag{13.62}$$

because $\cos^2(2^n q\pi) = 1$. Then

$$b(2^{n+1}q\pi) = \left\{\begin{array}{ll} \frac{2^2-1}{2} & \text{if } n = 0 \\ \frac{2^{2(n+1)}-1}{2} & \text{if } n > 0 \end{array}\right\} \tag{13.63}$$

and the theorem is proved by induction. ∎

The above theorem suggests that $b(\theta)$ grows at most quadratically. We prove something a little weaker, below; it is likely that the function really does grow only quadratically, but the following theorem suffices for the purposes of this paper.

Theorem 13.5. $|b(\theta)| < (\frac{3}{8} + \frac{25}{8}\lfloor \log_2 \theta \rfloor) \cdot \theta^2 + \frac{1}{2}$, *for all* $\theta > 1$. *That is,* $b(\theta)$ *grows moderately slowly for real* θ.

Proof. Notice that the series alternate in sign and decrease; hence, on $0 < \theta \leq 1$, $|b(\theta)| < \frac{1}{2} + \frac{3}{8} \cdot \theta^2$. This gives a base for the following induction.

Suppose the theorem is true on $0 < \theta \leq 2^\ell$. Now consider the functional equation, expressed in terms of 2θ:

$$b(2\theta) = 4\cos(\theta) \cdot b(\theta) + \frac{\theta \sin(2\theta)}{4} + \frac{3}{2}\cos^2(\theta)\,. \tag{13.64}$$

Taking absolute values and using the triangle inequality,

$$|b(2\theta)| < 4\left(\frac{1}{2} + \left(\frac{3}{8} + \frac{25\ell}{8}\right)\theta^2\right) + \frac{\theta}{4} + \frac{3}{2} \tag{13.65}$$

or

$$|b(2\theta)| < \left(\frac{3}{8} + \frac{25\ell}{8}\right)(2\theta)^2 + \frac{(2\theta)}{8} + 3 + \frac{1}{2}\,. \tag{13.66}$$

■

Since $2\theta > 1$, without loss of generality, $3/8 + 25/8(\ell+1) > 3/8 + 25/8\ell + 1/(8|2\theta|) + 3/(2\theta)^2$, and this establishes the truth of the theorem for $2^\ell < \theta \leq 2^{\ell+1}$.

13.5 Implications

We have now shown that

$$p_k(-2 + \frac{3}{2}\theta^2 \cdot 4^{-k}) = -\cos\theta + \left(-\frac{\theta^3 \sin\theta}{8}k + \theta^2 b(\theta)\right)4^{-k} + O(4^{-2k}) \tag{13.67}$$

(strictly speaking, we have not given a bound for the $O(4^{-2k})$ term; there is work that remains, here, but at this point there is little doubt), where $b(\theta)$ is entire and satisfies the functional equation (13.38), as well as a growth bound of the form $|b(\theta)| \leq M(\theta) \cdot \theta^2$. Thus, for a fixed large k, if $M(\theta) \cdot \theta^4 < 2^k$, say, then

$$\left| p_k(-2 + \frac{3}{2}\theta^2 \cdot 4^{-k}) + \cos\theta \right| < O(2^{-k})\,. \tag{13.68}$$

13.5.1 Several Largest Roots

This formula approximately locates several zeros of $p_k(z)$, near $\theta = (2\ell+1)\pi/2$, so long as $(2\ell+1)\pi/2 < O(2^{k/4})$, roughly; the larger k is, the more zeros we locate. Explicitly, the zeros are near

$$z_{k,\ell} = -2 + \frac{3}{2}\left(\frac{(2\ell+1)\pi}{2}\right)^2 4^{-k} + O(4^{-2k}) \tag{13.69}$$

because here $|p_k(z_{k,\ell})| < O(2^{-k})$.

Interestingly, this formula is already accurate for $k = 1$. Recall that $p_0(z) = 1$ and $p_1(z) = z + 1$, which has a root at $z = -1$. The formula predicts

$$\begin{aligned} z_{1,0} &= -2 + \frac{3}{2}\left(\frac{\pi}{2}\right)^2 4^{-1} + O(4^{-2}) \\ &= -2 + \frac{3\pi^2}{32} \\ &= -2 + 0.92527 \\ &= -1.0747, \end{aligned} \tag{13.70}$$

which is remarkably accurate for such small k. For $k = 2$, we have $p_2(z) = z(z+1)^2 + 1 = z(z^2 + 2z + 1) + 1 = z^3 + 2z^2 + z + 1$, and the largest magnitude zero is predicted to be

$$\begin{aligned} z_{2,0} &= -2 + \frac{3}{2}\left(\frac{\pi}{2}\right)^2 4^{-2} + O(4^{-4}) \\ &= -2 + 0.2313188 \\ &= -1.768681. \end{aligned} \tag{13.71}$$

The true largest root is (see also Fig. 13.1 and Table 13.2)

$$-1/6\sqrt[3]{100 + 12\sqrt{69}} - 2/3\frac{1}{\sqrt[3]{100 + 12\sqrt{69}}} - 2/3 \approx -1.75487766624670. \tag{13.72}$$

13.5.2 Newton-Like Improvements

We may improve the accuracy of these estimates by using Newton's method, as follows. Note that we will need $p_k'(z)$ at $z = z_{k,\ell}$, but this will be simple, using the formula ($\xi = -2 + \frac{3}{2}\theta^2 4^{-k}$) so $\xi'(\theta) = 3\theta 4^{-k}$,

Fig. 13.1 Error in approximating the largest three roots of $p_k(z)$. *Circles*, *squares*, and *crosses* are the largest, second largest, and third largest roots, respectively

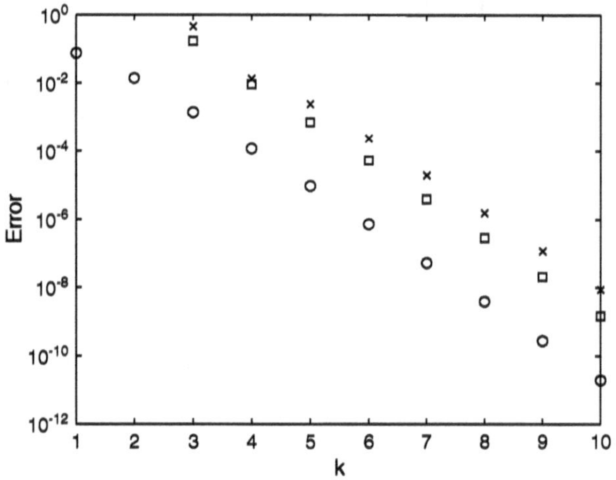

Table 13.2 Largest three roots of $p_k(z)$

k	z_0	z_1	z_2
1	−1	–	–
2	−1.7548776662466928	–	–
3	−1.9407998065294848	−1.3107026413368329	−1
4	−1.9854242530542053	−1.8607825222048549	−1.6254137251233037
5	−1.9963761377111938	−1.9667732163929287	−1.9072800910653020
6	−1.9990956823270185	−1.9918141725491222	−1.9771795870062574
7	−1.9997740486937273	−1.9979629155977143	−1.9943329667155349
8	−1.9999435217656740	−1.9994914380163981	−1.9985865888422069
9	−1.9999858811403921	−1.9998729117663291	−1.9996469177332729
10	−1.9999964703350087	−1.9999682317097476	−1.9999117502085037

$$\frac{d}{d\theta}p_k(\xi(\theta)) - \sin(\theta) = p_k'(\xi(\theta))\xi'(\theta) - \sin(\theta)$$
$$= \left(-\frac{1}{8}(3\theta^2\sin\theta + \theta^3\cos\theta)k + 2\theta b(\theta) + \theta^2 b'(\theta)\right)4^{-k} + O(4^{-2k}). \tag{13.73}$$

At $\theta = (2\ell+1)\pi/2$, the derivative of $p_k(z)$ is therefore

$$(-1)^\ell + \left(-\frac{1}{8}\left(\frac{3}{4}(2\ell+1)^2\pi^2\right)k + (2\ell+1)\pi b\left(\frac{(2\ell+1)\pi}{2}\right)\right.$$
$$\left. + \left(\frac{(2\ell+1)\pi}{2}\right)^2 b'\left(\frac{(2\ell+1)\pi}{2}\right)\right)4^{-k} + O(4^{-2k}). \tag{13.74}$$

Even without the $O(4^{-k})$ term, this gives us

$$p_k'(\xi(\theta)) = \frac{4^k}{3\theta}\sin\theta + O(1)$$
$$= \frac{4^k}{3(2\ell+1)\pi/2}(-1)^\ell + O(1), \tag{13.75}$$

which is enough to improve the accuracy of $z_{k,\ell}$ to $O(4^{-3k})$. If we work harder and include the next term (which requires us to compute $b'(\theta)$, not that this is so very hard), then we get something better, but with just this estimate $p_k'(-2+3\theta^2 4^{-k}/2) \approx -\sin\theta$, we get

$$z_{k,\ell}^{(1)} = z_{k,\ell} - \frac{p_k(z_{k,\ell})}{p_k'(z_{k,\ell})} \tag{13.76}$$

$$= -2 + \frac{3}{2}\theta^2 4^{-k} - \frac{(-\cos\theta + 4^{-k}(\frac{\theta^3}{8}\sin\theta k + \theta^2 b(\theta)))}{\sin\theta 4^k/3 + O(1)} \tag{13.77}$$

and if $\theta = (2\ell+1)\pi/2$,

$$= -2 + \frac{3}{2}\left(\frac{(2\ell+1)\pi}{2}\right)^2 4^{-k} - 3(-1)^\ell 4^{-2k}\left(\frac{\theta^3}{8}\sin\theta k + \theta^2 b(\theta)\right) + O(4^{-3k}). \tag{13.78}$$

For $k = 1$, this gives, with $\ell = 0$,

$$z_{1,0}^{(1)} = -2 + \frac{3}{2}\left(\frac{\pi}{2}\right)^2 4^{-1} - 3\left(-\frac{(\pi/2)^3}{8}\cdot 1 + \left(\frac{\pi}{2}\right)^2 b\left(\frac{\pi}{2}\right)\right)\cdot 4^{-2} \tag{13.79}$$

and, since $b(\pi/2) = 0.1285353$, this gives $z_{1,0}^{(1)} = -1.0434$, which is better than before. Of course, an accurate Newton step, using either $p_k'(z)$ directly or the formula with $b'(\theta)$ above, is better yet: also of course, this formula is better for larger k.

13.5.3 Interlacing

The largest roots, as predicted by this formula, have a curious interlacing property: between every root of $p_k(z)$ there are two roots of $p_{k+1}(z)$ and sometimes one of $p_{k-1}(z)$, as can be seen by graphing $-\cos(\theta/2)$, $-\cos(\theta)$, and $-\cos(2\theta)$ on the same graph.

Compare $p_7(-2+3/2\cos\theta^2 4^{-7}) = p_7(-2+3/2(2\theta)^2\cdot 4^{-8})$, $p_8(-2+3/2\theta^2\cdot 4^{-8})$ and $p_9(-2+3/2(\theta/2)^2\cdot 4^{-8})$. See Figs. 13.2 and 13.3. Furthermore, the largest root of $p_{k+1}(z)$ must lie between -2 and the largest root of $p_k(z)$.

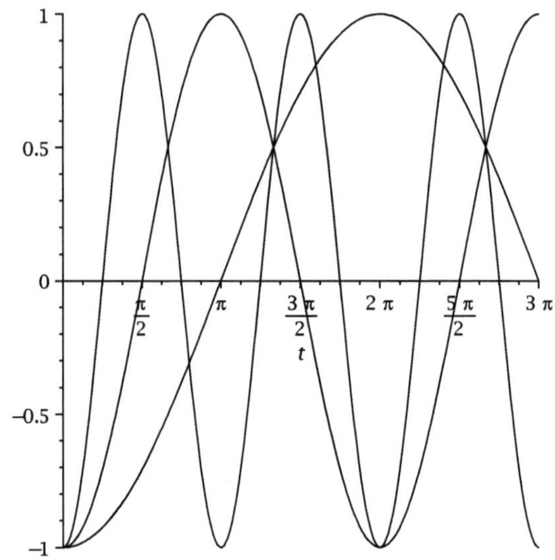

Fig. 13.2 $-\cos(\theta/2)$, $-\cos(\theta)$, and $-\cos(2\theta)$

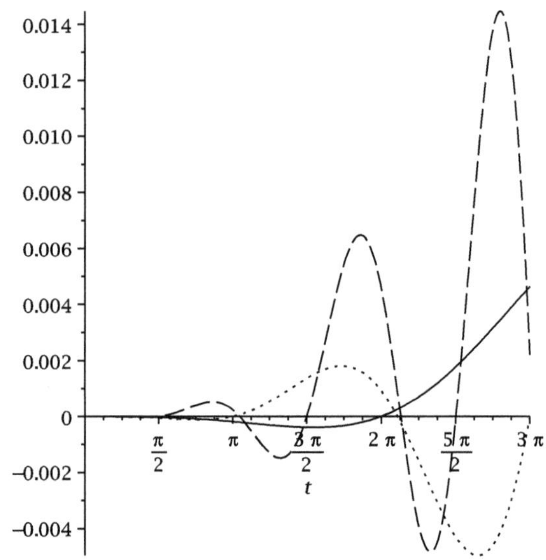

Fig. 13.3 The difference between $p_7(-2+3/2(2\theta)^2 4^{-8})$, $p_8(-2+3/2\theta^2 4^{-8})$, and $p_9(-2+3/2(\theta/2)^2 4^{-8})$ and the cosines in Fig. 13.2. The *solid line* is $k=7$, the *dotted line* is $k=8$, and the *dashed line* is $k=9$

13.5.4 Barycentric Weights

In [5] the authors consider eigenvalue methods based on Lagrange interpolation at the nodes $z=0$, $z=-2$, and $z=\xi_{k-1,j}$ for $1 \le j \le 2^{k-1}-1$, where $p_{k-1}(\xi_{k-1,j})=0$, and on Hermite interpolation at the same nodes. It turns out to matter greatly how widely varying the weights are: the condition number for evaluation (and rootfinding) is proportional to the ratio of the maximum barycentric weight to the minimum barycentric weight and inversely proportional to the minimum node separation raised to the maximum confluency. See [2–4].

Here, we can explicitly compute some of these barycentric weights, which are really just coefficients in partial fraction expansions. The expansions we need are

$$\frac{1}{zp_{k-1}(z)(z+2)} = \frac{\gamma_0}{z} + \sum_{j=1}^{2^{k-1}-1} \frac{\gamma_j}{(z-\xi_{k-1,j})} + \frac{\gamma_N}{(z+2)} \tag{13.80}$$

and

$$\frac{1}{zp_{k-1}^2(z)(z+2)} = \frac{\gamma_0}{z} + \sum_{j=1}^{2^{k-1}-1} \left(\frac{\gamma_{j,0}}{(z-\xi_{k-1,j})} + \frac{\gamma_{j,1}}{(z-\xi_{k-1,j})^2} \right) + \frac{\gamma_N}{(z+2)}. \tag{13.81}$$

It is easy to see that $\gamma_0 = \gamma_N = 1/2$, because $p_{k-1}(0) = 1$ and (for $k > 0$), p_{k-1} $(-2) = -1$.

Using our formulas for the roots $\xi_{k-1,\ell} \doteq -2+3((2\ell+1)/2)^2 4^{-k+1}$ (changing our notation somewhat), we can show that the weights in the Lagrange case are $-4/(2\ell+1)^2/\pi^2$ (for zeros near the maximum point -2), whereas for the Hermite case their ratios vary by a factor 4^k (which is not as bad as it seems—that is, after all, just the square of the degree of the polynomial, more or less).

This matters more because the minimum node separation also goes like 4^{-k}, and hence the condition number must be at least $O(4^{3k})$ for the Hermite case, whereas for the Lagrange case we have only shown $O(4^k)$. The other weights also matter, and experimental work (not given here) shows that their variation is somewhat greater but that the conclusion still holds: the Lagrange basis is better-conditioned.

13.6 More About the Functional Equation for $b(\theta)$

The functional equation (13.38) can be simplified. Note first that the homogeneous equation

$$b(\theta) = 4\cos\frac{\theta}{2} b\left(\frac{\theta}{2}\right) \tag{13.82}$$

has solution $H(\theta) = K\theta\sin\theta$:

$$K\theta\sin\theta = 4\cos\frac{\theta}{2}\cdot K\cdot\frac{\theta}{2}\cdot\sin\frac{\theta}{2} = K\cdot\theta\cdot 2\cdot\sin\frac{\theta}{2}\cdot\cos\frac{\theta}{2} = K\cdot\theta\cdot\sin\theta. \tag{13.83}$$

If we use the homogeneous solution, $\theta\sin\theta$, to try to identify $b(\theta)$ using variation of parameters, we are led to look at the change of variable $b(\theta) = \theta\sin\theta B(\theta)$. This gives

$$\theta\sin\theta B(\theta) = 4\cos\left(\frac{\theta}{2}\right)\left(\frac{\theta}{2}\right)\sin\left(\frac{\theta}{2}\right)B\left(\frac{\theta}{2}\right)+\frac{1}{8}\theta\sin\theta+\frac{3}{2}\cos^2\left(\frac{\theta}{2}\right)$$

$$= \theta\sin\theta\left(B\left(\frac{\theta}{2}\right)+1/8+\frac{3}{8}\frac{\cot\left(\frac{\theta}{2}\right)}{\left(\frac{\theta}{2}\right)}\right). \tag{13.84}$$

Dividing by $\theta\sin\theta$, this leads to the curious series

$$B(\theta) = \frac{N}{8}+\frac{3}{8}\left(\sum_{\ell=1}^{N}\frac{\cot(\theta/2^\ell)}{\theta/2^\ell}\right)+B\left(\frac{\theta}{2^N}\right). \tag{13.85}$$

One is tempted to let $N\to\infty$, but this series (obviously) diverges. But we know quite a bit about $b(\theta)$ for small θ, namely that its Taylor series begins $-1/2+3/8\theta^2-1/18\theta^4+\cdots$, and hence we know how $B(\theta/2^N)$ behaves for large N:

$$B(\theta/2^N) = \frac{1}{\theta/2^N\sin\theta/2^N}\left(-\frac{1}{2}+\frac{3}{8}\frac{\theta^2}{2^{2N}}-\frac{1}{18}\frac{\theta^4}{2^{4N}}+\cdots\right)$$

$$= -\frac{1}{2\theta^2}2^{2N}+\frac{7}{24}+O(2^{-2N}). \tag{13.86}$$

This is enough terms to identify and cancel all the singular parts as $N\to\infty$, in a rigorous form of renormalization, as we will see.

We also need the (well-known) series for the cotangent function:

$$\cot\theta = \sum_{m\geq 0}(-1)^m\frac{B_{2m}4^m}{(2m)!}\theta^{2m-1}, \tag{13.87}$$

where here B_{2m} represents the $2m$th Bernoulli number. Using this, we can expand $\cot(\theta/2^\ell)/(\theta/2^\ell)$ in Taylor series to get

$$\sum_{\ell=1}^{N}\frac{\cot(\theta/2^\ell)}{\theta/2^\ell} = \sum_{m=0}^{\infty}(-1)^m\frac{B_{2m}4^m}{(2m)!}\sum_{\ell=1}^{N}\left(\theta/2^\ell\right)^{2m-2}, \tag{13.88}$$

where we have interchanged the order of the (finite) first sum with the (absolutely convergent if $|\theta|<\pi$) second sum. But now the inner sum is a finite geometric series, which we can identify (even in the cases where the ratio is 1, when $m=1$ and the sum is just $-N/3$, and larger than 1, when $m=0$). If m is not 1,

$$\sum_{\ell=1}^{N}\left(\theta/2^\ell\right)^{2m-2} = \theta^{2m-2}\frac{\left(1-1/2^{N(2m-2)}\right)}{2^{2m-2}-1} \tag{13.89}$$

and putting this all together (not forgetting the factor $3/8$), we have

$$B(\theta) = \frac{N}{8} + \left(-\frac{N}{8} + \frac{1}{2}\frac{2^{2N}}{\theta^2} - \frac{1}{2\theta^2} + \frac{3}{8}\sum_{m=2}^{\infty} \frac{(-1)^m B_{2m} 4^m}{(2m)!(2^{2m-2}-1)}\theta^{2m-2} \right)$$
$$-\frac{1}{2}\frac{2^{2N}}{\theta^2} + \frac{7}{24} + O(4^{-N}), \tag{13.90}$$

which, as $N \to \infty$, gives us an explicit expression for the series coefficients of $B(\theta)$ and hence $b(\theta)$:

$$b(\theta) = -\frac{1}{2}\frac{\sin\theta}{\theta} + \frac{7}{24}\theta\sin\theta + \frac{3}{8}\theta\sin\theta \sum_{m\geq 2} \frac{(-1)^m B_{2m} 4^m}{(2m)!(2^{2m-2}-1)}\theta^{2m-2}. \tag{13.91}$$

This is still not a closed form, but somehow it is more satisfying than the recurrence relation. One could go a little further yet and write down an explicit finite sum containing factorials and Bernoulli numbers, for each Taylor coefficient of $b(\theta)$, by multiplying the known series for $\theta \sin\theta$ explicitly. But we have to lay the pen down somewhere, and it may as well be here.

13.7 Concluding Remarks

There remains quite a bit to be done. It seems obvious that we can compute many more terms in this expansion:

$$p_k(-2 + \frac{3}{2}\theta^2 4^{-k}) = -\cos\theta + \sum_{\ell\geq 1} v_k(\theta) 4^{-\ell k}, \tag{13.92}$$

where each $v_k(\theta)$ will be a polynomial of degree ℓ, in k, and give various linear inhomogeneous functional equations for the coefficients of those polynomials. This series might even be convergent.

One can ask about the *smallest* root, instead. Preliminary computations using a homotopy method to find the smallest root, from the previous iteration's smallest root, suggest (but only suggest, as we have only computed about 4 digits) that the smallest roots begin $s_k = 1/4 + \pi^2/k \pm i\beta/k^2 + \cdots$ for some real β near 20. The π^2 is very speculative and may well be completely wrong. Again, we leave that pursuit for another day.

However, the computations that we *have* done, using the OEIS, helped to discover an analytic formula for a family of roots of a nonlinear recurrence relation; more, a recurrence relation that has roots that approach a fractal boundary as $k \to \infty$. One wonders if any other interesting nonlinear equations are also susceptible of such a treatment.

Acknowledgements The discussion with Neil Calkin, during the course of the JonFest DownUnder conference, was very helpful. One might even say that had those discussions not taken place, RMC would not have been bitten (by the problem). Peter Borwein pointed out the work of Aczél on functional differential equations. This research work was partially supported by the Natural

Sciences and Engineering Research Council of Canada, the University of Western Ontario (and in particular by the Department of Applied Mathematics), the Australian National University (special thanks to the Coffee Gang, Joe Gani, Mike Osborne, David Heath, Ken Brewer, and the other regulars, for their interest), and by the University of Newcastle.

References

1. Aczél, J.: On history, applications and theory of functional equations. In: Functional Equations: History, Applications and Theory. D. Reidel, Dordrecht (1984)
2. Corless, R.M., Fillion, N.: A graduate introduction to numerical methods. Springer, to appear (2013)
3. Lawrence, P.W., Corless, R.M.: Stability of rootfinding for barycentric Lagrange interpolants. (submitted)
4. Lawrence, P.W., Corless, R.M.: Numerical stability of barycentric Hermite root-finding. In: Proceedings of the 4th International Workshop on Symbolic-Numeric Computation, pp. 147–148. San Jose, USA (2011)
5. Lawrence, P.W., Corless, R.M., Jeffrey, D.J.: Mandelbrot polynomials and matrices (2012) (in preparation)
6. Sloane, N.J.A.: An on-line version of the encyclopedia of integer sequences. Electron. J. Combin. **1**, 1–5 (1994). http://oeis.org/

Chapter 14
On the Fractal Distribution of Brain Synapses

Richard Crandall†

Abstract Herein we present mathematical ideas for assessing the fractal character of distributions of brain synapses. Remarkably, laboratory data are now available in the form of actual three-dimensional coordinates for millions of mouse-brain synapses (courtesy of Smithlab at Stanford Medical School). We analyze synapse datasets in regard to statistical moments and fractal measures. It is found that moments do not behave as if the distributions are uniformly random, and this observation can be quantified. Accordingly, we also find that the measured fractal dimension of each of two synapse datasets is 2.8 ± 0.05. Moreover, we are able to detect actual neural layers by generating what we call probagrams, paramegrams, and fractagrams—these are surfaces one of whose support axes is the y-depth (into the brain sample). Even the measured fractal dimension is evidently neural-layer dependent.

Key words: Brain • Fractals • Neural science • Synapses

Mathematics Subject Classifications (2010): Primary 28Axx, 28A80, 28A25; Secondary 65Cxx, 65C20.

COMMUNICATED BY HEINZ H. BAUSCHKE.

R. Crandall† (1947–2012) (✉)

D.H. Bailey et al. (eds.), *Computational and Analytical Mathematics*, Springer Proceedings in Mathematics & Statistics 50, DOI 10.1007/978-1-4614-7621-4_14,

14.1 Motivation

Those who study or delight in fractals know full well that often the fractal nature is underscored by structural rules. When the author was informed by colleagues[1] that 3D synapse data is now available in numerical form, it loomed natural that mathematical methods should be brought to bear.

Thus we open the discussion with the following disclaimer: The present paper is not a neurobiological treatise of any kind. It is a mathematical treatise. Moreover, there is no medical implication here, other than the possibility of using such measures as we investigate for creation of diagnostic tools.[2]

There is some precedent for this kind of mathematical approach. Several of many fractal studies on neurological structures and signals include [8–10]. on random point-clouds per se have even been suggested for the stringent testing of random-number generators [7]. Some researchers have attempted to attribute notions of context-dependent processing, or even competition to the activity within neural layers [1]. Indeed, it is known that dendrites—upon which synapses subsist—travel through layers. Some good rendition graphics are found in [16]. Again, our input datasets do not convey any information about dendritic structure; although, it could be that deeper analysis will ultimately be able to suggest dendritic presence [17].

14.2 Synapse Data for Mathematical Analysis

Our source data is in the section Appendix: Synapse datasets. It is important to note that said data consists exclusively of triples (x,y,z) of integers, each triple locating a single brain synapse, and we rescale to nanometers to yield physically realistic point-clouds. There is no neurological structure per se embedded in the data. This lack of structural information actually allows straightforward comparison to random point-clouds (Fig. 14.1).

To be clear, each synapse dataset has the form

$$x_0 \ y_0 \ z_0$$

$$x_1 \ y_1 \ z_1$$

[1]From Smithlab, of Stanford Medical School [15].

[2]Indeed, one motivation for high-level brain science in neurobiology laboratories is the understanding of such conditions as Alzheimer's syndrome. One should not rule out the possibility of "statistical" detection of some brain states and conditions—at least, that is our primary motive for bringing mathematics into play.

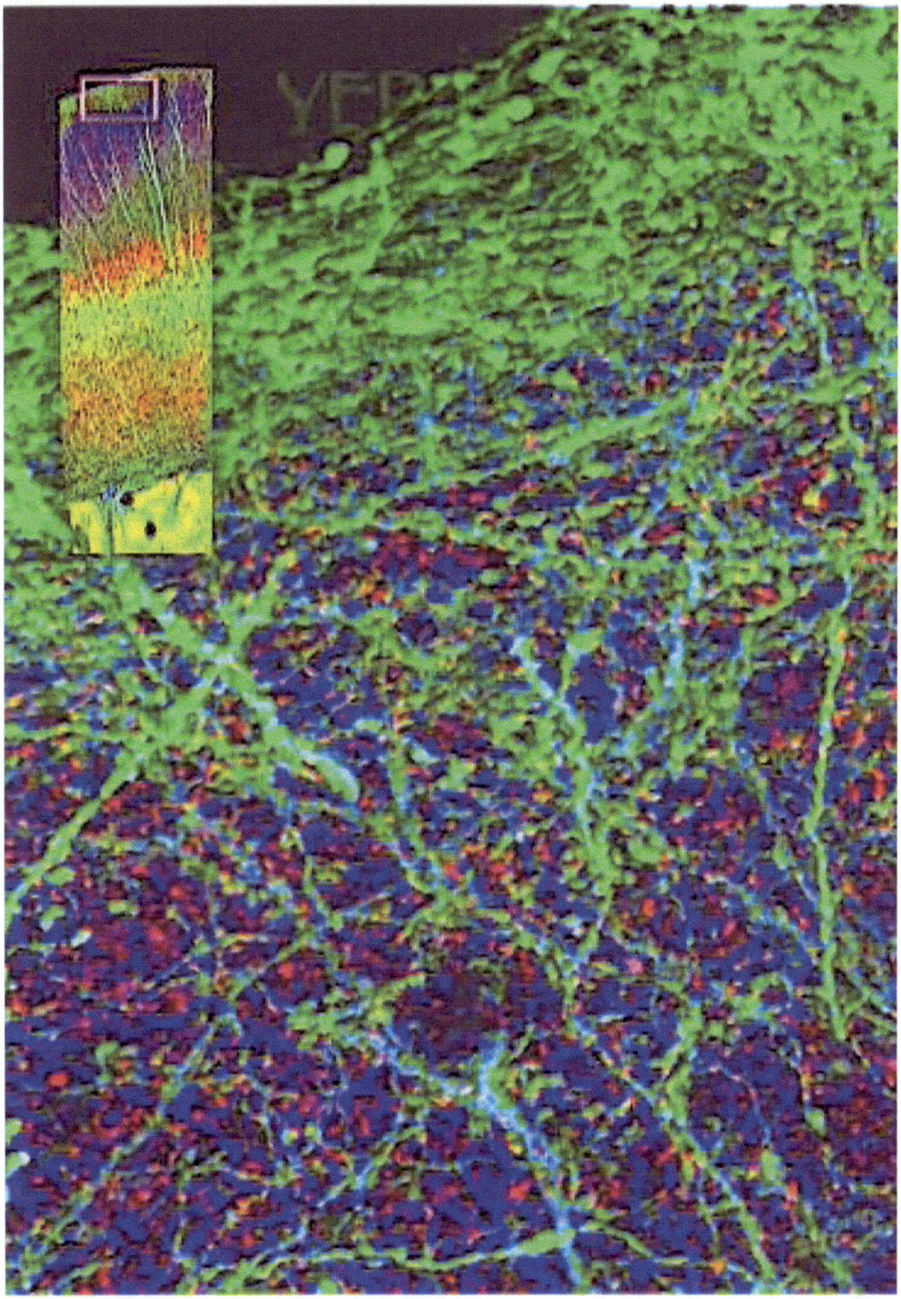

Fig. 14.1 Frame from video: The beginning (*top layer*, $y \sim 0$) of a mouse-brain section. Synapses (our present data of interest) are *red points*. The *vertical strip at upper left* represents the complete section—the *small light-pink rectangle* indicates the region we are currently seeing in the video (courtesy of Smithlab, Stanford medical school [15])

$$\cdots$$

$$x_j \ y_j \ z_j \qquad (= \mathbf{r})$$

$$\cdots$$

$$x_k \ y_k \ z_k \qquad (= \mathbf{q})$$

$$\cdots$$

$$x_{N-1} \ y_{N-1} \ z_{N-1},$$

where each x, y, z is an integer (Appendix 1 gives the nanometer quantization). There are N points, and we have indicated symbolically here that we envision some row as point $\mathbf{r}$ and some other row as point $\mathbf{q}$, for the purposes of statistical analysis (Fig. 14.2). (A point $\mathbf{r}$ may or may not precede a $\mathbf{q}$ on the list, although in our calculations we generally enforce $\mathbf{r} \neq \mathbf{q}$ to avoid singularities in some moments.)

14.3 The Modern Theory of Box Integrals

Box integrals—essentially statistical expectations, also called moments, over a unit box rather than over all of space—have a rich, decades-long history (see [2,3,5] and historical references therein). The most modern results involve such functions as

$$\Delta_n(s) := \langle |\mathbf{r} - \mathbf{q}| \rangle|_{\mathbf{r},\mathbf{q} \in [0,1]^n}$$

$$= \int_0^1 \cdots \int_0^1 \left(\sum_{k=1}^n (r_k - q_k)^2 \right)^{s/2} \mathrm{d}r_1 \mathrm{d}q_1 \mathrm{d}r_2 \mathrm{d}q_2 \cdots \mathrm{d}r_n \mathrm{d}q_n.$$

This can be interpreted physically as the expected value of v^s, where separation $v = |\mathbf{v}|$, $\mathbf{v} := \mathbf{r} - \mathbf{q}$ is the distance between two uniformly random points each lying in the unit n-cube (Fig. 14.3).

It is of theoretical interest that $\Delta_n(s)$ can be given a closed form for every integer s, in the cases $n = 1,2,3,4,5$ [5]. For example, the expected distance between two points in the unit 3-cube is given exactly by

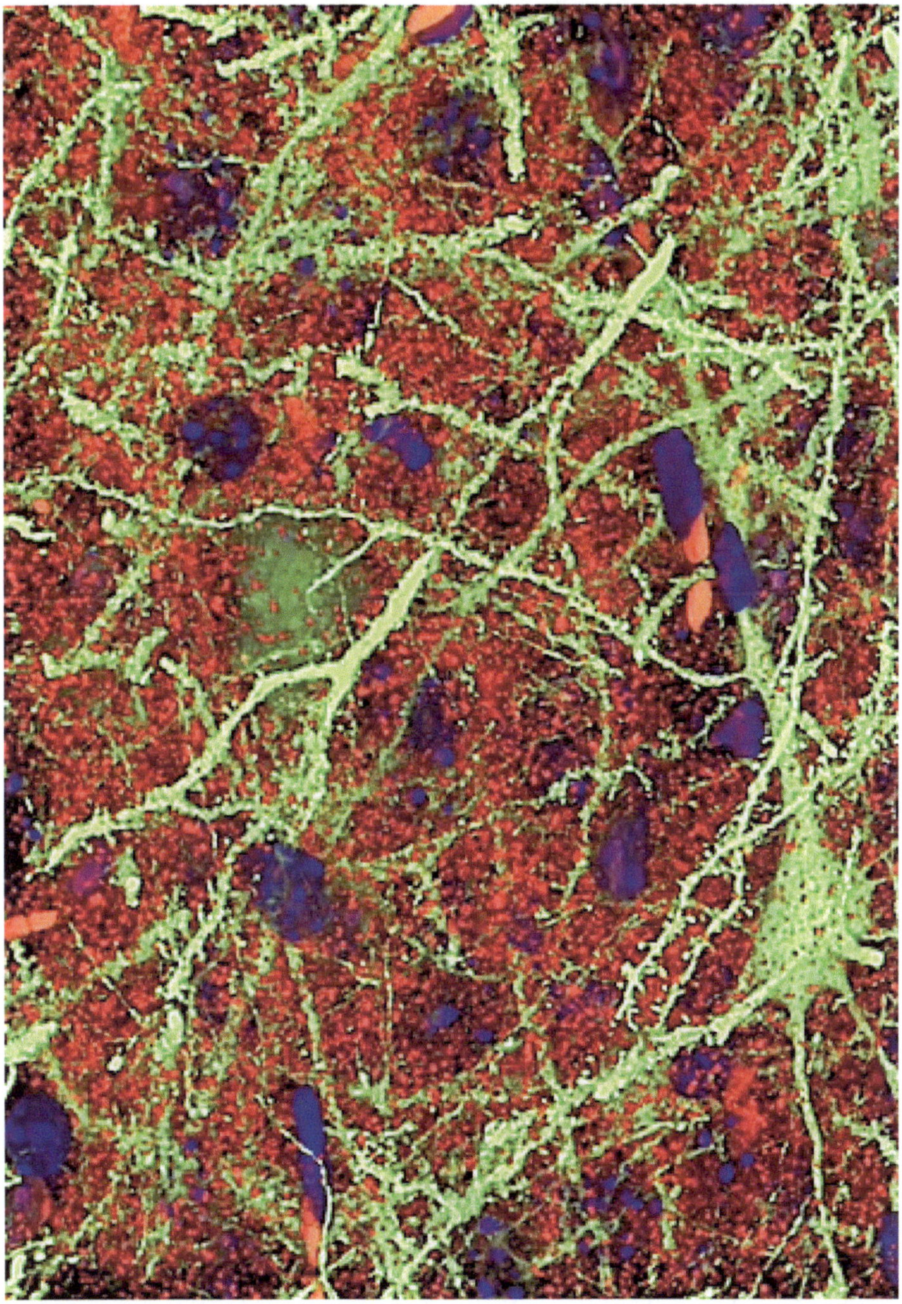

Fig. 14.2 A subsection in neural layer 5b. The chemical color-coding is as follows. *Green*: Thy1-H-YFP (layer 5B neuron subset); *Red*: Synapsin I (synapses); *Blue*: DAPI (DNA in all nuclei). All of our present analyses involve only the synapsin-detected synapses

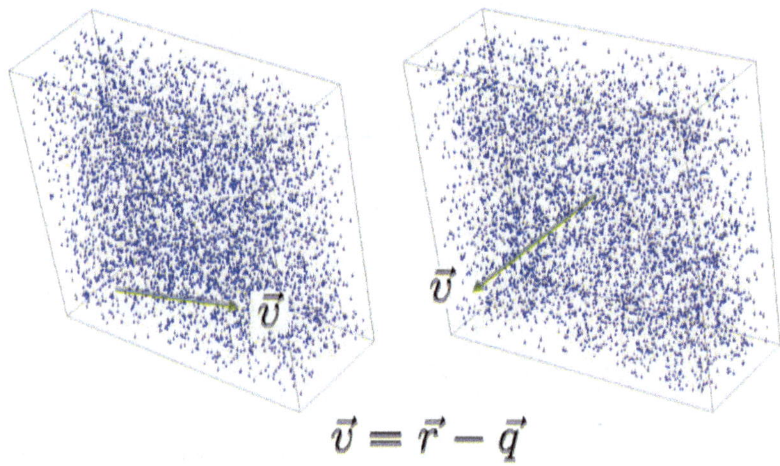

Fig. 14.3 Views of 5,000 random points (*left*) and 5,000 actual synapses (*right*) in a cuboid of given sides as follows (all in nanometers): $a = \Delta x \sim 103,300; b = \Delta y \sim 78,200; c = \Delta z \sim 11,400$, for *horizontal*, *vertical*, and *transverse* (angled into page), respectively. To convey an idea of scale, a millimeter is about 10x the horizontal span of either point-cloud. It is hard to see visual differences between the random points at *left* and the actual brain points at *right*. Nevertheless, sufficiently delicate statistical measures such as moments $\langle |\mathbf{v}|^s \rangle$ as well as fractal measurement do reveal systematic, quantifiable differences

$$\begin{aligned} \Delta_3(1) &= -\frac{118}{21} - \frac{2}{3}\pi + \frac{34}{21}\sqrt{2} - \frac{4}{7}\sqrt{3} + 2\log\left(1+\sqrt{2}\right) + 8\log\left(\frac{1+\sqrt{3}}{\sqrt{2}}\right) \\ &= 0.66170718226717623515583113324841358174640013579095\ldots. \end{aligned}$$

The exact formula allows a comparison between a given point-cloud and a random cloud: One may calculate the empirical expectation $\langle |\mathbf{r} - \mathbf{q}| \rangle$, where $\mathbf{r}, \mathbf{q}$ each runs over the point-cloud and compares with the exact expression $\Delta_3(1) \approx \ldots$. Similarly it is known that the expected inverse separation in the 3-cube is

$$\Delta_3(-1) := \left\langle \frac{1}{|\mathbf{r} - \mathbf{q}|} \right\rangle$$

$$= \frac{2}{5} - \frac{2}{3}\pi + \frac{2}{5}\sqrt{2} - \frac{4}{5}\sqrt{3} + 2\log\left(1+\sqrt{2}\right) + 12\log\left(\frac{1+\sqrt{3}}{\sqrt{2}}\right) - 4\log\left(2+\sqrt{3}\right)$$

$$= 1.88231264438966016010560083886836758785246288031070\ldots.$$

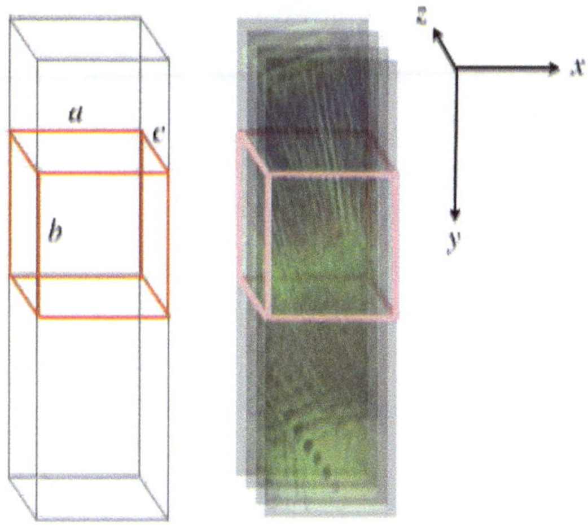

Fig. 14.4 Pictorial of the role of cuboid calculus in our analysis scenario. The *right-hand entity* pictorializes an array-tomography section of mouse brain (see Appendix: Synapse datasets for details). At the *left* is an idealized, *long cuboid* representing the full brain sample, inside of which is a chosen subsection as an (a, b, c)-cuboid. The idea is to statistically compare the synapse distribution within an (a, b, c)-cuboid against a random distribution having the same cuboid population. By moving the (a, b, c) cuboid downward, along the y-axis, one can actually detect neural layers

Such exact forms do not directly apply in our analysis of the brain data, because we need volume sections that are not necessarily cubical. For this reason, we next investigate a generalization of box integrals to cuboid volumes (Fig. 14.4).

14.4 Toward a Theory of Cuboid Integrals

In the present study we shall require a more general three-dimensional box integral involving a cuboid of sides (a, b, c).[3] Consider therefore an expectation for two points $\mathbf{r}, \mathbf{q}$ lying in the same cuboid (Fig. 14.5):

$$\Delta_3(s; a, b, c) := \langle |\mathbf{r} - \mathbf{q}| \rangle |_{\mathbf{r}, \mathbf{q} \in [0,a] \times [0,b] \times [0,c]}$$

$$= \frac{1}{a^2 b^2 c^2} \int_0^a \int_0^a \int_0^b \int_0^b \int_0^c \int_0^c |\mathbf{r} - \mathbf{q}|^s \, \mathrm{d}r_1 \, \mathrm{d}q_1 \, \mathrm{d}r_2 \, \mathrm{d}q_2 \, \mathrm{d}r_3 \, \mathrm{d}q_3.$$

This agrees with the standard box integral $\Delta_3(s)$ when $(a, b, c) = (1, 1, 1)$.

[3] A cuboid being a parallelepiped with all faces rectangular—essentially a "right parallelepiped."

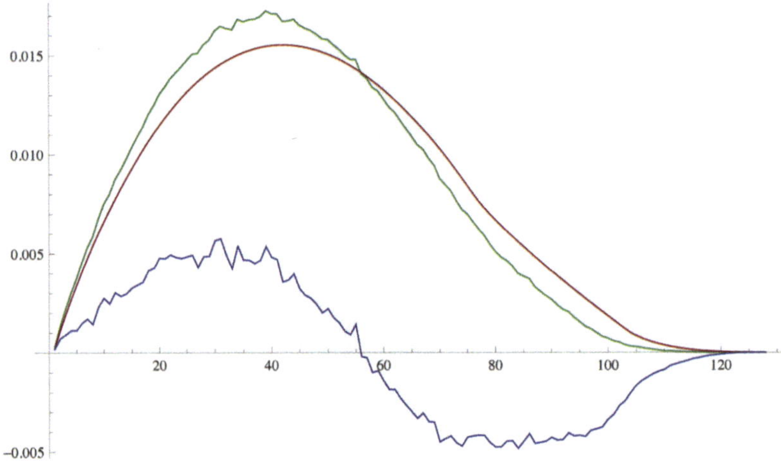

Fig. 14.5 Probability density curves for the separation $v = |\mathbf{r} - \mathbf{q}|$ (*horizontal axis*), taken over a cuboid of data, in the spirit of Fig. 14.4. The *green curve* (with highest peak) is extracted from subsegment 2 of dataset I, under the segmentation paradigm $\{12, 1, 128, \{1, 128\}\}$. The *red curve* (with rightmost peak) is theoretical—calculated from the Philip formula for $F_3(v; 146700, 107900, 2730)$. The *blue* "excess curve" is the point-wise curve difference (amplified $3\times$) and can be used in our "probagram" plots to show excess as a function of section depth y. The expected separations within this cuboid turn out to be $\langle v \rangle = 62018, 66789$ for brain, random, respectively

Figure 14.6 shows the result of empirical assessment of cuboid expectations for dataset I.

We introduce a generalized box integral, as depending on fixed parameters k, a_1, a_2, a_3 (we use a_i here rather than a, b, c just for economy of notation):

$$G_3(k; a_1, a_2, a_3) := \langle e^{-k|\mathbf{p}-\mathbf{q}|^2} \rangle$$

$$= \frac{1}{\prod a_i^2} \int_0^{a_1} \int_0^{a_1} \cdots \int_0^{a_3} \int_0^{a_3} e^{-k|\mathbf{r}-\mathbf{q}|^2} \, dr_1 \, dr_2 \, dr_3 \, dq_1 \, dq_2 \, dq_3,$$

which, happily, can be given a closed form

$$G_3(k; a_1, a_2, a_3) = \frac{1}{k^3} \prod_i \frac{e^{-a_i^2 k} + a_i \sqrt{\pi k} \, \mathrm{erf}\left(a_i \sqrt{k}\right) - 1}{a_i^2},$$

where $\mathrm{erf}(z) := 2/\sqrt{\pi} \int_0^z e^{-t^2} \, dt$ denotes the error function. The closed form here is quite useful, and by expanding the erf() in a standard series, we obtain for example a three-dimensional summation for G_3. The question is, can one write a summation that is of lower dimension? One possible approach is to expand the Gaussian in even

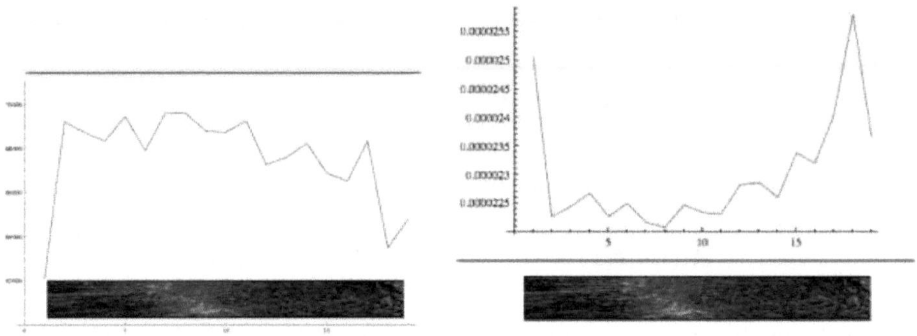

Fig. 14.6 Results for cuboid expectations of separation v and $1/v$ for cuboids of the type in Fig. 14.4, running over all y-depth. (The *dark green horizontal strip* represents the full sample, oriented *left-right* for these plots.) In both *left- and right-hand plots*, the *horizontal red line* is calculated from the exact formula for $\Delta_3(1;a,b,c)$. The segmentation paradigm here is $\{12,2,80,\{1,32\}\}$, dataset I

powers of $|\mathbf{p}-\mathbf{q}|$ and leverage known results in regard to box integrals Δ_n of Bailey et al. [2, 3]. Such dimensionality reduction remains an open problem.

Yet another expectation that holds promise for point-cloud analysis is what one might call a Yukawa expectation:

$$Y_3(k;a_1,a_2,a_3) \;:=\; \left\langle \frac{\mathrm{e}^{-k|\mathbf{r}-\mathbf{q}|}}{|\mathbf{r}-\mathbf{q}|} \right\rangle .$$

This is the expected Yukawa potential—of nuclear physics lore—between two points within the cuboid. The reason such potentials are of interest is that being "short-range" (just like nuclear forces) means that effects of closely clustered points will be amplified. Put another way: The boundary effects due to finitude of a cuboid can be rejected to some degree in this way.

14.4.1 Cuboid Statistics

Not just the exact expectation $\Delta_3(1;a,b,c)$ but the very probability density $F_3(v;a,b,c)$ has been worked out by Philip [14]. Both exact expressions in terms of a,b,c are quite formidable—see Appendix: Exact-density code for a programmatic way to envision the complexity. By probability density, we mean

$$\mathrm{Prob}\{|\mathbf{r}-\mathbf{q}| \in (v, v+\mathrm{d}v)\} \;=\; F_3(v;a,b,c)\,\mathrm{d}v;$$

hence we have a normalization integral with upper limit being the long cuboid diagonal:

$$\int_0^{\sqrt{a^2+b^2+c^2}} F_3(v;a,b,c)\,dv = 1.$$

More generally we can represent the moment Δ_3 in the form

$$\Delta_3(s;a,b,c) = \int_0^{\sqrt{a^2+b^2+c^2}} v^s F_3(v;a,b,c)\,dv.$$

The Philip density for separation v can also be used directly to obtain the density for a power of v, so

$$f_3(X := v^s;a,b,c) = \frac{1}{|s|} X^{\frac{1}{s}-1} F_3(X^{\frac{1}{s}};a,b,c).$$

For example, if we wish to plot the density of inverse separation $X := 1/v$ for a random point-cloud, we simply plot $X^{-2}F_3(1/X;a,b,c)$ for X running from $1/\sqrt{a^2+b^2+c^2}$ up to infinity; the area under this density will be 1.

14.5 Fractal Dimension

For the present research we used two fractal-measurement methods: The classical box-counting method, and a new, space-fill method. For a survey of various fractal-dimension definitions, including estimates for point-cloud data, see [6].

As for box-counting, we define a box dimension

$$\delta := \lim_{\varepsilon\to 0} \frac{\log\#(\varepsilon)}{-\log\varepsilon},$$

where for a given side ε of a microbox, $\#(\varepsilon)$ is the number of microboxes that minimally, collectively contain all the points of the point-cloud. Of course, our clouds are always finite, so the limit does not exist. But it has become customary to develop a #-vs.ε curve, such as the two curves atop Fig. 14.7, and report in some sense "best slope" as the measured box dimension.

There are two highly important caveats at this juncture: We choose to *redefine* the box-count number, as

$$\# \to \# \cdot \frac{1}{1-e^{-N\varepsilon^3}},$$

when the cloud has N total points. This statistical warp factor attempts to handle the scenario in which microboxes are so small that the finitude of points causes many empty microboxes. Put another way: The top curve of the top part of

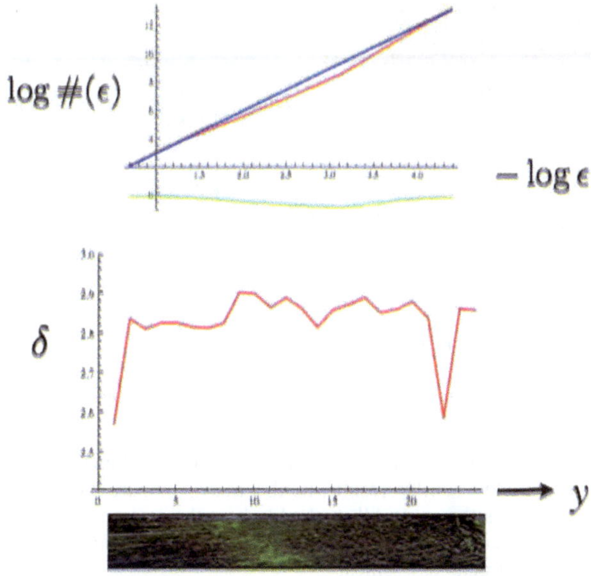

Fig. 14.7 Fractal-dimension measurement. Within a given cuboid we use the standard box-counting method, namely, in the *upper* figure is plotted log# vs. $\log(1/\varepsilon)$ for random points (*upper, blue curve*), then for the actual synapse points (*lower, red curve*), and with the excess as the *green* (*lowest*) plot. In the *bottom* figure, we use the excess to estimate fractal dimension for each cuboid in a segmentation paradigm $\{12, 2, 80, \{2, 80\}\}$. Evidently, the fractal dimension fluctuates depending on layer characteristics at depths *y*, with an average fractal dimension of ~ 2.8 for the whole of dataset I

Fig. 14.7—which curve should have slope 3 for N random points—stays straight and near slope 3 for a longer dynamic range because of the warp factor.

The second caveat is that we actually use not ε-microboxes but microcuboids. When the segment being measured is originally of sides (a, b, c), we simply rescale the cuboid to be in a unit box, which is equivalent to using a "microbrick" whose aspect ratios are that of the cuboid, and transform that microbrick to a cube of side $\varepsilon := (abc)^{1/3}$.

14.5.1 Space-Fill Method for Fractal Measurement

During this research, we observed that a conveniently stable fractal-measurement scheme exists for point-cloud datasets. We call this method the "space-fill" algorithm, which runs like so[4]:

---

[4]The present author devised this method in 1997, in an attempt to create "$1/f$" noise by digital means, which attempt begat the realization that fractal dimension could be measured with a Hilbert space-fill.

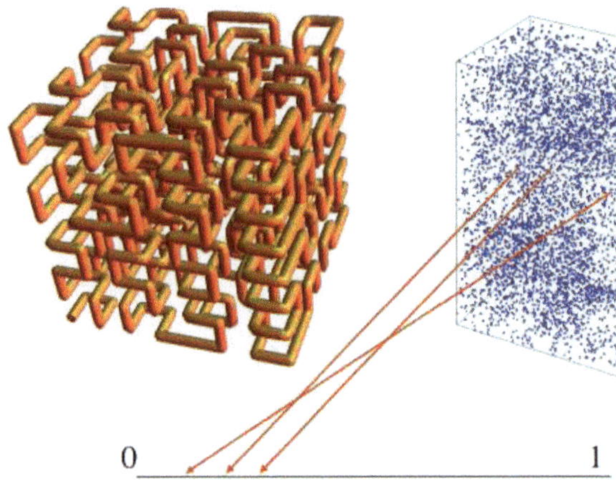

Fig. 14.8 The "space-fill" method for measuring point-cloud dimension. This algorithm as described in the text yields similar results to the more standard box-counting method, yet preliminary research reveals the space-fill method to be rather more stable with respect to graph noise. The basic idea is to create a set of pullbacks on the line $[0,1)$ and then use a quick sort and a simple one-dimensional fractal assessment

1. Assume a three-dimensional unit cube containing a point-cloud and construct a Hilbert space-filling curve, consisting of discrete visitation points $\mathbf{H}(t)$, where t runs over the integers in $[0, 2^{3b} - 1]$. (The resolution of this curve will be b binary bits per coordinate, therefore.)
2. Create a list of "pullback" rationals $t_k/2^{3b}$, corresponding to the points r_k of the point-cloud data.
3. Perform a one-dimensional sort on the set of pullbacks and measure the fractal dimension δ_1 using a simple interval counter.
4. Report fractal dimension of the point-cloud data as $\delta \;=\; 3 \cdot \delta$.

We do not report space-fill measurements herein—all of the results and figures employ the box-counting method—except to say (a) the space-fill method appears to be quite stable, with the fractagram surfaces being less noisy, and (b) the dimensions obtained in preliminary research with the space-fill approach are in good agreement with the box-counting method. Figure 14.8 pictorializes the space-fill algorithm.

14.6 Probagrams, Paramegrams, and Fractagrams

Our "grams" we have so coined to indicate their three-dimensional-embedding character.[5] Each 'gram is a surface, one of whose support dimensions is the section

[5] As in "sonogram"—which these days can be a medical ultrasound image, but originally was a moving spectrum, like a fingerprint of sound that would fill an entire sheet of strip-chart.

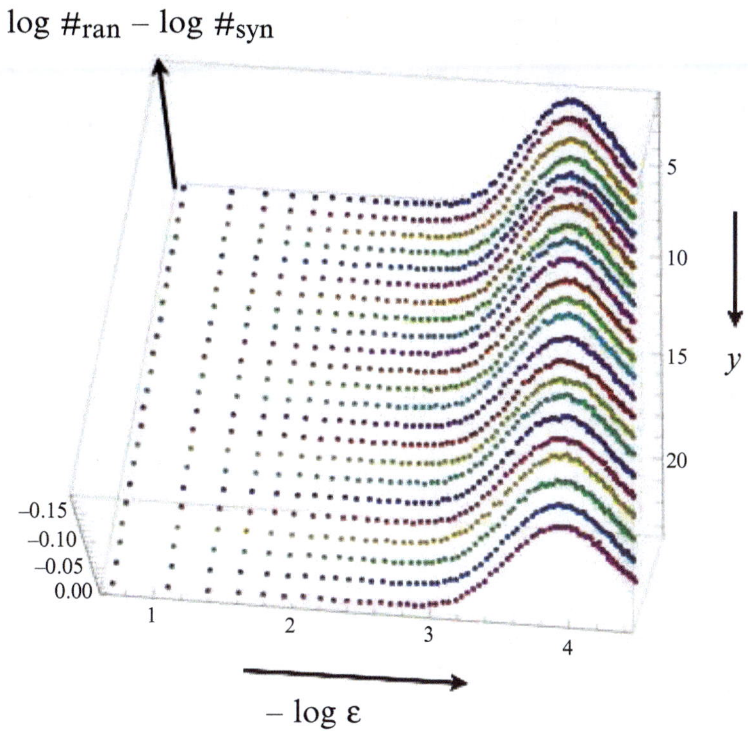

Fig. 14.9 The "fractagram" concept—which is similar for probagrams and paramegrams. For each cuboid in a given segmentation paradigm (here, paradigm $\{12, 2, 80, \{2, 80\}\}$) we generate the fractal-slope excess as in Fig. 14.7. The resulting "strands" of fractal data vs. y-depth in the dataset (here, dataset I) are much easier to interpret if plotted as a surface, which surface we then call a fractagram as pictured in Fig. 14.11

depth y. In our "grams", as in the original synapse datasets, $y = 0$ is the outside (pial) surface, while y increases into the brain sample. We typically have, in our "grams", *downward increasing* y, so that the top of a 'gram pictorial is the outside surface.

Precise definitions are:

- Probagram: Surface whose height is probability density of a given variable within a cuboid, horizontal axis is the variable, and the vertical axis is the y-depth into the sample.
- Paramegram: Surface whose height is a parameterized expectation (such as our function $G_3(k; a, b, c)$), horizontal axis is the parameter (such as k), and the vertical axis is y-depth.
- Fractagram: Surface whose height is the excess between the fractal-slope curve for a random cloud in a cuboid and the actual data cloud's fractal-slope curve, horizontal axis is $-\log \varepsilon$, and vertical axis is as before the y-depth (Fig. 14.9).

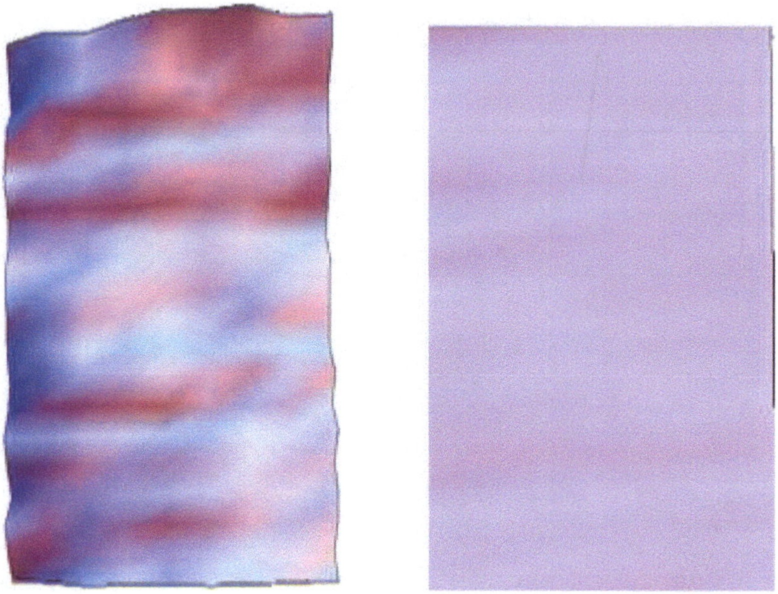

Fig. 14.10 A baseline experiment. At *left* is the probagram for dataset I and density $f_3(X := 1/v^2; a,b,c)$ under segmentation paradigm $\{12, 2, 16, \{2, 10\}\}$. At the *right* is the result of using the same number of points ($N = 1,119,299$) randomly placed within the full sample cuboid. This kind of experiment shows that the brain synapses are certainly *not* randomly distributed

In general, we display these "grams" looking down onto the surface or possible at a small tilt to be able to understand the surface visually.

What we shall call a segmentation paradigm is a set P of parameters that determine the precise manner in which we carve (a,b,c)-cuboids out of a full synapse dataset (Fig. 14.10). Symbolically,

$$P := \{M, G, H, \{b, e\}\},$$

where

- M is the "magnification" factor—the y-thickness of a cuboid divided into the full y-span of the dataset.
- G is the "grain"—which determines the oversampling; $1/G$ is the number of successively overlapping cuboids in one cuboid.
- H is the number of histogram bins in a 'gram plot, and we plot from bin b to bin e.

We generally use $g < 1$ to avoid possible alias effects at cuboid boundaries. The total number of cuboids analyzed in a 'gram thus turns out to be

$$S = 1 + G(M - 1).$$

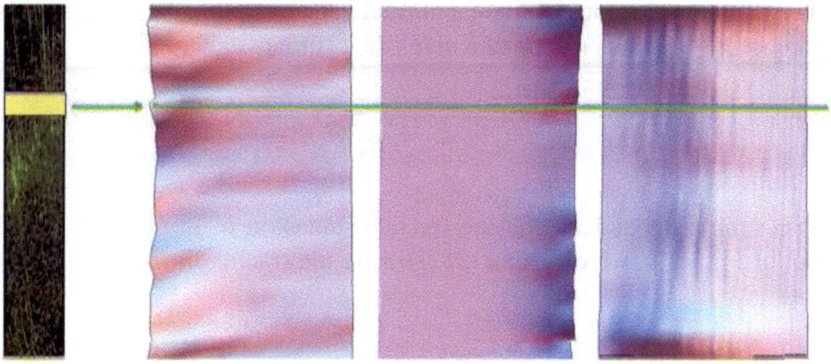

Fig. 14.11 Typical set of three "grams": At *far left* is a pictorialized a full-section sample, with a *small box* indicating a cuboid subsection. As said section is moved downward (increasing y), we obtain, *left-to-right* and for separation, $v := |\mathbf{r} - \mathbf{q}|$, the probagram for v^{-1}, then the paramegram for $\langle \exp(-kv^2) \rangle$, then the fractalgram. The phenomenon of neural layering is evident and qualitatively consistent (either correlated or anticorrelated) across all three "grams" for this sample (dataset I, detailed in Appendix: Synapse datasets)

For example, with grain $G = 3$ and $M = 10$, we calculate over a total of 28 cuboids. This is because there are generally $G = 3$ cuboids overlapping a given cuboid. In any case, one may take cuboid dimensions a, b, c as

$$a = x_{\max} - x_{\min};\ b = \frac{y_{\max} - y_{\min}}{M};\ c = z_{\max} - z_{\min},$$

where min, max coordinates are deduced from the data (Fig. 14.11). (In our "grams", we continually recompute the min, max for every cuboid to guard against such as corner holes in the data.)

14.7 How Do We Explain the Observed Fractal Dimension?

Let us give an heuristic argument for the interaction of cuboid expectations and fractal-dimension estimates. Whereas the radial volume element in 3-space is $4\pi r^2 dr$, imagine a point-cloud having the property that the number of points a distance r from a given point scales as $r^{\delta-1}$ where $\delta < 3$, say. Then, if the characteristic size of a point sample is R (here we are being rough, avoiding discussion of the nature of the region boundaries), we might estimate an expectation for point-separation v to the sth power as

$$\langle v^s \rangle \sim \frac{\int_0^R u^s u^{\delta-1}\, du}{\int_0^R u^{\delta-1}\, du}.$$

Note that we can avoid calculation of a normalization constant by dividing this way, to enforce $\langle v^0 \rangle = 1$. This prescription gives the estimate

$$\langle v^s \rangle \sim \frac{\delta}{s+\delta} R^s,$$

showing a simple dependence on the fractal dimension δ. In fact, taking the left-hand plot of Fig. 14.6, we can right off estimate the fractal dimension of the whole dataset as

$$\delta \sim 2.6,$$

not too off the mark from our more precise fractal measurements that we report as 2.8 ± 0.05.

So one way to explain our discovered fractal dimension $\sim 2.8 < 3$ for both datasets is to surmise that the distance metric is weighted in some nonuniform fashion (Fig. 14.12).

14.7.1 Generalized Cantor Fractals

One aspect undertaken during the present research was to attempt to fit the observed fractal properties of the datasets to some form of Cantor fractal. There is a way to define a generalized Cantor fractal in n dimensions so that virtually any desired fractal dimension in the interval $[n\frac{\log 2}{\log 3}, n]$ (see [4]).[6] Such generalized Cantor fractals were used to fine-tune our fractal measurement machinery.

Interestingly, the cuboid expectations for dataset II seem qualitatively resonant with the corresponding expectations for a certain generalized Cantor set called $C_3(\overline{33111111})$ having dimension $\delta = 2.795\ldots$. However, dataset I does *not* have similar expectations on typical cuboids. For one thing, the highest-peak curve in Fig. 14.5—which is from a cuboid within dataset I—shows $\langle v \rangle$ for the laboratory data being less than the same expectation for random data; yet, a Cantor fractal tends to have such expectation *larger* than random data.

We shall soon turn to a different fractal model that appears to encompass the features of both datasets. But first, a word is appropriate here as to the meaning of "holes" in a dataset. Clearly, holes in the laboratory point-clouds will be caused

---

[6]Mathematically, the available fractal dimensions for the generalized Cantor fractals are dense in said interval.

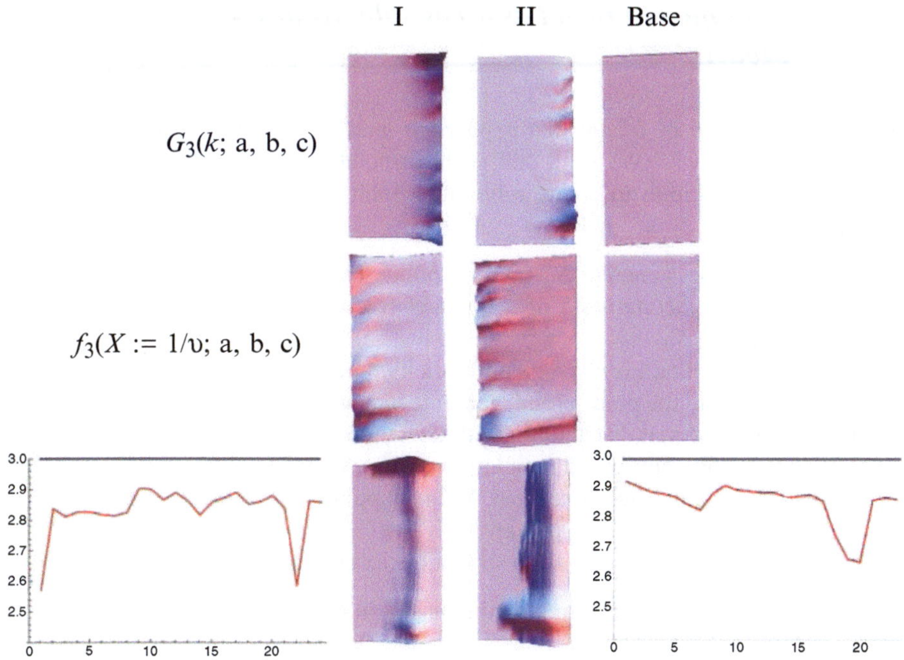

Fig. 14.12 The "grams" for the synapse-location datasets I, II. The *top row* shows G_3 paramegrams and baseline test for segmentation paradigm $\{12, 2, 32, \{1, 10\}\}$. The *second row* shows probagrams for inverse separation $1/v$, in the same segmentation paradigm. The two 3D plots at *bottom* are the fractagrams. At *far-left* and *far-right bottom* are graphical displays of the per-cuboid fractal-dimension estimate. Note that the baseline test here is for a randomly filled cuboid; the *horizontal lines* at dimension 3.0 really are less noisy than one pixel width. Thus the datasets I, II can be said both to have overall fractal dimension 2.8 ± 0.5, although the dimension is evidently neural-layer dependent

by the simple fact of synapses not subsisting within large bodies.[7] So, too, Cantor fractals can be created by successive removal of holes that scale appropriately. But here is the rub: The existence of holes *does not in itself necessarily alter fractal dimension*.[8] For example, take a random cloud and remove large regions, to create essentially a swiss-cheese structure in between whose holes are equidistributed points. The key is, fractal-measurement machinery will still give a dimension very close to $\delta = 3$.

[7]Synapses live on dendrites, exterior to actual neurons.

[8]Of course, the situation is different if hole existence is connected with microscopic synapse distribution, e.g., if synapses were to concentrate near surfaces of large bodies.

14.7.2 "Bouquet" Fractal as a Possible Synapse-Distribution Model

We did find a kind of fractal that appears to lend itself well to comparison with synapse distributions.[9] We shall call such artificial constructs "bouquet" fractals. A generating algorithm to create a bouquet point-cloud having N points runs as follows:

1. In a unit 3-cube, generate N_0 random points (N_0 and other parameters can be used to "tune" the statistics of a bouquet fractal). Thus the point-cloud starts with population N_0.
2. Choose an initial radius $r = r_0$, a multiplicity number m, and a scale factor $c < 1$.
3. For each point in the point-cloud, generate m new points a mean distance r away (using, say, a normal distribution with deviation r away from a given point). At this juncture the point-cloud population will be $N_0 \cdot m^k$ for k being the number of times this step 3 has been executed. If this population is $\geq N$, go to step 5.
4. Reduce r by $r = c \cdot r$ and go to step 3.
5. Prune the point-cloud population so that the exact population is achieved.

The bouquet fractal will have fractal dimension on the order of

$$\delta \sim \frac{\log m}{-\log c},$$

but this is an asymptotic heuristic; in practice, one should simply tune all parameters to obtain experimental equivalencies.[10] For example, our dataset I corresponds interestingly to bouquet parameters

$$\{N_0, r_0, m, c\} = \{1000, N_0^{-1/3}, 23, 1/3\}.$$

The measured fractal dimension of the resulting bouquet for population $N = 1{,}119{,}299$ is $\delta \sim 2.85$ and statistical moments also show some similarity.

Once again, something like a bouquet fractal may not convey any neurophysiological understanding of synapses locations, but there could be a diagnostic parameter set, namely that set for which chosen statistical measures come out quantitatively similar.

---

[9] Again, we are not constructing here a neurophysiological model; rather, a phenomenological model whose statistical measures have qualitative commonality with the given synapse data.

[10] The heuristic form of dimension δ here may not be met if there are not enough total points. This is because the fractal-slope paradigm has low-resolution box counts that depend also on parameters N_0, r.

14.7.3 Nearest-Neighbor Calculus

Another idea that begs for further research is to perform nearest-neighbor calculus on synapse cuboids. This is yet a different way to detect departure from randomness.

In an n-dimensional unit volume, the asymptotic behavior of the nearest-pair distance for N uniformly randomly placed points, namely

$$\mu_1 := \langle \min |\mathbf{r} - \mathbf{q}| \rangle_{\mathbf{r},\mathbf{q} \in V},$$

is given—in its first asymptotic term—by

$$\mu_1 \sim \Gamma\left(1 + \frac{1}{n}\right) \frac{2^{1/n}}{\sqrt{\pi}} \Gamma^{1/n}\left(1 + \frac{n}{2}\right) \frac{1}{N^{2/d}} + \dots.$$

In our $(n = 3)$-dimensional scenarios, we thus expect the nearest-pair separation to be

$$\mu_1 \sim \frac{\Gamma(4/3)}{\pi^{1/3}} \frac{1}{N^{2/3}} \approx \frac{0.6097}{N^{2/3}}.$$

It is interesting that this expression can be empirically verified with perhaps less inherent noise than one might expect.

Presumably a nearest-pair calculation on the synapse distributions will reveal once again significant departures from randomness. What we expect is a behavior like so

$$\mu_1 \sim \frac{\text{constant}}{N^{2/\delta}}$$

for fractal dimension δ. Probably the best research avenue, though, is to calculate the so-called k-nearest-pairs, meaning ordered k-tuples of successively more separate pairs, starting with the minimal pair, thus giving a list of expected ordered distances $\mu_1, \mu_2, \dots, \mu_k$.

Acknowledgments The author is grateful to S. Arch of Reed College, as well as N. Weiler, S. Smith, and colleagues of Smithlab at Stanford Medical School, for their conceptual and algorithmic contributions to this project. T. Mehoke aided this research via statistical algorithms and preprocessing of synapse files. Mathematical colleague T. Wieting supported this research by being a selfless, invaluable resource for the more abstract fractal concepts. D. Bailey, J, Borwein, and M. Rose aided the author in regard to experimental mathematics on fractal sets. This author benefitted from productive discussions with the Advanced Computation Group at Apple, Inc.; in particular, D. Mitchell provided clutch statistical tools for various of the moment analyses herein.

Appendix 1: Synapse Datasets

Referring to Table 14.1: Both datasets I, II are from adult-mouse "barrel cortex" which is a region of the somatosensory neocortex involved in processing sensation from the facial whiskers (one of the mouse's primary sensory modalities). The long y-axis of the volumes crosses all 6 layers of the neocortex (these are layers parallel to the cortical surface, and the long axis is perpendicular to the surface).

Neurophysiological considerations including array-tomography technology are discussed in [11–13] and web URL [15]; we give a brief synopsis:

Array tomography (AT) is a new high-throughput proteomic imaging method offering unprecedented capabilities for high-resolution imaging of tissue molecular architectures. AT is based on (1) automated physical, ultrathin sectioning of tissue specimens embedded in a hydrophilic resin, (2) construction of planar arrays of these serial sections on optical coverslips, (3) staining and imaging of these two-dimensional arrays, and (4) computational reconstruction into three dimensions, followed by (5) volumetric image analysis. The proteomic scope of AT is enhanced enormously by its unique amenability to high-dimensional immunofluorescence multiplexing via iterative cycles of antibody staining, imaging and antibody elution.

Appendix 2: Exact-Density Code

```
(* Evaluation of the exact Philip density F3[v,a,b,c]
 for an (a,b,c)-cuboid. *)

h11[u_, a_, b_, c_] := 1/(3 a^2 b^2 c^2) *
        If[u <= b^2, -3 Pi b c u + 4 b u^(3/2),
           If[u <= c^2,
                4 b^4 + 6 b^2 c Sqrt[u - b^2] -
                6 b c u ArcSin[b/Sqrt[u]],
                If[u <= b^2 + c^2, 4 b^4 + 6 b^2 c *
                      Sqrt[u - b^2] +
                      6 b c u (ArcCos[c/Sqrt[u]] -
                      ArcSin[b/Sqrt[u]]) -
```

Table 14.1 Synapse dataset characteristics

File, voxel nm×nm×nm	N	(x_{min}, x_{max})	(y_{min}, y_{max})	(z_{min}, z_{max})
I				
KDM-100824B 100×100×70	1,119,299	(2800, 151300)	(2300, 1298000)	(105, 2835)
II				
mMos3_Syn 100×100×200	1,732,051	(100, 103400)	(100, 1252600)	(105, 4095)

The point-cloud population N exceeds 10^6 for each dataset. The min, max parameters have been converted here to nm

```
                    2 b (2 u + c^2) Sqrt[u - c^2],
                    0
                ]
            ]
      ];

h12[u_, a_, b_, c_] := 1/(6 a^2 b^2 c^2) *
        If[u <= a^2,
           12 Pi a b c Sqrt[u] - 6 Pi a (b + c) u +
           8 (a + c) u^(3/2) - 3 u^2,
             If[u <= c^2,
                 5 a^4 - 6 Pi a^3 b +
                 12 Pi a b c Sqrt[u] +
                 8 c u^(3/2) - 12 Pi a b c *
                 Sqrt[u - a^2] - 8 c *
                 (u - a^2)^(3/2) -
                 12 a c u ArcSin[a/Sqrt[u]],
                 If[u <= a^2 + c^2,
                      5 a^4 - 6 Pi a^3 b +
                      6 Pi a b c^2 -
                      c^4 + 6 (Pi a b + c^2) u +
                      3 u^2 - 12 Pi a b c *
                      Sqrt[u - a^2] -
                      8 c (u - a^2)^(3/2) -
                      4 a (2 u + c^2)*
                      Sqrt[u - c^2] +
                      12 a c u *
              (ArcCos[c/Sqrt[u]]-ArcSin[a/Sqrt[u]]),
                 0
               ]
             ]
        ];

h22[u_, a_, b_, c_] := 1/(3 a^2 b^2 c^2) *
        If[u <= a^2, 0,
           If[u <= a^2 + b^2,
                3 Pi a^2 b (a + c) - 3 a^4 -
                6 Pi a b c Sqrt[u] +
                3 (a^2 + Pi b c) u +
                (6 Pi a b c - 2 (b + 3 c) a^2-4 b u)*
                Sqrt[u - a^2] -
                6 a b u ArcSin[a/Sqrt[u]],
                If[u <= a^2 + c^2,
                     3 a^2 b (Pi a - b) - 4 b^4-
                     12 a b c Sqrt[u]*
```

```
ArcSin[b Sqrt[u]/(Sqrt[a^2 + b^2] * Sqrt[u - a^2])]-
                    6 a c (a - Pi b) Sqrt[u - a^2]-
                    6 c (b^2 - a^2 +
                 2 a b ArcSin[a/Sqrt[a^2 + b^2]])*
                    Sqrt[u - a^2 - b^2] -
                    6 a b (a^2 + b^2)*
                    ArcSin[a/Sqrt[a^2 + b^2]] +
                    6 b c (a^2 + u) *
                    ArcSin[b/Sqrt[u - a^2]],
                    3 a^2 (a^2 - b^2 - c^2)-4 b^4-
                    3 a^2 u - 12 a b c  Sqrt[u]  *
(ArcSin[b Sqrt[u]/(Sqrt[a^2 + b^2] * Sqrt[u-a^2])]-
ArcCos[a c/(Sqrt[u - c^2] Sqrt[u - a^2])]) +
                    2 b (a^2 + c^2 + 2 u) *
                    Sqrt[u - a^2 - c^2] -
                    6 c *
(b^2 - a^2 + 2 a b ArcSin[a/Sqrt[a^2 + b^2]]) *
                    Sqrt[u - a^2 - b^2] -
                    6 a b (a^2 + b^2) *
                    ArcSin[a/Sqrt[a^2 + b^2]] +
                    6 b c (a^2 + u) *
(ArcSin[b/Sqrt[u - a^2]] - ArcCos[c/Sqrt[u - a^2]])+
                    6 a b (c^2 + u) *
                    ArcSin[a/Sqrt[u - c^2]]
              ]
            ]
        ];

h32[u_, a_, b_, c_] := h22[u, b, a, c];

h33[u_, a_, b_, c_] := 1/(6 a^2 b^2 c^2) *
        If[u <= b^2, 0,
           If[u <= a^2 + b^2,
                3 (2 Pi a b + b^2 + u) (u - b^2) -
                4 c (b^2 + 3 Pi a b + 2 u) *
                Sqrt[u - b^2],
                If[u <= b^2 + c^2, 3 (a^2 + b^2)^2 -
                   3 b^4 + 6 Pi a^3 b -
                   4 c (b^2 + 3 Pi a b + 2 u) *
                   Sqrt[u - b^2] +
                   4 c (a^2 + b^2 + 3 Pi a b + 2 u)*
                   Sqrt[u - a^2 - b^2],
                3 (a^2 + b^2)^2 + c^4 +
                   6 Pi a b (a^2 + b^2 - c^2) -
                   6 (Pi a b + c^2) u - 3 u^2 +
```

```
                    4 c (a^2 + b^2 + 3 Pi a b + 2 u)*
                    Sqrt[u - a^2 - b^2]
                ]
              ]
        ];

(* Next, the Philip density function for separation v.
    It must be arranged that a <= b <= c. *)

F3[v_, a_, b_, c_] :=
  2 v (h11[v^2, a, b, c] + h12[v^2, a, b, c] +
  h22[v^2, a, b, c] + h32[v^2, a, b, c] +
  h33[v^2, a, b, c]);
```

References

1. Adesnik, H., Scanziani, M.: Lateral competition for cortical space by layer-specific horizontal circuits. Nature **464**, 1155–1160 (2010)
2. Bailey, D., Borwein, J., Crandall, R.: Box integrals. J. Comput. Appl. Math. **206**, 196–208 (2007)
3. Bailey, D., Borwein, J., Crandall, R.: Advances in the theory of box integrals. Math. Comp. **79**, 1839–1866 (2010)
4. Bailey, D., Borwein, J., Crandall, R., Rose, M.: Expectations on fractal sets. Appl. Math. Comput. (2013)
5. Borwein, J., Chan, O-Y, Crandall, R.: Higher-dimensional box integrals. Exp. Math. **19**(3), 431–445 (2010)
6. Clarkson, K.J.: Nearest-neighbor searching and metric space dimensions. In: Shakhnarovich, G., et al. (eds.) Nearest-Neighbor Methods in Learning and Vision: Theory and Practice. MIT Press, Cambridge (2005)
7. Fischler, M.: Distribution of minimum distance among N random points in d dimensions. Technical Report FERMILAB-TM-2170 (2002)
8. Georgiev, S., et al.: EEG fractal dimension measurement before and after human auditory stimulation. Bioautomation **12**, 70–81 (2009)
9. Grski, A., Skrzat, J.: Error estimation of the fractal dimension measurements of cranial sutures. J. Anat. **208**(3), 353–359 (2006)
10. Lynnerup, N., Jacobsen, J.C.: Brief communication: age and fractal dimensions of human sagittal and coronal sutures. Am. J. Phys. Anthropol. **121**(4), 332–6 (2003)
11. Micheva, K.D., Smith, S.J.: Array tomography: A new tool for imaging the molecular architecture and ultrastructure of neural circuits. Neuron **55**, 25–36 (2007)
12. Micheva, K.D., Busse, B., Weiler, N.C., O'Rourke, N., Smith, S.J.: Single-synapse analysis of a diverse synapse population: proteomic imaging methods and markers. Neuron **68**(4), 639–53 (2010)
13. Micheva, K.D., O'Rourke, N., Busse, B., Smith, S.J.: Array tomography: high-resolution three-dimensional immunofluorescence. In: Imaging: A Laboratory Manual, Ch. 45, pp. 697–719, 3rd edn. Cold Spring Harbor Press, New York (2010)
14. Philip, J.: The probability distribution of the distance between two random points in a box. TRITA MAT 07 MA 10 (2007)
15. Smithlab: Array tomography. http://smithlab.stanford.edu/Smithlab/Array_Tomography.html

16. Spruston, N.: Pyramidal neurons: dendritic structure and synaptic integration. Nat. Rev. Neurosci. **9**, 206–221 (2008)
17. Teich, M., et al.: Fractal character of the neural spike train in the visual system of the cat. J. Opt. Soc. Am. A **14**(3), 529–546 (1997)

Chapter 15
Visible Points in Convex Sets and Best Approximation

Frank Deutsch, Hein Hundal, and Ludmil Zikatanov

Dedicated to Jonathan Borwein on the occasion of his 60th birthday

Abstract The concept of a *visible point* of a convex set relative to a given point is introduced. A number of basic properties of such visible point sets are developed. In particular, it is shown that this concept is useful in the study of best approximation, and it also seems to have potential value in the study of robotics.

Key words: Best approximation from convex sets • Visible points in convex sets

Mathematics Subject Classifications (2010): 41A65, 52A27.

15.1 Introduction

Unless explicitly stated otherwise, throughout this paper X will always denote a (real) normed linear space, and C a closed convex set in X. For any two distinct points x, v in X, we define interval notation analogous to that on the real line by

---

COMMUNICATED BY HEINZ H. BAUSCHKE.

F. Deutsch (✉) • L. Zikatanov
Department of Mathematics, The Pennsylvania State University, McAllister Building - MCL G5, University Park, PA 16802, USA
e-mail: deutsch@math.psu.edu; ludmil@psu.edu

H. Hundal
The Pennsylvania State University, 146 Cedar Ridge Drive, Port Matilda, PA 16870, USA
e-mail: hundalhh@yahoo.com

D.H. Bailey et al. (eds.), *Computational and Analytical Mathematics*, Springer Proceedings in Mathematics & Statistics 50, DOI 10.1007/978-1-4614-7621-4_15,

$$[x,v] := \{\lambda x + (1-\lambda)v \mid 0 \le \lambda \le 1\},$$
$$[x,v[:= \{\lambda x + (1-\lambda)v \mid 0 < \lambda \le 1\},$$
$$]x,v] := \{\lambda x + (1-\lambda)v \mid 0 \le \lambda < 1\} = [v,x[, \text{ and}$$
$$]x,v[:= \{\lambda x + (1-\lambda)v \mid 0 < \lambda < 1\}.$$

In other words, $[x,v]$ is just the closed line segment joining x and v, $[x,v[$ is the same line segment but excluding the end point v, and $]x,v[$ is the line segment $[x,v]$ with both end points x and v excluded.

Definition 15.1. Let $x \in X$. A point $v \in C$ is said to be **visible** to x with respect to C if and only if $[x,v] \cap C = \{v\}$ or, equivalently, $[x,v[\cap C = \emptyset$. The set of all visible points to x with respect to C is denoted by $V_C(x)$.

Thus

$$V_C(x) = \{v \in C \mid [x,v] \cap C = \{v\}\} = \{v \in C \mid [x,v[\cap C = \emptyset\}. \tag{15.1}$$

Geometrically, one can regard $V_C(x)$ as the "light" that would be cast on the set C if there were a light source at the point x emanating in all directions. Alternatively, one can regard the set C as an "obstacle" in X, a "robot" is located at a point $x \in X$, and the directions determined by the intervals $[x,v]$, where $v \in V_C(x)$, as directions to be avoided by the robot so as not to collide with the obstacle C.

In this paper we begin a study of visible sets. In Sect. 15.2, we will give some characterizations of visible sets (see Lemmas 15.4 and 15.10, and Theorem 15.15 below). We show that the visible set mapping V_C satisfies a translation property just like the well-known metric projection P_C (see Lemma 15.6 below). Recall that the generally set-valued *metric projection* (or nearest point mapping) P_C is defined on X by

$$P_C(x) := \{y \in C \mid \|x-y\| = \inf_{c \in C} \|x-c\|\}.$$

Those closed convex sets C such that the set of visible points to each point not in C is the whole set C are precisely the affine sets (Proposition 15.7). In Sect. 15.3 we study the connection between visible points and best approximations. Finally, in Sect. 15.4 we consider characterizing best approximations to a point in a Hilbert space from a polytope, i.e., the convex hull of a finite set of points.

15.2 Visibility from Convex Sets

The first obvious consequence of the definition of visibility is the following.

Lemma 15.2. *Let C be a closed convex set in X. If $x \in C$, then $V_C(x) = \{x\}$.*

This lemma shows that the most interesting case is when $x \in X \setminus C$ and the main results to follow actually require this condition as part of their hypotheses. Indeed, when $x \notin C$, there are additional useful criteria that characterize visible points. For any set C, let $\mathrm{bd}C$ denote the *boundary* of C.

Unlike the metric projection, the visibility operator is never empty-valued.

Lemma 15.3. *Let C be a closed convex set in X. Then*

1. *$V_C(x) \neq \emptyset$ for each $x \in X$, and*
2. *$V_C(x) \subset \mathrm{bd}C$ for each $x \in X \setminus C$.*

Proof.

1. Let $x \in X$. By Lemma 15.2 we may assume that $x \notin C$. Fix any $y \in C$. Then the interval $[x,y]$ contains points in C (e.g., y) and points not in C (e.g., x). Let

$$\lambda_0 := \sup\{\lambda \in [0,1] \mid \lambda x + (1-\lambda)y \in C\}.$$

Since C is closed, it follows that $v_0 := \lambda_0 x + (1-\lambda_0)y \in C$. Hence $\lambda_0 < 1$, and $[x,v_0] \cap C = \{v_0\}$. That is, $v_0 \in V_C(x)$.

2. Fix any $x \in X \setminus C$. To show that $v \in \mathrm{bd}C$ for each $v \in V_C(x)$. If not, then there exists some $v \in V_C(x)$ such that $v \in C \setminus \mathrm{bd}C$. Hence v is in the *interior* of C. Thus there must be a subinterval $[v_0, v]$ of the interval $[x,v]$ which lies in C. Hence $[x,v] \cap C \neq \{v\}$, a contradiction to $v \in V_C(x)$.

■

Lemma 15.4 (Characterization of visible points). *Let C be a closed convex set in X, $x \in X \setminus C$, and $v \in C$. Then the following statements are equivalent:*

1. *v is visible to x with respect to C.*
2. *$\lambda x + (1-\lambda)v \notin C$ for each $0 < \lambda \leq 1$.*
3. *$\max\{\lambda \in [0,1] \mid \lambda x + (1-\lambda)v \in C\} = 0$.*

Proof. (1) ⇒ (2): If (1) holds, then $[x,v[\cap C = \emptyset$. Since $[x,v[= \{\lambda x + (1-\lambda)v \mid 0 < \lambda \leq 1\}$, (2) follows.

(2) ⇒ (3): Since $v \in C$, (3) is an obvious consequence of (2).

(3) ⇒ (1): If (3) holds, then $[x,v[\cap C = \emptyset$. That is, $v \in V_C(x)$. ■

Simple examples in the Euclidean plane (e.g., a box) show that although C is convex, $V_C(x)$ is not convex in general. These simple examples also might seem to indicate that $V_C(x)$ is always closed. However, the following example in 3 dimensions shows that this is false in general.

Consider the subset of Euclidean 3-space $\ell_2(3)$ defined by

$$C := (1,0,0) + \mathrm{cone}\{(1,\alpha,\beta) \mid \alpha^2 + (\beta-1)^2 \leq 1\}. \tag{15.2}$$

Example 15.5. The set C defined by (15.2) is a closed convex subset of $\ell_2(3)$ such that $0 \notin C$ and $V_C(0)$ is not closed.

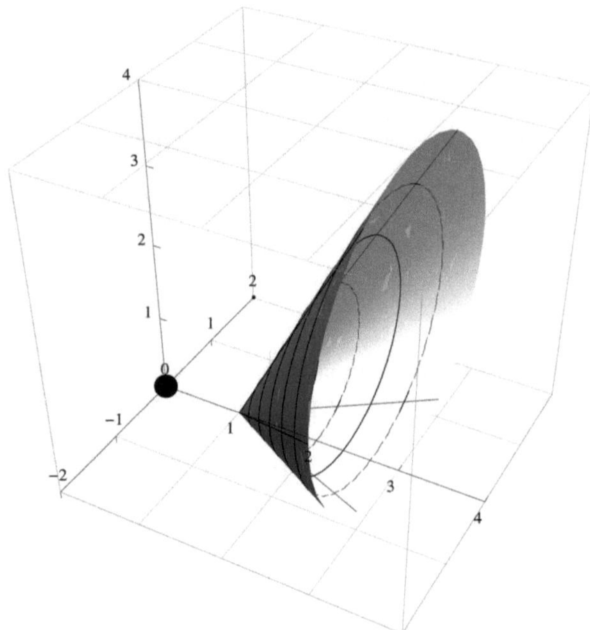

Fig. 15.1 The set C from Example 15.5

Proof. The result is geometrically obvious (see Fig. 15.1) by observing that the points $(2,\sin t, 1+\cos t)$ are in $V_C(0)$ for each $0<t<\pi$, but that the limit point $(2,0,0)$ (as $t\to\pi$) is not. However, the formal proof of this fact is a bit lengthy. Clearly, $0\notin C$ since the first component of any element of C is at least 1. We first verify the following claim.

Claim. *The points* $v(t):=(2,\,\sin t,\,1+\cos t)$ *are in* $V_C(0)$ *for each* $0<t<\pi$.

Using the classical trig identity $\sin^2 t+\cos^2 t=1$, it is clear that $v(t)\in C$ for each $0<t<\pi$. To complete the proof of the claim, it is enough to show that $[0,v(t)[\cap C=\emptyset$ for each $0<t<\pi$. By way of contradiction, suppose the claim is false. Then there exists $0<t_0<\pi$ such that $[0,v(t_0)[\cap C\neq\emptyset$. Since $0\notin C$, it follows that there exists $0<\lambda<1$ such that $\lambda v(t_0)\in C$. That is,

$$\begin{aligned}\lambda(2,\,\sin t_0,1+\cos t_0)\in C &= (1,0,0)+\operatorname{cone}\{(1,\alpha,\beta)\mid \alpha^2+(\beta-1)^2\le 1\}\\ &= (1,0,0)+\cup_{\rho\ge 0}\rho\{(1,\alpha,\beta)\mid \alpha^2+(\beta-1)^2\le 1\}.\end{aligned}$$

Since $\lambda\sin t_0\neq 0$, it follows that for some $\rho>0$,

$$\lambda(2,\,\sin t_0,1+\cos t_0)=(1,0,0)+\rho(1,\alpha,\beta) \tag{15.3}$$

for some α and β such that

$$\alpha^2+(\beta-1)^2\le 1. \tag{15.4}$$

By equating the corresponding components in (15.3), we obtain

$$2\lambda = 1+\rho \tag{15.5}$$

$$\lambda \sin t_0 = \rho\alpha \tag{15.6}$$

$$\lambda(\cos t_0 + 1) = \rho\beta \tag{15.7}$$

From (15.5) is deduced that $\rho = 2\lambda - 1 < 2 - 1 = 1$ and hence that

$$0 < \rho < 1. \tag{15.8}$$

Also, from (15.6) and (15.7) we deduce that $\alpha = \mu \sin t_0$ and $\beta = \mu(1+\cos t_0)$, where $\mu := \lambda/\rho$. Substituting these values for α and β into (15.4), we deduce after some algebra that $1 \geq 2\mu^2(1+\cos t_0) - 2\mu(1+\cos t_0) + 1$. Subtracting 1 from both sides of this inequality and then dividing both sides of the resulting inequality by the positive number $2\mu(1+\cos t_0)$, we obtain $\mu \leq 1$, i.e., $\lambda \leq \rho$. From (15.5), it follows that $\rho \geq 1$, which contradicts (15.8). This proves the claim.

It remains to note that the limit point $\lim_{t\to\pi} v(t) = v(\pi) = (2,0,0)$ is *not* in $V_C(0)$. For this, it is enough to note that $[0, v(\pi)[\cap C \neq \emptyset$. And for this, it suffices to show that $(3/4)v(\pi) \in C$. But

$$\frac{3}{4}v(\pi) = \left(\frac{6}{4},0,0\right) = (1,0,0) + \frac{1}{2}(1,0,0) \in C.$$

■

The following simple fact will be useful to us. It shows that the visible set mapping V_C satisfies a translation property that is also satisfied by the (generally set-valued) metric projection P_C.

Lemma 15.6. *Let C be a closed convex set and $x, y \in X$. Then*

$$V_C(x) = V_{C+y}(x+y) - y. \tag{15.9}$$

Proof. Let $v \in C$. Note that $v \in V_C(x) \Leftrightarrow [x, v[\cap C = \emptyset \Leftrightarrow [x+y, v+y[\cap (C+y) = \emptyset$ $\Leftrightarrow v+y \in V_{C+y}(x+y) \Leftrightarrow v \in V_{C+y}(x+y) - y$. ■

It is natural to ask which closed convex sets C have the property that $V_C(x) = C$ for each $x \notin C$. That is, for which sets is the whole set visible to any point outside the set? The next result shows that this is precisely the class of affine sets. Recall that a set A is *affine* if the line through each pair of points in A lies in A. That is, if the line $\operatorname{aff}\{a_1, a_2\} := \{\alpha_1 a_1 + \alpha_2 a_2 \mid \alpha_1 + \alpha_2 = 1\} \subset A$ for each pair $a_1, a_2 \in A$. Equivalently, A is affine if and only if $A = M + a$ for some (unique) linear subspace

M (namely, $M = A - A$) and (any) $a \in A$. Finally, the *affine hull* of a set C, $\operatorname{aff}(C)$, is the intersection of all affine sets which contain C. As is well known,

$$\operatorname{aff}(C) = \left\{ \sum_{j \in J} \alpha_j x_j \;\middle|\; J \text{ finite}, \sum_{j \in J} \alpha_j = 1, x_j \in C \right\}. \tag{15.10}$$

Proposition 15.7. *Let C be a closed convex set in X. Then the following statements are equivalent:*

1. *C is affine.*
2. *$V_C(x) = C$ for each $x \in X \setminus C$.*

Proof. (1) $\Rightarrow$ (2): Let us assume first that $C = M$ is actually a subspace, i.e., that $0 \in C$. Fix any $x \notin M$. Since $V_M(x) \subset M$, it suffices to show that $M \subset V_M(x)$. To this end, let $m \in M$. If $m \notin V_M(x)$, then $[x,m[\cap M \neq \emptyset$. Hence there exists $0 < \lambda < 1$ such that $\lambda x + (1-\lambda)m \in M$. Since $m \in M$, this implies that $\lambda x \in M$ and hence $x \in M$, a contradiction. This proves (2) in case C is a subspace.

In general, suppose C is affine. Then $C = M + c$ for some subspace M and $c \in C$. For any $x \in X \setminus C$, we see that $x - c \notin M$ and by the above proof and Lemma 15.6 we obtain

$$V_C(x) = V_{M+c}(x) = V_M(x-c) + c = M + c = C.$$

(2) $\Rightarrow$ (1): Assume (2) holds. If C is not affine, then there exist distinct points c_1, c_2 in C such that $\operatorname{aff}\{c_1,c_2\} \not\subset C$. Since C is closed convex and $\operatorname{aff}\{c_1,c_2\}$ is a line, it follows that either $\operatorname{aff}\{c_1,c_2\} \cap C = [y_1,y_2]$ or $\operatorname{aff}\{c_1,c_2\} \cap C = y_1 + \{\rho(y_2 - y_1) \mid \rho \geq 0\}$ for some distinct points y_1, y_2 in C. In either case, it is easy to verify that $x := 2y_1 - y_2 \notin C$. Also, $y_1 = \frac{1}{2}x + \frac{1}{2}y_2 \in [x, y_2[\cap C$, which proves that $y_2 \notin V_C(x)$ and hence contradicts the hypothesis that $V_C(x) = C$. Thus C must be affine. ■

Definition 15.8. Let C be a closed convex subset of X. For any point $y \in X$, we define the **translated cone** C_y of C by

$$C_y := \operatorname{cone}(C - y) + y.$$

Some basic facts about the translated cone follow.

Lemma 15.9. *Let C be a closed convex set in X. Then the following statements hold:*

1. *$C_y \supset C$ for each $y \in X$.*
2. *The set* $\operatorname{cone}(C - y)$, *and hence also C_y, is not closed in general.*
3. *If $y \in C$ and the set* $\operatorname{cone}(C - y)$ *is closed, then $C_y = T_C(y) + y$, where $T_C(y)$ is the* tangent cone *to C at y.*

Proof.

1. $C_y = \text{cone}(C-y)+y \supset C-y+y = C$.
2. Consider the closed ball C of radius one in the Euclidean plane centered at the point $(0,1)$ and let y denote the origin $(0,0)$. Then C_y is the open upper half-plane plus the origin, which is not closed.
3. This follows since the definition of the tangent cone to C at the point $y \in C$ is given by $T_C(y) = \overline{\text{cone}}(C-y)$ (see, e.g., [1, p. 100]).

■

One can also characterize the visible points via the translated cone.

Lemma 15.10. *Let C be a closed convex set in X, $x \in X \setminus C$, and $v \in C$. Then $v \in V_C(x)$ if and only if $x \notin C_v$. Equivalently, $v \notin V_C(x)$ if and only if $x \in C_v$.*

Proof. If $v \notin V_C(x)$, then $[x,v[\cap C \neq \emptyset$. Thus there exists $0 < \lambda < 1$ such that $y := \lambda x + (1-\lambda)v \in C$. Hence $x - v = (1/\lambda)(y-v) \in \text{cone}(C-v)$ and therefore $x \in C_v$.

Conversely, if $x \in C_v$, then there exist $\rho \geq 0$ and $y \in C$ such that $x = \rho(y-v)+v = \rho y + (1-\rho)v$. If $\rho \leq 1$, then x, being a convex combination of two points in C, must lie in C, a contradiction. It follows that $\rho > 1$ and $y = (1/\rho)x + ((\rho-1)/\rho)v \in [x,v[\cap C$. Thus $[x,v[\cap C \neq \emptyset$, and so $v \notin V_C(x)$ by (15.1). ■

The following proposition shows that the translated cones of C form the *external* building blocks for C.

Proposition 15.11. *Let C be a closed convex set in X. Then*

$$\bigcap_{y \in \text{bd}C} C_y = \bigcap_{y \in C} C_y = \bigcap_{y \in X} C_y = C.$$

Proof. By Lemma 15.9, $\cap_{y \in X} C_y \supset C$. Thus to complete the proof, it suffices to show that $\cap_{y \in \text{bd}C} C_y \subset C$. If not, then there exists $x \in \cap_{y \in \text{bd}C} C_y \setminus C$. Thus $x \in C_y \setminus C$ for each $y \in \text{bd}C$. By Lemma 15.10 $y \notin V_C(x)$ for all $y \in \text{bd}C$. But $V_C(x) \subset \text{bd}C$ by Lemma 15.3(2). This shows that $V_C(x) = \emptyset$, which contradicts Lemma 15.3(1). ■

A somewhat deeper characterization of visible points is available by using the strong separation theorem. Recall that two sets C_1 and C_2 in the normed linear space X can be *strongly separated* by a continuous linear functional $x^* \in X^*$ if

$$\sup_{y \in C_1} x^*(y) < \inf_{z \in C_2} x^*(z). \tag{15.11}$$

One can also interpret strong separation geometrically. Suppose C_1 and C_2 are strongly separated by the functional x^* such that (15.11) holds. Let b be any scalar such that

$$\sup_{y \in C_1} x^*(y) \leq b \leq \inf_{z \in C_2} x^*(z).$$

Define the hyperplane H and the (open) half-spaces H^+ and H^- by

$$H := \{y \in X \mid x^*(y) = b\}, \quad H^+ := \{y \in X \mid x^*(y) > b\}, \text{ and}$$
$$H^- := \{y \in X \mid x^*(y) < b\}.$$

(Note that H, H^-, and H^+ are disjoint sets such that $X = H \cup H^- \cup H^+$.) Then H is said to *strongly separate* the sets C_1 and C_2 in the sense that $C_1 \subset H \cup H^-$, $C_2 \subset H \cup H^+$, and (at least) one of the sets C_1 or C_2 is disjoint from H.

Fact 15.12 (Strong Separation Theorem; see [4, Theorem V.2.10, p. 417]). Let C_1 and C_2 be two disjoint closed convex sets in X, one of which is compact. Then the sets can be strongly separated by a continuous linear functional.

Definition 15.13. Let K be a convex subset of X. A point $e \in K$ is called an *extreme point* of K if $k_1 \in K$, $k_2 \in K$, $0 < \lambda < 1$, and $e = \lambda k_1 + (1-\lambda)k_2$ imply that $k_1 = k_2 = e$. The set of extreme points of K is denoted by $\operatorname{ext} K$.

The following fact is well known (see, e.g., [4, pp. 439–440]), and it will be needed in this section and the next.

Fact 15.14 (Krein–Milman). Let K be a nonempty compact convex subset of X. Then:

1. K has extreme points and K is the closed convex hull of its extreme points: $K = \overline{\operatorname{conv}}(\operatorname{ext} K)$.
2. If $x^* \in X^*$, then x^* attains its maximum (resp., minimum) value over K at an extreme point of K.

Theorem 15.15 (Another characterization of visible points). *Let C be a closed convex subset of X, $x \in X \setminus C$, and $v \in C$. Then the following statements are equivalent:*

1. *v is visible to x with respect to C.*
2. *For each point $y \in]x,v[$, there exists a functional $x^* \in X^*$ that strongly separates $[x,y]$ and C, and $x^*(y) = \max_{z\in[x,y]} x^*(z)$.*
3. *For each point $y \in]x,v[$, there exists a hyperplane $H = H_y$ that contains y and strongly separates $[x,y]$ and C.*

Proof. (1) $\Rightarrow$ (2): Suppose v is visible to x from C. Then $[x,v[\cap C = \emptyset$. In particular, for each $y \in [x,v[$, $[x,y] \cap C \subset [x,v[\cap C = \emptyset$. Thus $[x,y]$ and C are disjoint closed convex sets, and $[x,y]$ is compact. By Fact 15.12, there exists $x^* \in X^*$ such that

$$b := \sup_{z\in[x,y]} x^*(z) < \inf_{c\in C} x^*(c). \tag{15.12}$$

To verify (2), it remains to show that $x^*(y) = b$. If $x = y$, this is clear. Thus we may assume that $x \neq y$. Since $[x,y]$ is compact, the supremum on the left side of (15.12) is attained. Further, this maximum must be attained at an extreme point of $[x,y]$ by Fact 15.14(2). Since x and y are the only two extreme points of $[x,y]$, we must have $x^*(x) = b$ or $x^*(y) = b$.

Suppose $x^*(x) = b$. Since $v \in C$, we have $x^*(v) > b$ by (15.12). Since $y \in]x,v[$, there exists $0 < \lambda < 1$ such that $y = \lambda x + (1-\lambda)v$. Then

$$x^*(y) = \lambda x^*(x) + (1-\lambda)x^*(v) > \lambda b + (1-\lambda)b = b,$$

which contradicts the definition of b. Thus the condition $x^*(x) = b$ is not possible, and we must have that $x^*(y) = b$, which verifies (2).

(2) $\Rightarrow$ (3): Assume (2) holds. Let $y \in]x,v[$. Choose $x^* \in X^*$ as in (2), and define $H := \{z \in X \mid x^*(z) = b\}$, where $b = \max_{z \in [x,y]} x^*(z)$. Then H strongly separates $[x,y]$ and C, $x^*(y) = b$, and so $y \in H$. Thus (3) holds.

(3) $\Rightarrow$ (1): Suppose (3) holds but (1) fails. Then $[x,v[\cap C \neq \emptyset$. Choose any $y \in]x,v[\cap C$. By (3), there is a hyperplane H that strongly separates $[x,y]$ and C such that $y \in H$. Writing $H = \{z \in X \mid x^*(z) = b\}$, we see that $[x,y] \subset \{z \in X \mid x^*(z) \leq b\}$, $C \subset \{z \in X \mid x^*(z) > b\}$, and $x^*(y) = b$. But $y \in C$ and hence $x^*(y) > b$, which is a contradiction. ■

15.3 Visibility and Best Approximation

In this section we explore the connection between visibility and best approximation. The first such result states that the set of best approximations to x from C is always contained in the set of visible points to x with respect to C.

Lemma 15.16. *Let C be a closed convex subset of X. Then $P_C(x) \subset V_C(x)$ for each $x \in X$.*

Proof. The result is trivial if $P_C(x) = \emptyset$. If $x \in C$, then clearly $P_C(x) = \{x\}$ and $V_C(x) = \{x\}$ by Lemma 15.2.

Now suppose $x \in X \setminus C$ and let $x_0 \in P_C(x)$. Then $x_0 \in C$ so $x_0 \neq x$. If $[x,x_0[\cap C \neq \emptyset$, then there exists $0 < \lambda < 1$ such that $x_\lambda := \lambda x + (1-\lambda)x_0 \in C$. Hence

$$\|x - x_\lambda\| = \|(1-\lambda)(x - x_0)\| = (1-\lambda)\|x - x_0\| < \|x - x_0\|,$$

which is a contradiction to x_0 being a closest point in C to x. This shows that $[x,x_0[\cap C = \emptyset$ and hence that $x_0 \in V_C(x)$. ■

Recall that if X is a strictly convex reflexive Banach space, then each closed convex subset C is *Chebyshev* (see, e.g., [7]). That is, for each $x \in X$, there is a unique best approximation (i.e., nearest point) $P_C(x)$ to x from C. As is well known, the most important example of a strictly convex reflexive Banach space is a Hilbert space. It is convenient to use the following notation. If S is any subset of X, then the *convex hull* of S is denoted by $\mathrm{conv}(S)$ and the closed convex hull of S is denoted by $\overline{\mathrm{conv}}(S)$.

Another such relationship between visibility and best approximation is the following.

Lemma 15.17. *Let X be a strictly convex reflexive Banach space and C a closed convex subset of X. Then C is a Chebyshev set and if $x \in X \setminus C$, then*

$$P_C(x) = P_{V_C(x)}(x) = P_{\overline{\text{conv}}\, V_C(x)}(x). \tag{15.13}$$

Proof. By Lemma 15.16, $P_C(x) \in V_C(x)$. Since $V_C(x) \subset \overline{\text{conv}}\, V_C(x) \subset C$, it follows that $P_C(x) \in P_{V_C(x)}(x)$ and $P_C(x) = P_{\overline{\text{conv}}\, V_C(x)}(x)$. Thus $P_{V_C(x)}(x)$ is a singleton and (15.13) holds. ■

While the Krein–Milman theorem [Fact 15.14(1)] shows that the set of extreme points $\text{ext}\,C$ of a compact convex set C forms the *internal* building blocks of C, the next result shows that the sets C_e, where $e \in \text{ext}\,C$, form the *external* building blocks for C. It is a sharpening of Proposition 15.11 in the special case when the closed convex set C is actually compact.

Theorem 15.18. *Let C be a compact convex set in X. Then*

$$C = \bigcap\{C_e \mid e \in \text{ext}\,C\} = \bigcap\{C_y \mid y \in C\}. \tag{15.14}$$

Proof. Using Proposition 15.11, it suffices to show that $\cap\{C_e \mid e \in \text{ext}\,C\} \subset C$. If not, then there exists $x \in \cap\{C_e \mid e \in \text{ext}\,C\} \setminus C$. By Fact 15.12, there exists $x^* \in X^*$ such that

$$s := \sup_{c \in C} x^*(c) < x^*(x). \tag{15.15}$$

By compactness of C, the supremum of x^* over C is attained, i.e., there exists $c_0 \in C$ such that $x^*(c_0) = s$. As is easily verified, the set

$$\tilde{C} = C \cap \{y \in X \mid x^*(y) = s\} \tag{15.16}$$

is extremal in C and has extreme points (since it is a closed, hence compact, convex subset of C), and each extreme point of $\tilde{C}$ is an extreme point of C (see, e.g., [4, pp. 439–440]). Choose any extreme point $\tilde{c}$ in $\tilde{C}$. Then $\tilde{c} \in \text{ext}\,C$. Also, $x \in C_{\tilde{c}} = \text{cone}\,(C - \tilde{c}) + \tilde{c}$ implies that $x = \rho(c - \tilde{c}) + \tilde{c}$ for some $\rho > 0$ and $c \in C$ (see, e.g., [3, Theorem 4.4(5), p. 45]). Hence

$$s < x^*(x) = \rho[x^*(c) - x^*(\tilde{c})] + x^*(\tilde{c}) \leq x^*(\tilde{c}) = s,$$

which is impossible. This contradiction completes the proof. ■

Proposition 15.19. *Let C be a closed convex set in X, $x \in X \setminus C$, and let $x_0 \in C$ be a proper convex combination of points e_i in C. That is, $x_0 = \sum_1^k \lambda_i e_i$ for some $\lambda_i > 0$ with $\sum_1^k \lambda_i = 1$. If x_0 is visible to x with respect to C, then each e_i is also visible to x.*

Proof. If $k = 1$ the result is trivial. Assume that $k = 2$. (We will reduce the general case to this case.)

If the result were false, then we may assume without loss of generality that e_1 is not visible to x. Thus $]x, e_1[\cap C \neq \emptyset$. Hence there exists $0 < \mu < 1$ such that $x_1 := \mu x + (1-\mu)e_1 \in C$. It follows that

$$e_1 = \frac{1}{1-\mu}x_1 - \frac{\mu}{1-\mu}x. \tag{15.17}$$

Next consider, for each $\rho \in [0,1]$, the expression $x(\rho) := \rho x_1 + (1-\rho)e_2$. Clearly, $x(\rho) \in C$ for all such ρ since both x_1 and e_2 are in C and C is convex. Omitting some simple algebra, we deduce that

$$\begin{aligned} x(\rho) &= \rho[\mu x + (1-\mu)e_1] + (1-\rho)e_2 \\ &= \rho\mu x + (1-\rho\mu)x_0 + \rho(1-\mu)e_1 + (1-\rho)e_2 - (1-\rho\mu)x_0 \\ &= \rho\mu x + (1-\rho\mu)x_0 + [\rho(1-\mu+\lambda_1\mu) - \lambda_1]e_1 + [-\rho(1-\mu+\lambda_1\mu)+\lambda_1]e_2. \end{aligned}$$

In particular, if we choose

$$\tilde{\rho} := \frac{\lambda_1}{1-\mu+\lambda_1\mu}, \tag{15.18}$$

it is not hard to check that $0 < \tilde{\rho} < 1$. Thus $0 < \tilde{\rho}\mu < 1$ and

$$x(\tilde{\rho}) = \tilde{\rho}\mu x + (1-\tilde{\rho}\mu)x_0 \in C. \tag{15.19}$$

This proves that $x(\tilde{\rho}) \in]x, x_0[\cap C$, which contradicts the fact that x_0 is visible to x.

Finally, consider the case when $k \geq 3$. If the result were false, then without loss of generality, we may assume that e_1 fails to be visible to x. Write

$$x_0 = \lambda_1 e_1 + \mu \sum_{i=2}^{k} \frac{\lambda_i}{\mu} e_i,$$

where $\mu := \sum_2^k \lambda_i = 1 - \lambda_1$. Then $0 < \mu < 1$, $\lambda_1 = 1-\mu$, and $x_0 = (1-\mu)e_1 + \mu y$, where $y = \sum_2^k \frac{\lambda_i}{\mu} e_i \in C$ by convexity. By the case when $k = 2$ that we proved above, we get that e_1 (as well as y) is visible to x, which is a contradiction. ■

Remark 15.20. Simple examples in the plane (e.g., a triangle) show that the converse to Proposition 15.19 is *false*! That is, one could have a closed convex set C, a point $x \in X \setminus C$, points $e_i \in V_C(x)$ for $i = 1, 2, \ldots, k$, $k \geq 2$, but $x_0 = \frac{1}{k}\sum_1^k e_i \in C$ is not visible to x.

Theorem 15.21. *Let C be a closed and bounded convex set in an n-dimensional normed linear space X. Then*

$$C = \left\{ \sum_1^k \lambda_i e_i \;\middle|\; 1 \leq k \leq n+1,\ \lambda_i \geq 0,\ \sum_1^k \lambda_i = 1,\ e_i \in \operatorname{ext} C \right\}. \tag{15.20}$$

Further, let $x \in X \setminus C$. Then each point in $P_C(x)$ is a proper convex combination of no more than $n+1$ extreme points of C all of which are visible to x with respect to C. That is,

$$P_C(x) \subset \left\{ \sum_1^k \lambda_i e_i \;\middle|\; 1 \le k \le n+1,\ \lambda_i \ge 0,\ \sum_1^k \lambda_i = 1,\ e_i \in (\operatorname{ext} C) \cap V_C(x) \right\}. \tag{15.21}$$

Proof. By [6, Corollary 18.5.1], $C = \operatorname{conv}(\operatorname{ext} C)$. By Caratheodory's theorem (see, e.g., [2, p. 17]), each point in $\operatorname{conv}(\operatorname{ext} C)$ may be expressed as a convex combination of at most $n+1$ points of $\operatorname{ext} C$. That is,

$$\operatorname{conv}(\operatorname{ext} C) = \left\{ \sum_1^{n+1} \lambda_i e_i \;\middle|\; e_i \in \operatorname{ext} C, \lambda_i \ge 0, \sum_1^{n+1} \lambda_i = 1 \right\}. \tag{15.22}$$

This proves (15.20).

Now let $x \in X \setminus C$. By the first part, each point of $P_C(x)$ is in $\operatorname{conv}(\operatorname{ext} C)$. By Proposition 15.19 and Lemma 15.16, (15.21) follows. ∎

15.4 Best Approximation from a Simplex

In this section we investigate the problem of finding best approximations from a *polytope*, i.e., the convex hull of a finite number of points in a Hilbert space X. Such sets are compact (because they are closed and bounded in a finite-dimensional subspace).

Let $E := \{e_0, e_1, \ldots, e_n\}$ be a set of $n+1$ points in X that is *affinely independent*, i.e., $\{e_1 - e_0, e_2 - e_0, \ldots, e_n - e_0\}$ is linearly independent. This implies that each point in the convex hull $C = \operatorname{conv}\{e_0, e_1, \ldots, e_n\}$ has a unique representation as a convex combination of the points of E. In this case, C is also called an *n-dimensional simplex* with vertices e_i, since the dimension of the affine hull $\operatorname{aff}(C)$ of C is n. Further, the *relative interior* of C, that is, the interior of C relative to $\operatorname{aff}(C)$, is given by

$$\operatorname{ri}(C) := \left\{ \sum_{i=0}^n \lambda_i e_i \mid \lambda_i > 0,\ \sum_{i=0}^n \lambda_i = 1 \right\}. \tag{15.23}$$

It follows that the *relative boundary* of C, $\operatorname{rbd}(C) := C \setminus \operatorname{ri}(C)$, is given by

$$\operatorname{rbd}(C) = \left\{ \sum_{i=0}^n \lambda_i e_i \mid \lambda_i \ge 0,\ \sum_{i=0}^n \lambda_i = 1,\ \lambda_j = 0 \text{ for at least one } j \right\}. \tag{15.24}$$

(See [6, p. 44ff] and [5, p. 7ff] for more detail and proofs about the facts stated in this paragraph.)

We consider sets of affinely independent points, since this case captures the essence of our constructions and arguments. Convex hulls of n affinely dependent points (i.e., finite point sets that are not affinely independent) can be split into the union of a finite number of convex hulls of subsets of affinely independent points. Thus the problem of finding best approximation from the convex hull of an affinely dependent set of points can be reduced to a finite number of problems analogous to the one that we consider below in detail.

Under the above hypothesis that C is an n-dimensional simplex, we wish to compute $P_C(x)$ for any $x \in X$.

We give an explicit formula for $P_C(x)$ in the case when $n = 1$, that is, when $C = [e_0, e_1]$ is a line segment. Then, by a recursive argument, we will indicate how to compute $P_C(x)$ when C is an n-dimensional simplex for any $n \geq 2$. First we recall that the *truncation function* $[\cdot]_0^1$ is defined on the set of real numbers by

$$[\alpha]_0^1 = \begin{cases} 0 & \text{if } \alpha < 0, \\ \alpha & \text{if } 0 \leq \alpha \leq 1, \\ 1 & \text{if } \alpha > 1. \end{cases}$$

(Note that in the space $X = \mathbb{R}$, $[\alpha]_0^1 = P_{[0,1]}(\alpha)$ for all $\alpha \in \mathbb{R}$.)

Proposition 15.22. *Let $C = \operatorname{conv}\{e_0, e_1\} = [e_0, e_1]$ be a 1-dimensional simplex. Then, for each $x \in X$,*

$$P_C(x) = e_0 + \left[\frac{\langle x - e_0, e_1 - e_0\rangle}{\|e_1 - e_0\|^2}\right]_0^1 (e_1 - e_0). \tag{15.25}$$

Proof. Let $\alpha := \langle x - e_0, e_1 - e_0\rangle \|e_1 - e_0\|^{-2}$ and $c_0 := e_0 + [\alpha]_0^1(e_1 - e_0)$. Then $c_0 \in C$, and by the well-known characterization of best approximations from convex sets in Hilbert space (see, e.g., [3, p. 43]) it suffices to show that

$$\langle x - c_0, y - c_0\rangle \leq 0 \quad \text{for each } y \in C. \tag{15.26}$$

Let $y \in C$. Then $y = e_0 + \lambda(e_1 - e_0)$ for some $\lambda \in [0, 1]$. Hence

$$\begin{aligned} \langle x - c_0, y - c_0\rangle &= \langle x - e_0 - [\alpha]_0^1(e_1 - e_0), \lambda(e_1 - e_0) - [\alpha]_0^1(e_1 - e_0)\rangle \\ &= (\lambda - [\alpha]_0^1)[\langle x - e_0, e_1 - e_0\rangle - [\alpha]_0^1\|e_1 - e_0\|^2] \\ &= (\lambda - [\alpha]_0^1)\|e_1 - e_0\|^2 \left[\alpha - [\alpha]_0^1\right]. \end{aligned}$$

By considering the three possible cases: $\alpha < 0$, $\alpha \in [0, 1]$, and $\alpha > 1$, it is easy to see that the last expression is always ≤ 0. Hence (15.26) is verified. ∎

Before considering the cases when $n \geq 2$, let us first consider the problem of computing $P_A(x)$ for any $x \in X$, where $A = \operatorname{aff} C$.

Fact 15.23. Let $C = \operatorname{conv}\{e_0, e_1, \ldots, e_n\}$ be an n-dimensional simplex, and let $A = \operatorname{aff}(C)$. For any $x \in X$, we have

$$P_A(x) = e_0 + \sum_{j=1}^{n} \alpha_j (e_j - e_0), \tag{15.27}$$

where the scalars α_i satisfy the "normal" equations:

$$\sum_{j=1}^{n} \alpha_j \langle e_j - e_0, e_i - e_0 \rangle = \langle x - e_0, e_i - e_0 \rangle \qquad (i = 1, 2, \ldots, n). \tag{15.28}$$

The proof of this fact can be found e.g., in [1, p. 418] or [3, p. 215]. Moreover, the "reduction principle" that was established in [3, p. 80] (where it was stated in the particular case of a subspace) can be easily extended to affine sets as follows.

Fact 15.24 (Reduction principle). Let C be a closed convex set in the Hilbert space X and let $A = \overline{\operatorname{aff}}(C)$. Then $P_C = P_C \circ P_A$. That is, for each $x \in X$,

$$P_C(x) = P_C(P_A(x)) \qquad \text{and} \qquad d^2(x, C) = d^2(x, A) + d^2(P_A(x), C).$$

We are going to use the Reduction Principle as follows. We assume that it is straightforward to find the best approximation to any x in the set $A = \operatorname{aff} C$, where C is an n-dimensional simplex (since it involves only solving a linear system of n equations in n unknowns by Fact 15.23). The Reduction Principle states that (by replacing x with $P_A(x)$ if necessary) we may as well assume that our point x is in A to begin with, and we shall do this in what follows. We will see that the case when $n = 2$ can be reduced to the case when $n = 1$ (i.e., Proposition 15.22 above) for which there is an explicit formula.

Proposition 15.25. *Let $C = \operatorname{conv}\{e_0, e_1, e_2\}$ be a 2-dimensional simplex. Then for each $x \in \operatorname{aff}(C)$, either $x \in C$, in which case $P_C(x) = x$, or $x \notin C$, in which case*

$$P_C(x) = P_{[e_i, e_{i+1}]}(x) \quad \textit{for any } i \in \{0, 1, 2\} \textit{ that satisfies} \tag{15.29}$$

$$\|x - P_{[e_i, e_{i+1}]}(x)\| = \min_j \|x - P_{[e_j, e_{j+1}]}(x)\|. \tag{15.30}$$

(Here $e_3 := e_0$.)

Proof. If $x \in C$, then obviously $P_C(x) = x$. Thus we can assume that $x \in \operatorname{aff}(C) \setminus C$. It follows that $P_C(x)$ must lie on $\operatorname{rbd} C = \cup_{i=0}^{2} [e_i, e_{i+1}]$. That is, $P_C(x) \in [e_i, e_{i+1}]$ for some $i = 0, 1$, or 2.

Claim. *$P_C(x) = P_{[e_i,e_{i+1}]}(x)$ for each i such that $P_C(x) \in [e_i, e_{i+1}]$.*

To see this, we observe that since $P_C(x) \in [e_i, e_{i+1}]$, we have

$$\|x - P_C(x)\| = d(x,C) \leq d(x,[e_i,e_{i+1}]) \leq \|x - P_C(x)\|$$

which implies that $\|x - P_{[e_i,e_{i+1}]}(x)\| = d(x,[e_i,e_{i+1}]) = \|x - P_C(x)\|$. By uniqueness of best approximations from convex sets in Hilbert space, the claim is proved.

If k is any index such that $\|x - P_{[e_k,e_{k+1}]}(x)\| = \min_j \|x - P_{[e_j,e_{j+1}]}(x)\|$, then it is clear that we must have $P_C(x) = P_{[e_k,e_{k+1}]}(x)$. ∎

Now it appears to be straightforward to apply the idea of Proposition 15.25 to any n-dimensional simplex to describe how to determine $P_C(x)$.

Let $C = \text{conv}\{e_0, e_1, \ldots, e_n\}$ be an n-dimensional simplex in X and $x \in \text{aff}(C)$. If $x \in C$, we have $P_C(x) = x$. Thus we may assume that $x \in \text{aff}(C) \setminus C$. It follows that $P_C(x) \in \text{rbd}(C)$. From (15.24) we see

$$\text{rbd}(C) = \left\{ \sum_0^n \lambda_i e_i \mid \lambda_i \geq 0, \sum_0^n \lambda_i = 1, \lambda_j = 0 \text{ for some } j \right\}.$$

Since every $y \in \text{rbd}\, C$ is contained in (at least) one of the sets

$$C_j := \left\{ \sum_{i=0}^n \lambda_i e_i \mid \lambda_i \geq 0 \text{ for all } i, \lambda_j = 0, \text{ and } \sum_0^n \lambda_i = 1 \right\}, \tag{15.31}$$

it follows that

$$\text{rbd}\, C = \bigcup_{j=0}^n C_j.$$

Further, each C_j is a simplex of dimension $n-1$ in C, $P_C(x) \in C_j$ for at least one j, and for all such j, we have that

$$\|x - P_C(x)\| = d(x,C) \leq \|x - P_{C_j}(x)\| = d(x,C_j) \leq \|x - P_C(x)\|.$$

This implies that equality holds throughout these inequalities, and hence by the uniqueness of best approximations, we have $P_C(x) = P_{C_j}(x)$. If $J = \{j \mid \|x - P_{C_j}(x)\| = \min_i \|x - P_{C_i}(x)\|\}$, then clearly $P_C(x) = P_{C_j}(x)$ for each $j \in J$.

This discussion suggests the following recursive algorithm for computing $P_C(x)$ when $C = \text{conv}\{e_0, e_1, \ldots, e_n\}$ is an n-dimensional simplex. Let C_j be the $(n-1)$-dimensional simplices as defined in (15.31). Let $A = \text{aff}\, C$, $A_j = \text{aff}\, C_j$ for each $j = 0, 1, \ldots, n$, $x \in A \setminus C$, and $x_j = P_{C_j}(x_j)$ for all j. The algorithm below defines a function $P(n,x,C)$ which takes as input n and x and the set C and returns the best approximation $P_C(x)$.

Algorithm

1. If $n = 1$, then find $P(1, x, C)$ by using the formula given in Proposition 15.22.
2. If $n > 1$, then compute $x_j = P_{A_j}(x)$ and $P_{C_j}(x_j) = P(n-1, x_j, C_j)$ for $j = 0, 1, \ldots, n$.
3. Set $P_C(x) = P_{C_j}(x_j)$ for any $j \in \operatorname{argmin}_k \|x_k - P_{C_k}(x_k)\|$.

Acknowledgements We are greatly indebted to Heinz Bauschke who proofread this paper and made several suggestions that greatly improved the final version. Zikatanov was supported in part by the National Science Foundation DMS-1217142 and Department of Energy DE-SC0006903.

References

1. Bauschke, H.H., Combettes, P.L.: Convex Analysis and Monotone Operator Theory in Hilbert Spaces. Springer, New York (2011)
2. Cheney, E.W.: Introduction to Approximation Theory. McGraw-Hill, New York (1966)
3. Deutsch, F.: Best Approximation in Inner Product Spaces. Springer, New York (2001)
4. Dunford, N., Schwartz, J.T.: Linear Operators Part I: General Theory. Interscience Publ., New York (1958)
5. Holmes, R.B.: Geometric Functional Analysis and Its Applications. Springer, New York (1975)
6. Rockafellar, R.T.: Convex Analysis. Princeton University Press, Princeton (1970)
7. Singer, I.: Best Approximation in Normed Linear Spaces by Elements of Linear Subspaces. Springer, New York (1970)

Chapter 16
On Derivative Criteria for Metric Regularity

Asen L. Dontchev and Hélène Frankowska

Abstract We give a simple self-contained proof of the equality which links directly the graphical derivative and coderivative criteria for metric regularity. Then we present a sharper form of the criterion for strong metric regularity involving the paratingent derivative.

Key words: Set-valued mapping • Metric regularity • Strong metric regularity • Graphical derivative • Coderivative • Paratingent derivative

Mathematics Subject Classifications (2010): Primary 49J52; Secondary 49J53.

16.1 Introduction

In this paper we prove two theorems. The first one is as follows.

Theorem 16.1. *Let* $F : \mathbb{R}^n \rightrightarrows \mathbb{R}^m$ *be a set-valued map, let* $\bar{y} \in F(\bar{x})$*, and assume that* $\operatorname{gph} F$ *is locally closed at* $(\bar{x}, \bar{y})$*. Then*

COMMUNICATED BY HEINZ H. BAUSCHKE.

A.L. Dontchev (✉)
Mathematical Reviews, 416 Fourth Street, Ann Arbor, MI 48107-8604, USA
e-mail: ald@ams.org

H. Frankowska
CNRS, Institut de Mathématiques de Jussieu, Université Pierre et Marie Curie, 4 place Jussieu, 75252 Paris, France
e-mail: frankowska@math.jussieu.fr

D.H. Bailey et al. (eds.), *Computational and Analytical Mathematics*, Springer Proceedings in Mathematics & Statistics 50, DOI 10.1007/978-1-4614-7621-4_16,

$$\limsup_{\substack{(x,y)\to(\bar{x},\bar{y}),\\(x,y)\in\operatorname{gph}F}} \|DF(x\,|\,y)^{-1}\|^{-} = \|D^*F(\bar{x}\,|\,\bar{y})^{-1}\|^{+}. \tag{16.1}$$

The quantity on the left side of (16.1) involves the inner norm of the *graphical derivative* and the condition that it is finite is the so-called *derivative criterion for metric regularity*. The quantity on the right side is the outer norm of the *coderivative* and it is well known that F is metrically regular if and only if this quantity is finite. The graphical derivative and the coderivative are defined in further lines. In the case when F is metrically regular both quantities in (16.1) are equal to the regularity modulus of F. The reader can find these criteria and much more in books [2,5,8,9].

Recall that F is said to be *metrically regular* at $\bar{x}$ for $\bar{y}$ when $\bar{y} \in F(\bar{x})$ and there is a constant $\kappa > 0$ together with neighborhoods U of $\bar{x}$ and V of $\bar{y}$ such that

$$d(x, F^{-1}(y)) \le \kappa d(y, F(x)) \quad \text{for all } (x,y) \in U \times V.$$

The infimum of κ over all combinations of κ, U and V is called the regularity modulus and denoted by $\operatorname{reg}(F;\bar{x}\,|\,\bar{y})$.

Clearly, the equality (16.1) follows immediately from the combination of the derivative and coderivative criteria for metric regularity. In this paper we give a direct proof of (16.1) using a rather elementary duality argument without referring to metric regularity. This proof employs the approach used to prove basically the same result in [7]; however, the proof given here is simpler and, most importantly, self-contained. It may be used in an alternative proof of the coderivative criterion provided that derivative criterion is already proven, and vice versa.

Our second result is a derivative criterion for strong metric regularity. Recall that a mapping $F : \mathbb{R}^n \rightrightarrows \mathbb{R}^m$ is strongly metrically regular at $\bar{x}$ for $\bar{y}$ when there exist neighborhoods U of $\bar{x}$ and V of $\bar{y}$ such that the localization $V \ni y \mapsto F^{-1}(y) \cap U$ of the inverse mapping F^{-1} around $(\bar{y}, \bar{x})$ is a Lipschitz continuous function.

Theorem 16.2. *Consider a set-valued mapping $F : \mathbb{R}^n \rightrightarrows \mathbb{R}^m$ and $(\bar{x}, \bar{y}) \in \operatorname{gph} F$. If F is strongly metrically regular at $\bar{x}$ for $\bar{y}$, then*

$$\|PF(\bar{x}\,|\,\bar{y})^{-1}\|^{+} < \infty. \tag{16.2}$$

Furthermore, if the graph of F is locally closed at $(\bar{x}, \bar{y})$ and

$$\bar{x} \in \operatorname*{Liminf}_{y \to \bar{y}} F^{-1}(y), \tag{16.3}$$

then condition (16.2) *is also sufficient for strong metric regularity of F at $\bar{x}$ for $\bar{y}$. In this case the quantity on the left side of* (16.2) *equals* $\operatorname{reg}(F;\bar{x}\,|\,\bar{y})$.

Here $PF(x\,|\,y)$ denotes the *paratingent* derivative which we define below. Theorem 16.2 sharpens [9, Theorem 9.54], where it is assumed that the mapping F^{-1} has a local continuity property around $(\bar{y}, \bar{x})$ which is much stronger than (16.3). It also improves [8, Lemma 3.1], where another condition, again stronger than (16.3), is used.

Let us briefly introduce the notation and terminology used in the paper. The closed ball with center x and radius r is denoted by $\boldsymbol{B}_r(x)$; the closed unit ball is $\boldsymbol{B}$. We denote by $\|\cdot\|$ the Euclidean norm and by $\langle\cdot,\cdot\rangle$ the usual inner product. The Painlevé–Kuratowski lower and upper limits are denoted by Liminf and Limsup, respectively. A set C is said to be locally closed at $x \in C$ when there exists $r > 0$ such that the set $C \cap \boldsymbol{B}_r(x)$ is closed. For a set $C \subset \mathbb{R}^n$, a tangent vector v to C at $x \in C$, written $v \in T_C(x)$, is a vector for which there exist sequences $v_k \to v$ and $t_k \to 0_+$ such that $x + t_k v_k \in C$. The set of tangents, $T_C(x)$, is a closed cone, named the tangent cone. A paratingent vector w to C at $x \in C$, written $w \in P_C(x)$, is a vector for which there exist sequences $x_k \in C$, $x_k \to x$, $t_k \to 0_+$ and $v_k \to v$ such that $x_k + t_k v_k \in C$. Clearly, $T_C(x) \subset P_C(x)$. The polar K^* to the cone K consists of all vectors y such that $\langle y, x\rangle \leq 0$ for all $x \in K$. As is well known, $K^{**} = \operatorname{clco} K$; here and later "clco" means closed convex hull. The regular normal cone to a set C at a point $x \in C$, denoted $\hat{N}_C(x)$, is defined as the polar $T_C^*(x)$ to the tangent cone to C at x. A vector w is a generalized normal to C at x, written $w \in N_C(x)$, when there are sequences $u_k \to w$ and $x_k \to x$, $x_k \in C$ such that $u_k \in \hat{N}_C(x_k)$. The set of generalized normals $N_C(x)$ is the general normal cone to C at x. That is, $N_C(x) = \operatorname*{Limsup}_{y \to x,\, y \in C} \hat{N}_C(y) \supset \hat{N}_C(x)$.

Consider a mapping $F : \mathbb{R}^n \rightrightarrows \mathbb{R}^m$ and denote by $\operatorname{gph} F$ its graph defined by $\operatorname{gph} F := \{(x,y) \,|\, y \in F(x)\}$. For a pair (x,y) with $y \in F(x)$, recall that the *graphical (also called contingent) derivative* of F at x for y is the mapping $DF(x\,|\,y) : \mathbb{R}^n \rightrightarrows \mathbb{R}^m$ whose graph is the tangent cone $T_{\operatorname{gph} F}(x,y)$ to $\operatorname{gph} F$ at (x,y):

$$v \in DF(x\,|\,y)(u) \quad \Leftrightarrow \quad (u,v) \in T_{\operatorname{gph} F}(x,y).$$

The *coderivative* of F at x for y is the mapping $D^*F(x\,|\,y) : \mathbb{R}^m \rightrightarrows \mathbb{R}^n$ whose graph is defined by the general normal cone $N_{\operatorname{gph} F}(x,y)$ to $\operatorname{gph} F$ at (x,y) in the following way:

$$q \in D^*F(x\,|\,y)(p) \quad \Leftrightarrow \quad (q,-p) \in N_{\operatorname{gph} F}(x,y).$$

Finally, the *paratingent derivative* of F at x for y is the mapping $PF(x\,|\,y) : \mathbb{R}^n \rightrightarrows \mathbb{R}^m$ whose graph is the paratingent cone $P_{\operatorname{gph} F}(x,y)$ to $\operatorname{gph} F$ at (x,y):

$$v \in PF(x\,|\,y)(u) \quad \Leftrightarrow \quad (u,v) \in P_{\operatorname{gph} F}(x,y).$$

Both the tangent and the paratingent cones were introduced by Bouligand in 1930s. Further discussion on tangent cones and graphical derivatives can be found for instance in [2]. The paratingent derivative is called in [9] the *strict graphical derivative* and in [8] it is called *Thibault's limit set.* Directly from the definition we have $DF^{-1}(y\,|\,x) = DF(x\,|\,y)^{-1}$ and the same for the coderivative and the paratingent derivative.

A mapping $H : \mathbb{R}^n \rightrightarrows \mathbb{R}^m$ is said to be positively homogeneous if its graph is a cone with vertex at zero. For any positively homogeneous mapping H, the outer norm and the inner norm are defined, respectively, by

$$\|H\|^+ = \sup_{\|x\|\le 1} \sup_{y\in H(x)} \|y\| \quad \text{and} \quad \|H\|^- = \sup_{\|x\|\le 1} \inf_{y\in H(x)} \|y\|$$

with the convention $\inf_{y\in\emptyset} \|y\| = \infty$ and $\sup_{y\in\emptyset} \|y\| = -\infty$. The inner norm can be also defined as

$$\|H\|^- = \inf\Big\{ \kappa > 0 \,\Big|\, H(x) \cap \kappa \boldsymbol{B} \neq \emptyset \text{ for all } x \in \boldsymbol{B} \Big\}, \tag{16.4}$$

while the outer norm satisfies

$$\|H\|^+ = \inf\Big\{ \kappa > 0 \,\Big|\, y \in H(x) \Rightarrow \|y\| \le \kappa \|x\| \Big\}. \tag{16.5}$$

If H has closed graph, then furthermore $\|H\|^+ < \infty \;\Leftrightarrow\; H(0) = \{0\}$. The notation and terminology used in the paper are mainly from [5].

16.2 Proof of Theorem 16.1

In the proof of Theorem 16.1 we employ the following lemma whose proof is presented after the proof of the theorem.

Lemma 16.3. *Let C be a convex and compact set in $\mathbb{R}^d$, $K \subset \mathbb{R}^d$ be a closed set and $\bar{x} \in K$. Then $C \cap T_K(x) \neq \emptyset$ for all $x \in K$ near $\bar{x}$ if and only if $C \cap \operatorname{clco} T_K(x) \neq \emptyset$ for all $x \in K$ near $\bar{x}$.*

Proof (of Theorem 16.1). Since the graphical derivative and the coderivative are defined only locally around $(\bar{x}, \bar{y})$, we can assume without loss of generality that the graph of the mapping F is closed. We will show first that

$$\limsup_{\substack{(x,y)\to(\bar{x},\bar{y}),\\ (x,y)\in \operatorname{gph} F}} \|DF(x\,|\,y)^{-1}\|^- \ge \|D^*F(\bar{x}\,|\,\bar{y})^{-1}\|^+. \tag{16.6}$$

If the left side of (16.6) equals $+\infty$ there is nothing to prove. Let a positive constant c satisfy

$$c > \limsup_{\substack{(x,y)\to(\bar{x},\bar{y}),\\ (x,y)\in \operatorname{gph} F}} \|DF(x\,|\,y)^{-1}\|^-.$$

From (16.4) there exists $\delta > 0$ such that for all $(x,y) \in \operatorname{gph} F \cap (\boldsymbol{B}_\delta(\bar{x}) \times \boldsymbol{B}_\delta(\bar{y}))$ and for every $v \in \boldsymbol{B}$ there exists $u \in DF(x\,|\,y)^{-1}(v)$ such that $\|u\| < c$. Also, note that $(u,v) \in T_{\operatorname{gph} F}(x,y) \subset \operatorname{clco} T_{\operatorname{gph} F}(x,y) = T^{**}_{\operatorname{gph} F}(x,y)$.

Fix $(x,y) \in \operatorname{gph} F \cap (\boldsymbol{B}_\delta(\bar{x}) \times \boldsymbol{B}_\delta(\bar{y}))$ and let $v \in \boldsymbol{B} \subset \mathbb{R}^m$. Then there exists u with $(u,v) \in T^{**}_{\operatorname{gph} F}(x,y)$ such that $u = cw$ for some $w \in \boldsymbol{B}$. Let $(p,q) \in \hat{N}_{\operatorname{gph} F}(x,y) = T^*_{\operatorname{gph} F}(x,y)$. From the inequality $\langle u, p\rangle + \langle v, q\rangle \le 0$ we get

$$c\min_{w\in\boldsymbol{B}}\langle w,p\rangle+\langle v,q\rangle\le 0 \quad\text{which yields}\quad -c\|p\|+\langle v,q\rangle\le 0.$$

Since v is arbitrarily chosen in $\boldsymbol{B}$, we conclude that

$$\|q\|\le c\|p\| \quad \text{whenever} \quad (p,q)\in\hat{N}_{\operatorname{gph}F}(x,y). \tag{16.7}$$

Now, let $(p,q)\in N_{\operatorname{gph}F}(\bar{x},\bar{y})$; then there exist sequences $(x_k,y_k)\in\operatorname{gph}F$, $(x_k,y_k)\to(\bar{x},\bar{y})$ and $(p_k,q_k)\in\hat{N}_{\operatorname{gph}F}(x_k,y_k)$ such that $(p_k,q_k)\to(p,q)$. But then, from (16.7), $\|q_k\|\le c\|p_k\|$ and in the limit $\|q\|\le c\|p\|$. Thus, $\|q\|\le c\|p\|$ whenever $(-p,q)\in N_{\operatorname{gph}F}(\bar{x},\bar{y})$ and therefore we have $\|q\|\le c\|p\|$ whenever $(q,-p)\in N_{\operatorname{gph}F^{-1}}(\bar{y},\bar{x})$. By the definition of the coderivative,

$$\|q\|\le c\|p\| \quad \text{whenever} \quad q\in D^*F(\bar{x}\,|\,\bar{y})^{-1}(p).$$

This together with (16.5) implies that $c\ge\|D^*F(x,y)^{-1}\|^+$ and we obtain (16.6) by the arbitrariness of c.

For the converse inequality, it is enough to consider the case $\|D^*F(\bar{x}\,|\,\bar{y})^{-1}\|^+<\infty$. Let

$$c>\|D^*F(\bar{x}\,|\,\bar{y})^{-1}\|^+. \tag{16.8}$$

We first show that there exists $\delta>0$ such that for any $(x,y)\in\operatorname{gph}F\cap(\boldsymbol{B}_\delta(\bar{x})\times\boldsymbol{B}_\delta(\bar{y}))$ we have that

$$(0,v)\in\hat{N}_{\operatorname{gph}F}(x,y) \quad\Longrightarrow\quad v=0. \tag{16.9}$$

On the contrary, assume that there exist sequences $(x_k,y_k)\in\operatorname{gph}F$ with $(x_k,y_k)\to(\bar{x},\bar{y})$ and $v_k\in\mathbb{R}^m$ with $\|v_k\|=1$ such that $(0,v_k)\in\hat{N}_{\operatorname{gph}F}(x_k,y_k)$ for all k. But then there is $v\neq 0$ such that $(0,v)\in N_{\operatorname{gph}F}(\bar{x},\bar{y})$. Hence, there exists a nonzero v such that $v\in D^*F(\bar{x}\,|\,\bar{y})^{-1}(0)$. Taking into account (16.5), this contradicts (16.8).

Using (16.9), we will now prove a statement more general than (16.9) that there exists $\delta>0$ such that for any $(x,y)\in\operatorname{gph}F\cap(\boldsymbol{B}_\delta(\bar{x})\times\boldsymbol{B}_\delta(\bar{y}))$ we have

$$(v,-u)\in\hat{N}_{\operatorname{gph}F^{-1}}(y,x) \quad\Longrightarrow\quad \|v\|\le c\|u\|. \tag{16.10}$$

On the contrary, assume that there exists a sequence $(y_k,x_k)\to(\bar{y},\bar{x})$ such that for each k we can find $(v_k,-u_k)\in\hat{N}_{\operatorname{gph}F^{-1}}(y_k,x_k)$ satisfying $\|v_k\|>c\|u_k\|$. If $u_k=0$ for some k, then from (16.9) we get $v_k=0$, a contradiction. Thus, without loss of generality we assume that $\|u_k\|=1$. Let v_k be unbounded and let w be a cluster point of $\frac{1}{\|v_k\|}v_k$; then $\|w\|=1$. Since $(\frac{1}{\|v_k\|}v_k,-\frac{1}{\|v_k\|}u_k)\in\hat{N}_{\operatorname{gph}F^{-1}}(y_k,x_k)$, passing to the limit we get $(w,0)\in N_{\operatorname{gph}F^{-1}}(\bar{y},\bar{x})$ which contradicts (16.8) because of (16.5). Further, if v_k is bounded, then $(v_k,u_k)\to(v,u)$ for a subsequence, where $\|u\|=1$, $(v,-u)\in N_{\operatorname{gph}F^{-1}}(\bar{y},\bar{x})$, and $\|v\|\ge c$. This again contradicts (16.8). Thus, (16.10) holds for all $(y,x)\in\operatorname{gph}F^{-1}$ close to $(\bar{y},\bar{x})$.

Let $\delta > 0$ be such that (16.10) is satisfied for any $(x,y) \in \operatorname{gph} F \cap (\boldsymbol{B}_\delta(\bar{x}) \times \boldsymbol{B}_\delta(\bar{y}))$. Pick such (x,y). We will show that

$$(c\boldsymbol{B} \times \{w\}) \cap T^{**}_{\operatorname{gph} F}(x,y) \neq \emptyset \qquad \text{for every } w \in \boldsymbol{B}. \tag{16.11}$$

On the contrary, assume that there exists $w \in \boldsymbol{B}$ such that $(c\boldsymbol{B} \times \{w\}) \cap T^{**}_{\operatorname{gph} F}(x,y) = \emptyset$. Then, by the theorem on separation of convex sets, there exists a nonzero $(p,q) \in T^{*}_{\operatorname{gph} F}(x,y) = \hat{N}_{\operatorname{gph} F}(x,y)$ such that

$$\min_{u \in \boldsymbol{B}} \langle p, cu \rangle + \langle q, w \rangle > 0.$$

If $p = 0$, then $q \neq 0$ and then $(q,0) \in \hat{N}_{\operatorname{gph} F^{-1}}(y,x)$ in contradiction with (16.10). Hence, $p \neq 0$. Without loss of generality, let $\|p\| = 1$. Then $(q,p) \in \hat{N}_{\operatorname{gph} F^{-1}}(y,x)$ and

$$\langle q, w \rangle > \max_{u \in \boldsymbol{B}} \langle p, cu \rangle = c\|p\| = c. \tag{16.12}$$

By (16.5) and (16.10), $\|q\| \leq c$ and since $w \in \boldsymbol{B}$, this contradicts (16.12). Thus, (16.11) is satisfied.

By Lemma 16.3, for all $(x,y) \in \operatorname{gph} F$ sufficiently close to $(\bar{x},\bar{y})$, we have that (16.11) holds when the set $T^{**}_{\operatorname{gph} F}(x,y) = \operatorname{clco} T_{\operatorname{gph} F}(x,y)$ is replaced with $T_{\operatorname{gph} F}(x,y)$. This means that for any $w \in \boldsymbol{B}$ there exists $u \in DF(x|y)^{-1}(w)$ such that $\|u\| \leq c$. But then $c \geq \|DF(x|y)^{-1}\|^{-}$ for all $(x,y) \in \operatorname{gph} F$ sufficiently close to $(\bar{x},\bar{y})$. This combined with the arbitrariness of c in (16.8) implies the inequality opposite to (16.6) and hence the proof of the theorem if complete. ■

Proof (of Lemma 16.3). Clearly, $C \cap T_K(x) \neq \emptyset$ implies $C \cap \operatorname{clco} T_K(x) \neq \emptyset$. Assume that there exists an open neighborhood U of $\bar{x}$ such that $C \cap \operatorname{clco} T_K(x) \neq \emptyset$ for all $x \in K \cap U$. Let $\varepsilon > 0$ be such that $\boldsymbol{B}_\varepsilon(\bar{x}) \subset U$. Take any $x \in \boldsymbol{B}_{\varepsilon/3}(\bar{x})$ and let v be a projection of x on K. Then $\|v - x\| \leq \|\bar{x} - x\| \leq \varepsilon/3$ and hence,

$$\|v - \bar{x}\| \leq \|v - x\| + \|x - \bar{x}\| \leq \varepsilon/3 + \varepsilon/3 < \varepsilon.$$

Thus, there exists an open neighborhood W of $\bar{x}$ such that any metric projection of a point $x \in W$ on K belongs to $K \cap U$.

Fix $x \in K \cap W$. For all $t \geq 0$ define $\varphi(t) := \min\{\|u - v\| \mid u \in x + tC, v \in K\}$. The function φ is Lipschitz continuous. Indeed, for any $t_i \geq 0$, $i = 1,2$ there exist $c_i \in C$ and $k_i \in K$ such that $\varphi(t_i) = \|x + t_i c_i - k_i\|$, $i = 1,2$. Then

$$\begin{aligned} \varphi(t_1) - \varphi(t_2) &= \|x + t_1 c_1 - k_1\| - \|x + t_2 c_2 - k_2\| \\ &\leq \|x + t_1 c_2 - k_2\| - \|x + t_2 c_2 - k_2\| \leq \|c_2\| |t_1 - t_2|. \end{aligned}$$

Hence φ is absolutely continuous, that is, its derivative φ' exists almost everywhere and $\varphi(s) = \varphi(t) + \int_t^s \varphi'(\tau)\mathrm{d}\tau$ for all $s \geq t \geq 0$. We will prove next that

$$\varphi(t) = 0 \quad \text{for all sufficiently small} \quad t > 0. \tag{16.13}$$

If this holds, then for every small $t > 0$ there exists $v_t \in C$ such that $x + tv_t \in K$. Consider sequences $t_k \to 0+$ and $v_{t_k} \in C$ such that v_{t_k} converges to some v. Then $v \in T_K(x) \cap C$ and since $x \in K \cap W$ is arbitrary, we arrive at the claim of the lemma.

To prove (16.13), let $\gamma > 0$ be such that $x + [0,\gamma]C \subset W$. Assume that there exists $t_0 \in (0,\gamma]$ such that $\varphi(t_0) > 0$. Define $\bar{t} = \max\{t \mid \varphi(t) = 0 \text{ and } 0 \leq t < t_0\}$. Let $t \in (\bar{t}, t_0]$ be such that $\varphi'(t)$ exists. Then for some $v_t \in C$ and $x_t \in K$ we have $\varphi(t) = \|x + tv_t - x_t\| > 0$. Since x_t is a projection of $x + tv_t$ on K, by the observation in the beginning of the proof, we have $x_t \in K \cap U$. By assumption, there exists $w_t \in \operatorname{clco} T_K(x_t)$ such that $w_t \in C$. Then, for any $h > 0$ sufficiently small,

$$x + tv_t + hw_t = x + (t+h)\left(\frac{t}{t+h}v_t + \frac{h}{t+h}w_t\right) \in x + (t+h)C \subset W$$

because the set C is assumed convex. Thus

$$\varphi(t+h) - \varphi(t) \leq \|x + tv_t + hw_t - x_t\| - \|x + tv_t - x_t\|.$$

Dividing both sides of this inequality by $h > 0$ and passing to the limit when $h \to 0_+$, we get

$$\varphi'(t) \leq \left\langle \frac{x + tv_t - x_t}{\|x + tv_t - x_t\|}, w_t \right\rangle. \tag{16.14}$$

Recall that x_t is a projection of $x + tv_t$ on K and also the elementary fact that in this case $x + tv_t - x_t \in \hat{N}_K(x_t)$, see Proposition 4.1.2 in [2] or Example 6.16 in [9]. Since $w_t \in \operatorname{clco} T_K(x_t)$, we obtain from (16.14) that $\varphi'(t) \leq 0$. Having in mind that t is any point of differentiability of φ in $(\bar{t}, t_0)$, we get $\varphi(t_0) \leq \varphi(\bar{t}) = 0$. This contradicts the choice of t_0 according to which $\varphi(t_0) > 0$. Hence (16.13) holds and the lemma is proved. ∎

We note that more general versions of Lemma 16.3 are proved in [3, 4].

At the end of this section, we add the following corollary which is a simple consequence of the proof of Theorem 16.1 and recovers the last part of Theorem 1.2 in [6] and the first part of Theorem 4.3 in [1]. For a mapping $F : \mathbb{R}^n \rightrightarrows \mathbb{R}^m$ and a pair (x,y) with $y \in F(x)$, recall that the *convexified graphical derivative* of F at x for y is the mapping $\tilde{D}F(x|y) : \mathbb{R}^n \rightrightarrows \mathbb{R}^m$ whose graph is the closed convex hull of the tangent cone $T_{\operatorname{gph} F}(x,y)$ to $\operatorname{gph} F$ at (x,y):

$$v \in \tilde{D}F(x|y)(u) \quad \Leftrightarrow \quad (u,v) \in \operatorname{clco} T_{\operatorname{gph} F}(x,y).$$

Corollary 16.4. *Let $F : \mathbb{R}^n \rightrightarrows \mathbb{R}^m$ be a set-valued map, let $\bar{y} \in F(\bar{x})$ and assume that* gph F *is locally closed at $(\bar{x}, \bar{y})$. Then*

$$\limsup_{\substack{(x,y)\to(\bar{x},\bar{y}),\\ (x,y)\in \operatorname{gph} F}} \|\tilde{D}F(x\,|\,y)^{-1}\|^{-} = \|D^*F^{-1}(\bar{y}\,|\,\bar{x})\|^{+}.$$

Proof. Since $\tilde{D}F(x\,|\,y)^{-1}(v) \supset DF(x\,|\,y)^{-1}(v)$ we obtain $\|\tilde{D}F(x\,|\,y)^{-1}\|^{-} \le \|DF(x\,|\,y)^{-1}\|^{-}$. Thus, from (16.1),

$$\limsup_{\substack{(x,y)\to(\bar{x},\bar{y})\\ (x,y)\in \operatorname{gph} F}} \|\tilde{D}F(x\,|\,y)^{-1}\|^{-} \le \|D^*F^{-1}(\bar{y}\,|\,\bar{x})\|^{+}.$$

The converse inequality follows from the first part of the proof of Theorem 16.1, by limiting the argument to the convexified graphical derivative. ■

16.3 Proof of Theorem 16.2

Proposition 3G.1 in [5] says that a mapping F is strongly metrically regular at $\bar{x}$ for $\bar{y}$ if and only if it is metrically regular there and F^{-1} has a localization around $(\bar{y}, \bar{x})$ which is nowhere multi-valued. Furthermore, in this case for any $c > \operatorname{reg}(F;\bar{x}\,|\,\bar{y})$, there exists a neighborhood V of $\bar{y}$ such that F^{-1} has a localization around $(\bar{y}, \bar{x})$ which is a Lipschitz continuous function on V with constant c.

Let F be strongly metrically regular at $\bar{x}$ for $\bar{y}$, let $c > \operatorname{reg}(F;\bar{x}\,|\,\bar{y})$ and let U and V be open neighborhoods of $\bar{x}$ and $\bar{y}$, respectively, such that the localization $V \ni y \mapsto \varphi(y) := F^{-1}(y) \cap U$ is a Lipschitz continuous function on V with a Lipschitz constant c. We will show first that for any $v \in \mathbb{R}^m$ the set $PF(\bar{x}\,|\,\bar{y})^{-1}(v)$ is nonempty. Let $v \in \mathbb{R}^m$. Since $\operatorname{dom} \varphi \supset V$, we can choose sequences $t_k \to 0_+$ and u_k such that $\bar{x} + t_k u_k = \varphi(\bar{y} + t_k v)$ for large k. Then, from the Lipschitz continuity of φ with Lipschitz constant c, we conclude that $\|u_k\| \le c\|v\|$; hence u_k has a cluster point u which, by definition, is from $PF(\bar{x}\,|\,\bar{y})^{-1}(v)$. Now choose any $v \in \mathbb{R}^m$ and $u \in PF(\bar{x}\,|\,\bar{y})^{-1}(v)$; then there exist sequences $(x_k, y_k) \in \operatorname{gph} F$, $(x_k, y_k) \to (\bar{x}, \bar{y})$, $t_k \to 0_+$, $u_k \to u$, and $v_k \to v$ such that $y_k + t_k v_k \in V$, $x_k = \varphi(y_k)$, and $x_k + t_k u_k = \varphi(y_k + t_k v_k)$ for k sufficiently large. But then, again from the Lipschitz continuity of φ with Lipschitz constant c, we obtain that $\|u_k\| \le c\|v_k\|$. Passing to the limit we conclude that $\|u\| \le c\|v\|$ which implies that $\|PF(\bar{x}\,|\,\bar{y})^{-1}\|^{+} \le c$. Hence (16.2) is satisfied.

To prove the second statement, we first show that F^{-1} has a single-valued bounded localization, that is, there exist a bounded neighborhood U of $\bar{x}$ and a neighborhood V of $\bar{y}$ such that $V \ni y \mapsto F^{-1}(y) \cap U$ is single-valued. On the contrary, assume that for any bounded neighborhood U of $\bar{x}$ and any neighborhood V of $\bar{y}$ the intersection $\operatorname{gph} F^{-1} \cap (V \times U)$ is the graph of a multi-valued mapping. This means that there exist sequences $\varepsilon_k \to 0_+$, $x_k \to \bar{x}$, $x'_k \to \bar{x}$, $x_k \ne x'_k$ for all k such that $F(x_k) \cap F(x'_k) \cap \boldsymbol{B}_{\varepsilon_k}(\bar{y}) \ne \emptyset$ for all k. Let $t_k = \|x_k - x'_k\|$ and let $u_k = (x_k - x'_k)/t_k$. Then

$t_k \to 0$ and $\|u_k\| = 1$ for all k. Hence $\{u_k\}$ has a cluster point $u \neq 0$. Consider any $y_k \in F(x_k) \cap F(x'_k) \cap \boldsymbol{B}_{\varepsilon_k}(\bar{y})$. Then, $y_k + t_k 0 \in F(x'_k + t_k u_k)$ for all k. By the definition of the paratingent derivative, $0 \in PF(\bar{x},\bar{y})(u)$. Hence $\|PF(\bar{x},\bar{y})^{-1}\|^+ = \infty$ by (16.5), which contradicts (16.2). Thus, there exist neighborhoods U of $\bar{x}$ and V of $\bar{y}$ such that $\varphi(y) := F^{-1}(y) \cap U$ is at most single-valued on V and U is bounded. By (16.3), there exists a neighborhood $V' \subset V$ of $\bar{y}$ such that $F^{-1}(y) \cap U \neq \emptyset$ for any $y \in V'$; hence $V' \subset \operatorname{dom}\varphi$. Further, since $\operatorname{gph} F$ is locally closed at $(\bar{x},\bar{y})$ and φ is bounded, there exists an open neighborhood $V'' \subset V'$ of $\bar{y}$ such that φ is a continuous function on V''.

From the definition of the paratingent cone we deduce that the set-valued map $(x,y) \mapsto PF(x,y)$ has closed graph. We claim that condition (16.2) implies that

$$\limsup_{\substack{(x,y)\to(\bar{x},\bar{y}),\\ (x,y)\in \operatorname{gph} F}} \|PF(x\,|\,y)^{-1}\|^+ < \infty. \tag{16.15}$$

On the contrary, assume that there exist sequences $(x_k, y_k) \in \operatorname{gph} F$ converging to $(\bar{x},\bar{y})$, $v_k \in \boldsymbol{B}$ and $u_k \in PF(x_k\,|\,y_k)^{-1}(v_k)$ such that $\|u_k\| > k\|v_k\|$.

Case 1: There exists a subsequence $v_{k_i} = 0$ for all k_i. Since $\operatorname{gph} PF(x_{k_i}\,|\,y_{k_i})^{-1}$ is a cone, we may assume that $\|u_{k_i}\| = 1$. Let u be a cluster point of $\{u_{k_i}\}$. Then, passing to the limit we get $0 \neq u \in PF(\bar{x},\bar{y})^{-1}(0)$ which, combined with (16.5), contradicts (16.2).

Case 2: For all large k, $v_k \neq 0$. Since $\operatorname{gph} PF(x_k\,|\,y_k)^{-1}$ is a cone, we may assume that $\|v_k\| = 1$. Then $\lim_{k\to\infty} \|u_k\| = \infty$. Define $w_k := \frac{1}{\|u_k\|} u_k \in PF(x_k\,|\,y_k)^{-1}\left(\frac{1}{\|u_k\|} v_k\right)$ and let w be a cluster point of w_k. Then, passing to the limit we obtain $0 \neq w \in PF(\bar{x},\bar{y})^{-1}(0)$ which, combined with (16.5), again contradicts (16.2).

Hence (16.15) is satisfied. Therefore, there exists an open neighborhood $\tilde{V} \subset V''$ of $\bar{y}$ such that $\|PF(\varphi(y)\,|\,y)^{-1}\|^+ < \infty$ for all $y \in \tilde{V}$.

We will now prove that for every $(x,y) \in \operatorname{gph} F$ near $(\bar{x},\bar{y})$ and every $v \in \mathbb{R}^m$ we have that $DF(x,y)^{-1}(v) \neq \emptyset$. Fix $(x,y) \in \operatorname{gph} F \cap (U \times \tilde{V})$ and $v \in \mathbb{R}^m$, and let $h_k \to 0_+$; then there exists $u_k \in \mathbb{R}^n$ such that $x + h_k u_k = F^{-1}(y + h_k v) \cap U = \varphi(y + h_k v)$ for all large k and we also have that $h_k u_k \to 0$ by the continuity of φ. Assume that $\|u_k\| \to \infty$ for some subsequence (which is denoted in the same way without loss of generality). Set $t_k = h_k\|u_k\|$ and $w_k = \frac{1}{\|u_k\|} u_k$. Then $t_k \to 0_+$ and, for a further subsequence, $w_k \to w$ for some w with $\|w\| = 1$. Since $(y + t_k \frac{1}{\|u_k\|} v, x + t_k w_k) \in \operatorname{gph} F^{-1}$ we obtain that $w \in DF(x,y)^{-1}(0) \subset PF(x,y)^{-1}(0)$ for some $w \neq 0$. Thus $\|PF(x,y)^{-1}\|^+ = \infty$ contradicting the choice of $\tilde{V}$. Hence the sequence $\{u_k\}$ cannot be unbounded and since $y + h_k v \in F(x + h_k u_k)$ for all k, any cluster point u of $\{u_k\}$ satisfies $u \in DF(x,y)^{-1}(v)$. Hence $DF(x,y)^{-1}$ is nonempty-valued. From this and the inclusion $DF(x,y)^{-1}(v) \subset PF(x,y)^{-1}(v)$ we obtain

$$\|DF(x,y)^{-1}\|^- \leq \|PF(x,y)^{-1}\|^+. \tag{16.16}$$

Putting together (16.15) and (16.16), and utilizing the derivative criterion for metric regularity, that is, the fact that the finiteness of the expression on the left side of (16.1) implies metric regularity, we obtain that F is metrically regular at $\bar{x}$ for $\bar{y}$. But since F^{-1} has a single-valued localization around $(\bar{y},\bar{x})$ we conclude that F is strongly metrically regular at $\bar{x}$ for $\bar{y}$. The proof is complete.

Acknowledgements ALD was partially supported by the National Science Foundation Grant DMS-1008341 through the University of Michigan. HF was partially supported by the European Union under the 7th Framework Programme FP7-PEOPLE-2010-ITN. Grant agreement number 264735-SADCO.

References

1. Aubin, J.-P.: A viability approach to the inverse set-valued map theorem. J. Evol. Equ. **6**, 419–432 (2006)
2. Aubin, J.-P., Frankowska, H.: Set-Valued Analysis. Birkhäuser, Berlin (1990)
3. Aubin, J.-P., Frankowska, H.: Partial differential inclusions governing feedback controls. J. Convex Anal. **2**, 19–40 (1995)
4. Dal Maso, G., Frankowska, H.: Uniqueness of solutions to Hamilton-Jacobi equations arising in the calculus of variations. In: Menaldi, J., Rofman, E., Sulem, A. (eds.) Optimal Control and Partial Differential Equations, pp. 336–345. IOS Press, Amsterdam (2000)
5. Dontchev, A.L., Rockafellar, R.T.: Implicit Functions and Solution Mappings. Springer Monographs in Mathematics. Springer, Dodrecht (2009)
6. Dontchev, A.L., Quincampoix, M., Zlateva, N.: Aubin criterion for metric regularity. J. Convex Anal. **13**, 281–297 (2006)
7. Frankowska, H., Quincampoix, M.: Hölder metric regularity of set-valued maps. Math. Program. **132**, 333–354 (2012)
8. Klatte, D., Kummer, B.: Nonsmooth Equations in Optimization. Regularity, Calculus, Methods and Applications. Nonconvex Optimization and Its Applications, vol. 60. Kluwer Academic Publishers, Dordrecht (2002)
9. Rockafellar, R.T., Wets, R.J.-B.: Variational Analysis. Springer, Berlin (1998)

Chapter 17
Five Classes of Monotone Linear Relations and Operators

Mclean R. Edwards

Dedicated to Jonathan Borwein on the occasion of his 60th birthday

Abstract The relationships among five classes of monotonicity, namely 3^*-, 3-cyclic, strictly, para-, and maximal monotonicity, are explored for linear operators and linear relations in Hilbert space. Where classes overlap, examples are given; otherwise their relationships are noted for linear operators in $\mathbb{R}^2$, $\mathbb{R}^3$, and general Hilbert spaces. Along the way, some results for linear relations are obtained.

Key words: 3^*-monotone • Cyclic monotone • Cyclically monotone • Linear relations • Maximal monotone • Maximally monotone • Monotone operators • Paramonotone • Rectangular • Strictly monotone

Mathematics Subject Classifications (2010): Primary 47H05, 47A06; Secondary 47J20, 47N10.

17.1 Introduction

Monotone operators are multi-valued operators $T : X \to 2^X$ such that for all $x^* \in Tx$ and all $y^* \in Ty$,

$$\langle x-y, x^*-y^*\rangle \geq 0. \tag{17.1}$$

COMMUNICATED BY HEINZ H. BAUSCHKE.

M.R. Edwards (✉)
Department of Mathematics, University of British Columbia, Vancouver, BC V6T 1Z2, Canada
e-mail: mcleane@math.ubc.ca

D.H. Bailey et al. (eds.), *Computational and Analytical Mathematics*, Springer Proceedings in Mathematics & Statistics 50, DOI 10.1007/978-1-4614-7621-4_17,

They arise as a generalization of subdifferentials of convex functions and are used extensively in variational inequality (and by reformulation, equilibrium) theory.

Variational inequalities were first outlined in 1966 [23] and have since been used to model a large number of problems.

Definition 17.1 (Variational Inequality Problem). Given a nonempty closed convex set C and a monotone operator T acting on C, the *variational inequality problem*, $VIP(T,C)$, is to find an $\bar{x} \in C$ such that for some $\bar{x}^* \in T(\bar{x})$

$$\langle c - \bar{x}, \bar{x}^* \rangle \geq 0 \text{ for all } c \in C. \tag{17.2}$$

They provide a unified framework for, among others, constrained optimization, saddle point, Nash equilibrium, traffic equilibrium, frictional contact, and complementarity problems. For a good overview of sample problems and current methods used to solve them, see [19] and [20].

Monotone operators are also important for the theory of partial differential equations, where monotonicity both characterizes the vector fields of self-dual Lagrangians [21] and is crucial for the determination of equilibrium solutions (using a variational inequality) for elliptical and evolution differential equations and inclusions (see for instance [1]).

Over the years, various classes of monotone operators have been introduced in the exploration of their theory; however there have been few attempts to comprehensively compare those in use across disciplines.

Five special classes of monotone operators are studied here: strictly monotone, 3-cyclic monotone, 3^*-monotone, paramonotone, and maximal monotone. All possible relationships among these five properties are explored for linear operators in $\mathbb{R}^2$, $\mathbb{R}^n$, and in general Hilbert space, and the results are summarized in Tables 17.1 and 17.2 and in Figs. 17.1, 17.2, and 17.3.

Definition 17.2 (paramonotone). An operator $T : X \to 2^X$ is said to be *paramonotone* if T is monotone and for $x^* \in Tx, y^* \in Ty$, $\langle x - y, x^* - y^* \rangle = 0$ implies that $x^* \in Ty$ and $y^* \in Tx$.

A number of iterative methods for solving (17.2) have required paramonotonicity to converge. Examples include an interior point method using Bregman functions [15], an outer approximation method [14], and proximal point algorithms [2, 13]. Often, as in [8], with more work it is possible to show convergence with paramonotonicity where previously stronger conditions, such as strong monotonicity, were required. Indeed, the condition first emerged in this context [12] as a sufficient condition for the convergence of a projected-gradient-like method. For more on the theory of paramonotone operators and why this condition is important for variational inequality problems, see [24] and [31].

Definition 17.3 (strictly monotone). An operator $T : X \to 2^X$ is said to be *strictly monotone* if T is monotone and for all $(x,x^*),(y,y^*) \in \operatorname{gra} T$, $\langle x - y, x^* - y^* \rangle = 0$ implies that $x = y$.

Table 17.1 Monotone linear operators on $\mathbb{R}^2$: monotone class relationships

PM	SM	3CM	3*		
0	0	0	0	$\exists$	Example 17.41 ($R_{\pi/2}$)
0	*	*	1	$\emptyset$	Proposition 17.33
*	*	1	0	$\emptyset$	Fact 17.9
0	*	1	*	$\emptyset$	Proposition 17.10
0	1	*	*	$\emptyset$	Fact 17.8
1	*	*	0	$\emptyset$	Proposition 17.47
1	0	0	*	$\emptyset$	Remark 17.45
1	0	1	1	$\exists$	Example 17.43 ($A(x_1,x_2) := (x_1,0)$)
1	1	0	1	$\exists$	Example 17.41 (R_θ, $\pi/2 > \lvert\theta\rvert > \pi/3$)
1	1	1	1	$\exists$	*Id*

Where:
"PM" represents paramonotone
"SM" represents strictly monotone
"3CM" represents 3-cyclic monotone
"3*" represents 3*-monotone
1 represents that the property is present
0 represents an absence of that property
* represents that both 0/1 are covered by the result
$\exists$ represents that an example with these properties exists
$\emptyset$ represents that this combination of properties is impossible

Table 17.2 Monotone linear operators: monotone class relationships

PM	SM	3CM	3*	X		
0	0	0	0	$\mathbb{R}^2$	$\exists$	$R_{\pi/2}$
0	*	*	1	–	$\emptyset$	Proposition 17.33
*	*	1	0	–	$\emptyset$	Fact 17.9
0	*	1	*	–	$\emptyset$	Proposition 17.10
0	1	*	*	–	$\emptyset$	Fact 17.8
1	0	0	0	ℓ_2	$\exists$	Remark 17.51
1	0	0	1	$\mathbb{R}^2$	$\exists$	Example 17.49
1	0	1	1	$\mathbb{R}$	$\exists$	**0**
1	1	0	0	ℓ_2	$\exists$	Example 17.50
1	1	0	1	$\mathbb{R}^2$	$\exists$	Example 17.41 (R_θ, $\pi/2 > \lvert\theta\rvert > \pi/3$)
1	1	1	1	$\mathbb{R}$	$\exists$	*Id*

Where:
"PM" represents paramonotone
"SM" represents strictly monotone
"3CM" represents 3-cyclic monotone
"3*" represents 3* monotone
"X" represents the space the operator acts upon
1 represents that the property is present
0 represents an absence of that property
* represents that both 0/1 are covered by the result
$\exists$ represents that an example with these properties exists
$\emptyset$ represents that this combination of properties is impossible

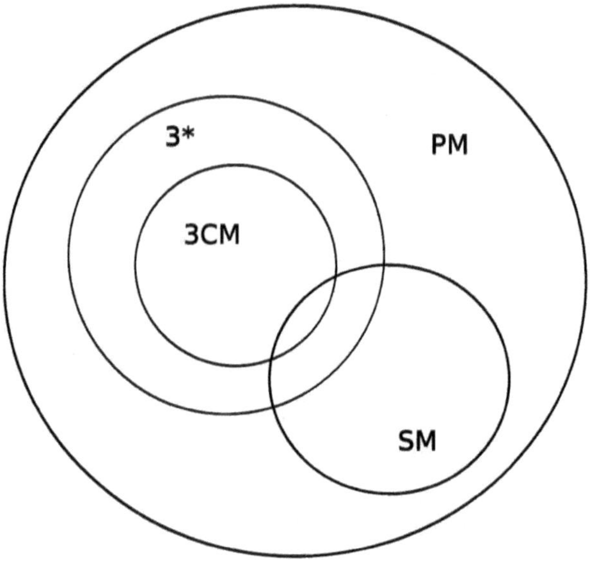

Fig. 17.1 Monotone linear operators: monotone class relationships. PM = paramonotone, SM = strictly monotone, 3CM = 3 cyclic monotone, 3* = 3^*-monotone

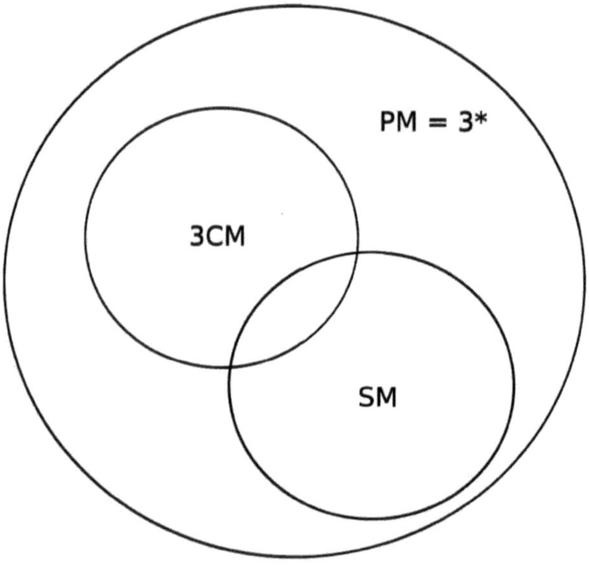

Fig. 17.2 Monotone linear operators on $\mathbb{R}^2$: monotone class relationships. PM = paramonotone, SM = strictly monotone, 3CM = 3 cyclic monotone, 3* = 3^*-monotone

Strict monotonicity is a stronger condition than paramonotonicity (Fact 17.8), and the strict monotonicity of an operator T guarantees the uniqueness of a solution to the variational inequality problem (see for instance [19]). These operators are somewhat analogous to the subdifferentials of strictly convex functions.

We adopt the notation of [32] and use the term 3^*-monotone, although this property was first introduced with no name. The property was first referenced simply by "$*$" [11] by Brézis and Haraux, and such operators were sometimes called

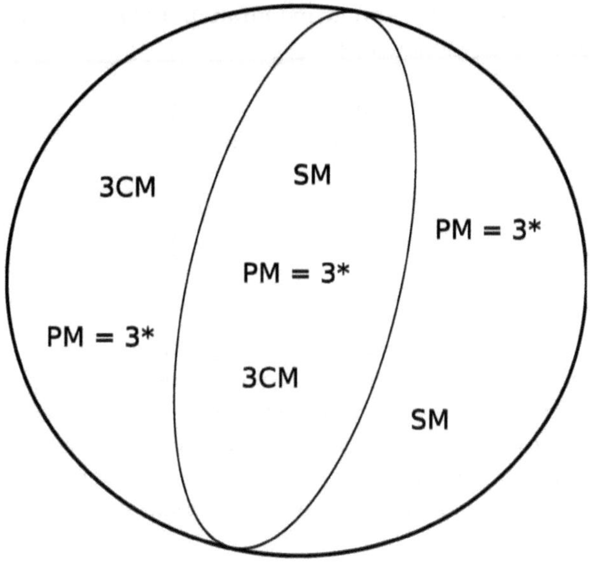

Fig. 17.3 Monotone linear operators on $\mathbb{R}^n$: monotone class relationships. PM = paramonotone, SM = strictly monotone, 3CM = 3 cyclic monotone, 3* = 3^*-monotone

(BH)-operators [16] in honour of these original authors. More recently the property has also taken on the name "rectangular" since the domain of the Fitzpatrick function of a monotone operator is rectangular precisely when the operator is 3^*-monotone [29].

Definition 17.4 (3^*-monotone). An operator $T : X \to 2^X$ is said to be 3^*-monotone if T is monotone and for all z in the domain of T and for all x^* in the range of T

$$\sup_{(y,y^*)\in \operatorname{gra} T} \langle z-y, y^*-x^*\rangle < +\infty. \tag{17.3}$$

3^*-monotonicity has the important property in that if T_1 and T_2 are 3^*-monotone, then as long as their sum is maximal monotone, the closure of the sum of their ranges is identical to the closure of the range of their sum. For instance, if two operators are 3^*-monotone, and one is surjective, then if the sum is maximal monotone, it is also surjective. Furthermore, if both are continuous monotone linear operators, and at least one is 3^*-monotone, then the kernel of the sum is the intersection of the kernels [3]. This property can be used, as shown in [11], to determine when solutions to $T^{-1}(0)$ exist by demonstrating that 0 is in the interior (or is not in the closure) of the sum of the ranges of an intelligent decomposition of a difficult to evaluate maximal monotone operator. It has also been shown for linear relations on Banach spaces that 3^*-monotonicity guarantees the existence of solutions to the primal-dual problem pairs in [27]. It should also be noted that operators with bounded range [32] and strongly coercive operators [11] are 3^*-monotone.

Definition 17.5 (n-cyclic monotone). Let $n \geq 2$. An operator $T : X \to 2^X$ is said to be n-cyclic monotone if

$$\left.\begin{array}{rl} (x_1, x_1^*) & \in \operatorname{gra} T \\ (x_2, x_2^*) & \in \operatorname{gra} T \\ \cdots & \in \operatorname{gra} T \\ (x_n, x_n^*) & \in \operatorname{gra} T \\ x_{n+1} & = x_1 \end{array}\right\} \Rightarrow \sum_{i=1}^{n} \langle x_i - x_{i+1}, x_i^* \rangle \geq 0. \tag{17.4}$$

A *cyclical monotone* operator is one that is n-cyclic monotone for all $n \in \mathbb{N}$.

Note that 2-cyclic monotonicity is equivalent to monotonicity. By substituting $(a_n, a_n^*) := (a_1, a_1^*)$, it easily follows from the definition that any n-cyclic monotone operator is $(n-1)$-cyclic monotone. 1-cyclic monotonicity is not defined, since the $n = 1$ case for (17.4) is trivial. 3-cyclic monotone operators serve to represent a special case of n-cyclic monotone operators that is also a stronger condition than 3^*-monotonicity. Of note, all subdifferentials of convex functions are cyclical monotone [28].

Definition 17.6 (maximality). An operator is *maximal n-cyclic monotone* if its graph cannot be extended while preserving n-cyclic monotonicity. A *maximal monotone* operator is a maximal 2-cyclic monotone operator. A *maximal cyclical monotone* operator is a cyclical monotone operator such that all proper graph extensions are not cyclical monotone.

There is a rich literature on the theory (see [9] for a good overview) and application (for instance [18]) of maximal monotone operators. Furthermore, it is well known that a maximal monotone operator T has the property that $T^{-1}(0)$ is convex, a property shared by paramonotone operators with convex domain (Proposition 17.11), and analogous to the fact that the minimizers of a convex function form a convex set. Maximal monotonicity is also an important property for general differential inclusions [10, 26].

Definition 17.7 (Five classes of monotone operator). An operator $T : X \to 2^X$ is said to be [Class] (with abbreviation [Code]) if and only if T is monotone and for every $(x, x^*), (y, y^*), (z, z^*)$ in gra T one has [Condition].

Code	Class	Condition (A)
	Monotone	$\langle x - y, x^* - y^* \rangle \geq 0$
PM	Paramonotone	$\langle x - y, x^* - y^* \rangle = 0 \Rightarrow (x, y^*), (y, x^*) \in \operatorname{gra} T$
SM	Strictly monotone	$\langle x - y, x^* - y^* \rangle = 0 \Rightarrow x = y$
3CM	3-cyclic monotone	$\langle x - y, x^* \rangle + \langle y - z, y^* \rangle + \langle z - x, z^* \rangle \geq 0$
MM	Maximal monotone	$(\forall a \in X)(\forall a^* \in X)$ $\langle x - a, x^* - a^* \rangle \geq 0 \Rightarrow (x, x^*) \in \operatorname{gra} T$
3*	3^*-monotone	$\sup_{(a,a^*) \in \operatorname{gra} T} \langle z - a, a^* - x^* \rangle < +\infty$

The order above, PM-SM-3CM-MM-3*, is fixed to allow a binary label of the classes to which an operator belongs. For instance, an operator with the label 10111 is paramonotone, not strictly monotone, 3-cyclic monotone, maximal monotone, and 3*-monotone.

After noting some general relationships among these classes in Sect. 17.2, we note in Sect. 17.3 that monotone operators belonging to particular combinations of these classes can be constructed in a product space.

Linear relations are a multi-valued extension of linear operators and are defined by those operators whose graph forms a vector space. This is a natural extension to consider as monotone operators are often multi-valued. We consider linear relations in Sect. 17.4 and explore their characteristics and structure. Of particular note, we fully explore the manner in which linear relations can be multi-valued and remark on a curious property of linear relations whose domains are not closed. Finally, we obtain a generalization to the fact that bounded linear operators that are 3*-monotone are also paramonotone (a corollary to a result in [11]), with conditions different from those in [7], and demonstrate by example that there is 3*-monotone linear relation that is not paramonotone.

In Sect. 17.5, we list various examples of linear operators satisfying or failing to satisfy the 5 properties defined above. The examples are chosen to have full domain, low dimension, and be continuous where possible. This is shown to yield a complete characterization of the dependence or independence of these five classes of monotone operator in $\mathbb{R}^2$, $\mathbb{R}^n$, and in a general Hilbert space X. One result of this section is that paramonotone and linear operators in $\mathbb{R}^2$ are exactly the symmetric or strictly monotone operators in $\mathbb{R}^2$.

We assume throughout that X is a real Hilbert space, with inner product $\langle\cdot,\cdot\rangle$. When an operator $T: X \to 2^X$ is such that for all $x \in X$, Tx contains at most one element, such operators are called *single-valued*. When T is single-valued, for brevity Tx is at times considered as a point rather than as a set (i.e., $x^* \in Tx$). The *orthogonal complement* of a set $C \subset X$ is denoted by $C^\perp$ and defined by

$$C^\perp := \{x \in X : \langle x, c\rangle = 0 \;\forall c \in C\}. \tag{17.5}$$

Note that for any set $C \subset X$, the set $C^\perp$ is closed in X. The operator P_V is the metric projection where V is a closed subspace of X. We use the convention that for set addition $A + \emptyset = \emptyset$, where $\emptyset$ is the empty set. A monotone extension $\tilde{T}: X \to 2^X$ of a monotone operator $T: X \to 2^X$ is a monotone operator such that $\operatorname{gra} T \subsetneq \operatorname{gra} \tilde{T}$, where $\operatorname{gra} T := \{(x, x^*) : x \in \operatorname{dom} T, x^* \in Tx\}$. An operator $T: X \to 2^X$ is said to be *locally bounded* if for every $x \in \operatorname{dom} T$, there is a neighbourhood V of x and an $M > 0$ such that for every $v \in V$, $\sup_{v^* \in Tv} \|v^*\| < M$. A selection of an operator $T: X \to 2^X$ is an operator $\tilde{T}$ such that $\operatorname{gra} \tilde{T} \subset \operatorname{gra} T$, and a single-valued selection of T is such an operator $\tilde{T}$ where $\tilde{T}: X \to X$.

17.2 Preliminaries

The following arises from the definitions of strict monotonicity and paramonotonicity.

Fact 17.8. Any strictly monotone operator $T : X \to 2^X$ is also paramonotone.

Two synonymous definitions of 3-cyclic monotonicity are worth explicitly stating. For an operator $T : X \to 2^X$ to be 3-cyclic monotone, every $(x,x^*),(y,y^*),(z,z^*) \in \operatorname{gra} T$ must satisfy

$$\langle x-y, x^*\rangle + \langle y-z, y^*\rangle + \langle z-x, z^*\rangle \geq 0, \tag{17.6}$$

or equivalently

$$\langle z-y, y^*-x^*\rangle \leq \langle x-z, x^*-z^*\rangle. \tag{17.7}$$

From (17.7), the following fact is obvious.

Fact 17.9. Any 3-cyclic monotone operator $T : X \to 2^X$ is also 3^*-monotone.

Another relationship among these classes of monotone operator was discovered in 2006 (Proposition 3.1 in [22]).

Proposition 17.10 ([22]). *If T is 3-cyclic monotone and maximal (2-cyclic) monotone, then T is paramonotone.*

Proof. Suppose that for some choice of $(x,x^*),(y,y^*) \in \operatorname{gra}(T)$, $\langle x-y, x^*-y^*\rangle = 0$, so $\langle y-x, x^*\rangle = \langle y-x, y^*\rangle$. Since T is 3-cyclic monotone, every $(z,z^*) \in \operatorname{gra}(T)$ satisfies

$$\begin{aligned}
0 &\geq \langle y-x, x^*\rangle + \langle z-y, y^*\rangle + \langle x-z, z^*\rangle \\
&= \langle -x, y^*\rangle + \langle z, y^*\rangle + \langle x-z, z^*\rangle \\
&= \langle z-x, y^*\rangle + \langle x-z, z^*\rangle \\
&= \langle x-z, z^*-y^*\rangle
\end{aligned}$$

and so

$$\langle x-z, y^*-z^*\rangle \geq 0 \quad \forall (z,z^*) \in \operatorname{gra}(T).$$

Since T is maximal monotone, $y^* \in Tx$. By exchanging the roles of x and y above, it also holds that $x^* \in T(y)$, and so T is paramonotone. ■

When finding the zeros of a monotone operator, it can be useful to know if the solution set is convex or not. It is well known that for a maximal monotone operator T, $T^{-1}(0)$ is a closed convex set (see for instance [4]). A similar result also holds for paramonotone operators.

Proposition 17.11. *Let $T : X \to 2^X$ be a paramonotone operator with convex domain. Then $T^{-1}(0)$ is a convex set.*

Proof. Suppose $T^{-1}(0)$ is nonempty. Let $x,y,z \in X$ such that $0 \in Tx$, $0 \in Tz$, and $y = \alpha x + (1-\alpha)z$ for some $\alpha \in]0,1[$. Then, $x-y = (1-\alpha)(x-z)$ and $y-z = \alpha(x-z)$, so $x-y = \frac{1-\alpha}{\alpha}(y-z)$. Since T has convex domain, $Ty \neq \emptyset$. By the monotonicity of T, for all $y^* \in Ty$

$$0 \leq \langle x-y, -y^* \rangle = \frac{1-\alpha}{\alpha}\langle y-z, -y^* \rangle \qquad \text{and} \qquad 0 \leq \langle y-z, y^* \rangle,$$

and so $\langle y-z, y^* \rangle = 0$. Therefore, by the paramonotonicity of T, $0 \in T(y)$, and so the set $T^{-1}(0)$ is convex. ■

However, if an operator is not maximal monotone, there is no guarantee that $T^{-1}(0)$ is closed, even if paramonotone, as the operator $T : \mathbb{R} \to \mathbb{R}$ below demonstrates:

$$Tx := \begin{cases} -1, & x \leq -1, \\ 0, & x \in]-1,1[, \\ 1, & x \geq 1. \end{cases} \tag{17.8}$$

17.3 Monotone Operators on Product Spaces

Let X_1 and X_2 be Hilbert spaces, and consider set valued operators $T_1 : X_1 \to 2^{X_1}$ and $T_2 : X_2 \to 2^{X_2}$. The product operator $T_1 \times T_2 : X_1 \times X_2 \to 2^{X_1 \times X_2}$ is defined as $(T_1 \times T_2)(x_1,x_2) := \{(x_1^*,x_2^*) : x_1^* \in T_1x_1 \text{ and } x_2^* \in T_2x_2\}$.

Proposition 17.12. *If both T_1 and T_2 are monotone, then the product operator $T_1 \times T_2$ is also monotone.*

Proof. For any points $((x_1,x_2),(x_1^*,x_2^*)), ((y_1,y_2),(y_1^*,y_2^*)) \in \operatorname{gra}(T_1 \times T_2)$,

$$\begin{aligned} &\langle (x_1,x_2)-(y_1,y_2),(x_1^*,x_2^*)-(y_1^*,y_2^*) \rangle \\ &= \langle x_1-y_1, x_1^*-y_1^* \rangle + \langle x_2-y_2, x_2^*-y_2^* \rangle \geq 0. \end{aligned}$$

Hence, $T_1 \times T_2$ is monotone. ■

Proposition 17.13. *If both T_1 and T_2 are paramonotone, then the product operator $T_1 \times T_2$ is also paramonotone.*

Proof. If $x_i^* \in T_ix_i$, $y_i^* \in T_iy_i$ for $i \in \{1,2\}$ and

$$\langle (x_1,x_2)-(y_1,y_2),(x_1^*,x_2^*)-(y_1^*,y_2^*) \rangle = 0,$$

then $\langle x_i - y_i, x_i^* - y_i^* \rangle = 0$ for $i \in \{1,2\}$ since both T_1 and T_2 are monotone. By the paramonotonicity of T_1 and T_2, $y_i^* \in T_i x_i$ and $x_i^* \in T_i y_i$ for $i \in \{1,2\}$, and so $(x_1^*, x_2^*) \in (T_1 \times T_2)(y_1, y_2)$ and $(y_1^*, y_2^*) \in (T_1 \times T_2)(x_1, x_2)$. ■

By following the same proof structure as Proposition 17.13, a similar result immediately follows for some other monotone classes.

Proposition 17.14. *If both T_1 and T_2 belong to the same monotone class, where that class is one of strict, n-cyclic, or 3^*-monotonicity, then so does their product operator $T_1 \times T_2$.*

Proposition 17.15. *If both T_1 and T_2 are maximal monotone, then the product operator $T_1 \times T_2$ is also maximal monotone.*

Proof. Suppose $T_1 \times T_2$ is not maximal monotone. Then there exists a point $((x_1, x_2), (x_1^*, x_2^*)) \notin \operatorname{gra}(T_1 \times T_2)$ such that for all $((y_1, y_2), (y_1^*, y_2^*)) \in \operatorname{gra}(T_1 \times T_2)$

$$\langle x_1 - y_1, x_1^* - y_1^* \rangle + \langle x_2 - y_2, x_2^* - y_2^* \rangle \geq 0, \tag{17.9}$$

and at least one of $(x_1, x_1^*) \notin \operatorname{gra} T_1$ or $(x_2, x_2^*) \notin \operatorname{gra} T_2$. Suppose without loss of generality that $(x_1, x_1^*) \notin \operatorname{gra} T_1$.

By the maximality of T_1, $\langle x_1 - z_1, x_1^* - z_1^* \rangle < 0$ for some $(z_1, z_1^*) \in \operatorname{gra} T_1$, and so by setting $(y_1, y_1^*) := (z_1, z_1^*)$ in (17.9), $\langle x_2 - y_2, x_2^* - y_2^* \rangle \geq 0$ for all $(y_2, y_2^*) \in \operatorname{gra} T_2$. Since T_2 is maximal monotone, it must be that $(x_2, x_2^*) \in \operatorname{gra} T_2$. Clearly, $((z_1, x_2), (z_1^*, x_2^*)) \in \operatorname{gra}(T_1 \times T_2)$, yet

$$\langle (x_1, x_2) - (z_1, x_2), (x_1^*, x_2^*) - (z_1^*, x_2^*) \rangle < 0.$$

This is a contradiction of (17.9), and so $T_1 \times T_2$ is maximal monotone. ■

Of course, if an operator $T_1 : X \to 2^X$ fails to satisfy the conditions for any of the classes of monotone operator here considered, then the product of that operator with any other operator $T_2 : Y \to 2^Y$, namely $T_1 \times T_2 : X \times Y \to 2^{X \times Y}$, will also fail the same condition. Simply consider the set of points P in the graph of T_1 which violate a particular condition in X, and instead consider the set of points $\tilde{P} := \{(p, a) \times (p^*, a^*) : p \in P\}$ for a fixed arbitrary point $(a, a^*) \in \operatorname{gra} T_2$. Clearly $\tilde{P} \subset \operatorname{gra} T_1 \times T_2$, and this set will violate the same conditions in $X \times Y$ that P violates for T_1 in X. For instance,

$$\langle (w, a) - (x, a), (y^*, a^*) - (z^*, a^*) \rangle = \langle w - x, y^* - z^* \rangle.$$

In this manner, the lack of a monotone class property (be it n-cyclic, para-, maximal, 3^*-, nor strict monotonicity) is dominant in the product space.

Taken together, the results of this section are that the product operator $T_1 \times T_2$ of monotone operators T_1 and T_2 operates with respect to monotone class inclusion as a logical AND operator applied to the monotone classes of T_1 and T_2. For instance, suppose that T_1 is paramonotone, not strictly monotone, 3-cyclic monotone, maximal monotone, and 3^*-monotone (with binary label 10111), and suppose that

T_2 is paramonotone, strictly monotone, not 3-cyclic monotone, maximal monotone, and not 3*-monotone (with binary label 11010). Then, $T_1 \times T_2$ is paramonotone, not strictly monotone, not 3-cyclic monotone, maximal monotone, and not 3*-monotone (with binary label 10010).

17.4 Linear Relations

Linear relations are the set-valued generalizations of linear operators, which we define using the nomenclature of R. Cross [17].

Definition 17.16 (linear relation). An operator $A : X \to 2^X$ is a *linear relation* if $\operatorname{dom} A$ is a linear subspace of X and for all $x, y \in \operatorname{dom} A$, $\lambda \in \mathbb{R}$

1. $\lambda Ax \subset A(\lambda x)$,
2. $Ax + Ay \subset A(x + y)$.

Equivalently, linear relations are exactly those operators $T : X \to 2^X$ whose graphs are linear subspaces of $X \times X$. The following results on linear relations are well known. Of note, Fact 17.17(1) and (2) are considered basic results and will not be cited in the work below.

Fact 17.17 ([30]). For any linear relation $A : X \to 2^X$,

(1) $\lambda Ax = A(\lambda x)$ for all $x \in \operatorname{dom} A$, $0 \neq \lambda \in \mathbb{R}$,
(2) $Ax + Ay = A(x + y)$ for all $x, y \in \operatorname{dom} A$,
(3) $A0$ is a linear subspace of X,
(4) $Ax = x^* + A0$ for all $(x, x^*) \in \operatorname{gra} A$,
(5) If A is single-valued at any point, it is single-valued at every point in its domain.

Proposition 17.18. *Suppose $A : X \to 2^X$ is a linear relation, and let $x \in \operatorname{dom} A$. Then, $P_{A0^\perp} Ax$ is a singleton and*

$$Ax \subset P_{A0^\perp} Ax + \overline{A0}. \tag{17.10}$$

If $A0$ is closed, then there is a unique $x_0^* \in Ax$ such that $x_0^* \in A0^\perp$, where $x_0^* = P_{A0^\perp} x^*$ for all $x^* \in Ax$.

Proof. Let $x \in \operatorname{dom} A$. Since $\overline{A0}$ and $A0^\perp$ are closed subspaces such that $\overline{A0} + A0^\perp = X$, then for all $x^* \in X$, $x^* = P_{\overline{A0}} x^* + P_{A0^\perp} x^*$. By Fact 17.17 (4), (17.10) holds and $P_{A0^\perp} Ax$ is a singleton. If $A0$ is closed, then for all $x^* \in Ax$,

$$Ax = x^* + A0 = P_{A0^\perp} x^* + A0.$$

Therefore, $P_{A0^\perp} y^* = P_{A0^\perp} x^*$ for all $y^* \in Ax$. Furthermore, since $0 \in A0$ always, $P_{A0^\perp} x^* \in Ax$. ■

Proposition 17.19. *Any monotone linear relation $A : X \to 2^X$ with full domain is maximal monotone and single-valued.*

Proof. Suppose that $A : X \to 2^X$ is a linear relation where $\operatorname{dom} A = X$. Let (z, z^*) be a point such that $\langle z - y, z^* - y^* \rangle \geq 0$ for all $(y, y^*) \in \operatorname{gra} A$. Choose an arbitrary $z_0^* \in Az$. Let $y = z - \varepsilon x$ for arbitrary $(x, x^*) \in \operatorname{gra} A$ and $\varepsilon > 0$, so that by linearity $-\varepsilon x^* \in A(-\varepsilon x)$. Therefore $z_0^* - \varepsilon x^* \in Ay$ and so $\langle \varepsilon x, z^* - z_0^* + \varepsilon x^* \rangle \geq 0$. Divide out the ε, and send $\varepsilon \to 0^+$ so that $\langle x, z^* - z_0^* \rangle \geq 0$ for all $x \in X$. Hence $z^* = z_0^*$ and T is single-valued and maximal monotone. ■

The following results appear respectively as Proposition 2.2(i) and Proposition 2.4 in [5].

Proposition 17.20 ([5]). *If $A : X \to 2^X$ is a monotone linear relation, then* $\operatorname{dom} A \subset (A0)^\perp$ *and* $A0 \subset (\operatorname{dom} A)^\perp$.

Corollary 17.21 ([5]). *If a linear relation $A : X \to 2^X$ is maximal monotone, then $(\operatorname{dom} A)^\perp = A0$, and so $\overline{\operatorname{dom} A} = (A0)^\perp$ and $A0$ is a closed subspace.*

This leads to a partial converse result to Proposition 17.19.

Corollary 17.22. *If a maximal monotone single-valued linear relation $A : X \to X$ is locally bounded, then it has full domain.*

Proof. Since A is single-valued, $A0 = 0$, and so by Corollary 17.21, $\overline{\operatorname{dom} A} = (A0)^\perp = X$. Choose any point $x \in X$. Since $\operatorname{dom} A$ is dense in X, there exist a sequence $(y_n, y_n^*)_{n \in \mathbb{N}} \subset \operatorname{gra} A$ such that $y_n \to x$. Since A is locally bounded, a subsequence $(y_{\phi(n)}^*)_{n \in \mathbb{N}}$ of $(y_n^*)_{n \in \mathbb{N}}$ weakly converges to some point $x^* \in X$. Therefore, for all $(z, z^*) \in \operatorname{gra} A$,

$$0 \leq \lim_{n \to +\infty} \langle y_{\phi(n)} - z, y_{\phi(n)}^* - z^* \rangle = \langle x - z, x^* - z^* \rangle.$$

Since A is maximal monotone, $(x, x^*) \in \operatorname{gra} A$, and so A has full domain. ■

The following fact appears in Proposition 2.2 in [5].

Fact 17.23 ([5]). Let $A : X \to 2^X$ be a monotone linear relation. For any $x, y \in \operatorname{dom} A$, the set

$$\{\langle y, x^* \rangle : x^* \in Ax\}$$

is a singleton, the value of which can be denoted simply by $\langle y, Ax \rangle$.

Proof. Let $x, y \in \operatorname{dom} A$ and suppose that $x_1^*, x_2^* \in Ax$. By Fact 17.17 (4), $x_2^* - x_1^* \in A0$. Now, by Proposition 17.20, $A0 \subset (\operatorname{dom} A)^\perp$, and so $x_2^* - x_1^* \in (\operatorname{dom} A)^\perp$. Since $y \in \operatorname{dom} A$, $\langle y, x_1^* \rangle = \langle y, x_2^* \rangle$. ■

Proposition 17.24 below demonstrates that multivalued linear relations are closely related to a number of single-valued linear relations. Note especially that $V = A0^\perp$ and $V = \overline{\operatorname{dom} A}$ both satisfy the conditions below.

Proposition 17.24 (dimension reduction). *Suppose that $A : X \to 2^X$ is a monotone linear relation. Let $V \subset X$ satisfy*

(1) V is a closed subspace of X,
(2) $\operatorname{dom} A \subset V$, and
(3) $A0 \subset V^\perp$.

Define the operator $\tilde{A} : V \to 2^V$, by $\tilde{A}x := P_V Ax$ on $\operatorname{dom} A$, and let $\tilde{A} = \emptyset$ when $x \notin \operatorname{dom} A$. Then, $\tilde{A}$ is a single-valued monotone linear relation and $\operatorname{dom} A = \operatorname{dom} \tilde{A}$. In the case where $V = A0^\perp$ and $A0$ is closed, the operator $\tilde{A}$ is a single-valued selection of A. If A is maximal monotone, then $V = A0^\perp = \overline{\operatorname{dom} A}$ is the only subspace satisfying conditions (17.24)–(17.24) above, and $\tilde{A}$ is a maximal monotone single-valued selection of A.

Proof. For any $x \in X$, $P_V(x) = P_V(P_{A0^\perp}x + P_{\overline{A0}}x) = P_V(P_{A0^\perp}x)$ as $\overline{A0} \subset V^\perp$. By Proposition 17.18, $\tilde{A}$ is always single-valued, and if $A0$ is closed, $P_{A0^\perp}x^* \in Ax$ for each $(x,x^*) \in \operatorname{gra} A$, and so if $V = A0^\perp$, then $\tilde{A}$ is a selection of A. Consider now arbitrary $(y,\tilde{y}^*), (z,\tilde{z}^*) \in \operatorname{gra} \tilde{A}$, and $\lambda \in \mathbb{R}$. Then, for $y^* \in Ay$ and $z^* \in Az$, we have that $P_V y^* = \tilde{y}^*$ and $P_V z^* = \tilde{z}^*$. Since A is a linear relation, $(y+\lambda z, y^*+\lambda z^*) \in \operatorname{gra} A$. Therefore, $(y+\lambda z, P_V(y^*+\lambda z^*)) \in \operatorname{gra} \tilde{A}$, and since P_V is itself a linear operator, $P_V(y^*+\lambda z^*) = \tilde{y}^*+\lambda\tilde{z}^*$, it follows that $\tilde{y}^*+\lambda\tilde{z}^* \in \tilde{A}(y+\lambda z)$. Since $\operatorname{dom} A = \operatorname{dom} \tilde{A}$, the operator $\tilde{A}$ is a linear relation. Finally, suppose that A is maximal monotone, and so from Corollary 17.21 we have that $A0^\perp = \overline{\operatorname{dom} A}$ and $A0$ is closed. The only subspace V satisfying the conditions in this case is $V = A0^\perp$. Suppose there exists a point (x,x^*) where $x \in V = A0^\perp$, that is monotonically related to $\operatorname{gra} \tilde{A}$. For all $(z,z^*) \in \operatorname{gra} A$, there is a $y \in A0$ such that $y + P_V z^* = z^*$. Then, by Fact 17.17 (4),

$$\langle x-z, x^*-z^* \rangle = \langle x-z, x^*-y-P_V z^* \rangle = \langle x-z, x^*-P_V z^* \rangle \geq 0.$$

Therefore, (x,x^*) is also monotonically related to A, and since A is maximal monotone, $(x,x^*) \in \operatorname{gra} A$. Since $x^* \in V$, $P_V x^* = x^*$, and so $(x,x^*) \in \operatorname{gra} \tilde{A}$. Therefore, $\tilde{A}$ is maximal monotone. ■

From the results in this section so far, we know that monotone linear relations $A : X \to 2^X$ can only be multi-valued such that $A0$ is a subspace of X, $Ax = x^* + A0$ for any $x^* \in Ax$, and $A0 \subset (\operatorname{dom} A)^\perp$. For the purposes of calculation by the inner product, for any $x, z \in \operatorname{dom} A$,

$$\langle x, Az \rangle = \langle x, \tilde{A}z \rangle, \tag{17.11}$$

where $\tilde{A}$ is the single-valued operator (a selection of A if $A0$ is closed) as calculated in Proposition 17.24 for $V = A0^\perp$. In the other direction, any single-valued monotone linear relation $\tilde{A} : X \to 2^X$ can be extended to a multivalued monotone linear relation $A : X \to 2^X$ by choosing any subspace $V \subset (\operatorname{dom} A)^\perp$ and setting $Ax := \tilde{A}x + V$.

Now, in the unbounded linear case, maximal monotone operators may not have a closed domain. The concept of a *halo* well captures this aspect.

Definition 17.25 (halo). The *halo* of a monotone linear relation $A : X \to 2^X$ is the set

$$\operatorname{halo} A := \{x \in X : (\exists M)(\forall (y,y^*) \in \operatorname{gra} A)\langle x-y, y^*\rangle \le M\|x-y\|\}. \tag{17.12}$$

The following is an amalgamation of Proposition 6.2 and Theorem 6.5 in [5].

Fact 17.26 ([5]). If $A : X \to 2^X$ is a monotone linear relation, then $\operatorname{dom} A \subset \operatorname{halo} A \subset (A0)^\perp$. Furthermore, A is maximal monotone if and only if $A0^\perp = \overline{\operatorname{dom} A}$ and $\operatorname{halo} A = \operatorname{dom} A$.

Now, if the domain of a linear relation is not closed, we have the following curious result. Below, A^m denotes the iterated operator composition, where for instance $A^3x = A(A(Ax))$. Note that if $\operatorname{dom} A$ is dense in X, the operator $P_V A$ is the same as A.

Proposition 17.27. *Suppose a maximal monotone linear relation $A : X \to 2^X$ is such that $\operatorname{dom} A$ is not closed, and let $V := \overline{\operatorname{dom} A}$. Then, there is a sequence $(z_n)_{n\in\mathbb{N}} \subset \operatorname{dom} A$ such that*

$$(P_V A)^m(z_n) \in \operatorname{dom} A,\ \forall 1 \le m < n, \tag{17.13}$$

$$(P_V A)^n(z_n) \notin \operatorname{dom} A, \tag{17.14}$$

where for all $z \in \operatorname{dom} A$, $P_V A z$ is a singleton set.

Proof. Since A is maximal monotone, $\operatorname{dom} A = \operatorname{halo} A \subsetneq \overline{\operatorname{dom} A}$, and by Corollary 17.21, $V = A0^\perp$. Therefore, by Proposition 17.18, $P_V A z \subset Az$ and is a singleton for every $z \in \operatorname{dom} A$. Choose any point $z_0 \in V$ such that $z_0 \notin \operatorname{dom} A$. We shall generate the sequence $(z_n)_{n\in\mathbb{N}} \subset \operatorname{dom} A$ iteratively as follows. For some $n \ge 0$, suppose that $z_n \in V$. By Minty's theorem [25], since A is maximal monotone, $\operatorname{ran}(\operatorname{Id} + A) = X$. Therefore, there exists a $z_{n+1} \in \operatorname{dom} A$ such that $z_n \in z_{n+1} + Az_{n+1}$. Since $z_n, z_{n+1} \in V$, $z_n \in z_{n+1} + P_V A z_{n+1}$, and so as $P_V A z_{n+1}$ is a singleton,

$$P_V A z_{n+1} = \{z_n - z_{n+1}\}.$$

Now, since both P_V and A are linear operators, if $n \ge 2$

$$\begin{aligned}(P_V A)^2 z_{n+1} &= P_V A(z_n - z_{n+1})\\ &= P_V A z_n - P_V A z_{n+1}\\ &= \{z_{n-1} - 2z_n + z_{n+1}\},\end{aligned} \tag{17.15}$$

a linear combination of the terms z_{n-1}, z_n, and z_{n+1}, with z_{n-1} appearing with coefficient 1. Similarly, if $n \ge 3$,

$$\begin{aligned}(P_V A)^3 z_{n+1} &= P_V A(z_{n-1} - 2z_n + z_{n+1}) \\ &= \{z_{n-2} - z_{n-1} - 2z_{n-1} + 2z_n + z_n - z_{n+1}\} \\ &= \{z_{n-2} - 3z_{n-1} + 3z_n - z_{n+1}\}.\end{aligned} \tag{17.16}$$

By iterative composition, $(P_V A)^m z_{n+1}$ is linear combination of the terms z_p for $n-m+1 \leq p \leq n+1$, with z_{n-m+1} appearing with coefficient 1, as long as $n-m+1 \geq 0$. Since $\operatorname{dom} A$ is a linear subspace of X, $(P_V A)^m z_{n+1} \subset \operatorname{dom} A$ if $n \geq m$. However, if $n+1 = m$, the single point in $(P_V A)^m z_{n+1}$ is not in $\operatorname{dom} A$ since $z_0 = x \notin \operatorname{dom} A$. ■

For any linear relation $A : X \rightarrow 2^X$ where $\operatorname{dom} A$ is not closed, sequences like those in Proposition 17.27 are plentiful. Every point $x \in \overline{\operatorname{dom} A}$ such that $x \notin \operatorname{dom} A$, including for instance the points λx for $\lambda > 0$, generates a different sequence $(z_n)_{n\in\mathbb{N}}$ using the method from the proof of Proposition 17.27.

To explore these concepts, consider the following example.

Example 17.28. Consider the infinite dimensional Hilbert space ℓ^2, the space of infinite sequences $\mathbf{x} = (x_k)_{k\in\mathbb{N}}$ such that $\sum_{k=1}^{+\infty} x_k^2 < +\infty$. Let $\mathbf{e}_k$ denote the kth standard unit vector (the kth element in the sequence is 1, and all other elements in the sequence are 0). Define the single-valued monotone relation $A : \ell_2 \rightarrow \ell_2$ for $\mathbf{x} \in \operatorname{dom} A$ by

$$A\mathbf{x} = A\Big(\sum_{k=1}^{+\infty} x_k \mathbf{e}_k\Big) := \sum_{k=1}^{+\infty} k x_k \mathbf{e}_k,$$

where

$$\operatorname{dom} A := \{x \in \ell_2 : \exists N \in \mathbb{N} \text{ s.t. } x_k = 0\ \forall k \geq N\}.$$

Considering the linear relation A in the example above, the point $\mathbf{x} := \sum_{k=1}^{+\infty} \frac{1}{k}\mathbf{e}_k$ is not in $\operatorname{halo} A$. This is because the sequence $(\mathbf{y}_n)_{n\in\mathbb{N}} \subset \operatorname{dom} A$ where $\mathbf{y}_n := \sum_{i=1}^{n} \frac{1}{2i}\mathbf{e}_i$ eventually violates (17.12) for any choice of $M > 0$ for a large enough n. (Therefore we know that A is not maximal monotone.) However, the point $\mathbf{z} := \sum_{i=1}^{+\infty} \frac{1}{i^2}\mathbf{e}_i$ is in $\operatorname{halo} A$, and $\operatorname{gra} A$ could be extended by the point $(\mathbf{z}, \mathbf{x})$ and remain monotone. Since $\mathbf{x} \in \overline{\operatorname{dom} A}$ but $\mathbf{x} \notin \operatorname{halo} A$, yet $\mathbf{x} = A\mathbf{z}$ and $\mathbf{z} \in \operatorname{halo} A$, we have the beginning of a sequence like those in Proposition 17.27 for any monotone extension of A containing $(\mathbf{z}, \mathbf{x})$ that is also a linear relation.

Finally, the following result is used later and appears in Proposition 4.6 in [6].

Proposition 17.29 ([6]). *Suppose that $A : X \rightarrow 2^X$ is a linear relation. Then A is maximal monotone and symmetric if and only if there exists a proper lower semicontinuous convex function $f : X \rightarrow \mathbb{R} \bigcup \{+\infty\}$ such that $A = \partial f$.*

17.5 Monotone Classes of Linear Relations

The recent result for paramonotonicity and 3*-monotonicity is a portion of the main result in [7].

Proposition 17.30 ([7]). *Suppose $A : X \to 2^X$ is a maximal monotone linear relation such that* $\operatorname{dom} A$ *and* $\operatorname{ran} A_+$ *are closed (A_+ is the symmetric part of A). Then, A is 3^*-monotone if and only if A is paramonotone.*

In this section we use a different approach to that used for Proposition 17.30, where we (while avoiding the use of the Fitzpatrick function) obtain results that apply to all monotone operators regardless maximal monotonicity. This is done by examining the density of $\operatorname{dom} A$ rather than its closure, further extending these results. First, we characterize paramonotonicity for linear relations with the following two facts.

Fact 17.31. Suppose $A : X \to 2^X$ is a monotone linear relation. Then, A is paramonotone if and only if for all $x \in X$

$$\langle x, Ax \rangle = 0 \Rightarrow Ax = A0. \tag{17.17}$$

Proof. Suppose that A is paramonotone and that for some $x \in \operatorname{dom} A$, $\langle x, Ax \rangle = 0$. Then, $\langle x - 0, Ax - A0 \rangle = 0$, since $A0 \subset (\operatorname{dom} A)^\perp$ (Proposition 17.20). Therefore, by paramonotonicity, every $x^* \in Ax$ is also in $A0$. By Fact 17.17 (3) and (4), $Ax = A0$.

Now, suppose that (17.17) holds for A and that for some $(y, y^*), (z, z^*) \in \operatorname{gra} A$,

$$\langle y - z, y^* - z^* \rangle = 0.$$

Let $x = y - z$. Since A is a linear relation, $y^* - z^* \in Ax$, and so $\langle x, Ax \rangle = 0$. Therefore, $Ax = A0$, and so $y^* - z^* \in A0$ and

$$y^* \in z^* + A0; \quad -z^* \in -y^* + A0.$$

By Fact 17.17 (1) and (4), $-y^* + A0 = -Ay$. Hence $y^* \in Az$ and $z^* \in Ay$, so A is paramonotone. ∎

Fact 17.32. Suppose $A : X \to 2^X$ is a monotone linear relation, and let $x \in X$. Then, $Ax = A0$ if and only if $0 \in Ax$ and if $0 \in Ax$, then $P_{A0^\perp} Ax = \{0\}$. If $A0$ is closed and $P_{A0^\perp} Ax = \{0\}$, then $0 \in Ax$.

Proof. Let $Ax = A0$. Since $A0$ is a linear subspace of X (Fact 17.17 (3)), $0 \in Ax$. Now, let $0 \in Ax$. Then, by Fact 17.17 (4), $Ax = A0$.

By Proposition 17.18, $P_{A0^\perp} Ax$ is a singleton, and since $0 \in A0^\perp$ by the definition of the orthogonal complement, $P_{A0^\perp} Ax = \{0\}$. Now, let $P_{A0^\perp} Ax = \{0\}$ and suppose that $A0$ is closed. Then, by Proposition 17.18, $0 \in Ax$. ∎

Proposition 17.33. *Suppose $A : X \to 2^X$ is a monotone linear relation such that* $\operatorname{dom} A$ *is dense in $A0^\perp$ and $A0$ is closed. If A is 3^*-monotone, then A is also paramonotone.*

Proof. Suppose that A is not paramonotone. Then, there exists an $x \in \operatorname{dom} A$ such that $\langle x, Ax\rangle = 0$ yet $Ax \neq A0$. Choose any $x^* \in Ax$, and let $x_0^* = P_{A0^\perp} x^*$. By Fact 17.32, $x_0^* \neq 0$ since $A0$ is closed. If $x_0^* \in \operatorname{dom} A$, let $w = \frac{1}{2}x_0^*$. If $x_0^* \notin \operatorname{dom} A$, there is a sequence $(y_n)_{n\in\mathbb{N}} \subset \operatorname{dom} A$ converging to x_0^* since $\operatorname{dom} A$ is dense in $A0^\perp$. In this case, let $w = y_n$ for some n such that

$$\langle w, Ax\rangle = \langle y_n, x_0^*\rangle \geq \frac{1}{2}\|x_0^*\|^2.$$

Let $v = \lambda x$ for some $\lambda > 0$ and let $u = 0$ so that

$$\langle w - v, Av - Au\rangle = \langle w - \lambda x, \lambda Ax\rangle \geq \frac{\lambda}{2}\|x_0^*\|^2$$

which is unbounded with respect to λ. Hence, A is not 3^*-monotone, yielding the contrapositive. ■

We therefore obtain by a different method Proposition 4.5 from [7].

Corollary 17.34 ([7]). *If the linear relation $A : X \to 2^X$ is maximal monotone and 3^*-monotone, then A is paramonotone.*

Proof. Follows directly from Proposition 17.33 and Corollary 17.21. ■

Corollary 17.35. *If the linear relation $A : X \to 2^X$ is 3^*-monotone, then the operator $\tilde{A} : X \to 2^X$ defined by*

$$\tilde{A}x := Ax + (\operatorname{dom} A)^\perp \tag{17.18}$$

is a linear relation and is a 3^*-monotone extension of A that is paramonotone.

Proof. The operator $\tilde{A}$ is a linear relation since A is a linear relation, since $\operatorname{dom}\tilde{A} = \operatorname{dom} A$, and since $(\operatorname{dom} A)^\perp$ is a linear subspace. (Recall that we are using the convention that $\emptyset + S = \emptyset$ for any set S.) More specifically, for all $x, y \in \operatorname{dom}\tilde{A} = \operatorname{dom} A$ and for all $\lambda \in \mathbb{R}$,

$$\lambda\tilde{A}x = \lambda Ax + \lambda(\operatorname{dom} A)^\perp \subset A(\lambda x) + (\operatorname{dom} A)^\perp = \tilde{A}(\lambda x),$$

and

$$\tilde{A}x + \tilde{A}y = Ax + (\operatorname{dom} A)^\perp + Ay \subset A(x+y) + (\operatorname{dom} A)^\perp = \tilde{A}(x+y).$$

By the definition of $(\operatorname{dom} A)^\perp$, for all $x, y, z \in \operatorname{dom}\tilde{A}$

$$\langle z - y, \tilde{A}y - \tilde{A}z\rangle = \langle z - y, Ay - Az\rangle.$$

Therefore, $\tilde{A}$ is monotone and 3*-monotone because A is monotone and 3*-monotone. Since by Proposition 17.20, $A0 \subset (\mathrm{dom}A)^{\perp}$, it follows from Fact 17.17 (4) that $\tilde{A}$ is a monotone extension of A and that $\tilde{A}0 = (\mathrm{dom}A)^{\perp}$. Therefore, $\tilde{A}0^{\perp} = \overline{\mathrm{dom}A}$, and so by Proposition 17.33 and since $\mathrm{dom}A = \mathrm{dom}\tilde{A}$, $\tilde{A}$ is paramonotone. ■

If the linear relation A from Proposition 17.33 is also a single-valued bounded linear operator, then Proposition 17.33 is a corollary to the stronger result of Proposition 2 in [11].

Proposition 17.36 ([11]). *Let $A : X \to X$ be a bounded monotone linear operator. Then, A is 3*-monotone if and only if there exists an $\alpha > 0$ such that*

$$\langle x, Ax\rangle \geq \alpha\langle Ax, Ax\rangle = \alpha\|Ax\|^2.$$

Corollary 17.37. *If $A : X \to X$ is a bounded linear 3*-monotone operator, then it is paramonotone.*

However, there are 3*-monotone linear relations that are not paramonotone.

Example 17.38. Let $X = \ell^2$ and define the operators $\tilde{A}, A : X \to 2^X$ for $\mathbf{x} = (x_1, x_2, \ldots) \in \ell_2$ by

$$\tilde{A}\mathbf{x} := \sum_{k=1}^{+\infty} x_{2k}\mathbf{e}_{2k} \tag{17.19}$$

and

$$A\mathbf{x} := x_1\mathbf{u} + \tilde{A}\mathbf{x} + A0, \tag{17.20}$$

where

$$\mathbf{u} := \left(\sum_{k=1}^{\infty} \frac{1}{k}\mathbf{e}_{2k+1}\right), \tag{17.21}$$

$$A0 := \{\mathbf{x} \in \ell_2 : \exists N \in \mathbb{N} \text{ s.t. } x_k = 0\ \forall k \geq N \text{ and } x_{2k+1} = 0\ \forall k \in \mathbb{N}\}, \tag{17.22}$$

and

$$\mathrm{dom}A = \mathrm{dom}\tilde{A} = \mathrm{span}\{\mathbf{e}_1, \mathbf{e}_2, \mathbf{e}_4, \mathbf{e}_6, \ldots\}. \tag{17.23}$$

Then, A is a 3*-monotone linear relation, but it is not paramonotone.

Proof. Both A and $\tilde{A}$ are by definition linear relations. Note that $\tilde{A} = \mathbf{0} \times J$ where J is a subgraph of Id. Therefore, $\tilde{A}$ is 3*-monotone as both Id and $\mathbf{0}$ are 3*-monotone. Also, $A0$ is a dense subspace of $\mathrm{span}\{\mathbf{e}_{2k+1} : k \in \mathbb{N}\}$, and so $A0^{\perp} = \mathrm{span}\{\mathbf{e}_{2k} : k \in \mathbb{N}\}$. Therefore, $P_{A0^{\perp}}A\mathbf{x} = \tilde{A}\mathbf{x}$ as $\mathbf{u} \in (\mathrm{dom}A)^{\perp}$. Since $\overline{A0} \subset (\mathrm{dom}A)^{\perp}$ (Proposition 17.20), for all $(\mathbf{x}, \mathbf{x}^*), (\mathbf{y}, \mathbf{y}^*), (\mathbf{z}, \mathbf{z}^*) \in \mathrm{gra}A$,

$$\langle \mathbf{z}-\mathbf{y},\mathbf{y}^*-\mathbf{x}^*\rangle = \langle \mathbf{z}-\mathbf{y},P_{A0^\perp}\mathbf{y}^*-P_{A0^\perp}\mathbf{x}^*\rangle = \langle \mathbf{z}-\mathbf{y},A\mathbf{y}-A\mathbf{x}\rangle,$$

and so A is also 3^*-monotone. Now,

$$A\mathbf{e}_1 = \mathbf{u}+A0 \not\subset A0,$$

and so $A\mathbf{e}_1 \neq A0$. However, $\langle \mathbf{e}_1,A\mathbf{e}_1\rangle = \langle \mathbf{e}_1,\tilde{A}\mathbf{e}_1\rangle = 0$. Therefore, A is not paramonotone. ■

17.6 Monotone Classes of Linear Operators

A *linear operator* is a single-valued linear relation with full domain, which is maximal monotone by Proposition 17.19. Although being single-valued and having full domain are restrictive conditions, when it comes to monotone classes, linear operators are highly characteristic of linear relations with closed domain.

If a monotone linear relation $A : X \to 2^X$ has closed domain, which is always the case if $X = \mathbb{R}^n$, then $\operatorname{dom} A$ is itself a Hilbert space and the results of Sects. 17.4 and 17.5 hold in their strongest form, as they do for all linear operators.

Let $\tilde{A} : \operatorname{dom} A \to 2^{\operatorname{dom} A}$ be the single-valued selection of A generated in the manner of Proposition 17.24 with $V = \operatorname{dom} A$. By Proposition 17.19, $\tilde{A}$ is a monotone linear operator. As the only difference between A and $\tilde{A}$ are elements perpendicular to the domain, for any $(x,x^*),(y,y^*) \in \operatorname{gra} A$,

$$\langle x-y,x^*-y^*\rangle = \langle x-y,\tilde{A}x-\tilde{A}y\rangle,$$

and so the monotone classes of each, while not necessarily equivalent, are highly correlated.

Below, we consider linear operators operating on $\mathbb{R}^2$, $\mathbb{R}^n$, and on Hilbert spaces of infinite dimension. Note that linear operators acting on $\mathbb{R}^n$ will be identified with their matrix representation in the standard basis, and recall from Proposition 17.29 that symmetric linear operators are the subdifferentials of a lower semicontinuous convex function.

17.6.1 Monotone Linear Operators on $\mathbb{R}^2$

In this section we consider linear operators $A : \mathbb{R}^2 \to \mathbb{R}^2$, which can be represented by the matrix

$$A = \begin{bmatrix} a & c \\ b & d \end{bmatrix}.$$

The operator A so defined is monotone if and only if $a+d \geq 0$ and $4ad \geq (b+c)^2$. We consider some simple examples, examine their properties, and provide some sufficient and necessary conditions for inclusion within various monotone classes.

Proposition 17.39 (3-cyclic monotone linear operators on $\mathbb{R}^2$). *If A is 3-cyclic monotone, then*

$$\max\{|b|,|c|\} - a - d \leq 0. \tag{17.24}$$

Proof. Choose $x = (0,0)$, $y = (1,0)$, and $z = (0,1)$; let $x^* = Ax = (0,0)$, $y^* = Ay = (a,b)$, and $z^* = Az = (c,d)$. If the mapping associated with A is 3-cyclic monotone, then

$$\begin{aligned} 0 &\leq \langle x-y, x^* \rangle + \langle y-z, y^* \rangle + \langle z-x, z^* \rangle \\ &= \langle (1,-1),(a,b) \rangle + \langle (0,1),(c,d) \rangle \\ &= a+d-b. \end{aligned}$$

Similarly, by choosing different y and z, the following conditions are also necessary for any matrix A as defined above:

$$0 \geq \begin{cases} b-a-d, & y=(1,0), z=(0,1), \\ -b-a-d, & y=(-1,0), z=(0,1), \\ c-a-d, & y=(0,1), z=(1,0), \\ -c-a-d, & y=(0,-1), z=(1,0). \end{cases} \tag{17.25}$$

In all cases, $x = (0,0)$. ■

There are many monotone linear operators in $\mathbb{R}^2$ that are not 3-cyclic monotone, and furthermore Examples 17.40 and 17.41 below demonstrate that 3-cyclic monotonicity does not follow from strict and maximal monotonicity.

Example 17.40. Consider the monotone linear operator $\tilde{R} : \mathbb{R}^2 \to \mathbb{R}^2$ defined by

$$\tilde{R} = \begin{bmatrix} 1 & -2 \\ 3 & 1 \end{bmatrix}. \tag{17.26}$$

The operator $\tilde{R}$ violates the necessary condition (17.24) for 3-cyclic monotonicity since $b-a-d > 0$ and $\tilde{R}$ satisfies the monotonicity conditions $(a+d) \geq 0$ and $4ad \geq (b+c)^2$, using the format $\tilde{R} = \begin{bmatrix} a & c \\ b & d \end{bmatrix}$ above. Note that $\langle x, \tilde{R}x \rangle = 0$ implies that $x = 0$, so $\tilde{R}$ is strictly monotone and therefore paramonotone. Hence, by Proposition 17.47, $\tilde{R}$ is also 3*-monotone. Finally, $\tilde{R}$ is maximal monotone by Proposition 17.19.

Example 17.41. Consider the rotation operator $R_\theta : \mathbb{R}^2 \to \mathbb{R}^2$ with matrix representation

$$R_\theta = \begin{bmatrix} \cos(\theta) & -\sin(\theta) \\ \sin(\theta) & \cos(\theta) \end{bmatrix}. \tag{17.27}$$

Note that R_θ is monotone if and only if $|\theta| \leq \pi/2$, since this is precisely when $\cos(\theta) \geq 0$. In this range, R_θ is maximal monotone by Proposition 17.19.

Now, R_θ is 3-cyclic monotone if and only if $|\theta| < \pi/3$ by Fact 17.42 below.

Therefore, for any $\theta \in]\pi/3, \pi/2[$, R_θ is maximal monotone and strictly monotone, but not 3-cyclic monotone.

Now, $\langle x, R_\theta x\rangle = 0$ implies that $x = 0$ unless $\theta = \pi/2$. Therefore, R_θ is strictly monotone and hence paramonotone when $|\theta| < \pi/2$. By Proposition 17.47, R_θ is 3*-monotone as well when $|\theta| < \pi/2$. When $\theta = \pi/2$, R_θ is not paramonotone, and therefore neither is it strictly monotone nor, by Proposition 17.33, is it 3*-monotone.

By the following fact (Proposition 7.1 in [3]), $\mathbb{R}^2$ is large enough to contain distinct instances of n-cyclic monotone operators for $n \geq 2$.

Fact 17.42 ([3]). Let $n \in \{2,3,\ldots\}$. Then R_θ is n-cyclic monotone if and only if $|\theta| \in [0, \pi/n]$.

Proof. See Example 4.6 in [3] for a detailed proof. ■

The zero operator yields trivial solutions to any associated variational inequality problem, and so the following, which shares the monotone classes of $\mathbf{0}$, is introduced in its stead.

Example 17.43. The orthogonal projection $A : \mathbb{R}^2 \to \mathbb{R}^2$ defined by $A(x_1,x_2) := (x_1,0)$ is maximal monotone, paramonotone, 3-cyclic monotone, and 3*-monotone.

Proof. Using the notation of Sect. 17.3, we have that $A = Id \times \mathbf{0}$, where $\mathbf{0} : \mathbb{R} \to \mathbb{R}$ is the zero operator, and $\mathrm{Id} : \mathbb{R} \to \mathbb{R}$ is the identity. The $\mathbf{0}$ operator is maximal monotone, paramonotone, 3-cyclic monotone, and 3*-monotone, as is Id, which is also strictly monotone, while $\mathbf{0}$ is not. The properties of A follow directly from the results in Sect. 17.3. ■

Finally, paramonotone linear operators in $\mathbb{R}^2$ are further restricted to be either strictly monotone or symmetric.

Proposition 17.44. *A linear operator $A : \mathbb{R}^2 \to \mathbb{R}^2$ is paramonotone if and only if it is strictly monotone or symmetric.*

Proof. Strictly monotone operators and symmetric linear operators are paramonotone by Facts 17.8 and 17.48, respectively. It remains to show that these are the only two possibilities. Assuming then that A is paramonotone, consider the general case, $A = \begin{bmatrix} a & c \\ b & d \end{bmatrix}$ and $A_+ = \begin{bmatrix} a & \frac{b+c}{2} \\ \frac{b+c}{2} & d \end{bmatrix}$. If $\ker(A_+) = \{0\}$, then A is strictly monotone by Fact 17.48. If $\ker(A_+) \neq \{0\}$, then by Fact 17.48 $\ker(A_+) \subseteq \ker(A)$, and so $\ker(A) \neq \{0\}$, from which $\det(A) = 0$ and $ad = bc$. Hence, since $\det(A_+) = 0$, $4bc = (b+c)^2$, so $(b-c)^2 = 0$ and $b = c$. Therefore A is symmetric. ■

Remark 17.45. The only paramonotone linear operators in $\mathbb{R}^2$ that are not strictly monotone are the symmetric linear operators $A := \begin{bmatrix} a & b \\ b & \frac{b^2}{a} \end{bmatrix}$ for $a > 0$ and $b \in \mathbb{R}$ and the zero operator $x \mapsto (0,0)$. By Proposition 17.29, since both examples of A are symmetric linear operators, they are also maximal monotone and maximal cyclical monotone, as they are subdifferentials of proper lower semicontinuous convex functions.

All relationships among the classes of monotone linear operators in $\mathbb{R}^2$ are now known completely and are summarized in Table 17.1. Recall that all monotone linear operators are assumed to have full domain and are therefore maximal monotone by Proposition 17.19.

17.6.2 Linear Operators on $\mathbb{R}^n$

On $\mathbb{R}^n$ the restriction that linear operators are single-valued is redundant as this also follows from having full domain.

Proposition 17.46. *A single-valued monotone linear relation* $A : \mathbb{R}^n \to \mathbb{R}^n$ *is maximal monotone if and only if* $\operatorname{dom} A = \mathbb{R}^n$.

Proof. In $\mathbb{R}^n$, all subspaces are closed, and so by Corollary 17.21, any maximal monotone single-valued linear relations have full domain. The converse follows from Proposition 17.19. ■

Since linear operators are maximal monotone, the following result is a consequence of Proposition 17.30 and appears in Remark 4.11 in [3].

Proposition 17.47 ([3]). *Given a monotone linear operator* $A : \mathbb{R}^n \to \mathbb{R}^n$, *A is* 3^*-*monotone if and only if A is paramonotone.*

In the following fact (from Proposition 3.2 in [24]), we denote by $A_+ := \frac{1}{2}(A + A^*)$ the symmetric part of a linear operator $A : \mathbb{R}^n \to \mathbb{R}^n$ and by $\ker A := \{x \in \mathbb{R}^n : Ax = 0\}$ the kernel of A.

Fact 17.48 ([24]). Let $A : \mathbb{R}^n \to \mathbb{R}^n$ be a linear operator. Then A is paramonotone if and only if A is monotone and $\ker(A_+) \subseteq \ker(A)$.

In Remark 17.45 we noted that the converse of Proposition 17.10 holds for monotone linear operators that are not strictly monotone operators on $\mathbb{R}^2$. We now demonstrate that this result does not generalize to $\mathbb{R}^3$.

Example 17.49. Let $T : \mathbb{R}^3 \to \mathbb{R}^3$ be the linear operator defined by

$$Tx := \begin{bmatrix} 1 & -2 & 1 \\ 3 & 1 & 3 \\ 1 & -2 & 1 \end{bmatrix} x. \tag{17.28}$$

The operator T is paramonotone and maximal monotone, but not strictly monotone. Further, T is not 3-cyclic monotone, but is 3^*-monotone.

Proof. The symmetric part of T is

$$T_+ := \begin{bmatrix} 1 & 1/2 & 1 \\ 1/2 & 1 & 1/2 \\ 1 & 1/2 & 1 \end{bmatrix}.$$

Since the eigenvalues of T_+, consisting of $\{0, \frac{1}{2}(3+\sqrt{3}), \frac{1}{2}(3-\sqrt{3})\}$, are nonnegative, T_+ is positive semidefinite, hence monotone, and so T is monotone.

An elementary calculation yields that $\ker T_+ = \{t(-1,0,1) : t \in \mathbb{R}\}$. Clearly, $\ker T = \ker T_+$, so by Fact 17.48, T is paramonotone. However, T is not strictly monotone since the kernel contains more than the zero element.

Furthermore, T is maximal monotone since it is linear and has full domain (Proposition 17.19). The operator T is not 3-cyclic monotone since the points $(0,0,0), (1,0,0)$, and $(0,1,0)$ do not satisfy the defining condition (17.6). (For a shortcut, call to mind Example 17.40 and Proposition 17.39.) Finally, since T is a linear operator in $\mathbb{R}^3$ that is paramonotone, it is 3^*-monotone by Proposition 17.47. ■

17.6.3 Monotone Linear Operators in Infinite Dimensions

Recall from Proposition 17.47 that linear paramonotone operators on $\mathbb{R}^n$ are 3^* monotone. Example 17.50 below demonstrates that larger spaces are more permissive. A similar example appears in [7].

Example 17.50. Let $\theta_k := \pi/2 - 1/k^4$ and let $A : \ell^2 \to \ell^2$ be the linear operator defined by

$$A\mathbf{x} \mapsto \sum_{k=1}^{+\infty} (\cos(\theta_k)x_{2k-1} - \sin(\theta_k)x_{2k})\,\mathbf{e}_{2k-1} + (\sin(\theta_k)x_{2k-1} + \cos(\theta_k)x_{2k})\,\mathbf{e}_{2k}. \tag{17.29}$$

The structure of A is such that every $\mathbf{x}^* = A\mathbf{x}$ obeys

$$\begin{bmatrix} x^*_{2k-1} \\ x^*_{2k} \end{bmatrix} = R_{\theta_k} \begin{bmatrix} x_{2k-1} \\ x_{2k} \end{bmatrix} \tag{17.30}$$

for all $\mathbf{x} \in \ell^2$ and $k \in \mathbb{N}$, where R_{θ_k} is the rotation matrix as introduced in Example 17.41. A is strictly monotone and maximal monotone, but not 3^*-monotone. It follows that A is also paramonotone but not 3-cyclic monotone.

Proof. The monotonicity of T is evident from (17.30). Suppose that $\mathbf{x} \in \ell^2$ is such that $\langle \mathbf{x}, A\mathbf{x} \rangle = 0$. Now,

$$\langle \mathbf{x}, A\mathbf{x} \rangle = \sum_{k=1}^{+\infty} \cos(\theta_k)(x_{2k-1}^2 + x_{2k}^2)$$

is equal to zero if and only if $\mathbf{x} = \mathbf{0}$, and so A is strictly monotone.

By Proposition 17.19, A is maximal monotone since it is linear and has full domain.

Let $\mathbf{x} = \mathbf{0}$, so that $A\mathbf{x} = \mathbf{0}$, and let $\mathbf{z} = \sum_{k=1}^{+\infty} \frac{1}{k}(\mathbf{e}_{2k-1} + \mathbf{e}_{2k})$. Define a sequence $\mathbf{y}_n \in \ell^2$ by $\mathbf{y}_n := n^2 \mathbf{e}_{2n-1}$, and so $A\mathbf{y}_n = n^2 \cos(\theta_n)\mathbf{e}_{2n-1} + n^2 \sin(\theta_n)\mathbf{e}_{2n}$. For all n, $0 < \cos(\theta_n) \leq 1/n^4$, and from the Taylor's series $\sin(\theta_n) \geq 1 - 1/(2n^8)$ for all large n. Considering the inequality (17.3) for 3*-monotonicity, we have

$$\begin{aligned} \langle \mathbf{z} - \mathbf{y}_n, A\mathbf{y}_n - A\mathbf{x} \rangle &= n(\cos(\theta_n) + \sin(\theta_n)) - n^4 \cos(\theta_n) \\ &\geq n(0 + 1 - 1/(2n^8)) - 1 \\ &\to +\infty, \qquad \text{as } n \to +\infty, \end{aligned} \tag{17.31}$$

and so A fails to be 3*-monotone. ∎

Remark 17.51. The operator A from Example 17.50 can be modified to lose its strict monotonicity property by using the zero function $\mathbf{0} : \mathbb{R} \to \mathbb{R}$ as a prefactor in the product space, yielding $T = \mathbf{0} \times A$. In this manner,

$$T\mathbf{x} := \sum_{k=1}^{+\infty} \begin{bmatrix} (\cos(\theta_k)x_{2k} - \sin(\theta_k)x_{2k+1})\,\mathbf{e}_{2k} \\ + (\sin(\theta_k)x_{2k} + \cos(\theta_k)x_{2k+1})\,\mathbf{e}_{2k+1} \end{bmatrix}. \tag{17.32}$$

Proof. The Hilbert space ℓ^2 can be written as a product space $\ell^2 = \mathbb{R} \times \ell^2$. More precisely, all of these spaces can be embedded in the larger space $\ell^2(\mathscr{Z})$ with standard unit vectors e_i for i in $\mathscr{Z}$, the set of integers. In this setting $\ell^2 = \text{span}\{e_i : i \in \mathbb{N}\}$, and let $V_0 = \text{span}\{e_0\}$ so that $\ell^2(\mathbb{N} \bigcup \{0\}) = V_0 \times \ell^2$. Let $T = \mathbf{0} \times A$, where A is the linear operator from Example 17.50. The operator $\mathbf{0} : V_0 \to V_0$ is paramonotone, maximal monotone, 3-cyclic monotone, and 3*-monotone, but not strictly monotone on $\mathbb{R}$. The operator $A : \ell^2 \to \ell^2$ from Example 17.50 is strictly monotone and maximal monotone, but not 3*-monotone. Therefore, by the results of Sect. 17.3, $T := \mathbf{0} \times A$ is paramonotone and maximal monotone and fails to be strictly monotone or 3*-monotone. ∎

Note that all linear operators are assumed to have full domain and are therefore maximal monotone by Proposition 17.19. Also, if a linear operator fails to be paramonotone, it fails to be 3*-monotone and 3-cyclic monotone as well. The monotone class characterizations for linear operators in a Hilbert space are now known completely, as summarized in Table 17.2 below.

Since the only linear operators on $\mathbb{R}$ are the constant operators, by the results shown in Table 17.1 and by Proposition 17.47, each example in Table 17.2 operates on a space with the lowest dimension for which its monotone class combination is possible. In particular, note how examples with binary label 1000, 1001, and 1100, although absent in Table 17.1, exist for spaces of higher dimension. Finally, for every operator T in Table 17.2, an operator with the same monotone class combination on any higher dimension can be constructed by a product space composition with Id ($T \times \mathrm{Id}$).

17.7 Summary

The relationships among the five classes of monotone linear operator considered, that is, maximal, para-, 3*-, 3-cyclic, and strictly monotone operators, are now fully understood in $\mathbb{R}^2$, $\mathbb{R}^n$, and in general Hilbert spaces. They are depicted in each case with the Venn diagrams below. Further, a sample linear monotone operator has been provided for every possible combination of monotone class. In Sect. 17.3, a method by which these examples can be combined and extended to create linear operators in higher dimension with a known monotone class configuration has been described. Various properties and monotone class relationships of linear relations have been explored above. Furthermore, only two monotone class relationships do not apply for linear relations: a linear relation may be 3*-monotone and not paramonotone (as in Example 17.38), and 3-cyclic monotone linear relations that are not paramonotone could exist, but they must not be maximal monotone.

Some of the results and examples in this paper were presented at a meeting of the Canadian Mathematical Society in Vancouver, Canada, on December 4, 2010, and a similar and complete characterization of these same monotone class relationships for nonlinear operators will appear shortly.

Acknowledgements The author would like to thank the professional and editorial support of Philip Loewen.

References

1. Attouch, H., Damlamian, A.: On multivalued evolution equations in Hilbert spaces. Israel J. Math. **12**(4), 373–390 (1972)
2. Auslender, A.A., Haddou, M.: An interior-proximal method for convex linearly constrained problems and its extension to variational inequalities. Math. Program. **71**, 77–100 (1995)
3. Bauschke, H.H., Borwein, J.M., Wang, X.: Fitzpatrick functions and continuous linear monotone operators. SIAM J. Optim. **18**, 789–809 (2007)
4. Bauschke, H.H., Combettes, P.L.: Convex Analysis and Monotone Operator Theory in Hilbert Spaces. Springer, New York (2011)
5. Bauschke, H.H., Wang, W., Yao, L.: Monotone linear relations: maximality and Fitzpatrick functions. J. Convex Anal. **25**, 673–686 (2009)
6. Bauschke, H.H., Wang, X., Yao, L.: On Borwein-Wiersma decompositions of monotone linear relations. SIAM J. Optim. **20**, 2636–2652 (2010)
7. Bauschke, H.H., Wang, X., Yao, L.: Rectangularity and paramonotonicity of maximally monotone operators. Optimization. In press

8. Bello Cruz, J.Y., Iusem, A.N.: Convergence of direct methods for paramonotone variational inequalities. Comput. Optim. Appl. **46**(2), 247–263 (2010)
9. Borwein, J.M.: Maximal monotonicity via convex analysis. J. Convex Anal. **13**(3), 561–586 (2006)
10. Bressan, A., Staicu, V.: On nonconvex perturbations of maximal monotone differential inclusions. Set-Valued Var. Anal. **2**, 415–437 (1994)
11. Brézis, H., Haraux, A.: Image d'une somme d'opérateurs monotones et applications. Israel J. Math. **23**(2), 165–186 (1976)
12. Bruck, R.E., Jr.: An iterative solution of a variational inequality for certain monotone operators in Hilbert space. Bull. Amer. Math. Soc. **81**(5), 890–892 (1975)
13. Burachik, R.S., Iusem, A.N.: An iterative solution of a variational inequality for certain monotone operators in a Hilbert space. SIAM J. Optim. **8**, 197–216 (1998)
14. Burachik, R.S., Lopes, J.O., Svaiter, B.F.: An outer approximation method for the variational inequality problem. SIAM J. Control Optim. **43**, 2071–2088 (2005)
15. Censor, Y., Iusem, A.N., Zenios, S.A.: An interior point method with Bregman functions for the variational inequality problem with paramonotone operators. Math. Program. **81**(3), 373–400 (1998)
16. Chu, L.-J.: On the sum of monotone operators. Michigan Math. J. **43**, 273–289 (1996)
17. Cross, R.: Monotone Linear Relations. M. Dekker, New York (1998)
18. Eckstein, J., Bertsekas, D.P.: On the Douglas-Rachford splitting method and the proximal point algorithm for maximal monotone operators. Math. Program. **55**, 293–318 (1992)
19. Facchinei, F., Pang, J.S.: Finite-dimensional Variational Inequalities and Complementarity Problems. Springer, New York (2003)
20. Ferris, M.C., Pang, J.S.: Engineering and economic applications of complementarity problems. SIAM Rev. **39**(4), 669–713 (December 1997)
21. Ghoussoub, N.: Selfdual partial differential systems and their variational principles. Springer Monographs in Mathematics, vol. 14. Springer, New York (2010)
22. Hadjisavvas, N., Schaible, S.: On a generalization of paramonotone maps and its application to solving the stampacchia variational inequality. Optimization **55**(5-6), 593–604 (October-December 2006)
23. Hartmann, P., Stampacchia, G.: On some non-linear elliptic differential-functional equations. Acta Math. **115**(1), 271–310 (1966)
24. Iusem, A.N.: On some properties of paramonotone operators. J. Convex Anal. **5**(2), 269–278 (1998)
25. Minty, G.J.: Monotone (nonlinear) operators in a Hilbert space. Duke Math. J. **29**, 341–346 (1962)
26. Papageorgiou, N.S., Shahzad, N.: On maximal monotone differential inclusions in RN. Acta Math. Hungar. **78**(3), 175–197 (1998)
27. Pennanen, T.: Dualization of monotone generalized equations, PhD thesis. University of Washington, Seattle, Washington (1999)
28. Rockafellar, R.T., Wets, R.J.-B.: Variational analysis. Grundlehren der Mathematischen Wissenschaften, vol. 317, 2nd edn. Springer, Berlin (2004)
29. Simons, S.: LC functions and maximal monotonicity. J. Nonlinear Convex Anal. **7**, 123–138 (2006)
30. Yagi, A.: Generation theorem of semigroup for multivalued linear operators. Osaka J. Math. **28**, 385–410 (1991)
31. Yamada, I., Ogura, N.: Hybrid steepest descent method for variational inequality problems over the fixed point set of certain quasi-nonexpansive mappings. Numer. Funct. Anal. Optim. **25**, 619–655 (2004)
32. Zeidler, E.: Nonlinear Functional Analysis and Its Applications, II/B - Nonlinear Monotone Operators. Springer, New York (1990)

Chapter 18
Upper Semicontinuity of Duality and Preduality Mappings

J.R. Giles

To celebrate Jonathan Borwein's 60th birthday

Abstract In their paper studying Hausdorff weak upper semicontinuity of duality and preduality mappings on the dual of a Banach space, Godefroy and Indumathi related these by an interesting geometrical property. This property actually characterises Hausdorff upper semicontinuity of the preduality mapping. When the duality mapping is Hausdorff upper semicontinuous with weakly compact image, we investigate how this same property persists with natural embedding into higher duals.

Key words: Duality and preduality mappings • Gateaux and Fréchet differentiability • Hausdorff upper semicontinuity

Mathematics Subject Classifications (2010): Primary: 46B20; Secondary: 54C60, 58C20

18.1 Introduction

In their paper [6] Godefroy and Indumathi studied weak upper semicontinuity of duality and preduality mappings on a Banach space. Given a Banach space X, the

---

COMMUNICATED BY JON D. VANDERWERFF.

J.R. Giles
School of Mathematical and Physical Sciences, The University of Newcastle,
New South Wales 2308, Australia
e-mail: John.Giles@newcastle.edu.au

D.H. Bailey et al. (eds.), *Computational and Analytical Mathematics*, Springer Proceedings in Mathematics & Statistics 50, DOI 10.1007/978-1-4614-7621-4_18,

subdifferential of the norm at $x \in S(X)$ is the non-empty weak* compact convex subset

$$\partial\|x\| \equiv \{f \in X^* : f(y) \leq \|x\|'_+(y) \text{ for all } y \in X\}.$$

The set-valued subdifferential mapping $x \mapsto \partial\|x\|$ is called the *duality mapping* on X. Denoting by τ the weak*, weak or norm topology on X^*, the duality mapping is *Hausdorff τ upper semicontinuous* at $x \in S(X)$ if given a τ neighbourhood N of O in X^* there exists $\delta > 0$ such that

$$\partial\|B(x,\delta)\| \subseteq \partial\|x\| + N;$$

equivalently, if and only if for each τ neighbourhood N of O in X^* there exists $\delta > 0$ such that

$$\{f \in B(X^*) : f(x) > 1 - \delta\} \subseteq \partial\|x\| + N \text{[3, Theorem 2.1, p. 102].}$$

The duality mapping $x \mapsto \partial\|x\|$ is always Hausdorff weak* upper semicontinuous [9, Proposition 2.5, p. 19]. The duality mapping $x \mapsto \partial\|x\|$ is Hausdorff norm upper semicontinuous at $x \in S(X)$ if and only if the norm is *strongly subdifferentiable* at x; that is, given $\varepsilon > 0$ there exists $\delta > 0$ such that

$$0 \leq \|x+y\| - \|x\| - \|x\|'_+(y) \leq \varepsilon\|y\| \text{ for all } \|y\| < \delta \text{ [5, Theorem 3.2, p. 28].}$$

The norm is *Gateaux differentiable* at $x \in S(X)$ if $\partial\|x\|$ is singleton, is *very smooth* at x if also the duality mapping $x \mapsto \partial\|x\|$ is Hausdorff weak upper semicontinuous at x and is *Fréchet differentiable* at x if also the duality mapping $x \mapsto \partial\|x\|$ is Hausdorff norm upper semicontinuous at x.

On the dual space X^*, given $f \in S(X^*)$, we call $\partial\|f\| \cap \hat{X}$ the *presubdifferential* of the dual norm at f, often considering $\partial\|f\| \cap \hat{X}$ as a subset of X. Of course this set could be empty. The mapping $f \mapsto \partial\|f\| \cap \hat{X}$ is called the *preduality mapping* on X^* and considering τ to be the weak or norm topology on X is said to be *Hausdorff τ upper semicontinuous* at f, provided $\partial\|f\| \cap \hat{X} \neq \emptyset$ and given a τ neighbourhood N of O in X there exists $\delta > 0$ such that

$$\partial\|B(f,\delta)\| \cap \hat{X} \subseteq (\partial\|f\| \cap \hat{X}) + \hat{N};$$

equivalently, if and only if for each τ neighbourhood N of O in X there exists $\delta > 0$ such that

$$\{x \in B(X) : f(x) > 1 - \delta\} \subseteq (\partial\|f\| \cap \hat{X}) + \hat{N}, \text{[6, Lemma 2.1, p. 319].}$$

Hausdorff weak upper semicontinuity of the duality and preduality mappings has significant characterisation by density properties.

Proposition 18.1. *Given a Banach space* X*,*

(i) The duality mapping $x \mapsto \partial\|x\|$ *on* X *is Hausdorff weak upper semicontinuous at* $x \in S(X)$ *if and only if*

$$\widehat{\partial\|x\|} \text{ is weak}^* \text{ dense } \sigma(X^{***}, X^{**}) \text{ in } \partial\|\hat{x}\|, \text{[3, Theorem 3.1, p. 103].}$$

(ii) Given $f \in S(X^*)$ *where* $\partial\|f\| \cap \hat{X} \neq \emptyset$*, the preduality mapping* $f \mapsto \partial\|f\| \cap \hat{X}$ *on* X^* *is Hausdorff weak upper semicontinuous at* f *if and only if*

$$\partial\|f\| \cap \hat{X} \text{ is weak}^* \text{ dense } \sigma(X^{**}, X^*) \text{ in } \partial\|f\|, \text{[6, Lemma 2.2, p. 320].}$$

In Sect. 18.2 we show that a geometrical property introduced by Godefroy and Indumathi actually characterises Hausdorff weak upper semicontinuity of the preduality mapping at $f \in S(X^*)$ where $\partial\|f\| \cap \hat{X} \neq \emptyset$. In Sect. 18.3 we investigate how Hausdorff upper semicontinuity properties of the duality mapping affect higher duals in particular when the subdifferential image is weakly compact.

18.2 The Geometrical Property of Godefroy and Indumathi

To establish links between duality and preduality mappings on the dual X^*, Godefroy and Indumathi introduced the following property. We say that a Banach space X has the *G-I Property* at $f \in S(X^*)$ if for any convex subset C of $B(X)$ such that $\sup_C f = 1$ we have $d(\hat{C}, \partial\|f\|) = 0$, [6, Fact 1, p. 321]. Following their argument [6, Theorem 2.3, p. 321] we determine significant implications of this property.

Lemma 18.2. *For a Banach space* X *with the G-I Property at* $f \in S(X^*)$

(i) $\partial\|f\| \cap \hat{X} \neq \emptyset$

(ii) For any $y \in B(X)$*,* $d(\hat{y}, \partial\|f\| \cap \hat{X}) = d(\hat{y}, \partial\|f\|)$

Proof. Given $y \in B(X)$ choose $r > d(\hat{y}, \partial\|f\|)$. Then there exists $F \in \partial\|f\|$ such that $\|\hat{y} - F\| < r$. Since the natural embedding of $B(y,r) \cap B(X)$ is weak* dense in $B^{**}(\hat{y}, r) \cap B(X^{**})$ there exists a net $\{x_\alpha\}$ in $B(y,r) \cap B(X)$ such that $\{\widehat{x_\alpha}\}$ is weak* convergent to F and so $f(x_\alpha) \to 1$. Then for $C \equiv B(y,r) \cap B(X)$ we have $\sup_C f = 1$ and so from the G-I Property given $\varepsilon > 0$ there exists $x_1 \in C$ such that $d(\widehat{x_1}, \partial\|f\|) < \varepsilon$. Again there exists $F_1 \in \partial\|f\|$ such that $\|\widehat{x_1} - F_1\| < \varepsilon$. Again using the weak* density of the natural embedding of $B(x_1, \varepsilon) \cap B(X)$ in $B^{**}(\widehat{x_1}, \varepsilon) \cap B(X^{**})$ we have for $C_1 \equiv B(x_1, \varepsilon) \cap B(X)$ with similar argument that there exists $x_2 \in C_1$ such that $d(\widehat{x_2}, \partial\|f\|) < \varepsilon/_2$. Continuing this argument we have a sequence $\{x_n\}$ such that

$$\|x_n - x_{n+1}\| < \varepsilon/_{2^{n-1}} \text{ and } d(\widehat{x_n}, \partial\|f\|) < \varepsilon/_{2^{n-1}}.$$

The sequence $\{x_n\}$ is Cauchy so converges to say z and we have $d(\hat{z}, \partial\|f\|) = 0$ which implies that $\hat{z} \in \partial\|f\| \cap \hat{X}$ so $\partial\|f\| \cap \hat{X} \neq \emptyset$. This completes the proof of (i).

Also

$$\begin{aligned}\|y-x_n\| &\le \|y-x_1\|+\|x_1-x_2\|+\ldots+\|x_{n-1}-x_n\|\\ &< r+\varepsilon+\frac{\varepsilon}{2}+\ldots+\frac{\varepsilon}{2^{n-2}}\\ &< r+2\varepsilon.\end{aligned}$$

As $\{x_n\}$ converges to z so $\|y-z\|\le r+2\varepsilon$ and since $r>\mathrm{d}(\hat{y},\partial\|f\|)$ and $\varepsilon>0$ were chosen arbitrarily and $z\in\partial\|f\|\cap\hat{X}$ we have

$$\mathrm{d}(\hat{y},\partial\|f\|\cap\hat{X})=d(\hat{y},\partial\|f\|).$$

This completes the proof of (ii). ■

Lemma 18.2(ii) has an immediate consequence.

Corollary 18.3. *Given a Banach space X with the G-I Property at $f\in S(X^*)$, for any convex subset C of $B(X)$ such that* $\sup_C f=1$ *we have* $\mathrm{d}(\hat{C},\partial\|f\|\cap\hat{X})=0$.

Lemma 18.2 enables us to give a characterisation of the G-I Property.

Theorem 18.4. *A Banach space X has the G-I Property at $f\in S(X^*)$ if and only if $\partial\|f\|\cap\hat{X}\neq\emptyset$ and the preduality mapping $f\mapsto\partial\|f\|\cap\hat{X}$ on X^* is Hausdorff weak upper semicontinuous at f.*

Proof. Suppose that the preduality mapping $f\mapsto\partial\|f\|\cap\hat{X}$ on X^* is not Hausdorff weak upper semicontinuous at f. Then by Proposition 18.1(ii), $\overline{\partial\|f\|\cap\hat{X}}^{\omega^*}\neq\partial\|f\|$. For $F\in\partial\|f\|\ \setminus\ \overline{\partial\|f\|\cap\hat{X}}^{\omega^*}$ there exists a convex weak* neighbourhood V of O in X^{**} such that

$$\overline{\partial\|f\|\cap\hat{X}}^{\omega^*}\cap(F+V)=\emptyset.$$

The convex set $\widehat{C_V}\equiv(F+V)\cap B(\hat{X})$ has $\sup_{C_V} f=1$ but

$$\mathrm{d}(\widehat{C_V},\partial\|f\|\cap\hat{X})>0.$$

So by Corollary 18.3, X does not have the G-I property at f.

Conversely, suppose there exists a convex subset C of $B(X)$ where $\sup_C f=1$ but $\mathrm{d}(\hat{C},\partial\|f\|)=r>0$. Then $(\partial\|f\|+rB(X^{**}))\cap\hat{C}=\emptyset$ which implies that the duality mapping $f\mapsto\partial\|f\|$ on X^* is not Hausdorff norm upper semicontinuous at f. But further, this implies that there exists an $\mathbf{F}\in X^{***}$ strongly separating $\hat{C}$ and $\partial\|f\|$ in X^{**}. So $\mathbf{F}\in X^{***}$ strongly separates $\overline{\hat{\hat{C}}}^{\omega^*}$ and $\overline{\widehat{\partial\|f\|}}^{\omega^*}$ in X^{****}. Since $\overline{\widehat{\partial\|f\|}}^{\omega^*}$ is weak* compact $\sigma(X^{****},X^{***})$ there exists a weak* open neighbourhood $W_{\mathbf{F}}$ of O in X^{****} such that $(\overline{\widehat{\partial\|f\|}}^{\omega^*}+W_{\mathbf{F}})\cap\overline{\hat{\hat{C}}}^{\omega^*}=\emptyset$. But then

$(\partial\|f\| + (W_{\mathbf{F}} \cap X^{**})) \cap \hat{C} = \emptyset$ which implies that the duality mapping $f \mapsto \partial\|f\|$ on X^* is not Hausdorff weak upper semicontinuous at f. But further, given that $\partial\|f\| \cap \hat{X} \neq \emptyset$, we have that

$$((\partial\|f\| \cap \hat{X}) + (W_{\mathbf{F}} \cap \hat{X})) \cap \hat{C} = \emptyset,$$

which implies that the preduality mapping $f \mapsto \partial\|f\| \cap \hat{X}$ on X^* is not Hausdorff weak upper semicontinuous at f. ■

In the proof of this characterisation we have revealed the following relation between duality and preduality mappings on the dual which are Hausdorff weak upper semicontinuous.

Corollary 18.5 ([6, Theorem 3.2, p. 321]). *Given a Banach space X if the duality mapping $f \mapsto \partial\|f\|$ on X^* is Hausdorff weak upper semicontinuous at $f \in S(X^*)$ then the preduality mapping $f \mapsto \partial\|f\| \cap \hat{X}$ on X^* is Hausdorff weak upper semicontinuous at f.*

Godefroy and Indumathi have pointed out that the converse of this result for Hausdorff weak upper semicontinuity does not hold [6, p. 322]. Nevertheless for Hausdorff norm upper semicontinuity we do have a reciprocal relation.

Theorem 18.6 ([7, Theorem 2.2, p. 400]). *On the dual X^* of a Banach space X, the duality mapping $f \mapsto \partial\|f\|$ is Hausdorff norm upper semicontinuous at $f \in S(X^*)$ if and only if the preduality mapping $f \mapsto \partial\|f\| \cap \hat{X}$ is Hausdorff norm upper semicontinuous at f.*

Proof. For the duality mapping $f \mapsto \partial\|f\|$ Hausdorff norm upper semicontinuous at f, given $\varepsilon > 0$ there exists $\delta > 0$ such that

$$\{F \in B(X^{**}) : F(f) > 1 - \delta\} \subseteq \partial\|f\| + \varepsilon B(X^{**}) \text{ so}$$

$$\{x \in B(X) : f(x) > 1 - \delta\}\hat{} \subseteq \partial\|f\| + \varepsilon B(X^{**}).$$

We saw in the proof of Theorem 18.4 that the duality mapping $f \mapsto \partial\|f\|$ on X^* being Hausdorff norm upper semicontinuous at f implies that the space X has the G-I Property at f. So from Lemma 18.2(ii) we have that for $z \in B(X)$

$$\hat{z} \in \partial\|f\| + \varepsilon B(X^{**}) \text{ if and only if } \hat{z} \in \partial\|f\| \cap \hat{X} + \varepsilon B(\hat{X}).$$

Then $\{x \in B(X) : f(x) > 1 - \delta\}\hat{} \subseteq \partial\|f\| \cap \hat{X} + \varepsilon B(\hat{X})$; that is, the preduality mapping $f \mapsto \partial\|f\| \cap \hat{X}$ on X^* is Hausdorff norm upper semicontinuous at f.

Conversely, for the preduality mapping $f \mapsto \partial\|f\| \cap \hat{X}$ on X^* Hausdorff norm upper semicontinuous at f, given $\varepsilon > 0$, there exists $\delta > 0$ such that

$$\begin{aligned}\{x \in B(X) : f(x) > 1 - \delta\}\hat{} &\subseteq (\partial\|f\| \cap \hat{X} + \varepsilon B(\hat{X})) \\ &\subseteq \partial\|f\| + \varepsilon B(X^{**}).\end{aligned}$$

Since $\partial\|f\|$ is weak* compact, $\partial\|f\|+\varepsilon B(X^{**})$ is weak* closed so

$$\begin{aligned}\{F\in B(X^{**}):F(f)>1-\delta\}&\subseteq\overline{\{x\in B(X):f(x)>1-\delta\}^{\wedge}}^{\omega^*}\\&\subseteq\partial\|f\|+\varepsilon B(X^{**})\end{aligned}$$

[3, Theorem 2.1, p. 102]; that is, the duality mapping $f\mapsto\partial\|f\|$ on X^* is Hausdorff norm upper semicontinuous at f. ■

Contreras and Payá proved that a Banach space X is reflexive if the duality mapping $f\mapsto\partial\|f\|$ on X^* is Hausdorff weak upper semicontinuous at every $f\in S(X^*)$ [1, Theorem 1.3, p. 453]. However from Lemma 18.2 and James Theorem [2, Corollary 3.56, p. 84] we deduce the improved result that X is reflexive if it satisfies the G-I Property at every $f\in S(X^*)$.

It follows from Proposition 18.1(ii) that if $\partial\|f\|\cap\hat{X}\neq\emptyset$ and the preduality mapping $f\mapsto\partial\|f\|\cap\hat{X}$ is Hausdorff weak upper semicontinuous residually on $S(X^*)$ then $\overline{\partial\|f\|\cap\hat{X}}^{\omega^*}=\partial\|f\|$ residually on $S(X^*)$ and that the dual norm is Fréchet differentiable residually on $S(X^*)$, [4, Theorem 1.3, p. 415].

18.3 Hausdorff Upper Semicontinuity and Higher Duals

Duality and preduality mappings are of particular interest when the subdifferential image is weakly compact.

Theorem 18.7. *Consider a Banach space X and $f\in S(X^*)$ where $\partial\|f\|\cap\hat{X}\neq\emptyset$. The preduality mapping $f\mapsto\partial\|f\|\cap\hat{X}$ on X^* is Hausdorff weak upper semicontinuous at f and $\partial\|f\|$ is weakly compact $\sigma(X^{**},X^{***})$ if and only if $\partial\|f\|\subseteq\hat{X}$.*

Proof. Given that the preduality mapping $f\mapsto\partial\|f\|\cap\hat{X}$ on X^* is Hausdorff weak upper semicontinuous at f we have from Proposition 18.1(ii) that

$$\overline{\partial\|f\|\cap\hat{X}}^{\omega^*}=\partial\|f\|.$$

But if $\partial\|f\|$ is weakly compact $\sigma(X^{**},X^{***})$ then so also is $\partial\|f\|\cap\hat{X}$. So $\partial\|f\|\cap\hat{X}$ is weak* closed $\sigma(X^{**},X^*)$ and

$$\partial\|f\|\cap\hat{X}=\partial\|f\|\text{ and then }\partial\|f\|\subseteq\hat{X}.$$

Conversely, the duality mapping $f\mapsto\partial\|f\|$ on X^* is Hausdorff weak* upper semicontinuous at f so given a weak open neighbourhood V of O in X and a weak* open neighbourhood W of O in X^{**} such that $\hat{V}=W\cap\hat{X}$ there exists a $\delta>0$ such that

$$\partial\|B(f,\delta)\|\subseteq\partial\|f\|+W.$$

As $\partial\|f\| \subseteq \hat{X}$ then

$$\partial\|B(f,\delta)\| \cap \hat{X} \subseteq (\partial\|f\| \cap \hat{X}) + \hat{V};$$

that is, the preduality mapping $f \mapsto \partial\|f\| \cap \hat{X}$ on X^* is Hausdorff weak upper semicontinuous at f. As $\partial\|f\|$ is weak* compact $\sigma(X^{**},X^*)$ but $\partial\|f\| \subseteq \hat{X}$ then $\partial\|f\|$ is weakly compact $\sigma(X^{**},X^{***})$. ∎

Hausdorff weak upper semicontinuity of duality and preduality mappings is linked across duals.

Theorem 18.8. *Given a Banach space X, if the preduality mapping $F \mapsto \partial\|F\| \cap \hat{X}^*$ on X^{**} is Hausdorff weak upper semicontinuous at $\hat{x} \in S(X^{**})$ then the duality mapping $x \mapsto \partial\|x\|$ on X is Hausdorff weak upper semicontinuous at x.*

Conversely, if the duality mapping $x \mapsto \partial\|x\|$ on X is Hausdorff weak upper semicontinuous at $x \in S(X)$ with weak compact image $\partial\|x\|$ then $\partial\|\hat{x}\| = \widehat{\partial\|x\|}$ and the preduality mapping $F \mapsto \partial\|F\| \cap \hat{X}^*$ on X^{**} is Hausdorff weak upper semicontinuous at $\hat{x} \in S(X^{**})$ with weak compact image $\partial\|\hat{x}\|$.

Proof. Since the preduality mapping $F \mapsto \partial\|F\| \cap \widehat{X^*}$ on X^{**} is Hausdorff weak upper semicontinuous at $\hat{x} \in S(X^{**})$, given a weak open neighbourhood V of O in X^* there exists $\delta > 0$ such that

$$\partial\|B(\hat{x},\delta)\| \cap \widehat{X^*} \subseteq (\partial\|\hat{x}\| \cap \widehat{X^*}) + \hat{V}.$$

But $\partial\|\widehat{B(x,\delta)}\| = \partial\|B(\hat{x},\delta)\| \cap \widehat{X^*}$ and $\widehat{\partial\|x\|} = \partial\|\hat{x}\| \cap \widehat{X^*}$ so $\partial\|B(x,\delta)\| \subseteq \partial\|x\| + \hat{V}$. Then the duality mapping $x \mapsto \partial\|x\|$ on X is Hausdorff weak upper semicontinuous at x.

Conversely, since the duality mapping $x \mapsto \partial\|x\|$ on X is Hausdorff weak upper semicontinuous at $x \in S(X)$ then by Proposition 18.1(i), we have $\partial\|\hat{x}\| = \overline{\widehat{\partial\|x\|}}^{\omega^*}$. The natural embedding of X^* into X^{***} maps $\partial\|x\|$ weakly compact $\sigma(X^*,X^{**})$ to $\widehat{\partial\|x\|}$ weakly compact $\sigma(X^{***},X^{****})$. Then $\widehat{\partial\|x\|}$ is weak* closed $\sigma(X^{***},X^{**})$, so $\partial\|\hat{x}\| = \widehat{\partial\|x\|} \subseteq \widehat{X^*}$. Theorem 18.7 implies that the preduality mapping $F \mapsto \partial\|F\| \cap \widehat{X^*}$ on X^{**} is Hausdorff weak upper semicontinuous at $\hat{x}$ with weak compact image $\partial\|\hat{x}\|$. ∎

Hausdorff norm upper semicontinuity of the duality mapping does have significant persistence properties for higher duals.

Theorem 18.9 ([3, Corollary 2.1, p. 103]). *Given a Banach space X, if the duality mapping $x \mapsto \partial\|x\|$ on X is Hausdorff norm upper semicontinuous at $x \in S(X)$ then the duality mapping $F \mapsto \partial\|F\|$ on X^{**} is Hausdorff norm upper semicontinuous at $\hat{x} \in S(X^{**})$ and through all even dual spaces. If also $\partial\|x\|$ is weakly compact $\sigma(X^*,X^{**})$ then $\partial\|\hat{x}\|$ is weakly compact $\sigma(X^{***},X^{****})$ and the subdifferential remains constant through all even dual spaces.*

Proof. Since the duality mapping $x \mapsto \partial\|x\|$ on X is Hausdorff norm upper semicontinuous at $x \in S(X)$ given $\varepsilon > 0$ there exists $\delta > 0$ such that

$$\{f \in B(X^*) : f(x) > 1-\delta\} \subseteq \partial\|x\| + \varepsilon B(X^*).$$

Then

$$\begin{aligned}\{\mathbf{F} \in B(X^{***}) : \mathbf{F}(\hat{x}) > 1-\delta\} &\subseteq \overline{\{f \in B(X^*) : f(x) > 1-\delta\}^{\wedge}}^{\omega^*}\\ &\subseteq \overline{(\partial\|x\| + \varepsilon B(X^*))^{\wedge}}^{\omega^*}\\ &\subseteq \partial\|\hat{x}\| + \varepsilon B(X^{***}), [3, \text{Theorem 2.1, p. 102}].\end{aligned}$$

So the duality mapping $F \mapsto \partial\|F\|$ on X^{**} is Hausdorff norm upper semicontinuous at $\hat{x}$. If also $\partial\|x\|$ is weakly compact $\sigma(X^*, X^{**})$ then from Theorem 18.8 we have that $\partial\|\hat{x}\| = \widehat{\partial\|x\|}$ and so $\partial\|\hat{x}\|$ is weakly compact $\sigma(X^{***}, X^{****})$. So applying Theorem 18.8 to duality mapping $F \mapsto \partial\|F\|$ on X^{**} we have

$$\partial\|\hat{\hat{x}}\| = \widehat{\partial\|\hat{x}\|} = \widehat{\widehat{\partial\|x\|}} \subseteq \widehat{\widehat{X^{**}}}. \qquad \blacksquare$$

This has a special case that if the norm of a Banach space X is Fréchet differentiable at $x \in S(X)$ then the norm of X^{**} is also Fréchet differentiable at $\hat{x} \in S(X^{**})$ and so on for all subsequent even dual spaces.

Such a persistence property does not hold generally for Hausdorff weak upper semicontinuity of the duality mapping. Godefroy and Indumathi give an example of a separable Banach space with X^{**}/X non-reflexive to show that even if the norm of X is very smooth at $x \in S(X)$ the norm of X^{**} is not necessarily very smooth at $\hat{x} \in S(X^{**})$ [6, Proposition 4.1, p. 326] (see also [8, Theorem 2.5, p. 1025]).

However, a similar persistence property does hold for Hausdorff weak upper semicontinuous duality mappings for a large class of Banach spaces, those which are an M-ideal in their second dual. A Banach space X which has $\hat{X}$ an M-ideal in X^{**} has $\hat{\hat{X}}$ an M-ideal in X^{****} [10, p. 1391] and this implies that for any element $\Phi \in X^{*****} = X^{\perp\perp\perp} \oplus X^{*\perp\perp}$ and $\Phi = \Phi_1 + \Phi_2$ where $\Phi_1 \in X^{\perp\perp\perp}$ and $\Phi_2 \in X^{*\perp\perp}$ we have $\|\Phi\| = \|\Phi_1\| + \|\Phi_2\|$, [6, p. 325]. Godefroy and Indumathi explored the persistence of smoothness properties to higher duals [6, Proposition 3.2, p. 324], but following their argument we have a more general result.

Lemma 18.10. *Given a Banach space X, if the duality mapping $x \mapsto \partial\|x\|$ on X is Hausdorff weak upper semicontinuous at $x \in S(X)$ with $\partial\|x\|$ weakly compact $\sigma(X^*, X^{**})$ then*

$$\partial\|\hat{\hat{x}}\|\,|_{X^{\perp\perp}} = \widehat{\partial\|\hat{x}\|}\,|_{X^{\perp\perp}}.$$

Proof. We are considering $\partial\|\hat{\hat{x}}\|$ and $\widehat{\partial\|\hat{x}\|}$ on $X^{****} = X^{*\perp} \oplus X^{\perp\perp}$. Since the duality mapping $x \mapsto \partial\|x\|$ on X is Hausdorff weak upper semicontinuous at $x \in S(X)$ and $\partial\|x\|$ is weakly compact $\sigma(X^*, X^{**})$ then by Theorem 18.8 we have $\partial\|\hat{x}\| = \widehat{\partial\|x\|}$. So $\widehat{\partial\|\hat{x}\|} \subseteq \widehat{X^*}$ and $\widehat{\partial\|\hat{x}\|}\,|_{X^{*\perp}} = 0$.

We denote by i_o the natural embedding of X into X^{**}. Consider $\varphi \in X^{\perp\perp} = i_o^{**}(X^{**}) \subseteq X^{****}$. Now there exists $F \in X^{**}$ such that $\varphi = i_o^{**}F$. Since the natural embedding of $B(X^{***})$ is weak* dense $\sigma(X^{*****}, X^{****})$ in $B(X^{*****})$, given $\Phi \in \partial\|\widehat{\widehat{x}}\|$ there exists a net $\{\mathbf{F}_\alpha\}$ in $B(X^{***})$ such that $\{\widehat{\mathbf{F}_\alpha}\}$ is weak* convergent to $\Phi \in B(X^{*****})$. Then $\widehat{\mathbf{F}_\alpha}(\widehat{\widehat{x}}) \to \Phi(\widehat{\widehat{x}}) = 1$ and so $i_o^{*}\mathbf{F}_\alpha(\hat{x}) \to 1$. Since the duality mapping $x \mapsto \partial\|x\|$ on X is Hausdorff weak upper semicontinuous at $x \in S(X)$ and $\partial\|x\|$ is weakly compact then $\{i_o^{*}\mathbf{F}_\alpha\}$ has a subnet $\{i_o^{*}\mathbf{F}_{\alpha_\beta}\}$ weakly convergent to some $f \in \partial\|x\|$ [3, Theorem 3.2, p. 104]. Then

$$\Phi(\varphi) = \lim \mathbf{F}_{\alpha_\beta}(\varphi) = \lim i_o^{**}F(\mathbf{F}_{\alpha_\beta}) = \lim F(i_o^{*}\mathbf{F}_{\alpha_\beta}) = F(f).$$

But also

$$\Phi(\hat{f}) = i_o^{**}F(f) = F(i_o^{*}\hat{f}) = F(f).$$

So for $\Phi \in \partial\|\widehat{\widehat{x}}\|\,|_{X^{\perp\perp}}$ we have $\Phi \in \widehat{\partial\|\hat{x}\|}\,|_{X^{\perp\perp}}$ remembering that $\partial\|\hat{x}\| = \widehat{\partial\|x\|}$. Since $\widehat{\partial\|\hat{x}\|} \subseteq \partial\|\widehat{\widehat{x}}\|$ we conclude that

$$\partial\|\widehat{\widehat{x}}\|\,|_{X^{\perp\perp}} = \widehat{\partial\|\hat{x}\|}\,|_{X^{\perp\perp}}.$$ ■

Theorem 18.11. *Given a Banach space X which is an M-ideal in X^{**}, if the duality mapping $x \mapsto \partial\|x\|$ on X is Hausdorff weak upper semicontinuous at $x \in S(X)$ and $\partial\|x\|$ is weakly compact $\sigma(X^{*}, X^{**})$ then*

$$\partial\|\widehat{\widehat{x}}\| = \widehat{\partial\|\hat{x}\|} = \widehat{\widehat{\partial\|x\|}} \subseteq \widehat{\widehat{X^{*}}}.$$

Further, all duality mappings on even duals are Hausdorff weak upper semicontinuous at embeddings of $x \in S(X)$ and the subdifferentials of these embeddings are all weakly compact and remain constant through all even duals.

Proof. Consider $\Phi \in \partial\|\widehat{\widehat{x}}\| \subseteq X^{*****} = X^{\perp\perp\perp} \oplus X^{*\perp\perp}$ and $\Phi = \Phi_1 + \Phi_2$ where $\Phi_1 \in X^{\perp\perp\perp}$ and $\Phi_2 \in X^{*\perp\perp}$. Since $\widehat{\widehat{X}}$ is an M-ideal in X^{****}, $\|\Phi\| = \|\Phi_1\| + \|\Phi_2\|$. Now $\Phi_1(\widehat{\widehat{x}}) = 0$ and $\Phi(\widehat{\widehat{x}}) = 1$ so $\Phi_2(\widehat{\widehat{x}}) = 1$. But $\|\Phi\| = 1$ so $\|\Phi_2\| \leq 1$, yet $\Phi_2(\widehat{\widehat{x}}) = 1$ so $\|\Phi_2\| = 1$. Then $\|\Phi_1\| = 0$ so $\Phi_1 = 0$. This means that $\Phi \in X^{*\perp\perp}$ then $\partial\|\widehat{\widehat{x}}\|\,|_{X^{*\perp}} = 0$. From Lemma 18.10 and Theorem 18.8 we conclude that

$$\partial\|\widehat{\widehat{x}}\| = \widehat{\partial\|\hat{x}\|} = \widehat{\widehat{\partial\|x\|}} \subseteq \widehat{\widehat{X^{*}}}.$$

From Theorem 18.7 we deduce that the preduality mapping $\Phi \mapsto \partial\|\Phi\| \cap (\widehat{X^{***}})$ on X^{****} is Hausdorff weak upper semicontinuous at $\widehat{\widehat{x}}$ with $\partial\|\widehat{\widehat{x}}\|$ weakly compact. By Theorem 18.8 we deduce that the duality mapping $F \mapsto \partial\|F\|$ on X^{**} is Hausdorff weak upper semicontinuous at $\hat{x}$ with $\partial\|\hat{x}\|$ weakly compact. Since X is an M-ideal in all duals of even order [10, Theorem 2, p. 1390] we can continue to establish our theorem. ■

This has as a special case that, if the norm of a Banach space X which is an M-ideal in X^{**} is very smooth at $x \in S(X)$, then the norm of X^{**} is very smooth at $\hat{x} \in S(X^{**})$ and so on for all subsequent even dual spaces.

Although Theorem 18.11 does not hold for Banach spaces in general, nevertheless it is worth noting that Theorems 18.7 and 18.8 imply that if a Banach space X has the property

$$\partial\|\hat{x}\| = \widehat{\partial\|x\|} \subseteq \widehat{X^*} \text{ at } x \in S(X)$$

then the duality mapping $x \mapsto \partial\|x\|$ on X is Hausdorff weak upper semicontinuous at x and $\partial\|x\|$ is weakly compact. So if the subdifferentials at $x \in S(X)$ and its natural embeddings remain constant on all even dual spaces then the duality mappings on all such dual spaces are Hausdorff weak upper semicontinuous at such embedding points and with weakly compact images.

Acknowledgements I would like to thank Scott Sciffer and Rebecca Smith for their assistance in the preparation of this paper.

References

1. Contreras, M.D., Payá, R.: On upper semicontinuity of duality mappings. Proc. Amer. Math. Soc. **121**, 451–459 (1994)
2. Fabian, M., Habala, P., Hájek, P., Montesinos Santalucía, V., Pelant, J., Zizler, V.: Functional analysis and infinite-dimensional geometry. CMS Books in Mathematics, Springer, New York (2001)
3. Giles, J.R, Gregory, D.A., Sims, B.: Geometrical implications of upper semicontinuity of the duality mapping on Banach space. Pacific J. Math. **79**, 99–109 (1978)
4. Giles, J.R., Kenderov, P.S., Moors, W.B., Sciffer, S.D.: Generic differentiability of convex functions on the dual of Banach space. Pacific J. Math. **172**, 413–431(1996)
5. Giles, J.R., Moors, W.B.: Generic continuity of restricted weak upper semicontinuous set-valued mappings. Set-Valued Anal. **4**, 25–39 (1996)
6. Godefroy, G., Indumathi, V.: Norm-to-weak upper semicontinuity of the duality and pre-duality mappings. Set-Valued Anal. **10**, 317–330 (2002)
7. Godefroy, G., Indumathi, V., Lust-Piquard, F.: Strong subdifferentiability of convex functionals and proximinality. J. Approx. Theory **116** 397–415 (2002)
8. Godefroy, G., Rao, T.S.S.R.K.: Renormings and extremal structures. Illinois J. Math. **48**, 1021–1029 (2004)
9. Phelps, Robert R.: Convex functions, monotone operators and differentiability. Springer Lecture Notes in Mathematics vol. 1364, 2nd edn. Springer, Berlin (1993)
10. Rao, T.S.S.R.K.: On the geometry of higher duals of a Banach space. Illinois J. Math. **45**, 1389–1392 (2001)

Chapter 19
Convexity and Variational Analysis

A.D. Ioffe

Dedicated to Jonathan Borwein on the occasion of his 60th birthday

In science things should be made as simple as possible but not any simpler.

A. Einstein

Abstract The paper focuses on the interplay between methods of general variational analysis, on the one hand, and convex analysis, on the other (with a special emphasis on the efficiency of the latter), in treating problems that come from the general variational analysis when applied in substantial or virtual presence of convexity. We discuss a number of relating problems: error bounds, metric regularity, linear openness, and stability as well as first-order necessary optimality conditions in semi-infinite programming.

Key words: Distance to infeasibility • Global error bound • Linear openness and Lipschitz stability • Metric regularity • Perfect regularity • Semi-infinite programming • Set-valued mapping

Mathematics Subject Classifications (2010): Primary 52A41, 58C06, 90C34; Secondary 49J52, 58C15, 90C48.

COMMUNICATED BY JON D. VANDERWERFF.

A.D. Ioffe (✉)
Department of Mathematics, Technion - Israel Institute of Technology, Haifa 32000, Israel
e-mail: ioffe@math.technion.ac.il

D.H. Bailey et al. (eds.), *Computational and Analytical Mathematics*, Springer Proceedings in Mathematics & Statistics 50, DOI 10.1007/978-1-4614-7621-4_19,

19.1 Introduction

The title of this paper resembles the name of the third chapter in the remarkable book of Borwein and Zhu [3]. But the emphasis is somewhat different. In [3] variational techniques directly "refer to proofs by way of establishing that an appropriate auxiliary function attains a minimum." This interpretation of variational analysis is close to its original meaning in the classical Morse's monograph [33]. But a more recent and also already classical monograph by Rockafellar and Wets [37] offers a much broader interpretation of the term which has become widely accepted by now. Pursuit of maximum generality is immanent in this interpretation and this is definitely an indication of the power and ambition of the emerging theory. But there are also some dangers.

Needless to say that general theories often offer good starting observation points to attack concrete problems. But equally true is that neglecting specific features and details may lead to heavy and awkward constructions poorly connected with the nature of the problem. These are rather trivial remarks, but there is a worrying tendency to indiscriminately use techniques of general variational analysis as the main tool to study very well structured problems for which a proper use of techniques taking specified structures into account could be much more efficient.[1]

My main intention in this paper is to demonstrate the interplay between methods of general variational analysis, on the one hand, and convex analysis, on the other (with more emphasis on the efficiency of the latter), in treating problems that come from the general variational analysis when applied in substantial or virtual presence of convexity.

We shall consider several problems connected either with (metric) regularity and perturbation stability of set-valued mappings or first-order optimality conditions. Not all of the problems are originally convex, but in either case the analysis involves convexity in one or another way. The paper basically surveys some results that recently have been or are about to be published, although there are some new results as well (those not supplied with references in the text or in the comments at the ends of each section), e.g., Theorem 19.28 containing an exact estimate for the modulus of Lipschitz stability of the set of solutions of a generalized equation with respect to joint variations of the right-hand part and linear perturbations of the set-valued mapping. For the majority of other results we just give new proofs, usually shorter than those available in the literature.

Now about the content of the paper. The next section contains necessary information from convex and modern variational analysis. In the third section we speak about calculation of global error bounds, for convex functions in the first part of the section and nonconvex l.s.c. functions in the second. This problem is an excellent example of the effect of specific structural properties of the objects and

[1] Some time ago I was literally horrified by an article by three acknowledged authors published in a respected journal in which a fairly elementary result, an error bound for a polyhedral function, was proved on more than ten pages using "advanced" technique of nonconvex subdifferentiation.

the concrete setting of the problem on the choice of technical instruments to solve the problem and on the result itself. We demonstrate this by considering first convex functions separately on Hilbert spaces, reflexive spaces, and arbitrary Banach spaces and then arbitrary lower semicontinuous functions on Banach spaces. The general results presented in the section can be found in Azé-Corvellec's papers [1, 2] and the monograph [46] by Zalinescu although our statements and proofs contain some new elements (more detailed bibliographic information is contained in comments concluding each of the last three sections). In the fourth section we consider set-valued mappings with convex graphs and systems of convex inequalities. The main questions discussed in the section relate to metric regularity (exact formulas for regularity moduli) and stability estimates of the regularity properties with respect to linear perturbation. We start with a new and a very short proof of the Robinson-Ursescu theorem substantially based on the general metric regularity criterion of [20]. Subsequent results are closely related to Canovas et al. [4–8], Ioffe-Sekiguchi [25], and Ioffe [23]. The first statement of the mentioned final result of the section Theorem 19.28 applies to all set-valued mappings with closed graph.

The main message of the last section is that, as far as the first-order necessary optimality conditions are concerned, smooth nonconvex inequality constraint or cost functions do not exist. By that I mean that if such a function appears in the statement of the problem, it can be replaced by a convex continuous local majorant with the same value and the same derivative at the point. This is the content of main lemma in the last section. This simple lemma does not seem to have appeared earlier, but a similar and even more elaborate idea was behind the Levitin-Milyutin-Osmolovski upper approximation in [27] (see also [19]). Using convex majorants offered by the lemma dramatically simplify the derivation of necessary optimality conditions.

A discussion with B. Mordukhovich after his talk (based on his joint paper with Nghia [32]) at J. Borwein anniversary conference in Vancouver revealed however that utility of such tricks may not seem to be obvious for everybody. So we show here that it is possible to furnish, based on the lemma, fairly elementary and short proofs for the main results of [32] and even for stronger versions of some of them. In fact, the message can be even strengthened: the convex majorant that appears after the application of the lemma is necessarily strictly differentiable, no matter whether the original function has this property or not. Thus, when we deal with first-order optimality conditions, for inequality constraint and cost functions the differentiability and strict differentiability assumptions make no difference. Of course, the difference is substantial for equality constraints.

In the text we supply with references mainly the results that are stated without proofs. The statements of known results supplied with proofs may slightly differ from the original statements, and we give relevant references in the comments at the ends of the sections.

Notation. Throughout the paper X, Y, etc. are Banach spaces, X^* is the topological dual of X, and $\langle x^*, x\rangle$ is the canonical pairing on $X^* \times X$. By $\mathrm{d}(x, Q)$ we denote the distance from x to Q. We shall always consider X^* with the weak*-topology and the symbol cl* means closure in this topology. The symbol $F : X \rightrightarrows Y$ is used for set-valued mappings from X into Y, F^{-1} is the inverse mapping: $F^{-1}(y) = \{x : y \in F(x)\}$.

If $Q \subset X$, then conv$Q$ is the convex hull of Q, cone$Q = \cup_{\lambda \geq 0} \lambda Q$ is the cone generated by Q, $S_Q(x)$ stands for the support function of Q, and Ind$_Q$ indicator of Q:

$$S_Q(x^*) = \sup_{x \in Q} \langle x^*, x \rangle; \qquad \text{IndQ} = \begin{cases} 0, & \text{if } x \in Q, \\ \infty, & \text{otherwise.} \end{cases}$$

19.2 Preliminaries

19.2.1 A Few Facts from Convex Analysis

Let X be a Banach space and f an extended-real-valued convex function on X. We call f *proper* if $f(x) > -\infty$. We define as usual the *domain* $\operatorname{dom} f = \{x : f(x) < \infty\}$ and *epigraph* $\operatorname{epi} f = \{(x, \alpha) \in X \times \mathbb{R} : \alpha \geq f(x)\}$ of f. It is said that f is *closed* if epi$f$ is a closed set or, equivalently, if f is lower semicontinuous. The *(Fenchel) conjugate* of f is

$$f^*(x^*) = \sup_u (\langle x^*, u \rangle - f(u)).$$

The *closure* cl$f$ of a convex function f is the greatest closed convex function majorized by f:

$$\operatorname{cl} f(x) = \liminf_{u \to x} f(u) = f^{**}(x) = \sup_{x^*} (\langle x^*, x \rangle - f^*(x^*)). \tag{19.1}$$

Observe that the lower bounds of a function and its closure coincide.

Recall that the *subdifferential* of f at $x \in \operatorname{dom} f$ is

$$\partial f(x) = \{x^* \in X^* : \langle x^*, h \rangle \leq f(x+h) - f(x), \quad \forall\, h \in X\}.$$

It is always a weak* closed set, bounded (hence weak*-compact) if f is continuous at x. Moreover, the set-valued mapping $x \mapsto \partial f(x)$ is norm-to-weak* upper semicontinuous; in particular, the function $\mathrm{d}(0, \partial f(\cdot))$ is lower semicontinuous.

If $x \in \operatorname{dom} f$, then the *directional derivative of f at x in the direction h* is

$$f'(x; h) = \lim_{t \to +0} t^{-1}(f(x+th) - f(x)).$$

This limit, finite or infinite, always exists, thanks to the following elementary fact: if $\varphi(t)$ is a convex function on a real line and $t_1 < t_2 \leq t_4$, $t_1 \leq t_3 < t_4$ belong to the domain of φ, then

$$\frac{\varphi(t_2)-\varphi(t_1)}{t_2-t_1} \le \frac{\varphi(t_4)-\varphi(t_3)}{t_4-t_3}. \tag{19.2}$$

As another immediate consequence, we get $f'(x;h) \ge -f'(x;-h)$.

The function $f'(x;\cdot)$ is *sublinear*, that is, convex and positively homogeneous of degree one: $f'(x;\lambda h) = \lambda f'(x;h)$ if $\lambda > 0$. It is continuous if f is continuous at x. Note that the closure of a sublinear function is also a sublinear function. The connection between the subdifferential and the directional derivative is described by the equality

$$\sup_{x^*\in\partial f(x)} \langle x^*, h\rangle = (\mathrm{cl}\, f'(x;\cdot))(h).$$

We also have that $f(x) = (\mathrm{cl}\, f)(x)$ and $\partial f(x) = \partial(\mathrm{cl}\, f)(x)$ if $\partial f(x) \neq \emptyset$.

The following proposition reveals the connection between the directional derivative and the subdifferential at the same point.

Proposition 19.1. *Let f be a proper closed convex function on a Banach space X. Then*

$$\mathrm{d}(0,\partial f(x)) = \sup_{\|h\|\le 1} (-f'(x;\cdot))(h) = \sup_{\|h\|\le 1} (-\mathrm{cl}\,(f'(x;\cdot)))(h).$$

(Here we adopt the standard conventions: $\sup\emptyset = -\infty$, $\inf\emptyset = \infty$, so that $d(x,\emptyset) = \infty$.)

Proof. If $\partial f(x) = \emptyset$, the equality holds by the standard convention. Furthermore, $0 \in \partial f(x)$ if and only if $f'(x;h) \ge 0$ for all h and the equality obviously holds in this case (just take $h = 0$). Assume now that $0 \notin \partial f(x) \neq \emptyset$. Then $d(0,\partial f(x)) > 0$. As $\partial f(x)$ is a weak* closed set and $\|\cdot\|$ is weak* l.s.c., there is an $x^* \in \partial f(x)$ such that $\|x^*\| = \mathrm{d}(0,\partial f(x))$. Take a small $\varepsilon > 0$ and set $Q_\varepsilon = (1-\varepsilon)\|x^*\| B_{X^*}$. This set is weak*-compact and does not meet $\partial f(x)$. Therefore there is an $h \in X$, $\|h\| = 1$ separating the sets, e.g.,

$$(1-\varepsilon)\|x^*\| = \sup_{\|u^*\|\le(1-\varepsilon)\|x^*\|} \langle u^*, h\rangle \le \inf_{u^*\in\partial f(x)} \langle u^*, h\rangle \le \langle x^*, h\rangle \le \|x^*\|.$$

On the other hand

$$\inf_{u^*\in\partial f(x)} \langle u^*, h\rangle = -\sup_{u^*\in\partial f(x)} \langle u^*, -h\rangle = -\mathrm{cl}\,(f'(x;\cdot))(-h)$$

and we get $|\mathrm{d}(0,\partial f(x)) - (-\mathrm{cl}\,(f'(x,\cdot))(-h)| \le \varepsilon \mathrm{d}(0,\partial f(x))$. ∎

Given a convex set $Q \subset X$ and an $\overline{x} \in Q$, the *tangent cone* to Q at $\overline{x}$ is

$$T(Q,\overline{x}) = \mathrm{cl}\,[\mathrm{cone}\,(Q-\overline{x})] = \mathrm{cl}\left(\bigcup_{\lambda>0} \lambda(T-\overline{x})\right)$$

and the *normal cone* to Q at $\overline{x}$ is

$$N(Q,\overline{x}) = \{x^* \in X^* : \langle x^*, x-\overline{x}\rangle \le 0, \ \forall\, x \in Q\}.$$

If Q_i, $i = 1,2$ are convex sets such that Q_1 meets the interior of Q_2 then for any $x \in Q_1 \cap Q_2$

$$N(Q_1 \cap Q_2, x) \subset N(Q_1, x) + N(Q_2, x).$$

If $F : X \rightrightarrows Y$ is a set-valued mapping with convex graph and $(\bar{x},\bar{y}) \in \operatorname{Graph} F$, then $T(\operatorname{Graph} F, (\bar{x},\bar{y}))$ can be considered the graph of the set-valued mapping $DF(\bar{x},\bar{y})$ defined by

$$DF(\bar{x},\bar{y})(h) = \{v \in Y : (h,v) \in T(\operatorname{Graph} F, (\bar{x},\bar{y}))\}.$$

This mapping is usually called the *derivative* (or *contingent derivative*) of F at $(\bar{x},\bar{y})$.

19.2.2 A Few Facts from General Variational Analysis

We need some facts from the local (metric) regularity theory of variational analysis. Given a set-valued mapping $F : X \rightrightarrows Y$ and an $(\bar{x},\bar{y}) \in \operatorname{Graph} F$, it is said that:

- F is *open* (or *covering*) *at a linear rate* at (or near) $(\bar{x},\bar{y})$ if there are $r > 0$ and $\varepsilon > 0$ such that

  $$B(v,tr) \cap B(\overline{y},\varepsilon) \subset F(B(x,t))$$

  whenever $\|x-\overline{x}\| < \varepsilon$, $0 \le t < \varepsilon$ and $v \in F(x)$. The upper bound of such r is called the *modulus* or *rate* of *surjection* or *openness* of F at $(\bar{x},\bar{y})$ and is denoted $\operatorname{sur} F(\overline{x}|\overline{y})$. If no such r and ε exist, we set $\operatorname{sur} F(\overline{x}|\overline{y}) = 0$.
- F is *metrically regular* at (or near) $(\bar{x},\bar{y})$ if there are $K > 0$ and $\varepsilon > 0$ such that

  $$\mathrm{d}(x, F^{-1}(y)) \le K\mathrm{d}(y, F(x))$$

  whenever $\|x-\overline{x}\| < \varepsilon$ and $\|y-\overline{y}\| < \varepsilon$. The lower bound of such K is called the *modulus* or *rate* of *metric regularity* of F at $(\bar{x},\bar{y})$ and is denoted $\operatorname{reg} F(\overline{x}|\overline{y})$. If no such K and ε exist, we set $\operatorname{reg} F(\overline{x}|\overline{y}) = \infty$.
- F^{-1} is said to be *pseudo-Lipschitz* or to have the *Aubin property* at (or near) $(\overline{y},\overline{x})$ if there are $K > 0$ and $\varepsilon > 0$ such that

  $$\mathrm{d}(x, F^{-1}(y)) \le K\mathrm{d}(y, v)$$

  whenever $\|x-\overline{x}\| < \varepsilon$, $\|y-\overline{y}\| < \varepsilon$ and $v \in F(x)$. The lower bound of such K is called the *Lipschitz modulus* or *rate* of F^{-1} at $(\overline{y}|\overline{x})$ and is denoted $\operatorname{lip} F^{-1}(\overline{y}|\overline{x})$. If no such K or ε exist, we set $\operatorname{lip} F^{-1}(\overline{y},\overline{x}) = \infty$.

Theorem 19.2 (equivalence theorem). *Under the convention that* $0 \cdot \infty = 1$, *for any set-valued mapping* $F : X \rightrightarrows Y$ *and any* $(\bar{x}, \bar{y}) \in \operatorname{Graph} F$

$$\operatorname{sur} F(\overline{x}|\overline{y}) \cdot \operatorname{reg} F(\overline{x}|\overline{y}) = 1; \qquad \operatorname{reg} F(\overline{x}|\overline{y}) = \operatorname{lip} F^{-1}(\overline{y}|\overline{x}).$$

We say that F is *regular* at $(\bar{x}, \bar{y})$ if the three properties are satisfied at $(\bar{x}, \bar{y})$.

Theorem 19.3 (criterion for local regularity). *Let* F *be a set-valued mapping whose graph is locally complete in the product metric, and let* $(\bar{x}, \bar{y}) \in \operatorname{Graph} F$. *Then* F *is regular near* $(\bar{x}, \bar{y})$ *if and only if there are* $\varepsilon > 0$ *and* $r > 0$ *such that for any* x, y, *and* v *satisfying* $d(x, \bar{x}) < \varepsilon$, $d(v, \bar{y}) < \varepsilon$, $y \in F(x)$, *and* $0 < d(y, v) < \varepsilon$, *there is a pair* $(u, z) \in \operatorname{Graph} F$, $(u, z) \neq (x, y)$, *and a* $\xi > 0$ *such that*

$$\|y - z\| \leq \|y - v\| - r d_{\xi}((x, y), (u, z)). \tag{19.3}$$

The upper bound of such r *coincides with* $\operatorname{sur} F(\overline{x}|\overline{y})$.

Moreover if F *is u.s.c. in the sense that the functions* $\varphi_y = d(y, F(\cdot))$ *are l.s.c. for any* y, *(19.3) can be replaced by*

$$\|y - z\| \leq \|y - v\| - r\|x - u\|.$$

Here d_{ξ} is the distance in $X \times Y$ associated with the norm $\max\{\|x - u\|, \xi\|y - z\|\}$. Note that the definitions and results extend without change (except for obvious replacement of the norms by distances) to arbitrary metric spaces. For the proofs of the theorems see, e.g., [20, 22].

The geometric meaning of the criterion is obvious: for any observation point (x, v) of the graph (close to $(\bar{x}, \bar{y})$) and any $y \neq v$ you can find a better observation position $(u, w) \in \operatorname{Graph} F$ such that the gain in the distance to y is proportional to the distance between the observation points. Less obvious is that the criterion is an excellent practical instrument, often better than more sophisticated means using slopes and subdifferentials.[2]

Calculation of regularity rates is typically a difficult task. But in certain cases it is sufficiently easy. Denote by $\operatorname{Epi} f$ the set-valued mapping $X \rightrightarrows \mathbb{R}$ whose graph coincides with the epigraph of f;

---

[2]To support this declaration we give below a proof of a set-valued version of the famous Milyutin's perturbation theorem [14] for the case when F is upper semicontinuous: *Let* Y *be a Banach space,* $F : X \rightrightarrows Y$ *with* $\overline{y} \in F(\overline{x})$, *and let* $g : X \to Y$ *be Lipschitz near* $\overline{x}$. *Then*

$$\operatorname{sur}(F + g)(\overline{x}, \overline{y} + g(\overline{x})) \geq \operatorname{sur} F(\bar{x}, \bar{y}) - \operatorname{lip} g(\overline{x}).$$

Proof. Set $\Phi(x) = F(x) + g(x)$, $\ell = \operatorname{lip} g(\overline{x})$. Then $v \in F(x) \Leftrightarrow w = v + g(x) \in \Phi(x)$. Take a $y \neq v$ and set $z = g(x) + y$. Then $\|y - v\| = \|z - w\|$. By the criterion (as F is regular) $\exists\, (x', v') \in \operatorname{Graph} F$ s.t. $(x', v') \neq (x, v)$ and

$$\|y - v'\| \leq \|y - v\| - r d(x, x').$$

Set $w' = g(x') + v' \in \Phi(x')$. Then

$$\operatorname{Epi} f(x) = \{\alpha \in \mathbb{R} : \alpha \geq f(x)\}.$$

Proposition 19.4. *Let f be a closed convex function on X. Then for any $x \in \operatorname{dom} f$*

$$\operatorname{sur}(\operatorname{Epi} f)(x, f(x)) = \liminf_{(u,f(u))\to(x,f(x))} d(0, \partial f(u)). \tag{19.4}$$

Proof. Note that for $\alpha \in \operatorname{Epi} f(x)$ the inclusion $B(\alpha,\varepsilon)=\alpha+[-\varepsilon,\varepsilon] \subset \operatorname{Epi} f(B(u,t))$ is equivalent to $\alpha+[-\varepsilon,\infty] \subset \operatorname{Epi} f(B(u,t))$. On the other hand, let $\partial f(u) \neq \emptyset$. Set $\rho = \mathrm{d}(0,\partial f(u))$. We have $f(u+th) \geq f(u)+t\langle u^*,h\rangle$ for any $h \in X$, any $u^* \in \partial f(u)$, and any $t > 0$. Together with Proposition 19.1 this implies that for any $\varepsilon > 0$

$$f(u)+t(1-\varepsilon)\rho[-1,\infty) \subset \operatorname{Epi} f(B(u,t)) \subset f(u)+t\rho[-1,\infty)$$

and (19.4) follows (as $d(0,\partial f(\cdot))$ is an l.s.c. function). ■

The *contingent* or *Bouligand tangent cone* to Q at $x \in Q$ is

$$T_B(Q,x) = \limsup_{\lambda\to+0} \lambda^{-1}(Q-x) = \{h : \liminf_{\lambda\to+0} \lambda^{-1}\mathrm{d}(x+\lambda h, Q) = 0\}.$$

The polar of $T_B(Q,x)$

$$N_{DH}(Q,x) = \{x^* \in X^* : \langle x^*, u-x\rangle \leq 0, \quad \forall\, u \in Q\}$$

is called the *Dini-Hadamard normal cone* to Q at x. Any normal cone to the graph of a set-valued mapping $X \rightrightarrows Y$ can be viewed as the graph of a set-valued mapping from Y^* into X^*. In particular, given $F : X \rightrightarrows Y$ and $(x,y) \in \operatorname{Graph} F$, the set-valued mapping

$$y^* \mapsto \{x^* : (x^*,-y^*) \in N_{DH}(\operatorname{Graph} F,(x,y))\} := D^*_{DH}F(x,y)(y^*)$$

is called the *Dini-Hadamard coderivative* of F at (x,y). We shall not need coderivatives associated with other types of normal cones.

The *Clarke tangent cone* $T_C(Q,x)$ to Q at x is the collection of $h \in X$ such that for any sequence $(x_n) \subset Q$ converging to x and any sequence (t_n) of positive numbers converging to zero there is a sequence (h_n) converging to h such that $x_n + t_n h_n \in Q$. This is always a closed convex cone. Its polar $N_C(Q,x)$ is called *Clarke's normal cone* to Q at x.

The inclusions $T_C(Q,x) \subset T_B(Q,x)$ and $N_{DH}(Q,x) \subset N_C(Q,x)$ always hold. If we actually have equalities, then Q is called *Clarke regular* at $x \in Q$.

$$\begin{aligned}\|z-w'\| &= \|z-g(x')-v'\| \leq \|z-g(x)-v'\| + \ell d(x,x')\\ &= \|y-v'\| + \ell d(x,x')\\ &\leq \|y-v\| - rd(x,x') + \ell d(x,x')\\ &= \|z-w\| - (r-\ell)d(x,x').\end{aligned}$$

The proof in the general case is almost equally simple.

19.3 Global Error Bounds

Let X be a metric space, and let f be an extended-real-valued function on X. We define the domain of f by $\operatorname{dom} f = \{x : |f(x)| < \infty\}$ and set $[f \le \alpha] = \{x \in \operatorname{dom} f : f(x) \le \alpha\}$, $[f = \alpha] = \{x \in \operatorname{dom} f : f(x) = \alpha\}$, etc.

Definition 19.5. Suppose $[f \le \alpha] \ne \emptyset$. A number $K \ge 0$ is called a *global error bound for f at level α* if

$$\mathrm{d}(x, [f \le \alpha]) \le K(f(x) - \alpha)^+, \quad \forall x \in X.$$

Clearly the set of all global error bounds has the minimal element. We shall denote by $K_f(\alpha)$ the smallest global error bound for f at level α. The reciprocal quantity $K_f(\alpha)^{-1}$ is sometimes called the *condition number* of f at the level α.

We shall look for estimates or exact expressions for global error bounds (condition numbers) that use only infinitesimal information about the function.

19.3.1 Convex Function on a Banach Space

As follows from the title of the subsection we shall consider here the case when X is a Banach space and f is a convex function. We assume throughout that

($\mathbf{A}_1$) f is a proper closed convex function and $[f \le \alpha] \ne \emptyset$.

In this subsection we prove the following theorem.

Theorem 19.6. *Let X be a Banach space and f a proper closed convex function on X satisfying ($\mathbf{A}_1$). Then for any α with $[f \le \alpha] \ne \emptyset$*

$$\begin{aligned} K_f(\alpha)^{-1} &= \inf_{x \in [f > \alpha]} \sup_{\|h\| \le 1} (-f'(x; h)) \\ &= \inf_{x \in [f > \alpha]} \mathrm{d}(0, \partial f(x)) \\ &= \inf_{x \in [f > \alpha]} \operatorname{sur}(\operatorname{Epi} f)(x, f(x)). \end{aligned} \tag{19.5}$$

Moreover, if X is a Hilbert space, then we also have

$$K_f(\alpha)^{-1} = \inf_{x \in [f = \alpha]} \inf_{h \in N([f \le \alpha], x), \|h\| = 1} f'(x; h). \tag{19.6}$$

Proof. The second and the third equalities follow from Propositions 19.1 and 19.4. So in order to prove the first statement we only have to show that $K_f(\alpha)^{-1}$ coincides with any of the quantities on the right. We shall do this separately for Hilbert spaces (along with (19.6)), reflexive spaces, and general Banach spaces. Set for brevity $S = [f \le \alpha]$, $S_0 = [f = \alpha]$.

1. **The Case of a Hilbert Space**. To begin with, consider a continuous convex function φ on the real segment $[0,T]$ which is equal to zero at 0 and strictly positive on $(0,T]$. Denote by $\varphi'(t\pm)$ the right and left derivatives of φ at t. We have $\varphi'(t+)+\varphi'(t-)\geq 0$ for all $t\in(0,T)$ and (by (19.2))

$$\varphi'(0+)=\lim_{t\to 0}\varphi'(t)=\lim_{t\to 0}(-\varphi'(t-))=\inf_{t>0}(-\varphi'(t-)). \tag{19.7}$$

It follows further from (19.2) and the mean value theorem that for any $t>0$ there is a $\tau\in(0,t)$ such that $-t\varphi'(t-)\geq\varphi(t)\geq-\tau\varphi'(\tau-)$. Together with (19.7) this implies that

$$\begin{aligned}\sup\{k\geq 0: kt\leq\varphi(t),\ \forall t\in[0,T]\} &= \varphi'(0+)\\ &= -\lim_{t\to 0}\varphi'(t-)=\inf_{t>0}(-\varphi'(t-)).\end{aligned} \tag{19.8}$$

Let $x\in(\operatorname{dom}f)\backslash S$, and $\bar{x}\in S$ be such that $\|x-\bar{x}\|=\mathrm{d}(x,S)$. As $f(x)<\infty$, we necessarily have $\bar{x}\in S_0$. Set $T=\|x-\bar{x}\|$, $h=T^{-1}(x-\bar{x})$ and let $\varphi(t)=f(\bar{x}+th)$, $t\in[0,T]$. It is clear that $\varphi'(t+)=f'(\bar{x}+th;h)$ and $\varphi'(t-)=f'(\bar{x}+th;-h)$. Note further that for any $\bar{x}\in S$ either $N(S,\bar{x})=\{0\}$ or $\bar{x}+h\notin\operatorname{dom}f$ for any nonzero $h\in N(S,\bar{x})$ in which case $f'(\bar{x};h)=\infty$ for such h, or finally there is an $h\in N(S,\bar{x})$, $\|h\|=1$ such that $f(\bar{x}+th)<\infty$ for some positive t. In the last case $\bar{x}$ is necessarily the closest to $\bar{x}+th$ element of S. Combining this with (19.8), we get

$$\begin{aligned}K_f(\alpha)^{-1} &= \sup\{k\geq 0:\ kd(x,S)\leq f(x)-\alpha,\ \forall\, x\notin S\}\\ &= \inf_{\bar{x}\in S_0}\ \inf_{h\in N(S,x),\|h\|=1} f'(x;h)=\inf_{x\notin S}\ \sup_{\|h\|\leq 1}(-f'(x;h)).\end{aligned}$$

This proves both (19.6) and the first equality in (19.5).

2. **The General Case: Proof that**

$$K_f(\alpha)^{-1}\leq\inf_{x\in[f>\alpha]}\ \sup_{\|h\|\leq 1}(-f'(x;h))=r. \tag{19.9}$$

Take an $x\in[f>\alpha]\cap\operatorname{dom}f$ and an $\bar{x}\in S_0$. Set $u=\bar{x}-x$ and $h=u/\|u\|$. We have from (19.2)

$$\frac{f(x)-f(\bar{x})}{\|x-\bar{x}\|}\leq -f'(x;h)$$

and (19.9) follows because $\bar{x}$ can be chosen to make $\|x-\bar{x}\|$ arbitrarily close to $\mathrm{d}(x,S)$. So it remains to prove the opposite inequality for which we can assume that $r>0$.

3. **The Case of a Reflexive Space: Completion of the Proof**. We have $\inf_{\|h\|\leq 1} f'(x;h)\leq -r$ for all $x\in[f>\alpha]$. This means that for any such x and any $r'<r$ there is a $t>0$ such that for some u we have

$$\|u-x\| \le t, \qquad f(u) \le f(x) - r't. \tag{19.10}$$

Fix x and ε, denote by $TU(x)$ the collection of pairs (t,u) satisfying (19.10), and consider the lower bound β of $f(u)$ over $TU(x)$.

We claim that there is a $(u,t) \in TU(x)$ such that $f(u) \le \alpha$. This is obvious if $\beta < \alpha$. For a $\beta > -\infty$, the set $TU(x)$ is convex and bounded by (19.10) and, as f is lower semicontinuous, it is a closed set. Since X is a reflexive space, then $TU(x)$ is weakly compact, so the lower bound is attained at some $(\overline{u},\overline{t})$. If we had assumed that $\beta > \alpha$ then there would be a pair $(u,t) \in TU(\overline{u})$ in which case $(u,t+\overline{t}) \in TU(x)$ and $f(u) < \beta$, a contradiction. Thus $f(\overline{u}) = \beta = \alpha$.

4. **General Case: Completion of the Proof**. The last argument does not work if X is not reflexive. In this case by Ekeland's variational principle for any $\delta > 0$ there is a pair $(\overline{u},\overline{t}) \in TU(x)$ such that $f(u) + \delta\|u-\overline{u}\|$ attains its minimum at $\overline{u}$. We take $\delta < r'$. If $f(\overline{u}) = \alpha$, we are done. If $f(\overline{u}) > \alpha$, there is an h with $\|h\| = 1$ such that $-f'(\overline{u};h) > r'$, that is, $f(\overline{u}+th) > f(\overline{u}) - r't$ for some $t > 0$. Set $u = \overline{u}+th$. Then $(u,t) \in TU(\overline{u})$ and we get a contradiction with the definition of $\overline{u}$, proving the claim. As x is an arbitrary point of $[f > \alpha]$, it follows that $K_f(\alpha) \le 1/r'$ and the desired inequality follows since r' can be arbitrarily close to r. ∎

Remark 19.7. Observe that for the cases of a Hilbert and reflexive X we only needed elementary convex analysis, whereas for a general case we have been compelled to invoke the variational principle of Ekeland. It would be interesting (albeit doubtful) to find a proof completely based on convex analysis also in the general case. An alternative simple proof that $K_f(\alpha)^{-1} \ge \inf_{x\in[f>\alpha]} d(0,\partial f(x))$, also based on Ekeland's principle, easily follows from Lemma 19.8 below (see also [26]).

19.3.2 General Results on Global Error Bounds

Here we consider the general case of an l.s.c. function on a complete metric space and the ways from these results to those of the preceding subsection. Recall [13] that the *slope* of f at x is

$$|\nabla f|(x) = \limsup_{u\to x, u\ne x} \frac{(f(x)-f(u))^+}{d(x,u)}$$

where $\alpha^+ = \max\{\alpha, 0\}$.

Lemma 19.8. *Let X be a complete metric space and f a lower semicontinuous function on X. Assume that for some $x \in \operatorname{dom} f$, and $\alpha < f(x)$, we have $|\nabla f|(u) \ge r > 0$ if $\alpha < f(u) \le f(x)$. Then $[f \le \alpha] \ne \emptyset$ and $d(x,[f\le\alpha]) \le r^{-1}(f(x)-\alpha)^+$.*

Proof. Set $g(u) = (f(u)-\alpha)^+$. By Ekeland's principle for any $\delta > 0$ there is a $\overline{u}$ such that $g(\overline{u}) \le g(x)$, $\mathrm{d}(\overline{u},x) \le g(x)/\delta$, and $g(u) + \delta \mathrm{d}(u,\overline{u}) \ge g(\overline{u})$ for all u. It follows that $|\nabla g|(\overline{u}) \le \delta$. If $\delta < r$, this can happen only if $g(\overline{u}) = 0$ for otherwise we would have $|\nabla g|(\overline{u}) = |\nabla f|(\overline{u}) \ge r$. Taking δ arbitrarily close to (and still smaller than) r, we prove the second statement. ∎

Denote by $K_f(\alpha,\beta)$ (where $\beta > \alpha$) the lower bound of K such that

$$d(x,[f \le \alpha]) \le K(f(x)-\alpha)^+ \text{ if } \alpha < f(x) \le \beta.$$

Clearly, $K_f(\alpha) = \lim_{\beta\to\infty} K_f(\alpha,\beta)$.

Theorem 19.9. *Let X be a complete metric space and f a lower semicontinuous function on X. If $[f \le \alpha] \ne \emptyset$, then*

$$\inf_{x\in[\alpha<f\le\beta]} |\nabla f|(x) = \inf_{\gamma\in[\alpha,\beta)} K_f(\gamma,\beta)^{-1}.$$

Proof. Set $r = \inf_{x\in[\alpha<f\le\beta]} |\nabla f|(x)$. The inequality $K_f(\gamma,\beta)^{-1} \ge r$ for $\alpha \le \gamma < \beta$ is immediate from Lemma 19.8. This proves that the left side of the equality cannot be greater than the quantity on the right. To prove the opposite inequality it is natural to assume that $K_f(\gamma,\beta)^{-1} \ge \xi > 0$ for all $\gamma \in [\alpha,\beta)$. For any $x \in [f > \alpha]$ and any $\varepsilon > 0$ such that $f(x) - \varepsilon > \alpha$, choose a $u = u(\varepsilon) \in [f \le f(x) - \varepsilon]$ such that $\mathrm{d}(x,u) \le (1+\varepsilon)\mathrm{d}(x,[f \le f(x)-\varepsilon]) \le (1+\varepsilon)\xi^{-1}\varepsilon$ and therefore $u \to x$ as $\varepsilon \to 0$. On the other hand, $\xi \mathrm{d}(x,u) \le f(x) - f(u)$ which (as $u \ne x$) implies that $\xi \le |\nabla f|(x)$, whence $\xi \le |\nabla f|(x)$, and the result follows. ■

As an immediate consequence we get

Corollary 19.10. *Under the assumption of the theorem*

$$K_f(\alpha)^{-1} \ge \inf_{x\in[f>\alpha]} |\nabla f|(x).$$

A trivial example of a function f having an isolated local minimum at a certain $\overline{x}$ and such that $\inf f < f(\overline{x})$ shows that the inequality can be strict. This may happen of course even if the slope is different from zero everywhere on $[f > \alpha]$. In this case an estimate of another sort can be obtained. Set (for $\beta > \alpha$)

$$d_f(\alpha,\beta) = \sup_{x\in[f\le\beta]} \mathrm{d}(x,[f \le \alpha])$$

and define the functions

$$\kappa_{f,\varepsilon}(t) = \sup\{\frac{1}{|\nabla f|(x)} : |f(x)-t| < \varepsilon\}; \quad \kappa_f(t) = \lim_{\varepsilon\to 0} \kappa_{f,\varepsilon}(t).$$

Proposition 19.11. *Let $\beta > \alpha$. Assume that $[f \le \alpha] \ne \emptyset$ and $|\nabla f|(x) \ge r > 0$ if $x \in [\alpha < f \le \beta]$. Then*

$$d_f(\alpha,\beta) \le \int_\alpha^\beta \kappa_f(t)\mathrm{d}t.$$

Proof. First we note that κ_f is measurable (so as it is nonnegative, the integral makes sense). Indeed, it is enough to verify that every $\kappa_{f,\varepsilon}$ is measurable. In fact the latter is even lower semicontinuous. Indeed, take a $\delta > 0$ and find an x with $|f(x)-t| < \varepsilon$ such that $|\nabla f|(x) > \kappa_{f,\varepsilon}(t) - \delta$. Take a positive $\gamma < \varepsilon - |f(x)-t|$. Then for any τ with $|t-\tau| < \gamma$ we have $|f(x)-\tau| < \varepsilon$ and therefore $\kappa_{f,\varepsilon}(\tau) \geq \kappa_{f,\varepsilon}(t) - \delta$.

Now fix an $\varepsilon > 0$ and let $\alpha = \tau_0 < \ldots < \tau_k = \beta$ be a partition of $[\alpha,\beta]$ with $(1/2)(\tau_{i+1}-\tau_i) = \varepsilon_i < \varepsilon$. Set $t_i = (\tau_i + \tau_{i-1})/2$, $i = 1,\ldots,k$. As follows from Theorem 19.9, $d_f(\tau_{i-1},\tau_i) \leq \kappa_{f,\varepsilon_i}(t_i)(\tau_{i+1}-\tau_i)$ and therefore

$$d_f(\alpha,\beta) \leq \sum_{i=1}^{k} \kappa_{f,\varepsilon_i}(t_i)(\tau_{i+1}-\tau_i) \leq \sum_{i=1}^{k} \kappa_{f,\varepsilon}(t_i)(\tau_{i+1}-\tau_i).$$

Passing to the limit over the net of all partitions of $[\alpha,\beta]$ we conclude that

$$d_f(\alpha,\beta) \leq \int_{\alpha}^{\beta} \kappa_{f,\varepsilon}(t)\mathrm{d}t.$$

The result now follows from the Lebesgue majorized convergence theorem as by the assumption $\kappa_{f,\varepsilon}(t) \leq r^{-1}$ for all t and ε if $t \in (\alpha,\beta]$. ∎

Returning to the case of a convex function on a Banach space, we first state the following elementary fact that serves as a bridge between the general and convex situations.

Proposition 19.12. *Let X be a convex function on a Banach space X, and let $x \in \operatorname{dom} f$. Then*

$$|\nabla f|(x) = \sup_{\|h\|\leq 1} (-f'(x;h)) = \mathrm{d}(0,\partial f(x)).$$

Proof. Clearly $|\nabla f|(x) = 0$ if and only $f'(x,h) \geq 0$ for all h and the equality holds with $h = 0$. If $\|h\| = 1$ and $u = x+th$, then $t = \|x-u\|$, so the equality $-f'(x;h) = \lim_{t\to 0}(f(x) - f(x+th))$ implies $-f'(x;h) \leq |\nabla f|(x)$. On the other hand, as $f'(x;h) \leq t^{-1}(f(x+th)-f(x))$ for all t and h, for a given $u \neq x$, we get $-f(x;h) \geq \|u-x\|$ if we set $t = \|u-x\|$ and $h = t^{-1}(u-x)$. ∎

Proposition 19.13. *Let f be a convex function on a Banach space X. Assume that $[f \leq \alpha] \neq \emptyset$. Let $\beta > \alpha$. Then for any $\gamma \in (\alpha,\beta)$*

$$K_f(\alpha,\beta) \geq K_f(\gamma,\beta).$$

Proof. We may assume that $K_f(\alpha,\beta) < \infty$ and $K_f(\gamma,\beta) > 0$ (which by definition means that $[f > \gamma] \cap \operatorname{dom} f \neq \emptyset$). Take an $x \in [f > \gamma] \cap \operatorname{dom} f$ and a $K > K_f(\alpha,\beta)$ and find a $u \in [f \leq \alpha]$ such that $\|x-u\| \leq K(f(x)-\alpha)$. As earlier, we may assume that $f(u) = \alpha$. As $\alpha < \gamma < f(x)$, there is a $t > 0$ such that $f(w) = \gamma$ for $w = tu+(1-t)x$. By convexity $t(f(x)-\alpha) \leq f(x)-\gamma$. We therefore have

$$\|x-w\| = t\|x-u\| \leq \|x-u\|\frac{f(x)-\gamma}{f(x)-\alpha} \leq K(f(x)-\gamma).$$

This is true for all $x \in [f > \gamma] \cap \operatorname{dom} f$ and all $K > K_f(\alpha,\beta)$, whence the result. ∎

Combining Theorem 19.9 and Propositions 19.12 and 19.13 we get still another proof of the first equality in Theorem 19.6.

19.3.3 Comments

Following the pioneering 1952 work by Hoffmann [18], error bounds, both for nonconvex and, especially, convex functions, are intensively studied, especially during last 2–3 decades, both theoretically, in connection with metric regularity, and also in view of their role in numerical analysis; see, e.g., [12, 17, 29, 31, 34, 39, 40, 44, 47, 48]. A finite dimensional version of (19.6) was proved in Lewis-Pang [29]. The equality can actually be extended to reflexive spaces (see Azé-Corvellec [1]). The equality $K_f(\alpha)^{-1} = \inf\{d(0,\partial f(x)) : x \in [f > \alpha]\}$ in Theorem 19.6 was proved by Zalinescu [45] (see also [46], Proposition 3.10.8, and for earlier results [11]). The first two equalities in the theorem can be found in [1, 2]. Theorem 19.9 and Proposition 19.13 were proved by Azé and Corvellec in [1]. The papers also contain sufficiently thorough bibliographic comments.

19.4 Convex Set-Valued Mappings

Let X and Y be Banach spaces and $F : X \rightrightarrows Y$. We shall say that F is a *convex mapping* if its graph Graph$F$ is a convex set. In this section we shall mainly discuss the regularity problems for convex mappings.

19.4.1 Theorem of Robinson–Ursescu

The standard statement of the Robinson-Ursescu theorem reads: *Let X and Y be Banach spaces, and let $F : X \rightrightarrows Y$ be a set-valued mapping with convex and locally closed graph. Assume that the image of X under F has a nonempty interior. Then F is regular at every $(\bar{x}, \bar{y})$ such that $\bar{y} \in \operatorname{int} F(X)$.*

This theorem can be rightfully viewed as an extension of the Banach-Schauder open mapping theorem. Moreover, the original proofs of the theorem followed the pattern of the classical proof of the open mapping theorem: first the Baire category theorem is applied to show that under the assumptions $\bar{y}$ belongs to the interior of the closure if the F-image of some ball around $\bar{x}$ and then basically the same iteration scheme as in the classical Banach proof is applied to show that the closure operation can be dropped and $\bar{y}$ belongs to the interior of the F-image (of the same ball) itself. Later it became clear that (as in many other results of the regularity theory) instead of the iteration procedure the variational principle of Ekeland can be used at the second stage of the proof. The latter is the basic fact behind the proof of the general regularity criterion of Theorem 19.3 quoted in the previous section.

The following elementary and short argument shows that the conclusion of the second part of the proof of the Robinson-Ursescu theorem, even in a more precise quantitative form, is a simple consequence of the general regularity criterion of Theorem 19.3. The only use of convexity in this argument is connected with the following obvious observation.

Proposition 19.14. *Let $F: X \rightrightarrows Y$ be a set-valued mapping with convex graph. If $\alpha > 0$ and $\beta > 0$ are such that $F(B(x,\alpha))$ is dense in $B(y,\beta)$ and $y \in F(x)$, then for any $\lambda \in (0,1)$ the F-image of $B(x,\lambda\alpha)$ is dense in $B(y,\lambda\beta)$.*

Passing to the proof of the Robinson-Ursescu theorem, we first find $\alpha > 0$ and $\beta > 0$ such that $B(\bar{y},\beta) \subset \mathrm{cl}\, F(B(\bar{x},\alpha))$ (whose existence as we have mentioned is proved through a standard application of the Baire category theorem) and then fix an $\varepsilon > 0$ such that $4\varepsilon < \min\{\alpha,\beta\}$. Let x,y,v satisfy $\|x-\bar{x}\| < \varepsilon$, $y \in F(x)$, $\|v-\bar{y}\| < \varepsilon$, $\|v-y\| < \varepsilon$. Then

$$B(y,\beta-2\varepsilon) \subset B(\bar{y},\beta) \subset \mathrm{cl}\, F(\bar{x},\alpha) \subset \mathrm{cl}\, F(\bar{x},\alpha+\varepsilon).$$

Setting $\xi = (\alpha+\varepsilon)/(\beta-2\varepsilon)$ we get from Proposition 19.14 that $B(y,t) \subset \mathrm{cl}\, F(x,\xi t)$ if, e.g., $t \in (0,\varepsilon)$. Let $z = \lambda v + (1-\lambda)y$ for some $\lambda \in (0,1)$. Then $t = \|z-y\| < \varepsilon$ and there is a u with $\|u-x\| \le \xi t$ such that $z \in F(u)$. We have

$$d_\xi((x,y),(u,v)) = \max\{\|x-u\|, \xi\|y-z\|\} = \xi\|y-z\|$$

and therefore $\|v-z\| = \|v-y\| - \|z-y\| \le \|v-y\| - r d_\xi((x,y),(u,z))$ if $r < \xi^{-1}$. A reference to Theorem 19.3 completes the proof. Moreover, we see that

$$\mathrm{sur}\, F(\bar{x}|\bar{y}) \ge \beta/\alpha.$$

19.4.2 Regularity Moduli of Convex Multifunctions

The next question we shall discuss in this section is how to compute regularity moduli of convex multifunction. As immediately follows from the arguments concluding the previous subsection (when $\varepsilon \to 0$), $\mathrm{sur}\, F(\bar{x}|\bar{y}) \ge \beta/\alpha$. A slight elaboration on this result gives a more precise conclusion.

Proposition 19.15. *Let $F: X \rightrightarrows Y$ have a convex and locally closed graph. Then*

$$\begin{aligned}\mathrm{sur}\, F(\bar{x}|\bar{y}) &= \lim_{\varepsilon\to 0} \sup\{r \ge 0:\ B(\bar{y},r\varepsilon) \subset \mathrm{cl}\, F(B(\bar{x},\varepsilon))\} \\ &= \lim_{\varepsilon\to 0} \varepsilon^{-1} \sup\{t \ge 0:\ B(\bar{y},t) \subset \mathrm{cl}\, F(B(\bar{x},\varepsilon))\}.\end{aligned}$$

Proof. The second equality is obvious (just take $r\varepsilon = t$). The first equality is trivial if $\bar{y}$ does not belong to $\mathrm{int\,cl}\,[F(B(\bar{x},\varepsilon))]$. For the case when $\bar{y}$ lies in the interior of $\mathrm{cl}\, F(B(\bar{x},\varepsilon))$, the inequality $\mathrm{sur}\, F(\bar{x}|\bar{y}) \ge \sup\{r \ge 0:\ B(\bar{y},r\varepsilon) \subset \mathrm{cl}\, F(B(\bar{x},\varepsilon))\}$

follows from the proof following Proposition 19.14: just take $\beta = r\varepsilon$ and $\alpha = \varepsilon$. And the opposite inequality is immediate from the definition of the modulus of surjection. ■

Although the formula can hardly be recommended for practical computation of the surjection moduli, it brings about a substantial simplification compare to the general case as there is no longer a need to verify similar inclusions for other points of the graph close to $(\bar{x},\bar{y})$. A duality-based working formula for the surjection modulus is offered by the following theorem.

Theorem 19.16. *Let $F : X \rightrightarrows Y$ be a set-valued mapping with convex and locally closed graph. If $\bar{y} \in F(\bar{x})$, then*

$$\operatorname{sur} F(\bar{x}|\bar{y}) = \lim_{\varepsilon \to +0} \inf_{\|y^*\|=1} \inf_{x^*} \Big(\|x^*\| + \frac{1}{\varepsilon} S_{\operatorname{Graph} F - (\bar{x},\bar{y})}(x^*,y^*) \Big).$$

Proof. To begin with we observe the following. Let $Q \subset Y$ be a closed convex set and $\bar{y} \in Q$. Then $B(\bar{y},r) \subset Q$ if and only if $\sup\{\langle y^*, y-\bar{y}\rangle : y \in Q\} \geq r$ for any y^* with $\|y^*\| = 1$. It follows that the lower bound of the supremum over the unit sphere in Y^* coincides with the upper bound of $r \geq 0$ such that $B(\bar{y},r) \subset Q$.

We have furthermore

$$\begin{aligned} &\sup\{r \geq 0 : B(\bar{y},r) \subset \operatorname{cl} F(B(\bar{x},\varepsilon))\} \\ &\quad = \inf_{\|y^*\|=1} \sup\{\langle y^*, y-\bar{y}\rangle : y \in F(\bar{x}+h),\ \|h\| \leq \varepsilon\} \\ &\quad = \inf_{\|y^*\|=1} \sup\{\langle y^*, v\rangle : (h,v) \in \operatorname{Graph} F - (\bar{x},\bar{y}),\ \|h\| \leq \varepsilon\} \\ &\quad = \inf_{\|y^*\|=1} (\operatorname{Ind}_{\operatorname{Graph} F - (\bar{x},\bar{y})} + \operatorname{Ind}_{\varepsilon B \times Y})^*(0,y^*). \end{aligned} \tag{19.11}$$

As $(0,0) \in (\operatorname{Graph} F - (\bar{x},\bar{y})) \cap \operatorname{int}(\varepsilon B \times Y)$, it follows from the standard duality between summation and infimal convolution:

$$\begin{aligned} &(\operatorname{Ind}_{\operatorname{Graph} F - (\bar{x},\bar{y})} + \operatorname{Ind}_{\varepsilon B \times Y})^*(0,y^*) \\ &\quad = \inf_{(x^*,v^*)} \{S_{\operatorname{Graph} F - (\bar{x},\bar{y})}(x^*,v^*) + S_{\varepsilon B \times Y}(-x^*, y^* - v^*)\} \\ &\quad = \inf_{x^*} \{S_{\operatorname{Graph} F - (\bar{x},\bar{y})}(x^*,y^*) + \varepsilon\|x^*\|\} \\ &\quad = \varepsilon \inf_{x^*} \{\|x^*\| + \varepsilon^{-1} S_{\operatorname{Graph} F - (\bar{x},\bar{y})}(x^*,y^*)\}. \end{aligned}$$

Together with (19.11) and Proposition 19.15 this completes the proof. ■

19.4.3 Systems of Convex Inequalities

This is the system of relations

$$\varphi_t(x) \le b_t, \quad t \in T, \tag{19.12}$$

where $x \in X$, X is a Banach space, T is a set of an arbitrary nature, and for any t, φ_t is a proper closed convex function on X and $b_t \in \mathbb{R}$. Set $b = (b_t)$ and let $\mathscr{S}(b)$ be the set of solutions of (19.12). Clearly, $\mathscr{S}(b)$ is a closed convex set (possibly empty). A natural question is about Lipschitz stability of the set-valued mapping $\mathscr{S}$ with respect to small perturbations of b near some nominal value $\overline{b}$.

Although we impose no a priori restrictions on elements of $\overline{b}$, there is no loss of generality in assuming that $\overline{b} = 0$. Otherwise, we can consider, instead of φ_t, the functions $\varphi_t - \overline{b}_t$. As perturbations of the right-hand side we shall consider arbitrary uniformly bounded real-valued functions on T, that is, elements of the space $\ell_\infty(T)$ with the standard uniform norm. As follows from the equivalence theorem, Lipschitz stability of solutions of (19.12) with $\overline{b} = 0$ is guaranteed by regularity at $(\overline{x}, 0)$ of the following set-valued mapping from X into $\ell_\infty(T)$:

$$F(x) = \{a = (a_t) \in \ell_\infty(T) : \ a_t \ge \varphi_t(x), \ \forall\, t \in T\}$$

and

$$\operatorname{lip} \mathscr{S}(0; \overline{x}) = (\operatorname{sur} F(\overline{x}|0))^{-1}.$$

Set

$$\Phi(x) = \sup_{t \in T} (\varphi_t(x) - \overline{b}_t).$$

Clearly, $\Phi(\overline{x}) \le 0$.

Theorem 19.17. *Let $\overline{x}$ be a solution of (19.12) with $b = \overline{b} = 0$. Then either* $\operatorname{sur} F(\overline{x}|\overline{b}) = \infty$ *or* $\Phi(\overline{x}) = 0$, $\partial \Phi(\overline{x}) \ne \emptyset$ *and*

$$\operatorname{sur} F(\overline{x}|\overline{b}) = \mathrm{d}(0, \partial \Phi(\overline{x})).$$

Thus the theorem effectively says that *Lipschitz stability of the solution map $\mathscr{S}$ at $(0, \overline{x})$ is equivalent to Lipschitz stability of the solution set of the single convex inequality*

$$\Phi(x) = \sup_{t \in T} \varphi_t(x) \le \alpha$$

at $(0, \overline{x})$ with the same Lipschitz modulus equal to $[\mathrm{d}(0, \partial \Phi(\overline{x})]^{-1}$.

Applying the theorem to the simplest case when T is a singleton, that is, when we deal with one convex function f and $f(\overline{x}) = \overline{\alpha}$, we conclude (again by virtue of the equivalence theorem) that

$$\mathrm{d}(x, [f \le \alpha]) \le K(f(x) - \alpha)^+$$

for all x and α close to $\overline{x}$ and $\overline{\alpha}$, respectively, with $K = (d(0, \partial f(\overline{x})))^{-1}$, provided $\partial f(\overline{x}) \ne \emptyset$. (Note that regularity of f in this sense is a stronger property than the existence of a local error bound at the level $\overline{\alpha}$.) We can now proceed with the proof of the theorem.

Proof. So we assume in the proof that $\overline{b}=0$. We may also harmlessly assume that φ_t are uniformly bounded from below (otherwise we can replace φ_t, say by $\max\{\varphi_t,-1\}$).

1. The cone $\mathscr{K}=\{a\in\ell_\infty:\ a_t\geq 0,\ \forall\, t\in T\}$ defines the standard order in $\ell_\infty(T)$. The dual cone $\mathscr{K}^*$ consists of all $p^*\in(\ell_\infty)^*$ such that $\langle p^*,a\rangle\geq 0$ if $a_t\geq 0$ for all t. We shall simply write $p^*\geq 0$ for elements of $\mathscr{K}^*$. For any $p^*\geq 0$, we define the function
$$(p^*\circ F)(x)=\inf\{\langle p^*,a\rangle:\ a\in F(x)\}$$
(clearly, the infimum is $-\infty$ if $p^*\notin\mathscr{K}^*$). This function is obviously convex. We claim that *for any x^* the function $p^*\mapsto(p^*\circ F)^*(x^*)$ on $(\ell_\infty)^*$ is convex and weak* lower semicontinuous on its domain.* Indeed, convexity follows from the obvious inequality:
$$\begin{aligned}\sup_x\big(\langle x^*,x\rangle-((\alpha p_1^*+(1-\alpha)p_2^*)\circ F)(x)\big)&\\ =\sup_x\big(\alpha(\langle x^*,x\rangle-(p_1^*\circ F)(x))+(1-\alpha)(\langle x^*,x\rangle-(p_2^*\circ F)(x))\big)&\\ \leq\alpha\sup_x(\langle x^*,x\rangle-(p_1^*\circ F)(x))+(1-\alpha)\sup_x(\langle x^*,x\rangle-(p_2^*\circ F)(x)).&\end{aligned}$$
On the other hand, if $a\in\ell_\infty$, then $p^*\mapsto\langle p^*,a\rangle$ is linear and weak*-continuous. It follows that for any x^* the function $p^*\mapsto(p^*\circ F)^*(x^*)$ is an upper bound of affine and weak*-continuous functions $\langle x^*,x\rangle-\langle p^*,a\rangle$ corresponding to $(x,a)\in\operatorname{Graph}F$.
2. Set $P^*=\{p^*\geq 0,\ \|p^*\|=1\}$. We shall show next that
$$\Phi(x)=\sup_{p^*\in P^*}(p^*\circ F)(x);\qquad \Phi^*(x^*)=\inf_{p^*\in P^*}(p^*\circ F)^*(x^*). \tag{19.13}$$
Indeed, the inequality $(p^*\circ F)(x)\geq\Phi(x)$ is obvious. The opposite inequality follows from the fact that $(\delta_t\circ F)(x)=\varphi_t(x)$, where δ_t is the "Dirac measure" at t: $\langle\delta_t,a\rangle=a_t$. This proves the first equality.
   As P^* is a convex and weak*-compact set, it follows, in view of the minimax theorem of Sion [38], that
$$\begin{aligned}\Phi^*(x^*)&=\sup_x(\langle x^*,x\rangle)-\sup_{p^*\in P^*}(p^*\circ F)(x))\\ &=\sup_x\inf_{p^*\in P^*}(\langle x^*,x\rangle-(p^*\circ F)(x))\\ &=\inf_{p^*\in P^*}\sup_x(\langle x^*,x\rangle-(p^*\circ F)(x))\\ &=\inf_{p^*\in P^*}(p^*\circ F)^*(x^*).\end{aligned}$$
   As the function $p^*\mapsto(p^*\circ F)(x^*)$ is weak* l.s.c., it follows that the infimum in the last expression is attained, so that $\Phi^*(x^*)=(p^*\circ F)^*(x^*)$ for some $p^*\in P^*$.

3. We have for $x^* \in X^*,\ p^* \in (\ell_\infty)^*$

$$s_{\mathrm{Graph}\,F-(\overline{x},0)}(x^*,-p^*) = \sup\{\langle x^*,x\rangle - \langle p^*,a\rangle : \ a_t \geq \varphi_t(x+\overline{x}),\ \forall\, t \in T\}.$$

If $s_{\mathrm{Graph}\,F-(\overline{x},0)}(x^*,-p^*) < \infty$, then necessarily $p^* \geq 0$ and

$$s_{\mathrm{Graph}\,F-(\overline{x},0)}(x^*,-p^*) = \sup_x \left(\langle x^*,x\rangle - (p^*\circ F)(x+\overline{x})\right) = (p^*\circ F)^*(x^*) - \langle x^*,\overline{x}\rangle.$$

Thus Theorem 19.16 along with (19.13) and the last equality gives

$$\mathrm{sur}\,F(\overline{x}|0) = \liminf_{\varepsilon\to 0}\inf_{x^*}\Big(\|x^*\| + \frac{1}{\varepsilon}(\Phi^*(x^*) - \langle x^*,\overline{x}\rangle)\Big).$$

If $\mathrm{sur}\,F(\overline{x}|0) = r < \infty$, then for any $\varepsilon >$ the infimum is attained at a certain $x^*(\varepsilon)$ with $\|x^*(\varepsilon)\| \leq r$ (indeed, $\Phi^*(x^*) - \langle x^*,\overline{x}\rangle \geq -\Phi(\overline{x}) \geq 0$ and the function in the parentheses is weak* lower semicontinuous and nondecreasing as $\varepsilon \to 0$).

Let x^* be a weak* limit point of $(x^*(\varepsilon))$ as $\varepsilon \to 0$. Then necessarily $\Phi^*(x^*) - \langle x^*,\overline{x}\rangle \leq 0$ which (as $\Phi(\overline{x}) \leq 0$) may happen only if $\Phi(\overline{x}) = 0$ and $x^* \in \partial\Phi(\overline{x})$. On the other hand, if $x^* \in \partial\Phi(\overline{x})$ and $\Phi(\overline{x}) = 0$, we get

$$\mathrm{sur}\,F(\overline{x}|0) = \inf\{\|x^*\| : \ x^* \in \partial\Phi(\overline{x})\}$$

and the proof is completed. ∎

Remark 19.18. It is to be emphasized that in no point in the proof the representation of elements of ℓ_∞ by finitely additive measures has been needed.

19.4.4 Perfect Regularity and Linear Perturbations

As follows from the formula for the modulus of surjection in Theorem 19.16, the value of the modulus is fully determined by the restriction of the support function to $\mathrm{Graph}\,F - (\bar{x},\bar{y})$ to the set on which it is smaller than $\varepsilon > 0$, no matter however small this ε is. But in general we cannot replace such sets by the zero level of the support function (which, as it is easy to see, is precisely the normal cone to $\mathrm{Graph}\,F$ at $(\bar{x},\bar{y})$).

Example 19.19. Let $X = Y = L^2[0,1]$ and $F : X \rightrightarrows Y$ is defined by $F(x) = x + K$ where K is the cone of nonnegative functions. Let $\overline{x}(t) \equiv -1$ and $y(t) \equiv 0$. Clearly, $\overline{y} \in F(\overline{x})$. Direct calculation gives

$$s_{\mathrm{Graph}\,F-(\bar{x},\bar{y})}(x^*,y^*) = \begin{cases} \|y^*\| + \displaystyle\int_0^1 |y^*(t)|dt, & \text{if } x^*+y^* = 0, y^*(t) \leq 0 \text{ a.e.} \\ \infty, & \text{otherwise.} \end{cases}$$

As the infimum of the L^1-norm on the unit sphere of L^2 is zero, it follows that $\mathrm{sur}\,F(\overline{x}|\overline{y}) = 1$.

On the other hand, the zero level set of the support function contains only the zero element, so by the standard convention ($\inf\emptyset = \infty$), we conclude that restricting the infimum in the formula of Theorem 19.16 only to this set (which does not meet the unit sphere) we get ∞, not 1.

Definition 19.20. We shall say that F is *perfectly regular* at $(\bar{x},\bar{y})$ if

$$\operatorname{sur} F(\overline{x}|\overline{y}) = \inf\{\|x^*\| : (x^*,y^*) \in N(\operatorname{Graph} F,(\bar{x},\bar{y})),\ \|y^*\| = 1\}.^3 \tag{19.14}$$

An example of infinite dimensional perfectly regular mapping is the F associated with the system of convex inequalities (19.12) in the concluding part of the previous section (see [23]).

It is possible to give a primal characterization of perfect regularity. Remind first that a *convex process* is a set-valued mapping whose graph is a convex cone and note that the surjection modulus of a convex process $A : X \rightrightarrows Y$ coincides with $\sup\{r \geq 0 : rB_Y \subset A(B_X)\}$. The latter is implicitly contained in [28]. It is a simple consequence of the fact that the inclusion $x+K \subset K$ holds for any point x of a convex cone K. Indeed, it follows that, given a convex process A, the inclusion $rB_Y \subset A(B_X)$ implies that for any $(x,y) \in \operatorname{Graph} A$ we have $y + rB_y \subset A(x+B_X)$.

If F is convex set-valued mapping and $(\bar{x},\bar{y}) \in \operatorname{Graph} F$, then the set-valued mapping $DF(\bar{x},\bar{y})$ whose graph is $T(\operatorname{Graph} F,(\bar{x},\bar{y}))$ is a convex process. It is clear that the tangent cone $T(\operatorname{Graph} F,(\bar{x},\bar{y}))$ to $\operatorname{Graph} F$ at $(\bar{x},\bar{y})$ contains $\operatorname{Graph} F - (\bar{x},\bar{y})$. Therefore

$$\begin{aligned}\operatorname{sur} F(\overline{x}|\overline{y}) &\leq \sup\{r \geq 0 : B(\overline{y},tr) \subset F(B(\overline{x},t))\}\\ &\leq \sup\{r \geq 0 : rB_Y \subset DF(\bar{x},\bar{y})(B_X)\} = \operatorname{sur} DF(\bar{x},\bar{y})(0|0).\end{aligned} \tag{19.15}$$

On the other hand the support function of $T(\operatorname{Graph} F,(\bar{x},\bar{y}))$ is precisely the indicator of $N(\operatorname{Graph} F,(\bar{x},\bar{y}))$ and therefore by Theorem 19.16 the right-hand side of the equality in the definition (19.20) is the modulus of surjection of $DF(\bar{x},\bar{y})$ at $(0,0)$. Thus

Proposition 19.21. *A convex mapping F is perfectly regular at $(\bar{x},\bar{y}) \in \operatorname{Graph} F$ if and only if the surjection moduli of F at $(\bar{x},\bar{y})$ and of the derivative of $DF(\bar{x},\bar{y})$ at the origin coincide.*

The following two propositions offer some sufficient conditions for perfect regularity.

Proposition 19.22 ([25], Proposition 5). *Let $F : X \rightrightarrows Y$ be a set-valued mapping with convex and locally closed graph. Suppose there is a weak-star closed convex subset Q^* of the unit sphere in Y^* such that for some $(\bar{x},\bar{y}) \in \operatorname{Graph} F$*

[3] The equality (19.14) can be used as the definition of perfect regularity for arbitrary set-valued mappings if as N we use limiting Fréchet or G-normal cones (depending on the geometry of the spaces).

$$s_{\mathrm{Graph}\,F-(\bar{x},\bar{y})}(x^*,y^*) < \infty \ \textit{and} \ \|y^*\| = 1 \quad \Rightarrow \quad y^* \in Q^*.$$

Then F is perfectly regular at $(\bar{x},\bar{y})$.

The simplest situation when the conditions of the last proposition are satisfied occurs when X is a space of continuous functions over a compact set T, Q^* is the set of probability measures on T, and K, the cone of nonnegative elements of X, is contained in $F(0)$.

The second proposition is an easy consequence of Theorem 19.16.

Proposition 19.23. *Let F be as above and $(\bar{x},\bar{y}) \in \mathrm{Graph}\,F$. For any $\varepsilon > 0$ set*

$$\mathscr{L}_\varepsilon = \{(x^*,y^*) : s_{\mathrm{Graph}\,F-(\bar{x},\bar{y})}(x^*,y^*) \le \varepsilon, \ \|x^*\| \le 1, \ \|y^*\| \le 1\}.$$

If the excess of $\mathscr{L}_\varepsilon$ over $N(\mathrm{Graph}\,F,(\bar{x},\bar{y}))$

$$\mathrm{ex}(\mathscr{L}_\varepsilon.N(\mathrm{Graph}\,F,(\bar{x},\bar{y}))) = \sup\{\mathrm{d}((x^*,y^*),N(\mathrm{Graph}\,F(\bar{x},\bar{y}))) : (x^*,y^*) \in \mathscr{L}_\varepsilon\}$$

goes to zero when $\varepsilon \to 0$, then F is perfectly regular at $(\bar{x},\bar{y})$.

Note that the condition of the proposition is automatically satisfied if both X and Y are finite dimensional.

Our main interest in this subsection is the effect of linear perturbations of F on regularity moduli. Specifically we shall consider mappings $F+A$ with A being a linear bounded operator from X into Y. We have (setting $y = v + Ax$)

$$\begin{aligned} s_{\mathrm{Graph}\,(F+A)-(\overline{x},\overline{y}+A\overline{x})}(x^*,y^*) &= \sup\{\langle x^*, x-\overline{x}\rangle + \langle y^*, y-(\overline{y}+A\overline{x})\rangle : \\ &\qquad y \in F(x)+Ax\} \\ &= \sup\{\langle x^*+A^*y^*, x-\overline{x}\rangle + \langle y^*, v-\overline{y}\rangle : v \in F(x)\} \\ &= s_{\mathrm{Graph}\,F-(\bar{x},\bar{y})}(x^*+A^*y^*,y^*). \end{aligned}$$

Theorem 19.16 now immediately gives

Proposition 19.24. *Let $F : X \rightrightarrows Y$ be a set-valued mapping with convex closed graph, let $(\bar{x},\bar{y}) \in \mathrm{Graph}\,F$, and let $A : X \to Y$ be a bounded linear operator. Then*

$$\mathrm{sur}\,(F(\overline{x}|\overline{y}+A\overline{x}) = \lim_{\varepsilon\to 0} \inf_{\|y^*\|=1} \inf_{x^*}(\|x^* - A^*y^*\| + \frac{1}{\varepsilon} s_{\mathrm{Graph}\,F-(\bar{x},\bar{y})}(x^*,y^*))$$

and consequently

$$\mathrm{sur}\,(F(\overline{x}|\overline{y}+A\overline{x}) \ge \mathrm{sur}\,F(\overline{x}|\overline{y}) - \|A\|.$$

This is a version of Milyutin's perturbation theorem [14, 16, 20] for the specific case of a convex mapping and a linear perturbation. A natural question that arises in connection with the last result is about the minimal norm of a linear perturbation which destroys regularity.

Definition 19.25 ([15]). The radius of regularity of $F : X \rightrightarrows Y$ at $(\bar{x},\bar{y}) \in \operatorname{Graph} F$ is the lower bound of norms of linear bounded operators $A : X \to Y$ such that $\operatorname{sur}(F + A)(\overline{x}|\overline{y} + A\overline{x}) = 0$. We shall denote it $\operatorname{rad} F(\overline{x}|\overline{y})$.

Theorem 19.26. *Let $F : X \rightrightarrows Y$ be a set-valued mapping with convex closed graph, and let $(\bar{x},\bar{y}) \in \operatorname{Graph} F$. Suppose that F is perfectly regular at $(\overline{x},\overline{y})$. Then*

$$\operatorname{rad} F(\overline{x}|\overline{y}) = \operatorname{sur} F(\overline{x}|\overline{y}).$$

Note that the condition of the theorem is satisfied under the assumptions of Propositions 19.22 and 19.23.

Proof. It follows from Proposition 19.24 that $\operatorname{rad} F(\overline{x}|\overline{y}) \geq \operatorname{sur} F(\overline{x}|\overline{y})$, so we have to prove the opposite inequality. Set $r := \operatorname{sur} F(\overline{x}|\overline{y})$. The theorem is obviously valid if $r = 0$. So we assume that $r > 0$. As F is perfectly regular at $(\bar{x},\bar{y})$, for any $\varepsilon > 0$ there are $(x^*_\varepsilon, y^*_\varepsilon) \in N(\operatorname{Graph} F, (\bar{x},\bar{y}))$ such that $\|y^*_\varepsilon\| = 1$, $\|x^*_\varepsilon\| \leq (1+\varepsilon)r$. In particular we have

$$S_{\operatorname{Graph} F - (\bar{x},\bar{y})}(x^*_\varepsilon, y^*_\varepsilon) = 0 \tag{19.16}$$

Let further $x_\varepsilon \in X$ and $y_\varepsilon \in Y$ satisfy

$$\|x_\varepsilon\| = \|y_\varepsilon\| = 1, \quad \langle x^*_\varepsilon, x_\varepsilon\rangle \geq (1-\varepsilon)\|x^*_\varepsilon\|, \quad \langle y^*_\varepsilon, y_\varepsilon\rangle \geq (1-\varepsilon).$$

We use these four vectors to define an operator $A_\varepsilon : X \to Y$ as follows:

$$A_\varepsilon x = \frac{\langle x^*_\varepsilon, x\rangle}{\langle y^*_\varepsilon, y_\varepsilon\rangle} x_\varepsilon.$$

Then $\|A_\varepsilon\| \leq \dfrac{1+\varepsilon}{1-\varepsilon} r$ and

$$A^*_\varepsilon y^* = \frac{\langle y^*, y_\varepsilon\rangle}{\langle y^*_\varepsilon, y_\varepsilon\rangle} x^*_\varepsilon.$$

In particular we see that $-x^*_\varepsilon = A^*_\varepsilon y^*_\varepsilon$. Combining this with Propositions 19.24 and (19.16) we get $\operatorname{sur}(F + A)(\overline{x}|\overline{y} + A\overline{x}) = 0$, that is, $\operatorname{rad} F(\bar{x},\bar{y}) \leq \|A_\varepsilon\| \to r$ as $\varepsilon \to 0$. ∎

Remark 19.27. 1. The perfect regularity condition is not necessary for the equality of the radius of regularity and the modulus of surjection. It can be easily verified that the equality holds in Example 19.19: just take A to be minus identity.

2. For mappings (even nonconvex) between finite dimensional spaces the equality holds [15]. It is also known that the inequality may fail to hold already for single-valued Lipschitz mappings from a Hilbert space into itself [21]. It would be interesting to find an example of a convex mapping for which the equality does not hold.

A natural related problem concerns stability of solutions of the inclusion

$$y \in F(x) + Ax \tag{19.17}$$

with respect to small variations of both y and A around some nominal values $\overline{y}$ and $\overline{A}$, given a nominal solution $\overline{x}$ corresponding to $\overline{y}$ and $\overline{A}$. This question moves us beyond the realm of convex problems (as $(x,A) \mapsto Ax$ is not a convex mapping).

Let $S(y,A)$ denote the set of solutions of (19.17). Our goal is to find the Lipschitz modulus of S at the nominal point. By the equivalence theorem, all we need is to find the modulus of surjection of the inverse mapping

$$\Phi(x) = \{(y,A) \in Y \times \mathscr{L}(X,Y) : \ y \in F(x) + Ax\}.$$

Here $\mathscr{L}(X,Y)$ is the space of linear bounded operators from X into Y with the standard operator norm. To correctly state the question we need to fix some norm in $Y \times \mathscr{L}(X,Y)$ (assuming the norms in X and Y are given). To this end, we take a norm ν in $\mathbb{R}^2$ and set

$$\|(y,A)\| = \nu(\|y\|, \|A\|).$$

We assume for convenience that $c \cdot \max\{\|y\|, \|A\|\} \geq \|(y,A)\| \geq \max\{\|y\|, \|A\|\}$ for some $c > 1$. By ν^* we denote the dual norm on $\mathbb{R}^2$.

Theorem 19.28. *Let $F : X \rightrightarrows Y$ be a set-valued mapping with closed graph. Let $\overline{A} \in \mathscr{L}(X,Y)$ and $(\bar{x},\bar{y}) \in \mathrm{Graph}\,(F + \overline{A})$ be given. Then*

$$\mathrm{sur}\,\Phi(\overline{x}|(\overline{y},\overline{A})) \geq \frac{1}{\nu^*(1, \|\overline{x}\|)} \mathrm{sur}\,(F + \overline{A})(\overline{x}|\overline{y} + \overline{A}\overline{x}). \tag{19.18}$$

The equality holds if F is a convex mapping which is perfectly regular at $(\bar{x},\bar{y})$. Therefore

$$\mathrm{lip}\, S((\overline{y},\overline{A})|\overline{x}) \leq \nu^*(1, \|\overline{x}\|)\mathrm{reg}\,(F + \overline{A})(\overline{x}|\overline{y} + \overline{A}\overline{x})$$

with the equality if F has the property specified above.

Proof. With no loss of generality we may assume in the proof that $\overline{y} = 0$ and $\overline{A} = 0$. Set $r = \mathrm{sur}\,F(\overline{x}|(0,0))$.

1. The inequality (19.18) automatically holds if $r = 0$, so we assume $r > 0$. Take a positive $\rho < r$. To prove the statement it will be sufficient to show that there is a $\delta > 0$ such that whenever

$$\|x - \overline{x}\| < \delta, \ \|y\| < \delta, \ \|A\| < \delta, \ (y,A) \in \Phi(x), \ t \in (0,\delta), \tag{19.19}$$

we have

$$B\big((y,A), \frac{\rho}{\nu^*(1, \|\overline{x}\|)} t\big) \subset \Phi(B(x,t)). \tag{19.20}$$

The definition of the modulus of surjection along with Milyutin's theorem imply that there is an $\varepsilon > 0$ such that for x, y, t, and A satisfying (19.19) with δ replaced by ε the inclusion

$$B(y,\rho t) \subset (F+A)(B(x,t)) \tag{19.21}$$

holds. Take a $\delta > 0$ satisfying

$$\delta < \frac{\varepsilon}{2}, \quad \delta(1+\delta+\|\overline{x}\|) < \varepsilon.$$

Let x, y, t, and A satisfy (19.19) and

$$(y',A') \in B\big((y,A), \frac{\rho}{\nu^*(1,\|\overline{x}\|)}t\big). \tag{19.22}$$

We have $y \in F(x)+Ax = F(x)+A'(x)+(A-A')x$, that is,

$$y-(A-A')x \in F(x)+A'x \tag{19.23}$$

and

$$\|y-(A-A')x\| \le \|y\|+\|A-A'\|\|x\| \le \delta+\delta(\|\overline{x}\|+\delta) < \varepsilon.$$

On the other hand, by (19.22)

$$\begin{aligned}\|y'-(y-(A-A')x)\| &\le \|y'-y\|+\|A'-A\|\|x\| \\ &\le \nu^*(1,\|x\|)\|(y'-y,A'-A)\| \le \rho t.\end{aligned}$$

By (19.21) and (19.23) there must be a u such that $\|u-x\| < t$ and $y' \in F(u)+A'u$ which means that $(y',A') \in \Phi(u)$ and (19.20) follows. This completes the proof of (19.18).

2. To prove the equality in case of a convex F which is perfectly regular at $(\overline{x},0)$, note first that for any $\Psi : X \rightrightarrows Y$ with $(\bar{x},\bar{y}) \in \operatorname{Graph}\Psi$

$$\operatorname{sur}\Psi(\overline{x}|\overline{y}) \le \sup\{r \ge 0 : \ B(\overline{y},tr) \subset \Psi(B(\overline{x},t))\} \tag{19.24}$$

for all sufficiently small $t \ge 0$. This is immediate from the definition.

Consider the operator $\Lambda : Y \times \mathscr{L}(X,Y) \to Y$ defined by $\Lambda(y,A) = y - A\overline{x}$. We claim that

$$\operatorname{ex}\big((\Lambda(\Phi(x) \cap tB_{Y\times\mathscr{L}(X,Y)}), F(x)\big) \le ct\|x-\overline{x}\|. \tag{19.25}$$

Here $\operatorname{ex}(Q,P)$ stands for the excess of Q over P:

$$\operatorname{ex}(Q,P) = \sup_{u\in Q} d(u,P).$$

Indeed, let $(y,A) \in \Phi(x)$ and $\|(y,A)\| \le t$. Then $y - Ax \in F(x)$, that is, $\Lambda(y,A) \in F(x) + \|A\|\,\|x-\bar{x}\|B_Y$. This means that $d(\Lambda(y,A),F(x)) \le \|A\|\,\|x-\bar{x}\|$ and (19.25) follows.

We observe furthermore that $F(x) \subset DF(\bar{x},0)(x-\bar{x})$ (as the graph of F is convex), so (19.25) implies that

$$\mathrm{ex}\big(\Lambda(\Phi(x) \cap tB_{Y\times\mathscr{L}(X,Y)}), (DF(\bar{x},0)(x-\bar{x}))\big) \le ct\|x-\bar{x}\|. \tag{19.26}$$

This along with (19.24) (applied to $\Psi = \Lambda \circ \Phi$) and (19.15) implies that

$$\mathrm{sur}\,(\Lambda \circ \Phi)(\bar{x},(0,0)) \le \mathrm{sur}\,DF(\bar{x},0). \tag{19.27}$$

The inequality

$$\mathrm{sur}\,(\Lambda \circ \Phi)(\bar{x}|0) \ge \mathrm{sur}\,\Lambda(0,0) \cdot \mathrm{sur}\,\Phi(\bar{x}|(0,0)) \tag{19.28}$$

is straightforward. Finally, as Λ is a linear bounded operator, its surjection modulus is the same at any point and

$$\mathrm{sur}\,\Lambda = \inf\{\|\Lambda^* y^*\| : \|y^*\| = 1\}. \tag{19.29}$$

We have

$$\langle y^*, y - A\bar{x}\rangle = \langle y^*, y\rangle - \langle y^* \otimes \bar{x}, A\rangle,$$

so $\Lambda^*(y^*) = (y^*, -y^* \otimes \bar{x})$ and

$$\|\Lambda^* y^*\| = \sup\{\langle y^*, y\rangle - \langle y^* \otimes \bar{x}, A\rangle : \nu(\|y\|,\|A\|) \le 1\} = \nu^*(\|y^*\|,\|y^* \otimes \bar{x}\|)$$

and therefore (as $\|y^* \otimes \bar{x}\| = \|y^*\|\|\bar{x}\|$) $\mathrm{sur}\,\Lambda = \nu^*(1,\|\bar{x}\|)$. Combining this with (19.27), (19.28), and (19.29) and taking into account that F is perfectly regular at $(\bar{x},0)$, we complete the proof. ∎

19.4.5 Comments

For original proofs of the Robinson-Ursescu theorem see [36, 42]. In both publications this was, as we have mentioned, a purely qualitative result. But the first estimate (an upper estimate for the modulus of regularity) for convex set-valued mappings was obtained by Robinson even earlier in [35] for so-called linear constraint systems using the Hörmander homogenization transform which with any convex set $Q \subset X$ associates a cone in $X \times \mathbb{R}$ generated by the set $Q \times \{1\}$. Robinson worked with the norm $\max\{\|x\|,|t|\}$ in $X \times \mathbb{R}$. In [25] we showed that the lower bound (over ε) of Robinson-type formulas corresponding to the norms $\max\{\|x\|,\varepsilon|t|\}$ in $X \times \mathbb{R}$ gives the exact value of the modulus of metric regularity. It was actually the first corollary of Theorem 19.16 also proved in [25]. Here the proof of the theorem has been substantially simplified. Systems of convex and linear inequalities have been thoroughly studied by Canovas et al. in [4–8]. My interest to the subject has been stimulated by some of their works. Theorem 19.17 was proved originally in [23]. The concept of perfect regularity was introduced in [25] Theorem 19.26. Its predecessor for arbitrary maps can be found in [23], but there it was assumed that $F+A$ is perfectly regular at $(\overline{x},\overline{y}+A\overline{x})$ for any A. Theorem 19.28 is also a new result as has been mentioned in the introduction. An earlier result of such sort was established in a recent paper [6] for systems of convex inequalities in $\mathbb{R}^n$ in which every inequality was independently perturbed by linear functions. Observe that the set-valued mappings associated with systems of convex inequalities are perfectly regular at all points of their graphs [23].

19.5 First-Order Necessary Conditions in Mathematical Programming

We start with the principal lemma.

Lemma 19.29 (Lemma on convex majorant). *Let f be a function on a Banach space X which is Fréchet differentiable at a certain $\overline{x}$. Then there is an $\varepsilon>0$ and a convex continuous function on X with the following properties:*

(a) $\varphi(\overline{x})=f(\overline{x})$ *and* $\varphi(x)\geq f(x)$ *if* $\|x-\overline{x}\|<\varepsilon$.
(b) φ *is strictly differentiable at* $\overline{x}$ *and* $\varphi'(\overline{x})=f'(\overline{x})$.
(c) $0\leq\varphi(x)-\varphi(\overline{x})-\langle\varphi'(\overline{x}),x-\overline{x}\rangle\leq 2\|x-\overline{x}\|$ *for all* x.

Proof. Let $x^*=f'(\overline{x})$. Then $|f(x)-f(\overline{x})-\langle x^*,x-\overline{x}\rangle|\leq r(\|x-\overline{x}\|)$, where $r(\lambda)\geq 0$ and $\lambda^{-1}r(\lambda)\to 0$ as $\lambda\to 0$. Without loss of generality we may assume that r is non-decreasing. As $r(\lambda=o(\lambda)$, there is an $\varepsilon>0$ such that $r(\lambda)\leq\lambda$ if $\lambda\leq\varepsilon$. Set

$$g(\xi)=\sup_{0<\lambda\leq\varepsilon} r(\lambda)\frac{2\xi-\lambda}{\lambda}.$$

Then g is a convex l.s.c. function on $\mathbb{R}$ as the upper envelop of a family of affine functions. We have $g(\xi)\leq 2\xi$ for all $\xi\geq 0$. Furthermore, $g(\xi)=0$ for $\xi\leq 0$ and $g(\xi)>0$ if $\xi>0$. Thus g is convex continuous on $[0,\infty]$. It is also clear that g is strictly increasing on $[0,\infty)$.

We notice next that for any ξ the function under the sign of supremum is nonnegative if and only if $\lambda\leq 2\xi$, so for any ξ we can take supremum over $(0,2\xi]$. It follows that

$$0\leq g(\xi)\leq\xi\sup_{0<\lambda\leq 2\xi}\frac{r(\lambda)}{\lambda}=o(\xi),\quad \text{as } \xi\to 0.$$

Define φ by

$$\varphi(x)=f(\overline{x})+\langle x^*,x-\overline{x}\rangle+g(\|x-\overline{x}\|).$$

Then (a) and (c) are obvious as well as differentiability of φ at $\bar{x}$ and the equality of derivatives of φ and f at $\bar{x}$. We have further for $\xi \geq \xi'$

$$g(\xi) - g(\xi') \leq \sup_{0<\lambda\leq 2\xi} \left[r(\lambda)\frac{2\xi - \lambda}{\lambda} - r(\lambda)\frac{2\xi' - \lambda}{\lambda} \right] \leq 2 \sup_{0<\lambda\leq \xi} \frac{r(\lambda)}{\lambda}(\xi - \xi')$$

which immediately implies (b) . ∎

Corollary 19.30. *Let T be a set and for any $t \in T$ let f_t be a function on X Fréchet differentiable at $\bar{x}$. Moreover, we assume that f_t are uniformly differentiable in the sense that there is an $r(\cdot)$ common for all f_t. Then there are functions φ_t ($t \in T$) and an $\varepsilon > 0$ such that*

(a) For any t the function φ_t is convex continuous on X, Fréchet differentiable at $\bar{x}$, and such that $\varphi_t(\bar{x}) = f_t(\bar{x})$, $\varphi_t'(\bar{x}) = f_t'(\bar{x})$ and $\varphi_t(x) \geq f_t(x)$ if $\|x - \bar{x}\| < \varepsilon$.
(b) φ_t are uniformly strictly differentiable at $\bar{x}$, that is,

$$\lim_{\delta\to 0} \sup\{(\|x-x'\|^{-1}|\varphi_t(x) - \varphi(x') - \langle\varphi'(\bar{x}), x-x'\rangle| : x,x' \in B(\bar{x},\delta),\ x \neq x'\} = 0.$$

(c) The inequality $0 \leq \varphi_t(x) - \varphi_t(\bar{x}) - \langle\varphi_t'(\bar{x}), x-\bar{x}\rangle \leq 2\|x-\bar{x}\|$ holds for all $x \in X$ and $t \in T$.

Proof. Just set $\varphi_t(x) = f_t(\bar{x}) + \langle f_t'(\bar{x}), x - \bar{x}\rangle + g(x - \bar{x})$ with the same g as in the lemma. ∎

An immediate consequence of the lemma is the practical trivialization of the procedure of developing first-order necessary optimality conditions when the cost function and the inequality constraint functions are Fréchet differentiable at the solution. Indeed, consider the problem:

(P$_1$) minimize $f_0(x)$ s.t. $f_t(x) \leq 0,\ t \in T;\ x \in Q$.

Here T is an arbitrary set, no matter finite or infinite. The nature of the constraint $x \in Q$ is not essential for a time being.

If f_0 and all f_t are Fréchet differentiable at $\bar{x}$, then we denote by φ_0 and φ_t convex functions obtained from f_0 and f_t using Lemma 19.29. Set further

$$\varphi(x) = \sup_{t\in T} \varphi_t(x).$$

Set $T_\varepsilon = \{t \in T :\ \varphi_t(\bar{x}) \geq -\varepsilon\} = \{t \in T :\ f_t(\bar{x}) \geq -\varepsilon\}$. We assume that

(A$_2$) there is an $\varepsilon > 0$ such that the set $\{f_t'(\bar{x}) :\ t \in T_\varepsilon\}$ is norm bounded in X^*.

By Corollary 19.30 φ is a convex continuous function satisfying for some $K > 0$

$$\varphi(x) \leq \varphi(\bar{x}) + K\|x - \bar{x}\|, \quad \forall x \in X. \tag{19.30}$$

For the subdifferential of φ at $\bar{x}$ we have the formula (see, e.g., [30, 43])

$$\partial\varphi(\bar{x}) = \bigcap_{\varepsilon>0} \mathrm{cl}^*\mathrm{conv}\Big(\bigcup_{t\in T_\varepsilon} \partial\varphi_t(\bar{x})\Big) = \bigcap_{\varepsilon>0} \mathrm{cl}^*\mathrm{conv}\Big(\bigcup_{t\in T_\varepsilon} \{f_t'(\bar{x})\}\Big). \tag{19.31}$$

Under additional conditions it is possible to guarantee the equality

$$\partial\varphi(\bar{x}) = \mathrm{cl}^*\mathrm{conv}\Big(\bigcup_{t\in T_0} f_t'(\bar{x})\Big),$$

for instance, if

($\mathbf{QC}_1$) T is a compact Hausdorff space and the function $t \mapsto \varphi_t(x)$ is upper semicontinuous for any x of a neighborhood of $\bar{x}$,

or

($\mathbf{QC}_2$) there is an $\varepsilon > 0$ such that the set $\mathrm{conv}\{(\varphi_t'(\bar{x}), \varphi_t(\bar{x})) : t \in T_\varepsilon\}$ is weak* closed.

Both conditions are well known and verification does not present any difficulty. Moreover in both cases it is possible to give precise representations for elements of $\partial\varphi(\bar{x})$: if ($\mathbf{QC}_1$) holds and X is a separable Banach space, then for any $x^* \in \partial\varphi(\bar{x})$ there is a probability measure μ on T_0 such that

$$x^* = \int_{T_0} \varphi_t'(\bar{x})\mathrm{d}\mu(t) \tag{19.32}$$

(see, e.g., [9, 24]. The same is obviously true if ($\mathbf{QC}_2$) holds, but here in this case we can consider only measures μ supported on finite sets so that the necessary conditions assume the form: for any $x^* \in \partial\varphi(\bar{x})$ there are $t_i \in T_0$, $i = 1,\dots,k$ and positive numbers α_i with $\sum\alpha_i = 1$ such that

$$x^* = \alpha_1\varphi_{t_1}'(\bar{x}) + \cdots + \alpha_k\varphi_{t_k}'(\bar{x}). \tag{19.33}$$

Let us return to the problem ($\mathbf{P}_1$) assuming that $\bar{x}$ is a solution. It is straightforward to see that $\bar{x}$ also solves

($\mathbf{P}_2$) minimize $\varphi_0(x)$ s.t. $\varphi(x) \le 0$, $x \in Q$.

Thus a problem with infinitely many inequality constraints reduces to a simple problem with one inequality constraint and both cost and constraint functions convex continuous. Further analysis depends of course on the structure of the constraint $x \in Q$. The simplest case occurs when Q is defined by the condition $F(x) = 0$ with $F : X \to Y$ strictly differentiable at $\bar{x}$ and the image of $F'(\bar{x})$ (i.e., $F'(\bar{x})(X)$) being a subspace of finite codimension. In this case ($\mathbf{P}_2$) is a standard problem for which the Lagrange multiplier rule holds: there are $\lambda_0 \ge 0$, $\lambda \ge 0$, and a $y^* \in Y^*$, not all equal to zero and such that

$$0 \in \lambda_0\varphi_0'(\bar{x}) + \lambda\partial\varphi(\bar{x}) + (F'(\bar{x}))^*y^*. \tag{19.34}$$

The standard constrained qualification condition. $F'(\overline{x})$ is surjective (that is to say, $F'(\overline{x})(X) = Y$) and there is a $h \in X$ such that ($F'(\overline{x})h = 0$ and $\varphi(\overline{x}) + h < 0$) guarantees that $\lambda_0 > 0$. In view of (19.31) the "Slater" part of this condition simply means that there are $\varepsilon > 0$ and $\delta > 0$ such that $\varphi_t(\overline{x}+h) \le -\delta$ for all $t \in T_\varepsilon$. We can further specify the necessary condition (19.34) under ($\mathbf{QC}_1$), ($\mathbf{QC}_2$), or alike. There is also no problem to express all these conclusions in terms of the original problem ($\mathbf{P}_1$). The standard constraint qualification condition gives now precisely the "perturbed Mangasarian-Fromovitz qualification condition"(PMFQC) of [32].

The conclusions of the last paragraph contain (with the exception of one theorem) all main results of [32].[4] But we can make a step further and easily get necessary conditions for more general types of the constraint $x \in Q$. The following proposition is straightforward.

Proposition 19.31. *Let f_0 and all f_t, $t \in T$ be Fréchet differentiable at $\overline{x}$. We assume that $f_0(\overline{x}) = 0$. Set*

$$\psi(x) = \max\{\varphi_0(x), \varphi(x)\}.$$

If $\overline{x}$ is a local solution of ($\mathbf{P}_1$), then it is a local solution of the problem

($\mathbf{P}_3$) *minimize $\varphi(x)$ s.t. $x \in Q$.*

In particular if ($\mathbf{A}_2$) holds then there is an $N > 0$ such that $\overline{x}$ is an unconditional local minimizer of $\psi(x) + N d(x, Q)$.

For the first statement see, e.g., [19] and for the second, e.g., Proposition 2.4.3 in [10]. Recall that φ is Lipschitz if $\{f_t'(\overline{x}),\ t \in T\}$ is a bounded set .

If the constrained $x \in Q$ is not convex, nonconvex subdifferentials of one or another sort become in principle necessary for further analysis. However reasonable optimality conditions can be obtained only under assumption that the behavior of Q near $\overline{x}$ is sufficiently regular which considerably simplifies the situation.

Proposition 19.32. *We assume that*

(a) ($\mathbf{A}_2$) holds
(b) Q is Clarke regular at $\overline{x}$

If $\overline{x}$ is a solution of ($\mathbf{P}_1$) then there is a $\lambda \in [0, 1]$ such that

$$0 \in \lambda \partial \varphi_0(\overline{x}) + (1-\lambda)\varphi(\overline{x}) + N_{DH}(Q, \overline{x}). \tag{19.35}$$

If moreover the Slater-type qualification condition

($\mathbf{QC}_3$) *there is an $h \in T_B(Q, \overline{x})$ such that $\psi(\overline{x}+h) < 0$,*

is satisfied, then $\lambda > 0$.

[4]The exception is Theorem 5.4 in which the cost function is assumed just lower semicontinuous.But in this case the problem of minimizing $f(x)$ s.t. $\varphi(x) \le 0$ and $F(x) = 0$ with convex continuous φ and strictly differentiable F is standard for nonsmooth mathematical programming and can be easily treated by already standard techniques using either the limiting Fréchet subdifferential (if X is an Asplund space) or the approximate G-subdifferential in the general case.

Proof. The first statement follows from Proposition 19.31 and standard calculus rules for Clarke's generalized gradient, as ψ is globally Lipschitz by (a). The second statement follows from (b) and, again, continuity of ψ as (19.35) cannot hold in this case with $\lambda = 0$. ■

The specification of the result for one or another structure of the constraint $x \in Q$ also does not present much difficulty. Let for instance $Q = \{x :\ 0 \in F(x)\}$, where $F : X \rightrightarrows Y$.

Proposition 19.33. *Let $F : X \rightrightarrows Y$, and let $(\bar{x}, \bar{y}) \in \operatorname{Graph} F$. Assume that F is metrically regular at $(\bar{x}, 0)$. If under this condition φ is a function on X which is Lipschitz near $\bar{x}$ and which attains a local minimum at $\bar{x}$ subject to $0 \in F(x)$, then there is a $K > 0$ such that the function $g(x,y) = \varphi(x) + K\|y\|$ attains a local minimum at $(\bar{x}, 0)$ subject to $(x,y) \in \operatorname{Graph} F$.*

Proof. Let ℓ be the Lipschitz constant of φ in a neighborhood of $\bar{x}$. Take $K > \ell \operatorname{reg} F(\bar{x}|0)$. By the equivalence theorem, for any $(x,y) \in \operatorname{Graph} F$ sufficiently close to $(\bar{x}, 0)$ there is a u such that $0 \in F(u)$ and $\ell\|x - u\| \le K\|y\|$. For such (x,y) and u, we have

$$\varphi(x) + K\|y\| \ge \varphi(x) + \ell\|x - u\| \ge \varphi(u) \ge \varphi(\bar{x})$$

as claimed. ■

Proposition 19.34. *Let F be metrically subregular at $(\bar{x}, 0) \in \operatorname{Graph} F$, that is (see, e.g., [16]),*

$$\mathrm{d}(x, F^{-1}(0)) \le K \mathrm{d}(0, F(x))$$

for all x of a neighborhood of $\bar{x}$. Then

$$T_B(Q, \bar{x}) = Pr_X\{h :\ (h, 0) \in T_B(\operatorname{Graph} F, (\bar{x}, 0))\}.$$

Proof. If $h \in T_B(Q, \bar{x})$, then $0 \in F(\bar{x} + t_n h_n)$ for some $t_n \to +0$, $h_n \to h$, that is, $(\bar{x}, 0) + t_n(h_n, 0) \in \operatorname{Graph} F$; hence $(h, 0) \in T_B(\operatorname{Graph} F, (\bar{x}, 0))$. Conversely, if the last inclusion holds then there are $t_n \to 0$ $h_n \to h$ and $v_n \to 0$ such that $t_n v_n \in F(\bar{x} + t_n h_n)$. By subregularity, $\mathrm{d}(\bar{x} + t_n h_n, F^{-1}(0)) \le t_n K\|v_n\|$ which means that there are $u_n \in X$ with $\|u_n\| \le 2K\|v_n\|$ such that $0 \in F(\bar{x} + t_n(h_n + u_n))$ but $h_n + u_n \to h$, whence $h \in T_B(Q, \bar{x})$. ■

Combining the last three propositions and taking into the account that metric regularity implies subregularity, we get

Proposition 19.35. *Consider ($\mathbf{P}_1$) with $Q = F^{-1}(0)$, where $F : X \rightrightarrows Y$ is a set-valued mapping with closed graph. Assume that*

(a) ($\mathbf{A}_2$) holds
(b) F is metrically regular at $(\bar{x}, 0)$
(b) $\operatorname{Graph} F$ is Clarke regular at $(\bar{x}, 0)$

If $\overline{x}$ is a solution of ($\mathbf{P}_1$) *then there are* $\lambda \in [0,1]$ *and* y^* *such that*

$$0 \in \lambda \varphi'(\overline{x}) + (1-\lambda)\partial\varphi(\overline{x}) + D^*_{DH}F(\overline{x},0)(y^*).$$

If moreover the Slater-type qualification condition

($\mathbf{QC}_3$) $\exists\, h \in X$ *such that* $(h,0) \in T_B(\operatorname{Graph} F, (\overline{x},0))$ *and* $\varphi(\overline{x}+h) < 0$,

is satisfied, then $\lambda > 0$.

Proof. The first statement is a consequence of the first part of Proposition 19.32 and 19.33, and the second is the consequence of the second part of Propositions 19.32 and 19.34. ∎

As above we can make the necessary condition more specific using either (19.32) or (19.33) under ($\mathbf{QC}_1$) or ($\mathbf{QC}_2$). Thus we get finally

Proposition 19.36. *We posit the assumptions of Proposition 19.35.*

(a) *If* ($\mathbf{QC}_1$) *holds with X being a separable Banach space, then there are* $\lambda \in [0,1]$ *and a probability measure* μ *supported on* T_0 *such that*

$$0 \in \lambda \varphi'(\overline{x}) + (1-\lambda)\int_T f_t'(\overline{x})\mathrm{d}\mu(t) + D^*_{DH}F(\overline{x},0)(y^*).$$

(b) *If* ($\mathbf{QC}_2$) *holds then there are* $t_i \in T_0$, $i = 1,\dots,k$ *and nonnegative* $\lambda, \lambda_1, \dots, \lambda_k$ *such that*

$$0 \in \lambda f_t'(\overline{x}) + \sum_{i=1}^{k} \lambda_i f_{t_i}'(\overline{x}) + D^*_{DH}F(\overline{x},0)(y^*).$$

In either case the condition

$\exists\, h \in X$ *such that* $\langle f_t'(\overline{x}), h\rangle < 0$ *for all* $t \in T$, $(h,0) \in T_B(\operatorname{Graph} F, (\overline{x},0))$, *implies that necessarily* $\lambda > 0$.

19.5.1 Comments

As have been mentioned in Introduction, this section has been written following the discussion with B. Mordukhovich at J. Borwein's 60th anniversary conference in Vancouver in May 2011.[5] My point was that generalized subdifferentiation is not an adequate tool to treat semi-infinite programming problems with differentiable data and the aim of the first part of the section was to demonstrate that convex analysis

[5] Video of B. Mordukhovich's talk and the subsequent discussion is available in the Internet: http://conferences.irmacs.sfu.ca/jonfest2011/talk/55,

offers a much more viable alternative. Another consequence of the discussion of this section is that, as far as first-order optimality conditions are concerned, semi-infinite programming with differentiable or convex inequalities and cost functions is not a particularly meaningful object to study: all results can be obtained from the standard results for problems with finitely many inequality constraints and convex subdifferential calculus. On the other hand, semi-infinite programming with non-differentiable inequality constraint functions remains rather Terra incognita and I am not aware of any results relating to the first-order necessary conditions for such problems. And of course for other problems, e.g., stability of solutions (or even feasible sets), the infinite number of constraints is an additional and serious challenge (see, e.g., [4–8, 23, 41]).

Acknowledgments I am thankful to V. Roschina for showing me a mistake in the original proof of Theorem 19.6.

References

1. Azé, D., Corvellec, J.-N.: On the sensitivity analysis of Hoffmann's constant for systems of linear inequalities. SIAM J. Optim. **12**, 913–927 (2002)
2. Azé, D., Corvellec, J.-N.: Characterization of error bounds for lower semicontinuous functions on metric spaces. ESAIM Control Optim. Calc. Var. **10**, 409–425 (2004)
3. Borwein, J.M., Zhu, J.: Techniques of Variational Analysis. CMS Books in Mathematics, vol. 20, Springer, Berlin (2006)
4. Cánovas, M.J., Gómez-Senent, F.J., Parra, J.: Stability of systems of convex equations and inequalities: distance to ill-posedness and metric regularity. Optimization **56**, 1–24 (2007)
5. Cánovas, M.J., Gómez-Senent, F.J., Parra, J.: Regularity modulus of arbitrarily perturbed linear inequality systems. J. Math. Anal. Appl. **343**, 315–327 (2008)
6. Cánovas, M.J., Gómez-Senent, F.J., Parra, J.: Linear regularity, equi-regularity and intersection mappings for convex semi-infinite inequality systems. Math. Program. Ser. B **123**, 33–60 (2010)
7. Cánovas, M.J., Klatte, D., Lopez, M.A., Parra, J.: Metric regularity of convex semi-infinite programming problem under convex perturbations. SIAM J. Optim. **18**, 717–732 (2007)
8. Cánovas, M.J., Lopez, M.A., Mordukhovich, B.S., Parra, J.: Variational analysis in linear semi-infinite and infinite programming, I: Stability of linear inequality system of feasible solutions. SIAM J. Optim. **20**, 1504–1526 (2009)
9. Castaing, C., Valadier, M.: Convex Analysis and measurable multifunctions. Lecture Notes in Mathematics, vol. 580, Springer, Berlin (1977)
10. Clarke, F.H.: Optimization and Nonsmooth Analysis. Wiley-Interscience, New York (1983)
11. Cornejo, O., Jourani, A., Zalinescu, C.: Conditioning and upper Lipschitz inverse subdifferential in nonsmooth optimization problems. J. Optim. Theory Appl. **95**, 127–148 (1997)
12. Coulibali, A., Crouzeix, J.-P.: Conditions numbers and error bounds in convex programming. Math. Program. **116**, Ser. B, 79–113 (2009)
13. De Giorgi, E., Marino, A., Tosques, M.: Problemi di evoluzione in spazi metrici e curve di massima pendenza. Atti Acad. Nat. Lincei, Rend. Cl. Sci. Fiz. Mat. Natur. **68**, 180–187 (1980)
14. Dmitruk, A.V., Milyutin, A.A., Osmolovskii, N.P.: Lyusternik theorem and the theory of extrema. Russian Math. Surveys **35**(6), 11–51 (1980)
15. Dontchev, A.L., Lewis, A.S., Rockafellar, R.T.: The radius of metric regularity. Trans. Amer. Math. Soc. **355**, 493–517 (2003)

16. Dontchev, A.L., Rockafellar, R.T.: Implicit Function and Solution Mapping. Springer, New York (2009)
17. Facchinei, F., Pang, J.S.: Finite-Dimensional Variational Inequalities and Complementarity Problems. Springer, New York (2003)
18. Hoffman, A.J.: On approximate solutions of systems of linear inequalities. J. Research Nat. Bur. Standards **49**, 263–265 (1952)
19. Ioffe, A.D.: Necessary and sufficient conditions for a local minimum, parts 1–3. SIAM J. Control Optim. **17**, 245–288 (1979)
20. Ioffe, A.D.: Metric regularity and subdifferential calculus. Uspehi Mat. Nauk **55**(3), 103–162 (2000) (in Russian). English translation: Russian Math. Surveys **55**(3), 501–558 (2000)
21. Ioffe, A.D.: On stability estimates for the regularity property of maps. In: Brezis, H., Chang, K.C., Li, S.J., Rabinowitz, P. (eds.) Topological Methods, Variational Methods and Their Applications, pp. 133–142. World Scientific, Singapore (2003)
22. Ioffe, A.D.: Regularity on fixed sets. SIAM J. Optim. **21**, 1345–1370 (2011)
23. Ioffe, A.D.: On stability of solutions to systems of convex inequalities. J. Convex Anal. **19**, 955–973 (2012)
24. Ioffe, A.D., Levin, V.L.: Subdifferentials of convex functions. Trans. Moscow Math. Soc. **26**, 1–72 (1972)
25. Ioffe, A.D., Sekiguchi, Y.: Regularity estimates for convex multifunctions. Math. Program. Ser. B **117**, 255–270 (2009)
26. Kruger, A., Ngai, N.V., Thera, M.: Stability of error bounds for convex constrained systems in Banach spaces. SIAM J. Optim. **20**, 3280–3296 (2010)
27. Levitin, E.S., Milyutin, A.A., Osmolovskii, N.P.: On conditions for a local minimum in a problemswith constraints. In: Mityagin, B.S.(ed.) Mathematical Economics and Functional Analysis. Nauka, Moscow (1974) (in Russian)
28. Lewis, A.S.: Ill-conditioned convex processes and conic linear systems. Math. Oper. Res. **24**, 829–834 (1999)
29. Lewis, A.S., Pang, J.S.: Error bounds for convex inequality systems. In: Crouzeix, J.P., Martinez-Legas, J.E., Volle, M. (eds.) Generalized Convexity, Generalized Monotonicity: Recent Results, pp. 75–110. Kluwer, Dordrecht (1998)
30. Lopez, M.A., Volle, M.: A formula for the set of optimal solutions of a relaxed minimization problem. Applications to subdifferential calculus. J. Convex Anal. **17**, 1057–1075 (2010)
31. Mangasarian, O.L.: A condition number for differentiable convex inequalities. Math. Oper. Res. **10**, 175–179 (1985)
32. Mordukhovich, B.S., Nghia, T.T.A.: Constraint qualification and optimality conditions for nonconvex semi-infinite and infinite programs. Math. Program **139**, 271–300 (2013)
33. Morse, M.: Variational Analysis: Critical Extremals and Sturmians Extensions. Wiley, New York (1973)
34. Ng, K.F., Yang, W.H.: Error bounds for some convex functions and distance composite functions. SIAM J. Optim. **15**, 1042–1056 (2005)
35. Robinson, S.M.: Stability theory for system of inequalities. Part 1: Linear systems. SIAM J. Numer. Anal. **12**, 754–769 (1975)
36. Robinson, S.M.: Regularity and stability for convex multivalued functions. Math. Oper. Res. **1**, 130–143 (1976)
37. Rockafellar, R.T., Wets, R.J.B.: Variational Analysis. Springer, New York (1998)
38. Sion, M.: On general minimax theorem. Pacific J. Math. **8**, 171–176 (1958)
39. Song, W.: Calmness and error bounds for convex constraint systems. SIAM J. Optim. **17**, 353–371 (2006)
40. Song, W.: Error bounds for convex constrained systems in Banach spaces. Control Cybernet. **36**, 775–792 (2007)
41. Stein, O.: Bi-level Strategies in Semi-Infinite Optimization. Kluwer, Boston (2003)
42. Ursescu, C.: Multifunctions with convex closed graphs. Czechoslovak Math. J. **25**, 438–411 (1975)

43. Volle, M.: Sous-différential d'une enveloppe supérieure de fonctions convexes. C.R. Acad. Sci. Paris , Sér. I **317** 845–849 (1993)
44. Wu, Z., Ye, J.: Sufficient conditions for error bounds. SIAM J. Optim. **12**, 421–435 (2001)
45. Zalinescu, C.: Weak sharp minima, well-behaving functions and global error bounds for convex inequalities in Banach spaces. In: Bulatov, V., Baturin, V. (eds.) Proceedings of the 12th Baikal International Conference on Optimization Methods and Their Applications pp. 272–284. Institute for System Dynamics Control Theory of SB RAS, Irkutsk (2001)
46. Zalinescu, C.: Convex Analysis in General Vector Spaces. World Scientific, New Jersey (2002)
47. Zhang, S.: Global error bounds for convex conic problems. SIAM J. Optim. **10**, 836–851 (2000)
48. Zheng, X.Y., Ng, K.F.: Metric regularity and constraint qualifications for convex inequalities in Banach spaces. SIAM J. Optim. **14**, 757–772 (2004)

Chapter 20
Generic Existence of Solutions and Generic Well-Posedness of Optimization Problems

P.S. Kenderov and J.P. Revalski

Dedicated to Jonathan Borwein on the occasion of his 60th birthday

Abstract We exhibit a large class of topological spaces in which the generic attainability of the infimum by the bounded continuous perturbations of a lower semicontinuous function implies generic well-posedness of the perturbed optimization problems. The class consists of spaces which admit a winning strategy for one of the players in a certain topological game and contains, in particular, all metrizable spaces and all spaces that are homeomorphic to a Borel subset of a compact space.

Key words: Baire category • Generic well-posedness of perturbation problems • Perturbed optimization problem • Variational principle • Well-posed problem

Mathematics Subject Classifications (2010): Primary: 49K27; Secondary: 49K40, 90C48

20.1 Introduction

Let X be a completely regular topological space and $f : X \to \mathbb{R} \cup \{+\infty\}$ be a fixed bounded from below lower semicontinuous function which is proper (the latter means that f has at least one finite value). We say that f *attains its infimum in* X, if there exists some $x \in X$ for which $f(x) = \inf_X f$. Denote by $C(X)$ the space

---

COMMUNICATED BY JON D. VANDERWERFF.

P.S. Kenderov • J.P. Revalski (✉)
Institute of Mathematics and Informatics, Bulgarian Academy of Sciences, Acad. G. Bonchev Street, block 8, 1113 Sofia, Bulgaria
e-mail: kenderovp@cc.bas.bg; revalski@math.bas.bg

D.H. Bailey et al. (eds.), *Computational and Analytical Mathematics*, Springer Proceedings in Mathematics & Statistics 50, DOI 10.1007/978-1-4614-7621-4_20,

of all bounded real-valued and continuous functions in X which we equip with the usual sup-norm $\|g\|_\infty = \sup\{|g(x)| : x \in X\}$, $g \in C(X)$. It has been shown in [17] that the set $E(f) = \{g \in C(X) : f+g$ attains its infimum in $X\}$ is dense in $C(X)$. We call the statement "$E(f)$ is dense in $C(X)$" a *variational principle for f with $C(X)$ as a set of perturbations*. The variational principle is called *generic* if the set $E(f)$ is residual in $C(X)$. Recall that $E(f)$ is *residual in* $C(X)$ if its complement is of the first Baire category in $C(X)$. Such a (or similar) setting, with different sets of perturbations, is present in several well-known variational principles–see, e.g., Ekeland [9], Stegall [22], Borwein and Preiss [3] and Deville, Godefroy, and Zizler [7,8] for the case of metric spaces X and [4,5] outside the case of metrizable spaces.

Our aim in this paper is to show that, for a very large class of spaces X, the residuality of $E(f)$ in $C(X)$ implies the residuality in the same space of the set $W(f) := \{g \in C(X) : f+g$ is well posed$\}$. Let us recall that a bounded from below function $h : X \to \mathbb{R} \cup \{+\infty\}$ (or more precisely, the problem to minimize h on X) is called *well posed* if every minimizing net $(x_\lambda)_\lambda \subset X$ for h has a cluster point. If h is lower semicontinuous and is well posed, then the set $M(h)$ of minimizers of h in X is a nonempty compact set in X and for every open $U \supset M(h)$ there exists $\varepsilon > 0$ for which $\{x \in X : h(x) < \inf_X h + \varepsilon\} \subset U$.

The spaces X for which we prove here that residuality of $E(f)$ (in $C(X)$) implies residuality of $W(f)$ are described by a topological game called *a determination game* and denoted by $DG(X)$. The reasons for this terminology will become clear later. Two players, Σ (who starts the game) and Ω, play by choosing at each step $n \geq 1$ nonempty sets A_n (the choices of Σ) and B_n (the choices of Ω) so that B_n is relatively open in A_n and $A_{n+1} \subset B_n \subset A_n$ for any n. Playing this way the players generate a sequence $p = \{A_n, B_n\}_{n\geq 1}$ which is called *a play*. The player Ω wins the play p if the intersection $\cap_n \overline{A}_n = \cap_n \overline{B}_n$ is either empty or a nonempty compact set such that, for each open set U containing $\cap_n \overline{B}_n$, there is some n with $B_n \subset U$. Otherwise, by definition, player Σ is declared to have won the play p. A *partial play* in the game $DG(X)$ is any finite sequence of the type $(A_1, B_1, \ldots, A_n)$ or $(A_1, B_1, \ldots, A_n, B_n)$, $n \geq 1$, where for $i = 1, \ldots, n$, the sets A_i and B_i are moves in $DG(X)$ of Σ and Ω correspondingly. A *strategy* ω for the player Ω is defined recursively and is a rule which to any possible partial play of the type $(A_1, \ldots, A_n)$, $n \geq 1$, puts into correspondence a nonempty set $B_n := \omega(A_1, \ldots, A_n) \subset A_n$ which is relatively open in A_n. If in a given play $\{A_n, B_n\}_{n\geq 1}$ of the game $DG(X)$ each choice B_n of Ω is obtained via the strategy ω, that is, $B_n = \omega(A_1, \ldots, A_n)$ for every $n \geq 1$, then this play p is called *an ω-play*. The strategy ω for the player Ω is called *winning* if the player Ω wins every ω-play in this game. The notions of strategy and winning strategy for the player Σ are introduced in a similar way. The term *the game is favorable (resp. unfavorable)* for some player means that the corresponding player has (resp. does not have) a winning strategy in the game.

In Theorem 20.3 we prove that if the player Ω has a winning strategy in the game $DG(X)$ and if for some proper bounded from below lower semicontinuous function f the set $E(f)$ is residual in $C(X)$, then the set $W(f)$ is also residual in $C(X)$. In other words, generic attainability of the infimum by the perturbations implies generic well-posedness of the perturbations. Let us mention that the class of spaces X for which Ω has a winning strategy for the game $DG(X)$ is quite large: it contains all metrizable spaces, all Borel subsets of compact spaces, a large class of fragmentable spaces, etc. (see the Concluding Remarks for more information about this class). There are spaces X however for which the phenomenon does not hold. In Example 20.4 we give a space X and a function f such that $E(f) = C(X)$ and $W(f) = \emptyset$.

The game $DG(X)$ has been used in [11] in order to give sufficient conditions when a semitopological group is, in fact, a topological group and in [12] to study the points of continuity of the so-called quasi-continuous mappings. Variants of $DG(X)$ have been used by Michael [19] (for the study of completeness properties of metric spaces), by Kenderov and Moors [13–15] (for characterization of the fragmentability of topological spaces), and by the authors in [6, 16, 17] (for proving the validity of generic variational principles).

20.2 Preliminary Results and Notions

Let X be a completely regular topological space and consider, as above, the Banach space $C(X)$ of all continuous and bounded functions in X equipped with its sup-norm. For a given function $f : X \to \mathbb{R} \cup \{+\infty\}$, the symbol $\mathrm{dom}(f)$ denotes the *effective domain* of f, which is the set of points $x \in X$ for which $f(x) \in \mathbb{R}$. For our further considerations we need the following statement:

Proposition 20.1 ([17], Lemma 2.1). *Let $f : X \to \mathbb{R} \cup \{+\infty\}$ be a lower semicontinuous function which is bounded from below and proper. Let $x_0 \in \mathrm{dom}(f)$ and $\varepsilon > 0$ be such that $f(x_0) < \inf_X f + \varepsilon$. Then, there exists a continuous function $g : X \to \mathbb{R}^+$ for which $\|g\|_\infty \leq \varepsilon$ and the function $f + g$ attains its infimum at x_0.*

In particular, this proposition shows that the set $E(f) = \{g \in C(X) : f + g$ attains its infimum in $X\}$ is dense in $C(X)$.

Further, any proper function $f : X \to \mathbb{R} \cup \{+\infty\}$ which is bounded from below defines a set-valued mapping $M_f : C(X) \rightrightarrows X$ as follows:

$$M_f(g) := \{x \in X : (f+g)(x) = \inf_X (f+g)\}, \quad g \in C(X),$$

which to each $g \in C(X)$ puts into correspondence the (possibly empty) set of minimizers in X of the perturbation $f + g$. It is known as the *solution mapping* determined by f.

We denote by $\mathrm{Gr}(M_f)$ the graph of M_f and by $\mathrm{Dom}(M_f)$ the set $\{g \in C(X) : M_f(g) \neq \emptyset\}$ which is called *effective domain of M_f*. The following properties are well known in the case when $f \equiv 0$. For an arbitrary proper bounded from below and lower semicontinuous f the proof of these properties is given in [6]. Recall that, for a set $A \subset X$, the symbol $\overline{A}$ denotes the closure of A in X.

Proposition 20.2 ([6], Proposition 2.4). *Let X be a completely regular topological space and let $f : X \to \mathbb{R} \cup \{+\infty\}$ be a proper bounded from below lower semicontinuous function. Then the solution mapping $M_f : C(X) \rightrightarrows X$ satisfies the following properties:*

(a) $\mathrm{Gr}(M_f)$ *is closed in the product topology in $C(X) \times X$.*
(b) $\mathrm{Dom}\,(M_f)$ *is dense in $C(X)$.*
(c) *M_f maps $C(X)$ onto* $\mathrm{dom}\,(f)$.
(d) *For any two open sets U of $C(X)$ and W of X such that $M_f(U) \cap W \neq \emptyset$ there is a nonempty open set $V \subset U$ such that $M_f(V) \subset W$.*
(e) *If $(V_n)_{n \geq 1}$ is a base of neighborhoods of $g_0 \in C(X)$ then $M_f(g_0) = \cap_n \overline{M(V_n)}$.*

The tool we use to show that a certain set is residual in a topological space is the well-known Banach-Mazur game. Given a topological space X and a set $S \subset X$, two players, denoted by α and β, play a game by choosing alternatively nonempty open sets U_n (the choices of β who starts the game) and V_n (the choices of α), $n \geq 1$, with the rule $U_{n+1} \subset V_n \subset U_n$. The player α wins the play $\{U_n, V_n\}_{n \geq 1}$ if $\cap_n U_n = \cap_n V_n \subset S$. Otherwise, β wins. The game is known as the Banach-Mazur game and is denoted by $BM(X,S)$. The notions of (winning) strategies for the players are defined as in the game $DG(X)$. It was proved by Oxtoby [20] that the player α has a winning strategy in $BM(X,S)$ if and only if the set S is residual in X.

20.3 Generic Well-Posedness of Perturbed Optimization Problems

In this section we formulate and prove our main result. Namely, we have the following

Theorem 20.3. *Let X be a completely regular topological space which admits a winning strategy for the player Ω in the determination game $DG(X)$. Suppose that for some proper bounded from below lower semicontinuous function $f : X \to \mathbb{R} \cup \{+\infty\}$ the set $E(f) = \{g \in C(X) : f + g$ attains its minimum in $X\}$ is residual in $C(X)$. Then the set $W(f) = \{g \in C(X) : f + g$ is well posed$\}$ is also residual in $C(X)$.*

Proof. Let X and f be as in the theorem. We will prove that the player α has a winning strategy in the Banach-Mazur game $B(C(X), W(f))$ played in $C(X)$ equipped with the sup-norm. According to the result of Oxtoby cited above this will imply that $W(f)$ is residual in $C(X)$.

First, knowing that $E(f)$ is residual in $C(X)$, let $(O_n)_n$ be a countable family of open and dense subsets of $C(X)$ such that $\cap_n O_n \subset E(f)$. Let us denote by ω the winning strategy in the game $DG(X)$ for the player Ω. We will construct now a winning strategy s for the player α in the game $BM(C(X), W(f))$.

To this end, let U_1 be an arbitrary nonempty open set of $C(X)$ which can be a legal move of the player β in this game. Take $A_1 := M_f(U_1)$ which is a nonempty set of

X, according to Proposition 20.2 (b). Consider this set as a first move of the player Σ in the determination game $DG(X)$. Then put $B_1 := \omega(A_1)$ to be the answer of the player Ω in the game $DG(X)$ according to his/her strategy ω. Since B_1 is relatively open subset of A_1 there is some open set $W_1 \subset X$ such that $B_1 = W_1 \cap A_1$. Now, by Proposition 20.2 (d), there is a nonempty open set V_1 of $C(X)$ for which $V_1 \subset U_1$ and $M_f(V_1) \subset W_1$. Thus $M_f(V_1) \subset W_1 \cap M_f(U_1) = W_1 \cap A_1 = B_1$. We may think, without loss of generality, that $V_1 \subset O_1$, $\overline{V}_1 \subset U_1$ and that in addition $\mathrm{diam}(V_1) < 1$. Define the value of the strategy s for the player α in the game $BM(C(X), W(f))$ at the set U_1 to be $s(U_1) := V_1$. Let further the nonempty open set $U_2 \subset V_1$ be an arbitrary legitimate choice of the player β in the game $BM(C(X), W(f))$ at the second step. Put $A_2 := M_f(U_2)$ which is a nonempty set of X according again to Proposition 20.2 (b). Since $A_2 = M_f(U_2) \subset M_f(V_1) \subset B_1$ the set A_2 can be a legal move of the player Σ in the game $DG(X)$ at the second step. Put $B_2 := \omega(A_1, B_1, A_2)$ to be the answer of the player Ω according to his/her strategy ω. The set B_2 is a nonempty relatively open subset of A_2; thus, there is some nonempty open set $W_2 \subset X$ such that $B_2 = W_2 \cap A_2$. Now, using once again Proposition 20.2 (d), there is some nonempty open subset V_2 of U_2 for which $M_f(V_2) \subset W_2$. Therefore, $M_f(V_2) \subset W_2 \cap M_f(U_2) = B_2$. Moreover, without loss of generality, we may think that $V_2 \subset O_2$, $\overline{V}_2 \subset U_2$ and that $\mathrm{diam}(V_2) < 1/2$. Define the value of the strategy s by $s(U_1, V_1, U_2) := V_2$.

Proceeding by induction we define a strategy s for the player α in the Banach-Mazur game $BM(C(X), W(f))$ such that for any s-play $\{U_n, V_n\}_{n \geq 1}$ in this game (i.e., $V_n = s(U_1, V_1, \ldots, U_n)$ for each $n \geq 1$) there exists an associated ω-play $\{A_n, B_n\}_{n \geq 1}$ in the game $DG(X)$ such that the following properties are satisfied for any $n \geq 1$:

(i) $A_n = M_f(U_n)$.
(ii) $M_f(V_n) \subset B_n$.
(iii) $V_n \subset O_n$.
(iv) $\overline{V}_{n+1} \subset U_{n+1} \subset V_n$.
(v) $\mathrm{diam}(V_n) < 1/n$.

Conditions (iv) and (v) ensure that the intersection $\cap_n V_n$ is a one point set, say $g \in C(X)$ and condition (iii) entails that $g \in \cap_n O_n \subset E(f)$. According to Proposition 20.2 (e) and taking into account (i) and (iv) we have

$$M_f(g) = \cap_n \overline{M_f(V_n)} = \cap_n \overline{M_f(U_n)} = \cap_n \overline{A}_n.$$

Since $g \in E(f)$, the set $M_f(g) = \cap_n \overline{A}_n$ is nonempty and therefore, because ω is a winning strategy for Ω in the determination game $DG(X)$, this set is compact and the family $(B_n)_n$ behaves like a base for $\cap_n \overline{A}_n = M_f(g)$, that is, for any open set U containing $M_f(g)$ there is some n such that $B_n \subset U$. We will show that $g \in W(f)$ and this will complete the proof. To show that the function $f + g$ is well posed let $(x_\lambda)_\lambda$ be a minimizing net for $f + g$, that is, $f(x_\lambda) + g(x_\lambda) \to \inf_X(f + g)$. We have to show that this net has a cluster point (necessarily lying in $M_f(g)$). For this, having in mind that the set of minima $M_f(g)$ for $f + g$ is nonempty and compact, it is enough to show that if U is an open subset of X so that $M_f(g) \subset U$, then $x_\lambda \in U$ eventually.

Fix $n \geq 1$ so large that $B_n \subset U$. Put $\varepsilon_\lambda := f(x_\lambda) + g(x_\lambda) - \inf_X(f+g) \geq 0$. We may think, without loss of generality, that $\varepsilon_\lambda > 0$ for every λ. By Proposition 20.1, for each λ, there is $g_\lambda \in C(X)$ with $\|g_\lambda\|_\infty < \varepsilon_\lambda$ and such that $x_\lambda \in M_f(g+g_\lambda)$. Since $(g_\lambda)_\lambda$ converges uniformly to zero, we have $g+g_\lambda \in V_n$ eventually. Thus, we have (using also (ii) above) $x_\lambda \in M_f(V_n) \subset B_n$ eventually (for λ). Therefore, $x_\lambda \in U$ eventually, and this completes the proof. ■

The next example shows that there are spaces in which we have generic attainment of the infimum by the perturbations, without having generic well-posedness of the perturbed optimization problems.

Example 20.4. Take Y to be the product of uncountably many copies of the unit interval $[0,1]$ with the usual product topology under which it is a compact topological space. Let X be the so-called sigma-product in Y, i.e., the subset of those $x \in Y$ for which only countable number of coordinates are different from zero. With the inherited topology X is a sequentially compact space which is not compact. Thus, for any proper bounded from below lower semicontinuous function $f : X \to \mathbb{R} \cup \{+\infty\}$ we will have $E(f) = C(X)$. In particular, this is so for any function $f \in C(X)$. Fix such a function f. In this case all the perturbations $f+g$, $g \in C(X)$ are continuous in X. On the other hand, it is easy to see that, for each value r of a continuous function h in X, the level set $h^{-1}(r) = \{x \in X : h(x) = r\}$ contains as a closed subset a copy of the sigma-product of uncountably many copies of the interval $[0,1]$. Hence, the set $h^{-1}(r)$ is not compact for $r = \inf_X h$ and thus $W(f) = \emptyset$.

20.4 Concluding Remarks

Some versions of the determination game $DG(X)$ have already been used for different purposes. We have in mind games in which the rules for selection of sets are as in $DG(X)$, but the rules for winning a play are different. We consider three of these versions here. In the first one, which is denoted by $G(X)$, Ω wins a play $\{A_n, B_n\}_{n\geq 1}$ if $\cap_n \overline{A}_n = \cap_n \overline{B}_n \neq \emptyset$. Otherwise Σ wins this play. The game $G(X)$ was used by Michael [19] for the study of completeness properties of metric spaces. It was also used by the authors in [6, 17] to show that the existence of a winning strategy for the player Ω in $G(X)$ ensures the validity of the following generic variational principle.

Theorem 20.5 ([17], Theorem 3.1). *If the player Ω has a winning strategy in the game $G(X)$, then, for any proper bounded from below lower semicontinuous function $f : X \to \mathbb{R} \cup \{+\infty\}$, the set $E(f) = \{g \in C(X) : f+g$ attains its minimum in $X\}$ is residual in $C(X)$.*

Note however that, for some particular functions f, the set $E(f)$ may be residual in $C(X)$ even if the space X does not admit a winning strategy for $G(X)$ (see, e.g., Example 5.2 from [6]).

In the second variant, denoted by $FG(X)$ and called *fragmenting game*, the player Ω wins a play $\{A_n, B_n\}_{n\geq 1}$ if $\cap_n \overline{A}_n = \cap_n \overline{B}_n$ is either empty or a one point set. Otherwise Σ wins this play. The game $FG(X)$ was used in [13–15] for the study of fragmentable spaces. Recall that a topological space X is called *fragmentable* (see Jayne and Rogers [10]) if there is a metric d in X such that for any nonempty set A of X and any $\varepsilon > 0$ there is a relatively open set B of A with the property d-diam$(B) < \varepsilon$, with d-diam(B) having the usual meaning of the diameter of the set B with respect to the metric d. Every metric space is fragmentable by its own metric. There are however interesting examples of nonmetrizable spaces which are fragmentable. For example every weakly compact subset of a Banach space is fragmented by the metric generated by the norm. Every bounded subset of the dual of an Asplund space is fragmented by the metric of the dual norm. The class of fragmentable spaces has proved its usefulness in the study of different problems in topology (e.g., single-valuedness of set-valued maps) and in the geometry of Banach spaces (e.g., differentiability of convex functions)–see [10, 13–15, 21] and the reference therein. It was proved in [13, 14] that

Theorem 20.6. *The space X is fragmentable if and only if the player Ω has a winning strategy in the fragmenting game $FG(X)$.*

Fragmentability is closely related to generic *Tykhonov well-posedness* of minimization problems. Tykhonov well posed are the problems which are well posed and have unique minimizer.

Theorem 20.7. *Let X be a topological space which is fragmented by a metric whose topology contains the original topology in X. Suppose that for some bounded from below and lower semicontinuous function $f : X \to \mathbb{R} \cup \{+\infty\}$ which is proper, the set $E(f)$ is residual in $C(X)$. Then the set $T(f) := \{g \in C(X) : f+g$ is Tykhonov well posed$\}$ is residual in $C(X)$ as well.*

Proof. Fragmentability by a metric whose topology contains the original topology in X is characterized by the fact that the player Ω possesses a special winning strategy ω in the determination game $DG(X)$ such that, for any ω-play $\{A_n, B_n\}_{n\geq 1}$, the set $\cap_n A_n$ is either empty or consists of just one point, say $\{x\}$, for which the family $(B_n)_n$ behaves like a base: for every open $U \ni x$ there is some $n \geq 1$ such that $B_n \subset U$. Further we proceed exactly as in the proof of Theorem 20.3 and construct a strategy s for the player α in the game $BM(C(X), T(f))$ such that, for any s-play $\{U_n, V_n\}_{n\geq 1}$, there exists an associate ω-play $\{A_n, B_n\}_{n\geq 1}$ in the game $DG(X)$ with the properties (i)-(v). As above $\cap_n U_n = \cap_n V_n$ is a one point set, say $g \in C(X)$, for which $g \in W(f)$. Since, in addition, we have that the target sets of the corresponding ω-plays are singletons, then we have, in fact, that $g \in T(f)$. And this completes the proof. ∎

Let us mention that if in the above theorem the metric which fragments X is also complete, then the set $T(f)$ is residual in $C(X)$–see [18], Theorem 2.3.

The third version of the game $DG(X)$ explains where the name "determination game" comes from. This version is played in a compactification bX of the completely regular topological space X. The moves of the players Σ and Ω are as in $DG(bX)$. The player Ω wins a play $p = \{A_n, B_n\}_{n\geq 1}$ in this new game if the *target set* $T(p) = \cap_n \overline{A}_n^{bX}$ (which is always nonempty in this setting) is either entirely in X or entirely in $bX \setminus X$. In the next statement the term *equivalent games*, for games with the same players, is used in the sense that the games are simultaneously favorable (unfavorable) for any of the players.

Proposition 20.8 ([12], Proposition 4). *Let X be a completely regular topological space and bX be any compactification of X. Then the game described above in bX and the game $DG(X)$ are equivalent. In particular, if some of the players Σ or Ω has a winning strategy in one compactification of X, then he/she has such a winning strategy in any other compactification of X.*

In other words, in the game $DG(X)$ the existence of a winning strategy for the payer Ω determines that, when using this strategy, the target sets of the corresponding plays in the compactification bX will be either entirely in X or entirely in the complement $bX \setminus X$. In a certain sense the game "determines" or "identifies" the space X.

Let us turn back to the game $DG(X)$ and denote by GD the class of *game determined* spaces X (for which the player Ω has a winning strategy in the game $DG(X)$). It has turned out that the class GD is rather large (for the following facts we refer to [12]): it includes all fragmentable spaces which are fragmented by a metric d whose topology contains the original topology in X; in particular, the class contains all metrizable spaces; the class GD contains also all p-spaces introduced by Arhangel'skii [1] and also all Moore spaces.

The class GD includes also the class of topological spaces introduced in [15] and called spaces with *countable separation*: the completely regular topological space X is said to have *countable separation* if for some compactification bX of X there is a countable family $(U_n)_n$ of open (in bX) sets such that for any two points x, y with $x \in X$ and $y \in bX \setminus X$ there is an element U_n of the family which contains exactly one of the points x and y. If X has countable separation then the latter property is satisfied in any compactification of X. Let us mention that each Borel set of a space with countable separation has again countable separation. Moreover, each set obtained by applying Souslin operations on subsets with countable separation has countable separation as well. The class GD also includes all spaces with *star separation* introduced in [2].

Acknowledgments The authors wish to thank an anonymous referee for the helpful suggestions to improve the presentation of the results. Both authors have been partially supported by the Bulgarian National Fund for Scientific Research, under grant DO02-360/2008.

References

1. Arhangel'skii, A.V.: On a class containing all metric and all locally bicompact spaces. Dokl. Akad. Nauk. SSSR **151**, 751–754 (1963)
2. Arhangel'skii, A.V., Čoban, M.M., Kenderov, P.S.: Topological games and continuity of group operations. Topology Appl. **157**, 2542–2552 (2010)
3. Borwein, J.M., Preiss, D.: A smooth variational principle with applications to subdifferentiability and differentiability of convex functions. Trans. Amer. Math. Soc. **303**, 517–527 (1987)
4. Čoban, M.M., Kenderov, P.S.: Dense Gâteaux differentiability of the sup-norm in $C^*(T)$ and the topological properties of T. Compt. Rend. Acad. Bulg. Sci. **38**, 1603–1604 (1985)
5. Čoban, M.M., Kenderov P.S., Revalski, J.P.: Generic well-posedness of optimization problems in topological spaces. Mathematika **36**, 301–324 (1989)
6. Čoban, M.M., Kenderov P.S., Revalski, J.P.: Variational principles and topological games. Topol. Appl., **159**(17), 3550–3562 (2012)
7. Deville, R., Godefroy G., Zizler, V.: A smooth variational principle with applications to Hamilton–Jacobi equations in infinite dimensions. J. Funct. Anal. **111**, 197–212 (1993)
8. Deville, R., Godefroy G., Zizler, V.: Smoothness and renormings in Banach spaces. Pitman monographs and Surveys in Pure and Applied Mathematics. Longman Scientific and Technical, Harlow (1993)
9. Ekeland, I.: On the variational principle. J. Math. Anal. Appl. **47**, 324–353 (1974)
10. Jayne, J.E., Rogers, C.A.: Borel selectors for upper semi-continuous set-valued maps. Acta Math. **155**, 41–79 (1985)
11. Kenderov, P.S., Kortezov, I.S., Moors, W.B.: Topological games and topological groups. Topol. Appl. **109**, 157–165 (2001)
12. Kenderov, P.S., Kortezov, I.S., Moors, W.B.: Continuity points of quasi-continuous mappings. Topol. Appl. **109**, 321–346 (2001)
13. Kenderov, P.S., Moors, W.B.: Game characterization of fragmentability of topological spaces. In: Mathematics and Education in Mathematics, Proceedings of the 25-th Spring conference of the Union of Bulgarian Mathematicians, April 1996, pp. 8–18, Kazanlak, Bulgaria (1996)
14. Kenderov, P.S., Moors, W.B.: Fragmentability of Banach spaces. Compt. Rend. Acad. Bulg. Sci. **49**(2), 9–12 (1996)
15. Kenderov, P.S., Moors, W.B.: Fragmentability and sigma-fragmentability of Banach spaces. J. London Math. Soc. **60**, 203–223 (1999)
16. Kenderov, P.S., Revalski, J.P.: The Banach-Mazur game and generic existence of solutions to optimization problems. Proc. Amer. Math. Soc. **118**, 911–917 (1993)
17. Kenderov, P.S., Revalski, J.P.: Dense existence of solutions of perturbed optimization problems and topological games. Compt. Rend. Acad. Bulg. Sci. **63**(7), 937–942 (2010)
18. Kenderov, P.S., Revalski, J.P.: Variational principles in non metrizable spaces. TOP. DOI: 10.1007/s11750-011-0243-3 (2011)
19. Michael, E.: A note on completely metrizable spaces. Proc. Amer. Math. Soc. **96**, 513–522 (1986)
20. Oxtoby, J.C.: The Banach-Mazur game and Banach Category Theorem. In: Contributions to the Theory of Games, Vol. III, Annals of Mathamatics Studies, PrincetonUniversity Press, Princeton **39**, 159–163 (1957)
21. Ribarska, N.K.: Internal characterization of fragmentable spaces. Mathematika **34**, 243–257 (1987)
22. Stegall, C.: Optimization of functions on certain subsets of Banach space. Math. Ann. **236**, 171–176 (1978)

Chapter 21
Legendre Functions Whose Gradients Map Convex Sets to Convex Sets

Alexander Knecht and Jon Vanderwerff

Dedicated to Jonathan Borwein on the occasion of his 60th birthday

Abstract We show that a differentiable function on a Banach space of dimension at least two has an affine gradient provided the gradient is continuous and one-to-one and maps convex sets to convex sets.

Key words: Affine • Convex function • Convex set • Gradient

Mathematics Subject Classifications (2010): Primary 52A41; Secondary 46T20, 49J50.

21.1 Introduction

We work in a real Banach space, usually denoted by X. We will say a nonempty $C \subset X$ is *convex* if $\lambda x + (1-\lambda)y \in C$ whenever $x, y \in C$ and $0 \le \lambda \le 1$. A function $f : X \to (-\infty, +\infty]$ is said to be *proper* provided $\mathrm{dom} f \neq \emptyset$ where $\mathrm{dom} f := \{x \in X \mid f(x) < \infty\}$. A proper function $f : X \to (-\infty, +\infty]$ is *convex* if

$$f(\lambda x + (1-\lambda)y) \le \lambda f(x) + (1-\lambda) f(y) \quad \text{for all } x, y \in \mathrm{dom} f,\ 0 \le \lambda \le 1.$$

In fact, in this note, we will focus on real-valued (convex) functions.

COMMUNICATED BY HEINZ H. BAUSCHKE.

A. Knecht • J. Vanderwerff (✉)
Department of Mathematics, La Sierra University, Riverside, CA 92515, USA
e-mail: akne224@lasierra.edu;jvanderw@lasierra.edu

D.H. Bailey et al. (eds.), *Computational and Analytical Mathematics*, Springer Proceedings in Mathematics & Statistics 50, DOI 10.1007/978-1-4614-7621-4_21,

A closed set in a Banach space is said to be a *Chebyshev set* if every point in the space has exactly one closest point in the set. It is a classical result that in Euclidean spaces a closed set is Chebyshev if and only if it is convex [11, 14]. The situation for infinite-dimensional Hilbert spaces remains a long-standing open problem. For more information on Chebyshev sets the reader may wish to see Borwein's survey paper [9] and Deutsch's book [12].

Recently, several authors have focused their attention on Chebyshev sets and related concepts with respect to Bregman distances [5, 7, 8]. The survey [5] also poses several open questions in this topic. In particular, [5, Question 3] asks whether there exists a convex function of Legendre type (which we will call *Legendre functions*) with full domain whose gradient maps convex sets to convex sets, and yet the gradient is not affine. We refer the reader to [16, Sect. 26] and [3, Sect. 5] for further information on fundamental properties of Legendre functions in Euclidean and general Banach spaces, respectively. For our purposes, it will be sufficient to know that the derivative of a Legendre function with full domain on $\mathbb{R}^n$ is continuous and one-to-one although more can be said (see [16, Theorem 26.5]), and in general Banach spaces, a Legendre function with full domain has a one-to-one (though not necessarily continuous) derivative [3].

The primary purpose of this note is to show that as long as the Banach space has dimension at least two, the answer to [5, Question 3] is negative. In addition, we have endeavored to use elementary techniques to the extent that only introductory analysis and linear algebra are needed in the event X is the Euclidean space $\mathbb{R}^n$. The interested reader can find additional relevant information on convex analysis in sources such as [4, 10, 15–17].

21.2 Main Result and Examples

We begin by outlining some natural elementary properties of mappings between Banach spaces.

Lemma 21.1. *Let X and Y be Banach spaces and suppose $T : X \to Y$ is a continuous and one-to-one mapping such that $T(0) = 0$ and T maps convex subsets of X into convex subsets of Y. Then*

(a) T maps lines in X into lines in Y, moreover given $x, y \in X$ we have

(i) $T(x + \mathbb{R}_+ y) \subset T(x) + \mathbb{R}_+[T(x+y) - T(x)]$
(ii) $T(x - \mathbb{R}_+ y) \subset T(x) - \mathbb{R}_+[T(x+y) - T(x)]$, where $\mathbb{R}_+ := [0, \infty)$

(c) When $y \neq 0$, $T(y) \neq 0$, and $T(-y) = \alpha T(y)$ for some $\alpha < 0$.
(d) $T(x)$ and $T(y)$ are linearly independent when x and y are linearly independent.
(e) For $\alpha, \beta \in \mathbb{R}$, $T(\alpha x + \beta y) \in \{tT(x) + sT(y) \mid s, t \in \mathbb{R}\}$.

Proof. (a) Both (i) and (ii) are trivial if $y = 0$ and so we assume $y \neq 0$. Now suppose

$$T(x + \mathbb{R}_+ y) \not\subset L \text{ where } L := T(x) + \mathbb{R}_+[T(x+y) - T(x)].$$

We fix $t_0 > 0$ such that $T(x+t_0y) \notin L$. Now fix a number $\beta > \max\{t_0, 1\}$ and let S be the line segment with endpoints $T(x)$ and $T(x+\beta y)$. Then $S \subset T(x+[0,\beta]y)$ because $T(x+[0,\beta]y)$ is a convex set. We can choose $0 < \alpha < \beta$ such that $T(x+\alpha y) \notin S$. Indeed, in the case $T(x+\beta y) \in L$, then $S \subset L$ because both endpoints of S are in the ray L. Then we let $\alpha = t_0$, and so $T(x+\alpha y) \notin S$ because $T(x+t_0y) \notin L$. In the case $T(x+\beta y) \notin L$ we let $\alpha = 1$. Because $T(x+\beta y) \notin L$, the line segment S meets the ray L only at $T(x)$; in particular $T(x+y) \in L \setminus S$ and so $T(x+\alpha y) \notin S$ in this case as well.

Now let $C_1 := T(x+[0,\alpha]y)$ and $C_2 := T(x+[\alpha,\beta]y)$. Then C_1, C_2 are convex and compact as continuous images of compact sets. Moreover, because T is one-to-one, $C_1 \cap C_2 = \{T(x+\alpha y)\}$. It follows that $C_1 \cap S$ and $C_2 \cap S$ are disjoint because $T(x+\alpha y) \notin S$. Thus we have two nonempty disjoint closed sets whose union is S. This contradicts that S is connected. Therefore $T(x+\mathbb{R}_+y) \subset L$ as desired.

For (ii), when $t \geq 0$, we use (i) to write

$$T(x) = T(x-ty+ty) = T(x-ty) + \alpha[T(x) - T(x-ty)], \text{ and similarly}$$
$$T(x+y) = T(x-ty+(t+1)y) = T(x-ty) + \beta[T(x) - T(x-ty)],$$

where the one-to-one property of T additionally ensures $\beta > \alpha$. Thus $T(x+y) - T(x) = (\beta - \alpha)[T(x) - T(x-ty)]$ from which (ii) follows.

Let us note that part (b) is a consequence of (a) and the injectivity of T. Now to prove (c), we suppose x and y are linearly independent in X, but $T(x)$ and $T(y)$ are not linearly independent. Because $T(x) \neq 0$ and $T(y) \neq 0$ we may write $T(x) = \lambda T(y)$ for some $\lambda \neq 0$. Then applying (a) and (b) we can find $\delta > 0$ so that $(-\delta, \delta)T(x) \subset T(\mathbb{R}x) \cap T(\mathbb{R}y)$. This contradicts the one-to-one property of T.

Part (d) can also be deduced in a natural fashion with the help of (a). ■

The following lemma is also elementary, but is crucial to our argument.

Lemma 21.2. *Let X and Y be any Banach spaces where X contains two linearly independent vectors. Suppose $T : X \to Y$ is a continuous and one-to-one mapping such that $T(0) = 0$ and T maps convex subsets of X into convex subsets of Y. Suppose x, y are linearly independent in X, then there exists $b \in \mathbb{R}$ such that $T(x+y) = T(x) + bT(y)$.*

Proof. First, Lemma 21.1(c) ensures $T(x)$ and $T(y)$ are linearly independent. Then using Lemma 21.1(d), we fix $a, b \in \mathbb{R}$ such that

$$T(x+y) - T(x) = aT(x) + bT(y). \tag{21.1}$$

We note $b \neq 0$ in (21.1) because otherwise $T(x+y) - (1+a)T(x) = 0$ which cannot occur since Lemma 21.1(c) ensures $T(x+y)$ and $T(x)$ are linearly independent. We now consider numbers $\lambda > 1$. Then Lemma 21.1(a)(i) implies there exists $s_\lambda > 0$ so that

$$T(x+\lambda y) = T(x) + s_\lambda[aT(x) + bT(y)]. \tag{21.2}$$

Now, Lemma 21.1(d) implies $T(y+\frac{1}{\lambda}x) = c_\lambda T(x) + d_\lambda T(y)$ for some $c_\lambda, d_\lambda \in \mathbb{R}$. Moreover, because 0, $\frac{1}{\lambda}(x+\lambda y)$, and $x+\lambda y$ are on the same line, Lemma 21.1(a) ensures that

$$T(x+\lambda y) = t_\lambda T\left(\frac{1}{\lambda}(x+\lambda y)\right) \quad \text{for some } t_\lambda \in \mathbb{R}.$$

Consequently,

$$t_\lambda(c_\lambda T(x) + d_\lambda T(y)) = T(x) + s_\lambda[aT(x) + bT(y)]. \tag{21.3}$$

Equating coefficients of $T(x)$ and $T(y)$ in (21.3) we obtain $t_\lambda c_\lambda = s_\lambda a + 1$ and $t_\lambda d_\lambda = s_\lambda b$. The quantities s_λ and b are both nonzero, so dividing and cancelling

$$\frac{c_\lambda}{d_\lambda} = \frac{s_\lambda a + 1}{s_\lambda b} = \frac{a + \frac{1}{s_\lambda}}{b}.$$

By the continuity of T, $\lim_{\lambda\to\infty} T(\frac{1}{\lambda}(x+\lambda y)) = T(y)$. Therefore, $\lim_{\lambda\to\infty} d_\lambda = 1$, and $\lim_{\lambda\to\infty} \frac{c_\lambda}{d_\lambda} = 0$. Therefore, $a + 1/s_\lambda \to 0$ and so $0 < 1/s_\lambda \to -a$ which means $a \le 0$.

To complete the proof, we will show $a \ge 0$ by proceeding along the lines above. Indeed, for $\lambda > 1$, similar to (21.2) we write

$$T(x-\lambda y) = T(x) + \sigma_\lambda[aT(x) + bT(y)]$$

where a and b are as in (21.1), and $\sigma_\lambda < 0$. Again, we can also find c_λ, d_λ, and τ_λ in $\mathbb{R}$ such that

$$T\left(\frac{1}{\lambda}x - y\right) = c_\lambda T(x) + d_\lambda T(y) \quad \text{and} \quad T(x-\lambda y) = \tau_\lambda T\left(\frac{1}{\lambda}(x-\lambda y)\right).$$

According to Lemma 21.1(b), $T(-y) = \alpha T(y)$ where $\alpha < 0$; consequently

$$\lim_{\lambda\to\infty} T\left(\frac{1}{\lambda}(x-\lambda y)\right) = T(-y) = \alpha T(y).$$

Proceeding as above, $\lim_{\lambda\to\infty} c_\lambda = 0$ and $\lim_{\lambda\to\infty} d_\lambda = \alpha$. Then as above

$$\lim_{\lambda\to\infty} \frac{\sigma_\lambda a + 1}{\sigma_\lambda b} = \lim_{\lambda\to\infty} \frac{a + \frac{1}{\sigma_\lambda}}{b} = 0.$$

Because $\sigma_\lambda < 0$, we deduce $a \ge 0$, and thus $a = 0$ as desired. ■

The following is our main result concerning general mappings.

Theorem 21.3. *Let X and Y be any Banach spaces where X contains two linearly independent vectors. Suppose $T : X \to Y$ is a continuous and one-to-one mapping such that T maps convex sets to convex sets. Then T is affine.*

Proof. Define $S(x) := T(x) - T(0)$ for $x \in X$. Then S has the assumed properties of T, and additionally, $S(0) = 0$. To prove the theorem, we will show that S is linear.

First, let x and y be any two linearly independent vectors in X. Lemma 21.2 implies there exist r and $s \in \mathbb{R}$ such that

$$S(x+y) = S(x) + rS(y) \quad \text{and} \quad S(y+x) = S(y) + sS(x).$$

Subtracting equations, $(r-1)S(y) - (s-1)S(x) = 0$. Now Lemma 21.1(c) implies $S(x)$ and $S(y)$ are linearly independent, and so $(r-1) = (s-1) = 0$. Consequently, $r = s = 1$, and thus

$$S(x+y) = S(x) + S(y) \text{ when } x \text{ and } y \text{ are linearly independent.} \tag{21.4}$$

We next show S preserves scalar multiplication. For this, we let $x \neq 0$ and choose any y linearly independent with x. By (21.4), we know for $n \in \mathbb{N}$ and $t > 0$, $S(x + nx + ty) = S(x) + S(nx + ty)$. Letting $t \to 0^+$, the continuity of S along with (21.4) implies

$$S(x+nx) = \lim_{t\to 0^+} S(x+nx+ty) = \lim_{t\to 0^+} S(x) + S(nx+ty) = S(x) + S(nx).$$

Using induction, it is easy to deduce that $S(kx) = kS(x)$ for any $k \in \mathbb{N}$.

Then for $k \in \mathbb{N}$, $S(x) = S(kk^{-1}x) = kS(k^{-1}x)$, and so $S(k^{-1}x) = k^{-1}S(x)$ for any $x \neq 0$, and $k \in \mathbb{N}$. Then for $m,n \in N$, $S(mn^{-1}x) = mS(n^{-1}x) = mn^{-1}S(x)$. That is, $S(rx) = rS(x)$ for any positive rational number r. Now for any $t \geq 0$ choose a sequence of positive rational numbers $r_n \to t$. Then the continuity of S implies $S(tx) = \lim_{n\to\infty} S(r_n x) = \lim_{n\to\infty} r_n S(x) = tS(x)$. Consequently, for $t > 0$,

$$0 = S(0) = S(tx - tx) = S(tx) + S(-tx) = tS(x) + S(-tx),$$

and so $S(-tx) = -tS(x)$ when $t > 0$. Therefore, $S(tx) = tS(x)$ for $x \in X$ and $t \in \mathbb{R}$. This with (21.4) implies S is a linear mapping as desired. ■

The following consequence of Theorem 21.3 answers [5, Question 3].

Corollary 21.4. *Suppose X is any Banach space that contains two linearly independent vectors. Let $f : X \to \mathbb{R}$ be a Legendre function such that $\nabla f(C)$ is convex whenever C is a convex subset of X. Then ∇f is affine.*

Proof. In the case X is finite dimensional, the result follows from Theorem 21.3 because ∇f is continuous and one-to-one when f is Legendre with full domain (in fact more is true, see [16, Theorem 26.5]).

Suppose X is infinite dimensional. We will assume ∇f is not affine and proceed by way of contradiction. Because ∇f is not affine, there exist $x, y \in X$ and $\lambda \in \mathbb{R}$ so that

$$\nabla f(\lambda x + (1-\lambda)y) \neq \lambda \nabla f(x) + (1-\lambda)\nabla f(y).$$

Now choose $u \in X$ so that

$$\langle \nabla f(\lambda x + (1-\lambda)y), u \rangle \neq \langle \lambda \nabla f(x) + (1-\lambda)\nabla f(y), u \rangle.$$

Let F be the linear span of x, y, and u (one can add an additional vector if necessary to the spanning set of F to ensure F has dimension at least two). Because f is Legendre with full domain, this means f is Gâteaux differentiable (see [3, Sect. 5] and [10, Chap. 7]), and so

$$\langle \nabla f(v), h \rangle = \lim_{t \to 0} \frac{f(v+th) - f(v)}{t}, \quad v, h \in X.$$

Thus defining the convex function $g : F \to \mathbb{R}$ by $g := f|_F$, it follows that $\nabla g(x) = \nabla f(x)|_F$, and for x, y, and u as above we have

$$\langle \nabla g(\lambda x + (1-\lambda)y), u \rangle \neq \langle \lambda \nabla g(x) + (1-\lambda)\nabla g(y), u \rangle.$$

Hence ∇g is not affine. However, because f is Legendre with full domain, g is a strictly convex differentiable function; see [3, Sect. 5]. Consequently ∇g is one-to-one and also continuous because F is finite dimensional [16, Corollary 25.5.1]. Finally, let C be any convex subset of F, and let $v, w \in C$ and $0 \leq \lambda \leq 1$. Because $\nabla f(C)$ is convex, there exists $z \in C$ such that

$$\nabla f(z) = \lambda \nabla f(v) + (1-\lambda)\nabla f(w).$$

Therefore, $\nabla f(z)|_F = \lambda \nabla f(v)|_F + (1-\lambda)\nabla f(w)|_F$. That is, $\nabla g(C)$ is convex. Then Theorem 21.3 ensures ∇g is affine which is a contradiction. ■

The following provides elementary examples that illustrate the necessity of various conditions in Theorem 21.3.

Example 21.5. (a) The function $t \mapsto t^4$ has a derivative that is continuous and one-to-one and maps convex sets to convex sets, but is not affine. So Theorem 21.3 fails on the one-dimensional Banach space $\mathbb{R}$.

(b) Let $f(t) = \mathrm{e}^{-|t|} + |t|$. Then the derivative of f is one-to-one, continuous, and bounded and maps convex sets to convex sets, but is not affine.

(c) Let $g : \mathbb{R}^2 \to \mathbb{R}$ be defined by $g(x,y) := x^4$. Then ∇g is continuous and maps convex sets into convex sets (convex subsets of $\mathbb{R} \times \{0\}$), but ∇g is not one-to-one, nor is it affine.

(d) Let $h(x,y) := x^4 + y^4$. Then h is a continuously differentiable convex function and ∇h is one-to-one. However, ∇h does not map convex sets to convex sets, nor is ∇h affine.

We close with some comments on how Corollary 21.4 relates to some other results in convex analysis.

Remark 21.6. The recent result [6, Theorem 4.2] shows a maximally monotone operator has an affine graph provided the graph is convex, and it is a now classical result that the gradient of a (proper lower-semi)continuous convex function is a maximally monotone operator (see e.g. [10, 17]). Thus it is natural to ask whether

[6, Theorem 4.2] could be used to deduce Corollary 21.4, and we thank L. Yao for bringing this question to our attention when the first author spoke about the subject of this note at the conference on *Analytical and Computational Mathematics* held in honor of Jonathan Borwein on the occasion of his 60th birthday in May 2011.

In this direction, observe that the graphs of the gradients of the convex functions in Example 21.5(a), (b), and (c) are not convex; thus a differentiable convex function may have a gradient that maps convex sets to convex sets, yet the graph of the gradient is not convex. Consequently, [6, Theorem 4.2] does not immediately provide the result of Corollary 21.4. However, this does not mean [6, Theorem 4.2] could not be used to provide an alternate or more efficient proof of Corollary 21.4 or extensions thereof. Indeed, the functions provided in Example 21.5(a),(b), and (c) are quite limited in that all have gradients whose ranges lie in a one-dimensional space, so it is a natural question to ask for weaker conditions than given in Theorem 21.3—or a characterization—for when maximally monotone operators (or other types of mappings between Banach spaces) are affine provided they map convex sets into convex sets.

Remark 21.7. We thank H.H. Bauschke for bringing the recent paper [13] to our attention and alerting us to its relevance to the subject of our note. In particular, [13, Corollary 3] shows that if V and W are arbitrary real vector spaces with $\dim(V) \geq 2$ and if $T : V \to W$ is one-to-one and maps segments to segments, then T is affine. The proof is built upon a similar result for finite-dimensional vector spaces as can be found in [1]; see also [2]. Further, it is important to note that no continuity assumption is needed on T, and this follows because affine mappings are continuous in finite-dimensional vector spaces.

Acknowledgments We are grateful to an anonymous referee for several suggestions that helped us improve the presentation of this note.

References

1. Artin, E.: Geometric Algebra. Interscience, New York (1957)
2. Artstein-Avidan, S., Milman, V.: The concept of duality in convex analysis, and the characterization of the Legendre transform. Ann. Math. **169**, 661–674 (2009)
3. Bauschke, H.H., Borwein, J.M., Combettes, P.L.: Essential smoothness, essential strict convexity, and Legendre functions in Banach spaces. Communications in Contemporary Mathematics **3**, 615–647 (2001)
4. Bauschke, H.H., Combettes, P.L.: Convex Analysis and Monotone Operator Theory in Hilbert Spaces. Springer, New York (2011)
5. Bauschke, H.H., Macklem, M.S., Wang, X.: Chebyshev Sets, Klee Sets, and Chebyshev Centers with respect to Bregman Distances: Recent Results and Open Problems. Preprint (2010). Available at arXiv:1003.3127
6. Bauschke, H.H., Wang, X., Yao, L.: Monotone linear relations: maximality and Fitzpatrick functions. J. Convex Anal. **16**, 673–686 (2009)
7. Bauschke, H.H., Wang, X., Ye, J., Yuan, X.: Bregman distances and Klee sets. J. Approx. Theory **158**, 170–183 (2009)

8. Bauschke, H.H., Wang, X., Ye, J., Yuan, X.: Bregman distances and chebyshev sets. J. Approx. Theory **159**, 3–25 (2009)
9. Borwein, J.M.: Proximality and Chebyshev sets. Optim. Lett. **1**, 21–32 (2007)
10. Borwein, J.M., Vanderwerff, J.D.: Convex Functions. Cambridge University Press, Cambridge (2010)
11. Bunt, L.N.H.: Bijdrage tot de theorie de convex puntverzamelingen, Thesis. University of Groningen (1932)
12. Deutsch, F.: Best Approximation in Inner Product Spaces. Springer, New York (2001)
13. Iusem, A.N, Reem, D., Svaiter, B.F: Order preserving and order reversing operators on the class of convex function in Banach spaces. arXiv:1212.1120v2
14. Motzkin, T.: Sur quelques propriétés caractérostistiques des ensembles convexes. Atti Acad. Naz. Lincei, Rend. VI Ser. **21**, 562–567 (1935)
15. Roberts, A.R., Varberg, D.E.: Convex Functions. Academic Press, New york (1973)
16. Rockafellar, R.T.: Convex Analysis. Princeton University Press, Princeton, New Jersey (1970)
17. Zălinescu, C.: Convex Analysis in General Vector Spaces. World Scientific, New Jersey (2002)

Chapter 22
On the Convergence of Iteration Processes for Semigroups of Nonlinear Mappings in Banach Spaces

W.M. Kozlowski and Brailey Sims

In tribute to Jonathan Borwein on his 60th birthday

Abstract Let C be a bounded, closed, convex subset of a uniformly convex Banach space X. We investigate the convergence of the generalized Krasnosel'skii-Mann and Ishikawa iteration processes to common fixed points of pointwise Lipschitzian semigroups of nonlinear mappings $T_t : C \to C$. Each of T_t is assumed to be pointwise Lipschitzian, that is, there exists a family of functions $\alpha_t : C \to [0, \infty)$ such that $\|T_t(x) - T_t(y)\| \leq \alpha_t(x)\|x - y\|$ for $x, y \in C$.

Key words: Asymptotic pointwise nonexpansive mapping • Common fixed point • Fixed-point • Fixed point iteration process • Ishikawa process • Krasnosel'skii-Mann process • Lipschitzian mapping • Mann process • Opial property • Pointwise Lipschitzian mapping • Semigroup of mappings • Uniformly convex Banach space

Mathematics Subject Classifications (2010): Primary 47H09; Secondary 47H10.

COMMUNICATED BY HEINZ H. BAUSCHKE.

W.M. Kozlowski (✉)
School of Mathematics and Statistics, University of New South Wales, Sydney, NSW 2052, Australia
e-mail: w.m.kozlowski@unsw.edu.au

B. Sims
School of Mathematical and Physical Sciences, The University of Newcastle, Callaghan, NSW 2308, Australia
e-mail: Brailey.Sims@newcastle.edu.au

D.H. Bailey et al. (eds.), *Computational and Analytical Mathematics*, Springer Proceedings in Mathematics & Statistics 50, DOI 10.1007/978-1-4614-7621-4_22,

22.1 Introduction

Let C be a bounded, closed, convex subset of a Banach space X. Let us consider a pointwise Lipschitzian semigroup of nonlinear mappings, that is, a family of mappings $T_t : C \to C$ satisfying the following conditions: $T_0(x) = x$, $T_{s+t}(x) = T_s(T_t(x))$, $t \mapsto T_t(x)$ is strong continuous for each $x \in C$, and each T_t is pointwise Lipschitzian. The latter means that there exists a family of functions $\alpha_t : C \to [0,\infty)$ such that $\|T_t(x) - T_t(y)\| \leq \alpha_t(x)\|x-y\|$ for $x,y \in C$ (see Definitions 22.1, 22.2, and 22.3 for more details). Such a situation is quite typical in mathematics and applications. For instance, in the theory of dynamical systems, the Banach space X would define the state space and the mapping $(t,x) \to T_t(x)$ would represent the evolution function of a dynamical system. Common fixed points of such a semigroup can be interpreted as points that are fixed during the state space transformation T_t at any given point of time t. Our results cater for both the continuous and the discrete time cases. In the setting of this paper, the state space may be an infinite-dimensional Banach space. Therefore, it is natural to apply these results not only to deterministic dynamical systems but also to stochastic dynamical systems.

The existence of common fixed points for families of contractions and nonexpansive mappings has been investigated since the early 1960s; see e.g., DeMarr [8], Browder [4], Belluce and Kirk [1, 2], Lim [22], and Bruck [5, 6]. The asymptotic approach for finding common fixed points of semigroups of Lipschitzian (but not pointwise Lipschitzian) mappings has been also investigated for some time; see e.g., Tan and Xu [33]. It is worthwhile mentioning the recent studies on the special case, when the parameter set for the semigroup is equal to $\{0,1,2,3,\ldots\}$ and $T_n = T^n$, the nth iterate of an asymptotic pointwise nonexpansive mapping, i.e. such a $T : C \to C$ that there exists a sequence of functions $\alpha_n : C \to [0,\infty)$ with $\|T^n(x) - T^n(y)\| \leq \alpha_n(x)\|x-y\|$ and $\limsup_{n\to\infty} \alpha_n(x) \leq 1$. Kirk and Xu [17] proved the existence of fixed points for asymptotic pointwise contractions and asymptotic pointwise nonexpansive mappings in Banach spaces, while Hussain and Khamsi extended this result to metric spaces [11] and Khamsi and Kozlowski to modular function spaces [14, 15]. Recently, Kozlowski proved existence of common fixed points for semigroups of nonlinear contractions and nonexpansive mappings in modular function spaces [20].

Several authors studied the generalizations of known iterative fixed-point construction processes like the Mann process (see e.g. [10, 23]) or the Ishikawa process (see, e.g., [12]) to the case of asymptotic and pointwise asymptotic nonexpansive mappings. There exists an extensive literature on the subject of iterative fixed-point construction processes for asymptotically nonexpansive mappings in Hilbert, Banach and metric spaces; see, e.g., [3, 7, 9, 11, 13, 24, 25, 27–37] and the works referred there. Schu [31] proved the weak convergence of the modified Mann iteration process to a fixed point of asymptotic nonexpansive mappings in uniformly convex Banach spaces with the Opial property and the strong convergence for compact asymptotic nonexpansive mappings in uniformly convex Banach spaces.

Tan and Xu [35] proved the weak convergence of the modified Mann and modified Ishikawa iteration processes for asymptotic nonexpansive mappings in uniformly convex Banach spaces satisfying the Opial condition or possessing Fréchet differentiable norm. Kozlowski [18] proved that – under some reasonable assumptions – the generalized Mann and Ishikawa processes converge weakly to a fixed point of an asymptotic pointwise nonexpansive mapping $T : C \to C$, where C is a bounded, closed, and convex subset of a uniformly convex Banach space X which satisfies the Opial condition.

Let us note that the existence of common fixed points for asymptotic pointwise nonexpansive semigroups has been recently proved by Kozlowski in [19]. However, the proof of this result does not provide a constructive method of finding such common fixed points. The aim of the current paper is to fill this gap. We prove that – under some reasonable assumptions – the generalized Krasnosel'skii-Mann and Ishikawa processes converge weakly, and – under additional assumptions – strongly, to a common fixed point of the asymptotic pointwise nonexpansive semigroups.

The paper is organized as follows:

(a) Section 22.2 provides the necessary preliminary material.
(b) Section 22.3 presents some technical results on approximate fixed-point sequences.
(c) Section 22.4 is devoted to proving the Demiclosedness Principle in a version relevant for this paper.
(d) Section 22.5 deals with the weak convergence of generalized Krasnosel'skii-Mann iteration processes to common fixed points of asymptotic pointwise nonexpansive semigroups.
(e) Section 22.6 deals with the weak convergence of generalized Ishikawa iteration processes to common fixed points of asymptotic pointwise nonexpansive semigroups.
(f) Section 22.7 presents the strong convergence result for both Krasnosel'skii-Mann and Ishikawa processes.

22.2 Preliminaries

Throughout this paper X will denote a Banach space, C a nonempty, bounded, closed, and convex subset of X, and J will be a fixed parameter semigroup of nonnegative numbers, i.e. a subsemigroup of $[0,\infty)$ with normal addition. We assume that $0 \in J$ and that there exists $t > 0$ such that $t \in J$. The latter assumption implies immediately that $+\infty$ is a cluster point of J in the sense of the natural topology inherited by J from $[0,\infty)$. Typical examples are: $J = [0,\infty)$ and ideals of the form $J = \{n\alpha : n = 0,1,2,3,\ldots\}$ for a given $\alpha > 0$. The notation $t \to \infty$ will mean that t tends to infinity over J.

Let us start with more formal definitions of pointwise Lipschitzian mappings and pointwise Lipschitzian semigroups of mappings and associated notational conventions.

Definition 22.1. We say that $T : C \to C$ is a pointwise Lipschitzian mapping if there exists a function $\alpha : C \to [0,\infty)$ such that

$$\|T(x) - T(y)\| \leq \alpha(x)\|x - y\| \text{ for all } x, y \in C. \tag{22.1}$$

If the function $\alpha(x) < 1$ for every $x \in C$, then we say that T is a pointwise contraction. Similarly, if $\alpha(x) \leq 1$ for every $x \in C$, then T is said to be a pointwise nonexpansive mapping.

Definition 22.2. A one-parameter family $\mathscr{F} = \{T_t; t \in J\}$ of mappings from C into itself is said to be a pointwise Lipschitzian semigroup on C if $\mathscr{F}$ satisfies the following conditions:

(i) $T_0(x) = x$ for $x \in C$
(ii) $T_{t+s}(x) = T_t(T_s(x))$ for $x \in C$ and $t, s \in J$
(iii) For each $t \in J$, T_t is a pointwise Lipschitzian mapping, i.e. there exists a function $\alpha_t : C \to [0,\infty)$ such that

$$\|T_t(x) - T_t(y)\| \leq \alpha_t(x)\|x - y\| \text{ for all } x, y \in C. \tag{22.2}$$

(iv) For each $x \in C$, the mapping $t \mapsto T_t(x)$ is strongly continuous.

For each $t \in J$ let $F(T_t)$ denote the set of its fixed points. Note that if $x \in F(T_t)$ then x is a periodic point (with period t) for the semigroup $\mathscr{F}$, i.e. $T_{kt}(x) = x$ for every natural k. Define then the set of all common fixed points for mappings from $\mathscr{F}$ as the following intersection:

$$F(\mathscr{F}) = \bigcap_{t \in J} F(T_t).$$

The common fixed points are frequently interpreted as the stationary points of the semigroup $\mathscr{F}$.

Definition 22.3. Let $\mathscr{F}$ be a pointwise Lipschitzian semigroup. $\mathscr{F}$ is said to be asymptotic pointwise nonexpansive if $\limsup_{t\to\infty} \alpha_t(x) \leq 1$ for every $x \in C$.

Denoting $a_0 \equiv 1$ and $a_t(x) = \max(\alpha_t(x), 1)$ for $t > 0$, we note that without loss of generality we can assume that $\mathscr{F}$ is asymptotically nonexpansive if

$$\|T_t(x) - T_t(y)\| \leq a_t(x)\|x - y\| \text{ for all } x, y \in C,\ t \in J, \tag{22.3}$$

$$\lim_{t\to\infty} a_t(x) = 1,\ a_t(x) \geq 1 \text{ for all } x \in C, \text{ and } t \in J. \tag{22.4}$$

Define $b_t(x) = a_t(x) - 1$. In view of (22.4), we have

$$\lim_{t\to\infty} b_t(x) = 0. \tag{22.5}$$

The above notation will be consistently used throughout this paper.

Definition 22.4. By $\mathscr{S}(C)$ we will denote the class of all asymptotic pointwise nonexpansive semigroups on C such that

$$M_t = \sup\{a_t(x) : x \in C\} < \infty, \quad \text{for every } t \in J, \tag{22.6}$$

$$\limsup_{t\to\infty} M_t = 1. \tag{22.7}$$

Note that we do not assume that all functions a_t are bounded by a common constant. Therefore, we do not assume that $\mathscr{F}$ is uniformly Lipschitzian.

Definition 22.5. We will say that a semigroup $\mathscr{F} \in \mathscr{S}(C)$ is equicontinuous if the family of mappings $\{t \mapsto T_t(x) : x \in C\}$ is equicontinuous at $t = 0$.

The following result of Kozlowski will be used in this paper to ensure existence of common fixed points.

Theorem 22.6. *[19] Assume X is uniformly convex. Let $\mathscr{F}$ be an asymptotically nonexpansive pointwise Lipschitzian semigroup on C. Then $\mathscr{F}$ has a common fixed point and the set $F(\mathscr{F})$ of common fixed points is closed and convex.*

The following elementary, easy to prove, lemma will be used in this paper.

Lemma 22.7. *[7] Suppose $\{r_k\}$ is a bounded sequence of real numbers and $\{d_{k,n}\}$ is a doubly indexed sequence of real numbers which satisfy*

$$\limsup_{k\to\infty}\limsup_{n\to\infty} d_{k,n} \le 0, \quad \textit{and} \quad r_{k+n} \le r_k + d_{k,n}$$

for each $k,n \ge 1$. Then $\{r_k\}$ converges to an $r \in \mathbb{R}$.

The notion of bounded away sequences of real numbers will be used extensively throughout this paper.

Definition 22.8. A sequence $\{c_n\} \subset (0,1)$ is called bounded away from 0 if there exists $0 < a < 1$ such that $c_n > a$ for every $n \in \mathbb{N}$. Similarly, $\{c_n\} \subset (0,1)$ is called bounded away from 1 if there exists $0 < b < 1$ such that $c_n < b$ for every $n \in \mathbb{N}$.

The following property of uniformly convex Banach spaces will play an important role in this paper.

Lemma 22.9. *[31,38] Let X be a uniformly convex Banach space. Let $\{c_n\} \subset (0,1)$ be bounded away from 0 and 1, and $\{u_n\}, \{v_n\} \subset X$ be such that*

$$\limsup_{n\to\infty} \|u_n\| \le a, \ \limsup_{n\to\infty} \|v_n\| \le a, \ \lim_{n\to\infty} \|c_n u_n + (1-c_n)v_n\| = a.$$

Then $\lim_{n\to\infty} \|u_n - v_n\| = 0$.

Using Kirk's result [16] (Proposition 2.1), Kozlowski [19] proved the following proposition.

Proposition 22.10. *Let $\mathscr{F}$ be a semigroup on C. Assume that all mappings $T_t \in \mathscr{F}$ are continuously Fréchet differentiable on an open convex set A containing C. Then $\mathscr{F}$ is asymptotic pointwise nonexpansive on C if and only if for each $x \in C$*

$$\limsup_{t\to\infty} \|(T_t)_x^{'}\| \leq 1. \tag{22.8}$$

This result, combined with Theorem 22.6, produces the following fixed-point theorem.

Theorem 22.11. *[19, Theorem 3.5] Assume X is uniformly convex. Let $\mathscr{F}$ be a pointwise Lipschitzian semigroup on C. Assume that all mappings $T_t \in \mathscr{F}$ are continuously Fréchet differentiable on an open convex set A containing C and for each $x \in C$*

$$\limsup_{t\to\infty} \|(T_t)_x^{'}\| \leq 1. \tag{22.9}$$

Then $\mathscr{F}$ has a common fixed point and the set $F(\mathscr{F})$ of common fixed points is closed and convex.

Because of the above, all the results of this paper can be applied to the semigroups of nonlinear mappings satisfying condition (22.9). This approach may be very useful for applications provided the Fréchet derivatives can be estimated.

22.3 Approximate Fixed-Point Sequences

The technique of approximate fixed-point sequences will play a critical role in proving fixed convergence to common fixed points for semigroups of mappings. Let us recall that given $T : C \to C$, a sequence $\{x_k\} \subset C$ is called an approximate fixed-point sequence for T if $\|T(x_k) - x_k\| \to 0$ as $k \to \infty$. We will also use extensively the following notion of a generating set.

Definition 22.12. A set $A \subset J$ is called a generating set for the parameter semigroup J if for every $0 < u \in J$ there exist $m \in \mathbb{N}$, $s \in A$, $t \in A$ such that $u = ms + t$.

Lemma 22.13. *Let C be a nonempty, bounded, closed, and convex subset of a Banach space X. Let $\mathscr{F} \in \mathscr{S}(C)$. If $\|T_s(x_n) - x_n\| \to 0$ for an $s \in J$ as $n \to \infty$ then for any $m \in \mathbb{N}$, $\|T_{ms}(x_n) - x_n\| \to 0$ as $n \to \infty$*

Proof. It follows from the fact that every a_t is a bounded function that there exists a finite constant $M > 0$ such that

$$\sum_{j=1}^{m-1} \sup\{a_{js}(x); x \in C\} \leq M. \tag{22.10}$$

It follows from

$$\|T_{ms}(x_n)-x_n\| \le \sum_{j=1}^{m-1} \|T_{(j+1)s}(x_n)-T_{js}(x_n)\| + \|T_s(x_n)-x_n\|$$

$$\le \|T_s(x_n)-x_n\|\Big(\sum_{j=1}^{m-1} a_{js}(x_n)+1\Big) \le (M+1)\|T_s(x_n)-x_n\| \tag{22.11}$$

that

$$\lim_{n\to\infty} \|T_{ms}(x_n)-x_n\| = 0, \tag{22.12}$$

which completes the proof. ■

Lemma 22.14. *Let C be a nonempty, bounded, closed, and convex subset of a Banach space X. Let $\mathscr{F} \in \mathscr{S}(C)$. If $\{x_k\} \subset C$ is an approximate fixed-point sequence for $T_s \in \mathscr{F}$ for any $s \in A$ where A is a generating set for J then $\{x_k\}$ is an approximate fixed-point sequence for T_s for any $s \in J$.*

Proof. Let $s,t \in A$ and $m \in \mathbb{N}$. We need to show that $\|T_{ms+t}(x_n)-x_n\| \to 0$ as $n \to \infty$. Indeed,

$$\|T_{ms+t}(x_n)-x_n\| \le \|T_{ms+t}(x_n)-T_{ms}(x_n)\| + \|T_{ms}(x_n)-x_n\|$$
$$\le a_{ms}(x_n)\|T_t(x_n)-x_n\| + \|T_{ms}(x_n)-x_n\|,$$

which tends to zero by boundedness of the function a_{ms} and by Lemma 22.13. ■

Lemma 22.15. *Let $\mathscr{F} \in \mathscr{S}(C)$ be equicontinuous and $\overline{B} = A \subset J$. If $\{x_k\} \subset C$ is an approximate fixed-point sequence for T_t for every $t \in B$ then $\{x_k\}$ is an approximate fixed-point sequence for T_s for every $s \in A$.*

Proof. Let $s \in A$, then there exists a sequence $\{s_n\} \subset B$ such that $s_n \to s$. Note that

$$\begin{aligned} \|T_s(x_k)-x_k\| &\le \|T_s(x_k)-T_{s_n}(x_k)\| + \|T_{s_n}(x_k)-x_k\| \\ &\le \sup_{x\in C} a_{\min(s,s_n)}(x) \sup_{x\in C} \|T_{|s-s_n|}(x)-x\| + \|T_{s_n}(x_k)-x_k\|. \end{aligned} \tag{22.13}$$

Fix $\varepsilon > 0$. By equicontinuity of $\mathscr{F}$ and by (22.6) there exists $n_0 \in \mathbb{N}$ such that

$$\sup_{x\in C} a_{\min(s,s_{n_0})}(x) \sup_{x\in C} \|T_{|s-s_{n_0}|}(x)-x\| < \frac{\varepsilon}{2}. \tag{22.14}$$

Since $\{x_k\}$ is an approximate fixed point for $T_{s_{n_0}}$ we can find $k_0 \in \mathbb{N}$ such that for every natural $k \ge k_0$

$$\|T_{s_{n_0}}(x_k)-x_k\| < \frac{\varepsilon}{2}. \tag{22.15}$$

By substituting (22.14) and (22.15) into (22.13) we get $\|T_s(x_k)-x_k\| < \varepsilon$ for large k. Hence $\{x_k\}$ is an approximate fixed point for T_s as claimed. ■

22.4 The Demiclosedness Principle

The following version of the Demiclosedness Principle will be used in the proof of our main convergence theorems. There exist several versions of the Demiclosedness Principle for the case of asymptotic nonexpansive mappings; see, e.g., Li and Sims [21], Gornicki [9], or Xu [37]. Recently, Kozlowski [18] proved a version of the Demiclosedness Principle for the asymptotic pointwise nonexpansive mappings, using the "parallelogram inequality" valid in the uniformly convex Banach spaces (Theorem 2 in [36]). For the completeness sake, we provide the proof for asymptotic pointwise nonexpansive semigroups.

Let us recall the definition of the Opial property which will play an essential role in this paper.

Definition 22.16. [26] A Banach space X is said to have the Opial property if for each sequence $\{x_n\} \subset X$ weakly converging to a point $x \in X$ (denoted as $x_n \rightharpoonup x$) and for any $y \in X$ such that $y \neq x$ there holds

$$\liminf_{n\to\infty} \|x_n - x\| < \liminf_{n\to\infty} \|x_n - y\|, \tag{22.16}$$

or equivalently

$$\limsup_{n\to\infty} \|x_n - x\| < \limsup_{n\to\infty} \|x_n - y\|. \tag{22.17}$$

Theorem 22.17. *Let X be a uniformly convex Banach space X with the Opial property. Let C be a nonempty, bounded, closed, and convex subset of X, and let $\mathcal{F} \in \mathcal{S}(C)$. Assume that there exists $w \in X$ and $\{x_n\} \subset C$ such that $x_n \rightharpoonup w$. Assume that there exists an $s \in J$ such that $\|T_s(x_n) - x_n\| \to 0$ as $n \to \infty$. Then $w \in F(T_{ks})$ for any natural k.*

Proof. Define a type $\varphi(x) = \limsup_{n\to\infty} \|x_n - x\|$ for $x \in C$. Let us fix $m \in \mathbb{N}$, $m > 2$ and observe that

$$\begin{aligned}\|T_{ms}(x_n) - x\| &\leq \sum_{i=1}^{m} \|T_{is}(x_n) - T_{(i-1)s}(x_n)\| + \|x_n - x\| \\ &\leq \|T_s(x_n) - x_n\| \Big(\sum_{i=2}^{m} a_{(i-1)s}(x_n) + 1 \Big) + \|x_n - x\|.\end{aligned}$$

Since all functions a_i are bounded and $\|T_s(x_n) - x_n\| \to 0$, it follows that

$$\limsup_{n\to\infty} \|T_{ms}(x_n) - x\| \leq \limsup_{n\to\infty} \|x_n - x\| = \varphi(x).$$

On the other hand, by Lemma 22.13, we have

$$\varphi(x) \leq \limsup_{n\to\infty} \|x_n - T_{ms}(x_n)\| + \limsup_{n\to\infty} \|T_{ms}(x_n) - x\| = \limsup_{n\to\infty} \|T_{ms}(x_n) - x\|.$$

Hence,

$$\varphi(x) = \limsup_{n\to\infty} \|T_{ms}(x_n) - x\|. \tag{22.18}$$

Because $\mathscr{F}$ is asymptotic pointwise nonexpansive, it follows that $\varphi\left(T_{ms}(x)\right) \leq a_{ms}(x)\varphi(x)$ for every $x \in C$. Applying this to w and passing with $m \to \infty$, we obtain

$$\lim_{m\to\infty} \varphi\left(T_{ms}(w)\right) \leq \varphi(w). \tag{22.19}$$

Since $x_n \rightharpoonup w$, by the Opial property of X, we have that for any $x \neq w$

$$\varphi(w) = \limsup_{n\to\infty} \|x_n - w\| < \limsup_{n\to\infty} \|x_n - x\| = \varphi(x),$$

which implies that $\varphi(w) = \inf\{\varphi(x) : x \in C\}$. This together with (22.19) gives us

$$\lim_{m\to\infty} \varphi\left(T_{ms}(w)\right) = \varphi(w). \tag{22.20}$$

By Proposition 3.4 in [17] (see also Theorem 2 in [36]) for each $d > 0$ there exists a continuous function $\lambda : [0,\infty) \to [0,\infty)$ such that $\lambda(t) = 0 \Leftrightarrow t = 0$, and

$$\|\alpha x + (1-\alpha)y\|^2 \leq \alpha\|x\|^2 + (1-\alpha)\|y\|^2 - \alpha(1-\alpha)\lambda(\|x-y\|), \tag{22.21}$$

for any $\alpha \in [0,1]$ and all $x, y \in X$ such that $\|x\| \leq d$ and $\|y\| \leq d$. Applying (22.21) to $x = x_n - w$, $y = x_n - T_{ms}(w)$ and $\alpha = \frac{1}{2}$ we obtain the following inequality:

$$\|x_n - \frac{1}{2}(w + T_{ms}(w))\|^2 \leq \frac{1}{2}\|x_n - w\|^2 + \frac{1}{2}\|x_n - T_{ms}(w)\|^2 - \frac{1}{4}\lambda\left(\|T_{ms}(w) - w\|\right).$$

Applying to both side $\limsup_{n\to\infty}$ and remembering that $\varphi(w) = \inf\{\varphi(x) : x \in C\}$ we have

$$\varphi(w)^2 \leq \frac{1}{2}\varphi(w)^2 + \frac{1}{2}\varphi\left(T_{ms}(w)\right)^2 - \frac{1}{4}\lambda\left(\|T_{ms}(w) - w\|\right),$$

which implies

$$\lambda\left(\|T_{ms}(w) - w\|\right) \leq 2\varphi\left(T_{ms}(w)\right)^2 - 2\varphi(w)^2.$$

Letting $m \to \infty$ and applying (22.20) we conclude that

$$\lim_{m\to\infty} \lambda\left(\|T_{ms}(w) - w\|\right) = 0.$$

By the properties of λ, we have $T_{ms}(w) \to w$. Fix any natural number k. Observe that, using the same argument, we conclude that $T_{(m+k)s}(w) \to w$. Note that

$$T_{ks}(T_{ms}(w)) = T_{(m+k)s}(w) \to w$$

By the continuity of T_{ks},

$$T_{ks}(T_{ms}(w)) \to T_{ks}(w)$$

and finally $T_{ks}(w) = w$ as claimed. ■

22.5 Weak Convergence of Generalized Krasnosel'skii-Mann Iteration Processes

Let us start with the precise definition of the generalized Krasnosel'skii-Mann iteration process for semigroups of nonlinear mappings.

Definition 22.18. Let $\mathscr{F} \in \mathscr{S}(\mathscr{C})$, $\{t_k\} \subset J$ and $\{c_k\} \subset (0,1)$. The generalized Krasnosel'skii-Mann iteration process gKM($\mathscr{F}, \{c_k\}, \{t_k\}$) generated by the semigroup $\mathscr{F}$, the sequences $\{c_k\}$ and $\{t_k\}$, is defined by the following iterative formula:

$$x_{k+1} = c_k T_{t_k}(x_k) + (1 - c_k)x_k, \text{ where } x_1 \in C \text{ is chosen arbitrarily,} \tag{22.22}$$

and

1. $\{c_k\}$ is bounded away from 0 and 1
2. $\lim_{k\to\infty} t_k = \infty$
3. $\sum_{n=1}^{\infty} b_{t_n}(x) < \infty$ for every $x \in C$

Definition 22.19. We say that a generalized Krasnosel'skii-Mann iteration process gKM($\mathscr{F}, \{c_k\}, \{t_k\}$) is well defined if

$$\limsup_{k\to\infty} a_{t_k}(x_k) = 1. \tag{22.23}$$

We will prove a series of lemmas necessary for the proof of the generalized Krasnosel'skii-Mann process convergence theorems.

Lemma 22.20. *Let C be a bounded, closed, and convex subset of a Banach space X. Let $\mathscr{F} \in \mathscr{S}(C)$, $w \in F(\mathscr{F})$, and let $\{x_k\}$ be a sequence generated by a generalized Krasnosel'skii-Mann process gKM($\mathscr{F}, \{c_k\}, \{t_k\}$). Then there exists an $r \in \mathbb{R}$ such that $\lim_{k\to\infty} \|x_k - w\| = r$.*

Proof. Let $w \in F(\mathscr{F})$. Since

$$\begin{aligned}
\|x_{k+1} - w\| &\le c_k\|T_{t_k}(x_k) - w\| + (1 - c_k)\|x_k - w\| \\
&= c_k\|T_{t_k}(x_k) - T_{t_k}(w)\| + (1 - c_k)\|x_k - w\| \\
&\le c_k(1 + b_{t_k}(w))\|x_k - w\| + (1 - c_k)\|x_k - w\| \\
&\le c_k b_{t_k}(w)\|x_k - w\| + \|x_k - w\| \\
&\le b_{t_k}(w)\operatorname{diam}(C) + \|x_k - w\|,
\end{aligned}$$

it follows that for every $n \in \mathbb{N}$

$$\|x_{k+n} - w\| \leq \|x_k - w\| + \operatorname{diam}(C) \sum_{i=k}^{k+n-1} b_{t_i}(w). \tag{22.24}$$

Denote $r_p = \|x_p - w\|$ for every $p \in \mathbb{N}$ and $d_{k,n} = \operatorname{diam}(C) \sum_{i=k}^{k+n-1} b_{t_i}(w)$. Observe that $\limsup_{k\to\infty} \limsup_{n\to\infty} d_{k,n} = 0$. By Lemma 22.7 then, there exists an $r \in \mathbb{R}$ such that $\lim_{k\to\infty} \|x_k - w\| = r$. ■

Lemma 22.21. *Let C be a bounded, closed, and convex subset of a uniformly convex Banach space X. Let $\mathscr{F} \in \mathscr{S}(C)$. Let $\{x_k\}$ be a sequence generated by a well-defined generalized Krasnosel'skii-Mann process $gKM(\mathscr{F}, \{c_k\}, \{t_k\})$. Then*

$$\lim_{k\to\infty} \|T_{t_k}(x_k) - x_k\| = 0 \tag{22.25}$$

and

$$\lim_{k\to\infty} \|x_{k+1} - x_k\| = 0. \tag{22.26}$$

Proof. By Theorem 22.6, $F(\mathscr{F}) \neq \emptyset$. Let us fix $w \in F(\mathscr{F})$. By Lemma 22.20, there exists an $r \in \mathbb{R}$ such that $\lim_{k\to\infty} \|x_k - w\| = r$. Because $w \in F(\mathscr{F})$, and the process is well defined, then there holds

$$\begin{aligned}
\limsup_{k\to\infty} \|T_{t_k}(x_k) - w\| &= \limsup_{k\to\infty} \|T_{t_k}(x_k) - T_{t_k}(w)\| \\
&\leq \limsup_{k\to\infty} a_{t_k}(x_k)\|x_k - w\| = r.
\end{aligned}$$

Observe that

$$\lim_{k\to\infty} \|c_k(T_{t_k}(x_k) - w) + (1 - c_k)(x_k - w)\| = \lim_{k\to\infty} \|x_{k+1} - w\| = r.$$

By Lemma 22.9 applied to $u_k = x_k - w$, $v_k = T_{t_k}(x_k) - w$,

$$\lim_{k\to\infty} \|T_{t_k}(x_k) - x_k\| = 0, \tag{22.27}$$

which by the construction of the sequence $\{x_k\}$ is equivalent to

$$\lim_{k\to\infty} \|x_{k+1} - x_k\| = 0. \tag{22.28}$$

■

Let us prove an important technical result which demonstrates that under suitable assumption the sequence $\{x_k\}$ generated by the generalized Krasnosel'skii-Mann iteration process becomes an approximate fixed-point sequence, which will provide a crucial step for proving the process convergence.

Lemma 22.22. *Let C be a bounded, closed, and convex subset of a uniformly convex Banach space X. Let $\mathscr{F} \in \mathscr{S}(\mathscr{C})$. Let the generalized Krasnosel'skii-Mann process $gKM(\mathscr{F}, \{c_k\}, \{t_k\})$ be well defined. Let $A \subset J$ be such that to every $s \in A$ there exists a strictly increasing sequence of natural numbers $\{j_k\}$ satisfying the following conditions:*

(a) $\|x_k - x_{j_k}\| \to 0$ *as* $k \to \infty$
(b) $\lim_{k\to\infty} \|T_{d_k}(x_{j_k}) - x_{j_k}\| = 0$, *where* $d_k = |t_{j_{k+1}} - t_{j_k} - s|$.

Then $\{x_k\}$ is an approximate fixed-point sequence for all mappings $\{T_{ms}\}$ where $s \in A$ and $m \in \mathbb{N}$, that is,

$$\lim_{k\to\infty} \|T_{ms}(x_k) - x_k\| = 0 \tag{22.29}$$

for every $s \in A$ and $m \in \mathbb{N}$. If, in addition, A is a generating set for J, then

$$\lim_{k\to\infty} \|T_t(x_k) - x_k\| = 0 \tag{22.30}$$

for any $t \in J$.

Proof. In view of Lemma 22.13, it is enough to prove (22.29) for $m = 1$. To this end, let us fix $s \in A$. Note that

$$\|x_{j_k} - x_{j_{k+1}}\| \to 0 \text{ as } k \to \infty. \tag{22.31}$$

Indeed,

$$\|x_{j_k} - x_{j_{k+1}}\| \le \|x_{j_k} - x_k\| + \|x_k - x_{k+1}\| + \|x_{k+1} - x_{j_{k+1}}\| \to 0, \tag{22.32}$$

in view of the above assumption (a) and of (22.26) in Lemma 22.21.

Observe that

$$\|x_{j_k} - T_s(x_{j_k})\| \to 0 \text{ as } k \to \infty. \tag{22.33}$$

Indeed,

$$\begin{aligned} \|x_{j_k} - T_s(x_{j_k})\| \le\ & \|x_{j_k} - x_{j_{k+1}}\| + \|x_{j_{k+1}} - T_{t_{j_{k+1}}}(x_{j_{k+1}})\| \\ & + \|T_{t_{j_{k+1}}}(x_{j_{k+1}}) - T_{t_{j_{k+1}}}(x_{j_k})\| + \|T_{t_{j_{k+1}}}(x_{j_k}) - T_{s+t_{j_k}}(x_{j_k})\| \\ & + \|T_{s+t_{j_k}}(x_{j_k}) - T_s(x_{j_k})\| \\ \le\ & \|x_{j_k} - x_{j_{k+1}}\| + \|x_{j_{k+1}} - T_{t_{j_{k+1}}}(x_{j_{k+1}})\| \end{aligned}$$

$$+a_{t_{j_{k+1}}}(x_{j_{k+1}})\|x_{j_{k+1}}-x_{j_k}\|+a_{s+t_{j_k}}(x_{j_k})\|T_{d_k}(x_{j_k})-x_{j_k}\|$$
$$+\sup_{x\in C}a_s(x)\|T_{t_{j_k}}(x_{j_k})-x_{j_k}\|,$$

which tends to the zero as $k\to\infty$ because of (22.31), Lemma 22.21, the fact that the process is well defined, assumptions (b) and (22.7), and the boundedness of each function a_s.

On the other hand,

$$\begin{aligned}\|x_k-T_s(x_k)\| &\le \|x_k-x_{j_k}\|+\|x_{j_k}-T_{t_{j_k}}(x_{j_k})\|+\|T_{t_{j_k}}(x_{j_k})-T_{s+t_{j_k}}(x_{j_k})\|\\ &\quad+\|T_{s+t_{j_k}}(x_{j_k})-T_s(x_{j_k})\|+\|T_s(x_{j_k})-T_s(x_k)\|\\ &\le \|x_k-x_{j_k}\|+\|x_{j_k}-T_{t_{j_k}}(x_{j_k})\|+a_{t_{j_k}}(x_{j_k})\|x_{j_k}-T_s(x_{j_k})\|\\ &\quad+a_s(x_{j_k})\|T_{t_{j_k}}(x_{j_k})-x_{j_k}\|+a_s(x_k)\|x_{j_k}-x_k\|\end{aligned}$$

which tends to the zero as $k\to\infty$ because of assumption (a), Lemma 22.21, the fact that the process is well defined, and the fact that the semigroup is asymptotic pointwise nonexpansive. If A is a generating set for J then by Lemma 22.14, $\{x_k\}$ is an approximate fixed point sequence for any T_s. This completes the proof of the Lemma. ∎

We will prove next a generic version of the weak convergence theorem for the sequences $\{x_k\}$ which are generated by the Krasnosel'skii-Mann iteration process and are at the same time approximate fixed-point sequences.

Theorem 22.23. *Let X be a uniformly convex Banach space X with the Opial property. Let C be a bounded, closed, and convex subset of a X. Let $\mathscr{F}\in\mathscr{S}(C)$. Assume that $gKM(\mathscr{F},\{c_k\},\{t_k\})$ is a well-defined Krasnosel'skii-Mann iteration process. If the sequence $\{x_k\}$ generated by $gKM(\mathscr{F},\{c_k\},\{t_k\})$ is an approximate fixed-point sequence for every $s\in A\subset J$ where A is a generating set for J, then $\{x_k\}$ converges weakly to a common fixed point $w\in F(\mathscr{F})$.*

Proof. Consider $y,z\in C$, two weak cluster points of the sequence $\{x_k\}$. Then there exist two subsequences $\{y_k\}$ and $\{z_k\}$ of $\{x_k\}$ such that $y_k\rightharpoonup y$ and $z_k\rightharpoonup z$. Fix any $s\in A$. Since $\{x_k\}$ is an approximate fixed-point sequence for s it follows that

$$\lim_{k\to\infty}\|T_s(x_k)-x_k\|=0. \tag{22.34}$$

It follows from the Demiclosedness Principle (Theorem 22.17) that $T_s(y)=y$ and $T_s(z)=z$. By Lemma 22.20 the following limits exist:

$$r_1=\lim_{k\to\infty}\|x_k-y\|,\ r_2=\lim_{k\to\infty}\|x_k-z\|. \tag{22.35}$$

We claim that $y = z$. Indeed, assume to the contrary that $y \neq z$. By the Opial property, we have

$$r_1 = \liminf_{k\to\infty} \|y_k - y\| < \liminf_{k\to\infty} \|y_k - z\| = r_2$$
$$= \liminf_{k\to\infty} \|z_k - z\| < \liminf_{k\to\infty} \|z_k - y\| = r_1. \quad (22.36)$$

The contradiction implies $y = z$ which means that the sequence $\{x_k\}$ has at most one weak cluster point. Since C is weakly sequentially compact, we deduce that the sequence $\{x_k\}$ has exactly one weak cluster point $w \in C$, which means that $x_k \rightharpoonup w$. Applying the Demiclosedness Principle again, we get $T_s(w) = w$. Since $s \in A$ was chosen arbitrarily and the construction of w did not depend on the selection of s, and A is a generating set for J, we conclude that $T_t(w) = w$ for any $t \in J$, as claimed. ∎

Let us apply the above result to some more specific situations. Let us start with a discrete case. First, we need to recall the following notions.

Definition 22.24. A strictly increasing sequence $\{n_i\} \subset \mathbb{N}$ is called quasi-periodic if the sequence $\{n_{i+1} - n_i\}$ is bounded, or equivalently if there exists a number $p \in \mathbb{N}$ such that any block of p consecutive natural numbers must contain a term of the sequence $\{n_i\}$. The smallest of such numbers p will be called a quasi-period of $\{n_i\}$.

Theorem 22.25. *Let X be a uniformly convex Banach space X with the Opial property. Let C be a bounded, closed and convex subset of a X. Let $\mathscr{F} \in \mathscr{S}(C)$ be a semigroup with a discrete generating set $A = \{\alpha_1, \alpha_2, \alpha_3 \ldots\}$. Assume that $gKM(\mathscr{F}, \{c_k\}, \{t_k\})$ is a well defined Krasnosel'skii-Mann iteration process. Assume that for every $m \in \mathbb{N}$, there exists a strictly increasing, quasi-periodic sequence of natural numbers $\{j_k(m)\}$, with a quasi-period p_m, such that for every $k \in \mathbb{N}$, $t_{j_{k+1}(m)} = \alpha_m + t_{j_k(m)}$. Then the sequence $\{x_k\}$ generated by $gKM(\mathscr{F}, \{c_k\}, \{t_k\})$ converges weakly to a common fixed point $w \in F(\mathscr{F})$.*

Proof. We will apply Lemma 22.22. Note that the assumption (b) of Lemma 22.22 is trivially satisfied since $t_{j_{k+1}(m)} - t_{j_k(m)} - \alpha_m = 0$. To prove (a), observe that by the quasi-periodicity of $\{j_k(m)\}$, to every positive integer k, there exists $j_k(m)$ such that $|k - j_k(m)| \leq p_m$. Assume that $k - p_m \leq j_k(m) \leq k$ (the proof for the other case is identical). Fix $\varepsilon > 0$. Note that by Lemma 22.21, $\|x_{k+1} - x_k\| < \dfrac{\varepsilon}{p_m}$ for k sufficiently large. Hence for k sufficiently large there holds

$$\|x_k - x_{j_k}\| \leq \|x_k - x_{k-1}\| + \ldots + \|x_{j_k(m)+1} - x_{j_k(m)}\| \leq p_m \frac{\varepsilon}{p_m} = \varepsilon. \quad (22.37)$$

This proves that (a) is also satisfied. Therefore, by Lemma 22.22, $\{x_k\}$ is an approximate fixed-point sequence for every T_s where $s \in \mathscr{J}$. By Theorem 22.23, $\{x_k\}$ converges weakly to a common fixed point $w \in F(\mathscr{F})$. ∎

Remark 22.26. Note that Theorem 4.1 in [18] is actually a special case of Theorem 22.25 with $A = \{1\}$.

Remark 22.27. It is easy to see that we can always construct a sequence $\{t_k\}$ with the properties specified in the assumptions of Theorem 22.25. When constructing concrete implementations of this algorithm, the difficulty will be to ensure that the constructed sequence $\{t_k\}$ is not "too sparse" in the sense that the Krasnosel'skii-Mann process $gKM(\mathscr{F}, \{c_k\}, \{t_k\})$ remains well defined (see Definition 22.19).

The following theorem is an immediate consequence of Theorem 22.23 and Lemmas 22.14, 22.22, and 22.15.

Theorem 22.28. *Let X be a uniformly convex Banach space X with the Opial property. Let C be a bounded, closed, and convex subset of X. Let $\mathscr{F} \in \mathscr{S}(C)$ be equicontinuous and $B \subset \overline{B} = A \subset J$ where A is a generating set for J. Let $\{x_k\}$ be generated by a well-defined Krasnosel'skii-Mann iteration process $gKM(\mathscr{F}, \{c_k\}, \{t_k\})$. If to every $s \in B$ there exists a strictly increasing sequence of natural numbers $\{j_k\}$ satisfying the following conditions:*

(a) $t_{j_{k+1}} - t_{j_k} \to s$ *as* $k \to \infty$
(b) $\|x_k - x_{j_k}\| \to 0$ *as* $k \to \infty$

then the sequence $\{x_k\}$ converges weakly to a common fixed point $w \in F(\mathscr{F})$.

Remark 22.29. Observe that the set B in Theorem 22.28 can be made countable. Hence by Remark 22.27 a sequence $\{t_k\}$ satisfying assumptions of Theorem 22.28 can be always constructed. Again, the main difficulty is in ensuring that the corresponding process $gKM(\mathscr{F}, \{c_k\}, \{t_k\})$ is well defined.

22.6 Weak Convergence of Generalized Ishikawa Iteration Processes

The two-step Ishikawa iteration process is a generalization of the one-step Krasnosel'skii-Mann process. The Ishikawa iteration process provides more flexibility in defining the algorithm parameters which is important from the numerical implementation perspective.

Definition 22.30. Let $\mathscr{F} \in \mathscr{S}(\mathscr{C})$, $\{t_k\} \subset J$. Let $\{c_k\} \subset (0,1)$, and $\{d_k\} \subset (0,1)$. The generalized Ishikawa iteration process generated by the semigroup $\mathscr{F}$, the sequences $\{c_k\}$, $\{d_k\}$, and $\{t_k\}$, is defined by the following iterative formula:

$$x_{k+1} = c_k T_{t_k}(d_k T_{t_k}(x_k) + (1 - d_k)x_k) + (1 - c_k)x_k, \tag{22.38}$$

where $x_1 \in C$ is chosen arbitrarily, and

1. $\{c_k\}$ is bounded away from 0 and 1, and $\{d_k\}$ is bounded away from 1
2. $\lim_{k\to\infty} t_k = \infty$
3. $\sum_{n=1}^{\infty} b_{t_n}(x) < \infty$ for every $x \in C$

Definition 22.31. We say that a generalized Ishikawa iteration process gI($\mathscr{F}, \{c_k\}, \{d_k\}, \{t_k\}$) is well defined if

$$\limsup_{k\to\infty} a_{t_k}(x_k) = 1. \tag{22.39}$$

Lemma 22.32. *Let C be a bounded, closed, and convex subset of a Banach space X. Let $\mathscr{F} \in \mathscr{S}(C)$, $w \in F(\mathscr{F})$, and let $gI(\mathscr{F}, \{c_k\}, \{d_k\}, \{t_k\})$ be a generalized Ishikawa process. Then there exists an $r \in \mathbb{R}$ such that $\lim_{k\to\infty} \|x_k - w\| = r$.*

Proof. Define $G_k : C \to C$ by

$$G_k(x) = c_k T_{t_k}\Big(d_k T_{t_k}(x) + (1-d_k)x\Big) + (1-c_k)x,\ x \in C. \tag{22.40}$$

It is easy to see that $x_{k+1} = G_k(x_k)$ and that $F(\mathscr{F}) \subset F(G_k)$ for every $k \geq 1$. Moreover, a straight calculation shows that each G_k satisfies

$$\|G_k(x) - G_k(y)\| \leq A_k(x)\|x-y\|, \tag{22.41}$$

where

$$A_k(x) = 1 + c_k a_{t_k}\Big(d_k T_{t_k}(x) + (1-d_k)x\Big)(1 + d_k a_{t_k}(x) - d_k) - c_k. \tag{22.42}$$

Note that $A_k(x) \geq 1$ which follows directly from the fact that $a_{t_k}(z) \geq 1$ for any $z \in C$. Using (22.42) and remembering that $w \in F(\mathscr{F})$ we have

$$B_k(w) = A_k(w) - 1 = c_k(1 + d_k a_{t_k}(w))(a_{t_k}(w) - 1) \leq (1 + a_{t_k}(w))b_{t_k}(w). \tag{22.43}$$

Fix any $M > 1$. Since $\lim_{k\to\infty} a_{t_k}(w) = 1$, it follows that there exists a $k_0 \geq 1$ such that for $k > k_0$, $a_{t_k}(w) \leq M$. Therefore, using the same argument as in the proof of Lemma 22.20, we deduce that for $k > k_0$ and $n > 1$

$$\begin{aligned}\|x_{k+n} - w\| &\leq \|x_k - w\| + \mathrm{diam}(C)\sum_{i=k}^{k+n-1} B_{t_i}(w)\\ &\leq \|x_k - w\| + \mathrm{diam}(C)(1+M)\sum_{i=k}^{k+n-1} b_{t_i}(w).\end{aligned} \tag{22.44}$$

Arguing like in the proof of Lemma 22.20, we conclude that there exists an $r \in \mathbb{R}$ such that $\lim_{k\to\infty} \|x_k - w\| = r$. ∎

Lemma 22.33. *Let C be a bounded, closed, and convex subset of a uniformly convex Banach space X. Let $\mathscr{F} \in \mathscr{S}(C)$. Let $gI(\mathscr{F}, \{c_k\}, \{d_k\}, \{t_k\})$ be a well-defined generalized Ishikawa iteration process. Then*

$$\lim_{k\to\infty} \|T_{t_k}(x_k) - x_k\| = 0 \tag{22.45}$$

and

$$\lim_{k\to\infty} \|x_{k+1} - x_k\| = 0. \tag{22.46}$$

Proof. By Theorem 22.6, $F(\mathscr{F}) \neq \emptyset$. Let us fix $w \in F(\mathscr{F})$. By Lemma 22.32, the limit $\lim_{k\to\infty} \|x_k - w\|$ exists. Let us denote it by r. Let us define

$$y_k = d_k T_{t_k}(x_k) + (1 - d_k)x_k. \tag{22.47}$$

Since $w \in F(\mathscr{F})$, $\mathscr{F} \in \mathscr{S}(C)$, and $\lim_{k\to\infty} \|x_k - w\| = r$, we have the following:

$$\begin{aligned}
\limsup_{k\to\infty} \|T_{t_k}(y_k) - w\| &= \limsup_{k\to\infty} \|T_{t_k}(y_k) - T_{t_k}(w)\| \\
&\leq \limsup_{k\to\infty} a_{t_k}(w)\|y_k - w\| \\
&= \limsup_{k\to\infty} a_{t_k}(w)\|d_k T_{t_k}(x_k) + (1 - d_k)x_k - w\| \\
&\leq \limsup_{k\to\infty} \left(d_k a_{t_k}(w)\|T_{t_k}(x_k) - w\| + (1 - d_k)a_{t_k}(w)\|x_k - w\|\right) \\
&\leq \lim_{k\to\infty} \left(d_k a_{t_k}^2(w)\|x_k - w\| + (1 - d_k)a_{t_k}(w)\|x_k - w\|\right) \leq r.
\end{aligned} \tag{22.48}$$

Note that

$$\begin{aligned}
&\lim_{k\to\infty} \|d_k(T_{t_k}(y_k) - w) + (1 - d_k)(x_k - w)\| \\
&= \lim_{k\to\infty} \|d_k T_{t_k}(y_k) + (1 - d_k)x_k - w\| = \lim_{k\to\infty} \|x_{k+1} - w\| = r.
\end{aligned} \tag{22.49}$$

Applying Lemma 22.9 with $u_k = T_{t_k}(y_k) - w$ and $v_k = x_k - w$, we obtain the equality $\lim_{k\to\infty} \|T_{t_k}(y_k) - x_k\| = 0$. This fact, combined with the construction formulas for x_{k+1} and y_k, proves (22.46).

Since

$$\begin{aligned}
\|T_{t_k}(x_k) - x_k\| &\leq \|T_{t_k}(x_k) - T_{t_k}(y_k)\| + \|T_{t_k}(y_k) - x_k\| \\
&\leq a_{t_k}(x_k)\|x_k - y_k\| + \|T_{t_k}(y_k) - x_k\| \\
&= d_k a_{t_k}(x_k)\|T_{t_k}(x_k) - x_k\| + \|T_{t_k}(y_k) - x_k\|,
\end{aligned} \tag{22.50}$$

it follows that

$$\|T_{t_k}(x_k) - x_k\| \leq (1 - d_k a_{t_k}(x_k))^{-1} \|T_{t_k}(y_k) - x_k\|. \tag{22.51}$$

The right-hand side of this inequality tends to zero because $\|T_{t_k}(y_k) - x_k\| \to 0$, $\limsup_{k\to\infty} a_{t_k}(x_k) = 1$ by the fact that the Ishikawa process is well defined, and $\{d_k\} \subset (0,1)$ is bounded away from 1. ■

We need the following technical result being the Ishikawa version of Lemma 22.22.

Lemma 22.34. *Let C be a bounded, closed, and convex subset of a uniformly convex Banach space X. Let $\mathscr{F} \in \mathscr{S}(\mathscr{C})$. Let the generalized Ishikawa process $gI(\mathscr{F}, \{c_k\}, \{d_k\}, \{t_k\})$ be well defined. Let $A \subset J$ be such that to every $s \in A$ there exists a strictly increasing sequence of natural numbers $\{j_k\}$ satisfying the following conditions:*

(a) $\|x_k - x_{j_k}\| \to 0$ *as* $k \to \infty$
(b) $\lim_{k\to\infty} \|T_{e_k}(x_{j_k}) - x_{j_k}\| = 0$, *where* $e_k = |t_{j_{k+1}} - t_{j_k} - s|$

Then $\{x_k\}$ is an approximate fixed-point sequence for all mappings $\{T_{ms}\}$ where $s \in A$ and $m \in \mathbb{N}$, that is,

$$\lim_{k\to\infty} \|T_{ms}(x_k) - x_k\| = 0 \tag{22.52}$$

for every $s \in A$ and $m \in \mathbb{N}$. If, in addition, A is a generating set for J then

$$\lim_{k\to\infty} \|T_t(x_k) - x_k\| = 0 \tag{22.53}$$

for any $t \in J$.

Proof. The proof is analogous to that of Lemma 22.22 with Lemma 22.20 replaced by Lemma 22.32 and Lemma 22.21 replaced by Lemma 22.33. ■

We are now ready to provide the weak convergence results for the Ishikawa iteration processes.

Theorem 22.35. *Let X be a uniformly convex Banach space X with the Opial property. Let C be a bounded, closed, and convex subset of a X. Let $\mathscr{F} \in \mathscr{S}(C)$. Assume that $gI(\mathscr{F}, \{c_k\}, \{d_k\}, \{t_k\})$ is a well-defined Ishikawa iteration process. If the sequence $\{x_k\}$ generated by $gI(\mathscr{F}, \{c_k\}, \{d_k\}, \{t_k\})$ is an approximate fixed-point sequence for every $s \in A \subset J$ where A is a generating set for J, then $\{x_k\}$ converges weakly to a common fixed point $w \in F(\mathscr{F})$.*

Proof. The proof is analogous to that of Theorem 22.23 with Lemma 22.22 replaced by Lemma 22.34 and Lemma 22.20 replaced by Lemma 22.32. ■

Similarly, it is easy to modify the proof of Theorems 22.25 and 22.28 to obtain the next two results.

Theorem 22.36. *Let X be a uniformly convex Banach space X with the Opial property. Let C be a bounded, closed, and convex subset of a X. Let* $\mathscr{F} \in \mathscr{S}(C)$ *be a semigroup with a discrete generating set* $A = \{\alpha_1, \alpha_2, \alpha_3 \dots\}$. *Assume that* $gI(\mathscr{F}, \{c_k\}, \{d_k\}, \{t_k\})$ *is a well-defined Ishikawa iteration process. Assume that for every* $m \in \mathbb{N}$ *there exists a strictly increasing, quasi-periodic sequence of natural numbers* $\{j_k(m)\}$, *with a quasi-period* p_m, *such that for every* $k \in \mathbb{N}$, $t_{j_{k+1}(m)} = \alpha_m + t_{j_k(m)}$. *Then the sequence* $\{x_k\}$ *generated by* $gI(\mathscr{F}, \{c_k\}, \{d_k\}, \{t_k\})$ *converges weakly to a common fixed point* $w \in F(\mathscr{F})$.

Theorem 22.37. *Let X be a uniformly convex Banach space X with the Opial property. Let C be a bounded, closed, and convex subset of X. Let* $\mathscr{F} \in \mathscr{S}(C)$ *be equicontinuous and* $B \subset \overline{B} = A \subset J$ *where A is a generating set for J. Let* $\{x_k\}$ *be generated by a well-defined Ishikawa iteration process* $gI(\mathscr{F}, \{c_k\}, \{d_k\}, \{t_k\})$. *If to every* $s \in B$ *there exists a strictly increasing sequence of natural numbers* $\{j_k\}$ *satisfying the following conditions:*

(a) $t_{j_{k+1}} - t_{j_k} \to s$ *as* $k \to \infty$
(b) $\|x_k - x_{j_k}\| \to 0$ *as* $k \to \infty$

then the sequence $\{x_k\}$ *converges weakly to a common fixed point* $w \in F(\mathscr{F})$.

22.7 Strong Convergence of Generalized Krasnosel'Skii-Mann and Ishikawa Iteration Processes

Lemma 22.38. *Let C be a compact subset of a Banach space X. Let* $\mathscr{F} \in \mathscr{S}(C)$ *and* $\{s_n\} \subset J$. *If* $s_n \to 0$ *as* $n \to \infty$ *then* $\mathscr{F}$ *is equicontinuous, that is,*

$$\lim_{n\to\infty} \sup_{x\in C} \|T_{s_n}(x) - x\| = 0. \tag{22.54}$$

Proof. Assume to the contrary that (22.54) does not hold. Then there exist $\{w_k\}$ a subsequence of $\{s_n\}$, a sequence $\{y_k\} \subset C$ and $\eta > 0$ such that for every $k \in \mathbb{N}$ there holds

$$\|T_{w_k}(y_k) - y_k\| > \eta > 0. \tag{22.55}$$

Using compactness of C and passing to a subsequence of $\{y_k\}$ if necessary we can assume that there exists $w \in C$ such that $\|y_k - w\| \to 0$ as $k \to \infty$.

$$
\begin{aligned}
0 < \eta &\leq \limsup_{k\to\infty} \|T_{w_k}(y_k) - y_k\| \\
&\leq \limsup_{k\to\infty} (\|T_{w_k}(y_k) - T_{w_k}(w)\| + \|T_{w_k}(w) - w\| + \|w - y_k\|) \\
&\leq \limsup_{k\to\infty} (a_{w_k}(w)\|y_k - w\| + \|T_{w_k}(w) - w\| + \|w - y_k\|) = 0 \qquad (22.56)
\end{aligned}
$$

since $\limsup_{k\to\infty} a_{w_k}(w) \leq 1$ and $t \mapsto T_t(w)$ is continuous, a contradiction. ∎

Theorem 22.39. *Let C be a compact, convex subset of a uniformly convex Banach space X. Let $\mathscr{F} \in \mathscr{S}(C)$ and $B \subset \overline{B} = A \subset J$ where A is a generating set for J. Let $\{x_k\}$ be generated by a well-defined Krasnosel'skii-Mann iteration process $gKM(\mathscr{F}, \{c_k\}, \{t_k\})$ (resp. generalized Ishikawa process $gI(\mathscr{F}, \{c_k\}, \{d_k\}, \{t_k\})$). If to every $s \in B$ there exists a strictly increasing sequence of natural numbers $\{j_k\}$ satisfying the following conditions:*

(a) $t_{j_{k+1}} - t_{j_k} \to s$ as $k \to \infty$
(b) $\|x_k - x_{j_k}\| \to 0$ as $k \to \infty$

then the sequence $\{x_k\}$ converges strongly to a common fixed point $x \in F(\mathscr{F})$.

Proof. We apply Lemma 22.22 (resp. Lemma 22.34) for the parameter set B. Note that condition (a) of Lemma 22.22 (resp. Lemma 22.34) is assumed. By Lemma 22.38 the semigroup $\mathscr{F}$ is equicontinuous and hence the assumption (b) of Lemma 22.22 (resp. Lemma 22.34) is satisfied. By Lemma 22.22 (resp. Lemma 22.34) then $\{x_k\}$ is an approximate fixed-point sequence for any T_t where $t \in B$. By Lemma 22.15 $\{x_k\}$ is an approximate fixed-point sequence for any T_t where $t \in A$. Since A is a generating set for J, it follows that $\{x_k\}$ is an approximate fixed-point sequence for any T_t where $t \in J$ (again, by Lemma 22.22 or respectively Lemma 22.34 for the Ishikawa case). Hence for every $t \in J$

$$\|T_t(x_k) - x_k\| \to 0 \text{ as } k \to \infty. \qquad (22.57)$$

Since C is compact there exist a subsequence $\{x_{p_k}\}$ of $\{x_k\}$ and $x \in C$ such that

$$\|T_t(x_{p_k}) - x\| \to 0 \text{ as } k \to \infty. \qquad (22.58)$$

Observe that

$$\|x_{p_k} - x\| \leq \|x_{p_k} - T_t(x_{p_k})\| + \|T_t(x_{p_k}) - x\|, \qquad (22.59)$$

which tends to zero as $k \to \infty$ by (22.57) and (22.58). Hence

$$\lim_{k\to\infty} \|x_{p_k} - x\| = 0. \qquad (22.60)$$

Finally

$$\begin{aligned} \|T_t(x)-x\| &\leq \|T_t(x)-T_t(x_{p_k})\|+\|T_t(x_{p_k})-x_{p_k}\|+\|x_{p_k}-x\| \\ &\leq a_t(x)\|x_{p_k}-x\|+\|T_t(x_{p_k})-x_{p_k}\|+\|x_{p_k}-x\|, \end{aligned} \tag{22.61}$$

which tends to zero as $k \to \infty$ by boundedness of the function a_t, by (22.60) and (22.57). Therefore, $T_t(x) = x$ for every $t \in J$, that is, x is a common fixed point for the semigroup $\mathscr{F}$. By Lemma 22.20 (resp. Lemma 22.32), $\lim_{k\to\infty} \|x_k - x\|$ exists which, via (22.60), implies that

$$\lim_{k\to\infty} \|x_k - x\| = 0 \tag{22.62}$$

completing the proof of the theorem. ∎

Acknowledgements The authors would like to thank the referee for the valuable suggestions to improve the presentation of the paper.

References

1. Belluce, L.P., Kirk, W.A.: Fixed-point theorems for families of contraction mappings. Pacific J. Math. **18**, 213–217 (1966)
2. Belluce, L.P., Kirk, W.A.: Nonexpansive mappings and fixed-points in Banach spaces. Illinois J. Math. **11**, 474–479 (1967)
3. Bose, S.C.: Weak convergence to the fixed point of an asymptotically nonexpansive map. Proc. Amer. Math. Soc. **68**, 305–308 (1978)
4. Browder, F.E.: Nonexpansive nonlinear operators in a Banach space. Proc. Nat. Acad. Sci. U.S.A. **54**, 1041–1044 (1965)
5. Bruck, R.E.: A common fixed point theorem for a commuting family of nonexpansive mappings. Pacific J. Math. **53**, 59–71 (1974)
6. Bruck, R.E.: A simple proof of the mean ergodic theorem for nonlinear contractions in Banach spaces. Israel J. Math. **32**, 107–116 (1979)
7. Bruck, R., Kuczumow, T., Reich, S.: Convergence of iterates of asymptotically nonexpansive mappings in Banach spaces with the uniform Opial property. Coll. Math. **65**(2), 169–179 (1993)
8. DeMarr, R.E.: Common fixed-points for commuting contraction mappings. Pacific J. Math. **13**, 1139–1141 (1963)
9. Gornicki, J.: Weak convergence theorems for asymptotically nonexpansive mappings in uniformly convex Banach spaces. Comment. Math. Univ. Carolin. **30**, 249–252 (1989)
10. Groetsch, C.W.: A note on segmenting Mann iterates. J. Math. Anal. Appl. **40**, 369–372 (1972)
11. Hussain, N., Khamsi, M.A.: On asymptotic pointwise contractions in metric spaces. Nonlinear Anal. **71**(10), 4423–4429 (2009)
12. Ishikawa, S.: Fixed points by a new iteration method. Proc. Amer. Math. Soc. **44**, 147–150 (1974)
13. Khamsi, M.A.: On Asymptotically Nonexpansive Mappings in Hyperconvex Metric Spaces. Proc. Amer. Math. Soc. **132**, 365–373 (2004)
14. Khamsi, M.A., Kozlowski, W.M.: On asymptotic pointwise contractions in modular function spaces. Nonlinear Anal. **73**, 2957–2967 (2010)
15. Khamsi, M.A., Kozlowski, W.M.: On asymptotic pointwise nonexpansive mappings in modular function spaces. J. Math. Anal. Appl. **380**(2), 697–708 (2011)
16. Kirk, W.A.: Mappings of generalized contractive type. J. Math. Anal. Appl. **32**, 567–572 (1974)

17. Kirk, W.A., Xu, H.K.: Asymptotic pointwise contractions. Nonlinear Anal. **69**, 4706–4712 (2008)
18. Kozlowski, W.M.: Fixed point iteration processes for asymptotic pointwise nonexpansive mappings in Banach spaces. J. Math. Anal. Appl. **377**, 43–52 (2011)
19. Kozlowski, W.M.: Common fixed points for semigroups of pointwise Lipschitzian mappings in Banach spaces. Bull. Austral. Math Soc. **84**, 353–361 (2011)
20. Kozlowski, W.M.: On the existence of common fixed points for semigroups of nonlinear mappings in modular function spaces. Comment. Math. **51**(1), 81–98 (2011)
21. Li, G., Sims, B.: τ-Demiclosedness principle and asymptotic behavior for semigroup of nonexpansive mappings in metric spaces. pp. 103–108, Yokohama Publisher, Yokohama (2008)
22. Lim, T.C.: A fixed point theorem for families of nonexpansive mappings. Pacific J. Math. **53**, 487–493 (1974)
23. Mann, W.R.: Mean value methods in iteration. Proc. Amer. Math. Soc. **4**, 506–510 (1953)
24. Nanjaras, B., Panyanak, B.: Demiclosed principle for asymptotically nonexpansive mappings in CAT(0) spaces. Fixed Point Theor. Appl. **2010**, 1–15 (2010)
25. Noor, M.A., Xu, B.: Fixed point iterations for asymptotically nonexpansive mappings in Banach spaces. J. Math. Anal. Appl. **267**, 444–453 (2002)
26. Opial, Z.: Weak convergence of the sequence of successive approximations for nonexpansive mappings. Bull. Amer. Math. Soc. **73**, 591–597 (1967)
27. Passty, G.B.: Construction of fixed points for asymptotically nonexpansive mappings. Proc. Amer. Math. Soc. **84**, 212–216 (1982)
28. Rhoades, B.E.: Fixed point iterations for certain nonlinear mappings. J. Math. Anal. Appl. **183**, 118–120 (1994)
29. Samanta, S.K.: Fixed point theorems in a Banach space satisfying Opial's condition. J. Indian Math. Soc. **45**, 251–258 (1981)
30. Schu, J.: Iterative construction of fixed points of asymptotically nonexpansive mappings,. J. Math. Anal. Appl. **158**, 407–413 (1991)
31. Schu, J.: Weak and strong convergence to fixed points of asymptotically nonexpansive mappings. Bull. Austral. Math. Soc. **43**, 153–159 (1991)
32. Tan, K-K., Xu, H-K.: The nonlinear ergodic theorem for asymptotically nonexpansive mappings in Banach spaces. Proc. Amer. Math. Soc. **114**, 399–404 (1992)
33. Tan, K-K., Xu, H-K.: A nonlinear ergodic theorem for asymptotically nonexpansive mappings. Bull. Austral. Math. Soc. **45**, 25–36 (1992)
34. Tan, K-K., Xu, H-K.: Approximating fixed points of nonexpansive mappings by the Ishikawa iteration process. J. Math. Anal. Appl. **178**, 301–308 (1993)
35. Tan, K-K., Xu, H-K.: Fixed point iteration processes for asymptotically nonexpansive mappings. Proc. Amer. Math. Soc. **122**, 733–739 (1994)
36. Xu, H-K.: Inequalities in Banach spaces with applications. Nonlinear Anal. **16**, 1127–1138 (1991)
37. Xu, H-K.: Existence and convergence for fixed points of asymptotically nonexpansive type. Nonlinear Anal. **16**, 1139–1146 (1991)
38. Zeidler, E.: Nonlinear Functional Analysis and Its Applications I, Fixed Points Theorems. Springer, New York (1986)

Chapter 23
Techniques and Open Questions in Computational Convex Analysis

Yves Lucet

Dedicated to Jonathan Borwein on the occasion of his 60th birthday

Abstract We present several techniques that take advantage of convexity and use optimal computational geometry algorithms to build fast (log-linear or linear) time algorithms in computational convex analysis. The techniques that have strong potential to be reused include: monotonicity of the argmax and injecting convexity to use that monotonicity, Lipschitzness of the argmin, exploiting various formulas in convex analysis, using a graph data structure to vectorize computation, and building a parametrization of the graph. We also point out the potential for parallelization. The techniques can be used as a check list on open problems to find an efficient algorithm. Finally, we list several currently open questions in computational convex analysis with links to computational geometry.

Key words: Computational Convex Analysis • Inf-convolution • Legendre-Fenchel Conjugate • Moreau-Yosida Approximalte • Proximal Mapping • Proximal Average • Convex Envelope • Convexity Test

Mathematics Subject Classifications (2010): Primary 52B55, 49M25; Secondary 90C25, 47H05, 47H09, 65Y05.

COMMUNICATED BY HEINZ H. BAUSCHKE.

Y. Lucet (✉)
ASC 350, Computer Science, Arts & Sciences 5, University of British Columbia Okanagan, 3333 University Way, Kelowna, BC V1V 1V7, Canada
e-mail: yves.lucet@ubc.ca

D.H. Bailey et al. (eds.), *Computational and Analytical Mathematics*, Springer Proceedings in Mathematics & Statistics 50, DOI 10.1007/978-1-4614-7621-4_23,

23.1 Introduction

Many important results and theories in convex optimization rely on convex analysis concepts like the conjugate function, the inf-convolution, the Moreau envelope, and the proximal mapping. Computational convex analysis (CCA) studies the computation of these transforms. Numerous applications of CCA are listed in [33]. Computer-aided convex analysis, where visualization of examples is essential, gives particular motivation to CCA for functions of one or two variables.

Algorithms to compute the conjugate, Moreau envelope, addition, and scalar multiplication of convex functions, when implemented, form a toolbox to manipulate convex functions [18, 19, 32]. New transforms, like the proximal average [2–5, 25], can be studied by composing these four transforms applied to convex functions. The toolbox can be extended to the composition of these four transforms applied to nonconvex functions with the addition of the convex envelope transform [17].

Computing a transform in CCA involves solving an optimization problem for each value of a parameter. It is a parametric optimization problem. To achieve acceptable computation times, CCA avoids the curse of dimensionality (storing graphs requires a number of points that increase exponentially with the dimension) by considering mostly functions of one (univariate) or two (bivariate) variables and rarely functions of more than two (multivariate) variables.

In such a low-dimensional framework, efficient algorithms are obtained by using combinations of ideas from computational geometry and convex analysis. The first suggestion of such an algorithm can be found in [35, Remark 5c, p. 282] where the author suggests to use the nonexpansiveness of the proximal operator to compute the Moreau envelope efficiently. The resulting nonexpansive prox (NEP) algorithm was studied much later in [32]. Other authors noted that convexity can be used to obtain more efficient algorithms to compute the conjugate, giving birth to the fast Legendre transform (FLT) algorithm [9, 36] that was studied in [13, 28]. Their log-linear worst-case time complexity was subsequently improved to linear time in [30] with the linear-time Legendre transform (LLT) algorithm.

The LLT algorithm linear-time complexity has been matched by the NEP algorithm [32] and the parabolic Moreau envelope (PE) algorithm [14, 16, 32]. All these algorithms belong to a family of algorithms named fast algorithms that approximate the input function with a piecewise linear model (linear spline). While the addition, scalar multiplication, and conjugate of a piecewise linear function are still piecewise linear, it is no longer the case for the Moreau envelope of even the simplest functions, e.g., the absolute value. Consequently, composing several transforms increases approximation errors and requires much technical skill to deduce the graph of the resulting operator. The proximal average operator is a prime example of such difficulties. Two new families of algorithms were introduced to solve that issue.

The piecewise linear-quadratic (PLQ) algorithms manipulate PLQ function (a univariate PLQ function is a quadratic spline). By allowing quadratic pieces, the class of PLQ functions is closed under all core convex operators, including the Moreau envelope. One can then compute proximal averages of PLQ functions as easily as proximal averages of piecewise linear functions. Moreover, a linear-time algorithm to compute the convex envelope of a PLQ function is available [17], which extends the PLQ toolbox to nonconvex functions.

A faster family of algorithms [19] for convex PLQ functions is based on Goebel's graph-matrix calculus [20]. Instead of storing the coefficients of quadratic polynomials, graph-matrix (GPH) algorithms store the (piecewise linear) graph of the subdifferential along with the value of the function at a finite set of points. Functions such as the conjugate are then computed as a matrix multiplication. Most of the GPH algorithms give rise to embarrassingly parallel algorithms. (An embarrassingly parallel algorithm can be split effortlessly into separate tasks that can be run concurrently.)

After introducing notations and definitions in Sect. 23.2, the present paper summarizes the techniques used to obtain efficient algorithms in CCA in Sect. 23.3. We list open questions in Sect. 23.4.

23.2 Preliminaries

We follow standard notations in convex analysis as found in [40] and [23]. We write the standard inner product either as $\langle x,y\rangle$, or in vector notation as $x^T y$. The associated Euclidean norm is noted $\|\cdot\|$. The identity operator is noted I.

We make the distinction between functions, operators, and transforms as follow. We call f a *function* if $f:\mathbb{R}^d \to \mathbb{R}\{-\infty,+\infty\}$ is always single valued (f may take the value $-\infty$ or $+\infty$). An *operator* $P:\mathbb{R}^d \rightrightarrows \mathbb{R}^d$ is an application from $\mathbb{R}^d$ to the sets of $\mathbb{R}^d$, i.e., $P(x)$ is a subset of $\mathbb{R}^d$. (Operators are also called multi-valued functions or set-valued functions in the literature.) A *transform* $\Gamma : X \to X$ is defined on the set of functions and takes images in the set of functions, i.e., $\Gamma(f)$ is a function. (Transforms are sometimes called functionals in the literature.)

As is standard in convex analysis, we say a function f is *proper* if for all $x \in \mathbb{R}^d$, $f(x) > -\infty$ and there exists $x \in \mathbb{R}^d$ with $f(x) < +\infty$. An important class of functions in convex analysis is the class of proper lower semicontinuous (lsc) convex functions.

We recall the main transforms and operators in convex analysis. The *subdifferential* operator $\partial f:\mathbb{R}^d \rightrightarrows \mathbb{R}^d$ of a function $f:\mathbb{R}^d \to \mathbb{R}\{-\infty,+\infty\}$ is defined as

$$\partial f(x) = \{s \in \mathbb{R}^d | f(y) \geq f(x) + \langle s, y-x\rangle, \forall y \in \mathbb{R}^d\}.$$

(When the function f is differentiable at x with gradient $\nabla f(x)$, we have $\partial f(x) \subseteq \{\nabla f(x)\}$ with equality when in addition the function is convex.)

The *Legendre-Fenchel transform* associates a function f with its *conjugate* function f^* defined as

$$f^*(s) = \sup_{x \in \mathbb{R}^d} \left[\langle s, x \rangle - f(x) \right].$$

The *Moreau-Yosida approximate* transform associates to a function f its *Moreau envelope*

$$f^\lambda(x) = \inf_{y \in \mathbb{R}^d} \left[f(y) + \frac{\|x-y\|^2}{2\lambda} \right].$$

The *proximal mapping* operator is the set of points where the Moreau envelope attains its minimum

$$P^\lambda(x) = \operatorname*{Argmin}_{y \in \mathbb{R}^d} \left[f(y) + \frac{\|x-y\|^2}{2\lambda} \right],$$

where $P^\lambda(x) = \emptyset$ if $f^\lambda(x)$ is not finite.

The *inf-convolution* transform takes two functions f_1 and f_2 and associates a new function

$$f_1 \oplus f_2(x) = \inf_{y \in \mathbb{R}^d} \left[f_1(y) + f_2(x-y) \right].$$

The *proximal average* transform associates to two functions f_0, f_1 and two numbers λ_0, λ_1 the proximal average function $\mathscr{P}(f_0, f_1; \lambda_0, \lambda_1)$ defined as a composition of Legendre-Fenchel transforms, additions, and scalar multiplications by

$$\mathscr{P}(f_0, f_1; \lambda_0, \lambda_1) = \left(\lambda_0 \left(f_0 + \frac{\|\cdot\|^2}{2} \right)^* + \lambda_1 \left(f_1 + \frac{\|\cdot\|^2}{2} \right)^* \right)^* - \frac{\|\cdot\|^2}{2}. \quad (23.1)$$

The *convex envelope* transform associates to a function f its convex envelope

$$\operatorname{co} f(x) = \inf \left\{ \sum_{i=1}^{d+1} \lambda_i f(x_i) | \sum_{i=1}^{d+1} \lambda_i x_i = x, \lambda_i \geq 0, \sum_{i=1}^{d+1} \lambda_i = 1 \right\}$$

which is the greatest convex function majorized by f (a function g is majorized by a function h if $g(x) \leq h(x)$ for all x).

We say that an operator P is *monotone* if for all x_1, x_2, and all $s_1 \in P(x_1)$, $s_2 \in P(x_2)$, we have $\langle s_1 - s_2, x_1 - x_2 \rangle \geq 0$. Subdifferentials of proper lsc convex functions are monotone operators.

23.3 Techniques

In this section, we list several techniques that, when applicable, allow one to write an efficient algorithm. The goal is to provide researchers with a starting point to compute new transforms. Unless otherwise specified, we restrict our attention to univariate functions.

23.3.1 Symbolic Computation

If we assume the supremum is attained, computing the conjugate amounts to finding its critical points x defined by

$$s \in \partial f(x). \tag{23.2}$$

When the function f is convex and differentiable everywhere, relation (23.2) becomes

$$s = \nabla f(x), \tag{23.3}$$

an equation that may be solved symbolically by inverting the gradient. The Maple SCAT package [7] is based on that idea. It can even handle multivariate functions by repeatedly applying the partial conjugate and is very useful to check manual computation.

The main drawback of solving (23.3) is there is no guarantee that a closed form formula exists. Moreover, a closed form formula that requires pages to write is of limited use.

23.3.2 Monotonicity of the Argmax/Argmin

Convexity brings monotonicity which can be used to build faster algorithms. The FLT algorithm was the first to exploit that idea in CCA. It was noted that for a univariate function f we have

$$s \mapsto \operatorname{Argmax}[sx - f(x)]$$

is a monotone operator for convex univariate functions f. Consequently, the computation of

$$s_j \mapsto \max_i [s_j x_i - f(x_i)]$$

can be reduced from a quadratic brute-force algorithm to a log-linear algorithm similar to the fast Fourier transform (FFT) [9, 13, 28, 36]. The technique applies to any transform that has a monotone argmax (or argmin).

A similar monotonicity property was used in the LLT algorithm to reduce the computation to merging two sorted sequences [30]. The key was to introduce convexity explicitly by first computing the convex envelope of the function. Then computing the conjugate amounts to finding the point where the line with slope s_j touches the epigraph of f. Since both s_j, $j = 1,\dots,m$ and the slopes of the convex envelope of the epigraph are nondecreasing sequences, we only need to merge the two sequences and extract where they intersect to obtain the conjugate in linear time.

The conjugate of multivariate functions can be computed in linear time by applying the partial conjugate repeatedly using the fact the dot product is a separable function.

23.3.3 Lipschitzness of the Argmin

The NEP algorithm exploits the nonexpansiveness of the proximal mapping to compute the Moreau envelope of a convex function $f : \mathbb{R} \mapsto \mathbb{R} \cup \{+\infty\}$. By considering a grid of points, computing the minimum is reduced to only one if statement, i.e., for each point x_i on the grid,

$$f^\lambda(x_i) = \min\left[f(x_j) + \frac{\|x_i - x_j\|^2}{2\lambda}, f(x_{j+1}) + \frac{\|x_i - x_{j+1}\|^2}{2\lambda}\right],$$

where $x_j = P^\lambda(x_i)$ is a point in the grid and $x_{j+1} = \mathrm{Succ}(x_j) = \mathrm{Succ}(P^\lambda(x_i))$.

Consequently, if the argmin (or argmax) of an operator is Lipschitz (not necessarily with constant 1), we obtain a linear-time algorithm.

Since the squared norm is a separable function, we obtain a linear-time algorithm for multivariate functions by repeatedly applying the algorithm to each dimension. (Computing the Moreau envelope of a function of d variables requires d application of the algorithm.)

23.3.4 Using a Formula

Operators that can be reduced to the composition of core operators immediately benefit from efficient algorithms. For example, the Moreau envelope of a (multivariate) convex function f can be computed by [31], [40, Example 11.26c]

$$f^\lambda(y) = \frac{\|y\|^2}{2\lambda} - \frac{1}{\lambda}\left(\lambda f + \frac{1}{2}\|\cdot\|^2\right)^\star(y). \tag{23.4}$$

Conversely, the conjugate can be computed using

$$f^{\star}(s) = \frac{\|s\|^2}{2} - \lambda\left(\frac{1}{\lambda}f - \frac{\|\cdot\|^2}{2\lambda}\right)^{\lambda}(s).$$

Similarly, the availability of an efficient algorithm to compute the conjugate, the addition, and the scalar multiplication transforms allow one to compute the proximal average of two convex lsc proper functions efficiently using Formula (23.1) and to compute the inf-convolution of two convex lsc proper functions f_1 and f_2 using the fact

$$f_1 \oplus f_2 = (f_1^* + f_2^*)^*.$$

Other convex analysis formulas can be used as the basis for a fast algorithm. For example, the formula [2, Theorem 6.7]

$$P^{\mu}(\mathscr{P}(f_0, f_1; \lambda_0, \lambda_1)) = \lambda_0 P^{\mu} f_0 + \lambda_1 P^{\mu} f_1$$

could be used to build a fast algorithm to compute the Moreau envelope of the proximal average.

23.3.5 Graph-Matrix Calculus

Considering the availability of matrix-based mathematical software like MATLAB and Scilab, which have optimized linear algebra subroutines, it is greatly advantageous to write an algorithm as a matrix multiplication for both ease of reading and performance. It turns out that computing core convex transforms amounts to a linear transformation of the subgradient and can be reduced to a matrix multiplication or vector operations.

First, we need to detail the data structure used in CCA algorithms. Instead of storing a PLQ univariate function $f : \mathbb{R} \to \mathbb{R} \cup \{+\infty\}$ defined by

$$f(x) = \begin{cases} a_1 x^2 + b_1 x + c_1 & \text{if } x \le x_1, \\ \vdots & \vdots \\ a_n x^2 + b_n x + c_n & \text{if } x_{n-1} \le x \le x_n, \\ a_{n+1} x^2 + b_{n+1} x + c_{n+1} & \text{if } x > x_n, \end{cases}$$

as a matrix (called the PLQ matrix)

$$\begin{bmatrix} x_1 & a_1 & b_1 & c_1 \\ \vdots & \vdots & \vdots & \vdots \\ x_n & a_n & b_n & c_n \\ +\infty & a_{n+1} & b_{n+1} & c_{n+1} \end{bmatrix},$$

where $a_i, b_i \in \mathbb{R}$ and $c_i \in \mathbb{R} \cup \{+\infty\}$, the GPH toolbox stores f in GPH matrix form

$$\begin{bmatrix} \xi_1, \cdots, \xi_m \\ s_1, \cdots, s_m \\ y_1, \cdots, y_m \end{bmatrix},$$

where $\xi_j, s_j, y_j \in \mathbb{R}$, $s_j \in \partial f(\xi_j)$ and $y_j = f(\xi_j)$. The PLQ matrix is easily converted into a GPH matrix using the formulas

$$\begin{aligned}
\xi_j &= x_{\lfloor \frac{j+1}{2} \rfloor}, & j &= 1, \ldots, 2n, \\
s_{2i-1} &= 2a_i x_i + b_i, & i &= 1, \ldots, n, \\
s_{2i} &= 2a_{i+1} x_i + b_{i+1}, & i &= 1, \ldots, n, \\
f_{2i} &= f_{2i-1} = a_i x_i^2 + b_i x_i + c_i, & i &= 1, \ldots, n.
\end{aligned}$$

(The values for $i = 1$ and $i = n+1$ are special cases; see [19].)

From the formulas representing a convex function f as $s \in \partial f(x)$ and $y = f(x)$, we get the representation of f^* as $x \in \partial f^*(s)$ and $y^* = sx - f(x)$ resulting in the following MATLAB/Scilab coded algorithm:

$$\begin{pmatrix} x^* \\ s^* \end{pmatrix} = \begin{pmatrix} s \\ x \end{pmatrix} = \begin{bmatrix} 0 & 1 \\ 1 & 0 \end{bmatrix} \begin{pmatrix} x \\ s \end{pmatrix},$$

$$y^* = s. * x - y,$$

where $.*$ represents the element-wise multiplication operator.

Similarly, the Moreau envelope is computed from the GPH matrix $[x; s; y]$ as the GPH matrix $[\xi; \sigma; m]$ with

$$\begin{pmatrix} \xi \\ \sigma \end{pmatrix} = \begin{pmatrix} x + \lambda s \\ s \end{pmatrix} = \begin{bmatrix} 1 & \lambda \\ 0 & 1 \end{bmatrix} \begin{pmatrix} x \\ s \end{pmatrix}$$

$$m = f + \frac{\lambda}{2} s.\hat{}2,$$

where $.\hat{}$ denotes the element-wise power operator.

Such vector formulas have three advantages: they provide a compact algorithm easily coded in MATLAB/Scilab, they immediately benefit from optimized matrix operations, and the resulting algorithm is embarrassingly parallel.

23.3.6 Parametrization

The parametric Legendre transform (PLT) algorithm [24] uses an explicit description of the graph of the conjugate to achieve its optimal linear-time complexity.

Indeed, the graph of f^* can be split in pieces where each piece is either linear and corresponds to a kink on the graph of f or is the inverse of the gradient of f. It can be parameterized by x as

$$\begin{aligned} s &\in \partial f(x), \\ f^*(s) &= sx - f(x). \end{aligned}$$

Since we want to compute the full graph when f is PLQ, the previous formula gives the PLT algorithm: for x in each piece, compute s and deduce $f^*(s)$.

The parametrization idea was also used in [19] to obtain a faster algorithm to compute the proximal average of two proper lsc convex univariate functions. Given f_1 (resp. f_2) as a GPH matrix $G_1 = [x_1; s_1; y_1]$ (resp. $G_2 = [x_2; s_2; y_2]$) and $\lambda_1 = 1 - \lambda$, $\lambda_2 = \lambda$ with $\lambda \in [0, 1]$, define the operator

$$P = (I + \partial f_2)^{-1}(I + \partial f_1),$$

where I is the identity. The operator P can be explicitly computed in GPH matrix form from G_1 and G_2. Then we compute

$$\begin{aligned} x &= \lambda_1 x_1 + \lambda_2 P x_1, \\ s &= x_1 + s_1 - s, \\ y_{Px_1} &= f_2(Px_1) \end{aligned}$$

to deduce the GPH matrix $[x; s; y]$ of the proximal average, where

$$y = \lambda_1 (y_1 + \frac{1}{2} x_1.\hat{\ }2) + \lambda_2 (y_{Px_1} + \frac{1}{2}(Px_1).\hat{\ }2 - \frac{1}{2} x.\hat{\ }2.$$

23.3.7 *Parallel Computing*

Using p processors, computing the maximum of n values on a CREW PRAM computer takes

$$O\left(\frac{n}{p} + \log p\right)$$

time by splitting the values into p chunks of size $\frac{n}{p}$ assigned to each processor and applying a max reduction operation (see, e.g., [39] for a presentation of computer models). When $p \geq n + 1$ we can improve the resulting $O(\log p)$ complexity to $O(1)$ on a common CRCW PRAM computer by taking $\varepsilon > 0$ such that $n^{1+\varepsilon} < n + 1$ and applying [41]. The $O(1)$ running time applies when using the formula

$$f^*(s_j) = \max_i s_j x_i - f(x_i),$$

for a univariate function f and does not use any convexity assumption. However, it is restricted to piecewise linear functions.

If convexity is assumed, consider the input as a GPH matrix $G = [x; s; y]$; then the conjugate has GPH matrix $[s; x; s.*x - y]$ and can be computed directly in $O(1)$ time on an EREW PRAM computer when $p \geq m$, where m is the number of columns of G. This strategy is much more efficient than using [41]: we do not even need to write an explicit parallel algorithm if we use a parallel linear algebra library.

When convexity is not assumed, computing the convex envelope is not easily parallelizable and one has to use Wang's technique [41] to achieve $O(1)$ worst-case running time to compute the conjugate. Note that optimal algorithms to compute the convex envelope, including output sensitive algorithms, are known in various computational models [6, 11, 21, 22, 34], but none achieve an $O(1)$ running time.

Since bivariate PLQ functions do not have a grid structure, no straightforward extension from the univariate case is available.

Remark 23.1. The complexity results give an insight into the difficulty of computing the conjugate in parallel. Considering that most computers have less than 8 processors today and that computing the maximum of 10^8 values using a serial program on a 2.4 GHz dual core computer in C takes 0.3 s, the sequential algorithm is expected to be used for many more years on the CPU. However, if massive conjugate computations are required, using a GPU with tens of thousands of threads may give more speedup.

23.4 Open Questions

We now list several open problems in computational convex analysis.

23.4.1 Nonconvex Inf-Convolution

Computing the inf-convolution of univariate PLQ convex functions can be done in linear time using

$$f_1 \oplus f_2 = (f_1^* + f_2^*)^*.$$

However, the nonconvex case is more challenging even for univariate functions. We first consider the case when one function is convex and the other nonconvex.

We first define some notations. For two univariate PLQ functions f and g, we note

$$\begin{aligned}
\varphi_x(y) &= f(x-y) + g(y), \\
H(x) &= \operatorname*{Argmin}_y \varphi_x(y), \\
h(x) &= \sup H(x).
\end{aligned}$$

For example, when $f \equiv 0$ and $g \equiv 1$, we have $f \oplus g \equiv 1$, $H \equiv \mathbb{R}$, and $h \equiv +\infty$. Note that in this case H is not monotone, which is why the sup is needed in the definition of h.

When $f = I_{[-8,2]}$ (the indicator function of the set $[-8,2]$, which is equal to 0 on that set and $+\infty$ outside) and $g = I_{[0,1]}$, we find $\operatorname{dom} f \oplus g = [-8,3]$. When $x \notin \operatorname{dom} f \oplus g$, $f \oplus g(x) = +\infty$, $H(x) = \emptyset$, and $h(x) = -\infty$. Otherwise, $x \in \operatorname{dom} f \oplus g$, $f \oplus g(x) = 0$, $H(x) = [x-2, x+8]$, and $h(x) = \max(x+8, 1)$. This example shows that h is not monotone on $\mathbb{R}$.

Another example to keep in mind is $f \equiv 0$ and $g(x) = -x^2$. Then $f \oplus g \equiv -\infty$, $H(x) \equiv \emptyset$, $h \equiv -\infty$, and $\operatorname{dom} f \oplus g = \emptyset$. In this case, $f \oplus g$ is not proper.

Remark 23.2. It is straightforward to check whether $f \oplus g$ is proper for univariate PLQ functions f and g. One has only to check the coefficients of the quadratic functions on the unbounded intervals and ensure that the function $\varphi_x(y)$ does not converge to $-\infty$ when y goes to either $-\infty$ or $+\infty$.

We are now ready to state our main result for the convex-nonconvex case.

Lemma 23.3. *Assume f and g are two univariate PLQ functions. In addition, assume f is convex. Then for all x_1, $x_2 \in \operatorname{dom} f \oplus g$, $x_1 < x_2 \Rightarrow h(x_1) \leq h(x_2)$.*

Proof. The proof is sketched in [13, p. 1549–1550] under different assumptions that ensure that h is always finite. We use the same core argument for PLQ functions and handle the infinite case.

The lemma is vacuously true if $\operatorname{dom} f \oplus g$ is empty or if it is a singleton (these cases are straightforward to handle numerically). We assume this is not the case and take $x_1, x_2 \in \operatorname{dom} f \oplus g$ with $x_1 < x_2$.

First, we assume that $h(x_1)$ is finite. Take $z < h(x_1)$. Then

$$x_1 - h(x_1) < x_2 - h(x_1) < x_2 - z$$

and

$$x_1 - h(x_1) < x_1 - z < x_2 - z.$$

Since f is convex, its slopes are nondecreasing as illustrated in Fig. 23.1. In other words,

$$\frac{f(x_1 - z) - f(x_1 - h(x_1))}{h(x_1) - z} \leq \frac{f(x_2 - z) - f(x_2 - h(x_1))}{h(x_1) - z}$$

so

$$f(x_1 - z) + f(x_2 - h(x_1)) \leq f(x_1 - h(x_1)) + f(x_2 - z).$$

But by definition of h we have

$$g(h(x_1)) + f(x_1 - h(x_1)) \leq g(z) + f(x_1 - z)$$

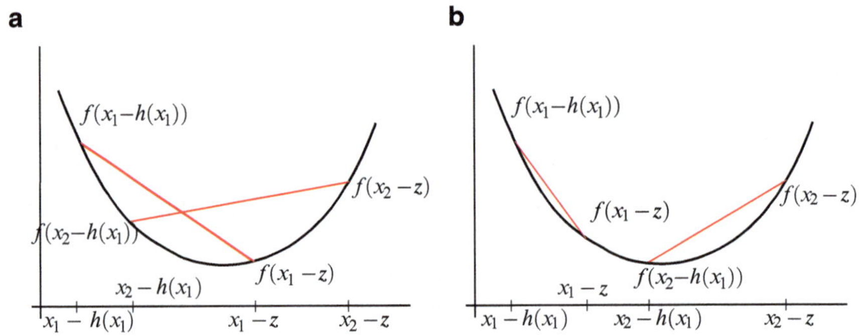

Fig. 23.1 Univariate convex functions have nondecreasing slopes. (**a**) Case $x_1 - h(x_1) \leq x_2 - h(x_1) \leq x_1 - z \leq x_2 - z$. (**b**) Case $x_1 - h(x_1) \leq x_1 - z \leq x_2 - h(x_1) \leq x_2 - z$

so we deduce

$$\begin{aligned}
g(h(x_1))+f(x_2-h(x_1)) &= g(h(x_1))+f(x_1-h(x_1))-f(x_1-h(x_1))+f(x_2-h(x_1)),\\
&\leq g(z)+f(x_1-z)+f(x_2-h(x_1))-f(x_1-h(x_1)),\\
&\leq g(z)+f(x_1-h(x_1))+f(x_2-z)-f(x_1-h(x_1)),\\
&= g(z)+f(x_2-z).
\end{aligned}$$

Since the inequality is true for all $z < h(x_1)$, we deduce $h(x_2) \geq h(x_1)$.

Now consider the case $h(x_1) = +\infty$. Then there exists a sequence h_1^k converging to $+\infty$ with $h_1^k \in H(x_1)$. Take $z < h_1^k$ and apply the previous argument to get

$$f \oplus g(x_2) \leq g(h_1^k) + f(x_1 - h_1^k) \leq g(z) + f(x_2 - z).$$

Since $x_2 \in \operatorname{dom} f \oplus g$ and f and g are PLQ functions, we deduce that for k large enough $\inf_{z<h_1^k}[g(z)+f(x_2-z)] = f \oplus g(x_2)$. Hence, $h_1^k \in H(x_2)$ which implies that $h(x_2) = +\infty$.

The only remaining case is $h(x_1) = -\infty$. We show that case cannot happen when $x_1 \in \operatorname{dom} f \oplus g$. Under our assumption we can always find x and y with $\varphi_x(y)$ finite. By definition of the inf, there is a sequence y_k with $\varphi_x(y_k) \to f \oplus g(x)$. Extracting subsequences if necessary, we have y_k converges to $\bar{y}$ either a finite or an infinite value. If the limit is finite, since f and g are continuous, we deduce $\varphi_x(\bar{y}) = f \oplus g(x)$ so $\bar{y} \in H(x_1)$ and $h(x_1) > -\infty$. Otherwise $|\bar{y}| = \infty$. Assume for simplicity $\bar{y} = +\infty$. Since φ_x is PLQ, there is a nonempty unbounded interval $[\bar{x}, +\infty)$ on which the function φ_x is quadratic. But the only way a quadratic function has a finite limit at infinity is when it is a constant function. In that case, $H(x_1)$ is nonempty and the proof is complete. ∎

Using the above lemma, a log-linear-time algorithm can be built (see Sect. 23.3.2). When none of the functions are convex, the problem is related to the all-pairs shortest-path problem [8] and a nontrivial subquadratic algorithm exists.

The nonconvex case raises the following open questions:

- Does a linear-time algorithm exist for the convex-nonconvex case?
- Does a log-linear algorithm exist for the nonconvex-nonconvex case?

23.4.2 Testing for Convexity

Given a set of points in $\mathbb{R}^3$ that corresponds to the sampling of a function, we can test whether there is a convex function interpolating the sample by checking a quadratic number of inequalities, i.e., by checking that all points are above any plane going through three neighboring points. A faster convexity test consists in computing the convex envelope and checking that all points are vertices of the graph of the convex envelope. Computing the convex envelope of a set of points in space can be performed in log-linear time [10] so convexity can be checked in log-linear time.

The following question is still open:

- Can we identify the log-linear number of inequalities that are required to check convexity?

A positive answer would give a faster algorithm to compute the closest convex function to a given nonconvex PLQ function. The resulting mathematical programming problem has been studied in [26, 27] in which several applications are listed. A different approach that approximates the convex envelope was proposed in [37, 38].

23.4.3 Bivariate Functions

Extending the CCA library [29] to bivariate functions is a work in progress [18]. The computation of the conjugate can be performed in log-linear time in the worst case and in linear time in the average-case. Moreover, a worst-case linear time algorithm exists when one restricts the desired precision for example for plotting purposes.

While the scalar multiplication of a bivariate PLQ function takes trivially linear time, the addition does not. In fact, adding two convex PLQ bivariate functions is known as the map overlay problem in computation geometry and has a quadratic output size in the worst case. The result can be refined using output sensitive algorithms [1, 12], and a linear-time algorithm can be achieved when one has some control on the domain of the functions to be added [15].

The Moreau envelope is directly deduced from the conjugate using Formula 23.4.

Consequently, we obtain a toolbox to manipulate convex PLQ bivariate functions. However, the class of convex PLQ bivariate function is not closed under the max operator. For univariate PLQ functions f_1 and f_2, the function $\max(f_1, f_2)$ (resp. $\min(f_1, f_2)$) is a univariate PLQ function and can be computed in linear time using a straightforward algorithm. This is no longer true for bivariate PLQ functions. Consider $f_1(x) = 1$ and $f_2(x) = \|x\|^2$. Then $\max(f_1, f_2)$ is not a PLQ function since its domain cannot be decomposed into the union of a finite number of polygons on which the function is quadratic.

The following questions are open:

- Is there a class of bivariate functions that is closed for the addition, scalar multiplication, conjugation, Moreau envelope, and maximum operators that still allows for linear-time algorithms? The question is open even in the convex case.
- Is there a worst-case linear-time algorithm to compute the conjugate of convex PLQ bivariate functions that does not need any precision assumption?

Acknowledgements Yves Lucet was partially supported by a Discovery Grant from the Natural Sciences and Engineering Research Council of Canada (NSERC). Part of the research was performed in the Computer-Aided Convex Analysis laboratory funded by the Canadian Foundation for Innovation. The author is very grateful to an anonymous referee for a very careful reading and for pointing out errors in a preliminary version of Lemma 23.3.

References

1. Balaban, I.J.: An optimal algorithm for finding segments intersections. In: Proceedings of the Eleventh Annual Symposium on Computational Geometry, SCG '95, pp. 211–219. ACM, New York (1995)
2. Bauschke, H.H., Goebel, R., Lucet, Y., Wang, X.: The proximal average: Basic theory. SIAM J. Optim. **19**, 768–785 (2008)
3. Bauschke, H.H., Lucet, Y., Trienis, M.: How to transform one convex function continuously into another. SIAM Rev. **50**, 115–132 (2008)
4. Bauschke, H.H., Lucet, Y., Wang, X.: Primal-dual symmetric intrinsic methods for finding antiderivatives of cyclically monotone operators. SIAM J. Control Optim. **46**, 2031–2051 (2007)
5. Bauschke, H.H., Moffat, S.M., Wang, X.: Self-dual smooth approximations of convex functions via the proximal average. In: Bauschke, H.H., Burachik, R.S., Combettes, P.L., Elser, V., Luke, D.R., Wolkowicz, H. (eds.) Fixed-Point Algorithms for Inverse Problems in Science and Engineering. Springer Optimization and Its Applications **49**, pp. 23–32. Springer, New York (2011)
6. Berkman, O., Schieber, B., Vishkin, U.: A fast parallel algorithm for finding the convex hull of a sorted point set. Internat. J. Comput. Geom. Appl. **6**, 231–241 (1996)
7. Borwein, J.M., Hamilton, C.H.: Symbolic computation of multidimensional Fenchel conjugates. In: ISSAC 2006, ACM, pp. 23–30. New York (2006)
8. Bremner, D., Chan, T.M., Demaine, E.D., Erickson, J., Hurtado, F., Iacono, J., Langerman, S., Taslakian, P.: Necklaces, convolutions, and $X + Y$. In: Algorithms—ESA 2006, Lecture Notes in Computer Science, vol. 4168, pp. 160–171. Springer, Berlin (2006)

9. Brenier, Y.: Un algorithme rapide pour le calcul de transformées de Legendre–Fenchel discrètes. C. R. Acad. Sci. Paris Sér. I Math. **308**, 587–589 (1989)
10. Chan, T.M.: Optimal output-sensitive convex hull algorithms in two and three dimensions. Discrete Comput. Geom. **16**, 361–368 (1996). Eleventh Annual Symposium on Computational Geometry (Vancouver, BC, 1995)
11. Chen, D.: Efficient geometric algorithms on the erew pram. Parallel and Distributed Systems, IEEE Transactions on **6**, 41 –47 (1995)
12. Chazelle, B., Edelsbrunner, H.: An optimal algorithm for intersecting line segments in the plane. J. ACM **39**, 1–54 (1992)
13. Corrias, L.: Fast Legendre–Fenchel transform and applications to Hamilton–Jacobi equations and conservation laws. SIAM J. Numer. Anal. **33**, 1534–1558 (1996)
14. Deniau, L., Blanc-Talon, J.: Fractal analysis with Hausdorff distance under affine transformations. Tech. report, ETCA-CREA-SP (1995)
15. Eppstein, D., Goodrich, M.T., Strash, D.: Linear-time algorithms for geometric graphs with sublinearly many crossings. In: Proceedings of the twentieth Annual ACM-SIAM Symposium on Discrete Algorithms, pp. 150–159. SIAM, Philadelphia, PA (2009)
16. Felzenszwalb, P.F., Huttenlocher, D.P.: Distance transforms of sampled functions. Tech. Report TR2004-1963, Cornell Computing and Information Science (Sept. 2004)
17. Gardiner, B., Lucet, Y.: Convex hull algorithms for piecewise linear-quadratic functions in computational convex analysis. Set-Valued Var. Anal. **18**, 467–482 (2010)
18. Gardiner, B., Lucet, Y.: Computing the conjugate of convex piecewise linear-quadratic bivariate functions. Math. Program. Ser. B **139**, 161–184 (2011)
19. Gardiner, B., Lucet, Y.: Graph-matrix calculus for computational convex analysis. In: Bauschke, H.H., Burachik, R.S., Combettes, P.L., Elser, V., Luke, D.R., Wolkowicz, H.(eds). Fixed-Point Algorithms for Inverse Problems in Science and Engineering . Springer Optimization and Its Applications, vol. 49, pp. 243–259. Springer, New York (2011)
20. Goebel, R.: Self-dual smoothing of convex and saddle functions. J. Convex Anal. **15**, 179–190 (2008)
21. Goodrich, M.T.: Finding the convex hull of a sorted point set in parallel. Inform. Process. Lett. **26**, 173–179 (1987)
22. Gupta, N., Sen, S.: Optimal, output-sensitive algorithms for constructing planar hulls in parallel. Comput. Geom. **8**, 151–166 (1997)
23. Hiriart-Urruty, J.-B., Lemaréchal, C.: Convex Analysis and Minimization Algorithms, Vol. I: Fundamentals, Vol. II: Advanced theory and bundle methods. Grundlehren der Mathematischen Wissenschaften [Fundamental Principles of Mathematical Sciences] vol. 305–306, Springer, Berlin (1993)
24. Hiriart-Urruty, J.-B., Lucet, Y.: Parametric computation of the Legendre–Fenchel conjugate. J. Convex Anal. **14**, 657–666 (2007)
25. Johnstone, J., Koch, V., Lucet, Y.: Convexity of the proximal average. J. Optim. Theory Appl. **148**, 107–124 (2011)
26. Lachand-Robert, T., Oudet, É: Minimizing within convex bodies using a convex hull method. SIAM J. Optim. **16**, 368–379 (2005) (electronic)
27. Lachand-Robert, T., Peletier, M.A.: The minimum of quadratic functionals of the gradient on the set of convex functions. Calc. Var. Part. Differ. Equat. **15**, 289–297 (2002)
28. Lucet, Y.: A fast computational algorithm for the Legendre–Fenchel transform. Comput. Optim. Appl. **6**, 27–57 (1996)
29. Lucet, Y.: Computational Convex Analysis library https://people.ok.ubc.ca/ylucet/cca.html (1996–2013)
30. Lucet, Y.: Faster than the Fast Legendre Transform, the Linear-time Legendre Transform. Numer. Algorithms **16**, 171–185 (1997)
31. Lucet, Y.: A linear Euclidean distance transform algorithm based on the Linear-time Legendre Transform. In: Proceedings of the Second Canadian Conference on Computer and Robot Vision (CRV 2005), pp. 262–267, IEEE Computer Society Press, Victoria BC (2005)

32. Lucet, Y.: Fast Moreau envelope computation, I: Numerical algorithms. Numer. Algorithms **43**, 235–249 (2006)
33. Lucet, Y.: What shape is your conjugate? A survey of computational convex analysis and its applications. SIAM Rev. **52**, 505–542 (2010)
34. Miller, R., Stout, Q.: Efficient parallel convex hull algorithms. IEEE Trans. Comput.**37**, 1605–1618 (1988)
35. Moreau, J.-J.: Proximité et dualité dans un espace Hilbertien. Bull. Soc. Math. France **93**, 273–299 (1965)
36. Noullez, A., Vergassola, M.: A fast Legendre transform algorithm and applications to the adhesion model. J. Sci. Comput. **9**, 259–281 (1994)
37. Oberman, A.M.: Computing the convex envelope using a nonlinear partial differential equation. Math. Models Methods Appl. Sci. **18**, 759–780 (2008)
38. Oberman, A.M.: The convex envelope is the solution of a nonlinear obstacle problem. Proc. Amer. Math. Soc. **135**, 1689–1694 (2007) (electronic)
39. Rauber, T., Rünger, G.: Parallel Programming: for Multicore and Cluster Systems. Springer Publishing Company Incorporated, New York (2010)
40. Rockafellar, R.T., Wets, R.J.-B.: Variational Analysis. Springer, Berlin (1998)
41. Wang, Y.-R., Horng, S.-J.: An O(1)time algorithm for the 3D Euclidean distance transform on the CRCW PRAM model. IEEE Trans. Parallel Distr. Syst. **14**, 973–982 (2003)

Chapter 24
Existence and Approximation of Fixed Points of Right Bregman Nonexpansive Operators

Victoria Martín-Márquez, Simeon Reich, and Shoham Sabach

Dedicated to Jonathan Borwein on the occasion of his 60th birthday

Abstract We study the existence and approximation of fixed points of right Bregman nonexpansive operators in reflexive Banach space. We present, in particular, necessary and sufficient conditions for the existence of fixed points and an implicit scheme for approximating them.

Key words: Bregman distance • Bregman firmly nonexpansive operator • Monotone mapping • Nonexpansive operator • Reflexive Banach space • Resolvent • Retraction • Totally convex function

Mathematics Subject Classifications (2010): 26B25, 46T99, 47H04, 47H05, 47H09, 47H10, 47J05, 47J25, 52A41, 54C15.

COMMUNICATED BY HEINZ H. BAUSCHKE.

V. Martín-Márquez
Department of Mathematical Analysis, University of Seville, 41012 Seville, Spain
e-mail: victoriam@us.es

S. Reich (✉) • S. Sabach
Department of Mathematics, The Technion – Israel Institute of Technology, 32000 Haifa, Israel
e-mail: sreich@tx.technion.ac.il; ssabach@tx.technion.ac.il

D.H. Bailey et al. (eds.), *Computational and Analytical Mathematics*, Springer Proceedings in Mathematics & Statistics 50, DOI 10.1007/978-1-4614-7621-4_24,

24.1 Introduction

The study of nonexpansive operators in Banach spaces has been an important topic in nonlinear functional analysis and optimization theory for almost 50 years now [3, 19–21]. There are several significant classes of nonexpansive operators which enjoy remarkable properties not shared by all such operators. We refer, for example, to firmly nonexpansive operators [11, 12]. These operators are of utmost importance in fixed-point, monotone mapping, and convex optimization theories in view of Minty's theorem regarding the correspondence between firmly nonexpansive operators and maximally monotone mappings [3, 19, 21, 27]. The largest class of nonexpansive operators comprises the quasi-nonexpansive operators. These operators still enjoy relevant fixed-point properties although nonexpansivity is only required about each fixed point [18].

In this paper we are concerned with certain analogous classes of operators which are, in some sense, nonexpansive not with respect to the norm, but with respect to Bregman distances [2, 9, 14, 17]. Since these distances are not symmetric in general, it seems natural to distinguish between left and right Bregman nonexpansive operators. Some left classes, so to speak, have already been studied and applied quite intensively [1, 4–6, 24, 31]. We have recently introduced and studied several classes of right Bregman nonexpansive operators in reflexive Banach spaces [22,23]. In these two papers we focused on the properties of their fixed-point sets. Our main aim in the present paper is to study the existence and approximation of fixed points of these operators.

Our paper is organized as follows In Sect. 24.2 we discuss several pertinent facts of convex analysis and Bregman operator theory. In Sect. 24.3 we present necessary and sufficient conditions for right quasi-Bregman nonexpansive operators to have (asymptotic) fixed points in general reflexive Banach spaces. Section 24.4 is devoted to a study of a Browder-type implicit algorithm [10] for computing fixed points of right Bregman firmly nonexpansive operators. Finally, in Sect. 24.5, we use the implicit method proposed in Sect. 24.4 to approximate zeroes of monotone mappings.

24.2 Preliminaries

All the results in this paper are set in a real reflexive Banach space X. The norms of X and X^*, its dual space, are denoted by $\|\cdot\|$ and $\|\cdot\|_*$, respectively. The pairing $\langle \xi, x\rangle$ is defined by the action of $\xi \in X^*$ at $x \in X$, that is, $\langle \xi, x\rangle := \xi(x)$. The set of all real numbers is denoted by $\mathbb{R}$ and $\overline{\mathbb{R}} = (-\infty, +\infty]$ is the extended real line, while $\mathbb{N}$ stands for the set of nonnegative integers. The closure of a subset K of X is denoted by $\overline{K}$. The (effective) *domain* of a convex function $f : X \to \overline{\mathbb{R}}$ is defined to be

$$\operatorname{dom} f := \{x \in X : f(x) < +\infty\}.$$

When $\operatorname{dom} f \neq \emptyset$ we say that f is *proper*. The *Fenchel conjugate* function of f is the convex function $f^* : X^* \to \overline{\mathbb{R}}$ defined by

$$f^*(\xi) = \sup\{\langle \xi, x\rangle - f(x) : x \in X\}.$$

It is not difficult to check that when f is proper and lower semicontinuous, so is f^*. The function f is called *cofinite* if $\operatorname{dom} f^* = X^*$.

In this section we present the basic notions and facts that are needed in the sequel. We divide this section into two parts in the following way. The first one (Sect. 24.2.1) is devoted to admissible functions, while the second (Sect. 24.2.2) concern, certain types of Bregman nonexpansive operators.

24.2.1 Admissible Functions

Let $x \in \operatorname{int}\operatorname{dom} f$, that is, let x belong to the interior of the domain of the convex function $f : X \to \overline{\mathbb{R}}$. For any $y \in X$, we define the *right-hand derivative* of f at the point x by

$$f^\circ(x,y) := \lim_{t \to 0^+} \frac{f(x+ty) - f(x)}{t}. \tag{24.1}$$

If the limit as $t \to 0$ in (24.1) exists for each y, then the function f is said to be *Gâteaux differentiable at* x. In this case, the *gradient* of f at x is the linear function $\nabla f(x)$, which is defined by $\langle \nabla f(x), y\rangle := f^\circ(x,y)$ for all $y \in X$ [25, Definition 1.3, p. 3]. The function f is called *Gâteaux differentiable* if it is Gâteaux differentiable at each $x \in \operatorname{int}\operatorname{dom} f$. When the limit as $t \to 0$ in (24.1) is attained uniformly for any $y \in X$ with $\|y\| = 1$, we say that f is *Fréchet differentiable* at x.

The function f is called *Legendre* if it satisfies the following two conditions:

(L1) $\operatorname{int}\operatorname{dom} f \neq \emptyset$ and the subdifferential ∂f is single-valued on its domain.
(L2) $\operatorname{int}\operatorname{dom} f^* \neq \emptyset$ and ∂f^* is single-valued on its domain.

The class of Legendre functions in infinite dimensional Banach spaces was first introduced and studied by Bauschke, Borwein, and Combettes in [4]. Their definition is equivalent to conditions (L1) and (L2) because the space X is assumed to be reflexive (see [4, Theorems 5.4 and 5.6, p. 634]). It is well known that in reflexive spaces, $\nabla f = (\nabla f^*)^{-1}$ (see [7, p. 83]). When this fact is combined with conditions (L1) and (L2), we obtain

$$\operatorname{ran}\nabla f = \operatorname{dom}\nabla f^* = \operatorname{int}\operatorname{dom} f^* \quad \text{and} \quad \operatorname{ran}\nabla f^* = \operatorname{dom}\nabla f = \operatorname{int}\operatorname{dom} f.$$

It also follows that f is Legendre if and only if f^* is Legendre (see [4, Corollary 5.5, p. 634]) and that the functions f and f^* are Gâteaux differentiable and strictly convex in the interior of their respective domains. When the Banach space X is smooth and strictly convex, in particular, a Hilbert space, the function $(1/p)\|\cdot\|^p$ with $p \in (1,\infty)$ is Legendre (cf. [4, Lemma 6.2, p. 639]). For examples and more information regarding Legendre functions, see, for instance, [1,4].

Throughout this paper, $f : X \to \overline{\mathbb{R}}$ is always an *admissible* function, that is, a proper, lower semicontinuous, convex, and Gâteaux differentiable function. Under these conditions we know that f is continuous in $\operatorname{int}\operatorname{dom} f$ (see [4, Fact 2.3, p. 619]).

The bifunction $D_f : \operatorname{dom} f \times \operatorname{int}\operatorname{dom} f \to [0,+\infty)$, which is defined by

$$D_f(y,x) := f(y) - f(x) - \langle \nabla f(x), y - x \rangle, \tag{24.2}$$

is called the *Bregman distance* (cf. [9, 16]).

The Bregman distance does not satisfy the well-known properties of a metric, but it does enjoy the following two important properties:

- The *three-point identity*: for any $x \in \operatorname{dom} f$ and $y, z \in \operatorname{int}\operatorname{dom} f$, we have

$$D_f(x,y) + D_f(y,z) - D_f(x,z) = \langle \nabla f(z) - \nabla f(y), x - y \rangle. \tag{24.3}$$

- The *four-point identity*: for any $y, w \in \operatorname{dom} f$ and $x, z \in \operatorname{int}\operatorname{dom} f$, we have

$$D_f(y,x) - D_f(y,z) - D_f(w,x) + D_f(w,z) = \langle \nabla f(z) - \nabla f(x), y - w \rangle. \tag{24.4}$$

According to [14, Sect. 1.2, p. 17] (see also [13]), the *modulus of total convexity of* f is the bifunction $\upsilon_f : \operatorname{int}\operatorname{dom} f \times [0,+\infty) \to [0,+\infty]$, which is defined by

$$\upsilon_f(x,t) := \inf\left\{D_f(y,x) : y \in \operatorname{dom} f,\ \|y - x\| = t\right\}.$$

The function f is called *totally convex at a point* $x \in \operatorname{int}\operatorname{dom} f$ if $\upsilon_f(x,t) > 0$ whenever $t > 0$. The function f is called *totally convex* when it is totally convex at every point $x \in \operatorname{int}\operatorname{dom} f$. This property is less stringent than uniform convexity (see [14, Sect. 2.3, p. 92]).

Examples of totally convex functions can be found, for instance, in [8, 14, 15]. We remark in passing that f is totally convex on bounded subsets if and only if f is uniformly convex on bounded subsets (see [15, Theorem 2.10, p. 9]).

24.2.2 Right Bregman Operators

Let $f : X \to \overline{\mathbb{R}}$ be admissible and let K be a nonempty subset of X. The *fixed-point set* of an operator $T : K \to X$ is the set $\{x \in K : Tx = x\}$. It is denoted by $\operatorname{Fix}(T)$. Recall that a point $u \in K$ is said to be an *asymptotic fixed point* [28] of T if there exists a sequence $\{x_n\}_{n\in\mathbb{N}}$ in K such that $x_n \rightharpoonup u$ (i.e., $\{x_n\}_{n\in\mathbb{N}}$ is weakly convergent to u) and $\|x_n - Tx_n\| \to 0$ as $n \to \infty$. We denote the asymptotic fixed-point set of T by $\widehat{\operatorname{Fix}}(T)$.

We first list significant types of nonexpansivity with respect to the Bregman distance.

Definition 24.1 (Right Bregman nonexpansivity). Let K and S be nonempty subsets of $\operatorname{dom} f$ and $\operatorname{int}\operatorname{dom} f$, respectively. An operator $T : K \to \operatorname{int}\operatorname{dom} f$ is said to be:

(i*) *Right Bregman firmly nonexpansive* (R-BFNE) if

$$\langle \nabla f(Tx) - \nabla f(Ty), Tx - Ty \rangle \leq \langle \nabla f(Tx) - \nabla f(Ty), x - y \rangle \tag{24.5}$$

for all $x, y \in K$ or, equivalently,

$$D_f(Tx, Ty) + D_f(Ty, Tx) + D_f(x, Tx) + D_f(y, Ty) \\ \leq D_f(x, Ty) + D_f(y, Tx). \tag{24.6}$$

(ii*) *Right quasi-Bregman firmly nonexpansive* (R-QBFNE) with respect to S if

$$0 \leq \langle \nabla f(p) - \nabla f(Tx), Tx - x \rangle \tag{24.7}$$

for all $x \in K$ and $p \in S$ or, equivalently,

$$D_f(Tx, p) + D_f(x, Tx) \leq D_f(x, p). \tag{24.8}$$

(iii*) *Right quasi-Bregman nonexpansive* (R-QBNE) with respect to S if

$$D_f(Tx, p) \leq D_f(x, p), \forall x \in K, p \in S. \tag{24.9}$$

(iv*) *Right Bregman strongly nonexpansive* (R-BSNE) with respect to S, if it is R-QBNE with respect to S and if whenever $\{x_n\}_{n\in\mathbb{N}} \subset K$ is bounded, $p \in S$, and

$$\lim_{n\to\infty} \left(D_f(x_n, p) - D_f(Tx_n, p)\right) = 0, \tag{24.10}$$

it follows that

$$\lim_{n\to\infty} D_f(x_n, Tx_n) = 0. \tag{24.11}$$

For the sake of completeness we also give here the definitions of left Bregman nonexpansivity.

Definition 24.2 (Left Bregman nonexpansivity). Let K and S be nonempty subsets of $\operatorname{int}\operatorname{dom} f$ and $\operatorname{dom} f$, respectively. An operator $T : K \to \operatorname{int}\operatorname{dom} f$ is said to be:

(i) *Left Bregman firmly nonexpansive* (L-BFNE) if

$$\langle \nabla f(Tx) - \nabla f(Ty), Tx - Ty \rangle \leq \langle \nabla f(x) - \nabla f(y), Tx - Ty \rangle \tag{24.12}$$

Table 24.1 Connections among types of right Bregman nonexpansivity

		Strictly R-QBFNE	$\Rightarrow$	Strictly R-BSNE	$\Rightarrow$	Strictly R-QBNE
		$\Downarrow$		$\Downarrow$		$\Downarrow$
R-BFNE	$\Rightarrow$	Properly R-QBFNE	$\Rightarrow$	Properly R-BSNE	$\Rightarrow$	Properly R-QBNE

for any $x,y \in K$ or, equivalently,

$$D_f(Tx,Ty)+D_f(Ty,Tx)+D_f(Tx,x)+D_f(Ty,y) \leq D_f(Tx,y)+D_f(Ty,x). \tag{24.13}$$

(ii) *Left quasi-Bregman firmly nonexpansive* (L-QBFNE) with respect to S if

$$0 \leq \langle \nabla f(x) - \nabla f(Tx), Tx - p \rangle \tag{24.14}$$

for any $x \in K$ and $p \in S$, or equivalently,

$$D_f(p,Tx)+D_f(Tx,x) \leq D_f(p,x). \tag{24.15}$$

(iii) *Left quasi-Bregman nonexpansive* (L-QBNE) with respect to S if

$$D_f(p,Tx) \leq D_f(p,x) \;\; \forall x \in K,\, p \in S. \tag{24.16}$$

(iv) *Left Bregman strongly nonexpansive* (L-BSNE) with respect to S if it is L-QBNE with respect to S and if whenever $\{x_n\}_{n\in\mathbb{N}} \subset K$ is bounded, $p \in S$, and

$$\lim_{n\to\infty} \left(D_f(p,x_n) - D_f(p,Tx_n)\right) = 0, \tag{24.17}$$

it follows that

$$\lim_{n\to\infty} D_f(Tx_n,x_n) = 0. \tag{24.18}$$

Remark 24.3 (Types of Bregman nonexpansivity with respect to S). As in [24], we distinguish between two types of Bregman nonexpansivity, depending on the set S, in such a way that if $S = \operatorname{Fix}(T)$ we say that T is properly Bregman nonexpansive, whereas if $S = \widehat{\operatorname{Fix}}(T)$ we say that T is strictly Bregman nonexpansive, according to the different notions of Bregman nonexpansivity. The connections among all these classes of right Bregman nonexpansive operators are presented in Table 24.1.

The following result [22] is essential for the proof of our approximation result in Sect. 24.4. It shows that the operator $I - T$ has a certain demiclosedness property. Before formulating this result, we recall that a mapping $B : X \to X^*$ is said to be *weakly sequentially continuous* if the weak convergence of $\{x_n\}_{n\in\mathbb{N}} \subset X$ to x implies the weak* convergence of $\{Bx_n\}_{n\in\mathbb{N}}$ to Bx.

Proposition 24.4 (Asymptotic fixed-point set of R-BFNE operators). *Let $f : X \to \mathbb{R}$ be Legendre and uniformly continuous on bounded subsets of X, and let*

∇f be weakly sequentially continuous. Let K be a nonempty subset of $\operatorname{dom} f$ *and let* $T : K \to \operatorname{int}\operatorname{dom} f$ *be an R-BFNE operator. Then* $\operatorname{Fix}(T) = \widehat{\operatorname{Fix}}(T)$.

In [22] we studied properties of several classes of right Bregman nonexpansive operators from the point of view of their fixed-point sets. A useful tool for such a study is the following operator.

Definition 24.5 (Conjugate operator). Let $f : X \to \overline{\mathbb{R}}$ be Legendre and let $T : K \subset \operatorname{int}\operatorname{dom} f \to \operatorname{int}\operatorname{dom} f$ be an operator. We define the *conjugate operator* associated with T by

$$T_f^* := \nabla f \circ T \circ \nabla f^* : \nabla f(K) \to \operatorname{int}\operatorname{dom} f^*.$$

When there is no danger of confusion we use the notation T^* for T_f^*. We also denote $\left(T_f^*\right)_{f^*}^*$ by T^{**}. It is very natural to ask what the connections between left and right classes of Bregman nonexpansivity are. This question can be answered by using the following [22, Proposition 2.7].

Proposition 24.6 (Properties of the conjugate operator). *Let* $f : X \to \overline{\mathbb{R}}$ *be Legendre and let* $T : K \subset \operatorname{int}\operatorname{dom} f \to \operatorname{int}\operatorname{dom} f$ *be an operator. Then the following properties hold:*

(i) $\operatorname{dom} T^* = \nabla f(\operatorname{dom} T)$ *and* $\operatorname{ran} T^* = \nabla f(\operatorname{ran} T)$.
(ii) T is R-BFNE if and only if T^* is L-BFNE.
(iii) $\operatorname{Fix}(T) = \nabla f^*(\operatorname{Fix}(T^*))$.
(iv) T is R-QBFNE (R-QBNE or R-BSNE) if and only if T^* is L-QBFNE (L-QBNE or L-BSNE).
(v) $T^{**} = T$.
(vi) If, in addition, ∇f and ∇f^* are uniformly continuous on bounded subsets of $\operatorname{int}\operatorname{dom} f$ *and* $\operatorname{int}\operatorname{dom} f^*$*, respectively, then*

$$\widehat{\operatorname{Fix}}(T^*) = \nabla f\left(\widehat{\operatorname{Fix}}(T)\right).$$

This connection between left and right Bregman nonexpansive operators allows us to get properties of right Bregman nonexpansive operators from their left counterparts (cf. [22]). The following result is an example of this.

Proposition 24.7 ($\nabla f(\operatorname{Fix}(T))$ of an R-QBNE operator is closed and convex). *Let $f : X \to \overline{\mathbb{R}}$ be a Legendre function and let K be a nonempty subset of* $\operatorname{int}\operatorname{dom} f$ *such that $\nabla f(K)$ is closed and convex. If $T : K \to \operatorname{int}\operatorname{dom} f$ is an R-QBNE operator, then $\nabla f(\operatorname{Fix}(T))$ is closed and convex.*

Proof. Since T is R-QBNE, the conjugate operator T^* is L-QBNE with respect to f^* [see Proposition 24.6(iv)]. Moreover, f^* is Legendre, and the domain of T^* is $\nabla f(K)$, which is closed and convex by assumption. Applying [31, Lemma 15.5, p. 307] and Proposition 24.6(iii), we get that $\operatorname{Fix}(T^*) = \nabla f(\operatorname{Fix}(T))$ is closed and convex, as asserted. ∎

The *right Bregman projection* cf. [6,22]) with respect to f of $x \in \operatorname{int}\operatorname{dom} f$ onto a nonempty, closed, and convex set $K \subset \operatorname{int}\operatorname{dom} f$ is defined by

$$\overrightarrow{\operatorname{proj}}_K^f(x) := \operatorname*{argmin}_{y\in K}\{D_f(x,y)\} = \{z \in K : D_f(x,z) \le D_f(x,y)\ \forall y \in K\}. \quad (24.19)$$

It is not clear a priori that the right Bregman projection is well defined because D_f is not convex in its second variable. However, Bauschke et al. (cf. [6, Proposition 7.1, p. 9]) proved that

$$\overrightarrow{\operatorname{proj}}_K^f = \nabla f^* \circ \overleftarrow{\operatorname{proj}}_{\nabla f(K)}^{f^*} \circ \nabla f, \quad (24.20)$$

where $\overleftarrow{\operatorname{proj}}_K^f$ stands for the left Bregman projection onto K with respect to f (see [14,15] for more information). As a consequence, one is able to prove that the right Bregman projection with respect to functions with admissible and totally convex conjugates has a variational characterization (cf. [22, Proposition 4.10]) as long as $\nabla f(K)$ is closed and convex.

Proposition 24.8 (Characterization of the right Bregman projection). *Let $f : X \to \mathbb{R}$ be a function such that f^* is admissible and totally convex. Let $x \in X$ and let K be a subset in $\operatorname{int}\operatorname{dom} f$ such that $\nabla f(K)$ is closed and convex. If $\hat{x} \in K$, then the following conditions are equivalent:*

(i) The vector $\hat{x}$ is the right Bregman projection of x onto K with respect to f.
(ii) The vector $\hat{x}$ is the unique solution of the variational inequality

$$\langle \nabla f(z) - \nabla f(y), z - x\rangle \ge 0 \quad \forall y \in K.$$

(iii) The vector $\hat{x}$ is the unique solution of the inequality

$$D_f(z,y) + D_f(x,z) \le D_f(x,y) \quad \forall y \in K.$$

Given two subsets $K \subset C \subset X$, an operator $R : C \to K$ is said to be a *retraction of C onto K* if $Rx = x$ for each $x \in K$. A retraction $R : C \to K$ is said to be *sunny* (see [21,26]) if

$$R(Rx + t(x - Rx)) = Rx$$

for each $x \in C$ and any $t \ge 0$, whenever $Rx + t(x - Rx) \in C$.

Under certain conditions on f, it turns out that the right Bregman projection is the unique sunny R-QBNE retraction of X onto its range (cf. [22, Corollary 4.6]).

Proposition 24.9 (Properties of the right Bregman projection). *Let $f : X \to \mathbb{R}$ be a Legendre, cofinite, and totally convex function, and assume that f^* is totally convex. Let K be a nonempty subset of X.*

(i) If $\nabla f(K)$ is closed and convex, then the right Bregman projection,

$$\overrightarrow{\text{proj}}_K^f = \nabla f^* \circ \overleftarrow{\text{proj}}_{\nabla f(K)}^{f^*} \circ \nabla f,$$

is the unique sunny R-QBNE retraction of X onto K.

(ii) If K is a sunny R-QBNE retract of X, then $\nabla f(K)$ is closed and convex, and $\overrightarrow{\text{proj}}_K^f$ is the unique sunny R-QBNE retraction of X onto K.

The previous result yields the fact that the fixed-point set of any R-QBNE operator is a sunny R-QBNE retract of X and the corresponding retraction is uniquely defined by the right Bregman projection onto the fixed-point set (cf. [22, Corollary 4.7]).

Proposition 24.10 (Fix (T) **is a sunny R-QBNE retract).** *Let $f : X \to \mathbb{R}$ be Legendre, cofinite, and totally convex, with a totally convex conjugate f^*. If $T : X \to X$ is an R-QBNE operator, then there exists a unique sunny R-QBNE retraction of X onto* Fix (T), *and this is the right Bregman projection onto* Fix (T).

24.3 Existence of Fixed Points

In this section we obtain necessary and sufficient conditions for R-QBNE operators to have (asymptotic) fixed points in general reflexive Banach spaces. We begin with a necessary condition for a strictly R-QBNE operator to have an asymptotic fixed point.

Proposition 24.11 (Necessary condition for $\widehat{\text{Fix}}(T)$ to be nonempty). *Let $f : X \to \overline{\mathbb{R}}$ be an admissible and totally convex function. Let $T : K \subset \operatorname{int}\operatorname{dom} f \to K$ be an operator. The following assertions hold:*

(i) If T is strictly R-QBNE and $\widehat{\text{Fix}}(T)$ is nonempty or
(ii) If T is properly R-QBNE and Fix (T) *is nonempty,*

then $\{T^n x\}_{n\in\mathbb{N}}$ is bounded for each $x \in K$.

Proof.

(i) We know from (24.9) that

$$D_f(Tx,p) \le D_f(x,p)$$

for any $p \in \widehat{\text{Fix}}(T)$ and $x \in K$. Therefore

$$D_f(T^n x,p) \le D_f\left(T^{n-1}x,p\right) \le \cdots \le D_f(x,p)$$

for any $p \in \widehat{\text{Fix}}(T)$ and $x \in K$. This inequality shows that the nonnegative sequence $\{D_f(T^n x,p)\}_{n\in\mathbb{N}}$ is bounded. Now the boundedness of the sequence $\{T^n x\}_{n\in\mathbb{N}}$ follows from [30, Lemma 3.1, p. 31].

(ii) This result is a consequence of the arguments in assertion (i) when $p \in \widehat{\text{Fix}}(T)$ is replaced with $p \in \text{Fix}(T)$. ■

A left variant of Proposition 24.11(ii) has already been proved in [31, Theorem 15.7, p. 307]. Note that this left variant result can be rewritten as follows, where the conditions on f, T, and K are somewhat different.

Proposition 24.12 (Necessary condition for $\mathrm{Fix}(T)$ **to be nonempty (left variant)).** *Let $f: X \to \overline{\mathbb{R}}$ be an admissible function and assume that ∇f^* is bounded on bounded subsets of* $\mathrm{int}\,\mathrm{dom}\, f^*$. *Let $T: K \subset \mathrm{int}\,\mathrm{dom}\, f \to K$ be a properly L-QBNE operator. If* $\mathrm{Fix}(T)$ *is nonempty, then $\{T^n x\}_{n\in\mathbb{N}}$ is bounded for each $x \in K$.*

Using this result and the properties of the conjugate operator, we can now obtain a variant of Proposition 24.11(ii) under different assumptions on f.

Proposition 24.13 (Necessary condition for $\mathrm{Fix}(T)$ **to be nonempty (second version)).** *Let $f: X \to \overline{\mathbb{R}}$ be a function such that f^* is admissible, and assume that ∇f and ∇f^* are bounded on bounded subsets of* $\mathrm{int}\,\mathrm{dom}\, f$ *and* $\mathrm{int}\,\mathrm{dom}\, f^*$, *respectively. Let $T: K \subset \mathrm{int}\,\mathrm{dom}\, f \to K$ be a properly R-QBNE operator. If* $\mathrm{Fix}(T)$ *is nonempty, then $\{T^n x\}_{n\in\mathbb{N}}$ is bounded for each $x \in K$.*

Proof. Since T is a properly R-QBNE operator with $\mathrm{Fix}(T) \neq \emptyset$, it follows from Proposition 24.6(iii) and (iv) that

$$T^* := \nabla f \circ T \circ \nabla f^* : \nabla f(K) \to \nabla f(K) \tag{24.21}$$

is a properly L-QBNE operator with respect to f^* with $\mathrm{Fix}(T^*) = \nabla f(\mathrm{Fix}(T)) \neq \emptyset$. Since the assumptions of Proposition 24.12 hold, the sequence $\{(T^*)^n \xi\}_{n\in\mathbb{N}}$ is bounded for each $\xi \in \nabla f(K)$.

Next we note that

$$(T^*)^n = T^* \circ \cdots \circ T^* = \nabla f \circ T^n \circ \nabla f^* = (T^n)^*. \tag{24.22}$$

Therefore $\{(T^n)^* \xi\}_{n\in\mathbb{N}}$ is bounded for each $\xi \in \nabla f(K)$, which means that the sequence $\{\nabla f(T^n x)\}_{n\in\mathbb{N}}$ is bounded for each $x \in K$. Now the desired result follows because ∇f^* is bounded on bounded subsets of $\mathrm{int}\,\mathrm{dom}\, f^*$. ■

Given an operator $T: K \subset \mathrm{int}\,\mathrm{dom}\, f \to K$, we let

$$S_n^f(z) := (1/n) \sum_{k=1}^{n} \nabla f\left(T^k z\right), \quad z \in K. \tag{24.23}$$

Using these f-averages, we now present a sufficient condition for R-BFNE operators to have a fixed point. We start by proving this result directly.

Proposition 24.14 (Sufficient condition for $\mathrm{Fix}(T)$ **to be nonempty).** *Let $f: X \to \overline{\mathbb{R}}$ be an admissible function. Let K be a nonempty subset of* $\mathrm{int}\,\mathrm{dom}\, f$ *such that $\nabla f(K)$ is closed and convex, and let $T: K \to K$ be an R-BFNE operator. If there exists $x \in K$ such that $\left\|S_n^f(x)\right\| \nrightarrow \infty$ as $n \to \infty$, then* $\mathrm{Fix}(T)$ *is nonempty.*

Proof. Assume there exists $x \in K$ such that $\left\|S_n^f(x)\right\| \nrightarrow \infty$ as $n \to \infty$. Let $y \in K$, $k \in \mathbb{N}$, and $n \in \mathbb{N}$ be given. Since T is R-BFNE, we have [see (24.6)]

$$D_f\left(T^{k+1}x,Ty\right)+D_f\left(Ty,T^{k+1}x\right)\le D_f\left(y,T^{k+1}x\right)+D_f\left(T^kx,Ty\right), \quad (24.24)$$

where $T^0=I$, the identity operator. From the three-point identity [see (24.3)] and (24.24) we get

$$\begin{aligned}D_f\left(T^{k+1}x,Ty\right)+D_f\left(Ty,T^{k+1}x\right)\le{}& D_f\left(T^kx,Ty\right)+D_f\left(Ty,T^{k+1}x\right)\\&+D_f(y,Ty)\\&+\left\langle\nabla f\left(T^{k+1}x\right)-\nabla f(Ty),Ty-y\right\rangle.\end{aligned}$$

This implies that

$$\begin{aligned}0\le{}& D_f(y,Ty)+D_f\left(T^kx,Ty\right)-D_f\left(T^{k+1}x,Ty\right)\\&+\left\langle\nabla f\left(T^{k+1}x\right)-\nabla f(Ty),Ty-y\right\rangle.\end{aligned}$$

Summing up these inequalities with respect to $k=0,1,\ldots,n-1$, we now obtain

$$\begin{aligned}0\le{}& nD_f(y,Ty)+D_f(x,Ty)-D_f(T^nx,Ty)\\&+\left\langle\sum_{k=0}^{n-1}\nabla f\left(T^{k+1}x\right)-n\nabla f(Ty),Ty-y\right\rangle.\end{aligned}$$

Dividing this inequality by n, we get

$$\begin{aligned}0\le{}& D_f(y,Ty)+\frac{1}{n}\left[D_f(x,Ty)-D_f(T^nx,Ty)\right]\\&+\left\langle\frac{1}{n}\sum_{k=0}^{n-1}\nabla f\left(T^{k+1}x\right)-\nabla f(Ty),Ty-y\right\rangle\end{aligned}$$

and hence

$$0\le D_f(y,Ty)+\frac{1}{n}D_f(x,Ty)+\left\langle S_n^f(x)-\nabla f(Ty),Ty-y\right\rangle. \quad (24.25)$$

Since $\left\|S_n^f(x)\right\|\nrightarrow\infty$ as $n\to\infty$ by assumption, we know that there exists a subsequence $\left\{S_{n_k}^f(x)\right\}_{k\in\mathbb{N}}$ of $\left\{S_n^f(x)\right\}_{n\in\mathbb{N}}$ such that $S_{n_k}^f(x)\rightharpoonup\xi\in X^*$ as $k\to\infty$. Substituting n_k for n in (24.25) and letting $k\to\infty$, we obtain

$$0\le D_f(y,Ty)+\langle\xi-\nabla f(Ty),Ty-y\rangle. \quad (24.26)$$

Since $\nabla f(K)$ is closed and convex, we know that $\xi\in\nabla f(K)$. Therefore there exists $p\in K$ such that $\nabla f(p)=\xi$ and from (24.26) we obtain

$$0 \le D_f(y, Ty) + \langle \nabla f(p) - \nabla f(Ty), Ty - y \rangle. \tag{24.27}$$

Setting $y = p$ in (24.27), we get from the four-point identity [see (24.4)] that

$$\begin{aligned} 0 &\le D_f(p, Tp) + \langle \nabla f(p) - \nabla f(Tp), Tp - p \rangle \\ &= D_f(p, Tp) + D_f(p, p) - D_f(p, Tp) - D_f(Tp, p) + D_f(Tp, Tp) \\ &= -D_f(Tp, p). \end{aligned}$$

Hence $D_f(Tp, p) \le 0$ and so $D_f(Tp, p) = 0$. It now follows from [4, Lemma 7.3(vi), p. 642] that $Tp = p$. That is, $p \in \operatorname{Fix}(T)$. ■

At this point we recall the left variant of this result [31, Theorem 15.8, p. 310], where

$$S_n(z) := (1/n) \sum_{k=1}^{n} T^k z, \quad z \in K. \tag{24.28}$$

Proposition 24.15 (Sufficient condition for $\operatorname{Fix}(T)$ **to be nonempty (left variant)).** *Let $f : X \to \overline{\mathbb{R}}$ be an admissible function. Let K be a nonempty, closed, and convex subset of* $\operatorname{int}\operatorname{dom} f$, *and let $T : K \to K$ be an L-BFNE operator. If there exists $x \in K$ such that $\|S_n(x)\| \nrightarrow \infty$ as $n \to \infty$, then* $\operatorname{Fix}(T)$ *is nonempty.*

Using this result, we obtain a second version of Proposition 24.14 under different assumptions on the function f.

Proposition 24.16 (Sufficient condition for $\operatorname{Fix}(T)$ **to be nonempty (second version)).** *Let $f : X \to \overline{\mathbb{R}}$ be a function such that f^* is admissible. Let K be a nonempty subset of* $\operatorname{int}\operatorname{dom} f$ *such that $\nabla f(K)$ is closed and convex, and let $T : K \to K$ be an R-BFNE operator. If there exists $x \in K$ such that $\left\| S_n^f(x) \right\| \nrightarrow \infty$ as $n \to \infty$, then* $\operatorname{Fix}(T)$ *is nonempty.*

Proof. Since T is an R-BFNE operator, we obtain from Proposition 24.6(ii) that T^* is an L-BFNE operator. In addition, from (24.22), we get the following connection between the f-average operator S_n^f [see (24.23)] and the operator S_n (defined by (24.28) for the operator T) with respect to the conjugate operator T^*, which here we denote by $S_n^{T^*}$. Given $x \in K$ and $\xi := \nabla f(x) \in \nabla f(K)$,

$$\begin{aligned} S_n^f(x) &= \frac{1}{n} \sum_{k=1}^{n} \nabla f\left(T^k x\right) = \frac{1}{n} \sum_{k=1}^{n} \nabla f\left(T^k\left(\nabla f^*(\xi)\right)\right) \\ &= \frac{1}{n} \sum_{k=1}^{n} \left(\nabla f \circ T \circ \nabla f^*(\xi)\right)^k = \frac{1}{n} \sum_{k=1}^{n} \left(T^*(\xi)\right)^k := S_n^{T^*}(\xi). \end{aligned}$$

Hence the assumption that there exists $x \in K$ such that $\left\| S_n^f(x) \right\| \nrightarrow \infty$ as $n \to \infty$ is equivalent to the assumption that there exists $\xi \in \nabla f(K)$ such that $\left\| S_n^{T^*}(\xi) \right\| \nrightarrow \infty$ as $n \to \infty$. Now we apply Proposition 24.15 to f^* and T^* on $\nabla f(K)$, which is assumed to be closed and convex, and get that $\mathrm{Fix}(T^*)$ is nonempty. From Proposition 24.6(iii) we obtain that $\mathrm{Fix}(T)$ is nonempty too. ∎

From Propositions 24.14 and 24.16 we deduce the following result which says that every nonempty set K such that $\nabla f(K)$ is bounded, closed, and convex has the fixed-point property for R-BFNE self-operators.

Corollary 24.17. *Let $f : X \to \overline{\mathbb{R}}$ be either an admissible function or a function such that f^* is admissible. Let K be a nonempty subset of* $\mathrm{int}\,\mathrm{dom}\, f$ *such that $\nabla f(K)$ is bounded, closed, and convex, and let $T : K \to K$ be an R-BFNE operator. Then* $\mathrm{Fix}(T)$ *is nonempty.*

24.4 Approximation of Fixed Points

In this section we study the convergence of a Browder-type implicit algorithm [10] for computing fixed points of R-BFNE operators with respect to a Legendre function f.

Theorem 24.18 (Implicit method for approximating fixed points). *Let $f : X \to \mathbb{R}$ be a Legendre and positively homogeneous function of degree $\alpha > 1$, which is uniformly continuous on bounded subsets of X. Assume that ∇f is weakly sequentially continuous and f^* is totally convex. Let K be a nonempty and bounded subset of X such that $\nabla f(K)$ is bounded, closed, and convex with $0^* \in \nabla f(K)$, and let $T : K \to K$ be an R-BFNE operator. Then the following two assertions hold:*

(i) For each $t \in (0,1)$, there exists a unique $u_t \in K$ satisfying $u_t = tTu_t$.
(ii) The net $\{u_t\}_{t\in(0,1)}$ converges strongly to $\overrightarrow{\mathrm{proj}}^f_{\mathrm{Fix}(T)}(0)$ as $t \to 1^-$.

Proof.

(i) Fix $t \in (0,1)$ and let S_t be the operator defined by $S_t = tT$. Note that, since ∇f is positively homogeneous of degree $\alpha - 1 > 0$, we have $\nabla f(0) = 0^* \in \nabla f(K)$. This implies that S_t is an operator from K into K. Indeed, it is easy to see that for any $x \in K$, since $t^{\alpha-1} \in (0,1)$ and $\nabla f(K)$ is convex, we have

$$\nabla f^*\left(t^{\alpha-1}\nabla f(Tx) + \left(1 - t^{\alpha-1}\right)\nabla f(0)\right) \in K.$$

On the other hand,

$$\begin{aligned}\nabla f^*\left(t^{\alpha-1}\nabla f(Tx) + \left(1 - t^{\alpha-1}\right)\nabla f(0)\right) &= \nabla f^*\left(t^{\alpha-1}\nabla f(Tx)\right)\\ &= \nabla f^*\left(\nabla f(tTx)\right)\\ &= tTx.\end{aligned}$$

Hence $S_t x \in K$ for any $x \in K$. Next we show that S_t is an R-BFNE operator. Given $x, y \in K$, since T is R-BFNE, we have

$$\begin{aligned}\langle \nabla f(S_t x) - \nabla f(S_t y), S_t x - S_t y\rangle &= t^{\alpha}\langle \nabla f(Tx) - \nabla f(Ty), Tx - Ty\rangle \\ &\leq t^{\alpha}\langle \nabla f(Tx) - \nabla f(Ty), x - y\rangle \\ &= t\langle \nabla f(S_t x) - \nabla f(S_t y), x - y\rangle \\ &\leq \langle \nabla f(S_t x) - \nabla f(S_t y), x - y\rangle .\end{aligned}$$

Thus S_t is indeed R-BFNE. Since $\nabla f(K)$ is bounded, closed, and convex, it follows from Corollary 24.17 that S_t has a fixed point. Furthermore, $\mathrm{Fix}(S_t)$ consists of exactly one point. Indeed, if $u, u' \in \mathrm{Fix}(S_t)$, then it follows from the right Bregman firm nonexpansivity of S_t that

$$\begin{aligned}\left\langle \nabla f(u) - \nabla f(u'), u - u'\right\rangle &= \left\langle \nabla f(S_t u) - \nabla f(S_t u'), S_t u - S_t u'\right\rangle \\ &\leq \left\langle \nabla f(S_t u) - \nabla f(S_t u'), u - u'\right\rangle \\ &= t^{\alpha-1}\left\langle \nabla f(u) - \nabla f(u'), u - u'\right\rangle ,\end{aligned}$$

which means that

$$\left\langle \nabla f(u) - \nabla f(u'), u - u'\right\rangle \leq 0.$$

Since f is Legendre, we know that f is strictly convex and therefore ∇f is strictly monotone. Hence $u = u'$. Thus there exists a unique point $u_t \in K$ such that $u_t = S_t u_t$.

(ii) Note that, since T is R-BFNE, it follows from Corollary 24.17 that $\mathrm{Fix}(T)$ is nonempty. Furthermore, since T is R-QBNE (see Table 24.1), from Proposition 24.7, we know that $\nabla f(\mathrm{Fix}(T))$ is closed and convex. Therefore Proposition 24.8 shows that $\overrightarrow{\mathrm{proj}}^{f}_{\mathrm{Fix}(T)}$ is well defined and has a variational characterization. Let $\{t_n\}_{n\in\mathbb{N}}$ be an arbitrary sequence in the real interval $(0,1)$ such that $t_n \to 1^-$ as $n \to \infty$. Denote $x_n = u_{t_n}$ for all $n \in \mathbb{N}$. It suffices to show that $x_n \to \overrightarrow{\mathrm{proj}}^{f}_{\mathrm{Fix}(T)}(0)$ as $n \to \infty$. Since K is bounded, there is a subsequence $\{x_{n_k}\}_{k\in\mathbb{N}}$ of $\{x_n\}_{n\in\mathbb{N}}$ such that $x_{n_k} \rightharpoonup v$ as $k \to \infty$. From the definition of x_n, we see that $\|x_n - Tx_n\| = (1-t_n)\|Tx_n\|$ for all $n \in \mathbb{N}$. So, we have $\|x_n - Tx_n\| \to 0$ as $n \to \infty$ and hence $v \in \widehat{\mathrm{Fix}}(T)$. Proposition 24.4 now implies that $v \in \mathrm{Fix}(T)$. We next show that $x_{n_k} \to v$ as $k \to \infty$. Fix $n \in \mathbb{N}$. Since T is properly R-QBFNE (see Table 24.1), we have

$$0 \leq \langle \nabla f(Tx_n) - \nabla f(v), x_n - Tx_n\rangle .$$

Since $x_n - Tx_n = (t_n - 1)Tx_n$, we also have

$$0 \leq \langle \nabla f(Tx_n) - \nabla f(v), (t_n - 1)Tx_n\rangle .$$

This yields

$$0 \leq \langle \nabla f(Tx_n) - \nabla f(v), -Tx_n\rangle \tag{24.29}$$

and

$$\langle \nabla f(Tx_n) - \nabla f(v), Tx_n - v \rangle \leq \langle \nabla f(Tx_n) - \nabla f(v), -v \rangle. \tag{24.30}$$

Since $x_{n_k} \rightharpoonup v$ and $\left\| x_{n_k} - Tx_{n_k} \right\| \to 0$ as $k \to \infty$, it follows that $Tx_{n_k} \rightharpoonup v$. From the weak sequential continuity of ∇f we obtain that $\nabla f\left(Tx_{n_k}\right) \overset{*}{\rightharpoonup} \nabla f(v)$ as $k \to \infty$. Hence it follows from the monotonicity of ∇f and from (24.30) that

$$\begin{aligned} 0 &\leq \liminf_{k \to \infty} \left\langle \nabla f\left(Tx_{n_k}\right) - \nabla f(v), Tx_{n_k} - v \right\rangle \\ &\leq \limsup_{k \to \infty} \left\langle \nabla f\left(Tx_{n_k}\right) - \nabla f(v), -v \right\rangle \\ &= 0. \end{aligned} \tag{24.31}$$

Thus

$$\lim_{k \to \infty} \left\langle \nabla f\left(Tx_{n_k}\right) - \nabla f(v), Tx_{n_k} - v \right\rangle = 0.$$

Since

$$D_f\left(v, Tx_{n_k}\right) + D_f\left(Tx_{n_k}, v\right) = \left\langle \nabla f\left(Tx_{n_k}\right) - \nabla f(v), Tx_{n_k} - v \right\rangle,$$

it follows that

$$\lim_{k \to \infty} D_f\left(v, Tx_{n_k}\right) = \lim_{k \to \infty} D_f\left(Tx_{n_k}, v\right) = 0.$$

From [32, Proposition 2.2, p. 3] we get that $\left\| Tx_{n_k} - v \right\| \to 0$ as $k \to \infty$. Finally, we claim that $v = \overrightarrow{\mathrm{proj}}^f_{\mathrm{Fix}(T)}(0)$. Indeed, note that inequality (24.29) holds when we replace v with any $p \in \mathrm{Fix}(T)$. Then, since $\nabla f\left(Tx_{n_k}\right) \overset{*}{\rightharpoonup} \nabla f(v)$ and $Tx_{n_k} \to v$ as $k \to \infty$, letting $k \to \infty$ in this inequality, we get

$$0 \leq \langle \nabla f(v) - \nabla f(p), -v \rangle$$

for any $p \in \mathrm{Fix}(T)$. In other words,

$$0 \leq \langle \nabla f(v) - \nabla f(p), 0 - v \rangle$$

for any $p \in \mathrm{Fix}(T)$. Now we obtain from Proposition 24.8 that $v = \overrightarrow{\mathrm{proj}}^f_{\mathrm{Fix}(T)}(0)$, as asserted. ■

Here is the left variant of this result [31].

Proposition 24.19 (Implicit method for approximating fixed point (left variant)). *Let $f : X \to \mathbb{R}$ be a Legendre and totally convex function, which is positively homogeneous of degree $\alpha > 1$, uniformly Fréchet differentiable, and bounded on bounded subsets of X. Let K be a nonempty, bounded, closed, and* *convex subset of X with* $0 \in K$*, and let* $T : K \to K$ *be an L-BFNE operator. Then the following two assertions hold:*

(i) For each $t \in (0,1)$*, there exists a unique* $u_t \in K$ *satisfying* $u_t = tTu_t$.
(ii) The net $\{u_t\}_{t\in(0,1)}$ *converges strongly to* $\overleftarrow{\mathrm{proj}}^f_{\mathrm{Fix}(T)}(\nabla f^*(0^*))$ *as* $t \to 1^-$.

Again using the left variant and the conjugation properties, we can obtain a right variant under somewhat different conditions.

Theorem 24.20 (Implicit method for approximating fixed points (second version)). *Let* $f : X \to \overline{\mathbb{R}}$ *be a Legendre and cofinite function. Assume that* f^* *is totally convex, positively homogeneous of degree* $\alpha > 1$*, and uniformly Fréchet differentiable and bounded on bounded subsets of* X^**. Let K be a nonempty subset of* $\operatorname{int}\operatorname{dom} f$ *such that* $\nabla f(K)$ *is bounded, closed, and convex with* $0^* \in \nabla f(K)$*. Let* $T : K \to K$ *be an R-BFNE operator. Then the following two assertions hold:*

(i) For each $t \in (0,1)$*, there exists a unique* $u_t \in K$ *satisfying* $u_t = tTu_t$.
(ii) The net $\{u_t\}_{t\in(0,1)}$ *converges strongly to* $\overrightarrow{\mathrm{proj}}^f_{\mathrm{Fix}(T)}(0)$ *as* $t \to 1^-$.

Proof.

(i) Since T is an R-BFNE operator, we obtain from Proposition 24.6(ii) that the conjugate operator $T^* : \nabla f(K) \to \nabla f(K)$ is an L-BFNE operator with respect to f^*. Now we apply Proposition 24.19(i) to T^* and get that for each $t \in (0,1)$, there exists a unique $\xi_t \in \nabla f(K)$ satisfying $\xi_t = tT^*\xi_t$. Denote $u_t = \nabla f^*(\xi_t) \in K$. Then from the definition of conjugate operators we get

$$\begin{aligned}\xi_t = tT^*\xi_t &\Leftrightarrow \nabla f(u_t) = tT^*\nabla f(u_t)\\ &\Leftrightarrow \nabla f(u_t) = t(\nabla f \circ T \circ \nabla f^*)(\nabla f(u_t))\\ &\Leftrightarrow \nabla f(u_t) = t\nabla f(Tu_t).\end{aligned}$$

Note that, since ∇f^* is positively homogeneous of degree $\alpha - 1 > 0$, the gradient ∇f is positively homogeneous of degree $1/(\alpha - 1) > 0$. Hence

$$\nabla f(u_t) = \nabla f\left(t^{\alpha-1}Tu_t\right).$$

So, for each $t \in (0,1)$, there exists a unique $u_t \in K$ satisfying $u_t = t^{\alpha-1}Tu_t$, which yields assertion (i) because $\alpha - 1 > 0$ and $0 < t < 1$.

(ii) From the positive homogeneity, we deduce that $\nabla f^*(0^*) = 0$. Therefore, applying Proposition 24.19(ii) to f^* and the conjugate operator T^* on $\nabla f(K)$, we get that the net $\{\xi_t\}_{t\in(0,1)}$ converges strongly to

$$\overleftarrow{\mathrm{proj}}^{f^*}_{\mathrm{Fix}(T^*)}(\nabla f(0)) = \overleftarrow{\mathrm{proj}}^{f^*}_{\mathrm{Fix}(T^*)}(0^*)$$

as $t \to 1^-$. Now, since $u_t = \nabla f^*(\xi_t) \in K$ for all $t \in (0,1)$, it follows from (24.20) that

$$\lim_{t\to 1^-} \nabla f(u_t) = \overleftarrow{\mathrm{proj}}^{f^*}_{\mathrm{Fix}(T^*)}(0^*)$$
$$= \nabla f\left(\overrightarrow{\mathrm{proj}}^{f}_{\mathrm{Fix}(T)}(\nabla f^*(0^*))\right)$$
$$= \nabla f\left(\overrightarrow{\mathrm{proj}}^{f}_{\mathrm{Fix}(T)}(0)\right). \tag{24.32}$$

Since f^* is uniformly Fréchet differentiable and bounded on bounded subsets of $\mathrm{int\,dom} f^*$, we know that ∇f^* is uniformly continuous on bounded subsets of X^* [29, Proposition 2.1]. Since $\{\xi_t = \nabla f(u_t)\}_{t\in(0,1)}$ is bounded as a convergent sequence, it now follows from (24.32) that $\{u_t\}_{t\in(0,1)}$ converges strongly to $\overrightarrow{\mathrm{proj}}^{f}_{\mathrm{Fix}(T)}(0)$ as $t \to 1^-$. ■

Remark 24.21. Under the hypotheses of Theorem 24.20, since $\nabla f(K)$ is closed and convex, if we assume, in addition, that f is totally convex, then Proposition 24.9 implies that the right Bregman projection onto $\mathrm{Fix}(T)$ is the unique sunny R-QBNE retraction of X onto $\mathrm{Fix}(T)$. In other words, the sequence $\{u_t\}_{t\in(0,1)}$ converges strongly to the value of the unique sunny R-QBNE retraction of X onto $\mathrm{Fix}(T)$ at the origin. In the setting of a Hilbert space, when $f = (1/2)\|\cdot\|^2$, this fact recovers the result of Browder [10], which shows that, for a nonexpansive mapping T, the approximating curve $x_t = (1-t)u + tTx_t$ generates the unique sunny nonexpansive retraction onto $\mathrm{Fix}(T)$ when $t \to 1^-$, in the particular case where $u = 0$.

24.5 Zeroes of Monotone Mappings

Let $A : X \to 2^{X^*}$ be a set-valued mapping. Recall that the (effective) *domain* of the mapping A is the set $\mathrm{dom}A = \{x \in X : Ax \neq \emptyset\}$. We say that A is *monotone* if for any $x, y \in \mathrm{dom}A$, we have

$$\xi \in Ax \text{ and } \eta \in Ay \quad \Longrightarrow \quad 0 \leq \langle \xi - \eta, x - y \rangle. \tag{24.33}$$

A monotone mapping A is said to be *maximal* if the graph of A is not a proper subset of the graph of any other monotone mapping.

A problem of great interest in optimization theory is that of finding zeroes of set-valued mappings $A : X \to 2^{X^*}$. Formally, the problem can be written as follows:

$$\text{Find } x \in X \text{ such that } 0^* \in Ax. \tag{24.34}$$

This problem occurs in practice in various forms. For instance, minimizing a lower semicontinuous and convex function $f : X \to \overline{\mathbb{R}}$, a basic problem of optimization, amounts to finding a zero of the mapping $A = \partial f$, where $\partial f(x)$ stands for the subdifferential of f at the point $x \in X$. Finding solutions of some classes of differential equations can also be reduced to finding zeroes of certain set-valued mappings $A : X \to 2^{X^*}$.

In the case of a Hilbert space $\mathscr{H}$, one of the most important methods for solving (24.34) consists of replacing it with the equivalent fixed-point problem for the classical resolvent $R_A : \mathscr{H} \to 2^{\mathscr{H}}$ of A, defined by

$$R_A := (I + A)^{-1}.$$

In this case, provided that A satisfies some monotonicity conditions, the resolvent R_A is single-valued, nonexpansive, and even firmly nonexpansive. When X is not a Hilbert space, the classical resolvent R_A is of limited interest and other operators should be employed. For example, in several papers (see, for instance, [5, 31]), the f-resolvent Res_A^f was used for finding zeroes of monotone mappings A in general reflexive Banach spaces. More precisely, given a set-valued mapping $A : X \to 2^{X^*}$, the *f-resolvent* of A is the operator $\mathrm{Res}_A^f : X \to 2^X$ which is defined by

$$\mathrm{Res}_A^f := (\nabla f + A)^{-1} \circ \nabla f. \tag{24.35}$$

In this paper we consider another variant of the classical resolvent for general reflexive Banach spaces, namely, the conjugate resolvent of a mapping A [22].

Definition 24.22 (Conjugate f-resolvent). Let $A : X \to 2^{X^*}$ be a set-valued mapping. The *conjugate resolvent* of A with respect to f, or the conjugate f-resolvent, is the operator $\mathrm{CRes}_A^f : X^* \to 2^{X^*}$ defined by

$$\mathrm{CRes}_A^f := (I + A \circ \nabla f^*)^{-1}. \tag{24.36}$$

The conjugate resolvent satisfies the following properties [22].

Proposition 24.23 (Properties of conjugate f-resolvents). *Let $f : X \to \overline{\mathbb{R}}$ be an admissible function and let $A : X \to 2^{X^*}$ be a mapping such that* $\operatorname{int}\operatorname{dom} f \cap \operatorname{dom} A \neq \emptyset$. *The following statements hold:*

(i) $\operatorname{dom} \mathrm{CRes}_A^f \subset \operatorname{int}\operatorname{dom} f^*$.
(ii) $\operatorname{ran} \mathrm{CRes}_A^f \subset \operatorname{int}\operatorname{dom} f^*$.
(iii) $\nabla f^* \left(\operatorname{Fix} \left(\mathrm{CRes}_A^f \right) \right) = \operatorname{int}\operatorname{dom} f \cap A^{-1}(0^*)$.
(iv) Suppose, in addition, that A is a monotone mapping. Then the following assertions also hold:

  (a) If $f|_{\operatorname{int}\operatorname{dom} f}$ is strictly convex, then the operator CRes_A^f *is single-valued on its domain and R-BFNE.*
  (b) If $f : X \to \mathbb{R}$ is such that $\operatorname{ran} \nabla f \subset \operatorname{ran}(\nabla f + A)$, *then* $\operatorname{dom} \mathrm{CRes}_A^f = \operatorname{int}\operatorname{dom} f^*$.

According to Proposition 24.23(iii) and (iv)(a), we can apply Theorem 24.18 in the dual space X^* to the conjugate resolvent CRes_A^f and obtain an implicit method for approximating zeroes of monotone mappings.

Theorem 24.24 (Implicit method for approximating zeroes). *Let $f : X \to \mathbb{R}$ be a Legendre and totally convex function such that f^* is positively homogeneous of degree $\alpha > 1$ and uniformly continuous on bounded subsets of X^*. Assume that ∇f^* is weakly sequentially continuous. Let K^* be a nonempty and bounded subset of X^* such that $\nabla f^*(K^*)$ is bounded, closed, and convex with $0 \in \nabla f^*(K^*)$. Let λ be any positive real number and let $A : X \to 2^{X^*}$ be a monotone mapping such that $\nabla f(\mathrm{dom} A) \subset K^* \subset \mathrm{ran}(I + \lambda A \circ \nabla f^*)$. Then the following two assertions hold:*

(i) For each $t \in (0,1)$, there exists a unique $\xi_t \in K^*$ satisfying $\xi_t = t\mathrm{CRes}^f_{\lambda A}\xi_t$.
(ii) The net $\{\xi_t\}_{t\in(0,1)}$ converges strongly to $\overrightarrow{\mathrm{proj}}^f_{\nabla f(A^{-1}(0^*))}(0^*)$ as $t \to 1^-$.

Acknowledgements The first author was partially supported by DGES (Grant MTM2009-10696-C02-01) and by the Junta de Andalucía (Grant FQM-127); the second and third authors were partially supported by the Israel Science Foundation (Grant 647/07), the Graduate School of the Technion, the Fund for the Promotion of Research at the Technion and by the Technion VPR Fund. Part of the research pertaining to this paper was carried out when the authors were visiting each other's institutions. They thank their colleagues for their kind hospitality. The authors are also grateful to an anonymous referee for several helpful comments and suggestions.

References

1. Bauschke, H.H., Borwein, J.M.: Legendre functions and the method of random Bregman projections. J. Convex Anal. **4**, 27–67 (1997)
2. Bauschke, H.H., Borwein, J.M.: Joint and separate convexity of the Bregman distance. In: Inherently Parallel Algorithms in Feasibility and Optimization and their Applications. Studies in Computational Mathematics, vol. 8, pp. 23–36. North-Holland, Amsterdam (2001)
3. Bauschke, H.H., Combettes, P.L.: Convex Analysis and Monotone Operator Theory in Hilbert Spaces. Springer, New York (2011)
4. Bauschke, H.H., Borwein, J.M., Combettes, P.L.: Essential smoothness, essential strict convexity, and Legendre functions in Banach spaces. Comm. Contemp. Math. **3**, 615–647 (2001)
5. Bauschke, H.H., Borwein, J.M., Combettes, P.L.: Bregman monotone optimization algorithms. SIAM J. Contr. Optim. **42**, 596–636 (2003)
6. Bauschke, H.H., Wang, X., Ye, J., Yuan, X.: Bregman distances and Chebyshev sets. J. Approx. Theory **159**, 3–25 (2009)
7. Bonnans, J.F., Shapiro, A.: Perturbation Analysis of Optimization Problems. Springer, New York (2000)
8. Borwein, J.M., Reich, S., Sabach, S.: A characterization of Bregman firmly nonexpansive operators using a new monotonicity concept. J. Nonlinear Convex Anal. **12**, 161–184 (2011)
9. Bregman, L.M.: The relaxation method of finding the common point of convex sets and its application to the solution of problems in convex programming. USSR Comput. Math. Math. Phys. **7**, 200–217 (1967)
10. Browder, F.E.: Convergence of approximants to fixed points of nonexpansive nonlinear mappings in Banach spaces. Arch. Rational Mech. Anal. **24**, 82–90 (1967)
11. Bruck, R.E.: Nonexpansive projections on subsets of Banach spaces. Pacific J. Math. **47**, 341–355 (1973)
12. Bruck, R.E., Reich, S.: Nonexpansive projections and resolvents of accretive operators in Banach spaces. Houston J. Math. **3**, 459–470 (1977)

13. Butnariu, D., Censor, Y., Reich, S.: Iterative averaging of entropic projections for solving stochastic convex feasibility problems. Comput. Optim. Appl. **8**, 21–39 (1997)
14. Butnariu, D., Iusem, A.N.: Totally Convex Functions for Fixed Points Computation and Infinite Dimensional Optimization. Kluwer Academic Publishers, Dordrecht (2000)
15. Butnariu, D., Resmerita, E.: Bregman distances, totally convex functions and a method for solving operator equations in Banach spaces. Abstr. Appl. Anal. **2006**, 1–39 (2006) (Art. ID 84919)
16. Censor, Y., Lent, A.: An iterative row-action method for interval convex programming. J. Optim. Theory Appl. **34**, 321–353 (1981)
17. Censor, Y., Zenios, S.A.: Parallel Optimization. Oxford University Press, New York (1997)
18. Dotson, W.G., Jr.: Fixed points of quasi-nonexpansive mappings. J. Austral. Math. Soc. **13**, 167–170 (1972)
19. Goebel, K., Kirk, W.A.: Topics in metric fixed point theory. In: Cambridge Studies in Advanced Mathematics, vol. 28. Cambridge University Press, Cambridge (1990)
20. Goebel, K., Kirk, W.A.: Classical theory of nonexpansive mappings. In: Handbook of Metric Fixed Point Theory, pp. 49–91. Kluwer Academic Publishers, Dordrecht (2001)
21. Goebel, K., Reich, S.: Uniform Convexity, Hyperbolic Geometry, and Nonexpansive Mappings. Dekker, New York (1984)
22. Martín-Márquez, V., Reich, S., Sabach, S.: Right Bregman nonexpansive operators in Banach spaces. Nonlinear Anal. **75**, 5448–5465 (2012)
23. Martín-Márquez, V., Reich, S., Sabach, S.: Bregman strongly nonexpansive operators in Banach spaces. J. Math. Anal. Appl. **400**, 597–614 (2013)
24. Martín-Márquez, V., Reich, S., Sabach, S.: Iterative methods for approximating fixed points of Bregman nonexpansive operators. Discrete Contin. Dyn. Syst. Ser. S **6**, 1043–1063 (2013)
25. Phelps, R.R.: Convex Functions, Monotone Operators, and Differentiability, 2nd edn. In: Lecture Notes in Mathematics, vol. 1364. Springer, Berlin (1993)
26. Reich, S.: Asymptotic behavior of contractions in Banach spaces. J. Math. Anal. Appl. **44**, 57–70 (1973)
27. Reich, S.: Extension problems for accretive sets in Banach spaces. J. Func. Anal. **26**, 378–395 (1977)
28. Reich, S.: A weak convergence theorem for the alternating method with Bregman distances. In: Theory and Applications of Nonlinear Operators of Accretive and Monotone Type, pp. 313–318. Dekker, New York (1996)
29. Reich, S., Sabach, S.: A strong convergence theorem for a proximal-type algorithm in reflexive Banach spaces. J. Nonlinear Convex Anal. **10**, 471–485 (2009)
30. Reich, S., Sabach, S.: Two strong convergence theorems for a proximal method in reflexive Banach spaces. Numer. Funct. Anal. Optim. **31**, 22–44 (2010)
31. Reich, S., Sabach, S.: Existence and approximation of fixed points of Bregman firmly nonexpansive operators in reflexive Banach spaces. In: Fixed-Point Algorithms for Inverse Problems in Science and Engineering, Optimization and Its Applications, vol. 49, pp. 301–316. Springer, New York (2011)
32. Resmerita, E.: On total convexity, Bregman projections and stability in Banach spaces. J. Convex Anal. **11**, 1–16 (2004)

Chapter 25
Primal Lower Nice Functions and Their Moreau Envelopes

Marc Mazade and Lionel Thibault

Dedicated to Jonathan Borwein on the occasion of his 60th birthday

Abstract This paper studies two equivalent definitions of primal lower nice functions and some subdifferential characterizations of such functions. Various regularity properties of the associated Moreau envelopes and proximal mappings are also provided.

Key words: Borwein–Preiss principle • Infimum convolution • Moreau envelopes • Primal lower nice functions • Proximal mappings • Semiconvex functions • Subdifferentials

Mathematics Subject Classifications (2010): Primary 49J52, 49J53; Secondary 34A60

25.1 Introduction

The class of primal lower nice (pln for short) functions covers a large class of functions with an underlying subsmooth structure. Among these functions, we quote all lower semicontinuous (lsc) proper convex functions, lower-$\mathscr{C}^2$ functions, and

COMMUNICATED BY HEINZ H. BAUSCHKE.

M. Mazade (✉) • L. Thibault
Université de Montpellier II, Place Eugène Bataillon, Case Courrier 51, 34095 Montpellier, France
e-mail: mmazade@math.univ-montp2.fr; thibault@math.univ-montp2.fr

D.H. Bailey et al. (eds.), *Computational and Analytical Mathematics*, Springer Proceedings in Mathematics & Statistics 50, DOI 10.1007/978-1-4614-7621-4_25,

qualified convexly composite functions. The first definition of pln functions was given in [22] by Poliquin in the finite-dimensional setting. Poliquin began the study of this class of functions and their properties, as the coincidence of their proximal and Clarke subdifferentials and a first subdifferential characterization. He also obtained the following integration theorem: if two lsc functions f and g are pln at some point x of their effective domain and have the same subgradients on a neighborhood of x, then $g = f + \alpha$ near x, where α is a constant. Carrying on this way, Levy, Poliquin, and Thibault showed in [15] that the subdifferential characterization for pln functions is still valid in the context of a general Hilbert space, as well as the coincidence of proximal and Clarke subdifferentials. Then Thibault and Zagrodny [25] extended to the infinite-dimensional Hilbert setting Poliquin's integration theorem for pln functions. In [4], Bernard, Thibault, and Zagrodny developed some additional results of integration for this class of functions.

One of the interests in studying regularity properties of pln functions is due to the strong connection between this class of functions and differentiability properties of Moreau envelopes. Such properties arise in the study of existence of solutions of nonconvex dynamical differential inclusions (see, e.g., [9, 16] and the references therein) and also in the development of proximal-like algorithms for some nonconvex optimization problems. Several authors have developed a series of local properties of Moreau envelopes, first Poliquin and Rockafellar [23], concerning prox-regular functions in the finite-dimensional setting. In [3], Bernard and Thibault carried on working on local regularity properties of Moreau envelopes and the related proximal mappings of prox-regular and pln functions. Some years later, Marcellin and Thibault [16] obtained some existence results for differential inclusions associated with subdifferentials of pln functions, studying the corresponding differential equations governed by the gradients of the Moreau envelopes. Recently, Bačák, Borwein, Eberhard, and Mordukhovich [1] established some new subdifferential properties of Moreau envelopes of prox-regular functions in Hilbert spaces.

Pln functions have been currently involved in optimal controlled problems. Serea and Thibault obtained new results in [24] for pln properties of value functions of Mayer-type control problems. In that paper, the authors also revisited the historical Poliquin's definition of pln functions, giving a new one which is equivalent.

All these facts motivated us to study the class of pln functions and their associated Moreau envelopes through the new definition given in [24]. Our approach involves for the subdifferential characterization of pln functions some ideas in [2] and for differential properties of Moreau envelopes of such functions some recent techniques used in [1] for the study of Moreau envelopes of prox-regular functions. The paper is organized as follows. Section 25.2 is devoted to some preliminaries and to a first comparison between two definitions of pln functions. In Sect. 25.3, we establish a subdifferential characterization of pln functions as in [15, 22] with the new definition. Section 25.4 studies some regularity properties of Moreau envelopes of pln functions, considering again the equivalent definition.

25.2 Pln Functions and First Properties

Let us recall some fundamental definitions. Throughout all the paper, unless otherwise stated, $(X, \| \, \|)$ is a Banach space and X^* is its topological dual endowed with the dual norm $\| \, \|_*$ that we will denote by $\| \, \|$ for convenience. When X will be a Hilbert space, we will identify, as usual, X^* with X through the Riesz isometry. The open (resp. closed) ball of X centered at $\bar{x}$ with radius ε is denoted by $B(\bar{x}, \varepsilon)$ (resp. $B[\bar{x}, \varepsilon]$), and we will set $\mathbb{B}_X := B[0,1]$. For a set-valued mapping $M : X \rightrightarrows Y$ from X into a Banach space Y, we will denote by $\operatorname{Dom} M$ and $\operatorname{gph} M$ its effective domain and graph, respectively, that is,

$$\operatorname{Dom} M := \{x \in X : M(x) \neq \emptyset\} \text{ and } \operatorname{gph} M := \{(x,y) \in X \times Y : y \in M(x)\}.$$

For a set S of X and $\bar{x} \in S$, the Clarke tangent cone of S at $\bar{x}$ is defined as the Painlevé–Kuratowski limit inferior of the set-differential quotient

$$T^C(S;\bar{x}) := \operatorname*{Lim\,inf}_{t \downarrow 0; x \xrightarrow[S]{} \bar{x}} \frac{1}{t}(S - x),$$

that is, a vector $h \in T^C(S;\bar{x})$ if for any sequence $(x_n)_n$ in S converging to $\bar{x}$ and any sequence $(t_n)_n$ of positive numbers converging to 0 there exists a sequence $(h_n)_n$ in X converging to h such that

$$x_n + t_n h_n \in S \text{ for all } n \in \mathbb{N}.$$

The Clarke tangent cone $T^C(S;\bar{x})$ is known to be closed and convex (see [6]). The Clarke normal cone $N^C(S;\bar{x})$ of S at $\bar{x}$ is the negative polar $(T^C(S;\bar{x}))^0$ of the Clarke tangent cone, that is,

$$N^C(S;\bar{x}) := \{\zeta \in X^* : \langle \zeta, h \rangle \leq 0 \quad \forall h \in T^C(S;\bar{x})\}.$$

Let $f : X \to \mathbb{R} \cup \{+\infty\}$ be an extended real-valued function and let $\bar{x} \in \operatorname{dom} f$, that is, $f(\bar{x}) < +\infty$. Through the Clarke normal cone, one defines the Clarke subdifferential $\partial_C f(\bar{x})$ of the function f at $\bar{x}$ as

$$\partial_C f(\bar{x}) := \{\zeta \in X^* : (\zeta, -1) \in N^C(\operatorname{epi} f; (x, f(x)))\},$$

where $\operatorname{epi} f := \{(x;r) \in X \times \mathbb{R} : f(x) \leq r\}$ is the epigraph of f in $X \times \mathbb{R}$. One also puts $\partial_C f(\bar{x}) = \emptyset$ when $f(\bar{x}) = +\infty$. When f is lsc on an open set $\mathcal{O}$, then $\mathcal{O} \cap \operatorname{Dom} \partial_C f$ is dense in $\mathcal{O} \cap \operatorname{dom} f$.

If f is Lipschitz continuous near $\bar{x}$, one has

$$\partial_C f(\bar{x}) = \{\zeta \in X^* : \langle \zeta, h \rangle \leq f^o(\bar{x}; h) \; \forall h \in X\},$$

where

$$f^o(\bar{x};h) := \limsup_{t\downarrow 0, x\to\bar{x}} t^{-1}[f(x+th)-f(x)].$$

For f Lipschitz continuous (resp. $\mathscr{C}^1$) near $\bar{x}$ one has

$$\partial_C(f+g)(\bar{x}) \subset \partial_C f(\bar{x}) + \partial_C g(\bar{x}) \quad \big(\text{resp. } \partial_C(f+g)(\bar{x}) = Df(\bar{x}) + \partial_C g(\bar{x})\big) \quad (25.1)$$

for any extended real-valued function $g : X \to \mathbb{R}\cup\{+\infty\}$.

Besides the Clarke subdifferential, some other subdifferentials will be crucial in the development of the paper. First, we recall that the Fréchet subdifferential of f at $\bar{x}$ is the set

$$\partial_F f(\bar{x}) := \left\{ \zeta \in X^* : \liminf_{x\to\bar{x}} \frac{f(x)-f(\bar{x})-\langle\zeta, x-\bar{x}\rangle}{\|x-\bar{x}\|} \geq 0 \right\}.$$

In other words, $\zeta \in \partial_F f(\bar{x})$ provided that for each $\varepsilon > 0$, there exists $\eta > 0$ such that for all $x \in B(\bar{x},\eta)$,

$$\langle\zeta, x-\bar{x}\rangle \leq f(x) - f(\bar{x}) + \varepsilon\|x-\bar{x}\|.$$

When $f(\bar{x}) = +\infty$, by convention, $\partial_F f(\bar{x}) = \emptyset$.

If X is an Asplund space (i.e., the topological dual of any separable subspace of X is separable) the Mordukhovich limiting subdifferential (see [17]) of f at $\bar{x}$ is then given by

$$\partial_L f(\bar{x}) := \{w^* - \lim \zeta_n : \zeta_n \in \partial_F f(x_n), x_n \to_f \bar{x}\},$$

where $w^* - \lim \zeta_n$ is the weak star limit of ζ_n and where $x_n \to_f \bar{x}$ means $\|x_n - \bar{x}\| \to 0$ with $f(x_n) \to f(\bar{x})$. (Of course, any reflexive Banach space is Asplund.) Under the Asplund property of X one has

$$\partial_F f(\bar{x}) \subset \partial_L f(\bar{x}) \subset \partial_C f\bar{x}).$$

The proximal subdifferential is known to be an efficient tool for many variational studies. Below, it will be involved in the differential study of the Moreau envelope. An element $\zeta \in X^*$ belongs to the proximal subdifferential $\partial_P f(\bar{x})$ of f at $\bar{x}$ whenever there exist $\eta > 0$ and $r > 0$ such that

$$\langle\zeta, x-\bar{x}\rangle \leq f(x) - f(\bar{x}) + r\|x-\bar{x}\|^2, \text{ for all } x \in B(\bar{x},\eta).$$

By convention, one sets $\partial_P f(\bar{x}) = \emptyset$ when $f(\bar{x}) = +\infty$. We obviously see that $\partial_P f(\bar{x}) \subset \partial_F f(\bar{x})$. We also recall (see [17]) that, when X is a Hilbert space, one has

$$\partial_L f(\bar{x}) := \{w - \lim \zeta_n : \zeta_n \in \partial_P f(x_n), x_n \to_f \bar{x}\}.$$

Like for the Clarke subdifferential if f is lsc on an open set $\mathcal{O}$ and if X is an Asplund space (resp. a Hilbert space), then $\mathcal{O} \cap \operatorname{Dom} \partial_F f$ (resp. $\mathcal{O} \cap \operatorname{Dom} \partial_P f$) is dense in $\mathcal{O} \cap \operatorname{dom} f$ (see [17]).

Now we define the concept of pln functions in a quantified way.

Definition 25.1. Let $f : X \to \mathbb{R} \cup \{+\infty\}$ be a function defined on the Banach space X. For an open convex subset $\mathcal{O}$ of X with $\mathcal{O} \cap \operatorname{dom} f \neq \emptyset$, the function f is said to be pln on $\mathcal{O}$ provided that f is lsc on $\mathcal{O}$ and there exists some real number $c \geq 0$ such that for all $x \in \mathcal{O} \cap \operatorname{Dom} \partial_C f$ and for all $\zeta \in \partial_C f(x)$ we have

$$f(y) \geq f(x) + \langle \zeta, y - x \rangle - c(1 + \|\zeta\|)\|y - x\|^2 \tag{25.2}$$

for each $y \in \mathcal{O}$. The real $c \geq 0$ will be called a pln constant for f over $\mathcal{O}$ and we will say that f is c-pln on $\mathcal{O}$. For $\bar{x} \in \operatorname{dom} f$, we say that f is pln at $\bar{x}$ whenever it is pln on some open convex set containing the point $\bar{x}$.

Remark 25.2. The definition above is an adaptation of the one in [24]. For convenience, we use in the definition above $\partial_C f(x)$ instead of $\partial_F f(x)$, but it will be seen in Theorem 25.10 that requiring (25.2) with $\partial_C f(x)$ is equivalent to requiring it with $\partial_F f(x)$.

Using the arguments in Proposition 2.2 in [24], it is easily seen that the definition above is equivalent to the pioneering definition of pln functions, introduced by Poliquin in [22], if in the definition in [22] we invoke the Clarke subdifferential in place of the proximal one. We state this in the following proposition.

Proposition 25.3. *A function $f : X \to \mathbb{R} \cup \{+\infty\}$ is pln at $\bar{x} \in \operatorname{dom} f$ in the sense of Definition 25.1 if and only if it is pln at $\bar{x}$ in the pioneering sense of Poliquin, that is, there are positive constant real numbers s_0, c_0, Q_0, such that f is lsc on an open set containing $B[\bar{x}, s_0]$ and for all $x \in B[\bar{x}, s_0]$, for all $q \geq Q_0$, and for all $\zeta \in \partial_C f(x)$ with $\|\zeta\| \leq c_0 q$, one has*

$$f(y) \geq f(x) + \langle \zeta, y - x \rangle - \frac{q}{2}\|y - x\|^2 \tag{25.3}$$

for each $y \in B[\bar{x}, s_0]$.

The proof is given in detail in [24].

Our first result below will establish a link between continuous pln functions and semiconvex functions. Recall that a function $f : X \to \mathbb{R} \cup \{+\infty\}$ is said to be (linearly) semiconvex on a convex set $\mathcal{O} \subset X$ with constant $c \geq 0$ whenever for all reals $\lambda \in]0, 1[$

$$f(\lambda x + (1 - \lambda)y) \leq \lambda f(x) + (1 - \lambda)f(y) + c\lambda(1 - \lambda)\|x - y\|^2.$$

When f is semiconvex near each point of $\mathcal{O}$, we will say that f locally semiconvex on $\mathcal{O}$.

The link between pln and semiconvex continuous functions will use the following proposition.

Proposition 25.4. *Assume that X is a Hilbert space and let $f : \mathcal{O} \subset X \to \mathbb{R} \cup \{+\infty\}$ be a function on a convex set $\mathcal{O}$ of X. The following are equivalent:*

(a) the function f is semiconvex on $\mathcal{O}$ with constant $c \geq 0$.
(b) the function $f + c\| \|^2$ is convex on $\mathcal{O}$.

Proof. For all $\lambda \in]0,1[$, all $x, y \in \mathcal{O}$, the following inequalities are equivalent:

$$f(\lambda x+(1-\lambda)y)+c\|\lambda x+(1-\lambda)y\|^2 \leq \lambda f(x)+c\lambda\|x\|^2+(1-\lambda)f(y) + c(1-\lambda)\|y\|^2$$

$$f(\lambda x+(1-\lambda)y) \leq \lambda f(x)+c\lambda\|x\|^2+(1-\lambda)f(y) + c(1-\lambda)\|y\|^2 - c\|\lambda x+(1-\lambda)y\|^2$$

$$f(\lambda x+(1-\lambda)y) \leq \lambda f(x)+(1-\lambda)f(y)+c\lambda\|x\|^2+c(1-\lambda)\|y\|^2 - c\lambda^2\|x\|^2-2c\lambda(1-\lambda)\langle x,y\rangle - c(1-\lambda)^2\|y\|^2$$

$$f(\lambda x+(1-\lambda)y) \leq \lambda f(x)+(1-\lambda)f(y)+c\lambda(1-\lambda)\|x\|^2 + c(1-\lambda)\lambda\|y\|^2-2c\lambda(1-\lambda)\langle x,y\rangle$$

and hence the convexity of $f + c\|\cdot\|^2$ is equivalent to the inequality

$$f(\lambda x+(1-\lambda)y) \leq \lambda f(x)+(1-\lambda)f(y)+c\lambda(1-\lambda)\|x-y\|^2,$$

for all $\lambda \in]0,1[, x, y \in \mathcal{O}$. ■

Now we can establish the link between pln and semiconvex functions in showing that a continuous function on $\mathcal{O}$ is pln at each point of $\mathcal{O}$ if and only if it is locally semiconvex on $\mathcal{O}$. So, we extend a result proved in [24] in the finite-dimensional setting. (The general result in the context of any Banach space will appear in a forthcoming work.)

Proposition 25.5. *Assume that X is a Hilbert space and $f : \mathcal{O} \to \mathbb{R} \cup \{+\infty\}$ is a proper lsc function, defined on an open set $\mathcal{O}$ of X. The following are equivalent :*

(a) f is locally semiconvex, finite, and locally Lipschitz continuous on $\mathcal{O}$.
(b) f is locally semiconvex, finite and continuous on $\mathcal{O}$.
(c) f is locally bounded from above on $\mathcal{O}$ and pln at any point of $\mathcal{O}$.

Proof. Clearly, (a) implies (b). Suppose now f is locally semiconvex, finite, and continuous on $\mathcal{O}$. Let $\bar{x} \in \mathcal{O}$. There exist $\eta > 0$ and $c \geq 0$ such that $f + c\|\cdot\|^2$ is convex on $B(\bar{x}, \eta)$. As f is continuous, we may suppose without loss of generality that f is bounded from above on $B(\bar{x}, \eta)$. Choose $x \in B(\bar{x}, \eta)$ and $\zeta \in \partial_C f(x)$. Note that

$$\zeta + 2cx \in \partial_C f(x) + \nabla(c\|\cdot\|^2)(x) = \partial_C(f + c\|\cdot\|^2)(x)$$

since $c\|\cdot\|^2$ is $\mathscr{C}^1$ on X. Thus, for all $x' \in B(\bar{x}, \eta)$,

$$\langle \zeta + 2cx, x' - x\rangle \leq f(x') + c\|x'\|^2 - f(x) - c\|x\|^2$$
$$\langle \zeta, x' - x\rangle \leq f(x') - f(x) + c\|x' - x\|^2$$

and we get f is pln at $\bar{x}$. So, (b) $\Rightarrow$ (c) since the local boundedness property obviously follows from the continuity of f on $\mathscr{O}$.

Now suppose that (c) holds and let $\bar{x} \in \mathscr{O}$. Since f is lsc, there exist some positive number η_0 and some $\gamma \in \mathbb{R}$ such that $f(x) \geq \gamma$ for all $x \in B(\bar{x}, \eta_0)$. By assumption, we can find $0 < \eta_1 < \eta_0$ and β_1 such that $f(x) \leq \beta_1$ for all $x \in B(\bar{x}, \eta_1)$. So, for $\beta := \beta_1 + |\gamma|$ we have $|f(x)| \leq \beta$ for all $x \in B(\bar{x}, \eta_1)$. Since f is pln at $\bar{x}$, there exist $0 < \eta < \eta_1$ and $c \geq 0$ such that for all $x \in B(\bar{x}, \eta)$ and all $\zeta \in \partial_C f(x)$ the inequality

$$\langle \zeta, x' - x\rangle \leq f(x') - f(x) + c(1 + \|\zeta\|)\|x' - x\|^2 \tag{25.4}$$

is valid for all $x' \in B(\bar{x}, \eta)$. Choose some real number $r > 0$ such that $cr < 1$ and $B(\bar{x}, 2r) \subset B(\bar{x}, \eta)$. Considering $x \in B(\bar{x}, r) \cap \operatorname{Dom} \partial_C f$ and $\zeta \in \partial_C f(x)$, for $x' = x + rb$, where $b \in \mathbb{B}$, we obtain from (25.4)

$$\langle \zeta, rb\rangle \leq 2\beta + c(1 + \|\zeta\|)r^2.$$

Hence

$$\|\zeta\| \leq r^{-1}(1 - cr)^{-1}(2\beta + cr^2) =: k. \tag{25.5}$$

This entails that f is Lipschitz continuous on $B(\bar{x}, r)$ (see [25]) which ensures the desired local Lipschitz property of f on $\mathscr{O}$. It then remains to establish the semiconvexity of f near $\bar{x}$. To get that we first observe by (25.4) and (25.5) that for any $\zeta \in \partial_C f(x)$, with $x \in B(\bar{x}, r)$,

$$\langle \zeta, x' - x\rangle \leq f(x') - f(x) + c(1 + k)\|x' - x\|^2$$

for all $x' \in B(\bar{x}, r)$. So, if we set $c' := c(1 + k)$,

$$\langle (\zeta_1 + 2c'x_1) - (\zeta_2 + 2c'x_2), x_1 - x_2\rangle \geq 0$$

for all $\zeta_i \in \partial_C f(x_i)$ with $x_i \in B(\bar{x}, r)$. The latter inequality is equivalent (see [7,8,21]) to the convexity of $f + c'\|\cdot\|^2$ on $B(\bar{x}, r)$. Consequently, (c) $\Rightarrow$ (a) is established.

■

25.3 Subdifferential Characterizations

Recall that a set-valued mapping $T : X \rightrightarrows X^*$ is hypomonotone on a subset $\mathscr{O}$ of X if there exists $\sigma > 0$ such that

$$\langle \zeta_1 - \zeta_2, x_1 - x_2 \rangle \geq -\sigma \|x_1 - x_2\|^2$$

whenever $\zeta_i \in T(x_i), x_i \in \mathscr{O}$. The pln behavior of a function f is characterized by some local linear hypomonotonicity of $\partial_P f$ or $\partial_C f$ in [15,21]. Using Definition 25.1, as the main result of the paper, we establish in the next theorem a similar characterization with a variant of the hypomonotonicity property of the subdifferential $\partial_C f$. Before stating Theorem 25.7, we prove the following lemma which is an adaptation of Lemma 4.2 in [25] and ideas in [12].

Lemma 25.6 (see [25]). *Let X be a normed vector space and let $f : X \to \mathbb{R} \cup \{+\infty\}$ be a function with $f(\bar{x}) < +\infty$. Let s be a positive number such that f is bounded from below over $B[\bar{x}, s]$ by some real α. Let $\beta \in \mathbb{R}$ and θ be a nonnegative number. For each real $c \geq 0$, let*

$$F_{\beta,c}(\zeta, x, y) := f(y) + \beta \langle \zeta, x - y \rangle + c(1 + \|\zeta\|)\|x - y\|^2 \quad \textit{for all } x, y \in X,\ \zeta \in X^*.$$

Let any real $c_0 \geq \frac{16|\beta|}{3s}$ such that $c_0 > \frac{16}{3s^2}(\theta + f(\bar{x}) - \alpha)$. Then, for any real $c \geq c_0$, for any $\zeta \in X^*$, and for any $x \in B[\bar{x}, \frac{s}{4}]$, any point $u \in B[\bar{x}, s]$ such that

$$F_{\beta,c}(\zeta, x, u) \leq \inf_{y \in B[\bar{x}, s]} F_{\beta,c}(\zeta, x, y) + \theta \tag{25.6}$$

must belong to $B(\bar{x}, \frac{3s}{4})$.

Proof. Fix $x \in B[\bar{x}, s/4]$ and $\zeta \in X^*$ and fix also any real $c \geq c_0$. Take any $y \in B[\bar{x}, s]$ with $\|y - \bar{x}\| > 3s/4$. Since $\|x - y\| \geq \|\bar{x} - y\| - \|x - \bar{x}\| \geq s/2$, we observe that

$$\|x - y\|^2 - \|x - \bar{x}\|^2 \geq \frac{s^2}{4} - \frac{s^2}{16} = \frac{3s^2}{16}.$$

Then, for $F(y) := F_{\beta,c}(\zeta, x, y)$, we have

$$\begin{aligned} F(y) - F(\bar{x}) - \theta &\geq \alpha - f(\bar{x}) - \theta + \beta \langle \zeta, \bar{x} - y \rangle + c(1 + \|\zeta\|)(\|x - y\|^2 - \|x - \bar{x}\|^2) \\ &\geq \alpha - f(\bar{x}) - \theta - s|\beta|\|\zeta\| + c(1 + \|\zeta\|)\frac{3s^2}{16} \\ &= \left(\alpha - f(\bar{x}) - \theta + c\frac{3s^2}{16}\right) + s\left(c\frac{3s}{16} - |\beta|\right)\|\zeta\|, \end{aligned}$$

so, for $\eta := \alpha - f(\bar{x}) - \theta + c\frac{3s^2}{16} > 0$, we obtain $F(y) - \eta \geq F(\bar{x}) + \theta$, finishing the proof of the lemma. ∎

Additional properties of function of type (25.6) will be developed in Sect. 25.4.

Theorem 25.7. *Let $f : X \to \mathbb{R} \cup \{+\infty\}$ be a function on the Banach space X which is finite at $\bar{x} \in X$ and lsc near $\bar{x}$. The following are equivalent :*

(a) f is pln at $\bar{x}$.
(b) There exist $\varepsilon > 0$ and $c \geq 0$ such that

$$\langle \zeta_1 - \zeta_2, x_1 - x_2 \rangle \geq -c(1 + \|\zeta_1\| + \|\zeta_2\|)\|x_1 - x_2\|^2$$

whenever $\zeta_i \in \partial_C f(x_i)$ and $\|x_i - \bar{x}\| \leq \varepsilon$.

Proof. We first show that (a) implies (b). Since f is pln at $\bar{x}$, there exist $\eta_0 > 0$ and $c \geq 0$ such that the inequality

$$f(x') \geq f(x) + \langle \zeta, x' - x \rangle - c(1 + \|\zeta\|)\|x' - x\|^2$$

holds true for all $x' \in X$ with $\|x' - \bar{x}\| \leq \eta_0$, $x \in \operatorname{Dom} \partial_C f$ with $\|x' - \bar{x}\| \leq \eta_0$, and $\zeta \in \partial_C f(x)$. If $\zeta_i \in \partial_C f(x_i)$, with $\|x_i - \bar{x}\| \leq \eta_0$, then $f(x_1)$ and $f(x_2)$ are finite and

$$f(x_1) \geq f(x_2) + \langle \zeta_2, x_1 - x_2 \rangle - c(1 + \|\zeta_2\|)\|x_1 - x_2\|^2$$
$$f(x_2) \geq f(x_1) + \langle \zeta_1, x_2 - x_1 \rangle - c(1 + \|\zeta_1\|)\|x_2 - x_1\|^2.$$

Adding these inequalities yields

$$\langle \zeta_1 - \zeta_2, x_1 - x_2 \rangle \geq -c(2 + \|\zeta_1\| + \|\zeta_2\|)\|x_1 - x_2\|^2 \tag{25.7}$$

and then

$$\langle \zeta_1 - \zeta_2, x_1 - x_2 \rangle \geq -2c(1 + \|\zeta_1\| + \|\zeta_2\|)\|x_1 - x_2\|^2.$$

Now, using some ideas in [2], we show that (b) implies (a). Let $\varepsilon > 0$ and $c \geq 0$ for which the inequality of the assertion (b) holds and such that f is lsc on $B(\bar{x}, \varepsilon)$. Fix $0 < \varepsilon' < \min\{\varepsilon, \frac{1}{c}\}$ (with the convention $\frac{1}{c} = +\infty$ if $c = 0$) such that $\alpha := \inf_{B[\bar{x}, \varepsilon']} f$ is finite according to the lsc property of f. Following the lemma above fix $c_0 \geq \frac{16}{3\varepsilon'}$ satisfying $c_0 > \frac{16}{3(\varepsilon')^2}(1 + f(\bar{x}) - \alpha)$ and fix also any real $c' > \max\{c_0, \frac{c}{1 - c\varepsilon'}\}$. Let $x \in B(\bar{x}, \frac{\varepsilon'}{4}) \cap \operatorname{Dom} \partial_C f$ and $\zeta \in \partial_C f(x)$. Put

$$\varphi(y) := f(y) + \langle \zeta, x - y \rangle + c'(1 + \|\zeta\|)\|y - x\|^2 \text{ for all } y \in X$$

and

$$\bar{\varphi}(y) := \begin{cases} \varphi(y) & \text{if } y \in B[\bar{x}, \varepsilon'] \\ +\infty & \text{otherwise,} \end{cases}$$

and note that the function $\bar{\varphi}$ is lsc on X.

Let $(\varepsilon_n)_n$ be a sequence of real numbers converging to 0 with $0 < \varepsilon_n < \min\{1, (\varepsilon'/4)^2\}$. For each integer n, take $u_n \in X$ such that

$$\bar{\varphi}(u_n) \leq \inf_X \bar{\varphi} + \varepsilon_n.$$

By Lemma 25.6 applied with $\beta = \theta = 1$, we have $\{u_n\} \subset B(\bar{x}, \frac{3\varepsilon'}{4})$. By the Ekeland variational principle (see [10]), there exists $(x_n)_n$ such that

$$\bar{\varphi}(x_n) < \inf_X \bar{\varphi} + \varepsilon_n,\ \|x_n - u_n\| < \sqrt{\varepsilon_n} \text{ and } \bar{\varphi}(x_n) = \inf_{u \in X} \{\bar{\varphi}(u) + \sqrt{\varepsilon_n}\|u - x_n\|\}.$$

We deduce $\|x_n - \bar{x}\| < \varepsilon'$ and

$$0 \in \partial_C(\varphi + \sqrt{\varepsilon_n}\|\cdot - x_n\|)(x_n);$$

hence

$$0 \in \partial_C f(x_n) - \zeta + c'(1 + \|\zeta\|)\partial_C(\|\cdot - x\|^2)(x_n) + \sqrt{\varepsilon_n}\mathbb{B}_{X^*}.$$

This gives $\zeta_n \in \partial_C f(x_n)$ and $\xi_n \in -\zeta + c'(1 + \|\zeta\|)\partial_C(\|\cdot - x\|^2)(x_n)$ such that

$$\|\zeta_n + \xi_n\| \leq \sqrt{\varepsilon_n}. \tag{25.8}$$

Putting $\xi_n' = \dfrac{\xi_n + \zeta}{c'(1 + \|\zeta\|)}$, we have $\xi_n' \in \partial_C(\|\cdot - x\|^2)(x_n)$, so

$$\langle \xi_n', x_n - x\rangle = 2\|x_n - x\|^2 \text{ and } \|\xi_n'\| = 2\|x_n - x\|. \tag{25.9}$$

For n large enough, say $n \geq n_0$, we have $\|x_n - x\| < \varepsilon'$ since

$$\|x_n - x\| \leq \|x_n - u_n\| + \|u_n - \bar{x}\| + \|\bar{x} - x\| < \sqrt{\varepsilon_n} + \frac{3\varepsilon'}{4} + \|\bar{x} - x\|$$

$$\text{and} \quad \frac{3\varepsilon'}{4} + \|\bar{x} - x\| < \varepsilon'.$$

Then, for $n \geq n_0$, we have $\|\xi_n'\| = 2\|x_n - x\| < 2\varepsilon'$ and from the equality $\xi_n = -\zeta + c'(1 + \|\zeta\|)\xi_n'$ we get

$$\|\xi_n\| \leq \|\zeta\| + c'(1 + \|\zeta\|)\|\xi_n'\| \leq \|\zeta\| + 2c'\varepsilon'(1 + \|\zeta\|),$$

which yields by (25.8)

$$\|\zeta_n\| = \|\zeta_n + \xi_n - \xi_n\| \leq \|\zeta_n + \xi_n\| + \|\xi_n\| \leq \sqrt{\varepsilon_n} + \|\zeta\| + 2c'\varepsilon'(1 + \|\zeta\|). \tag{25.10}$$

We note by the assertion (b) applied with $\zeta_2 = \zeta$ and $\zeta_1 = \zeta_n$ that

$$\langle \zeta - \zeta_n, x - x_n\rangle \geq -c(1 + \|\zeta\| + \|\zeta_n\|)\|x - x_n\|^2. \tag{25.11}$$

On the other hand, (25.8) and (25.9) entail that

$$\begin{aligned}\langle \zeta - \zeta_n, x - x_n\rangle &= \langle c'(1+\|\zeta\|)\xi_n' - \xi_n - \zeta_n, x - x_n\rangle \\ &= \langle c'(1+\|\zeta\|)\xi_n', x - x_n\rangle + \langle -\xi_n - \zeta_n, x - x_n\rangle \\ &= -2c'(1+\|\zeta\|)\|x - x_n\|^2 + \langle -\xi_n - \zeta_n, x - x_n\rangle \\ &\le -2c'(1+\|\zeta\|)\|x - x_n\|^2 + \sqrt{\varepsilon_n}\|x - x_n\|,\end{aligned}$$

and concerning the last member above, (25.11) ensures

$$-2c'(1+\|\zeta\|)\|x - x_n\|^2 + \sqrt{\varepsilon_n}\|x - x_n\| \ge -c(1+\|\zeta\|+\|\zeta_n\|)\|x - x_n\|^2.$$

This leads to

$$\left(2c'(1+\|\zeta\|) - c(1+\|\zeta\|+\|\zeta_n\|)\right)\|x - x_n\| \le \sqrt{\varepsilon_n}.$$

Through the inequality (25.10) we have the estimation

$$\begin{aligned}2c'(1+\|\zeta\|) - c(1+\|\zeta\|+\|\zeta_n\|) &\ge 2c'(1+\|\zeta\|) - c(1+\|\zeta\|) \\ &\quad - c(\sqrt{\varepsilon_n} + \|\zeta\| + 2c'\varepsilon'(1+\|\zeta\|)) \\ &> 2c'(1+\|\zeta\|) - c(1+\|\zeta\|) \\ &\quad - c(1+\|\zeta\| + 2c'\varepsilon'(1+\|\zeta\|)) \\ &= 2(c' - c - cc'\varepsilon')(1+\|\zeta\|).\end{aligned}$$

Consequently,

$$2(c' - c - cc'\varepsilon')(1+\|\zeta\|)\|x - x_n\| \le \sqrt{\varepsilon_n}$$

and the inequality $c' > \frac{c}{1-c\varepsilon'}$ (in the choice above of c') guarantees that $c' - c - cc'\varepsilon' > 0$. It follows that $x_n \to x$ as $n \to +\infty$ and $u_n \to x$. Recall that u_n satisfies

$$f(u_n) + \langle \zeta, x - u_n\rangle + c'(1+\|\zeta\|)\|u_n - x\|^2 \le \inf_{y \in B[\bar{x},\varepsilon']} \{f(y) + \langle \zeta, x - y\rangle + c'(1+\|\zeta\|)\|y - x\|^2\} + \varepsilon_n.$$

Thus, according to the lower semicontinuity of f, we obtain

$$f(x) \le \liminf_n f(u_n) \le \inf_{y \in B[\bar{x},\varepsilon']} \{f(y) + \langle \zeta, x - y\rangle + c'(1+\|\zeta\|)\|y - x\|^2\},$$

which entails that

$$f(y) \ge f(x) + \langle \zeta, y - x\rangle - c'(1+\|\zeta\|)\|y - x\|^2$$

for all $y \in B(\bar{x}, \frac{\varepsilon'}{4})$, which means that f is c'-pln on $B(\bar{x}, \frac{\varepsilon'}{4})$. ∎

Remark 25.8. With this proposition, when X is a Hilbert space, writing (a) implies (b) for a c-pln function f, we get

$$\langle \zeta_1 - \zeta_2, x_1 - x_2 \rangle \geq -c(2 + \|\zeta_1\| + \|\zeta_2\|)\|x_1 - x_2\|^2 \qquad \text{(HYPM)}$$

for all $\zeta_i \in \partial_C f(x_i)$, with $\|x_i - \bar{x}\| \leq \bar{\eta}$. For some $c' > c$, we set $c_0 := \frac{1}{4c'}$, $q_0 := 4c'$. Then for all $x_i \in B[\bar{x}, \bar{\eta}]$, for all $q \geq q_0$, and for all $\zeta_i \in \partial_C f(x_i)$, such that $\|\zeta_i\| \leq c_0 q$ we have

$$\langle \zeta_1 - \zeta_2, x_1 - x_2 \rangle \geq -q\|x_1 - x_2\|^2.$$

Indeed, if $\|\zeta_1\| + \|\zeta_2\| \leq 2$, putting this in (25.7), we obtain

$$\langle \zeta_1 - \zeta_2, x_1 - x_2 \rangle \geq -4c'\|x_1 - x_2\|^2 \geq -q\|x_1 - x_2\|^2.$$

Else, if $\|\zeta_1\| + \|\zeta_2\| > 2$,

$$\langle \zeta_1 - \zeta_2, x_1 - x_2 \rangle \geq -c(2 + \|\zeta_1\| + \|\zeta_2\|)\|x_1 - x_2\|^2 \geq -4c(\|\zeta_1\| + \|\zeta_2\|)\|x_1 - x_2\|^2;$$

hence the inequality $\|\zeta_i\| \leq c_0 q$ ensures

$$\langle \zeta_1 - \zeta_2, x_1 - x_2 \rangle \geq -4c'c_0 q\|x_1 - x_2\|^2 \geq -q\|x_1 - x_2\|^2.$$

As a result, the operator $qI + T_{\frac{q}{4c'}}$ is monotone for all $q \geq 4c'$, where $T_{\frac{q}{4c'}}$ is the truncation of $\partial_C f$ at $\bar{x}$ whose graph is defined by

$$\operatorname{gph} T_{\frac{q}{4c'}} := \left\{ (x, \zeta) \in \operatorname{gph} \partial_C f : \|x - \bar{x}\| \leq \bar{\eta}, \|\zeta\| \leq \frac{q}{4c'} \right\}.$$

Note that when $c > 0$, we can take $c' = c$.

The next proposition is a slight modification of Proposition 1.6 in [16], which gives a closure property concerning the graph of the (proximal) subdifferential of a pln function (see also Lemma 2.4 in [13]).

Proposition 25.9. *Let $\mathscr{O}$ be an open convex set of the Banach space X and $f : X \to \mathbb{R} \cup \{+\infty\}$ be a function which is lsc on $\mathscr{O}$ with $\mathscr{O} \cap \operatorname{dom} f \neq \emptyset$. Let ∂f be any of the subdifferentials $\partial_P f, \partial_F f, \partial_C f$. Assume that the inequality* (25.2) *holds for all $x' \in \mathscr{O} \cap \operatorname{Dom} \partial f$ and all $\zeta \in \partial f(x')$. Let $x \in \mathscr{O} \cap \operatorname{Dom} \partial f$ and $(x_j)_{j \in J}$ be a net converging strongly to x in X, and let $(\zeta_j)_{j \in J}$ be a bounded net of X^* converging weakly star to some ζ in X^* with $\zeta_j \in \partial f(x_j)$. Then*

$$\zeta \in \partial_P f(x) \subset \partial f(x) \quad \text{and} \quad \lim_{j \in J} f(x_j) = f(x);$$

hence in particular $\partial f(x)$ is weakly star closed in X^*.

Proof. Take $c \geq 0$ such that

$$f(x') \geq f(x'') + \langle \zeta, x' - x'' \rangle - c(1 + \|z\|)\|x' - x''\|^2$$

whenever $x', x'' \in \mathscr{O}$ and $\zeta \in \partial f(x'')$. Choose a real $\gamma > 0$ such that $\|\zeta_j\| \leq \gamma$ for all $j \in J$ (according to the boundedness assumption) and choose also some $j_0 \in J$ such that $x_j \in \mathscr{O}$ for all $j \succeq j_0$ (since $x_j \to x$). Fixing $x' \in \mathscr{O}$, this yields, for all $j \succeq j_0$,

$$\begin{aligned} f(x') &\geq f(x_j) + \langle \zeta_j, x' - x_j \rangle - c(1 + \|\zeta_j\|)\|x' - x_j\|^2 \\ &\geq f(x_j) + \langle \zeta_j, x' - x_j \rangle - c(1 + \gamma)\|x' - x_j\|^2. \end{aligned} \tag{25.12}$$

Taking the limit inferior, it follows that

$$f(x') \geq f(x) + \langle \zeta, x' - x \rangle - c(1 + \gamma)\|x' - x\|^2,$$

which means $\zeta \in \partial_P f(x)$. Further, putting $x' = x$ in (25.12), we obtain

$$f(x) \geq f(x_j) + \langle \zeta_j, x - x_j \rangle - c(1 + \gamma)\|x - x_j\|^2$$

for $j \succeq j_0$, which allows us to write

$$f(x) \geq \limsup_{j \in J} f(x_j) \geq \liminf_{j \in J} f(x_j) \geq f(x).$$

It remains to show that $\partial f(x)$ is weak star closed. From the definition of the Clarke subdifferential, this is obvious for $\partial_C f(x)$. So, consider ∂f as $\partial_F f$ (resp. $\partial_P f$) if X is Asplund (resp. Hilbert). Then the set $\partial f(x)$ being convex, the Krein–Šmulian theorem guarantees that $\partial f(x)$ is weakly star closed since the arguments above ensure that the weak star limit of any bounded net of $\partial f(x)$ remains in $\partial f(x)$. ■

A characterization result for pln functions (due to Poliquin [22] in finite dimension and Levy–Poliquin–Thibault [15] in Hilbert space) ensures that if a function $f : X :\to \mathbb{R} \cup \{+\infty\}$ is pln at $\bar{x} \in \operatorname{dom} f$, then for all x in a neighborhood of $\bar{x}$, the proximal subdifferential of f at x agrees with the Clarke subdifferential of f at x. Here, in the same vein of [15, 22], we get the same result at any point of $\mathscr{O}$ with Definition 25.1 for a pln function on $\mathscr{O}$.

Theorem 25.10. *Let $\mathscr{O}$ be an open convex set of the Banach space X and $f : X \to \mathbb{R} \cup \{+\infty\}$ be a function which is lsc on $\mathscr{O}$ with $\mathscr{O} \cap \operatorname{dom} f \neq \emptyset$. The following hold:*

(a) If f is pln on $\mathscr{O}$, then for all $x \in \mathscr{O}$, we have

$$\partial_P f(x) = \partial_F f(x) = \partial_L f(x) = \partial_C f(x).$$

(b) *Assume that X is an Asplund space (resp. a Hilbert space) and let $c \geq 0$. The function f is c-pln on $\mathcal{O}$ if and only if for all $x \in \mathcal{O} \cap \mathrm{Dom}\, \partial_F f$ (resp. $x \in \mathrm{Dom}\, \partial_P f$) and for all $\zeta \in \partial_F f(x)$ (resp. $\zeta \in \partial_P f(x)$) one has*

$$f(y) \geq f(x) + \langle \zeta, y - x \rangle - c(1 + \|\zeta\|)\|y - x\|^2, \tag{25.13}$$

for each $y \in \mathcal{O}$.

Proof. To prove (a), fix any $x \in \mathcal{O}$. First we always have $\partial_P f(x) \subset \partial_C f(x)$. Let $\zeta \in \partial_C f(x)$. According to the definition of a pln function, there exists $c \geq 0$ such that

$$f(y) \geq f(x) + \langle \zeta, y - x \rangle - c(1 + \|\zeta\|)\|y - x\|^2$$

for all $y \in \mathcal{O}$; hence $\zeta \in \partial_P f(x)$.
(b): Assume that X is an Asplund space (resp. a Hilbert space). The implication $\Rightarrow$ being obvious, let us show the converse one. The arguments of the proof are similar to [15]. Let $c \geq 0$ be as in Definition 25.1, that is, for each $(x, \zeta) \in \mathrm{gph}\, \partial_F f$ (resp. $\zeta \in \mathrm{gph}\, \partial_P f$) with $x \in \mathcal{O}$

$$f(y) \geq f(x) + \langle \zeta, y - x \rangle - c(1 + \|\zeta\|)\|y - x\|^2 \text{ for all } y \in \mathcal{O}. \tag{25.14}$$

Fix any point $x \in \mathcal{O}$. Since the subdifferential $\partial_F f(x)$ [resp. $\partial_P f(x)$] is included in the Clarke subdifferential $\partial_C f(x)$, we may suppose that $\partial_C f(x) \neq \emptyset$. From Mordukhovich and Shao (see [18]), we have

$$\partial_C f(x) = \overline{\mathrm{co}}^* [V + V_0],$$

where $\overline{\mathrm{co}}^*$ denotes the weak star closed convex hull in X^*, and for $\partial f(u) := \partial_F f(u)$ [resp. $\partial f(u) = \partial_P f(u)$],

$$V := \{w - \lim \zeta_n : \zeta_n \in \partial f(x_n), x_n \to_f \bar{x}\} = \partial_L f(\bar{x}),$$

where $x_n \to_f \bar{x}$ means $\|x_n - \bar{x}\| \to 0$ with $f(x_n) \to f(\bar{x})$ and

$$V_0 := \{w - \lim \sigma_n \zeta_n : \zeta_n \in \partial f(x_n), x_n \to_f \bar{x}, \sigma_n \downarrow 0\} =: \partial_L^\infty f(\bar{x}).$$

Let $\zeta \in V$. There exist x_n, converging strongly to x, and $\zeta_n \in \partial f(x_n)$, with (ζ_n) converging weakly star to ζ, and $(f(x_n))$ converging to $f(x)$. The weak star convergence of (ζ_n) gives some $\gamma > 0$ such that $\|\zeta_n\| \leq \gamma$ for all n. Further, for n large enough $x_n \in \mathcal{O}$, so by (25.14), we have, for any $y \in \mathcal{O}$,

$$f(y) \geq f(x_n) + \langle \zeta_n, y - x_n \rangle - c(1 + \|\zeta_n\|)\|y - x_n\|^2.$$

Taking the limit as $n \to \infty$, we get

$$f(y) \geq f(x) + \langle \zeta, y - x \rangle - c(1 + \gamma)\|y - x\|^2 \tag{25.15}$$

for all $y \in \mathcal{O}$.

Now consider $\zeta_0 \in V_0$, i.e., there exist (x_n) converging strongly to x, $\sigma_n \downarrow 0$, and $\zeta_n \in \partial f(x_n)$ such that $f(x_n)$ converges to $f(x)$ and $\sigma_n \zeta_n$ converges weakly star to ζ_0. Since $\sigma_n \zeta_n$ converges weakly star, there exists some $\delta > 0$ such that $\|\sigma_n \zeta_n\| \leq \delta$ for all n. From (25.14) again, for each $y \in \mathscr{O} \cap \operatorname{dom} f$, we get for n large enough

$$f(y) \geq f(x_n) + \langle \zeta_n, y - x_n \rangle - c(1 + \|\zeta_n\|)\|y - x_n\|^2$$

and then

$$\sigma_n f(y) \geq \sigma_n f(x_n) + \langle \sigma_n \zeta_n, y - x_n \rangle - \sigma_n c(1 + \|\zeta_n\|)\|y - x_n\|^2$$

which entails

$$\sigma_n f(y) \geq \sigma_n f(x_n) + \langle \sigma_n \zeta_n, y - x_n \rangle - \sigma_n c\|y - x_n\|^2 - c\delta\|y - x_n\|^2.$$

Taking the limit, we get

$$0 \geq \langle \zeta_0, y - x \rangle - c\delta\|y - x\|^2 \tag{25.16}$$

for all $y \in \mathscr{O} \cap \operatorname{dom} f$.

Finally, we show $V + V_0 \subset \partial_P f(x)$. Indeed, for any $\zeta \in V$ and $\zeta_0 \in V_0$, combining (25.15) and (25.16), we have for any $y \in \mathscr{O} \cap \operatorname{dom} f$

$$f(y) \geq f(x) + \langle \zeta + \zeta_0, y - x \rangle - c(1 + \gamma + \delta)\|y - x\|^2,$$

and the latter obviously still holds for any $y \in \mathscr{O}$. Thus, the set $\partial_P f(x)$ being weak star closed according to the proposition above, we conclude that

$$\partial_C f(x) = \overline{\text{co}}^*[V + V_0] \subset \overline{\text{co}}^* \partial_P f(x) = \partial_P f(x) \subset \partial_C f(x).$$

So the inequality (25.14) holds for each $(x, \zeta) \in \operatorname{gph} \partial_C f$, which means that f is pln in the sense of Definition 25.1. ∎

25.4 Regularity Properties of Moreau Envelopes

This section is devoted to properties of Moreau envelopes, also known as infimal convolutions. For two extended real-valued functions $f, g : X \to \mathbb{R} \cup \{+\infty\}$ on a normed space X, the Moreau infimum convolution of f and g is defined by

$$(f \square g)(x) = \inf_{y \in X} \{f(y) + g(y - x)\} \quad \text{for all } x \in X.$$

The particular important case of $\frac{1}{2\lambda}\|\cdot\|^2$ as function g yields to the concepts of Moreau envelope and proximal mapping.

Definition 25.11 (see [19,20]). Let X be a normed vector space and let $f : X \to \mathbb{R} \cup \{+\infty\}$ be a proper function. Consider two positive numbers λ and ε. The Moreau envelope of f with index λ is the function from X into $[-\infty, +\infty]$ defined by

$$e_\lambda f(x) := \inf_{y \in X} \left\{ f(y) + \frac{1}{2\lambda} \|x - y\|^2 \right\}, \quad \text{for all } x \in X,$$

and the corresponding proximal mapping is defined by

$$P_\lambda f(x) := \operatorname*{arg\,min}_{y \in X} \left\{ f(y) + \frac{1}{2\lambda} \|x - y\|^2 \right\}, \quad \text{for all } x \in X.$$

The local Moreau envelope of f associated with λ and ε is defined by

$$e_{\lambda,\varepsilon} f(x) := \inf_{\|y\| \le \varepsilon} \left\{ f(y) + \frac{1}{2\lambda} \|x - y\|^2 \right\}, \quad \text{for all } x \in X,$$

and the corresponding local proximal mapping is defined by

$$P_{\lambda,\varepsilon} f(x) := \operatorname*{arg\,min}_{\|y\| \le \varepsilon} \left\{ f(y) + \frac{1}{2\lambda} \|x - y\|^2 \right\}, \quad \text{for all } x \in X.$$

For any $\lambda > 0$, the function $e_\lambda f$ is the infimum convolution of f and $\frac{1}{2\lambda}\| \cdot \|^2$ (see [20]), so we can write $e_\lambda f = f \Box \frac{1}{2\lambda}\| \cdot \|^2$. Given any $\lambda, \varepsilon > 0$, we have

$$e_{\lambda,\varepsilon} f(x) = e_\lambda (f + \delta_{B[0,\varepsilon]})(x)$$

and

$$P_{\lambda,\varepsilon} f(x) = P_\lambda (f + \delta_{B[0,\varepsilon]})(x),$$

where δ_C is the indicator function of the subset C, that is, $\delta_C(x) = 0$ if $x \in C$ and $\delta_c(x) = +\infty$ otherwise.

The following lemma is of great importance to prove some regularity properties of $e_\lambda f$.

Lemma 25.12 (from Correa-Jofré-Thibault [8]). *Let X be a normed space and $f, g : X \to \mathbb{R} \cup \{+\infty\}$. Let $\bar{x} \in X$ be a point where $(f \Box g)(\bar{x})$ is finite. If the infimum convolution at $\bar{x}$*

$$(f \Box g)(\bar{x}) = \inf_{y \in H} \{ f(y) + g(\bar{x} - y) \}$$

is exact, that is, if the preceding infimum is attained at some $\bar{y}$, then

$$\partial_P (f \Box g)(\bar{x}) \subset \partial_P f(\bar{y}) \cap \partial_P g(\bar{x} - \bar{y}).$$

Proof. Let $\zeta \in \partial_P(f\Box g)(\bar{x})$, $\varepsilon > 0$, and $c > 0$ such that

$$\langle \zeta, x-\bar{x}\rangle \leq (f\Box g)(x) - (f\Box g)(\bar{x}) + c\|x-\bar{x}\|^2 \quad \text{for all} \, x \in B(\bar{x},\varepsilon).$$

Then for all $y \in B(\bar{y},\varepsilon)$

$$\begin{aligned}\langle \zeta, y-\bar{y}\rangle &= \langle \zeta, y-\bar{y}+\bar{x}-\bar{x}\rangle \\ &\leq (f\Box g)(y-\bar{y}+\bar{x}) - (f\Box g)(\bar{x}) + c\|y-\bar{y}\|^2 \\ &\leq f(y) + g(\bar{x}-\bar{y}) - f(\bar{y}) - g(\bar{x}-\bar{y}) + c\|y-\bar{y}\|^2 \\ &= f(y) - f(\bar{y}) + c\|y-\bar{y}\|^2,\end{aligned}$$

so $\zeta \in \partial_P f(\bar{y})$. Analogously for all $y \in B(\bar{x}-\bar{y},\varepsilon)$

$$\begin{aligned}\langle \zeta, y-(\bar{x}-\bar{y})\rangle &= \langle \zeta, y-\bar{x}+\bar{y}\rangle \\ &\leq (f\Box g)(y+\bar{y}) - (f\Box g)(\bar{x}) + c\|y+\bar{y}-\bar{x}\|^2 \\ &\leq f(\bar{y}) + g(y) - f(\bar{y}) - g(\bar{x}-\bar{y}) + c\|y+\bar{y}-\bar{x}\|^2 \\ &= g(y) - g(\bar{x}-\bar{y}) + c\|y+\bar{y}-\bar{x}\|^2;\end{aligned}$$

hence $\zeta \in \partial_P g(\bar{x}-\bar{y})$. ■

Throughout the remaining of the paper, X is assumed to be a Hilbert space endowed with the inner product $\langle\cdot,\cdot\rangle$ and the associated norm $\|x\| = \sqrt{\langle x,x\rangle}$.

Let us recall the link between the $\mathscr{C}^1$ property of $e_\lambda f$ and the Lipschitz continuity of $P_\lambda f$. We refer to [26, Example 3.14] for some complements about C^1 property of $e_\lambda f$ and continuity property of $P_\lambda f$.

Proposition 25.13 (see [3]). *Let X be a Hilbert space and $f : X \to \mathbb{R}\cup\{+\infty\}$ be a proper lsc function minorized by a quadratic function on X. Consider $\lambda > 0$ and let U be an open subset of the Hilbert space X. The following properties are equivalent:*

(a) $e_\lambda f$ is $\mathscr{C}^1$ on U.
(b) $P_\lambda f$ is nonempty, single-valued, and continuous on U.

When these properties hold, $\nabla e_\lambda f = \lambda^{-1}(I - P_\lambda f)$ on U.

We can now establish, for a pln function f, a list of properties of the Moreau envelope and proximal mapping of $f + \delta(\cdot,B)$ for an appropriate closed ball B. We proceed, involving some arguments of [1], to developing a proof avoiding the use (made in [16]) of the truncation of the graph of ∂_P. In [11], the authors observed that a real-valued function is locally $\mathscr{C}^{1,1}$ if and only if it is simultaneously semiconvex and semiconcave. We state this result in the next theorem.

Theorem 25.14. *Let X be a Hilbert space and $f : X \to \mathbb{R}$ be a function which is continuous on $B(\bar{x},\gamma)$. Then, f is locally $\mathscr{C}^{1,1}$ around $\bar{x} \in X$ if and only if it is simultaneously locally semiconvex and locally semiconcave around $\bar{x}$. More precisely, if there exists $c > 0$ such that $f + (c/2)\|\cdot\|^2$ and $(c/2)\|\cdot\|^2 - f$ are convex on $B(0,\gamma)$, then f is $\mathscr{C}^{1,1}$ on $B(0,\gamma)$, with ∇f c-Lipschitz continuous on $B(\bar{x},\gamma)$ and conversely.*

See [11] or [1] for an alternative proof. Theorem 25.14 is also a consequence of Theorem 6.1 in [14] concerning a $\psi(\cdot)$-paraconvex function on a normed space.

Proposition 25.15. *Let X be a Hilbert space and $F : X \to \mathbb{R} \cup \{+\infty\}$ be an extended real-valued lsc function with $F(0) = 0$ and such that there exists $\sigma > 0$ for which*

$$F(x) \geq -\frac{\sigma}{2}\|x\|^2 \quad \text{for all } x \in X. \tag{25.17}$$

Let $\lambda_0 \in]0, \frac{1}{\sigma}[$ and $\eta > 0$. Let $\beta, \gamma > 0$ be positive real numbers such that

$$(1 + 2(1 - \lambda_0\sigma)^{-1})\gamma + \sqrt{2\lambda_0(1 - \lambda_0\sigma)^{-1}\beta} < \eta. \tag{25.18}$$

Then for all $\lambda \in]0, \lambda_0[$, the following hold:

(a) For all $x \in B(0,\gamma)$ and $x' \in P_\lambda F(x)$ one has $\|x'\| < \eta$, $\|x - x'\| < \eta$.
(b) For any $x \in B(0,\gamma)$

$$e_\lambda F(x) = \inf_{x' \in B(0,\eta)} \left\{ F(x') + \frac{1}{2\lambda}\|x - x'\|^2 \right\}.$$

(c) The function $e_\lambda F$ is Lipschitz continuous on $B(0,\gamma)$, with Lipschitz modulus η/λ.

The proof requires first the following lemma due to Poliquin–Rockafellar (see [23, Lemma 4.1]). For completeness we reproduce their proof in the Hilbert setting.

Lemma 25.16 (see [23]). *Let X be a Hilbert space and $F : X \to \mathbb{R} \cup \{+\infty\}$ be an extended real-valued function with $F(0) = 0$ and satisfying (25.17). Let $\lambda \in]0, 1/\sigma[$, $\rho \geq 0$, and $x, x' \in X$. If*

$$F(x') + \frac{1}{2\lambda}\|x' - x\|^2 \leq e_\lambda F(x) + \rho, \tag{25.19}$$

we have the estimate

$$\|x'\| \leq 2(1 - \lambda\sigma)^{-1}\|x\| + \sqrt{2\lambda(1 - \lambda\sigma)^{-1}\rho}.$$

Proof. First with $x' = 0$ we observe for all $x \in X$ the inequality

$$e_\lambda F(x) \le F(0) + \frac{1}{2\lambda}\|0-x\|^2 = \frac{1}{2\lambda}\|x\|^2.$$

So for any x, x' satisfying (25.19) we have according to (25.17)

$$-\frac{\sigma}{2}\|x'\|^2 + \frac{1}{2\lambda}\|x'-x\|^2 \le F(x') + \frac{1}{2\lambda}\|x'-x\|^2 \le e_\lambda F(x) + \rho \le \frac{1}{2\lambda}\|x\|^2 + \rho;$$

hence

$$-\frac{\sigma}{2}\|x'\|^2 + \frac{1}{2\lambda}\|x'-x\|^2 \le \frac{1}{2\lambda}\|x\|^2 + \rho.$$

Computing this, we get

$$\left(\frac{1}{2\lambda} - \frac{\sigma}{2}\right)\|x'\|^2 \le \frac{1}{\lambda}\langle x', x\rangle + \rho$$

$$(1-\lambda\sigma)\|x'\|^2 \le 2\langle x', x\rangle + 2\lambda\rho.$$

Putting $\alpha := (1-\lambda\sigma)^{-1}$, we obtain

$$\|x'\|^2 \le 2\alpha\langle x', x\rangle + 2\lambda\alpha\rho \le 2\alpha\|x'\|\,\|x\| + 2\lambda\alpha\rho;$$

hence

$$(\|x'\| - \alpha\|x\|)^2 \le \alpha^2\|x\|^2 + 2\lambda\alpha\rho,$$

which yields

$$\|x'\| \le \alpha\|x\| + \sqrt{\alpha^2\|x\|^2 + 2\lambda\alpha\rho} \le 2\alpha\|x\| + \sqrt{2\lambda\alpha\rho}. \qquad \blacksquare$$

The proof of the next lemmas uses strong ideas from [5].

Lemma 25.17. *Let X be a Hilbert space and $F : X \to \mathbb{R}\cup\{+\infty\}$ be a proper lsc function. Assume that there exist $\beta, \gamma, \sigma \ge 0$ such that*

$$F(y) \ge -\frac{\sigma}{2}\|y\|^2 - \beta\|y\| - \gamma \quad \textit{for all } y \in X.$$

Let $x \in \operatorname{Dom}\partial_P e_\lambda F$. Then, for all $\lambda \in]0, \frac{1}{\sigma}[$ (with convention $\frac{1}{\sigma} = +\infty$ when $\sigma = 0$), $P_\lambda F(x) \neq \emptyset$ and, for all $x' \in P_\lambda F(x)$, the following hold:

$$\partial_P e_\lambda F(x) = \{\lambda^{-1}(x-x')\} \quad \text{and} \quad \lambda^{-1}(x-x') \in \partial_P F(x').$$

Consequently, $\partial_P e_\lambda F(x)$ and $P_\lambda F(x)$ are singleton sets whenever $x \in \operatorname{Dom}\partial_P e_\lambda F$.

Proof. Let $x \in X$ be such that $\partial_P e_\lambda F(x) \neq \emptyset$ and let $\zeta \in \partial_P e_\lambda F(x)$. Fix a sequence of positive numbers $(t_n)_n$ with $t_n \downarrow 0$ and a sequence $(y_n)_n$ such that

$$F(y_n) + \frac{1}{2\lambda}\|x - y_n\|^2 \leq e_\lambda F(x) + t_n^2.$$

Since

$$F(y) \geq -\frac{\sigma}{2}\|y\|^2 - \beta\|y\| - \gamma \quad \text{for all } y \in X,$$

we get

$$-\frac{\sigma}{2}\|y_n\|^2 - \beta\|y_n\| - \gamma + \frac{1}{2\lambda}\|x - y_n\|^2 \leq e_\lambda F(x) + t_n^2$$

$$\left(\frac{1}{2\lambda} - \frac{\sigma}{2}\right)\|y_n\|^2 - \beta\|y_n\| - \frac{1}{\lambda}\langle x, y_n\rangle + \frac{1}{2\lambda}\|x\|^2 \leq e_\lambda F(x) + t_n^2 + \gamma$$

$$\left(\frac{1}{2\lambda} - \frac{\sigma}{2}\right)\|y_n\|^2 - \beta\|y_n\| - \frac{1}{\lambda}\|x\|\|y_n\| \leq e_\lambda F(x) + t_n^2 + \gamma - \frac{1}{2\lambda}\|x\|^2$$

$$\left(\frac{1}{2\lambda} - \frac{\sigma}{2}\right)\|y_n\|^2 - (\beta + \frac{1}{\lambda}\|x\|)\|y_n\| \leq e_\lambda F(x) + t_n^2 + \gamma - \frac{1}{2\lambda}\|x\|^2;$$

hence

$$\left(\frac{1}{2\lambda} - \frac{\sigma}{2}\right)\left(\|y_n\| - \frac{\beta + \frac{1}{\lambda}\|x\|}{2\left(\frac{1}{2\lambda} - \frac{\sigma}{2}\right)}\right)^2 \leq e_\lambda F(x) + t_n^2 + \gamma - \frac{1}{2\lambda}\|x\|^2 + \left(\frac{1}{2\lambda} - \frac{\sigma}{2}\right)\left(\frac{\beta + \frac{1}{\lambda}\|x\|}{2\left(\frac{1}{2\lambda} - \frac{\sigma}{2}\right)}\right)^2.$$

Since $\lambda < \frac{1}{\sigma}$, that is, $(\frac{1}{2\lambda} - \frac{\sigma}{2}) > 0$, the sequence (y_n) must be bounded and there exists a subsequence that we will not relabel converging weakly to some $\bar{y}$ and such that $\|x - y_n\| \to \alpha$.

We will prove that $\|x - \bar{y}\| = \alpha$ and we will conclude that (y_n) converges strongly to $\bar{y}$ and hence $\bar{y} \in P_\lambda F(x)$. The inequality $\|x - \bar{y}\| \leq \alpha$ being a consequence of the weak lower semicontinuity of the norm, let us prove that $\|x - \bar{y}\| \geq \alpha$. Since $\zeta \in \partial_P F(x)$, there is some $r > 0$ such that for n large enough,

$$\begin{aligned}
\langle \zeta, y_n - x\rangle &\leq t_n^{-1}\big(e_\lambda F(x - t_n(x - y_n)) - e_\lambda F(x)\big) + t_n r\|x - y_n\|^2 \\
&\leq t_n^{-1}\left(F(y_n) + \frac{1}{2\lambda}\|(1 - t_n)(x - y_n)\|^2 - F(y_n) - \frac{1}{2\lambda}\|x - y_n\|^2 + t_n^2\right) \\
&\quad + t_n r\|x - y_n\|^2 \\
&= -\frac{1}{\lambda}\|x - y_n\|^2 + \frac{t_n}{2\lambda}\|x - y_n\|^2 + t_n + t_n r\|x - y_n\|^2,
\end{aligned}$$

and taking the limit, we obtain

$$\langle \zeta, \bar{y} - x \rangle \le -\frac{1}{\lambda}\alpha^2 \quad \text{and then } \frac{\alpha^2}{\lambda} \le \|\zeta\| \|x - \bar{y}\|. \tag{25.20}$$

Analogously, for any $z \in X$ and n large enough, we have

$$\begin{aligned}
\langle \zeta, z \rangle &\le t_n^{-1}\left(e_\lambda F(x + t_n z) - e_\lambda F(x)\right) + t_n r\|x\|^2 \\
&\le t_n^{-1}\left(F(y_n) + \frac{1}{2\lambda}\|x - y_n + t_n z\|^2 - F(y_n) - \frac{1}{2\lambda}\|x - y_n\|^2 + t_n^2\right) + t_n r\|x\|^2 \\
&\le \frac{1}{\lambda}\|x - y_n\|\|z\| + \frac{t_n}{2\lambda}\|z\|^2 + t_n + t_n r\|x\|^2.
\end{aligned}$$

Taking the limit, we obtain

$$\langle \zeta, z \rangle < \frac{\alpha}{\lambda}\|z\| \quad \text{and then } \|\zeta\| \le \frac{\alpha}{\lambda}. \tag{25.21}$$

From (25.20) and (25.21), we obtain that $\|x - \bar{y}\| \ge \alpha$; hence $\alpha = \|x - \bar{y}\|$ as desired. The nonemptiness of $P_\lambda F(x)$ is fulfilled. According to Lemma 25.12, for any $x' \in P_\lambda F(x)$, we have

$$\partial_P e_\lambda F(x) \subset \partial_P F(x') \cap \{(1/\lambda)(x' - x)\}$$

which is exactly what we expected. ■

Proof.

(a) Fix λ_0, η, β, and γ as in Proposition 25.15. Let $\lambda \in]0, \lambda_0[$ and $x \in B(0, \gamma)$. Given $x' \in P_\lambda F(x)$, according to Lemma 25.16 with $\rho = 0$, one obtains

$$\|x'\| \le 2(1 - \lambda\sigma)^{-1}\|x\| < 2(1 - \lambda_0\sigma)^{-1}\gamma < \eta.$$

Moreover

$$\|x - x'\| \le \|x\| + \|x'\| \le \gamma + 2(1 - \lambda\sigma)^{-1}\gamma < \gamma + 2(1 - \lambda_0\sigma)^{-1}\gamma < \eta.$$

(b) Suppose that some $x' \in X$ satisfies the inequality

$$F(x') + \frac{1}{2\lambda}\|x' - x\|^2 \le e_\lambda F(x) + \beta.$$

Applying Lemma 25.16 we have

$$\begin{aligned}
\|x'\| &\le 2(1 - \lambda\sigma)^{-1}\gamma + \sqrt{2\lambda(1 - \lambda\sigma)^{-1}\beta} \\
&< 2(1 - \lambda_0\sigma)^{-1}\gamma + \sqrt{2\lambda_0(1 - \lambda_0\sigma)^{-1}\beta} < \eta,
\end{aligned}$$

which justifies our representation in (b) .

(c) Observe that for any $x \in B(0,\gamma)$ we have on one hand

$$e_\lambda F(x) \le F(0) + \frac{1}{2\lambda}\|x\|^2 = \frac{1}{2\lambda}\|x\|^2$$

and on the other hand the inequalities, for any $y \in B(0,\eta)$,

$$F(y) + \frac{1}{2\lambda}\|x-y\|^2 \ge -\frac{\sigma}{2}\|y\|^2 + \frac{1}{2\lambda}\|x-y\|^2 \ge \frac{-\sigma\eta^2}{2},$$

give through (b) that $e_\lambda F(x) \ge \frac{-\sigma\eta^2}{2}$. This says in particular that $e_\lambda F(\cdot)$ is finite over $B(0,\gamma)$.

So, putting $\Phi_{x'}(x) := F(x') + \frac{1}{2\lambda}\|x'-x\|^2$ for each $x' \in D_{F,\eta} := B(0,\eta) \cap \operatorname{dom} F$ and noting that the family of functions $(\Phi_{x'})_{x' \in D_{F,\eta}}$ is equi-Lipschitzian on $B(0,\gamma)$, we obtain that $e_\lambda F$ is Lipschitz continuous on $B(0,\gamma)$. Let us estimate the Lipschitz constant of $e_\lambda F(\cdot)$ over $B(0,\gamma)$. For any $x \in \operatorname{Dom} \partial_P e_\lambda F$ with $\|x\| < \gamma$ and any $\zeta \in \partial_P e_\lambda F(x)$ we have by Lemma 25.17

$$\zeta = \frac{1}{\lambda}(x - x') \quad \text{for } x' = P_\lambda F(x);$$

hence according to the assertion (a)

$$\|\zeta\| \le \frac{\eta}{\lambda}.$$

The function $e_\lambda F$ being lsc on $B(0,\gamma)$ we deduce (see [25, Theorem 2.1]) that $e_\lambda F$ is $\frac{\eta}{\lambda}$-Lipschitz continuous on $B(0,\gamma)$. ■

The proof of the next proposition is an adaptation of the arguments of [1, Theorem 4.9].

Proposition 25.18. *Let X be a Hilbert space and $F : X \to \mathbb{R} \cup \{+\infty\}$ be a function which is c-pln on an open convex set containing $B[0,s_0]$ for a positive constant c satisfying $s_0 < 1/2c$. Assume $F(0) = 0$, $0 \in \partial_C F(0)$ and define $\bar{F}(\cdot) = F(\cdot) + \delta(\cdot, B[0,s_0])$. Let $\lambda_0 \in]0, \frac{1-2cs_0}{2c}[$. Then for any $\beta, \gamma > 0$ satisfying*

$$(1 + 2(1-2\lambda_0 c)^{-1})\gamma + \sqrt{2\lambda_0(1-2\lambda_0 c)^{-1}\beta} < s_0, \tag{25.22}$$

the following properties hold:

(a) For all $\lambda \in]0,\lambda_0[$, the function $e_\lambda \bar{F}(\cdot)$ is $\mathscr{C}^{1,1}$ on $B(0,\gamma)$ with $\nabla e_\lambda \bar{F}$ a_λ-Lipschitz continuous on $B(0,\gamma)$ for $a_\lambda := \frac{1}{\lambda(1-2c\lambda(1+s_0/\lambda))}$.

(b) $P_\lambda \bar{F}(\cdot)$ is nonempty, single-valued, and $(1 + \frac{1}{1-2c\lambda_0(1+s_0)})$-Lipschitz continous on $B(0,\gamma)$.

Proof. (a) By assumption, for all $x \in B[0,s_0]$, all $\zeta \in \partial_C F(x)$,

$$F(y) - F(x) \geq \langle \zeta, y - x \rangle - c(1 + \|\zeta\|)\|y - x\|^2$$

for all $y \in B[0, s_0]$. Since $0 \in \partial_C F(0)$ and $F(0) = 0$, the latter inequality yields

$$F(x) \geq -c\|x\|^2 \text{ for all } x \in B[0, s_0].$$

Consequently

$$\bar{F}(x) \geq -c\|x\|^2 \text{ for all } x \in X.$$

So $\bar{F}(\cdot)$ satisfies (25.17) with $\sigma = 2c$ and $\bar{F}$ is lsc on X since F is by assumption lsc at each point of $B[0, s_0]$. Fix $\lambda_0 \in]0, \frac{1-2cs_0}{2c}[$ and $\eta = s_0$. Take, for the function $\bar{F}$, any positive real numbers β, γ satisfying (25.18) in Proposition 25.15 for $\sigma = 2c$, that is, satisfying (25.22). For any $\lambda \in]0, \lambda_0[$, let $x \in B(0, \gamma)$ where $e_\lambda \bar{F}$ is ∂_P-subdifferentiable and let $\zeta \in \partial_P e_\lambda \bar{F}(x)$. We know by Lemma 25.17 that $P_\lambda \bar{F}(x)$ is a singleton and that for $x' = P_\lambda \bar{F}(x)$ we have $\zeta = \lambda^{-1}(x - x')$ and $\zeta \in \partial_P \bar{F}(x')$. Moreover, according to (a) and (c) in Proposition 25.15, we have

$$\|x'\| < s_0, \ \|x - x'\| < s_0 \text{ and } \|\zeta\| < s_0/\lambda;$$

hence in particular $\zeta \in \partial_P \bar{F}(x') = \partial_P F(x')$. Since F is c-pln on an open convex set containing $B[0, s_0]$, we get

$$F(z) - F(x') \geq \langle \zeta, z - x' \rangle - c(1 + \|\zeta\|)\|z - x'\|^2$$

for all $z \in B(0, s_0)$, which yields for any $y \in X$

$$\begin{aligned} F(z) + \frac{1}{2\lambda}\|z - y\|^2 - \left(F(x') + \frac{1}{2\lambda}\|x' - x\|^2\right) \\ \geq \frac{1}{2\lambda}(\|z - y\|^2 - \|x' - x\|^2) + \langle \zeta, z - x' \rangle - c(1 + \|\zeta\|)\|z - x'\|^2 =: h(z). \end{aligned} \tag{25.23}$$

Note that the equality $x' = P_\lambda \bar{F}(x)$ entails $e_\lambda \bar{F}(x) = \bar{F}(x') + \frac{1}{2\lambda}\|x - x'\|^2 = F(x') + \frac{1}{2\lambda}\|x - x'\|^2$ and due to (b) of Proposition 25.15 applied with the function $\bar{F}$

$$e_\lambda \bar{F}(y) = \inf_{z \in B(0, s_0)} \left\{ F(z) + \frac{1}{2\lambda}\|z - y\|^2 \right\} \tag{25.24}$$

for all $y \in B(0, \gamma)$. According to the inequality $\|\zeta\| < s_0/\lambda$, we observe that

$$\begin{aligned} \frac{1}{2\lambda}(\|z - y\|^2 - \|x' - x\|^2) + \langle \zeta, z - x' \rangle - c(1 + \|\zeta\|)\|z - x'\|^2 \\ \geq \frac{1}{2\lambda}(\|z - y\|^2 - \|x' - x\|^2) + \langle \zeta, z - x' \rangle - c\left(1 + \frac{s_0}{\lambda}\right)\|z - x'\|^2. \end{aligned}$$

Since $\lambda_0 < \frac{1-2cs_0}{2c}$, we have $(\frac{1}{2c} - s_0)\frac{1}{\lambda_0} > 1$; hence

$$\frac{1}{2\lambda} - c\left(1+\frac{s_0}{\lambda}\right) = c\left(\left(\frac{1}{2c} - s_0\right)\frac{1}{\lambda} - 1\right) > c\left(\left(\frac{1}{2c} - s_0\right)\frac{1}{\lambda_0} - 1\right) > 0.$$

We deduce that

$$\lim_{\|z\|\to+\infty} \frac{1}{2\lambda}(\|z-y\|^2 - \|x'-x\|^2) + \langle\zeta, z-x'\rangle - c\left(1+\frac{s_0}{\lambda}\right)\|z-x'\|^2 = +\infty.$$

The function h in the right-hand side in (25.23) is then a coercive lsc convex function and hence attains its minimum on X at some point $\bar{z} \in X$. Writing $\nabla h(\bar{z}) = 0$ and using the equality $\zeta = \lambda^{-1}(x-x')$ yield

$$0 = \frac{1}{\lambda}(\bar{z}-y) + \frac{1}{\lambda}(x-x') - 2c(1+\|\zeta\|)(\bar{z}-x'),$$

which gives

$$\bar{z} = x' - \frac{1}{1-2c\lambda(1+\|\zeta\|)}(x-y).$$

So

$$\begin{aligned}
\inf_X h &= h(\bar{z}) \\
&= \frac{1}{2\lambda}\left(\|x'-y-\frac{1}{1-2c\lambda(1+\|\zeta\|)}(x-y)\|^2 - \|x'-x\|^2\right) \\
&\quad + \langle\zeta, -\frac{1}{1-2c\lambda(1+\|\zeta\|)}(x-y)\rangle \\
&\quad - \frac{2c(1+\|\zeta\|)}{2}\|\frac{1}{1-2c\lambda(1+\|\zeta\|)}(x-y)\|^2 \\
&= \frac{1}{2\lambda}\left(\|x'-x+(1-\frac{1}{1-2c\lambda(1+\|\zeta\|)})(x-y)\|^2\right. \\
&\quad \left.- \|x'-x\|^2\right) + \frac{1}{1-2c\lambda(1+\|\zeta\|)}\langle\zeta, y-x\rangle \\
&\quad - \frac{2c(1+\|\zeta\|)}{2}\frac{1}{(1-2c\lambda(1+\|\zeta\|))^2}\|(x-y)\|^2;
\end{aligned}$$

hence

$$\begin{aligned}
\inf_X h &= \frac{1}{2\lambda}\Big(-\frac{4c\lambda(1+\|\zeta\|)}{1-2c\lambda(1+\|\zeta\|)}\langle x'-x,x-y\rangle\Big) \\
&\quad +\left(\frac{2c\lambda(1+\|\zeta\|)}{1-2c\lambda(1+\|\zeta\|)}\right)^2 \|x-y)\|^2\Big) \\
&\quad +\frac{1}{1-2c\lambda(1+\|\zeta\|)}\langle\zeta,y-x\rangle-\frac{2c(1+\|\zeta\|)}{2(1-2c\lambda(1+\|\zeta\|))^2}\|x-y\|^2 \\
&= \left(\frac{1}{1-2c\lambda(1+\|\zeta\|)}-\frac{2c\lambda(1+\|\zeta\|)}{1-2c\lambda(1+\|\zeta\|)}\right)\langle\zeta,y-x\rangle \\
&\quad -\Big(\frac{2c(1+\|\zeta\|)}{2(1-2c\lambda(1+\|\zeta\|))^2}-\frac{1}{2\lambda}\left(\frac{2c\lambda(1+\|\zeta\|)}{1-2c\lambda(1+\|\zeta\|)}\right)^2\Big)\|x-y\|^2 \\
&= \langle\zeta,y-x\rangle-\frac{2c(1+\|\zeta\|)}{2(1-2c\lambda(1+\|\zeta\|))}\|x-y\|^2 \\
&= \langle\zeta,y-x\rangle-\frac{c(1+\|\zeta\|)}{(1-2c\lambda(1+\|\zeta\|))}\|x-y\|^2.
\end{aligned}$$

The latter equality combined with the inequality (25.23) and the equality (25.24) gives

$$\begin{aligned}
e_\lambda\bar{F}(y)-e_\lambda\bar{F}(x) &\geq \langle\zeta,y-x\rangle-\frac{c(1+\|\zeta\|)}{1-2c\lambda(1+\|\zeta\|)}\|x-y\|^2 \\
&\geq \langle\zeta,y-x\rangle-\frac{c}{1-2c\lambda(1+s_0/\lambda)}(1+s_0/\lambda)\|x-y\|^2
\end{aligned}$$

for all $y\in B(0,\gamma)$. Setting $k_\lambda := \frac{c}{1-2c\lambda(1+s_0/\lambda)}(1+s_0/\lambda)$, we see through the equality $\|x-y\|^2=\|y\|^2-\|x\|^2-2\langle x,y-x\rangle$ that, for all $x\in B(0,\gamma)\cap \operatorname{Dom}\partial_P\bar{F}$ and $\zeta\in\partial_P\bar{F}(x)$,

$$e_\lambda\bar{F}(y)+k_\lambda\|y\|^2-e_\lambda\bar{F}(x)-k_\lambda\|x\|^2\geq\langle\zeta+2k_\lambda x,y-x\rangle \text{ for all } y\in B(0,\gamma),$$

which means by [8] that the function $e_\lambda\bar{F}(\cdot)+k_\lambda\|\cdot\|^2$ is convex on $B(0,\gamma)$. We deduce that $e_\lambda\bar{F}(\cdot)+k_\lambda\|\cdot\|^2+\frac{1}{2\lambda}\|\cdot\|^2$ is also convex on $B(0,\gamma)$.

Writing, for each $x\in B(0,\gamma)$,

$$\begin{aligned}
-\left(e_\lambda\bar{F}(x)-\frac{1}{2\lambda}\|x\|^2\right) &= -\left(\inf_{y\in X}\{\bar{F}(y)+\frac{1}{2\lambda}\|x-y\|^2\}-\frac{1}{2\lambda}\|x\|^2\right) \\
&= \sup_{y\in X}\left\{\frac{1}{\lambda}\langle x,y\rangle-(\frac{1}{2\lambda}\|y\|^2+\bar{F}(y))\right\},
\end{aligned}$$

we get that $-(e_\lambda \bar{F}(\cdot) - \frac{1}{2\lambda}\|\cdot\|^2)$ is a pointwise supremum of affine functions and hence convex, so $e_\lambda \bar{F}(\cdot) - \frac{1}{2\lambda}\|\cdot\|^2$ is concave. Thus we get the concavity of $e_\lambda \bar{F}(\cdot) - k_\lambda\|\cdot\|^2 - \frac{1}{2\lambda}\|\cdot\|^2$ on $B(0,\gamma)$. We observe that

$$\begin{aligned} k_\lambda + \frac{1}{2\lambda} &= \frac{1}{1-2c\lambda(1+s_0/\lambda)}\left(c(1+s_0/\lambda) + \frac{1}{2\lambda}(1-2c\lambda(1+s_0/\lambda))\right) \\ &= \frac{1}{2\lambda(1-2c\lambda(1+s_0/\lambda))}. \end{aligned}$$

Therefore $e_\lambda \bar{F}(\cdot)$ is simultaneously semiconvex and semiconcave on $B(0,\gamma)$, or more precisely, $e_\lambda \bar{F}(\cdot) + (a_\lambda/2)\|\cdot\|^2$ and $(a_\lambda/2)\|\cdot\|^2 - e_\lambda \bar{F}(\cdot)$ are both convex for

$$a_\lambda := \lambda^{-1}(1-2c\lambda(1+s_0/\lambda))^{-1}. \tag{25.25}$$

So, the function $e_\lambda \bar{F}(\cdot)$ being continuous on $B(0,\gamma)$ according to Proposition 25.15 we conclude via Theorem 25.14 that $e_\lambda \bar{F}(\cdot)$ is $\mathscr{C}^{1,1}$ on $B(0,\gamma)$ with $\nabla e_\lambda \bar{F}(\cdot)$ a_λ-Lipschitz continuous on $B(0,\gamma)$. The assertion (a) of the proposition is then established.

It remains to prove the assertion (b). Since $e_\lambda \bar{F}(\cdot)$ is $\mathscr{C}^{1,1}$ on $B(0,\gamma)$, according to Proposition 25.13, we have

$$\nabla e_\lambda \bar{F}(\cdot) = \lambda^{-1}(I - P_\lambda \bar{F}(\cdot)) \text{ on } B(0,\gamma).$$

Then for all $x, y \in B(0,\gamma)$, we have

$$\begin{aligned} \|P_\lambda \bar{F}(y) - P_\lambda \bar{F}(x)\| &= \|(I-P_\lambda \bar{F})(x) - (I-P_\lambda \bar{F})(y) - (x-y)\| \\ &\le \|(I-P_\lambda \bar{F})(x) - (I-P_\lambda \bar{F})(y)\| + \|x-y\| \\ &\le (1+\lambda a_\lambda)\|x-y\|. \end{aligned}$$

Computing $1+\lambda a_\lambda$, according to (25.25), we get

$$\begin{aligned} 1+\lambda a_\lambda &= 1 + \frac{1}{(1-2c\lambda(1+s_0/\lambda))} \\ &\le 1 + \frac{1}{1-2c(\lambda_0+s_0)}, \end{aligned}$$

hence $P_\lambda \bar{F}(\cdot)$ is $(1+\frac{1}{1-2c(\lambda_0+s_0)})$-Lipschitz continous on $B(0,\gamma)$. ■

Now we are in a position to establish the second main result of the paper. It concerns the $\mathscr{C}^{1,1}$-property of the Moreau envelope of a pln function.

Theorem 25.19. *Let X be a Hilbert space and $f : X \to \mathbb{R} \cup \{+\infty\}$ be an extended real-valued function. Assume that f is c-pln on an open convex set containing*

$B[u_0,s_0]$ for a positive constant c satisfying $s_0 < (1/2c)$. Let $(x_0,y_0) \in \operatorname{gph} \partial_C f$ with $\|x_0 - u_0\| < \frac{s_0}{18}$ and let

$$\bar{f}(\cdot) = f(\cdot) + \delta\left(\cdot, B\left[x_0, \frac{s_0}{2}\right]\right).$$

Then there exists some threshold λ_0 such that for any $\lambda \in]0,\lambda_0[$:

(a) $e_\lambda \bar{f}$ is $\mathscr{C}^{1,1}$ on $B(u_0,\frac{s_0}{18})$ with $\nabla e_\lambda \bar{f}$ d_λ-Lipschitz continuous on $B(u_0,\frac{s_0}{18})$ for

$$d_\lambda := \frac{1}{\lambda(1 - 2c(\lambda(1+\|y_0\|) + s_0/2))} + \|y_0\|.$$

(b) $P_\lambda \bar{f}$ is nonempty, single-valued, and k_0-Lipschitz continuous on $B(u_0,\frac{s_0}{18})$ for

$$k_0 := 1 + \lambda_0 + \frac{1}{1 - 2c(\lambda_0 + s_0/2)}.$$

(c) $P_\lambda \bar{f}(x_0 + \lambda y_0) = x_0$.
(d) $\nabla e_\lambda \bar{f} = \lambda^{-1}(I - P_\lambda \bar{f})$ on $B(u_0,\frac{s_0}{18})$.
(e) $\|\nabla e_\lambda \bar{f}(x_0)\| \le k_0 \|y_0\|$.
(f) $P_\lambda \bar{f}(B(u_0,\frac{s_0}{18})) \subset B(u_0,\frac{7}{16}s_0)$.

Further, given any $x \in B(u_0,\frac{s_0}{18})$:

(g) $\nabla e_\lambda \bar{f}(x) \in \partial_P f(P_\lambda \bar{f}(x))$.
(h) $\|x - x_0\| \ge (1 - c\lambda(2 + \|\nabla e_\lambda \bar{f}(x)\| + k_0\|y_0\|))\|P_\lambda \bar{f}(x) - P_\lambda \bar{f}(x_0)\|$.

Proof. Let $\mathscr{O}$ be an open convex set containing $B[u_0,s_0]$ such that f is c-pln on $\mathscr{O}$. (a): For all $x \in X$, put

$$F(x) := \frac{1}{1+\|y_0\|}\left(g(x + x_0) - f(x_0) - \langle y_0, x\rangle\right),$$

where $g(x) := f(x) + \delta(x, B[u_0,s_0])$, and put also

$$\bar{F}(x) := F(x) + \delta(x, B[0,\varepsilon]) \quad \text{with} \quad \varepsilon = s_0/2.$$

The function $F(\cdot)$ is lsc on X, $F(0) = 0$ and for $\zeta \in \partial_P F(x)$ with $x \in B(u_0 - x_0, \varepsilon)$ we have $(1 + \|y_0\|)\zeta + y_0 \in \partial_P f(x + x_0)$.

The c-pln property of f on the open convex set $\mathscr{O}$ containing $B[u_0,s_0]$ ensures that F is c-pln on $\mathscr{O}_0 := B(0, 17s_0/18)$. Indeed, we have for all $\zeta \in \partial_P F(x)$ with $x \in \mathscr{O}_0$

$$\begin{aligned} f(x' + x_0) - f(x + x_0) \ge{}& \langle (1 + \|y_0\|)\zeta + y_0, x' - x\rangle \\ & - c(1 + (1 + \|y_0\|)\|\zeta\| + \|y_0\|)\|x' - x\|^2 \end{aligned}$$

for all $x' \in \mathcal{O}_0 = B(0, 17s_0/18)$, because for such x' we have $x' + x_0 \in B(u_0, s_0) \subset \mathcal{O}$. Then

$$F(x') - F(x) \geq \langle \zeta, x' - x \rangle - \frac{c}{1 + \|y_0\|}(1 + (1 + \|y_0\|)\|\zeta\| + \|y_0\|)\|x' - x\|^2,$$
$$F(x') - F(x) \geq \langle \zeta, x' - x \rangle - c(1 + \|\zeta\|)\|x' - x\|^2,$$

for all $x' \in \mathcal{O}_0$, which means that F is c-pln on $\mathcal{O}_0 = B(0, 17s_0/18)$. Further, $B(0, 17s_0/18) \supset B[0, \varepsilon]$ and $0 \in \partial_P F(0)$ since $y_0 \in \partial_P f(x_0)$. Consequently, $\bar{F}(x) \geq -c\|x\|^2$, for all $x \in X$. As regards the Moreau envelopes, for any $x \in X$ and $\lambda > 0$, we can write

$$\begin{aligned}
& e_{\lambda(1+\|y_0\|)}\bar{F}(x) \\
&= \inf_{u \in X}\{F(u) + \delta(u, B[0, \varepsilon]) + \frac{1}{2\lambda(1 + \|y_0\|)}\|x - u\|^2\} \\
&= \inf_{u \in X}\{\frac{g(u + x_0) - f(x_0) - \langle y_0, u \rangle}{1 + \|y_0\|} + \delta(u, B[0, \varepsilon]) + \frac{1}{2\lambda(1 + \|y_0\|)}\|x - u\|^2\} \\
&= \frac{-f(x_0) + \langle y_0, x_0 \rangle}{1 + \|y_0\|} + \inf_{u \in X}\Big\{\frac{g(u + x_0) - \langle y_0, u + x_0 \rangle}{1 + \|y_0\|} + \delta(u + x_0, B[x_0, \varepsilon]) \\
&\qquad + \frac{1}{2\lambda(1 + \|y_0\|)}\|x - u\|^2\Big\} \\
&= \frac{-f(x_0) + \langle y_0, x_0 \rangle}{1 + \|y_0\|} + \inf_{z \in X}\Big\{\frac{g(z) - \langle y_0, z \rangle}{1 + \|y_0\|} + \delta(z, B[x_0, \varepsilon]) \\
&\qquad + \frac{1}{2\lambda(1 + \|y_0\|)}\|x + x_0 - z\|^2\Big\} \\
&= \frac{-f(x_0) + \langle y_0, x_0 \rangle}{1 + \|y_0\|} + \frac{1}{1 + \|y_0\|}\inf_{z \in X}\Big\{f(z) - \langle y_0, z \rangle + \delta(z, B[u_0, s_0] \cap B[x_0, \varepsilon]) \\
&\qquad + \frac{1}{2\lambda}\|x + x_0 - z\|^2\Big\}.
\end{aligned}$$

Since $\|x_0 - u_0\| < \frac{s_0}{18}$ and $\varepsilon = \frac{s_0}{2}$, we have $B[u_0, s_0] \cap B[x_0, \varepsilon] = B[x_0, \varepsilon]$ and hence

$$\begin{aligned}
e_{\lambda(1+\|y_0\|)}\bar{F}(x) =& \frac{-f(x_0) + \langle y_0, x_0 \rangle}{1 + \|y_0\|} + \frac{1}{1 + \|y_0\|}\Big(-\frac{\lambda}{2}\|y_0\|^2 - \langle y_0, x \rangle - \langle y_0, x_0 \rangle \\
& + \inf_{z \in X}\{f(z) + \delta(z, B[x_0, \varepsilon]) + \frac{1}{2\lambda}\|x + x_0 + \lambda y_0 - z\|^2\}\Big).
\end{aligned}$$

Thus

$$e_{\lambda(1+\|y_0\|)}\bar{F}(x) = \frac{-f(x_0) - \langle y_0, x \rangle - \frac{\lambda}{2}\|y_0\|^2}{1 + \|y_0\|} + \frac{1}{1 + \|y_0\|}e_\lambda \bar{f}(x + x_0 + \lambda y_0).$$

So, for any $x \in X$ and any $\lambda > 0$,

$$e_\lambda \bar{f}(x) = (1+\|y_0\|)e_{\lambda(1+\|y_0\|)}\bar{F}(x-(x_0+\lambda y_0)) \\ +f(x_0)+\langle y_0, x-(x_0+\lambda y_0)\rangle + \frac{\lambda}{2}\|y_0\|^2. \tag{25.26}$$

Let $\lambda_0 \in]0, \min\{\frac{1-2cs_0}{2c}, \frac{3}{14c}\}[$. Since the inequality $\lambda_0 < \frac{3}{14c}$ yields

$$1+\frac{2}{1-2c\lambda_0} < \frac{9}{2} \quad \text{or equivalently} \quad (1+2(1-2c\lambda_0)^{-1})^{-1}\frac{s_0}{2} > \frac{s_0}{9},$$

we can choose $\gamma > 0$ such that

$$\frac{s_0}{9} < \gamma < \frac{s_0}{2}\left(1+\frac{2}{1-2c\lambda_0}\right)^{-1}.$$

Then there exists $\beta > 0$ such that

$$(1+2(1-2c\lambda_0)^{-1})\gamma + \sqrt{2\lambda_0(1-2c\lambda_0)^{-1}\beta} < s_0/2.$$

Applying Proposition 25.18 to F, c, and $\varepsilon < 1/2c$, we get that for all $\lambda \in]0, \lambda_0[$, $e_\lambda \bar{F}(\cdot)$ is $\mathscr{C}^{1,1}$ on $B(0,\gamma)$ with $\nabla e_\lambda \bar{F}$ b_λ-Lipschitz continuous on $B(0,\gamma)$, for $b_\lambda := \frac{1}{\lambda(1-2c\lambda(1+\varepsilon/\lambda))}$. Define $\bar{\lambda}_0 = \min\{\lambda_0, \gamma - s_0/9\}$. We deduce according to (25.26) that, for all $\lambda \in]0, \frac{\bar{\lambda}_0}{1+\|y_0\|}[$, $e_\lambda \bar{f}(\cdot)$ is $\mathscr{C}^{1,1}$ on $B(x_0+\lambda y_0, \gamma)$. Given any $0 < \lambda < \frac{\bar{\lambda}_0}{1+\|y_0\|}$, if x belongs to $B[u_0, s_0/18]$, then

$$\begin{aligned}\|x-(x_0+\lambda y_0)\| &\le \|x-u_0\| + \|x_0-u_0\| + \|\lambda y_0\| \\ &\le s_0/18 + s_0/18 + \frac{\bar{\lambda}_0\|y_0\|}{1+\|y_0\|} \\ &< s_0/9 + \gamma - s_0/9 = \gamma,\end{aligned}$$

so $B[u_0, s_0/18] \subset B(x_0+\lambda y_0, \gamma)$; hence $e_\lambda \bar{f}(\cdot)$ is $\mathscr{C}^{1,1}$ on $B(u_0, s_0/18)$ and according to (25.26), $\nabla e_\lambda \bar{f}$ is d_λ-Lipschitz continuous on $B(u_0, s_0/18)$ for

$$\begin{aligned} d_\lambda &:= (1+\|y_0\|)b_{\lambda(1+\|y_0\|)} + \|y_0\| \\ &= \frac{(1+\|y_0\|)}{\lambda(1+\|y_0\|)(1-2c\lambda(1+\|y_0\|)(1+\frac{\varepsilon}{\lambda(1+\|y_0\|)}))} + \|y_0\| \\ &= \frac{1}{\lambda(1-2c(\lambda(1+\|y_0\|)+\varepsilon))} + \|y_0\|.\end{aligned}$$

This finishes the proof of (a).

(b), (c), and (d): As in Proposition 25.18, thanks to Proposition 25.13, we deduce that $P_\lambda \bar{f}$ is nonempty, single-valued, and continuous on $B(u_0, \frac{s_0}{18})$, and

$$\nabla e_\lambda \bar{f}(\cdot) = \lambda^{-1}(I - P_\lambda \bar{f})(\cdot) \quad \text{on} \quad B(u_0, s_0/18). \tag{25.27}$$

So, $P_\lambda \bar{f}$ is $(1+\lambda d_\lambda)$-Lipschitz continuous on $B(u_0, \frac{s_0}{18})$. Computing, we get that for all $\lambda \in]0, \frac{\bar{\lambda}_0}{1+\|y_0\|}[$

$$1+\lambda d_\lambda = 1+\lambda\|y_0\| + \frac{1}{1-2c(\lambda(1+\|y_0\|)+\varepsilon)},$$

$$1+\lambda d_\lambda < 1+\frac{\bar{\lambda}_0\|y_0\|}{1+\|y_0\|} + \frac{1}{1-2c(\bar{\lambda}_0+\varepsilon)} < 1+\bar{\lambda}_0+\frac{1}{1-2c(\bar{\lambda}_0+\varepsilon)};$$

hence $P_\lambda \bar{f}$ is k_0-Lipschitz continuous on $B(u_0, s_0/18)$ for $k_0 := 1+\bar{\lambda}_0+(1-2c(\bar{\lambda}_0+\varepsilon))^{-1}$. Furthermore, given any $\lambda \in]0, \bar{\lambda}_0/(1+\|y_0\|)[$, combining with (25.26), we get

$$x_0 + P_{\lambda(1+\|y_0\|)}\bar{F}(u) = P_\lambda \bar{f}(u+x_0+\lambda y_0) \tag{25.28}$$

for all $u \in B(0, \varepsilon/4)$, both mappings being nonempty and single-valued on this ball. Indeed let $u \in B(0, \varepsilon/4)$ and $p \in X$ such that

$$e_{\lambda(1+\|y_0\|)}\bar{F}(u) = \bar{F}(p) + \frac{1}{2\lambda(1+\|y_0\|)}\|u-p\|^2.$$

Then, according to (25.26) and the definition of $\bar{F}$, we have

$$\begin{aligned}
e_\lambda \bar{f}(u+x_0+\lambda y_0) &= (1+\|y_0\|)(\bar{F}(p) + \frac{1}{2\lambda(1+\|y_0\|)}\|u-p\|^2) \\
&\quad + f(x_0) + \langle y_0, u\rangle + \frac{\lambda}{2}\|y_0\|^2 \\
&= f(x_0+p) + \delta(x_0+p, B[u_0, s_0]) + \delta(p, B[0,\varepsilon]) \\
&\quad + \frac{1}{2\lambda}\|u-p\|^2 - \langle y_0, p\rangle + \langle y_0, u\rangle + \frac{\lambda}{2}\|y_0\|^2 \\
&= f(x_0+p) + \delta(x_0+p, B[u_0, s_0] \cap B[x_0, \varepsilon]) \\
&\quad + \frac{1}{2\lambda}\|u+x_0+\lambda y_0 - (x_0+p)\|^2,
\end{aligned}$$

and since $B[u_0, s_0] \cap B[x_0, \varepsilon] = B[x_0, \varepsilon]$ we obtain

$$e_\lambda \bar{f}(u+x_0+\lambda y_0) = f(x_0+p) + \delta(x_0+p, B[x_0, \varepsilon])$$

$$+\frac{1}{2\lambda}\|u+x_0+\lambda y_0-(x_0+p)\|^2$$
$$=\bar{f}(x_0+p)+\frac{1}{2\lambda}\|u+x_0+\lambda y_0-(x_0+p)\|^2,$$

which justifies the desired equality (25.28).

Using the inequality

$$\bar{F}(x) \geq -c\|x\|^2 \text{ for all } x \in X,$$

we get that for all $x \in X$

$$\bar{F}(x)+\frac{1}{2\lambda(1+\|y_0\|)}\|x\|^2 \geq \left(\frac{1}{2\lambda(1+\|y_0\|)}-c\right)\|x\|^2 \geq 0$$
$$\left(\text{since } \lambda(1+\|y_0\|) < \bar{\lambda}_0 < \frac{1}{2c}\right)$$

so we deduce that $0 \in P_{\lambda(1+\|y_0\|)}\bar{F}(0)$. Since $P_{\lambda(1+\|y_0\|)}\bar{F}(\cdot)$ is single-valued on $B(0,\gamma)$ we get that $P_{\lambda(1+\|y_0\|)}\bar{F}(0)=0$. Putting this in (25.28) with $u=0$ we obtain

$$P_\lambda \bar{f}(x_0+\lambda y_0)=x_0 \text{ whenever } \lambda \in]0,\bar{\lambda}_0/(1+\|y_0\|)[. \tag{25.29}$$

So (b), (c), and (d) are established.
(e): To get (e), we observe thanks to (25.29), (25.27), and the inequality $1+\lambda d_\lambda < k_0$ that

$$\begin{aligned}\|\nabla e_\lambda \bar{f}(x_0)\| &= \|\lambda^{-1}(x_0-P_\lambda \bar{f}(x_0))\| \\ &= \|\lambda^{-1}(P_\lambda \bar{f}(x_0+\lambda y_0)-P_\lambda \bar{f}(x_0))\| \\ &\leq \lambda^{-1}(1+\lambda d_\lambda)\|x_0+\lambda y_0-x_0\| = (1+\lambda d_\lambda)\|y_0\| \leq k_0\|y_0\|.\end{aligned}$$

(f), (g): The arguments to get (f) and (g) are similar to those in the proof of Proposition 2.8 in [16]. We give them in detail for completeness. As f is bounded from below on $B[x_0,\frac{s_0}{2}]$ with $f(x_0) \in \mathbb{R}$, Lemma 25.6 (applied with $\beta=0$, $\theta=0$ and $\zeta=0$) provides some real number $\lambda_0'>0$ such that for all $\lambda \in]0,\lambda_0'[$, $P_\lambda \bar{f}(B[x_0,\frac{s_0}{8}]) \subset B[x_0,\frac{3s_0}{8}]$. We set $\bar{\lambda}_1 := \min\{\frac{\bar{\lambda}_0}{1+\|y_0\|},\lambda_0'\}$. As $B[u_0,\frac{s_0}{18}] \subset B[x_0,\frac{s_0}{8}]$ and $B[x_0,\frac{3s_0}{8}] \subset B(u_0,\frac{7s_0}{16})$, we get (f) for each $\lambda \in]0,\bar{\lambda}_1[$.

Let $x \in B(u_0,\frac{s_0}{18})$ and $\lambda \in]0,\bar{\lambda}_1[$ be fixed in the remaining of the proof. We first show that $\nabla e_\lambda \bar{f}(x) \in \partial_P f(P_\lambda \bar{f}(x))$. By the assertion (f) proved above we have $P_\lambda \bar{f}(x) \in B(u_0,\frac{7s_0}{16}) \subset B(x_0,\frac{s_0}{2})$. The functions f and $\bar{f}$ are lsc and coincide on an open neighborhood of $P_\lambda \bar{f}(x)$. It follows that

$$\partial_P \bar{f}(P_\lambda \bar{f}(x)) = \partial_P f(P_\lambda \bar{f}(x)). \tag{25.30}$$

Combining Lemma 25.12 and the single valuedness of $P_\lambda \bar{f}$ on $B(u_0, \frac{s_0}{18})$, we get $\nabla e_\lambda \bar{f}(x) \in \partial_P \bar{f}(P_\lambda \bar{f}(x))$, so (g) is true.
(h): To get finally (h), since $x, x_0 \in B(u_0, s_0/18) \subset \mathscr{O}$, we use (g) and the hypomonotonicity-like property of $\partial_P f$ on $\mathscr{O}$ [see the inequality (25.7)] to see that

$$\begin{aligned}\langle \nabla e_\lambda \bar{f}(x) - \nabla e_\lambda \bar{f}(x_0), P_\lambda \bar{f}(x) - P_\lambda \bar{f}(x_0)\rangle \\ \geq -c(2 + \|\nabla e_\lambda \bar{f}(x)\| + \|\nabla e_\lambda \bar{f}(x_0)\|)\|P_\lambda \bar{f}(x) - P_\lambda \bar{f}(x_0)\|^2 \\ \geq -c(2 + \|\nabla e_\lambda \bar{f}(x)\| + k_0\|y_0\|)\|P_\lambda \bar{f}(x) - P_\lambda \bar{f}(x_0)\|^2,\end{aligned}$$

where the latter inequality is due to the assertion (e) proved above. This entails, according to the equality (25.27),

$$\begin{aligned}\langle x - x_0, P_\lambda \bar{f}(x) - P_\lambda \bar{f}(x_0)\rangle \\ \geq (1 - c\lambda(2 + \|\nabla e_\lambda \bar{f}(x)\| + k_0\|y_0\|))\|P_\lambda \bar{f}(x) - P_\lambda \bar{f}(x_0)\|^2,\end{aligned}$$

and this yields (h) and finishes the proof of the theorem. ■

References

1. Bačák, M., Borwein, J.M., Eberhard, A., Mordukhovich, B.: Infimal convolutions and Lipschitzian properties of subdifferentials for prox regular functions in Hilbert spaces. J. Convex Anal. **17**, 737–763 (2010)
2. Bernard, F., Thibault, L.: Prox-regularity of functions and sets in Banach spaces. Set-Valued Anal. **12**, 25–47 (2004)
3. Bernard, F., Thibault, L.: Prox-regular functions in Hilbert spaces. J. Math. Anal. Appl. **303**, 1–14 (2005)
4. Bernard, F., Thibault, L., Zagrodny, D.: Integration of primal lower nice functions in Hilbert spaces. J. Optim. Theory Appl. **124**, 561–579 (2005)
5. Borwein, J.M., Giles, J.R.: The proximal normal formula in Banach space. Trans. Am. Math. Soc. **302**, 371–381 (1987)
6. Clarke, F.H.: Optimization and Nonsmooth Analysis. Wiley, New York (1983)
7. Correa, R., Jofre, A., Thibault, L.: Characterization of lower semicontinuous convex functions. Proc. Am. Math. Soc. **116**, 67–72 (1992)
8. Correa, R., Jofre, A., Thibault, L.: Subdifferential characterization of convexity. In: Du, D.-Z, et al. (eds.) Recent Advances in Nonsmooth Optimization, pp. 18–23. World Scientific, Singapore (1994)
9. Degiovanni, M., Marino, A., Tosques, M.: Evolution equations with lack of convexity. Nonlinear Anal. **9**, 1401–1443 (1985)
10. Ekeland, I.: Nonconvex minimization problems. Bull. Am. Math. Soc. **1**, 531–536 (1979)
11. Hiriart-Urruty, J.B., Plazanet, Ph.: Moreau's decomposition theorem revisited. Analyse non linéaire (Perpignan, 1987). Ann. Inst. H. Poincaré Anal. Non Linéaire **6**, suppl, 325–338 (1989)
12. Ivanov, M., Zlateva, N.: On primal lower nice property. C.R. Acad. Bulgare Sci. **54**(11), 5–10 (2001)
13. Ivanov, M., Zlateva, N.: Subdifferential characterization of primal lower nice functions on smooth spaces. C.R. Acad. Bulgare Sci. **57**, 13–18 (2004)

14. Jourani, A., Thibault, L., Zagrodny, D.: $\mathscr{C}^{1,\omega(\cdot)}$-regularity and Lipschitz-like properties of subdifferential. Proc. London Math. Soc. **105**(1), 189–223 (2012)
15. Levy, A.B., Poliquin, R.A., Thibault, L.: Partial extensions of Attouch's theorem with applications to proto-derivatives of subgradient mappings. Trans. Am. Math. Soc. **347**, 1269–1294 (1995)
16. Marcellin, S., Thibault, L.: Evolution problems associated with primal lower nice functions. J. Convex Anal. **13**(2), 385–421 (2006)
17. Mordukhovich, B.S.: Variational analysis and generalized differentiation I and II. Grundlehren der Mathematischen Wissenschaften [A Series of Comprehensive Studies in Mathematics], vol. 330 and 331. Springer, Berlin (2005)
18. Mordukhovich, B.S., Shao, Y.H.: Nonsmooth analysis in Asplund spaces. Trans. Am. Math. Soc. **348**, 1235–1280 (1996)
19. Moreau, J.J.: Proximité et dualité dans un espace Hilbertien. Bull. Soc. Math. France **93**, 273–299 (1965)
20. Moreau, J.J.: Fonctionnelles Convexes, 2nd edn. Facoltà di Ingegneria Università di Roma "Tor Vergata" (2003)
21. Poliquin, R.A.: Subgradient monotonicity and convex functions. Nonlinear. Anal. **14**, 305–317 (1990)
22. Poliquin, R.A.: Integration of subdifferentials of nonconvex functions. Nonlinear Anal. **17**, 385–398 (1991)
23. Poliquin, R.A., Rockafellar, R.T.: Prox-regular functions in variational analysis. Trans. Am. Math. Soc. **348**, 1805–1838 (1996)
24. Serea, O.S., Thibault, L.: Primal lower nice property of value functions in optimization and control problems. Set-Valued Var. Anal. **18**, 569–600 (2010)
25. Thibault, L., Zagrodny, D.: Integration of subdifferentials of lower semicontinuous functions on Banach Spaces. J. Math. Anal. Appl. **189**, 33–58 (1995)
26. Wang, X.: On Chebyshev functions and Klee functions. J. Math. Anal. Appl. **368**, 293–310 (2010)

Chapter 26
Bundle Method for Non-Convex Minimization with Inexact Subgradients and Function Values

Dominikus Noll

Abstract We discuss a bundle method to minimize locally Lipschitz functions which are both nonconvex and nonsmooth. We analyze situations where only inexact subgradients or function values are available. For suitable classes of such nonsmooth functions we prove convergence of our algorithm to approximate critical points.

Key words: Convergence • Inexact function values • Inexact subgradients • Lower C^1 functions • Nonconvex bundle method

Mathematics Subject Classifications (2010): Primary 90C56; Secondary 49J52, 65K05, 65K10.

26.1 Introduction

We consider optimization programs of the form

$$\min_{x \in \mathbb{R}^n} f(x), \tag{26.1}$$

where $f : \mathbb{R}^n \to \mathbb{R}$ is locally Lipschitz but neither differentiable nor convex. We present a bundle algorithm which converges to a critical point of (26.1) if exact function and subgradient evaluation of f are provided and to an approximate critical

---

COMMUNICATED BY HEINZ H. BAUSCHKE.

D. Noll (✉)
Institut de Mathématiques, Université Paul Sabatier, Toulouse, France
e-mail: noll@mip.ups-tlse.fr

D.H. Bailey et al. (eds.), *Computational and Analytical Mathematics*, Springer Proceedings in Mathematics & Statistics 50, DOI 10.1007/978-1-4614-7621-4_26,

point if subgradients or function values are inexact. Here $\bar{x} \in \mathbb{R}^n$ is approximate critical if

$$\operatorname{dist}(0, \partial f(\bar{x})) \leq \varepsilon, \tag{26.2}$$

where $\partial f(x)$ is the Clarke subdifferential of f at x.

The method discussed here extends the classical bundle concept to the nonconvex setting by using downshifted tangents as a substitute for cutting planes. This idea was already used in the 1980s in Lemaréchal's M2FC1 code [32] or in Zowe's BT codes [48, 54]. Its convergence properties can be assessed by the model-based bundle techniques [6, 7, 40, 42]. Recent numerical experiments using the downshift mechanism are reported in [8, 19, 50]. In the original paper of Schramm and Zowe [48] downshift is discussed for a hybrid method combining bundling, trust region, and line-search elements.

For convex programs (26.1) bundle methods which can deal with inexact function values or subgradients have been discussed at least since 1985; see Kiwiel [26, 28]. More recently, the topic has been revived by Hintermüller [22], who presented a method with exact function values but inexact subgradients $g \in \partial_\varepsilon f(x)$, where ε remains unknown to the user. Kiwiel [30] expands on this idea and presents an algorithm which deals with inexact function values and subgradients, both with unknown errors bounds. Kiwiel and Lemaréchal [31] extend the idea further to address column generation. Incremental methods to address large problems in stochastic programming or Lagrangian relaxation can be interpreted in the framework of inexact values and subgradients; see, e.g., Emiel and Sagastizábal [15, 16] and Kiwiel [29]. In [39] Nedic and Bertsekas consider approximate functions and subgradients which are in addition affected by deterministic noise.

Nonsmooth methods without convexity have been considered by Wolfe [52], Shor [49], Mifflin [38], Schramm and Zowe [48], and more recently by Lukšan and Vlček [35], Noll and Apkarian [41], Fuduli et al. [17, 18], Apkarian et al. [6], Noll et al. [42], Hare and Sagastizábal [21], Sagastizábal [47], Lewis and Wright [33], and Noll [40]. In the context of control applications, early contributions are Polak and Wardi [44], Mayne and Polak [36, 37], Kiwiel [27], Polak [43], Apkarian et al. [1–7], and Bompart et al. [9]. All these approaches use exact knowledge of function values and subgradients.

The structure of the paper is as follows. In Sect. 26.2 we explain the concept of an approximate subgradient. Section 26.3 discusses the elements of the algorithm, acceptance, tangent program, aggregation, cutting planes, recycling, and the management of proximity control. Section 26.4 presents the algorithm. Section 26.5 analyzes the inner loop in the case of exact function values and inexact subgradients. Section 26.6 gives convergence of the outer loop. Section 26.7 extends to the case where function values are also inexact. Section 26.8 uses the convergence theory of Sects. 26.5–26.7 to derive a practical stopping test. Section 26.9 concludes with a motivating example from control.

26.2 Preparation

Approximate subgradients in convex bundle methods refer to the ε-subdifferential [24]:

$$\partial_\varepsilon f(x) = \{g \in \mathbb{R}^n : g^\top (y-x) \le f(y) - f(x) + \varepsilon \text{ for all } y \in \mathbb{R}^n\}, \tag{26.3}$$

whose central property is that $0 \in \partial f_\varepsilon(\bar{x})$ implies ε-minimality of $\bar{x}$, i.e., $f(\bar{x}) \le \min f + \varepsilon$. Without convexity we cannot expect a tool with similar global properties. We shall work with the following very natural approximate subdifferential

$$\partial_{[\varepsilon]} f(x) = \partial f(x) + \varepsilon B, \tag{26.4}$$

where B is the unit ball in some fixed Euclidian norm and $\partial f(x)$ is the Clarke subdifferential of f. The present section motivates this choice.

The first observation concerns the optimality condition (26.2) arising from the choice (26.4). Namely $0 \in \partial_{[\varepsilon]} f(\bar{x})$ can also be written as $0 \in \partial(f + \varepsilon\| \cdot - x\|)(x)$, meaning that a small perturbation of f is critical at x.

We can also derive a weak form of ε-optimality from $0 \in \partial_{[\varepsilon]} f(x)$ for composite functions $f = g \circ F$ with g convex and F smooth or, more generally, for lower C^2 functions (see [45]) which have such a representation locally.

Lemma 26.1. *Let $f = g \circ F$ where g is convex and F is of class C^2, and suppose $0 \in \partial_{[\varepsilon]} f(x)$. Fix $r > 0$, and define*

$$c_r := \max_{\|d\|=1} \max_{\|x'-x\| \le r} \max_{\phi \in \partial g(F(x))} \phi^\top D^2 F(x')[d,d].$$

Then x is $(r\varepsilon + r^2 c_r/2)$-optimal on the ball $B(x,r)$.

Proof. We have to prove $f(x) \le f(x^+) + r\varepsilon + r^2 c_r/2$ for every $x^+ \in B(x,r)$. Write $x^+ = x + td$ for some $\|d\| = 1$ and $t \le r$. Since $0 \in \partial_{[\varepsilon]} f(x)$, and since $\partial f(x) = DF(x)^* \partial g(F(x))$, there exists $\phi \in \partial g(F(x))$ such that $\|DF(x)^* \phi\| \le \varepsilon$. In other words, $\|\phi^\top DF(x)d\| \le \varepsilon$ because $\|d\| = 1$. By the subgradient inequality we have

$$\phi^\top (F(x+td) - F(x)) \le g(F(x+td)) - g(F(x)) = f(x^+) - f(x). \tag{26.5}$$

Second-order Taylor expansion of $t \mapsto \phi^\top F(x+td)$ at $t = 0$ gives

$$\phi^\top F(x+td) = \phi^\top F(x) + t\phi^\top DF(x)d + \tfrac{t^2}{2}\phi^\top D^2F(x_t)[d,d]$$

for some x_t on the segment $[x, x+td]$. Substituting this into (26.5) and using the definition of c_r give

$$f(x) \le f(x^+) + t\|\phi^\top DF(x)d\| + \tfrac{t^2}{2}\|\phi^\top D^2F(x_t)[d,d]\| \le f(x^+) + r\varepsilon + \tfrac{r^2}{2}c_r,$$

hence the claim. ■

Remark 26.2. For convex f we can try to relate the two approximate subdifferentials in the sense that

$$\partial_\varepsilon f(x) \subset \partial_{[\varepsilon']} f(x)$$

for a suitable $\varepsilon' = \varepsilon'(x,\varepsilon)$. For a convex quadratic function $f(x) = \frac{1}{2}x^\top Qx + q^\top x$ it is known that $\partial_\varepsilon f(x) = \{\nabla f(x) + Q^{1/2}z : \frac{1}{2}\|z\|^2 \leq \varepsilon\}$, [24], so that $\partial_\varepsilon f(x) \subset \partial f(x) + \varepsilon' B = \partial_{[\varepsilon']} f(x)$ for $\varepsilon' = \sup\{\|Q^{1/2}z\| : \frac{1}{2}\|z\|^2 \leq \varepsilon\}$, which means that $\varepsilon'(x,\varepsilon)$ is independent of x and behaves as $\varepsilon' = \mathcal{O}(\varepsilon^{1/2})$. We expect this type of relation to hold as soon as f has curvature information around x. On the other hand, if $f(x) = |x|$, then $\partial f_\varepsilon(x) = \partial f(x) + \frac{\varepsilon}{|x|}B$ for $x \neq 0$ (and $\partial_\varepsilon f(0) = \partial f(0)$), which means that the relationship $\varepsilon' = \varepsilon/|x|$ is now linear in ε for fixed $x \neq 0$. In general it is difficult to relate ε to ε'. See Hiriart-Urruty and Seeger [23] for more information on this question.

Remark 26.3. For composite functions $f = g \circ F$ with g convex and F of class C^1 we can introduce

$$\partial_\varepsilon f(x) = DF(x)^* \partial_\varepsilon g(F(x)),$$

where $\partial_\varepsilon g(y)$ is the usual convex ε-subdifferential (26.3) of g and $DF(x)^*$ is the adjoint of the differential of F at x. Since the corresponding chain rule is valid in the case of an affine F, $\partial_\varepsilon f(x)$ is consistent with (26.3). Without convexity $\partial f_\varepsilon(x)$ no longer preserves the global properties of (26.3). Yet, for composite functions $f = g \circ F$, a slightly more general version of Lemma 26.1 combining $\partial_{[\sigma]} f$ and $\partial_\varepsilon f$ can be proved along the lines of [41, Lemma 2]. In that reference the result is shown for the particular case $g = \lambda_1$, but an extension can be obtained by reasoning as in Lemma 26.1.

Remark 26.4. For convex f the set $\partial_{[\varepsilon]} f(x)$ coincides with the Fréchet ε-sub differential $\partial_\varepsilon^F f(x)$. According to [34, Corollary 3.2] the same remains true for approximate convex functions. For the latter see Sect. 26.5.

26.3 Elements of the Algorithm

26.3.1 Local Model

Let x be the current iterate of the outer loop. The inner loop with counter k generates a sequence y^k of trial steps, one of which is eventually accepted to become the new serious step x^+. At each instant k we dispose of a convex working model $\phi_k(\cdot,x)$, which approximates f in a neighborhood of x. We suppose that we know at least one approximate subgradient $g(x) \in \partial_{[\varepsilon]} f(x)$. The affine function

$$m_0(\cdot,x) = f(x) + g(x)^\top(\cdot - x)$$

will be referred to as the exactness plane at x. For the moment we assume that it gives an exact value of f at x, but not an exact subgradient. The algorithm assures $\phi_k(\cdot,x) \geq m_0(\cdot,x)$ at all times k, so that $g(x) \in \partial\phi_k(x,x)$ for all k. In fact we construct $\phi_k(\cdot,x)$ in such a way that $\partial\phi_k(x,x) \subset \partial_{[\varepsilon]}f(x)$ at all times k.

Along with the first-order working model $\phi_k(\cdot,x)$ we also consider an associated second-order model of the form

$$\Phi_k(y,x) = \phi_k(y,x) + \tfrac{1}{2}(y-x)^\top Q(x)(y-x),$$

where $Q(x)$ depends on the serious iterate x, but is fixed during the inner loop k. We allow $Q(x)$ to be indefinite.

26.3.2 Cutting Planes

Suppose y^k is a null step. Then model $\Phi_k(\cdot,x)$ which gave rise to y^k was not rich enough and we have to improve it at the next inner loop step $k+1$ in order to perform better. We do this by modifying the first-order part. In convex bundling one includes a cutting plane at y^k into the new model $\phi_{k+1}(\cdot,x)$. This remains the same with approximate subgradients and values (cf. [22, 30]) as soon as the concept of cutting plane is suitably modified. Notice that we have access to $g_k \in \partial_{[\varepsilon]}f(y^k)$, which gives us an approximate tangent

$$t_k(\cdot) = f(y^k) + g_k^\top(\cdot - y^k)$$

at y^k. Since f is not convex, we cannot use $t_k(\cdot)$ directly as cutting plane. Instead we use a technique originally developed in Schramm and Zowe [48] and Lemaréchal [32], which consists in shifting $t_k(\cdot)$ downwards until it becomes useful for $\phi_{k+1}(\cdot,x)$. Fixing $c > 0$ once and for all, we call

$$s_k := [t_k(x) - f(x)]_+ + c\|y^k - x\|^2 \tag{26.6}$$

the downshift and introduce

$$m_k(\cdot,x) = t_k(\cdot) - s_k,$$

called the downshifted tangent.

We sometimes use the following more stringent notation, where no reference to the counter k is made. The approximate tangent is $t_{y,g}(\cdot) = f(y) + g^\top(\cdot - y)$, bearing a reference to the point y where it is taken and to the specific approximate subgradient $g \in \partial_{[\varepsilon]}f(y)$. The downshifted tangent is then $m_{y,g}(\cdot,x) = t_{y,g}(\cdot) - s$, where $s = s(y,g,x) = [t_{y,g}(x) - f(x)]_+ + c\|y - x\|^2$ is the downshift. Since this notation is fairly heavy, we will try to avoid it whenever possible and switch to the former, bearing in mind that $t_k(\cdot)$ depends both on y^k and the subgradient $g_k \in \partial_{[\varepsilon]}f(y^k)$. Similarly, the downshifted tangent plane $m_k(\cdot,x)$ depends on y^k, g^k, and on x, as does the downshift s_k. We use $m_k(\cdot,x)$ as a substitute for the classical cutting plane. For convenience we continue to call $m_k(\cdot,x)$ a cutting plane.

The cutting plane satisfies $m_k(x,x) \leq f(x) - c\|y^k - x\|^2$, which assures that it does not interfere with the subdifferential of $\phi_{k+1}(\cdot,x)$ at x. We build $\phi_{k+1}(\cdot,x)$ in such a way that it has $m_k(\cdot,x)$ as an affine minorant.

Proposition 26.5. *Let* $\phi_{k+1}(\cdot,x) = \max\{m_\nu(\cdot,x) : \nu = 0,\dots,k\}$. *Then* $\partial\phi_{k+1}(x,x) \subset \partial_{[\varepsilon]} f(x)$.

Proof. As all the downshifts s_k are positive, $\phi_{k+1}(y,x) = m_0(y,x)$ in a neighborhood of x; hence $\partial\phi_{k+1}(x,x) = \partial m_0(x,x) = \{g(x)\} \subset \partial_{[\varepsilon]} f(x)$. ■

26.3.3 Tangent Program

Given the local model $\Phi_k(\cdot,x) = \phi_k(\cdot,x) + \frac{1}{2}(\cdot - x)^\top Q(x)(\cdot - x)$ at serious iterate x and inner loop counter k, we solve the tangent program

$$\min_{y \in \mathbb{R}^n} \Phi_k(y,x) + \tfrac{\tau_k}{2}\|y - x\|^2. \tag{26.7}$$

We assume that $Q(x) + \tau_k I \succ 0$, which means (26.7) is strictly convex and has a unique solution y^k, called a trial step. The optimality condition for (26.7) implies

$$(Q(x) + \tau_k I)(x - y^k) \in \partial\phi_k(y^k,x). \tag{26.8}$$

If $\phi_k(\cdot,x) = \max\{m_\nu(\cdot,x) : \nu = 0,\dots,k\}$, with $m_\nu(\cdot,x) = a_\nu + g_\nu^\top(\cdot - x)$, then we can find $\lambda_0 \geq 0,\dots,\lambda_k \geq 0$, summing up to 1, such that

$$g_k^* := (Q(x) + \tau_k I)(x - y^k) = \sum_{\nu=0}^{k} \lambda_\nu g_\nu.$$

Traditionally, g_k^* is called the aggregate subgradient at y^k. We build the aggregate plane

$$m_k^*(\cdot,x) = a_k^* + g_k^{*\top}(\cdot - x),$$

where $a_k^* = \sum_{\nu=1}^{k} \lambda_\nu a_\nu$. Keeping $m_k^*(\cdot,x)$ as an affine minorant of $\phi_{k+1}(\cdot,x)$ allows to drop some of the older cutting planes to avoid overflow. As $\partial\phi_k(y^k,x)$ is the subdifferential of a max-function, we know that $\lambda_\nu > 0$ precisely for those $m_\nu(\cdot,x)$ which are active at y^k. That is, $\sum_{\nu=1}^{k} \lambda_\nu m_\nu(y^k,x) = \phi_k(y^k,x)$. Therefore the aggregate plane satisfies

$$m_k^*(y^k,x) = \phi_k(y^k,x). \tag{26.9}$$

As our algorithm chooses ϕ_{k+1} such that $m_k^*(\cdot,x) \leq \phi_{k+1}(\cdot,x)$, we have $\phi_k(y^k,x) \leq \phi_{k+1}(y^k,x)$. All this follows the classical line originally proposed in Kiwiel [25]. Maintaining a model $\phi_k(\cdot,x)$ which contains aggregate subgradients from previous sweeps instead of *all* the older g_ν, $\nu = 0,\ldots,k$ does not alter the statement of Proposition 26.5 nor of formula (26.9).

26.3.4 Testing Acceptance

Having computed the kth trial step y^k via (26.7), we have to decide whether it should be accepted as the new serious iterate x^+. We compute the test quotient

$$\rho_k = \frac{f(x) - f(y^k)}{f(x) - \Phi_k(y^k,x)}.$$

Fixing constants $0 < \gamma < \Gamma < 1$, we call y^k *bad* if $\rho_k < \gamma$ and *good* if $\rho_k \geq \Gamma$. If y^k is not bad, meaning $\rho_k \geq \gamma$, then it is accepted to become x^+. We refer to this as a serious step. Here the inner loop ends. On the other hand, if y^k is bad, then it is rejected and referred to as a null step. In this case the inner loop continues.

26.3.5 Management of τ in the Inner Loop

The most delicate point is the management of the proximity control parameter during the inner loop. Namely, it may turn out that the trial steps y^k proposed by the tangent program (26.7) are too far from the current x, so that no decrease below $f(x)$ can be achieved. In the convex case one relies entirely on the mechanism of cutting planes. Indeed, if y^k is a null step, then the convex cutting plane, when added to model $\phi_{k+1}(\cdot,x)$, will cut away the unsuccessful y^k, paving the way for a better y^{k+1} at the next sweep.

The situation is more complicated without convexity, where cutting planes are no longer tangents to f. In the case of downshifted tangents the information stored in the ideal set of all theoretically available cutting planes may *not* be sufficient to represent f correctly when y^k is far away from x. This is when we have to force smaller steps by increasing τ, i.e., by tightening proximity control. As a means to decide when this has to happen, we use the parameter

$$\tilde{\rho}_k = \frac{f(x) - M_k(y^k,x)}{f(x) - \Phi_k(y^k,x)}, \tag{26.10}$$

where $m_k(\cdot,x)$ is the new cutting plane drawn for y^k as in Sect. 26.3.1 and $M_k(\cdot,x) = m_k(\cdot,x) + \frac{1}{2}(\cdot - x)^\top Q(x)(\cdot - x)$. We fix a parameter $\tilde{\gamma}$ with $\gamma < \tilde{\gamma} < 1$ and make the following decision.

$$\tau_{k+1} = \begin{cases} 2\tau_k & \text{if } \rho_k < \gamma \text{ and } \tilde{\rho}_k \geq \tilde{\gamma}, \\ \tau_k & \text{if } \rho_k < \gamma \text{ and } \tilde{\rho}_k < \tilde{\gamma}. \end{cases} \tag{26.11}$$

The idea in (26.11) can be explained as follows. The quotient $\tilde{\rho}_k$ in (26.10) can also be written as $\tilde{\rho}_k = \left(f(x) - \Phi_{k+1}(y^k,x)\right) / \left(f(x) - \Phi_k(y^k,x)\right)$, because the cutting plane at stage k will be integrated into model Φ_{k+1} at stage $k+1$. If $\tilde{\rho}_k \approx 1$, we can therefore conclude that adding the new cutting plane at the null step y^k hardly changes the situation. Put differently, had we known the cutting plane before computing y^k, the result would not have been much better. In this situation we decide to force smaller trial steps by increasing the τ-parameter. If on the other hand $\tilde{\rho}_k \ll 1$, then the gain of information provided by the new cutting plane at y^k is substantial with regard to the information already stored in Φ_k. Here we continue to add cutting planes and aggregate planes only, hoping that we will still make progress *without* having to increase τ. The decision $\tilde{\rho}_k \approx 1$ versus $\tilde{\rho}_k \ll 1$ is formalized by the rule (26.11).

Remark 26.6. By construction $\tilde{\rho}_k \geq 0$, because aggregation assures that $\phi_{k+1}(y^k,x) \geq \phi_k(y^k,x)$. Notice that in contrast ρ_k may be negative. Indeed, $\rho_k < 0$ means that the trial step y^k proposed by the tangent program (26.7) gives no descent in the function values, meaning that it is clearly a bad step.

26.3.6 Management of τ in the Outer Loop

The proximity parameter τ will also be managed dynamically between serious steps $x \to x^+$. In our algorithm we use a memory parameter $\tau_j^\sharp$, which is specified at the end of the $(j-1)$st inner loop and serves to initialize the jth inner loop with $\tau_1 = \tau_j^\sharp$.

A first rule which we already mentioned is that we need $Q(x^j) + \tau_k I \succ 0$ for all k during the jth inner loop. Since τ is never decreased during the inner loop, we can assure this if we initialize $\tau_1 > -\lambda_{\min}(Q(x^j))$.

A more important aspect is the following. Suppose the $(j-1)$st inner loop ended at inner loop counter k_{j-1}, i.e., $x^j = y^{k_{j-1}}$ with $\rho_{k_{j-1}} \geq \gamma$. If acceptance was good, i.e., $\rho_{k_{j-1}} \geq \Gamma$, then we can trust our model, and we account for this by storing a smaller parameter $\tau_j^\sharp = \frac{1}{2}\tau_{k_{j-1}} < \tau_{k_{j-1}}$ for the jth outer loop. On the other hand, if acceptance of the $(j-1)$st step was neither good nor bad, meaning $\gamma \leq \rho_{k_{j-1}} \leq \Gamma$, then there is no reason to decrease τ for the next outer loop, so we memorize $\tau_{k_{j-1}}$, the value we had at the end of the $(j-1)$st inner loop. Altogether

$$\tau_j^\sharp = \begin{cases} \max\{\frac{1}{2}\tau_{k_{j-1}}, -\lambda_{\min}(Q(x^j)) + \zeta\} & \text{if } \rho_{k_{j-1}} \geq \Gamma, \\ \max\{\tau_{k_{j-1}}, -\lambda_{\min}(Q(x^j)) + \zeta\} & \text{if } \gamma \leq \rho_{k_{j-1}} < \Gamma, \end{cases} \tag{26.12}$$

where $\zeta > 0$ is some small threshold fixed once and for all.

26.3.7 Recycling of Planes

In a convex bundle algorithm one keeps in principle all cutting planes in the model, using aggregation to avoid overflow. In the nonconvex case this is no longer possible. Cutting planes are downshifted tangents, which links them to the value $f(x)$ of the current iterate x. As we pass from x to a new serious iterate x^+, the cutting plane $m_{z,g}(\cdot,x) = a + g^\top(\cdot - x)$ with $g \in \partial_{[\varepsilon]} f(z)$ for some z cannot be used as such, because we have no guarantee whether $a + g^\top(x^+ - x) \leq f(x^+)$. But we can downshift it again if need be. We recycle the plane as

$$m_{z,g}(\cdot,x^+) = a - s^+ + g^\top(\cdot - x), \quad s^+ = [m_{z,g}(x^+,x) - f(x^+)]_+ + c\|x^+ - z\|^2.$$

In addition one may also apply a test whether z is too far from x^+ to be of interest, in which case the plane should simply be removed from the stock.

26.4 Algorithm

Algorithm (Proximity control algorithm for (26.1)).

Parameters: $0 < \gamma < \Gamma < 1$, $\gamma < \tilde{\gamma} < 1$, $0 < q < \infty$, $q < T < \infty$, $\tilde{\varepsilon} > 0$.

1: **Initialize outer loop**. Choose initial guess x^1 and an initial matrix $Q_1 = Q_1^\top$ with $-qI \preceq Q_1 \preceq qI$. Fix memory control parameter $\tau_1^\sharp$ such that $Q_1 + \tau_1^\sharp I \succ 0$. Put $j = 1$.

2: **Stopping test**. At outer loop counter j, stop if $0 \in \partial_{[\tilde{\varepsilon}]} f(x^j)$. Otherwise go to inner loop.

3: **Initialize inner loop**. Put inner loop counter $k = 1$ and initialize τ-parameter using the memory element, i.e., $\tau_1 = \tau_j^\sharp$. Choose initial convex working model $\phi_1(\cdot,x^j)$, possibly recycling some planes from previous sweep $j-1$, and let $\Phi_1(\cdot,x^j) = \phi_1(\cdot,x^j) + \frac{1}{2}(\cdot - x^j)^\top Q_j(\cdot - x^j)$.

4: **Trial step generation.** At inner loop counter k solve tangent program

$$\min_{y \in \mathbb{R}^n} \Phi_k(y,x^j) + \frac{\tau_k}{2}\|y - x^j\|^2.$$

The solution is the new trial step y^k.

5: **Acceptance test.** Check whether

$$\rho_k = \frac{f(x^j) - f(y^k)}{f(x^j) - \Phi_k(y^k, x^j)} \geq \gamma.$$

If this is the case put $x^{j+1} = y^k$ (serious step), quit inner loop, and go to step 8. If this is not the case (null step) continue inner loop with step 6.

6: **Update proximity parameter**. Compute a cutting plane $m_k(\cdot, x^j)$ at x^j for the null step y^k. Let $M_k(\cdot, x^j) = m_k(\cdot, x^j) + \frac{1}{2}(\cdot - x^j)^\top Q_j(\cdot - x^j)$ and compute secondary control parameter

$$\tilde{\rho}_k = \frac{f(x^j) - M_k(y^k, x^j)}{f(x^j) - \Phi_k(y^k, x^j)}.$$

Put $\tau_{k+1} = \begin{cases} \tau_k, & \text{if } \tilde{\rho}_k < \tilde{\gamma} \quad \text{(bad)} \\ 2\tau_k, & \text{if } \tilde{\rho}_k \geq \tilde{\gamma} \quad \text{(too bad)} \end{cases}$

7: **Update working model**. Build new convex working model $\phi_{k+1}(\cdot, x^j)$ based on null step y^k by adding the new cutting plane $m_k(\cdot, x^j)$ (and using aggregation to avoid overflow). Keep exactness plane in the working model. Then increase inner loop counter k and continue inner loop with step 4.

8: **Update Q_j and memory element**. Update matrix $Q_j \to Q_{j+1}$, respecting $Q_{j+1} = Q_{j+1}^\top$ and $-qI \preceq Q_{j+1} \preceq qI$. Then store new memory element

$$\tau_{j+1}^\sharp = \begin{cases} \tau_k, & \text{if } \gamma \leq \rho_k < \Gamma \quad \text{(not bad)} \\ \frac{1}{2}\tau_k, & \text{if } \rho_k \geq \Gamma \quad \text{(good)} \end{cases}$$

Increase $\tau_{j+1}^\sharp$ if necessary to ensure $Q_{j+1} + \tau_{j+1}^\sharp I \succ 0$. If $\tau_{j+1}^\sharp > T$ then reset $\tau_{j+1}^\sharp = T$. Increase outer loop counter j by 1 and loop back to step 2.

26.5 Analysis of the Inner Loop

In this section we analyze the inner loop and show that there are two possibilities. Either the inner loop terminates finitely with a step $x^+ = y^k$ satisfying $\rho_k \geq \gamma$ or we get an infinite sequence of null steps y^k which converges to x. In the latter case, we conclude that $0 \in \partial_{[\tilde{\varepsilon}]} f(x)$, i.e., that x is approximate optimal.

Suppose the inner loop turns forever. Then there are two possibilities. Either τ_k is increased infinitely often, so that $\tau_k \to \infty$, or τ_k is frozen, $\tau_k = \tau_{k_0}$ for some k_0 and all $k \geq k_0$. These scenarios will be analyzed in Lemmas 26.9 and 26.11. Since the matrix $Q(x)$ is fixed during the inner loop, we write it simply as Q.

To begin with, we need an auxiliary construction. We define the following convex function:

$$\phi(y,x) = \sup\{m_{z,g}(y,x) : z \in B(0,M), g \in \partial_{[\varepsilon]} f(y)\}, \tag{26.13}$$

where $B(0,M)$ is a fixed ball large enough to contain x and all trial steps encountered during the inner loop. Recall that $m_{z,g}(\cdot,x)$ is the cutting plane at z with approximate subgradient $g \in \partial_{[\varepsilon]} f(z)$ with respect to the serious iterate x. Due to boundedness of $B(0,M)$, $\phi(\cdot,x)$ is defined everywhere.

Lemma 26.7. *We have $\phi(x,x) = f(x)$, $\partial\phi(x,x) = \partial_{[\varepsilon]} f(x)$, and ϕ is jointly upper semicontinuous. Moreover, if $y^k \in B(0,M)$ for all k, then $\phi_k(\cdot,x) \le \phi(\cdot,x)$ for every first-order working model ϕ_k.*

Proof. (1) The first statement follows because every cutting plane drawn at some $z \neq x$ and $g \in \partial_{[\varepsilon]} f(z)$ satisfies $m_{z,g}(x,x) \le f(x) - c\|x-z\|^2 < f(x)$, while cutting planes at x obviously have $m_{x,g}(x,x) = f(x)$.

(2) Concerning the second statement, let us first prove $\partial_{[\varepsilon]} f(x) \subset \partial\phi(x,x)$. We consider the set of limiting subgradients

$$\partial^l f(x) = \{\lim_{k\to\infty} \nabla f(y^k) : y^k \to x, f \text{ is differentiable at } y^k\}.$$

Then $\operatorname{co}\partial^l f(x) = \partial f(x)$ by [13]. It therefore suffices to show $\partial^l f(x) + \varepsilon B \subset \partial\phi(x,x)$, because $\partial\phi(x,x)$ is convex and we then have $\partial\phi(x,x) \supset \operatorname{co}(\partial^l f(x) + \varepsilon B) = \operatorname{co}\partial^l f(x) + \varepsilon B = \partial f(x) + \varepsilon B$.

Let $g_a \in \partial^l f(x) + \varepsilon B$. We have to show $g_a \in \partial\phi(x,x)$. Choose $g \in \partial^l f(x)$ such that $\|g - g_a\| \le \varepsilon$. Pick a sequence $y^k \to x$ and $g_k = \nabla f(y^k) \in \partial f(y^k)$ such that $g_k \to g$. Let $g_{a,k} = g_k + g_a - g$ and then $g_{a,k} \in \partial_{[\varepsilon]} f(y^k)$ and $g_{a,k} \to g_a$. Let $m_k(\cdot,x)$ be the cutting plane drawn at y^k with approximate subgradient $g_{a,k}$, then $m_k(y^k,x) \le \phi(y^k,x)$. By the definition of the downshift process

$$m_k(y,x) = f(y^k) + g_{a,k}^\top(y - y^k) - s_k,$$

where s_k is the downshift (26.6). There are two cases, $s_k = c\|y^k - x\|^2$, and $s_k = t_k(x) - f(x) + c\|y^k - x\|^2$ according to whether the term $[\ldots]_+$ in (26.6) equals zero or not.

Let us start with the second case, where $t_k(x) > f(x)$. Then $s_k = f(y^k) + g_{a,k}^\top(x - y^k) - f(x) + c\|y^k - x\|^2$ and

$$\begin{aligned} m_k(y,x) &= f(y^k) + g_{a,k}^\top(y - y^k) - f(y^k) - g_{a,k}^\top(x - y^k) + f(x) - c\|y^k - x\|^2 \\ &= f(x) + g_{a,k}^\top(y - x) - c\|y^k - x\|^2. \end{aligned}$$

Therefore

$$\phi(y,x) - \phi(x,x) \ge m_k(y,x) - f(x) = g_{a,k}^\top(y - x) - c\|y^k - x\|^2.$$

Passing to the limit using $y^k \to x$ and $g_{a,k} \to g_a$ proves $g_a \in \partial\phi(x,x)$.

It remains to discuss the first case, where $t_k(x) \le f(x)$, so that $s_k = c\|y^k - x\|^2$. Then

$$m_k(\cdot,x) = f(y^k) + g_{a,k}^\top(\cdot - y^k) - c\|y^k - x\|^2.$$

Therefore

$$\begin{aligned}\phi(y,x) - \phi(x,x) &\ge m_k(y,x) - f(x)\\ &= f(y^k) - f(x) + g_{a,k}^\top(y - y^k) - c\|y^k - x\|^2\\ &= f(y^k) - f(x) + g_{a,k}^\top(x - y^k) + g_{a,k}^\top(y - x) - c\|y^k - x\|^2.\end{aligned}$$

As y is arbitrary, we have $g_{a,k} \in \partial_{|\zeta_k|}\phi(x,x)$, where $\zeta_k = f(y^k) - f(x) + g_{a,k}^\top(x - y^k) - c\|y^k - x\|^2$. Since $\zeta_k \to 0$, $y^k \to x$ and $g_{a,k} \to g_a$, we deduce again $g_a \in \partial\phi(x,x)$. Altogether for the two cases $[\dots]_+ = 0$ and $[\dots]_+ > 0$ we have shown $\partial^l f(x) + \varepsilon B \subset \partial\phi(x,x)$.

(3) Let us now prove $\partial\phi(x,x) \subset \partial f(x) + \varepsilon B$. Let $g \in \partial\phi(x,x)$ and $m(\cdot,x) = f(x) + g^\top(\cdot - x)$ the tangent plane to the graph of $\phi(\cdot,x)$ at x associated with g. By convexity $m(\cdot,x) \le \phi(\cdot,x)$. We fix $h \in \mathbb{R}^n$ and consider the values $\phi(x+th,x)$ for $t > 0$. According to the definition of $\phi(\cdot,x)$ we have $\phi(x+th,x) = m_{z_t,g_t}(x+th,x)$, where $m_{z_t,g_t}(\cdot,x)$ is a cutting plane drawn at some $z_t \in B(0,M)$ with $g_t \in \partial_{[\varepsilon]} f(z_t)$. The slope of the cutting plane along the ray $x + \mathbb{R}_+ h$ is $g_t^\top h$. Now the cutting plane passes through $\phi(x+th,x) \ge m(x+th,x)$, which means that its value at $x+th$ is above the value of the tangent. On the other hand, according to the downshift process, the cutting plane satisfies $m_{z_t,g_t}(x,x) \le f(x) - c\|x - z_t\|^2$. Its value at x is therefore below the value of $m(x,x) = f(x)$. These two facts together tell us that $m_{z_t,g_t}(\cdot,x)$ is steeper than $m(\cdot,x)$ along the ray $x + \mathbb{R}_+ h$. In other words, $g^\top h \le g_t^\top h$. Next observe that $\phi(x+th,x) \to \phi(x,x) = f(x)$ as $t \to 0^+$. That implies $m_{z_t,g_t}(x+th,x) \to f(x)$. Since by the definition of downshift $m_{z_t,g_t}(x+th,x) \le f(x) - c\|x - z_t\|^2$, it follows that we must have $\|x - z_t\|^2 \to 0$, i.e., $z_t \to x$ as $t \to 0^+$. Passing to a subsequence, we may assume $g_t \to \hat{g}$ for some $\hat{g}$. With $z_t \to x$ it follows from upper semicontinuity of the Clarke subdifferential that $\hat{g} \in \partial_{[\varepsilon]} f(x)$. On the other hand, $g^\top h \le g_t^\top h$ for all t implies $g^\top h \le \hat{g}^\top h$. Therefore $g^\top h \le \sigma_K(h) = \max\{\tilde{g}^\top h : \tilde{g} \in K\}$, where σ_K is the support function of $K = \partial_{[\varepsilon]} f(x)$. Given that h was arbitrary, and as K is closed convex, this implies $g \in K$ by Hahn–Banach.

(4) Upper semicontinuity of ϕ follows from upper semicontinuity of the Clarke subdifferential. Indeed, let $x_j \to x$, $y_j \to y$. Using the definition (26.13) of ϕ, find cutting planes $m_{z_j,g_j}(\cdot,x_j) = t_{z_j}(\cdot) - s_j$ at serious iterate x_j, drawn at z_j with $g_j \in \partial_{[\varepsilon]} f(z_j)$, such that $\phi(y_j,x_j) \le m_{z_j,g_j}(y_j,x_j) + \varepsilon_j$ and $\varepsilon_j \to 0$. We have $t_{z_j}(y) = f(z_j) + g_j^\top(y - z_j)$. Passing to a subsequence, we may assume $z_j \to z$ and $g_j \to g \in \partial_{[\varepsilon]} f(z)$. That means $t_{z_j}(\cdot) \to t_z(\cdot)$, and since $y_j \to y$ also $t_{z_j}(y_j) \to t_z(y)$. In order to conclude for the $m_{z_j,g_j}(\cdot,x_j)$ we have to see how the downshift behaves. We have indeed $s_j \to s$, where s is the downshift at z with respect to the

approximate subgradient g and serious iterate x. Therefore $m_{z,g}(\cdot,x) = t_z(\cdot) - s$. This shows $m_{z_j,g_j}(\cdot,x_j) = t_{z_j}(\cdot) - s_j \to t_z(\cdot) - s = m_{z,g}(\cdot,x)$ as $j \to \infty$, and then also $m_{z_j,g_j}(y_j,x_j) = t_{z_j}(y_j) - s_j \to t_z(y) - s = m_{z,g}(y,x)$, where uniformity comes from boundedness of the g_j. This implies $\lim m_{z_j,g_j}(y_j,x_j) = m_{z,g}(y,x) \leq \phi(y,x)$ as required.

(5) The inequality $\phi_k \leq \phi$ is clear, because $\phi_k(\cdot,x)$ is built from cutting planes $m_k(\cdot,x)$, and all these cutting planes are below the envelope $\phi(\cdot,x)$. ■

Remark 26.8. In [40, 42] the case $\varepsilon = 0$ is discussed and a function $\phi(\cdot,x)$ with the properties in Lemma 26.7 is called a first-order model of f at x. It can be understood as a generalized first-order Taylor expansion of f at x. Every locally Lipschitz function f has the standard or Clarke model $\phi^\sharp(y,x) = f(x) + f^0(x,y-x)$, where $f^0(x,d)$ is the Clarke directional derivative at x. In the present situation it is reasonable to call $\phi(\cdot,x)$ an ε-model of f at x.

Following [34] a function f is called ε-convex on an open convex set U if $f(tx + (1-t)y) \leq tf(x) + (1-t)f(y) + \varepsilon t(1-t)\|x-y\|$ for all $x,y \in U$ and $0 \leq t \leq 1$. Every ε-convex function satisfies $f'(y,x-y) \leq f(x) - f(y) + \varepsilon\|x-y\|$; hence for $g \in \partial f(y)$,

$$g^\top(x-y) \leq f(x) - f(y) + \varepsilon\|x-y\|. \tag{26.14}$$

A function f is called approximate convex if for every x and $\varepsilon > 0$ there exists $\delta > 0$ such that f is ε-convex on $B(x,\delta)$. Using results from [14, 34] one may show that approximate convex functions coincide with lower C^1 function in the sense of Spingarn [51].

Lemma 26.9. *Suppose the inner loop turns forever and $\tau_k \to \infty$.*

1. *If f is ε'-convex on a set containing all y^k, $k \geq k_0$, then $0 \in \partial_{[\tilde\varepsilon]} f(x)$, where $\tilde\varepsilon = \varepsilon + (\varepsilon' + \varepsilon)/(\tilde\gamma - \gamma)$.*
2. *If f is lower C^1, then $0 \in \partial_{[\alpha\varepsilon]} f(x)$, where $\alpha = 1 + (\tilde\gamma - \gamma)^{-1}$.*

Proof.

(i) The second statement follows from the first, because every lower C^1 function is approximate convex, hence ε'-convex on a suitable neighborhood of x. We therefore concentrate on the first statement.

(ii) By assumption none of the trial steps is accepted, so that $\rho_k < \gamma$ for all $k \in \mathbb{N}$. Since τ_k is increased infinitely often, there are infinitely many inner loop instances k where $\tilde\rho_k \geq \tilde\gamma$. Let us prove that under these circumstances $y^k \to x$. Recall that $g_k^* = (Q + \tau_k I)(x - y^k) \in \partial\phi_k(y^k,x)$. By the subgradient inequality this gives

$$g_k^{*\top}(x-y^k) \leq \phi_k(x,x) - \phi_k(y^k,x). \tag{26.15}$$

Now use $\phi_k(x,x) = f(x)$ and observe that $m_0(y^k,x) \leq \phi_k(y^k,x)$, where $m_0(\cdot,x)$ is the exactness plane. Since $m_0(y,x) = f(x) + g(x)^\top(y-x)$ for some $g(x) \in \partial_{[\varepsilon]} f(x)$, expanding the term on the left of (26.15) gives

$$(x-y^k)^\top(Q+\tau_k I)(x-y^k) \leq g(x)^\top(x-y^k) \leq \|g(x)\|\|x-y^k\|. \quad (26.16)$$

Since $\tau_k \to \infty$, the term on the left-hand side of (26.16) behaves asymptotically like $\tau_k\|x-y^k\|^2$. Dividing (26.16) by $\|x-y^k\|$ therefore shows that $\tau_k\|x-y^k\|$ is bounded by $\|g(x)\|$. As $\tau_k \to \infty$, this could only mean $y^k \to x$.

(iii) Let us use $y^k \to x$ and go back to formula (26.15). Since the left hand side of (26.15) tends to 0 and $\phi_k(x,x) = f(x)$, we see that the limit superior of $\phi_k(y^k,x)$ is $f(x)$. On the other hand, $\phi_k(y^k,x) \geq m_0(y^k,x)$, where $m_0(\cdot,x)$ is the exactness plane. Since clearly $m_0(y^k,x) \to m_0(x,x) = f(x)$, the limit inferior is also $f(x)$, and we conclude that $\phi_k(y^k,x) \to f(x)$.

Keeping this in mind, let us use the subgradient inequality (26.15) again and subtract a term $\frac{1}{2}(x-y^k)^\top Q(x-y^k)$ from both sides. That gives the estimate

$$\tfrac{1}{2}(x-y^k)^\top Q(x-y^k) + \tau_k\|x-y^k\|^2 \leq f(x) - \Phi_k(y^k,x).$$

Fix $0 < \zeta < 1$. Using $\tau_k \to \infty$ we have

$$(1-\zeta)\tau_k\|x-y^k\| \leq \|g_k^*\| \leq (1+\zeta)\tau_k\|x-y^k\|$$

and also

$$\tfrac{1}{2}(x-y^k)^\top Q(x-y^k) + \tau_k\|x-y^k\|^2 \geq (1-\zeta)\tau_k\|x-y^k\|^2$$

for sufficiently large k. Therefore,

$$f(x) - \Phi_k(y^k,x) \geq \tfrac{1-\zeta}{1+\zeta}\|g_k^*\|\|x-y^k\| \quad (26.17)$$

for k large enough.

(iv) Now let $\eta_k := \operatorname{dist}\left(g_k^*, \partial\phi(x,x)\right)$. We argue that $\eta_k \to 0$. Indeed, using the subgradient inequality at y^k in tandem with $\phi(\cdot,x) \geq \phi_k(\cdot,x)$, we have for all $y \in \mathbb{R}^n$

$$\phi(y,x) \geq \phi_k(y^k,x) + g_k^{*\top}(y-y^k).$$

Here our upper envelope function (26.13) is defined such that the ball $B(0,M)$ contains x and all trial points y^k at which cutting planes are drawn.

Since the subgradients g_k^* are bounded by part (ii), there exists an infinite subsequence $\mathcal{N} \subset \mathbb{N}$ such that $g_k^* \to g^*$, $k \in \mathcal{N}$, for some g^*. Passing to the limit $k \in \mathcal{N}$ and using $y^k \to x$ and $\phi_k(y^k,x) \to f(x) = \phi(x,x)$, we have $\phi(y,x) \geq \phi(x,x) + g^{*\top}(y-x)$ for all y. Hence $g^* \in \partial\phi(x,x)$, which means $\eta_k = \operatorname{dist}(g_k^*, \partial\phi(x,x)) \leq \|g_k^* - g^*\| \to 0$, $k \in \mathcal{N}$, proving the argument.

(v) Using the definition of η_k, choose $\tilde{g}_k \in \partial\phi(x,x)$ such that $\|g_k^* - \tilde{g}_k\| = \eta_k$. Now let $\mathrm{dist}(0, \partial\phi(x,x)) = \eta$. Then $\|\tilde{g}_k\| \geq \eta$ for all $k \in \mathscr{N}$. Hence $\|g_k^*\| \geq \eta - \eta_k > (1-\zeta)\eta$ for $k \in \mathscr{N}$ large enough, given that $\eta_k \to 0$ by (iv). Going back with this to (26.17) we deduce

$$f(x) - \Phi_k(y^k,x) \geq \tfrac{(1-\zeta)^2}{1+\zeta}\eta\|x-y^k\| \tag{26.18}$$

for $k \in \mathscr{N}$ large enough.

(vi) We claim that $f(y^k) \leq M_k(y^k,x) + (1+\zeta)(\varepsilon' + \varepsilon)\|x - y^k\|$ for all k sufficiently large. Indeed, we have $m_k(\cdot,x) = t_k(\cdot) - s_k$, where s_k is the downshift of the approximate tangent $t_k(\cdot)$ at y^k, $g_{\varepsilon k} \in \partial_{[\varepsilon]} f(y^k)$, with regard to the serious iterate x. There are two cases. Assume first that $t_k(x) > f(x)$. Then

$$\begin{aligned} m_k(y,x) &= f(y^k) + g_{\varepsilon k}^\top (y - y^k) - s_k \\ &= f(y^k) + g_{\varepsilon k}^\top (y - y^k) - c\|x - y^k\|^2 - t_k(x) + f(x) \\ &= f(x) + g_{\varepsilon k}^\top (y - x) - c\|x - y^k\|^2. \end{aligned}$$

In consequence

$$\begin{aligned} f(y^k) - m_k(y^k,x) &= f(y^k) - f(x) - g_{\varepsilon k}^\top (y^k - x) + c\|x - y^k\|^2 \\ &= f(y^k) - f(x) - g_k^\top (y^k - x) + (g_k - g_{\varepsilon k})^\top (x - y^k) + c\|x - y^k\|^2. \end{aligned}$$

Now since f is ε'-convex, estimate (26.14) is valid under the form

$$g_k^\top (x - y^k) \leq f(x) - f(y^k) + \varepsilon'\|x - y^k\|.$$

We therefore get

$$f(y^k) - m_k(y^k,x) \leq (\varepsilon' + \varepsilon)\|x - y^k\| + c\|x - y^k\|^2.$$

Subtracting a term $\frac{1}{2}(x - y^k)^\top Q(x - y^k)$ on both sides gives

$$f(y^k) - M_k(y^k,x) \leq (\varepsilon' + \varepsilon + \nu_k)\|x - y^k\|,$$

where $\nu_k := c\|x - y^k\|^2 - \frac{1}{2}(x - y^k)^\top Q(x - y^k) \to 0$ and $M_k(y,x) = m_k(y,x) + \frac{1}{2}(y - x)^\top Q(y - x)$. Therefore

$$f(y^k) - M_k(y^k,x) \leq (1+\zeta)(\varepsilon' + \varepsilon)\|x - y^k\| \tag{26.19}$$

for k large enough.

Now consider the second case $t_k(x) \leq f(x)$. Here we get an even better estimate than (26.19), because $s_k = c\|x - y^k\|^2$, so that $f(y^k) - m_k(y^k,x) = c\|x - y^k\|^2 \leq \varepsilon\|x - y^k\|$ for k large enough.

(vii) To conclude, using (26.18) and (26.19) we expand the coefficient $\tilde{\rho}_k$ as

$$\tilde{\rho}_k = \rho_k + \frac{f(y^k) - M_k(y^k,x)}{f(x) - \Phi_k(y^k,x)}$$
$$\leq \rho_k + \frac{(1+\zeta)^2(\varepsilon'+\varepsilon)\|x-y^k\|}{(1-\zeta)^2\eta\|x-y^k\|} = \rho_k + \frac{(1+\zeta)^2(\varepsilon'+\varepsilon)}{(1-\zeta)^2\eta}.$$

This shows

$$\eta < \frac{(1+\zeta)^2(\varepsilon'+\varepsilon)}{(1-\zeta)^2(\tilde{\gamma}-\gamma)}.$$

For suppose we had $\eta \geq \frac{(1+\zeta)^2(\varepsilon'+\varepsilon)}{(1-\zeta)^2(\tilde{\gamma}-\gamma)}$, then $\tilde{\rho}_k \leq \rho_k + (\tilde{\gamma}-\gamma) \leq \tilde{\gamma}$ for all k, contradicting $\tilde{\rho}_k > \tilde{\gamma}$ for infinitely many k. As $0 < \zeta < 1$ was arbitrary, we have the estimate $\eta \leq \frac{\varepsilon'+\varepsilon}{\tilde{\gamma}-\gamma}$. Since $\partial\phi(x,x) = \partial f(x) + \varepsilon B$ by Lemma 26.7, we deduce $0 \in \partial\phi(x,x) + \eta B \subset \partial f(x) + (\varepsilon+\eta)B$, and this is the result claimed in statement 1. ■

Remark 26.10. Suppose we choose γ very small and $\tilde{\gamma}$ close to 1, then $\alpha = 2+\xi$ for some small ξ, so roughly $\alpha \approx 2$.

Lemma 26.11. *Suppose the inner loop turns forever and τ_k is frozen. Then $y^k \to x$ and $0 \in \partial_{[\varepsilon]} f(x)$.*

Proof.

(i) The control parameter is frozen from counter k_0 onwards, and we put $\tau := \tau_k$, $k \geq k_0$. This means that $\rho_k < \gamma$ and $\tilde{\rho}_k < \tilde{\gamma}$ for all $k \geq k_0$.

(ii) We prove that the sequence of trial steps y^k is bounded. Notice that

$$g_k^{*\top}(x-y^k) \leq \phi_k(x,x) - \phi_k(y^k,x)$$

by the subgradient inequality at y^k and the definition of the aggregate subgradient. Now observe that $\phi_k(x,x) = f(x)$ and $\phi_k(y^k,x) \geq m_0(y^k,x)$. Therefore, using the definition of g_k^*, we have

$$(x-y^k)^\top(Q+\tau I)(x-y^k) \leq f(x) - m_0(y^k,x) = g(x)^\top(x-y^k) \leq \|g(x)\|\|x-y^k\|.$$

Since the τ-parameter is frozen and $Q+\tau I \succ 0$, the expression on the left is the square $\|x-y^k\|^2_{Q+\tau I}$ of the Euclidean norm derived from $Q+\tau I$. Since both norms are equivalent, we deduce after dividing by $\|x-y^k\|$ that $\|x-y^k\|_{Q+\tau I} \leq C\|g(x)\|$ for some constant $C > 0$ and all k. This proves the claim.

(iii) Let us introduce the objective function of tangent program (26.7) for $k \geq k_0$:

$$\psi_k(\cdot,x) = \phi_k(\cdot,x) + \tfrac{1}{2}(\cdot - x)^\top(Q+\tau I)(\cdot - x).$$

Let $m_k^*(\cdot,x)$ be the aggregate plane, then $\phi_k(y^k,x) = m_k^*(y^k,x)$ by (26.9) and therefore also

$$\psi_k(y^k,x) = m_k^*(y^k,x) + \tfrac{1}{2}(y^k - x)^\top(Q+\tau I)(y^k - x).$$

We introduce the quadratic function $\psi_k^*(\cdot,x) = m_k^*(\cdot,x) + \frac{1}{2}(\cdot - x)^\top(Q+\tau I)(\cdot - x)$. Then

$$\psi_k(y^k,x) = \psi_k^*(y^k,x) \tag{26.20}$$

by what we have just seen. By construction of model $\phi_{k+1}(\cdot,x)$ we have $m_k^*(y,x) \leq \phi_{k+1}(y,x)$, so that

$$\psi_k^*(y,x) \leq \psi_{k+1}(y,x). \tag{26.21}$$

Notice that $\nabla\psi_k^*(y,x) = \nabla m_k^*(y,x) + (Q+\tau I)(y-x) = g_k^* + (Q+\tau I)(y-x)$, so that $\nabla\psi_k^*(y^k,x) = 0$ by (26.8). We therefore have the relation

$$\psi_k^*(y,x) = \psi_k^*(y^k,x) + \tfrac{1}{2}(y-y^k)^\top(Q+\tau I)(y-y^k), \tag{26.22}$$

which is obtained by Taylor expansion of $\psi_k^*(\cdot,x)$ at y^k. Recall that step 8 of the algorithm assures $Q+\tau I \succ 0$, so that the quadratic expression defines the Euclidean norm $\|\cdot\|_{Q+\tau I}$.

(iv) From the previous point (iii) we now have

$$\begin{aligned}
\psi_k(y^k,x) &\leq \psi_k^*(y^k,x) + \tfrac{1}{2}\|y^k - y^{k+1}\|_{Q+\tau I}^2 && [\text{using (26.20)}]\\
&= \psi_k^*(y^{k+1},x) && [\text{using (26.22)}]\\
&\leq \psi_{k+1}(y^{k+1},x) && [\text{using (26.21)}]\\
&\leq \psi_{k+1}(x,x) && (y^{k+1} \text{ minimizer of } \psi_{k+1})\\
&= \phi_{k+1}(x,x) = f(x).
\end{aligned} \tag{26.23}$$

We deduce that the sequence $\psi_k(y^k,x)$ is monotonically increasing and bounded above by $f(x)$. It therefore converges to some value $\psi^* \leq f(x)$.

Going back to (26.23) with this information shows that the term $\frac{1}{2}\|y^k - y^{k+1}\|_{Q+\tau I}^2$ is squeezed in between two convergent terms with the same limit, ψ^*, which implies $\frac{1}{2}\|y^k - y^{k+1}\|_{Q+\tau I}^2 \to 0$. Consequently, $\|y^k - x\|_{Q+\tau I}^2 - \|y^{k+1} - x\|_{Q+\tau I}^2$ also tends to 0, because the sequence of trial steps y^k is bounded by part (ii).

Recalling $\phi_k(y,x) = \psi_k(y,x) - \frac{1}{2}\|y-x\|_{Q+\tau I}^2$, we deduce, using both convergence results, that

$$\begin{aligned}
&\phi_{k+1}(y^{k+1},x) - \phi_k(y^k,x)\\
&\quad = \psi_{k+1}(y^{k+1},x) - \psi_k(y^k,x) - \tfrac{1}{2}\|y^{k+1} - x\|_{Q+\tau I}^2 + \tfrac{1}{2}\|y^k - x\|_{Q+\tau I}^2 \to 0.
\end{aligned} \tag{26.24}$$

(v) We want to show that $\phi_k(y^k,x) - \phi_{k+1}(y^k,x) \to 0$ and then of course also $\Phi_k(y^k,x) - \Phi_{k+1}(y^k,x) \to 0$.

Recall that by construction the cutting plane $m_k(\cdot,x)$ is an affine support function of $\phi_{k+1}(\cdot,x)$ at y^k. By the subgradient inequality this implies

$$g_k^\top(y-y^k) \leq \phi_{k+1}(y,x) - \phi_{k+1}(y^k,x) \tag{26.25}$$

for all y. Therefore

$$\begin{aligned} 0 &\leq \phi_{k+1}(y^k,x) - \phi_k(y^k,x) && \text{(using aggregation)} \\ &= \phi_{k+1}(y^k,x) + g_k^\top(y^{k+1}-y^k) - \phi_k(y^k,x) - g_k^\top(y^{k+1}-y^k) \\ &\leq \phi_{k+1}(y^{k+1},x) - \phi_k(y^k,x) + \|g_k\|\|y^{k+1}-y^k\| && \text{[using (26.25)]} \end{aligned}$$

and this term converges to 0, because of (26.24), because the g_k are bounded, and because $y^k - y^{k+1} \to 0$ according to part (iv) above. Boundedness of the g_k follows from boundedness of the trial steps y^k shown in part (ii). Indeed, $g_k \in \partial f(y^k) + \varepsilon B$, and the subdifferential of f is uniformly bounded on the bounded set $\{y^k : k \in \mathbb{N}\}$. We deduce that $\phi_{k+1}(y^k,x) - \phi_k(y^k,x) \to 0$. Obviously, that also gives $\Phi_{k+1}(y^k,x) - \Phi_k(y^k,x) \to 0$.

(vi) We now proceed to prove $\Phi_k(y^k,x) \to f(x)$ and then also $\Phi_{k+1}(y^k,x) \to f(x)$. Assume this is not the case, then $\limsup_{k\to\infty} f(x) - \Phi_k(y^k,x) =: \eta > 0$. Choose $\delta > 0$ such that $\delta < (1-\tilde{\gamma})\eta$. It follows from (v) above that there exists $k_1 \geq k_0$ such that

$$\Phi_{k+1}(y^k,x) - \delta \leq \Phi_k(y^k,x)$$

for all $k \geq k_1$. Using $\tilde{\rho}_k \leq \tilde{\gamma}$ for all $k \geq k_0$ then gives

$$\tilde{\gamma}\Big(\Phi_k(y^k,x) - f(x)\Big) \leq \Phi_{k+1}(y^k,x) - f(x) \leq \Phi_k(y^k,x) + \delta - f(x).$$

Passing to the limit implies $-\tilde{\gamma}\eta \leq -\eta + \delta$, contradicting the choice of δ. This proves $\eta = 0$.

(vii) Having shown $\Phi_k(y^k,x) \to f(x)$ and therefore also $\Phi_{k+1}(y^k,x) \to f(x)$, we now argue that $y^k \to x$. This follows from the definition of ψ_k, because

$$\Phi_k(y^k,x) \leq \psi_k(y^k,x) = \Phi_k(y^k,x) + \tfrac{\tau}{2}\|y^k - x\|^2 \leq \psi^* \leq f(x).$$

Since $\Phi_k(y^k,x) \to f(x)$ by part (vi), we deduce $\frac{\tau}{2}\|y^k - x\|^2 \to 0$ using a sandwich argument, which also proves *en passant* that $\psi^* = f(x)$ and $\phi_k(y^k,x) \to f(x)$.

To finish the proof, let us now show $0 \in \partial_{[\varepsilon]} f(x)$. Remember that by the necessary optimality condition for (26.7) we have $(Q + \tau I)(x - y^k) \in \partial\phi_k(y^k,x)$. By the subgradient inequality,

$$\begin{aligned}(x-y^k)^\top(Q+\tau I)(y-y^k) &\leq \phi_k(y,x)-\phi_k(y^k,x)\\ &\leq \phi(y,x)-\phi_k(y^k,x),\end{aligned}$$

where ϕ is the upper envelope (26.13) of all cutting planes drawn at $z \in B(0,M)$, $g \in \partial_{[\varepsilon]}f(z)$, which we choose large enough to contain the bounded set $\{x\} \cup \{y^k : k \in \mathbb{N}\}$, a fact which assures $\phi_k(\cdot,x) \leq \phi(\cdot,x)$ for all k (see Lemma 26.7). Passing to the limit, observing $\|x-y^k\|^2_{Q+\tau I} \to 0$ and $\phi_k(y^k,x) \to f(x) = \phi(x,x)$, we obtain

$$0 \leq \phi(y,x)-\phi(x,x)$$

for all y. This proves $0 \in \partial\phi(x,x)$. Since $\partial\phi(x,x) \subset \partial_{[\varepsilon]}f(x)$ by Lemma 26.7, we have shown $0 \in \partial_{[\varepsilon]}f(x)$. ■

26.6 Convergence of the Outer Loop

In this section we prove subsequence convergence of our algorithm for the case where function values are exact and subgradients are in $\partial_{[\varepsilon]}f(y^k)$. We write $Q_j = Q(x^j)$ for the matrix of the second-order model, which depends on the serious iterates x^j.

Theorem 26.12. *Let x^1 be such that $\Omega = \{x \in \mathbb{R}^n : f(x) \leq f(x^1)\}$ is bounded. Suppose f is ε'-convex on Ω and that subgradients are drawn from $\partial_{[\varepsilon]}f(y)$, whereas function values are exact. Then every accumulation point $\bar{x}$ of the sequence of serious iterates x^j satisfies $0 \in \partial_{[\tilde{\varepsilon}]}f(\bar{x})$, where $\tilde{\varepsilon} = \varepsilon + (\varepsilon'+\varepsilon)/(\gamma-\tilde{\gamma})$.*

Proof.

(i) From the analysis in Sect. 26.5 we know that if we apply the stopping test in step 2 with $\tilde{\varepsilon} = \varepsilon + (\varepsilon'+\varepsilon)/(\gamma-\tilde{\gamma})$, then the inner loop ends after a finite number of steps k with a new x^+ satisfying the acceptance test in step 5, unless we have finite termination due to $0 \in \partial_{[\tilde{\varepsilon}]}f(x)$. Let us exclude this case, and let x^j denote the infinite sequence of serious iterates. We assume that at outer loop counter j the inner loop finds a serious step at inner loop counter $k = k_j$. In other words, $y^{k_j} = x^{j+1}$ passes the acceptance test in step 5 of the algorithm and becomes a serious iterate, while the y^k with $k < k_j$ are null steps. That means

$$f(x^j)-f(x^{j+1}) \geq \gamma\left(f(x^j)-\Phi_{k_j}(x^{j+1},x^j)\right). \tag{26.26}$$

Now recall that $(Q_j + \tau_{k_j} I)(x^j - x^{j+1}) \in \partial \phi_{k_j}(x^{j+1}, x^j)$ by optimality of the tangent program (26.7). The subgradient inequality for $\phi_{k_j}(\cdot, x^j)$ at x^{j+1} therefore gives

$$\begin{aligned}(x^j - x^{j+1})^\top (Q_j + \tau_{k_j} I)(x^j - x^{j+1}) &\le \phi_{k_j}(x^j, x^j) - \phi_{k_j}(x^{j+1}, x^j)\\ &= f(x^j) - \phi_{k_j}(x^{j+1}, x^j),\end{aligned}$$

using $\phi_{k_j}(x^j, x^j) = f(x^j)$. With $\Phi_k(y, x^j) = \phi_k(y, x^j) + \frac{1}{2}(y - x^j)^\top Q_j (y - x^j)$ we have

$$\tfrac{1}{2}\|x^{j+1} - x^j\|^2_{Q_j + \tau_{k_j} I} \le f(x^j) - \Phi_{k_j}(x^{j+1}, x^j) \le \gamma^{-1}\left(f(x^j) - f(x^{j+1})\right), \tag{26.27}$$

using (26.26). Summing (26.27) from $j = 1$ to $j = J$ gives

$$\sum_{j=1}^{J} \|x^{j+1} - x^j\|^2_{Q_j + \tau_{k_j} I} \le \gamma^{-1} \sum_{j=1}^{J} \left(f(x^j) - f(x^{j+1})\right) = \gamma^{-1}\left(f(x^1) - f(x^{J+1})\right).$$

Here the right-hand side is bounded above because our method is of descent type in the serious steps and Ω is bounded. Consequently the series on the left is summable, and therefore $\|x^{j+1} - x^j\|^2_{Q_j + \tau_{k_j} I} \to 0$ as $j \to \infty$. Let $\bar{x}$ be an accumulation point of the sequence x^j. We have to prove $0 \in \partial_{[\tilde{\varepsilon}]} f(\bar{x})$. We select a subsequence $j \in J$ such that $x^j \to \bar{x}$, $j \in J$. There are now two cases. The first is discussed in part (ii); the second is more complicated and will be discussed in (iii)–(ix).

(ii) Suppose there exists an infinite subsequence J' of J such that $g_j := (Q_j + \tau_{k_j} I)\left(x^j - x^{j+1}\right)$ converges to 0, $j \in J'$. We will show that in this case $0 \in \partial_{[\tilde{\varepsilon}]} f(\bar{x})$.

In order to prove this claim, notice first that since $\Omega = \{x \in \mathbb{R}^n : f(x) \le f(x^1)\}$ is bounded by hypothesis, and since our algorithm is of descent type in the serious steps, the sequence x^j, $j \in \mathbb{N}$ is bounded. We can therefore use the convex upper envelope function ϕ of (26.13), where $B(0, M)$ contains Ω and also *all* the trial points y^k visited during all inner loops j.

Indeed, the set of x^j being bounded, so are the $\|g(x^j)\|$, where $g(x^j) \in \partial_{[\varepsilon]} f(x^j)$ is the exactness subgradient of the jth inner loop. From (26.16) we know that $\|x^j - y^k\|_{Q_j + \tau_k I} \le \|g(x^j)\|$ for every j and every trial step y^k arising in the jth inner loop at some instant k. From the management of the τ-parameter in the outer loop (26.12) we know that $Q_j + \tau_k I \succ \zeta I$ for some $\zeta > 0$, so $\|x^j - y^k\| \le \zeta^{-1}\|g(x^j)\| \le C < \infty$, meaning the y^k are bounded. During the following the properties of ϕ obtained in Lemma 26.7 will be applied at every $x = x^j$.

Since g_j is a subgradient of $\phi_{k_j}(\cdot,x^j)$ at $x^{j+1}=y^{k_j+1}$, we have for every test vector h

$$\begin{aligned} g_j^\top h &\le \phi_{k_j}(x^{j+1}+h,x^j)-\phi_{k_j}(x^{j+1},x^j)\\ &\le \phi(x^{j+1}+h,x^j)-\phi_{k_j}(x^{j+1},x^j) \qquad [\text{using } \phi_{k_j}(\cdot,x^j)\le\phi(\cdot,x^j)]. \end{aligned}$$

Now $y^{k_j}=x^{j+1}$ was accepted in step 5 of the algorithm, which means

$$\gamma^{-1}\left(f(x^j)-f(x^{j+1})\right)\ge f(x^j)-\Phi_{k_j}(x^{j+1},x^j).$$

Combining these two estimates for a fixed test vector h gives

$$\begin{aligned} g_j^\top h &\le \phi(x^{j+1}+h,x^j)-f(x^j)+f(x^j)-\phi_{k_j}(x^{j+1},x^j)\\ &= \phi(x^{j+1}+h,x^j)-f(x^j)+f(x^j)-\Phi_{k_j}(x^{j+1},x^j)\\ &\quad +\tfrac{1}{2}(x^j-x^{j+1})^\top Q_j(x^j-x^{j+1})\\ &\le \phi(x^{j+1}+h,x^j)-f(x^j)+\gamma^{-1}\left(f(x^j)-f(x^{j+1})\right)\\ &\quad +\tfrac{1}{2}(x^j-x^{j+1})^\top Q_j(x^j-x^{j+1})\\ &= \phi(x^{j+1}+h,x^j)-f(x^j)+\gamma^{-1}\left(f(x^j)-f(x^{j+1})\right)\\ &\quad +\tfrac{1}{2}(x^j-x^{j+1})^\top (Q_j+\tau_{k_j}I)(x^j-x^{j+1})-\tfrac{\tau_{k_j}}{2}\|x^j-x^{j+1}\|^2\\ &\le \phi(x^{j+1}+h,x^j)-f(x^j)+\gamma^{-1}\left(f(x^j)-f(x^{j+1})\right)\\ &\quad +\tfrac{1}{2}(x^j-x^{j+1})^\top (Q_j+\tau_{k_j}I)(x^j-x^{j+1}). \end{aligned}$$

Now fix $h'\in\mathbb{R}^n$. Plugging $h=x^j-x^{j+1}+h'$ in the above estimate gives

$$\tfrac{1}{2}\|x^j-x^{j+1}\|^2_{Q_j+\tau_{k_j}I}+g_j^\top h'\le\phi(x^j+h',x^j)-f(x^j)+\gamma^{-1}\left(f(x^j)-f(x^{j+1})\right). \tag{26.28}$$

Passing to the limit $j\in J'$ and using, in the order named, $\|x^j-x^{j+1}\|^2_{Q_j+\tau_{k_j}I}\to 0$, $g_j\to 0$, $x^j\to\bar{x}$, $f(x^j)\to f(\bar{x})=\phi(\bar{x},\bar{x})$ and $f(x^j)-f(x^{j+1})\to 0$, we obtain

$$0\le\phi(\bar{x}+h',\bar{x})-\phi(\bar{x},\bar{x}). \tag{26.29}$$

In (26.28) the rightmost term $f(x^j)-f(x^{j+1})\to 0$ converges by monotonicity, convergence of the leftmost term was shown in part (i), and $g_j\to 0$ is the working hypothesis. Now the test vector h' in (26.29) is arbitrary, which shows $0\in\partial\phi(\bar{x},\bar{x})$. By Lemma 26.7 we have $0\in\partial_{[\varepsilon]}f(\bar{x})\subset\partial_{[\bar{\varepsilon}]}f(\bar{x})$.

(iii) The second more complicated case is when $\|g_j\| = \|(Q_j + \tau_{k_j}I)(x^j - x^{j+1})\| \geq \mu > 0$ for some $\mu > 0$ and every $j \in J$. The remainder of this proof will be entirely dedicated to this case.

We notice first that under this assumption the τ_{k_j}, $j \in J$, must be unbounded. Indeed, assume on the contrary that the τ_{k_j}, $j \in J$, are bounded. By boundedness of Q_j and boundedness of the serious steps, there exists then an infinite subsequence $j \in J'$ of J such that Q_j, τ_{k_j}, and $x^j - x^{j+1}$ converge respectively to $\bar{Q}$, $\bar{\tau}$, and $\delta\bar{x}$ as $j \in J'$. This implies that the corresponding subsequence of g_j converges to $(\bar{Q} + \bar{\tau}I)\delta\bar{x}$, where $\|(\bar{Q} + \bar{\tau}I)\delta\bar{x}\| \geq \mu > 0$. Similarly, $(x^j - x^{j+1})^\top(Q_j + \tau_{k_j}I)(x^j - x^{j+1}) \to \delta\bar{x}^\top(\bar{Q} + \bar{\tau}I)\delta\bar{x}$. By part (i) of the proof we have $g_j^\top(x^{j+1} - x^j) = \|x^{j+1} - x^j\|^2_{Q_j + \tau_{k_j}I} \to 0$, which means $\delta\bar{x}^\top(\bar{Q} + \bar{\tau}I)\delta\bar{x} = 0$. Since $\bar{Q} + \bar{\tau}I$ is symmetric and $\bar{Q} + \bar{\tau}I \succeq 0$, we deduce $(\bar{Q} + \bar{\tau}I)\delta\bar{x} = 0$, contradicting $\|(\bar{Q} + \bar{\tau}I)\delta\bar{x}\| \geq \mu > 0$. This argument proves that the τ_{k_j}, $j \in J$, are unbounded.

(iv) Having shown that the sequence τ_{k_j}, $j \in J$ is unbounded, we can without loss assume that $\tau_{k_j} \to \infty$, $j \in J$, passing to a subsequence if required. Let us now distinguish two types of indices $j \in J$. We let J^+ be the set of those $j \in J$ for which the τ-parameter was increased at least once during the jth inner loop. The remaining indices $j \in J^-$ are those where the τ-parameter remained unchanged during the jth inner loop. Since the jth inner loop starts at $\tau_j^\sharp$ and ends at τ_{k_j}, we have

$$J^+ = \{j \in J : \tau_{k_j} < \tau_j^\sharp\} \text{ and } J^- = \{j \in J : \tau_{k_j} = \tau_j^\sharp\}.$$

We claim that the set J^- must be finite. For suppose J^- is infinite, then $\tau_{k_j} \to \infty$, $j \in J^-$. Hence also $\tau_j^\sharp \to \infty$, $j \in J^-$. But this contradicts the rule in step 8 of the algorithm, which forces $\tau_j^\sharp \leq T < \infty$. This contradiction shows that J^+ is cofinal in J.

(v) Remember that we are still in the case whose discussion started in point (iii). We are now dealing with an infinite subsequence $j \in J^+$ of $j \in J$ such that $\tau_{k_j} \to \infty$, $\|g_j\| \geq \mu > 0$, and such that the τ-parameter was increased at least once during the jth inner loop. Suppose this happened for the last time at stage $k_j - \nu_j$ for some $\nu_j \geq 1$. Then

$$\tau_{k_j} = \tau_{k_j - 1} = \cdots = \tau_{k_j - \nu_j + 1} = 2\tau_{k_j - \nu_j}. \tag{26.30}$$

According to step 6 of the algorithm, the increase at counter $k_j - \nu_j$ is due to the fact that

$$\rho_{k_j - \nu_j} < \gamma \text{ and } \tilde{\rho}_{k_j - \nu_j} \geq \tilde{\gamma}. \tag{26.31}$$

This case is labelled *too bad* in step 6 of the algorithm.

(vi) Condition (26.31) means that there are infinitely many $j \in J^+$ satisfying

$$\rho_{k_j-\nu_j} = \frac{f(x^j) - f(y^{k_j-\nu_j})}{f(x^j) - \Phi_{k_j-\nu_j}(y^{k_j-\nu_j}, x^j)} < \gamma$$

and

$$\tilde{\rho}_{k_j-\nu_j} = \frac{f(x^j) - M_{k_j-\nu_j}(y^{k_j-\nu_j}, x^j)}{f(x^j) - \Phi_{k_j-\nu_j}(y^{k_j-\nu_j+1}, x^j)} \geq \tilde{\gamma}.$$

Notice first that as $\tau_{k_j} \to \infty$ and $\tau_{k_j} = 2\tau_{k_j-\nu_j}$, boundedness of the subgradients $\tilde{g}_j := (Q_j + \frac{1}{2}\tau_{k_j} I)(x^j - y^{k_j-\nu_j}) \in \partial\phi_{k_j-\nu_j}(y^{k_j-\nu_j}, x^j)$ shows $y^{k_j-\nu_j} \to \bar{x}$. Indeed, boundedness of the $\tilde{g}_j$ follows from the subgradient inequality

$$\begin{aligned}(x^j - y^{k_j-\nu_j})^\top (Q_j + \tau_{k_j-\nu_j} I)(x^j - y^{k_j-\nu_j}) &\leq \phi_{k_j-\nu_j}(x^j, x^j) - \phi_{k_j-\nu_j}(y^{k_j-\nu_j}, x^j) \\ &\leq f(x^j) - m_0(y^{k_j-\nu_j}, x^j) \\ &= g(x^j)^\top (x^j - y^{k_j-\nu_j}) \\ &\leq \|g(x^j)\| \|x^j - y^{k_j-\nu_j}\|, \qquad (26.32)\end{aligned}$$

where $m_0(\cdot, x^j) = f(x^j) + g(x^j)^\top(\cdot - x^j)$ is the exactness plane at x^j. As $\tau_{k_j} \to \infty$, we have $\tau_{k_j-\nu_j} = \frac{1}{2}\tau_{k_j} \to \infty$, too, so the left-hand side of (26.32) behaves asymptotically like constant times $\tau_{k_j-\nu_j}\|x^j - y^{k_j-\nu_j}\|^2$. On the other hand the $x^j \in \Omega$ are bounded, hence so are the $g(x^j)$. The right-hand side therefore behaves asymptotically like constant times $\|x^j - y^{k_j-\nu_j}\|$. This shows boundedness of $\tau_{k_j-\nu_j}\|x^j - y^{k_j-\nu_j}\|$, and therefore $x^j - y^{k_j-\nu_j} \to 0$, because $\tau_{k_j-\nu_j} \to \infty$.

(vii) Recall that $x^j \to \bar{x}$, $j \in J$. By (vi) we know that $y^{k_j-\nu_j} \to \bar{x}$, $j \in J$. Passing to a subsequence J' of J, we may assume $\tilde{g}_j \to \tilde{g}$ for some $\tilde{g}$. We show $\tilde{g} \in \partial\phi(\bar{x}, \bar{x})$.

For a test vector h and $j \in J'$,

$$\begin{aligned}\tilde{g}_j^\top h &\leq \phi_{k_j-\nu_j}(y^{k_j-\nu_j} + h, x^j) - \phi_{k_j-\nu_j}(y^{k_j-\nu_j}, x^j) \\ &\leq \phi(y^{k_j-\nu_j} + h, x^j) - \phi_{k_j-\nu_j}(y^{k_j-\nu_j}, x^j). \qquad (26.33)\end{aligned}$$

Using the fact that $\tilde{\rho}_{k_j-\nu_j} \geq \tilde{\gamma}$, we have

$$f(x^j) - \Phi_{k_j-\nu_j}(y^{k_j-\nu_j}, x^j) \leq \tilde{\gamma}^{-1}\left(f(x^j) - M_{k_j-\nu_j}(y^{k_j-\nu_j}, x^j)\right).$$

Adding $\frac{1}{2}(y^{k_j-\nu_j} - x^j)^\top Q_j (y^{k_j-\nu_j} - x^j)$ on both sides gives

$$\begin{aligned}&f(x^j) - \phi_{k_j-\nu_j}(y^{k_j-\nu_j}, x^j) \\ &\leq \tilde{\gamma}^{-1}\left(f(x^j) - M_{k_j-\nu_j}(y^{k_j-\nu_j}, x^j)\right) + \tfrac{1}{2}(y^{k_j-\nu_j} - x^j)^\top Q_j (y^{k_j-\nu_j} - x^j).\end{aligned}$$

Combining this and estimate (26.33) gives

$$\tilde{g}_j^\top h \leq \phi(y^{k_j-\nu_j}+h,x^j) - f(x^j) + \tilde{\gamma}^{-1}\left(f(x^j) - M_{k_j-\nu_j}(y^{k_j-\nu_j},x^j)\right)$$
$$+\tfrac{1}{2}(y^{k_j-\nu_j}-x^j)^\top Q_j(y^{k_j-\nu_j}-x^j). \tag{26.34}$$

As we have seen $y^{k_j-\nu_j}-x^j \to 0$, hence the rightmost term in (26.34) converges to 0 by boundedness of Q_j. Moreover, we claim that $\lim f(x^j) - M_{k_j-\nu_j}(y^{k_j-\nu_j},x^j) = 0$, so the term $\tilde{\gamma}^{-1}(\dots)$ on the right-hand side of (26.34) converges to 0. Indeed, to see this claim, notice first that it suffices to show $f(x^j) - m_{k_j-\nu_j}(y^{k_j-\nu_j},x^j) \to 0$, because the second-order term converges to 0. Since $m_{k_j-\nu_j}(\cdot,x^j)$ is a cutting plane at x^j, we have $m_{k_j-\nu_j}(y^{k_j-\nu_j},x^j) \leq f(y^{k_j-\nu_j})$ by definition of the downshift. So it suffices to show $\liminf m_{k_j-\nu_j}(y^{k_j-\nu_j},x^j) \geq f(\bar{x})$. Now this follows from the definition of the downshift s_j at $y^{k_j-\nu_j}$ with regard to x^j. Recall that for the tangent $t_{k_j-\nu_j}(\cdot)$ at $y^{k_j-\nu_j}$, approximate subgradient $\tilde{g}_j$, and serious iterate x^j, we have

$$s_j = [t_{k_j-\nu_j}(x^j) - f(x^j)]_+ + c\|y^{k_j-\nu_j}-x^j\|^2.$$

We can clearly concentrate on proving $t_{k_j-\nu_j}(x^j) - f(x^j) \to 0$. Now $t_{k_j-\nu_j}(x^j) - f(x^j) = f(y^{k_j-\nu_j}) - f(x^j) + \tilde{g}_j^\top(x^j - y^{k_j-\nu_j})$, and since $y^{k_j-\nu_j} \to \bar{x}$, $x^j \to \bar{x}$, and the $\tilde{g}_j$ are bounded, our claim follows.

Going back to (26.34) with the information $\tilde{g}_j^\top h \to \tilde{g}^\top h$, it remains to prove $\limsup \phi(y^{k_j-\nu_j}+h,x^j) \leq \phi(\bar{x}+h,\bar{x})$. Indeed, once this is proved, passing to the limit in (26.34) shows $\tilde{g}^\top h \leq \phi(\bar{x}+h,\bar{x}) - f(\bar{x}) = \phi(\bar{x}+h,\bar{x}) - \phi(\bar{x},\bar{x})$. This proves $\tilde{g} \in \partial\phi(\bar{x},\bar{x})$, and then $\tilde{g} \in \partial_{[\varepsilon]} f(\bar{x})$ by Lemma 26.7.

What remains to be shown is obviously joint upper semicontinuity of ϕ at $(\bar{x}+h,\bar{x})$, and this follows from Lemma 26.7; hence our claim $\tilde{g} \in \partial_{[\varepsilon]} f(\bar{x})$ is proved.

(viii) Let $\eta := \operatorname{dist}(0,\partial\phi(\bar{x},\bar{x}))$. Then $\|\tilde{g}\| \geq \eta$ by (vii) above. Let us fix $0 < \zeta < 1$; then, as $\tilde{g}_j \to \tilde{g}$, we have $\|\tilde{g}_j\| \geq (1-\zeta)\eta$ for $j \in J'$ large enough.

Now, assuming first $[\dots]_+ > 0$ in the downshift, we have

$$\begin{aligned} m_{k_j-\nu_j}(\cdot,x^j) &= f(y^{k_j-\nu_j}) + \tilde{g}_j^\top(\cdot - y^{k_j-\nu_j}) - s_j \\ &= f(y^{k_j-\nu_j}) + \tilde{g}_j^\top(\cdot - y^{k_j-\nu_j}) - c\|y^{k_j-\nu_j}-x^j\|^2 - t_{k_j-\nu_j}(x^j) + f(x^j) \\ &= f(x^j) + \tilde{g}_j^\top(\cdot - x^j) - c\|y^{k_j-\nu_j}-x^j\|^2, \end{aligned}$$

for $\tilde{g}_j \in \partial_{[\varepsilon]} f(y^{k_j-\nu_j})$ as above. Pick $g_j \in \partial f(y^{k_j-\nu_j})$ such that $\|g_j - \tilde{g}_j\| \leq \varepsilon$. Then

$$f(y^{k_j-\nu_j}) - m_{k_j-\nu_j}(y^{k_j-\nu_j},x^j) = f(y^{k_j-\nu_j}) - f(x^j) - \tilde{g}_j^\top(y^{k_j-\nu_j}-x^j)$$
$$+c\|y^{k_j-\nu_j}-x^j\|^2$$

$$= f(y^{k_j-\nu_j}) - f(x^j) - g_j^\top(y^{k_j-\nu_j} - x^j)$$
$$+(\tilde{g}_j - g_j)(y^{k_j-\nu_j} - x^j) + c\|y^{k_j-\nu_j} - x^j\|^2.$$

Since f is ε'-convex, we have $g_j^\top(y^{k_j-\nu_j} - x^j) \leq f(x^j) - f(y^{k_j-\nu_j}) + \varepsilon'\|y^{k_j-\nu_j} - x^j\|$. Substituting this we get

$$f(y^{k_j-\nu_j}) - m_{k_j-\nu_j}(y^{k_j-\nu_j}, x^j) \leq (\varepsilon' + \varepsilon)\|y^{k_j-\nu_j} - x^j\| + c\|y^{k_j-\nu_j} - x^j\|^2. \tag{26.35}$$

In the case $[\dots]_+ = 0$ an even better estimate is obtained, so that (26.35) covers both cases. Subtracting a term $\frac{1}{2}(y^{k_j-\nu_j} - x^j)^\top Q_j(y^{k_j-\nu_j} - x^j)$ on both sides of (26.35) and using $y^{k_j-\nu_j} - x^j \to 0$, we get

$$f(y^{k_j-\nu_j}) - M_{k_j-\nu_j}(y^{k_j-\nu_j}, x^j) \leq (\varepsilon' + \varepsilon + \nu_j)\|y^{k_j-\nu_j} - x^j\|,$$

where $\nu_j \to 0$. In consequence

$$f(y^{k_j-\nu_j}) - M_{k_j-\nu_j}(y^{k_j-\nu_j}, x^j) \leq (1+\zeta)(\varepsilon' + \varepsilon)\|y^{k_j-\nu_j} - x^j\| \tag{26.36}$$

for j large enough. Recall that $\tilde{g}_j = (Q_j + \frac{1}{2}\tau_{k_j} I)(x^j - y^{k_j-\nu_j}) \in \partial\phi_{k_j-\nu_j}(y^{k_j-\nu_j}, x^j)$ by (26.8) and (26.30). Hence by the subgradient inequality

$$\tilde{g}_j^\top(x^j - y^{k_j-\nu_j}) \leq \phi_{k_j-\nu_j}(x^j, x^j) - \phi_{k_j-\nu_j}(y^{k_j-\nu_j}, x^j).$$

Subtracting a term $\frac{1}{2}(x^j - y^{k_j-\nu_j})^\top Q_j(x^j - y^{k_j-\nu_j})$ from both sides gives

$$\tfrac{1}{2}(x^j - y^{k_j-\nu_j})^\top Q_j(x^j - y^{k_j-\nu_j}) + \tfrac{1}{2}\tau_{k_j}\|x^j - y^{k_j-\nu_j}\|^2 \leq f(x^j) - \Phi_{k_j-\nu_j}(y^{k_j-\nu_j}, x^j). \tag{26.37}$$

As $\tau_{k_j} \to \infty$, we have

$$(1-\zeta)\tfrac{1}{2}\tau_{k_j}\|x^j - y^{k_j-\nu_j}\| \leq \|\tilde{g}_j\| \leq (1+\zeta)\tfrac{1}{2}\tau_{k_j}\|x^j - y^{k_j-\nu_j}\| \tag{26.38}$$

and

$$\tfrac{1}{2}(x^j - y^{k_j-\nu_j})^\top Q_j(x^j - y^{k_j-\nu_j}) + \tfrac{1}{2}\tau_{k_j}\|x^j - y^{k_j-\nu_j}\|^2 \geq (1-\zeta)\tfrac{1}{2}\tau_{k_j}\|x^j - y^{k_j-\nu_j}\|^2 \tag{26.39}$$

both for j large enough. Therefore, plugging (26.38) and (26.39) into (26.37) gives

$$f(x^j) - \Phi_{k_j-\nu_j}(y^{k_j-\nu_j}, x^j) \geq \tfrac{1-\zeta}{1+\zeta}\|\tilde{g}_j\|\|x^j - y^{k_j-\nu_j}\|$$

for j large enough. Since $\|\tilde{g}_j\| \geq (1-\zeta)\eta$ for j large enough, we deduce

$$f(x^j) - \Phi_{k_j-\nu_j}(y^{k_j-\nu_j}, x^j) \geq \tfrac{(1-\zeta)^2}{1+\zeta}\eta \|x^j - y^{k_j-\nu_j}\|. \tag{26.40}$$

(ix) Combining (26.36) and (26.40) gives the estimate

$$\begin{aligned}\tilde{\rho}_{k_j-\nu_j} &= \rho_{k_j-\nu_j} + \frac{f(y^{k_j-\nu_j}) - M_{k_j-\nu_j}(y^{k_j-\nu_j}, x^j)}{f(x^j) - \Phi_{k_j-\nu_j}(y^{k_j-\nu_j}, x^j)} \\ &\leq \rho_{k_j-\nu_j} + \frac{(1+\zeta)^2(\varepsilon' + \varepsilon)\|y^{k_j-\nu_j} - x^j\|}{(1-\zeta)^2\eta \|y^{k_j-\nu_j} - x^j\|}.\end{aligned} \tag{26.41}$$

This proves

$$\eta \leq \frac{(1+\zeta)^2(\varepsilon' + \varepsilon)}{(1-\zeta)^2(\tilde{\gamma} - \gamma)}.$$

For suppose we had $\eta > \frac{(1+\zeta)^2(\varepsilon'+\varepsilon)}{(1-\zeta)^2(\tilde{\gamma}-\gamma)}$, then $\frac{(1+\zeta)^2(\varepsilon'+\varepsilon)}{(1-\zeta)^2\eta} < \tilde{\gamma} - \gamma$, which gave $\tilde{\rho}_{k_j-\nu_j} \leq \rho_{k_j-\nu_j} + \tilde{\gamma} - \gamma < \tilde{\gamma}$ for all j, contradicting $\tilde{\rho}_{k_j-\nu_j} \geq \tilde{\gamma}$ for infinitely many $j \in J$.

Since ζ in the above discussion was arbitrary, we have shown $\eta \leq \frac{\varepsilon'+\varepsilon}{\tilde{\gamma}-\gamma}$. Recall that $\eta = \operatorname{dist}\left(0, \partial_{[\varepsilon]} f(\bar{x})\right)$. We therefore have shown $0 \in \partial_{[\tilde{\varepsilon}]} f(\bar{x})$, where $\tilde{\varepsilon} = \varepsilon + \eta$. This is what is claimed. ■

Corollary 26.13. *Suppose $\Omega = \{x \in \mathbb{R}^n : f(x) \leq f(x^1)\}$ is bounded and f is lower C^1. Let approximate subgradients be drawn from $\partial_{[\varepsilon]} f(y)$, whereas function values are exact. Then every accumulation point $\bar{x}$ of the sequence of serious iterates x^j satisfies $0 \in \partial_{[\alpha\varepsilon]} f(\bar{x})$, where $\alpha = 1 + (\tilde{\gamma} - \gamma)^{-1}$.*

Remark 26.14. At first glance one might consider the class of lower C^1 functions used in Corollary 26.13 as too restrictive to offer sufficient scope. This misapprehension might be aggravated, or even induced, by the fact that lower C^1 functions are *approximately convex* [14, 34], an unfortunate nomenclature which erroneously suggests something close to a convex function. We therefore stress that lower C^1 is a large class which includes all examples we have so far encountered in practice. Indeed, applications are as a rule even lower C^2, or *amenable* in the sense of Rockafellar [45], a much smaller class, yet widely accepted as of covering all applications of interest.

Recent approaches to nonconvex nonsmooth optimization like [21, 33, 47] all work with composite (and therefore lower C^2) functions. This is in contrast with our own approach [19, 20, 40, 42], which works for lower C^1 and is currently the only one I am aware of that *has* the technical machinery to go beyond lower C^2. On second glance one will therefore argue that it is rather the class of lower C^2 functions which does not offer sufficient scope to justify the development of a new theory, because the chapter on nonsmooth composite convex functions $f = g \circ F$ in

[46] covers this class nicely and leaves little space for new contributions and because one *can* do things for lower C^1.

26.7 Extension to Inexact Values

In this section we discuss what happens when we have not only inexact subgradients but also inexact function values. In the previous sections we assumed that for every approximate subgradient g_a of f at x, there exists an exact subgradient $g \in \partial f(x)$ such that $\|g_a - g\| \leq \varepsilon$. Similarly, we will assume that approximate function values $f_a(x)$ satisfy $|f_a(x) - f(x)| \leq \bar{\varepsilon}$ for a fixed error tolerance $\bar{\varepsilon}$. We do not assume any link between ε and $\bar{\varepsilon}$.

Let us notice the following fundamental difference between the convex and the nonconvex case, where it is often reasonable to assume $f_a \leq f$; see, e.g., [30, 31]. Suppose f is convex, x is the current iterate, and an approximate value $f(x) - \bar{\varepsilon} \leq f_a(x) \leq f(x)$ is known. Suppose y^k is a null step, so that we draw an approximate tangent plane $t_k(\cdot) = f_a(y^k) + g_k^\top(\cdot - y^k)$ at y^k with respect to $g_k \in \partial_{[\varepsilon]} f(y^k)$. If we follow [30, 31], then $t_k(\cdot)$, while not a support plane, is still an affine minorant of f. It may then happen that $t_k(x) = f_a(y^k) + g_k^\top(x - y^k) > f_a(x)$, because $f_a(x), f_a(y^k)$ are approximations only. Now the approximate cutting plane gives us viable information as to the fact that the true value $f(x)$ satisfies $f(x) \geq t_k(x) > f_a(x)$. We shall say that *we can trust the value* $t_k(x) > f_a(x)$.

What should we do if we find a value $t_k(x)$ in which we can trust and which reveals our estimate $f_a(x)$ as too low? Should we correct $f_a(x)$ and replace it by the better estimate now available? If we do this we create trouble. Namely, we have previously rejected trial steps y^k during the inner loop at x based on the incorrect information $f_a(x)$. Some of these steps might have been acceptable, had we used $t_k(x)$ instead. But on the other hand, x was accepted as serious step in the inner loop at x^- because $f_a(x)$ was sufficiently below $f_a(x^-)$. If we correct the approximate value at x, then acceptance of x may become unsound as well. For short, correcting values as soon as better estimates arrive is not a good idea, because we might be forced to go repeatedly back all the way through the history of our algorithm.

In order to avoid this backtracking, Kiwiel [30] proposes the following original idea. If $f_a(x)$, being too low, still allows progress in the sense that x^+ with $f_a(x^+) < f_a(x)$ can be found, then why waste time and correct the value $f_a(x)$? After all, there is still progress! On the other hand, if the underestimation $f_a(x)$ is so severe that the algorithm will stop, then we should be sure that no further decrease within the error tolerances $\bar{\varepsilon}, \varepsilon$ is possible. Namely, if this is the case, then we can stop in all conscience. To check this, Kiwiel progressively relaxes proximity control in the inner loop, until it becomes clear that the model of all possible approximate cutting planes itself does not allow to descend below $f_a(x)$ and, therefore, does not allow to descend more than $\bar{\varepsilon}$ below $f(x)$.

The situation outlined is heavily based on convexity and does not appear to carry over to nonconvex problems. The principal difficulty is that without convexity we cannot trust values $t_{y,g}(x) > f_a(x)$ even in the case of *exact* tangent planes, $g \in \partial f(y)$. We know that tangents have to be downshifted, and without the exact knowledge of $f(x)$, the only available reference value to organize the downshift is $f_a(x)$. Naturally, as soon as we downshift with reference to $f_a(x)$, cutting planes $m_{y,g}(\cdot,x)$ satisfying $m_{y,g}(x,x) > f_a(x)$ can no longer occur. This removes one of the difficulties. However, it creates, as we shall see, a new one.

In order to proceed with inexact function values, we will need the following property of the cutting plane $m_k(\cdot,x) := t_k(\cdot) - s_k$ at null step y^k and approximate subgradient $g_k \in \partial_{[\varepsilon]} f(y^k)$. We need to find $\tilde{\varepsilon} > 0$ such that $f_a(y^k) \leq m_k(y^k,x) + \tilde{\varepsilon}\|x - y^k\|$. More explicitly, this requires

$$f_a(y^k) \leq f_a(x) + g_k^\top (y^k - x) + \tilde{\varepsilon}\|x - y^k\|.$$

If f is ε'-convex, then

$$\begin{aligned} f(y^k) &\leq f(x) + g^\top (y^k - x) + \varepsilon'\|x - y^k\| \\ &\leq f(x) + g_k^\top (y^k - x) + (\varepsilon' + \varepsilon)\|x - y^k\| \end{aligned}$$

for $g \in \partial f(y^k)$ and $\|g - g_k\| \leq \varepsilon$. That means

$$f(y^k) - (f(x) - f_a(x)) \leq f_a(x) + g_k^\top (y^k - x) + (\varepsilon + \varepsilon')\|x - y^k\|.$$

So what we need in addition is something like

$$f_a(y^k) \leq f(y^k) - (f(x) - f_a(x)) + \varepsilon''\|x - y^k\|,$$

because then we get the desired relation with $\tilde{\varepsilon} = \varepsilon + \varepsilon' + \varepsilon''$. The condition can still be slightly relaxed to make it more useful in practice. The axiom we need is that there exist $\delta_k \to 0^+$ such that

$$f(x) - f_a(x) \leq f(y^k) - f_a(y^k) + (\varepsilon'' + \delta_k)\|x - y^k\| \tag{26.42}$$

for every $k \in \mathbb{N}$. Put differently, as $y^k \to x$, the error we make at y^k by underestimating $f(y^k)$ by $f_a(y^k)$ is larger than the corresponding underestimation error at x, up to a term proportional to $\|x - y^k\|$. The case of exact values $f = f_a$ corresponds to $\varepsilon'' = 0, \delta_k = 0$.

Remark 26.15. As f is continuous at x, condition (26.42) implies upper semi-continuity of f_a at serious iterates, i.e., $\limsup f_a(y^k) \leq f_a(x)$.

We are now ready to modify our algorithm and then run through the proofs of Lemmas 26.9 and 26.11 and Theorem 26.12 and see what changes need to be made to account for the new situation. As far as the algorithm is concerned, the changes are easy. We replace $f(y^k)$ and $f(x)$ by $f_a(y^k)$ and $f_a(x)$. The rest of the procedure is the same.

We consider the same convex envelope function $\phi(\cdot,x)$ defined in (26.13). We have the following.

Lemma 26.16. *The upper envelope model satisfies $\phi(x,x) = f_a(x)$, $\phi_k \leq \phi$. ϕ is jointly upper $2\bar{\varepsilon}$-semicontinuous, and $\partial\phi(x,x) \subset \partial_{[\varepsilon]}f(x) \subset \partial_{2\bar{\varepsilon}}\phi(x,x)$, where $\partial_{2\bar{\varepsilon}}\phi(x,x)$ is the $2\bar{\varepsilon}$-subdifferential of $\phi(\cdot,x)$ at x in the usual convex sense.*

Proof.

(1) Any cutting plane $m_{z,g}(\cdot,x)$ satisfies $m_{z,g}(x,x) \leq f_a(x) - c\|x-z\|^2$. This shows $\phi(x,x) \leq f_a(x)$, and if we take $z = x$, we get equality $\phi(x,x) = f_a(x)$.

(2) We prove $\partial_{[\varepsilon]}f(x) \subset \partial_{2\bar{\varepsilon}}\phi(x,x)$. Let $g \in \partial f(x)$ be a limiting subgradient, and choose $y^k \to x$, where f is differentiable at y^k with $g_k = \nabla f(y^k) \in \partial f(y^k)$ such that $g_k \to g$. Let g_a be an approximate subgradient such that $\|g - g_a\| \leq \varepsilon$. We have to prove $g_a \in \partial_{2\bar{\varepsilon}}\phi(x,x)$. Putting $g_{a,k} := g_k + g_a - g \in \partial_{[\varepsilon]}f(y^k)$ we have $g_{a,k} \to g_a$. Let $m_k(\cdot,x)$ be the cutting plane drawn at y^k with approximate subgradient $g_{a,k}$. That is, $m_k(\cdot,x) = m_{y^k,g_{a,k}}(\cdot,x)$. Then

$$m_k(y,x) = f_a(y^k) + g_{a,k}^\top(y - y^k) - s_k,$$

where $s_k = [f_a(x) - t_k(x)]_+ + c\|x - y^k\|^2$ is the downshift and where $t_k(\cdot)$ is the approximate tangent at y^k with respect to $g_{a,k}$. There are two cases, $s_k = c\|x - y^k\|^2$ and $s_k = f_a(x) + t_k(x) + c\|x - y^k\|^2$, according to whether $[\dots]_+ = 0$ or $[\dots]_+ > 0$. Let us start with the case $t_k(x) > f_a(x)$. Then

$$s_k = f_a(y^k) + g_{a,k}^\top(x - y^k) + c\|x - y^k\|^2$$

and

$$m_k(y,x) = f_a(y^k) + g_{a,k}^\top(y - y^k) - f_a(y^k) - g_{a,k}^\top(x - y^k) + f_a(x) - c\|x - y^k\|^2.$$

Therefore

$$\phi(y,x) - \phi(x,x) \geq m_k(y^k,x) - f_a(x) = g_{a,k}^\top(y - x) - c\|x - y^k\|^2.$$

Passing to the limit $k \to \infty$ proves $g_a \in \partial\phi(x,x)$, so in this case a stronger statement holds.

Let us next discuss the case where $t_k(x) \leq f_a(x)$, so that $s_k = c\|x - y^k\|^2$. Then

$$m_k(y,x) = f_a(y^k) + g_{a,k}^\top(y - y^k) - c\|x - y^k\|^2.$$

Therefore

$$\begin{aligned}\phi(y,x)-\phi(x,x) &\geq m_k(y^k,x)-f_a(x)\\ &= f_a(y^k)-f_a(x)+g_{a,k}^\top(y-y^k)-c\|x-y^k\|^2\\ &= f_a(y^k)-f_a(x)+g_{a,k}^\top(x-y^k)-c\|x-y^k\|^2+g_{a,k}^\top(y-x).\end{aligned}$$

Put $\zeta_k := g_{a,k}^\top(x-y^k)-c\|x-y^k\|^2+(g_{a,k}-g_a)^\top(y-x)$ then

$$\phi(y,x)-\phi(x,x)\geq f_a(y^k)-f_a(x)+\zeta_k+g_a^\top(y-x).$$

Notice that $\lim \zeta_k = 0$, because $g_{a,k}\to g_a$ and $y^k\to x$. Let $F_a(x):=\liminf_{k\to\infty} f_a(y^k)$, then we obtain

$$\phi(y,x)-\phi(x,x)\geq F_a(x)-f_a(x)+g_a^\top(y-x).$$

Putting $\varepsilon(x):=[f_a(x)-F_a(x)]_+$, we therefore have shown

$$\phi(y,x)-\phi(x,x)\geq -\varepsilon(x)+g_a^\top(y-x),$$

which means $g_a\in\partial_{\varepsilon(x)}\phi(x,x)$. Since approximate values f_a are within $\bar\varepsilon$ of exact values f, we have $|f_a(x)-F_a(x)|\leq 2\bar\varepsilon$, hence $\varepsilon(x)\leq 2\bar\varepsilon$. That shows $g_a\in\partial_{\varepsilon(x)}\phi(x,x)\subset\partial_{2\bar\varepsilon}\phi(x,x)$.

(3) The proof of $\partial\phi(x,x)\subset\partial_{[\varepsilon]}f(x)$ remains the same, after replacing $f(x)$ by $f_a(x)$.

(4) If a sequence of planes $m_r(\cdot)$, $r\in\mathbb{N}$, contributes to the envelope function $\phi(\cdot,x)$ and if $m_r(\cdot)\to m(\cdot)$ in the pointwise sense, then $m(\cdot)$ also contributes to $\phi(\cdot,x)$, because the graph of $\phi(\cdot,x)$ is closed. On the other hand, we may expect discontinuities as $x_j\to x$. We obtain $\limsup_{j\to\infty}\phi(y_j,x_j)\leq\phi(y,x)+\bar\varepsilon$ for $y_j\to y$, $x_j\to x$. ■

Remark 26.17. If approximate function values are underestimations, $f_a\leq f$, as is often the case, then $|F_a-f_a|\leq\bar\varepsilon$ and the result holds with $\partial\phi(x,x)\subset\partial_{[\varepsilon]}f(x)\subset\partial_{\bar\varepsilon}\phi(x,x)$.

Corollary 26.18. *Under the hypotheses of Lemma 26.16, if x is a point of continuity of f_a, then $\partial\phi(x,x)=\partial_{[\varepsilon]}f(x)$ and ϕ is jointly upper semicontinuous at (x,x).*

Proof. Indeed, as follows from part (2) of the proof above, for a point of continuity x of f_a, we have $\varepsilon(x)=0$. ■

Lemma 26.19. *Suppose the inner loop at serious iterate x turns forever and $\tau_k\to\infty$. Suppose f is ε'-convex on a set containing all y^k, $k\geq k_0$, and let (26.42) be satisfied. Then $0\in\partial_{[\tilde\varepsilon]}f(x)$, where $\tilde\varepsilon=\varepsilon+(\varepsilon''+\varepsilon'+\varepsilon)/(\tilde\gamma-\gamma)$.*

Proof. We go through the proof of Lemma 26.9 and indicate the changes caused by using approximate values $f_a(y^k)$, $f_a(x)$. Part (ii) remains the same, except that

$\phi(x,x) = f_a(x)$. The exactness subgradient has still $g(x) \in \partial_{[\varepsilon]} f(x)$. Part (iii) leading to formula (26.17) remains the same with $f_a(x)$ instead of $f(x)$. Part (iv) remains the same, and we obtain the analogue of (26.18) with $f(x)$ replaced by $f_a(x)$.

Substantial changes occur in part (v) of the proof leading to formula (26.19). Indeed, consider without loss the case where $t_k(x) > f_a(x)$. Then

$$\begin{aligned} m_k(y,x) &= f_a(y^k) + g_{\varepsilon k}^\top (y - y^k) - s_k \\ &= f_a(x) + g_{\varepsilon k}^\top (y - x) - c\|x - y^k\|^2, \end{aligned}$$

as in the proof of Lemma 26.9, and therefore

$$f_a(y^k) - m_k(y^k,x) = f_a(y^k) - f_a(x) - g_k^\top (y^k - x) + (g_k - g_{\varepsilon k})^\top (x - y^k) + c\|x - y^k\|^2.$$

Since f is ε'-convex, we have $g_k^\top (x - y^k) \leq f(x) - f(y^k) + \varepsilon'\|x - y^k\|$. Hence

$$f_a(y^k) - m_k(y^k,x) \leq f(x) - f_a(x) - \left(f(y^k) - f_a(y^k)\right) + (\varepsilon' + \varepsilon + \nu_k)\|x - y^k\|,$$

where $\nu_k \to 0$. Now we use axiom (26.42), which gives

$$f_a(y^k) - m_k(y^k,x) \leq (\varepsilon'' + \varepsilon' + \varepsilon + \delta_k + \nu_k)\|x - y^k\|,$$

for $\delta_k, \nu_k \to 0$. Subtracting the usual quadratic expression on both sides gives $f_a(y^k) - M_k(y^k,x) \leq (\varepsilon'' + \varepsilon' + \varepsilon + \delta_k + \tilde{\nu}_k)\|x - y^k\|$ with $\delta_k, \tilde{\nu}_k \to 0$. Going back with this estimation to the expansion $\tilde{\rho}_k \leq \rho_k + \frac{\varepsilon'' + \varepsilon' + \varepsilon}{\eta}$ shows $\eta < \frac{\varepsilon'' + \varepsilon' + \varepsilon}{\tilde{\gamma} - \gamma}$ as in the proof of Lemma 26.9, where $\eta = \operatorname{dist}(0, \partial\phi(x,x))$. Since $\partial\phi(x,x) \subset \partial_{[\varepsilon]} f(x)$ by Lemma 26.16, we have $0 \in \partial_{[\varepsilon+\eta]} f(x)$. This proves the result. ■

Lemma 26.20. *Suppose the inner loop turns forever and τ_k is frozen from some counter k onwards. Then $0 \in \partial_{[\varepsilon]} f(x)$.*

Proof. Replacing $f(x)$ by $f_a(x)$, the proof proceeds in exactly the same fashion as the proof of Lemma 26.11. We obtain $0 \in \partial\phi(x,x)$ and use Lemma 26.16 to conclude $0 \in \partial_{[\varepsilon]} f(x)$. ■

As we have seen, axiom (26.42) was necessary to deal with the case $\tau_k \to \infty$ in Lemma 26.19, while Lemma 26.20 gets by without this condition. Altogether, that means we have to adjust the stopping test in step 2 of the algorithm to $0 \in \partial_{[\tilde{\varepsilon}]} f(x^j)$, where $\tilde{\varepsilon} = \varepsilon + (\varepsilon'' + \varepsilon' + \varepsilon)/(\tilde{\gamma} - \gamma)$. As in the case of exact function values, we may delegate the stopping test to the inner loop, so if the latter halts due to insufficient progress, we interpret this as $0 \in \partial_{[\tilde{\varepsilon}]} f(x^j)$, which is the precision we can hope for. Section 26.8 below gives more details.

Let us now scan through the proof of Theorem 26.12 and see what changes occur through the use of inexact function values $f_a(y^k)$, $f_a(x^j)$.

Theorem 26.21. *Let x^1 be such that $\Omega' = \{x \in \mathbb{R}^n : f(x) \leq f(x^1) + 2\bar{\varepsilon}\}$ is bounded. Suppose f is ε'-convex on Ω, that subgradients are drawn from $\partial_{[\varepsilon]} f(y)$, and that* *inexact function values* $f_a(y)$ *satisfy* $|f(y) - f_a(y)| \leq \bar{\varepsilon}$. *Suppose axiom* (26.42) *is satisfied. Then every accumulation point* $\bar{x}$ *of the sequence* x^j *satisfies* $0 \in \partial_{[\tilde{\varepsilon}]} f(\bar{x})$, *where* $\tilde{\varepsilon} = \varepsilon + (\varepsilon'' + \varepsilon' + \varepsilon)/(\tilde{\gamma} - \gamma)$.

Proof. Notice that $\tilde{\varepsilon}$ used in the stopping test has a different meaning than in Theorem 26.21. Replacing $f(x^j)$ by $f_a(x^j)$ and $f(y^{k_j})$ by $f_a(y^{k_j})$, we follow the proof of Theorem 26.12. Part (i) is still valid with these changes. Notice that $\Omega = \{x : f_a(x) \leq f_a(x^1)\} \subset \Omega'$ and Ω' is bounded by hypothesis, so Ω is bounded.

As in the proof of Theorem 26.12 the set of all trial points $y^1, \ldots, y^{k_j}$ visited during all the inner loops j is bounded. However, a major change occurs in part (ii). Observe that the accumulation point $\bar{x}$ used in the proof of Theorem 26.12 is neither among the trial points nor the serious iterates. Therefore, $f_a(\bar{x})$ is never called for in the algorithm. Now observe that the sequence $f_a(x^j)$ is decreasing and by boundedness of Ω converges to a limit $F_a(\bar{x})$. We redefine $f_a(\bar{x}) = F_a(\bar{x})$, which is consistent with the condition $|f_a(\bar{x}) - f(\bar{x})| \leq \bar{\varepsilon}$, because $f_a(x^j) \geq f(x^j) - \bar{\varepsilon}$, so that $F_a(\bar{x}) \geq f(\bar{x}) - \bar{\varepsilon}$.

The consequences of the redefinition of $f_a(\bar{x})$ are that the upper envelope model ϕ is now jointly upper semicontinuous at $(\bar{x}, \bar{x})$, and that the argument leading to formula (26.29) remains unchanged, because $f_a(x^j) \to \phi(\bar{x}, \bar{x})$.

Let us now look at the longer argument carried out in parts (iii)–(ix) of the proof of Theorem 26.12, which deals with the case where $\|g_j\| \geq \mu > 0$ for all j. Parts (iii)–(vii) are adapted without difficulty. Joint upper semicontinuity of ϕ at $(\bar{x}+h, \bar{x})$ is used at the end of (vii), and this is assured as a consequence of the redefinition $f_a(\bar{x}) = F_a(\bar{x})$ of f_a at $\bar{x}$.

Let us next look at part (viii). In Theorem 26.12 we use ε'-convexity. Since the latter is in terms of exact values, we need axiom (26.42) for the sequence $y^{k_j - \nu_j} \to \bar{x}$, similarly to the way it was used in Lemma 26.16. We have to check that despite the redefinition of f_a at $\bar{x}$ axiom (26.42) is still satisfied. To see this, observe that $y^{k_j - \nu_j}$ is a trial step which is rejected in the jth inner loop, so that its approximate function value is too large. In particular, $f_a(y^{k_j - \nu_j}) \geq f_a(x^{j+1})$, because x^{j+1} is the first trial step accepted. This estimate shows that (26.42) is satisfied at $\bar{x}$.

Using (26.42) we get the analogue of (26.36), which is

$$f_a(y^{k_j - \nu_j}) - M_{k_j - \nu_j}(y^{k_j - \nu_j}, x^j) \leq (\varepsilon'' + \varepsilon' + \nu_j + \delta_j)\|y^{k_j - \nu_j} - x^j\|$$

for certain $\nu_j, \delta_j \to 0$. Estimate (26.40) remains unchanged, so we can combine the two estimates to obtain the analogue of (26.41) in part (ix), which is

$$\tilde{\rho}_{k_j - \nu_j} \leq \rho_{k_j - \nu_j} + \frac{(1+\zeta^2)(\varepsilon'' + \varepsilon' + \varepsilon)}{(1-\zeta)^2 \eta}.$$

Using the same argument as in the proof of Theorem 26.12, we deduce

$$\eta \leq \frac{(1+\zeta)^2(\varepsilon'' + \varepsilon' + \varepsilon)}{(1-\zeta)^2(\tilde{\gamma} - \gamma)}$$

for $\eta = \text{dist}(0, \partial\phi(x,x))$. Since $0 < \zeta < 1$ was arbitrary, we obtain $\eta \leq \frac{\varepsilon''+\varepsilon'+\varepsilon}{\tilde{\gamma}-\gamma}$. Now as $\bar{x}$ is a point of continuity of f_a, Corollary 26.18 tells us that $\eta = \text{dist}(0, \partial_{[\varepsilon]} f(\bar{x}))$. Therefore $0 \in \partial_{[\varepsilon+\eta]} f(\bar{x})$. Since $\varepsilon + \eta = \tilde{\varepsilon}$, we are done. ■

26.8 Stopping

In this section we address the practical problem of stopping the algorithm. The idea is to use tests which are based on the convergence theory developed in the previous sections.

In order to save time, the stopping test in step 2 of the algorithm is usually delegated to the inner loop. This is based on Lemmas 26.9 and 26.11 and the following.

Lemma 26.22. *Suppose tangent program* (26.7) *has the solution* $y^k = x$. *Then* $0 \in \partial_{[\varepsilon]} f(x)$.

Proof. From (26.8) we have $0 \in \partial\phi_k(x,x) \subset \partial\phi(x,x) \subset \partial_{[\varepsilon]} f(x)$ by Lemma 26.16. ■

In [20] we use the following two-stage stopping test. Fixing a tolerance level tol > 0, if x^+ is the serious step accepted by the inner loop at x, and if x^+ satisfies

$$\frac{\|x - x^+\|}{1 + \|x\|} < \text{tol},$$

then we stop the outer loop and accept x^+ as the solution, the justification being Lemma 26.22. On the other hand, if the inner loop at x fails to find x^+ and either exceeds a maximum number of allowed inner iterations or provides three consecutive trial steps y^k satisfying

$$\frac{\|x - y^k\|}{1 + \|x\|} < \text{tol},$$

then we stop the inner loop and the algorithm and accept x as the final solution. Here the justification comes from Lemmas 26.9 and 26.11.

Remark 26.23. An interesting aspect of inexactness theory with unknown precisions $\varepsilon, \varepsilon', \varepsilon''$ are the following two scenarios, which may require different handling. The first is when functions and subgradients are inexact or noisy, but we do not take this into account and proceed as if information were exact. The second scenario is when we deliberately use inexact information in order to gain speed or deal with problems of very large size. In the first case we typically arrange all elements of the algorithm like in the exact case, including situations where we are not even aware that information is inexact. In the second case we might introduce new elements which make the most of the fact that data are inexact.

As an example of the latter, in [30] where f is convex, the author does not use downshift with respect to $f_a(x)$, and as a consequence one may have $\phi_k(x,x) > f_a(x)$, so that the tangent program (26.7) may fail to find a predicted descent step y^k at x. The author then uses a sub-loop of the inner loop, where the τ-parameter is decreased until *either* a predicted descent step is found *or* optimality within the allowed tolerance of function values is established.

26.9 Example from Control

Optimizing the H_∞-norm [4,7,19,20] is a typical application of (26.1) where inexact function and subgradient evaluations may arise. The objective function is of the form

$$f(x) = \max_{\omega \in \mathbb{R}} \overline{\sigma}\left(G(x, j\omega)\right), \tag{26.43}$$

where $G(x,s) = C(x)\left(sI - A(x)\right)^{-1} B(x) + D(x)$ is defined on the open set $S = \{x \in \mathbb{R}^n : A(x) \text{ stable}\}$ and where $A(x)$, $B(x)$, $C(x)$, $D(x)$ are matrix-valued mappings depending smoothly on $x \in \mathbb{R}^n$. In other words, for $x \in S$ each $G(x,s)$ is a stable real-rational transfer matrix.

Notice that f is a composite function of the form $f = \|\cdot\|_\infty \circ \mathscr{G}$, where $\|\cdot\|_\infty$ is the H_∞-norm, which turns the Hardy space $\mathscr{H}_\infty$ of functions G which are analytic and bounded in the open right-half plane [53, p. 100] into a Banach space,

$$\|G\|_\infty = \sup_{\omega \in \mathbb{R}} \overline{\sigma}\left(G(j\omega)\right),$$

and $\mathscr{G} : S \to \mathscr{H}_\infty$, $x \mapsto G(x,\cdot) = C(x)(\cdot I - A(x))^{-1}B(x) + D(x) \in \mathscr{H}_\infty$ is a smooth mapping, defined on the open subset $S = \{x \in \mathbb{R}^n : A(x) \text{ stable}\}$. Since composite functions of this form are lower C^2, and therefore also lower C^1, we are in business. For the convenience of the reader we also include a more direct argument proving the same result:

Lemma 26.24. *Let f be defined by* (26.43)*, then f is lower C^2, and therefore also lower C^1, on the open set $S = \{x \in \mathbb{R}^n : A(x)$ stable$\}$.*

Proof. Recall that $\overline{\sigma}(G) = \max_{\|u\|=1} \max_{\|v\|=1} \operatorname{Re} uGv^H$, so that

$$f(x) = \max_{\omega \in \mathbb{S}^1} \max_{\|u\|=1} \max_{\|v\|=1} \operatorname{Re} u G(x, j\omega) v^H.$$

Here, for $x \in S$, the stability of $G(x,\cdot)$ assures that $G(x,s)$ is analytic in s on a band $\mathscr{B}$ on the Riemann sphere $\mathbb{C} \cup \{\infty\}$ containing the zero meridian $j\mathbb{S}^1$ with $\mathbb{S}^1 = \{\omega : \omega \in \mathbb{R} \cup \{\infty\}\}$, a compact set homeomorphic to the real 1-sphere. This shows that f is lower C^2 on the open set S. Indeed, $(x,\omega,u,v) \mapsto F(x,\omega,u,v) := \operatorname{Re} u G(x, j\omega) v^H$ is jointly continuous on $S \times \mathbb{S}^1 \times \mathbb{C}^m \times \mathbb{C}^p$ and smooth in x, and $f(x) = \max_{(\omega,u,v) \in K} F(x,\omega,u,v)$ for the compact set $K = \mathbb{S}^1 \times \{u \in \mathbb{C}^m : \|u\| = 1\} \times \{v \in \mathbb{C}^p : \|v\| = 1\|\}$. ■

The evaluation of $f(x)$ is based on the iterative bisection method of Boyd et al. [10]. Efficient implementations use Boyd and Balakrishnan [11] or Bruisma and Steinbuch [12] and guarantee quadratic convergence. All these approaches are based on the Hamiltonian test from [10], which states that $f(x) > \gamma$ if and only if the Hamiltonian

$$H(x,\gamma) = \begin{bmatrix} A(x) & 0 \\ 0 & -A(x)^\top \end{bmatrix} - \begin{bmatrix} 0 & B(x) \\ C(x)^\top & 0(x) \end{bmatrix} \begin{bmatrix} \gamma I & D(x) \\ D(x)^\top & \gamma I \end{bmatrix}^{-1} \begin{bmatrix} C(x) & 0 \\ 0 & -B(x)^\top \end{bmatrix} \tag{26.44}$$

has purely imaginary eigenvalues $j\omega$. The bundle method of [7], which uses (26.44) to compute function values, can now be modified to use approximate values $f_a(y^k)$ for unsuccessful trial points y^k. Namely, if the trial step y^k is to become the new serious iterate x^+, its value $f(y^k)$ has to be below $f(x)$. Therefore, as soon as the Hamiltonian test (26.44) certifies $f(y^k) > f(x)$ even before the exact value $f(y^k)$ is known, we may dispense with the exact computation of $f(y^k)$. We may stop the Hamiltonian algorithm at the stage where the first γ with $f(y^k) > \gamma \geq f(x)$ occurs, compute the intervals where $\omega \mapsto \overline{\sigma}(G(x,j\omega))$ is above γ, take the midpoints of these intervals, say $\omega_1,\dots,\omega_r$, and pick the one where the frequency curve is maximum. If this is ω_ν, then $f_a(y^k) = \overline{\sigma}(G(x,j\omega_\nu))$. The approximate subgradient g_a is computed via the formulas of [4] with ω_ν replacing an active frequency. This procedure is trivially consistent with (26.42), because $f(x) = f_a(x)$ and $f_a(y) \leq f(y)$.

If we wish to allow inexact values not only at trial points y but also at serious iterates x, we can use the termination tolerance of the Hamiltonian algorithm [11]. The algorithm works with estimates $f_l(x) \leq f(x) \leq f_u(x)$ and terminates when $f_u(x) - f_l(x) \leq 2\eta_x F(x)$, returning $f_a(x) := (f_l(x) + f_u(x))/2$, where we have the choice $F(x) \in \{f_l(x), f_u(x), f_a(x)\}$. Then $|f(x) - f_a(x)| \leq 2\eta_x |F(x)|$. As η_x is under control, we can arrange that $\eta_x|F(x)| \leq \eta_y|F(y)| + o(\|x-y\|)$ in order to assure condition (26.42).

Remark 26.25. The outlined method applies in various other cases in feedback control where function evaluations use iterative procedures, which one may stop short to save time. We mention IQC-theory [2], which uses complex Hamiltonians, [7] for related semi-infinite problems, or the multidisk problem [3], where several H_∞-criteria are combined in a progress function. The idea could be used quite naturally in the ε-subgradient approaches [36, 37] or in search methods like [1].

Acknowledgements The author acknowledges funding by Fondation d'Entreprise EADS under grant *Technicom* and by Fondation de Recherche pour l'Aéronautique et l'Espace under grant *Survol*.

References

1. Apkarian, P., Noll, D.: Controller design via nonsmooth multidirectional search. SIAM J. Contr. Appl. **44**(6), 1923–1949 (2006)
2. Apkarian, P., Noll, D.: IQC analysis and synthesis via nonsmooth optimization. Syst. Contr. Lett. **55**(12), 971–981 (2006)
3. Apkarian, P., Noll, D.: Nonsmooth optimization for multidisk H_∞-synthesis. Eur. J. Contr. **12**(3), 229–244 (2006)
4. Apkarian, P., Noll, D.: Nonsmooth H_∞-synthesis. Trans. Automat. Contr. **51**(1), 229–244 (2006)
5. Apkarian, P., Noll, D.: Nonsmooth optimization for multiband frequency domain control design. Automatica **43**(4), 724–731 (2007)
6. Apkarian, P., Noll, D., Prot, O.: A trust region spectral bundle method for non-convex eigenvalue optimization. SIAM J. Optim. **19**(1), 281–306 (2008)
7. Apkarian, P., Noll, D., Prot, O.: A proximity control algorithm to minimize non-smooth and non-convex semi-infinite maximum eigenvalue functions. J. Convex Anal. **16**, 641–666 (2009)
8. Apkarian, P., Ravanbod-Hosseini, L., Noll, D.: Time-domain constrained structured H_∞-synthesis. Int. J. Robust Nonlinear Contr. **21**(2), 197–217 (2010)
9. Bompart, V., Noll, D., Apkarian, P.: Second order nonsmooth optimization for H_∞-synthesis. Numerische Mathematik **107**(3), 433–454 (2007)
10. Boyd, S., Balakrishnan, V.: A regularity result for the singular values of a transfer matrix and a quadratically convergent algorithm to compute its L_∞-norm. Syst. Contr. Lett. **15**, 1–7 (1990)
11. Boyd, S., Balakrishnan, V., Kabamba, P. : A bisection method for computing the H_∞-norm of a transfer matrix and related problems. Math. Contr. Signals Syst. **2**, 207–219 (1989)
12. Bruisma, N., Steinbuch, M.: A fast algorithm to compute the H_∞ norm of a transfer function. Syst. Contr. Lett. **14**, 287–293 (1990)
13. Clarke, F.: Optimization and Nonsmooth Analysis. SIAM Publications, Philadelphia (1983)
14. Daniilidis, A., Georgiev, P.: Approximate convexity and submonotonicity. J. Math. Anal. Appl. **291**, 117–144 (2004)
15. Emiel, G., Sagastizábal, C.: Incremental-like bundle methods with application to energy planning. Comput. Optim. Appl. **46**(2), 305–332 (2010)
16. Emiel, G., Sagastizábal, C.: Dynamic subgradient methods. Optimization online: http://www.optimization-online.org/DB_HTML/2008/08/2077.html
17. Fuduli, A., Gaudioso, M., Giallombardo, G.: A DC piecewise affine model and a bundling technique in nonconvex nonsmooth optimization. Optim. Methods Softw. **19**, 89–102 (2004)
18. Fuduli, A., Gaudioso, M., Giallombardo, G.: Minimizing nonconvex nonsmooth functions via cutting planes and proximity control. SIAM J. Optim. **14**, 743–756 (2004)
19. Gabarrou, M., Alazard, D., Noll, D.: Structured flight control law design using non-smooth optimization. In: Proceedings of the 18th IFAC Symposium on Automatic Control in AeroSpace (ACA). Nara, Japan (2010)
20. Gabarrou, M., Alazard, D., Noll, D.: Design of a flight control architecture using a non-convex bundle method. Math. Contr. Signals Syst. **25**(2), 257–290 (2013)
21. Hare, W.L., Sagastizábal, C.: Redistributed proximal bundle method for nonconvex optimization. SIAM J. Optim. **20**(5), 2442–2473 (2010)
22. Hintermüller, M.: A proximal bundle method based on approximate subgradients. Comput. Optim. Appl. **20**, 245–266 (2001)
23. Hiriart-Urruty, J.-B., Seeger, A.: The second-order subdifferential and the Dupin indicatrix of a non-differentiable convex function. Proc. London Math. Soc. **58**, 351–365 (1989)
24. Hiriart-Urruty, J.-B., Lemaréchal, C.: Convex Analysis and Minimization Algorithms II. Springer, Berlin (1993)
25. Kiwiel, K.C.: An aggregate subgradient method for non-smooth convex minimization. Math. Programming **27**, 320–341 (1983)

26. Kiwiel, K.C.: An algorithm for nonsmooth convex minimization with errors. Math. Comput. **45**, 171–180 (1985)
27. Kiwiel, K.C.: A linearization algorithm for computing control systems subject to singular value inequalities. IEEE Trans. Autom. Contr. **AC-31**, 595–602 (1986)
28. Kiwiel, K.C.: Approximations in proximal bundle methods and decomposition of convex programs. J. Optim. Theory Appl. **84**(3), 529–548 (1995)
29. Kiwiel, K.C.: Convergence of approximate and incremental subgradient methods for convex optimization. SIAM J. Optim. **14**(3), 807–840 (2003)
30. Kiwiel, K.C.: A proximal bundle method with approximate subgradient linearizations. SIAM J. Optim. **16**, 1007–1023 (2006)
31. Kiwiel, K.C., Lemaréchal, C.: An inexact bundle variant suited to column generation. Math. Programming, Ser. A **118**(1), 177–206 (2007)
32. Lemaréchal, C., Strodiot, J.-J., Bihain, A.: On a bundle algorithm for non-smooth optimization. In: Mangasarian, O.L., Meyer, R.R., Robinson, S.M. (eds.) Nonlinear Programming, vol. 4, pp. 245–282. Academic, New York (1981)
33. Lewis, A., Wright, S.: A proximal method for composite minimization arXiv:0812.0423 [math.oc]
34. Luc, D.T., Van Ngai, H., Théra, M.: On ε-monotonicity and ε-convexity. In: Calculus of Variations and Differential Equations (Haifa 1998). CRC Research Notes in Mathematics, vol. 410, pp. 82–100. Chapman & Hall, Boca Raton (2000)
35. Lukšan, L., Vlček, J.: A bundle-Newton method for nonsmooth unconstrained minimization. Math. Programming **83**, 373–391 (1998)
36. Mayne, D., Polak, E.: Algorithms for the design of control systems subject to singular value inequalities. Math. Programming Stud. **18**, 112–134 (1982)
37. Mayne, D., Polak, E., Sangiovanni, A.: Computer aided design via optimization. Automatica **18**(2), 147–154 (1982)
38. Mifflin, R.: A quasi-second-order proximal bundle algorithm. Math. Programming, Ser. A **73**(1), 51–72 (1996)
39. Nedić, A., Bertsekas, D.P.: The effect of deterministic noise in subgradient methods. Math. Programming, Ser. A **125**(1), 75–99 (2010)
40. Noll, D.: Cutting plane oracles to minimize non-smooth non-convex functions. Set-Valued Variational Anal. **18**(3–4), 531–568 (2010)
41. Noll, D., Apkarian, P.: Spectral bundle method for non-convex maximum eigenvalue functions: first-order methods. Math. Programming, Ser. B **104**, 701–727 (2005)
42. Noll, D., Prot, O., Rondepierre, A.: A proximity control algorithm to minimize non-smooth and non-convex functions. Pacific J. Optim. **4**(3), 569–602 (2008)
43. Polak, E.: On the mathematical foundations of nondifferentiable optimization in engineering design. SIAM Rev. **29**, 21–89 (1987)
44. Polak, E., Wardi, Y.: A nondifferentiable optimization algorithm for the design of control systems subject to singular value inequalities over the frequency range. Automatica **18**, 267–283 (1982)
45. Rockafellar, R.T., Wets, R.: Variational Analysis. Springer, Berlin (2004)
46. Ruszczyński, A.: Nonlinear Optimization. Princeton University Press, Princeton (2006)
47. Sagastizábal, C.: Composite proximal bundle method. Math. Programming **138**(1) (2013)
48. Schramm, H., Zowe, J.: A version of the bundle idea for minimizing nondifferentiable functions: conceptual idea, convergence analysis, numerical results. SIAM J. Optim. **2**, 121–151 (1992)
49. Shor, N.: Minimization Methods for Nondifferentiable Functions. Springer, Berlin (1985)
50. Simões, A., Apkarian, P., Noll, D.: Non-smooth multi-objective synthesis with applications. Contr. Eng. Pract. **17**(11), 1338–1348 (2009)
51. Spingarn, J.E.: Submonotone subdifferentials of Lipschitz functions. Trans. Am. Math. Soc. **264**, 77–89 (1981)

52. Wolfe, P.: A method of conjugate subgradients for minimizing nondifferentiable functions. In: Balinski, M.L., Wolfe, P. (eds.) Nondifferentiable Optimization. Mathematical Programming Studies, vol. 3, pp. 145–173. North-Holland, Amsterdam (1975)
53. Zhou, K., Doyle, J.C., Glover, K.: Robust and Optimal Control. Prentice Hall, Upper Saddle River (1996)
54. Zowe, J.: The BT-algorithm for minimizing a non-smooth functional subject to linear constraints. In: Clarke, F.H., Demyanov, V.F., Gianessi, F. (eds.) Non-smooth Optimization and Related Topics. Plenum Press, New York (1989)

Chapter 27
Convergence of Linesearch and Trust-Region Methods Using the Kurdyka–Łojasiewicz Inequality

Dominikus Noll and Aude Rondepierre

Abstract We discuss backtracking linesearch and trust-region descent algorithms for unconstrained optimization and prove convergence to a critical point if the objective is of class C^1 and satisfies the Kurdyka–Łojasiewicz condition. For linesearch we investigate in which way an intelligent management memorizing the stepsize should be organized. For trust-regions we present a new curvature-based acceptance test which ensures convergence under rather weak assumptions.

Key words: Backtracking • Descent method • Kurdyka–Łojasiewicz inequality • Linesearch • Memorized steplength • Nonlinear optimization • Trust-region

Mathematics Subject Classifications (2010): Primary 65K05; Secondary 90C30, 90C53, 03C64.

COMMUNICATED BY HENRY WOLKOWICZ.

D. Noll (✉)
Institut de Mathématiques de Toulouse, Université Paul Sabatier, 118 route de Narbonne, 31062 Toulouse, France
e-mail: Dominikus.Noll@math.univ-toulouse.fr

A. Rondepierre
Institut de Mathématiques de Toulouse, INSA de Toulouse, 135 avenue de Rangueil 31077 Toulouse, France
e-mail: Aude.Rondepierre@math.univ-toulouse.fr

D.H. Bailey et al. (eds.), *Computational and Analytical Mathematics*, Springer Proceedings in Mathematics & Statistics 50, DOI 10.1007/978-1-4614-7621-4_27,

27.1 Introduction

Global convergence for linesearch descent methods traditionally only assures subsequence convergence to critical points (see, e.g., [4, Proposition 1.2.1] or [13, Theorem 3.2]), while convergence of the entire sequence of iterates is not guaranteed. Similarly, subsequence convergence in trust-region methods is established by relating the progress of trial points to the minimal progress achieved by the Cauchy point. These results are usual proved for $C^{1,1}$ or C^2-functions; see [8, Theorem 6.4.6] or [13, Theorem 4.8].

Recently Absil et al. [1] proved convergence of iterates of descent methods to a single limit point for analytic objective functions, using the fact that this class satisfies the so-called Łojasiewicz inequality [11,12]. Here we prove convergence of linesearch and trust-region descent methods to a single critical point for C^1 functions satisfying the Kurdyka–Łojasiewicz (KL) inequality [10], a generalization of the Łojasiewicz inequality. This is motivated by recent convergence results based on this condition in other fields; see, e.g., [2,3,5,6].

For linesearch methods we prove convergence for C^1 functions, and we show that it is allowed to memorize the accepted steplength between serious steps if the objective is of class $C^{1,1}$. This option may be of interest for large-scale applications, where second-order steps are not practical, and restarting each linesearch at $t = 1$ may lead to unnecessary and costly backtracking.

For trust-region methods we discuss acceptance tests which feature conditions on the curvature of the objective along the proposed step, in tandem with the usual criteria relating the achieved progress to the minimal progress guaranteed by the Cauchy point.

The paper is organized as follows. Section 27.2 presents the Kurdyka–Łojasiewicz inequality. Sections 27.3–27.5 are devoted to the convergence of backtracking linesearch for functions satisfying the KL inequality. In Sect. 27.6 convergence for trust-region methods under the KL condition is discussed and new conditions to guarantee convergence in practice are investigated.

27.2 The Kurdyka–Łojasiewicz Condition

In 1963 Łojasiewicz [11, 12] proved that a real analytic function $f : \mathbb{R}^n \to \mathbb{R}$ has the following property, now called the Łojasiewicz property. Given a critical point $\bar{x} \in \mathbb{R}^n$ of f, there exists a neighborhood U of $\bar{x}$, $c > 0$ and $\frac{1}{2} \leq \theta < 1$ such that

$$|f(x) - f(\bar{x})|^\theta \leq c\|\nabla f(x)\|$$

for all $x \in U$. In 1998 K. Kurdyka presented a more general construction which applies to differentiable functions definable in an o-minimal structure [10]. The following extension to nonsmooth functions has been presented in [5]:

Definition 27.1. A proper lower semicontinuous function $f : \mathbb{R}^n \to \mathbb{R} \cup \{+\infty\}$ has the Kurdyka–Łojasiewicz property (for short KL-property) at $\bar{x} \in \operatorname{dom} \partial f = \{x \in \mathbb{R}^n : \partial f(x) \neq \emptyset\}$ if there exist $\eta > 0$, a neighborhood U of $\bar{x}$, and a continuous concave function $\varphi : [0, \eta] \to [0, +\infty)$ such that:

1. $\varphi(0) = 0$, φ is C^1 on $(0, \eta)$, and $\varphi' > 0$ on $(0, \eta)$.
2. For every $x \in U \cap \{x \in \mathbb{R}^n : f(\bar{x}) < f(x) < f(\bar{x}) + \eta\}$,

$$\varphi'\left(f(x) - f(\bar{x})\right) \operatorname{dist}(0, \partial f(x)) \geq 1. \tag{27.1}$$

The Łojasiewicz inequality or property is a special case of the KL-property when $\varphi(s) = s^{1-\theta}$, $\theta \in [\frac{1}{2}, 1)$. It is automatically satisfied for noncritical points, so (27.1) is in fact a condition on critical points. We will need the following preparatory result.

Lemma 27.2. *Let $K \subset \mathbb{R}^n$ be compact. Suppose f is constant on K and has the KL-property at every $\bar{x} \in K$. Then there exist $\varepsilon > 0$, $\eta > 0$, and a continuous concave function $\varphi : [0, \eta] \to [0, \infty)$, which is C^1 on $(0, \eta)$ and satisfies $\varphi(0) = 0$, $\varphi' > 0$ on $(0, \eta)$, such that*

$$\varphi'(f(x) - f(\bar{x}))\operatorname{dist}(0, \partial f(x)) \geq 1$$

for every $\bar{x} \in K$ and every x such that $\operatorname{dist}(x, K) < \varepsilon$ and $f(\bar{x}) < f(x) < f(\bar{x}) + \eta$.

Proof. The proof is a slight extension of a similar result in [2] for functions having the Łojasiewicz property.

For every $\bar{x} \in K$ pick a neighborhood $B(\bar{x}, \varepsilon_{\bar{x}})$ of $\bar{x}$ and $\eta_{\bar{x}} > 0$ in tandem with a function $\varphi_{\bar{x}}$ as in Definition 27.1. Since K is compact, there exist finitely many $\bar{x}_i \in K$, $i = 1, \ldots, N$ such that $K \subset \bigcup_{i=1}^N B(\bar{x}_i, \frac{1}{2}\varepsilon_{\bar{x}_i})$. Write for simplicity $\varepsilon_i := \varepsilon_{\bar{x}_i}$, $\eta_i := \eta_{\bar{x}_i}$, $\varphi_i := \varphi_{\bar{x}_i}$. Then put

$$\eta = \min_{i=1\ldots N} \eta_i > 0 \quad \text{and} \quad \varepsilon = \min_{i=1,\ldots,N} \tfrac{1}{2}\varepsilon_i > 0.$$

It follows immediately that $\{x \in \mathbb{R}^n : \operatorname{dist}(x, K) < \varepsilon\} \subset \bigcup_{i=1}^{N} B(\bar{x}_i, \varepsilon_i)$.

Suppose $f(x) = \underline{f}$ for every $x \in K$. Then (27.1) holds uniformly on K in the sense that given any x with $\operatorname{dist}(x, K) < \varepsilon$ and $\underline{f} < f(x) < \underline{f} + \eta$, there exists $i(x) \in \{1, \ldots, N\}$ such that

$$\varphi'_{i(x)}(f(x) - \underline{f}) \operatorname{dist}(0, \partial f(x)) \geq 1.$$

To conclude the proof, it remains to define the function $\varphi : [0, \eta] \to [0, \infty)$. We let

$$\varphi(t) = \int_0^t \max_{i=1\ldots N} \varphi'_i(\tau)\, d\tau, \qquad t \in [0, \eta].$$

Observe that $\tau \mapsto \max_{i=1\ldots N} \varphi'_i(\tau)$ is continuous on $(0, \eta)$ and decreasing on $[0, \eta]$. Then φ is well defined and continuous on $[0, \eta]$ and of class C^1 on $(0, \eta)$. We also easily check $\varphi(0) = 0$, φ concave on $[0,\eta]$ and strictly increasing on $(0,\eta)$. Finally we have

$$\begin{aligned}\varphi'(f(x)-f(\bar{x}))\,\mathrm{dist}\,(0,\partial f(x)) &= \varphi'(f(x)-\underline{f})\,\mathrm{dist}\,(0,\partial f(x))\\ &\geq \varphi'_{i(x)}(f(x)-f(\bar{x}))\,\mathrm{dist}\,(0,\partial f(x)) \geq 1\end{aligned}$$

for all $\bar{x} \in K$ and all $x \in \mathbb{R}^n$ such that $\mathrm{dist}(x,K) < \varepsilon$ and $\underline{f} < f(x) < \underline{f} + \eta$. ■

Next we address convergence of linesearch methods assuming f of class C^1 and having the (KL) property. We will need the following technical lemma, whose proof can be found, e.g., in [7]:

Lemma 27.3. *Let f be of class C^1 and $x_j \to x$, $y_j \to x$. Then*

$$\frac{f(y_j)-f(x_j)-\nabla f(x_j)^\top (y_j-x_j)}{\|y_j-x_j\|} \to 0.$$

27.3 Linesearch Without Memory

Descent methods which attempt second-order steps usually start the linesearch at the steplength $t = 1$. We refer to this as *memory-free*. The challenge is to prove convergence for C^1 functions.

The algorithm discussed hereafter uses the following well-known definition:

Definition 27.4. A sequence d^j of descent directions chosen by a descent algorithm at points x^j is called *gradient oriented* if there exists $0 < c < 1$ such that the angle $\phi_j := \angle\left(d^j, -\nabla f(x^j)\right)$ satisfies

$$\forall j \in \mathbb{N}, 0 < c \leq \cos\phi_j. \tag{27.2}$$

Algorithm (Linesearch descent method without memory).

Parameters: $0 < \gamma < 1$, $0 < \underline{\theta} < \overline{\theta} < 1$, $\tau > 0$, $0 < c < 1$.

1: **Initialize**. Choose initial guess x^1. Put counter $j = 1$.

2: **Stopping test**. Given iterate x^j at counter j, stop if $\nabla f(x^j) = 0$. Otherwise compute a gradient oriented descent direction d^j with $\cos\phi_j \geq c$ and goto linesearch.

3: **Initialize linesearch**. Put linesearch counter $k = 1$ and initialize steplength t_1 such that:

$$t_1 \geq \tau \frac{\|\nabla f(x^j)\|}{\|d^j\|}.$$

4: **Acceptance test.** At linesearch counter k and steplength $t_k > 0$ check whether

$$\rho_k = \frac{f(x^j) - f(x^j + t_k d^j)}{-t_k \nabla f(x^j)^\top d^j} \geq \gamma.$$

If $\rho_k \geq \gamma$, put $x^{j+1} = x^j + t_k d^j$, quit linesearch, increment counter j, and go back to step 2.
On the other hand, if $\rho_k < \gamma$, reduce steplength such that $t_{k+1} \in [\underline{\theta} t_k, \overline{\theta} t_k]$, increment linesearch counter k, and continue linesearch with step 4.

Lemma 27.5. *Suppose f is differentiable and $\nabla f(x^j) \neq 0$ and let d^j be a descent direction at x^j. Then the linesearch described in Algorithm 27.3 needs a finite number of backtracks to find a steplength t_k such that $x^j + t_k d^j$ passes the acceptance test $\rho_k \geq \gamma$.*

Proof. The proof is straightforward. Suppose the linesearch never ends, then $\rho_k < \gamma$ for all k and $t_k \to 0$. Since $f'(x^j, d^j) = \nabla f(x^j)^\top d^j < 0$, $\rho_k < \gamma$ transforms into

$$\frac{f(x^j + t_k d^j) - f(x^j)}{t_k} > \gamma \nabla f(x^j)^\top d^j = \gamma f'(x^j, d^j),$$

and the left-hand side converges to $f'(x^j, d^j)$. This leads to $0 > f'(x^j, d^j) \geq \gamma f'(x^j, d^j)$, contradicting $0 < \gamma < 1$. ■

Having proved that an acceptable steplength is found in a finite number of backtracks, we now focus on convergence of the whole algorithm. The proof of Theorem 27.6 below first establishes stationarity of limit points, generalizing well-known results for gradient methods (see, e.g., [4, Proposition 1.2.1]), and then proves the convergence of the iterates using the Kurdyka–Łojasiewicz condition.

Theorem 27.6. *Let $\Omega = \{x \in \mathbb{R}^n : f(x) \leq f(x^1)\}$ be bounded. Suppose f is of class C^1 and satisfies the Kurdyka–Łojasiewicz condition. Then the sequence of iterates x^j generated by Algorithm 27.3 is either finite and ends with $\nabla f(x^j) = 0$, or it converges to a critical point $\bar{x}$ of f.*

Proof.

(1) We can clearly concentrate on the case of an infinite sequence x^j. Consider the following normalized sequence of descent directions $\tilde{d}^j = (\|\nabla f(x^j)\| / \|d^j\|) d^j$. Then the directions $\tilde{d}^j$ are also gradient oriented and $\|\tilde{d}^j\| = \|\nabla f(x^j)\|$. A trial step $x^j + t d^j$ can then also be written as $x^j + \tilde{t} \tilde{d}^j$, where the stepsizes t, $\tilde{t}$ are in one-to-one correspondence via $\tilde{t} = (\|d^j\| / \|\nabla f(x^j)\|) t$. Neither the backtracking rule in step 4 nor the acceptance test is affected if we write steps $x^j + t d^j$ as $x^j + \tilde{t} \tilde{d}^j$. The initial condition in step 3 becomes $\tilde{t} \geq \tau$. Switching back to the notation $x^j + t d^j$, we may therefore assume $\|d^j\| = \|\nabla f(x^j)\|$ and that the linesearch is initialized at $t_1 \geq \tau$. The gradient oriented direction d^j now satisfies

$$\|\nabla f(x^j)\|^2 \geq -\nabla f(x^j)^\top d^j \geq c \|d^j\| \|\nabla f(x^j)\| = c \|\nabla f(x^j)\|^2. \tag{27.3}$$

(2) From Lemma 27.5 we know that the linesearch ends after a finite number of backtracks, let us say with steplength $t_{k_j} > 0$. So $x^{j+1} = x^j + t_{k_j} d^j$. From the acceptance test $\rho_{k_j} \geq \gamma$ we know that

$$\begin{aligned} f(x^j) - f(x^{j+1}) &\geq -\gamma \nabla f(x^j)^\top (x^{j+1} - x^j), \\ &\geq -\gamma t_{k_j} \nabla f(x^j)^\top d^j \\ &\geq c\gamma t_{k_j} \|\nabla f(x^j)\|^2 \quad \text{[according to (27.3)].} \qquad (27.4) \end{aligned}$$

By construction we have $t_{k_j} = \|x^{j+1} - x^j\| / \|d^j\| = \|x^{j+1} - x^j\| / \|\nabla f(x^j)\|$, so that

$$f(x^j) - f(x^{j+1}) \geq c\gamma \|\nabla f(x^j)\| \|x^{j+1} - x^j\|, \qquad (27.5)$$

in which we recognize the so-called strong descent condition in [1]. Summing (27.5) from $j = 1$ to $j = m - 1$ gives

$$\sum_{j=1}^{m-1} \|\nabla f(x^j)\| \|x^{j+1} - x^j\| \leq (c\gamma)^{-1} \sum_{j=1}^{m-1} f(x^j) - f(x^{j+1}) = (c\gamma)^{-1} \left(f(x^1) - f(x^m) \right).$$

Since the algorithm is of descent type, the right-hand side is bounded above, so the series on the left is summable. In particular, $\|\nabla f(x^j)\| \|x^{j+1} - x^j\| \to 0$, or equivalently $t_{k_j} \|\nabla f(x^j)\|^2 \to 0$.

(3) Fix an accumulation point $\bar{x}$ of x^j and select a subsequence $j \in J$ such that $x^j \to \bar{x}$, $j \in J$. To show that $\bar{x}$ is critical, it suffices to find a subsequence $j' \in J'$ such that $\nabla f(x^{j'}) \to 0$.

Suppose on the contrary that no such subsequence exists, so that $\|\nabla f(x^j)\| \geq \mu > 0$ for some $\mu > 0$ and all $j \in J$. To obtain a contradiction, we will focus on the last step before acceptance.

(3.1) First note that we must have $t_{k_j} \to 0$, $j \in J$. Indeed using $\|\nabla f(x^j)\| \|x^{j+1} - x^j\| \geq \mu \|x^{j+1} - x^j\|$, $j \in J$ in tandem with the results from part (2), we see that $\|x^{j+1} - x^j\| \to 0$, $j \in J$. Then, knowing that

$$t_{k_j} = \|x^{j+1} - x^j\| / \|\nabla f(x^j)\| \leq \mu^{-1} \|x^{j+1} - x^j\|,$$

we deduce $t_{k_j} \to 0$ and by boundedness of the x^j also $t_{k_j} \|\nabla f(x^j)\| \to 0$, $j \in J$.

(3.2) We now claim that there exists an infinite subsequence J' of J such that (i) $\|\nabla f(x^j)\| \geq \mu > 0$, $j \in J'$, (ii) $t_{k_j} \to 0$, $j \in J'$, and (iii) $k_j \geq 2$ for $j \in J'$, i.e., for $j \in J'$, there was at least one backtrack during the jth linesearch. Item (iii) is a consequence of the initial condition $t_1 \geq \tau$ in step 3 of the algorithm. Namely, in tandem with $t_{k_j} \to 0$, $j \in J$, this condition says that the set $J' = \{j \in J : k_j \geq 2\} = \{j \in J : t_{k_j} < t_1\}$ cannot be finite.

This sequence $j \in J'$ satisfies $\rho_{k_j} \geq \gamma$, $\rho_{k_j-1} < \gamma$, $t_{k_j} \to 0$, $\|\nabla f(x^j)\| \geq \mu > 0$. Because of the backtracking rule, we then also have $t_{k_j-1} \to 0$. Putting $y^{k_j-1} = x^j + t_{k_j-1} d^j$, given that $x^j \to \bar{x}$, $t_{k_j}\|\nabla f(x^j)\| \to 0$, $j \in J'$, and $t_{k_j-1}\|d^j\| = t_{k_j-1}\|\nabla f(x^j)\| \leq \underline{\theta}^{-1} t_{k_j}\|\nabla f(x^j)\|$, we have $y^{k_j-1} \to \bar{x}$, $j \in J'$.

Note that d^j is gradient oriented so that $y^{k_j-1} - x^j$ is also gradient oriented and

$$-\nabla f(x^j)^\top (y^{k_j-1} - x^j) \geq c\|\nabla f(x^j)\|\|y^{k_j-1} - x^j\| \geq c\mu\|y^{k_j-1} - x^j\|. \tag{27.6}$$

(3.3) Now we expand

$$\begin{aligned}\rho_{k_j-1} &= \frac{f(x^j)-f(y^{k_j-1})}{-\nabla f(x^j)^\top (y^{k_j-1}-x^j)} = 1-\frac{f(y^{k_j-1})-f(x^j)-\nabla f(x^j)^\top (y^{k_j-1}-x^j)}{-\nabla f(x^j)^\top (y^{k_j-1}-x^j)}\\ &=: 1-R_j.\end{aligned}$$

Using (27.6) gives

$$\begin{aligned}|R_j| &= \frac{|f(y^{k_j-1}) - f(x^j) - \nabla f(x^j)^\top (y^{k_j-1} - x^j)|}{-\nabla f(x^j)^\top (y^{k_j-1} - x^j)}\\ &\leq \frac{|f(y^{k_j-1}) - f(x^j) - \nabla f(x^j)^\top (y^{k_j-1} - x^j)|}{c\mu\|y^{k_j-1} - x^j\|}.\end{aligned}$$

Since f is of class C^1 and since $x^j \to \bar{x}$, $y^{k_j-1} \to \bar{x}$, Lemma 27.3 guarantees the existence of a sequence $\varepsilon_j \to 0$ such that

$$\left|f(y^{k_j-1}) - f(x^j) - \nabla f(x^j)^\top (y^{k_j-1} - x^j)\right| \leq \varepsilon_j\|y^{k_j-1} - x^j\|.$$

We deduce $|R_j| \leq \varepsilon_j/(c\mu) \to 0$; hence $\rho_{k_j-1} \to 1$ contradicting $\rho_{k_j-1} < \gamma$. This proves that $\|\nabla f(x^j)\| \geq \mu > 0$ for all $j \in J$ was impossible. Therefore $\bar{x}$ is critical, and so are all the accumulation points of x^j.

(4) By boundedness of the sequence x^j the set K of its accumulation points $\bar{x}$ is bounded and consists of critical points of f. It is also closed, as can be shown by a diagonal argument. Hence K is compact. Since the algorithm is of descent type, f has constant value on K.

Since f satisfies the Kurdyka–Łojasiewicz condition at every $\bar{x} \in K$, Lemma 27.2 gives us $\varepsilon > 0$, $\eta > 0$, and a continuous concave function $\varphi : [0,\eta] \to [0,\infty)$ with $\varphi(0) = 0$ and $\varphi' > 0$ on $(0,\eta)$ such that for every $\bar{x} \in K$ and every x with $\text{dist}(x,K) < \varepsilon$ and $f(\bar{x}) < f(x) < f(\bar{x}) + \eta$ we have

$$\varphi'\left(f(x) - f(\bar{x})\right)\|\nabla f(x)\| \geq 1. \tag{27.7}$$

(5) Assume without loss of generality that $f(\bar{x}) = 0$ on K. Then $f(x^j) > 0$ for all j, because our algorithm is of descent type. Concavity of φ implies

$$\varphi\left(f(x^j)\right) - \varphi\left(f(x^{j+1})\right) \geq \varphi'\left(f(x^j)\right)\left(f(x^j) - f(x^{j+1})\right). \quad (27.8)$$

Using $f(\bar{x}) = 0$, the Kurdyka–Łojasiewicz estimate (27.7) gives

$$\varphi'\left(f(x^j)\right) = \varphi'\left(f(x^j) - f(\bar{x})\right) \geq \|\nabla f(x^j)\|^{-1}. \quad (27.9)$$

Hence by (27.8)

$$\begin{aligned}\varphi\left(f(x^j)\right) - \varphi\left(f(x^{j+1})\right) &\geq \|\nabla f(x^j)\|^{-1}\left(f(x^j) - f(x^{j+1})\right) \\ &\geq c\gamma\|x^{j+1} - x^j\| \qquad \text{[using (27.5)].}\end{aligned}$$

Summing from $j = 1$ to $j = m-1$ gives

$$c\gamma\sum_{j=1}^{m-1}\|x^j - x^{j+1}\| \leq \varphi\left(f(x^1)\right) - \varphi\left(f(x^m)\right),$$

and since the term on the right-hand side is bounded, the series on the left converges. This shows that x^j is a Cauchy sequence, which converges therefore to some $\bar{x} \in K$, proving that $K = \{\bar{x}\}$ is singleton. ■

27.4 Memorizing the Steplength

In Newton-type descent schemes it is standard to start the linesearch at steplength $t = 1$. However, if a first-order method is used, a different strategy may be more promising. To avoid unnecessary backtracking, we may decide to start the $(j+1)$st linesearch where the jth ended. Such a concept may be justified theoretically if f is of class $C^{1,1}$.

Standard proofs for backtracking linesearch algorithms use indeed $C^{1,1}$ functions. The Lipschitz constant of ∇f on Ω allows a precise estimation of the Armijo stepsize

$$t_\gamma = \sup\{t > 0 : f(x+td) - f(x) < \gamma t\nabla f(x)^\top d\}.$$

As long as the linesearch starts with large steps, $t > t_\gamma$, backtracking $t_{k+1} \in [\underline{\theta}t_k, \overline{\theta}t_k]$ will lead to an acceptable steplength t^* such that $\underline{\theta}t_\gamma \leq t^* \leq t_\gamma$. This mechanism guarantees that the accepted steplength is not too small and replaces the usual conditions against small stepsizes. However, what we plan to do in this section is to memorize the last accepted steplength. So the above argument will not work, because our linesearch may already start small, and we will have no guarantee to end up in the interval $[\underline{\theta} t_\gamma, t_\gamma]$. In that situation the safeguard against too small steps is more subtle to assure. We propose the following:

Algorithm (Descent method with memorized steplength).

Parameters: $0 < \gamma < \Gamma < 1, 0 < c < 1, 0 < \underline{\theta} < \overline{\theta} < 1, \Theta > 1$.

1: **Initialize**. Choose initial guess x^1. Fix memory steplength $\tau_1 = 1$. Put counter $j = 1$.

2: **Stopping test**. Given iterate x^j at counter j, stop if $\nabla f(x^j) = 0$. Otherwise compute descent direction d^j with $\|d^j\| = \|\nabla f(x^j)\|$ and $\cos\phi_j \geq c$ and goto linesearch.

3: **Initialize linesearch**. Put linesearch counter $k = 1$ and use memory steplength τ_j to initialize linesearch at steplength $t_1 = \tau_j$.

4: **Acceptance test.** At linesearch counter k and steplength $t_k > 0$ check whether

$$\rho_k = \frac{f(x^j) - f(x^j + t_k d^j)}{-t_k \nabla f(x^j)^\top d^j} \geq \gamma.$$

If $\rho_k \geq \gamma$ put $x^{j+1} = x^j + t_k d^j$, quit linesearch and goto step 5. On the other hand, if $\rho_k < \gamma$ backtrack by reducing steplength to $t_{k+1} \in [\underline{\theta} t_k, \overline{\theta} t_k]$ and continue linesearch with step 4.

5: **Update memory steplength**. Define the new memory steplength τ_{j+1} as

$$\tau_{j+1} = \begin{cases} t_k & \text{if } \gamma \leq \rho_k < \Gamma \\ \Theta t_k & \text{if } \rho_k \geq \Gamma \end{cases},$$

where t_k is the accepted steplength in step 4. Increment counter j and go back to step 2.

Theorem 27.7. *Let $\Omega = \{x \in \mathbb{R}^n : f(x) \leq f(x^1)\}$ be bounded, and suppose f satisfies the Kurdyka–Łojasiewicz condition and is of class $C^{1,1}(\Omega)$. Let x^j be the sequence of steps generated by the descent Algorithm 27.4. Then either $\nabla f(x^j) = 0$ for some j or x^j converges to a critical point of f.*

Proof.

(1) As in the proof of Theorem 27.6 we concentrate on the case where the sequence x^j is infinite. As required by Algorithm 27.4, the sequence d^j is already normalized to $\|d^j\| = \|\nabla f(x^j)\|$. We now follow the proof of Theorem 27.6 until the end of part (2), where $t_{k_j}\|\nabla f(x^j)\|^2 \to 0$ is proved.

(2) We wish to prove $\nabla f(x^j) \to 0$, $j \in \mathbb{N}$. Assume on the contrary that there exists an infinite set $J \subset \mathbb{N}$ such that $\|\nabla f(x^j)\| \geq \mu > 0$ for all $j \in J$. Then we must have $t_{k_j} \to 0$, $j \in J$. This is shown precisely as in part (3.1) of the proof of Theorem 27.6.

(3) Using the sequence $j \in J$ which satisfies $\|\nabla f(x^j)\| \geq \mu$ and $t_{k_j} \to 0$, $j \in J$, we now have the first substantial modification. We construct another infinite

sequence $J' \subset \mathbb{N}$ such that $t_{k_j} \to 0$, $j \in J'$, and such that in addition for every $j \in J'$ the jth linesearch did at least one backtrack. In other words, $k_j \geq 2$ for every $j \in J'$. In contrast with Theorem 27.6 we do not claim that J' is a subsequence of J. Neither do we have any information as to whether $\|\nabla f(x^j)\| \geq \mu$ for $j \in J'$, and we therefore cannot use such an estimate, as we did in the proof of Theorem 27.6.

Now J' can be constructed as follows. Put

$$j'(j) = \min\{j' \in \mathbb{N} : j' \geq j, k_{j'} \geq 2\}, \quad \text{and} \quad J' = \{j'(j) : j \in J\}.$$

We claim that $j'(j) < \infty$ for every $j \in J$. Suppose there exists $j \in J$ such that $k_{j'} = 1$ for all $j' \geq j$. Then no backtracking is done in any of the linesearches j' following j. Since the stepsize t is not decreased between linesearches, it is not decreased at all, so it cannot become arbitrarily small any more. This contradicts $t_{k_j} \to 0$, $j \in J$. This argument shows $j \leq j'(j) < \infty$ for all $j \in J$, so J' is an infinite set.

For the indices $j \in J'$ we have $k_j \geq 2$ and

$$t_{k_j} \text{ accepted}, \qquad t_{k_j-1} \text{ rejected}, \qquad \underline{\theta} t_{k_j-1} \leq t_{k_j} \leq \overline{\theta} t_{k_j-1}.$$

In particular, $\rho_{k_j-1} < \gamma$, $\rho_{k_j} \geq \gamma$. Moreover, $t_{k_j-1} \to 0$, $j \in J'$. Writing $y^{k_j-1} = x^j + t_{k_j-1} d^j$, we see that $x^j - y^{k_j-1} \to 0$, $j \in J'$. Now we expand

$$\begin{aligned} \rho_{k_j-1} &= \frac{f(x^j) - f(y^{k_j-1})}{-t_{k_j-1} \nabla f(x^j)^\top d^j} = 1 - \frac{f(y^{k_j-1}) - f(x^j) - \nabla f(x^j)^\top (y^{k_j-1} - x^j)}{-t_{k_j-1} \nabla f(x^j)^\top d^j} \\ &=: 1 + R_j. \end{aligned}$$

Since f is of class $C^{1,1}$ and since the sequences x^j and y^{k_j-1} are bounded and $x^j - y^{k_j-1} \to 0$, there exists a constant $L > 0$ (the Lipschitz constant of ∇f on Ω) such that

$$\left| f(y^{k_j-1}) - f(x^j) - \nabla f(x^j)(y^{k_j-1} - x^j) \right| \leq \tfrac{L}{2} \|y^{k_j-1} - x^j\|^2 = \tfrac{L}{2} t_{k_j-1}^2 \|d^j\|^2$$

for all $j \in J'$. Gradient orientedness of d^j implies $|\nabla f(x^j)^\top d^j| \geq c\|d^j\|^2$, so the residual term R_j may be estimated as

$$|R_j| \leq \frac{\frac{L}{2} t_{k_j-1}^2 \|d^j\|^2}{c t_{k_j-1} \|d^j\|^2} = (L/2c)\, t_{k_j-1} \to 0 \quad (j \in J').$$

That shows $\rho_{k_j-1} \to 1$, $(j \in J')$, contradicting $\rho_{k_j-1} < \gamma$. This argument proves $\nabla f(x^j) \to 0$, $j \to \infty$. In consequence, every accumulation point $\bar{x}$ of the sequence x^j is a critical point.

(4) The remainder of the proof is now identical with (4)–(5) in the proof of Theorem 27.6, and the conclusion is the same. ∎

27.5 A Practical Method

In Algorithm 27.4 we cannot a priori exclude the possibility that τ_j becomes arbitrarily small, even though it has in principle the possibility to recover if good steps are made (see step 5 of Algorithm 27.4). Let us see what happens if $d^j = -P_j^{-1}\nabla f(x^j)$, where P_j is the Hessian of f or a quasi-Newton substitute of the Hessian. The crucial question is, will this method eventually produce good steps $\rho_k \geq \Gamma$, so that the memorized steplength increases to reach $\tau_j = 1$, from whereon the full Newton step is tried first?

Theorem 27.8. *Let $0 < \gamma < \Gamma < \frac{1}{2}$. Let the Newton steps $d^j = -\nabla^2 f(x^j)^{-1}\nabla f(x^j)$ at x^j form a sequence of gradient oriented descent directions. Let $\bar{x}$ be a local minimum of f satisfying the second-order sufficient optimality condition.*

Then there exists a neighborhood V of $\bar{x}$ such that as soon as $x^j \in V$, the iterates stay in V, the first trial step $x^{j+1} = x^j + t_1 d^j$ is accepted with $\rho_1 \geq \Gamma$, so that the memory steplength is increased from τ_j to $\tau_{j+1} = \min\{\Theta\tau_j, 1\}$, until it reaches 1 after a finite number of steps. From that moment on the full Newton step is tried and accepted, and the method converges quadratically to $\bar{x}$.

Proof. This theorem is similar to Theorem 6.4 from [9] with the following differences: the step t_k does not necessarily satisfy the second Wolfe condition, and the sequence x^j of iterates is not assumed to converge toward $\bar{x}$. Instead we have to use the hypothesis of gradient orientedness and the backtracking process of the linesearch to prove the same result.

Since the local minimum $\bar{x}$ satisfies the second-order sufficient optimality condition, the Hessian of f at $\bar{x}$ is positive definite, and we have $\mu := \lambda_{\min}(\nabla^2 f(\bar{x})) > 0$. Using a well-known result on Newton's method (see, e.g., [9, Theorem 2.1]), there exists an open neighborhood U of the local minimum $\bar{x}$, where the Newton iterates are well defined, remain in U, and converge to $\bar{x}$ and

$$\lambda_{\min}(\nabla^2 f(x)) \geq \frac{\mu}{2} \quad \text{and} \quad \lambda_{\max}(\nabla^2 f(x)) \leq K < \infty \tag{27.10}$$

for every $x \in U$.

Assume now that the iterates x^j reach U. We first prove that the Newton step is acceptable in the sense that $f(x^j + d^j) - f(x^j) < \gamma \nabla f(x^j)^\top d^j$ because of $\gamma < \frac{1}{2}$. Indeed, as in the proof of Theorem 6.4 in [9], the combined use of the mean value theorem, gradient orientedness, and hypothesis (27.10) implies that for all j with $x^j \in U$, the Newton iterate $x^j + d^j$ is accepted by any Armijo parameter $< \frac{1}{2}$, so that it even passes the acceptance test with the larger constant Γ instead of γ due to $0 < \gamma < \Gamma < \frac{1}{2}$. Note that the same is then true for every damped Newton step, namely as soon as $t = 1$ passes the acceptance test, so does any $t < 1$.

The last point is to prove that if the iterates x^j enter U with $\tau_j < 1$, then our algorithm starts to increase τ until the Newton step is actually made. Indeed, even though at the beginning a smaller step $x^j + td^j$ with $t < 1$ is made, according to what was previously shown, this step is accepted at once with $\rho_1 > \Gamma$ and remains in U.

We then update $\tau_{j+1} = \Theta \tau_j$ (with a fixed $\Theta > 1$), meaning that τ_j is increased until it hits 1 after a finite number of iterations j. From that moment onward the Newton step is tried first and then accepted at once, and quadratic convergence prevails. ■

Remark 27.9. This result indicates that Γ should be only slightly larger than γ, at least near the second-order minimum.

Remark 27.10. The following modification of the update rule of τ seems interesting. Fix $1 < \Theta < \Xi$ and put

$$\tau_{j+1} = \begin{cases} t_{k_j} & \text{if } \gamma \leq \rho_{k_j} < \Gamma, \\ \Theta t_{k_j} & \text{if } \rho_{k_j} \geq \Gamma \text{ and } k_j \geq 2, \\ \Xi t_{k_j} & \text{if } \rho_{k_j} \geq \Gamma \text{ and } k_j = 1. \end{cases}$$

This accelerates the increase of τ if acceptance is immediate and helps to get back to $\tau = 1$ faster if the neighborhood of attraction of Newton's method is reached. Our convergence analysis covers this case as well.

27.6 Convergence of Trust-Region Methods for Functions of Class C^1

The idea of memorizing the steplength in a linesearch method is paralleled by the trust-region strategy. The basic trust-region algorithm uses a quadratic model

$$m(y,x^j) = f(x^j) + \nabla f(x^j)^\top (y - x^j) + \frac{1}{2}(y - x^j)^\top B_j (y - x^j)$$

to approximate the objective function f within the trust-region $\{x \in \mathbb{R}^n : \|y - x^j\| \leq \Delta_k\}$ around the current iterate x^j, where $\Delta_k > 0$ is the trust-region radius and B_j an approximation of the Hessian at x^j. One then computes an approximate solution y^{k+1} of the tangent program

$$\min\{m(y,x^j) : \|y - x^j\| \leq \Delta_k, y \in \mathbb{R}^n\}. \tag{27.11}$$

Instead of minimizing the trust-region model, the step y^{k+1} is only supposed to achieve a decrease of $m(\cdot, x^j)$, which is at least a given percentage of the reduction obtained by the Cauchy point x_C^{j+1}. This means y^{k+1} satisfies

$$f(x^j) - m(y^{k+1}, x^j) \geq c \left[f(x^j) - m(x_C^{j+1}, x^j) \right], \tag{27.12}$$

where $0 < c < 1$ is fixed once and for all and where the Cauchy point x_C^{j+1} is defined as the solution of the one-dimensional problem:

$$\min\left\{m\left(x^j - t\frac{\nabla f(x^j)}{\|\nabla f(x^j)\|}, x^j\right) : t \in \mathbb{R}, 0 \le t \le \Delta_k\right\}. \tag{27.13}$$

Here we follow the line of Conn et al. [8], who determine a step y^{k+1} satisfying the weaker condition

$$f(x^j) - m(y^{k+1}, x^j) \ge c\|\nabla f(x^j)\| \min\left(\Delta_k, \frac{\|\nabla f(x^j)\|}{1 + \|B_j\|}\right). \tag{27.14}$$

It can be shown that (27.12) implies (27.14) and that the exact solution of (27.11) satisfies (27.14). With these preparations we can now state our algorithm.

Algorithm (Trust-region method).

Parameters: $0 < \gamma < \Gamma < 1$, $0 < \underline{\theta} < \overline{\theta} < 1$, $\tau > 0$.

1: **Initialize**. Choose initial guess x^1 and initial trust-region radius $\Delta_1^\sharp > 0$. Put counter $j = 1$.

2: **Stopping test**. Given iterate x^j at counter j, stop if $\nabla f(x^j) = 0$. Otherwise goto step 3.

3: **Model definition.** Define a model $m(\cdot, x^j)$ of f in $\{x \in \mathbb{R}^n : \|x - x^j\| \le \Delta_j^\sharp\}$:

$$m(y, x^j) = f(x^j) + \nabla f(x^j)^\top (y - x^j) + \tfrac{1}{2}(y - x^j)^\top B_j (y - x^j).$$

4: **Initialize inner loop.** Put counter $k = 1$ and $\Delta_1 = \Delta_j^\sharp$.

5: **Tangent program.** At inner loop counter k let y^{k+1} be an approximate solution of

$$\min\{m(y, x^j) : \|y - x^j\| \le \Delta_k, y \in \mathbb{R}^n\}$$

in the sense of (27.12).

6: **Acceptance test.** At counter k, check whether

$$\rho_k = \frac{f(x^j) - f(y^{k+1})}{f(x^j) - m(y^{k+1}, x^j)} \ge \gamma. \tag{27.15}$$

- If $\rho_k \ge \gamma$ put $x^{j+1} = y^{k+1}$, and update:

$$\Delta_{j+1}^\sharp \in \begin{cases} [\Delta_k, +\infty[& \text{if } \rho_k > \Gamma \text{ and } \|y^{k+1} - x^j\| = \Delta_k \\ [\bar{\theta}\Delta_k, \Delta_k] & \text{otherwise.} \end{cases}$$

  Increment outer counter j, and go back to step 2.
- If $\rho_k < \gamma$, then: $\Delta_{k+1} \in [\underline{\theta}\Delta_k, \overline{\theta}\Delta_k]$. Increment inner counter k and go to step 5.

The trial point y^{k+1} computed in step 5 of the algorithm is called a *serious step* if accepted as a new iterate x^{j+1} and a *null step* if rejected. To decide whether a step y^{k+1} is accepted, we compute the ratio

$$\rho_k = \frac{f(x^j) - f(y^{k+1})}{f(x^j) - m(y^{k+1}, x^j)},$$

which reflects the agreement between f and its model at y^{k+1}. If the model $m(\cdot, x^j)$ is a good approximation of f at y^{k+1}, we expect $\rho_k \approx 1$, so here y^{k+1} is a good point and should be accepted. If $\rho_k \ll 1$, y^{k+1} is bad and we reject it. Step 6 of the algorithm formalizes this decision.

The proof of the global convergence of the trust-region algorithm for functions of class C^1 in the sense of subsequences can be found in, e.g., [13, Theorem 4.8]. One first proves finiteness of the inner loop and then global convergence of Algorithm 27.6.

Our issue here is to prove convergence of the sequence, which requires the Kurdyka–Łojasiewicz condition and the so-called strong descent condition in [1]:

Theorem 27.11. *Let $\Omega = \{x \in \mathbb{R}^n : f(x) \leq f(x^1)\}$ be bounded. Suppose f is of class C^1 and satisfies the Kurdyka–Łojasiewicz condition. Let the Hessian matrices B_j be uniformly bounded. If the sequence x^j, $j \in \mathbb{N}$, of iterates of Algorithm 27.6 satisfies the strong descent condition*

$$f(x^j) - f(x^{j+1}) \geq \sigma \|\nabla f(x^j)\| \|x^{j+1} - x^j\|, \tag{27.16}$$

then it is either finite and ends with $\nabla f(x^j) = 0$ or it converges to a critical point $\bar{x}$ of f.

Proof. Let K be the set of the accumulation points of the sequence x^j, $j \in \mathbb{N}$. As in the proof of Theorem 27.6 we prove compactness of K and show that f is constant on K. Then the Kurdyka–Łojasiewicz condition gives

$$\begin{aligned} \varphi(f(x^j)) - \varphi(f(x^{j+1})) &\geq \varphi'(f(x^j)) \left(f(x^j) - f(x^{j+1})\right) \\ &\geq \|\nabla f(x^j)\|^{-1} \left(f(x^j) - f(x^{j+1})\right). \end{aligned}$$

Assuming the strong descent condition $f(x^j) - f(x^{j+1}) \geq \sigma \|\nabla f(x^j)\| \|x^{j+1} - x^j\|$ as in [1] now yields

$$\varphi(f(x^j)) - \varphi(f(x^{j+1})) \geq \sigma \|x^{j+1} - x^j\|.$$

Using the series argument from Theorem 27.6 proves convergence of the sequence of iterates x^j to some $\bar{x} \in K$, and then $K = \{\bar{x}\}$. ■

Now we have to give practical criteria which imply the strong descent condition (27.16). Several easily verified conditions for the iterates of the trust-region algorithm are given in [1]. Here we focus on conditions involving the curvature of the model along the search direction. Let $\omega(y, x^j)$ denote the curvature of the model $m(\cdot,x^j)$ between x^j and y^{k+1}, namely,

$$\omega(y^{k+1},x^j) = \frac{(y^{k+1}-x^j)^\top B_j(y^{k+1}-x^j)}{\|y^{k+1}-x^j\|^2}.$$

Note that the curvature along the Cauchy point direction is

$$\omega(x_C^{j+1},x^j) = \frac{\nabla f(x^j)^\top B_j \nabla f(x^j)}{\|\nabla f(x^j)\|^2}.$$

We propose the following modified tangent program in Algorithm 27.6:

Fix $0 < \mu < 1$.

5': **Tangent program.** Compute an approximate solution y^{k+1} of

$$\min\{m(y,x^j) : \|y-x^j\| \leq \Delta_k, y \in \mathbb{R}^n\}$$

in the sense of (27.12), such that in addition

$$\omega(y^{k+1},x^j) \geq \mu\,\omega(x_C^{j+1},x^j) \geq 0 \tag{27.17}$$

as soon as the Cauchy point lies in the interior of the trust-region, i.e., if $\|\nabla f(x^j)\| \leq \Delta_k \omega(x_C^{j+1},x^j)$.

This modified step (5') in the algorithm has a solution y^{k+1}, because the Cauchy point satisfies the two conditions (27.12) and (27.17). We have to prove the convergence of the modified trust-region algorithm, which we will achieve by proving the strong descent condition. We will need the following preparatory:

Lemma 27.12. *When y^{k+1} is a descent step of the model $m(\cdot,x^j)$ away from x^j, then it satisfies*

$$\|\nabla f(x^j)\| \geq \frac{1}{2}\omega(y^{k+1},x^j)\|y^{k+1}-x^j\|.$$

Each serious step x^{j+1} generated by Algorithm 27.6 satisfies

$$\|\nabla f(x^j)\| \geq \tfrac{1}{2}\omega(x^{j+1},x^j)\|x^{j+1}-x^j\|.$$

Proof. By definition every descent step y^{k+1} of the model $m(\cdot,x^j)$ at the current iterate x^j has to verify $-\nabla f(x^j)^\top(y^{k+1}-x^j) > 0$ and $f(x^j) - m(y^{k+1},x^j) \geq 0$, so that

$$-\nabla f(x^j)^\top(y^{k+1}-x^j) \geq \tfrac{1}{2}(y^{k+1}-x^j)^\top B_j(y^{k+1}-x^j).$$

Using the Cauchy–Schwarz inequality $\|\nabla f(x^j)\|\|y^{k+1}-x^j\| \geq -\nabla f(x^j)^\top(y^{k+1}-x^j)$, we obtain

$$\|\nabla f(x^j)\| \geq \frac{1}{2}\frac{(y^{k+1}-x^j)^\top B_j(y^{k+1}-x^j)}{\|y^{k+1}-x^j\|} = \frac{1}{2}\omega(y^{k+1},x^j)\|y^{k+1}-x^j\|.$$

According to the acceptance test, any serious step is also a descent step of the model at the current iterate, which proves the second part of the lemma. ■

Note that the previous result is only useful when the curvature is positive.

Proposition 27.13. *The iterates x^j generated by the Algorithm 27.6 with step 5' replacing the original step 5 satisfy the strong descent condition* (27.16).

Proof. The idea here is to show that the Cauchy step is bounded below by a fraction of the step, i.e., there exists $\eta \in (0,1)$ such that

$$\|x_C^{j+1}-x^j\| \geq \eta\|x^{j+1}-x^j\|. \tag{27.18}$$

Indeed, the sufficient decrease condition (27.12) together with (27.18) gives strong descent (see Theorem 4.4 from [1]). By the definition of the Cauchy point we have

$$\|x_C^{j+1}-x^j\| = \begin{cases} \dfrac{\|\nabla f(x^j)\|}{\omega(x_C^{j+1},x^j)} & \text{if } \|\nabla f(x^j)\| \leq \Delta_{k_j}\omega(x_C^{j+1},x^j), \\ \Delta_{k_j} & \text{otherwise.} \end{cases}$$

In the first case, that is, when $\|\nabla f(x^j)\| \leq \Delta_{k_j}\omega(x_C^{j+1},x^j)$, the curvature condition (27.17) gives

$$\|x_C^{j+1}-x^j\| = \frac{\|\nabla f(x^j)\|}{\omega(x_C^{j+1},x^j)} \geq \mu\frac{\|\nabla f(x^j)\|}{\omega(x^{j+1},x^j)} \geq \frac{\mu}{2}\|x^{j+1}-x^j\|$$

according to Lemma 27.12. In the second case we have $\|x_C^{j+1}-x^j\| = \Delta_{k_j} \geq \|x^{j+1}-x^j\|$, since x^{j+1} has to belong to the trust-region. Thus (27.18) is satisfied in both case with $\eta = \frac{\mu}{2}$. ■

In the last part of the paper we present yet another version (5") of the tangent program based on condition (27.14) from Conn et al. [8], which allows to prove convergence and yet is weaker than the sufficient decrease condition. Note that this condition is at least satisfied by the Cauchy point and the exact solution of the tangent program.

Now with (5") each serious step satisfies

$$f(x^j)-m(x^{j+1},x^j) \geq c\|\nabla f(x^j)\|\min\left(\Delta_{k_j},\frac{\|\nabla f(x^j)\|}{\|B_j\|}\right)$$

5": **Tangent program.** Compute an approximate solution y^{k+1} of

$$\min\{m(y,x^j) : \|y-x^j\| \leq \Delta_k, y \in \mathbb{R}^n\}$$

in the sense of (27.14), i.e., $f(x^j) - m(y^{k+1},x^j) \geq c\|\nabla f(x^j)\| \min\left(\Delta_k, \frac{\|\nabla f(x^j)\|}{1+\|B_j\|}\right)$.

$$\begin{aligned}
&\geq c\|\nabla f(x^j)\| \min\left(\|x^{j+1}-x^j\|, \frac{\|\nabla f(x^j)\|}{\|B_j\|}\right) \\
&\geq c\min\left(1, \frac{\|\nabla f(x^j)\|}{\|B_j\|\|x^{j+1}-x^j\|}\right)\|\nabla f(x^j)\|\|x^{j+1}-x^j\|.
\end{aligned} \tag{27.19}$$

To infer the strong descent condition (27.16), the question is how to guarantee that $\frac{\|\nabla f(x^j)\|}{\|B_j\|\|x^{j+1}-x^j\|}$ remains bounded away from 0. Let us first consider the simpler case when the matrix B_j is positive.

Proposition 27.14. *Consider the following conditions:*

(H_1) *B_j is positive definite and there exists a $\kappa \geq 1$ such that*

$$\operatorname{cond}(B_j) := \|B_j\|\|B_j^{-1}\| \leq \kappa \quad \textit{(using the matrix 2-norm).}$$

(H_2) *There exists $\bar{\sigma} > 0$ and $\underline{\sigma} > 0$ such that $\bar{\sigma} I \succ B_j \succeq \underline{\sigma} I \succ 0$.*

Then $(H_2) \Rightarrow (H_1)$. Moreover condition (H_1) in tandem with the acceptance condition (27.14) *used within the modified step (5") of Algorithm 27.6 guarantees strong descent.*

Proof. Clearly (H_2) implies (H_1). Now for the second part assume that the matrix B_j is positive definite. Then the curvature of the model $m(\cdot,x^j)$ is also positive and by (27.19) and Lemma 27.12:

$$\begin{aligned}
f(x^j) - m(x^{j+1},x^j) &\geq c\min\left(1, \frac{\|\nabla f(x^j)\|}{\|B_j\|\|x^{j+1}-x^j\|}\right)\|\nabla f(x^j)\|\|x^{j+1}-x^j\| \\
&\geq c\min\left(1, \frac{1}{2}\frac{\omega(x^{j+1},x^j)}{\|B_j\|}\right)\|\nabla f(x^j)\|\|x^{j+1}-x^j\|.
\end{aligned}$$

Note that $\dfrac{\omega(x^{j+1},x^j)}{\|B_j\|} \leq 1$; therefore

$$f(x^j) - m(x^{j+1},x^j) \geq \frac{c}{2}\frac{\omega(x^{j+1},x^j)}{\|B_j\|}\|\nabla f(x^j)\|\|x^{j+1}-x^j\|.$$

Condition (H_1) clearly guarantees that $\omega(x^{j+1},x^j)/\|B_j\|$ stays bounded away from 0; hence we have strong descent (27.16). ■

In order to cover also those cases where B_j is not positive, we propose to replace the acceptance test (27.15) by the following. Fix $0<\mu<1$. The trial step y^{k+1} is accepted to become x^{j+1} if it satisfies

$$\rho_k = \frac{f(x^j)-f(y^{k+1})}{f(x^j)-m(y^{k+1},x^j)} \geq \gamma \quad \text{and} \quad \|\nabla f(x^j)\| \geq \mu \|B_j\| \|x^{j+1}-x^j\|. \quad (27.20)$$

The following result shows that condition (27.20) is eventually satisfied by the trial steps y^{k+1} according to (5"). Convergence of the trust-region algorithm with the tangent program (5") follows then with the same method of proof.

Proposition 27.15. *Let $x \in \mathbb{R}^n$ be the current iterate. Suppose f is differentiable and $\nabla f(x) \neq 0$. Then the inner loop of the trust-region algorithm with condition* (27.14) *and acceptance test* (27.20)*, finds a serious iterate after a finite number of trial steps.*

Proof. Suppose on the contrary that the inner loop turns forever. Then $\Delta_k \to 0$ and $y^{k+1} \to x$ $(k \to \infty)$. Now we expand

$$\rho_k = \frac{f(x)-f(y^{k+1})}{f(x)-m(y^{k+1},x)} = 1 - \frac{f(y^{k+1})-m(y^{k+1},x)}{f(x)-m(y^{k+1},x)}.$$

By condition (27.14) at each inner iteration k we have

$$\begin{aligned} f(x)-m(y^{k+1},x) &\geq c\|\nabla f(x)\| \min\left(\frac{\|\nabla f(x)\|}{1+\|B\|},\Delta_k\right) \\ &\geq c\|\nabla f(x)\|\Delta_k && \text{for sufficiently large } k. \\ &\geq c\|\nabla f(x)\|\,\|y^{k+1}-x\| && \text{for sufficiently large } k. \end{aligned}$$

On the other hand we also have

$$\begin{aligned} |f(y^{k+1})-m(y^{k+1},x)| &\leq |f(y^{k+1})-f(x)-\nabla f(x)^\top(y^{k+1}-x)| \\ &\quad +\tfrac{1}{2}|(y^{k+1}-x)^\top B(y^{k+1}-x)| \\ &\leq \|y^{k+1}-x\|\varepsilon_k + \tfrac{1}{2}\|B\|\|y^{k+1}-x\|^2, \end{aligned}$$

where the existence of $\varepsilon_k \to 0$ follows from Lemma 27.3. Combining the previous inequalities, we obtain

$$\begin{aligned} \left|\frac{f(y^{k+1})-m(y^{k+1},x)}{f(x)-m(y^{k+1},x)}\right| &\leq \frac{\|y^{k+1}-x\|\varepsilon_k + \frac{1}{2}\|B\|\,\|y^{k+1}-x\|^2}{c\|\nabla f(x)\|.\|y^{k+1}-x\|} \\ &\leq \frac{\varepsilon_k + \frac{1}{2}\|B\|\,\|y^{k+1}-x\|}{c\|\nabla f(x)\|} \to 0 \quad (k \to \infty), \end{aligned}$$

which implies $\rho_k \to 1$ $(k \to \infty)$. By our working hypothesis the acceptance test (27.20) fails. Since it requires two conditions and since the first of these two conditions, $\rho_k \geq \gamma$, is satisfied for large k, the second condition must eventually fail, i.e., there must exist $K \in \mathbb{N}$ such that

$$\|\nabla f(x)\| < \mu \|B\| \|y^{k+1} - x\|$$

for all $k \geq K$. But from $y^{k+1} \to x$ $(k \to \infty)$ we deduce $\nabla f(x) = 0$, a contradiction.■

References

1. Absil, P.A., Mahony, R., Andrews, B.: Convergence of the iterates of descent methods for analytic cost functions. SIAM J. Optim. **16**(2), 531–547 (2005)
2. Attouch, H., Bolte, J.: On the convergence of the proximal algorithm for nonsmooth functions involving analytic features. Math. Program. Ser. B, **116**(1–2), 5–16 (2009)
3. Attouch, H., Bolte, J., Redont, P., Soubeyran, A.: Proximal alternating minimization and projection methods for nonconvex problems: an approach based on the Kurdyka-Łojasiewicz inequality. Math. Oper. Res. **35**(2), 438–457 (2010)
4. Bertsekas, D.P.: Nonlinear Programming, 2nd edn. Optimization and Neural Computation Series. Athena Scientific, Belmont (1999)
5. Bolte, J., Daniilidis, A., Lewis, A., Shiota, M.: Clarke subgradients of stratifiable functions. SIAM J. Optim. **18**(2), 556–572 (2007) (electronic)
6. Bolte, J., Daniilidis, A., Ley, O., Mazet, L.: Characterizations of Łojasiewicz inequalities: subgradient flows, talweg, convexity. Trans. Am. Math. Soc. **362**(6), 3319–3363 (2010)
7. Cartan, H.: Calcul diffÉrentiel. Hermann, Paris (1967)
8. Conn, A.R., Gould, N.I.M., Toint, Ph.L.: Trust-Region Methods. MPS/SIAM Series on Optimization. Society for Industrial and Applied Mathematics (SIAM), Philadelphia (2000)
9. Dennis, J.E., Moré, J.J.: Quasi-Newton methods, motivation and theory. SIAM Rev. **19**(1), 46–89 (1977)
10. Kurdyka, K.: On gradients of functions definable in o-minimal structures. Ann. Inst. Fourier (Grenoble) **48**, (3), 769–783 (1998)
11. Łojasiewicz, S.: Une propriété topologique des sous-ensembles analytiques réels. In: Les Équations aux Dérivées Partielles (Paris, 1962), pp. 87–89. Éditions du Centre National de la Recherche Scientifique, Paris (1963)
12. Łojasiewicz, S.: Sur les ensembles semi-analytiques. In: Actes du Congrès International des Mathématiciens (Nice, 1970), Tome 2, pp. 237–241. Gauthier-Villars, Paris (1971)
13. Nocedal, J., Wright, S.J.: Numerical Optimization, 2nd edn. Springer Series in Operations Research and Financial Engineering. Springer, New York (2006)

Chapter 28
Strong Duality in Conic Linear Programming: Facial Reduction and Extended Duals

Gábor Pataki

Dedicated to Jonathan Borwein on the occasion of his 60th birthday

Abstract The facial reduction algorithm (FRA) of Borwein and Wolkowicz and the extended dual of Ramana provide a strong dual for the conic linear program

$$\sup\{\,\langle c,x\rangle\,|\,Ax \le_K b\,\} \tag{P}$$

in the absence of any constraint qualification. The FRA solves a sequence of auxiliary optimization problems to obtain such a dual. Ramana's dual is applicable when (P) is a semidefinite program (SDP) and is an explicit SDP itself. Ramana, Tunçel, and Wolkowicz showed that these approaches are closely related; in particular, they proved the correctness of Ramana's dual using certificates from a facial reduction algorithm. Here we give a simple and self-contained exposition of facial reduction, of extended duals, and generalize Ramana's dual:

- We state a simple FRA and prove its correctness.
- Building on this algorithm we construct a family of extended duals when K is a *nice* cone. This class of cones includes the semidefinite cone and other important cones.

Key words: Conic linear programming • Strong duality • Semidefinite programming • Facial reduction • Extended duals • Nice cones

COMMUNICATED BY HEINZ H. BAUSCHKE.

G. Pataki (✉)
Department of Statistics and Operations Research, University of North Carolina at Chapel Hill, Chapel Hill, NC 27599-3216, USA
e-mail: gabor@unc.edu

D.H. Bailey et al. (eds.), *Computational and Analytical Mathematics*, Springer Proceedings in Mathematics & Statistics 50, DOI 10.1007/978-1-4614-7621-4_28,

Mathematics Subject Classifications (2010): Primary: 90C46, 49N15, 90C22, 90C25; secondary: 52A40, 52A41

28.1 Introduction

28.1.1 Conic Linear Programs

Conic linear programs generalize ordinary linear programming, as they require membership in a closed convex cone in place of the usual nonnegativity constraint. Conic LPs share some of the duality theory of linear optimization: weak duality always holds in a primal-dual pair, and assuming a suitable constraint qualification (CQ), their objective values agree and are attained.

When a CQ is lacking and the underlying cone is not polyhedral, pathological phenomena can occur: nonattainment of the optimal values, positive gaps, and infeasibility of the dual even when the primal is bounded. All these pathologies appear in semidefinite programs (SDPs), second-order cone programs, and p-order conic programs, arguably the most important and useful classes of conic LPs [1–4, 7, 15, 16, 32, 34].

28.1.2 Facial Reduction and Extended Duals

Here we study two fundamental approaches to duality in conic linear programs that work without assuming any CQ. The first approach is the facial reduction algorithm (FRA) of Borwein and Wolkowicz [5, 6], which constructs a so-called minimal cone of a conic linear system. Using this minimal cone one can always ensure strong duality in a primal-dual pair of conic LPs.

The second approach is Ramana's extended dual for SDPs [26]. (Ramana named his dual an extended Lagrange–Slater dual, or ELSD dual. We use the shorter name for simplicity.) The extended dual is an explicit SDP with a fairly large number of (but polynomially many) variables and constraints. It has the following desirable properties: it is feasible if and only if the primal problem is bounded; and when these equivalent statements hold, it has the same value as the primal and attains it.

Though these approaches at first sight look quite different, Ramana, Tunçel, and Wolkowicz in [27] showed that they are closely related: in the case of semidefinite programming, they proved the correctness of Ramana's dual using certificates from the algorithm of [5, 6].

The goal of our paper is to give a simple and self-contained exposition of facial reduction, of extended duals, study their connection, and give simple proofs of generalizations of Ramana's dual. We will use ideas from the paper of Ramana, Tunçel, and Wolkowicz [27], although our development is different. We will state an FRA and prove its correctness using only elementary results from the duality theory of conic LPs and convex analysis. We build on this algorithm and generalize Ramana's dual: we construct a family of extended duals for (*P*) when K is a *nice* cone. This class of cones includes the semidefinite cone and other important cones, as p-order, in particular, second-order cones.

Next we present our framework in more detail. A conic linear program can be stated as

$$\begin{aligned} \sup \;\; & \langle c,x \rangle \\ s.t. \; & Ax \leq_K b, \end{aligned} \tag{P}$$

where $A : X \to Y$ is a linear map between finite dimensional Euclidean spaces X and Y and $c \in X$, $b \in Y$. The set $K \subseteq Y$ is a closed, convex cone, and we write $Ax \leq_K b$ to mean $b - Ax \in K$. We naturally associate a dual program with (*P*). Letting A^* be the adjoint operator of A, and K^* the dual cone of K, i.e.,

$$K^* = \{y \,|\, \langle y,x \rangle \geq 0 \forall x \in K\},$$

the dual problem is

$$\begin{aligned} \inf \; & \langle b,y \rangle \\ s.t. \; & y \geq_{K^*} 0 \\ & A^* y = c. \end{aligned} \tag{D}$$

When (*P*) is feasible, we say that *strong duality holds between (P) and (D)* if the following conditions are satisfied:

- Problem (*P*) is bounded, if and only if (*D*) is feasible.
- When these equivalent conditions hold, the optimal values of (*P*) and (*D*) agree and the latter is attained.

We say that (*P*) is *strictly feasible*, or *satisfies Slater's condition*, if there is an $x \in X$ such that $b - Ax$ is in the relative interior of K. When (*P*) is strictly feasible, it is well known that strong duality holds between (*P*) and (*D*).

The facial reduction algorithm of Borwein and Wolkowicz constructs a suitable face of K, called the *minimal cone of (P)*, which we here denote by $F_{\min}$. The minimal cone has two important properties:

- The feasible set of (*P*) remains the same if we replace its constraint set by

$$Ax \leq_{F_{\min}} b.$$

- The new constraint set satisfies Slater's condition.

Thus, if we also replace K^* by $F_{\min}^*$ in (*D*), strong duality holds in the new primal-dual pair. The algorithm in [5, 6] constructs a decreasing chain of faces starting with K and ending with $F_{\min}$, in each step solving a pair of auxiliary conic linear programs.

28.1.3 Contributions of the Paper

We first state a simplified FRA and prove its correctness. Building on this algorithm, and assuming that cone K is *nice,* i.e., the set $K^* + F^\perp$ is closed for all F faces of K, we show that the dual of the minimal cone has a representation

$$\begin{aligned} F^*_{\min} = \{ & u_{\ell+1} + v_{\ell+1} : \\ & (u_0, v_0) = (0,0) \\ & (A,b)^*(u_i + v_i) = 0,\ i = 1,\ldots,\ell, \\ & (u_i, v_i) \in K^* \times \tan(u_0 + \cdots + u_{i-1}, K^*),\ i = 1,\ldots,\ell+1 \}, \end{aligned} \tag{28.1}$$

where $\tan(u, K^*)$ denotes the tangent space of the cone K^* at $u \in K^*$ and ℓ is a suitable integer. Plugging this expression for $F^*_{\min}$ in place of K^* in (D) we obtain a dual with the properties of Ramana's dual. We show the correctness of several representations of $F^*_{\min}$, each leading to a different extended dual. We note that the results of [27] already imply that such a representation is possible, but this is not stated there explicitly.

The cone of positive semidefinite matrices is nice (and also self-dual), so in this case the representation of (28.1) is valid. In this case we can also translate the tangent space constraint into an explicit semidefinite constraint and recover variants of Ramana's dual.

We attempted to simplify our treatment of the subject as much as possible: as background we use only the fact that strong duality holds in a primal-dual pair of conic LPs, when the primal is strictly feasible, and some elementary facts in convex analysis.

28.1.4 Literature Review

Borwein and Wolkowicz originally presented their FRA in the two papers [5, 6]. Their algorithm works for a potentially nonlinear conic system of the form $\{x \,|\, g(x) \in K\}$. Luo et al. in [20] studied a so-called *conic expansion method* which finds a sequence of increasing sets starting with K^* and ending with $F^*_{\min}$: thus their algorithm can be viewed as a dual variant of facial reduction. Their paper also contains an exposition of facial reduction and Ramana's dual. Sturm in [33] introduced an interesting and novel application of facial reduction: deriving error bounds for semidefinite systems that lack a strictly feasible solution. Luo and Sturm in [19] generalized this approach to mixed semidefinite and second-order conic systems. Lewis in [18] used facial reduction to derive duality results without a CQ assumption in partially finite convex programming. Tunçel in his recent book [35] constructed an SDP instance with n by n semidefinite matrices that requires $n-1$ iterations of the facial reduction algorithm to find the minimal cone and thus showed that the theoretical worst case is essentially attainable.

Waki and Muramatsu in [36] also described an FRA, rigorously showed its equivalence to the conic expansion approach of Luo et al, and presented computational results on SDPs. A preliminary version of this paper appeared in [21]. Pólik and Terlaky in [25] used the results of [21] to construct strong duals for conic LPs over homogeneous cones. Wang et al. in [37] presented an FRA for nonsymmetric semidefinite least squares problems.

Tunçel and Wolkowicz in [30] described a connection between the lack of strict complementarity in the homogeneous primal and dual systems and positive duality gaps in SDPs: in particular, they proved that when strict complementarity in the homogeneous problems fails in a certain minimal sense, one can generate instances with an arbitrary positive duality gap. Cheung et al. in [8] developed a relaxed version of an FRA, in which one can allow an error in the solution of the auxiliary conic LPs, and applied their method to SDPs, in particular, to instances generated according to the results of [30].

Nice cones appear in other areas of optimization as well. In [22] we studied the question of when the linear image of a closed convex cone is closed and described necessary and sufficient conditions. These lead to a particularly simple and exact characterization when the dual of the cone in question is nice. We call a conic linear system *well behaved* if for all objective functions the resulting conic linear program has strong duality with its dual and *badly behaved*, otherwise. In related work, [23], we described characterizations of well- and badly behaved conic linear systems. These become particularly simple when the underlying cone is nice and yield combinatorial type characterizations for semidefinite and second-order conic systems.

Chua and Tunçel in [10] showed that if a cone K is nice, then so is its intersection with a linear subspace. Thus, all homogeneous cones are nice, since they arise as the slice of a semidefinite cone with a suitable subspace, as proven independently by Chua in [9] and by Faybusovich in [11]. In [10] the authors also proved that the preimage of a nice cone under a linear map is also nice and in [24] we pointed out that this result implies that the intersection of nice cones is also nice. In [24] we gave several characterizations of nice cones and proved that they must be facially exposed; facial exposedness with a mild additional condition implies niceness and conjectured that facially exposed and nice cones are actually the same class of cones. However, this conjecture was proven false by Roshchina [31].

Most articles on strong duality deal with instances with a fixed right-hand side. Schurr et al. in [29] obtained characterizations of *universal duality*, i.e., of the situation when strong duality holds for all right-hand sides, and objective functions.

Klep and Schweighofer in [17] derived a strong dual for SDPs that also works without assuming any constraint qualification. Their dual resembles Ramana's dual, but interestingly, it is derived using concepts from algebraic geometry, whereas all other references known to us use convex analysis.

Recently Gouveia et al. in [14] studied the following fundamental question: can a convex set be represented as the projection of an affine slice of a suitable closed, convex cone? They gave necessary and sufficient conditions for such a *lift* to exist and showed that some known lifts from the literature are in the lowest dimension possible. The representation of (28.1) is related in spirit, as we also represent the set $F_{\min}^*$ as the projection of a conic linear system in a higher dimensional space.

28.1.5 Organization of the Paper and Guide to the Reader

In Sect. 28.2 we fix notation, review preliminaries, and present two motivating examples. The reader familiar with convex analysis can skip the first part of this section and go directly to the examples. In Sect. 28.3 we present a simple FRA, prove its correctness, and show how $F_{\min}^*$ can be written as the projection of a nonlinear conic system in a higher dimensional space.

Assuming that K is nice, in Sect. 28.4 we arrive at the representation in (28.1), i.e., show that $F_{\min}^*$ is the projection of a conic *linear* system, and derive an extended dual for conic LPs over nice cones. Here we obtain our first Ramana-type dual for SDPs which is an explicit SDP itself, but somewhat different from the dual proposed in [26].

In Sect. 28.5 we describe variants of the representation in (28.1), of extended duals, and show how we can exactly obtain Ramana's dual. In Sect. 28.6 we show that the minimal cone $F_{\min}$ itself also has a representation similar to the representation of $F_{\min}^*$ in (28.1) and discuss some open questions.

The paper is organized to arrive at an explicit Ramana-type dual for SDP as quickly as possible. Thus, if a reader is interested in only the derivation of such a dual, it suffices for him/her to read only Sects. 28.2–28.4.

28.2 Preliminaries

28.2.1 Matrices and Vectors

We denote operators by capital letters and matrices (when they are considered as elements of a Euclidean space and not as operators) and vectors by lowercase letters. The ith component of vector x is denoted by x_i and the (i, j)th component of matrix z by z_{ij}. We distinguish vectors and matrices of similar type with lower indices, i.e., writing $x_1, x_2, \ldots$ The jth component of vector x_i is denoted by $x_{i,j}$. This notation is somewhat ambiguous, as x_i may denote a vector, or the ith component of the vector x, but the context will make it clear which one is meant.

28.2.2 Convex Sets

For a set C we write $\mathrm{cl}\, C$ for its closure, $\mathrm{lin}\, C$ for its linear span, and $C^{\perp}$ for the orthogonal complement of its linear span. For a convex set C we denote its relative interior by $\mathrm{ri}\, C$. For a one-element set $\{x\}$ we abbreviate $\{x\}^{\perp}$ by $x^{\perp}$. The open line segment between points x_1 and x_2 is denoted by (x_1, x_2).

For a convex set C, and F, a convex subset of C, we say that F is a face of C if $x_1, x_2 \in C$ and $(x_1, x_2) \cap F \neq \emptyset$ implies that x_1 and x_2 are both in F. For $x \in C$ there is a unique minimal face of C that contains x, i.e., the face that contains x in its relative interior: we denote this face by $\mathrm{face}(x, C)$. For $x \in C$ we define the set of feasible directions and the tangent space at x in C as

$$\mathrm{dir}(x,C) = \{\, y \,|\, x + ty \in C \text{ for some } t > 0 \,\},$$
$$\tan(x,C) = \mathrm{cl}\, \mathrm{dir}(x,C) \cap -\mathrm{cl}\, \mathrm{dir}(x,C).$$

28.2.3 Cones

We say that a set K is a cone, if $\lambda x \in K$ holds for all $x \in K$ and $\lambda \geq 0$, and define the dual of cone K as

$$K^* = \{\, z \,|\, \langle z, x \rangle \geq 0 \text{ for all } x \in K \,\}.$$

For an F face of a closed convex cone K and $x \in \mathrm{ri}\, F$ the complementary or conjugate face of F is defined alternatively as (the equivalence is straightforward)

$$F^{\triangle} = K^* \cap F^{\perp} = K^* \cap x^{\perp}.$$

The complementary face of a face G of K^* is defined analogously and denoted by $G^{\triangle}$. A closed convex cone K is facially exposed, i.e., all faces of K arise as the intersection of K with a supporting hyperplane iff for all F faces of K we have $(F^{\triangle})^{\triangle} = F$. For brevity we write $F^{\triangle *}$ for $(F^{\triangle})^*$ and $F^{\triangle \perp}$ for $(F^{\triangle})^{\perp}$.

For a closed convex cone K and $x \in K$ we have

$$\tan(x,K) = \mathrm{face}(x,K)^{\triangle \perp}, \tag{28.2}$$

as shown in [23, Lemma 1].

28.2.4 The Semidefinite Cone

We denote the space of n by n symmetric and the cone of n by n symmetric, positive semidefinite matrices by $\mathcal{S}^n$, and $\mathcal{S}^n_+$, respectively. The space $\mathcal{S}^n$ is equipped with the inner product

$$\langle x, z \rangle := \sum_{i,j=1}^{n} x_{ij} z_{ij},$$

and $\mathscr{S}^n_+$ is self-dual with respect to it. For $y \in \mathscr{S}^n$ we write $y \succeq 0$ to denote that y is positive semidefinite. Using a rotation $v^T(.)v$ by a full-rank matrix v any face of $\mathscr{S}^n_+$ and its conjugate face can be brought to the form

$$F = \left\{ \begin{pmatrix} x & 0 \\ 0 & 0 \end{pmatrix} | x \in \mathscr{S}^r_+ \right\}, F^\triangle = \left\{ \begin{pmatrix} 0 & 0 \\ 0 & y \end{pmatrix} | y \in \mathscr{S}^{n-r}_+ \right\}, \tag{28.3}$$

where r is a nonnegative integer.

For a face of this form and related sets we use the shorthand

$$F = \begin{pmatrix} \oplus & 0 \\ 0 & 0 \end{pmatrix}, F^\triangle = \begin{pmatrix} 0 & 0 \\ 0 & \oplus \end{pmatrix}, F^{\triangle\perp} = \begin{pmatrix} \times & \times \\ \times & 0 \end{pmatrix}, \tag{28.4}$$

when the sizes of the blocks in the partition are clear from the context. The $\oplus$ sign denotes a positive semidefinite submatrix and the sign $\times$ stands for a submatrix with arbitrary elements.

For an x positive semidefinite matrix we collect some expressions for $\tan(x, \mathscr{S}^n_+)$ below: these play an important role when constructing explicit duals for SDPs. The second part of Proposition 28.1 is based on Lemma 1 in [27].

Proposition 28.1. *The following statements hold.*

(1) Suppose $x \in \mathscr{S}^n_+$ is of the form

$$x = \begin{pmatrix} I_r & 0 \\ 0 & 0 \end{pmatrix}, \tag{28.5}$$

and $F = \mathrm{face}(x, \mathscr{S}^n_+)$. Then F, $F^\triangle$, and $F^{\triangle\perp}$ are as displayed in (28.4), with the upper left block r by r, and

$$\tan(x, \mathscr{S}^n_+) = F^{\triangle\perp} = \begin{pmatrix} \times & \times \\ \times & 0 \end{pmatrix}. \tag{28.6}$$

(2) For an arbitrary $x \in \mathscr{S}^n_+$ we have

$$\tan(x, \mathscr{S}^n_+) = \left\{ w + w^T \;\middle|\; \begin{pmatrix} x & w \\ w^T & \beta I \end{pmatrix} \succeq 0 \text{ for some } \beta \in \mathbb{R} \right\}. \tag{28.7}$$

Proof. Statement (1) is straightforward from the form of x and the expression for the tangent space given in (28.2) with $K = \mathscr{S}^n_+$.

To see (2) first assume that x is of the form as in (28.5); then our claim follows from part (1).

Suppose now that $x \in \mathscr{S}^n_+$ is arbitrary and let q be a matrix of suitably scaled eigenvectors of x with eigenvectors corresponding to nonzero eigenvalues coming first. Let us write $T(x)$ for the set on the right-hand side of (28.7). Then one easily checks $\tan(q^T xq, \mathscr{S}^n_+) = q^T \tan(x, \mathscr{S}^n_+)q$ and $T(q^T xq) = q^T T(x)q$, so this case reduces to the previous case. ■

28.2.5 Conic LPs

An ordinary linear program is clearly a special case of (P). If we choose $X = \mathbb{R}^m, Y = \mathscr{S}^n$, and $K = \mathscr{S}^n_+$, then problem (P) becomes an SDP. Since K is self-dual, the dual problem (D) is also an SDP. The operator A and its adjoint are defined via symmetric matrices $a_1, \ldots, a_m$ as

$$Ax = \sum_{i=1}^{m} x_i a_i \text{ and } A^* y = (\langle a_1, y \rangle, \ldots, \langle a_m, y \rangle)^T.$$

We use the operator Feas() to denote the feasible set of a conic system.

28.2.6 The Minimal Cone

Let us choose $x \in \operatorname{ri} \operatorname{Feas}(P)$. We define the minimal cone of (P) as the unique face of K that contains $b - Ax$ in its relative interior and denote this face by $F_{\min}$.

For an arbitrary $y \in \operatorname{Feas}(P)$ there is $z \in \operatorname{Feas}(P)$ such that $x \in (y, z)$. Hence $b - Ax \in (b - Ay, b - Az)$, so $b - Ay$ and $b - Az$ are in $F_{\min}$, and (P) is equivalent to

$$Ax \leq_{F_{\min}} b,$$

and this constraint system satisfies Slater's condition.

28.2.7 Nice Cones

We say that a closed convex cone K is nice if

$$K^* + F^\perp \text{ is closed for all } F \text{ faces of } K.$$

Most cones appearing in the optimization literature, such as polyhedral, semidefinite, p-order, in particular second-order cones, are nice: see, e.g., [5, 6, 22].

Example 28.2. In the linear inequality system

$$\begin{pmatrix} 1 & 0 & 0 \\ 0 & -1 & 1 \\ 0 & 1 & 0 \\ 0 & 0 & -1 \\ 0 & 0 & 1 \end{pmatrix} \begin{pmatrix} x_1 \\ x_2 \\ x_3 \end{pmatrix} \leq \begin{pmatrix} 0 \\ 0 \\ 0 \\ 0 \\ 0 \end{pmatrix}, \tag{28.8}$$

all feasible solutions satisfy the last four inequalities at equality, and for, say, $x = (-1,0,0)^T$ the first inequality is strict. So the minimal cone of this system is

$$F_{\min} = \mathbb{R}^1_+ \times \{0\}^4.$$

In linear programs strong duality holds even without strict feasibility, so this example illustrates only the concept of the minimal cone.

Example 28.3. In the SDP

$$\begin{aligned} &\sup x_1 \\ &s.t.\ x_1 \begin{pmatrix} 0\,1\,0 \\ 1\,0\,0 \\ 0\,0\,0 \end{pmatrix} + x_2 \begin{pmatrix} 0\,0\,1 \\ 0\,1\,0 \\ 1\,0\,0 \end{pmatrix} \preceq \begin{pmatrix} 1\,0\,0 \\ 0\,0\,0 \\ 0\,0\,0 \end{pmatrix}, \end{aligned} \tag{28.9}$$

a feasible positive semidefinite slack z must have all entries equal to zero, except for z_{11}, and there is a feasible slack with $z_{11} > 0$. So the minimal cone and its dual are

$$F_{\min} = \begin{pmatrix} \oplus\,0\,0 \\ 0\,0\,0 \\ 0\,0\,0 \end{pmatrix}, F^*_{\min} = \begin{pmatrix} \oplus \times \times \\ \times \times \times \\ \times \times \times \end{pmatrix}. \tag{28.10}$$

The optimal value of (28.9) is clearly zero. Writing y for the dual matrix, the dual program is equivalent to

$$\begin{aligned} &\inf y_{11} \\ &s.t. \begin{pmatrix} y_{11} & 1/2 & -y_{22}/2 \\ 1/2 & y_{22} & y_{23} \\ -y_{22}/2 & y_{23} & y_{33} \end{pmatrix} \succeq 0. \end{aligned} \tag{28.11}$$

The dual has an unattained 0 minimum: y_{11} can be an arbitrarily small positive number, at the cost of making y_{22} and in turn y_{33} large; however, y_{11} cannot be made 0, as y_{12} is $1/2$.

Suppose that in (28.11) we replace the constraint $y \succeq 0$ by $y \in F^*_{\min}$. Then we can set y_{11} to zero, so with this modification, the dual attains.

We will return to these examples later to illustrate our FRA and extended duals.

We assume throughout the paper that (P) is feasible. It is possible to remove this assumption and modify the facial reduction algorithm of sect. 28.3 either to prove the infeasibility of (P) or to find the minimal cone in finitely many steps: such an algorithm was described by Waki and Muramatsu in [36].

28.3 A Simple Facial Reduction Algorithm

We now state a simplified FRA that is applicable when K is an arbitrary closed convex cone. We prove its correctness and illustrate it on Examples 28.2 and 28.3.

Let us recall that $F_{\min}$ denotes the minimal cone of (P) and for brevity define the subspace L as

$$L = \mathcal{N}((A,b)^*). \tag{28.12}$$

Lemma 28.4. *Suppose that an F face of K satisfies $F_{\min} \subseteq F$. Then the following hold:*

(1) For all $y \in F^* \cap L$ we have

$$F_{\min} \subseteq F \cap y^{\perp} \subseteq F. \tag{28.13}$$

(2) There exists $y \in F^* \cap L$ such that the second containment in (28.13) is strict, iff $F_{\min} \neq F$. We can find such a y or prove $F = F_{\min}$ by solving a pair of auxiliary conic linear programs.

Proof. To prove (1) suppose that x is feasible for (P) and let $y \in F^* \cap L$. Then $b - Ax \in F_{\min} \subseteq F$, hence $\langle b - Ax, y \rangle = 0$, which implies the first containment; the second is obvious.

In statement (2) the "only if" part is obvious. To see the "if" part, let us fix $f \in \operatorname{ri} F$ and consider the primal-dual pair of conic linear programs that we call reducing conic LPs below:

$$\begin{array}{lllll} & \sup & t & \inf & \langle b,y\rangle & \\ (R-P) & s.t. & Ax + ft \leq_F b & s.t. & y \geq_{F^*} 0 & (R-D) \\ & & & & A^*y = 0 & \\ & & & & \langle f,y\rangle = 1. & \end{array}$$

First let us note

$$\begin{aligned} F_{\min} = F &\Leftrightarrow \exists x \text{ s.t.} \, b - Ax \in \operatorname{ri} F \\ &\Leftrightarrow \exists x \text{ and} \, t > 0 \, \text{s.t.} \, b - Ax - ft \in F. \end{aligned}$$

Here in the first equivalence the direction $\Rightarrow$ is obvious from the definition of the minimal cone. To see the direction $\Leftarrow$ assume $b - Ax \in \operatorname{ri} F$. Then $\operatorname{ri} F \cap F_{\min} \neq \emptyset$ and $F_{\min}$ is a face of K, so Theorem 18.1 in [28] implies $F \subseteq F_{\min}$, and the reverse containment is already given. The second equivalence is obvious.

Fig. 28.1 The facial reduction algorithm

FACIAL REDUCTION ALGORITHM	
Initialization:	Let $y_0 = 0$, $F_0 = K$, $i = 1$.
repeat	
	Choose $y_i \in L \cap F_{i-1}^*$.
	Let $F_i = F_{i-1} \cap y_i^\perp$.
	Let $i = i + 1$.
end repeat	

Therefore, $F_{\min} \neq F$ iff the optimal value of $(R-P)$ is 0. Note that $(R-P)$ is strictly feasible, with some x such that $b - Ax \in F$ and some $t < 0$.

Hence $F_{\min} \neq F$ iff $(R-D)$ has optimal value 0 and attains it, i.e., iff there is $y \in F^* \cap L$ with $\langle f, y \rangle = 1$. Such a y clearly must satisfy $F \cap y^\perp \subsetneq F$. ■

Based on Lemma 28.4 we now state a simple FRA in Fig. 28.1.

The algorithm of Fig. 28.1 may not terminate in general, as it allows the choice of a y_i in iteration i such that $F_i = F_{i-1}$; it even allows $y_i = 0$ for all i. Based on this general algorithm, however, it will be convenient to construct a representation of $F_{\min}^*$.

We call an iteration of the FRA reducing if the y_i vector found therein satisfies $F_i \subsetneq F_{i-1}$; we can make sure that an iteration is reducing or that we have found the minimal cone by solving the pair of conic linear programs $(R-P)-(R-D)$. It is clear that after a sufficient number of reducing iterations the algorithm terminates.

Let us define the quantities

$$\begin{aligned} \ell_K &= \text{the length of the longest chain of faces in } K, \\ \ell &= \min\{\ell_K - 1, \dim L\}. \end{aligned} \tag{28.14}$$

We prove the correctness of our FRA and an upper bound on the number of reducing iterations in Theorem 28.5:

Theorem 28.5. *Suppose that the FRA finds $y_0, y_1, \ldots,$ and corresponding faces $F_0, F_1, \ldots$ Then the following hold:*

(1) $F_{\min} \subseteq F_i$ for $i = 0, 1, \ldots$.
(2) After a sufficiently large number of reducing iterations the algorithm finds $F_{\min} = F_t$ in some iteration t. Furthermore,

$$F_{\min} = F_i$$

holds for all $i \geq t$.
(3) The number of reducing iterations in the FRA is at most ℓ.

Proof. Let us first note that the face F_i found by the algorithm is of the form

$$F_i = K \cap y_0^\perp \cap \cdots \cap y_i^\perp, i = 0, 1, \ldots$$

Statement (1) follows from applying repeatedly part (1) of Lemma 28.4.

In (2) the first part of the claim is straightforward; in particular, the number of reducing iterations cannot exceed $\ell_K - 1$. Suppose $i \geq t$. Since $F_t = F_{\min}$, we have

$$F_{\min} \subseteq F_i = F_t \cap y_{t+1}^{\perp} \cap \cdots \cap y_i^{\perp} \subseteq F_{\min}, \tag{28.15}$$

so equality holds throughout in (28.15), which proves $F_i = F_{\min}$.

To prove (3) let us denote by k the number of reducing iterations. It remains to show that $k \leq \dim L$ holds, so assume to the contrary $k > \dim L$. Suppose that $y_{i_1}, \ldots, y_{i_k}$ are the vectors found in reducing iterations, where $i_1 < \cdots < i_k$. Since they are all in L, they must be linearly dependent, so there is an index $r \in \{1, \ldots, k\}$ such that

$$y_{i_r} \in \operatorname{lin}\{y_{i_1}, \ldots, y_{i_{r-1}}\} \subseteq \operatorname{lin}\{y_0, y_1, \ldots, y_{i_r - 1}\}.$$

For brevity let us write $s = i_r$. Then $y_0^{\perp} \cap \cdots \cap y_{s-1}^{\perp} \subseteq y_s^{\perp}$, so

$$F_s = F_{s-1},$$

i.e., the sth step is not reducing, which is a contradiction. ∎

Next we illustrate our algorithm on the examples of Sect. 28.2.

Examples 28.2 and 28.3 continued Suppose we run our algorithm on the linear system (28.8). The y_i vectors below, with corresponding faces shown, are a possible output:

$$\begin{aligned} y_0 &= 0,\, F_0 = \mathbb{R}^5_+, \\ y_1 &= (0,0,0,1,1)^T,\, F_1 = \mathbb{R}^3_+ \times \{0\}^2, \\ y_2 &= (0,1,1,0,-1)^T,\, F_2 = F_{\min} = \mathbb{R}^1_+ \times \{0\}^4. \end{aligned} \tag{28.16}$$

The algorithm may also finish in one step, by finding, say, $y_0 = 0$, and

$$y_1 = (0,1,1,2,1)^T. \tag{28.17}$$

Of course, in linear systems, there is always a reducing certificate that finds the minimal cone in one step, i.e., $F_{\min} = K \cap y_1^{\perp}$ for some $y_1 \geq 0$; this is straightforward from LP duality.

When we run our algorithm on the instance of (28.9), the y_i matrices below, with corresponding F_i faces, are a possible output:

$$\begin{aligned} y_0 &= 0,\, F_0 = \mathcal{S}^3_+, \\ y_1 &= \begin{pmatrix} 0 & 0 & 0 \\ 0 & 0 & 0 \\ 0 & 0 & 1 \end{pmatrix},\, F_1 = \begin{pmatrix} \multicolumn{2}{c}{\oplus} & 0 \\ & & 0 \\ 0 & 0 & 0 \end{pmatrix}, \\ y_2 &= \begin{pmatrix} 0 & 0 & -1 \\ 0 & 2 & 0 \\ -1 & 0 & 0 \end{pmatrix},\, F_2 = F_{\min} = \begin{pmatrix} \oplus & 0 & 0 \\ 0 & 0 & 0 \\ 0 & 0 & 0 \end{pmatrix}. \end{aligned} \tag{28.18}$$

Indeed it is clear that the y_i are orthogonal to all the constraint matrices in problem (28.9) and that $y_i \in F_{i-1}^*$ for $i = 1,2$.

Let us now consider the conic system

$$\left.\begin{aligned} y_0 &= 0 \\ y_i &\in F_{i-1}^*, \text{ where} \\ F_{i-1} &= K \cap y_0^\perp \cap \cdots \cap y_{i-1}^\perp, i = 1,\ldots,\ell+1 \\ y_i &\in L, i = 1,\ldots,\ell \end{aligned}\right\} \qquad (EXT)$$

that we call an extended system.

We have the following representation theorem:

Theorem 28.6. $F_{\min}^* = \{y_{\ell+1} \,|\, (y_i)_{i=0}^{\ell+1}$ *is feasible in (EXT)*$\}$.

Before proving Theorem 28.6 we make some remarks. First, the two different ranges for the i indices in the constraints of (EXT) are not accidental: the sequence $y_0,\ldots,y_\ell$ is a possible output of our FRA, iff with some $y_{\ell+1}$ it is feasible in (EXT), and the variable $y_{\ell+1}$ represents the dual of the minimal cone. It also becomes clearer now why we allow nonreducing iterations in our algorithm: in the conic system (EXT) some y_i correspond to reducing iterations, but others do not.

The extended system (EXT) is not linear, due to how the y_i vectors depend on the previous y_j, and in general we also don't know how to describe the duals of faces of K. Hence the representation of Theorem 28.6 is not yet immediately useful. However, in the next section we state an equivalent conic linear system to represent $F_{\min}^*$ when K is nice and arrive at the representation of (28.1) and at an extended dual of (P).

Proof of Theorem 28.6. Let us write G for the set on the right-hand side. Suppose that $(y_i)_{i=0}^{\ell+1}$ is feasible in (EXT) with corresponding faces $F_0,\ldots,F_\ell$. By part (1) in Theorem 28.5 we have

$$F_{\min} \subseteq F_\ell, \text{ hence } F_{\min}^* \supseteq F_\ell^*. \qquad (28.19)$$

Since $y_{\ell+1} \in F_\ell^*$ in G, the containment $F_{\min}^* \supseteq G$ follows.

By part (2)–(3) in Theorem 28.5 there exists $(y_i)_{i=0}^{\ell+1}$ that is feasible in (EXT), with corresponding faces $F_0,\ldots,F_\ell$ such that equality holds in (28.19). This proves the inclusion $F_{\min}^* \subseteq G$. ■

28.4 When K Is Nice: An Extended Dual and an Explicit Extended Dual for Semidefinite Programs

From now on we make the following assumption:

K is nice.

Let us recall the definition of L from (28.12) and consider the conic system

$$\left.\begin{array}{l}(u_0,v_0) = (0,0) \\ (u_i,v_i) \in K^* \times \tan(u_0+\cdots+u_{i-1},K^*),\, i=1,\ldots,\ell+1 \\ u_i+v_i \in L,\, i=1,\ldots,\ell\end{array}\right\}. \qquad (EXT_{\text{nice}})$$

This is a conic linear system, since the set

$$\{\,(u,v)\,|\,u \in K^*,\, v \in \tan(u,K^*)\,\}$$

is a convex cone, although it may not be closed (e.g., if $K^* = \mathbb{R}^2_+$, then $(\varepsilon,1)$ is in this set for all $\varepsilon > 0$, but $(0,1)$ is not).

Theorem 28.7. $\text{Feas}(EXT) = \{\,(u_i+v_i)_{i=0}^{\ell+1} : (u_i,v_i)_{i=0}^{\ell+1} \in \text{Feas}(EXT_{\text{nice}})\,\}$.

Proof. To see the inclusion $\subseteq$ suppose that $(y_i)_{i=0}^{\ell+1}$ is feasible in (EXT), with faces

$$F_{i-1} = K \cap y_0^{\perp} \cap \cdots \cap y_{i-1}^{\perp},\, i=1,\ldots,\ell+1. \qquad (28.20)$$

For $i=1,\ldots,\ell+1$ we have $y_i \in F_{i-1}^*$, and K is nice, so we can write $y_i = u_i+v_i$ for some $u_i \in K^*$ and $v_i \in F_{i-1}^{\perp}$. Also, let us set $u_0 = v_0 = 0$, then of course $y_0 = u_0+v_0$.

We show that $(u_i,v_i)_{i=0}^{\ell+1}$ is feasible in (EXT_{nice}). To do this, it is enough to verify

$$F_{i-1}^{\perp} = \tan(u_0+\cdots+u_{i-1},K^*) \qquad (28.21)$$

for $i=1,\ldots,\ell+1$. Equation (28.21) will follow if we prove

$$F_{i-1} = K \cap (u_0+\cdots+u_{i-1})^{\perp} \qquad (28.22)$$

for $i=1,\ldots,\ell+1$; indeed, from (28.22) we directly obtain

$$F_{i-1} = \text{face}(u_0+\cdots+u_{i-1},K^*)^{\triangle},$$

hence

$$\begin{aligned} F_{i-1}^{\perp} &= \text{face}(u_0+\cdots+u_{i-1},K^*)^{\triangle\perp} \\ &= \tan(u_0+\cdots+u_{i-1},K^*), \end{aligned}$$

where the second equality comes from (28.2).

So it remains to prove (28.22). It is clearly true for $i=1$. Let i be a nonnegative integer at most $\ell+1$ and assume that (28.22) holds for $1,\ldots,i-1$. We then have

$$\begin{aligned} F_{i-1} &= F_{i-2} \cap y_{i-1}^{\perp} \\ &= F_{i-2} \cap (u_{i-1}+v_{i-1})^{\perp} \\ &= F_{i-2} \cap u_{i-1}^{\perp} \\ &= K \cap (u_0+\cdots+u_{i-2})^{\perp} \cap u_{i-1}^{\perp} \\ &= K \cap (u_0+\cdots+u_{i-2}+u_{i-1})^{\perp}. \end{aligned}$$

Here the second equation follows from the definition of (u_{i-1}, v_{i-1}), the third from $v_{i-1} \in F_{i-2}^{\perp}$, the fourth from the inductive hypothesis, and the last from all u_j being in K^*.

Thus the proof of the containment $\subseteq$ is complete.

To prove the opposite inclusion let us choose $(u_i, v_i)_{i=0}^{\ell+1}$ to be feasible in (EXT_{nice}) and define $y_i = u_i + v_i$ for all i and the faces $F_0, \ldots, F_\ell$ as in (28.20). Repeating the previous argument verbatim, (28.21) holds, so we have

$$y_i \in K^* + F_{i-1}^{\perp} = F_{i-1}^*, \, i = 1, \ldots, \ell+1.$$

Therefore $(y_i)_{i=0}^{\ell+1}$ is feasible in (EXT) and this completes the proof. ∎

We now arrive at the representation of $F_{\min}^*$ that we previewed in (28.1) and at an extended dual of (P):

Corollary 28.8. *The dual of the minimal cone of (P) has a representation*

$$F_{\min}^* = \{u_{\ell+1} + v_{\ell+1} : (u_i, v_i)_{i=0}^{\ell+1} \text{ is feasible in} (EXT_{\text{nice}})\}, \tag{28.23}$$

and the extended dual

$$\begin{array}{ll} \inf & \langle b, u_{\ell+1} + v_{\ell+1} \rangle \\ s.t. & A^*(u_{\ell+1} + v_{\ell+1}) = c \\ & (u_i, v_i)_{i=0}^{\ell+1} \text{ is feasible in} (EXT_{\text{nice}}) \end{array} \tag{D_{ext}}$$

has strong duality with (P).

In particular, if (P) is an SDP with m variables, independent constraint matrices, and $K = \mathscr{S}_+^n$, then the problem

$$\begin{array}{lrl} \inf & & \langle b, u_{\ell+1} + v_{\ell+1} \rangle \\ s.t. & & A^*(u_{\ell+1} + v_{\ell+1}) = c \\ & & (A,b)^*(u_i + v_i) = 0, \, i = 1, \ldots, \ell \\ & & u_i \succeq 0, \, i = 1, \ldots, \ell+1 \\ & (*) & \begin{pmatrix} u_0 + \cdots + u_{i-1} & w_i \\ w_i^T & \beta_i I \end{pmatrix} \succeq 0, i = 1, \ldots, \ell+1 \\ & & v_i = w_i + w_i^T, \, i = 1, \ldots, \ell+1 \\ & & w_i \in \mathbb{R}^{n \times n}, \, i = 1, \ldots, \ell+1 \\ & & \beta_i \in \mathbb{R}, \, i = 1, \ldots, \ell+1 \\ & & (u_0, v_0) = (0, 0), \end{array} \tag{$D_{\text{ext,SDP}}$}$$

where

$$\ell = \min\{n, n(n+1)/2 - m - 1\}, \tag{28.24}$$

has strong duality with (P).

Proof. The representation (28.23) follows from combining Theorems 28.6 and 28.7. The second statement of the theorem follows, since replacing K^* by $F^*_{\min}$ in (D) yields a strong dual for (P).

Suppose now that (P) is an SDP with $K = \mathcal{S}^n_+$, with m variables, and with independent constraint matrices. The length of the longest chain of faces in $\mathcal{S}^n_+$ is $n+1$ and the dimension of the subspace $\mathcal{N}((A,b)^*)$ is $n(n+1)/2 - m - 1$. Hence we can choose ℓ as in (28.24) to obtain a correct extended dual.

Let $v_i \in \mathcal{S}^n$ and $u_0, \dots, u_{i-1} \in \mathcal{S}^n_+$, where $i \in \{1, \dots, \ell+1\}$. The representation of the tangent space in $\mathcal{S}^n_+$ in (28.7) implies that $v_i \in \tan(u_0 + \cdots + u_{i-1}, K^*)$ holds, iff $v_i, u_0, \dots, u_{i-1}$ with some w_i (possibly nonsymmetric) matrices and β_i scalars satisfies the ith constraint of $(D_{\text{ext,SDP}})$ marked by (*). This proves the correctness of the extended dual $(D_{\text{ext,SDP}})$. ■

For the reducing certificates found for the linear system (28.8) and displayed in (28.16) the reader can easily find the decomposition whose existence we showed in Theorem 28.7.

Example 28.3 continued Recall that when we run our FRA on the SDP instance (28.9), matrices y_0, y_1, y_2 shown in equation (28.18) are a possible output.

We illustrate their decomposition as proved in Theorem 28.7, in particular, as $y_i = u_i + v_i$ with $u_i \in K^*$ and $v_i \in \tan(u_0 + \cdots + u_{i-1}, K^*)$ for $i = 1, 2$:

$$\begin{aligned} u_0 &= 0,\ v_0 = 0, \\ u_1 &= \begin{pmatrix} 0&0&0 \\ 0&0&0 \\ 0&0&1 \end{pmatrix},\ v_1 = 0, \\ u_2 &= \begin{pmatrix} 0&0&0 \\ 0&2&0 \\ 0&0&0 \end{pmatrix},\ v_2 = \begin{pmatrix} 0&0&-1 \\ 0&0&0 \\ -1&0&0 \end{pmatrix}. \end{aligned} \tag{28.25}$$

We can check $v_2 \in \tan(u_1, \mathcal{S}^3_+)$ by using the tangent space formula (28.6).

To illustrate the correctness of the extended dual $(D_{\text{ext,SDP}})$, we first note that $n = m = 3$, so by formula (28.24) we can choose $\ell = 2$ to obtain a correct extended dual. Recall that $y \in F^*_{\min}$ is an optimal dual solution if and only if it is of the form

$$y = \begin{pmatrix} 0 & 1/2 & -y_{22}/2 \\ 1/2 & y_{22} & y_{23} \\ -y_{22}/2 & y_{32} & y_{33} \end{pmatrix}. \tag{28.26}$$

Consider the $(u_i, v_i)_{i=0}^2$ sequence shown in (28.25); we prove that any y optimal matrix satisfies

$$y \in \mathcal{S}^3_+ + \tan(u_0 + u_1 + u_2, \mathcal{S}^3_+). \tag{28.27}$$

Indeed, $\tan(u_0+u_1+u_2,\mathscr{S}_+^3)$ is the set of 3 by 3 matrices with the component in the $(1,1)$ position equal to zero, and the other components arbitrary, and this proves (28.27).

In fact, considering the expression for $F_{\min}^*$ in (28.10), it follows that any $y \in F_{\min}^*$ satisfies (28.27).

28.5 Variants of Extended Duals

So far we proved the correctness of an extended dual of (*P*), which is itself an explicit SDP when (*P*) is. Ramana's original dual is somewhat different from $(D_{\text{ext,SDP}})$ though. Here we describe several variants of extended duals for (*P*) and show how to derive Ramana's dual.

First let us define a simplified extended system

$$\left.\begin{aligned} (u_0,v_0) &= (0,0) \\ (u_i,v_i) &\in K^* \times \tan(u_{i-1},K^*),\ i=1,\ldots,\ell+1 \\ u_i+v_i &\in L,\ i=1,\ldots,\ell. \end{aligned}\right\} \qquad (EXT_{\text{nice,simple}})$$

We prove that this system works just as well as (EXT_{nice}) when constructing extended duals.

Corollary 28.9. *The dual of the minimal cone of (P) has a representation*

$$F_{\min}^* = \{u_{\ell+1}+v_{\ell+1} : (u_i,v_i)_{i=0}^{\ell+1} \textit{ is feasible in} (EXT_{\text{nice,simple}})\}, \qquad (28.28)$$

and the extended dual

$$\begin{aligned} &\inf\ \langle b, u_{\ell+1}+v_{\ell+1}\rangle \\ &s.t.\ \ A^*(u_{\ell+1}+v_{\ell+1}) = c \\ &\qquad (u_i,v_i)_{i=0}^{\ell+1} \textit{ is feasible in} (EXT_{\text{nice,simple}}), \end{aligned} \qquad (D_{\text{ext,simple}})$$

where ℓ is defined in (28.14), has strong duality with (P).

In particular, if (P) is an SDP as described in Corollary 28.8, then the problem obtained from $(D_{\text{ext,SDP}})$ by replacing the constraint (*) by

$$(**)\ \begin{pmatrix} u_{i-1} & w_i \\ w_i^T & \beta_i I \end{pmatrix} \succeq 0,\ i=1,\ldots,\ell+1,$$

has strong duality with (P).

Proof. It is enough to prove the representation in equation (28.28); given this, the rest of the proof is analogous to the proof of the second and third statements in Corollary 28.8.

We will use the representation of $F_{\min}^*$ in (28.23). Let us denote by G the set on the right-hand side of equation (28.28); we will prove $G = F_{\min}^*$.

To show $G \subseteq F_{\min}^*$ suppose $u_{\ell+1} + v_{\ell+1} \in G$, where $(u_i, v_i)_{i=0}^{\ell+1}$ is feasible in $(EXT_{\text{nice,simple}})$. Then it is also feasible in (EXT_{nice}), since applying the tangent space formula (28.2) with K^* in place of K implies that

$$\tan(u_{i-1}, K^*) \subseteq \tan(u_0 + \cdots + u_{i-1}, K^*)$$

holds for $i = 1, \ldots, \ell+1$.

To prove $G \supseteq F_{\min}^*$ suppose that $u_{\ell+1} + v_{\ell+1} \in F_{\min}^*$, where $(u_i, v_i)_{i=0}^{\ell+1}$ is feasible in (EXT_{nice}). Again, by (28.2), the sets $\tan(u_0, K^*), \ldots, \tan(u_0 + \cdots + u_{i-1}, K^*)$ are all contained in $\tan(u_0 + \cdots + u_{i-1}, K^*)$ for $i = 1, \ldots, \ell$. Hence

$$v_1 + \cdots + v_i \in \tan(u_0 + \cdots + u_{i-1}, K^*),\ i = 1, \ldots, \ell \tag{28.29}$$

holds, and we also have

$$v_{\ell+1} \in \tan(u_0 + \cdots + u_\ell, K^*). \tag{28.30}$$

Let us define

$$(u_i', v_i') = (u_0 + \cdots + u_i, v_0 + \cdots + v_i),\ i = 1, \ldots, \ell.$$

By (28.29) and (28.30) it follows that $(u_{\ell+1}, v_{\ell+1})$ with $(u_i', v_i')_{i=0}^{\ell}$ is feasible for $(EXT_{\text{nice,simple}})$, so the inclusion follows. ■

Let us now consider another extended system

$$\left.\begin{array}{l} (u_0, v_0) = (0,0) \\ (u_i, v_i) \in K^* \times \tan'(u_{i-1}, K^*),\ i = 1, \ldots, \ell+1 \\ u_i + v_i \in L,\ i = 1, \ldots, \ell \end{array}\right\}, \qquad (EXT'_{\text{nice,simple}})$$

where the set $\tan'(u, K^*)$ satisfies the following two requirements for all $u \in K^*$:

1. $\tan'(u, K^*) \subseteq \tan(u, K^*)$.
2. For all $v \in \tan(u, K^*)$ there exists $\lambda_v > 0$ such that $v \in \tan'(\lambda_v u, K^*)$.

Corollary 28.10. *The dual of the minimal cone of (P) has the representation*

$$F_{\min}^* = \{u_{\ell+1} + v_{\ell+1} : (u_i, v_i)_{i=0}^{\ell+1} \textit{ is feasible in } (EXT'_{\text{nice,simple}})\}, \tag{28.31}$$

and the extended dual

$$\begin{array}{ll} \inf & \langle b, u_{\ell+1} + v_{\ell+1} \rangle \\ s.t. & A^*(u_{\ell+1} + v_{\ell+1}) = c \\ & (u_i, v_i)_{i=0}^{\ell+1} \textit{ is feasible in} (EXT'_{\text{nice,simple}}), \end{array} \qquad (D'_{\text{ext,simple}})$$

where ℓ is defined in (28.14), has strong duality with (P).

In particular, if (P) is an SDP as described in Corollary 28.8, then the problem obtained from ($D_{\text{ext,SDP}}$) by replacing the constraint (*) by

$$(***) \begin{pmatrix} u_{i-1} & w_i \\ w_i^T & I \end{pmatrix} \succeq 0, \; i = 1, \ldots, \ell+1,$$

and dropping the β_i variables, has strong duality with (P).

Proof. We use the representation of $F_{\min}^*$ in (28.28). Let us denote by G the set on the right-hand side of equation (28.31). We will prove $G = F_{\min}^*$.

It is clear that $G \subseteq F_{\min}^*$, since if $(u_i, v_i)_{i=0}^{\ell+1}$ is feasible in ($EXT'_{\text{nice,simple}}$), then by the first property of the operator tan', it is also feasible in ($EXT_{\text{nice,simple}}$).

To show the opposite inclusion, suppose $u_{\ell+1} + v_{\ell+1} \in F_{\min}^*$, where $(u_i, v_i)_{i=0}^{\ell+1}$ is feasible in ($EXT_{\text{nice,simple}}$). Let us choose $\lambda_\ell, \lambda_{\ell-1}, \ldots, \lambda_1$ positive reals such that

$$\begin{aligned} v_{\ell+1} &\in \text{tan'}(\lambda_\ell u_\ell, K^*), \\ \lambda_\ell v_\ell &\in \text{tan'}(\lambda_{\ell-1} u_{\ell-1}, K^*), \\ &\vdots \\ \lambda_2 v_2 &\in \text{tan'}(\lambda_1 u_1, K^*), \end{aligned} \tag{28.32}$$

and for completeness, set $\lambda_0 = 0$. Then $(u_{\ell+1}, v_{\ell+1})$ with $(\lambda_i u_i, \lambda_i v_i)_{i=0}^{\ell}$ is feasible in ($EXT'_{\text{nice,simple}}$), and this proves $F_{\min}^* \subseteq G$. ∎

We finally remark that in the extended duals for semidefinite programming it is possible to eliminate the v_i variables and use the w_i matrices directly in the constraints; thus one can exactly obtain Ramana's dual. We leave the details to the reader.

28.6 Conclusion

We gave a simple and self-contained exposition of a FRA and of extended duals: both approaches yield strong duality for a conic linear program, without assuming any constraint qualification. We generalized Ramana's dual: we proved that when K is a nice cone, the set $F_{\min}^*$ has an extended formulation, i.e., it is the projection of the feasible set of a conic linear system in a higher dimensional space. The only nontrivial constraints in this system are of the form $u \geq_{K^*} 0$, and $v \in \tan(u, K^*)$.

This formulation leads to an extended, strong dual of (P), when K is nice. When $K = K^*$ is the semidefinite cone, by writing the tangent space constraint as a semidefinite constraint, we obtain an extended strong dual, which is an SDP itself, and thus recover variants of Ramana's dual.

One may wonder whether there is an extended formulation for $F_{\min}$ itself. Suppose that K is an arbitrary closed convex cone. When a fixed $\bar{s} \in \operatorname{ri} F_{\min}$ is given, then obviously

$$F_{\min} = \{ s \,|\, 0 \leq_K s \leq_K \alpha \bar{s} \text{ for some } \alpha \geq 0 \}.$$

The minimal cone can also be represented without such an $\bar{s}$, since

$$F_{\min} = \{ s \,|\, 0 \leq_K s \leq_K \alpha b - Ax \text{ for some } x, \text{ and } \alpha \geq 0 \}. \tag{28.33}$$

This representation was obtained by Freund [12], based on the article by himself and Roundy and Todd [13].

It is also natural to ask whether there are other nice cones, for which the set

$$\{ (u,v) \,|\, u \in K^*, \, v \in \tan(u,K^*) \}$$

has a formulation in terms of K^*; e.g., is this true for the second-order cone? Conic linear programs over such cones would also have Ramana-type (i.e., expressed only in terms of K^*) extended duals.

Acknowledgements I would like to thank an anonymous referee and Minghui Liu for their helpful comments.

References

1. Alizadeh, F., Goldfarb, D.: Second-order cone programming. Math. Program. Ser. B **95** 3–51 (2003)
2. Andersen, E.D., Roos, C., Terlaky, T.: Notes on duality in second order and p-order cone optimization. Optimization **51**, 627–643 (2002)
3. Ben-Tal, A., Nemirovskii, A.: Lectures on Modern Convex Optimization. MPS/SIAM Series on Optimization. SIAM, Philadelphia (2001)
4. Bonnans, F.J., Shapiro, A.: Perturbation Analysis of Optimization Problems. Springer Series in Operations Research. Springer, New York (2000)
5. Borwein, J.M., Wolkowicz, H.: Facial reduction for a cone-convex programming problem. J. Aust. Math. Soc. **30**, 369–380 (1981)
6. Borwein, J.M., Wolkowicz, H.: Regularizing the abstract convex program. J. Math. Anal. App. **83**, 495–530 (1981)
7. Boyd, S., Vandenberghe, L.: Convex Optimization. Cambridge University Press, Cambridge (2004)
8. Cheung, V., Wolkowicz, H., Schurr, S.: Preprocessing and regularization for degenerate semidefinite programs. Technical Report, University of Waterloo (2012)
9. Chua, C.-B.: Relating homogeneous cones and positive definite cones via T-algebras. SIAM J. Optim. **14**, 500–506 (2003)
10. Chua, C.-B., Tunçel, L.: Invariance and efficiency of convex representations. Math. Program. B **111**, 113–140 (2008)
11. Faybusovich, L.: On Nesterov's approach to semi-definite programming. Acta Appl. Math. **74**, 195–215 (2002)

12. Freund, R.M.: Talk at the University of Waterloo (1994)
13. Freund, R.M., Roundy, R., Todd, M.J.: Identifying the set of always-active constraints in a system of linear inequalities by a single linear program. Technical Report Sloan W.P. No. 1674–1685, Sloan School of Business, MIT (1985)
14. Gouveia, J., Parrilo, P., Thomas, R.: Lifts of convex sets and cone factorizations. Math. Oper. Res. (2013) (to appear)
15. Güler, O.: Foundations of Optimization. Graduate Texts in Mathematics. Springer, New York (2010)
16. Helmberg, C.: The semidefinite programming webpage. http://www-user.tu-chemnitz.de/~semidef.html
17. Klep, I., Schweighofer, M.: An exact duality theory for semidefinite programming based on sums of squares. Technical Report, Universität Konstanz. http://arxiv.org/abs/1207.1691 (2012)
18. Lewis, A.S.: Facial reduction in partially finite convex programming. Math. Program. B **65**, 123–138 (1994)
19. Luo, Z.-Q., Sturm, J.: Error analysis. In: Saigal, R., Vandenberghe, L., Wolkowicz, H. (eds.) Handbook of Semidefinite Programming. Kluwer Academic Publishers, Dordrecht (2000)
20. Luo, Z.-Q., Sturm, J., Zhang, S.: Duality results for conic convex programming. Technical Report 9719/A, Econometric Institute, Erasmus University, Rotterdam (1997)
21. Pataki, G.: A simple derivation of a facial reduction algorithm and extended dual systems. Technical Report, Columbia University (2000)
22. Pataki, G.: On the closedness of the linear image of a closed convex cone. Math. Oper. Res. **32**, 395–412 (2007)
23. Pataki, G.: Bad semidefinite programs: they all look the same. Technical Report, University of North Carolina at Chapel Hill (2010). Available from http://arxiv.org/abs/1112.1436 (under review)
24. Pataki, G.: On the connection of facially exposed and nice cones. J. Math. Anal. App. **400**, 211–221 (2013)
25. Pólik, I., Terlaky, T.: Exact duality for optimization over symmetric cones. Technical Report, Lehigh University (2009)
26. Ramana, M.V.: An exact duality theory for semidefinite programming and its complexity implications. Math. Program. Ser. B **77**, 129–162 (1997)
27. Ramana, M.V., Tunçel, L., Wolkowicz, H.: Strong duality for semidefinite programming. SIAM J. Opt. **7**, 641–662 (1997)
28. Rockafellar, T.R.: Convex Analysis. Princeton University Press, Princeton (1970)
29. Schurr, S.P., Tits, A.L., O'Leary, D.P.: Universal duality in conic convex optimization. Math. Program. **109**, 69–88 (2007)
30. Tunçel, L., Wolkowicz, H.: Strong duality and minimal representations for cone optimization. Comput. Optim. Appl. **53**, 619–648
31. Roshchina, V.: Facially exposed cones are not nice in general. Technical Report, University of Ballarat (2013). http://arxiv.org/abs/1301.1000
32. Saigal, R., Vandenberghe, L., Wolkowicz, H. (eds.): Handbook of Semidefinite Programming. Kluwer Academic Publishers, Dordrecht (2000)
33. Jos Sturm: Error bounds for linear matrix inequalities. SIAM J. Optim. **10**, 1228–1248 (2000)
34. Todd, M.J.: Semidefinite Optimization. Acta Numer. **10**, 515–560 (2001)
35. Tunçel, L.: Polyhedral and Semidefinite Programming Methods in Combinatorial Optimization. Fields Institute Monographs. American Mathematical Society, Providence (2011)
36. Waki, H., Muramatsu, M.: Facial reduction algorithms for conic optimization problems. Technical Report CS-09-01, Department of Computer Science, The University of Electro-Communications (2009)
37. Wang, Y., Xiu, N., Luo, Z.: A regularized strong duality for nonsymmetric semidefinite least squares problem. Optim. Lett. **5**, 665–682 (2011)

Chapter 29
Towards a New Era in Subdifferential Analysis?

Jean-Paul Penot

Dedicated to Jon Borwein on the occasion of his 60th birthday

Abstract We give some new attention to the foundations of nonsmooth analysis. We endeavour to delineate the common features of usual subdifferentials. In particular, we stress calculus rules and properties linked with order. Our objective is to give the possibility of using subdifferentials without dealing with specific constructions.

Key words: Calculus rules • Nonsmooth analysis • Subdifferential • Subgradient

Mathematics Subject Classifications (2010): Primary 49J52; Secondary 26A27, 90C56.

29.1 Introduction

During several decades nonsmooth analysis has been viewed by some authors and most users as a field of disorder, a "ménagerie" to take the expression in the preface of [1]. On the contrary, it appears to some other authors that nonsmooth analysis has joint features, in particular the passages from sets to functions via indicator functions and distance functions and the reverse passage using epigraphs. It is the

---

COMMUNICATED BY MICHEL THÉRA.

J.-P. Penot (✉)
Laboratoire Jacques-Louis Lions, Université Pierre et Marie Curie,
4 place Jussieu 75252 Paris Cedex 05, France
e-mail: penot@ann.jussieu.fr

D.H. Bailey et al. (eds.), *Computational and Analytical Mathematics*, Springer Proceedings in Mathematics & Statistics 50, DOI 10.1007/978-1-4614-7621-4_29,

purpose of the present paper to show that many properties are shared by the different concepts of subdifferential. As a consequence, the user may avoid the specific constructions and just retain the rules needed for the application in view. That does not mean that for all problems an abstract subdifferential should be used. We admit that some problems are best dealt with by using a specific subdifferential.

It is not the first time an axiomatic approach to subdifferentials is adopted. On the contrary, such attempts abound. However, our aim, as described above, is different from the ones we are aware of in the literature. In [15] A. Ioffe explores different possibilities of defining tangent cones and generalized derivatives. In [20] (resp. [26]) he gives conditions aiming at showing minimality (resp. uniqueness) of his subdifferential. A similar objective is present in Sect. 2.5.1 of the monograph [31]. D. Aussel et al., A. Ioffe, M. Lassonde, and J. Zhu in [2, 23, 29, 39] consider a list of axioms as reduced as possible in order to show the equivalence of several crucial properties of subdifferentials such as sum rules and mean value theorems. R. Correa et al. [10] adopt conditions (which exclude some subdifferentials) in order to characterize convexity; see [11, 38] for more general conditions including a sum rule. In [34] a first exploration of conditions ensuring some compatibility with order is undertaken. It is carried on here with the observation that subdifferential analysis is much used for optimization problems and such problems involve order. Thus, order properties should be given more attention than what they received in the literature.

Of course, the properties we adopt as axioms are suggested by the well-known properties of differential calculus and by the properties of the main specific subdifferentials. But we also detect properties linked with order that have not been exhibited. Since the main applications of subdifferentials concern optimization questions, it is natural to stress such order properties.

After a presentation of the conditions (or axioms) we select in the next section, we deduce some consequences of these general properties. Then, in Sect. 29.2.3, we display some variants because the user may prefer to deal with simple versions or on the contrary may prefer to dispose of more powerful properties. The definitions of the most usual subdifferentials are reminded in Sect. 29.2.4 and it is checked that the conditions we selected are satisfied by them. Section 29.3 is devoted to extension questions. The main one concerns the extension of subdifferentials from the class $\mathscr{L}$ of locally Lipschitzian functions to the class $\mathscr{I}$ of lower semicontinuous functions. Thus, our main result reduces the proof of the expected properties to the case of the class $\mathscr{L}$ for which the constructions are often simpler. Finally, we turn to some additional properties which enable to develop a full calculus.

Such a general approach cannot replace the study of specific subdifferentials, because all the main subdifferentials have particular properties of interest not shared by the other ones. But it shows that the family of subdifferentials has much in common and is a genuine family.

The notation we use is the notation of [36], hence is essentially compatible with the ones of [7, 9]. In particular, if X and Y are two normed spaces, $L(X,Y)$ stands for the set of continuous linear maps from X into Y and $B(\overline{x},r)$ (resp. $B[\overline{x},r]$) denotes the open (resp. closed) ball with center $\overline{x}$ and radius r in X. For a function f on X

finite at $\overline{x}$, $B(\overline{x},r,f)$ is the set of $x \in B(\overline{x},r)$ such that $|f(x) - f(\overline{x})| < r$. Here, as in [36], we say that a map $g : X \to Y$ between two normed spaces is *circa-differentiable* (or strictly differentiable) *at* $\overline{x} \in X$ if there exists some $A \in L(X,Y)$ such that for all $\varepsilon > 0$ there exists some $\delta > 0$ such that $g - A$ is Lipschitzian with rate ε on the ball $B(\overline{x},\delta)$. That is the case when g is of class C^1 at (resp. around) $\overline{x}$ in the sense that g is differentiable on a neighborhood of $\overline{x}$ and its derivative g' is continuous at $\overline{x}$ (resp. on a neighborhood of $\overline{x}$). We always endow a product $X \times Y$ of normed spaces with a *product norm*, i.e., a norm on $X \times Y$ for which the projections are nonexpansive and the insertions $x \mapsto (x,0)$ and $y \mapsto (0,y)$ are continuous.

29.2 General Properties of Subdifferentials

29.2.1 An Axiomatic Approach

Below is a list of essential properties shared by all interesting subdifferentials besides global subdifferentials that are outside the realm of infinitesimal analysis (see [33] for instance for subdifferentials adapted to generalized convexity). Taking these properties as axioms we can devise several other properties. Of course, it is desirable to dispose of as many properties as possible. On the other hand, it is convenient to make this list as short as possible, while keeping these properties. We do not look for independence of these conditions, but for a list which is natural enough and as efficient as possible for its uses. Several properties have some variants. The strongest ones are more difficult to check for a specific subdifferential. The coarsest ones may not reflect the full power of subdifferentials. Thus, in general we choose the one whose expression is the simplest one while being general enough, and we mention possible variants. We observe that it is possible to rule out all subdifferentials but one by requiring some particular conditions. Again, on the contrary, we want to encompass all usual subdifferentials. Thus, the list we give is the result of a compromise. Depending on the problem at hand or the needs one may have, it can be shortened or completed. In the course of the paper we will examine some additional properties which may be of great importance.

Some versatility is obtained by restricting the attention to some particular class of spaces $\mathscr{X}$ or some particular class of functions $\mathscr{F}$, as it is known that important properties of some subdifferentials are valid only in finite dimensional spaces or Asplund spaces. We assume $\mathscr{X}$ is stable by products and contains $\mathbb{R}$. By a class of functions, we mean that for all X in $\mathscr{X}$ we are given a set $\mathscr{F}(X)$ of extended real-valued lower semicontinuous functions on X. The main classes of functions we consider are the class $\mathscr{L}$ of locally Lipschitzian functions and the class $\mathscr{I}$ of (extended real-valued proper) lower semicontinuous functions. For instance, we can take the class of functions which are sums of a lower semicontinuous convex function and a function of class C^1. Here we assume that the class of functions contains at least the class $\mathscr{L}$. Of course, the conditions we impose require that the subdifferential ∂ is local and coincides with the Fenchel–Moreau subdifferential ∂_{FM} when applied to a convex function f, where

$$\partial_{\mathrm{FM}} f(x) = \{x^* : \langle x^*, w\rangle \leq f(x+w) - f(x), \ \forall\, w \in X\}.$$

Definition 29.1. Given a class $\mathscr{X}$ of Banach spaces and a class $\mathscr{F}$ of functions, by *subdifferential* or subdifferential of classical type, we mean a mapping which associates with any $X \in \mathscr{X}$, any $f \in \mathscr{F}(X)$, and any $\overline{x} \in \operatorname{dom} f = \{x : |f(x)| < \infty\}$ a set $\partial f(\overline{x}) \subset X^*$ in such a way that the following properties are satisfied whenever V, W, X, Y, Z are members of $\mathscr{X}$, $\overline{x} \in X$, $\overline{y} \in Y$:

– *Localizability:*

(S1) If $f, g \in \mathscr{F}(X)$ coincide in a neighborhood of $\overline{x}$, then $\partial f(\overline{x}) = \partial g(\overline{x})$.

– *Contiguity:*

(S2) If $f \in \mathscr{F}(X)$ is convex, then $\partial f(\overline{x}) = \partial_{\mathrm{FM}} f(\overline{x})$.

– *Optimality:*

(S3) If $f \in \mathscr{F}(X)$ attains a local minimum at $\overline{x} \in \operatorname{dom} f$, then $0 \in \partial f(\overline{x})$.

– *Calculability:*

(S4) If for some $g : X \to Y$ of class C^1 at $\overline{x} \in X$ with $A(X) = Y$ for $A := g'(\overline{x})$, $\lambda > 0$, $\ell \in X^*$, $c \in \mathbb{R}$, and $h \in \mathscr{F}(Y)$, one has $f(x) = \lambda h(g(x)) + \langle \ell, x\rangle + c$, then $\partial f(\overline{x}) = \lambda A^T \partial h(g(\overline{x})) + \ell$.

(S4b) For $m \in \mathbb{N}\backslash\{0\}$, $g_1 \in \mathscr{F}(X_1), \ldots, g_m \in \mathscr{F}(X_m)$, $X := X_1 \times \cdots \times X_m$, $g := g_1 \times \cdots \times g_m : X \to \mathbb{R}^m$, $j : \mathbb{R}^m \to \mathbb{R}$ of class C^1 around $\overline{r} := g(\overline{x}) := (g_1(\overline{x}_1), \ldots, g_m(\overline{x}_m))$ and nondecreasing in each of its m arguments, with $D_i j(\overline{r}) \neq 0$ for $i = 1, \ldots, m$, if $f := j \circ g$, then $\partial f(\overline{x}) \subset j'(\overline{r}) \circ (\partial g_1(\overline{x}_1) \times \cdots \times \partial g_m(\overline{x}_m))$.

– *Consistency*:

(S5) If $f(x,y) := \max(g(x), h(y))$, $(\overline{x}^*, \overline{y}^*) \in \partial f(\overline{x}, \overline{y})$ with $g \in \mathscr{F}(X)$, continuous $h \in \mathscr{F}(Y)$ of class C^1 around $\overline{y}$, $g(\overline{x}) = h(\overline{y})$, $\overline{y}^* \neq h'(\overline{y}) \neq 0$, then $(\overline{x}^*, \overline{y}^*) \in \{\lambda \partial g(\overline{x}) \times (1-\lambda)\{h'(\overline{y})\} : \lambda \in]0,1]\}$.

– *Pseudo-homotonicity* (or order compatibility)*:*

(S6a) If $A \in L(V,W)$ with $W = A(V)$, $\overline{w} \in W$, $M \subset A^{-1}(\overline{w})$, $\varphi \in \mathscr{F}(V)$, $p \in \mathscr{F}(W)$ are such that $p \circ A \leq \varphi$ and that for every sequences $(\alpha_n) \to 0_+$, $(w_n) \to \overline{w}$ with $(p(w_n)) \to p(\overline{w}) \in \mathbb{R}$, $w_n \in A(\operatorname{dom} \varphi)$ for all n, one can find $\overline{v} \in M$, an infinite subset N of $\mathbb{N}$, a sequence $(v_n)_{n \in N} \to \overline{v}$ such that $A(v_n) = w_n$ and $\varphi(v_n) \leq p(w_n) + \alpha_n$ for all $n \in N$, then one has

$$A^T(\partial p(\overline{w})) \subset \bigcup_{\overline{v} \in M} \partial \varphi(\overline{v}). \tag{29.1}$$

(S6b) If $A \in L(V,W)$ with $W = A(V)$, if E is a closed subset of W, $\overline{w} \in E$, $\varphi \in \mathscr{F}(V)$, $p := d_E$, the distance function to E, $M \subset A^{-1}(\overline{w})$ are such that $p \circ A \leq \varphi$, and if for every sequences $(\alpha_n) \to 0_+$, $(w_n) \to \overline{w}$ with $w_n \in E$ for all n, one can find $\overline{v} \in M$, an infinite subset N of $\mathbb{N}$, and a sequence $(v_n)_{n \in N} \to \overline{v}$ such that $A(v_n) = w_n$ and $\varphi(v_n) \leq \alpha_n$ for all $n \in N$, then relation (29.1) holds.

If necessary, we make clearer the definition of ∂ by writing $(\partial, \mathscr{X}, \mathscr{F})$. It may be convenient to extend the definition of ∂ to any triple (X,f,x) with X in $\mathscr{X}$, $f \in \mathscr{F}(X)$, $x \in X$ by requiring the following condition:

– *Substantiality:*

(S0) $\partial f(x) = \varnothing$ if $x \in X \backslash \mathrm{dom} f$.

The terminology we use is mostly due to A. Ioffe. It is clear, but for condition (S5), "consistency" can be explained by the fact that (S5) serves to show the agreement of a geometrical device with the present analytical approach, as proved in Proposition 29.23.

The preceding set of properties can be supplemented by two other sets: one less demanding and one slightly more exacting. Both versions are close to the classical one above.

Definition 29.2. Given a class $\mathscr{X}$ of Banach spaces and a class $\mathscr{F}$ of functions, by *subdifferential of alleviated type* we mean a mapping $\partial : (X,f,x) \mapsto \partial f(x)$ satisfying the conditions of the preceding definition but (S4c) and (S5c) replaced with:

(S4a) If for some $A \in L(X,Y)$ with $A(X) = Y$, $\lambda > 0$, $b \in Y$, $\ell \in X^*$, $c \in \mathbb{R}$ $h \in \mathscr{F}(Y)$, one has $f(x) = \lambda h(A(x)+b) + \langle \ell, x\rangle + c$, then $\partial f(\overline{x}) = \lambda A^T \partial h(A(\overline{x})+b) + \ell$.

(S5a) If $f(x,y) := \max(g(x), h(y))$ with $g \in \mathscr{F}(X)$, $h \in Y^* \backslash \{0\}$, $g(\overline{x}) = h(\overline{y})$, $(\overline{x}^*, \overline{y}^*) \in \partial f(\overline{x}, \overline{y})$, $\overline{y}^* \neq h$, then $(\overline{x}^*, \overline{y}^*) \in \{\lambda \partial g(\overline{x}) \times (1-\lambda)\{h'(\overline{y})\} : \lambda \in]0,1]\}$.

Definition 29.3. Given a class $\mathscr{X}$ of Banach spaces and a class $\mathscr{F}$ of functions, by *subdifferential of deepened type*, we mean a mapping $\partial : (X,f,x) \mapsto \partial f(x)$ satisfying the conditions of Definition but (S4c) and (S5c) replaced with:

(S4d) If for some $g : X \to Y$ circa-differentiable at $\overline{x} \in X$ with $A(X) = Y$ for $A := g'(\overline{x})$, $\lambda > 0$, $b \in Y$, $\ell \in X^*$, $c \in \mathbb{R}$, $h \in \mathscr{F}(Y)$ one has $f(x) = \lambda h(g(x)+b) + \langle \ell, x\rangle + c$, then $\partial f(\overline{x}) = \lambda A^T \partial h(g(\overline{x})+b) + \ell$.

(S5d) If $f(x,y) := \max(g(x), h(y))$, $(\overline{x}^*, \overline{y}^*) \in \partial f(\overline{x}, \overline{y})$ with $g \in \mathscr{F}(X)$, $h \in \mathscr{F}(Y)$ circa-differentiable at $\overline{y}$, $g(\overline{x}) = h(\overline{y})$, $\overline{y}^* \neq h'(\overline{y}) \neq 0$, then $(\overline{x}^*, \overline{y}^*) \in \{\lambda \partial g(\overline{x}) \times (1-\lambda)\{h'(\overline{y})\} : \lambda \in]0,1]\}$.

In the sequel (S4) and (S5) are sometimes relabelled (S4c) and (S5c), respectively, in order to put in light the analogies with the other conditions of the same type; (S6) stands for the conjunction of (S6a) and (S6b).

Simple consequences of the preceding conditions are displayed in the next subsection.

29.2.2 *Some Simple Consequences*

We shall show that the conditions expounded above have some interesting direct consequences and are a starting point for a rich set of nontrivial calculus rules provided some further properties are added. They will be introduced in the last section.

Clearly, condition (S4a) ensures invariance by isomorphisms; in particular ∂ is independent of the choice of a compatible norm. Correspondingly, (S4c) (or (S4d)) implies invariance under diffeomorphisms: if $f = h \circ G$, where G is a diffeomorphism, then $\partial f(x) = G'(x)^T \partial h(G(x))$, so that subdifferentials of functions on differential manifolds can be introduced. Conversely, if this invariance property holds, then a weaken version of (S4c) is satisfied, as shown by the submersion theorem: given $f = h \circ G$ with $h \in \mathscr{F}(Y)$, $G : X \to Y$ of class C^1 around $\overline{x} \in X$, then $\partial f(\overline{x}) = A^T \partial h(G(\overline{x}))$ provided the kernel Z of $A := G'(\overline{x})$ has a topological supplement. Moreover, invariance by diffeomorphism implies that in (S6a), (S6b) A can be replaced with a submersion.

A tight contiguity property follows from conditions (S4c) and (S4d).

Proposition 29.4

(a) Under conditions (S2) and (S4c) (resp. (S4d)), if f is of class C^1 at $\overline{x}$ (resp. circa-differentiable at $\overline{x}$), then $\partial f(\overline{x}) = \{f'(\overline{x})\}$.

(b) If $f(x,y) := g(x) + h(y)$, where $g \in \mathscr{F}(X)$, $h \in \mathscr{F}(Y)$ being circa-differentiable at $\overline{y}$, then $\partial f(\overline{x},\overline{y}) \subset \partial g(\overline{x}) \times \{h'(\overline{y})\}$.

A special case of condition (S4b) concerns the sum of m functions of independent variables. It is obtained by taking $j : \mathbb{R}^m \to \mathbb{R}$ given by $j(r_1,\ldots,r_m) = r_1 + \cdots + r_m$. In view of its importance, we state it for $m = 2$ in the following form.

(S4s) If $f(x,y) = g(x) + h(y)$ with $f \in \mathscr{F}(X \times Y)$, $g \in \mathscr{F}(X)$, $h \in \mathscr{F}(Y)$, then $\partial f(\overline{x},\overline{y}) \subset \partial g(\overline{x}) \times \partial h(\overline{y})$.

The general case follows from an easy induction on m.

A particular case of (S5a) is the following condition that has interesting consequences:

(S5o) If $f(x,y) := \max(g(x),h(y))$, $(\overline{x}^*,0) \in \partial f(\overline{x},\overline{y})$ with $g \in \mathscr{L}(X)$, $h \in Y^* \backslash \{0\}$, $g(\overline{x}) = h(\overline{y})$, then $\overline{x}^* \in \partial g(\overline{x})$.

A simplified form of condition (S6a) consists in taking for M a singleton:

(S6s) If $A \in L(V,W)$ with $W = A(V)$, $\overline{v} \in V$, $\overline{w} := A\overline{v}$, $\varphi \in F(V)$, $p \in F(W)$ are such that $p \circ A \leq \varphi$, $\varphi(\overline{v}) = p(\overline{w})$ and that for every sequences $(\alpha_n) \to 0_+$, $(w_n) \to \overline{w}$ with $(p(w_n)) \to p(\overline{w})$ one can find a sequence $(v_n) \to \overline{v}$ such that $A(v_n) = w_n$ and $\varphi(v_n) \leq p(w_n) + \alpha_n$ for all $n \in N$ large enough, then one has $A^T(\partial p(\overline{w})) \subset \partial \varphi(\overline{v})$.

A similar simplification of (S6b) can be given.

Conditions (S6a) and (S6b) entail exact forms that are often convenient. In order to formulate them we recall that a multimap $S : W \rightrightarrows V$ between two Banach (or metric) spaces is lower semicontinuous at $\overline{w} \in W$ if $S(\overline{w})$ is nonempty and if for all $\overline{v} \in S(\overline{w})$ one has $d(\overline{v}, S(w)) \to 0$ as $w \to \overline{w}$.

(S6e) If $A \in L(V,W)$, $\varphi \in \mathscr{F}(V)$, $p \in \mathscr{F}(W)$, $\overline{w} \in W$ are such that $p \circ A \leq \varphi$, $W = A(V)$ and for some neighborhood W_0 of $\overline{w}$ and some lower semicontinuous multimap $S : W_0 \rightrightarrows V$ satisfying $A(S(w)) = w$, $\varphi(v) = p(w)$ for all $w \in W_0$, $v \in S(w)$, then, for all $\overline{v} \in S(\overline{w})$, one has $A^T(\partial p(\overline{w})) \subset \partial \varphi(\overline{v})$.

(S6f) If $A \in L(V,W)$, $\varphi \in \mathscr{F}(V)$, E is a closed subset of W, $p = d_E$, $\overline{w} \in E$ are such that $p \circ A \leq \varphi$, $W = A(V)$ and for some neighborhood W_0 of $\overline{w}$ and some lower semicontinuous multimap $S : W_0 \cap E \rightrightarrows V$ satisfying $A(v) = w$, $\varphi(v) = p(w)$ for all $w \in W_0 \cap E$, $v \in S(w)$, then, for all $\overline{v} \in S(\overline{w})$, one has $A^T(\partial p(\overline{w})) \subset \partial \varphi(\overline{v})$.

A number of other direct consequences can be drawn from the axioms presented above. An immediate one concerns the case that a function is independent of one of its variables. A second one deals with an intertwined sum.

Proposition 29.5. *Let X, Y in $\mathscr{X}$, $g \in \mathscr{F}(X)$ be such that the function f given by $f(x,y) = g(x)$ belongs to $\mathscr{F}(X \times Y)$. Then $\partial f(x,y) = \partial g(x) \times \{0\}$.*

Proof. The assertion derives from (S4a) by taking for A the canonical projection and $\ell := 0$. ∎

Let us note that since $g(\cdot) = \inf\{f(\cdot,y) : y \in Y\}$ and since for any sequence $(x_n) \to x$ one has $f(x_n,y) = g(x_n)$, condition (S6a) (or (S6e) with $S(\cdot) := \{(\cdot,y)\}$ or $S(x) := \{x\} \times Y$) implies that for all $(x,y) \in X \times Y$ one has $\partial g(x) \times \{0\} \subset \partial f(x,y)$. More generally (S6a) implies the inclusion $A^T(\partial h(Ax)) \subset \partial f(x)$ when $f := h \circ A$ as in condition (S4a). Although (S4a) and (S6a) could be merged into a single statement adding to the conclusion of (S6) the equality $A^T(\partial p(\overline{w})) = \partial \varphi(\overline{v})$ when $p = \varphi \circ A$, we prefer to state them separately.

Proposition 29.6. *Let ∂ be a subdifferential of alleviated type. Let W, X in $\mathscr{X}$, $g \in \mathscr{F}(W)$, $h \in \mathscr{F}(X)$, $B \in L(W,X)$ such that f given by $f(w,x) := g(w) + h(Bw + x)$ belongs to $\mathscr{F}(W \times X)$. Then for any $(w^*,x^*) \in \partial f(w,x)$ one has $w^* - B^T x^* \in \partial g(w)$, $x^* \in \partial h(Bw + x)$, i.e., $(w^*,x^*) = (u^* + B^T x^*, x^*)$ for some $u^* \in \partial g(w)$, while $x^* \in \partial h(Bw + x)$.*

If ∂ is a classical subdifferential and if for some map $j : W \to X$ of class C^1 one has $f(w,x) := g(w) + h(j(w) + x)$, then, for any $(w^*,x^*) \in \partial f(w,x)$, setting $B := Dj(w)$ one has $w^* - B^T x^* \in \partial g(w)$, $x^* \in \partial h(j(w) + x)$.

Proof. For the second assertion, setting $G(w',x') := (w', j(w') + x')$, observing that G is a C^1-diffeomorphism, that $f = k \circ G$, where $k(u,v) := g(u) + h(v)$, and applying (S4c) and (S4s), for $(w^*,x^*) \in \partial f(w,x)$, one has $(w^*,x^*) = A^T(u^*,v^*)$ for $A := DG(w,x)$, some $(u^*,v^*) \in \partial k(w, j(w) + x)$, so that $u^* \in \partial g(w)$, $v^* \in \partial h(j(w) + x)$. Since $A^T(u^*,v^*) = (u^* + B^T v^*, v^*)$, one gets $v^* = x^*$, $u^* = w^* - B^T x^*$. The first assertion is obtained similarly, using (S4a) instead of (S4c). ∎

A boundedness property can be easily derived from the above conditions.

Proposition 29.7. *If $f \in \mathscr{F}(X)$ is Lipschitzian with rate r near $\overline{x} \in X$, then for any subdifferential ∂, one has $\partial f(\overline{x}) \subset rB_{X^*}$.*

Proof. By (S1), we may suppose f is globally Lipschitzian with rate r. Then, for all $w \in X$ we have

$$f(w) = \inf\{\varphi(w,u) : u \in X\} \quad \text{for } \varphi(w,u) := f(u) + r\|u - w\|.$$

Moreover, one has $f = \varphi \circ S$ with $S(w) := \{(w,w)\}$. Then (S6), or even (S6e), ensures that for all $\overline{x}^* \in \partial f(\overline{x})$ one has $(\overline{x}^*, 0) \in \partial \varphi(\overline{x}, \overline{x})$. Using Proposition 29.6 and (S2), we get some $\overline{u}^* \in \partial f(\overline{x})$ such that $(\overline{x}^*, 0) = (\overline{x}^*, \overline{u}^* - \overline{x}^*)$, with $\overline{u}^* \in r\partial \|\cdot\|(0) = rB_{X^*}$. Thus $\overline{x}^* \in rB_{X^*}$. ∎

Corollary 29.8. *If $f \in \mathscr{F}(X)$ is circa-differentiable at $\overline{x}$ and if ∂ is a subdifferential of alleviated type, then $\partial f(\overline{x}) \subset \{f'(\overline{x})\}$.*

Proof. If $f \in \mathscr{F}(X)$ is circa-differentiable at $\overline{x}$ with derivative ℓ, setting $g := f - \ell$, for every $\varepsilon > 0$ one can find a neighborhood of $\overline{x}$ on which g is Lipschitzian with rate less than ε. By (S4a) and the preceding proposition one gets

$$\partial f(\overline{x}) - \ell = \partial g(\overline{x}) \subset \varepsilon B_{X^*}.$$

It follows that $\partial f(\overline{x}) \subset \{\ell\}$. ■

The case $m = 1$, $h : r \mapsto r^p$ and $r \mapsto r^{1/p}$ of condition (S4b) yields the following special case of Leibniz rule.

Proposition 29.9. *If $f, g \in \mathscr{F}(X)$ are positive and such that, for some $p > 0$, $f = g^p$ around some x satisfying $g(x) > 0$, then $\partial f(x) = pg(x)^{p-1}\partial g(x)$.*

The usual Leibniz rule stems from the case $m = 2$ in (S4b), using the product $(r, s) \mapsto rs$.

Proposition 29.10. *If $g, h \in \mathscr{L}(X)$, $f := gh$ satisfy $g(\overline{x}) > 0$, $h(\overline{x}) > 0$ then $\partial f(\overline{x}) \subset g(\overline{x})\partial h(\overline{x}) + h(\overline{x})\partial g(\overline{x})$.*

29.2.3 Some Variants

It may be useful to detect some variants of the preceding set of conditions and to make some comments. First, one has to note that making a slight modification of the previous axioms one can eliminate any of the usual subdifferentials. For instance, it has been observed by M. Lassonde that changing the inclusion $\partial f(x,y) \subset \partial g(x) \times \partial h(y)$ of condition (S4s) for $f(x,y) := g(x) + h(y)$ into an equality excludes the Clarke subdifferential. Other variants considered in the present subsection exclude the limiting subdifferential or the Fréchet or the (Dini-) Hadamard subdifferentials.

There are no serious reasons to change conditions (S0)–(S3).

One may prefer to (S5c) or (S5d) the more intrinsic condition:

(S5i) If $g \in \mathscr{F}(X)$, $h \in \mathscr{F}(Y)$, $f(x,y) := \max(g(x), h(y))$, $g(\overline{x}) = h(\overline{y})$, with $0 \notin \partial g(\overline{x})$, $0 \notin \partial h(\overline{y})$, then $\partial f(\overline{x},\overline{y}) \subset \{(1-t)\partial g(\overline{x}) \times t\partial h(\overline{y}) : t \in [0,1]\}$.

An easy induction enables to reformulate this condition for m functions:

If $m \in \mathbb{N}\backslash\{0,1\}$, $f(x_1,\ldots,x_m) := \max(g_1(x_1),\ldots,g_m(x_m))$ with $g_1 \in \mathscr{F}(X_1),\ldots,g_m \in \mathscr{F}(X_m)$, $g_1(\overline{x}_1) = \cdots = g_m(\overline{x}_m)$, with $0 \notin \partial g_i(\overline{x}_i)$, $i = 1,\ldots,m$, then $\partial f(\overline{x}) \subset \{r_1^*\partial g_1(\overline{x}) \times \cdots \times r_m^*\partial f_m(\overline{x}) : r_1^*,\ldots,r_m^* \in \mathbb{R}_+,\ r_1^* + \cdots + r_m^* = 1\}$.

However, it is not clear whether (S5i) is satisfied by the firm (or Fréchet) subdifferential or the directional (or Dini–Hadamard) subdifferential.

On the other hand, (S4b) and (S5i) are special cases of the following chain rule in which $p_i : X := X_1 \times \cdots \times X_m \to X_i$ denotes the canonical projection, $\overline{x} := (\overline{x}_1,\ldots,\overline{x}_m) \in X$, $\overline{r} := (g_1(\overline{x}_1),\ldots,g_m(\overline{x}_m))$:

(C) If $f := j \circ (g_1 \circ p_1, \ldots, g_m \circ p_m)$ with $g_1 \in \mathscr{L}(X_1), \ldots, g_m \in \mathscr{L}(X_m)$, $j \in \mathscr{L}(\mathbb{R}^m)$ nondecreasing, then $\partial f(\overline{x}) \subset \{r_1^* \partial g_1(\overline{x}_1) \times \cdots \times r_m^* \partial g_m(\overline{x}_m) : (r_1^*, \ldots, r_m^*) \in \partial j(\overline{r})\}$.

For (S4b) that follows from the choice $j(r_1, \ldots, r_m) = r_1 + \cdots + r_m$ and for (S5i) that stems from the choice $j(r_1, \ldots, r_m) = \max(r_1, \ldots, r_m)$ for $(r_1, \ldots, r_m) \in \mathbb{R}^m$, j being convex with $\partial j(r, \ldots, r) = \{(r_1^*, \ldots, r_m^*) \in \mathbb{R}_+^m : r_1^* + \cdots + r_m^* = 1\}$.

Obviously, (S5) or (S5o) entails the following special case in which $h = I_{\mathbb{R}}$:

(S5m) If $g \in \mathscr{L}(X)$, if $k(x,r) := g(x) \vee r := \max(g(x), r)$ for $(x,r) \in X \times \mathbb{R}$, and if $(\overline{x}^*, 0) \in \partial k(\overline{x}, \overline{r})$ with $\overline{r} := g(\overline{x})$, then $\overline{x}^* \in \partial g(\overline{x})$.

Now, setting $m(x,r) := \max(g(x) - r, 0) = k(x,r) - r$, condition (S4) ensures that

$$(\overline{x}^*, -1) \in \partial m(\overline{x}, \overline{r}) \Longleftrightarrow (\overline{x}^*, 0) \in \partial k(\overline{x}, \overline{r}). \tag{29.2}$$

Thus, condition (S5m) is seen to be equivalent to $\overline{x}^* \in \partial g(\overline{x})$ whenever $(\overline{x}^*, -1) \in \partial m(\overline{x}, \overline{r})$ and $\overline{r} := g(\overline{x})$. Now, $m(x,r) = d((x,r), \operatorname{epi} g)$ for (x,r) close to $(\overline{x}, \overline{r})$ with $\overline{r} = g(\overline{x})$ when $X \times \mathbb{R}$ is endowed with an appropriate norm. This fact and Definition 29.24 below justify our terminology.

Let us note that in fact (S5m) is equivalent to (S5o) in view of (S4): given $g \in \mathscr{L}(X)$, $h \in Y^* \backslash \{0\}$, f with $f(x,y) := \max(g(x), h(y))$, $(\overline{x}, \overline{y}) \in X \times Y$ with $\overline{r} := h(\overline{y}) = g(\overline{x})$ as in (S5), introducing $A := I_X \times h \in L(X \times Y, X \times \mathbb{R})$ which is surjective, one has $A^T(u^*, r^*) = (u^*, r^* h)$, $f = k \circ A$, hence $\partial f(\overline{x}, \overline{y}) = A^T(\partial k(\overline{x}, h(\overline{y})))$ and $(\overline{x}^*, 0) \in \partial f(\overline{x}, \overline{y})$ if and only if $(\overline{x}^*, 0) \in \partial k(\overline{x}, \overline{r})$, so that $\overline{x}^* \in \partial g(\overline{x})$ when (S5m) holds.

In view of the equivalence (29.2), (S5m) is a consequence of the inclusion $\partial m(\overline{x}, \overline{r}) \subset [0,1](\partial g(\overline{x}) \times \{-1\})$. In turn, this inclusion is a consequence of (S4) and of the inclusion $\partial k^+(\overline{z}) \subset [0,1] \partial k(\overline{z})$ for $k \in \mathscr{L}(X \times \mathbb{R})$, with $k^+ := \max(k, 0)$. That inclusion motivates some other variants below. However, it is not satisfied by the directional (or Dini–Hadamard) or the firm (or Fréchet) subdifferential (take $k : \mathbb{R} \to \mathbb{R}$ given by $k(x) = \min(x, 0)$). Thus, we do not retain this inclusion.

If one adopts condition (S4c), condition (S5a) is equivalent to condition (S5). In fact, if f, g, h, $\overline{x}$, $\overline{y}$, $\overline{x}^*$ are as in (S5c), the submersion theorem ensures that there exist open neighborhoods U, V, W of $\overline{r} := h(\overline{y})$, 0, $\overline{y}$ in $\mathbb{R}$, Z, Y, where $Z := h'(\overline{y})^{-1}(0)$ and a bijection $\varphi : W \to U \times V$ of class C^1 at $\overline{y}$ such that $h(\varphi^{-1}(r, v)) = r$ for all $(r, v) \in U \times V$. Then, $k(x, r, v) := f(x, \varphi^{-1}(r, v)) = \max(g(x), r)$. Since $(x, r, v) \mapsto (x, \varphi^{-1}(r, v))$ is a bijection of class C^1 at $(\overline{x}, \overline{r}, 0)$, condition (S4c) ensures that $(\overline{x}^*, 0, 0) \in \partial k(\overline{x}, \overline{r}, 0)$. Then, using the linear form $(r, v) \mapsto r$ on $\mathbb{R} \times Z$ and (S5a), we get $\overline{x}^* \in \partial g(\overline{x})$, so that (S5c) is satisfied.

Condition (S6) is a weakening of the *homotonicity* property:

(H) If $f, g \in \mathscr{F}(X)$ are such that $f \geq g$ and $f(\overline{x}) = g(\overline{x})$ for some $\overline{x} \in X$, then $\partial g(\overline{x}) \subset \partial f(\overline{x})$.

Note that when (S4a) holds, condition (H) is equivalent to the following condition:

(S6h) If $A \in L(V, W)$ with $W = A(V)$, $\overline{v} \in V$, $\overline{w} := A\overline{v}$, $\varphi \in \mathscr{F}(V)$, $p \in \mathscr{F}(W)$ are such that $p \circ A \leq \varphi$ and $\varphi(\overline{v}) = p(\overline{w})$, then one has $A^T(\partial p(\overline{w})) \subset \partial \varphi(\overline{v})$.

Condition (H) is satisfied by the elementary subdifferentials and the viscosity subdifferentials, but not by the Clarke subdifferential nor the limiting subdifferential, so that we eschew it but we retain the weaker condition (S6a) or one of its variants below.

Note that condition (S6a) can be reformulated as follows, denoting by $\mathrm{epi}_s f$ the strict epigraph of a function f: $\mathrm{epi}_s f := \{(v,r) \in V \times \mathbb{R} : r > f(v)\}$:

(S6o) If $A \in L(V,W)$ with $W = A(V)$, $\overline{v} \in V$, $\overline{w} := A\overline{v}$, $\varphi \in \mathscr{F}(V)$, $p \in \mathscr{F}(W)$ are such that $p \circ A \leq \varphi$ and that the map $A \times I_{\mathbb{R}} : (v,r) \mapsto (A(v),r)$ is open at $(\overline{v}, \varphi(\overline{v}))$ from $\mathrm{epi}_s \varphi \cup \{(\overline{v}, \varphi(\overline{v}))\}$ to $\mathrm{epi}_s p \cup \{(\overline{w}, p(\overline{w}))\}$, then one has $A^T(\partial p(\overline{w})) \subset \partial \varphi(\overline{v})$.

For most purposes it suffices to consider the case V is a product $V := W \times X$ and A is the canonical projection from $W \times X$ onto W. Note that in such a case the conclusion is $\partial p(\overline{w}) \times \{0\} \subset \partial \varphi(\overline{w}, \overline{x})$. A basic variant of (S6b) called quasi-homotonicity in [34] has some interest for the constructions we shall devise, besides its analogy with condition (H):

(QH) If E is a subset of X and if $f \in \mathscr{F}(X)$ is such that $d_E \leq f$ and $f = 0$ on E, then $\partial d_E(x) \subset \partial f(x)$ for all $x \in E$.

The list of conditions we have given is not minimal: the pseudo-homotonicity condition (S6a) entails the optimality condition (S3): if $f \in \mathscr{F}(X)$ attains its minimum at $\overline{x}$, setting $V := X$, $W := \{0\}$, $p(0) := f(\overline{x})$, $\varphi := f$, the assumption of (S6a) is clearly satisfied and we get $0 \in \partial f(\overline{x})$. Also, there is an analogy between conditions (S6a) and (S4a) obtained by taking $\varphi := f$, $p := h$ when $f := h \circ A$ as in (S4a); however (S6) only yields the inclusion $A^T \partial h(A\overline{x}) \subset \partial(h \circ A)(\overline{x})$ and not the equality. But equality could be added in (S6a) in the case $\varphi = p \circ A$ and then it would be redundant to state (S4a).

29.2.4 Checking the Conditions

For the sake of brevity, we just take a sample of subdifferentials, recalling briefly their definitions. We refer to [26, 36] for other subdifferentials.

We first define the *tangent cone* (or directional tangent cone or contingent cone) to a subset E of a normed space X at $a \in E$ as the set $T^D(E,a)$ of $v \in X$ such that there exist sequences $(t_n) \to 0_+$, $(v_n) \to v$ satisfying $a + t_n v_n \in E$ for all $n \in \mathbb{N}$. The *circa* (or Clarke) *tangent cone* to E at a is the set $T^C(E,a)$ of $v \in X$ such that for any sequences $(t_n) \to 0_+$, $(e_n) \to a$ in E there exists a sequence $(v_n) \to v$ satisfying $e_n + t_n v_n \in E$ for all $n \in \mathbb{N}$ [8]. Given a function $f : X \to \overline{\mathbb{R}} := \mathbb{R} \cup \{-\infty, +\infty\}$ finite at $\overline{x} \in X$ we define the subderivates of f at $\overline{x}$ in the direction $u \in X$ as

$$f^D(\overline{x}, u) := \inf\{s \in \mathbb{R} : (u,s) \in T^D(E,a)\},$$

$$f^C(\overline{x}, u) := \inf\{s \in \mathbb{R} : (u,s) \in T^C(E,a)\}$$

with $a := (\overline{x}, f(\overline{x}))$, $E := \mathrm{epi} f := \{(x,r) \in X \times \mathbb{R} : r \geq f(x)\}$. Then the *directional* (or Dini–Hadamard, or contingent) subdifferential and the circa (or Clarke) subdifferential of f at $\overline{x}$ are given, respectively, by

$$\partial_D f(\overline{x}) := \{x^* \in X^* : x^*(\cdot) \leq f^D(\overline{x}, \cdot)\},$$

$$\partial_C f(\overline{x}) := \{x^* \in X^* : x^*(\cdot) \leq f^C(\overline{x}, \cdot)\}.$$

The *firm* or Fréchet subdifferential of f at $\overline{x}$ is the set $\partial_F f(\overline{x})$ of $x^* \in X^*$ such that for some remainder r (i.e., a function $r : X \to \overline{\mathbb{R}}$ such that $r(0) = 0$ and $r(x)/\|x\| \to 0$ as $x \to 0$, $x \neq 0$) one has

$$\forall x \in X \qquad f(\overline{x}+x) - f(\overline{x}) - \langle \overline{x}^*, x\rangle \geq -r(x).$$

The *limiting* (firm) subdifferential of f at $\overline{x}$ is the set $\partial_L f(\overline{x})$ of $x^* \in X^*$ such that there exist sequences $(\varepsilon_n) \to 0_+$, $(x_n) \to \overline{x}$, $(x_n^*) \overset{*}{\to} x^*$ (i.e., $(x_n^*) \to x^*$ for the weak* topology) satisfying $(f(x_n)) \to f(\overline{x})$, $x_n^* \in \partial^{\varepsilon_n} f(x_n)$ for all $n \in \mathbb{N}$, where for $x \in f^{-1}(\mathbb{R})$, $\varepsilon > 0$ one sets

$$\partial^{\varepsilon} f(x) := \left\{ x^* \in X^* : \liminf_{w \to 0,\ w \neq 0} \frac{1}{\|w\|}(f(x+w) - f(x) - \langle x^*, w\rangle) \geq -\varepsilon \right\}.$$

It can be shown that conditions (S0)–(S3) are satisfied by the firm (or Fréchet) and directional (or Dini–Hadamard) subdifferentials, the viscosity subdifferentials, the limiting subdifferential, and the Clarke subdifferential and its variants (see [36] for instance). In order to check (S4a) and its variants we need the Lyusternik–Graves Theorem (see [36, Theorem 2.67] for instance).

Lemma 29.11. *Let X and Y be Banach spaces, let W be an open subset of X, and let $g : W \to Y$ be circa-differentiable at $\overline{x} \in W$ with a surjective derivative. Then there exist ρ, σ, $\kappa > 0$ such that $B(\overline{x}, \rho) \subset W$ and for all $w \in B(\overline{x}, \rho)$, $y \in B(g(\overline{x}), \sigma)$ there exists $x \in W$ satisfying $g(x) = y$ and $\|x - w\| \leq \kappa \|g(w) - y\|$.*

Proposition 29.12. *The directional subdifferential, the firm subdifferential, and the Clarke subdifferential satisfy (S4d), hence (S4a) and (S4c) for the class $\mathcal{I}$ of lower semicontinuous functions. The limiting subdifferential satisfies (S4c).*

Proof. Let us first consider the case of the Clarke subdifferential. The case of (S4c) for Lipschitzian functions is given in [9, Theorem 3.2, p. 79]. Using the definition of $\partial_C f$ in terms of the normal cone to the epigraph of f, (S4d) for lower semicontinuous functions is equivalent to the assertion:

if X and Y are Banach spaces, if W is an open subset of X and $g : W \to Y$ is circa-differentiable at $\overline{x} \in W$ with $g'(\overline{x})(X) = Y$, then for a closed subset H of Y containing $\overline{y} := g(\overline{x})$ and $F := g^{-1}(H)$ one has $T^C(F, \overline{x}) = g'(\overline{x})^{-1}(T^C(H, \overline{y}))$ or equivalently $N^C(F, \overline{x}) = g'(\overline{x})^T(N^C(H, \overline{y}))$.

Given $u \in T^C(F, \overline{x})$, let $(t_n) \to 0_+$, $(y_n) \to \overline{y}$ in H. Since g is open at $\overline{x}$ by the preceding lemma, one can find a sequence $(x_n) \to \overline{x}$ such that $g(x_n) = y_n$ for all n large enough. Then $x_n \in F$ and for some sequence $(u_n) \to u$ one has $x_n + t_n u_n \in F$. Since g is circa-differentiable at $\overline{x}$ and since $A := g'(\overline{x})$ is open, one can find sequences $(z_n) \to 0$ in Y, $(w_n) \to 0$ in X such that $g(x_n + t_n u_n) = g(x_n) + t_n A(u_n) + t_n z_n$ and $A(w_n) = z_n$ for all n. Then, since $g(x_n + t_n u_n) \in H$ and $(u_n + w_n) \to u$, one gets $A(u) \in T^C(H, \overline{y})$.

Conversely, let $u \in A^{-1}(T^C(H,\overline{y}))$. Let $v := A(u)$ and let $(t_n) \to 0_+$, $(x_n) \to \overline{x}$ in F. Then $(y_n) := (g(x_n)) \to \overline{y}$ in H, so that there exists a sequence $(v_n) \to v$ satisfying $y_n + t_n v_n \in H$ for all n. The preceding lemma yields some $\kappa > 0$ and some sequence (w_n) such that $g(w_n) = g(x_n) + t_n v_n$ and for n large enough one has

$$\|x_n + t_n u_n - w_n\| \leq \kappa \|g(x_n) + t_n v_n - g(x_n + t_n u_n)\|.$$

Then, setting $u'_n := t_n^{-1}(w_n - x_n)$, $g(x_n + t_n u_n) = g(x_n) + t_n A(u_n) + t_n z_n$, one gets $\|u'_n - u_n\| \leq \kappa \|A(u_n) + z_n - v_n\|$, hence $(u'_n - u_n) \to 0$ and $x_n + t_n u'_n = w_n \in g^{-1}(g(x_n) + t_n v_n) \subset g^{-1}(H) = F$. Thus $u \in T^C(F,\overline{x})$.

A similar result holds for the directional normal cone and the firm normal cone (see [36, Theorem 2.111]). Assertion (S4c) for ∂_L follows from a passage to the limit. ■

Proposition 29.13. *The directional subdifferential, the firm subdifferential, and the Clarke subdifferential satisfy (S4b) for the class $\mathcal{F} = \mathcal{I}$. The limiting subdifferential satisfies (S4b) for the class $\mathcal{F} = \mathcal{L}$.*

Proof. For the sake of simplicity, we suppose $m = 2$ and $f = j \circ (g \times h)$, where $g \in \mathcal{F}(X)$, $h \in \mathcal{F}(Y)$, $j : \mathbb{R}^2 \to \mathbb{R}$ is Lipschitzian, nondecreasing, of class C^1 near $(\overline{q}, \overline{r})$ with $a := D_1 j(\overline{q}, \overline{r}) > 0$, $b := D_2 j(\overline{q}, \overline{r}) > 0$, $\overline{q} := g(\overline{x})$, $\overline{r} := h(\overline{y})$. Given $(x^*, y^*) \in \partial f(\overline{x}, \overline{y})$ let us show that $x^*/a \in \partial g(\overline{x})$, the inclusion $y^*/b \in \partial h(\overline{y})$ being similar. Let F (resp. G) be the epigraph of f (resp. g) and let $\overline{p} := f(\overline{x}, \overline{y})$. Let us first suppose ∂ is the Clarke subdifferential ∂_C. Let us show that for all $(u,s) \in T^C(G, (\overline{x}, \overline{q}))$, the Clarke tangent cone to G at $(\overline{x}, \overline{q})$, one has $(u/a, 0, s) \in T^C(F, (\overline{x}, \overline{y}, \overline{p}))$. That will prove that $x^*/a \in \partial g(\overline{x})$ since then $\langle x^*/a, u \rangle - s = \langle (x^*, y^*, -1), (u/a, 0, s) \rangle \leq 0$. Let $(t_n) \to 0_+$, $((x_n, y_n, p_n)) \to (\overline{x}, \overline{y}, \overline{p})$ in F. Since $\liminf_n g(x_n) \geq g(\overline{x})$, $\liminf_n h(y_n) \geq h(\overline{y})$, since j is continuous, nondecreasing in each of its arguments and increasing near $\overline{q}$, $\overline{r}$, respectively, and since $p_n \geq j(g(x_n), h(y_n))$, we see that $(q_n) := (g(x_n)) \to \overline{q} := g(\overline{x})$. By definition of $T^C(G, (\overline{x}, \overline{q}))$, we can find a sequence $((u_n, s_n)) \to (u, s)$ such that $(x_n, q_n) + t_n(u_n, s_n) \in G$, i.e., $g(x_n + t_n u_n) \leq q_n + t_n s_n$ for all n and, for some sequence $(a_n) \to a$,

$$f(x_n + t_n u_n, y_n) \leq j(g(x_n) + t_n s_n, h(y_n)) = j(g(x_n), h(y_n)) + t_n a_n s_n \leq p_n + t_n a_n s_n.$$

Thus $(u, 0, as) \in T^C(F, (\overline{x}, \overline{y}, \overline{p}))$ and $(u/a, 0, s) \in T^C(F, (\overline{x}, \overline{y}, \overline{p}))$.

A similar (but simpler) proof holds for the directional subdifferential ∂_D since for every $(u,s) \in T(G, (\overline{x}, \overline{q}))$, the usual tangent cone to G at $(\overline{x}, \overline{q})$, one has $(u/a, 0, s) \in T(F, (\overline{x}, \overline{y}, \overline{p}))$.

Let us consider the case of the firm subdifferential ∂_F. Without loss of generality, we suppose $g(\overline{x}) = 0$, $h(\overline{y}) = 0$. Suppose $x^*/a \notin \partial_F g(\overline{x})$: there exist $\varepsilon \in]0,1]$ and a sequence $(x_n) \to 0$ such that $g(\overline{x} + x_n) < \langle x^*/a, x_n \rangle - \varepsilon \|x_n\|$ for all n. Then, given $\alpha > 0$, for n large enough, one has

$$\begin{aligned} f(\overline{x} + x_n, \overline{y}) &\leq j(\langle x^*/a, x_n \rangle - \varepsilon \|x_n\|, \overline{r}) \\ &\leq D_1 j(\overline{q}, \overline{r})(\langle x^*/a, x_n \rangle - \varepsilon \|x_n\|) + \alpha(\|x^*/a\| + \varepsilon) \|x_n\| \\ &\leq \langle x^*, x_n \rangle + (-a\varepsilon + \alpha \|x^*/a\| + \alpha\varepsilon) \|x_n\|. \end{aligned}$$

Taking α such that $\alpha(\|x^*/a\|+\varepsilon) < a\varepsilon$, we get a contradiction with $(x^*,y^*) \in \partial_F f(\overline{x},\overline{y})$.

The case of the limiting firm subdifferential is obtained by using compactness and a passage to the limit. ∎

Now let us consider condition (S5) and its variants.

Proposition 29.14. *Condition (C) (hence conditions (S5), (S5d)) is satisfied by the Clarke subdifferential ∂_C in the class of all Banach spaces and by the limiting subdifferential ∂_L in the class of Asplund spaces for $\mathscr{F} = \mathscr{L}$.*

Proof. For $\partial := \partial_L$ see [31, Theorem 3.41]. Let us consider the case $\partial := \partial_C$. The Chain Rule ([9, Theorem 2.5 p. 76]) for $f := j \circ (g_1 \circ p_1, \ldots, g_m \circ p_m)$, $\overline{r} := (g_1(\overline{x}_1), \ldots, g_m(\overline{x}_m))$:

$$\partial f(\overline{x}) \subset \overline{\mathrm{co}}^*\{\partial(r_1^* g_1 \circ p_1 + \cdots + r_m^* g_m \circ p_m)(\overline{x}) : (r_1^*, \ldots, r_m^*) \in \partial j(\overline{r})\},$$

the Sum Rule, and the relations $\partial(r_i^* g_i \circ p_i)(\overline{x}) = r_i^* p_i^T(\partial g_i(\overline{x}_i))$ for $i \in \mathbb{N}_m := \{1, \ldots m\}$, $r_i^* \in \mathbb{R}_+$ show that $\partial f(\overline{x}) \subset \overline{\mathrm{co}}^*(A)$, where

$$A := \{r_1^* \partial g_1(\overline{x}_1) \times \cdots \times r_m^* \partial g_m(\overline{x}_m) : (r_1^*, \ldots, r_m^*) \in \partial j(\overline{r})\}.$$

Now, since $\partial g_i(\overline{x})$ and $\partial j(\overline{r})$ are weak* compact and since the map

$$(x_1^*, \ldots, x_m^*, r_1^*, \ldots, r_m^*) \mapsto r_1^* x_1^* + \cdots + r_m^* x_m^*$$

is continuous for the weak* topologies, A is weak* compact. Let us show that A is convex. Let $t \in]0,1[$, $a^* := (r_1^* u_1^*, \ldots, r_m^* u_m^*) \in A$, $b^* := (s_1^* v_1^*, \ldots, s_m^* v_m^*)$ with u_i^*, $v_i^* \in \partial g_i(\overline{x}_i)$ for $i \in \mathbb{N}_m$ $r^* := (r_1^*, \ldots, r_m^*) \in \partial j(\overline{r})$, $s^* := (s_1^*, \ldots, s_m^*) \in \partial j(\overline{r})$. Let $t^* := (1-t)r^* + ts^* \in \partial j(\overline{r})$, $t^* := (t_1^*, \ldots, t_m^*)$. If $t_i^* := (1-t)r_i^* + ts_i^* > 0$, let $x_i^* := (1/t_i^*)((1-t)r_i^* u_i^* + ts_i^* v_i^*) \in \partial g_i(\overline{x}_i)$; if $t_i^* = 0$, let us pick $x_i^* \in \partial g_i(\overline{x}_i)$ arbitrary and note that $r_i^* = 0$, $s_i^* = 0$, so that $(1-t)r_i^* u_i^* + ts_i^* v_i^* = t_i^* x_i^*$ for all $i \in \mathbb{N}_m$. Then, in all cases

$$(1-t)a^* + tb^* := (t_1^* x_1^*, \ldots, t_m^* x_m^*) \in A.$$

∎

Remark 29.15. A direct proof of condition (S5) can be given for ∂_C. Using support functions, it suffices to show that for all $(u,v) \in X \times Y$ one has

$$f^C((\overline{x},\overline{y}),(u,v)) \leq \max(g^C(\overline{x},u), h^C(\overline{y},v))$$

where $g^C(\overline{x},u)$ is the Clarke derivative of g at $\overline{x}$ given by

$$g^C(\overline{x},u) := \inf\{r \in \mathbb{R} : (u,r) \in T^C(\mathrm{epi}\, g, (\overline{x}, g(\overline{x})))\},$$

$T^C(\text{epi}\, g, e)$ being the Clarke tangent cone to the set $G := \text{epi}\, g$ at $e := (\overline{x}, g(\overline{x})) \in \text{epi}\, g$. By definition of this cone, given $s \geq \max(g^C(\overline{x}, u), h^C(\overline{y}, v))$, we have to prove that for any sequence $(t_n) \to 0_+$ and any sequence $((x_n, y_n, p_n))$ in $\text{epi}\, f$ with limit $(\overline{x}, \overline{y}, f(\overline{x}, \overline{y})$ one can find a sequence $((u_n, v_n, s_n)) \to (u, v, s)$ such that $(x_n, y_n, p_n) + t_n(u_n, v_n, s_n) \in \text{epi}\, f$ for all n. Since $((x_n, p_n)) \to (\overline{x}, g(\overline{x})$ in $\text{epi}\, g$,) we can find a sequence $((u_n, q_n)) \to (u, s)$ such that $(x_n, p_n) + t_n(u_n, q_n) \in \text{epi}\, g$ for all n. Similarly, one can find a sequence $((v_n, r_n)) \to (v, s)$ such that $(y_n, p_n) + t_n(v_n, r_n) \in \text{epi}\, h$ for all n. Then, taking $s_n := \max(q_n, r_n)$, we get the required sequence.

Proposition 29.16. *Condition (S5d) (hence condition (S5)) is satisfied by the directional subdifferential ∂_D and the firm subdifferential ∂_F on the class of lower semicontinuous functions.*

Proof. Let f, g, h, $\overline{x}$, $\overline{y}$ be as in (S5d) and let $(\overline{x}^*, \overline{y}^*) \in \partial f(\overline{x}, \overline{y})$ with $g(\overline{x}) = h(\overline{y})$, $\overline{y}^* \neq h'(\overline{y}) \neq 0$, assuming without loss of generality that $\overline{x} = 0, \overline{y} = 0, g(\overline{x}) = h(\overline{y}) = 0$. Let $v \in Y$ be such that $h'(\overline{y})v = 1$. Let us first consider the case $\partial = \partial_D$. For $t > 0$ we have $f(0, tv) = h(tv)$, hence $1 = h'(\overline{y})v = \lim_{t \to 0_+} (1/t) f(0, tv) \geq \langle \overline{y}^*, v \rangle$. Similarly, for all $w \in Y$ such that $h'(\overline{y})w > 0$, we have $h'(\overline{y})w \geq \langle \overline{y}^*, w \rangle$. The same is true if $h'(\overline{y})w \geq 0$ as follows by taking a sequence $(w_n) \to w$ such that $h'(\overline{y})w_n > 0$ for all n. Thus there exists $\lambda \geq 0$ such that $v^* - \overline{y}^* = \lambda v^*$ for $v^* := h'(\overline{y})$ and $\overline{y}^* = (1 - \lambda)v^*$. The assumption $\overline{y}^* \neq h'(\overline{y})$ yields $\lambda > 0$. Observing that $f(0, -tv) \leq 0$ for t small enough, we get $\langle \overline{y}^*, -v \rangle \leq 0$, hence $1 - \lambda = (1 - \lambda)\langle v^*, v \rangle = \langle \overline{y}^*, v \rangle \geq 0$. Now, given $u \in X$, $s \geq g^D(\overline{x}, u)$, let us show that $\lambda s \geq \langle \overline{x}^*, u \rangle$. Taking a sequence $((s_n, t_n, u_n)) \to (s, 0_+, u)$ such that $t_n s_n \geq g(t_n u_n)$, setting $s'_n := t_n^{-1} h(t_n s v)$, we note that $(s'_n) \to s$ and $(s''_n) \to s$ for $s''_n := \max(s_n, s'_n)$. Since $f(t_n u_n, t_n s v) \leq t_n s''_n$ for all n, we get $\langle \overline{x}^*, u \rangle + \langle \overline{y}^*, sv \rangle \leq s$ or $\langle \overline{x}^*, u \rangle \leq \lambda s$. Then $\overline{x}^* = 0$, hence $\lambda = 0$ cannot occur. Thus one gets $u^* := \overline{x}^*/\lambda \in \partial g(\overline{x})$,

$$(\overline{x}^*, \overline{y}^*) = (\lambda u^*, (1 - \lambda)v^*) \in \lambda \partial g(\overline{x}) \times (1 - \lambda)h'(\overline{y}). \tag{29.3}$$

Now let us consider the case of the Fréchet subdifferential ∂_F, assuming again that $\overline{x} = 0$, $\overline{y} = 0$, $g(\overline{x}) = h(\overline{y}) = 0$. As above, we have $\overline{y}^* = (1 - \lambda)v^*$ for $v^* := h'(\overline{y})$, $\lambda \in [0, 1]$ and we cannot have $\lambda = 0$. Suppose $\overline{x}^*/\lambda \notin \partial_F g(\overline{x})$: there exist $\alpha > 0$ and a sequence $(u_n) \to 0$ such that $g(u_n) < s_n := \langle \overline{x}^*/\lambda, u_n \rangle - \alpha \|u_n\|$. Since $(h(s_n v)/s_n) \to 1$, there exists a sequence $(\sigma_n) \to 0$ in $\mathbb{R}$ such that $f(u_n, s_n v) \leq (1 + \sigma_n)s_n$. Then, for some sequence $(\varepsilon_n) \to 0_+$, one gets

$$\begin{aligned}(1 + \sigma_n)s_n &\geq \langle \overline{x}^*, u_n \rangle + \langle \overline{y}^*, s_n v \rangle - \varepsilon_n(\|u_n\| + s_n \|v\|) \\ &\geq \lambda s_n + \alpha\lambda \|u_n\| + (1 - \lambda)s_n - \varepsilon_n(\|u_n\| + s_n \|v\|).\end{aligned}$$

Then one has

$$(|\sigma_n| + \varepsilon_n \|v\|) |s_n| \geq (\sigma_n + \varepsilon_n \|v\|)s_n \geq (\alpha\lambda - \varepsilon_n) \|u_n\|$$

and $|s_n| \leq (\|\overline{x}^*\|/\lambda + \alpha) \|u_n\|$, a contradiction since $(|\sigma_n| + \varepsilon_n \|v\|) \to 0$. ∎

Remark 29.17. When $\mathscr{F}$ is the class $\mathscr{L}$ of locally Lipschitzian functions, for all $u \in X$, there exists some $s \in \mathbb{R}$ such that $s \geq g^D(\overline{x}, u)$, so that the preceding proof shows that $\overline{x}^* = 0$ if $\lambda = 0$. Thus the following slightly more general condition is satisfied:

(S5') If $f(x,y) := \max(g(x), h(y))$, $(\overline{x}^*, \overline{y}^*) \in \partial f(\overline{x}, \overline{y})$ with $g \in \mathscr{F}(X)$, $h \in \mathscr{F}(Y)$ of class C^1 around $\overline{y}$, $g(\overline{x}) = h(\overline{y})$, $(\overline{x}^*, \overline{y}^*) \neq (0, h'(\overline{y}))$, $h'(\overline{y}) \neq 0$, then $(\overline{x}^*, \overline{y}^*) \in \{\lambda \partial g(\overline{x}) \times (1-\lambda)\{h'(\overline{y})\} : \lambda \in]0,1]\}$.

Remark 29.18. The conclusion $\overline{x}^* \in [0,1]\partial g(\overline{x})$ cannot hold in general when $\overline{y}^* = h'(\overline{y})$, $0 \notin \partial g(\overline{x})$ since we may have $\partial g(\overline{x}) = \varnothing$ and $(0, h'(\overline{y})) \in \partial f(\overline{x}, \overline{y})$ as $f(x,y) \geq h(y)$ for all $(x,y) \in X \times Y$ and $f(\overline{x}, \overline{y}) = h(\overline{y})$.

Proposition 29.19. *Conditions (S6a) and (S6b) are satisfied by the directional subdifferential ∂_D, the firm subdifferential ∂_F, the viscosity subdifferentials, the Clarke subdifferential on the class of all Banach spaces, and by the limiting subdifferential ∂_L on the class of Asplund spaces.*

Proof. In fact, one has $A^T(\partial p(\overline{w})) \subset \partial \varphi(\overline{v})$ for any $\overline{v} \in A^{-1}(\overline{w})$ satisfying $\varphi(\overline{v}) = p(\overline{w})$, and one can find such a $\overline{v}$ in M since φ is lower semicontinuous. In the case M is a singleton, it is proved in [34, 36] that the Clarke subdifferential ∂_C satisfies (S6a), (S6b) on the class of all Banach spaces; the general case is similar. The proof for the limiting subdifferential ∂_L on the class of Asplund spaces is given in [36, Proposition 6.21]. ■

29.2.5 Normal Cones

The notion of subdifferential has a bearing on geometrical concepts. Two situations may occur: either $\mathscr{F}(X)$ coincides with the class $\mathscr{I}(X)$ of lower semicontinuous functions or $\mathscr{F}(X)$ just contains the class $\mathscr{L}(X)$ of locally Lipschitzian functions.

Definition 29.20. When $\mathscr{F}(X)$ contains $\mathscr{I}(X)$, the normal cone to a closed subset E of a member X of $\mathscr{X}$ at $x \in E$ is the set

$$N(E,x) = \partial \iota_E(x),$$

where ι_E is the *indicator function* of E (given by $\iota_E(x) = 0$ for $x \in E$, $\iota_E(x) = +\infty$ for $x \in X \backslash E$).

Assuming $\mathscr{F}(X)$ contains the space $\mathscr{L}(X)$ of locally Lipschitzian functions on X, the metric normal cone to a subset E of X at $x \in E$ is the cone $N^m(E,x)$ generated by $\partial d_E(x)$:

$$N^m(E,x) := \mathbb{R}_+ \partial d_E(x).$$

By (S6m), one always has $N^m(E,x) \subset N(E,x)$ for all $x \in E$. Moreover, $N^m(E,x)$ is independent of the choice of the norm among the norms inducing the topology of X. By (S0) $N(E,x)$ is nonempty only if $x \in E$.

Proposition 29.21. *For every $x \in E$, $N(E,x)$ is a cone containing 0. If $\overline{x}$ is an interior point of E, then $N^m(E,\overline{x}) = N(E,\overline{x}) = \{0\}$.*

Proof. We have $0 \in N^m(E,x)$ for every $x \in E$ by (S3). Since $r\iota_E = \iota_E$ for all $r > 0$, $N(E,x)$ is a cone. Let $\overline{x} \in \operatorname{int} E$ and let $r > 0$ be such that $B[\overline{x},r] \subset E$ and let g be the indicator of this closed ball. This is a convex function and $\partial g(\overline{x}) = \{0\}$ by (S2). Now (S1) with $f = \iota_E$ implies the statement. ■

Of course, it does not follow from the proposition that $N(E,x)$ contains nonzero elements if x is a boundary point of E. This is not always the case even if E is convex.

Some properties of normal cones can be derived from conditions (S1)–(S6). Let us prove one of them which will be used soon.

Proposition 29.22. *Let V, W in $\mathscr{X}$, $A \in L(V,W)$ and let $\overline{v} \in F \subset V$, $G \subset W$ be such that $A(F) \subset G$. Suppose that for every sequence $(w_n) \overset{G}{\to} \overline{w} := A\overline{v}$ there exists a sequence $(v_n) \overset{F}{\to} \overline{v}$ such that $Av_n = w_n$ for all $n \in \mathbb{N}$ large enough. Then*

$$A^T(N(G,\overline{w})) \subset N(F,\overline{v}) \tag{29.4}$$

$$A^T(N^m(G,\overline{w})) \subset N^m(F,\overline{v}). \tag{29.5}$$

In particular, these relations hold when $F = A^{-1}(G)$ and $W = A(V)$.

These relations are equalities when A is an isomorphism. Thus the metric normal cone does not depend on the choice of the norm in an equivalence class.

Proof. Let ι_F and ι_G be the indicator functions of F and G, respectively. Since $A(F) \subset G$, one has $\iota_G \circ A \leq \iota_F$. Our assumption allows to apply (S6e) which yields (29.4). In order to prove (29.5), setting $\varphi := \|A\|\, d_F$, $p := d_G$, let us observe that, for all $v \in V$, we have

$$d_G(Av) \leq \inf_{u \in V}(\|Av - Au\| + \iota_F(u)) \leq \inf_{u \in V}(\|A\|\, \|u - v\| + \iota_F(u)) = \varphi(v).$$

Under our assumption condition (S6m) ensures that for all $\overline{w}^* \in \partial d_G(\overline{w})$ one has $A^T(\overline{w}^*) \in \|A\|\, \partial d_F(\overline{v})$. Inclusion (29.5) ensues.

The last but one assertion is obtained by interchanging F and G and changing A into A^{-1}. Taking for A the identity map from X endowed with some norm to X endowed with another norm, the last assertion ensues. ■

The next proposition explains in details the choice of the term "consistency" we gave to (S5).

Proposition 29.23. *For every subdifferential $(\partial, \mathscr{X}, \mathscr{F})$ and every X in $\mathscr{X}$, $f \in \mathscr{L}(X)$, $\overline{x} \in X$ one has*

$$\partial f(\overline{x}) = \{x^* \in X^* : (x^*, -1) \in N^m(\operatorname{epi} f, (\overline{x}, f(\overline{x})))\}.$$

Proof. Let $f \in \mathscr{L}(X)$ with X in $\mathscr{X}$, $\overline{x} \in X$ and let E be the epigraph of f. Without loss of generality we may assume f is globally Lipschitzian with rate $c > 0$ and even that $c = 1$ (since we can change the norm of X to the norm $c\|\cdot\|$ and the metric normal cone to epi f will remain unchanged). Then, endowing $X \times \mathbb{R}$ with the sum norm, we have $d_E(x,r) = (f(x) - r)^+$ for all $(x,r) \in X \times \mathbb{R}$, as easily checked. Thus,

$$\forall (x,r) \in X \times \mathbb{R} \qquad f(x) \leq \varphi(x,r) := d_E(x,r) + r$$

and the map $S : X \to X \times \mathbb{R}$ given by $S(x) := (x, f(x))$ is continuous and satisfies $S(\overline{x}) = (\overline{x}, f(\overline{x}))$. Thus, by (S6e) and (S4a) in which we take the linear form $\ell : (x,r) \mapsto r$, for all $\overline{x}^* \in \partial f(\overline{x})$, one has $(\overline{x}^*, 0) \in \partial\varphi(\overline{x}, f(\overline{x})) = \partial d_E(\overline{x}, f(\overline{x})) + (0,1)$, or $(\overline{x}^*, -1) \in \partial d_E(\overline{x}, f(\overline{x}))$. As noted above, the converse is ensured by condition (S5) in view of the relation $d_E(x,r) = (f(x) - r)^+$. ■

29.3 Extension of Subdifferentials

29.3.1 The Metric Extension

Quite often a subdifferential ∂ is easy to define on the space $\mathscr{L}(X)$ of locally Lipschitzian functions on X. Then it is of interest to extend ∂ to a larger set, for instance, the whole set $\mathscr{S}(X)$ of l.s.c. functions on X. That can be done by setting $\partial f(x) := \partial^m f(x)$ for $f \in \mathscr{S}(X)$ and $x \in \mathrm{dom} f$, where

$$\partial^m f(x) := \{x^* \in X^* : (x^*, -1) \in N^m(E, x_f)\}, \tag{29.6}$$

$E := E_f$ being the epigraph of f and $x_f := (x, f(x))$. Of course, for $x \in X \backslash \mathrm{dom} f$, we set $\partial f(x) = \varnothing$. Proposition 29.23 shows that this definition is coherent when $f \in \mathscr{L}(X)$, so that the terminology "consistency" for (S5) is justified. More is required in the next definition.

Definition 29.24. A subdifferential $\partial := (\partial, \mathscr{X}, \mathscr{F})$ is said to be geometrically consistent if the equality $\partial f(x) = \partial^m f(x)$ holds for all $f \in \mathscr{F}(X)$, $x \in \mathrm{dom} f$.

Proposition 29.23 ensures that every subdifferential $\partial := (\partial, \mathscr{X}, \mathscr{F})$ with $\mathscr{F} = \mathscr{L}$ is geometrically consistent. Moreover, let us note that, by construction, when ∂ is defined on $\mathscr{L}(X)$, its extension to $\mathscr{S}(X)$ we have just defined is geometrically consistent. It is easy to show that the Fréchet subdifferential and the Clarke subdifferential are geometrically consistent. Of course, geometric consistency is desirable in order to avoid confusions. When it is not satisfied, one has to be careful when using ∂ on $\mathscr{S}(X) \backslash \mathscr{L}(X)$.

For the moment, we note an easy consequence of geometric consistency.

Proposition 29.25. *Suppose ∂ is geometrically consistent on the class $\mathscr{S}$ of lower semicontinuous functions. Let E be a closed subset of X and let the normal cone $N(E,x)$ to E at $x \in E$ be defined as above: $N(E,x) := \partial\iota_E(x)$. Then one has*

$$N(E,x) = [1, +\infty)\partial d_E(x) = N^m(E,x).$$

Moreover, for every $f \in \mathscr{I}(X)$ and every $x \in \operatorname{dom} f$, one has

$$\partial f(x) = \{x^* \in X^* : (x^*, -1) \in N(\operatorname{epi} f, x_f)\}.$$

Proof. Since $\mathbb{R}_+ \partial d_E(x) =: N^m(E,x) \subset N(E,x) := \partial \iota_E(x)$, to prove the first assertion, it suffices to show the inclusion $\partial \iota_E(x) \subset [1, +\infty)\partial d_E(x)$. Since the epigraph F of ι_E is just $E \times \mathbb{R}_+$, taking the sum norm on $X \times \mathbb{R}$, one has $d_F(u,r) = d_E(u) + r^-$, where $r^- := \max(-r, 0)$. By (S4s) and (S2) one gets

$$\partial d_F(x,0) \subset \partial d_E(x) \times [-1,0].$$

Since by Definition 29.24, for all $x^* \in \partial \iota_E(x)$, one has $(x^*, -1) \in \mathbb{R}_+ \partial d_F(x,0)$, one can find $c \in \mathbb{R}_+$, $w^* \in \partial d_E(x)$, $r \in [-1,0]$ such that $(x^*, -1) = (cw^*, cr)$, hence $c \geq 1$ and $x^* = cw^* \in c\partial d_E(x)$.

The last assertion follows from the relations $\partial f(x) = \partial^m f(x)$ and $N(\operatorname{epi} f, x_f) = N^m(\operatorname{epi} f, x_f)$. ■

Theorem 29.26. *Let $\partial = (\partial, \mathscr{X}, \mathscr{L})$ be a subdifferential of alleviated type on the class $\mathscr{L}$ of locally Lipschitzian functions. Then the extension ∂^m of ∂ to the class $\mathscr{I}$ of lower semicontinuous functions is a subdifferential of alleviated type.*

Proof. For the sake of brevity, for $f \in \mathscr{I}(X)$ and $x \in \operatorname{dom} f$, we set $x_f := (x, f(x))$. Given functions f, g, and h, we denote by F, G, and H, respectively, their epigraphs.

(S0) is obtained by construction

(S1) follows from the fact that if f and g coincide on a ball $B(x,\varepsilon)$, then their epigraphs F and G are such that $F \cap B(e,\varepsilon) = G \cap B(e,\varepsilon)$, where $e = x_f = x_g = (x, f(x))$. Then $d_F = d_G$ on $B(e, \varepsilon/2)$ and $\partial d_F(e) = \partial d_G(e)$, so that $\partial^m f(x) = \partial^m g(x)$.

(S2) If f is convex, then its epigraph F is convex; hence $d_F := d(\cdot, F)$ is convex. Since d_F is Lipschitzian, $\partial d_F = \partial_{\text{FM}} d_F$, where ∂_{FM} is the subdifferential in the sense of convex analysis. Now $\mathbb{R}_+ \partial_{\text{FM}} d_F(x_f)$ is the normal cone $N_{\text{FM}}(F, x_f)$ to F at x_f in the sense of convex analysis. Thus, (S2) for ∂^m follows from the equivalence $x^* \in \partial_{\text{FM}} f(x)$ iff $(x^*, -1) \in \mathbb{R}_+ \partial_{\text{FM}} d_F(x_f)$.

(S3) If f attains a local minimum at $\overline{x}$, modifying f outside some closed ball and subtracting $f(\overline{x})$, we may assume that $\overline{x}$ is a global minimizer of f and $f(\overline{x}) = 0$. Then the epigraph F of f is contained in $X \times \mathbb{R}_+$, so that for all $(w,r) \in X \times \mathbb{R}$, one has

$$\varphi(w,r) := d((w,r),F) + r \geq d((w,r), X \times \mathbb{R}_+) + r = (-r)^+ + r \geq f(\overline{x}),$$

and $\varphi(\overline{x}, f(\overline{x})) = f(\overline{x})$: φ attains its minimum at $\overline{x}_f := (\overline{x}, f(\overline{x}))$. By (S3) and (S4a) we get $(0,0) \in \partial d_F(\overline{x}_f) + (0,1)$ or $(0,-1) \in \partial d_F(\overline{x}_f)$; hence $0 \in \partial^m f(\overline{x})$.

(S4a) Let us check this condition step by step. Suppose first that $f := h \circ A$, where $A \in L(X,Y)$ is surjective. Without loss of generality we may suppose $\|y\| = \inf\{\|x\| : x \in A^{-1}(y)\}$ for all $y \in Y$. Then $(z,s) \in H$ if and only if there exists $u \in A^{-1}(z)$ such that $(u,s) \in F$; hence for all $(x,r) \in X \times \mathbb{R}$

$$d_H(Ax,r) = \inf_{(z,s)\in H}(\|Ax-z\|+|r-s|) = \inf_{(u,s)\in F}(\|x-u\|+|r-s|) = d_F(x,r)$$

or $d_F = d_H \circ (A \times I)$.Thus, by (S4a) for ∂, $x^* \in X^*$ is in $\partial^m f(\overline{x})$, i.e., $(x^*,-1) \in t\partial d_F(\overline{x}, f(\overline{x}))$ for some $t > 0$, if and only if one has $x^* = A^T y^*$ for some $y^* \in Y^*$ such that $(y^*,-1) \in t\partial d_H(\overline{y}, h(\overline{y}))$ for some $t > 0$, if and only if $x^* = A^T y^*$ for some $y^* \in \partial^m h(\overline{y})$.

When $f = g + c$ with $c \in \mathbb{R}$, we have $F = G + (0,c)$, so that $d_F(x,r) = d_G(x,r-c)$ and $\partial^m f(x) = \partial^m g(x)$ by (S4a) for ∂. Similarly, when $f(x) := g(x+b)$, one has $F = G - (b,0)$, so that $d_F(x,r) = d_G(x+b,r)$ and $\partial^m f(x) = \partial^m g(x)$.

Suppose $f = \lambda g$ with $\lambda > 0$. The proof that $\partial^m f(x) = \lambda \partial^m g(x)$ is given in [34, Proposition 7] and in [26]. For the sake of completeness we present a proof using the fact that ∂^m satisfies the composition (S4a). Let $h : x \mapsto g(\lambda x)$ and let F, G, H be the epigraphs of f, g, h, respectively. For $(x,r) \in X \times \mathbb{R}$ one has

$$d_F(x,r) = \lambda \inf\{\|x/\lambda - w/\lambda\| + |r/\lambda - s| : (w,s) \in G\} = \lambda d_H(x/\lambda, r/\lambda),$$

or $d_F = \lambda d_H \circ A$ with $A(x,r) := (x/\lambda, r/\lambda)$. By (S4a) for ∂ we get $\partial d_F(x, f(x)) = \lambda \partial (d_H \circ A)(x, \lambda g(x)) = \partial d_H(x/\lambda, g(x)) = \partial d_H(x/\lambda, h(x/\lambda))$. Thus, given $x^* \in \partial^m f(x)$, one has $x^* \in \partial^m h(x/\lambda)$. Since ∂^m satisfies the above composition rule, noting that $h = g \circ B$ with $B(x) := \lambda x$, we get $x^* \in \lambda \partial^m g(x)$. Observing that $g = \lambda^{-1} f$, we obtain $\partial f(x) = \lambda \partial g(x)$.

Now suppose that $f := g + \ell$, where ℓ is linear and continuous. For $(x,r) \in X \times \mathbb{R}$, since $h := h_{x,r} : (u,q) \mapsto \|u-x\| + |q + \ell(u) - r|$ is Lipschitzian with rate $c := 1 + \|\ell\|$, the Penalization Lemma [9, Proposition 6.3, p. 50], [36, Proposition 1.120] yields that

$$\begin{aligned} d_F(x,r) &= \inf_{(u,q)\in G}\{\|u-x\| + (q+\ell(u)-r)^+\} \\ &= \inf_{(u,q)\in X\times\mathbb{R}}\{\|u-x\| + (q+\ell(u)-r)^+ + c d_G(u,q)\}. \end{aligned}$$

For $(x,r,u,q) \in (X \times \mathbb{R})^2$, setting $\varphi(x,r,u,q) := \|u-x\| + (q+\ell(u)-r)^+ + c d_G(u,q)$ and $T(x,r) := (x, r - \ell(x))$, we have $T(x,r) \in G$ when $(x,r) \in F$; hence $\varphi(S(x,r)) = 0$ for $S(x,r) := (x,r,T(x,r))$. Thus, using (S6e) and Proposition 29.6 with $B(u,q) = (-u, -q - \ell(u))$, for every $x^* \in \partial^m f(\overline{x})$, $t > 0$ such that $(x^*,-1) \in t\partial d_F(\overline{x}, f(\overline{x}))$, we can find $u^* \in \partial \|\cdot\|(0)$, $s^* \in \partial |\cdot|(0)$, $(w^*, q^*) \in c\partial d_G(\overline{x}, g(\overline{x}))$ satisfying

$$(t^{-1}x^*, -t^{-1}, 0, 0) = (u^*, s^*, w^* - u^* - s^*\ell, q^* - s^*),$$

i.e., $x^* = tu^*$, $q^* = s^* = -t^{-1}$, $w^* = t^{-1}x^* + s^*\ell$. Then $tw^* \in \partial^m g(\overline{x})$ and $x^* = tw^* + \ell$. Let us note that we may also apply (S6b), since for any sequences $(\alpha_n) \to 0_+$, $((x_n, r_n)) \to (\overline{x}, f(\overline{x}))$, we can pick $(u_n, s_n) \in F$ such that $\|(u_n, s_n) - (x_n, r_n)\| \le$

$d_F(x_n,r_n)+\alpha_n$. Then, setting $q_n := s_n - \ell(u_n)$, we have $(u_n,q_n) \in G$,; hence $\|u_n - x_n\| + |q_n + \ell(u_n) - r_n| + cd_G(u_n,q_n) \le d_F(x_n,r_n)+\alpha_n$.

In order to check (S4b), we first present a proof of (S4s), as it is more intuitive. Let $f \in \mathscr{I}(X\times Y)$, $g \in \mathscr{I}(X)$, $h \in \mathscr{I}(Y)$ be such that $f(x,y) = g(x)+h(y)$ for $(x,y) \in X\times Y$ and let $(\overline{x}^*,\overline{y}^*) \in \partial^m f(\overline{x},\overline{y})$ for some $(\overline{x},\overline{y}) \in \mathrm{dom} f$. Let G, H, F be the epigraphs of g, h, f, respectively. Observing that for a, b, $r \in \mathbb{R}$ one has

$$\inf\{|s+t-r| : s \ge a,\ t \ge b\} = (a+b-r)^+$$
$$= \min_{s,t\in\mathbb{R}}\{(a-s)^+ + (b-t)^+ :\ s+t = r\},$$

the minimum being attained for $s := a-(1/2)(a+b-r)^+$, $t := b-(1/2)(a+b-r)^+$, we get, for $(x,y,r) \in X\times Y\times\mathbb{R}$,

$$\begin{aligned} d_F(x,y,r) &= \inf\{\|x-u\|+\|y-v\|+|s+t-r| : u\in X,\ v\in Y,\ s\ge g(u),\ t\ge h(v)\} \\ &= \inf\{\|x-u\|+\|y-v\|+(g(u)-s)^+ + (h(v)-t)^+ : u\in X,\ v\in Y,\ s+t=r\} \\ &= \inf\{d_G(x,s)+d_H(y,t) :\ s,t\in\mathbb{R},\ s+t=r\}. \end{aligned}$$

Thus, setting $A(x,y,s,t) := (x,y,s+t)$, $\varphi(x,y,s,t) := d_G(x,s)+d_H(y,t)$, one has

$$d_F(x,y,r) = \inf\{\varphi(x,y,s,t) : A(x,y,s,t) := (x,y,r)\}.$$

Moreover, given sequences $(\alpha_n) \to 0_+$, $((x_n,y_n,r_n)) \to (\overline{x},\overline{y},\overline{r}) := (\overline{x},\overline{y},f(\overline{x},\overline{y}))$, picking $(u_n,v_n,q_n) \in F$ such that $\|(u_n,v_n,q_n)-(x_n,y_n,r_n)\| \le d_F(x_n,y_n,r_n)+\alpha_n$, setting

$$\begin{aligned} s_n &:= (1/2)(g(u_n)-h(v_n)+r_n), & t_n &:= (1/2)(-g(u_n)+h(v_n)+r_n), \\ s_n' &:= s_n+(1/2)(q_n-r_n), & t_n' &:= t_n+(1/2)(q_n-r_n), \end{aligned}$$

one has $s_n+t_n = r_n$, $(s_n)\to g(\overline{x})$, $(t_n)\to h(\overline{y})$ since $(u_n)\to\overline{x}$, $(v_n)\to\overline{y}$, $(q_n)\to\overline{r}$, g and h are l.s.c., and $(u_n,s_n') \in G$, $(v_n,t_n') \in H$, hence

$$\varphi(x_n,y_n,s_n,t_n) \le \|x_n-u_n\|+\|y_n-v_n\|+|r_n-q_n| \le d_F(x_n,y_n,r_n)+\alpha_n$$

for all n. Thus, the assumption of (S6s) is satisfied. Let us note that when $(x_n,y_n,r_n)\in F$ for all n we can take $(u_n,v_n,q_n) = (x_n,y_n,r_n)$ and get $\varphi(x_n,y_n,s_n,t_n) = 0$ with $(s_n,t_n) \in T(x_n,y_n,r_n)$ where

$$T(x,y,r) := \{(s,t)\in\mathbb{R}^2 : s\ge g(x),\ t\ge h(y),\ s+t=r\}$$

so that T has nonempty values on F and is lower semicontinuous at $((\overline{x},\overline{y},\overline{r}),(g(\overline{x}),h(\overline{y})))$: the assumption of (S6b) is satisfied. Thus, taking $c>0$ such that

$(\overline{x}^*,\overline{y}^*,-1) \in c\partial d_F(\overline{x},\overline{y},\overline{r})$, using (S6a) or (S6b) and the relation $A^T(x^*,y^*,r^*) = (x^*,y^*,r^*,r^*)$, one gets

$$(c^{-1}\overline{x}^*,c^{-1}\overline{y}^*,-c^{-1},-c^{-1}) \in \partial\varphi(\overline{x},\overline{y},g(\overline{x}),h(\overline{y})),$$

and since $(x,s) \mapsto d_G(x,s)$ and $(y,t) \mapsto d_H(y,t)$ are Lipschitzian with independent variables, $(\overline{x}^*,-1) \in c\partial d_G(\overline{x},g(\overline{x}))$, $(\overline{y}^*,-1) \in c\partial d_H(\overline{y},h(\overline{y}))$ by (S4s), so that $(\overline{x}^*,\overline{y}^*) \in \partial^m g(\overline{x}) \times \partial^m h(\overline{y})$.

We are ready to check (S4b). For the sake of simplicity of notation, we just present the case $m = 2$, the general case being similar. We adopt a notation close to the case (S4s), replacing the addition with a Lipschitzian map $j : \mathbb{R}^2 \to \mathbb{R}$ of class C^1 with $D_1 j(g(\overline{x}),h(\overline{y})) > 0$, $D_2 j(g(\overline{x}),h(\overline{y})) > 0$ and setting $(\overline{s},\overline{t}) := (g(\overline{x}),h(\overline{y}))$, $\overline{r} := j(\overline{s},\overline{t})$, $f(x,y) := j(g(x),h(y))$,

$$\varphi(x,y,r,u,v,s,t) := \|x-u\| + \|y-v\| + (j(s,t)-r)^+ + \lambda d_G(u,s) + \lambda d_H(v,t),$$

where λ is the Lipschitz rate of $(u,v,s,t) \mapsto \|x-u\| + \|y-v\| + (j(s,t)-r)^+$ and F, G, H are again the epigraphs of f, g, h. Since j is continuous and increasing in its two variables around $(\overline{s},\overline{t}) := (g(\overline{x}),h(\overline{y}))$ (modifying j off a neighborhood of $(g(\overline{x}),h(\overline{y}))$, alternatively, one may suppose j is increasing and such that $j(\overline{s}+s,\overline{t}) \to +\infty$ as $s \to +\infty$), for some neighborhood U of $(\overline{x},\overline{y},\overline{r})$, one has $F \cap U = \{(u,v,j(s,t)) : (u,s) \in G,\ (v,t) \in H\} \cap U$, so that, for (x,y,r) near $(\overline{x},\overline{y},\overline{r})$,

$$\begin{aligned} d_F(x,y,r) &= \inf\{\|x-u\| + \|y-v\| + (j(s,t)-r)^+ : (u,s) \in G,\ (v,t) \in H\} \\ &= \inf\{\varphi(x,y,r,u,v,s,t) : (u,v,s,t) \in X \times Y \times \mathbb{R}^2\} \end{aligned}$$

in view of the Penalization Lemma. Let us define the multimap $T : F \rightrightarrows \mathbb{R}^2$ by

$$T(x,y,r) := \{(s,t) : s \geq g(x),\ t \geq h(y),\ j(s,t) \leq r\}.$$

It has nonempty values on F since for all $(x,y,r) \in F$ one has $(g(x),h(y)) \in T(x,y,f(x,y)) \subset T(x,y,r)$. Moreover, for all $(x,y,r) \in F$ and $(s,t) \in T(x,y,r)$, one has $\varphi(x,y,r,x,y,s,t) = 0$. Let us show that T is lower semicontinuous at $((\overline{x},\overline{y},\overline{r}),(\overline{s},\overline{t}))$ on F. Let $((x_n,y_n,r_n))$ be a sequence in F with limit $(\overline{x},\overline{y},\overline{r})$. Since $r_n \geq j(g(x_n),h(y_n))$ for all n, since g and h are lower semicontinuous and for every $\varepsilon > 0$ one has $\min(j(\overline{s}+\varepsilon,\overline{t}), j(\overline{s},\overline{t}+\varepsilon) > \overline{r} = \lim r_n$, the set

$$N_\varepsilon := \{n \in \mathbb{N} : g(x_n) \geq g(\overline{x}) + \varepsilon\} \cup \{n \in \mathbb{N} : h(y_n) \geq h(\overline{y}) + \varepsilon\}$$

is finite. Thus, by lower semicontinuity of g and h, we have $(s_n) := (g(x_n)) \to g(\overline{x})$, $(t_n) := (h(y_n)) \to h(\overline{y})$. Since $r_n \geq j(g(x_n),h(y_n))$, we have $(s_n,t_n) \in T(x_n,y_n,r_n)$ for all n. Thus T is lower semicontinuous at $((\overline{x},\overline{y},\overline{r}),(\overline{s},\overline{t}))$ on F. Given $(x^*,y^*) \in \partial^m f(\overline{x},\overline{y})$, $\overline{s} := g(\overline{x})$, $\overline{t} := h(\overline{y})$ and $c > 0$ such that $(x^*,y^*,-1) \in c\partial d_F(\overline{x},\overline{y},\overline{r})$, $\overline{r} := f(\overline{x},\overline{y})$, applying (S6f), we get that $(x^*,y^*,-1,0,0,0,0) \in c\partial\varphi(\overline{x},\overline{y},\overline{r},\overline{x},\overline{y},\overline{s},\overline{t})$.

Using Proposition 29.6 and (S4b) we can find $(u^*,s^*) \in \lambda\partial d_G(\overline{x},\overline{s})$, $(v^*,t^*) \in \lambda\partial d_H(\overline{y},\overline{t})$, $w^* \in B_{X^*}$, $z^* \in B_{Y^*}$, $r^* \in [-1,1]$ such that

$$(x^*,y^*,-1,0,0,0,0) = c(w^*,z^*,r^*,u^*-w^*,v^*-z^*,s^*-r^*D_1j(\overline{s},\overline{t}),t^*-r^*D_2j(\overline{s},\overline{t})).$$

Thus $r^* = -1/c$, $s^* = -(1/c)D_1j(\overline{s},\overline{t})$, $t^* = -(1/c)D_2j(\overline{s},\overline{t})$, $x^* = cw^* = cu^* = D_1j(\overline{s},\overline{t})(-u^*/s^*)$, with $-u^*/s^* \in \partial^m g(\overline{x})$, $y^* = cz^* = cv^* = D_2j(\overline{s},\overline{t})(-v^*/t^*)$, with $-v^*/t^* \in \partial^m h(\overline{y})$; hence $(x^*,y^*) \in Dj(\overline{s},\overline{t}) \circ (\partial^m g(\overline{x}) \times \partial^m h(\overline{y}))$, as expected.

Let us turn to condition (S5). Let $g \in \mathscr{F}(X)$, h of class C^1 around $\overline{y}$, $f(x,y) := \max(g(x),h(y))$, $(\overline{x},\overline{y}) \in X \times Y$ with $\overline{r} := g(\overline{x}) = h(\overline{y})$, $(\overline{x}^*,\overline{y}^*) \in \partial^m f(\overline{x},\overline{y})$ with $\overline{x}^* \neq 0$ or $\overline{y}^* \neq h'(\overline{y})$. Let $c > 0$ be such that $(\overline{x}^*,\overline{y}^*,-1) \in c\partial d_F(\overline{x},\overline{y},\overline{r})$. Let us first check that for all $(x,y,r) \in X \times Y \times \mathbb{R}$ we have

$$d_F(x,y,r) \le 2(d_G(x,r) \vee d_H(y,r)). \tag{29.7}$$

Given $m > d_G(x,r) \vee d_H(y,r)$ we can find $(u,s) \in G$, $(v,t) \in H$ such that $\|u-x\| + (s-r)^+ < m$, $\|v-y\| + (t-r)^+ < m$, so that $(u,v,s \vee t) \in F$

$$d_F(x,y,r) \le \|u-x\| + \|v-y\| + (s \vee t - r)^+ < 2m$$

and inequality (29.7) holds. Since φ given by $\varphi(x,y,r) := d_G(x,r) \vee d_H(y,r)$ is null on F, condition (S6b) ensures that $(\overline{x}^*,\overline{y}^*,-1) \in 2c\partial\varphi(\overline{x},\overline{y},\overline{r})$. Now, since there is no loss of generality in assuming that h is Lipschitzian with rate 1, we have

$$\varphi(x,y,r) = d_G(x,r) \vee (h(y)-r)^+ = d_G(x,r) \vee (h(y)-r) = (d_G(x,r)+r) \vee h(y) - r.$$

Let $\psi((x,r),y) := \varphi(x,y,r) + r$, so that $(\overline{x}^*/2c, 1-1/2c, \overline{y}^*/2c) \in \partial\psi(\overline{x},\overline{r},\overline{y})$ and one cannot have $(\overline{x}^*/2c, 1-1/2c) = (0,0)$ and $\overline{y}^*/2c = h'(\overline{y})$ as these relations imply $2c = 1$, $\overline{x}^* = 0$, $\overline{y}^*/2c = h'(\overline{y})$. Conditions (S4a) and (S5) for ∂ yield some $\lambda \in]0,1]$ such that $(\overline{x}^*/2c, 1-1/2c) \in \lambda(\partial d_G(x,r) + (0,1))$ and $\overline{y}^*/2c = (1-\lambda)h'(\overline{y})$ or $(\overline{x}^*,-\mu) \in 2\lambda c\partial d_G(\overline{x},\overline{r})$ for $\mu := 1-2c(1-\lambda) \in]0,1]$ since $\lambda \le 1$, $\partial d_G(\overline{x},\overline{r}) \subset X^* \times \mathbb{R}_-$ and since $\mu = 0$ implies $\overline{y}^* = h'(\overline{y})$. Thus $\overline{x}^*/\mu \in \partial^m g(\overline{x})$, $(\overline{x}^*,\overline{y}^*) \in \mu\partial^m g(\overline{x}) \times (1-\mu)h'(\overline{y})$: (S5) is satisfied.

Now, let us check condition (S6a). Let $A : V \to W$ be a surjective continuous linear map between two members of $\mathscr{X}$; let $\varphi \in \mathscr{I}(V)$ and $p \in \mathscr{I}(W)$ be such that $\varphi \ge p \circ A$. Suppose that for some $\overline{w} \in W$, $M \subset A^{-1}(\overline{w})$ and every sequences $(\alpha_n) \to 0_+$, $(w_n) \to \overline{w}$ with $(p(w_n)) \to \overline{r} := p(\overline{w})$ one can find $\overline{v} \in M$, an infinite subset N of $\mathbb{N}$, a sequence $(v_n)_{n\in N} \to \overline{v}$ such that $A(v_n) = w_n$ and $\varphi(v_n) \le p(w_n) + \alpha_n$ for all $n \in N$. Denoting by P (resp. F) the epigraph of p (resp. φ), let us prove that a similar property holds with $V \times \mathbb{R}$, $W \times \mathbb{R}$, $B := A \times I_{\mathbb{R}}$, d_F, d_P, $(\overline{w},\overline{r})$, $M \times \{\overline{r}\}$ substituted to V, W, A, φ, p, $\overline{w}$, M, respectively. Since A is open, we may endow W with the norm given by $\|w\| := \inf\{\|v\| : Av = w\}$. Then $\|A\| \le 1$. We have $d_P \circ B \le d_F$: given $(u,r) \in V \times \mathbb{R}$, since $(Av,s) \in P$ whenever $(v,s) \in F$, we obtain

$$d_P(B(u,r)) = \inf\{\|Au - w\| + |r - s| : (w,s) \in P\}$$
$$\leq \inf\{\|A\| \, \|u - v\| + |r - s| : (v,s) \in F\} \leq d_F(u,r).$$

Let $((w_n, s_n)) \to (\overline{w}, \overline{r})$, $(\alpha_n) \to 0_+$ and let $(\delta_n) := (d_P(w_n, s_n)) \to d_P(\overline{w}, p(\overline{w})) = 0$. There exists a sequence $((w'_n, s'_n))$ in P such that $\|(w'_n, s'_n) - (w_n, s_n)\| \leq (1 + \alpha_n)\delta_n$. Then $(w'_n) \to \overline{w}$, $(s'_n) \to p(\overline{w})$, so that

$$p(\overline{w}) \leq \liminf_n p(w'_n) \leq \limsup_n p(w'_n) \leq \lim_n s'_n = p(\overline{w}),$$

and $(p(w'_n)) \to p(\overline{w})$. By our assumption, there exist $\overline{v} \in M$, an infinite subset N of $\mathbb{N}$, a sequence $(v'_n) \to \overline{v}$ such that $A(v'_n) = w'_n$ and $\varphi(v'_n) \leq p(w'_n) + \alpha_n/2 \leq s'_n + \alpha_n/2$ for all $n \in N$. By the choice of the norm on W, there exists a sequence (v_n) such that $\|v_n - v'_n\| \leq (1 + \alpha_n)\|w_n - w'_n\|$ and $A(v_n) = w_n$. Thus the sequence $((v_n, s_n))$ is such that $((v_n, s_n)) \to (\overline{v}, \overline{r})$, $B(v_n, s_n) = (w_n, s_n)$, and

$$d_F(v_n, s_n) \leq \left\|(v'_n, s'_n + \alpha_n/2) - (v_n, s_n)\right\| = \left\|v'_n - v_n\right\| + \left|s'_n + \alpha_n/2 - s_n\right|$$
$$\leq (1 + \alpha_n)\left\|w_n - w'_n\right\| + \left|s'_n - s_n\right| + \alpha_n/2 \leq (1 + \alpha_n)^2\delta_n + \alpha_n/2,$$

since $\|w_n - w'_n\| \leq (1 + \alpha_n)\delta_n$. Then, since we may suppose $\alpha_n(2 + \alpha_n)\delta_n \leq \alpha_n/2 + \delta_n$ for all $n \in N$, we get $d_F(v_n, s_n) \leq d_P(w_n, s_n) + \alpha_n + \delta_n$ for all $n \in N$. Given $\overline{w}^* \in \partial^m p(\overline{w})$, let $t \in \mathbb{R}_+$ be such that $(\overline{w}^*, -1) \in t\partial d_P(\overline{w}, p(\overline{w}))$. Applying (S6a) to the Lipschitzian functions d_F,d_P we get that $t^{-1}(A^T w^*, -1) = t^{-1}B^T(w^*, -1) \in \partial d_F(\overline{v}, \varphi(\overline{v}))$ and $A^T w^* \in \partial^m \varphi(\overline{v})$, as required.

The proof of (S6b) is similar. Keeping the preceding notation, with $p := d_E$, we just observe that for any sequences $(\alpha_n) \to 0_+$, $(w_n) \to (\overline{w})$ in E, we have $((w_n, 0)) \to (\overline{w}, 0)$ with $(w_n, 0) \in P$ and if $N \subset \mathbb{N}$, $\overline{v} \in M$, $(v_n)_{n \in N} \to \overline{v}$ are such that $A(v_n) = w_n$, $\varphi(v_n) \leq \alpha_n$ for all $n \in N$, we have $B(v_n, 0) = (w_n, 0)$ and $d_F(v_n, 0) \leq \alpha_n$, so that we can use (S6b) for d_P and d_F instead of p and φ and we get the inclusion $A^T(\partial^m d_E(\overline{w})) \in \partial^m \varphi(\overline{v})$ as above. ∎

Remark 29.27. Another proof of the relation $\partial^m(\lambda g) = \lambda \partial^m g$ uses (S6b). Keeping the same notation, using the Penalization Lemma and the fact that $h_{x,r} : (u,q) \mapsto \|x - u\| + |r - \lambda q|$ is Lipschitzian with rate $\mu := \max(\lambda, 1)$, we have

$$d_F(x,r) = \inf\{h_{x,r}(u,q) : (u,q) \in G\} = \inf\{h_{x,r}(u,q) + \mu d_G(u,q) : (u,q) \in X \times \mathbb{R}\}.$$

Setting $\varphi(x,r,u,q) := h_{x,r}(u,q) + \mu d_G(u,q)$, $S(x,r) := (x, r/\lambda)$ for $(x,r,u,q) \in (X \times \mathbb{R})^2$, we have $\varphi(x,r,S(x,r)) = 0 = d_F(x,r)$ when $(x,r) \in F$. Applying (S6b) and Proposition 29.6, for every $x^* \in \partial^m f(x)$, $t > 0$ such that $(x^*, -1) \in t\partial d_F(x, f(x))$, we can find $u^* \in \partial \|\cdot\|(0)$, $s^* \in \partial |\cdot|(0)$, $(w^*, q^*) \in \mu \partial d_G(\overline{x}, g(\overline{x}))$ satisfying

$$(t^{-1}x^*, -t^{-1}, 0, 0) = (u^*, s^*, u^* - w^*, q^* - \lambda s^*),$$

i.e., $x^* = tu^* = tw^*$, $s^* = -t^{-1}$, $q^* = -\lambda t^{-1}$. Setting $y^* := \lambda^{-1}tw^*$, we get $(y^*, -1) \in \lambda^{-1}t\mu\partial d_G(\overline{x}, g(\overline{x}))$ or $y^* \in \partial^m g(\overline{x})$ and $x^* = \lambda y^*$. We can get the same conclusion by showing that the assumption of (S6a) is satisfied. For every sequences $(\alpha_n) \to 0_+$, $((x_n, r_n)) \to (\overline{x}, f(\overline{x}))$, let $(u_n, s_n) \in F$ be such that $\|(u_n, s_n) - (x_n, r_n)\| \leq d_F(x_n, r_n) + \alpha_n$. Then, setting $q_n := \lambda^{-1}s_n$, we have $(u_n, q_n) \in G$; hence $\|x_n - u_n\| + |r_n - \lambda q_n| + \mu d_G(u_n, q_n) \leq d_F(x_n, r_n) + \alpha_n$.

29.3.2 Limiting Subdifferentials

In order to make a multimap $M : X \rightrightarrows X^*$ from a Banach space to its dual X^* more robust, a general procedure is available, setting

$$\overline{M}(x) := \bigcup_{r>0} w^*\text{-}\limsup_{x' \to x}(M(x') \cap rB_{X^*}).$$

A variant consists in taking the sequential limsup, the symbol $\overset{*}{\to}$ denoting weak* convergence:

$$\overline{M}^s(x) := \{x^* : \exists (x_n) \to x,\ (x_n^*) \overset{*}{\to} x^*,\ \forall n \in \mathbb{N}\ x_n^* \in M(x_n)\}.$$

When ∂ is a subdifferential and $f \in \mathscr{F}(X)$, we set $\overline{\partial} f(x) := \overline{M}(x)$, $\overline{\partial}^s f(x) := \overline{M}^s(x)$, with $M := \partial f$, replacing the convergence $x' \to x$ (resp. $(x_n) \to x$) by the convergence $x' \to_f x$, i.e., $(x', f(x')) \to (x, f(x))$ (resp. $(x_n) \to_f x$). Here we take $x \in \operatorname{dom} f$ and we set $\overline{\partial} f(x) = \overline{\partial}^s f(x) = \varnothing$ for $x \in X \backslash \operatorname{dom} f$.

Theorem 29.28. *Let $\partial := (\mathscr{X}, \mathscr{F}, \partial)$ be a subdifferential satisfying (S6e) rather than (S6a), (S6b). Then $\overline{\partial} := (\mathscr{X}, \mathscr{F}, \overline{\partial})$ is a subdifferential satisfying (S6e) rather than (S6a), (S6b). If the spaces in $\mathscr{X}$ are such that their dual unit balls are weak* sequentially compact, the same assertion holds for $\overline{\partial}^s := (\mathscr{X}, \mathscr{F}, \overline{\partial}^s)$.*

Proof. Conditions (S1) and (S2) are obviously satisfied by $\overline{\partial}$ and $\overline{\partial}^s$ since for every bounded subset B^* of the dual of a Banach space X the coupling function $\langle \cdot, \cdot \rangle$ is continuous on $X \times B^*$ when B^* is endowed with the weak* convergence. Condition (S3) is a consequence of the inclusions $\partial f \subset \overline{\partial}^s f \subset \overline{\partial} f$. Let us show that property (S4c) is obtained by a passage to the limit. Let $g : X \to Y$ be of class C^1 around $\overline{x}$ and $f = h \circ g$ with $h \in \mathscr{F}(Y)$. For $y^* \in \overline{\partial} h(\overline{y})$ with $\overline{y} := g(\overline{x})$, one can find a net $((y_i, y_i^*))_{i \in I}$ in the graph of ∂h with $(y_i) \to_h \overline{y}$, $(y_i^*) \overset{*}{\to} y^*$, (y_i^*) being bounded. Since g is open at $\overline{x}$, there exists a net $(x_i)_{i \in I} \to \overline{x}$ such that $y_i = A(x_i)$ for all $i \in I$. Then $(f(x_i))_{i \in I} = (h(y_i))_{i \in I} \to h(\overline{y}) = f(\overline{x})$, $(x_i^*) := (g'(x_i)^T y_i^*) \overset{*}{\to} g'(\overline{x})^T y^*$, the mapping $(\ell, y^*) \mapsto \ell^T(y^*) := y^* \circ \ell$ being continuous on bounded sets when $L(X,Y)$ is endowed with the norm topology and Y^* is provided with the weak* topology. Since $g'(x_i)^T y_i^* \in \partial f(x_i)$ one gets $g'(\overline{x})^T y^* \in \overline{\partial} f(x)$. The reverse inclusion $\overline{\partial} f(\overline{x}) \subset g'(\overline{x})^T(\overline{\partial} h(\overline{y}))$ uses the same continuity argument and the fact that there exists some

$c > 0$ and a neighborhood U of $\overline{x}$ such that $\inf\{\|x\| : x \in g'(u)^{-1}(y)\} \leq c\,\|y\|$ for all $u \in U$, $y \in Y$, so that if $(x_i^*) = (y_i^* \circ g'(x_i))$ is bounded, with $(x_i) \to \overline{x}$, then (y_i^*) is bounded, hence has a weak* converging subnet.

In order to check **(S5)**, i.e., consistency of $\overline{\partial}$, let $g \in \mathscr{F}(X)$, $h \in \mathscr{F}(Y)$ of class C^1 around $\overline{y}$, $g(\overline{x}) = h(\overline{y})$, $(\overline{x}^*, \overline{y}^*) \in \overline{\partial} f(\overline{x}, \overline{y})$ for f given by $f(x,y) := \max(g(x), h(y))$. Suppose $\overline{y}^* \neq h'(\overline{y}) \neq 0$ and let $((x_i, y_i, x_i^*, y_i^*))_{i\in I}$ be in the graph of ∂f, with $((x_i, y_i)) \to_f (\overline{x}, \overline{y})$, $((x_i^*, y_i^*)) \overset{*}{\to} (\overline{x}^*, \overline{y}^*)$ and bounded. We cannot have $g(x_i) < h(y_i)$ for all i in a cofinal subset J of I because otherwise we would have $f(x,y) = h(y)$ for (x,y) near (x_i, y_i) with $i \in J$; hence $(x_i^*, y_i^*) = (0, h'(y_i))$ and, passing to the limit, $\overline{y}^* = h'(\overline{y})$, a contradiction with our assumption. If $g(x_i) > h(y_i)$ for all i in a cofinal subset K of I we have $f(x,y) = g(x)$ for (x,y) near (x_i, y_i) with $i \in K$; hence $x_i^* \in \partial g(x_i)$, $y_i^* = 0$ by (S4a) and $x^* \in \overline{\partial} g(x)$, $\overline{y}^* = 0$. If for some cofinal subset L of I we have $g(x_i) = h(y_i)$ for all $i \in L$, then, since $(x_i^*, y_i^*) \in \partial f(x_i, y_i)$ and for $i \in L$ large enough $y_i^* \neq h'(y_i) \neq 0$, by consistency of ∂ we get some $\lambda_i \in]0,1]$, $u_i^* \in \partial g(x_i)$ such that $(x_i^*, y_i^*) = (\lambda_i u_i^*, (1-\lambda_i) h'(y_i))$. Taking $\lambda \in [0,1]$ and a subnet such that $(\lambda_i) \to \lambda$, we cannot have $\lambda = 0$ since $\overline{y}^* \neq h'(\overline{y})$; hence $(u_i^*) \to u^* := \lambda^{-1} \overline{x}^*$ and $u^* \in \overline{\partial} g(\overline{x})$, so that $(\overline{x}^*, \overline{y}^*) = (\lambda u^*, (1-\lambda) h'(\overline{y}))$ with $\lambda \in]0,1]$.

Now, let V, W, A, p, φ, $\overline{w}$, S be as in (S6e) and let $\overline{w}^* \in \overline{\partial} p(\overline{w})$. Let $((w_i, w_i^*, p(w_i)))_{i\in I}$ be a net such that $((w_i, p(w_i))) \to (\overline{w}, p(\overline{w}))$, $(w_i^*) \overset{*}{\to} \overline{w}^*$, (w_i^*) being bounded with $w_i^* \in \partial p(w_i)$ for all $i \in I$. By assumption, given $\overline{v} \in S(\overline{w})$, there exist $\overline{i} \in I$ and a net $(v_j)_{j\in J} \to \overline{v}$ for $J := \{i \in I : i \geq \overline{i}\}$ such that $v_j \in S(w_j)$ for all $j \in J$. Since ∂ satisfies (S6e), and since we may suppose S is lower semicontinuous at w_j for all $j \in J$, we have $A^T(w_j^*) \in \partial\varphi(v_j)$ for all $j \in J$. Since A^T is weak* continuous and bounding (i.e., maps bounded sets into bounded sets) and since $(\varphi(v_j)) = (p(w_j)) \to p(\overline{w}) = \varphi(\overline{v})$, we get $A^T(\overline{w}^*) \in \overline{\partial}\varphi(\overline{v})$. The sequential case is similar. ∎

29.4 Further Properties of Subdifferentials

Now let us deal with properties subdifferentials may have or may not have.

29.4.1 Subdifferentiability Spaces and Variational Subdifferentials

The following notion reflects the fact that it is natural to deal with subdifferentials that have sufficiently many points of nonemptiness.

Definition 29.29 ([13, 14]). Given a subdifferential $\partial := (\mathscr{X}, \mathscr{F}, \partial)$, a Banach space X in $\mathscr{X}$ is called a ∂-subdifferentiability space (or just a subdifferentiability space if there is no risk of confusion) if for any $f \in \mathscr{F}(X)$ the set $\{(x, f(x)) : x \in X,\ \partial f(x) \neq \varnothing\}$ is dense in the graph of f.

Then, for any $f \in \mathscr{F}(X)$, the domain of ∂f is dense in the domain of f. In fact, for every $\overline{x} \in \mathrm{dom} f$ and every $\varepsilon > 0$, there exists $x \in \mathrm{dom}\partial f \cap B(\overline{x}, \varepsilon, f)$. A criterion for such a property appears with the next definition inspired by variational principles.

Definition 29.30. A subdifferential $\partial := (\mathscr{X}, \mathscr{F}, \partial)$ is *variational* if for any X in $\mathscr{X}$, $f \in \mathscr{F}(X)$, $\overline{x} \in \mathrm{dom} f$, $\varepsilon > 0$ such that $f(\overline{x}) < \inf f(X) + \varepsilon$ and for any γ, $\delta > 0$ satisfying $\gamma\delta \geq \varepsilon$, there exist $y \in B(\overline{x}, \delta)$ and $y^* \in \partial f(y)$ such that $\|y^*\| < \gamma$, $|f(y) - f(\overline{x})| < \varepsilon$.

In [24] the term "β-variational subdifferential" is used for the viscosity subdifferential associated with a bornology β. Since no bornology is present here, the risk of confusion is low. Variational subdifferentials provide nontrivial concepts.

Proposition 29.31. *Let ∂ be a variational subdifferential on the class $\mathscr{I}$ of lower semicontinuous functions. Then all X in $\mathscr{X}$ are ∂-subdifferentiability spaces.*

Proof. Let $f \in \mathscr{I}(X)$, $\overline{x} \in \mathrm{dom}\ f$, and $\varepsilon > 0$ be given. Let $\rho \in (0, \varepsilon]$ be such that $f(x) \geq f(\overline{x}) - \varepsilon$ for all $x \in B := B(\overline{x}, \rho)$. Then $(f + \iota_B)(\overline{x}) \leq \inf(f + \iota_B)(X) + \varepsilon$ where ι_B is the indicator function of B given by $\iota_B(x) = 0$ if $x \in B$, $+\infty$ else. Let $\delta \in]0, \rho[$. Since ∂ is variational, there exist $y \in B(\overline{x}, \delta)$ and $y^* \in \partial(f + \iota_B)(y)$ such that $|f(y) - f(\overline{x})| < \varepsilon$. Since $y \in \mathrm{int} B$, f and $f + \iota_B$ coincide around y; hence $y^* \in \partial f(y)$. ∎

Not all subdifferentials are variational. For any Banach space X that is not an Asplund space the Fréchet subdifferential is not variational since there is a concave continuous function whose points of subdifferentiability are not dense in X. In particular, if S is a connected compact topological space not reduced to a singleton and $X := C(S)$ is the space of continuous functions on S endowed with the supremum norm $\|\cdot\|_\infty$ the function f given by $f(x) := -\sup_{s \in S} x(s)$ is such that for every $x \in X$ one has $\partial_F f(x) = \varnothing$.

In spaces whose powers are ∂-subdifferentiability spaces, one can give a (weak) approximate rule for finite sums of lower semicontinuous functions at points where they attain their infima. A stronger rule is considered in the next subsection.

29.4.2 *Reliable and Trustworthy Subdifferentials*

Let us introduce a class of subdifferentials useful for optimization theory. We shall see that the two properties of the following definition are in fact equivalent. They lead to fuzzy calculus rules of utmost importance [12, 13, 31, 37]. Moreover, they are the basis of the equivalences established in [23, 29, 39] including a Mean Value Theorem.

Definition 29.32 ([32]). A subdifferential $\partial := (\mathscr{X}, \mathscr{F}, \partial)$ is said to be *reliable* if for any $\varepsilon > 0$, for any X in $\mathscr{X}$, $f \in \mathscr{F}(X)$, g convex Lipschitzian on X such that $f+g$ attains a local minimum at some $\overline{x} \in \mathrm{dom} f$ there exist $(y, y^*) \in \partial f$, $(z, z^*) \in \partial g$ with $y, z \in B(\overline{x}, \varepsilon)$, $|f(y) - f(\overline{x})| < \varepsilon$ and $\|y^* + z^*\| < \varepsilon$.

If such a property holds whenever $g \in \mathscr{L}(X)$, then ∂ is said to be *trustworthy*.

Taking for $\mathscr{F}$ the class $\mathscr{I}$ of l.s.c. functions, the Clarke subdifferential ∂_C and the Ioffe subdifferential ∂_G are trustworthy on the class of all Banach spaces; the Fréchet subdifferential and the limiting subdifferential are reliable on the class of Asplund spaces. Reliability is a property stronger than the ones studied in the preceding section as the next proposition shows.

Proposition 29.33. *A reliable subdifferential is variational.*

Proof. Let $\partial := (\mathscr{X}, \mathscr{F}, \partial)$ be a reliable subdifferential. Let X in $\mathscr{X}$, $f \in \mathscr{F}(X)$, $\overline{x} \in \mathrm{dom} f$ be such that $f(\overline{x}) < \inf f(X) + \varepsilon$ and let γ, $\delta > 0$ satisfying $\gamma\delta \geq \varepsilon$. Let $\beta \in]0,1[$ be such that $f(\overline{x}) < \inf f(X) + \varepsilon\beta$. The Ekeland's variational principle yields some $w \in B(\overline{x}, \delta)$ such that $f(w) \leq f(\overline{x})$ and

$$\forall x \in X \qquad f(w) \leq f(x) + \beta\gamma\|x - w\|.$$

Since ∂ is reliable, and since w is a minimizer of $f+g$ for g given by $g(x) := \beta\gamma\|x - w\|$, setting $\varepsilon' := \min(\delta - d(w, \overline{x}), (1-\beta)\gamma, \varepsilon)$, we get $y, z \in B(w, \varepsilon') \subset B(\overline{x}, \delta)$, $y^* \in \partial f(y)$, $z^* \in \partial g(z) \subset \beta\gamma B_{X^*}$ such that $\|y^* + z^*\| < (1-\beta)\gamma$, $\|f(y) - f(w)\| < \varepsilon'$. Then $\|y^*\| < \gamma$ and $f(\overline{x}) - \varepsilon < f(y) < f(w) + \varepsilon' \leq f(\overline{x}) + \varepsilon$. ■

Theorem 29.34. *Any reliable subdifferential $\partial := (\mathscr{X}, \mathscr{F}, \partial)$ is trustworthy if for any member X of $\mathscr{X}$ and any $m \in \mathbb{N} \backslash \{0\}$, X^m is a member of $\mathscr{X}$.*

Proof. That follows from the implication (R1)$\Rightarrow$(R5) in [29, Theorem 3.1], observing that if ∂ is reliable, then for all X in $\mathscr{X}$, condition (R1) in [29, Theorem 3.1] is satisfied. ■

Reliable subdifferentials provide nontrivial geometric concepts as well as analytic concepts.

Proposition 29.35. *Let ∂ be a reliable subdifferential on $\mathscr{I}(X)$. If E is a closed subset of X and if $\overline{x} \in E$ is a boundary point of E, then for any $\varepsilon > 0$ there is some $x \in E$ with $\|x - \overline{x}\| < \varepsilon$ such that $N(E, x)$ contains a nonzero element.*

Proof. Since $\overline{x}$ is a boundary point of E, given $\varepsilon \in]0, 1/2[$, we can find $w \in X \backslash E$ such that $\|w - \overline{x}\| < \varepsilon^2/2$. By the Ekeland's variational principle there is some $z \in E$ with $\|z - \overline{x}\| < \varepsilon/2$ such that the function g given by

$$g(x) := \|w - x\| + \iota_E(x) + \varepsilon\|x - z\|$$

attains its minimum at z. Since ∂ is reliable, there are $x \in E$ and $u \in X$ such that $\|x - z\| < \varepsilon/2$, $\|u - z\| < \|w - z\|$ and some $x^* \in N(E, x)$, $u^* \in \partial(\|w - \cdot\| + \varepsilon\| \cdot - z\|)(u)$ such that $\|u^* + x^*\| < \varepsilon$. As $u \neq w$, we conclude that $\|u^*\| > 1 - \varepsilon$ and $\|x^*\| > 1 - 2\varepsilon > 0$. ■

Let us provide a criterion for reliability. For such a purpose, we recall from [2,4] the notion of ∂-differentiability: a function $f \in \mathcal{F}(X) \cap (-\mathcal{F}(X))$ is said to be *∂-differentiable* at $x \in \mathrm{dom} f \cap \mathrm{dom}(-f)$ if $\partial f(x)$ and $\partial(-f)(x)$ are nonempty. As in [36], a function f on X is said to be of *class* D^1 if it is directionally differentiable and if the map $(x,u) \mapsto f'(x)u$ is continuous. A function k on X is said to be *forcing* if $(x_n) \to 0$ whenever $(k(x_n)) \to 0$. A space X in $\mathcal{X}$ is said to be *∂-smooth* if there exists a Banach space $W \subset \mathcal{F}(X)$ containing the set of Lipschitzian ∂-differentiable convex functions and a ∂-differentiable forcing function such that for all $f \in \mathcal{F}(X)$, $g \in W$, $x \in X$, one has

$$\sup_{x \in X} \sup\{\|x^*\| : x^* \in \partial g(x) \cup \partial(-g)(x)\} \leq \|g\|_W,$$

$$\partial(f+g)(x) \subset \partial f(x) + \partial g(x).$$

Any Banach space X is ∂_C-smooth as one can take for W the space of Lipschitzian functions on X with the norm given by $\|k\|_W = |k(0)| + \sup_{x \neq x' \in X} |k(x) - k(x')| / \|x - x'\|$. If X has a Lipschitzian forcing function of class C^1 (resp. of class D^1), then it is ∂_F-smooth (resp. ∂_D-smooth) as one can take for W the space of Lipschitzian functions k of class C^1 (resp. D^1) with the norm $\|\cdot\|_W$ given by $\|k\|_W = |k(0)| + \sup_{x \in X} \|k'(x)\|$.

Proposition 29.36. *Let $\partial := (\partial, \mathcal{X}, \mathcal{F})$ be a subdifferential. If X is ∂-smooth, then ∂ is reliable on $\mathcal{F}(X)$.*

Proof. It is an adaptation of the classical decoupling method for the minimization of the sum of two functions using the Deville–Godefroy–Zizler Variational Principle. ∎

Another criterion follows the line of [29]. It uses the following concept.

Definition 29.37. A subdifferential ∂ is said to be ∂-reliable if for any X in $\mathcal{X}$, any $f \in \mathcal{F}(X)$, g convex, continuous and ∂-differentiable at $\overline{x} \in X$, one has $0 \in \partial f(\overline{x}) + \partial g(\overline{x})$ whenever $\overline{x}$ is a minimizer of $f+g$.

As in [2, 3, 29] we say that the norm of a Banach space $(X, \|\cdot\|)$ is *∂-smooth* if for any converging sequence (x_n) of X and any sequence (t_n) of $\mathbb{R}_+$ with sum 1, the function $x \mapsto \sum_{n \geq 0} t_n \|x - x_n\|^2$ is ∂-differentiable on X.

Proposition 29.38. *Let ∂ be a ∂-reliable subdifferential on a class $\mathcal{X}$. Suppose that for all X in $\mathcal{X}$ and all convex continuous functions g, h on X that are ∂-differentiable on X the function $h+g$ is ∂-differentiable on X. If for some X in $\mathcal{X}$ the norm of X is ∂-smooth, then ∂ is reliable on X.*

We note that the assumption on the sum of two ∂-differentiable functions is satisfied by the directional subdifferential, the firm subdifferential, the Clarke subdifferential, the Ioffe subdifferential, and the limiting subdifferential.

Let us sketch the proof, following the arguments of [29, Theorem 4.4]. In view of the equivalence of conditions (R2) and (R3) in [29], it suffices to prove the following assertion.

Given X in $\mathscr{X}$, $f \in \mathscr{F}(X)$, a closed linear subspace W of X such that f attains a robust minimum at some $\overline{x} \in W \cap \mathrm{dom} f$, we have to prove that there exist sequences $(x_n) \to \overline{x}$, (x_n^*), such that $(x_n^* \mid_W) \to 0$, $(f(x_n)) \to f(\overline{x})$, $x_n^* \in \partial f(x_n)$ for all $n \in \mathbb{N}$.

Here $\overline{x}$ is a *robust* (or uniform) minimizer of f on W if $f(\overline{x}) = \sup_{\delta \to 0} \inf f(B[W, \delta])$, where $B[W,\delta] := \{x \in X : d_W(x) \leq \delta\}$, with $d_W(x) := \inf_{w \in W} \|w - x\|$. Moreover, using the arguments of [29, Theorem 4.4], we may assume W is proximinal and that d_W^2 is ∂-differentiable. Let $\varepsilon \in]0,1[$ be given; we may suppose that ε is so small that $m_\varepsilon := \inf f(B[\overline{x}, \varepsilon]) > -\infty$. Let $\delta > 0$ be such that

$$\inf f(B[W,\delta]) > f(\overline{x}) - \varepsilon^2/2.$$

We chose $c > 0$ such that $c\delta^2 > f(\overline{x}) - \varepsilon^2/2 - m_\varepsilon$; hence

$$\forall x \in B[\overline{x},\varepsilon] \backslash B[W,\delta] \qquad f(x) + c d_W^2(x) > f(\overline{x}) - \varepsilon^2/2.$$

Then $\overline{x}$ is an $\varepsilon^2/2$ approximate minimizer of $f + c d_W^2$ on $B[\overline{x}, \varepsilon]$:

$$\forall x \in B[\overline{x},\varepsilon] \qquad f(x) + c d_W^2(x) > f(\overline{x}) - \varepsilon^2/2.$$

The Borwein–Preiss' variational principle [7, 2.5.2] yields some $x_\varepsilon \in X$ and some ∂-differentiable function g on X such that $\|x_\varepsilon - \overline{x}\|^2 < \varepsilon^2$, $f(x_\varepsilon) + c d_W^2(x_\varepsilon) + g(x_\varepsilon) \leq f(\overline{x})$ and x_ε is a minimizer of $f + c d_W^2 + g$ on $B[\overline{x}, \varepsilon]$, g being given by $g(x) := \sum_{n \geq 0} t_n \|x - x_n\|^2$ with $t_0 = 1/2$, $\sum_{n \geq 0} t_n = 1$, $x_n \in B[\overline{x}, \varepsilon]$ for all n. Then g is Lipschitzian with rate 2ε on $B[\overline{x}, \varepsilon]$ and since $c d_W^2 + g$ is ∂-differentiable and ∂ is ∂-reliable, we have $0 \in \partial f(x_\varepsilon) + \partial(c d_W^2 + g)(x_\varepsilon)$. Using the sum rule for convex continuous functions we can find $x_\varepsilon^* \in \partial f(x_\varepsilon)$, $y_\varepsilon^* \in c \partial d_W^2(x_\varepsilon)$, $z_\varepsilon^* \in \partial g(x_\varepsilon)$ such that $x_\varepsilon^* + y_\varepsilon^* + z_\varepsilon^* = 0$. Then we have $\|z_\varepsilon^*\| \leq 2\varepsilon$, $y_\varepsilon^* \mid_W = 0$ and $f(x_\varepsilon) \leq f(\overline{x})$. Since f is lower semicontinuous at $\overline{x}$, replacing ε with the general term ε_n of a sequence with limit 0_+ and observing that $\|x_{\varepsilon_n}^* \mid_W\| \leq 2\varepsilon_n$, we get the announced claim.

As a conclusion, let us give a nonconvex approximate separation theorem.

Theorem 29.39. *Let $\partial := (\partial, \mathscr{X}, \mathscr{F})$ be a reliable subdifferential and let C, D be two closed subsets of a member X of $\mathscr{X}$. Suppose C is convex with a nonempty interior, $D \cap \mathrm{int} C = \varnothing$ and $a \in C \cap D$. Then, for every $\varepsilon > 0$, there exists $x \in C \cap B(a,\varepsilon)$, $y \in D \cap B(a,\varepsilon)$, $x^* \in N(C,x)$, $y^* \in N(D,y)$ such that $\|x^*\| \geq 1$ and $\|x^* + y^*\| \leq \varepsilon$.*

Proof. Given $\varepsilon \in]0,1/2[$, without loss of generality, using a translation and a dilation if necessary, we may suppose $0 \in \mathrm{int} C$ and $\|a\| \leq 1 - 2\varepsilon$. Let $\rho \in]0,1[$ be such that $\rho B_X \subset C$ and let j be the Minkowki gauge of C, so that $C = j^{-1}([0,1])$, $\mathrm{int} C = j^{-1}([0,1[)$, and j is continuous and sublinear, hence Lipschitzian. Since $D \cap \mathrm{int} C = \varnothing$, a is a minimizer of j on D. Thus, there exists $x \in C \cap B(a, \varepsilon\rho)$, $y \in D \cap B(a,\varepsilon)$, $x^* \in \partial j(x)$, $y^* \in N(D,y)$ such that $\|x^* + y^*\| \leq \varepsilon$. Taking $w := x - a \in \varepsilon\rho B_X \subset \varepsilon C$ we have

$$\langle x^*, -a \rangle = \langle x^*, w - x \rangle \leq j(w) - j(x) \leq \varepsilon - (j(a) - j(a - x)) \leq 2\varepsilon - 1,$$

and $\|x^*\| (1 - 2\varepsilon) \geq \|x^*\| . \|a\| \geq \langle x^*, a \rangle \geq 1 - 2\varepsilon$, hence $\|x^*\| \geq 1$. ∎

Acknowledgments The author gratefully acknowledges the influence of numerous exchanges with A.D. Ioffe during the last decades and the useful criticisms of an anonymous referee.

References

1. Aubin, J.-P., Frankowska, H.: Set-Valued Analysis. Birkhaüser, Boston (1990)
2. Aussel, D., Corvellec, J.-N., Lassonde, M.: Mean value property and subdifferential criteria for lower semicontinuous functions. Trans. Am. Math. Soc. **347**, 4147–4161 (1995)
3. Aussel, D., Corvellec, J.-N., Lassonde, M.: Nonsmooth constrained optimization and multidirectional mean value inequalities. SIAM J. Optim. **9**, 690–706 (1999)
4. Borwein, J., Preiss, D.: A smooth variational principle with applications to subdifferentiability and differentiability of convex functions. Trans. Am. Math. Soc. **303**, 517–537 (1987)
5. Borwein, J., Zhu, Q.J.: Viscosity solutions and viscosity subderivatives in smooth Banach spaces with applications to metric regularity. SIAM J. Control Optim. **34**, 1568–1591 (1996)
6. Borwein, J., Zhu, Q.J.: A survey of subdifferential calculus with applications. Nonlinear Anal. **38**, 687–773 (1999)
7. Borwein, J.M., Zhu, Q.J.: Techniques of Variational Analysis. CMS Books in Mathematics, vol. 20. Springer, New York (2005)
8. Clarke, F.H.: Optimization and Nonsmooth Analysis. Wiley, New York (1983)
9. Clarke, F.H., Ledyaev, Y.S., Stern, R.J., Wolenski, P.R.: Nonsmooth Analysis and Optimal Control Theory. Springer, New York (1998)
10. Correa, R., Jofre, A., Thibault, L.: Subdifferential monotonicity as characterization of convex functions. Numer. Funct. Anal. Optim. **15**(5–6), 531–535 (1994)
11. Correa, R., Jofre, A., Thibault, L.: Subdifferential characterization of convexity. In: Du, D.Z., Qi L., Womersley R.S. (eds.). Recent Advances in Nonsmooth Optimization, pp. 18–23. World Scientific, Singapore (1995)
12. Fabian, M.: Subdifferentiability and trustworthiness in the light of the new variational principle of Borwein and Preiss. Acta Univ. Carol. Math. Phys. **30**, 51–56 (1989)
13. Ioffe, A.: On subdifferentiability spaces. Ann. N.Y. Acad. Sci. **410**, 107–119 (1983)
14. Ioffe, A.D.: Subdifferentiability spaces and nonsmooth analysis. Bull. Am. Math. Soc. **10**, 87–89 (1984)
15. Ioffe, A.D.: On the theory of subdifferential. Fermat Days 85: Mathematics for optimization (Toulouse, 1985), North-Holland Math. Stud., vol. 129, pp. 183–200. North-Holland, Amsterdam (1986)
16. Ioffe, A.D.: Calculus of Dini subdifferentials of functions and contingent coderivatives of set-valued maps. Nonlinear Anal. **8**, 517–539 (1984)
17. Ioffe, A.: Approximate subdifferentials and applications, I. Trans. Am. Math. Soc. **281**, 389–416 (1984)
18. Ioffe, A.D.: Approximate subdifferentials and applications, II. Mathematika **33**, 111–128 (1986)
19. Ioffe, A.D.: On the local surjection property. Nonlinear Anal. **11**, 565–592 (1987)
20. Ioffe, A.: Approximate subdifferentials and applications, III: The metric theory. Mathematika **36**, 1–38 (1989)
21. Ioffe, A.D.: Proximal analysis and approximate subdifferentials. J. London Math. Soc. **41**(2), 175–192 (1990)
22. Ioffe, A.: Separable reduction theorem for approximate subdifferentials. C. R. Acad. Sci. Paris **323**(Série I), 107–112 (1996)

23. Ioffe, A.: Fuzzy principles and characterization of trustworthiness. Set-Valued Anal. **6**, 265–276 (1998)
24. Ioffe, A.D.: Codirectional compactness, metric regularity and subdifferential calculus. In: Thera M. (ed.). Constructive, Experimental and Nonlinear Analysis. CMS Conference proceeding, vol. 27, pp. 123–165 (2000)
25. Ioffe, A.D.: Metric regularity and subdifferential calculus. Uspehi Mat. Nauk **55**(3), 103–162 (2000) (in Russian). English translation: Russian Math. Surveys, **55**(3), 501–558 (2000)
26. Ioffe, A.D.: On the theory of subdifferentials. Adv. Nonlinear Anal. **1**, 47–120 (2012)
27. Ioffe, A.D., Penot, J.-P.: Subdifferentials of performance functions and calculus of coderivatives of set-valued mappings. Serdica Math. J. **22**, 359–384 (1996)
28. Jules, F.: Sur la somme de sous-différentiels de fonctions continues inférieurement. Dissertationes Mathematicae 423, Polska Akad. Nauk, Warsaw (2003)
29. Lassonde, M.: First-order rules for nonsmooth constrained optimization. Nonlinear Anal. Ser. A Theory Methods **44**(8), 1031–1056 (2001)
30. Mordukhovich, B.S.: Metric approximations and necessary conditions for optimality for general classes of nonsmooth optimization problems. Dokl. Acad. Nauk SSSR **254**, 1072–1076 (1980)
31. Mordukhovich, B.S.: Variational Analysis and Generalized Differentiation. Springer, Berlin (2005)
32. Penot, J.-P.: Miscellaneous incidences of convergence theories in optimization and nonsmooth analysis, II: applications to nonsmooth analysis. In: Du, D.Z., Qi L., Womersley, R.S., (eds.). Recent Advances in Nonsmooth Optimization, pp. 289–321. World Scientific Publishers, Singapore (1995)
33. Penot, J.-P.: What is quasiconvex analysis? Optimization **47**, 35–110 (2000)
34. Penot, J.-P.: The compatibility with order of some subdifferentials. Positivity **6**, 413–432 (2002)
35. Penot, J.-P.: Semigroups of relations and subdifferentials of the minimal time function and of the geodesic distance. SIAM J. Control Optim. **51**(4), 2839–2868 (2013)
36. Penot, J.-P.: Calculus Without Derivatives. Graduate Texts in Mathematics. 266 Springer, New York (2013)
37. Rockafellar, R.T., Wets, R.J.-B.: Variational Analysis. Springer, Berlin (1998)
38. Thibault, L.: A note on the Zagrodny mean value theorem. Optimization **35**, 127–130 (1995)
39. Zhu, Q.J.: The equivalence of several basic theorems Please cite it for subdifferentials. Set-Valued Anal. **6**, 171–185 (1998)

Chapter 30
Modular Equations and Lattice Sums

Mathew Rogers and Boonrod Yuttanan

Abstract We highlight modular equations due to Ramanujan and Somos and use them to prove new relations between lattice sums and hypergeometric functions. We also discuss progress towards solving Boyd's Mahler measure conjectures. Finally, we conjecture a new formula for $L(E,2)$ of conductor 17 elliptic curves.

Key words: Hypergeometric series • Lattice sums • Mahler measure • Modular equations.

Mathematics Subject Classifications (2010): Primary 11R06, 33C20; Secondary 11F03, 33C75, 33E05.

30.1 Introduction

Modular equations appear in a variety of number-theoretic contexts. Their connection to formulas for $1/\pi$ [15], Ramanujan constants such as $e^{\pi\sqrt{163}}$ [22], and elliptic curve cryptography is well established. In the classical theory of modular forms, an

---

COMMUNICATED BY FRANK GARVAN.

M. Rogers (✉)
Department of Mathematics and Statistics, Université de Montréal, CP 6128 succ. Centre-ville, Montréal QC, Canada, H3C 3J7
e-mail: mathewrogers@gmail.com

B. Yuttanan
Department of Mathematics and Statistics, Prince of Songkla University, Songkhla 90112, Thailand,
e-mail: boonrod_y@hotmail.com

D.H. Bailey et al. (eds.), *Computational and Analytical Mathematics*, Springer Proceedings in Mathematics & Statistics 50, DOI 10.1007/978-1-4614-7621-4_30,

nth-degree modular equation is an algebraic relation between $j(\tau)$ and $j(n\tau)$, where $j(\tau)$ is the j-invariant. For our purposes a modular equation is simply a nontrivial algebraic relation between theta or eta functions. In this paper we use modular equations to study four-dimensional lattice sums. The lattice sums are interesting because they arise in the study of Mahler measures of elliptic curves.

There are many hypothetical relations between special values of L-series of elliptic curves and Mahler measures of two-variable polynomials. The Mahler measures $m(\alpha)$, $n(\alpha)$, and $g(\alpha)$ are defined by

$$m(\alpha) := \int_0^1 \int_0^1 \log\left|y+y^{-1}+z+z^{-1}+\alpha\right| \mathrm{d}\theta_1 \mathrm{d}\theta_2, \tag{30.1}$$

$$n(\alpha) := \int_0^1 \int_0^1 \log\left|y^3+z^3+1-\alpha yz\right| \mathrm{d}\theta_1 \mathrm{d}\theta_2, \tag{30.2}$$

$$g(\alpha) := \int_0^1 \int_0^1 \log\left|(y+1)(z+1)(y+z)-\alpha yz\right| \mathrm{d}\theta_1 \mathrm{d}\theta_2, \tag{30.3}$$

where $y=e^{2\pi i\theta_1}$ and $z=e^{2\pi i\theta_2}$. Boyd conjectured that for all integral values of $k\neq 4$ [6]:

$$m(k) \stackrel{?}{=} \frac{q}{\pi^2} L(E,2),$$

where E is an elliptic curve, q is rational, and both E and q depend on k. He also discovered many formulas involving $g(\alpha)$ and $n(\alpha)$. In cases where E has a small conductor, it is frequently possible to express $L(E,2)$ in terms of four-dimensional lattice sums. Thus many of Boyd's identities can be regarded as series acceleration formulas. The main goal of this paper is to prove new formulas for the lattice sum $F(b,c)$, defined in (30.13). There are at least 18 instances where $F(b,c)$ is known (or conjectured) to reduce to integrals of elementary functions. The modular equations of Ramanujan and Somos are the main tools in our analysis.

30.2 Eta Function Product Identities

Somos discovered thousands of new modular equations by searching for linear relations between products of Dedekind eta functions. Somos refers to these formulas as *eta function product identities*. The existence of eta function product identities follows from the fact that $j(\tau)$ equals a rational expression involving eta functions. One can transform classical modular equations into eta function product identities by simply clearing denominators. Somos's experimental approach turned up many surprisingly simple identities. In order to give an example, first consider the eta function with respect to q:

$$\eta(q) = q^{1/24}\prod_{n=1}^{\infty}(1-q^n) = \sum_{n=-\infty}^{\infty}(-1)^n q^{(6n+1)^2/24},$$

and adopt the shorthand notation

$$e_j = \eta(q^j).$$

Formula (30.4) is the smallest eta function product identity in Somos's list [20]:

$$e_2e_6e_{10}e_{30} = e_1e_{12}e_{15}e_{20} + e_3e_4e_5e_{60}. \tag{30.4}$$

Notice that all three monomials are products of four eta functions and are essentially weight-two modular forms. No identities are known if the eta products have weight less than two, and (30.4) appears to be the only three-term linear relation between products of four eta functions. Many additional identities are known if the number of terms is allowed to increase or if eta products of higher weight are considered. For additional examples see formulas (30.16), (30.24), (30.25), (30.28), and (30.31) below.

Identities such as (30.4) can be proved almost effortlessly with the theory of modular forms. A typical proof involves checking that the first few Fourier coefficients of an identity vanish. Sturm's Theorem furnishes an upper bound on the number of coefficients that need to be examined [14]. We note that it is often possible, but usually more difficult, to prove such identities via q-series methods. Ramanujan derived hundreds of modular equations with q-series techniques. We conclude this section by proving (30.4). This proof fills in a gap in the literature of Ramanujan–Berndt-style proofs.

Theorem 30.1. *The identity* (30.4) *is true.*

Proof. Let us denote the usual theta functions by

$$\varphi(q) := \sum_{n=-\infty}^{\infty} q^{n^2}, \qquad \psi(q) := \sum_{n=0}^{\infty} q^{n(n+1)/2}. \tag{30.5}$$

Furthermore define u_j and z_j by

$$u_j := 1 - \frac{\varphi^4(-q^j)}{\varphi^4(q^j)}, \qquad z_j := \varphi^2(q^j).$$

Ramanujan uses a slightly different notation [2]. He typically sets $\alpha = u_1$ and says that "β has degree j over α" when $\beta = u_j$. Certain values of the eta function can be expressed in terms of u_1 and z_1 [2, p. 124]. We have

$$\eta(q) = 2^{-1/6}u_1^{1/24}(1-u_1)^{1/6}\sqrt{z_1}, \tag{30.6}$$

$$\eta(q^2) = 2^{-1/3}\left\{u_1(1-u_1)\right\}^{1/12}\sqrt{z_1}, \tag{30.7}$$

$$\eta(q^4) = 2^{-2/3}u_1^{1/6}(1-u_1)^{1/24}\sqrt{z_1}. \tag{30.8}$$

Now we prove (30.4). By (30.7) the left-hand side of the identity becomes

$$e_2e_6e_{10}e_{30}=2^{-4/3}\{u_1u_3u_5u_{15}(1-u_1)(1-u_3)(1-u_5)(1-u_{15})\}^{1/12}\sqrt{z_1z_3z_5z_{15}}.$$

By (30.6) and (30.8), the right-hand side of the identity becomes

$$\begin{aligned}e_1e_{12}e_{15}e_{20}+e_3e_4e_5e_{60}\\ =2^{-5/3}\Big(&\{u_3u_5(1-u_1)(1-u_{15})\}^{1/6}\{u_1u_{15}(1-u_3)(1-u_5)\}^{1/24}\\ &+\{u_1u_{15}(1-u_3)(1-u_5)\}^{1/6}\{u_3u_5(1-u_1)(1-u_{15})\}^{1/24}\Big)\sqrt{z_1z_3z_5z_{15}}.\end{aligned}$$

Combining the last two formulas shows that (30.4) is equivalent to

$$\begin{aligned}&2^{1/3}\{u_1u_3u_5u_{15}(1-u_1)(1-u_3)(1-u_5)(1-u_{15})\}^{1/24}\\ &\quad=\{u_3u_5(1-u_1)(1-u_{15})\}^{1/8}+\{u_1u_{15}(1-u_3)(1-u_5)\}^{1/8}.\end{aligned}\tag{30.9}$$

It is sufficient to show that (30.9) can be deduced from Ramanujan's modular equations.

The first modular equation we require can be recovered by multiplying entries 11.1 and 11.2 in [2, p. 383]:

$$\left((u_1u_{15})^{1/8}+\{(1-u_1)(1-u_{15})\}^{1/8}\right)\left((u_3u_5)^{1/8}+\{(1-u_3)(1-u_5)\}^{1/8}\right)=1.$$

Rearranging yields an identity for the right-hand side of (30.9):

$$\begin{aligned}&\{u_3u_5(1-u_1)(1-u_{15})\}^{1/8}+\{u_1u_{15}(1-u_3)(1-u_5)\}^{1/8}\\ &\quad=1-\{u_1u_3u_5u_{15}\}^{1/8}-\{(1-u_1)(1-u_3)(1-u_5)(1-u_{15})\}^{1/8}.\end{aligned}\tag{30.10}$$

By [2, p. 385, Entry 11.14], it is clear that

$$\begin{aligned}&1-\{u_1u_3u_5u_{15}\}^{1/8}-\{(1-u_1)(1-u_3)(1-u_5)(1-u_{15})\}^{1/8}\\ &\quad=2^{1/3}\{u_1u_3u_5u_{15}(1-u_1)(1-u_3)(1-u_5)(1-u_{15})\}^{1/24}.\end{aligned}\tag{30.11}$$

The theorem follows from combining (30.10) and (30.11) to recover (30.9). ■

We find it slightly surprising that Ramanujan overlooked (30.4). He possessed a tremendous ability to derive modular equations, and he discovered all of the necessary intermediate results. Perhaps it is simply not obvious *why* identities such as (30.4) exist. We are unable to offer much insight, beyond pointing out that there are many additional formulas in Somos's tables. A conceptual proof of (30.4) might lead to a systematic method for generating more identities. In the next section, we demonstrate that this is an important topic in the study of lattice sums.

30.3 Lattice Sums

In this section we investigate four-dimensional lattice sums. Many of these sums are related to L-functions of elliptic curves. Let us define

$$F(a,b,c,d) := (a+b+c+d)^2 \times \sum_{n_i=-\infty}^{\infty} \frac{(-1)^{n_1+n_2+n_3+n_4}}{(a(6n_1+1)^2+b(6n_2+1)^2+c(6n_3+1)^2+d(6n_4+1)^2)^2}.$$

The four-dimensional series is not absolutely convergent, so it is necessary to employ summation by cubes [5]. Notice that Euler's pentagonal number theorem can be used to represent $F(a,b,c,d)$ as an integral

$$F(a,b,c,d) = -\frac{(a+b+c+d)^2}{24^2}\int_0^1 \eta(q^a)\eta(q^b)\eta(q^c)\eta(q^d)\log q\frac{\mathrm{d}q}{q}. \tag{30.12}$$

We also use the shorthand notation

$$F(b,c) := F(1,b,c,bc), \tag{30.13}$$

since we are primarily interested in cases where $a=1$, $d=bc$, and b and c are rational.

The interplay between values of $F(b,c)$, Boyd's Mahler measure conjectures, and the Beilinson conjectures is outlined in [17]. If $(b,c)\in\mathbb{N}^2$ and $(1+b)(1+c)$ divides 24, then $F(b,c)=L(E,2)$ for an elliptic curve E. Formulas are now proved relating each of those eight L-values to Mahler measures [23]. Mahler measures often reduce to generalized hypergeometric functions, so many of Boyd's identities can be regarded as series transformations [13, 16]. It is known that

$$\begin{aligned}
m(\alpha) &= \operatorname{Re}\left[\log(\alpha) - \frac{2}{\alpha^2}{}_4F_3\left({\tfrac{3}{2},\tfrac{3}{2},1,1 \atop 2,2,2};\frac{16}{\alpha^2}\right)\right], \text{if } \alpha\neq 0,\\
n(\alpha) &= \operatorname{Re}\left[\log(\alpha) - \frac{2}{\alpha^3}{}_4F_3\left({\tfrac{4}{3},\tfrac{5}{3},1,1 \atop 2,2,2};\frac{27}{\alpha^3}\right)\right], \text{ if } |\alpha| \text{ is sufficiently large,}\\
3g(\alpha) &= n\left(\frac{\alpha+4}{\alpha^{2/3}}\right) + 4n\left(\frac{\alpha-2}{\alpha^{1/3}}\right), \text{ if } |\alpha| \text{ is sufficiently large.}
\end{aligned}$$

The function $m(\alpha)$ also reduces to a ${}_3F_2$ function if $\alpha\in\mathbb{R}$ [12, 17]. Rogers and Zudilin [18] recently proved that

$$F(3,5) = \frac{4\pi^2}{15}m(1) = \frac{\pi^2}{15}{}_3F_2\left({\tfrac{1}{2},\tfrac{1}{2},\tfrac{1}{2} \atop 1,\tfrac{3}{2}}\middle|\frac{1}{16}\right). \tag{30.14}$$

Equation (30.14) is equivalent to a formula that Deninger conjectured [9]. The same formula helped motivate Boyd's seminal paper [6]. It is also possible to prove formulas for values such as $F(1,4)$ and $F(2,2)$ [17]. These lattice sums are not related to elliptic curve L-values in an obvious way, so it was conjectured that it should be possible to "sum up" $F(b,c)$ for arbitrary values of b and c.

30.3.1 Lacunary Cases

In general, the difficulty of dealing with a lattice sum depends on whether it is *lacunary* or *non-lacunary*. Lacunary examples are usually much easier to work with. We say that a lattice sum is lacunary if it equals the Mellin transform of a lacunary modular form. Modular forms are called lacunary whenever their Fourier series coefficients have zero arithmetic density. To detect lacunary eta products, first expand the eta product in a series

$$\eta(q^a)\eta(q^b)\eta(q^c)\eta(q^d) = q^{(a+b+c+d)/24}\left(a_0 + a_1 q + a_2 q^2 + \cdots\right), \tag{30.15}$$

and then check that $a_n = 0$ for almost all n. It seems to be an open problem to classify quadruples (a,b,c,d) which make (30.15) lacunary. While cusp forms associated with CM elliptic curves are always lacunary, only three of those cusp forms actually equal products of four eta functions [14]. Less is known if an eta product is not obviously related to an elliptic curve. Empirically, it appears that many values of $e_a e_b e_c e_d$ can be expressed as linear combinations of two-dimensional theta functions. Expansions such as

$$\eta^2(q)\eta^2(q^2) = \sum_{\substack{k=-\infty \\ n\geq 0}}^{\infty} (-1)^{k+n}(2n+1)q^{\frac{(2k)^2+(2n+1)^2}{4}}$$

imply an eta product is lacunary, because subsequences of integers generated by quadratic forms have zero density. Unfortunately, it is not clear if every lacunary value of $e_a e_b e_c e_d$ possesses such an expansion. There is also insufficient evidence to conjecture how often $e_a e_b e_c e_d$ is lacunary. This stems from the fact that it is difficult to detect the property numerically. It requires thousands of q-series coefficients to convincingly demonstrate that (easy) cases like e_1^4 are lacunary. The calculations become much worse for more complicated examples.

The lattice sums $F(1,1)$, $F(1,2)$, and $F(1,3)$ equal L-values of CM elliptic curves. Therefore they are lacunary. These examples, and additional values such as $F(1,4)$ and $F(2,2)$, reduce to two-dimensional sums via classical theta series results. Less obvious lacunary sums include $F(2,9)$ and $F(4,7,7,28)$. These cases require eta function product identities. A result of Ramanujan [3, p. 210, Entry 56], shows that

$$3e_1e_2e_9e_{18} = -e_1^2e_2^2 + e_1^3\frac{e_{18}^2}{e_9} + e_2^3\frac{e_9^2}{e_{18}}. \tag{30.16}$$

Substituting classical theta expansions for e_1^3, e_2^2/e_1, and e_1^2/e_2 [11, pp. 114–117] leads to

$$\begin{aligned}3\eta(q)\eta(q^2)\eta(q^9)\eta(q^{18}) =& -\sum_{\substack{n=0\\k=0}}^{\infty}(-1)^n(2n+1)q^{\frac{(2n+1)^2+(2k+1)^2}{8}}\\ &+\sum_{\substack{n=0\\k=0}}^{\infty}(-1)^n(2n+1)q^{\frac{(2n+1)^2+9(2k+1)^2}{8}}\\ &+\sum_{\substack{n=0\\k=-\infty}}^{\infty}(-1)^{n+k}(2n+1)q^{\frac{(2n+1)^2+9(2k)^2}{4}}.\end{aligned} \tag{30.17}$$

The eta product equals a finite linear combination of two-dimensional theta functions. Therefore it is lacunary. Formula (30.17) is the main ingredient needed to relate $F(2,9)$ to hypergeometric functions and Mahler measures.

Theorem 30.2. *Let $t=\sqrt[4]{12}$, then the following identity is true:*

$$\begin{aligned}\frac{144}{25\pi^2}F(2,9) =& -3m(4i)+2m\left(\frac{1}{\sqrt{2}}\left(4-2t-2t^2+t^3\right)\right)\\ &+m\left(4i\left(7+4t+2t^2+t^3\right)\right).\end{aligned} \tag{30.18}$$

Proof. The most difficult portion of the calculation is to find a two-dimensional theta series for $e_1e_2e_9e_{18}$. This task has been accomplished via an eta function product identity. The remaining calculations parallel those carried out in [17]. Integrating (30.17) leads to

$$\begin{aligned}\frac{3}{25}F(2,9)+F(1,2) =& 4\sum_{\substack{n=0\\k=0}}^{\infty}\frac{(-1)^n(2n+1)}{((2n+1)^2+9(2k+1)^2)^2}\\ &+\sum_{\substack{n=0\\k=-\infty}}^{\infty}\frac{(-1)^{n+k}(2n+1)}{((2n+1)^2+9(2k)^2)^2}.\end{aligned} \tag{30.19}$$

There are two possible formulas for $F(1,2)$ [16]:

$$F(1,2)=\frac{\pi^2}{8}m\left(2\sqrt{2}\right)=\frac{\pi^2}{16}m(4i). \tag{30.20}$$

By the formula for $F_{(1,2)}(3)$ in [17, Eq. 115], we also have

$$\sum_{\substack{n=0\\k=-\infty}}^{\infty}\frac{(-1)^{n+k}(2n+1)}{((2n+1)^2+9(2k)^2)^2}=\frac{\pi^2}{48}m\left(4i\left(7+4t+2t^2+t^3\right)\right), \tag{30.21}$$

where $t = \sqrt[4]{12}$. Next we evaluate the remaining term in (30.19). Notice that for $x > 0$

$$\sum_{\substack{n=0\\k=0}}^{\infty} \frac{(-1)^n(2n+1)}{((2n+1)^2+x(2k+1)^2)^2}$$
$$= \frac{\pi^2}{16}\int_0^{\infty} u \left(\sum_{n=0}^{\infty}(-1)^n(2n+1)e^{-\pi(n+1/2)^2 u}\right)\left(\sum_{k=0}^{\infty} e^{-\pi x(k+1/2)^2 u}\right) du.$$

By the involution for the weight-$3/2$ theta function

$$\sum_{n=0}^{\infty}(-1)^n(2n+1)e^{-\pi(n+1/2)^2 u} = \frac{1}{u^{3/2}}\sum_{n=0}^{\infty}(-1)^n(2n+1)e^{-\pi(n+1/2)^2\frac{1}{u}},$$

this becomes

$$\sum_{\substack{n=0\\k=0}}^{\infty} \frac{(-1)^n(2n+1)}{((2n+1)^2+x(2k+1)^2)^2}$$
$$= \frac{\pi^2}{16}\sum_{\substack{n=0\\k=0}}^{\infty}(-1)^n(2n+1)\int_0^{\infty} u^{-1/2}e^{-\pi\left((n+1/2)^2\frac{1}{u}+x(k+1/2)^2 u\right)} du$$
$$= \frac{\pi^2}{16\sqrt{x}}\sum_{\substack{n=0\\k=0}}^{\infty}(-1)^n\frac{(2n+1)}{(2k+1)}e^{-\frac{\pi\sqrt{x}}{2}(2n+1)(2k+1)}$$
$$= \frac{\pi^2}{16\sqrt{x}}\sum_{n=0}^{\infty}(-1)^n(2n+1)\log\left(\frac{1+e^{-\pi\sqrt{x}(n+1/2)}}{1-e^{-\pi\sqrt{x}(n+1/2)}}\right).$$

Applying formulas (1.6), (1.7), and (2.9) in [13], we have

$$= \frac{\pi^2}{32\sqrt{x}}\left(m\left(\frac{4}{\sqrt{\alpha_{x/4}}}\right) - m\left(\frac{4i\sqrt{1-\alpha_{x/4}}}{\sqrt{\alpha_{x/4}}}\right)\right)$$
$$= \frac{\pi^2}{32\sqrt{x}} m\left(4\left(\frac{1-\sqrt{1-\alpha_{x/4}}}{1+\sqrt{1-\alpha_{x/4}}}\right)\right),$$

where α_x is the singular modulus (recall that $\alpha_x = 1-\varphi^4(-e^{-\pi\sqrt{x}})/\varphi^4(e^{-\pi\sqrt{x}})$). The second-degree modular equation shows that

$$\frac{1-\sqrt{1-\alpha_{x/4}}}{1+\sqrt{1-\alpha_{x/4}}} = \sqrt{\alpha_x},$$

and hence we obtain

$$\sum_{\substack{n=0\\k=0}}^{\infty} \frac{(-1)^n(2n+1)}{((2n+1)^2+x(2k+1)^2)^2} = \frac{\pi^2}{32\sqrt{x}} m\left(4\sqrt{\alpha_x}\right). \tag{30.22}$$

It is well known that α_n can be expressed in terms of class invariants if $n \in \mathbb{Z}$:

$$\alpha_n = \frac{1}{2}\left(1-\sqrt{1-1/G_n^{24}}\right).$$

The values of G_n have been extensively tabulated [4, p. 188]. Setting $n = 9$ yields

$$\begin{aligned}\alpha_9 &= \frac{1}{2}\left(1-\sqrt{1-\left(\frac{\sqrt{2}}{\sqrt{3}+1}\right)^8}\right)\\ &= \frac{1}{2}\left(1-4t+t^3\right)\\ &= \frac{\left(4-2t-2t^2+t^3\right)^2}{32},\end{aligned}$$

where $t = \sqrt[4]{12}$. Formula (30.22) reduces to

$$\sum_{\substack{n=0\\k=0}}^{\infty} \frac{(-1)^n(2n+1)}{((2n+1)^2+9(2k+1)^2)^2} = \frac{\pi^2}{96} m\left(\frac{1}{\sqrt{2}}\left(4-2t-2t^2+t^3\right)\right). \tag{30.23}$$

The proof of (30.18) follows from combining (30.19), (30.20), (30.21), and (30.23). ∎

We have chosen to exclude the explicit formula for $F(4,7,7,28)$ from this paper.[1] It suffices to say that the sum reduces to an extremely unpleasant expression involving hypergeometric functions and Meijer G-functions. The key modular equation is due to Somos [21, Entry $q_{28,9,35}$]:

$$28e_4e_7^2e_{28} = -7e_1e_7^3 - \frac{e_1^5}{e_2^2}\frac{e_{14}^2}{e_7} + 8\frac{e_2^5}{e_1^2}e_{14}. \tag{30.24}$$

By classical theta expansions [11, p. 114–117], the eta product becomes

[1] The formula is available upon request.

$$28\eta(q^4)\eta^2(q^7)\eta(q^{28}) = -7\sum_{\substack{n=-\infty\\k=0}}^{\infty}(-1)^{n+k}(2k+1)q^{\frac{(6n+1)^2+21(2k+1)^2}{24}} - \sum_{\substack{n=-\infty\\k=0}}^{\infty}(6n+1)q^{\frac{(6n+1)^2+21(2k+1)^2}{24}} + 8\sum_{n,k=-\infty}^{\infty}(-1)^{n+k}(3n+1)q^{\frac{4(3n+1)^2+7(6k+1)^2}{12}}.$$

As a result it is easy to see that $e_4e_7^2e_{28}$ is lacunary.

We believe that there are additional lacunary values of $F(a,b,c,d)$. It might be interesting to try to detect them numerically. Another possible extension of this research involves looking at linear combinations of lattice sums. One can prove that certain linear combinations of lattice sums reduce to Mahler measures. As an example, briefly consider the following modular equation [21, Entry $x_{50,6,81}$]:

$$5e_1e_2e_{25}e_{50} + 2e_1^2e_2e_{50} + 2e_1e_2^2e_{25} = -e_1^2e_2^2 + e_1^3\frac{e_{50}^2}{e_{25}} + e_2^3\frac{e_{25}^2}{e_{50}}. \tag{30.25}$$

All three eta quotients on the right-hand side of (30.25) have two-dimensional theta series expansions. As a result we can prove that

$$\frac{5}{13^2}F(2,25) + \frac{2}{9^2}F(1,1,2,50) + \frac{2}{5^2}F(1,2,2,25) = \frac{\pi^2}{80}\left(-5m(4i) + 2m(4\sqrt{\alpha_{25}}) + m\left(4i\sqrt{\frac{1-\alpha_{25}}{\alpha_{25}}}\right)\right), \tag{30.26}$$

where $\alpha_{25} = \frac{1}{2^{13}}\left(\sqrt{5}-1\right)^8\left(\sqrt[4]{5}-1\right)^8$. There are many additional results like (30.26), which we will not discuss here.

30.3.2 Non-lacunary Cases

The calculations become far more difficult when $F(a,b,c,d)$ *does not* reduce to a two-dimensional sum. The recent proofs of formulas for $F(1,5)$, $F(2,3)$, and $F(3,5)$ are all based upon new types of q-integral transformations [18, 19]. The fundamental transformation for $F(2,3)$ is

$$\int_0^1 q^{1/2}\psi(q)\psi(q^3)\varphi(-q^x)\varphi(-q^{3x})\log q\frac{dq}{q} = \frac{2\pi}{3x}\operatorname{Im}\int_0^1 \omega q\psi^4\left(\omega^2q^2\right)\log\left(4q^{3x}\frac{\psi^4(q^{12x})}{\psi^4(q^{6x})}\right)\frac{dq}{q},$$

where $\omega = e^{2\pi i/3}$. When $x = 1$ the left-hand side equals $-4F(2,3)$ (to see this use $q^{1/8}\psi(q) = \eta^2(q^2)/\eta(q)$ and $\varphi(-q) = \eta^2(q)/\eta(q^2)$), and the right-hand side becomes an extremely complicated elementary integral. The most difficult portion of the calculation is to reduce the elementary integral to hypergeometric functions,

$$F(2,3) = \frac{\pi^2}{6}m(2) = \frac{\pi^2}{12}{}_3F_2\left({\frac{1}{2},\frac{1}{2},\frac{1}{2} \atop 1,\frac{3}{2}};\frac{1}{4}\right).$$

Boyd's numerical work was instrumental in the calculation, because it allowed the final formula to be anticipated in advance.

Non-lacunary lattice sums reduce to intractable integrals quite frequently. We recently used the method from [18] to find a formula for $F(1,8)$:

$$F(1,8) = \frac{9\pi\sqrt[4]{2}}{128}\int_0^1 \frac{(1-k)^2+2\sqrt{2(k+k^3)}}{(1+k)(k+k^3)^{3/4}}\log\left(\frac{1+2k-k^2+2\sqrt{k-k^3}}{1+k^2}\right)\mathrm{d}k. \tag{30.27}$$

The proof of (30.27) is long and complicated, so we verified this monstrous identity to 100 decimal places by calculating $F(1,8)$ with (30.12). We speculate that the integral should reduce to something along the lines of (30.18).

Eta function identities occasionally provide shortcuts for avoiding integrals like (30.27). We have already demonstrated that linear dependencies exist between lattice sums [see (30.26)]. In certain cases it is possible to relate new lattice sums to well-known examples. Consider a forty-fifth-degree modular equation due to Somos [21, Entry $x_{45,4,12}$]:

$$6e_1e_5e_9e_{45} = -e_1^2e_5^2 - 2e_3^2e_{15}^2 - 9e_9^2e_{45}^2 + e_3^4 + 5e_{15}^4. \tag{30.28}$$

We were unable to prove (30.28) by elementary methods. Integrating (30.28) leads to a linear dependency between three lattice sums. We have

$$9F(5,9) = 45F(1,1) - 50F(1,5). \tag{30.29}$$

Both $F(1,1)$ and $F(1,5)$ equal values of hypergeometric functions [16, 18], so we easily obtain the following theorem.

Theorem 30.3. *Recall that $n(\alpha)$ is defined in* (30.2). *We have*

$$\frac{108}{5\pi^2}F(5,9) = 8n\left(3\sqrt[3]{2}\right) - 9n\left(2\sqrt[3]{4}\right). \tag{30.30}$$

Boyd's Mahler measure conjectures imply various additional formulas. A proof of Boyd's conductor 30 conjectures would lead to closed forms for both $F(2,15)$ and $F(2,5/3)$. To make this explicit we use two relations. First consider a four-term modular equation due to Somos [20]:

$$e_1e_3e_5e_{15} + 2e_2e_6e_{10}e_{30} = e_1e_2e_{15}e_{30} + e_3e_5e_6e_{10}. \tag{30.31}$$

Integrating (30.31), and then using the evaluation $F(3,5) = 4\pi^2 m(1)/15$ from [19], leads to

$$F(2,15) + 4F\left(2, \frac{5}{3}\right) = \frac{8\pi^2}{5} m(1). \tag{30.32}$$

Next we require an unproven relation. Boyd conjectured[2] that for a conductor 30 elliptic curve

$$L(E_{30}, 2) \stackrel{?}{=} \frac{2\pi^2}{15} g(3),$$

where $g(\alpha)$ is defined in (30.3). The modularity theorem guarantees that $L(E_{30}, 2) = L(f_{30}, 2)$, where $f_{30}(e^{2\pi i \tau})$ is a weight-two cusp form on $\Gamma_0(30)$. Somos has calculated a basis for the space of cusp forms on $\Gamma_0(30)$. It follows that the cusp form associated with conductor 30 elliptic curves is

$$f_{30}(q) = \eta(q^3)\eta(q^5)\eta(q^6)\eta(q^{10}) - \eta(q)\eta(q^2)\eta(q^{15})\eta(q^{30}).$$

Upon integrating $f_{30}(q)$, Boyd's conjecture becomes

$$F\left(2, \frac{5}{3}\right) - \frac{1}{4} F(2,15) \stackrel{?}{=} \frac{2\pi^2}{15} g(3). \tag{30.33}$$

Combining (30.32) and (30.33) leads to a pair of conjectural evaluations.

Conjecture 30.4. Recall that $m(\alpha)$ and $g(\alpha)$ are defined in (30.1) and (30.3). The following equivalent formulas are numerically true:

$$\frac{15}{4\pi^2} F(2,15) \stackrel{?}{=} 3m(1) - g(3), \tag{30.34}$$

$$\frac{15}{\pi^2} F\left(2, \frac{5}{3}\right) \stackrel{?}{=} 3m(1) + g(3). \tag{30.35}$$

Tracking backwards shows that a proof of either (30.34) or (30.35) would settle Boyd's conductor 30 Mahler measure conjectures. Eisenstein series identities due to Berkovich and Yesilyurt could be of use here [1].

[2]See Table 2 in [6]. In our notation, Boyd's entries correspond to values of $g(2-k)$.

30.4 Conclusion: Conductor 17 Elliptic Curves

An important connection exists between lattice sums and Mahler measures; however this relationship has limitations. Even if we could "sum up" $F(b,c)$ for arbitrary values of b and c, this would only settle a few of Boyd's conjectures [6]. Conductor 17 curves are the first cases in Cremona's list [8], where $L(E,2)$ probably does not reduce to values of $F(b,c)$. If we let E_{17} denote a conductor 17 curve (we used $y^2+xy+y=x^3-x^2-x$), then

$$\frac{17}{2\pi^2}L(E_{17},2) \stackrel{?}{=} m\left(\frac{(1+\sqrt{17})^2}{4}\right) - m\left(\sqrt{17}\right). \tag{30.36}$$

We discovered (30.36) via numerical experiments involving elliptic dilogarithms.[3] The cusp form associated with conductor 17 curves is stated in [10]. We have

$$f_{17}(q) = \frac{\eta(q)\eta^2(q^4)\eta^5(q^{34})}{\eta(q^2)\eta(q^{17})\eta^2(q^{68})} - \frac{\eta^5(q^2)\eta(q^{17})\eta^2(q^{68})}{\eta(q)\eta^2(q^4)\eta(q^{34})}. \tag{30.37}$$

Since $L(E_{17},2) = L(f_{17},2)$, formula (30.36) can be changed into a complicated elementary identity. There does not seem to be an easy way to relate $L(E_{17},2)$ to Mahler measures of rational polynomials. This probably explains why conductor 17 curves never appear in Boyd's paper [6]. After examining $f_{17}(q)$ in detail, we feel reasonably confident that $L(E_{17},2)$ is linearly independent from values of $F(b,c)$ over $\mathbb{Q}$.

Acknowledgments The authors thank David Boyd, Wadim Zudilin, and Bruce Berndt for their useful comments and encouragement. The authors are especially grateful to Michael Somos for the useful communications and for providing his list of modular equations. Mat Rogers thanks the Max Planck Institute of Mathematics for their hospitality. The authors thank the referee for many helpful suggestions which improved the exposition of the paper.

References

1. Berkovich, A., Yesilyurt, H.; Ramanujan's identities and representation of integers by certain binary and quaternary quadratic forms. Ramanujan J. **20**, 375–408 (2009)
2. Berndt, B.C.: Ramanujan's Notebooks, Part III. Springer-Verlag, New York (1991)
3. Berndt, B.C.: Ramanujan's Notebooks, Part IV. Springer-Verlag, New York (1994)
4. Berndt, B.C.: Ramanujan's Notebooks, Part V. Springer-Verlag, New York (1998)
5. Borwein, D., Borwein, J.M., Taylor, K.F.: Convergence of lattice sums and Madelung's constant. J. Math. Phys. **26**(11), 2999–3009 (1985)

[3] Brunault recently informed us that he can prove (30.36) with a method based upon Beilinson's theorem [7].

6. Boyd, D.W.: Mahler's measure and special values of L-functions. Experiment. Math. **7**, 37–82 (1998)
7. Brunault, F.: Version explicite du théorème de Beilinson pour la courbe modulaire $X_1(N)$. C. R. Math. Acad. Sci. Paris **343**(8), 505–510 (2006)
8. Cremona, J.E.: Algorithms for modular elliptic curves. Available at: http://www.warwick.ac.uk/~masgaj/ftp/data/
9. Deninger, C.: Deligne periods of mixed motives, K-theory and the entropy of certain $\mathbb{Z}^n$-actions. J. Am. Math. Soc. **10**(2), 259–281 (1997)
10. Finch, S.: Primitive cusp forms. Preprint (2009)
11. Köhler, G.: Eta products and theta series identities. Springer, Heidelberg (2011)
12. Kurokawa, N., Ochiai, H.: Mahler measures via crystalization. Comment. Math. Univ. St. Pauli **54**, 121–137 (2005)
13. Lalín, M.N., Rogers, M.D.: Functional equations for Mahler measures of genus-one curves. Algebra Number Theory **1**(1), 87–117 (2007)
14. Ono, K.: The Web of Modularity: Arithmetic of the Coefficients of Modular Forms and q-series. American Mathematical Society, Providence (2004)
15. Ramanujan, S.: Modular equations and approximations to π. Quart. J. Math. **45**, 350–372 (1914). Collected papers of Srinivasa Ramanujan, 23–29, AMS Chelsea Publ., Providence, RI (2000)
16. Rodriguez-Villegas, F.: Modular Mahler measures I. In: Topics in number theory (University Park, PA, 1997), Math. Appl. vol. 467, pp. 17–48. Kluwer Acad. Publ., Dordrecht (1999)
17. Rogers, M.: Hypergeometric formulas for lattice sums and Mahler measures. Intern. Math. Res. Not. **17**, 4027–4058 (2011)
18. Rogers, M., Zudilin, W.: From L-series of elliptic curves to Mahler measures. Compositio Math. **148**(2), 385–414 (2012)
19. Rogers, M., Zudilin, W.: On the mahler measure of $1+X+X^{-1}+Y+Y^{-1}$. Preprint, arXiv: 1102.1153 [math.NT] (2011)
20. Somos, M.: A Remarkable eta-product Identity. Preprint (2008)
21. Somos, M.: Dedekind eta function product identities. Available at: http://eta.math.georgetown.edu/
22. Weisstein, E.W.: Ramanujan Constant. From MathWorld–A Wolfram Web Resource. http://mathworld.wolfram.com/RamanujanConstant.html
23. Zudilin, W.: Arithmetic hypergeometric series. Russian Math. Surveys **66**(2) 369–420 (2011)

Chapter 31
An Epigraph-Based Approach to Sensitivity Analysis in Set-Valued Optimization

Douglas E. Ward and Stephen E. Wright

Abstract In this paper, we obtain estimates for the contingent and adjacent derivatives of the epigraph of the marginal multifunction in parametric set-valued optimization. These estimates generalize some sensitivity results from scalar-valued optimization and provide new information in the setting of multiobjective nonlinear programming.

Key words: Adjacent derivative • Basic normal cone • Contingent derivative • Sensitivity analysis • Set-valued optimization • Tangent cone.

Mathematics Subject Classifications (2010): Primary 49J53; Secondary 90C31.

31.1 Introduction

Let W, X, and Y be real normed spaces, and let $P \subset Y$ be a nonempty cone; i.e., $\lambda p \in P$ for each scalar $\lambda \geq 0$ and each $p \in P$. Let $H : W \rightrightarrows X$ and $G : W \times X \rightrightarrows Y$ be set-valued mappings. We consider the parametrized family of set-valued optimization problems defined by

$$Q(w) := \mathrm{Min}_P F(w), \tag{31.1}$$

COMMUNICATED BY HEINZ H. BAUSCHKE.

D.E. Ward (✉)
Department of Mathematics, Miami University, 204 Bachelor Hall, Oxford, OH 45056, USA
e-mail: wardde@MiamiOH.edu

S.E. Wright
Department of Statistics, Miami University, Oxford, OH 45056, USA
e-mail: wrightse@MiamiOH.edu

D.H. Bailey et al. (eds.), *Computational and Analytical Mathematics*, Springer Proceedings in Mathematics & Statistics 50, DOI 10.1007/978-1-4614-7621-4_31,

where

$$F(w) := \{y \in G(w,x) \mid x \in H(w)\} \tag{31.2}$$

and

$$\mathrm{Min}_P F(w) := \{y \in F(w) \mid (\{y\} - P) \cap F(w) \subset \{y\} + P\}.$$

Here we interpret W as the parameter space, X as the decision space, and Y as the objective space, with the cone P defining an ordering on objective values in Y. For each value of the parameter w, $H(w)$ defines a feasible set in X, while $G(w,x)$ gives a set of objective values in Y. The set $F(w)$ comprises the graph of the objective values over the entire feasible region of the optimization problem, whereas $Q(w)$ selects those objective values that cannot be improved relative to the P-ordering. (The elements of $Q(w)$ are known as "efficient points" of $F(w)$.) Note that this model generalizes parametric vector optimization—the case where G is single-valued—and parametric scalar optimization, where in addition $Y = \mathbb{R}$ and $P = \mathbb{R}_+ := \{y \in \mathbb{R} \mid y \geq 0\}$.

The theory of set-valued optimization is a subject that is attracting increasing interest (see [13] and its references). In particular, there is a growing body of work on sensitivity analysis in parametric multiobjective optimization, where generalized derivative concepts are employed to estimate how the sets $F(w)$ and $Q(w)$ vary with changes in w. The generalized derivatives used for this purpose include both "primal" constructions based on tangent cones [1, 2, 5, 11, 15–19, 21, 25, 26, 28–31] and "dual" constructions based on normal cones [4, 12, 20].

One major branch of this sensitivity analysis literature, beginning with the seminal papers of Tanino [28, 29], makes extensive use of the contingent derivative [1, 2], which is defined in terms of the contingent cone. In these studies [11, 15–18, 25, 26, 28, 29], epigraphs of set-valued maps play an important role.

Definition 31.1. Let $M : X \rightrightarrows Y$ be a set-valued mapping.

(a) The *graph* of M is the set $\mathrm{gph}\, M := \{(x,y) \mid y \in M(x)\}$.
(b) For a cone $P \subset Y$, the epigraph mapping (or profile map) $(M+P) : X \rightrightarrows Y$ is defined by $(M+P)(x) = M(x) + P$. The *epigraph* of M is the set $\mathrm{epi}\, M = \mathrm{gph}(M+P)$.

These studies include detailed investigation of the relationship between the contingent derivative of $M+P$ and the epigraph of the contingent derivative of M, two quantities that coincide for scalar-valued functions but may differ more generally. Also considered are relationships among the contingent derivatives of $F, F+P, Q, Q+P$ for various concepts of minimization with respect to P, with the goal of estimating the contingent derivatives of F and Q. These discussions make significant progress in clarifying relationships that are easily verified for scalar-valued functions but can be much more problematic in a multiobjective setting.

Notably absent from this analysis, however, are results relating generalized derivatives of $F+P$ and $G+P$. We would argue that such results merit a prominent position in this theory, based on what is known about sensitivity analysis in scalar-valued optimization. In the scalar-valued case, one key type of sensitivity estimate gives bounds on directional derivatives of the *marginal function* Q in terms of directional derivatives of G and local approximations to the graph of H (see [3,6,8–10,21,32,34,35] and their references). Since directional derivatives of scalar-valued functions can be defined via tangent cones to epigraphs, these estimates relate generalized derivatives of $Q+P$ and $G+P$ (and generalized derivatives of $F+P$ and $G+P$, since $Q+P$ and $F+P$ coincide in the scalar-valued case). One advantage of these estimates is that they are valid under rather mild hypotheses on the problem data—in particular, milder hypotheses than those required to guarantee that tangent cone approximations to the *graphs* of Q and F are well behaved.

In the present paper, we apply the techniques developed in [34] to produce estimates for the contingent and adjacent derivatives of $F+P$ in terms of those for $G+P$ and H. Our estimates, which generalize some known results for contingent derivatives of F, give sensitivity information for a number of problems in which these earlier results are not applicable (see Example 31.27).

To set the stage for these developments, we review relevant aspects of tangent and normal cone calculus in Sect. 31.2. We present our main theorems in Sect. 31.3, giving inclusions for contingent and adjacent derivatives of $F+P$ that generalize inequalities from [34]. Then in Sect. 31.4, we focus on the case where G is single-valued and strictly differentiable and derive an extension of [28, Theorem 4.1].

We conclude this section by setting the basic terminology and notation. Throughout, $P \subset Y$ denotes a general cone—any further hypotheses on P are given in the statements of results requiring them. The cone P is said to be *pointed* if $P \cap -P = \{0\}$. The *recession cone* $\hat{P} := \cap_{y \in P}(P-y)$ is a convex cone for which $\hat{P} \subset P = P + \hat{P}$. Note that $P = \hat{P}$ if and only if the cone P is convex.

For $\varepsilon > 0$ and $x \in X$, we define open and closed balls, respectively, as

$$B(x,\varepsilon) := \{z \in X \mid \|z-x\| < \varepsilon\} \quad \text{and} \quad \bar{B}(x,\varepsilon) := \{z \in X \mid \|z-x\| \leq \varepsilon\}.$$

We say that $S \subset X$ is *locally closed* around $\bar{x} \in S$ if there exists $\varepsilon > 0$ such that $S \cap \bar{B}(\bar{x},\varepsilon)$ is closed. By $u \to_S x$, we mean $u \to x$ with $u \in S$. We denote the closure of S by $\operatorname{cl} S$ and the interior of S by $\operatorname{int} S$. The *cone generated by* S is the set $\operatorname{cone} S := \{ts \mid t \geq 0,\ s \in S\}$.

31.2 Tangent Cones and Intersection Theorems

In this paper we work with generalized derivatives of multifunctions defined via tangent cones to their graphs. Intuitively, we can think of a tangent cone R as a multifunction that assigns, to each set S and point $x \in S$, a cone $R(S,x)$ giving a local approximation to S near x. For our purposes, two tangent cones are particularly useful:

Definition 31.2. Let S be a subset of X and $x \in S$.

(a) The contingent cone to S at x is defined by

$$T(S,x) := \left\{ z \in X \;\middle|\; \exists \{(t_j, z^j)\} \to (0^+, z) \text{ such that } x + t_j z^j \in S \right\}.$$

(b) The adjacent cone to S at x is defined by

$$A(S,x) := \left\{ z \in X \;\middle|\; \forall \{t_j\} \to 0^+, \exists \{z^j\} \to z \text{ such that } x + t_j z^j \in S \right\}.$$

Basic properties of the contingent and adjacent cones are listed in [2, Chap. 4]. In particular, both are always closed cones containing the origin, and the inclusion $A(S,x) \subset T(S,x)$ is always satisfied. One important class of sets for which the contingent and adjacent cones coincide is the class of convex sets. Specifically, if S is convex and $\bar{x} \in S$, then

$$T(S,\bar{x}) = A(S,\bar{x}) = \operatorname{cl}\operatorname{cone}(S - \bar{x}).$$

Given a concept of tangent cone, we define an associated generalized derivative as follows:

Definition 31.3. Let $M : X \rightrightarrows Y$ be a set-valued mapping, and let R be a tangent cone. For $\bar{y} \in M(\bar{x})$, define

$$R(M,(\bar{x},\bar{y})) := R(\operatorname{gph} M,(\bar{x},\bar{y})).$$

The *R-derivative* of M at $(\bar{x},\bar{y})$ is the mapping $\mathbf{D}^R M : X \rightrightarrows Y$ defined by

$$\mathbf{D}^R M(\bar{x},\bar{y})(x) := \{ y \,|\, (x,y) \in R(M,(\bar{x},\bar{y})) \}.$$

In particular, the T-derivative and A-derivative are known, respectively, as the contingent and adjacent derivatives.

To give an idea of how the generalized derivatives in Definition 31.3 "generalize the derivative," we mention an important special case. Let $f : X \to Y$ and $\bar{x} \in X$ be such that the Hadamard directional derivative

$$f'(\bar{x};x) := \lim_{t \downarrow 0, v \to x} \frac{f(\bar{x} + tv) - f(\bar{x})}{t}$$

exists. Then it is easy to show that

$$\mathbf{D}^T f(\bar{x}, f(\bar{x}))(x) = \mathbf{D}^A f(\bar{x}, f(\bar{x}))(x) = \{ f'(\bar{x};x) \}.$$

In particular, when f is Fréchet differentiable at $\bar{x}$ with derivative $\nabla f(\bar{x})$, we have

$$\mathbf{D}^T f(\bar{x}, f(\bar{x}))(x) = \mathbf{D}^A f(\bar{x}, f(\bar{x}))(x) = \{\nabla f(\bar{x})x\}, \quad \forall x \in X. \tag{31.3}$$

Properties of tangent cones can be used to develop the calculus of generalized derivatives [2, 22–24, 33]. Especially important for us is the question of when an intersection of tangent cones of sets is contained in a tangent cone of the intersection of those sets. One type of assumption under which "intersection theorems" can be established involves the Clarke tangent cone and interior Clarke tangent cone.

Definition 31.4. Let $S \subset X$ and $x \in S$.

(a) The *Clarke tangent cone* to S at x is defined by

$$C(S,x) := \{z \in X \mid \forall x^j \to_S x, \forall t_j \to 0^+, \exists \{z^j\} \to z \text{ with } x^j + t_j z^j \in S\}.$$

(b) The *interior Clarke tangent cone* to S at x is defined by

$$\mathrm{IC}(S,x) := \{z \in X \mid \exists \varepsilon > 0 \text{ with } (B(x,\varepsilon) \cap S) + (0,\varepsilon)B(z,\varepsilon) \subset S\}.$$

Remark 31.5

(a) Sets S such that $\mathrm{IC}(S,x) \neq \emptyset$ are said to be *epi-Lipschitzian* at x (see [7, 22–24] for further discussion).
(b) It follows easily from Definition 31.4 that both the Clarke tangent cone and interior Clarke tangent cone preserve Cartesian products of sets. Specifically, if $S_1 \subset X_1$, $S_2 \subset X_2$, and $(x_1,x_2) \in S_1 \times S_2$, then

$$C(S_1 \times S_2, (x_1,x_2)) = C(S_1,x_1) \times C(S_2,x_2) \tag{31.4}$$

and

$$\mathrm{IC}(S_1 \times S_2, (x_1,x_2)) = \mathrm{IC}(S_1,x_1) \times \mathrm{IC}(S_2,x_2). \tag{31.5}$$

With hypotheses involving the Clarke and interior Clarke tangent cones, the following intersection theorem can be obtained (see [33, Proposition 2.6]):

Proposition 31.6. *Let $S_i \subset X$, $i = 1,\ldots,n$, and consider $x \in \cap_{i=1}^n S_i$. If*

$$C(S_1,x) \cap \left[\bigcap_{i=2}^n \mathrm{IC}(S_i,x)\right] \neq \emptyset, \tag{31.6}$$

then

$$T(S_1,x) \cap \left[\bigcap_{i=2}^n A(S_i,x)\right] \subset T\left(\bigcap_{i=1}^n S_i, x\right) \tag{31.7}$$

and

$$\bigcap_{i=1}^{n} A(S_i,x) = A\left(\bigcap_{i=1}^{n} S_i,x\right). \tag{31.8}$$

When X is finite-dimensional and each S_i is (locally) closed, (31.7) and (31.8) are satisfied under assumptions involving the basic normal cone that are less demanding than (31.6). We state this finite-dimensional intersection theorem after reviewing the definition of the basic normal cone [20].

Definition 31.7. Let X be finite-dimensional and S a nonempty subset of X. Let $< \cdot,\cdot >$ denote the Euclidean inner product on X.

(a) For $x \in S$ and $\varepsilon \geq 0$, the set of ε-*normals* to S at x is defined by

$$\hat{N}_\varepsilon(S,x) := \left\{ z \in X \;\middle|\; \limsup_{u \to_S x} \frac{< z, u-x >}{\|u-x\|} \leq \varepsilon \right\}.$$

(b) The *basic normal cone* to S at $\bar{x} \in S$ is defined by

$$N(S,\bar{x}) := \{ z \in X \mid \exists \{z^j\} \to z, \exists \{\varepsilon_j\} \to 0^+, \exists \{x^j\} \to_S \bar{x} \text{ with } z^j \in \hat{N}_{\varepsilon_j}(S,x^j) \}.$$

(c) For set-valued mappings $M : X \rightrightarrows Y$ with $\bar{y} \in M(\bar{x})$, we set

$$N(M,(\bar{x},\bar{y})) := N(\operatorname{gph} M,(\bar{x},\bar{y})).$$

The normal cone, like the Clarke and interior Clarke tangent cones, preserves Cartesian products [20, Proposition 1.2]. For $S_1 \subset X_1$, $S_2 \subset X_2$, and $(x_1,x_2) \in S_1 \times S_2$, we have

$$N(S_1 \times S_2,(x_1,x_2)) = N(S_1,x_1) \times N(S_2,x_2). \tag{31.9}$$

Moreover, the following intersection theorem is valid:

Theorem 31.8 ([20, 34]). *Let X be finite-dimensional and $S_i \subset X$, $i = 1,\dots,n$ be locally closed around $x \in \cap_{i=1}^n S_i$. Suppose that*

$$\sum_{i=1}^{n} z_i = 0,\; z_i \in N(S_i,x) \quad \textit{implies} \quad z_1 = z_2 = \cdots = z_n = 0. \tag{31.10}$$

Then (31.7) and (31.8) are satisfied, along with

$$N\left(\bigcap_{i=1}^{n} S_i,x\right) \subset \sum_{i=1}^{n} N(S_i,x). \tag{31.11}$$

Remark 31.9. In the finite-dimensional setting of Theorem 31.8, condition (31.6) implies, but is not implied by, condition (31.10). Condition (31.10) is also a less demanding condition than

$$\forall v_1,\dots,v_n,\quad \bigcap_{i=1}^{n}(C(S_i,x)-v_i)\neq\emptyset,$$

the hypothesis under which (31.7) and (31.8) are derived in Corollaries 4.3.5 and 4.3.6 of [2].

For example, let $X=\mathbb{R}^2$, $S_1:=\{(x_1,x_2)\,|\,x_2=|x_1|\}$, and $S_2:=\{0\}\times\mathbb{R}$. Then $N(S_1,(0,0))=S_1\cup\{(x_1,x_2)\,|\,x_2\leq -|x_1|\}$ and $N(S_2,(0,0))=\mathbb{R}\times\{0\}$, so that (31.10) holds. However, $\mathrm{IC}(S_1,(0,0))=\mathrm{IC}(S_2,(0,0))=\emptyset$, implying that both $C(S_1,(0,0))\cap\mathrm{IC}(S_2,(0,0))$ and $\mathrm{IC}(S_1,(0,0))\cap C(S_2,(0,0))$ are empty. Moreover, $C(S_1,(0,0))=\{(0,0)\}$ and $C(S_2,(0,0))=S_2$, so that

$$(C(S_1,(0,0))-(1,0))\cap(C(S_2,(0,0))-(0,0))=\emptyset.$$

Since (31.10) is less stringent than these other conditions, we will we able to deduce stronger results with the help of Theorem 31.8 than are obtainable via the theory developed in [2].

31.3 Generalized Derivatives of the Epigraph of the Objective Multifunction

In this section, we apply Proposition 31.6 and Theorem 31.8 to derive inclusions relating the contingent and adjacent derivatives of the epigraphs of G and F. These inclusions are valid in some fairly general circumstances, in particular the case when G is single-valued and locally Lipschitzian. We also identify some situations where the inclusions are satisfied as equations.

Theorem 31.10. *In (31.2), let $\bar{x}\in H(\bar{w})$ and $\bar{y}\in G(\bar{w},\bar{x})$.*

(a) If

$$\mathrm{IC}(G+P,(\bar{w},\bar{x},\bar{y}))\cap(C(H,(\bar{w},\bar{x}))\times Y)\neq\emptyset, \tag{31.12}$$

then for all $w\in W$

$$\bigcup_{\{x\,|\,(w,x)\in T(H,(\bar{w},\bar{x}))\}} \boldsymbol{D}^A(G+P)(\bar{w},\bar{x},\bar{y})(w,x)\subset \boldsymbol{D}^T(F+P)(\bar{w},\bar{y})(w) \tag{31.13}$$

and

$$\bigcup_{\{x \,|\, (w,x) \in A(H,(\bar{w},\bar{x}))\}} \boldsymbol{D}^A(G+P)(\bar{w},\bar{x},\bar{y})(w,x) \subset \boldsymbol{D}^A(F+P)(\bar{w},\bar{y})(w). \tag{31.14}$$

Similarly, if

$$C(G+P,(\bar{w},\bar{x},\bar{y})) \cap (\mathrm{IC}(H,(\bar{w},\bar{x})) \times Y) \neq \emptyset, \tag{31.15}$$

then (31.14) holds for all $w \in W$, as does

$$\bigcup_{\{x \,|\, (w,x) \in A(H,(\bar{w},\bar{x}))\}} \boldsymbol{D}^T(G+P)(\bar{w},\bar{x},\bar{y})(w,x) \subset \boldsymbol{D}^T(F+P)(\bar{w},\bar{y})(w). \tag{31.16}$$

(b) Suppose that W, X, Y are finite-dimensional, gph$H$ *is locally closed near $(\bar{w},\bar{x})$, and* epi$G$ *is locally closed near $(\bar{w},\bar{x},\bar{y})$. If*

$$-N(H,(\bar{w},\bar{x})) \cap \{(w^*,x^*) \,|\, (w^*,x^*,0) \in N(G+P,(\bar{w},\bar{x},\bar{y}))\} = \{(0,0)\}, \tag{31.17}$$

then (31.13), (31.14), and (31.16) are satisfied.

Proof. Suppose that (31.12) holds. To prove (31.13), let

$$y \in \mathbf{D}^A(G+P)(\bar{w},\bar{x},\bar{y})(w,x) \quad \text{with} \quad (w,x) \in T(H,(\bar{w},\bar{x})).$$

Then

$$(w,x,y) \in A(\operatorname{epi} G,(\bar{w},\bar{x},\bar{y})) \cap T(\operatorname{gph} H \times Y,(\bar{w},\bar{x},\bar{y})).$$

Since $C(Y,\bar{y}) = Y$, (31.4) and (31.12) guarantee that (31.6) holds with $n = 2$, $S_1 :=$ gph$H \times Y$, and $S_2 :=$ epi$G$. Applying Proposition 31.6, we obtain

$$(w,x,y) \in T(\operatorname{epi} G \cap (\operatorname{gph} H \times Y),(\bar{w},\bar{x},\bar{y})).$$

By the definition of the contingent cone, there then exist sequences $\{t_j\} \to 0^+$ and $\{(w^j,x^j,y^j)\} \to (w,x,y)$ such that

$$(\bar{w},\bar{x},\bar{y}) + t_j(w^j,x^j,y^j) \in \operatorname{epi} G \cap (\operatorname{gph} H \times Y).$$

In other words,

$$(\bar{w}+t_jw^j,\bar{x}+t_jx^j) \in \operatorname{gph} H$$

and

$$(\bar{w}+t_jw^j,\bar{x}+t_jx^j,\bar{y}+t_jy^j) \in \operatorname{epi} G,$$

which implies that

$$(\bar{w}+t_j w^j, \bar{y}+t_j y^j) \in \operatorname{epi} F.$$

Hence $y \in \mathbf{D}^T(F+P)(\bar{w},\bar{y})(w)$, establishing (31.13). The proof of (31.14) under assumption (31.12) is analogous to this one, as are the proofs of (31.14) and (31.16) under assumption (31.15).

Now suppose that W, X, Y are finite-dimensional, $\operatorname{gph} H$ is locally closed near $(\bar{w},\bar{x})$, $\operatorname{epi} G$ is locally closed near $(\bar{w},\bar{x},\bar{y})$, and (31.17) holds. By (31.9) and the fact that $N(Y,\bar{y}) = \{0\}$, (31.17) implies that

$$-N(\operatorname{gph} H \times Y, (\bar{w},\bar{x},\bar{y})) \cap N(G+P, (\bar{w},\bar{x},\bar{y})) = \{0\}.$$

We can then prove (31.13), (31.14), and (31.16) exactly as above, applying Theorem 31.8 instead of Proposition 31.6. ■

Example 31.11. The hypotheses of Theorem 31.10 are satisfied for a wide variety of problems, including many situations where the data are non-Lipschitzian. For example, suppose $X = Y = W = \mathbb{R}$, $P = \mathbb{R}_+$, and let $G(w,x) := \{|x|^{1/2}\}$, $H(w) := [w, +\infty)$, and $\bar{w} = \bar{x} = \bar{y} = 0$. In this example

$$N(H,(0,0)) = \{(w,x) \mid w \geq 0, x = -w\}$$

and $N(G+P,(0,0,0)) = \{0\} \times \mathbb{R} \times (-\infty, 0]$, so that (31.17) holds. Condition (31.15) is also satisfied, since

$$C(G+P,(0,0,0)) = \mathbb{R} \times \{0\} \times \mathbb{R}_+$$

and $\mathrm{IC}(H,(0,0)) = \{(w,x) \mid x > w\}$. By Theorem 31.10, (31.13), (31.14), and (31.16) hold. Indeed, one can verify that both the left-hand side and right-hand side of these inclusions give $\mathbb{R}_+$ when $w \leq 0$ and $\emptyset$ when $w > 0$.

In the scalar-valued case ($Y := \mathbb{R}$, $P := \mathbb{R}_+$), one situation in which (31.12), (31.15), and (31.17) are all satisfied is that in which G is a locally Lipschitzian function. If $G : W \times X \to \mathbb{R}$ is Lipschitzian near $(\bar{w},\bar{x})$, then

$$\{(w^*,x^*) \mid (w^*,x^*,0) \in N(G+P,(\bar{w},\bar{x},\bar{y}))\} = \{(0,0)\} \qquad (31.18)$$

(see [20, Corollary 1.81]), so that (31.17) holds. More generally, Bao and Mordukhovich [4, Proposition 1] have shown that (31.18) holds when $G : W \times X \rightrightarrows Y$ is an epi-Lipschitz-like set-valued mapping, a class of mappings that includes the graphs of locally Lipschitzian functions.

When $G : W \times X \to \mathbb{R}$ is Lipschitzian near $(\bar{w},\bar{x})$, it is also true that

$$\{(w,x) \mid \exists y \text{ such that } (w,x,y) \in \mathrm{IC}(\operatorname{epi} G, (\bar{w},\bar{x},G(\bar{w},\bar{x})))\} = W \times X, \qquad (31.19)$$

which implies (31.12). Equation (31.19) can be extended to vector-valued functions if the recession portion of the ordering cone P has nonempty interior or, equivalently, if P can be expressed as the sum of a general cone and a convex cone with nonempty interior. To demonstrate such an extension, we first need the following fact.

Lemma 31.12. *Let Y be a real normed space with unit ball $B := \bar{B}(0,1)$. Consider a cone $P \subset Y$ with nonempty interior and a bounded set $S \subset Y$. Then there exists a point $y \in Y$ and a scalar $\varepsilon > 0$ such that $B(y,\varepsilon) \subset z + P$ for every $z \in S$.*

Proof. By hypothesis, there exists a point $v \in Y$ along with scalars $r > 0$ and $\varepsilon > 0$ such that $v + \varepsilon B \subset P$ and $S \subset rB$. Define $u := (r/\varepsilon)v$, $y := u + v$, and $K := \text{cone}(v + \varepsilon B) \subset P$. Now consider any $z \in S$. First, we observe that

$$u - z = \frac{r}{\varepsilon}v - z \in \frac{r}{\varepsilon}v - S \subset \frac{r}{\varepsilon}(v + \varepsilon B) \subset \frac{r}{\varepsilon}K \subset K.$$

Hence we have $u \in z + K$. By the convexity (hence additivity) of the cone K, this yields

$$u + K \subset z + K \subset z + P.$$

In particular, we see that

$$B(y,\varepsilon) = y + \varepsilon B = u + v + \varepsilon B \subset u + K \subset z + P.$$

This holds for any $z \in S$, verifying our assertion. ■

Making use of Lemma 31.12, we can now derive an extension of (31.19).

Proposition 31.13. *Let $f : X \to Y$ be Lipschitzian near $\bar{x} \in X$. Suppose Y is ordered by a cone P for which $\hat{P}$ has nonempty interior. Then*

$$\{x \mid \exists y \textit{ such that } (x,y) \in \text{IC}(\text{epi}\, f, (\bar{x}, f(\bar{x})))\} = X.$$

Proof. Let $x \in X$. Since f is Lipschitzian near $\bar{x}$, there exist $L > 0$ and $\varepsilon > 0$ such that

$$\|f(u) - f(u')\| \le L\|u - u'\| \quad \forall u, u' \in B(\bar{x}, \varepsilon).$$

Let $\varepsilon_0 \in (0,\varepsilon)$ such that

$$B(\bar{x}, \varepsilon_0) + (0, \varepsilon_0)B(x, \varepsilon_0) \subset B(\bar{x}, \varepsilon).$$

Then for all $u \in B(\bar{x}, \varepsilon_0)$, $t \in (0, \varepsilon_0)$, and $x' \in B(x, \varepsilon_0)$,

$$\frac{\|f(u + tx') - f(u)\|}{t} \le L\|x'\| \le L(\|x\| + \varepsilon_0).$$

Since $\hat{P}$ has nonempty interior, by Lemma 31.12 there exist $y \in Y$ and $\varepsilon_1 \in (0,\varepsilon_0)$ such that for each $y' \in B(y,\varepsilon_1)$ and each z in the bounded set

$$S := \left\{ \frac{f(u+tx') - f(u)}{t} \,\middle|\, u \in B(\bar{x},\varepsilon_1), t \in (0,\varepsilon_1), x' \in B(x,\varepsilon_1) \right\},$$

we have $y' \in z + \hat{P}$.

We now show that $(x,y) \in \mathrm{IC}(\mathrm{epi}\, f, (\bar{x}, f(\bar{x})))$. Let $(u,w) \in B((\bar{x}, f(\bar{x})),\varepsilon_1) \cap \mathrm{epi}\, f$, $t \in (0,\varepsilon_1)$, $(x',y') \in B((x,y),\varepsilon_1)$. Then $w = f(u) + p_1$ for some $p_1 \in P$ and

$$y' = \frac{f(u+tx') - f(u)}{t} + p_2$$

for some $p_2 \in \hat{P}$. It follows that

$$f(u+tx') = ty' + f(u) - tp_2 = ty' + w - tp_2 - p_1.$$

Because $P + \hat{P} = P$, we have $ty' + w \in f(u+tx') + P$, or $(u,w) + t(x',y') \in \mathrm{epi}\, f$. Hence $(x,y) \in \mathrm{IC}(\mathrm{epi}\, f, (\bar{x}, f(\bar{x})))$, as asserted. ∎

Proposition 31.13 shows that condition (31.12) in Theorem 31.10 is satisfied if $\mathrm{int}\hat{P} \neq \emptyset$ and G is single-valued and locally Lipschitzian near $(\bar{w},\bar{x})$. To see this, note that $(0,0) \in C(\mathrm{gph}\, H, (\bar{w},\bar{x}))$. By Proposition 31.13, there exists y with $(0,0,y) \in \mathrm{IC}(\mathrm{epi}\, G, (\bar{w},\bar{x},\bar{y}))$, so the intersection of sets in (31.12) must be nonempty. If W, X, and Y are finite-dimensional, this means that (31.17) also holds, since (31.12) implies (31.17) as mentioned in Remark 31.9.

In order to guarantee equality in the inclusions of Theorem 31.10, some additional condition must be satisfied. We next consider one such condition.

Definition 31.14 ([5]). Let $M : X \rightrightarrows Y$ with $\bar{y} \in M(\bar{x})$. M is said to be *directionally compact* at $(\bar{x},\bar{y})$ in the direction $x \in X$ if for all sequences $\{t_j\} \to 0^+$ and $\{x^j\} \to x$, every sequence $\{y^j\}$ with $\bar{y} + t_j y^j \in M(\bar{x} + t_j x^j)$ has a convergent subsequence.

Remark 31.15. Directional compactness holds, in particular, when M is single-valued and $M'(\bar{x};x)$ exists. In this case, M is directionally compact at $(\bar{x}, M(\bar{x}))$ in the direction x. To see this, suppose that $t_j \downarrow 0$, $x^j \to x$ and $M(\bar{x}) + t_j y^j \in M(\bar{x} + t_j x^j)$. Then

$$y^j = \frac{M(\bar{x} + t_j x^j) - M(\bar{x})}{t_j},$$

so that $y^j \to M'(\bar{x};x)$.

Proposition 31.16. *In (31.2), let $\bar{x} \in H(\bar{w})$ and $\bar{y} \in G(\bar{w},\bar{x})$. Suppose that the mapping $M : W \times Y \rightrightarrows X$ defined by*

$$M(u,v) := \{d \,|\, d \in H(u), v \in (G+P)(u,d)\} \tag{31.20}$$

is directionally compact at $(\bar{w},\bar{y},\bar{x})$ *in the direction* (w,y)*, and assume that* $y \in \mathbf{D}^T(F+P)(\bar{w},\bar{y})(w)$*. Then there exists* $x \in X$ *such that*

$$y \in \mathbf{D}^T(G+P)(\bar{w},\bar{x},\bar{y})(w,x) \quad \text{with} \quad (w,x) \in T(H,(\bar{w},\bar{x})).$$

Proof. By our assumption that $y \in \mathbf{D}^T(F+P)(\bar{w},\bar{y})(w)$, there exist sequences $\{(w^j,y^j)\} \to (w,y)$ and $\{t_j\} \to 0^+$ such that

$$\bar{y} + t_j y^j \in (F+P)(\bar{w}+t_j w^j).$$

It follows from (31.2) that there exists a sequence $\{d^j\}$ with

$$d^j \in H(\bar{w}+t_j w^j) \quad \text{and} \quad \bar{y}+t_j y^j \in (G+P)(\bar{w}+t_j w^j, d^j).$$

Define $x^j := (d^j - \bar{x})/t_j$. Then $d^j = \bar{x}+t_j x^j$ and

$$\bar{x}+t_j x^j \in M(\bar{w}+t_j w^j, \bar{y}+t_j y^j).$$

Since M is directionally compact at $(\bar{w},\bar{y},\bar{x})$ in the direction (w,y), we may assume, taking a subsequence if necessary, that $x^j \to x$ for some $x \in X$. Therefore

$$y \in \mathbf{D}^T(G+P)(\bar{w},\bar{x},\bar{y})(w,x) \quad \text{with} \quad (w,x) \in T(H,(\bar{w},\bar{x})),$$

as asserted. ■

The following example illustrates Theorem 31.10 and Proposition 31.16.

Example 31.17. Let $W = \mathbb{R}$, $X = Y = \mathbb{R}^2$, and $P = \mathbb{R}^2_+$. Define

$$H(w) := \{(x_1,x_2) \in {\mathbb{R}^2}_+ \,|\, x_1 x_2 = w\}$$

and $G(w,x_1,x_2) := \{(x_1,x_2)\}$. Let $\bar{w} = 1$, $\bar{x} = (1,1)$, and $\bar{y} = (1,1)$. Then

$$M(u,v_1,v_2) = \{(d_1,d_2) \,|\, d_1 d_2 = u, v_1 \geq d_1, v_2 \geq d_2\},$$

and one can verify that for any $(w,y_1,y_2) \in \mathbb{R}^3$ with $w \leq y_1 + y_2$, M is directionally compact at $(\bar{w},\bar{y},\bar{x})$ in the direction (w,y). To see this, suppose that sequences $t_j \to 0^+$, $w^j \to w$, $(y_1^j,y_2^j) \to (y_1,y_2)$, and (x_1^j,x_2^j) satisfy

$$(1,1)+t_j(x_1^j,x_2^j) \in M((1,1,1)+t_j(w^j,y_1^j,y_2^j)).$$

Then $x_1^j + x_2^j \to w$, $x_1^j \leq y_1^j$, and $x_2^j \leq y_2^j$, which together imply that $\{(x_1^j,x_2^j)\}$ is bounded and therefore has a convergent subsequence.

We note that G is Lipschitzian, so that (31.12) and (31.17) hold, and we may apply Theorem 31.10. Moreover, one can calculate that

$$T(H,(\bar{w},\bar{x})) = A(H,(\bar{w},\bar{x})) = \{(w,x_1,x_2) \mid x_1 + x_2 = w\}.$$

It follows from (31.16) and Proposition 31.16 that for all $w \in \mathbb{R}$

$$\bigcup_{\{x \mid (w,x) \in T(H,(\bar{w},\bar{x}))\}} \mathbf{D}^T(G+P)(\bar{w},\bar{x},\bar{y})(w,x) = \mathbf{D}^T(F+P)(\bar{w},\bar{y})(w).$$

Indeed, one can check that both sides of this equation reduce to the set

$$\{(w,y_1,y_2) \mid y_1 + y_2 \geq w\}.$$

Remark 31.15 mentions one instance in which directional compactness is satisfied. We next identify another class of mappings with this property.

Definition 31.18 (see [1,28]). A set-valued mapping $M : X \rightrightarrows Y$ is said to be upper locally Lipschitz at $\bar{x} \in X$ if there exist $\varepsilon > 0$ and $L > 0$ such that

$$M(x) \subset M(\bar{x}) + L\|x - \bar{x}\|\bar{B}(0,1) \quad \text{for all} \quad x \in B(\bar{x},\varepsilon).$$

Theorem 31.19. *In (31.2), suppose that X is finite-dimensional, and let $\bar{x} \in H(\bar{w})$ and $\bar{y} \in G(\bar{w},\bar{x})$. Suppose that the mapping $M : W \times Y \rightrightarrows X$ defined in (31.20) is upper locally Lipschitz at $(\bar{w},\bar{y})$ and $M(\bar{w},\bar{y}) = \{\bar{x}\}$. Then for all $w \in W$,*

$$\mathbf{D}^T(F+P)(\bar{w},\bar{x})(w) \subset \bigcup_{\{x \mid (w,x) \in T(H,(\bar{w},\bar{x}))\}} \mathbf{D}^T(G+P)(\bar{w},\bar{x},\bar{y})(w,x). \tag{31.21}$$

Proof. Let $w \in W$ and $y \in \mathbf{D}^T(F+P)(\bar{w},\bar{x})(w)$. By Proposition 31.16, it suffices to show that the mapping M defined in (31.20) is directionally compact at $(\bar{w},\bar{y},\bar{x})$ in the direction (w,y). To that end, suppose that $\{(w^j,y^j)\} \to (w,y)$, $\{t_j\} \to 0^+$, and $\{x^j\} \subset X$ satisfy

$$\bar{x} + t_j x^j \in M(\bar{w} + t_j w^j, \bar{y} + t_j y^j).$$

Since M is upper locally Lipschitz at $(\bar{w},\bar{y})$ and $M(\bar{w},\bar{y}) = \{\bar{x}\}$, there exists $L > 0$ such that for all j large enough

$$\|\bar{x} + t_j x^j - \bar{x}\| \leq L\|(\bar{w} + t_j w^j, \bar{y} + t_j y^j) - (\bar{w},\bar{y})\|,$$

so that

$$\|x^j\| \leq L\|(w^j,y^j)\|.$$

Hence $\{x^j\}$ is bounded. Since X is finite-dimensional, it follows that $\{x^j\}$ has a convergent subsequence, establishing the fact that M is directionally compact at $(\bar{w},\bar{y},\bar{x})$ in the direction (w,y). By Proposition 31.16, there exists $x \in X$ such that

$$y \in \mathbf{D}^T(G+P)(\bar{w},\bar{x},\bar{y})(w,x) \quad \text{with} \quad (w,x) \in T(H,(\bar{w},\bar{x})).$$

Therefore (31.21) holds. ■

In the scalar-valued case ($Y = \mathbb{R}, P = \mathbb{R}_+$) of problem (31.1), we have

$$F(w)+P = Q(w)+P \tag{31.22}$$

for all w such that $Q(w)$ is nonempty. More generally, Eq. (31.22) does not always hold (see [28, Example 3.1]). But when (31.22) does hold for all w in some neighborhood of $\bar{w}$, Theorem 31.10 gives information about contingent and adjacent derivatives of the epigraph of Q.

Theorem 31.20. *In (31.1) and (31.2), let $\bar{x} \in H(\bar{w})$ and $\bar{y} \in G(\bar{w},\bar{x})$ with $\bar{y} \in Q(\bar{w})$, and suppose that (31.22) holds for all w in some neighborhood of $\bar{w}$.*

(a) If (31.12) is satisfied, then for all $w \in W$,

$$\bigcup_{\{x\,|\,(w,x)\in T(H,(\bar{w},\bar{x}))\}} \mathbf{D}^A(G+P)(\bar{w},\bar{x},\bar{y})(w,x) \subset \mathbf{D}^T(Q+P)(\bar{w},\bar{y})(w) \tag{31.23}$$

and

$$\bigcup_{\{x\,|\,(w,x)\in A(H,(\bar{w},\bar{x}))\}} \mathbf{D}^A(G+P)(\bar{w},\bar{x},\bar{y})(w,x) \subset \mathbf{D}^A(Q+P)(\bar{w},\bar{y})(w). \tag{31.24}$$

Similarly, if (31.15) is satisfied, then (31.24) holds for all $w \in W$, as does

$$\bigcup_{\{x\,|\,(w,x)\in A(H,(\bar{w},\bar{x}))\}} \mathbf{D}^T(G+P)(\bar{w},\bar{x},\bar{y})(w,x) \subset \mathbf{D}^T(Q+P)(\bar{w},\bar{y})(w). \tag{31.25}$$

(b) Suppose that W, X, Y are finite-dimensional, gph$H$ *is locally closed near $(\bar{w},\bar{x})$, and* epi$G$ *is locally closed near $(\bar{w},\bar{x},\bar{y})$. If (31.17) holds, then (31.23)–(31.25) are satisfied.*

Proof. Since (31.22) holds for all w in a neighborhood of $\bar{w}$, we have

$$\mathbf{D}^R(Q+P)(\bar{w},\bar{y}) = \mathbf{D}^R(F+P)(\bar{w},\bar{y})$$

for $R := T, A$. The assertions then follow immediately from Theorem 31.10. ■

Remark 31.21. Equation (31.22) holds, in particular, when P is convex and

$$F(w) \subset Q(w) + P. \tag{31.26}$$

To see this, note that $Q(w) \subset F(w)$ by definition, so that by (31.26),

$$Q(w) + P \subset F(w) + P \subset Q(w) + P + P = Q(w) + P.$$

When (31.26) is true for all w in some neighborhood of $\bar{w}$, F is said to be *dominated* by Q near $\bar{w}$. This domination property is known to be satisfied for some large classes of problems [27]. We note in particular two cases mentioned in [14, Lemma 3.1]:

- If P is convex with compact base, $(F+P)(w)$ is closed and convex, and $Q(w) \neq \emptyset$, then (31.26) holds.
- The cone P is said to be *regular* if every sequence in Y that is P-decreasing and P-lower bounded converges to an element of P. (A *P-decreasing* sequence $\{y^j\}$ satisfies $y^m - y^n \in P$ whenever $m \leq n$. The sequence $\{y^j\}$ is *P-lower bounded* if there exists $\bar{y} \in Y$ with $y^m - \bar{y} \in P$ for all m.) If P is convex and regular and $F(w)$ is closed and P-lower bounded, then $Q(w) \neq \emptyset$ and (31.26) holds.

Remark 31.22. For $f : X \to \mathbb{R}$ and a tangent cone R, the *R-epiderivative* of f at $x \in X$ in the direction $y \in X$ is defined by

$$f^R(x;y) := \inf\{r \,|\, (y,r) \in R(\operatorname{epi} f, (x, f(x)))\}.$$

For $R := T, A$, this definition implies that

$$\operatorname{epi} f^R(x;\cdot) = R(\operatorname{epi} f, (x, f(x))).$$

In the case where $Y := \mathbb{R}$, $P := \mathbb{R}_+$, Theorem 31.20(b) essentially reduces to [34, Theorem 3.1], a result on epiderivatives of Q. In this scalar-valued setting, inclusion (31.23), for example, gives the inequality

$$Q^T(\bar{w};w) \leq \inf_x \{G^A((\bar{w},\bar{x});(w,x)) \,|\, (w,x) \in T(H,(\bar{w},\bar{x}))\}.$$

31.4 Applications to Multiobjective Nonlinear Programming

In this section we explore the implications of the results in Sect. 31.3 for the case in which G is single-valued and smooth. In this case contingent and adjacent cones to graphs and epigraphs often coincide, as was mentioned in Sect. 31.2. In particular, the following fact is useful in interpreting Theorems 31.10 and 31.20.

Lemma 31.23. *Suppose P is closed and $f : X \to Y$ is Hadamard directionally differentiable at $\bar{x}$ in the direction x; i.e., $f'(\bar{x};x)$ exists. Then*

$$\boldsymbol{D}^T(f+P)(\bar{x},f(\bar{x}))(x) \;=\; \boldsymbol{D}^A(f+P)(\bar{x},f(\bar{x}))(x) \;=\; f'(\bar{x};x)+P. \qquad (31.27)$$

Proof. As was shown in Remark 31.15, f is directionally compact at $(\bar{x},f(\bar{x}))$ in the direction x. It follows from [5, Proposition 5] that

$$\mathbf{D}^T(f+P)(\bar{x},f(\bar{x}))(x) = \mathbf{D}^T f(\bar{x},f(\bar{x}))(x)+P. \qquad (31.28)$$

We next observe that

$$\mathbf{D}^A f(\bar{x},f(\bar{x}))(x)+P \subset \mathbf{D}^A(f+P)(\bar{x},f(\bar{x}))(x). \qquad (31.29)$$

Indeed, suppose $y \in \mathbf{D}^A f(\bar{x},f(\bar{x}))(x)$ and $p \in P$. Let $\{t_j\} \to 0^+$. There exists $\{(x^j,y^j)\} \to (x,y)$ such that $f(\bar{x})+t_jy^j = f(\bar{x}+t_jx^j)$, which implies that

$$f(\bar{x})+t_j(y^j+p) \in (f+P)(\bar{x}+t_jx^j).$$

Therefore $y+p \in \mathbf{D}^A(f+P)(\bar{x},f(\bar{x}))(x)$, establishing (31.29). Combining (31.28) and (31.29), we conclude that

$$\begin{aligned}
\mathbf{D}^T(f+P)(\bar{x},f(\bar{x}))(x) &= \mathbf{D}^T f(\bar{x},f(\bar{x}))(x)+P \\
&= f'(\bar{x};x)+P \\
&= \mathbf{D}^A f(\bar{x},f(\bar{x}))(x)+P \\
&\subset \mathbf{D}^A(f+P)(\bar{x},f(\bar{x}))(x) \\
&\subset \mathbf{D}^T(f+P)(\bar{x},f(\bar{x}))(x).
\end{aligned}$$

Therefore (31.27) holds. ■

In the remainder of this section, we are primarily concerned with the case where $G: W\times X \to Y$ is *strictly differentiable* at $(\bar{w},\bar{x})$.

Definition 31.24 ([7]). The function $f: X \to Y$ is said to be *strictly differentiable* at $x \in X$ if there exists a linear mapping $\nabla f(x): X \to Y$ such that for all $y \in X$,

$$\nabla f(x)y = \lim_{(w,v,t)\to(x,y,0^+)} \frac{f(w+tv)-f(w)}{t}.$$

In finite dimensions, under the assumption that G is strictly differentiable, Theorem 31.20 and Proposition 31.16 yield the following result:

Theorem 31.25. *In (31.1), suppose that W, X, and Y are finite-dimensional and $G: W\times X \to Y$ is strictly differentiable at $(\bar{w},\bar{x})$, where $\bar{x} \in H(\bar{w})$ and $\bar{y} = G(\bar{w},\bar{x}) \in Q(\bar{w})$. Assume that P is closed and that $\hat{P}$ has nonempty interior. If (31.22) holds for all w in some neighborhood of $\bar{w}$, then*

$$\{\nabla_w G(\bar{w},\bar{x})w+\nabla_x G(\bar{w},\bar{x})x \,|\, (w,x)\in T(H,(\bar{w},\bar{x}))\}+P \subset \boldsymbol{D}^T(Q+P)(\bar{w},\bar{y})(w) \tag{31.30}$$

and

$$\{\nabla_w G(\bar{w},\bar{x})w+\nabla_x G(\bar{w},\bar{x})x \,|\, (w,x)\in A(H,(\bar{w},\bar{x}))\}+P \subset \boldsymbol{D}^A(Q+P)(\bar{w},\bar{y})(w). \tag{31.31}$$

If, in addition, the mapping M defined in (31.20) is directionally compact at $(\bar{w},\bar{y},\bar{x})$ in the direction (w,y) for each $y\in\boldsymbol{D}^T(Q+P)(\bar{w},\bar{y})(w)$, then

$$\{\nabla_w G(\bar{w},\bar{x})w+\nabla_x G(\bar{w},\bar{x})x \,|\, (w,x)\in T(H,(\bar{w},\bar{x}))\}+P = \boldsymbol{D}^T(Q+P)(\bar{w},\bar{y})(w). \tag{31.32}$$

Moreover, if (31.32) holds and P is pointed, then

$$Min_P\{\nabla_w G(\bar{w},\bar{x})w+\nabla_x G(\bar{w},\bar{x})x \,|\, (w,x)\in T(H,(\bar{w},\bar{x}))\} \subset \boldsymbol{D}^T Q(\bar{w},\bar{y})(w). \tag{31.33}$$

Proof. Since G is strictly differentiable at $(\bar{w},\bar{x})$, G is also Lipschitzian near $(\bar{w},\bar{x})$ [7, Proposition 2.2.1]. By Proposition 31.13, (31.12) is satisfied, and thus (31.17) holds as well. We can then apply Theorem 31.20. By Lemma 31.23,

$$\begin{aligned}\mathbf{D}^A(G+P)(\bar{w},\bar{x},\bar{y})(w,x) &= \mathbf{D}^T(G+P)(\bar{w},\bar{x},\bar{y})(w,x)\\ &= \nabla_w G(\bar{w},\bar{x})w+\nabla_x G(\bar{w},\bar{x})x+P,\end{aligned}$$

and so (31.23) and (31.24) imply (31.30) and (31.31). Equation (31.32) follows from (31.30) and Proposition 31.16. Finally, to obtain (31.33) from (31.32), note that

$$\begin{aligned}&\mathrm{Min}_P\{\nabla_w G(\bar{w},\bar{x})w+\nabla_x G(\bar{w},\bar{x})x \,|\, (w,x)\in T(H,(\bar{w},\bar{x}))\}+P\\ &= \mathrm{Min}_P\{\nabla_w G(\bar{w},\bar{x})w+\nabla_x G(\bar{w},\bar{x})x \,|\, (w,x)\in T(H,(\bar{w},\bar{x}))\}\end{aligned}$$

by [13, Lemma 4.7] and that

$$\mathrm{Min}_P\mathbf{D}^T(Q+P)(\bar{w},\bar{y})(w) \subset \mathbf{D}^T Q(\bar{w},\bar{y})(w)$$

by [28, Theorem 2.1]. ∎

It is instructive to compare Theorem 31.25 with a previous result from [28]:

Theorem 31.26. *([28, Theorem 4.1]) In (31.1) and (31.2), suppose that W, X, and Y are finite-dimensional, P is convex, closed, and pointed with nonempty interior,* *and* $G : W \times X \to Y$ *is continuously differentiable. Let* $\bar{x} \in H(\bar{w})$ *be such that* $G(\bar{w},\bar{x})$ *is a Benson properly minimal point of* $F(\bar{w})$*; i.e.,* $\mathrm{cl\,cone}(F(\bar{w}) - G(\bar{w},\bar{x})) \cap -P = \{0\}$*. In addition, assume that*

(a) H *is upper locally Lipschitz at* $\bar{w}$*.*
(b) $H(w)$ *is compact for each* w *in some neighborhood of* $\bar{w}$*.*
(c) The mapping $\tilde{M}(u,v) := \{d \in H(u) \mid v = G(u,d)\}$ *is upper locally Lipschitz at* $(\bar{w}, G(\bar{w},\bar{x}))$*.*
(d) $\tilde{M}(\bar{w}, G(\bar{w},\bar{x})) = \{\bar{x}\}$*.*

Then (31.33) holds.

In comparing Theorems 31.25 and 31.26, it is helpful to keep in mind that assumption (b) in Theorem 31.26 is made solely in order to guarantee that F is dominated by Q near $\bar{w}$. Assumption (b) may in fact be replaced by

($\hat{\text{b}}$) F is dominated by Q near $\bar{w}$.

It is also worth noting that since $G(\bar{w},\bar{x}) \in Q(\bar{w})$, $\tilde{M}(\bar{w}, G(\bar{w},\bar{x})) = M(\bar{w}, G(\bar{w},\bar{x}))$ for M defined in (31.20), so that assumption (d) in Theorem 31.26 may be replaced by

($\hat{\text{d}}$) $M(\bar{w}, G(\bar{w},\bar{x})) = \{\bar{x}\}$.

Remembering these observations along with Theorem 31.19, we see that Theorems 31.25 and 31.26 both give (31.33) under rather similar assumptions. However, Theorem 31.25 also yields (31.30) and (31.31), inclusions that hold under mild assumptions and reduce to inequalities for the contingent and adjacent epiderivatives of Q in the scalar-valued case (as noted in Remark 31.22). It is easy to find examples in which Theorem 31.26 is not applicable but for which Theorem 31.25 provides sensitivity information.

Example 31.27 ([9]). Let $W = X = \mathbb{R}^2$, $Y = \mathbb{R}$, and $P = \mathbb{R}_+$. Take $G(w_1,w_2,x_1,x_2) = -x_2$, and let

$$H(w_1,w_2) = \{(x_1,x_2) \mid {x_1}^2 + x_2 \le w_1,\ -{x_1}^2 + x_2 \le w_2\}$$

and $\bar{w} = (0,0)$. Here $Q(\bar{w}) = \{0\}$, P has nonempty interior, and (31.22) is satisfied, so (31.30) and (31.31) hold in this example. Since

$$H(w_1,w_2) = \{(x_1,x_2) \mid g_1(x_1,x_2) \le w_1,\ g_2(x_1,x_2) \le w_2\}$$

for $g_1(x_1,x_2) := {x_1}^2 + x_2$ and $g_2(x_1,x_2) := -{x_1}^2 + x_2$, and ${g_1}'((0,0);(x_1,x_2))$ and ${g_2}'((0,0);(x_1,x_2))$ exist, one can calculate (as in [34, Theorem 4.1]) that

$$T(H,(\bar{w},\bar{x})) = A(H,(\bar{w},\bar{x})) = \{(w_1,w_2,x_1,x_2) \mid x_2 \le w_1,\ x_2 \le w_2\}.$$

Inclusions (31.30) and (31.31) thus reduce to

$$[\max(-w_1,-w_2),+\infty) \subset \mathbf{D}^R(Q+P)(0,0)(w_1,w_2) \tag{31.34}$$

for $R := T, A$. One can readily verify that

$$Q(w_1,w_2) = \begin{cases} -w_1 & \text{if } w_2 \geq w_1, \\ -(w_1+w_2)/2 & \text{if } w_2 < w_1, \end{cases}$$

which means that

$$\mathbf{D}^R(Q+P)(0,0)(w_1,w_2) = \begin{cases} [-w_1,+\infty) & \text{if } w_2 \geq w_1, \\ [-(w_1+w_2)/2,+\infty) & \text{if } w_2 < w_1, \end{cases}$$

consistent with (31.34). Note, however, that (31.32) does not hold in this example. For the mapping M in (31.20), we have $M((0,0),0) = \{(0,0)\}$, so it must be that M is not upper locally Lipschitz at $((0,0),0)$. Indeed, it turns out that for all $w > 0$

$$\tilde{M}((w,w),0) = [-\sqrt{w},\sqrt{w}] \times \{0\} \subset M((w,w),0),$$

so that neither M nor $\tilde{M}$ is upper locally Lipschitz at $((0,0),0)$. This means that hypothesis (c) of Theorem 31.26 is not satisfied, and therefore Theorem 31.26 gives no information for this example. In fact, inclusion (31.33) is not always satisfied, since the left-hand side of (31.33) is equal to $\max(-w_1,-w_2)$, while

$$\mathbf{D}^T Q((0,0),0)(w_1,w_2) = \begin{cases} -w_1 & \text{if } w_2 \geq w_1, \\ -(w_1+w_2)/2 & \text{if } w_2 < w_1. \end{cases}$$

31.5 Conclusion

In this paper, we have established inclusions that relate the contingent and adjacent derivatives of the epigraph of the marginal multifunction Q to those of the epigraph of the objective mapping G. These inclusions, which are derived via the calculus of contingent and adjacent cones, are valid for a large class of nonsmooth optimization problems and give new information even in the case where G is single-valued.

Since our analysis is based on tangent cone intersection theorems that are special cases of more general intersection theorems for second-order tangent sets ([33, Proposition 2.6], [34, Theorem 2.4]), the methods of this paper can also be used to build a theory of second-order sensitivity analysis for parametric set-valued optimization. We hope to pursue this topic in future work.

Acknowledgments We are grateful for the helpful comments of a referee.

References

1. Aubin, J.-P., Ekeland, I.: Applied Nonlinear Analysis. Wiley, New York (1984)
2. Aubin, J.-P., Frankowska, H.: Set-valued analysis. Birkhäuser, Boston (1990)
3. Auslender, A.: Differentiable stability in nonconvex and nondifferentiable programming. Math. Program. Study **10**, 29–41 (1979)
4. Bao, T.Q., Mordukhovich, B.S.: Variational principles for set-valued mappings with applications to multiobjective optimization. Control Cybernet. **36**, 531–562 (2007)
5. Bednarczuk, E.M., Song, W.: Contingent epiderivative and its applications to set-valued optimization. Control Cybernet. **27**, 375–386 (1998)
6. Bonnans, J.F., Shapiro, A.: Perturbation Analysis of Optimization Problems. Springer, New York (2000)
7. Clarke, F.H.: Optimization and Nonsmooth Analysis. Wiley, New York (1983)
8. Fiacco, A.V.: Introduction to Sensitivity and Stability Analysis in Nonlinear Programming. Academic Press, New York (1983)
9. Gauvin, J.: Theory of Nonconvex Programming. Les Publications CRM, Université de Montréal, Montreal (1993)
10. Gauvin, J., Tolle, J.W.: Differential stability in nonlinear programming. SIAM J. Control Optim. **15**, 294–311 (1977)
11. Göpfert, A., Riahi, H., Tammer, C., Zalinescu, C.: Variational Methods in Partially Ordered Spaces. Springer, New York (2003)
12. Huy, N.Q., Mordukhivich, B.S., Yao, J.C.: Coderivatives of frontier maps and solution maps in parametruc multiobjective optimization. Taiwanese J. Math. **12**, 2083–2111 (2008)
13. Jahn, J.: Vector Optimization: Theory, Appplications, and Extensions. Springer, Berlin (2004)
14. Jahn, J., Khan, A.A.: Existence theorems and characterizations of generalized contingent epiderivatives. J. Nonlinear Convex Anal. **3**, 315–330 (2002)
15. Khan, A.A., Ward, D.E.: Toward second-order sensitivity analysis in set-valued optimization. J. Nonlinear Convex Anal. **13**, 65–83 (2012)
16. Klose, J.: Sensitivity analysis using the tangent derivative. Numer. Funct. Anal. Optim. **13**, 143–153 (1992)
17. Kuk, K., Tanino, T., Tanaka, M.: Sensitivity analysis in vector optimization. J. Optim. Theory Appl. **89**, 713–730 (1996)
18. Kuk, K., Tanino, T., Tanaka, M.: Sensitivity analysis in parametrized convex vector optimization. J. Math. Anal. Appl. **202**, 511–522 (1996)
19. Lee, G.M., Huy, N.Q.: On sensitivity analysis in vector optimization. Taiwanese J. Math. **11**, 945–958 (2007)
20. Mordukhovich, B.S.: Variational Analysis and Generalized Differentiation I: Basic Theory. Springer, Berlin (2006)
21. Penot, J.-P.: Differentiability of relations and differential stability of perturbed optimization problems. SIAM J. Control Optim. **22**, 529–551 (1984)
22. Rockafellar, R.T.: Directionally lipschitzian functions and subdifferential calculus. Proc. London Math. Soc. **39**, 331–355 (1979)
23. Rockafellar, R.T.: Generalized directional derivatives and subgradients of nonconvex functions. Can. J. Math. **32**, 157–180 (1980)
24. Rockafellar, R.T.: The Theory of Subgradients and Its Applications to Problems of Optimization: Convex and Nonconvex Functions. Heldermann Verlag, Berlin (1981)
25. Shi, D.S.: Contingent derivative of the perturbation map in multiobjective optimization. J. Optim. Theory Appl. **70**, 385–396 (1991)
26. Shi, D.S.: Sensitivity analysis in convex vector optimization. J. Optim. Theory Appl. **77**, 145–159 (1993)
27. Sonntag, Y., Zalinescu, C.: Comparison of existence results for efficient points. J. Optim. Theory Appl. **105**, 161–188 (2000)

28. Tanino, T.: Sensitivity analysis in multiobjective optimization. J. Optim. Theory Appl. **56**, 479–499 (1988)
29. Tanino, T.: Stability and sensitivity analysis in convex vector optimization. SIAM J. Control Optim. **26**, 521–536 (1988)
30. Tanino, T.: Stability and sensitivity analysis in multiobjective nonlinear programming. Ann. Oper. Res **27**, 97–114 (1990)
31. Tanino, T.: Sensitivity analysis in MCDM. In: Gal, T., Stewart T.J., Hanne, T. (eds.) Multicriteria Decision Making: Advances in MCDM Models, Algorithms, Theory, and Applications. Kluwer, Boston (1999)
32. Ward, D.E.: Differential stability in non-Lipschitzian optimization. J. Optim. Theory Appl. **73**, 101–120 (1992)
33. Ward, D.E.: Calculus for parabolic second-order derivatives. Set-Valued Anal. **1**, 213–246 (1993)
34. Ward, D.E.: Epiderivatives of the marginal function in nonsmooth parametric optimization. Optimization **31**, 47–61 (1994)
35. Ward, D.E.: Dini derivatives of the marginal function of a non-Lipschitzian program. SIAM J. Optim. **6**, 198–211 (1996)